Relativity of Evolution

Markus Knoflacher

Relativity of Evolution

Translated by Robert D. Martin

 Springer

Markus Knoflacher
Club of Vienna
Wien, Austria

Translated by
Robert D. Martin
Negaunee Integrative Research Center, The Field Museum
Chicago, IL, USA

ISBN 978-3-662-69422-0   ISBN 978-3-662-69423-7   (eBook)
https://doi.org/10.1007/978-3-662-69423-7

Translation from the German language edition: "Relativität der Evolution" by Markus Knoflacher, © Der/die
Herausgeber bzw. der/die Autor(en), exklusiv lizenziert an Springer-Verlag GmbH, DE, ein Teil von Springer
Nature 2022. Published by Springer Berlin Heidelberg. All Rights Reserved.

Cover: © Author Cover design: deblik, Berlin

This Springer imprint is published by the registered company Springer-Verlag GmbH, DE, part of Springer Nature.
The registered company address is: Heidelberger Platz 3, 14197 Berlin, Germany

If disposing of this product, please recycle the paper.

Foreword by Professor Elmar Heinzle

Where do we come from? Where are we going? These are two very big questions that have always preoccupied human beings—especially the researchers among us. With their traditional creation stories, religions provide us with explanations regarding the first question. Ever since Charles Darwin, however, evolution has occupied the core of our attempts to account for our origins. Fossil remains and structural patterns of plants and animals, together with behaviour and adaptations of animals and plants, all support this theory. If you take a closer look, as Markus Knoflacher does in his book *Relativity of Evolution*, you soon realize that our knowledge has expanded enormously—especially with support from molecular biology, ecology, and systems biology—and continues to grow at a breathtaking pace. This has allowed us to broaden our perspective on evolutionary processes. However, there are still many unanswered questions, particularly with respect to decisive stages, and new ones are constantly arising. Such questions include: How did the first organic molecules that are now the building blocks of organisms come into being? How did the chirality of biologically important molecules such as amino acids and carbohydrates, as we know them from Nature, come about? How did thermodynamically disadvantageous polymers develop from low molecular building blocks? How did polymers with an ordered sequence of building blocks such as in DNA, RNA, and proteins come into being? How did it come to be that these molecules can replicate and modify themselves, e.g. catalysed by RNA or DNA? How did formation of compartments enclosed by membranes come into existence? How did systems develop to provide the energy for these endogenous processes? How did self-replicating cells develop? How did eukaryotes with their inner compartments evolve? What role do loose cell assemblies play in the development of multicellular organisms? How did sexual reproduction evolve? What part does epigenetics play in evolution? What role do viruses play in evolution, where do they come from and how were they able to survive in an evolutionary context without having their own energy-supply systems? In contrast to higher animals and plants, the evolution of these processes cannot be studied using fossils. However, the different stages of emergence can be studied in organisms and systems that are alive today. In addition, certain evolutionary processes can also be simulated in laboratory experiments, e.g. enzyme evolution or evolution of microorganisms for improved degradation of xenobiotics.

The above list of unresolved questions could be extended almost indefinitely right up to the development of human language and social structures. In his book, Markus Knoflacher focuses on the processes that play a role in the evolution of individual cells through to multicellular organisms, as well as in important interactions with the environment. A basic tenor of the book is the importance of networks at various levels for the metabolic provision of energy, in the regulation of cellular processes and in regulating the behaviour of cell clusters.

Markus Knoflacher begins his presentation with a description of quantum-physical phenomena and processes that can play a part in the differentiation of molecules. These processes are usually studied and intensively applied particularly in physics, but also for technical applications. Just think of the quantum computers currently under development with their completely novel data-processing approaches. In biology, it is mainly photons, electrons, and protons that play a role that is increasingly discussed in quantum physics. In the foreground is light, which plays a central role in transmission of information, but also in generation of energy. In enzyme catalysis, proton and electron tunnelling is important in overcoming energy barriers. Protons play a core role in the production of the universal energy carrier ATP at membranes. It has been hypothesized that in all organisms the fundamental regulatory processes are of a quantum-physical nature.

Markus Knoflacher describes important fundamental characteristics of life: information processes, biological energy conversion, and nonequilibrium systems. The capacity to synthesize the required building blocks and provide the necessary energy is the basis for life. This includes efficient regulation of the processes involved so that a stable metabolic state is achieved, i.e. homeostasis. Similarly, life requires inherent growth and death— homeostasis at the level of a population. This is true at all levels of life. All life is based on more-or-less stable genetic information, on the one hand, and allows flexible adaptation to changing environmental conditions, on the other. In the three domains of organisms— archaeans, bacteria, and eukaryotes—the great majority live as single cells. Because of the vast number of individuals, an enormous diversity is possible. Moreover, due to their very small spatial dimensions, no special transportation or complex control systems are required. Considerable space is devoted to description of genetic and epigenetic processes in prokaryotes and eukaryotes. In addition to inheritance of genetic information by subsequent generations, combined with mutations of various kinds, the so-called horizontal gene transfer plays a major role in evolution—e.g. through viruses or direct gene exchange with neighbouring cells. To a certain extent, information about environmental influences can also be passed on by means of epigenetic processes. Description of the significantly greater diversity in genomes of prokaryotes compared to eukaryotes is also an interesting aspect. Markus Knoflacher throws considerable doubt on the paradigm of directed evolution from simple genetic endowments to more differentiated ones. Instead, it seems that long periods of increasing genetic impoverishment are followed by shorter periods with rapid augmentation of genetic differentiation. Active persistence and enduring further evolution of a very large number of species of organisms is also important. Of particular interest—in line with the emphasis on interacting systems—is the call for abandonment of the strict family tree paradigm in favour of a network concept. In addition to the biosphere, this network includes the interacting abiotic realm.

Unicellular microorganisms, i.e. eubacteria, archaebacteria, and eukaryotic protozoans, play a central role in the discussion of evolutionary diversification. In all living systems, compartmentalization by means of membranes is essential. In all cell types, the provision of energy is crucial for biosynthetic processes, e.g. for cell growth and cell reproduction as well as for processes of communication with the environment. The immense diversity of microorganisms with their more-or-less specialized capabilities allows them to colonize and interact in and with a huge variety of environmental conditions, ranging from hot deep-sea springs to the human digestive tract. It is assumed that the characteristic inner compartments of eukaryotes were formed by inclusion of bacteria—the chloroplasts of phototrophic cells from cyanobacteria, and mitochondria, found in almost all eukaryotic cells, from α-proteobacteria. Microbiomes, which can be found in a wide variety of forms, notably in biofilms, are now seen as an intermediate step towards multicellular organisms. A transition to multicellular forms can already be observed in microorganisms that are typically unicellular, such as myxobacteria. In more complex multicellular systems, apoptosis (programmed cell death) plays an important part in continuous renewal of tissues or elimination of unwanted cells such as cancer cells. All of those life forms can be observed side-by-side today and have not been displaced by the so-called higher development in later-emerging species.

Markus Knoflacher poses the question of whether the long-established and successful concept of species and their arrangement in phylogenetic trees can still be maintained in any meaningful way, pointing in particular to the interconnectedness of systems. Enormous diversity and attempts to deal with it effectively are also reflected by increasing difficulties in attaining a comprehensible, clear nomenclature for microorganisms (see "Microbial Taxonomy Run Amok" in Sanford et al., Trends in Microbiology, 2021). As a general rule, categorization of organism types according to genetic criteria now predominates. The performance of sequencing techniques and associated bioinformatic procedures has undergone enormous expansion. However, there is an ever-increasing gap in knowledge of the physiological potential and ecological significance of these organisms, for whose characterization no comparably potent tools are available. Most methods are well suited to pure cultures, but those are practically non-existent in nature.

A decisive characteristic of life and a prerequisite thereof is successful handling of information of different kinds at various levels. Information is available intracellularly or extracellularly. Diverse signals provide data for biological algorithms, which in turn generate new information for a wide range of actions. The complexity of these processes increases from viruses to higher organisms. While intracellular regulatory systems primarily serve to maintain homeostasis, but also cell growth and cell proliferation, extracellular signals of an abiotic and biotic nature are required to adapt to environmental conditions or to cope with changes in the environment. Microorganisms can already communicate with each other in various ways, e.g. for food intake or for defence against external threats. Transmission of signals, for instance in the form of alarmons, is finely tuned to the environment and corresponding possibilities for diffusing those signals. With the formation of aggregates from single cells, and even more with the development of multicellularity, the effort expended for coordination increases and with it the complexity of information acquisition and processing. In this context, it is also interesting to note that many developments evidently first occurred under marine conditions. Only thereafter did colonization of the land take place. In fungi as well as in plants and animals, intercellular communication can be observed both within and between individuals and also in interaction with the environment. In most cases, this occurs by means of a variety of chemical signalling substances. Only in animals did neural systems—a prerequisite for faster coordinated movements—develop. The final stage was centralization of the nervous system.

For the essential aspect of energy conversion, Markus Knoflacher provides a wide-ranging discussion, beginning with the initial biochemical processes for synthesis of polymers and extending to the activity of nervous systems. Within every cell, a universal energy currency that can be employed everywhere exists in the form of ATP and allied compounds. It is assumed that energy conversion developed first through chemical processes and subsequently through conversion of solar energy in the form of light. In order to maintain its vitality, every cell must be kept in a state of dynamic equilibrium. In ecospheres, such as biofilms, which are usually colonized by many different organisms, conversion of compounds for energy production involves a wide variety of interactions. At the same time, close proximity favours the exchange of information with the help of signal molecules and genetic information. While chemolithotrophic nutrition is dependent on the local occurrence of energy-rich inorganic compounds, phototrophic nutrition permits organisms to disperse over the entire surface of earth. Development of photosynthetic mechanisms, first in cyanobacteria and subsequently in plants, probably began under anoxygenic conditions. Only later was sufficient oxygen released into the atmosphere, until today's oxygen concentration was eventually reached. It is assumed that horizontal gene transfer played a major role in the development of photosynthetic mechanisms, but many questions remain unanswered. It cannot be determined with certainty whether exclusively autotrophic processes took place at the outset, i.e. production of organic carbon exclusively from carbon dioxide, or whether heterotrophic lifestyles were already established in parallel. Today, the diversity of heterotrophic cells outweighs that of autotrophic cells, e.g. algae and plants. Transportation of substances from the environment to cells and also between cells takes place predominantly by diffusion in unicellular organisms. In aquatic environments, transportation can be enhanced by convection. Multicellular organisms have developed pumping systems to transport more nutrients. Plants use capillary systems to transport nutrients and water between roots and leaves, while animals use circulatory systems to transport nutrients between individual tissues. Here, too, Markus Knoflacher uses many aptly assembled examples to show how the evolution of such processes can be inferred.

A separate chapter is dedicated to nonequilibrium systems, which can be described as a prerequisite for life. Such systems require a constant supply of energy and nutrients of various kinds, such as minerals, carbon dioxide, or carbohydrates. The greatest efficiency in energy production is achieved through the process of oxidation permitted by oxygen. However, this always requires the availability of oxidizable compounds. Oxidation with oxygen on a large scale only became possible following the increase in the atmospheric oxygen concentration that occurred around 2.5 billion years ago. This change is associated with a pronounced increase in biodiversity. Large-scale glaciations also exerted a major influence on organismal diversity. Overall, it must be recognized that various interactions between the lithosphere and hydrosphere, as well as the atmosphere have a major influence on the biosphere, but that the latter can also influence the former. The input of minerals from land into the seas generally favours growth of algae, which in turn releases oxygen into the atmosphere and leads to sedimentation as a result of die-off.

Developments in marine habitats evidently took place before those on land, favoured by broadly constant conditions in the seas. Especially in epipelagic zones, the evolution of photosynthesis was facilitated, thus providing a nutrient base for many heterotrophic organisms. Spatial niches are of great importance for evolution, as they provide the foundation for development of cell aggregates and multicellular organisms. For higher marine organisms, the opportunity arose to conquer the land.

Developments in the terrestrial realm are generally better documented as a result of the evolution of larger animals. Here, too, the interaction of different systems is important for the existence and evolution of organisms. Their diversity extends far beyond the reasonably well-known interactions of mycorrhizae on the roots of plants. However, the early phases of terrestrial colonization are still largely obscure. Diversity in the terrestrial realm also showed oscillations, which are particularly noticeable in the case of vertebrates. In addition to interactions of protozoans and fungi with plants, interactions between plants and animals, e.g. large vertebrates, are also decisive factors in evolution.

If evolution proceeds according to clear laws, it should in fact be possible to reproduce the process experimentally. At first, this may seem completely unrealistic because of the extensive timespan required. To a certain extent, however, experiments can be designed in such a way that partial processes become apparent: the so-called primordial soup with formation of more-or-less complex organic molecules, polymerisation with the participation of polyphosphate, evolutionary improvement of degradation of xenobiotics by microorganisms, or evolutionary generation of enzymes.

Ultimately, the findings compiled and organized by Markus Knoflacher lead us back to one of the core questions stated at the beginning: Where are we going? Where is earth—along with all of its systems relevant for us—heading? The question arises as to what extent we can foresee, predict, or at least anticipate future developments. What short- and long-term effects will increasingly intensive human activities have on the future, interacting (eco-) system earth? One can imagine the enormous difficulties concerned in any attempt to find answers to these questions if the overwhelming diversity of individual evolutionary events and even more so of all relevant interactions in the overall system are to be taken into account.

What is particularly striking about Markus Knoflacher's book is the abundance of cited, competently categorized, and interpreted literature—it contains over 5000 citations. A huge wealth of information has been meticulously compiled and is presented to the reader in a novel, easy-to-read form, in which the notion of networked, permanently interacting systems predominates. This very well-organized wealth of information provides many interesting ideas for teachers at various levels. In general, the book can certainly be recommended to anyone who wishes to gain a deeper insight into evolutionary processes and seeks a different approach to understanding them. I also recommend this book to anyone who is concerned about the future of our natural, technical, and social environment. This excellent book clearly demonstrates that we need some understanding of the interconnectedness of these processes in order to think judiciously about future developments. Focussing exclusively on the increase in atmospheric carbon dioxide will not be enough!

Technische Biochemie   Elmar Heinzle
Universität des Saarlandes
Saarbrücken, Germany

Foreword by Professor Robert D. Martin

I was delighted to receive the invitation to translate Markus Knoflacher's magnificent treatise from the German original (entitled *Relativität der Evolution* and published in 2022) into this English version, similarly entitled *Relativity of Evolution*. This venture was initiated just over three years ago, when—after accepting a request to provide a foreword for the German edition—I read through the entire manuscript with steadily mounting enthusiasm. After doing so, I duly prepared my foreword and sent in to Markus. In my accompanying message I expressed not only my great admiration for his work but also my strong conviction that he should arrange to have his book translated into English as soon as possible. This is a milestone volume in evolutionary biology that surely deserves to reach the widest possible readership! In due course, I was (as I had quietly hoped) invited to perform the translation myself, and I unhesitatingly accepted. Thereafter, I spent almost a year painstakingly carrying out an adaptation into my mother tongue that I sincerely hope accurately conveys both the content and the spirit of his *magnum opus*.

It is rare indeed for a new book about evolutionary biology to inspire fundamental rethinking of long-held assumptions. But that is precisely what Markus's treatise *Relativity of Evolution* does. It elegantly and convincingly demonstrates that the core concept of relativity has implications throughout biology in much the same way as its more fundamental *alter ego* revolutionized physics. And Markus is admirably qualified to present and discuss the detailed and encyclopaedic evidence that he has assembled to support his well-argued case.

Markus started out as an undergraduate studying zoology and botany at the University of Vienna in Austria, subsequently progressing to acquire a doctorate degree in biology. From there, he embarked on an extended research career in extra-university institutions, including the Austrian Academy of Sciences, Joanneum Research, and the Austrian Institute of Technology. In the course of his professional activities, Markus devoted his attention particularly to interdisciplinary questions tackled from the special perspective of systems theory. From 2013 onwards—in a liberating transition that fruitfully permitted him to focus more on his primary interests—he has continued to conduct research as an independent scientist. Markus has published prolifically throughout his career, and his output includes numerous edited books, the most recent being *Herausforderungen der evolutionären Komplexität* (*Challenges of Evolutionary Complexity*). That volume, published in 2017, includes a chapter contributed by Markus that neatly prepared the ground for this single-authored masterpiece: *Relativity of Evolution*. In parallel, he has been notably active in co-organizing broad-based scientific meetings to promote interdisciplinary endeavours. Indeed, as one outcome of my flourishing connection with Markus, I was invited to participate in a splendid conference entitled *Wechselwirkungen und Zufall in der Evolution* (*Interaction and Chance in Evolution*) held at the Natural History Museum in Vienna on 6–7 October, 2022. Appropriately reflecting the breadth of Markus's vision, the programme included a dozen excellent cameo contributions, ranging from the origin of the universe through the evolution of viruses, bacteria, fungi, and lichens and on to invertebrates and then vertebrates. The programme ended with a presentation by Markus of his recently published book *Relativität der Evolution*. Altogether a feast for broad-based biologists!

Before proceeding any further, however, I must now confess to a deeply unsettling outcome of reading the original German text of *Relativity of Evolution*. It dramatically opened my eyes to the fact that my own approach to evolutionary biology, pursued over a 60-year academic career entirely devoted to exploration of primate evolution, has been too narrow. This unseen limitation arose despite the fact that from the outset I deliberately set out to take a very broad approach to my subject matter, with the ultimate goal of clarifying human evolution on a very broad basis. As Markus notes, Man's place in Nature has been a key question ever since the Darwin/Wallace theory of evolution first became public. To me, it seemed self-evident starting out that—to contribute meaningfully to an understanding of human origins and evolution—a wide-ranging perspective was essential. As a species, we are unique in crucial ways, most notably with respect to the great size and complexity of our brain, our highly unusual pattern of bipedal locomotion, and our radically modified jaws and teeth. Comparisons restricted to our closest relatives, the great apes, are simply inadequate to generate testable hypotheses regarding the evolution of such unusual human features. Instead, comprehensive investigations are required to recognize general principles that we can confidently apply to interpret the challenging case of human evolution.

Across the sciences, increasing specialization and compartmentalization since the time of Darwin and Wallace has brought many benefits, but has also given rise to substantial and ever-growing obstacles to interdisciplinary communication. Keenly aware of this, I vowed to avoid specialization as far as I possibly could. Having started out as a zoologist, I decided to study as many different primate species as possible. I began with the relatively primitive prosimians (lemurs, lorises, and tarsiers), proceeded to the relatively more advanced simians (monkeys and apes), and then at last began to focus on the human species. As regards subject matter, my approach was equally wide-ranging. Beginning with classical morphology—first in extant primates and then in their fossil relatives as well—I progressively explored behaviour, ecology, reproductive biology, physiology, and eventually molecular biology as sources of information. Moreover, I soon realized that reliable reconstruction of primate evolution actually demands at least a basic understanding of mammal evolution, starting some 200 million years ago. So I expanded my range even further. After decades of research along these widely ranging lines, I thought that I had pretty much mastered my chosen sector of the Tree of Life. I naively thought that I had managed to carve out a realm of biology that I could effectively consider in splendid isolation.

As, with steadily growing admiration, I read through Markus's arresting and absorbing treatise in the German original, my complacency was seriously rattled. It became increasingly clear to me that the origin and diversification of life on planet Earth is a far more complex and challenging topic than I had ever imagined. As I gradually absorbed the importance and reach of Markus's narrative, the supposed breadth of my lifelong research into primate evolution slowly withered before my eyes. Beginning with basic physical principles that influence all living organisms, he conducts a broad sweep across domains and themes with an authoritative command of hard-won knowledge at every level. Slowly but surely, it becomes utterly clear that all things in biology are so closely and intimately interdependent that isolated consideration of any single sector of the Tree of Life must inevitably lead to major omissions in our understanding.

Markus's account is distinctive in that it focusses mainly on evolutionary *processes*, rather than on the mere facts of evolutionary *history* (although he actually provides a very informative overview of key developments in the Tree of Life). Throughout the book, he displays a truly impressive mastery of the literature relating to the evolution of life on Earth—as witnessed by an extensive, remarkably up-to-date bibliography comprising over 5000 references, systematically flagged in footnotes in the German version for ease of reading. Topics covered include the morphology and physiology of living organisms, fossil evidence of their origins and evolutionary radiation, and molecular evidence indicating their relationships. However, instead of just painting a picture of the Tree of Life as presently reconstructed, his primary concern is to identify systematic principles that apply across all living systems.

Markus identifies a pervasive, deeply-rooted problem with discussions of evolutionary biology in the general tendency to perceive life forms as "primitive" or "advanced" and to imply progressive trends towards "higher", more complex, levels of organization. In fact, this kind of interpretation of life on Earth existed long before the theory of evolution emerged and became enshrined in the philosophical notion of the *Scala naturae*—also known as the Great Chain of Being. Within this framework, various life forms are arranged on an ascending ladder of supposedly increasing "perfection". When I began my own research into primate evolution, this model was explicitly applied to the array of modern species. On the bottom rung of the ladder were tree-shrews (whose relationship to primates has since been shown to be very remote at best), while humans were inevitably placed on the uppermost rung. As it happened, I specifically rejected this approach to primate evolutionary history in 1968, in one of the first papers arising from my PhD thesis on the evolutionary relationships of tree-shrews, pointing out its highly misleading influence. A half century has since elapsed, and I had begun to believe that the notion of an evolutionary stepladder had been largely abandoned. Admittedly, a common tendency to refer to prosimians and simians as "lower" and "higher" primates, respectively, persists and is often excused (even by me) as a short-hand description for popular consumption. But Markus's account clearly demonstrates that embedded notions of "primitive" and "advanced" organisms, reflecting a perceived general trend towards increasing complexity, are still seriously impeding a realistic assessment of evolutionary biology. The very fact that vast numbers of single-cell organisms still exist around the globe today shows that they have been just as "successful"—in objective biological terms—as the lineage leading to modern humans. Indeed, the current COVID-19 pandemic clearly shows that viruses, which lack any independent metabolic machinery and hence do not qualify for identification as living organisms, can be extraordinarily successful in crude biological terms. No living organism is "primitive" in the sense of remaining exactly the same as its initial ancestor. Extant life forms may retain primitive characters, such as single-celled status, to a greater or lesser degree; but there is no such thing as a "frozen ancestor".

Another striking example of a permeating biological principle comes from the many mutually beneficial cooperative arrangements between organisms—*symbioses*—that have arisen countless times virtually since the very beginning of life on Earth. The earliest cells at the bacterial level (*prokaryotes*) had no nucleus to isolate the genetic material (DNA in a circular or linear array) from the cell cytoplasm. A notable evolutionary development accompanying the origin of ancestral cells that eventually gave rise to protists, fungi, plants, and animals (collectively known as *eukaryotes*) was the development of a nucleus containing DNA packaged into linear chromosomes. The latter feature is closely linked to the presence of a special form of cell division—*meiosis*—to produce sex cells for reproduction. Eukaryote cells are also characterized by the presence in their cytoplasm of *mitochondria*, tiny organelles that play an important part in energy turnover. As Markus notes, the earliest fossil evidence for eukaryotic cells (presumably possessing all of those distinctive features) comes from deposits dated at around 1,800,000,000 years old.

Several decades ago, molecular biological investigations revealed that mitochondria—which possess a circular genome like many prokaryotes—actually originated from free-living bacteria. It seems likely that ancestral eukaryote cells had a symbiotic relationship, which probably evolved just once, with bacteria present in their cytoplasm. In the meantime, molecular evidence has clearly revealed that mitochondria are related to free-living bacteria belonging to a particular class—Alphaproteobacteria. Initial tree-building indicated that mitochondria are direct relatives of *Rickettsia* (tick-borne typhus), but within the past few years new evidence has revealed that the relationship is actually rather more remote. [Inclusion of this very recent correction, which updated my own understanding, is just one of many indications of Markus's supreme command of the relevant literature.]

In fact, during the early radiation of eukaryotes, the last common ancestor of all plants must have been characterized by a *double* symbiosis. Their cells contain not only mitochondria but also *chloroplasts*, which were subsequently independently derived from a different group of bacteria—Cyanobacteria—presumably by means of ingestion (phagocytosis). By virtue of the chlorophyll that they contain, chloroplasts provide the biological mechanism for *photosynthesis,* a defining feature of all plants. Available evidence indicates that inclusion of chloroplasts in early eukaryotes dates back over 1,500,000,000 years. As with mitochondria, this was possibly a single event, but it may have developed separately in a number of lineages, for instance in certain algae.

All living protists, fungi, plants, and animals hence owe their existence to ancient cooperative arrangements with bacteria that have persisted to the present day. But that is just one particularly striking example of mutually beneficial associations between different organisms. As Markus's book shows, such associations—which blur the boundaries between individual life forms—are extremely widespread in Nature. One long-overlooked example that has recently been received increasing attention in medical circles is the gut biome in humans and other vertebrates. Some, but not all, bacteria that live in the digestive tract are beneficial symbionts, with wide-ranging effects that influence human health.

Another prominent example, which does not directly involve humans and other vertebrates is horizontal gene transfer, in which genes may be passed from one species to another. It has gradually become apparent that in relatively simple organisms, primarily single-celled species, genes can "jump" from one organism to another. This is so common for such species, in fact, that it is often impossible to construct neatly branching evolutionary trees using standard methods based on molecular similarities. Indeed, the resulting confused picture has thwarted attempts to distinguish clearly between "species". However, this problem is prevalent only in single-celled species and has not been reported for vertebrates and other multicellular animals. It is therefore easy to assume (as I did) that "jumping genes" are not directly relevant to vertebrates, such as the primates that I have studied so intensively. However, horizontal gene transfer remains a common feature of microorganisms and can have important spin-off effects. In human medicine, for instance, it has now been shown that gene transfer between microbes within the human body can play a part in the more rapid development of the scary phenomenon of antibiotic resistance.

Another long-standing and very widespread relationship that is often mutually beneficial (i.e. symbiotic) is the association between fungi and the roots of plants known as a mycorrhiza. The fungus receives organic products of photosynthesis such as sugars from the plant, while the fungus supplies the plant with minerals (e.g. phosphorus) and water. Mycorrhizal associations, which have an extended evolutionary history, make major contributions to plant nutrition and to the biology and chemistry of soils. In addition to their evolutionary significance they are also of considerable economic importance. A more recent, symbiotic development was an association between bacteria in the family Rhizobiaceae and roots of leguminous plants, endowing the pivotal capacity for nitrogen fixation.

The examples presented above provide only a tiny sample of the multifarious interactions provided in Markus's "big picture" survey. But they suffice to show how a multitude of interconnections among organisms, and between them and their environments, have shaped evolution ever since life first emerged on our planet over 3,000,000,000 years ago. They are also sufficient to demonstrate that in the study of evolutionary biology reductionist approaches that consider only a limited number of factors (for example, molecular processes alone) are bound to be deficient. In the memorable words of H.L. Mencken: "For every complex problem there is an answer that is clear, simple, and wrong".

Markus's ultimate aim in pursuing his research and writing this book was to derive lessons to guide future human actions. He emphasizes the many dangers of current human activities, not just with respect to the increasingly dramatic effects of climate change associated with other factors such as increasing mechanization and automation. In addition, he shows that the enormous complexity of interconnecting life processes demands far greater caution in attempts to modify biological systems. Such intervention is almost bound to have unexpected, and perhaps extremely unwelcome, side-effects. Moreover, there are clear lessons for conservation biology. It is now widely accepted that a single-species approach to preservation of biological diversity is simply inappropriate and inadequate. Many conservation programmes now recognize the general importance of ecological factors driving entire biomes. However, the amazingly extensive networks that Markus reveals indicate that an even more comprehensive approach will be necessary for successful conservation over the longer term.

I have long believed that politicians and civil servants should be required to be educated with a basic training in biological sciences, not least to establish respect for the complexity of biological systems and their potential fragility in the face of exploitation of natural resources and galloping technological innovation. Partly because of a widespread inclination in the humanities to reject any relevance of evolutionary biology to their disciplines, there has been little movement towards that goal. *The Relativity of Evolution* eloquently underlines the need for comprehensive biological understanding at the highest levels of human society.

Emeritus Curator, Science & Education   Robert D. Martin
Negaunee Integrative Research Center, The Field Museum
Chicago, IL, USA

Academic Guest, Institute for Evolutionary Medicine
University of Zürich
Zürich, Switzerland

Preface

This is a revised version of the original German publication *Relativität der Evolution*, published in 2022. In this English version—expertly translated by Professor Robert Martin—more recent findings have been incorporated as far as possible, individual errors corrected, and illustrations revised.

The thematic focus on fundamental developmental trends in biological evolution remains the same. In terms of content, a prospective approach has been chosen. In contrast to retrospective approaches, comparative analyses are based on the—so far known—contextual conditions of the particular geological phases under consideration. Prerequisites for this are analyses of characteristics in potentials of biological processes (Chaps. 2–7) and of abiotic conditions (Chap. 8). Establishment of these foundations permitted comparative analyses of evolutionary conditions and processes in marine (Chap. 9) and terrestrial (Chap. 10) environments.

All analyses were based on published sources that had been evaluated in an interdisciplinary context according to criteria of quality and consistency. The prospective approach and the methodological procedure opened up new perspectives regarding evolutionary processes and many novel questions. Those issues are addressed from an anthropological point of view in Professor Martin's foreword and from a biochemical point of view in Professor Heinzle's foreword. Let it be noted that omission of a glossary was an intentional decision, made with the aim of promoting contextual thinking. Although this may make it more difficult to access certain information rapidly, it will hopefully encourage engaged readers to delve more deeply into the interactions between social and evolutionary processes.

Inevitably, the question of the origins of biological evolution—which may possibly reside in quantum phenomena—has remained unanswered. It is, however, a fascinating fact that biological self-organization—manifested in living organisms—has been able to maintain itself continuously over billions of years in an abiotic environment characterized by exactly contrary—entropic—directions of change over time. All organisms—ranging from tiny single-celled organisms to the largest-bodied multicellular organisms—share the capacities of self-reproduction, information processing, and autonomous conversion of energy. Viruses differ from organisms primarily in lacking the latter capacity.

Limits for the evolution of all organisms are determined energetically by conversion of abiotic energy potentials into biologically utilizable energy potentials (autotrophic energy conversion). Within this dynamic framework, abiotic processes, viruses, and interactions with other organisms continuously promote functional differentiation of life forms. Biological evolution develops through direct interactions between organisms and abiotic conditions. From this it can be inferred that evolutionary processes are influenced by local abiotic conditions and autotrophic transformation rates as well as large-scale processes of heterarchical interaction. In simplified terms, biological evolution can be characterized by ongoing adaptations to spatially heterogeneous framework conditions with differing dynamics.

Connections are most clearly recognized through comparisons of biological evolution between open oceans and terrestrial habitats. With the first, the existence of all organisms is still based on autotrophic energy conversion by predominantly unicellular organisms. Interactions between organisms are to a large extent determined by randomness in dynamic water currents. With the second—much more recent—case, multicellular plants dominate autotrophic energy conversion. They interact with other organisms in a variety of ways and also modulate abiotic factors such as air currents or the water balance. In terrestrial ecosystems, interactions can be found resembling those in microbial biofilms, which are probably the oldest organismal communities in Earth's history. Such biofilms can develop both on a chemoautotrophic basis—for example, in submarine hydrothermal springs—and on a photoautotrophic basis—for instance, as stromatolites.

Communities of organisms permit a broader functional spectrum of biological outputs than individual species and thus augment opportunities for evolution within abiotic framework conditions. Opportunities for evolution can be characterized only to a small degree by the biomass of the organisms involved. Structural differentiation and "infrastructure benefits" across different spatial dimensions are far more important. Examples of structural differentiation can be found in coral reefs or tropical rainforests; examples of infrastructure benefits are reduction of losses of energy and material through the extracellular matrix in biofilms or through fungi, animals, and microorganisms in terrestrial vegetation. Among other things, these benefits are attained by coordinating energy conversion processes, exchanging information, or horizontal gene transfer. As regularity and diversity of interactions increase, the risk of collapse of organismal communities is also reduced by the disposition—common to all organisms—towards unlimited population growth.

Within the group containing great apes and hominids, evolutionary developments in the genus *Homo* provide an example—from the perspective of an individual species—of effects of surmounting restrictive framework conditions. Increasing differentiation of tools, active use of fire, construction of dwellings, and modification of ecosystems all provide evidence of this. Most of the time during human evolutionary history, anthropogenic modifications were limited by the transformation potential of ecosystems. Not only food but also combustible materials and fuels for thermal processes were of organic origin. Under these conditions, society had to endure repeated famines and—owing to a lack of knowledge about bacterial life forms and viruses—deadly epidemics. With increasing scientific knowledge and access to fossil fuels, in theory options for sustainable development of human societies became accessible. Theoretically, decreasing mortality rates enabled by medical knowledge permitted falling birth rates to stabilize population sizes. To compensate for crop failures, new options presumably became available for temporary increases in production on existing farmland or for rapid transportation of food from areas with surplus production.

In reality, options have been exploited for global warfare and increasingly accelerated global expansion of land settled by humans. Ongoing global modification of living environments—such as expansion of urban landscapes —as well as manufacturing and economic processes has led human society to embark on a new developmental trajectory. Increasingly, continuing population growth, and above all disproportionately rising material and energy demands, can be met only through utilization of fossil fuels and nuclear energy. Ultimately, the human social system is drifting into an entropy trap that is no longer manageable.

Reasons for this state of affairs reside not in scientific findings that permit previously unknown freedoms, but evolutionarily developed individual and social behaviour patterns of human beings. Indeed, resulting motivations of greed for material possessions and acquisition of social power lead to abusive decisions concerning the application of acquired knowledge. Repression and suppression of discussions of these inherent characteristics of human behaviour facilitate the creation of authoritarian social structures, on the one hand, and hinder scientific research into the associated phenomena, on the other. The processes involved are increasingly accelerated by directly perceptible, constantly decreasing individual performance requirements as a result of ever-increasing technological substitutions.

Critical comments in the text regarding limits to scientific knowledge and to measures for "climate protection" are intended to highlight risks and problems in a systemic context. This is by no means a fundamental negation of scientific working methods or a denial of anthropogenic influences on climate change. Instead, criticism is very much aimed at ill-considered transference of findings from studies conducted in closed systems to processes in open systems and at overestimation of the controllability of global environmental systems.

Wien, Austria   Markus Knoflacher

Acknowledgements

I owe my deepest gratitude to my wife, who has borne the personal and material burdens associated with years of intensive investigation! I would also like to thank the Club of Vienna for financing the translation! Without the enthusiastic and thorough work of Professor Robert Martin, the translation would not have been of such high quality, for which I thank him warmly! I also thank him for summarizing his impressions (including comments on the translation) in the accompanying foreword, along with Professor Elmar Heinzle, whose foreword is based on his intensive work on the manuscript for the original German version of the book. I would also like to thank Mr Rick Lumpkin (NOAA/AOML) for his gracious and swift provision of the overview map of global ocean currents. Last but not least, I would like to thank the countless scientists who made this treatise possible through their research and resulting publications of the results.

Contents

About the Author

Markus Knoflacher studied Zoology and Botany at the University of Vienna in Austria and acquired a doctoral degree in philosophy. His professional career in extra-university institutions was focussed on interdisciplinary research tackled from the perspective of systems theory. After retirement he conducts research as an independent scientist.

Why This Book?

Much ink has been expended on and about the subject of evolution. Leading topics have been the presentation of evolutionary phenomena, their interpretation, or questions of faith as to whether the implementation of a plan of creation is discernible therein. Fundamentally, the contents of this book belong in the first two categories. In fact, only a small portion of the book is devoted to describing the historical course of evolution. The primary focus is on investigating the processes involved. And this is applied with the optimistic aim of deriving insights for meaningful human action over the long term. It is optimistic in the sense that, to all appearances, human society is feverishly engaged in its own extinction. Here, I allude not to the omnipresent issue of climate change, but to far more subtle but intensively conducted developments. One of these developments is increasing mechanization and automation of human activities. Originally promoted to augment human labour, the effects of displacing humans from economic processes are becoming increasingly apparent. Whereas in the beginning it was mainly insufficiently educated people who were affected, in the meantime more and more highly qualified people are being displaced from socioeconomic processes. The ultimate aim of endeavouring to create "artificial intelligence", or genetically tailoring "perfect humans" beyond genuine therapeutic needs, is the abolition of contemporary humanity. Indirectly, this is an admission that we humans cannot cope with being human.

We seek solutions outside our existing abilities instead of within them. What is striking here is the resemblance to the religious-culturally dominant notions of an ideal world beyond the present, which form the essential foundations of both Abrahamic[1] and Buddhist religions. The cultural scope of these notions is revealed by comparisons with the contents and effects of Confucian philosophy (Bauer 2009). Because of the mis-match between the resulting increasing changes in living environments and actual action permitted by human

---

[1] Judaic, Christian and Moslem religions.

© The Author(s), under exclusive license to Springer-Verlag GmbH, DE, part of Springer Nature 2024
M. Knoflacher, *Relativity of Evolution*,
https://doi.org/10.1007/978-3-662-69423-7_1

capacities, new challenges are constantly generated that overtax our ability to find solutions. The truth of this is in fact confirmed by statements that the world is becoming increasingly complex. This is true only in terms of our own perception, whereas it is quite wrong in the context of evolution. Perhaps that part of the creation story connected with the expulsion from paradise led to loss of a crucial sentence: "If you believe you were made in my image, go forth and prove it."

Long before the first known traces of human existence, the world was far more complex than we can possibly imagine. This becomes particularly evident when the "forgotten" 99.999… per cent of all life forms—single-celled organisms such as archaeans and bacteria—are taken into account. In their interplay with viruses—which are not classified as living organisms—they have influenced biological evolution from its beginning and will do so until to its future end. Multicellular organisms—the core of classical evolutionary theories—were able to develop only within the framework of microbiological processes. Through them, however, microbial processes were able to unfold in otherwise unattainable physical dimensions and could therefore also modify abiotic processes to a greater extent in certain realms. Increasingly, this complexity became apparent only through replacement of processes brought about by evolution by large-scale application of technologies. Given the orders of magnitude involved, however, it would be utterly presumptuous to infer from this that we are increasingly able to control evolutionary processes.

One final note regarding the gender issue: As a rule, both genders are meant throughout the text. In the interests of readability and avoidance of repetition, however, only one gender form is generally used.

Reference

Bauer W (2009) Geschichte der chinesischen Philosophie. Beck, Munich

Darwin's Long Shadow

2

Since Charles Darwin's first publication on the origin of species appeared in 1859, discussions about biological evolution have almost always culminated in the question of man's status within the domain of organisms. Even the co-founder of the theory of evolution—Alfred Russel Wallace (1823–1913)—spurned the association of his name with the term. Subsequently, in polemics against Wallace, advocates of Darwinian primacy often cited this attitude. Across his life trajectory we find both significant contributions to development of the theory of evolution and increasing distancing from it during his later years (Glaubrecht 2013). A decisive motivation for this shift can be found in Wallace's social stance. Because he had already experienced at first hand the misery of repressed social classes during his formative years, he apparently perceived an element of danger inherent in unconditional transfer of the principle of selection to the rules governing human society. It remains unclear why he sought extra-societal reasons for the justification of social equality. Perhaps it seemed too feeble to him to argue that social equivalence cannot be determined from natural laws but only through society itself.

Wallace's intellectual life perfectly characterizes the zones of conflict between the scientifically anchored contents of the theory of evolution and human claims to a special status in the universe. Instead of recognizing the importance of evolutionary processes for human emergence and deriving from them insights concerning personal responsibility for action, escape routes are sought in ideological notions. As a result, factual discussions rapidly drift into emotional arguments between different persuasions of faith. Advances in knowledge over the last one-hundred-and-fifty years pass unheeded.

This date of affairs stems from interactions between a variety of factors and processes: In everyday life, the topic of biological evolution carries no special significance—because it is predominantly understood in the sense of a historical description. In this context, some people even see a need for action to "overcome evolution". Advocates of genetic

M. Knoflacher, *Relativity of Evolution*, https://doi.org/10.1007/978-3-662-69423-7_2

engineering are particularly vocal in this regard. Ironically, the technologies concerned employ "tools" that have bacteria and viruses have used for billions of years to manipulate genes. Slightly modified and labelled with mystical abbreviations—CRISPR,[1] for example—their use drives aspirations and generates huge earnings (Kozubek 2016). Profits generated through exploitation of evolved behaviour patterns using digital information-processing technologies—so-called "social media"—are even greater.

Underestimation of the all-embracing importance of evolutionary processes for social evolution also contributes to public disinterest. Roots of such disinterest are to be found in a collective misuse of phenomena of evolution in different scientific disciplines. Collective misuse was particularly evident in the horrific atrocities of fascism. Less obviously—but also with grave social consequences—traces of ideological misuse can be found in the notions of neoliberalism. As an example, consider the central credo of "survival of the fittest".

Scientific conceits complicate, and may even prohibit, open discussion of the importance of evolutionary processes for society. Disciplines In the humanities often deny any connection between social and evolutionary processes. Dissenting opinions are denigrated as "biologisms" and promptly assigned to the shadowy realm of fascist ideologies. Overlooked here is the fact that in some cases arguments analogous to those embedded in the theories of T.D. Lyssenko (1898–1976) are employed (Roll-Hansen 2006; Gording 2012). The postulated independence of social processes from evolutionary factors is reminiscent of the claim—based on genes—that crops have unrestricted adaptability. Dogmatic implementation of the theory of "Lysenkoism" in the Soviet Union during the Stalinist era not only dramatically exacerbated famines, but also led to imprisonment, or even murder, of dissenters.

The theory of evolution is generally accepted across biological disciplines, but marked differences of opinion nevertheless exist with respect to the meaning of relevant processes. Greatly simplified, two different viewpoints can be outlined: The first assumes that evolution of organisms is autonomous—"blind" to the environment. Changes are brought about by mutation and random selection driven by environmental factors. From this viewpoint, the hierarchical order of nature (*Scala naturae*) formulated by Aristotle (384–322 B.C.) and the static system of species definition by Carl von Linné (1707–1778) can be reconciled—at least superficially. The paradigm of goal-directed evolution from "lower" to "higher" organisms lurks within portrayals of many "phylogenetic trees" (Fig. 2.1a). Implicitly, it is assumed that "higher" organisms subsume and surpass all capacities of the relevant "lower" organisms. In particular, capacities of organisms at the "roots" of phylogenetic trees are held to be superseded by those of "higher (better)" organisms. As a result, it is implied that "lower" organisms have largely lost any importance in evolutionary processes.

---

[1] Clustered Regularly Interspaced Palindromic Repeats. For anyone interested, a brief explanation is available on Wikipedia.

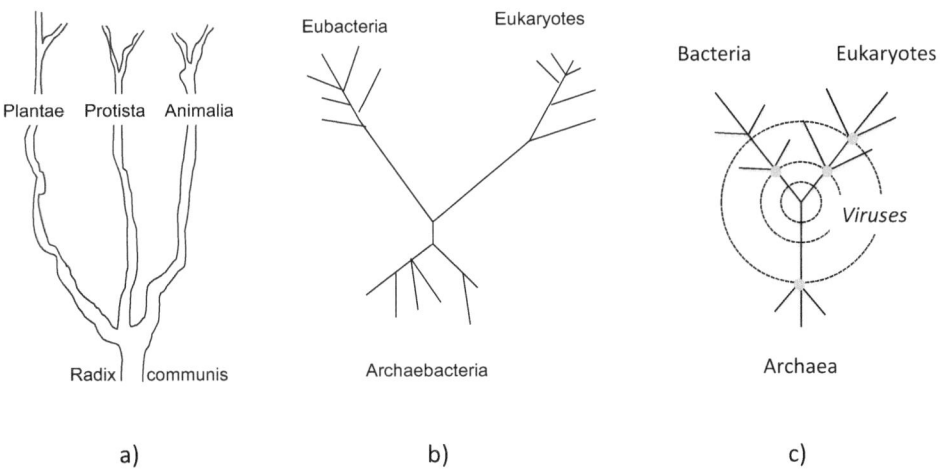

Fig. 2.1 Comparison of interpretations of biological evolution according to Haeckel (Haeckel Ernst (1834—1919) German zoologist, physician and philosopher) 1866 (**a**), the cladistic approach of Woese (1987) (**b**) and current knowledge (**c**)

The opposing viewpoint, by contrast, grants individual organisms the capacity to adapt to variable environmental conditions. This also raises questions about the boundaries of species and the evolutionary relationships between species, as well as their grouping into superordinate units. The first viewpoint seemingly derived support from the decoding of DNA, which was cited in a comprehensive justification by one of its most prominent representatives—the evolutionary biologist Ernst Mayr (1904–2005) (Mayr 1984). In that context, the second viewpoint—founded by the biologist Willi Hennig (1913–1976)—was extensively questioned (Hennig 1966). The cladistic approach of the microbiologist Carl Woese (1928–2012), which led to a radical change in the classification of organisms (Fig. 2.1b), was also rejected (Woese 1987, 1998; Mayr 1998; Ho et al. 2013).

A new classification proposed by Woese and Fox (1977) was based on structural comparisons of ribosomal RNA (rRNA) from a total of 13 organisms from three major groups (domains). It was not the cell's own gene sequences in the DNA or morphological features that were used, but the "regulatory information" of tiny cell organelles (ribosomes). Through them, the genetic information of the cell is "translated" into the synthesis of proteins. To perform this function, the tiny organelles[2] are found in large numbers in all cells. The method employed was explained in detail in a subsequent publication (Olsen and Woese 1993). The main reasons underlying the method are the greater diversity of characteristics at the genetic level compared to the phenotype, the more precise detection of genetic characteristics and the identification of characteristics that cannot be detected phenotypically. The rationale cited for the specific use of ribosomal RNA is based on its

---

[2] The diameter is about one four hundred thousandth of a millimeter.

ubiquitous occurrence in cells, sufficient variability in relation to evolutionary developments, and functional location between inherent genetic traits and environmental factors.

Analyses using this method do not identify an explicit "root"—the three distinct groups
of organisms (domains) are related at the same level. Two groups of organisms (Eubacteria
and Archaebacteria) include apparently "primitive" unicellular organisms. All animals and
plants in Ernst Haeckel's representation, on the other hand, are grouped—together with
the fungi—in the Eukaryotes.[3] Woese's representation thus deviates from the prevailing
view of evolutionary processes. The usual picture of deterministic evolution from "simple" unicellular organisms to "complex" multicellular organisms is here reinterpreted as a
multidimensional, non-directional process. In archaeans and bacteria—as well as in
eukaryotes—diverse genetic networks are found without the framework of a phylogenetic
tree. This in fact led various authors to present antagonistic arguments against the classical
paradigm of evolution (McBreen and Lockhart 2006; Doolittle and Bapteste 2007). Quite
apart from such arguments it has been discussed whether instead of three domains only
two might be justified. Such discussions were triggered by increasing evidence of features
shared by cells of archaeans (Asgardarchaeota) and eukaryotes. Based on the—presumed—temporal sequence in the origin of archaeans and eukaryotes, the latter accordingly lose their status as a separate domain. Only the two domains Archaea and Bacteria
would remain (Williams et al. 2019). This ignores the fact that characteristics of bacteria
and archaeans are identifiable in eukaryote cells. Emergence of these properties was made
possible by horizontal gene transfer (HGT), which is mainly due to viruses (see Sect.
5.2.3). (Fig. 2.1c). In the meantime, it has also been demonstrated that even the properties
of ribosomes can be exchanged between bacteria and eukaryotes by viruses (Colussi et al.
2015). Despite the widely demonstrated capacities of viruses to cross domain boundaries,
definition of a fourth domain—at least for giant viruses (Nucleocytoplasmic Large
Viruses—NCLDV)—is also being considered (Boyer et al. 2010).

Such arguments can easily be questioned on the basis of classical empirical analyses of
evolutionary processes. After all, evolutionary lineages in the framework of phylogenetic
trees can be constructed from chronological sequences of fossils of plants and animals. In
an analogous fashion, imitation of female insects by plant flowers and resulting copulation
by male insects do not result in crosses between insects and plants. The only advantages
that ensue are for plant reproduction thanks to more effective pollination of flowers (Ellis
and Johnson 2010). After all, the diversity of genetically modified plants and animals
shows how easily genes can be transferred between different life forms. The technological
potential of such modifications—including in humans—often extends far outside ethically
and socially acceptable limits (Giraldo and Montoliu 2001; Niemann and Kues 2007;
Sanchis 2011; Ahmad et al. 2012; Bagle et al. 2012; Chandler and Sanchez 2012; Napier
et al. 2015; Abiri et al. 2016; Herrmann and Jayne 2019; Mbah et al. 2020).

---

[3] Eukaryotes should be grouped within Archaebacteria according to the opinion of some scientists
(Williams et al. 2013).

These examples confirm the possibilities that exist for genetic networking between different organisms, but also reveal the huge gap between evolutionary processes in the laboratory and in open ecosystems. In the first context, genetic modifications are directed at a specific performance trait of target organisms and occur through direct intervention in genetic structures. Morphological and physiological barriers to reproduction are thereby circumvented. Organisms created in this way can usually survive over the long term only under anthropogenically maintained conditions, in extreme cases in protected laboratory environments and in the simplest case under controlled cultivation. In open ecosystems evolutionary processes take place under dynamic conditions involving random interactions between a wide variety of organisms. Under such conditions, longer-term evolution depends on adaptations of a large number of different traits to specific environmental conditions. With increasing specialization—forced enhancement of a single capacity at the expense of all others—the risk of detriment increases both for the species concerned and for the systems affected by it. The effects of genetically modified organisms in open systems, which cannot be properly assessed, also constitute the core problem of their release.

Functional relationships within ecosystems can perhaps be illustrated with the example of music. Because the tonal possibilities of any individual instrument are limited, it can never achieve the sonority of an orchestra. In a simplified analogy, scores represent the energetic and material framework conditions of ecosystems, while conducting stands for the different dynamic expressions of abiotic parameters. In a simplified manner, the latter can be characterized by the cycles of solar irradiation and their modification by atmospheric processes. The more extensive an orchestra's instrumentation and the more effectively the orchestra members are attuned to each other, the better it is able to meet the demands of the score and of conducting. Variations in interpretation arise over the short term from conductors, and over the medium term from age-related changes in orchestra membership. Well-rehearsed orchestras fill vacancies only with highly qualified newcomers. Any initially unrecognized weaknesses of newcomers are detected quite quickly and such orchestra members are replaced by more qualified musicians. Understaffed and inadequately qualified orchestras are scarcely capable of such corrections. As new additions are made, they can drift in completely different directions or end up in cacophony. In the long run—in analogy to large-scale abiotic changes—cultural developments modify the selection of preferred scores. To be clear: In this analogy we humans are also only orchestra members and never conductors!

Whereas a focus on multicellular organisms was still understandable in Darwin's time because of the limited possibilities for studying microscopically small prokaryotes (Eubacteria and Archaebacteria), it has lost all justification today. Advanced development of methods for investigation now permits insights into details of the morphology and physiology of the smallest cells down to single atoms (Celler et al. 2013). It is becoming increasingly clear that the purely morphological classification of organisms into species provides an inadequate picture of the true extent of biological diversity. Even in the case of very large and conspicuous animals—for example orangutans or Chinese giant salamanders—new species were detectable only through genetic analyses (Stokstad 2017; Yan et al. 2018). The never-ending appearance of publications announcing newly "discovered" species is therefore not at all surprising. In scientific terms, the phenomenon of so-called cryptic species reveals that evolutionary changes do not necessarily manifest themselves in morphological changes (Fernandez et al. 2006; Šlapeta et al. 2006; Amato et al. 2007; Trontelj and Fišer 2009; Poulin 2010; Lumbsch and Leavitt 2011; Funk et al. 2012; Hawksworth 2012; Carstens and Satler 2013). In social terms, reports regarding "newly discovered species" tend to obscure the genuinely dramatic losses of biodiversity brought about by rapid expansion and intensification of anthropogenic (human) interventions (Waldron et al. 2017).

With genetic detection methods, evidence of prokaryotes can be found even in supposedly hostile localities—for instance, in hot springs or inside rock. Despite the diversity of life forms discovered so far, it is estimated that only about 0.1 to 1 percent of all microorganisms have as yet been studied in laboratories (Solden et al. 2016). Investigation of DNA sequences found freely in the environment using metagenomic methods permits detection of gene sets of non-culturable microorganisms. According to the results available to date, organisms would again have to be grouped into two superordinate groups—with bacteria separated from a common group consisting of eukaryotes and archaeans (Williams et al. 2013; Hug et al. 2016). Regardless, prokaryotes have an evolutionary history about twice as long as that for eukaryotes (Dodd et al. 2017). Alternative hypotheses question the chronological sequence and propose renaming prokaryotes to akaryotes. Eukaryotes and akaryotes would hence be placed on an equal footing (Penny et al. 2014; Harish and Kurland 2017). This raises the legitimate question as to whether these are no more than "marginal phenomena" of evolutionary processes?

It is remarkable, however, that in most treatises on evolutionary theory organisms are considered detached from all environmental influences. Even in the most recent works on evolution, there are only minimal references to the substantial requirements of living organisms and the inevitably associated interactions with their environment. Aspects of metabolism and energy balance or the interactions between different organisms are largely left to the disciplines of physiology or ecology. Exceptions to this are global upheavals of abiotic conditions, which have manifested themselves paleontologically in the disappearance of large numbers of organisms—so-called mass extinctions. Inevitably, systemic interrelationships are lost when evolutionary processes are considered. As a result, essential foundations for investigating the significance of evolutionary processes for human society are absent.

Accordingly, in this book the attempt is made to embrace functional relationships in evolution and ultimately to examine their significance for society. No claim can be made to completeness, because on the one hand the necessary foundations are scattered over many disciplines and on the other many processes remain insufficiently researched. It is hoped, however, that this approach will open up new perspectives and stimulate new areas of investigation.

References

Abiri R, Valdiani A, Maziah M, Shaharuddin NA, Sahebi M et al (2016) A critical review of the concept of transgenic plants: insights into pharmaceutical biotechnology and molecular farming. Curr Issues Mol Biol 18:21–42

Ahmad P, Ashraf M, Younis M, Hu X, Kumar A et al (2012) Role of transgenic plants in agriculture and biopharming. Biotechnol Adv 30(3):524–640

Amato A, Kooistra WHCF, Ghiron JHL, Mann DG, Pröschold T, Montresor M (2007) Reproductive isolation among sympatric cryptic species in marine diatoms. Protist 158(2):193–207

Bagle TT, Kunkulol RR, Baig MS, More SY (2012) Transgenic animals and their application in medicine. Int J Med Res Health Sci 2(1):107–116

Boyer M, Madoui M-A, Gimenez G, La Scola B, Raoult D (2010) Phylogenetic and phyletic studies of informational genes in genomes highlight existence of a 4 domain of life including giant viruses. PLoS One 5(12):e15530

Carstens BC, Satler JD (2013) The carnivorous plant described as *Sarracenia alata* contains two cryptic species. Biol J Linn Soc 109:737–746

Celler K, Koning RI, Koster AJ, van Wezel GP (2013) Multidimensional view of bacterial cytoskeleton. J Bacteriol 195(8):1627–1636

Chandler SF, Sanchez C (2012) Genetic modification; the development of transgenic ornamental plant varieties. Plant Biotechnol J 10:891–903

Colussi TM, Costantino DA, Zhu J, Donohue JP, Korostelev AA et al (2015) Initiation of translation in bacteria by a structured eukaryotic IRES RNA. Nature 519:110–113

Dodd MS, Papineau D, Grenne T, Slack JF, Rittner M et al (2017) Evidence for early life in Earth's oldest hydrothermal vent precipitates. Nature 543:60–64

Doolittle WF, Bapteste E (2007) Pattern pluralism and the tree of life hypothesis. PNAS 104(7):2043–2049

Ellis AG, Johnson SD (2010) Floral mimicry enhances pollen export: the evolution of pollination by sexual deceit outside of the orchidaceae. Am Nat 176(5):E143–E151

Fernandez CC, Shevock JR, Glazer AN, Thompson JN (2006) Cryptic species within the cosmopolitan desiccation-tolerant moss *Grimmia laevigata*. PNAS 103(3):637–642

Funk WC, Caminer M, Ron SR (2012) High levels of cryptic species diversity uncovered in Amazonian frogs. Proc R Soc B 279:1806–1814

Giraldo P, Montoliu L (2001) Size matters: use of YACs, BaCs and PACs in transgenic animals. Transgenic Res 10:83–103

Glaubrecht M (2013) Am Ende des Archipels. Alfred Russel Wallace, Galiani Berlin

Gording MD (2012) How Lysenkoism Became Pseudoscience: Dobzhansky to Velikovsky. J Hist Biol 45:443–468

Harish A, Kurland CG (2017) Akaryotes and Eukaryotes are independent descendants of a universal common ancestor. Biochimie 138:168–183

Hawksworth DL (2012) Global species numbers of fungi: are tropical studies and molecular approaches contributing to a more robust estimate. Biodivers Conserv 21:2425–2433

Hennig W (1966) Phylogenetic systematics. University of Illinois Press, Urbana

Herrmann K, Jayne K (eds) (2019) Animal experimentation: working towards a paradigm change. Brill, Leiden

Ho C-C, Lau SKP, Woo PCY (2013) Romance of the three domains: how cladistics transformed the classification of cellular organisms. Protein Cell 4(9):664–676

Hug LA, Baker BJ, Anantharaman K, Brown CT, Probst AJ et al (2016) A new view of the tree of life. Nat Microb Lett. https://doi.org/10.1038/nmicrobiol.2016.48

Kozubek J (2016) Modern prometheus. Cambridge University Press, Cambridge

Lumbsch HT, Leavitt SD (2011) Goodbye morphology? A paradigm shift in the delimitation of species in lichenized fungi. Fungal Divers. https://doi.org/10.1007/s13225-011-0123-z

Mayr E (1984) Die Entwicklung der biologischen Gedankenwelt. Springer, Berlin

Mayr E (1998) Two empires of three? PNAS 95:9729–9723

Mbah DA, Tawah CL, Guewo-Fokeng M (2020) Genetic modification of animals: potential benefits and concerns. J Cameroon Acad Sci 15(3):163–174

McBreen K, Lockhart PL (2006) Reconstructing reticulate evolutionary histories of plants. Trends Plant Sci 11(8):398–404

Napier JA, Usher S, Haslam RP, Ruiz-Lopez N, Sayanova O (2015) Transgenic plants as a sustainable, terrestrial source of fish oils. Eur J Lipid Sci 117:1317–1324

Niemann H, Kues WA (2007) Transgenic farm animals: an update. Reprod Fertil Dev 19:762–770

Olsen GJ, Woese CR (1993) Ribosomal RNA: a key to phylogeny. FASEB J 7:113–123

Penny D, Collins LJ, Daly TK, Cox SJ (2014) The relative ages of eukaryotes and akaryotes. J Mol Evol 79:228–239

Poulin R (2010) Uneven distribution of cryptic diversity among higher taxa of parasitic worms. Biol Lett. https://doi.org/10.1098/rsbl.2010.0640

Roll-Hansen N (2006) The Lysenko effect: undermining the autonomy of science. Endeavour 29(4):143–147

Sanchis V (2011) From microbial sprays to insect-resistant transgenic plants: history of the biospesticide *Bacillus thuringiensis*. A review. Agron Sustain Dev 31:217–231

Šlapeta J, López-García P, Moreira D (2006) Global dispersal and ancient cryptic species in the smallest marine eukaryotes. Mol Biol Evol 23(1):23–29

Solden L, Lloyd K, Wrighton K (2016) The bright side of microbial dark matter: lessons learned from the uncultivated majority. Curr Opin Microb 31:217–226

Stokstad E (2017) New ape found, sparking fears for its survival. Science 358:572–573

Trontelj P, Fišer C (2009) Perspectives: cryptic species diversity should not be trivialised. Syst Biodivers 7(1):1–3

Waldron A, Miller DC, Redding D, Mooers A, Kuhn TS et al (2017) Reductions in global biodiversity loss predicted from conservation spending. Nature 551:364–367

Williams TA, Foster PG, Cox CJ, Embley TM (2013) An archaeal origin of eukaryotes supports only two primary domains of life. Nature 504:231–236

Williams TA, Cox CJ, Foster PG, Szöllösi GJ, Embley M (2019) Phylogenomics provides robust support for a two-domains tree of life. Nat Ecol Evol. https://doi.org/10.1038/s41559-019-1040-x

Woese CE (1987) Bacterial evolution. Microbiol Rev 51(2):221–271

Woese CR (1998) Default taxonomy: Ernst Mayr's view of the microbial world. PNAS 95:11043–11046

Woese CR, Fox GE (1977) Phylogenetic structure of the prokaryotic domain: the primary kingdoms. PNAS 74(11):5088–5090

Yan F, Lü J, Zhang B, Yuan Z, Zhao H et al (2018) The Chinese giant salamander exemplifies the hidden extinction of cryptic species. Curr Biol 28:R581–R598

Where Can We Find Approaches to Understanding Evolutionary Processes?

3.1   The Fuzzy Boundaries of Life

Answers to this question will vary depending on the scientific discipline. In their work, bacteriologists confronted with the increasing resistance of pathogens to drugs tackle processes completely different from those confronting palaeontologists, who compare fragments of fossils from various sites and time periods. Virologists, who observe the emergence of virus strains every year in order to prepare suitable vaccines for influenza vaccination in a timely fashion, also set themselves different priorities.

Biologically trained people will sense a drift in topic in the last sentence, because viruses do not possess all characteristics of living organisms. Nevertheless, they influence the evolutionary processes of living organisms to a great extent. Their influences also differ from those of an inanimate nature. At this point, an extensive discussion of the question of the characteristics of life might begin, ultimately leading to the statement that life can be defined only by a combination of different characteristics (Moreira and López-Garcia 2009). The following examples are intended to illustrate this statement:

The distinction between life and non-life becomes challenging when it comes to the question of when vital organs may be removed from a human body. Only functional—in the broadest sense living—organs are suitable for organ donation, but their donor must by definition be dead. This example alone reveals the fluent transitions between life and non-life, which become more diverse with increasing consideration of other life forms (Cleland and Chyba 2002; Bitbol and Luisi 2004; Ruiz-Mirazo et al. 2004; Benner 2010; Forterre 2010; Gayon 2010; Tessera 2011). Use of simple characteristics—for example, active metabolism—does not suffice for demarcation. In plant seeds or dormant stages of bacteria, metabolism is extremely reduced, but they are nevertheless able to activate it under suitable conditions. Similar difficulties are created by isolated reliance on the criterion of

© The Author(s), under exclusive license to Springer-Verlag GmbH, DE, part of
Springer Nature 2024
M. Knoflacher, *Relativity of Evolution*,
https://doi.org/10.1007/978-3-662-69423-7_3

self-replication, which permits development of new identical forms. Viruses also have this capacity for self-replication, but no metabolic activity can be detected in them. In certain viruses, however, genes for forming structures that can convert energy can be found.

Scientific methods quickly reach their performance limits when confronted with the multidimensional natures of life phenomena, as neither clear causal relationships nor directed developments can be demonstrated. Serious engagements with this discomfiting fact are usually avoided by reducing the number of parameters considered (Fig. 3.1). Only partial aspects of life can be grasped from the perspectives of any individual scientific discipline, and they are difficult to unify into a common picture. In general, there is nothing wrong with this reductionist approach, as long as it is not accompanied by a claim to comprehensive explanation. Otherwise, life very quickly becomes equated with the confines of the respective discipline, be it genetics, physics or chemistry (Pross 2011). As comforting as such views may seem, they do not contribute to expanding knowledge about the phenomenon of life and its evolution. This is not at all the same as claiming that the laws of physics and chemistry are suspended in living systems. Rather, the question to be examined is whether and which specific rules of the game operate in biological systems. This opens up novel questions that have generally been ignored in previous approaches to

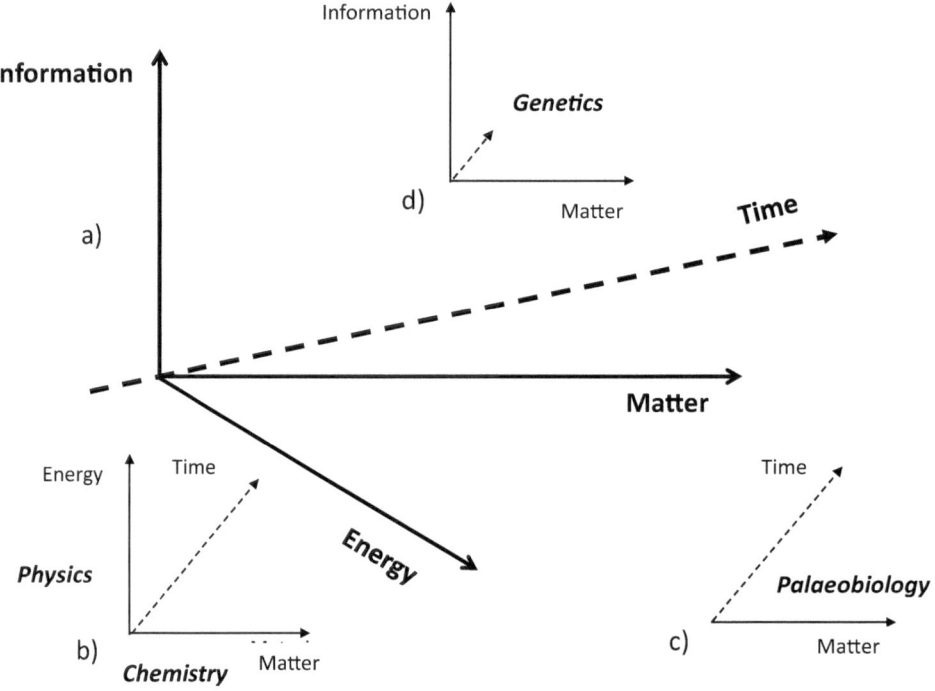

Fig. 3.1 (**a**) Schematic diagram of at least four characteristic dimensions of life (matter, energy, information, time) and their reduction from viewpoints of the scientific disciplines cited as examples, (**b**) chemistry/physics, (**c**) palaeobiology, (**d**) genetics

evolutionary research. The primary focus of investigation was and still is the recording and description of structures—ranging from morphological to molecular dimensions. However, living organisms are dynamic systems with specific functional properties and variable system states. Given that interpretation of functional properties from structural characteristics quickly reaches its limits (Schwenk 2000), it is all the more difficult to derive dynamic life processes from them. Precisely such knowledge is, however, necessary for an improved understanding of the differences between biological and abiotic evolutionary processes. Yet that can be gained from studies of living organisms alone. In combination with carefully collected observational data, formal methods—implemented, for instance, in models—can also contribute to increasing knowledge. Without qualitatively sufficient data, even the most elaborate computer models—including self-learning algorithms, so-called artificial intelligence—remain just as speculative as non-formal methods (Hut et al. 2006; Tegmark 2007; Chaitin 2010; Chen 2019).

3.2   Beginning of Life or Beginning of Evolution?

While defining life unambiguously it already challenging, it is fundamentally impossible to portray the origin of life with certainty. This obstacle is immediately followed by the—equally unanswerable—question regarding the initial steps of independent conversion of inorganic matter in the direction of biological evolution. Nonetheless, in recent decades much attention has been devoted to this question under the heading of astrobiology (Wong and Laczano 2009). The impetus for this has been society's aspirations to conquer new habitats in outer space, combined with a search for extraterrestrial life. Relevant investigations and discussions are aimed at establishing whether the impetus for biological evolution can be found on Earth alone or also comes from other planets, in tandem with a quest to identify planets with suitable living conditions (Hanslmeier 2014). Fans of science fiction may well be disappointed that this latter aspect is not addressed here. As regards the first issue, the suspense-creating factor is intimately linked to the question of when the evolutionary paths of inanimate and animate matter began to diverge.

Although the questions are fundamentally unanswerable, hypothetical levels of potential separation processes are outlined in Fig. 3.1. The diagrammatic presentation is certainly not intended as an attempt at explanation, but rather as an aid to orientation for discussion of the extensive literature on this topic. The presentation is also designed to show that, if the first impulses[1] were already evident during the formation of the universe some 14 billion years ago, vast periods of time might lie between a possible beginning of (biological) evolution and the beginning of life. This possibility makes it futile to discuss a possible importation of seeds of life—for example through meteorites—because the potential for the evolution of life would have to be present throughout the entire universe. But such a development would be possible only under certain conditions—for example,

---

[1] Here called "Quantum physical differentiation".

during the earliest history of the Earth. Before people with biological interests rush to dismiss such hypotheses as nonsense, it should be noted that a well-known property of organisms and organic molecules is already found at the level of quantum physical processes. Among the four fundamental forces of nature (Gravitation, Electromagnetic Force, Strong Interaction, Weak Interaction), under certain conditions, Weak Interaction causes an unequal distribution of the orbital angular momentum of the radiated electrons. Specifically, the proportion of "left-rotated" electrons increases to around 70 percent (Bleck-Neuhaus 2013). Such mirror-image asymmetries—also known as chirality (handedness)—determine functional properties of the molecules involved in biological processes (Berg et al. 2013; George and Korolev 2018). Taken in isolation, this comment has little persuasive power; after all, the first process takes place near absolute zero and during nuclear decay, while processes of the second type take place in biological cells at higher temperatures between around 270 Kelvin (K) and 370 K. The frequently used phrase "in an aqueous environment" has been deliberately avoided here, as many molecules also have water-repellent (hydrophobic) regions. However, the increasing evidence for quantum physical processes in organisms surely gives pause for thought (Davies 2004; Cline 2005; Hunter 2006; Arndt et al. 2009; Fleming et al. 2011; McClintock 2011; Jarvis 2012; Lambert et al. 2013; Trixler 2013; Mohseni et al. 2014; Mender 2015). Moreover, quantum physical properties would provide more conclusive explanations for various biological processes than is currently possible. Because of the scanty evidence, such highly speculative considerations must be deferred for future discussion.

The second group of scenarios for the origin of life—here called "thermodynamic differentiation"—attracts the greatest attention in scientific and social discussions. Formation of organic molecules and the framework conditions needed for this are illuminated from a wide variety of perspectives. The spectrum ranges from mathematical and physical models to chemical reaction chains (Eigen and Schuster 1978; Ebeling et al. 1990; Haken 1990; Kauffman 1993; Segré et al. 2000; Scheuring et al. 2003; Miller and Cleaves 2007; Schuster 2007; Nowak and Ohtsuki 2008; Moreno and Ruiz-Mirazo 2009; Horowitz et al. 2010; Lukeš et al. 2011; Derr et al. 2012). Accordingly, there is a wide variety of concepts regarding the sequence of formation of complex organic molecules, for example proteins, ribonucleic acids (RNA), deoxyribonucleic acids (DNA) or lipids (Lazcano 1986; Joyce 1989, 2002; Cech 1993; Hager et al. 1996; Poole et al. 1999; Segré and Lancet 2000; Segré et al. 2001; Orgel 2004, 2008; Staub et al. 2004; Caetano-Anollés and Caetano-Anollés 2005, 2013; Forterre 2005; Koonin 2006; Wang et al. 2006, 2011; Yu and Thorne 2006; Di Giulio 2008; Shabalina and Koonin 2008; Takeuchi et al. 2008; Caetano-Anollés et al. 2009, 2011; Koonin and Novozhilov 2009; Penny et al. 2009; Wang and Caetano-Anollés 2009; Conway Morris 2010; Koonin and Wolf 2010; Lane and Darst 2010; Yarus 2010; Benner et al. 2012; Bernhardt 2012; Caetano-Anollés and Nasir 2012; Kim and Caetano-Anollés 2012; Mittenthal et al. 2012; Noller 2012; Robertson and Joyce 2012; Nasir et al. 2014; Becker et al. 2018). Because of the absence of fossil evidence, it is also impossible to establish definitively whether viruses originated by reduction from cells or directly from molecular building blocks (indicated by question marks in Fig. 3.2) (Moya et al. 2000;

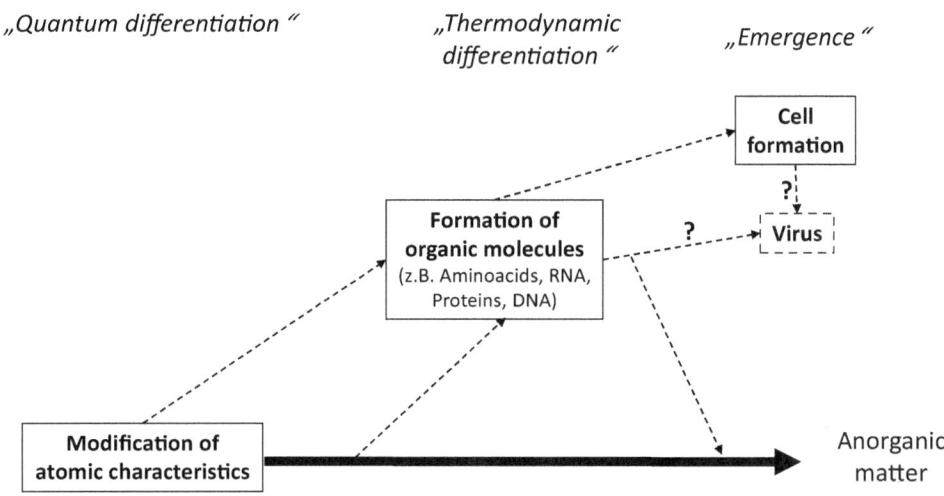

Fig. 3.2 Schematic representation of hypothetical starting points for biological evolution. (Explanations in the text)

Claverie 2006; Hedges et al. 2006; Koonin et al. 2006; Dimmock et al. 2007; Claverie and Ogata 2009; Holmes 2011; Nasir and Caetano-Anollés 2015). The variety of scenarios is further augmented by differing assumptions regarding the conditions of origin, for example in outer space, in the primordial oceans, in the vicinity of hydrothermal sources or in ice (Cronin and Pizzarello 1983; Canfield 1998; Line 2002; Miyakawa et al. 2002; Russell and Martin 2004; Trinks et al. 2005; Martin and Russell 2007; Pudritz et al. 2007; Dobretsov et al. 2008; Pizzarello et al. 2008; Mulkidjanian 2009; Mulkidjanian and Galperin 2009; Atmanspacher 2004; Gargaud et al. 2011; Nitschke and Russell 2013; Becker et al. 2016; Priye et al. 2017). However, presumed extremely high temperatures in the early phase of the Earth's formation limit the possible time available for continuous evolution of organisms to around 3.5 to 4 billion years (Stacey and Davis 2008).

On closer inspection, a relatively uniform and narrow basic issue is found in the vast majority of scenarios—the emergence of complex structures from simple elements. In a few studies, the incompleteness of explanations is acknowledged or the dynamics and self-structuring of inorganic matter or their interactions with the organic world are taken into account (Eigen and Schuster 1978; Prigogine 1985; Orgel 2000; Santosh 2010; Deamer 2011; Russell et al. 2013). Enthusiasm regarding the proof of the initiation of autonomously operating reaction chains overshadows their occurrence in abiotic realms, for example in natural nuclear reactors of early Earth history (Evins et al. 2005). In living organisms, however, such processes run in coordination with the associated demand and do not proceed without inhibition until the reaction partners are exhausted. For organisms, self-structuring from chaos also does not end in the solidified order of mineral crystals, but in a dynamic management of internal order. Living organisms are in kinetic equilibrium far removed from thermodynamic equilibrium (Pross 2003, 2004, 2005, 2009, 2011; Pross and Khodorkovsky 2004; Williams and Fraústo da Silva 2006; Pascal 2012; Pross and Pascal 2013; Pascal et al. 2013). For this, organisms need an independent energy balance and the capacity to process information.

Linear explanations for formation of functional cells end in an extended chicken-and-egg problem, usually circumvented by focussing on sub-aspects (Russell et al. 1994; Woese 2002; Becerra et al. 2007; Martin 2012; Koonin and Wolf 2009). If, for example, primary evolution of regulatory functions is assumed, the question arises as to how everything could function without an energy supply. One attempt to explain self-structuring at the beginning of evolution is based on metabolic processes in hyperthermal prokaryotes that operate without enzymatic regulation (Ralser 2018). If, on the other hand, energy supply is granted primacy, the question arises as to the regulation of energy conversion. Equally difficult is clarification of the question regarding the systemic state in which molecular assemblies were enclosed by cell membranes. It is no coincidence that, from a creationist perspective, the unresolved questions offer ideal starting points for critiques of the theory of evolution (Behe 2006). What is omitted here, however, is a comprehensive discussion of biological complexity. Although hypothetical evolutionary trajectories leading to complex life forms under natural selection can be constructed at a macroscopic level, such approaches fail on a molecular level. After all, the most complex forms of protein molecules, for example in ATP synthase for energy conversion (see Sect. 4.5), are already found in Archaea and Bacteria. Stated simply, this complex molecule can be compared to an electric motor in which (electrical) energy is converted into mechanical energy via a rotor. In ATP synthase, however, mechanical energy is used for the regeneration of cellular energy potentials. Archaea and Bacteria, with constructively similar molecules, employ mechanical energy for the rotary drive of their filamentous flagella for locomotion (see Sect. 5.1.5). Since ATP synthases are fundamentally important for living cells, this example shows that complex structures had already been developed during the early stages of biological evolution.

But biological evolutionary processes cannot have occurred in isolation in individual phylogenetic lineages of organisms. This can be explained using the example of model calculations for evolution of the structural and functional diversity of proteins. In the detailed presentation in the original study (Caetano-Anollés et al. 2007), it is assumed that the first prokaryotes appeared after about three quarters of evolutionary time had elapsed (Fig. 3.3b). If the age of the Earth is taken as the basis for the total period, this would have occurred around 1.15 billion years ago. According to more recent findings, however, the first traces of microbial processes can be traced as far back as some 3.7 billion years ago (Nutman et al. 2016). If, on the other hand, the first signals of independent protein differentiation are interpreted in connection with cell formation (Fig. 3.3a), this would result in better agreement with the evidence of the first traces of life. In connection with evidence of the first traces of life, regardless of the divergences cited as examples, the question

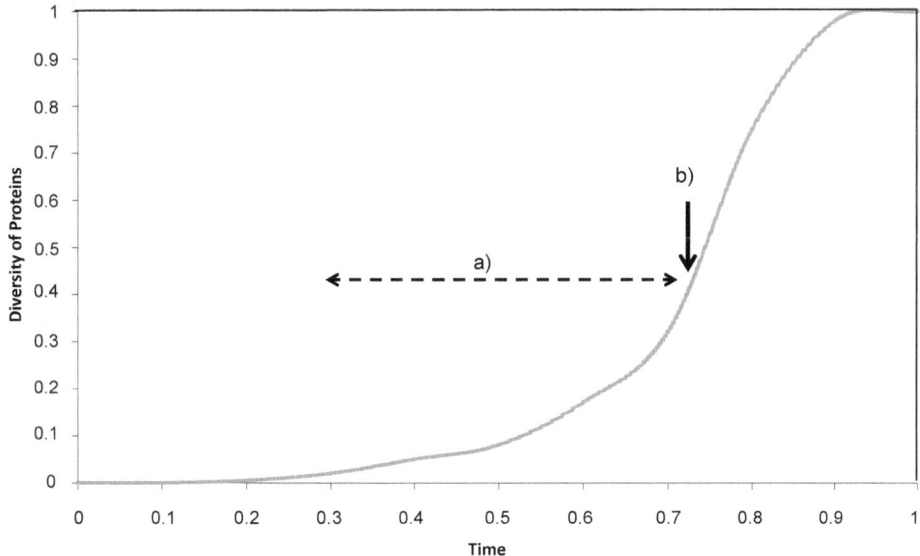

Fig. 3.3 Example for discussion of interpretation problems for prebiotic evolution in the developmental model of protein diversity advocated by Caetano-Anollés et al. (2007). (**a**) Indications of first independent developmental lines of protein diversity. (**b**) Beginning of cellular life according to representations in the original study

remains open as to how complex organic life was able to evolve from inorganic matter in the relatively short period of around 1 billion years. Here, a concluding thought experiment is provided, in which the absolute length of the time axis in Fig. 3.3 is estimated. As an assumption, point (b) serves as the beginning of life and the currently estimated time between its occurrence and the present is around 3.7 billion years. With these assumptions, the differentiation of proteins would have begun around 14 billion years ago—approximately at the same time as the beginning of the universe (Hanslmeier 2014).

The example discussed is based on the widespread paradigm of directed biological evolution. Starting from current life forms or gene sets, a retrospective search is made for possible paths of their emergence (Boussau and Gouy 2012). From the large number of different possible changes, variants with the smallest number of steps (parsimony) are usually taken to account for evolutionary relationships. In this way, evolutionary processes are inevitably assumed to have proceeded in a directional manner into the life forms that currently exist. In view of the huge number of extinct life forms and unpredictable changes in abiotic framework conditions, such an evolutionary principle is extremely unlikely. Far more likely—especially in the case of far-reaching changes in living conditions—are flexible, adaptive search processes. In real terms, it is above all the still existing life forms with their genetic make-up that are available to us for study. Through the processes of horizontal gene transfer and the multiplication of gene sets described below, a wide variety of "experimental prototypes" can develop and "test" their survivability under prevailing conditions. This does not necessarily mean that organisms with the smallest number of modifications of their gene sets will emerge as survivors, but rather those with the functionally best adapted gene sets. This may well include life forms with more extensive modifications both of their gene sets and of their morphological characteristics. As theoretical considerations also show, biological evolutionary processes were significantly influenced by functional interactions between organisms (Szappanos et al. 2016). The diversity of interactions—especially among microorganisms—extends far beyond simple food chains and can be described only incompletely in this monograph.

References

Arndt M, Juffmann T, Vedral V (2009) Quantum physics meets biology. HFSP J 3(6):386–400

Atmanspacher H (2004) Quantum theory and consciousness: an overview with selected examples. Discrete Dyn Nat Soc 1:51–73

Becerra A, Delaye L, Islas S, Lazcano A (2007) The very early stages of biological evolution and the nature of the last common ancestor of the three major cell domains. Annu Rev Ecol Syst 38:161–179

Becker S, Schneider C, Okamura H, Crisp A, Amatov T (2018) Wet-dry cycles enable the parallel origin of canonical and non-canonical nucleosides by continuous synthesis. Nat Commun 9:163

Becker S, Thoma I, Deutsch A, Gehrke T, Mayer P et al (2016) A high-yielding, strictly regioselective prebiotic purine nucleoside formation pathway. Science 352(6287):833–836

Behe MJ (2006) Darwins Black Box. Free Press, New York

Benner SA (2010) Defining life. Astrobiology 10(10):1021–1030

Benner SA, Kim H-J, Yang Z (2012) Setting the stage: the history, chemistry, and geobiology behind RNA. Cold Spring Harb Perspect Biol 4:a003541

Berg JM, Tymoczko JL, Stryer L (2013) Biochemie. Springer Spektrum, Berlin

Bernhardt HS (2012) The RNA world hypothesis: the worst theory of the early evolution of life (except for all the others). Biol Direct 7:23

Bitbol M, Luisi PL (2004) Autopoiesis with or without cognition: defining life at its edge. J R Soc Interface 1:99–107

Bleck-Neuhaus J (2013) Elementare Teilchen. Springer Spektrum, Berlin

Boussau B, Gouy M (2012) What genomes have to say about the evolution of the Earth. Gondwana Res 21(2–3):483–494

Caetano-Anollés G, Caetano-Anollés D (2005) Universal sharing patterns in proteomes and evolution of protein fold architecture and life. J Mol Evol 60:484–498

Caetano-Anollés K, Caetano-Anollés G (2013) Structural phylogenomics reveals gradual evolutionary replacement of abiotic chemistries by protein enzymes in purine metabolism. PLoS One 8(3:e59300

Caetano-Anollés G, Kim HS, Mittenthal JE (2007) The origin of modern metabolic networks inferred from phylogenomic analysis of protein architecture. PNAS 104(22):9358–9363

Caetano-Anollés D, Kim KM, Mittenthal JE, Caetano-Anollés G (2011) Proteome evolution and the metabolic origins of translation and cellular life. J Mol Evol 72:14–33

Caetano-Anollés G, Nasir A (2012) Benefits of using molecular structure and abundance in phylogenomic analysis. Front Genet 3:172. https://doi.org/10.3389/fgene.2012.00172

Caetano-Anollés G, Wang M, Caetano-Anollés D, Mittenthal JE (2009) The origin, evolution and structure of the protein world. Biochem J 417:621–637

Canfield DE (1998) A new model for proterozoic ocean chemistry. Nature 396:450–453

Cech TR (1993) The efficiency and versatility of catalytic RNA: implications for the RNA world. Gene 135:33–36

Chaitin G (2010) To a mathematical theory of evolution and biological creativity. CDMTCS Research Report Series 291

Chen EK (2019): Time's arrow in a quantum universe: on the status of statistical mechanical probabilities. arXiv:1902.04564v1

Claverie J-M (2006) Viruses take center stage in cellular evolution. Genome Biology 7:110

Claverie J-M, Ogata H (2009) Ten good reasons not to exclude giruses from the evolutionary picture. Nat Rev Microb 7(8):615

Cleland CE, Chyba CF (2002) Defining life. Orig Life Evol Biosph 32:387–393

Cline DB (2005) On the physical origin of the homochirality of life. Eur Rev 13(2):49–59

Conway Morris S (2010) Evolution: like any other science it is predictable. Philos Trans R Soc B 365:133–145

Cronin JR, Pizzarello S (1983) Amino acids in meteorites. Adv Space Res 3(9):5–18

Davies PCW (2004) Does quantum mechanics play a non-trivial role in life? BioSystems 78:69–79

Deamer D (2011) First life. University of California Press, Berkeley

Derr J, Manapat ML, Rajamani S, Leu K, Xulvi-Brunet R et al (2012) Prebiotically plausible mechanisms increase compositional diversity of nucleic acid sequences. Nucleic Acids Res 2012:1–12

Di Giulio M (2008) An extension of the coevolution theory of the origin of the genetic code. Biol Direct 3:27

Dimmock NJ, Easton AJ, Leppard KN (2007) Introduction to modern virology. Blackwell Publishing, Malden

Dobretsov NL, Kolchanov NA, Suslov VV (2008) On important stages of geosphere and biosphere evolution. In: Dobretsov NL, Kolchanov NA, Rozanov A (eds) Biosphere origin and evolution. Springer

Ebeling W, Engel A, Feistel R (1990) Physik der Evolutionsprozesse. Akademie Verlag, Berlin

Eigen M, Schuster P (1978) The hypercylce. Naturwissenschaften 65:7–41

Evins LZ, Jensen KA, Ewing RC (2005) Uraninite recrystallization and Pb loss in the Oklo and Bangombé natural fission reactors, Gabon. Geochim Cosmochim Acta 69(6):1589–1606

Fleming GR, Scholes GD, Cheng Y-C (2011) Quantum effects in biology. Procedia Chem 3:38–57

Forterre P (2005) The two ages of the RNA world, and the transition to the DNA world: a story of viruses and cells. Biochimie 87:793–803

Forterre P (2010) Defining life: the virus viewpoint. Orig Life Evol Biosph 40(2):151–160

Gargaud M, López-García P, Martin H (2011) Origins and evolution of life. Cambridge University Press, Cambridge

Gayon J (2010) Defining life: synthesis and conclusions. Orig Life Evol Biosph 40:231–244

George AB, Korolev KS (2018) Chirality provides a direct fitness advantage and facilitates intermixing in cellular aggregates. PLoS Comput Biol 14(12):e1006645

Hager AJ, Pollard JD, Szostak JW (1996) Ribozymes: aiming at RNA replication and protein synthesis. Chem Biol 3:717–725

Haken H (1990) Synergetik. Springer, Berlin

Hanslmeier A (2014) Einführung in Astronomie und Astrophysik. Springer Spektrum, Berlin

Hedges SB, Battistuzzi FU, Blair JE (2006) Molecular timescale of evolution in the proterozoic. In: Xiao S, Kaufman AJ (eds) Neoproterozoic geobiology and paleobiology. Springer, New York, pp 199–229

Holmes EC (2011) What does virus evolution tell us about virus origin? J Virol 85(11):5247–5251

Horowitz ED, Engelhart AE, Chen MC, Quarles KA, Smith MW et al (2010) Intercalation as a means to suppress cyclization and promote polymerization of base-pairing oligonucleotides in a prebiotic world. PNAS 107(12):5288–5293

Hunter P (2006) A quantum leap in biology. EMBO Rep 7(10):971–974

Hut P, Alford M, Tegmark M (2006) On math, matter and mind. arXiv:physics/0510188v2

Jarvis PD (2012) Quantum physics meets biology: a case study. In: Finlayson TR (ed) Preserving the Humboldt tradition of scholarship in Australia. Australian Association of Humboldt Fellows, Melbourne

Joyce GF (1989) RNA evolution and the origins of life. Nature 338:217–224

Joyce GF (2002) The antiquity of RNA-based evolution. Nature 418:214–221

Kauffman SA (1993) The origins of order. Oxford University Press, Oxford

Kim KM, Caetano-Anollés G (2012) The evolutionary history of protein fold families and proteomes confirms that the archaeal ancestor is more ancient than the ancestors of other superkingdoms. BMC Evol Biol 12:13

Koonin EV (2006) Temporal order of evolution of DNA replication systems inferred by comparison of cellular and viral DNA polymerases. Biol Direct 1:39

Koonin EV, Novozhilov AS (2009) Origin and evolution of the genetic code: the universal enigma. IUBMB Life 61(2):99–111

Koonin EV, Senkevich TG, Dolja VV (2006) The ancient virus world and evolution of cells. Biol Direct 1:29

Koonin EV, Wolf YI (2009) The fundamental units, processes and patterns of evolution, and the tree of life conundrum. Biol Direct 4:33

Koonin EV, Wolf YI (2010) The common ancestry of life. Biol Direct 5:64

Lambert N, Chen Y-N, Cheng Y-C, Li C-M, Chen G-Y, Nori F (2013) Quantum biology. Nat Phys 9:10–18

Lane WJ, Darst SA (2010) Molecular evolution of multi-subunit RNA polymerases: sequence analysis. J Mol Biol 395(5):671

Lazcano A (1986) Prebiotic evolution and the origins of cells. Treb Soc Cat Biol 39:73–103

Line MA (2002) The enigma of the origin of life and its timing. Microbiology (Reading) 148:21–27

Lukeš J, Archibald JM, Keeling PJ, Doolittle WF, Gray MW (2011) How a neutral evolutionary ratchet can build cellular complexity. IUBMB Life 63(7):528–537

Martin WF (2012) Hydrogen, metals, bifurcating electrons, and proton gradients: the early evolution of biological energy conservation. FEBS Lett 586:485–493

Martin W, Russell MJ (2007) On the origin of biochemistry at an alkaline hydrothermal vent. Philos Trans R Soc B 362:1887–1925

McClintock PVE (2011) Quantum aspects of life. Contemp Phys 52(1):71–73

Mender D (2015) From quantum photosynthesis to the sentient brain. NeuroQuantology 13(4):420–425

Miller SL, Cleaves HJ (2007) Prebiotic chemistry on the primitive earth. In: Rigoutsos I, Stephanopoulos G (eds) Systems biology: Volume I: Genomics. Oxford University Press, Oxford

Mittenthal J, Caetano-Anollés D, Caetano-Anollés G (2012) Biphasic patterns of diversification and the emergence of modules. Front Genet 3:147

Miyakawa S, Yamanashi H, Kobayashi K, Cleaves HJ, Miller SL (2002) Prebiotic synthesis from CO atmospheres: implications for the origins of life. PNAS 99(23):14628–14631

Mohseni M, Omar Y, Engel GS, Plenio MB (2014) Quantum effects in biology. Cambridge University Press, Cambridge

Moreira D, López-Garcia P (2009) Ten reasons to exclude viruses from the tree of life. Nature 7:306–311

Moreno A, Ruiz-Mirazo K (2009) The problem of the emergence of functional diversity in prebiotic evolution. Biol Philos 24:585–605

Moya A, Elena SF, Bracho A, Miralles R, Barrio E (2000) The evolution of RNA viruses: a population genetics view. PNAS 97(13):6967–6973

Mulkidjanian AY (2009) On the origin of life in the Zinc world: 1. Photosynthesizing, porous edifices built of hydrothermally precipitated zinc sulphide as cradles of life on Earth. Biol Direct 4:26

Mulkidjanian AY, Galperin MY (2009) On the origin of life in the Zinc world: 2. Validation of the hypothesis on the photosynthesizing zinc sulphide edifices as cradles of life on Earth. Biol Direct 4:27

Nasir A, Caetano-Anollés G (2015) A phylogenomic data-driven exploration of viral origins and evolution. Sci Adv 1:e1500527

Nasir A, Kim KM, Caetano-Anollés G (2014) A phylogenomic census of molecular functions identifies modern thermophilic archaea as the most ancient form of cellular life. Archaea 2014:706468

Nitschke W, Russell MJ (2013) Beating the acetyl coenzyme A-pathway to the origin of life. Philos Trans R Soc B 368:20120258

Noller HF (2012) Evolution of protein synthesis from an RNA world. Cold Spring Harb Perspect Biol 4:a003681

Nowak MA, Ohtsuki H (2008) Prevolutionary dynamics and the origin of evolution. PNAS 105(39):14924–14927

Nutman AP, Bennett VC, Friend CRL, Van Kranendonk MJ, Chivas AR (2016) Rapid emergence of life shown by discovery of 3700-million-year-old microbial structures. Nature 537:535–538

Orgel LE (2000) Self-organizing biochemical cycles. PNAS 97(23):12503–12507

Orgel LE (2004) Prebiotic chemistry and the origin of the RNA world. Biochem Mol Biol 39:99–123

Orgel LE (2008) The implausibility of metabolic cycles on the prebiotic earth. PLoS Biol 6(1):e18

Pascal R (2012) Suitable energetic conditions for dynamic chemical complexity and the living state. J Syst Chem 3:3

Pascal R, Pross A, Sutherland JD (2013) Towards an evolutionary theory of the origin of life based on kinetics and thermodynamics. Open Biol 3:130156

Penny D, Hoeppner MP, Poole AM, Jeffares DC (2009): An overview of the introns-first theory. J Mol Evol. doi: https://doi.org/10.1007/s00239-009-9279-5.

Pizzarello S, Huang H, Alexandre MR (2008) Molecular asymmetry in extraterrestrial chemistry: insights from a pristine meteorite. PNAS 105(10):3700–3704

Poole A, Jeffares D, Penny D (1999) Early evolution: prokaryotes, the new kids on the block. BioEssay 21(10):880–889

Prigogine I (1985) Vom Sein zum Werden. Piper, Munich

Priye A, Yu Y, Hassam YA, Ugaz VM (2017) Synchronized chaotic targeting and acceleration of surface chemistry in prebiotic hydrothermal microenvironments. PNAS 114(6):1275–1280

Pross A (2003) The driving force for life's emergence: kinetic and thermodynamic considerations. J Theor Biol 220:393–400

Pross A (2004) Causation and the origins of life. Metabolism or replication first? Orig Life Evol Biosph 34:307–321

Pross A (2005) Stability in chemistry and biology: life as a kinetic state of matter. Pure Appl Chem 77(11):1905–1921

Pross A (2009) Seeking the chemical roots of Darwinism: bridging between chemistry and biology. Chem Eur J 15:8374–8381

Pross A (2011) Toward a general theory of evolution: extending Darwinian theory to inanimate matter. J Syst Chem 2:1

Pross A, Khodorkovsky V (2004) Extending the concept of kinetic stability: toward a paradigm for life. J Phys Org Chem 17:312–316

Pross A, Pascal R (2013) The origin of life: what we know, what we can know and what we will never know. Open Biol 3:120190

Pudritz R, Higgs P, Stone J (eds) (2007) Planetary systems and the origins of life. Cambridge University Press, Cambridge

Ralser M (2018) An appeal to magic? The discovery of a non-enzymatic metabolism and its role in the origins of life. Biochem J 475:2577–2592

Robertson MP, Joyce GF (2012) The origins of the RNA world. Cold Spring Harb Perspect Biol 4:a003608

Ruiz-Mirazo K, Pereto J, Moreno A (2004) A universal definition of life: autonomy and open-ended evolution. Orig Life Evol Biosph 34:323–346

Russell MJ, Daniel RM, Hall AJ, Sherringham JA (1994) A hydrothermally precipitated catalytic iron sulphide membrane as a first step toward life. J Mol Evol 39:231–243

Russell MJ, Martin W (2004) The rocky roots of the acetyl-CoA pathway. Trends Biochem Sci 29(7):358–363

Russell MJ, Nitschke W, Branscomb E (2013) The inevitable journey of being. Philos Trans R Soc Lond B Biol Sci 368:20120254

Santosh M (2010) Supercontinent tectonics and biogeochemical cycle: a matter of 'life and death'. Geosci Front 1:21–30

Scheuring I, Czárán T, Szabó P, Károlyi G, Toroczkai Z (2003) Spatial models of prebiotic evolution: soup before pizza? Orig Life Evol Biosph 33:319–355

Schuster P (2007) Nonlinear dynamics from physics to biology. Complexity 12(4):9–11

Schwenk K (ed) (2000) Feeding—form, function, and evolution in tetrapod vertebrates. Academic Press, San Diego

Segré D, Ben-eli D, Deamer DW, Lancet D (2001) The lipid world. Orig Life Evol Biosph 31:119–145

Segré D, Ben-eli D, Lancet D (2000) Compositional genomes: prebiotic information transfer in mutually catalytic noncovalent assemblies. PNAS 97(8):4112–4117

Segré D, Lancet D (2000) Composing life. EMBO Rep 1(3):217–222

Shabalina SA, Koonin EV (2008) Origins and evolution of eukaryotic RNA interference. Trends Ecol Evol 23(10):578–587

Stacey FD, Davis PM (2008) Physics of the Earth. Cambridge University Press, Cambridge

Staub E, Fiziev P, Rosenthal A, Hinzmann B (2004) Insights into the evolution of the nucleolus by an analysis of its protein domain repertoire. Bioessays 26(5):567–581

Szappanos B, Fritzemeier J, Csörgö B, Lázár V, Lu X et al (2016) Adaptive evolution of complex innovations through stepwise metabolic niche expansion. Nat Commun. https://doi.org/10.1038/ncomms11607

Takeuchi N, Salazar L, Poole AM, Hogeweg P (2008) The evolution of strand preference in simulated RNA replicators with strand displacement: implications for the origin of transcription. Biol Direct 3:33

Tegmark M (2007): The mathematical universe. arXiv:0704.0646v2

Tessera M (2011) Origin of evolution versus origin of life: a shift of paradigm. Int J Mol Sci 12:3445–3458

Trinks H, Schröder W, Biebricher CK (2005) Ice and the origin of life. Orig Life Evol Biosph 35:429–445

Trixler F (2013) Quantum tunneling to the origin and evolution of life. Curr Org Chem 17:1758–1770

Wang M, Boca SM, Kalelkar R, Mittenthal JE, Caetano-Anollés G (2006) A phylogenomic reconstruction of the protein world based on a genomic census of protein fold architecture. Complexity 12(1):27–40

Wang M, Caetano-Anollés G (2009) The evolutionary mechanics of domain organization in proteomes and the rise of modularity in the protein world. Structure 17:66–78

Wang M, Kurland CG, Caetano-Anollés G (2011) Reductive evolution of proteomes and protein structures. PNAS 108(29):11954–11958

Williams RJP, Fraústo da Silva JJR (2006) The chemistry of evolution. Elsevier, Amsterdam

Woese CR (2002) On the evolution of cells. PNAS 99(13):8742–8747

Wong JT-F, Laczano A (eds) (2009) Prebiotic evolution and astrobiology. Landes Bioscience, Austin

Yarus M (2010) Getting past the RNA world: the initial Darwinian ancestor. Cold Spring Harb Perspect Biol 1:a003590

Yu J, Thorne JL (2006) Dependence among sites in RNA evolution. Mol Biol Evol 23(8):1525–1537

Perplexing Cats and Demons: Pointers to the Quantum-Physical Foundations of Life

In the first third of the twentieth century, Niels Bohr (1885–1962), Pasqual Jordan (1902–1980) and Max Delbrück (1906–1981), among others, tackled the question of connections between life processes and quantum physical phenomena. Under the political pressure of fascism and the military conflicts of the Second World War, these approaches were increasingly forgotten. Only the image of a cat—both alive and dead in a box—from Erwin Schrödinger's[1] thought experiment provides clues to that early preoccupation with the above-mentioned question. This imagery also brings ideas about quantum properties closer to those circles of people for whom the word "quantum"triggers head-shaking at best. At the same time, for decades this parable has rendered the question of possible connections between biological and quantum-physical processes if not impossible to answer— although E. Schrödinger himself tried to respond to the question of life in lectures (Schrödinger 1943).

Additional obstacles were present in the paradigms of the disciplines recruited for discussion. Among the biological disciplines, morphology—structures and composition of organs—has been at the center of evolutionary debates ever since Darwin, although it is now overshadowed by the genetic perspective. In physiology and biochemistry, chemical interpretation of biological processes obviates the need for a closer examination of quantum physical phenomena. Because of notable successes in semiconductor and laser technology, interest in quantum physics has concentrated mainly on technical applications. For outsiders, this subject is accessible only with difficulty because—with few exceptions— pre-linguistic quantum physical phenomena are predominantly represented by mathematical formulae (Prigogine 1985; Feynman 1993; Kiefer 2012; Bleck-Neuhaus 2013; Clegg 2015; Susskind and Friedman 2015). As a result, a differentiated approach to the "quantum world", which comprises various particles with different characteristics, has also

---

[1] Austrian physicist and Nobel Prize laureate 1887–1961.

M. Knoflacher, *Relativity of Evolution*,
https://doi.org/10.1007/978-3-662-69423-7_4

failed to materialize. Biologically relevant elementary particles—electrons and protons, together with photons—have diverse properties. The first two belong to the fermions, in which individual particles always assume different states. With bosons—to which the photons belong—many particles can be in the same state (Bleck-Neuhaus 2013; Clegg 2015). From a classical evolutionary point of view, it is difficult to imagine why subatomic particles might be of biological importance. In recent years, however, there has been increasing reference in the literature to overcoming intellectual barriers between the disciplines, from which impulses towards new insights can also be expected to emerge (Kohen and Klinman 1999; Davies 2004; Hunter 2006; Abbott et al. 2008a, b; Arndt et al. 2009; Sergi 2009; Romero-Isart et al. 2010; Fleming et al. 2011a; McClintock 2011; Chang 2012; Rieper 2011; Chin et al. 2012; Lambert et al. 2013; Trixler 2013; Mohseni et al. 2014; Vattay et al. 2014; Al-Khalili and McFadden 2015; Asano et al. 2015; Maruani 2021).

One pertinent example is the hypothesis of "Quantum Darwinism"in which—in contrast to classical macroscopic concepts—universal quantum physical properties of the universe are assumed (Zurek 2009, 2022; Korbicz 2021). In this context, macroscopic structures become the environment for the quantum processes. Quantum information is transmitted in those macroscopic subsets of the environment which possess selectively adjusted "receivers". Within a dynamic context, in each case the subsets with the best adapted "receivers" prevail—the name bestowed on the hypothesis also refers to these processes. Furthermore, with this hypothesis it is notable that the properties of time and space also differ from the background of the macroscopic world.

Another approach to quantum phenomena in biology is embodied in the "Quantum Probability Theory" in the field of human psychology (Busemeyer et al. 2015; Pothos and Busemeyer 2022). Simply expressed, psychological responses are analyzed comparatively with classical and quantum physical probability functions. In many cases, better explanations for various psychological responses emerged with this approach.

4.1   Information and Energy from Light

Before you ask yourself in exasperation what all this has to do with evolution, you should close your eyes for at least a minute. With this exercise you can directly experience the importance of two sense organs—operating in accordance with quantum physical principles—for the perception of your environment. For the time being, let us stick with light, which occurs in very diverse forms in the natural environment, for example as lightning, in the northern lights, during volcanic eruptions, or as fire or from celestial bodies. Of particular importance is light from the sun, which shines in essentially regular cycles.

For us it is natural to perceive structures and colors with our eyes and not just the intensity of a stimulus as, for example, with the heat receptors in the skin, or special sense organs in some reptile species (Goris 2011). Colloquially, we associate the first case with the term "light" and the second with the term "heat"—although, when we lie on a beach in the sun, both phenomena are triggered by the same source. From a physical point of view, we perceive only a minor—but for us essential—sector of the huge spectrum of electromagnetic radiation. Approximately the same spectrum is also used by many plants for conversion of energy in photosynthesis (Fig. 4.1). In this sector of the electromagnetic spectrum, different organisms convert **identical frequency ranges** in different ways! Here, a decisive role is played by the particular molecular and structural organization of their "receiving systems". In other words: Organisms can selectively use different properties of quanta.

We are inevitably confronted here with the quantum physical dualism of particles and waves (Feynman 1985, 1993; Johnsen 2012). Although wave representations predominate in representations of classical optics and also of the physiology of vision, only a concern with particles—the photons—permits a deeper insight into biological processes. These ghost particles have no mass of their own, but an impact energy, which depends on their oscillation frequency. Simply put, impact energy increases with increasing oscillation frequency—or decreasing wavelength (Fig. 4.1). Physically, the double-slit experiment is often used to explain the dualism. In this—depending on the experimental set-up—the wave or particle properties of light are revealed. More directly, the light mills on offer in many furniture stores—vacuum-sealed glass spheres with a four-bladed rotor inside—demonstrate the impact energy of photons. As soon as such spheres are exposed to light, the rotors begin to turn as if by magic. In concrete terms, this demonstrates how light or photon energy is converted into motion.

The quantum property of light—as with all electromagnetic radiation—is also evident in the measurement of decreasing light intensity. If light consisted only of waves the measurement signal would necessarily decrease continuously with decreasing light intensity (Fig. 4.2a). Instead, the average number of pulses measured decreases, while the strength of the pulses consistently remains the same (Fig. 4.2b). Light thus comes in small

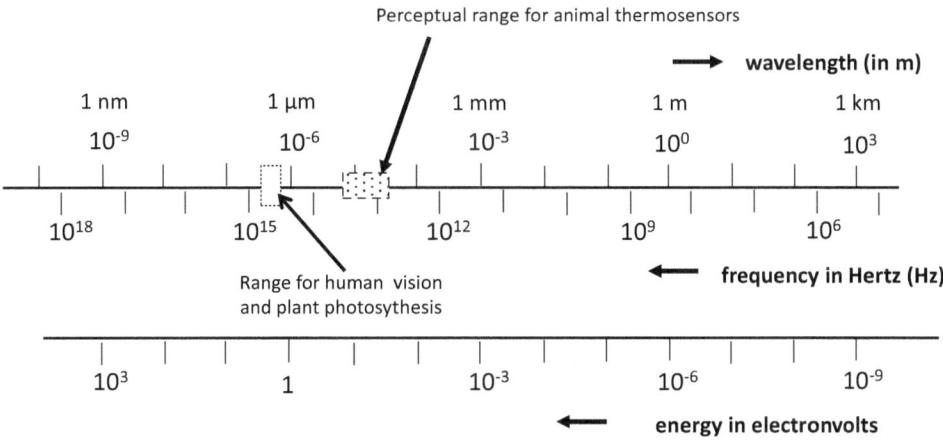

Fig. 4.1 Sector of the spectrum of electromagnetic radiation indicating the areas of perception of light and heat rays. Data sources: Vogel (1974), Goris (2011), Terakita et al. (2012)

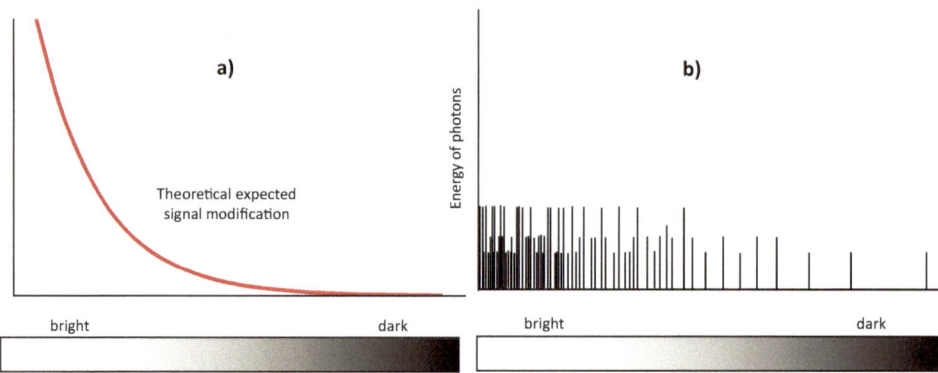

Fig. 4.2 Schematic representation of metrological proof of the quantum properties of light. (**a**) If light were to show only wave properties, a continuous decrease of the observed signal would be expected. (**b**) In fact, however, the number of single pulses measured decreases with decreasing light intensity, while the strength of a single pulse always remains constant—depending on its particular color (represented by different line lengths)

"packets"—in quanta (Land and Osorio 2003; Kelber and Roth 2006; Kelber and Osorio 2010; Warrant 2007, 2015). Physiologically, single photons can also be detected by the rods in the human retina—which are relevant for twilight vision (Tinsley et al. 2016). However, optical perception depends not only on existing irradiation conditions but also on sensory-physiological processing of photon signals. Among insects, it proved possible to demonstrate greater optical resolution in various fly species attained through augmented frequencies of photon detection combined with more rapid signal processing (Juusola et al. 2017; Song et al. 2016; Juusola and Song 2017). Morphological adaptations of the eyes enable nocturnal butterflies and hymenopterans to perceive specific flower colours in the dark through bundled detection of incoming photons (Somanathan et al. 2008; Warrant 2008; Warrant and Dacke 2011; Warrant and Somanathan 2022).

In simplified terms, during light perception photons striking photoreceptor molecules (rhodopsin) in our sensory organs deform them, but without any further utilization of the energy released. Within the frequency range detected, the number of incoming photons is "counted"—a basic prerequisite for color perception. In contrast to our technologically influenced way of thinking, detection takes place not in clearly delimited zones, but in the form of "bell curves" that also show overlap. In other words, perception of a particular color is created only during downstream processing of the acquired signals, while enabling the sensation of a continuous color spectrum (Fig. 4.3). This is an example of diffuse sensing—predominant in biological information processing—which is best represented in the formal framework of fuzzy logic (Bothe 1993; Biewer 1997).

The example of the human eye shows only one of the multitudinous methods for extracting information from light. Demonstration of 19 state-spaces with up to 55 local dimensions in quantum entanglements of photons provides some idea of the diverse possibilities for information extraction from light by organisms (Valencia et al. 2020). These

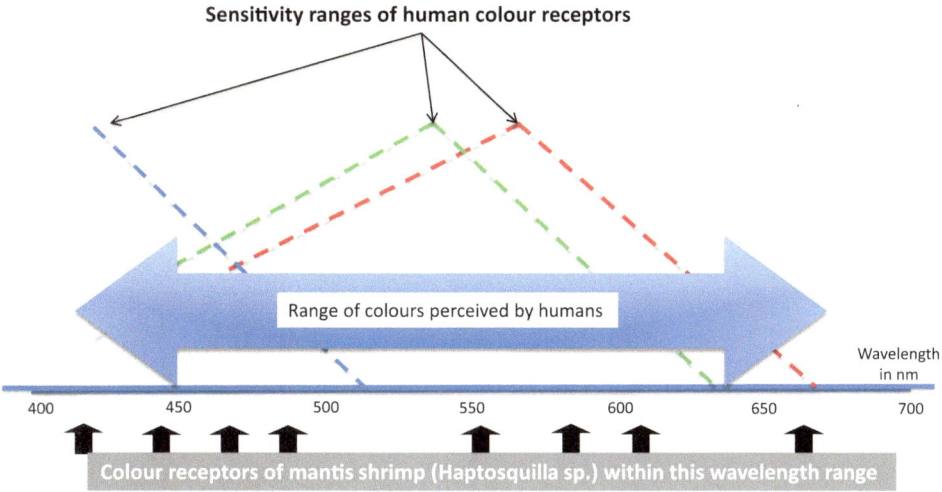

Fig. 4.3 Schematic comparison of the perceptual ranges and numbers of color sensors in the visual sense organs of humans and the mantis shrimp (*Haptosquilla trispinosa*). Data sources: Bowmaker and Dartnall (1980), Thoen et al. (2014)

possibilities range from various organelles in bacteria to perception of ultraviolet light in insects and vertebrates (Gowardovskii et al. 2000; Yokoyama and Shi 2000; Groenhof et al. 2004; Carvalho et al. 2011; Gehring 2012; Sia et al. 2014). In contrast, light receptors of nocturnal animals or deep-sea fish—including coelacanths—exhibit a narrow reception bandwidth around a wavelength maximum at about 480 nanometers (Yokoyama et al. 1999, 2008; Yokoyama 2008). In individual species of deep-sea fish, light perception is shifted far into the red range by means of coupling of rhodopsin with modified chlorophyll molecules (Douglas et al. 1998). With an increasing number of different color sensors, more differentiated detection of different wavelengths—and hence improvement in interpolation over the detectable light spectrum—becomes possible. How closely color perception in animals is linked to evolutionary adaptations to environmental conditions can be seen, for instance, in sea snakes. In various species—originally descended from land-living ancestors—a spectrum of color sensors comparable to that in diurnal primates inhabiting tropical rainforests has evolved in the colorful world of coral reefs (Carvalho et al. 2017; Simões et al. 2020).

Extreme examples in this regard are mantis shrimps with twelve or even more different color sensors (Fig. 4.3) (Thoen et al. 2014; Porter et al. 2020). Moreover, their light receptors—as in other animals—permit detection of linearly polarized light and the direction of rotation of circularly polarized light (Fig. 4.4) (Hawryshyn and McFarland 1987; Shashar et al. 1996; Cronin et al. 2003; Chiou et al. 2008; Kleinlogel and White 2008; Mäthger et al. 2009b; Brady and Cummings 2010; Warrant 2010; Saba et al. 2014). Such versatile sensory-physiological equipment is understandable given the special optical conditions of aquatic habitats. Both spectral composition and polarization of incident light change as a

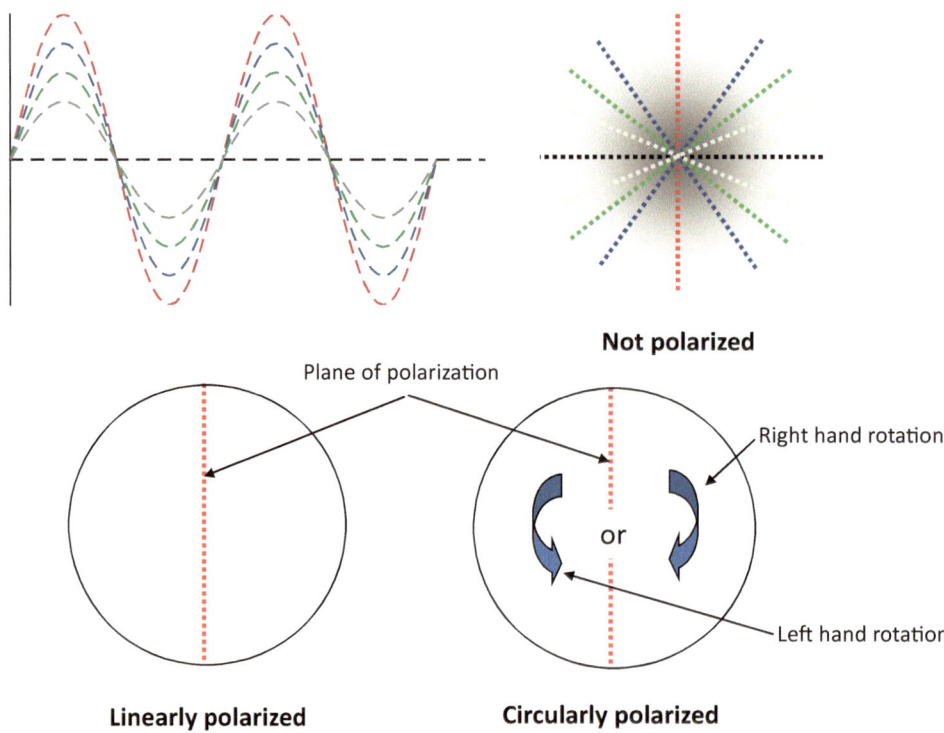

Fig. 4.4 Schematic representation of differences between the "wave cloud" of non-polarized light, polarized light, and circularly polarized light

result of superimposed influences of various factors, for instance, position of the sun, water depth, turbulence or constitution and content of suspended matter (Johnsen 2012; Kirk 2011). Many organisms "camouflage" themselves in water with "invisible" bodies. Specifically, transparent tissues scatter incident light in a fashion resembling that in the surrounding water. Various crustacean larvae—living in open water—can even camouflage their—otherwise visible—eyes with adaptable physiological filters (Bagge 2019; Feller et al. 2019; McDonald et al. 2022; Feller and Porter 2023; Shavit et al. 2023). Thanks to their special sensory apparatus, however, mantis shrimps—adapted to the relevant spectral ranges—can in fact recognize transparent organisms because of differences in light polarization (Johnsen and Widder 2001; Johnsen 2012, 2014; Gagnon et al. 2007; Johnsen et al. 2011; Cronin et al. 2014; Patel et al. 2021).

An important property of the quantum world becomes recognizable:

– The incidence of photons is distributed randomly and not deterministically.

And an important feature of the information perception of organisms hence emerges:

– Reception and processing of signals takes place in the context of the sense organs and downstream information processing.

Organisms themselves reveal the energy content of **photons** in multiple ways. The best known example is the energetic basis of all multicellular life—oxidative photosynthesis with absorption of light by chlorophyll molecules. In various forms—ranging from bacteria (Cyanobacteria) to flowering plants—photon energy is converted into chemical energy and employed to build biomass (Ritz et al. 2002; Vermaas 2002; Mulkidjanian et al. 2006; Kiang et al. 2007; Johnston et al. 2009; Nobel 2009; Neilson and Durnford 2010; Engel 2011; Hohmann-Marriott and Blankenship 2011; Renger 2011: Scholes et al. 2011; van Grondelle and Novoderezhkin 2011). An initial indication of the complexity of the challenges overcome in this process is evident, for instance, from the fact that huge time differences in the process sequences must be overcome between the conversion of photon energy and the storage of that energy in organic molecules. Whereas quantum physical processes operate at orders of magnitude of femtoseconds (10^{-15}), biochemical processes take milliseconds (10^{-3})—i.e. take place billions of times more slowly (Cupellini et al. 2020). Organisms can actively adapt their photosynthetic systems to changes in light irradiation. Under the dynamic conditions of marine habitats (see Sect. 8.6.3), Cyanobacteria, for example, chromatically adapt their receiving systems to the spectral range available at different water depths. The requisite remodeling of molecular structures is completed in about a week (Stomp et al. 2007; Wiltbank and Kehoe 2019; Villafani et al. 2020). In terrestrial plants, photosynthetic systems can adapt rapidly to changes in light intensities by modifying the spatial positioning of their organelles (chloroplasts) and molecular structures in the receiving systems (Capretti et al. 2019; Kong and Wada 2016; Wada and Kong 2018; Mascoli et al. 2020).

Less widely known is direct, non-photosynthetic use of photon energy for cellular metabolism through proteorhodopsins in archaeans and bacteria, and possibly in viruses as well (Fig. 4.5) (Bratanov et al. 2019; Needham et al. 2019). Photon energy is employed in a variety of ways to transport ions into or out of cells (Schäfer et al. 1999; Béjà et al. 2000, 2001; Man et al. 2003; McCarren and DeLong 2007; Kouyama et al. 2010; Palovaara et al. 2014; Kandori 2015; Govorunova et al. 2017). Under extreme conditions, Cyanobacteria also switch to energy production by means of different rhodopsins (Hasegawa et al. 2020; Astashkin et al. 2022). Modified forms of these molecules—operating according to quantum physical principles (Logunov and Schulten 1996; Sudo and Spudich 2006; Terakita et al. 2012)—are used in other bacterial groups, by photosynthetic algae and by certain animals for light perception and associated regulation of their physiological processes and spatial orientation. All known types of biological photoreceptors are found in fungi. Hence, they can not only adapt their life cycles or growth of their fruiting bodies, but also—by adapting their physiological performance—colonize extreme locations, for example in lichens or on surfaces of plants (Spudich 2006; McCarren and DeLong 2007; Kianianmomeni and Hallmann 2014; Spudich et al. 2014; Fischer et al. 2016; Muggia et al. 2016; Yu et al. 2021). In addition, there is evidence for

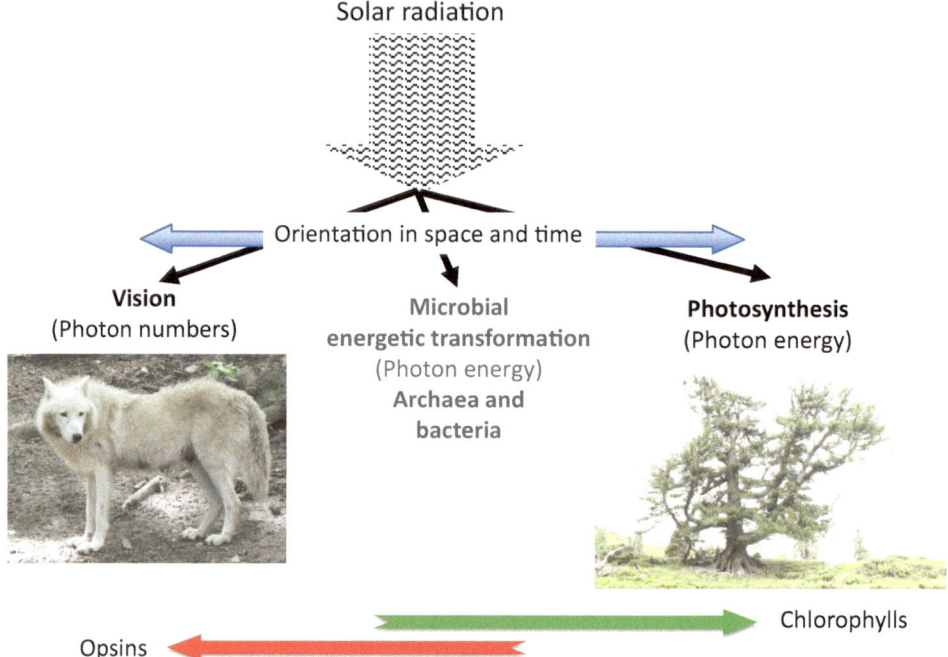

Fig. 4.5 Schematic representation of the different biological uses of sunlight

proteorhodopsin genes in viruses, which are presumably used for sensory manipulation of infected organisms (Yutin and Koonin 2012; Bratanov et al. 2019).

In animals—including humans—ultraviolet radiation with a wavelength near 280 nm (UV-B) is physiologically important for synthesis of vitamin D. Studies of cold-blooded (poikilothermous) animals such as reptiles have revealed that heat radiation within the infrared range is also needed for complete synthesis of this vitamin (Holick et al. 1995). Among other things, deficiency in the supply of vitamin D in humans leads to disturbances in calcium metabolism, which manifest themselves in skeletal malformations (rickets) in children and in aching bones and fatigue fractures in adults. Because intense ultraviolet radiation can also damage DNA and folic acid (vitamin B_9) and cause skin cancer, the degree of radiation exposure required is still a subject of scientific debate (Rangel et al. 2009; Bikle 2014; Holick 2014; Lucock et al. 2014). In evolutionary terms, however, clear correlations between regionally varying intensities of ultraviolet radiation and skin pigmentation have been shown in humans (Jablonski and Chaplin 2013).

In contrast to sensory perception, energetic use of light converts the impulse energy of photons into biochemical energy. In the biological quantum world, the challenges involved are comparable to harnessing energy from rain rather than from a steam engine. Technically, rain falling on a landscape can be used efficiently for energy conversion in a hydroelectric power plant only if adequate quantities of water are supplied from higher altitudes. A supply of water from lower-lying areas, on the other hand, requires an energy input that can only be covered—without long-term disadvantages—by energy generated elsewhere that cannot be used in some other way. An example in technological systems would be the use of "surplus electricity" from wind energy to fill reservoirs in pumped storage hydro-power stations. Biological examples of this can be found in various purple bacteria, which also use longer-wave light and—up to thermodynamically permissible limits—supply energy "uphill" to the photosynthetic centre (Trissl et al. 1999).

In the photosynthesis of land plants and green algae, only vibrational energy from two narrow wavelength ranges (680 and 700 nm) can be used directly for conversion into bio-chemical energy (Falkowski et al. 2004; Hohmann-Marriott and Blankenship 2011). If only the photon energy of these wavelengths were to be converted—as is the case with vision—most of the available radiant energy would remain unused. In photosynthetic complexes, as in the technological example provided above, antenna systems capture radi-ant energy at shorter wavelengths—i.e. higher energy levels—and feed it to reaction cen-tres. The processes involved have not been completely studied as yet, but they clearly exhibit quantum physical properties (Ritz et al. 2002; Hammes-Schiffer 2011; Scholes et al. 2011; van Grondelle and Novoderezhkin 2011; Romero et al. 2014). Indications of this include almost loss-free transmission and an extremely high transmission speed in the femtosecond range.[2] Simply stated, the largest possible fraction of incoming radiant energy is captured during photosynthesis and conducted in aggregate form to the reaction centres—the chlorophylls (Chl) and bacteriochlorophylls (BChl) (indicated in Fig. 4.6 by "range of 'antennas' ").

To interpret Fig. 4.6, it should be noted that the energetic irradiation curves apply only to ideal conditions and, in the case of water, only to the uppermost layer of pure water. In pure water, the energy of individual wavelengths is absorbed by water molecules. Furthermore, with increasing water depth dissolved substances and suspended particles modify and absorb the radiation spectrum (Buiteveld et al. 1994; Stomp et al. 2007; Kirk 2011). For these reasons, no quantitative values have been indicated on the vertical axis.

Figure 4.6 shows the differences between direct energetic conversion of photon energy by bacteriorhodopsins (BR) and by complex photosynthetic systems (Chl and BChl). The former are found in energetically optimal wavelength ranges—for aquatic conditions—which can also be used up to the limits of light penetration depths (Pinhassi et al. 2016). Energetic conversion ranges of reaction centres for oxygenic (Chl) and anoxygenic (BChl) photosynthesis lie at the upper range limit of irradiation in aquatic habitats, but with their antenna systems are in each case able to exploit the entire bandwidth of available wave-lengths (Heldt and Piechulla 2008; Niedzwiedzki et al. 2020). Because of the great adapt-ability of photosynthetic systems, no reliable conclusions can be drawn about their evolutionary origin. Evidence of obligate photosynthetic bacteria in deep-sea hydrother-mal vents suggests that the—non-thermal—wavelengths present there (see insert in Fig. 4.6) can also be used for energy conversion (Beatty et al. 2005). The temporal sequence for the origin of anoxygenic and oxygenic photosynthetic systems is being

---

[2] 1 Femtosecond = 10^{-15} s.

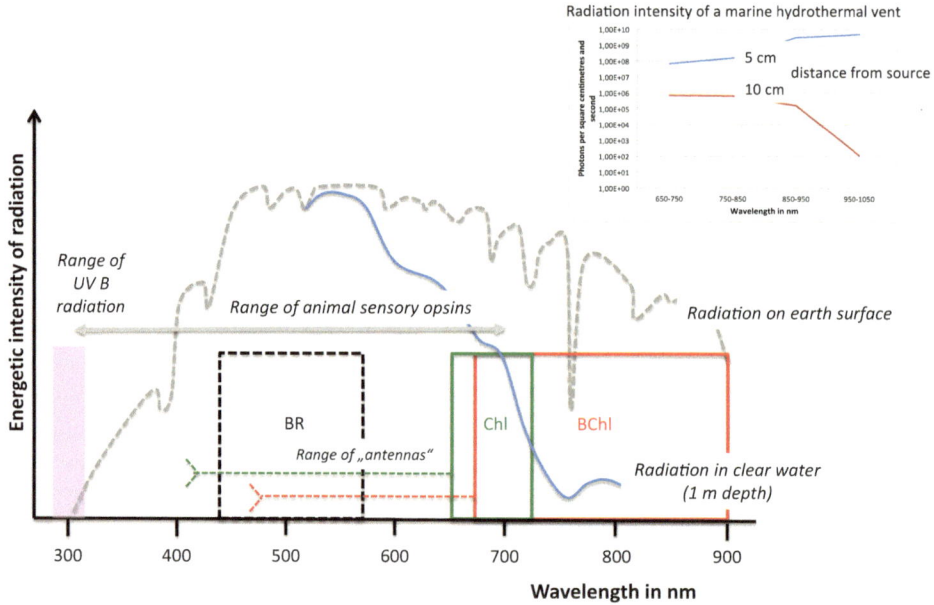

Fig. 4.6 Schematic representation of wavelength ranges used for energy conversion by bacteriorho-dopsins (BR), chlorophylls (Chl) and bacteriochlorophylls (BChl). For the latter, the total wavelength ranges used by antenna systems (range of "antennas") are indicated. The bandwidth of visual perception of animal opsins and the relative intensities of solar irradiation at Earth's surface and in clear water are shown for comparison. Data sources: Wetzel (1975), Stomp et al. (2007), Heldt and Piechulla (2008), Pinhassi et al. (2016), Niedzwiedzki et al. (2020), Vöcking et al. (2022). Insert: Measured photon density in the wavelength range between 650 and 1050 nm at distances of 5 and 10 cm in a deep-sea source with an elevated temperature of about 350 degrees. Data source: Van Dover et al. (1996)

questioned increasingly (Cardona 2019; Ward and Shih 2021; Sephus et al. 2022). It is also unclear where the first photosynthetic systems evolved. Potentially, aquatic surface zones and deep-sea hydrothermal vents might be considered. In the former, the damaging effects of ultraviolet radiation must be taken into account, while in the latter high temperatures in the vicinity of sources are problematic (Boomer et al. 2000; Cockell 2000; Madigan et al. 2003; Olson and Blankenship 2004; Okubo et al. 2006; Alboresi et al. 2010; Rozhnov 2013; Djokic et al. 2017).

Cyanobacteria and photoautotrophic eukaryotes would be doomed if they were unable to perceive light as well. In heterogeneous natural environments, photosynthesis would not be possible over the long term without orientation to prevailing light distributions. Photosensitive organs for orientation to the incidence and wavelengths of light have evolved along different pathways in both cyanobacteria and eukaryotic algae. Functionally, the variety of organs ranges from arrays of a few photosensitive molecules to differentiated compound lens eyes (Jékely 2009; Schuergers et al. 2016; Colley and Nilsson 2016; Gavelis et al. 2017). Algae must adapt to changes in the spectral composition of light in aquatic habitats. From a sensory point of view, the diverse requirements involved are met by phytochromes covering the entire visible spectrum (Rockwell et al. 2014). Regarding regulation, the perceived light signals are used not only for orientation of the photosynthetic organs, but also for adaptation of photosynthetic processes to the relevant spectral conditions (Raven 2009; Kehoe 2010; Hirose et al. 2013). In the short-wave range, biliproteins serve the tasks of both photoreception and support of photosynthesis (Ting et al. 2001; Ma et al. 2012; Duanmu et al. 2013). Figure 4.6 also shows that the spectral range of animal light perception is predominantly consistent with aquatic habitat conditions. Here it can be suggested that physiological capacities for obtaining information from light were determined by primary evolution of animals in aquatic habitats. This inference is strengthened by the fact that the greatest diversity for perception of different light qualities among vertebrates is found in bony fishes and among arthropods in mantis shrimps (Stomatopoda) (Cronin et al. 2014; Koyanagi and Terakita 2014; Carleton and Yourick 2020; Porter et al. 2020; Musilova et al. 2021).

4.2   Modification of Light

When looking at multicoloured insects or flowers, we are rarely aware of the physical processes that make such perception possible. We are even less aware of those processes when we fail to perceive "camouflaged" animals. These two examples are intended only to illustrate briefly the role that modification of light can play in interactions between various organisms. We perceive only a small sector of light modification by organisms—colours in the visible range of light. Many phenomena beyond our perception or in zones that we can access only with difficulty, such as the deep sea, remain forever hidden from the average observer. From an evolutionary perspective, on the other hand, the primary question that arises concerns processes involved in the emergence of phenomena that are already detectable in animal fossils over 500 million years old (Parker 1998).

Physically, light is modified by pigments and microscopic (photonic) structures. This is always the result of modifications that remain evident, be it the "remainder" of the colour spectrum in the case of pigments or polarization of remaining radiation (Fig. 4.7). In the case of pigment colours, the brilliance of the colour impression is influenced both by the proportion of surface reflection (reflected solid line in Fig. 4.7a) and by reflection conditions in the colour layer. Structural colours are created, for example, by modulation of light in ultra-thin layers with different material densities, multi-dimensional light-transmitting structures—so-called photonic crystals—or surface structures of various shapes (Fig. 4.7b). In addition to optical effects—for example the suppression of reflections—surface structures can also have self-cleaning properties. Examples of this can be found in the dark scales of the Gabon viper, in the "black" plumage of birds of paradise or in deep-sea fish (Spinner et al. 2013, 2014; McCoy et al. 2018; Davis et al. 2020a). Because of the narrow bandwidth of the reflected light, structural colours usually appear more brilliant compared to pigment colours. Both basic principles of light modification can occur

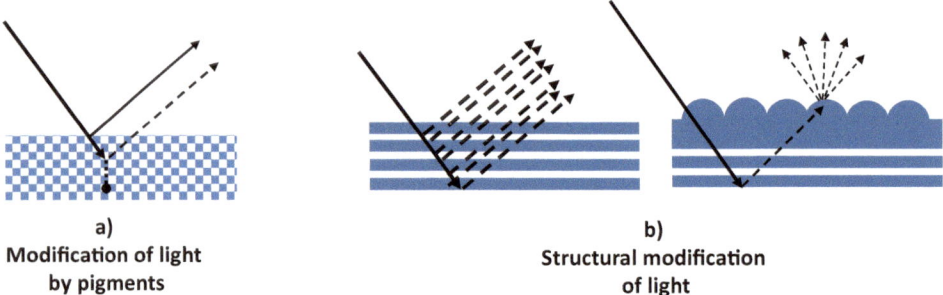

a)
**Modification of light
by pigments**

b)
**Structural modification
of light**

Fig. 4.7 Simplified schemes of light modification. (**a**) Colour pigments absorb a part of the colour spectrum (vertical dotted line) and trigger the impression of colour across the remaining part of the spectrum (dashed line); (**b**) In structures, light is modified by complex physical processes in multiple material layers of different densities, triggering, for example, particularly brilliant colour effects or maintaining the colour impression constant from all viewing directions (right diagram)

either in isolation or in various combinations in organisms (Vukusic and Sambles 2003; Kinoshita et al. 2008; Ghiradella and Butler 2009; Shawkey et al. 2009; Johnsen 2012; Sun et al. 2013; Wiersma 2013; Arwin et al. 2014; Hernández-Jiménez et al. 2014; Stavenga 2014; Wilts et al. 2014a, b).

The quantum physical processes of colour formation is most easily illustrated with the example of pigment colours (Fig. 4.7a). As already mentioned, "subtraction" from the light spectrum results from absorption of specific wavelengths by pigments. During absorption, photons disappear but transfer their energy to pigment molecules. In the molecules, the energy is absorbed by elementary particles, the electrons—which are crucial for biological processes. These attain a higher energy level, which in the case of absorption is "decomposed" in small steps (Fig. 4.8b). If, however, the amount of energy released in a degradation step is higher than the required excitation level of a photon, light emission occurs. Because of losses in energy conversion, the frequency of emitted photon is lower. Because of the longer wavelengths resulting, the colour of the secondarily emitted light shifts towards the red range (see Fig. 4.3). Conventionally, two different terms are used for the described process (Valeur and Berberan-Santos 2011; Johnsen 2012): If secondary photon emission occurs immediately (in the range of 10^{-8} s) and only during illumination, it is **fluorescence** (Fig. 4.8b). If, on the other hand, secondary photon emission is subject to delay (between 10^{-3} and 1 s) and persists for some time after primary illumination has ended, it is **phosphorescence** (Fig. 4.8b). The reason for this effect is temporary change in electron properties, which delays the release of energy.

The phenomena described so far are based on material properties and material structures that occur in a barely comprehensible variety in many groups of organisms. Caution is needed in biological interpretations of the observed phenomena, as they can be associated with a wide diversity of effects. In the simplest case, the cause is structure formation, for example, when optical effects can be visualized only by processing materials in the laboratory. Equally difficult is biological interpretation of structural colours in bacterial

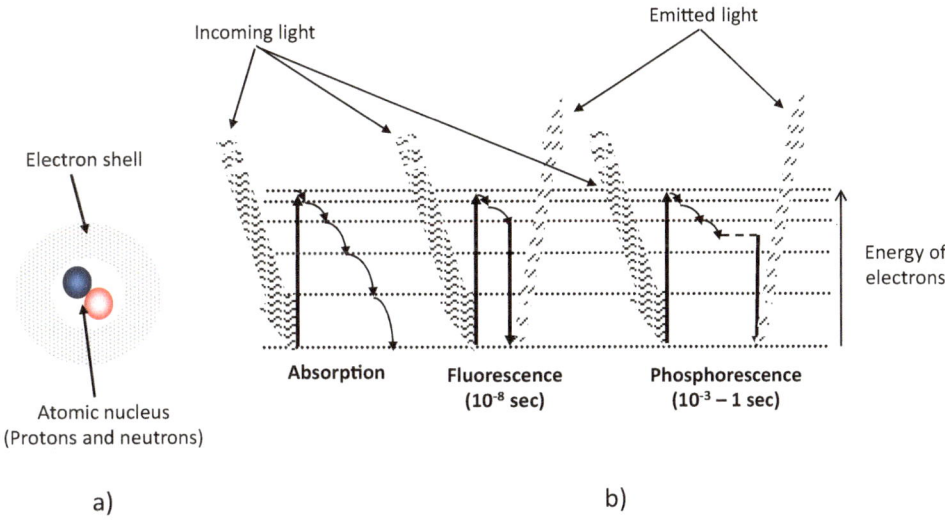

Fig. 4.8 Diagrammatic comparison between different processes of light conversion on surfaces. (**a**) Diagram of an atom with one proton and one neutron in the nucleus, along with the electron shell, in which the electron can occupy different energy levels; (**b**) Comparison of the different effects of decay in the increased energy content of electrons in the shell induced by excitation by light

films, whereas pigment colours are usually associated with certain physiological functions (Kientz et al. 2012, 2016; De Carvalho et al. 2014). According to currently available literature, a developmentally differentiated variety of light-modifying pigments and structures can be found in plants. For red algae and liverworts, it has proven possible to detect pigments that are presumably used to regulate surface structures serving irradiation by light (Duckett and Ligrone 2006; Chandler et al. 2015). It will be left open here whether the striking differences between gymnosperms and angiosperms (flowering plants) that is evident with respect to visually effective structures reflects a natural distinction of some kind or merely of biased research interest (Strout et al. 2013; Vignolini et al. 2013; Diah et al. 2014). Among flowering plants, which have been relatively intensively studied, there is clearly a more extensive colour spectrum for flowers than for fruits (Stournaras et al. 2013). In published accounts, evidence for visually effective structures in flowers also predominates over that for fruits (Whitney et al. 2009; Glover and Whitney 2010; Vignolini et al. 2012). Functionally, this may be linked to various influences, such as signals for pollinating insects or protection from ultraviolet radiation, as with the Edelweiss blossom (van der Kooi et al. 2014; Vigneron et al. 2014). One of the most widely cited examples of functionally disparate effects of nanostructures is the so-called "lotus effect". Water-repelling and self-cleaning effects of nanostructures originally discovered on leaves of the lotus plant (*Nelumbo* sp.) have now been identified in a wide variety of plants, contributing to the diversity of structural colours (Neinhuis and Barthlott 1997; Gu et al. 2003; Bhushan and Jung 2008; Zhou et al. 2018).

Turning to animals, even human observers are struck by the great variety of colours in insects—especially butterflies and beetles—spiders and birds (Vukusic et al. 2001; Prum et al. 2004, 2006, 2009; Stavenga et al. 2004; Santos et al. 2006; Kemp and Rutowski 2007; Welch and Vigneron 2007; Gaillot et al. 2008; Michielsen and Stavenga 2008; Doucet and Meadows 2009; Poladian et al. 2009; Sharma et al. 2009; Vukusic et al. 2009; Wilts et al. 2009, 2015; Dong et al. 2010; Stavenga et al. 2010, 2011, 2014; Saranathan et al. 2010; Biró and Vigneron 2011; Stoddard and Prum 2011; Yoshioka et al. 2011; Yin et al. 2012; Foelix et al. 2013; Schenk et al. 2013; Arwin et al. 2014; del Rio et al. 2014; Hernández-Jiménez et al. 2014; Saade and Ogryzko 2014; Sharma et al. 2014; Thomé et al. 2014). In most cases the impression of colour is generated by combinations of pigments and structures, although the array of emitted light usually extends far beyond the spectrum of human perception in terms of colour diversity and polarization patterns. The functional significance of the various light modifications has so far been determined only in individual cases, because for this purpose—in addition to qualities of the emitted light—the sensory performance of potential interaction partners must also be recorded. One example is a competitive advantage observed in the mating ritual of males of a jumping spider species (Lim and Li 2013). Apart from visual effects, nanostructures are also associated with a wide range of functional influences in animals. Examples include water-repelling or water-binding effects, self-cleaning of the skin surface and use of molecular electrostatic effects to increase retention forces independent of the effects of gravity on a wide variety of surfaces. In everyday life, such effects can be observed with flies on window panes or ceilings and, at southerly latitudes, with geckos (Beutel and Gorb 2001; Federle 2006; Sukontason et al. 2006; Bhushan 2009; Watson et al. 2015, 2017).

Photonic structures on body surfaces can facilitate camouflage protection from potential predators in open-water marine fish or enhance the colouration of, for example, neon tetras (Lythgoe and Shand 1989; Levy-Lior et al. 2010; Yoshioka et al. 2011; Johnsen 2012; Brady et al. 2015). Structural colours are present relatively rarely on the body surfaces of mammals—for example, in male mandrills or in shrew opossums. In this case, a link between competitive strength and colour intensity is suspected (Prum and Torres 2004).

In addition to static presentation of colours, individual species can actively modify their colouration, for example by altering the fluid content of photonic structures in the Panamanian leaf beetle *Charidotella egregia* (Vigneron et al. 2007). The better-known colour changes shown by squids and octopuses are achieved by combined modification of pigment-containing cells and reflective protein structures underlying them. In addition, this may involve reflection of polarized light (Crookes et al. 2004; Mäthger et al. 2009a; DeMartini et al. 2015). In chameleons, colour changes are enhanced by layers of photonic crystals overlying the pigment cells (Teyssier et al. 2015).

Fluorescence is one of the protective mechanisms that plants possess to prevent cell damage under strong solar radiation (Krause and Weis 1991; Horton et al. 1994; Campbell et al. 1998). In this process, the differential between incoming and currently photosynthetically convertible energy is converted back into light and emitted (Fig. 4.8b). These phenomena are not observable macroscopically, but are routinely exploited in a wide variety of investigations and experiments in the microscopic range. According to reports so far available, in animals fluorescence phenomena are connected with a variety of functional contexts. Fluorescence in sponges with a silicate skeleton is likely to be related to signal processing (Wang et al. 2012; Müller et al. 2013). Fluorescence of corals, by contrast, is probably related to protection against strong solar radiation (Alieva et al. 2008; Roth et al. 2010; Roth and Deheyn 2013; Smith et al. 2013). Green fluorescent proteins have been found in the lancelet *Amphioxus* (Cephalochordata) (Bomati et al. 2014). In shells of the mussel *Pinctada vulgaris*, on the other hand, red fluorescence is generated by haemoglobin-like porphyrins (Arma et al. 2014). The signalling effect of fluorescence has been observed in a hydrozoan and certain fish (Gerlach et al. 2014; Wucherer and Michiels 2014; Haddock and Dunn 2015). The functional significance of fluorescence in scorpions, many fish and sea turtles remains obscure (Stachel et al. 1999; Michiels et al. 2008; Sparks et al. 2014; Gruber and Sparks 2015). In butterflies, mantis shrimps and parrots, fluorescence is likely to be associated with intraspecific signalling (Arnold et al. 2002; Mazel et al. 2003; Vukusic and Hooper 2005).

4.3   Light Production: Bioluminescence

Should you meander through a meadow or forest on a warm summer night, you are highly likely to come across the "flying light spots" of male glowworms—members of the firefly family Lampyridae. This single example, however, scarcely does justice to the manifold forms of light production that can be found among organisms, above all in the oceans (Haddock et al. 2010; Widder 2010; Martini and Haddock 2017).

Bioluminescence differs from fluorescence and phosphorescence in the energy source for the primary excitation of electrons. Whereas electrons are excited by light in fluorescence and phosphorescence, in bioluminescence excitation is achieved with chemical energy (Fig. 4.9a). The light-emitting process—in which luciferins react with luciferases—requires energy from cell metabolism, for example in the form of adenosine triphosphate (ATP) (Tsuji and Leisman 1981; Tsuji 1985; Haddock et al. 2010; Widder and Falls 2013). Studies on insects indicate that the energy requirement for such light production is approximately one third of the total energy requirement at rest: about 37 percent in fireflies (Lewis and Cratsley 2008) and about 26 percent in the larva of the Australian fungus gnat *Arachnocampa flava* (Willis et al. 2011). Luciferins and luciferases occur in organisms in different chemical forms (Day et al. 2004; Haddock et al. 2010; Widder 2010; Brodl et al. 2018). Among marine organisms, the luciferin coelenterazine is the most widespread and is found in almost all bioluminescent multicellular organisms. Unicellular dinoflagellates and multicellular luminescent shrimps, however, possess a dinoflagellate luciferin (Nicolas et al. 1987; Widder 2010; Valiadi and Iglesias-Rodriguez 2013). Over the course of evolution, the capacity to produce light is thought to have evolved separately up to 50 times (Haddock et al. 2010; Widder 2010; Schnitzler et al. 2012; Davis et al. 2016). Note, however, that the specific colour of the emitted light can be determined either by bioluminescence alone or through secondarily triggered fluorescence (Fig. 4.9b) (Widder et al. 1983; Ohmiya and Hirano 1996; Haddock and Case 1999;

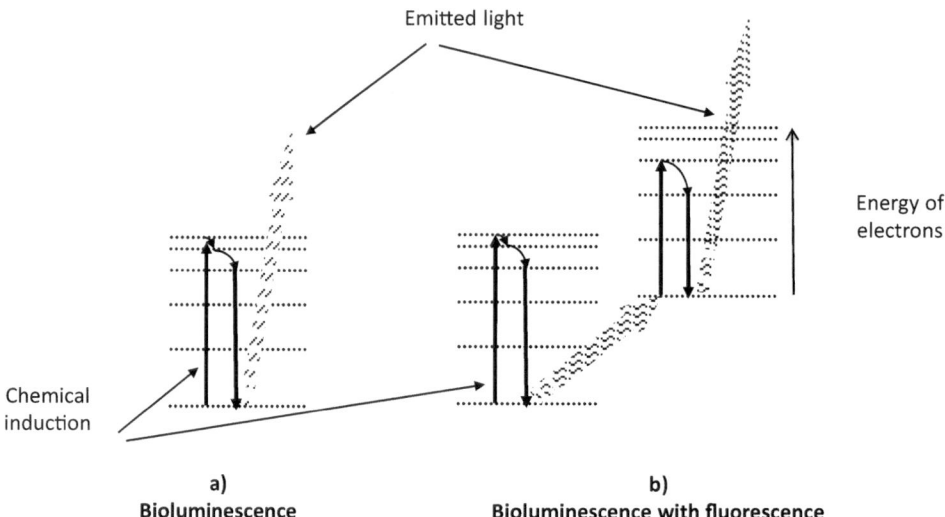

Fig. 4.9 Schematic representation of the process of bioluminescence. (**a**) As a stand-alone process; (**b**) in combination with fluorescence

Kudryasheva et al. 2002; Gurskaya et al. 2003; Nakatsu et al. 2006; Nemtseva and Kudryasheva 2007; Widder 2010; Navizet et al. 2011; Johnsen et al. 2012).

Luminescent bacteria occur in vast numbers in oceans, and to a limited extent in other habitats—for example on decaying material. The spectrum of their habitats extends from open water of oceans to the inside of organisms in the form of symbioses or parasitism (Nealson and Hastings 1979; Meighen 1991; Kozakiewicz et al. 2005; Heath-Heckman et al. 2013). The evolutionary roots and biological significance of bioluminescence in bacteria remain enigmatic (Węgrzyn and Czyż 2002; Stabb 2005).

As with bacteria, the majority of all bioluminescent eukaryotes—an estimated 80 per cent—live in oceans (Morin 1986; Deheyn et al. 2000; Deheyn and Wilson 2010; Thuesen et al. 2010; Haddock et al. 2005, 2010; Widder 1998, 2010; Gouveneaux and Mallefet 2013; Fishelson et al. 2005). Biological luminescent systems in marine habitats are correspondingly diverse. Their functional importance is known only partially, but ranges across attracting potential prey (Widder 1998; Fishelson et al. 2005; Haddock et al. 2005), illumination in hunting (Widder 2001; Robison et al. 2003; Kubodera et al. 2007), distracting or repelling predators (Osborn et al. 2009; Jones and Mallefet 2010, 2013; Cusick and Widder 2014), recognizing individuals of the same species (Davis et al. 2014), and camouflage with backlighting (McFall-Ngai and Morin 1991; Rees et al. 1998; Jones and Nishiguchi 2004; Johnsen 2005, 2014; Davis et al. 2020b; Paitio et al. 2020). Only about one fifth of all eukaryotes use luminescent bacteria for light production; the majority possess independent organs for this purpose. Both forms of light production have various advantages and disadvantages. Symbioses with luminescent bacteria reduce the energy needed for light production, but activities of the bacteria are more difficult to regulate. For example, the dwarf squid *Euprymna scolopes* controls the light emissions of symbiotic bacteria by means of special light sensors in its luminescent organ (Tong et al. 2009). This permits elimination of inactive bacteria and regulation of the light intensity of the luminescent organ (Ruby and McFall-Ngai 1992; Ruby and Lee 1998; Anetzberger et al. 2009). According to its light requirements, the deep-sea fish *Photostomias guernei* can fold its luminescent organ in or out (Widder 2001). Independent luminescent organs require more energy, but they are easier to control by means of hormonal or neuronal pathways, as is the case with deep-sea sharks, e.g. lantern sharks *Etmopterus* sp. (Claes and Mallefet 2011; Claes et al. 2011, 2012, 2015; Straube et al. 2015; Gruber et al. 2016; Duchatelet et al. 2021; Mizuno et al. 2021; Tomita et al. 2023).

Even taking into account habitats in tropical regions and in caves, bioluminescence is not as widespread outside the oceans as in the deep sea (Viviani 2009; Viviani et al. 2010). The diversity of luminescent systems in fungi that was originally discussed has now been refuted by more recent studies and attributed to a uniform principle (Bermudes et al. 1992; Weitz 2004; Desjardin et al. 2005, 2008; Deheyn and Latz 2007; Bondar et al. 2011; Hayashi et al. 2012; Oliveira et al. 2012; Bechara 2015). The functional significance of their often relatively weak luminescence is still under discussion, because experimental investigations of individual hypotheses have also led to contradictory results—for example, attraction of insects for spore dispersal (Sivinski 1998; Weinstein et al. 2016). Among terrestrial animals, bioluminescence is so far known to occur in individual species of earthworms (Oligochaeta), snails (Pulmonata), centipedes and millipedes (Chilopoda and Diplopoda), springtails (Collembola) and flies (Diptera) along with a large number of firefly species (Lampyridae) (Isobe et al. 1991; Sivinski 1998; Day et al. 2004; Ghiradella and Schmidt 2004; Marek et al. 2011; Meyer-Rochow and Moore 2009; Viviani 2009; Oba et al. 2011; Marek and Moore 2015; Kusy et al. 2020). The larvae of fungus gnats (*Arachnocampa* sp.), which occur in caves in Australia and New Zealand, are well known as a tourist attraction due to their combination of bioluminescence with construction of sticky threads for catching insects (Willis et al. 2011; Vršanský et al. 2012; von Byern et al. 2016). Differentiation of light signals in fireflies is probably related to a variety of factors (Lewis and Cratsley 2008; Lall et al. 2010), for example the colonization of different habitats—including termite burrows—or the presence of toxic substances (De Cock and Matthysen 2001, 2003; Moosman et al. 2009; Costa and Vanin 2010). For reasons that are as yet unknown, larvae of some species, such as the "railroad worm" (*Phrixotrix* sp.), exhibit conspicuous colour patterns with red-and-green bioluminescence (Viviani et al. 2006, 2013; Arnoldi et al. 2010).

4.4   Strong Pointers but No Definitive Proof: Magnetic Field and Molecules Perceived

Various bacteria (e.g. *Magnetospirillum* sp.) orient themselves to the Earth's magnetic field with the aid of intrinsically formed iron crystals, which are stored in an independent cell organelle—the magnetosome. It has been assumed that the iron particles were originally stored for physiological protection against oxygen free radicals and only secondarily came to stimulate signal processing. The number of crystals—each of which measures approximately 37 nanometers (nm)—varies, and in some species up to a thousand can be detected (Schüler 2008; Naresh et al. 2011; Lin et al. 2014, 2020; Amor et al. 2016). The bacteria utilize the obvious magnetosensitive orientation of the crystals both in isolation and in combination with other indicators—for example oxygen concentrations or light intensity—to detect optimal conditions for survival (Scheffel et al. 2008; Chen et al. 2014; Zhang et al. 2012; Bennet et al. 2014; Bradlaugh et al. 2023).

It is much more difficult to confirm a physiological basis for magnetic orientation in other organisms, such as birds or insects. In addition to light perception, it is suspected that the retina of migratory birds may perceive certain features of the geomagnetic field. Electron pairs in cryptochrome molecules excited by blue light (radical pairs) can be considered as potential sensors (Gegear et al. 2008; Rodgers and Hore 2009). Results of recent investigations suggest that electromagnetic induction in the inner ear of pigeons is a signal generator for orientation to the magnetic field. Human studies have revealed that brain waves are influenced by magnetic fields, but it is unclear whether spatial orientation is also possible (Nimpf et al. 2019; Wang et al. 2019).

Widely distributed cryptochromes count among the photolyase enzymes that add or remove chemical double bonds using photon energy (Heldt and Piechulla 2008; Berg et al. 2013). In bacteria, cryptochromes catalyze DNA repair (Weber 2005), while in plants they are involved—among other things—in regulating growth and alignment with the direction of light radiation (phototropism) (Heldt and Piechulla 2008). In animals, they also serve as regulators of the day/night cycle (circadian rhythm) (Rivera et al. 2012).

According to findings so far reported, the incident angle (inclination) of magnetic field lines is used for orientation in the magnetic field—as is also the case in bacteria with magnetite bodies (magnetosomes) (Kirschvink et al. 2001; Richter et al. 2007; Lefèvre and Bazylinski 2013; Chulliat et al. 2015; Amor et al. 2016). A similar sensory system for magnetic orientation has been assumed to be present in insect antennas (Gegear et al. 2010; Wajnberg et al. 2010). The best known example is that of the monarch butterfly (*Danaus plexippus*), which migrates between summer ranges in the eastern United States and wintering grounds in the Michoacán Mountains of Mexico (Guerra et al. 2014). Although there are many indirect indications of the existence of this phenomenon, no direct experimental confirmation has yet been achieved (Wiltschko et al. 2002; Mouritsen and Ritz 2005; Wiltschko and Wiltschko 2005; Liedvogel et al. 2007; Efimova and Hore 2008; Maeda et al. 2008; Ritz et al. 2009; Gegear et al. 2010; Ritz et al. 2010; Winklhofer 2010; Hein et al. 2011; Bandyopadhyay et al. 2012; Lau et al. 2012; Solovyov et al. 2014; Hiscock et al. 2016). This is due to a variety of factors, in the same way as many quantum physical phenomena are not directly observable in experiments. In behaviour-based experiments, various—and non-linearly interacting—factors can influence the results, for example in studies on sea turtles (Lohmann et al. 2007; Gould 2011; Brothers and Lohmann 2015). Moreover, no definitive confirmation has emerged for competing theories of orientation by birds, either with magnetic bodies (magnetosomes) in the beak or with sensors in the inner ear (Diebel et al. 2000; Gagliardo et al. 2009; Falkenberg et al. 2010; Wu and Dickman 2011).

A similar kind of discussion is observed with regard to the process of differential perception of chemical substances—the first stage in evolution of the sense of smell. Here, the question is how molecules in the olfactory receptors recognize differences between externally impinging molecules—without forming strong bonds. One hypothesis assumes a lock-and-key principle in which structures of impinging molecules form loose bonds with matching structures of receptor molecules, thus triggering the primary odour signal (Block et al. 2015). This hypothesis was challenged by observations in which organisms showed different responses to structurally completely identical molecules with different molecular weights. In these experiments, the lighter hydrogen atoms in the molecules were exchanged for the isotope deuterium, which is about twice as heavy. From this it was concluded that not only the structure but also the mass of external molecules is detected by the receptor molecules through vibrations (Turin 1996). In the hypothesis based on this finding, it is assumed that electrons in the receptor molecule pick up vibrational properties from the attached external molecule is, and that the resulting change in electron state is registered by the receptor molecule. In various theoretical approaches, the quantum physical process of "electron tunneling" has been studied, but no definitive evidence has been obtained (Kay and Stopfer 2006; Brookes 2010; Franco et al. 2011; Bittner et al. 2012; Gane et al. 2013; Stoneham et al. 2014; Tirandaz et al. 2015, 2017). The main reasons for this are the diversity of chemoreceptors and continuing inadequate knowledge of their properties (Solovyov et al. 2012; Derby et al. 2016).

4.5   Ubiquitous, but Not Perceivable, in Basic Life Processes

The preceding chapters may give the impression that evidence for quantum physical processes is to be found primarily in multicellular organisms. Such an impression is simply the outcome of attempts to present abstract quantum physical principles in conjunction with known, observable biological phenomena. This approach is by no means intended to open the way to any interpretation—utterly misguided in my view—according to which "higher development" of organisms, however that may be conceived, could be attributable to increasing permeation by quantum physical processes. The exact opposite view—that quantum physical processes determine all basic life processes—is far more consistent with observed results at cellular and subcellular levels.

Maxwell's demon provides a starting point for argumentation. In the thought experiment formulated by the physicist J.W. Maxwell (1831–1875), a tiny demon sorts molecules in a container into two different chambers according to their velocity. Despite the small amount of movement required to open and close a sliding partition, the demon consumes some energy. This illustrates the statement of the Second Law of Thermodynamics, according to which every preservation of order involves energy expenditure. According to the states of order in a living cell, an incalculable number of demons would be expected to intervene continuously to ensure regulation. Instead of demons, we find regulatory systems—often conforming to quantum-physical principles—that maintain dynamic order in cells with minimal energy expenditure. From a classical chemical perspective, the counter-argument could be offered that specially structured chemical catalysts—enzymes—regulate biochemical processes within cells (Berg et al. 2013). Detailed investigations of the reaction processes themselves have increasingly shown that the catalytic properties involved are primarily based on multi-step dynamic processes in time ranges between femtoseconds (10^{-15} s) and milliseconds (10^{-3} s). In fact, quantum physical processes ensure the precision and selectivity of the catalytic processes involved (Agarwal et al. 2002; Masgrau et al. 2006; Nagel and Klinman 2006, 2010; Pu et al. 2006; Klinman 2009; Schwartz and Schramm 2009; Nashine et al. 2010; Boekelheide et al. 2011; Wang et al. 2014).

Development of concepts regarding the self-regulatory processes of cells is difficult even for professional specialists. Here, one biochemistry textbook uses the everyday example of using traffic lights to control movement of vehicles as an analogy for regulation of metabolic processes (Berg et al. 2013). In fact, the example of a traffic roundabout would be far more suitable, as it incorporates both features of structural preconditions and local regulation depending on inflows from the connecting roads. Equally misleading is symbolic description of cellular metabolic processes with the term "burn". The only useful point here is the shared feature that energy is converted both in cells and in combustion processes, such that both systems are wholly subject to the empirical statement in the Second Law of Thermodynamics:

> No continuously operating machine does nothing but produce mechanical work and cool a thermal reservoir. (Vogel 1974)

Key differences, however, reside in the processes of energy conversion and the realms of work performance. In a technological context, the statement of the Second Law can be easily illustrated by a steam engine, in which heat is generated with a continuously supplied fuel source and the steam thus generated releases energy in the form of mechanical work through a suitable device. In the realm of biology, on the other hand, a not inconsiderable part of the converted energy is expended to maintain internal order. At the same time, in order to maintain viability, an organism must be able to balance differences between external energy supply and internal and external energy demand. Such requirements can be met only if each individual transformation step can be activated or arrested as needed. One prerequisite for this is structural differentiation of biomolecules into "energy reservoirs", which can be tapped only by overcoming the lowest barrier in each case (Fig. 4.10b). Therefore, activation of reaction processes requires catalytic energy that can be recovered as they proceed. Abiotic reactions, on the other hand, run spontaneously under suitable conditions—until the potential energy has been completely dissipated (Fig. 4.10a).

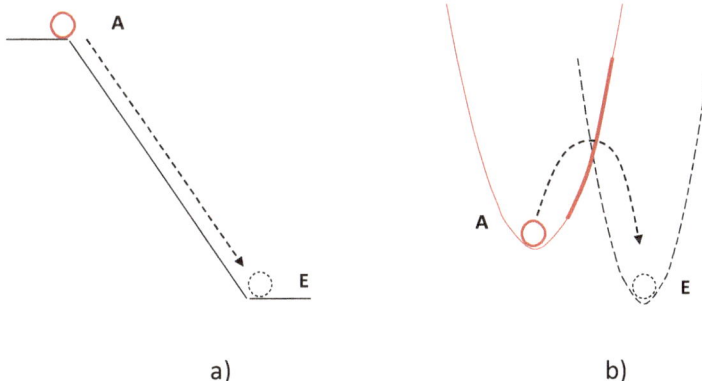

a)   b)

Fig. 4.10 Schematic comparison between the conditions of energy conversion in abiotic and biotic processes. (**a**) From initial states (A) to final states (E); (**b**) biological processes with controlled intermediate steps in "energy reservoirs"

In many biochemical processes, the energy needed to overcome the energy barrier in cells is reduced through quantum physics by the so-called tunneling effect (Fig. 4.11). Originally, the term was used for simplified description of overcoming energy barriers in the radiation of alpha particles from atomic nuclei (Bleck-Neuhaus 2013). In the context of biochemistry, however, it is used for overcoming barriers in biochemical processes involving both protons and electrons (Page et al. 1999; Lin et al. 2005; Hayashi and Stuchebrukhov 2010; Fleming et al. 2011a, b; Schreiner et al. 2011; Choi et al. 2012; Klinman and Kohen 2013; Trixler 2013). But this terminology easily leads to confusion when it is understood in connection with everyday experience, because the process concerned has nothing whatsoever to do with physically tunneling under mountains. In analogy with Förster's energy resonance theory (Lerner et al. 2018), radio waves provide a better indication from daily experience—for example, with mobile phones—as they can also penetrate walls. In the quantum-physical tunneling effect, however, the wave form is only an intermediate stage between the excited particles of the starting molecule and the final molecule. As an example, remember that electrons originate from the electron shells (Fig. 4.8a), which are no longer assigned to individual atoms in biochemical molecules—conjugated polyenes (Johnson et al. 1955; Berg et al. 2013). In a biological context, protons, on the other hand, are hydrogen nuclei without electron shells, the transfer of which changes the electrical charge of the molecules involved.

Discussion of the significance of tunneling effects in biological processes clearly reveals the different approaches of participants. From a laboratory perspective involving isolated reactions in glass vessels (*in vitro*)—in contrast to an approach involving investigation of processes in active cells (*in vivo*)—no need to consider quantum physical processes is recognized (Phillips and Quake 2006; Kohen 2012). This difference in viewpoints also provides clues to the role of quantum physical processes in living cells. This becomes particularly clear in the cells of archaeans and bacteria, which usually have no internal

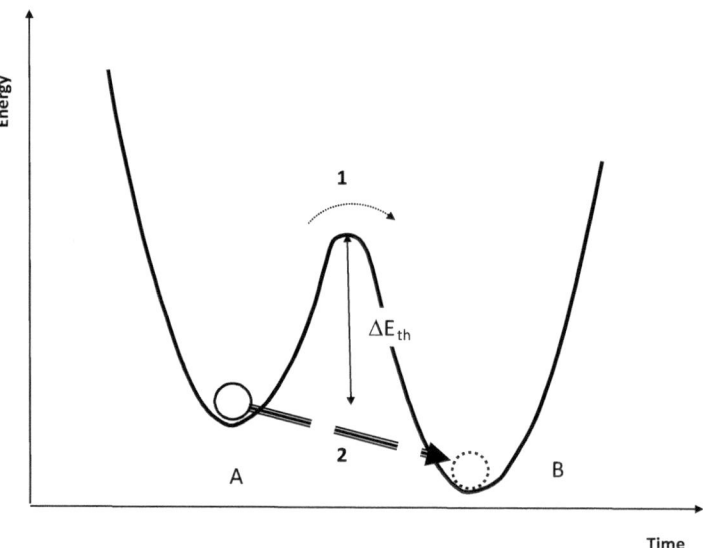

Fig. 4.11 Schematic diagram of the quantum-physical tunneling effect. As a result of wave-particle duality, quanta—electrons or protons—can penetrate energetic barriers (ΔE_{th}) between the states A and B with a lower energy expenditure (path 2) than in non-quantum physical processes (path 1)

differentiation through membranes and yet—often under extreme environmental conditions—are the most successful organisms in evolutionary terms. With classical laboratory methods, all physiological reactions would collapse within a very short time and the outcome would be an unstructured mush of differently sized molecules.

In the living cell, by contrast, through dynamic variation in their structures, protein molecules provide the framework conditions for an orderly course of reactions through quantum-physical tunneling effects. Transfer takes place at a higher velocity between proteins than in water or in a vacuum; moreover, greater transfer distances are feasible (Gray and Winkler 2005; Guo et al. 2016; Richardson et al. 2016). According to empirical and theoretical investigations so far conducted, electrons can tunnel over greater distances (Moser et al. 1992, 2008; Langen et al. 1995; Winkler 2000; Tezcan et al. 2001; Gray and Winkler 2005; Li et al. 2006; Hayashi and Stuchebrukhov 2010, 2011), and coupled tunnel sections are also possible (Gray and Winkler 2009; Choi et al. 2012). Note that the logarithmic scale of the vertical time axis must be borne in mind when viewing the schematic illustration. Furthermore, the energy content of the quanta decreases with increasing transmission time. At the orders of magnitude considered, long-range structural modifications of proteins influence the tunneling of electrons (Borgis et al. 1989; Nagel and Klinman 2006; Wang et al. 2006; Iyengar et al. 2008; Hay et al. 2009; Klinman 2009; Boekelheide et al. 2011; Klinman and Kohen 2013).

Because of the smaller tunneling distances of protons, their transmission is greatly influenced by small-scale modifications of protein structures (Bahnson et al. 1997;

Masgrau et al. 2006; Meyer et al. 2008). Presumably, interactions between proteins and DNA are also based on tunneling effects (Löwdin 1963). Of great importance for the distribution of organisms in extremely cold or hot habitats is the limited temperature-dependence of certain protein reactions involving tunnel processes (Agarwal et al. 2002; Knapp and Klinman 2002; Masgrau et al. 2004; Limbach et al. 2006; Hay et al. 2008).

Tunneling effects also play an essential role in energy transfer, which has already been mentioned in connection with photosynthesis. Electrons excited by photons transfer energy from the antennae to the reaction centres. Intensive investigations of bacteriochlorophyll systems—named Fenna-Matthews-Olson complexes (FMO) after their discoverers—have revealed efficiencies reaching up to almost 100 percent (Permentier et al. 2000; Engel et al. 2007; Müh et al. 2007; Mohseni et al. 2008; Plenio and Huelga 2008; Read et al. 2008; Caruso et al. 2009; Cheng and Fleming 2009; Rebentrost et al. 2009; Hoyer et al. 2011; Shim et al. 2012; Wong et al. 2012; Chenu and Scholes 2015). In addition, evidence has been found for an influence of memory effects on the transmission efficiency attained (Fujita et al. 2012).

As already mentioned, proteorhodopsins can also transfer photon energy directly to protons (Serrano et al. 2004; DeLong and Béja 2010; Kouyama et al. 2010; Slamovits et al. 2011). This enables ongoing regeneration of the central energy carrier of cells—adenosine triphosphate (ATP)—in archaeans. Conversion of proton energy into the biologically usable energy of ATP takes place in what are probably the smallest "motors" in the world, with an efficiency approaching 100 percent. In archaeans and bacteria, the motors are located in cell membranes; in eukaryotic cells, they are housed in the membranes of vacuoles and mitochondria, the organelles responsible for energy metabolism (Fig. 4.12). The extremely variable "motors" are driven by the proton motor force, the sum of chemical and electrical gradients between the regions separated by a membrane (Lolkema et al. 2003; Mulkidjanian et al. 2007; Berg et al. 2013; Schlegel et al. 2012; Vargová et al. 2021). This ensures separation of the end product from the catalyst—the ATPase—after the reaction of adenosine diphosphate (ADP) with phosphate to form ATP (Dimroth et al. 1998; Sambongi et al. 1999; Stock et al. 1999, 2000; Weber and Senior 2003; Capaldi and Aggeler 2002; Mulkidjanian et al. 2007; Berg et al. 2013; Veshaguri et al. 2016).

An even smaller motor—consisting of three protein components—is employed by viruses during reproduction to insert DNA into the already completed envelopes (capsids). Current knowledge indicates that this motor has no rotating parts, but uses electrostatic forces for transportation. Viruses "construct" the motor, while its "components" and the energy required are taken from the parasitized bacterial cell (Sun et al. 2008, 2010; Bustamante et al. 2011; Casjens 2011; Allemand et al. 2012).

It was not until 2008 that one of the presumably most ancient forms of energy regulation with quantum physical characteristics in cells was described for the first time (Li et al. 2008). In an oxygen-free (anaerobic) environment, archaeans and bacteria balance energy between a reaction with energy excess (exergonic) and a reaction with energy requirement (endergonic) through an electron pair. In sum, both reactions must result in at least a small

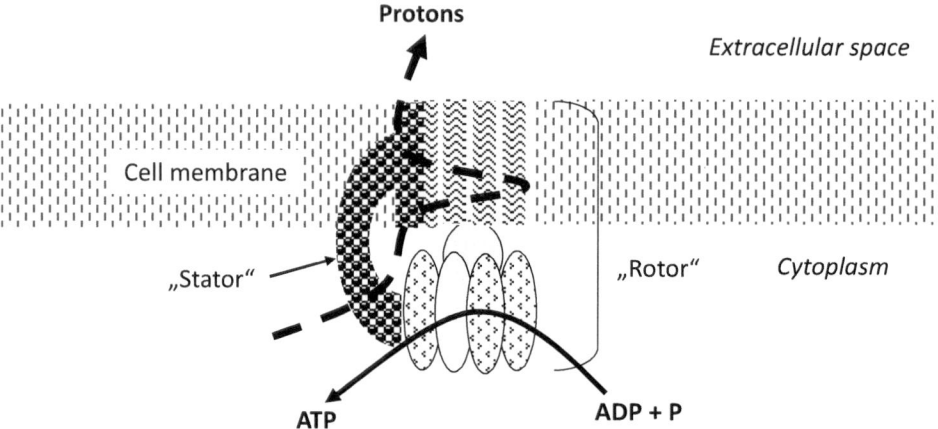

Fig. 4.12 Simplified diagram of the proton motor for regenerating adenosine triphosphate (ATP). The "rotor" is driven by transmission of the proton motor force in its upper part during transfer of protons from inside to outside. Regeneration is regulated by contact between ATPase and the reaction partners ADP and phosphate, with subsequent separation of the ATP in the lower part of the rotor. Greatly simplified version of information from different sources; see text

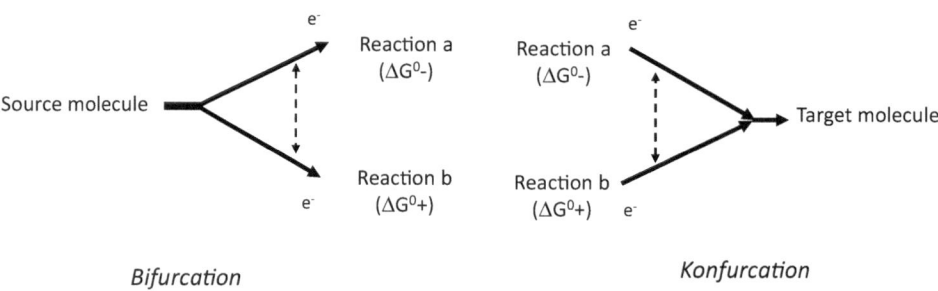

Fig. 4.13 Simplified overview diagrams of the two forms of electron bifurcation in archaeans and bacteria. Abbreviations and explanation of signs: e = electrons; ΔG^0 = changes in free enthalpy; dashed line symbolizes the energy balance between the two reactions. Note: The signs symbolize spontaneously possible ("downhill") reactions with ($-$) and reactions dependent on energy input ("uphill") reactions (+)

surplus of free energy. Although the electron pair can be transferred from one donor molecule to two reactions (bifurcation) as well as from two reactions to one recipient molecule (confurcation), the term electron bifurcation has become commonplace (Fig. 4.13) (Sieber et al. 2010; Kaster et al. 2011; Nitschke and Russell 2011; Huang et al. 2012; Martin-Delgado 2012; Poehlein et al. 2012; Schuchmann and Müller 2012; Branscomb and Russell 2013; Buckel and Thauer 2013; Chowdhury et al. 2014; Okamoto et al. 2014; Demmer et al. 2015; Hess et al. 2016; Metcalf 2016; Peters et al. 2016). Electron bifurcation plays a central role in energy conversion processes of all organisms, under both oxygenic and anoxygenic conditions. The catalytic effect is based on a combination of efficient energy transfer and adaptive self-regulation—presumably through interactions between the two electrons. Model calculations underpin the quantum physical features of these processes (Tacchino et al. 2019; Wise et al. 2021; Zuchan et al. 2021).

4.6   Boundaries Between Quantum Physical and Macroscopic Phenomena: Or, Why All Organisms Consist of Cells

References to quantum physical phenomena in biological processes raise the question of central differences between biotic and abiotic developments. The answer is readily provided by reference to the capacity for self-organization. However, the dimensions on which this ability should be present currently remain open. Ideal protection for avoiding answers is provided by "building block arguments", whether they be based on amino acids, proteins or genes—best illustrated by the so-called "central dogma" of genetics. From this perspective, all biological phenomena beyond the building blocks disappear, along with any need for further discussion. It is therefore unsurprising that in the ranks of physics or chemistry the opinion is held that biological processes are nothing more than physical or chemical processes. Physical and chemical regularities naturally also play a role in biological processes—interestingly, most strongly where they would be least expected according to the classical paradigm of evolutionary science—as will be justified in later chapters. Yet the sciences themselves provide the most detailed evidence that they work for components of organisms but not with biological processes. *To be scientifically acceptable, any investigation in physics or chemistry must be able to demonstrate connections between causes and effects and yield predictions regarding experimental results. This means that they are able describe with great accuracy cause-effect relationships of detailed parts of biological processes. However, they cannot provide explanations for the countless organized dynamic interactions in living biological systems.*

Given that no solutions for complex biological systems can be expected from causal approaches, an approximation will be attempted from a phenomenological perspective. It can be deduced from the previously mentioned phenomenon of self-organization that this goal can be achieved only if at least some of the factors influencing biological processes exhibit a minimal degree of independence from environmental factors. In the search for answers, a first approach is provided by knowledge of the bandwidths of environmental conditions used by living organisms. Among organisms generally, the largest ranges are found in archaeans and bacteria. Here, for example, the energetically particularly relevant temperature range extends from a few degrees Celsius below zero to around 120 °C (Amend and Shock 2001; Madigan et al. 2003; Lever et al. 2015). The liquid state of surrounding water is crucial for maintaining vital functions in cells—only a few organisms can endure desiccation, for limited periods of time (Alpert 2005). Interestingly, improvements in methods for physical investigation of the liquid state of water are providing increasing evidence of the internal dynamics of molecules in time dimensions ranging from femto- (10^{-15}) to picoseconds (10^{-12}). In these time dimensions, ion distributions and molecular structures change due to the bonding dynamics of hydrogen atoms. The hydrogen ion concentration (pH value[3]), which has been recognized since the beginning of the twentieth century as a measure of acidic or basic properties, describes only the static average state of solutions. The quantum physical properties of water molecules become even more apparent in interactions with other molecules and ions (Geissler et al. 2001; Fayer 2012; Kühne and Khaliullin 2013; Hassanali et al. 2014; Richardson et al. 2016; Wolke et al. 2016). Given the very incomplete state of current evidence and knowledge, a serious assessment of the significance of the physical properties of water for terrestrial biological evolution it is not yet possible.

At the lower limits of the temperature range, organisms can remain active only in solutions such as seawater. At the upper limits, organisms remain active only if sufficient water pressure prevents evaporation—for example in the vicinity of deep-sea hydrothermal vents. From the wide range observed, it is evident that biological processes are largely independent of the thermal conditions of the environment. In case you happen to disagree because of a heat or cold wave that you have just experienced, it should be noted that the given range is applicable not to a single species but across all known organisms. For an organism with a temperature optimum of around 90 °C, 80 °C may already be suboptimal. Extreme values given for eukaryotes (Møberg et al. 2011), on the other hand, usually refer to resting stages that are maintained until favourable conditions are attained.

Free-living organisms can achieve relative independence from environmental temperature conditions only through pre-existing internal conditions. An indication of this is provided by the publication of Phillips and Quake (Phillips and Quake 2006), who argue that the bonding energy of electrons below the order of nanometres (10^{-9} m) is higher than the thermal energy. This is also the range in which the prevalent chemical bonds of biological molecules are found. At sizes below the critical threshold, the structures of biochemical molecules achieve relative independence from external temperature conditions. As comparisons between different organisms show, the specific temperature resistance of proteins depends on their respective structures. As an example, consider a comparison between effects of boiling water on vegetables or hyperthermophilic archaeans from geysers—in the former case, the cell contents decompose, while in the latter case the organisms reside in their comfort zone.

Here, a question also arises regarding the dimensions at which quantum-physical processes merge into processes of classical mechanics and thus become subject to entropy. Theoretical approaches to this have been developed on the basis of quantum information—specifically quantum entanglement. In very simplified terms, this is based on the assumption that in entangled quantum system the state is defined only by the overall system constituted. These properties are also assumed to be central characteristics of biological processes. In a mathematical model, environmental factors and the size of the systems involved, among other things, were used as prerequisites for quantum entanglement. The

---

[3] Given as the decadic logarithm of the hydrogen ion concentration.

model described quantum entanglements for electrons and protons. Because of the multiple interactions between the factors taken into account in the model, it was not possible to determine absolute sizes for the transition ranges (Zurek and Paz 1994; Zurek 2002). It is even more difficult to estimate size dimensions for the multi-layered processes in living cells (McFadden 2000)—which cannot be described by the model. Nevertheless, it can be hypothetically assumed that in all living cells signal transmission, information processing and storage are based on quantum physical processes.

This line of reasoning is based on a number of pointers: In the meantime, quantum-physical phenomena have been experimentally proven and theoretically justified for signal transmission between photons, between photons and solids and between solids (Barrett et al. 2004; Riebe et al. 2004; Sherson et al. 2006; Nalbach et al. 2010; Pfaff et al. 2014). Experimentally, even at room temperatures, short-term information storage—lasting as long as milliseconds—can be demonstrated on a quantum-physical basis with accuracies of up to 98 percent (Chen et al. 2008; Zhao et al. 2008; Cooper 2011; Hosseini et al. 2015). An additional group of indications results indirectly from findings of molecular biological studies of signal transmission in cells. With the chosen approaches to investigation, changes and connections between proteins can be identified in many signal transmissions. However, the reaction times that can be described in this way lie in the range of seconds and higher. This does not explain the decisive regulatory processes for phenomena at the molecular level and experimentally observable variations (noise) in signal transmissions (Bray 1995; Raser and O'Shea 2005; Cai et al. 2006; Newman et al. 2006; Dehmelt and Bastiaens 2010; Alberts et al. 2015). The third group of circumstantial evidence relates to observed reactions and signal processing capacities in supposedly "simple" cells, especially bacteria. Synchronous processing of different signals, the capacity to learn and communication between different cells have all been documented. In some publications, the observed responses have even been interpreted as bacterial "intelligence" (Jiang et al. 1997; Wolf and Arkin 2003; Jacob et al. 2004; van der Horst et al. 2007; Mole et al. 2007; Trevors and Mason 2011; Majumdar and Pal 2017). *In sum, it can be postulated that the fundamental processes of information processing and storage in all organisms are of a quantum-physical nature—but that their macroscopically observable expression is conditioned by prevailing environmental conditions.*

Here, an additional question arises as to how the random properties of quantum physical processes can be transformed into the deterministic properties of the macroscopic world. The random properties of quantum physical processes are often characterized by the double-slit experiment. In this, photons are transmitted from a light source through two closely spaced slits and the spatial distribution of incoming photons is determined. With a sufficiently large number of photons, a density distribution of the arrival points can be measured, but this does not permit any conclusions to be drawn regarding the passage of photons through the individual slits. Nor can it be predicted through which slit the next photon will pass (Feynman 1985). Usually, the results of the double-slit experiment are explained by the wave properties of light (Freudenstein and Kulisch 2016). According to all known research findings, quanta—electrons and photons—determine the processes of biological processes, so it makes sense to start from a quantum-related interpretation of the double-slit experiment when searching for transitions between the quantum-physical and the macroscopic worlds.

In discussions of quantum physical features of biological processes (Ball 2011), the essential feature of *functional redundancy* at the molecular level is usually forgotten. In contrast to designs of laboratory experiments with single, isolated components, many components always operate in parallel in cells. Even in the smallest cells, the number of components—of proteins, for example—can be several hundreds or thousands (Cohen and Golden 2015)—a fundamental prerequisite for coupling quantum physical processes with those of classical physics. With large numbers of processes acting in parallel, the inherently random results of quantum physical processes (Feynman 1985; Wilde 2017) may be interpreted by means of a representative value and can provide unambiguous signals for subsequent chemical and physical processes or information (Fig. 4.14). Because of the random properties involved, each repetition of the processes provides new input values and thus also slightly modifies characteristic values. This makes it possible to explain the

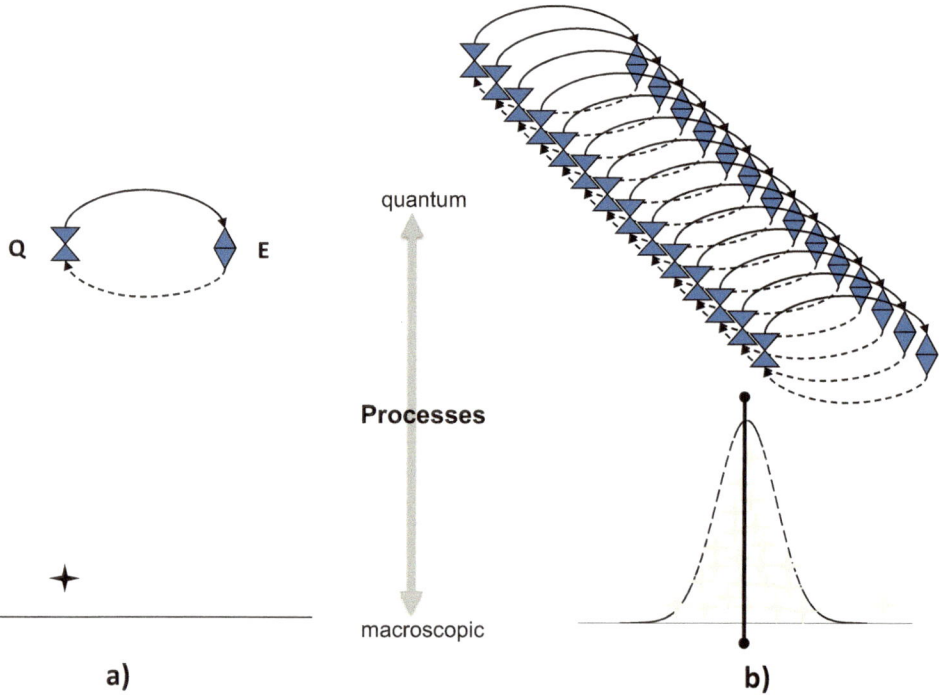

Fig. 4.14 Schematic diagram of the effect of multipartite redundant transmission systems in the case of results for random transference in the transition between quantum physical processes and macroscopic processes. (**a**) In the case of a simple system, the random value is adopted; (**b**) for redundant systems, the random individual values can be subsumed by a representative value (e.g. mean value—sketched with the vertical line)

specific properties of biological processes that exhibit a high degree of self-similarity—but can never be precisely predicted in concrete individual cases. It also makes it easier to understand why monocausal approaches to investigation with singular, isolated systems cannot provide adequate answers regarding fundamental properties of life functions.

The effects of the processes concerned can be observed under laboratory conditions in variations in phenotypic characteristics of clonal bacterial populations. Although all bacteria have identical gene sets and are exposed to the same, constant environmental conditions, subpopulations with different physiological characteristics arise randomly—without genetic mutations (Kirschner et al. 2000; Balaban et al. 2004; Veening et al. 2008).

Functionally redundant systems— within their performance bandwidths—are also more robust against disturbance by external influences. Strongly deviating individual values temporarily influence the "representative value", which once again approaches the normal value in subsequent transmission processes. In addition, filtering processes, for example, can suppress perception of inappropriate signals and only allow signals above a certain threshold value to pass through, as in the case of stochastic resonance (Hänggi 2002).

A prerequisite for an orderly flow of the processes described is their integration into structured conditions, such as those offered by biological cells, that are adequately shielded from external influences. For macroscopic dimensions, the quantum physical hypothesis for the foundations of life (Fig. 4.16a) requires that a single cell must suffice for self-preservation of all life functions. In concrete terms, this shows that multicellularity cannot be an inevitable end-point but only a special case of evolutionary processes.

4.7   How Does Time Play Its Part?

Achievement of independence from thermal energy enables the maintenance of internal order in cells of organisms. However, its dynamic maintenance requires the coordination of molecular processes in the cells and supplementation by other forms of energy (Mori et al. 2015). The various forms of quantum physical energy transfer are preferred candidates to meet this need. In conjunction with potentials for the dynamic structural modification of biological molecules, a theoretical basic system for maintenance of functions in all living cells can be postulated. In fact, this also makes it easier to understand why living organisms—despite their globally small share of thermal energy flows (Fig. 4.15)—have been able to maintain their functionality over billions of years and have avoided subjection to abiotic decay processes.

The preservation of basic life functions in the maelstrom of abiotic processes provides the strongest evidence for their quantum-physical foundations. The consistency of the self-reproduction capacity of all organisms can be understood most plausibly in light of the time reversal or time invariance of quantum systems (Feynman 1993; Bleck-Neuhaus 2013; Susskind and Friedman 2015; Schiansky et al. 2023). According to this principle, original states of order can be restored as often as desired and do not necessarily decay over the course of time—as would be required by the Second Law of Thermodynamics

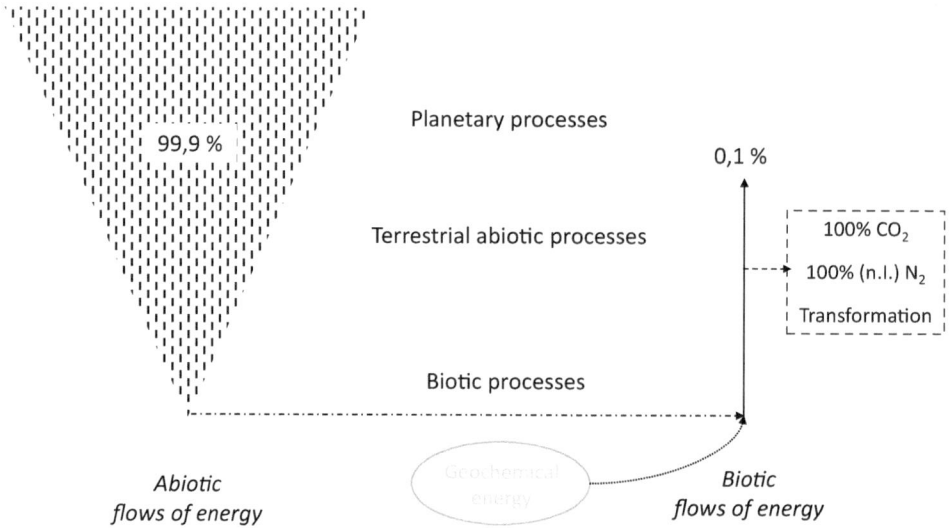

Fig. 4.15 Schematic diagram of the current relationships between average global energy fluxes in abiotic and biotic systems. In the latter, contributions to conversion of atmospheric carbon dioxide (CO_2) and non-industrial (n.I.) conversion of atmospheric nitrogen are shown as examples. Further explanations in the text

(Vogel 1974). Life apparently functions without Maxwellian demons. Possible indications of this are provided by the phenomena of electron bifurcation (see Sect. 4.5).

The principle of time invariance also makes it easier to understand that life in its general manifestation has been able to renew itself continuously for billions of years. Autonomous, continuous repairs—for example of not uncommon breaks in DNA—also ensure preservation of the functionality of living cells (Kozubek 2016). Only in this way can the state of genetic information be controlled and its implementation monitored. Analogies to self-correcting quantum codes also provide evidence here for quantum physical properties of fundamental biological processes (Laflamme et al. 1996; Bennett et al. 1998). Under constantly changing framework conditions, the postulated *quantum-physical reproductive system* in the form of a *reprosome* ensures the continuous reproductive capacity of organisms. Because of its time-independence, this system can also have no goal orientation. However, its central role is to be seen in continuous stimulation of a search for suitable environmental conditions to maintain reproductive capacity (Fig. 4.16a). Evidence for physical implementation can be found among viruses, which at a minimum require nucleic acids and proteins to maintain the capacity to self-replicate (Dimmock et al. 2007). The suggested existence of a central set of genes that is essential for self-replication—so-called "ultraconservative elements"—has been invalidated in laboratory experiments (McCormack et al. 2012). Following experimental removal of sections of the ultra-conservative elements, mice remained healthy and were also able to reproduce without any problem (Ahituv et al. 2007).

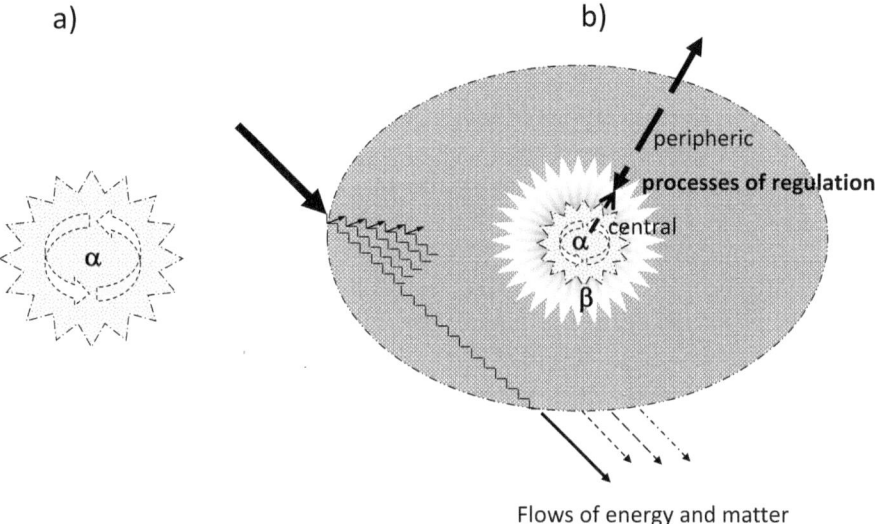

a)   b)

peripheric

processes of regulation

central

α

β

Flows of energy and matter

Fig. 4.16 (**a**) Schematic representation of the hypothetical reprosome, or temporally invariant quantum-physical reproductive system; (**b**) which is subject to temporal changes due to physical conversion into matter (molecular systems)

With physical conversion into matter, the reproductive system becomes subject to the conditions of the Second Law of Thermodynamics and therefore requires the regular transformation of free energy and substances to maintain its functionality (Fig. 4.16b). Accordingly, every organic system is also subject to temporal changes.

Because of variability in environmental conditions, the specific forms of transformation must be able to adapt to those conditions—within the respective biological bandwidths. To maintain the capacity to reproduce, the reprosome must be protected from deleterious changes caused by environmental factors (Ogryzko 2008; Saade and Ogryzko 2014). This requires a peripheral regulatory system (β in Fig. 4.16b), by means of which the requirements of the central reproductive system (α in Fig. 4.16b) can be adjusted to the actual environmental potentials of abiotic and biotic nature. In relation to the explanations for Fig. 4.14, an environmental context thus comes into play for interpretation of the signals concerned. This can be understood both in terms of changes in the principle of signal interpretation for individual processes as well as with regard to the relative weighting of signals from different processes.

In order to achieve a high probability of reproduction, the "material" of transformation systems must fulfill specific and sometimes contradictory requirements. For example, the "basic building blocks" of the material—chemical elements—must be ubiquitously available and, under the environmental conditions for active organisms, appropriate for synthesis into suitable molecules. Molecules with a high selectivity in their reactions are suitable for fulfilling the requirements mentioned. Appropriate molecules should be reactive on the one hand and inhibitory on the other. They should react in close proximity with other molecules and only with the relevant required partners. In the context of the transformation systems involved, the uptake, storage and release of information, energy and substances should be able to occur in a controlled manner. For this, transformation systems of the individual units must be able to seal themselves off from unsuitable environmental factors, yet remain open to appropriate environmental factors. The multiple requirements concerned are met through use of a limited number of chemical elements—first and foremost carbon, oxygen and hydrogen—as basic components (Mason and Moore 1985; Williams and Fraústo da Silva 2006). Biological macromolecules are formed mainly through combination of simple basic molecules and structural differentiation; additional chemical elements are used to a limited extent (Berg et al. 2013). The orders of magnitude involved can be outlined using amino acids and proteins as examples (Beiko et al. 2005; Belitz et al. 2008; Buddrus 2011; Berg et al. 2013). The spatially structured molecules of amino acids usually contain atoms of four chemical elements—carbon, hydrogen, oxygen and nitrogen—and occasionally sulphur as well. Of a few hundred known amino acids, 20 form the basic building blocks for several hundred thousand different proteins. Decisive for the diversity achieved are the structures of the molecules, in terms of both the number of atoms incorporated and their spatial arrangement. In the macromolecules of proteins, up to four levels of structural formation can be distinguished. Although their structures can be represented by biochemical methods, their dynamic behaviour can only be adequately explained with quantum physical methods (Phillips and Quake 2006; Ditzler et al. 2009).

With this functional repertoire, all organisms can pursue the central goal of all biological processes—the preservation of reproductive capacity—adapted to prevailing environmental conditions in the most varied and non-deterministic ways. I might have chosen the formulation "preservation of life" instead of "preservation of reproductive capacity". However, this would have inadmissibly narrowed the perspective, since viruses are also subject to this principle.

Viruses have presumably used archaeans and bacteria for their replication since the earliest stages of their evolution, thus maintaining a dynamic of attack and defence strategies. However, their long-term conservation is only ensured if they do not completely destroy populations of their host organisms. Genome analyses of viruses show that their distribution is also influenced by abiotic environmental factors, because their genome patterns reflect the habitats of prokaryotes (Krupovic et al. 2011; Datsenko et al. 2012).

Lack of determinism in biological responses is most clearly demonstrated by interactions between bacteria and multicellular organisms. In connection with various influencing factors, many groups of bacteria can live with multicellular organisms, either in symbiosis or triggering diseases. Interpretation of environmental conditions by the epigenetic systems of bacteria[4] in conjunction with social coordination (quorum sensing) within bacterial populations is decisive for the corresponding effect (Casadesús and Low 2006; Adam et al. 2008; Rumbaugh et al. 2009; de Jong et al. 2011; Rainey et al. 2011: Bierne et al. 2012; Bondarev et al. 2013; Sachs 2013). This reveals the multi-stability of

---

[4] The term "epigenetics" was defined by Waddington; van Speybroeck (2002).

interactions between organisms and the relative meaning of *a priori* classifications such as "symbiont"or "pathogen".

4.8   An Attempted Conclusion

Using directly observable phenomena, the initial sections of this chapter showed the extent to which quantum physical and classical physical principles interact. Because of the need for clarity and various interdependences, however, structures for morphological implementation—eyes, for example—have not mentioned and will be dealt with only at an appropriate point elsewhere. Also completely omitted was a discussion of quantum physical interpretations of neural processes and quantum entanglements with interindividual effects, possibly in connection with certain forms of mimicry (Hameroff et al. 2002; Woolf and Hameroff 2001; Atmanspacher 2004; Schwartz et al. 2005; Tarlaci 2011; Chang 2012; Briegel and Popescu 2014). Similarly not discussed—because they lie outside the scope of this work—were questions of similarities or differences between quantum physical and quantum chemical approaches (Logunov and Schulten 1996; Damjanović et al. 2001; Frutos et al. 2007; Khrenova et al. 2010). However, a hypothesis was presented to the effect that the basic regulatory processes of all organisms are of a quantum-physical nature.

Overall, available evidence indicates that quantum physical basic principles are of central importance for the origins and subsequent evolution of organisms. It can be assumed that the greatest degrees of freedom for self-preservation and modification are to be found in the effective dimensions of quantum physical processes. In the transition from the quantum physical to the macroscopic world, a key role is probably played by *functional redundancy* of molecular processes in cells. Summarized interpretation of random individual signals from a generally very large number of processes running in parallel permits derivation of a representative parameter. This makes it easier to understand why biological forms and processes show a high degree of self-similarity but not full correspondence. However, it also becomes more understandable why all organisms consist of cells and why unicellularity is the dominant form of life. From this perspective, multicellularity is to be seen as an optional—but by no means necessary—special case of evolution.

If these conclusions are correct, a number of consequences can be recognized. Of lesser importance is the fact that discussion of whether life arose on Earth in or space is rendered superfluous—the first evolutionary steps would have to have already taken place during the formation phase of the universe. It is furthermore conceivable that the quantum-physical reproductive system manifests itself in alternative life forms on other planets under suitable framework conditions. Of far greater consequence, however, will be any conclusions that answer the question regarding the fundamental characteristics of life. As mentioned in connection with quantum physics, many phenomena lack sufficient means of comparison to put them into words. It will be all the more difficult to proceed beyond this "pre-lingustistic" stage to the far more complex quantum biological processes. Perhaps we have to pay this price in order to achieve a better understanding of our limits and possibilities?

It can also be deduced from interactions between quantum physical and classical physical principles (Fig. 4.16), that the phenomena of evolution—observable with respect to organisms—are not based on any purposeful design principle. Characteristics and properties of organisms can evolve anew, but can also be lost again. Observable evolutionary processes are the result of multi-layered interactions of self-organizing systems with environmental factors, driven by maintenance of self-reproduction. The deliberately chosen formulation of the previous sentence is intended to make it clear that viruses—which are not usually counted as organisms—nevertheless play an essential role in evolutionary processes.

References

Abbott D, Davies PCW, Pati AK (eds) (2008a) Quantum aspects of life. Imperial College Press, London

Abbott D, Gea-Banacloche J, Davies PCW, Hameroff S, Zeilinger A et al (2008b) Plenary debate: quantum effects in biology: trivial or not? Fluct Noise Lett 8(1):C5–C26

Adam M, Murali B, Glenn NO, Potter SS (2008) Epigenetic inheritance based evolution of antibiotic resistance in bacteria. BMC Evol Biol 8:52

Agarwal PK, Billeter SR, Rajagopalan PTR, Benkovic SJ, Hammes-Schiffer S (2002) Network of coupled promoting motions in enzyme catalysis. PNAS 99(5):2794–2799

Ahituv N, Zhu Y, Visel A, Holt A, Afzal V et al (2007) Deletion of ultraconserved elements yields viable mice. PLoS Biol 5(9):e234

Alberts B, Johnson A, Lewis J, Morgan D, Raff M et al (2015) Molecular biology of the cell. Garland Science, New York

Alboresi A, Gerotto C, Giacometti GM, Bassi R, Morosinotto T (2010) *Physcomitrella patens* mutants affected on heat dissipation clarify the evolution of photoprotection mechanisms upon land colonization. PNAS 107(24):11128–11133

Alieva VO, Konzen KA, Field SF, Meleshkevitch EA, Hunt ME et al (2008) Diversity and evolution of coral fluorescent proteins. PLoS One 3(7):e2680

Al-Khalili J, McFadden J (2015) Der Quantenbeat des Lebens. Ulstein, Berlin

Allemand J-F, Maier B, Smith DE (2012) Molecular motors for DNA translocation in prokaryotes. Curr Opin Biotechnol 23(4):503–509

Alpert P (2005) The limits and frontiers of desiccation-tolerant life. Integr Comp Biol 45:685–695

Amend JP, Shock EL (2001) Energetics of overall metabolic reactions of thermophilic and hyperthermophilic archaea and bacteria. FEMS Microbiol Rev 25:175–243

Amor M, Busigny V, Louvat P, Gélabert A, Cartigny P et al (2016) Mass-dependent and -independent signature of Fe isotopes in magnetotactic bacteria. Science 352:705–708

Anetzberger C, Pirch T, Jung K (2009) Heterogeneity in quorum sensing-regulated bioluminescence of *Vibrio harveyi*. Mol Microbiol 73(2):267–277

Arma LH, Saitoh A, Ishibashi Y, Asahi T, Sueoka Y et al (2014) Red fluorescence lamellae in calcitic prismatic layer of Pinctada vulgaris shell (Mollusc, bivalvia). Opt Mater Express 4:9. https://doi.org/10.1364/OME.4.001813

Arndt M, Juffmann T, Vedral V (2009) Quantum physics meets biology. HFSP J 3(6):386–400

Arnold KE, Owens IPF, Marshall NJ (2002) Fluorescent signaling in parrots. Science 295:92

Arnoldi FGC, da Silva Neto AJ, Viviani VR (2010) Molecular insights on the evolution of the lateral and head lantern luciferases and bioluminescence colors in Mastinocerini railroad-worms (Coleoptera: Phengodidae). Photochem Photobiol Sci 9:87–92

Arwin H, Magnusson R, del Río LF, Åkerlind C, Muñoz-Pineda E et al (2014) Exploring optics of beetle cuticles with Mueller-matrix ellipsometry. Mater Tod Proc 18:155–160

Asano M, Khrennikov A, Ohya M, Tanaka Y, Yamato I (2015) Quantum adaptivity in biology: from genetics to cognition. Springer, Dordrecht

Astashkin R, Kovalev K, Bukhdruker S, Vaganova S, Kuzmin A et al (2022) Structural insights into light driven anion pumping in cyanobacteria. Nat Commun 13:6460

Atmanspacher H (2004) Quantum theory and consciousness: an overview with selected examples. In: Discrete dynamics in nature and society, vol 1. Hindawi, pp 51–73

Bagge LE (2019) Not as clear as it may appear: challenges associated with transparent camouflage in the ocean. Integr Comp Biol 59(6):1653–1663

Bahnson BJ, Colby TD, Chin JK, Goldstein BM, Klinman JP (1997) A link between protein structure and enzyme catalyzed hydrogen tunneling. PNAS 94:12797–12802

Balaban NQ, Merrin J, Chait R, Kowalik L, Leibler S (2004) Bacterial persistence as a phenotypic switch. Science 305:1622–1625

Ball P (2011) The dawn of quantum biology. Nature 474:272–274

Bandyopadhyay JN, Paterek T, Kszlikowski D (2012) Quantum coherence and sensitivity of avian magnetoreception. arXiv:1204.6528v2

Barrett MD, Chlaverini J, Schaetz T, Britton J, Itano WM et al (2004) Deterministic quantum teleportation of atomic qubits. Nature 429:737–739

Beatty JT, Overmann J, Lince MT, Manske AK, Lang AS et al (2005) An obligately photosynthetic bacterial anaerobe from a deep-sea hydrothermal vent. PNAS 102(26):9306–9310

Bechara EJH (2015) Bioluminescence: a fungal nightlight with an internal timer. Curr Biol 25:R269–R293

Beiko RG, Harlow TJ, Ragan MA (2005) Highways of gene sharing in prokaryotes. PNAS 102(40):14332–14337

Béjà O, Aravind L, Koonin EV, Suzuki MT, Hadd A et al (2000) Bacterial rhodopsin: evidence for a new type of phototrophy in the sea. Science 289:1902–1906

Béjà O, Spudich EN, Spudich JL, Leclerc M, DeLong EF (2001) Proteorhodopsin phototrophy in the ocean. Nature 411:786–789

Belitz H-D, Gosch W, Schieberle P (2008) Lehrbuch der Lebensmittelchemie. Springer, Berlin

Bennet M, McCarthy A, Fix D, Edwards MR, Repp F et al (2014) Influence of magnetic fields on magneto-aerotaxis. PLoS One 9(7):e101150

Bennett CH, Gács P, Li M, Vitányi PMB, Zurek WH (1998) Information distance. IEEE Trans Inf Theory 44(4):1407–1423

Berg JM, Tymoczko JL, Stryer L (2013) Biochemie. Springer Spektrum, Berlin

Bermudes D, Petersen RH, Nealson KH (1992) Low-level bioluminescence detected in *Mycena haematopus* basidiocarps. Mycologia 84(5):799–802

Beutel RG, Gorb SN (2001) Ultrastructure of attachment specializations of hexapods (Arthropoda): evolutionary patterns inferred from a revised ordinal phylogeny. J Zool Syst Evol Res 39:177–207

Bhushan B (2009) Biomimetics: lessons from nature—an overview. Philos Trans R Soc A 367:1445–1486

Bhushan B, Jung YC (2008) Wetting, adhesion and friction of superhydrophobic and hydrophilic leaves and fabricated micro/nanopatterned surfaces. J Phys Condens Matter 20:225010

Bierne H, Hamon M, Cossart P (2012) Epigenetics and bacterial infections. Cold Spring Harb Perspect Med 2:a010272

Biewer B (1997) Fuzzy-Methoden. Springer, Berlin

Bikle DD (2014) Vitamin D metabolism, mechanism of action, and clinical applications. Chem Biol 21:319–329

Biró LP, Vigneron J-P (2011) Photonic nanoarchitectures in butterflies and beetles: valuable sources for bioinspiration. Laser Photonics Rev 5(1):27–51

Bittner ER, Madalan A, Czader A, Roman G (2012) Quantum origins of molecular recognition and olfaction in drosophila. J Chem Phys 137:22A551

Bleck-Neuhaus J (2013) Elementare Teilchen. Springer Spektrum, Berlin

Block E, Jang S, Matsunami H, Sekharan S, Dethier B et al (2015) Implausibility of the vibrational theory of olfaction. PNAS 112:E2766–E2774

Boekelheide N, Salomón-Ferrer R, Miller TF (2011) Dynamics and dissipation in enzyme catalysis. PNAS 108(39):16159–16163

Bomati EK, Haley JE, Noel JP, Deheyn DD (2014) Spectral and structural comparison between bright and dim green fluorescent proteins in Amphioxus. Sci Rep 4:5469

Bondar VS, Puzyr AP, Purtov KV, Medvedeva SE, Rodicheva EK, Gitelson JI (2011) The luminescent system of the luminous fungus *Neonothopanus nambi*. Dokl Biochem Biophys 438:138–140

Bondarev V, Richter M, Romano S, Piel J, Schwedt A, Schulz-Vogt HN (2013) The genus *Pseudovibrio* contains metabolically versatile bacteria adapted for symbiosis. Environ Microbiol 15(7):2095–2113

Boomer SM, Pierson BK, Austinhirst R, Castenholz RW (2000) Characterization of novel bacteriochlorophyll-a-containing red filaments from alkaline hit springs in Yellowstone National Park. Arch Microbiol 174:152–161

Borgis DC, Lee S, Hynes JT (1989) A dynamical theory of nonadiabatic proton and hydrogen atom transfer reaction rates in solution. Chem Phys Lett 162(1–2):19–26

Bothe H-H (1993) Fuzzy logic. Springer, Berlin

Bowmaker JK, Dartnall HJA (1980) Visual pigments of rods and cones in a human retina. J Physiol 298:501–511

Bradlaugh AA, Fedele G, Munro AL, Hansen CN, Hares JM et al (2023) Essential elements of radical pair magnetosensitivity in Drosophila. Nature 615:111–116

Brady P, Cummings M (2010) Differential response to circularly polarized light by the jewel scarab beetle *Chrysina gloriosa*. Am Nat 175(5):614–620

Brady PC, Gilerson AA, Kattawar GW, Sullivan JM, Twardowski MS et al (2015) Open-ocean fish reveal an omnidirectional solution to camouflage in polarized environments. Science 350(6263):965–969

Branscomb E, Russell MJ (2013) Turnstiles and bifurcators: The disequilibrium converting engines that put metabolism on the road. Biochim Biophys Acta 1827:62–78

Bratanov D, Kovalev K, Machtens J-P, Astashkin R, Chizhov I et al (2019) Unique structure and function of viral rhodopsins. Nat Commun. https://doi.org/10.1038/s4167-019-12718-0

Bray D (1995) Protein molecules as computational elements in living cells. Nature 376:307–312

Briegel HJ, Popescu S (2014) A perspective on possible manifestations of entanglement in biological systems. In: Mohseni M, Omar Y, Engel GS, Plenio MB (eds) Quantum effects in biology. Cambridge University Press, Cambridge, pp 277–310

Brodl E, Winkler A, Macheroux P (2018) Molecular mechanisms of bacterial bioluminescence. Comput Struct Biotechnol J 16:551–564

Brookes JC (2010) Science is perception: what can our sense of smell tell us about ourselves and the world around us? Philos Trans R Soc A 368:3491–3502

Brothers JR, Lohmann KJ (2015) Evidence for geomagnetic imprinting and magnetic navigation in the natal homing of sea turtles. Curr Biol 25:392–396

Buckel W, Thauer RK (2013) Energy conservation via electron bifurcating ferredoxin reduction and proton/Na$^+$ translocating ferredoxin oxidation. Biochim Biophys Acta 1827:94–113

Buddrus J (2011) Grundlagen der Organischen Chemie. De Gruyter, Berlin

Buiteveld H, Hakvoort JHM, Douze M (1994) The optical properties of pure water. In: Jaffe JS (ed) SPIE proceedings on ocean optics XII, vol 2258, pp 174–183

Busemeyer JR, Wang Z, Townsend JT, Eidels A (2015) The Oxford handbook of computational mathematical psychology. Oxford University Press, Oxford

Bustamante C, Cheng W, Mejia YX (2011) Revisiting the central dogma one molecule at a time. Cell 144:480–497

Cai L, Friedman N, Xie XS (2006) Stochastic protein expression in individual cells at the single molecule level. Nature 440:358–362

Campbell D, Hurry V, Clarke AK, Gustafsson P, Öquist G (1998) Chlorophyll fluorescence analysis of cyanobacterial photosynthesis and acclimation. Microb Mol Biol Rev 62(3):667–683

Capaldi RA, Aggeler R (2002) Mechanism of the F_1F_0-type ATP synthase, a biological rotary motor. Trends Biochem Sci 27(3):154–160

Capretti A, Ringsmuth AK, van Velzen JF, Rosnik A, Croce R, Gregorkiewicz T (2019) Nanophotonics of higher-plant photosynthetic membranes. Light Sci Appl 8:5

Cardona T (2019) Thinking twice about the evolution of photosynthesis. Open Biol 9:180426

Carleton KL, Yourick MR (2020) Axes of visual adaptation in the ecologically diverse family *Cichlidae*. Semin Cell Dev Biol 106:43–52

Caruso F, Chin AW, Datta A, Huelga SF, Plenio MB (2009) Highly effective energy excitation transfer in light-harvesting complexes: the fundamental role of noise-assisted transport. arXiv:0901.4454v2

Carvalho LS, Knott B, Berg ML, Bennett ATD, Hunt DM (2011) Ultraviolet-sensitive vision in long lived birds. Proc Biol Sci 278:107–114

Carvalho LS, Pessoa DMA, Mountfort JK, Davies WIL, Hunt DM (2017) The genetic and evolutionary drives behind primate color vision. Front Ecol Evol 5:34

Casadesús J, Low D (2006) Epigenetic gene regulation in the bacterial world. Microbiol Mol Biol Rev 70(3):830–856

Casjens SR (2011) The DNA-packaging nanomotor of tailed bacteriophages. Nature 9:647–657

Chandler CJ, Wilts BD, Vignolini S, Brodie J, Steiner U et al (2015) Structural colour in *Chondrus crispus*. Sci Rep 5:11645

Chang Y-F (2012) Extensive quantum biology, application of nonlinear biology and nonlinear mechanisms of memory. NeuroQuantology 10(2):183–189

Chen Y-A, Chen S, Yuan Z-S, Zhao B, Chuu C-S et al (2008) Memory-built-in quantum teleportation with photonic and atomic qubits. Nat Phys 4:103–107

Chen Y-R, Zhang R, Du H-J, Pan H-M, Zhang W-Y et al (2014) A novel species of ellipsoidal multicellular magnetotactic prokaryotes from Lake Yuehu in China. Environ Microbiol. https://doi.org/10.1111/1462-2920.12480

Cheng Y-C, Fleming GR (2009) Dynamics of light harvesting in photosynthesis. Annu Rev Phys Chem 60:241–262

Chenu A, Scholes GD (2015) Coherence in energy transfer and photosynthesis. Annu Rev Phys Chem 66:69–96

Chin AW, Huelga SF, Plenio MB (2012) Coherence and decoherence in biological systems: principles of noise-assisted transport and the origin of long-lived coherences. Philos Trans R Soc A 370:3638–3657

Chiou T-H, Kleinlogel S, Cronin T, Caldwell R, Loeffler B et al (2008) Circular polarization vision in a stomatopod crustacean. Curr Biol 10:429–434

Choi M, Shin S, Davidson VL (2012) Characterization of electron tunneling and hole hopping reactions between different forms of MauG and methylamine dehydrogenase within a natural protein complex. Biochemistry 51(35):6942–6949

Chowdhury NP, Mowafy AM, Demmer JK, Upadhyay V, Koelzer S et al (2014) Studies on the mechanism of electron bifurcation catalyzed by electron transferring flavoprotein (Etf) and butyryl-CoA Dehydrogenase (Bcd) of *Acidaminococcus fermentans*. J Biol Chem 289(8):5145–5157

Chulliat A, Alken P, Nair M, Woods A, Maus S, et al (2015) The US/UK world magnetic model for 2015–2020. Technical report, National Geophysical Data Center, NOAA

Claes JM, Mallefet J (2011) Control of luminescence from lantern shark (*Etmopterus spinax*) photophores. Commun Integr Biol 4(3):251–253

Claes JM, Sato K, Mallefet J (2011) Morphology and control of photogenic structures in a rare dwarf pelagic lantern shark (*Etmopterus splendidus*). J Exp Mar Biol Ecol 406:1–5

Claes JM, Ho H-C, Mallefet J (2012) Control of luminescence from pygmy shark (*Squaliolus aliae*) photophores. J Exp Biol 215:1691–1699

Claes JM, Nilsson D-E, Mallefet J, Straube N (2015) The presence of lateral photophores correlates with increased speciation in deep-sea bioluminescent sharks. R Soc Open Sci 2:150219

Clegg B (2015) Quantentheorie in 30 Sekunden. Librero, Kerkdriel

Cockell CS (2000) Ultraviolet radiation and the photobiology of earth's early oceans. Orig Life Evol Biosph 30:467–499

Cohen SE, Golden SS (2015) Circadian rhythms in cyanobacteria. Microbiol Mol Biol Rev 79(4):373–385

Colley NJ, Nilsson D-E (2016) Photoreception in phytoplankton. Integr Comp Biol. https://doi.org/10.1093/icb/icw037

Cooper WG (2011) The molecular clock in terms of quantum information processing of coherent states, entanglement and replication of evolutionarily selected decohered isomers. Interdiscip Sci 3(2):91–109

Costa C, Vanin SA (2010) Coleoptera larval fauna associated with termite nests (isoptera) with emphasis on the "bioluminescent termite nests" from Central Brazil. Psyche. https://doi.org/10.1155/2010/23947

Cronin TW, Shashar N, Caldwell RL, Marshall J, Cheroske AG, Chiou T-H (2003) Polarization vision and its role in biological signaling. Integr Comp Biol 43:549–558

Cronin TW, Bok MJ, Marshall NJ, Caldwell RJ (2014) Filtering and polychromatic vision in mantis shrimps: themes in visible and ultraviolet vision. Philos Trans R Soc B 369:20130032

Crookes WJ, Ding L-L, Huang QL, Kimbell JR, Horwitz J, McFall-Ngai MJ (2004) Reflectins: the unusual proteins of squid reflective tissues. Science 308:235–238

Cupellini L, Bondanza M, Nottoli M, Mennucci B (2020) Successes and challenges in the atomistic modeling of light-harvesting and its photoregulation. Biochim Biophys Acta Bioenerg 1861:148049

Cusick KD, Widder EA (2014) Intensity differences in bioluminescent dinoflagellates impact foraging efficiency in a nocturnal predator. Bull Mar Sci 90(3):797–811

Damjanović A, Kosztin I, Schulten K (2001) Excitons in a photosynthetic light-harvesting system: a combined molecular Dynamics, quantum chemistry, and polaron model study. arXiv:physics/0107064v1

Datsenko KA, Pougach J, Tikhonov A, Wanner BL, Severinov K, Semenova E (2012) Molecular memory of prior infections activates the CRISPR/Cas adaptive bacterial immunity system. Nat Commun 3:945. https://doi.org/10.1038/ncomms1937

Davies PCW (2004) Does quantum mechanics play a non-trivial role in life? BioSystems 78:69–79

Davis MP, Holcroft NI, Wiley EO, Sparks JS, Smith WL (2014) Species-specific bioluminescence facilitates speciation in the deep sea. Mar Biol 161:1139–1148

Davis MP, Sparks JS, Smith WL (2016) Repeated and widespread evolution of bioluminescence in marine fishes. PLoS One 11(6):e0155154

Davis AL, Thomas KN, Goetz FE, Robison BH, Johnsen S, Osborn KJ (2020a) Ultra-black camouflage in deep-sea fishes. Curr Biol 30:3470–3476

Davis AL, Sutton TT, Kier WM, Johnsen S (2020b) Evidence that eye-facing photophores serve as a reference for counterillumination in an order of deep-sea fishes. Proc Biol Soc 287:20192918

Day JC, Tisi LC, Bailey MJ (2004) Evolution of beetle bioluminescence: the origin of beetle luciferin. Luminescence 19:8–20

De Carvalho JC, Cardoso LC, Ghiggi V, Woiciechowski AL, Vandenberghe LPS, Soccol CR (2014) Microbial pigments. In: Brar SK, Dhillon GS, Soccol CR (eds) Biotransformation of waste biomass into high value biochemicals. Springer, New York

De Cock R, Matthysen E (2001) Aposematism and bioluminescence: experimental evidence from glow-worm larvae (Coleoptera: Lampyridae). Evol Ecol 13:619–639

De Cock R, Matthysen E (2003) Glow-worm larvae bioluminescence (Coleoptera: Lampyridae) operates as an aposematic signal upon toads (Bufo bufo). Behav Ecol 14(1):103–108

de Jong IG, Haccou P, Kuipers OP (2011) Bet hedging or not? A guide to proper classification of microbial survival strategies. Bioessays 33:215–223

Deheyn DD, Latz ML (2007) Bioluminescence characteristics of a tropical terrestrial fungus (Basidiomycetes). Luminescence 22:462–467

Deheyn DD, Wilson NG (2010) Bioluminescent signals spatially amplified by wavelength-specific diffusion through the shell of a marine snail. Proc Biol Sci. https://doi.org/10.1098/rspb.2010.2203

Deheyn DD, Mallefet J, Jangoux M (2000) Evidence from polychromatism and bioluminescence that the cosmopolitan ophiuroid *Amphipholis squamata* might not represent a unique taxon. Life Sci 323:499–509

Dehmelt L, Bastiaens PIH (2010) Spatial organization of intracellular communication: insights from imaging. Nat Rev Mol Cell Biol 11:440–452

del Rio LF, Arwin H, Järrendahl K (2014) Polarization of light reflected from Chrysina gloriosa under various illuminations. Mater Today Proc 1S:172–176

DeLong EF, Béja O (2010) The light-driven proton pump proteorhodopsin enhances bacterial survival during tough times. PLoS Biol 8(4):e1000359. https://doi.org/10.1371/journal.pbio.1000359

DeMartini DG, Izumi M, Weaver AT, Pandolfi E, Morse DE (2015) Structures, organization, and function of reflectin proteins in dynamically tunable reflective cells. J Biol Chem 290(4):15238–15249

Demmer JK, Huang H, Wang S, Demmer U, Thauer RK, Ermler U (2015) Insights into flavin-based electron bifurcation via the NADH-dependent reduced ferredoxin:NADP oxidoreductase structure. J Biol Chem 290(36):21985–21995

Derby CD, Kozma MT, Senatore A, Schmidt M (2016) Molecular mechanisms of reception and perireception in crustacean chemoreception: a comparative review. Chem Senses 41:381–398

Desjardin DE, Capelari M, Stevani CV (2005) A new bioluminescent agaric from São Paulo, Brazil. Fungal Divers 18:9–14

Desjardin DE, Oliveira AG, Stevani CV (2008) Fungi bioluminescence revisited. Photochem Photobiol Sci 7:170–182

Diah SZM, Karman SB, Gebeshuber IC (2014) Nanostructural colouration in malaysian plants: lessons for biomimetics and biomaterials. J Nanomater 2014:878409

Diebel CE, Proksch R, Green CR, Nellson P, Walker MM (2000) Magnetite defines a vertebrate magnetoreceptor. Nature 406:299–302

Dimmock NJ, Easton AJ, Leppard KN (2007) Introduction to modern virology. Blackwell Publishing, Malden

Dimroth P, Kaim G, Matthey U (1998) The motor of the ATP synthase. Biochim Biophys Acta 1365:87–92

Ditzler MA, Otyepka M, Šponer J, Walter NG (2009) Molecular dynamics and quantum mechanics of RNA: conformational and chemical change we can believe in. Acc Chem Res 43(1):40–45

Djokic T, Van Kranendonk MJ, Campbell KA, Walter MR, Ward CR (2017) Earliest signs of life on land preserved in ca. 3.5 Ga hot spring deposits. Nat Commun. https://doi.org/10.1038/ncomms15263

Dong BQ, Liu XH, Zhan TR, Jiang LP, Yin HW, Liu F, Zi J (2010) Structural coloration and photonic pseudogap in natural random close-packing photonic structures. Opt Express 18(14):14430

Doucet SM, Meadows MG (2009) Iridescence: a functional perspective. J R Soc Interface 6:S115–S132

Douglas RH, Partridge JC, Dulai K, Hunt D, Mullineaux CW, Tauber AY, Hynninen PH (1998) Dragon fish see using chlorophyll. Nature 393:423–424

Duanmu D, Casero D, Dent RM, Gallaher S, Yang W et al (2013) Retrograde bilin signaling enables *Chlamydomonas* greening and phototrophic survival. PNAS 110(9):3621–3626

Duchatelet L, Claes JM, Delroisse J, Flammang P, Mallefet J (2021) Glow on sharks: state of the art on bioluminescence research. Oceans 2:822–842

Duckett JG, Ligrone R (2006) *Cyathodium* Kunze (Cyathodiaceae: Marchantiales), a tropical liverwort genus and family new to Europe, in Southern Italy. J Bryol 28:88–96

Efimova O, Hore PJ (2008) Role of exchange and dipolar interactions in the radical pair model of the avian magnetic compass. Biophys J 94:1565–1574

Engel GS (2011) Quantum coherence in photosynthesis. Procedia Chem 3:222–231

Engel GS, Calhoun TR, Read EL, Ahn T-K, Mančal T et al (2007) Evidence of wavelike energy transfer through quantum coherence in photosynthetic systems. Nature 446:782–786

Falkenberg G, Fleissner G, Schuchardt K, Kuehnbacher M, Thalau P et al (2010) Avian magnetoreception: elaborate iron mineral containing dendrites in the upper beak seem to be a common feature of birds. PLoS One 5(2):e9231

Falkowski PG, Schofield O, Katz ME, Van de Schootbrugge B, Knoll AH (2004) Why is the land green and the ocean red? In: Thierstein HR, Young JR (eds) Coccolithophores. Springer, Berlin, pp 429–453

Fayer MD (2012) Dynamics of water interacting with interfaces, molecules, and ions. Acc Chem Res 45(1):3–14

Federle W (2006) Why are so many adhesive pads hairy? J Exp Biol 209:2611–2621

Feller K, Porter M (2023) Photonic tinkering in the open ocean. Science 379:643–644

Feller KD, Wilby D, Jacucci G, Vignolini S, Mantell J et al (2019) Long-wavelength reflecting filters found in the larval retinas of one mantis shrimp family (*Nannosquillidae*). Curr Biol 29:3101–3108

Feynman RP (1985) QED: the strange theory of light and matter. Princeton University Press, Princeton

Feynman RP (1993) Vom Wesen physikalischer Gesetze. Piper, Munich

Fischer R, Aguirre J, Herrera-Estrella A, Corrochano LM (2016) The complexity of fungal vision. Microbiol Spectr 4(6). https://doi.org/10.1128/microbiolspec.FUNK-0020-2016

Fishelson L, Gon O, Goren M, Ben-David-Zaslow R (2005) The oral cavity and bioluminescent organs of the cardinal fish species *Siphamia permutata* and *S. cephalotes* (Perciformes, Apogonidae). Mar Biol 147:603–609

Fleming GR, Scholes GD, Cheng Y-C (2011a) Quantum effects in biology. Procedia Chem 3:38–57

Fleming LE, Kirkpatrick B, Backer LC, Walsh CJ, Nierenberg K et al (2011b) Review of Florida red tides and human health effects. Harmful Algae 10(2):224–233

Foelix RF, Erb B, Hill DE (2013) Structural colors in spiders. In: Nentwig W (ed) Spider ecophysiology. Springer, Berlin, pp 333–347

Franco MI, Turin L, Mershin A, Skoulakis EMC (2011) Molecular vibration-sensing component in *Drosophila melanogaster* olfaction. PNAS 108(9):3797–3802

Freudenstein R, Kulisch W (2016) Quantenmechanik. Wiley, Weinheim

Frutos LM, Andruniów T, Santoro F, Ferré N, Olivucci M (2007) Tracking the excited-state time evolution of the visual pigment with multiconfigurational quantum chemistry. PNAS 104(19):7764–7765

Fujita T, Brookes JC, Saikin SK, Aspuru-Guzik A (2012) Memory-assisted excitation diffusion in the chlorosome light-harvesting antenna of green sulphur bacteria. arXiv:1206.5812v1

Gagliardo A, Iolè P, Savini M, Wild M (2009) Navigational abilities of adult and experienced homing pigeons deprived of olfactory or trigeminally mediated magnetic information. J Exp Biol 212:3119–3124

Gagnon YL, Shashar N, Warrant EJ, Johnsen SJ (2007) Light scattering by selected zooplankton from the Gulf of Aqaba. J Exp Biol 210:3728–3735

Gaillot DP, Deparis O, Welch V, Wagner BK, Vigneron JP, Summers CJ (2008) Composite organic-inorganic butterfly scales: production of photonic structures with atomic layer deposition. Phys Rev E 78:031922

Gane S, Georganakis D, Maniati K, Vamvakias M, Ragoussis N et al (2013) Molecular vibration-sensing component in human olfaction. PLoS One 8(1):e55780

Gavelis GS, Keeling PJ, Leander BS (2017) How exaptations facilitated photosensory evolution: seeing the light by accident. Bioessays 39:1600266

Gegear RJ, Casselman A, Waddell S, Reppert SM (2008) Cryptochrome mediates light-dependent magnetosensitivity in Drosophila. Nature 454:1014–1018

Gegear RJ, Foley LE, Casselman A, Reppert SM (2010) Animal cryptochromes mediate magnetoreception by an unconventional photochemical mechanism. Nature 463:804–907

Gehring WJ (2012) The evolution of vision. Wiley Interdiscip Rev Dev Biol. https://doi.org/10.1002/wdev.96

Geissler PL, Dellago C, Chandler D, Hutter J, Parrinello M (2001) Autoionization in liquid water. Science 291:2121–2124

Gerlach T, Sprenger D, Michiels NK (2014) Fairy wrasses perceive and respond to their deep red fluorescent coloration. Proc Biol Sci 281:20140787

Ghiradella HT, Butler MW (2009) Many variations on a few themes: a broader look at development of iridescent scales (and feathers). J R Soc Interface 6:S243–S251

Ghiradella H, Schmidt JT (2004) Fireflies at one hundred plus: a new look at flash control. Integr Comp Biol 44:203–212

Glover BJ, Whitney HM (2010) Structural colour and iridescence in plants: the poorly studied relations of pigment colour. Ann Bot 105:505–511

Goris RC (2011) Infrared organs of snakes: an integral part of vision. J Herpetol 45(1):2–14

Gould JL (2011) Animal navigation: longitude at last. Curr Biol 21(6):R225–R227

Gouveneaux A, Mallefet J (2013) Physiological control of bioluminescence in a deep-sea planktonic worm, *Tomopteris helgolandica*. J Exp Biol 216:4285–4289

Govorunova EG, Sineshchekov OA, Li H, Spudich JL (2017) Microbial rhodopsins: diversity, mechanisms, and optogenetic applications. Annu Rev Biochem 86:845–872

Gowardovskii VI, Fyhrquist N, Reuter T, Kuzmin DG, Donner K (2000) In search of the visual pigment template. Vis Neurosci 17:509–528

Gray HB, Winkler JR (2005) Long-range electron transfer. PNAS 102(10):3534–3539

Gray HB, Winkler JR (2009) Electron flow through proteins. Chem Phys Lett 24(483):1–9

Groenhof G, Bouxin-Cademartory M, Hess B, de Visser SP, Berendsen HJC et al (2004) Photoactivation of the photoactive yellow protein: why photon absorption triggers a trans-to-cis isomerization of the chromophore in the protein. J Am Chem Soc 126:4228–4233

Gruber DF, Sparks JS (2015) First observation of fluorescence in marine turtles. Am Mus Novit 3845

Gruber DF, Loew ER, Deheyn DD, Akkaynak D, Gaffney JP et al (2016) Biofluorescence in Catsharks (Scyliorhinidae): fundamental description and relevance for elasmobranch visual ecology. Sci Rep 6:24751

Gu Z-Z, Uetsuka H, Takahashi K, Nakajima R, Onishi H et al (2003) Structural color and the lotus effects. Angew Chem 42(8):894–897

Guerra PA, Gegear RJ, Reppert SM (2014) A magnetic compass aids monarch butterfly migration. Nat Commun 5:4164

Guo J, Lü J-T, Feng Y, Chen J, Peng J et al (2016) Nuclear quantum effects of hydrogen bonds probed by tip-enhanced inelastic electron tunneling. Science 352(6283):321–325

Gurskaya NG, Fradkov AF, Pounkova NI, Staroverov DB, Bulina ME et al (2003) A colourless green fluorescent protein homologue from the non-fluorescent hydromedusa *Aequorea coerulescens* and its fluorescent mutants. Biochem J 373:403–408

Haddock SHD, Case JF (1999) Bioluminescence spectra of shallow and deep-sea gelatinous zooplankton: ctenophores, medusae and siphonophores. Mar Biol 133:571–582

Haddock SHD, Dunn CW (2015) Fluorescent proteins function as a prey attractant: experimental evidence from the hydromedusa *Olindias formosus* and other marine organisms. Biol Open 4:1094–1104

Haddock SHD, Dunn CW, Pugh PR, Schnitzler CE (2005) Bioluminescent and red-fluorescent lures in a deep-sea siphonophore. Science 309:263–264

Haddock SHD, Moline MA, Case JF (2010) Bioluminescence in the sea. Ann Rev Mar Sci 2:443–493

Hameroff S, Nip A, Porter M, Tuszynski J (2002) Conduction pathways in microtubules, biological quantum computation, and consciousness. Biosystems 64(1–3):149–168

Hammes-Schiffer S (2011) When electrons and protons get excited. PNAS 108(21):8531–8532

Hänggi P (2002) Stochastic resonance in biology. Chemphyschem 3:285–290

Hasegawa M, Hosaka T, Kojima K, Nishimura Y, Nakajima Y et al (2020) A unique clade of light-driven proton-pumping rhodopsins evolved in the cyanobacterial lineage. Sci Rep 10:16752

Hassanali AA, Cuny J, Verdolino V, Parrinello M (2014) Aqueous solutions: state of the art in *ab initio* molecular dynamics. Philos Trans R Soc A 372:20120482

Hawryshyn CW, McFarland W (1987) Cone photoreceptor mechanisms and the detection of polarized light in fish. J Comp Physiol A 160:439–465

Hay S, Pudney C, Hothi P, Johanissen LO, Masgrau L et al (2008) Atomistic insight into the origin of the temperature-dependence of kinetic isotope effects and H-tunnelling in enzyme systems is revealed through combined experimental studies and biomolecular simulation. Biochem Soc Trans 36(1):16–21

Hay S, Pudney CR, Scrutton NG (2009) Structural and mechanistic aspects of flavoproteins: probes of hydrogen tunnelling. FEBS J 276:3930–3941

Hayashi T, Stuchebrukhov AA (2010) Electron tunnelling in respiratory complex I. PNAS 107(45):19157–19162

Hayashi T, Stuchebrukhov AA (2011) Quantum electron tunneling in respiratory complex I. J Phys Chem B 115(18):5354–5364

Hayashi S, Fukushima R, Wada N (2012) Extraction and purification of a luminiferous substance from the luminous mushroom *Mycena chlorophos*. Biophysics 8:111–114

Heath-Heckman EAC, Peyer SM, Whistler CA, Apicella MA, Goldman WE, McFall-Ngai MJ (2013) Bacterial bioluminescence regulates expression of a host cryptochrome gene in the squid-vibrio symbiosis. mBio 4(2):e00167–e00113

Hein CM, Engels S, Kishkiner D, Mouritsen H (2011) Robins have a magnetic compass in both eyes. Nature 471:E11–E13

Heldt HW, Piechulla B (2008) Pflanzenbiochemie. Springer Spektrum, Heidelberg

Hernández-Jiménez M, Azofeifa DE, Libby E, Barboza-Aguilar C, Solis Á et al (2014) Qualitative correlation between structural chirality through the cuticle of *Chrysina aurigans* scarabs and left-handed circular polarization of the reflected light. Opt Mater Express 4(12). https://doi.org/10.1364/OME.4.002632

Hess V, Gallegos R, Jones JA, Barquera B, Malamy MH, Müller V (2016) Occurrence of ferredoxin:NAD$^+$ oxidoreductase activity and its ion specificity in several gram-positive and gram-negative bacteria. Peer J. https://doi.org/10.7717/peerj.1515

Hirose Y, Rockwell NC, Nishiyama K, Narikawa R, Ukaji Y et al (2013) Green/red cyanobacteriochromes regulate complementary chromatic acclimation via a protochromic photocycle. PNAS 110(13):4974–4979

Hiscock HG, Worster S, Kattnig DR, Steers C, Jin Y et al (2016) The quantum needle of the avian magnetic compass. PNAS 113(17):4634–4639

Hohmann-Marriott MF, Blankenship RE (2011) Evolution of photosynthesis. Annu Rev Plant Biol 62:515–548

Holick MF (2014) Cancer, sunlight and vitamin D. J Clin Transl Endocrinol 1:179–186

Holick MF, Tian XO, Allen M (1995) Evolutionary importance for the membrane enhancement of the production of vitamin D$_3$ in the skin of poikilothermic animals. PNAS 92:3124–3126

Horton P, Ruhan AV, Walters RG (1994) Regulation of light harvesting in green plants. Plant Physiol 106:415–420

Hosseini M, Campbell G, Sparkes BM, Lam PK, Buchler BC (2015) Unconditional room temperature quantum memory. arXiv:1412.8235v2

Hoyer S, Ishizaki A, Whaley KB (2011) Propagating quantum coherence for a biological advantage. arXiv:1106.2911v1

Huang H, Wang S, Moll J, Thauer RK (2012) Electron bifurcation involved in the energy metabolism of the acetogenic bacterium *Moorella thermoacetica* growing on glucose or H$_2$ plus CO$_2$. J Bacteriol 194(14):3689–3699

Hunter P (2006) A quantum leap in biology. EMBO Rep 7(10):971–974

Isobe M, Uyakul D, Sigurdsson JB, Goto T, Lam TJ (1991) Fluorescent substances in the luminous land snail, *Dyakia striata*. Agric Biol Chem 55(8):1947–1951

Iyengar SS, Summer I, Jakowski J (2008) Hydrogen tunneling in an enzyme active site: a quantum wavepacket dynamical perspective. J Phys Chem B 112:7601–7613

Jablonski NG, Chaplin G (2013) Epidermal pigmentation in the human lineage is an adaptation to ultraviolet radiation. J Hum Evol 65:671–675

Jacob EB, Becker I, Shapira Y, Levine H (2004) Bacterial linguistic communication and social intelligence. Trends Microbiol 12(8):366–372

Jékely G (2009) Evolution of phototaxis. Philos Trans R Soc Lond B Biol Sci 364:2795–2808

Jiang Z-Y, Gest H, Bauer CE (1997) Chemosensory and photosensory perception in purple photosynthetic bacteria utilize common signal transduction components. J Bacteriol 179(18):5720–5727

Johnsen S (2005) The red and the black: bioluminescence and the color of animals in the deep sea. Integr Comp Biol 45:234–246

Johnsen S (2012) The optics of life. Princeton University Press, Princeton

Johnsen S (2014) Hide and seek in the open sea: pelagic camouflage and visual countermeasures. Annu Rev Mar Sci 6(9):1–24

Johnsen S, Widder EA (2001) Ultraviolet absorption in transparent zooplankton and its implications for depth distribution and visual predation. Mar Biol 138:717–730

Johnsen S, Marshall NJ, Widder EA (2011) Polarization sensitivity as a contrast enhancer in pelagic predators: lessons from *in situ* polarization imaging of transparent zooplankton. Philos Trans R Soc Lond B Biol Sci 366:655–670

Johnsen S, Frank TM, Haddock SHD, Widder EA, Messing CG (2012) Light and vision in the deep-sea benthos: I. Bioluminescence at 500-1000 m depth in the Bahamian Islands. J Exp Biol 215:3335–3343

Johnson FH, Eyring H, Polissar MJ (1955) The kinetic basis of molecular biology. Wiley, New York

Johnston DT, Wolfe-Simon F, Pearson A, Knoll AH (2009) Anoxygenic photosynthesis modulated Proterozoic oxygen and sustained Earth's middle age. PNAS 106(40):16925–16929

Jones A, Mallefet J (2010) Aposematic use of bioluminescence in *Ophiopsila aranea* (Ophiuroidea, Echinodermata). Luminescence 25:155–156

Jones A, Mallefet J (2013) Why do brittle stars emit light? Behavioural and evolutionary approaches of bioluminescence. Cah Biol Mar 54:729–734

Jones BW, Nishiguchi MK (2004) Counterillumination in the Hawaiian bobtail squid, *Euprymna scolopes* Berry (Mollusca: Cephalopoda). Mar Biol 144:1151–1155

Juusola M, Song Z (2017) How a fly photoreceptor samples light information in time. J Physiol 16:5427–5437

Juusola M, Dau A, Song Z, Solanki N, Rien D et al (2017) Microsaccadic sampling of moving image information provides Drosophila hyperacute vision. eLife 6:e26117

Kandori H (2015) Ion-pumping microbial rhodopsins. Front Mol Biosci 2:52

Kaster A-K, Moll J, Parey K, Thauer RK (2011) Coupling of ferredoxin and heterodisulfide reduction via electron bifurcation in hydrogenotrophic methanogenic archaea. PNAS 108(7):2981–2986

Kay LM, Stopfer M (2006) Information processing in the olfactory systems of insects and vertebrates. Cell Environ Biol 17(4):433–442

Kehoe DM (2010) Chromatic adaptation and the evolution of light color sensing in cyanobacteria. PNAS 107(20):9029–9030

Kelber A, Osorio D (2010) From spectral information to animal colour vision: experiments and concepts. Proc Biol Sci 277:1617–1625

Kelber A, Roth LSV (2006) Nocturnal colour vision—not as rare as we might think. J Exp Biol 209:781–788

Kemp DJ, Rutowski RL (2007 Jan) Condition dependence, quantitative genetics, and the potential signal content of iridescent ultraviolet butterfly coloration. Evolution 61:168–183

Khrenova MG, Nemukhin AV, Grigorenko BL, Krylov AI, Domratcheva TM (2010) Quantum chemistry calculations provide support to the mechanism of the light-induced structural changes in the Flavin-binding photoreceptor proteins. J Chem Theor Comput 6:2293–2302

Kiang NY, Siefert J, Govindjee, Blankenship RE (2007) Spectral signatures of photosynthesis. I. Review of earth organisms. Astrobiology 7(1):222–251

Kianianmomeni A, Hallmann A (2014) Algal photoreceptors: in vivo functions and potential applications. Planta 239:1–26

Kiefer C (2012) Quantentheorie. Fischer, Frankfurt

Kientz B, Ducret A, Luke S, Vukusic P, Mignot T, Rosenfeld E (2012) Glitter-Like iridescence within the bacteroidetes especially *Cellulophaga* spp.: optical properties and correlation with gliding motility. PLoS One 7(12):e52900

Kientz B, Luke S, Vukusic Péteri R, Beaudry C, Renault T et al (2016) A unique self-organization of bacterial sub-communities creates iridescence in *Cellulophaga lytica* colony biofilms. Sci Rep 6:19906

Kinoshita S, Yoshioka S, Miyazaki J (2008) Physical structural colors. Rep Prog Phys. https://doi.org/10.1088/0034-4995/71/076401

Kirk JTO (2011) Light and photosynthesis in aquatic ecosystems. Cambridge University Press, Cambridge

Kirschner M, Gerhart J, Mitchinson T (2000) Molecular "vitalism". Cell 100:79–88

Kirschvink JL, Walker MM, Diebel CE (2001) Magnetite-based magnetoreception. Curr Opin Neurobiol 11:462–467

Kleinlogel S, White AG (2008) The secret world of shrimps: polarisation vision at its best. PLoS One 3(5):e2190

Klinman JP (2009) An integrated model for enzyme catalysis emerges from studies of hydrogen tunneling. Chem Phys Lett 471(4–6):179–193

Klinman JP, Kohen A (2013) Hydrogen tunneling links protein dynamics to enzyme catalysis. Annu Rev Biochem 82:471–496

Knapp MJ, Klinman JP (2002) Environmentally coupled hydrogen tunneling. Eur J Biochem 269:3113–3121

Kohen A (2012) Enzyme dynamics: consensus and controversy. J Biocatal Biotransformation 1:1

Kohen A, Klinman JP (1999) Hydrogen tunneling in biology. Chem Biol 6:R191–R198

Kong S-G, Wada M (2016) Molecular basis of chloroplast photorelocation movement. J Plant Res 129:159–166

Korbicz JK (2021) Roads to objectivity: quantum Darwinism, spectrum broadcast structures, and strong quantum Darwinism—a review. arXiv:2007.04276v2

Kouyama T, Kanada S, Takeguchi Y, Narusawa A, Murakami M, Ihara K (2010) Crystal structure of the light-driven chloride pump Halorhodopsin from *Natronomonas pharaonis*. J Mol Biol 396:564–579

Koyanagi M, Terakita A (2014) Diversity of animal opsin-based pigments and their optogenetic potential. Biochim Biophys Acta 1937:710–716

Kozakiewicz J, Gajewska M, Łyżeń R, Czyż A, Węgrzyn G (2005) Bioluminescence-mediated stimulation of photoreactivation in bacteria. FEMS Microb Lett 250:105–110

Kozubek J (2016) Modern prometheus. Cambridge University Press, Cambridge

Krause GH, Weis E (1991) Chlorophyll fluorescence and photosynthesis: the basics. Annu Rev Plant Physiol 42:313–349

Krupovic M, Prangishvili D, Hendrix RW, Bamford DH (2011) Genomics of bacterial and archaeal viruses: dynamics within the prokaryotic virosphere. Microbiol Mol Biol Rev 75(4):610–635

Kubodera T, Koyama Y, Mori K (2007) Observations of wild hunting behaviour and bioluminescence of a large deep-sea, eight-armed squid, *Taningia danae*. Proc Biol Sci. https://doi.org/10.1098/rspb.2006.0236

Kudryasheva NS, Nemtseva EV, Sizykh AG, Kratasyuk VA, Visser AJWG (2002) Estimation of energy of the upper electron-excited states of the bacterial bioluminescent emitter. J Photochem Photobiol B Biol 68:88–92

Kühne TD, Khaliullin RZ (2013) Electronic signature of the instantaneous asymmetry in the first coordination shell of liquid water. Nat Commun. https://doi.org/10.1038/ncomms2459

Kusy D, He J-W, Bybee SM, Motyka M, Bi W-X et al (2020) Phylogenomic relationships of bioluminescent elateroids define the 'lampyroid' clade with clicking Sinopyrophoridae as its earliest member. Syst Entomol. https://doi.org/10.1111/syen.12451

Laflamme R, Miquel C, Paz JP, Zurek WH (1996) Prefect quantum error correction code. arXiv:quant-ph/9602019v1

Lall AB, Cronin TW, Carvalho AA, de Souza JM, Barros MP et al (2010) Vision in click beetles (Coleoptera: Elateridae): pigments and spectral correspondence between visual sensitivity and species bioluminescence emission. J Comp Physiol A Neuroethol Sens Neural Behav Physiol. https://doi.org/10.1007/s00359-010-0549-x

Lambert N, Chen Y-N, Cheng Y-C, Li C-M, Chen G-Y, Nori F (2013) Quantum biology. Nat Phys 9:10–18

Land MF, Osorio DC (2003) Colour vision: colouring the dark. Curr Biol 13:R83–R85

Langen R, Chang I-J, Germanas JP, Richards JH, Winkler JR, Gray HB (1995) Electron tunneling in proteins: coupling through a β strand. Science 268:1733–1735

Lau JCS, Rodgers CT, Hore PJ (2012) Compass magnetoreception in birds arising from photo-induced radical pairs in rotationally disordered cryptochromes. J R Soc Interface 9:3329–3337

Lefèvre CT, Bazylinski DA (2013) Ecology, diversity, and evolution of magnetotactic bacteria. Microbiol Mol Biol Rev 77(3):497–526

Lerner E, Cordes T, Ingagiola A, Alhadid Y, Chung SY et al (2018) Toward dynamic structural biology: two decades of single-molecule Förster resonance energy transfer. Science 359:eaan1133

Lever MA, Rogers KL, Lloyd KG, Overmann J, Schink B et al (2015) Life under extreme energy limitation: a synthesis of laboratory- and field-based investigations. FEMS Microbiol Rev 39(5):688–728

Levy-Lior A, Shimoni E, Schwartz O, Gavish-Regev E, Oron D et al (2010) Guanine-based biogenic photonic-crystal arrays in fish and spiders. Adv Funct Mater 20:320–329

Lewis SM, Cratsley CK (2008) Flash signal evolution, mate choice, and predation in fireflies. Annu Rev Entomol 53:293–321

Li N, Xu J-Z, Yao H, Zhu J-J, Chen H-Y (2006) The direct electron transfer of myoglobin based on the electron tunneling in proteins. J Phys Chem B 110:11561–11565

Li F, Hinderberger J, Seedorf H, Zhang J, Buckel W, Thauer RK (2008) Coupled ferredoxin and crotonyl coenzyme A (CoA) reduction with NADH catalyzed by the butyryl-CoA dehydrogenase/Etf complex from *Clostridium kluyveri*. J Bacteriol 190(3):843–850

Liedvogel M, Maeda K, Henbest K, Schleicher E, Simon T et al (2007) Chemical magnetoreception: bird cryptochrome 1a is excited by blue light and forms long-lived radical-pairs. PLoS One 2(10):e1106

Lim MLM, Li D (2013) UV-green iridescence predicts male quality during jumping spider contests. PLoS One 8(4):e59774

Limbach H-H, Lopez JM, Kohen A (2006) Arrhenius curves of hydrogen transfers: tunnel effects, isotope effects and effects of pre-equilibria. Philos Trans R Soc Lond B Biol Sci 361:1399–1415

Lin J, Balabin IA, Beratan DN (2005) The nature of aqueous tunneling pathways between electron-transfer proteins. Science 310(5752):1311–1313

Lin W, Bazylinsku DA, Xiao T, Wu L-F, Pan Y (2014) Life with compass: diversity and biogeography of magnetotactic bacteria. Environ Microbiol 16(9):2646–2658

Lin W, Kirschvink JI, Paterson GA, Bazylinski DA, Pan Y (2020) On the origin of microbial magnetoreception. Natl Sci Rev 7:472–479

Logunov I, Schulten K (1996) Quantum chemistry: molecular dynamics study of the dark-adaptation process in bacteriorhodopsin. J Am Chem Soc 118:9727–9735

Lohmann KJ, Lohmann CMF, Putman NF (2007) Magnetic maps in animals: nature's GPS. J Exp Biol 210:3697–3705

Lolkema JS, Chaban Y, Boekema EJ (2003) Subunit composition, structure, and distribution of bacterial V-type ATPases. J Bioenerg Biomembr 35(4):323–335

Löwdin P-O (1963) Proton tunneling in DNA and its biological implications. Rev Mod Phys 35(3):724–732

Lucock M, Yates Z, Martin C, Choi J-H, Boyd L et al (2014) Vitamin D, folate, and potential early lifecycle environmental origin of significant adult phenotypes. Evol Med Public Health 2014(1):69–91

Lythgoe JN, Shand J (1989) The structural basis for iridescent colour changes in dermal and corneal iridophores in fish. J Exp Biol 141:313–325

Ma Q, Hua H-H, Chen Y, Liu B-B, Krämer AL et al (2012) A rising tide of blue-absorbing bilipro-
   tein photoreceptors—characterization of seven such bilin-binding GAF domains in *Nostoc* sp.
   PCC7120. FEBS J 279:4095–4108
Madigan MT, Martinko JM, Parker J (2003) Brock biology of microorganisms. Prentice Hall, Upper
   Saddle River, NJ
Maeda K, Henbest KB, Cintolesi F, Kuprov I, Rodgers CT et al (2008) Chemical compass model of
   avian magnetoreception. Nature 453:387–390
Majumdar S, Pal S (2017) Bacterial intelligence: imitation games, time-sharing, and long-range
   quantum coherence. J Cell Commun Signal 11:281–284
Man D, Wang W, Sabehi G, Aravind L, Post AF et al (2003) Diversification and spectral tuning in
   marine proteorhodopsins. EMBO J 22(8):1725–1731
Marek P, Moore W (2015) Discovery of a glowing millipede in California and the gradual evolution
   of bioluminescence in Diplopoda. PNAS 112(20):6419–6424
Marek P, Papaj D, Yeager J, Mollna S, Moore W (2011) Bioluminescent Aposematism in millipedes.
   Curr Biol 21(18):R680–R681
Martin-Delgado MA (2012) On quantum effects in a theory of biological evolution. Sci Rep 2:302
Martini S, Haddock SHD (2017) Quantification of bioluminescence from the surface to the deep sea
   demonstrates its predominance as an ecological trait. Sci Rep 7:45750
Maruani J (2021) Can quantum theory concepts shed light on biological evolution processes? In:
   Mammino L, Ceresoli D, Maruani J, Brändas E (eds) Advances in quantum systems in chem-
   istry, physics, and biology; progress in theoretical chemistry and physics 32. Springer Nature,
   Switzerland, pp 437–465
Mascoli V, Novoderezhkin V, Liguori N, Xu P, Croce R (2020) Design principles of solar light
   harvesting in plants: functional architecture of the monomeric antenna CP29. BBA Bioenerg
   1861:148156
Masgrau L, Basran J, Hothi P, Sutcliffe MJ, Scrutton NS (2004) Hydrogen tunneling in quinopro-
   teins. Arch Biochem Biophys 428:41–51
Masgrau L, Roujeinikova A, Johanissen LO, Hothi P, Basran J et al (2006) Atomic description of an
   enzyme reaction dominated by proton tunneling. Science 312:237–241
Mason B, Moore CB (1985) Grundzüge der Geochemie. Enke, Stuttgart
Mäthger LM, Denton EJ, Marshall NJ, Hanlon RT (2009a) Mechanisms and behavioural functions
   of structural coloration in cephalopods. J R Soc Interface 6:S149–S163
Mäthger LM, Shashar N, Hanlon RT (2009b) Do cephalopods communicate using polarized light
   reflections from their skin? J Exp Biol 212:2133–2140
Mazel CH, Cronin TW, Caldwell RL, Marshall NJ (2003) Fluorescent enhancement of signaling in
   a Mantis Shrimp. Science 303(5654):51
McCarren J, DeLong EF (2007) Proteorhodopsin photosystem gene clusters exhibit co-evolutionary
   trends and shared ancestry among diverse marine microbial phyla. Environ Microbiol
   9(4):846–858
McClintock PVE (2011) Quantum aspects of life. Contemp Phys 52(1):71–73
McCormack JE, Faircloth BC, Crawford NG, Gowaty PA, Brumfield RT, Glenn TC (2012)
   Ultraconserved elements are novel phylogenomic markers that resolve placental mammal phy-
   logeny when combined with species-tree analysis. Genome Res 22:746–754
McCoy DE, Feo T, Harvey TA, Prum RO (2018) Structural absorption by barbule microstructures of
   super black bird of paradise feathers. Nat Commun. https://doi.org/10.1038/s41467-017-02088-w
McDonald MS, Palecanda S, Cohen JH, Porter ML (2022) Ultraviolet vision in larval *Neogonodactylus*
   oerstedii. J Exp Biol. https://doi.org/10.1242/jeb.243256
McFadden J (2000) Quantum evolution. Norton & Company, New York

McFall-Ngai M, Morin JG (1991) Camouflage by disruptive illumination in Leiognathids, a family of shallow-water, bioluminescent fishes. J Exp Biol 156:119–137

Meighen EA (1991) Molecular biology of bacterial bioluminescence. Microbiol Rev 55(1):123–142

Metcalf WW (2016) Classic spotlight: electron bifurcation, a unifying concept for energy conservation in anaerobes. J Bacteriol 198(9):1358

Meyer MP, Tomchick DR, Klinman JP (2008) Enzyme structure and dynamics affect hydrogen tunneling: the impact of a remote side chain (I553) in soybean lipoxygenase-1. PNAS 105(4):1146–1151

Meyer-Rochow VB, Moore S (2009) Hitherto unreported aspects of the ecology and anatomy of a unique gastropod: the bioluminescent freshwater pulmonate *Latia neritoides*. In: Meyer-Rochow VB (ed) Bioluminescence in focus—a collection of illuminating essays. Research Signpost, Thiruvananthapuram, Kerala, pp 85–104

Michiels NK, Anthes N, Hart NS, Herler J, Meixner AJ et al (2008) Red fluorescence in reef fish: a novel signalling mechanism? BMC Ecol. https://doi.org/10.1186/1472-6785-8-16

Michielsen K, Stavenga DG (2008) Gyroid cuticular structures in butterfly wing scales: biological photonic crystals. J R Soc Interface 5:85–94

Mizuno G, Yano D, Paitio J, Endo H, Oba Y (2021) *Etmopterus* lantern sharks use coelenterazine as the substrate for their luciferin-luciferase bioluminescence system. Biochem Biophys Res Commun 577:139–145

Møberg N, Halberg KA, Jørgensen A, Persson D, Bjørn M et al (2011) Survival in extreme environments—On the current knowledge of adaptations in tardigrades. Acta Physiol 202:409–420

Mohseni M, Rebentrost P, Lloyd S, Aspuru-Guzik A (2008) Environment-assisted quantum walks in photosynthetic energy transfer. arXiv:0805.2741v2

Mohseni M, Omar Y, Engel GS, Plenio MB (2014) Quantum effects in biology. Cambridge University Press, Cambridge

Mole BM, Baltrus DA, Dangl JL, Grant SR (2007) Global virulence regulation networks in phytopathogenic bacteria. Trends Microbiol 15(8):363–371

Moosman PR Jr, Cratsley CK, Lehto SD, Thomas HH (2009) Do courtship flashes of fireflies (Coleoptera: Lampyridae) serve as aposematic signals to insectivorous bats? Anim Behav 78:1019–1025

Mori T, Mchaourab H, Johnson CH (2015) Circadian clocks: unexpected biochemical cogs. Curr Biol 25:R842–R844

Morin JG (1986) "Firefleas" of the sea: luminescent signaling in marine ostracode crustaceans. Insect Behav Ecol 69(1):105–121

Moser CC, Keske JM, Warncke K, Farid RS, Dutton PL (1992) Nature of biological electron transfer. Nature 355:796–802

Moser CC, Chobot SE, Page CC, Dutton PL (2008) Distance metrics for heme protein electron tunneling. Biochim Biophys Acta 1777:1032–1037

Mouritsen H, Ritz T (2005) Magnetoreception and its use in bird navigation. Curr Opin Neurobiol 15:406–414

Muggia L, Fleischhacker A, Kopun T, Grube M (2016) Extremotolerant fungi from alpine rock lichens and their phylogenetic relationships. Fungal Divers 76:119–142

Müh F, Madjet ME-A, Adolphs J, Abdurrahman A, Rabenstein B et al (2007) α-Helices direct excitation energy flow in the Fenna-Matthews-Olson protein. PNAS 104(43):16862–16867

Mulkidjanian AY, Koonin EV, Makarova KS, Mekhedov SL, Sorokin A et al (2006) The cyanobacterial genome core and the origin of photosynthesis. PNAS 103(35):13126–13131

Mulkidjanian AY, Makarova KS, Galperin MY, Koonin EV (2007) Inventing the dynamo machine: the evolution of the F-type and Y-type ATPases. Nat Rev Microbiol 5:892–899

Müller WEG, Schröder HC, Pisignano D, Markl JS, Wang X (2013) Metazoan circadian rhythm: toward an understanding of a light-based zeitgeber in sponges. Integr Comp Biol. https://doi.org/10.1093/icb/ict001

Musilova Z, Salzburger W, Cortesi F (2021) The visual opsin gene repertoires of teleost fishes: evolution, ecology and function. Annu Rev Cell Dev Biol 37:441–468

Nagel ZD, Klinman JP (2006) Tunneling and dynamics of enzymatic hydride transfer. Chem Rev 106:3095–3118

Nagel ZD, Klinman JP (2010) Update 1 of: tunneling and dynamics in enzymatic hydride transfer. Chem Rev 110(12):PR41–PR67

Nakatsu T, Ichiyama S, Hiratake J, Saldanha A, Kobashi N et al (2006) Structural basis for the spectral difference in luciferase bioluminescence. Nature 440:372–376

Nalbach P, Eckel J, Thorwart M (2010) Quantum coherent biomolecular energy transfer with spatially correlated fluctuations. arXiv:1003.3857v1

Naresh M, Sharma M, Mittla A (2011) Intracellular magneto-spatial organization of magnetic organelles inside intact bacterial cells. J Biomed Nanotechnol 7:572–577

Nashine VC, Hammes-Schiffer S, Benkovic SJ (2010) Coupled motions in enzyme catalysis. Curr Opin Chem Biol 14(5):644–651

Navizet I, Liu Y-J, Ferré N, Roca-Sanyuán D, Lindh R (2011) The chemistry of bioluminescence: an analysis of chemical functionalities. Chem Phys Chem 12:1064–1076

Nealson KH, Hastings JW (1979) Bacterial bioluminescence: its control and ecological significance. Microbiol Rev 43(4):496–518

Needham DM, Yoshizawa S, Hosaka T, Poirier C, Choi CJ et al (2019) A distinct lineage of giant viruses brings a rhodopsin photosystem to unicellular marine predators. PNAS 116(41):20574–20583

Neilson JAD, Durnford DG (2010) Structural and functional diversification of the light-harvesting complexes in photosynthetic eukaryotes. Photosynth Res 106:57–71

Neinhuis C, Barthlott W (1997) Characterization and distribution of water-repellent, self-cleaning plant surfaces. Ann Bot 79:667–677

Nemtseva EV, Kudryasheva NS (2007) The mechanism of excitation in the bacterial bioluminescent reaction. Russ Chem Rev 761:91–100

Newman JRS, Ghaemmaghami S, Ihmels J, Breslow DK, Noble M et al (2006) Single-cell proteomic analysis of S. cerevisiae reveals the architecture of biological noise. Nature 441:840–846

Nicolas M-T, Nicolas G, Johnseon CH, Bassot J-M, Hastings JW (1987) Characterization of the bioluminescent organelles in Gonyaulax polyedra (Dinoflagellates) after fast-freeze fixation and antiluciferase immunogold staining. J Cell Biol 105:723–735

Niedzwiedzki DM, Swainsbury DJK, Canniffe DP, Hunter CN, Hitchcock A (2020) A photosynthetic antenna complex foregoes unity carotenoid-to-bacteriochlorophyll energy transfer efficiency to ensure photoprotection. PNAS 117(12):6502–6508

Nimpf S, Nordmann GC, Kagerbauer D, Malkemper EP, Landler L et al (2019) A putative mechanism for magnetoreception by electromagnetic induction in the pigeon inner ear. Curr Biol 29:4052–4059

Nitschke W, Russell MJ (2011) Redox bifurcations: mechanisms and importance to life now, and at its origin: a widespread means of energy conversion in biology unfolds.... Bioessays 34:106–109

Nobel PS (2009) Physiochemical and environmental plant physiology. Academic Press, Oxford

Oba Y, Branham MA, Fukatsu T (2011) The terrestrial bioluminescent animals of Japan. Zool Sci 28:771–769

Ogryzko VV (2008) Erwin Schroedinger, Francis Crick and epigenetic stability. Biol Direct. https://doi.org/10.1186/1745-6150-3-15

Ohmiya Y, Hirano T (1996) Shining the light: the mechanism of the bioluminescence reaction of calcium-binding photoproteins. Chem Biol 3:337–347

Okamoto A, Hashimoto K, Nealson KH (2014) Flavin redox bifurcation as a mechanism for controlling the direction of electron flow during extracellular electron transfer. Angew Chem. https://doi.org/10.1002/anie.201407004

Okubo Y, Futamata H, Hiraishi A (2006) Characterization of phototrophic purple nonsulfur bacteria forming colored microbial mats in a Swine Wastewater Ditch. Appl Environ Microbiol 72(9):6225–6233

Oliveira AG, Desjardin DE, Perry BA, Stevani CV (2012) Evidence that a single bioluminescent system is shared by all known bioluminescent fungal lineages. Photochem Photobiol Sci 11:848–852

Olson JM, Blankenship RE (2004) Thinking about the evolution of photosynthesis. Photosynth Res 80:373–386

Osborn KJ, Haddock SHD, Pleijel F, Madin LP, Rouse GW (2009) Deep-sea, swimming worms with luminescent "bombs". Science 325:964

Page CP, Moser CC, Chen X, Dutton L (1999) Natural engineering principles of electron tunneling in biological oxidation-reduction. Nature 402:47–52

Paitio J, Yano D, Muneyama E, Takei S, Asada H et al (2020) Reflector of the body photophore in lanternfish is mechanistically tuned to project the biochemical emission in photocytes for counterillumination. Biochem Biophys Res Commun 521:821–826

Palovaara J, Akram N, Baltar F, Bunse C, Forsberg J et al (2014) Stimulation of growth by proteorhodopsin phototrophy involves regulation of central metabolic pathways in marine planktonic bacteria. PNAS 111:E3650–E3658

Parker AR (1998) Colour in Burgess Shale animals and the effects of light on evolution in the Cambrian. Proc R Soc Lond B 265:967–972

Patel RN, Khil V, Abdurahmonova L, Driscoli H, Patel S et al (2021) Mantis shrimp identify an object by its shape rather than its color during visual recognition. J Exp Biol 224. https://doi.org/10.1242/jeb.242256

Permentier HP, Neerken S, Schmidt KA, Overmann J, Amesz J (2000) Energy transfer and charge separation in the purple non-sulfur bacterium *Roseospirillum parvum*. Biochim Biophys Acta 1460:338–345

Peters JW, Miller A-F, Jones AK, King PW, Adams MWW (2016) Electron bifurcation. Curr Opin Chem Biol 31:146–152

Pfaff W, Hensen B, Bernien H, van Dam SB, Blok MS, et al (2014) Unconditional quantum teleportation between distant solid-state qubits. arXiv:1404.4369v3

Phillips R, Quake SR (2006) The biological frontiers of physics. Phys Today 2006:38–43

Pinhassi J, DeLong EF, Béjà O, González JM, Pedrós-Alió C (2016) Marine bacterial and archaeal ion-pumping rhodopsins: genetic diversity, physiology, and ecology. Microbiol Mol Biol Rev 80(4):929–954

Plenio MB, Huelga SF (2008) Dephasing assisted transport: Quantum networks and biomolecules. arXiv:0807.4902v1

Poehlein A, Schmidt S, Kaster A-K, Goenrich M, Vollmers J et al (2012) An ancient pathway combining carbon dioxide fixation with the generation and utilization of a sodium ion gradient for ATP synthesis. PLoS One 7(3):e33439. https://doi.org/10.1371/journal.pone.0033439

Poladian L, Wickham S, Lee K, Large MCJ (2009) Iridescence from photonic crystals and its suppression in butterfly scales. J R Soc Interface 6:S233–S242

Porter ML, Awata H, Bok MJ, Cronin TW (2020) Exceptional diversity of opsin expression patterns in *Neogonodactylus oerstedii* (Stomatopoda) retinas. PNAS 117(16):8948–8957

Pothos EM, Busemeyer JR (2022) Quantum cognition. Annu Rev Psychol 73:749–778

Prigogine I (1985) Vom Sein zum Werden. Piper, Munich

Prum RO, Torres RH (2004) Structural colouration of mammalian skin: convergent evolution of coherently scattering dermal collagen arrays. J Exp Biol 207:2157–2172

Prum RO, Cole JA, Torres RH (2004) Blue integumentary structural colours in dragonflies (Odonata) are not produced by incoherent Tyndall scattering. J Exp Biol 207:3999–4009

Prum RO, Quinn T, Torres RH (2006) Anatomically diverse butterfly scales all produce structural colours by coherent scattering. J Exp Biol 209:748–765

Prum RO, Dufresne ER, Quinn T, Waters K (2009) Development of colour-producing β-keratin nanostructures in avian feather barbs. J R Soc Interfaces 6:S253–S265

Pu J, Gao J, Truhlar DG (2006) Multidimensional tunneling, recrossing, and the transmission coefficient for enzymatic reactions. Chem Rev 106(8):3140–3169

Rainey PB, Beaumont HJE, Ferguson GC, Gallie J, Kost C et al (2011) The evolutionary emergence of stochastic phenotype switching in bacteria. Microb Cell Fact 10(S1):S14

Rangel NL, Williams KS, Seminario JM (2009) Light-activated molecular conductivity in the photoreactions of vitamin D. J Phys Chem 113:6740–6744

Raser JM, O'Shea EK (2005) Noise in gene expression: origins, consequences, and control. Science 309:2010–2013

Raven JA (2009) Functional evolution of photochemical energy transformation in oxygen-producing organisms. Funct Plant Biol 36:505–515

Read EL, Schlau-Cohen GS, Engel GS, Wen J, Blankenshio RE, Fleming GR (2008) Visualization of excitonic structure in the Fenna-Matthews-Olson photosynthetic complex by polarization-dependent two-dimensional electronic spectroscopy. Biophys J 95:847–856

Rebentrost P, Mohseni M, Kassal I, Lloyd S, Aspuru-Guzik A (2009) Environment-assisted quantum transport. arXiv:0807.0929v2

Rees J-F, de Wergifosse B, Noiset O, Dubuisson M, Janssens B, Thompson EM (1998) The origins of marine bioluminescence: turning oxygen defence mechanisms into deep-sea communication tools. J Exp Biol 201:1211–1221

Renger T (2011) Modeling of photosynthetic light-harvesting: from structure to function. Procedia Chem 3:236–247

Richardson JO, Pérez C, Lobsiger S, Reid AA, Temelso B et al (2016) Concerted hydrogen-bond breaking by quantum tunneling in the water hexamer prism. Science 351(6279):1310–1313

Richter M, Kube M, Bazylinski DA, Lombardot T, Glöckner FO et al (2007) Comparative genome analysis of four magnetotactic bacteria reveals a complex set of group-specific genes implicated in magnetosome biomineralization and function. J Bacteriol 189(13):4899–4910

Riebe M, Häffner H, Roos CF, Hänsel W, Benhelm J et al (2004) Deterministic quantum teleportation with atoms. Nature 429:734–737

Rieper E (2011) Quantum coherence in biological systems. Dissertation, National University of Singapore

Ritz T, Damjanovic A, Schulten K (2002) The quantum physics of photosynthesis. ChemPhysChem 3:243–248

Ritz T, Wiltschko R, Hore PJ, Rodgers CT, Stapput K et al (2009) Magnetic compass of birds is base on a molecule with optimal directional sensitivity. Biophys J 96:3451–3457l

Ritz T, Ahmad M, Mouritsen H, Wiltschko R, Wiltschko W (2010) Photoreceptor-based magnetoreception: optimal design of receptor molecules, cells, and neuronal processing. J R Soc Interface 7:S135–S146

Rivera AS, Ozturk N, Fahey B, Plachetzki DC, Degnen BM et al (2012) Blue-light-receptive Cryptochrome is expressed in a sponge eye lacking neurons and opsin. J Exp Biol 215:1278–1286

Robison BH, Reisenbichler KR, Hunt JH, Haddock SHD (2003) Light production by the arm tips of the deep-sea cephalopod *Vampyroteuthis infernalis*. Biol Bull 205:102–109

Rockwell NC, Duanmu D, Martin SS, Bachy C, Price DC et al (2014) Eukaryotic algal phytochromes span the visible spectrum. PNAS 111(10):3871–3876

Rodgers CT, Hore PJ (2009) Chemical magnetoreception in birds: the radical pair mechanism. PNAS 106(2):353–360

Romero E, Augulis R, Novoderezhkin VI, Ferretti M, Thieme J et al (2014) Quantum coherence in photosynthesis for efficient solar energy conversion. Nat Phys 10(9):676–682

Romero-Isart O, Juan ML, Quidant R, Cirac JI (2010) Toward quantum superposition of living organisms. New J Phys 1:033015

Roth MS, Deheyn DD (2013) Effects of cold stress and heat stress on coral fluorescence in reef-building corals. Sci Rep 3:1421

Roth MS, Latz MI, Goericke R, Deheyn DD (2010) Green fluorescent protein regulation in the coral *Acropora yongei* during photoacclimation. J Exp Biol 213:3644–3655

Rozhnov SV (2013) At the dawn of the aerobic biosphere: the effect of oxygen on the development of biota in the proterozoic and early paleozoic. Paleontol J 47(9):961–972

Ruby EG, Lee K-H (1998) The *Vibrio fischeri—Euprymna scolopes* light organ association: current ecological paradigms. Appl Environ Microbiol 64(3):805–812

Ruby EG, McFall-Ngai MJ (1992) A squid that glows in the night: development of an animal-bacterial mutualism. J Bacteriol 174(15):4865–4870

Rumbaugh KP, Diggle SP, Watters CM, Ross-Gillespie A, Griffin AS, West SA (2009) Quorum sensing and the social evolution of bacterial virulence. Curr Biol 19:341–345

Saade E, Ogryzko VV (2014) Epigenetics: what it is about? Biopolym Cell 30(1):3–9

Saba M, Wilts BD, Hielscher J, Schröder-Turk GE (2014) Absence of circular polarization in reflections of butterfly wing scales with chiral Gyroid structure. Mater Today Proc 2014:193–208

Sachs JL (2013) Origins, evolution, and breakdown of bacterial symbiosis. In: Levin SA (ed) Encyclopedia of biodiversity. Academic Press, Amsterdam, pp 637–644

Sambongi Y, Iko Y, Tanabe M, Omote H, Iwamoto-Kihara A et al (1999) Mechanical rotation of the c subunit oligomer in ATP synthase (F_0F_1): direct observation. Science 286:1722–1724

Santos SICO, Elward B, Lumeij JT (2006) Sexual dichromatism in the blue-fronted Amazon parrot (*Amazona aestiva*) revealed by multiple-angle spectrometry. J Avian Med Surg 20(1):8–14

Saranathan V, Osuji CO, Mochrie SGJ, Noh H, Narayanan S et al (2010) Structure, function, and self-assembly of single network gyroid ($I4_132$) photonic crystals in butterfly wing scales. PNAS 107(26):11676–11681

Schäfer G, Engelhard M, Müller V (1999) Bioenergetics of the archaea. Microbiol Mol Biol Rev 63(3):570–620

Scheffel A, Gärdes A, Grünberg K, Wanner G, Schüler D (2008) The major magnetosome proteins MamGFDC are not essential for magnetite biomineralization in *Magnetospirillum gryphiswaldense* but regulate the size of magnetosome crystals. J Bacteriol 190(1):377–386

Schenk F, Wilts BD, Stavenga DG (2013) The Japanese jewel beetle: a painter's challenge. Bioinspir Biomim. https://doi.org/10.1088/1748-3182/4/045002

Schiansky P, Strömberg T, Trillo D, Saggio V, Dive B et al (2023) Demonstration of universal time-reversal for qubit processes. Optica. https://doi.org/10.1354/OPTICA.469109

Schlegel K, Leone V, Faraldo-Gómez JD, Müller V (2012) Promiscuous archaeal ATP synthase concurrently coupled to Na^+ and H^+ translocation. PNAS 109(3):947–952

Schnitzler CE, Pang K, Powers ML, Reitzel AM, Ryan JF et al (2012) Genomic organization, evolution, and expression of photoprotein and opsin genes in *Mnemiopsis leidyi*: a new view of ctenophore photocytes. BMC Biol 10:107

Scholes GD, Fleming GR, Olaya-Castro A, van Grondelle R (2011) Lessons from nature about solar light harvesting. Nat Chem 3:763–774

Schreiner PR, Reisenauer HP, Ley D, Gerbig D, Wu C-H, Allen WD (2011) Methylhydroxycarbene: tunneling control of a chemical reaction. Science 332:1300–1303

Schrödinger E (1943) What is life? Trinity College, Dublin

Schuchmann K, Müller V (2012) A bacterial electron-bifurcating hydrogenase. J Biol Chem 287(37):31165–31171

Schuergers N, Lenn T, Kampmann R, Meissner MV, Esteves T et al (2016) Cyanobacteria use micro-optics to sense light direction. eLife. https://doi.org/10.7554/eLife.12620

Schüler D (2008) Genetics and cell biology of magnetosome formation in magnetotactic bacteria. FEMS Micorbiol Rev 32:654–672

Schwartz SD, Schramm VL (2009) Enzymatic transition states and dynamic motion in barrier crossing. Nat Chem Biol 5(8):551–558

Schwartz JM, Stapp HP, Beauregard M (2005) Quantum physics in neuroscience and psychology: a neurophysical model of mind-brain interaction. Philos Trans R Soc Lond B Biol Sci 360:1309–1327

Sephus CD, Fer E, Garcia AK, Adam ZR, Schwieterman EW, Kacar B (2022) Earliest photic zone niches probed by ancestral microbial rhodopsins. Mol Biol Evol 39(5):msac 100

Sergi A (2009) Quantum biology. Atti della Accademia Peloritana dei Pericolanti LXXXVII:C1C0901001

Serrano A, Perez-Castiñeira JR, Baltscheffsky H, Baltscheffsky M (2004) Proton-pumping inorganic pyrophosphatases in some archaea and other extremophilic prokaryotes. J Bioenerg Biomembr 36(1):127–133

Sharma V, Crne M, Park JO, Srinivasarao M (2009) Structural origin of circularly polarized iridescence in jeweled beetles. Science 325:449–451

Sharma V, Crne M, Park JO, Srinivasarao M (2014) Bouligand structures underlie circularly polarized iridescence of scarab beetles: a closer view. Mater Today Proc 2014:161–171

Shashar N, Rutledge PS, Cronin TW (1996) Polarization vision in cuttlefish—a concealed communication channel? J Exp Biol 199:2077–2084

Shavit K, Wagner A, Schertel L, Farstey V, Akkaynak D et al (2023) A tunable reflector enabling crustaceans to see but not be seen. Science 379:695–700

Shawkey MD, Morehouse NL, Vukusic P (2009) A protean palette: colour materials and mixing in birds and butterflies. J R Soc Interface 6:S221–S231

Sherson JF, Krauter H, Olsson RK, Julsgaard B, Hammerer K et al (2006) Quantum teleportation between light and matter. Nature 443:557–560

Shim S, Rebentrost P, Valleau S, Aspuru-Guzik A (2012) Atomistic study of the long-lived quantum coherences in the Fenna-Matthews-Olson complex. Biophys J 102:649–660

Sia PI, Luiten AN, Stace TM, Wood JPM, Casson RJ (2014) Quantum biology of the retina. Clin Exp Ophthalmol 42:582–589

Sieber JR, Sims DR, Han C, Kim E, Lykidis A et al (2010) The genome of *Syntrophomonas wolfei*: new insights into syntrophic metabolism and biohydrogen production. Environ Microbiol 12(8):2289–2301

Simões BF, Gower DJ, Rasmussen AR, Sarker MAR, Fry GC et al (2020) Spectral diversification and trans-species allelic polymorphism during the land-to-sea transition in snakes. Curr Biol 30:1–8

Sivinski JM (1998) Phototropism, bioluminescence, and the Diptera. Fla Entomol 81(3):282–292

Slamovits CH, Okamoto N, Burri L, James ER, Keeling PJ (2011) A bacterial proteorhodopsin proton pump in marine eukaryotes. Nat Commun. https://doi.org/10.1038/ncomms1188

Smith EG, D'Angelo C, Salih A, Wiedenmann J (2013) Screening by coral green fluorescent protein (GFP)-like chromoproteins supports a role in photoprotection of zooxanthellae. Coral Reefs. https://doi.org/10.1007/s00338-012-0994-9

Solovyov IA, Chang P-Y, Schulten K (2012) Vibrationally assisted electron transfer mechanism of olfaction: myth or reality? Phys Chem Chem Phys 14(40):13861–13871

Solovyov IA, Ritz T, Schulten K, Hore PJ (2014) A chemical compass for bird navigation. In: Mohseni M, Omar Y, Engel GS, Plenio MB (eds) Quantum effects in biology. Cambridge University Press, Cambridge, pp 218–236

Somanathan H, Borges RM, Warrant EJ, Kelber A (2008) Nocturnal bees learn landmark colours in starlight. Curr Biol 18(21):R996–R997

Song Z, Zhou Y, Juusola M (2016) Random photon absorption model elucidates how early gain cotrol in fly photoreceptors arises from quantal sampling. Front Comput Neurosci 10:61

Sparks JS, Schelly RC, Smith WL, Davis MP, Tchernov D et al (2014) The covert world of fish biofluorescence: a phylogenetically widespread and phenotypically variable phenomenon. PLoS One 9(1):e83259

Spinner M, Kovalev A, Gorb SN, Westhoff G (2013) Snake velvet black: hierarchical micro- and nanostructure enhances dark colouration in *Bitis rhinoceros*. Sci Rep 3:1846

Spinner M, Gorb SN, Balmert A, Bleckmann H, Westhoff G (2014) Non-contaminating camouflage: multifunctional skin microornamentation in the west african gaboon viper (*Bitis rhinoceros*). PLoS One 9(3):e91087

Spudich JL (2006) The multitalented microbial sensory rhodopsins. Trends Microbiol 14(11):480–487

Spudich JL, Sineshchekov OA, Govorunova EG (2014) Mechanism divergence in microbial rhodopsins. Biochim Biophys Acta 1837:546–552

Stabb EV (2005) Shedding light on the bioluminescence "paradox". ASM News 71(5):223–229

Stachel SJ, Stockwell SA, Van Vranken DL (1999) The fluorescence of scorpions and cataractogenesis. Chem Biol 6:531–539

Stavenga DG (2014) Thin film and multilayer optics cause structural colors of many insects and birds. Mater Today Proc 1S:109–121

Stavenga DG, Stowe S, Siebke K, Zeil J, Arikawa K (2004) Butterfly wing colours: scale beads make white pierid wings brighter. Proc Biol Sci 271:1577–1584

Stavenga DG, Leertouwer HL, Marshall NJ, Osorio D (2010) Dramatic colour changes in a bird of paradise caused by uniquely structured breast feather barbules. Proc Biol Sci. https://doi.org/10.1098/rspb.2010.2293

Stavenga DG, Wilts BD, Leertouwer HL, Haryiyama T (2011) Polarized iridescence of the multilayered elytra of the Japanese jewel beetle, *Chrysochroa fulgidissima*. Philos Trans R Soc B 366:709–723

Stavenga DG, Leertouwer HL, Wilts BD (2014) Coloration principles of nymphaline butterflies—thin films, melanin, ommochromes and wing scale stacking. J Exp Biol 217:2171–2180

Stock D, Leslie AGW, Walker JE (1999) Molecular architecture of the rotary motor in ATP synthase. Science 286:1700–1705

Stock D, Gibbons C, Arechaga I, Leslie AGW, Walker JE (2000) The rotary mechanism of ATP synthase. Curr Opin Struct Biol 10:672–679

Stoddard MJ, Prum RO (2011) How colorful are birds? Evolution of the avian plumage color gamut. Behav Ecol. https://doi.org/10.1093/beheco/arr088

Stomp M, Huisman J, Stal LJ, Matthijs HCP (2007) Colorful niches of phototrophic microorganisms shaped by vibrations of the water molecule. ISME J 1:271–282

Stoneham AM, Turin L, Brookes JC, Horsfield AP (2014) Quantum vibrational effects of sense in smell. In: Mohseni M, Omar Y, Engel GS, Plenio MB (eds) Quantum effects in biology. Cambridge University Press, Cambridge, pp 264–276

Stournaras KE, Lo E, Böhning-Gaese K, Cazetta E, Dehling DM et al (2013) How colourful are fruits? Limited color diversity in fleshy fruits on local and global scles. New Phytol 198:617–629

Straube N, Li C, Claes JM, Corrigan S, Naylor GJP (2015) Molecular phylogeny of squaliformes and first occurrence of bioluminescence in sharks. BMC Evol Biol 15:162

Strout G, Russell SD, Pulsifer DP, Erten S, Lakhtakia A, Lee DW (2013) Silica nanoparticles aid in structural leaf coloration in the Malaysian tropical rainforest understorey herb *Mapania caudata*. Ann Bot. https://doi.org/10.1093/aob/mct172

Sudo Y, Spudich JL (2006) Three strategically placed hydrogen-bonding residues convert a proton pump into a sensory receptor. PNAS 103(44):16129–16134

Sukontason KL, Bunchu N, Methanitikorn R, Chaiwong T, Kuntalue B, Sukontason K (2006) Ultrastructure of adhesive device in fly in families calliphoridae, muscidae and sarcophygidae, and their implication as mechanical carriers of pathogens. Parasitol Res 98:477–481

Sun S, Kondabagil K, Draper B, Alam TI, Bowman VD et al (2008) The structure of the phage T4 DNA packaging motor suggests a mechanism dependent on electrostatic forces. Cell 135:1251–1262

Sun S, Rao VB, Rossmann MG (2010) Genome packaging in viruses. Curr Opin Struct Biol 20(1):114–120

Sun J, Bhushan B, Tong J (2013) Structural coloration in nature. RSC Adv 3:14862–14889

Susskind L, Friedman A (2015) Quantum mechanics. Penguin Books

Tacchino F, Succurro A, Ebenhöh O, Gerace D (2019) Optimal efficiebcy of the Q-cycle mechanism around physiological temperatures from an open quantum system approach. Sci Rep 9:16657

Tarlaci S (2011) Quantum physics in living matter: from quantum biology to quantum neurobiology. NeuroQuantology 9(4):692–701

Terakita A, Kawano-Yamashita E, Koyanagi M (2012) Evolution and diversity of opsins. In: Wiley interdisciplinary reviews: membrane transport and signaling, vol 1. Wiley, pp 104–111

Teyssier J, Saenko SV, van der Marel D, Milinkovitch MC (2015) Photonic crystals cause active colour change in chameleons. Nat Commun. https://doi.org/10.1938/ncomms7368

Tezcan FA, Crane BR, Winkler JR, Gray HB (2001) Electron tunneling in protein crystals. PNAS 98(9):5002–5006

Thoen HH, How MJ, Chiou T-H, Marshall J (2014) A different form of color vision in Mantis shrimp. Science 343:411–413

Thomé M, Nicole L, Berthier S (2014) Multiscale replication of iridescent butterfly wings. Mater Today Proc 1S:221–224

Thuesen EV, Goetz FE, Haddock SHD (2010) Bioluminescent organs of two deep-sea arrow worms, *Eukrohnia fowleri* and *Caecosagitta macrocephala*, with further observations on bioluminescence in chaetognaths. Biol Bull 219:100–111

Ting CS, Rocap G, King J, Chisholm SW (2001) Phycobiliprotein genes of the marine photosynthetic prokaryote prochlorococcus: evidence for rapid evolution of genetic heterogeneity. Microbiology 147:3171–3182

Tinsley JN, Molodtsov MI, Prevedel R, Wartmann D, Espigulé-Pons J et al (2016) Direct detection of a single photon by humans. Nat Commun. https://doi.org/10.1038/ncomms12172

Tirandaz A, Ghahramani FT, Shafiee A (2015) Dissipative vibrational model for chiral recognition in olfaction. Phys Rev E 92:032724

Tirandaz A, Ghahramani FT, Salari V (2017) Validity examination of the dissipative quantum model of olfaction. arXiv:1701.01050v1

Tomita T, Toda M, Kaneko A, Murakumo K, Miyamoto K, Sato K (2023) Successful delivery of viviparous lantern shark from an artificial uterus and the self-production of lantern shark luciferin. PLoS One 18(9):e0291224

Tong D, Rozas NS, Oakley TH, Mitchell J, Colley NJ, McFall-Ngai M (2009) Evidence for light perception in a bioluminescent organ. PNAS 106(24):9836–9841

Trevors JT, Mason L (2011) Quantum microbiology. Curr Issues Mol Biol 13:43–49

Trissl H-W, Law CJ, Cogdell RJ (1999) Uphill energy transfer in LH2-containing purple bacteria at room temperature. Biochim Biophys Acta 1412:149–172

Trixler F (2013) Quantum tunneling to the origin and evolution of life. Curr Org Chem 17:1758–1770

Tsuji FI (1985) ATP-dependent bioluminescence in the firefly squid, *Watasenia scintillans*. PNAS 82:4629–4632

Tsuji FI, Leisman GB (1981) K⁺/Na⁺-triggered bioluminescence in the oceanic squid *Symplectotheuthis oualaniensis*. PNAS 78(11):6719–6723

Turin L (1996) A spectroscopic mechanism for primary olfactory reception. Chem Senses 21:773–791

Valencia NH, Srivastav V, Pivoluska M, Huber M, Friis N, et al (2020) High-dimensional pixel entanglement: efficient generation and certification. arXiv:2004.04994v4

Valeur B, Berberan-Santos MN (2011) A brief history of fluorescence and phosphorescence before the emergence of quantum theory. J Chem Educ 88:731–738

Valiadi M, Iglesias-Rodriguez D (2013) understanding bioluminescence in dinoflagellates—how far have we come? Microorganisms 1:3–25

van der Horst MA, Key J, Hellingwerf KJ (2007) Photosensing in chemotrophic, non-phototrophic bacteria: let there be light sensing too. Trends Microbiol 15(12):554–562

van der Kooi CJ, Wilts BD, Leertouwer HL, Staal M, Elzenga JTM, Stavenga DG (2014) Iridescent flowers? Contribution of surface structures to optical signalling. New Phytol 203:667–673

Van Dover CL, Reynolds GT, Chave AD, Tyson JA (1996) Light at deep-sea hydrothermal vent. Geophys Res Lett 23(16):2049–2052

van Grondelle R, Novoderezhkin VI (2011) Quantum effects in photosynthesis. Procedia Chem 3:198–210

Van Speybroeck L (2002) From epigenesis to epigenetics: the case of C. H. Waddington. Ann N Y Acad Sci 981:61–81

Vargová R, Wideman JG, Derelle R, Klimeš V, Kahn RA et al (2021) An eukaryote-wide perspective on the diversity and evolution of the ARF GTPase protein family. Genome Biol Evol 13:8

Vattay G, Kauffman S, Niiranen S (2014) Quantum biology on the edge of quantum chaos. PLoS One 9(3):e89017

Veening J-W, Smits WK, Kuipers OP (2008) Bistability, epigenetics, and bet-hedging in bacteria. Annu Rev Microbiol 62:193–210

Vermaas WFJ (2002) Evolution of photosynthesis. In: Encyclopedia of life science. Wiley. www.els.net

Veshaguri S, Christensen SM, Kemmer GC, Ghale G, Møller MP et al (2016) Direct observation of proton pumping by a eukaryotic P-type ATPase. Science 351(6280):1469–1473

Vigneron JP, Pasteels JM, Windsor DM, Vértesy Z, Rassart M et al (2007) Switchable reflector in the Panamanian tortoise beetle *Charidotella egregia* (Chrysomelidae: Cassidinae). Phys Rev E 76:031907

Vigneron JP, Rassart M, Vértesy Z, Kertesz K, Sarrasin M, et al (2014) Optical structure and function of the white filamentary hair covering the edelweiss bracts. arXiv:0710.2695v1

Vignolini S, Rudall PJ, Rowland AV, Reed A, Moyoroud E et al (2012) Pointillist structural color in *Pollia* fruit. PNAS 109(39):15712–15715

Vignolini S, Moyroud E, Glover BJ, Steiner U (2013) Analysing photonic structures in plants. J R Interface 10:20130394

Villafani Y, Yang HW, Park Y-I (2020) Color sensing and signal transmission diversity of cyanobacterial phytochromes and cyanobacteriochromes. Mol Cells 43(6):509–516

Viviani V (2009) Terrestrial bioluminescence. In: Smith KC (ed) Photobiological sciences online. American Society for Photobiology. http://www.photobiology.info/

Viviani V, Arnoldi FGC, Venkatesh B, Neto AJS, Ogawa FGT et al (2006) Active-site properties of *Phrixotrix* railroad worm green and red bioluminescence-eliciting luciferases. J Biochem 140:467–474

Viviani V, Rocha MY, Hagen O (2010) Fauna de besouros bioluminescentes (Coleoptera: Elateroidea: Lampyridae; Phengodidae, Elateridae) nos municípios de Campinas, Sorocaba-Votorantim e Rio Claro-Limeira (SP, Brasil): biodiversidade e influência da urbanização. Biota Neotrop 10(2) http://www.biotaneotropica.org.br

Viviani V, Amaral DT, Neves DR, Simões A, Arnoldi FGC (2013) The luciferin binding site residues C/T311 (S314) influence the bioluminescence color of beetle luciferases through main-chain interaction with oxyluciferin phenolate. Biochemistry 52(1):19–27

Vöcking O, Macias-Muñoz A, Jaeger SJ, Oakley TH (2022) Deep diversity: extensive variation in the components of complex visual systems across animals. Cell 11:3966

Vogel H (1974) Physik. Springer, Berlin

von Byern J, Dorrer V, Merritt DJ, Chandler P, Stringer I et al (2016) Characterization of the Fishing Lines in Titiwai (=*Arachnocampa luminosa* Skuse, 1890) from New Zealand and Australia. PLoS One 11(12):e0162687

Vršanský P, Chorvát D, Fritzsche I, Hain M, Ševčik R (2012) Light-mimicking cockroaches indicate tertiary origin of recent terrestrial luminescence. Naturwissenschaften. https://doi.org/10.1007/s00114-112-0956-7

Vukusic P, Hooper I (2005) Directionally controlled fluorescence emission in butterflies. Science 310:1151

Vukusic P, Sambles JR (2003) Photonic structures in biology. Nature 424:852–855

Vukusic P, Sambles JR, Larence CR, Wootton RJ (2001) Limited-view iridescence in the butterfly *Ancyluris meliboeus*. Proc Biol Sci 269:7–14

Vukusic P, Kelly R, Hooper I (2009) A biological sub-micron thickness optical broadband reflector characterized using both light and microwaves. J R Soc Interface 6:S193–S201

Wada M, Kong S-G (2018) Actin-mediated movement of chloroplasts. J Cell Sci 131. https://doi.org/10.1242/jcs.210310

Wajnberg E, Acosta-Avalos D, Alves OC, de Oliveira JF, Srygley RB, Esquivel DMS (2010) Magnetoreception in eusocial insects: an update. J R Soc Interface 7:S207–S225

Wang L, Goodey NM, Benkovic SJ, Kohen A (2006) The role of enzyme dynamics and tunnelling in catalysing hydride transfer: studies of distal mutants of dihydrofolate reductase. Philos Trans R Soc Lond B Biol Sci. https://doi.org/10.1098/rstb.3006.1871

Wang Y, Labandeira CC, Shih C, Ding Q, Wang C et al (2012) Jurassic mimicry between a hanging-fly and a gingko from China. PNAS 109(50):20514–20519

Wang L, Fried SD, Boxer SG, Markland TE (2014) Quantum delocalization of protons in the hydrogen-bond network of an enzyme active site. PNAS 111(52):18454–18459

Wang CX, Hilburn IA, Wu D-A, Mizuhara Y, Cousté CP et al (2019) Transduction of the geomagnetic field as evidenced from alpha-band activity in the human brain. eNeuro 6(2). https://doi.org/10.1523/ENEURO.0483-18.2019

Ward LM, Shih PM (2021) Granick revisited: synthesizing evolutionary and ecological evidence for the late origin of bacteriochlorophyll via ghost lineages and horizontal gene transfer. PLoS One 16(1):e0239248

Warrant EJ (2007) Nocturnal bees. Curr Biol 17(23):R991–R992

Warrant EJ (2008) Seeing in the dark: vision and visual behaviour in nocturnal bees and wasps. J Exp Biol 211:1737–1746

Warrant EJ (2010) Polarisation vision: beetles see circularly polarised light. Curr Biol 20(14):R610–R612

Warrant EJ (2015) Visual tracking in the dead of night. Science 348(6420):1212–1213

Warrant EJ, Dacke M (2011) Vision and visual navigation in nocturnal insects. Annu Rev Entomol 56:239–254

Warrant E, Somanathan H (2022) Colour vision in nocturnal insects. Philos Trans R Soc Lond B Biol Sci 377:20210285

Watson GS, Schwarzkopf L, Cribb BW, Myhra S, Gellender M, Watson JA (2015) Removal mechanisms of dew via self-propulsion off the gecko skin. J R Soc Interface 12:20141396

Watson GS, Watson JA, Cribb BW (2017) Diversity of cuticular micro- and nanostructures on insects: properties, functions, and potential applications. Annu Rev Entomol 62:185–205

Weber S (2005) Light-driven enzymatic catalysis of DNA repair: a review of recent biophysical studies on photolyases. Biochim Biophys Acta 1707:1–23

Weber J, Senior AE (2003) ATP synthesis driven by proton transport in F_1F_0-ATP synthase. FEBS Lett 545:61–70

Węgrzyn G, Czyż A (2002) How do marine bacteria produce light, why are they luminescent, and can we employ bacterial bioluminescence in aquatic biotechnology? Oceanologia 44(3):291–305

Weinstein P, Delean S, Wood T, Austin AD (2016) Bioluminescence in the ghost fungi *Omphalotus nidiformis* does not attract potential spore dispersing insects. Fungus 7(2):229–234

Weitz HJ (2004) Naturally bioluminescent fungi. Mycologist 18(1):4–5

Welch VL, Vigneron J-P (2007) Beyond butterflies—the diversity of biological photonic crystals. Opt Quant Electron 39:295–303

Wetzel RG (1975) Limnology. WB Saunders Company Philadelphia London Toronto. ISBN: 7216-9240-0.

Whitney HM, Kolle M, Andrew P, Chittka L, Steiner U, Glover BJ (2009) Floral iridescence, produced by diffractive optics, acts as a cue for animal pollinators. Science 323:130–133

Widder EA (1998) A predatory use of counterillumination by the squaloid shark, *Isistius brasiliensis*. Environ Biol Fishes 53:267–273

Widder EA (2001) Marine bioluminescence. Biosci Explained 1(1):1–9

Widder EA (2010) Bioluminescence in the ocean: origins of biological, chemical, and ecological diversity. Science 328:704–708

Widder EA, Falls B (2013) Review of bioluminescence for engineers and scientists in biophotonics. IEEE J Sel Top Quantum Electron. https://doi.org/10.1109/JSTQE.2013.2284434

Widder EA, Latz MI, Case JF (1983) Marine bioluminescence spectra measured with an optical multichannel detection system. Biol Bull 165:791–810

Wiersma DS (2013) Disordered photonics. Nat Photonics 7:188–196

Wilde MM (2017) Quantum information theory. Cambridge University Press, Cambridge

Williams RJP, Fraústo da Silva JJR (2006) The chemistry of evolution. Elsevier, Amsterdam

Willis RE, White CR, Merritt DJ (2011) Using light as a lure is an efficient predatory strategy in Arachnocampa flava, an Australian glowworm. J Comp Physiol B 181:477–486

Wiltbank LB, Kehoe DM (2019) Diverse light responses of cyanobacteria mediated by phytochrome superfamily photoreceptors. Nat Rev Microbiol 17:37–50

Wilts BD, Leertouwer HL, Stavenga DG (2009) Imaging scatterometry and microspectrophotometry of lycaenid butterfly wing scales with perforated multilayers. J R Soc Interface 6:S185–S192

Wilts BD, Michelsen K, De Raedt H, Stavenga DG (2014a) Sparkling feather reflections of a bird-of-paradise explained by finite-difference time-domain modelling. PNAS 111(12):4363–4368

Wilts BD, Whitney HM, Glover BJ, Steiner U, Vignolini S (2014b) Natural helicoidal structures: morphology, self-assembly and optical properties. Mater Today Proc 1S:177–185

Wilts BD, Matsushita A, Arikawa K, Stavenga DG (2015) Spectrally tuned structural and pigmentary coloration of birdwing butterfly wing scales. J R Soc Interface 12:20150717

Wiltschko W, Wiltschko R (2005) Magnetic orientation and magnetoreception in birds and other animals. J Comp Physiol A 191:675–693

Wiltschko W, Traudt J, Güntürkün O, Prior H, Wiltschko R (2002) Lateralization of magnetic compass orientation in migratory birds. Nature 419:467–470

Winkler JR (2000) Electron tunneling pathways in proteins. Curr Opin Chem Biol 4:192–198

Winklhofer M (2010) Magnetoreception. J R Soc Interface 7:S131–S134

Wise CE, Ledinina AE, Yuly JL, Artz JH, Lubner CE (2021) The role of thermodynamic features on the functional activity of electron bifurcating enzymes. BBA Bioenerg 1862(4):148377

Wolf DM, Arkin AP (2003) Motifs, modules and games in bacteria. Curr Opin Microbiol 6:125–134

Wolke CT, Fournier JA, Dzugan LC, Fagiani MR, Odbadrakh TT et al (2016) Spectroscopic snapshots of the proton-transfer mechanism in water. Science 354:1131–1135

Wong CY, Alvey RM, Turner DB, Wilk KE, Bryant DA et al (2012) Electronic coherence lineshapes reveal hidden excitonic correlations in photosynthetic light harvesting. Nat Chem 4:396–404

Woolf NJ, Hameroff SR (2001) A quantum approach to visual consciousness. Trends Cogn Sci 5(11):472–478

Wu L-Q, Dickman JD (2011) Magnetoreception in an avian brain in part mediated by inner ear lagena. Curr Biol 21:418–423

Wucherer MF, Michiels NK (2014) Regulation of red fluorescent light emission in a cryptic marine fish. Front Zool 11:1

Yin H, Dong B, Liu Y, Zhan Y, Shi L et al (2012) Amorphous diamond-structured photonic crystal in the feather barbs of the scarlet macaw. PNAS 109(27):10798–10801

Yokoyama S (2008) Evolution of dim-light and color vision pigments. Annu Rev Genomics Hum Genet 9:259–282

Yokoyama S, Shi Y (2000) Genetics and evolution of ultraviolet vision in vertebrates. FEBS Lett 486:167–172

Yokoyama S, Zhang H, Radlwimmer FB, Blow NS (1999) Adaptive evolution of color vision of the Comoran coelacanth (*Latimeria chalumnae*). PNAS 96:6278–6284

Yokoyama S, Tada T, Zhang H, Britt L (2008) Elucidation of phenotypic adaptations: molecular analyses of dim-light vision proteins in vertebrates. PNAS 105(36):13480–13485

Yoshioka S, Matsushana B, Tanaka S, Inouye Y, Oshima N, Kinoshita S (2011) Mechanism of variable structural colour in neon tetra: quantitative evaluation of the Venetian blind model. J R Soc Interface 8:56–66

Yu Z, Streng C, Seibeld RF, Igbalajobi OA, Leister K et al (2021) Genome-wide analyses of light-regulated genes in Aspergillus nidulans reveal a complex interplay between different photoreceptors and novel photoreceptor functions. PLoS Genet 17(10):e1009845

Yutin N, Koonin EV (2012) Proteorhodopsin genes in giant viruses. Biol Direct 7:34

Zhang W-J, Santini C-L, Bernadac A, Ruan J, Zhang S-D et al (2012) Complex spatial organization and flagellin composition of flagellar propeller from marine magnetotactic ovoid strain MO-1. J Mol Biol 416:558–570

Zhao B, Chen Y-A, Bao X-H, Strassel T, Chu C-S, et al (2008) A millisecond quantum memory for scalable quantum networks. arXiv:0807.5064v1

Zhou C-J, Tian D, He J-H (2018) What factors affect lotus effect? Therm Sci 22(4):1737–1743

Zuchan K, Baymann F, Baffert C, Brugna M, Nitschke W (2021) The dyad of the Y-junction- and a flavin module unites diverse redox enzymes. BBA Bioenerg 1862(6):148401

Zurek WH (2002) Decoherence and the transition from quantum to classical—revisited. Los Alamos Sci 27

Zurek W H (2009) Quantum Darwinism. arXiv:0903.5082v1

Zurek WH (2022) Quantum theory of the classical: einselection, envariance, quantum Darwinism and extantons. Entropy 24:1520

Zurek WH, Paz JP (1994) Decoherence, chaos, and the second law. arXiv:gr-qc/9402006v2

General Characteristics and Properties of Organic Life

5.1   The Basic Unit of All Organisms: The Cell(s)

Despite all of the differences in size and diversity among life forms, all organisms share a common feature in that they originate from a single cell. They contain the entire specific potential for development of a complete organism, be it a bacterium, a sequoia or a human being. Whereas the majority of all organisms remain permanently unicellular, multicellular organisms pass through this stage at least at the beginning of their life cycles. Seen from this perspective, the diversity of functional processes in a single cell and its significance for the entire course of biological evolution become especially apparent. Although our life functions in their entirety depend on cells and single-celled organisms, we cannot directly perceive the processes involved. Only with elaborate technical aids has it become increasingly possible to penetrate the spatial and temporal dimensions of the molecular processes behind them. Because of prevailing paradigms—still founded on observations of the macroscopic world—the evolutionary significance of cellular processes is only gradually becoming recognized. One obstacle to understanding such processes is the lack of structural differentiation in a system for information processing—as in nervous systems in (most) animals. On the other hand, a wide variety of receptor molecules on cell membranes for the detection of diverse signals from the environment has been identified. This means, for example, that chemical or electrical signals can be perceived and also employed for communication between cells. Some groups of bacteria can orient themselves to the terrestrial magnetic field with the help of stored iron compounds in a special cell organelle (magnetosome). Various receptor systems enable many life forms to orient themselves to light signals (Visick and Fuqua 2005; Bassler and Losick 2006; van der Horst et al. 2007; Baraquet et al. 2009; Naresh et al. 2011; Ellison et al. 2017). Internal and external cell

M. Knoflacher, *Relativity of Evolution*, https://doi.org/10.1007/978-3-662-69423-7_5

structures enable many unicellular life forms to move across surfaces or in liquids (Ounjai et al. 2013; Zhao et al. 2014).

For billions of years, archaeans and bacteria have managed the conversion of elements derived from inorganic compounds into complex organic compounds (Fig. 5.1). Classical chemical approaches that focus primarily on the concentration ratios of a few elements in nutrient solutions—mostly carbon (C), nitrogen (N) and phosphorus (P)—do not yield a useful explanation for biological transformation processes. However, when additional elements are taken into account, stoichiometry provides important clues to various external chemical conditions over the course of biological evolution. An easier approach to understanding cellular processes is provided by considering the electrochemical characteristics of chemical elements and compounds—for example protons (H^+) or electrons (e^-)—in biochemical transformation processes. As explained in the previous chapter, these characteristics enable the conversion of both material compounds and energy—as well as information processing. The latter was explained in more detail in the context of conversion of photon energy into biochemical energy during the processes of photosynthesis or extraction of information from light.

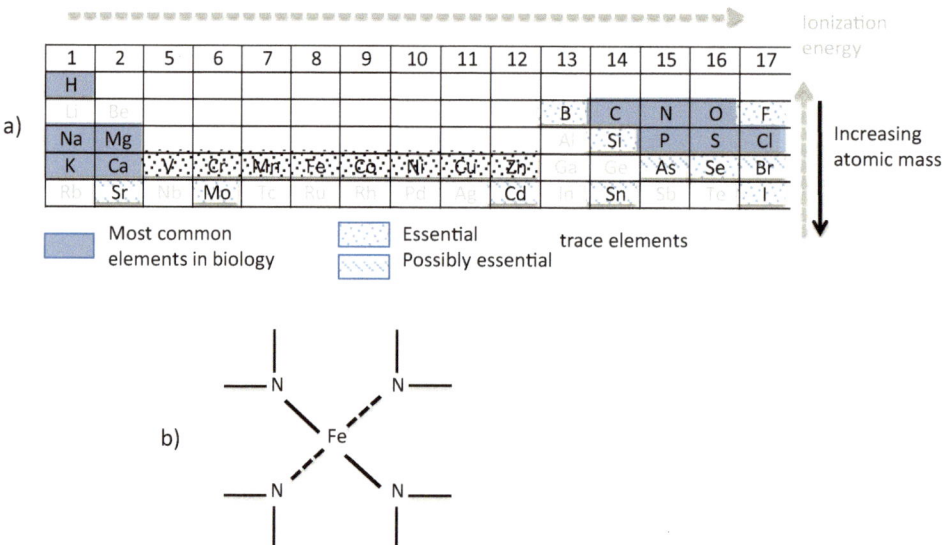

Fig. 5.1 (a) Organisms use simplified lighter chemical elements (common elements) to form biochemical structures and individual heavier elements (trace elements) to develop functional processes; (b) Greatly simplified illustration (central region of a haem molecule) showing the importance of multidimensional structural connections between different atoms for the functionality of biological molecules. Explanations: Biologically unused elements in grey symbols; numbers = group classification of chemical element according to IUPAC (IUPAC = International Union of Pure and Applied Chemistry) (3, 4 and 18 omitted along with elements of periods 6 and 7 in the periodic table). Sources: Modified according to Fraústo da Silva and Williams (2001), Berg et al. (2013)

Relevant processes can be abstractly summarized in the—mutually interdependent—groups of energy conversion, synthesis and information processing—including genetic information processes (Fig. 5.2). The calcium ion (Ca^{2+}), which is involved (among other things) in signal transmission, metabolism and the processes of cell division or cell movement, serves as an example of a chemical element. In general, all benefits involve requirements for maintaining physiologically favourable calcium concentrations in the solutions (cytosols) that individual cells contain. Marked variation in benefits is related to the diversity of regulators of the passive influx of calcium ions into cells and energetically costly exportation from cells (Isayenkov et al. 2010; Kazmierczak et al. 2013). In almost all organisms—from prokaryotes to plants, fungi and animals—calcium ions regulate various physiological processes as well as the transmission and storage of information (Michiels et al. 2001; Verret et al. 2010; Bose et al. 2011; Cai and Clapham 2011; Cai et al. 2015; Dominguez et al. 2015; Plattner 2015; Beckmann et al. 2016; Kudla et al. 2018).

Research results obtained over recent decades have provided increasing evidence for the importance of calcium ions for information processing in vertebrates—including humans. The decisive factor here was a focus of sensory physiological research on the morphology and electrical signals of nerve cells. The numerous glial cells—especially in central nervous systems—have primarily been allocated supporting functions, including during the development of nervous systems. It was only because of more detailed investigations that increasing evidence emerged showing that glial cells are also directly involved in the transmission and storage of information—based on calcium ions—in brains

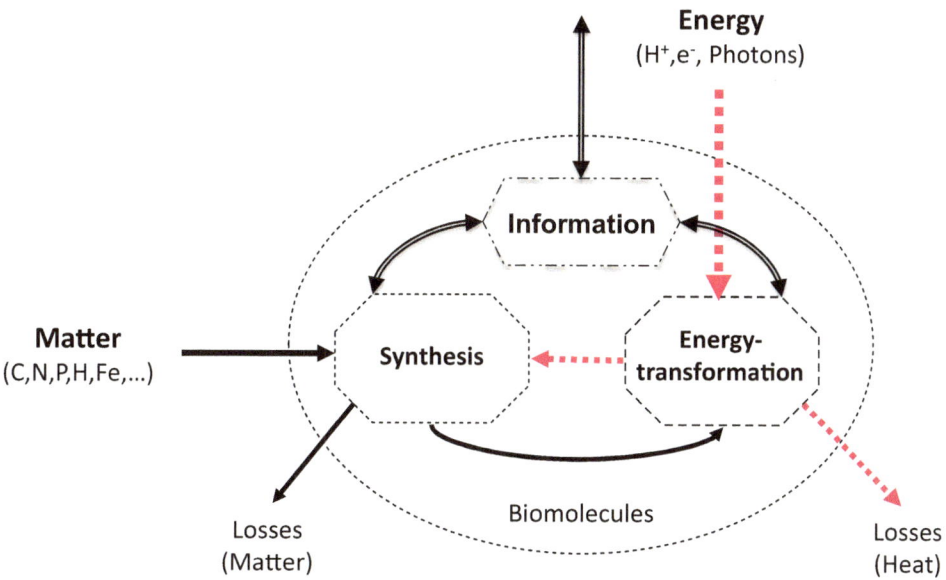

Fig. 5.2 Generalized representation of central processes for cell maintenance and growth

(Kettenmann and Verkhratsky 2008; Tasker et al. 2012; Araque et al. 2014; Falk and Götz 2017).

The importance of unicellular organisms for biological evolution can be seen macroscopically from evidence of early environmental modifications, for example changes in atmospheric oxygen content and deposition of iron or calcium compounds (Konhauser 2007; Knoll et al. 2016; Brocks 2018; Lenton and Daines 2018; van Maldegem et al. 2019). In an exclusively retrospective view of events, it is easy to overlook the fact that unicellular organisms, because of their large numbers and ubiquitous distribution, continue to exert a major influence on both the dynamics of biological evolution and the global framework (Cho and Azam 1999; Kappler et al. 2005; Parkes et al. 2005; Jørgensen 2006; Dupraz et al. 2009; Miot et al. 2009; Konhauser et al. 2011; Kamennaya et al. 2012; Sarmento 2012). Deviating from classical notions of biological evolution, archaeans and bacteria, with their genetic endowments and physiological capacities, continuously adapt to altered framework conditions. Unlike multicellular organisms, the emerging functional groups cannot be divided into species using the usual methodological approaches of systematics (Doolittle 2009a, b; Polz et al. 2013). Delimitation of species among fungi is also difficult and partially unresolved, among other things as a result of their particular formation of cell structures and genetic variability (Giraud et al. 2008; Groenewald et al. 2012; Hawksworth 2012).

5.1.1 Challenges of Maintaining Cellular Functionality

Any given development takes place autonomously, regulated by internal processes but depending on certain environmental conditions. Here, we are already faced with an essential process of evolution—*the interplay between autonomous endogenous development and variable environmental conditions*. In the context of the central property of organic life—reproduction (Sect. 4.7)—this interplay must be dynamically balanced on at least two levels, that of the individual organism and that of the population (Fig. 5.3). Reproduction leads to opposing effects at these two levels. For individual organisms, reproduction is associated with material and energetic costs, sometimes directly linked with the death of organisms in the parent generation—for example in *Myxococcus* (see Fig. 6.11), octopuses or male spiders. At the population level, reproduction leads to an increased number of organisms, thus contributing to long-term maintenance of the life form concerned (Fig. 5.3). At this level, reproduction can be interpreted as self-organization within populations. It is safeguarded by the endogenous stimulation of reproduction, which is common to all life forms. As far as can be determined, this is the only evolutionary strategy—common to all organisms—for maintaining specific life forms. By contrast, there are no known common strategies for avoiding population losses, either due to endogenous factors—for example, overpopulation or violent conflicts between subpopulations—or resulting from external influences—for example, epidemics or changes in abiotic conditions.

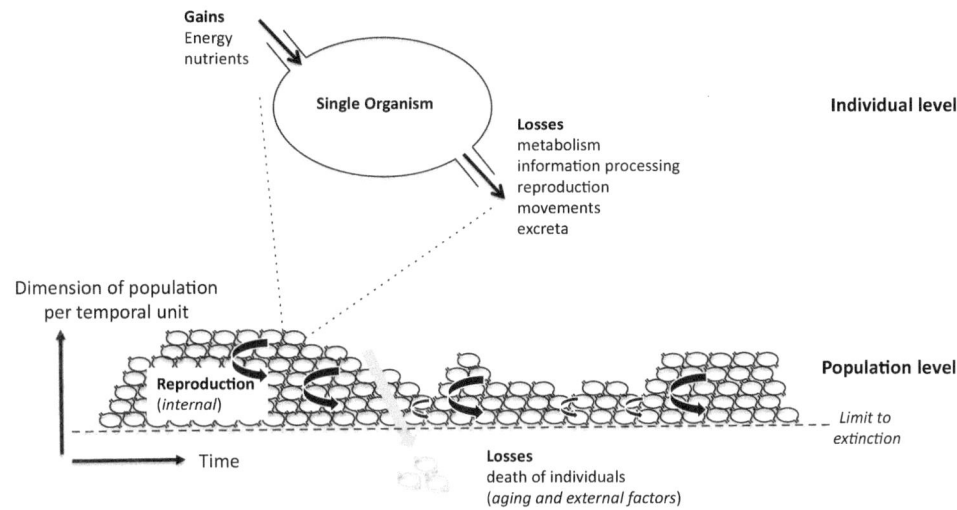

Fig. 5.3 Simplified representation of requirements for the maintenance of life functions in a single organism and in populations (For discussion of the concept see Waples and Gaggiotti 2006) (further explanations in the text)

The reasons for this reside in differing properties at the two system levels. At the individual level are selectively closed systems in the form of single cells or multicellular organisms. Their dynamic equilibrium (homeostasis) is maintained by internal regulation of all transformation processes over an organism's lifespan. At the population level are open systems in which changes can theoretically trigger the most diverse effects over arbitrarily large distances and long periods of time. In reality, such effects are dampened for thermodynamic reasons alone and are superimposed by a diversity of simultaneously occurring processes (see Chap. 8).

Processes in open systems cannot be completely understood for any single life form. For reasons underlying decision-making capacity and thus viability alone, organisms can detect processes only within content-related, spatial and dynamic boundaries (see Chap. 6). Information from these evolutionarily developed "perceptual spaces" accordingly also dominates individual decisions and actions in all organisms. For example, chemical signals at distances of micrometres are relevant for bacterial cells, whereas humans cannot even visually perceive such dimensions, although they do perceive processes at distances between centimetres and several hundred metres.

This will surely trigger objections on the grounds that people using technical devices can observe dimensions ranging from atomic to astronomical. These are the privileges of a vanishingly small minority in global terms. But even these privileged few, at least during their private everyday lives, experience their environments using the senses and observation possibilities of all human beings within certain—evolutionarily predetermined—limits. Yet social notions and interests are fed from this realm of perception—in interaction with the environmental conditions encountered on a daily basis. All else is not directly accessible to individual experience and can—depending on individual attitudes—be believed, ignored or rejected across a wide transitional bandwidth.

Because only information from their specific perceptual space is relevant for organisms belonging to a single life form, for enhancement of biological evolutionary conditions abiotic processes can be modified only through interactions with complementary perceptual spaces and functional capacities (see Chaps. 9 and 10).

In principle, every organism—whether consciously or unconsciously—must make decisions regarding the distribution of energy costs in the present for an unknown future. At the same time, the latitude available for this is limited by the energy-conversion potentials of the present and recent past. The long-term "memory" of its genetic constitution and the short-term epigenetic memory of an individual organism form the basis for the decision-making strategies. These challenges can be met with different strategies—always in interrelation with prevailing environmental conditions (Fig. 5.4). If there are very low probabilities for subsequent generations to find suitable habitats for reproduction, a strategy with large numbers of offspring but limited effort for individual offspring will prove successful. By contrast, a strategy of extended reproductive cycles with large organisms and few offspring can be successful only under dynamically stable environmental conditions.

Compared to organisms, viruses have significantly lower weights, very small genomes and no independent energy metabolism. Among viruses, virophages—viruses that prey on

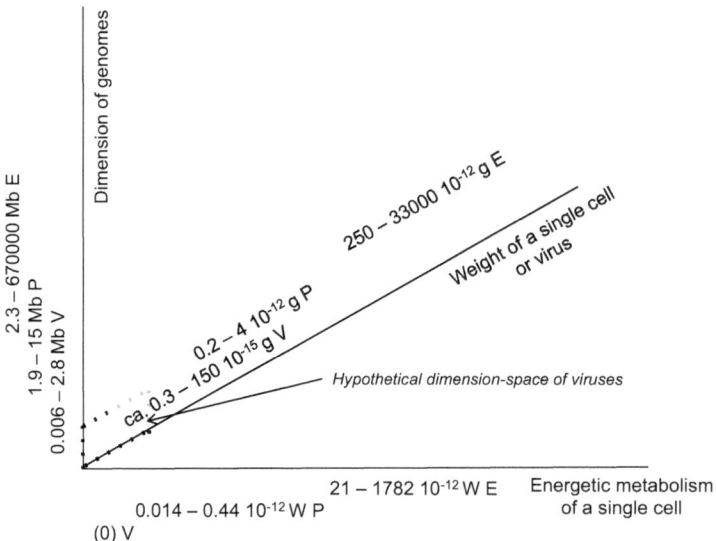

Fig. 5.4 Comparison of genome sizes, weights and energy turnover of cells of prokaryotes (P) and unicellular eukaryotes (E), supplemented by genome sizes and estimated weights of viruses (V). Further abbreviations: Mb = megabases (= 10^6 bases), g = grams. Data sources: Gupta et al. (2004), McGrath and Katz (2004), Keeling et al. (2005), Gregory et al. (2007), Schneiker et al. (2007), Lane and Martin (2010), Han et al. (2013), Aherfi et al. (2016), Bekliz et al. (2016)

viruses—have the smallest weights and genomes. On the other hand, their hosts—giant viruses—have the largest and weights genomes. With their genomes, they are able to set up independent reproductive systems—virus factories—in the cytoplasm of their host cells: unicellular and multicellular eukaryotes (Boughalmi et al. 2013; Fischer 2013). This means that they require only substances and energy from host cells for reproduction, but are at the same time exposed to attacks by virophages.

As a general rule, parasitic life forms are also found in the lower ranges of size scales among prokaryotes and eukaryotes. With few exceptions, however, they are also confronted with attacks by viruses and therefore require a genetic repertoire for their defence (Cohen et al. 2019). The genomes of soil-dwelling myxobacteria permit a variable and versatile metabolism, for example including conversion of cellulose (Schneiker et al. 2007). In bacteria, the extent and configuration of genomes vary depending on their particular habitats—an essential factor with respect to the problem already mentioned of classification of bacterial species. Instead, systematic units of bacteria are defined with the help of pan-genomes—consisting of a matching fraction of genes, the core genome and "peripheral" genes (Medini et al. 2005; Hendrickson 2009; Lapierre and Gogarten 2009; Vernikos et al. 2013). Inevitably, therefore, differences in the stated genomes of "species" of bacteria are to be expected. Within their systematic units, eukaryotes show far less variation in genome sizes. However, no correlations between genome size and multicellularity exist in eukaryotes. The two extreme values in Fig. 5.4 are found in unicellular life forms; the minimal value is found in endoparasitic microsporidia (unicellular fungi) and the maximal value in free-living amoebae. The genome of unicellular amoebae contains about 200 times as many base pairs as that of humans (McGrath and Katz 2004; Keeling et al. 2005; Mohr 2018). The smallest genomes of eukaryotes with about 0.4 Mb (megabases) are found in the residual cell nuclei (nucleomorphs) of algae (*Guillardia*) living in secondary endosymbiosis (Douglas et al. 2001; Curtis et al. 2012).

Data on cell weights and energy turnover should provide some idea of comparative differences between prokaryotes and unicellular eukaryotes. Especially in the maximum ranges of both groups, giant forms such as bacterial cells with a diameter of around 750 μm (*Thiomargarita* sp.) or unicellular eukaryotes (protists, Xenophyophores) up to several decimetres in size can be found on deep sea beds (Levin and Gooday 1992; Lecroq et al. 2009; Viswanathan 2012; Gooday et al. 2017). In both cases, the cells also serve as "food stores", rendering the measurement of physiologically relevant biomass and energy turnover considerably more difficult. Estimates for *Thiomargarita* give a cell weight of about $1 \cdot 10^{-6}$ grams (g) and an energy turnover of about $500 \cdot 10^{-12}$ watts (W) (Lane and Martin 2012). For Xenophyophores no data have yet been reported.

5.1.2   Dominance of Unicellular Life Forms

Among the three domains of organisms (archaeans, bacteria and eukaryotes; Fig. 2.1), archaeans show the greatest potential for survival under extreme environmental conditions. Particularly striking is the temperature tolerance of individual life forms—which can continue to grow at around 120 °C but can also survive at −140 °C—as well as the ability of various archaeans to adapt to minimal energy flow (Rothschild and Mancinelli 2001; Madigan et al. 2003; Lewalter and Müller 2006; Stetter 2006; Schoepp-Cothenet et al. 2013; Mayer and Müller 2014; Lever et al. 2015). An essential prerequisite for survival under extreme conditions is the special structure of the cell membrane, which greatly reduces uncontrolled energy losses (Valentine 2007; Sojo et al. 2014). Bacteria, by comparison, require more favourable energetic conditions (Kim and Gadd 2008). Many life forms show great flexibility in the use of different energy potentials and can colonize an almost unimaginable variety of habitats in community with other prokaryotes. Examples are hydrothermal vents (Inskeep et al. 2005; Teske and Reysenbach 2015), hypersaline waters (Des Marais 2003), oxygen-free zones (Kiørboe and Jackson 2001; Treude et al. 2005; Ghosh and Dam 2009), inclusions in ice (Staley and Gosink 1999), microfissures in soils and rocks (Hallberg et al. 2006; MacLean et al. 2007; Vos et al. 2013; Osburn et al. 2014)—along with all surfaces and digestive systems of multicellular eukaryotes (Walter and Ley 2011; Belkaid and Segre 2014). Representatives of both domains can, under suitable conditions, remain viable in the form of resting stages for up to a hundred million years (Vreeland et al. 2000; Stan-Lotter et al. 2002; Jaakkola et al. 2012; Morono et al. 2020). These are capacities that in all probability safeguarded the reproductive functions of organisms in the early stages of biological evolution (Nisbet and Sleep 2001).

Thanks to continual improvements in methods of investigations, differentiation of cell structures in prokaryotes is increasingly detectable. In terms of cell physiology, temporarily formed microcompartments surrounded by proteins fulfil a wide variety of functions in bacteria, for example the fixation of CO_2 in autotrophs or of organic compounds in heterotrophs (Kerfeld et al. 2018). In addition to the previously known double cell membranes in Gram-negative bacteria and the magnetosomes—used for orientation—new findings have included highly differentiated cell organelles for photosynthesis, endosymbiotic organelles and nucleus-like structures in bacteria (Löwe and Amos 2009; Murat et al. 2010; Diekmann and Pereira-Leal 2013), as well as membrane-enveloped extracellular regions (intramembrane compartments) in archaeans from hot springs (Huber et al. 2012). Archaeans with double cell membranes (Perras et al. 2014) and tiny life forms living symbiotically with other archaeans (Nanoarchaeota) have also been discovered (Huber et al. 2003; Wurch et al. 2016).

Archaeans and bacterial cells possess various cell appendages, some of which consist of protrusions of the cell membrane and many of which are mobile flagella with a wide range of different functions. Individual archaeans have cell appendages with anchor-like ends, presumably for attachment to surfaces (Moissl et al. 2005). Cell protrusions mainly serve to Improve the exchange of substances with the environment (Wagner et al. 2006; Shetty et al. 2011). In all unicellular life forms, small bubbles (membrane vesicles) for the transport of different molecules—for instance enzymes, signalling substances, toxins—can be formed by cell protrusions (Kulp and Kuehn 2010; Deatherage and Cookson 2012; Brown et al. 2015). Preservation of the original substance concentration by the envelope is advantageous for attaining the relevant effect—e.g. enrichment of substances, perception of signals or destruction of other cells. Dependence of physical transport on the randomness of environmental conditions is disadvantageous. Flagella enable electron exchange with the environment or other prokaryotes, penetration of the cell membranes of other organisms for gene exchange or infusion of toxic substances, and—most importantly— locomotion (Schäfer et al. 1999; Lovley 2008; Haiko and Westerlund-Wikström 2013; Jarrell et al. 2013; Reardon and Mueller 2013; Bhowmick and Tripathy 2014; Malvankar and Lovley 2014).

Prevailing physical conditions at microscopic dimensions, in conjunction with naturally occurring nutrient concentrations, also influence the development of cell sizes in prokaryotes. In contrast to eukaryotes, they can only absorb their nutrients through osmosis. Moreover, according to current knowledge, they do not possess any structures in the cell interior for nutrient transport. Optimal cell sizes are below 0.3 micrometres (µm) for very unfavourable nutrient conditions, but between 5 and 30 µm for more favourable conditions (Schulz and Jørgensen 2001). Proposed alternative explanations for the anatomically determined specific energy availability per gene (Lane and Martin 2012) encounter obstacles as soon as very large prokaryote life forms are considered (Bresler et al. 1998; Angert 2006; Mußmann et al. 2007; Koonin 2015; Lynch and Marinov 2015). An example is provided by the bacterial genus *Thiomargarita* sp. with a cell diameter of around 750 µm, which has a large number of identical gene sets that are all distributed on the outer cell membrane. Its particular cell size enables this—only passively moving—life form to store nitrate for survival under adverse conditions. Other bacterial large life forms actively use the transition zones between oxygenated and oxygen-free sectors by influencing currents or distributing cell attachments across the different zones (Schulz and Jørgensen 2001).

Discussions about the origin of eukaryotic cells address all features that differ from prokaryotic cells, such as the cell nucleus, membranes, chloroplasts and mitochondria (Fig. 5.5a). Argumentation is particularly hindered by a lack of fossil evidence from the early periods of eukaryote evolution and multiple changes in the genome due to loss and acquisition of various genes (Martin et al. 2003; Kurland et al. 2006; Anantharaman et al. 2007; Keeling and Palmer 2008; Heiss et al. 2011; Shih and Matzke 2013; Guy et al. 2014; Ku et al. 2015; Raymann et al. 2015; Eme et al. 2017). Different hypotheses have been proposed regarding the origin of the membrane-enveloped nucleus and the intracellular membrane system (Martin 1999, 2005; Gould et al. 2016; Rout and Field 2017; Hartman and Fedorov 2002; López-Garcia and Moreira 2020; Imachi et al. 2020; Mills et al. 2022). On the basis of comparative protein analyses, one hypothesis for development of the first eukaryotic cell assumes an original cell type (chronocyte) in which the cell nucleus and the intracellular membrane system were able to evolve. Other hypotheses propose evolution of eukaryotic cells from symbiotic communities of archaeans and bacteria, and close evolutionary connections between eukaryotes and Asgard archaeans are postulated (Martin et al. 2003; Kurland et al. 2006; Anantharaman et al. 2007; Keeling and Palmer 2008; Heiss et al. 2011; Shih and Matzke 2013; Guy et al. 2014; Ku et al. 2015; Raymann et al. 2015; Eme et al. 2017).

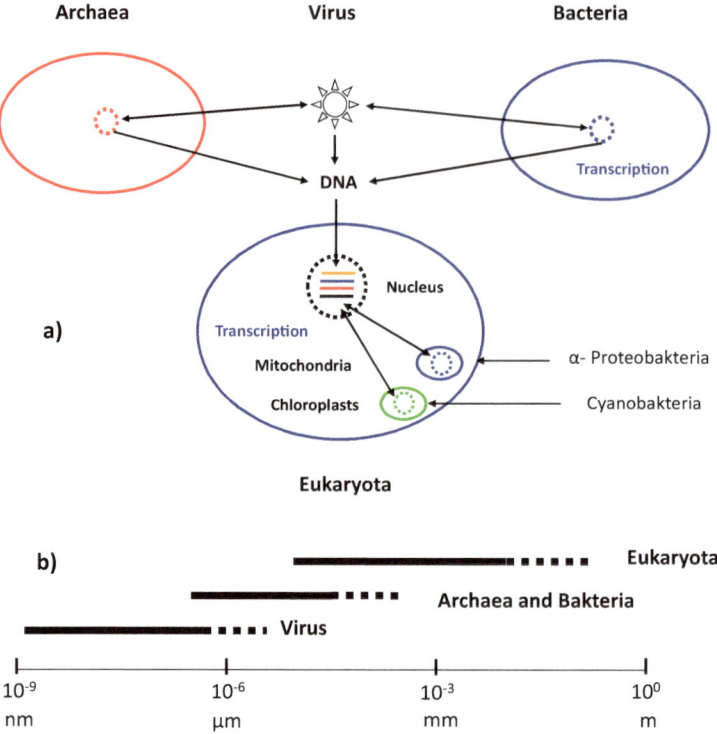

Fig. 5.5 Simplified representation of dimensions and composition of eukaryotic cells. (**a**) Composition of eukaryotic cells from different elements of prokaryotes—especially DNA—and the evolutionary influence of viruses. Mitochondria—presumably formed by symbioses—determine energy metabolism in all eukaryotes, while in photosynthetically active eukaryotes chloroplasts additionally govern the conversion of solar radiation. (**b**) Comparison of the linear dimensions of viruses and cells; dashed areas symbolise extreme sizes of individual life forms. Explanations in the text

Broad agreement exists regarding the frequency of acquisition and origin of mitochondria and chloroplasts. Mitochondria probably originated from α-proteobacteria and were probably acquired only once. However, their phylogenetic classification among the α-proteobacteria remains open. An oft-suspected relationship with endoparasitic Rickettsiae—pathogens causing various diseases, such as typhus—has not been confirmed by recent studies (Martin 1999, 2005; Gould et al. 2016; Rout and Field 2017; Wang and Wu 2015; Martijn et al. 2018). Genetically and structurally, mitochondria show great diversity—in the context of prevailing functional environmental conditions in the host cell (Lane and Archibald 2008; Vafai and Mootha 2012; Huynen et al. 2013; Gray 2014; Stairs et al. 2015). Opinions about the mode of acquisition can be simplified into two viewpoints. One of them assumes acquisition by phagocytosis (Emelyanov 2003; de Duve 2007; Pittis and Gabaldon 2016; Zachar and Szathmáry 2017). Central weaknesses of this standpoint lie in the facts that no prokaryotic life form with phagocytosis has yet been discovered and that the process—with archaeans as host cells—would be energetically unmanageable (Martin et al. 2017). The second standpoint assumes a syntrophic (see Sect. 7.3) precursor between the host cell and an α-proteobacterium (López-Garcia and Moreira 1999; Dyall et al. 2004; Nowack and Melkonian 2010; Degli Esposti et al. 2014; Degli Esposti 2014; Martin et al. 2016). The question as to why this process should have occurred only once in evolutionary history remains unanswered. Chloroplasts presumably originated from cyanobacteria and were probably acquired by phagocytosis (Gould et al. 2008; Archibald 2009; Hohmann-Marriott and Blankenship 2011; Jensen and Leister 2014; Stiller et al. 2014). Indications of this are provided on the one hand by similarities between the photosynthetic mechanisms of cyanobacteria and eukaryotes, and on the other hand by multiple repetition of the process over the course of evolution.

Comparisons of cell sizes (Fig. 5.5b) show that most unicellular organisms—despite their often considerable size—are subject to the specific conditions of the microscopic world mentioned in the text. The largest viruses, with lengths of up to 1500 nm (*Pithovirus* sp.), attain the sizes of prokaryotes, but are incapable of active movement (Abergel et al. 2015; Leslie 2017). The largest single cells in prokaryotes have so far been found in bacteria with a diameter of around 750 μm (*Thiomargarita* sp.). Large bacterial cells (Beggiatoaceae) are themselves colonized by different microorganisms (constituting the microbiome), with which they interact in a manner that is not yet fully understood in functional terms (Viswanathan 2012; Ionescu and Bizic 2019; Flood et al. 2021). Approximations to conditions in the macroscopic world can be achieved only through multicellularity. For example, archaeans in mangrove swamps form filaments up to 30 millimetres long with a diameter of about 10 μm by aggregating some 1500 cells (Muller et al. 2010). On deep sea beds, unicellular eukaryotes (Xenophyophores) can reach dimensions of up to 25 cm, but a large fraction of the cell serves to store food (Levin and Gooday 1992; Gage and Tyler 1999).

Although multicellularity has evolved numerous times separately in eukaryotes, large-bodied life forms have developed only in plants, fungi and animals (King 2004; Parfrey and Lahr 2013; Du et al. 2015). In fact, the great diversity of unicellular eukaryotes so far discovered is reflected by differentiated presentation of a recent classification based on molecular analyses (Fig. 5.6) (Burki et al. 2020). In addition to this classification, there are several classifications with partially differing labels. The technical form of presentation was chosen because each classification is based on interpretations of the research group concerned and can hence undergo subsequent specific changes. The relative significance of phylogenetic analyses becomes even clearer in comparison with the current revised version of eukaryote classification (Adl et al. 2019), which is not considered in detail here. Furthermore, it should be noted that classifications based on morphological methods alone inevitably lead to diverging results (Trontelj and Fišer 2009; Ruggiero et al. 2015).

Cells with very large volumes usually have high proportions of stored substances to bridge periods of deficiency or to permit autonomous development of organisms—for example in the eggs of birds and reptiles (Sauria) (Flindt 1986; Horner 2000; Fenchel and Finlay 2006; Finkel et al. 2010; Levin and Angert 2015). Apart from these special features,

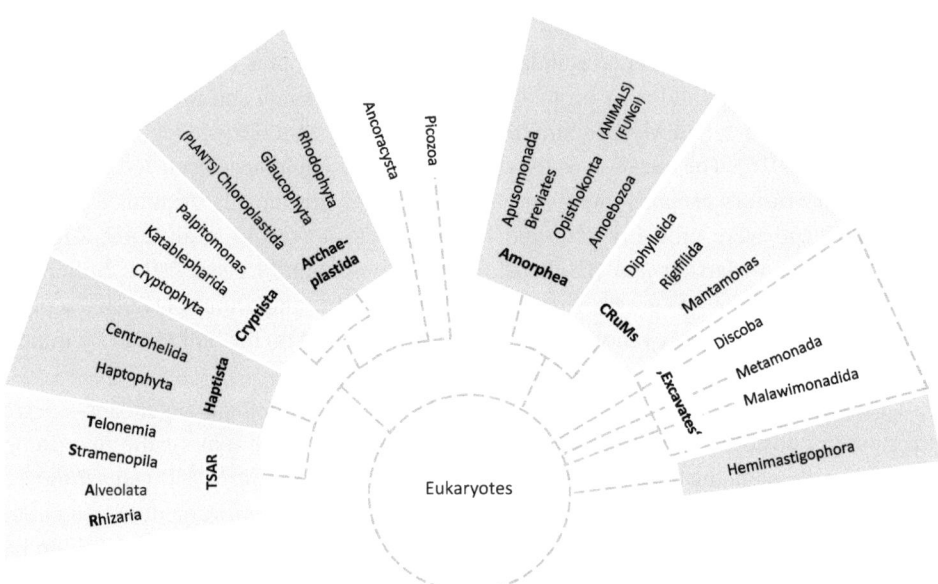

Fig. 5.6 A characteristic feature of cladograms for eukaryotes is the dominance of unicellular life forms. At this systematic level, multicellular life forms cannot be adequately represented. For orientation, multicellular organisms have been indicated with capital letters next to the relevant groups. Abbreviations: CRuMS = collodictyonids + *Rigifilida* + *Mantamonas*; TSAR = Telonemia + Stramenopiles + Alveolata + Rhizaria. Systematic designations according to Burki et al. (2020)

maximal sizes of cells are limited by a variety of external and internal factors (Mergaert et al. 2006; Ferrell et al. 2011; Ferrezuelo et al. 2012; Lloyd 2013; López-Sandoval et al. 2014; Taheri-Araghi et al. 2015). Influential external factors are, for instance, differing diffusion gradients of substances in air and water, availability of nutrients or convertible energy, competitive pressure from other organisms, or physical barriers in the environment (Pedrós-Alió et al. 2000; Stewart 2003; Vogel 2003; Pernthaler 2005; Schmidt et al. 2006; Beardall et al. 2009; Nürnberg et al. 2015; Amodeo and Skotheim 2016). Internally, the framework conditions for cell development are determined by the genome and the concrete regulation by related epigenetic processes, in interaction with environmental conditions, substances in air and water, availability of nutrients or convertible energy, competitive pressure from other organisms or physical barriers in the environment. Internally, the framework conditions for cell development are determined by the genome and specific regulation by epigenetic processes in interaction with environmental conditions (Gregory 2001; Jorgensen et al. 2004; Noble 2012; Amodeo and Skotheim 2016; Noble et al. 2019; Owen et al. 2023).

Here, examples of various unicellular plankton species also reveal flexible adaptability to different combinations of factors. As mentioned above, under nutrient-poor conditions in near-surface zones in many species only small cells develop. At the same time, however, the largest cells of eukaryotic diatoms are found under these conditions. This contradiction only becomes understandable when the morphology and lifestyle of these species is examined: The cells of diatoms are enclosed in a hard silicate shell and are therefore heavier than salt water. As is known, the ratio between cell surface and cell volume decreases with increasing cell size. This also increases the proportion of the weight of the outer shell to the total weight. Accordingly, large cells are lighter than smaller cells compared to salt water and can therefore change their buoyancy through the proportion of carbohydrates in the cell contents. During time spent in the nutrient-poor surface zones, that proportion increases as a result of photosynthesis, causing the cells to sink into deeper, nutrient-rich zones. There, because of the absence of photosynthesis, cells mainly take up nitrogen compounds and the content of carbohydrates is reduced. As a result, the cells become lighter and rise once more to near-surface zones (Moore and Villareal 1996; Villareal et al. 1999; Sunda and Hardison 2010).

The minimal sizes of cells, on the other hand, are primarily determined by the morphological and functional characteristics of the organism species concerned. Because of their simpler internal morphology, free-living cells of archaeans or bacteria are smaller than the smallest free-living cells of eukaryotes (Derelle et al. 2006). However, cells of endoparasitic life forms can also be smaller than cells of related free-living species.

5.1.3   Cell Reproduction in Prokaryotes and Eukaryotes

As the growth of a single cell is limited for the reasons mentioned, organisms can only reproduce or—in the case of multicellular life forms—grow by means of cell division. The prerequisite for maintaining viability is the transfer of all genetic and epigenetic information to the newly formed cell (O'Donnell et al. 2013; Alberts et al. 2015). This is achieved by duplicating the DNA as an information carrier prior to cell division. In prokaryotes, the DNA resides in a circular structure (prokaryote chromosome) in the cytoplasm. In eukaryotes, segments of DNA each wrap around a group of eight histone proteins to form a nucleosome. These are connected by DNA links (linker DNA) in a bead-like fashion to form a chromosome (Fig. 5.7 below). The best known processes—known as the cell cycle—are observed with eukaryotic chromosomes between the formation of a new cell and the subsequent cell division. An initial phase without any recognizable changes in the chromosomes (G1—G for "gap"), is followed by a phase involving production of copies of the DNA sets (S—synthesis phase). This is succeeded by another phase without recognizable changes in the chromosomes (G2). In the final phase—mitosis (M)—preparation for cell division occurs through separation of the copies and their spatial relocation to opposite poles of the cell (Fig. 5.7) (Alberts et al. 2015).

In prokaryotes, the general sequence is simpler than in eukaryotes (Fig. 5.7). After an initial phase (B) without any recognizable changes, a phase (C) follows in which the DNA is copied and the two circular structures are separated. Cell division occurs only after a subsequent phase (D) without recognizable changes to the DNA. In archaeans and bacteria, the overall cycles differ in the lengths of their phases (Bernander 1998). In reality, a

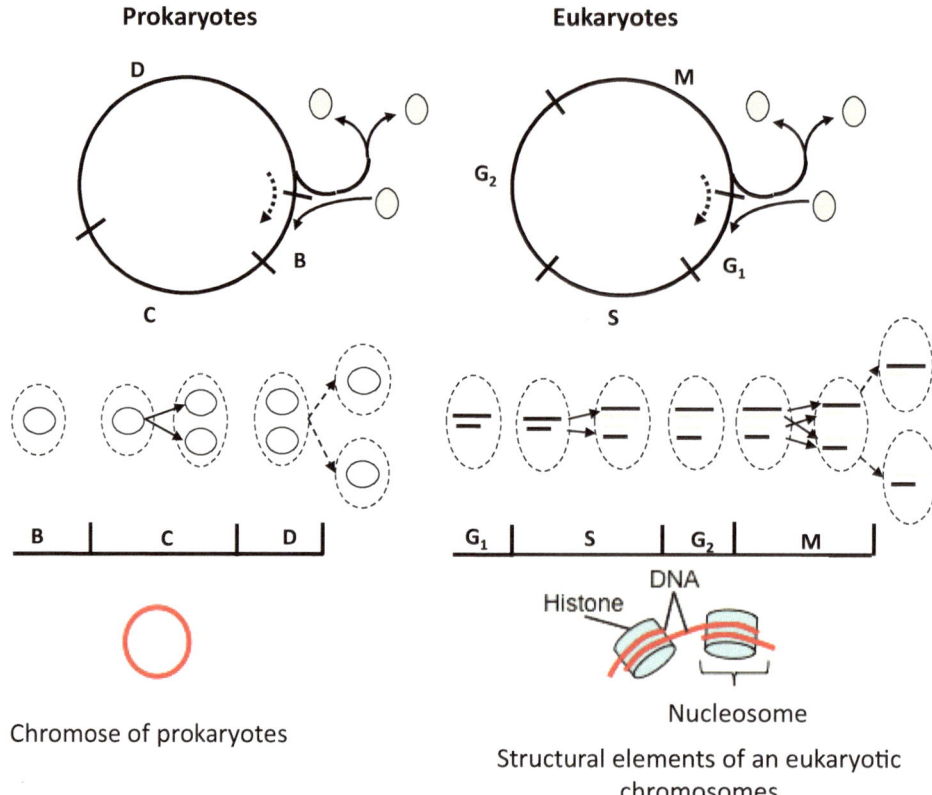

Fig. 5.7 Comparative representation of cell cycles in prokaryotes and eukaryotes. Top: Comparative, simplified representation of transmission of genetic information during cell division (cell cycle) in prokaryotes and eukaryotes. The chronological sequence of processes, beginning with the single cell symbolized as a grey oval, runs clockwise. Dotted arrows inside the circle symbolize possible repetitions of genome duplication without cell division. Below: Schematic comparison of DNA structures in prokaryotes (DNA rings) and basic units of a eukaryotic chromosome (nucleosomes). Compiled from different sources. Further explanations in the text

wide range of variations in genome reproduction can be observed, such as shifted phases in the case of multiple DNA structures, multiple repetitions of the cycle or asymmetrical cell divisions in the formation of spores (Binder and Chisholm 1995; Bernander and Poplawski 1997; Draper and Gober 2002; Egan et al. 2005; Margolin 2005; Skerker et al. 2005; Breuert et al. 2006; Gladfelter et al. 2006; Woldringh and Nanninga 2006; Lundgren and Bernander 2007; Rasmussen et al. 2007; Lindås et al. 2008; Lundgren et al. 2008; Slater et al. 2009: Makarova et al. 2010; Reyes-Lamothe et al. 2012). During mitosis, breaks in the DNA can be repaired and exchange of genes between chromosomes (crossing-over) can take place (Allers and Mevarech 2005; Matos et al. 2011; White 2011; Matos and West 2014).

Reproduction of the extrachromosomal plastid genomes of prokaryotic cells has not been taken into account here. Also not shown is the ever-present combination of prokaryotic and eukaryotic genome reproductions in eukaryotes. This is because reproduction of both the eukaryotic cell's genome and the prokaryotic genomes of mitochondria and chloroplasts is essential.

In methodological terms, automated analyses of nucleotide sequences of DNAs and RNAs opened up novel possibilities for rapid investigation of vast amounts of genetic material. Using various computer-assisted procedures (algorithms), information about potential genes or potential species, for instance, can be obtained. Simply stated, when seeking to identify potential genes, triplets (codons) of the genetic code are examined. Identified codon sequences that begin with a start codon and end with a stop codon (open reading frames—ORFs) provide the database for further investigation. When considering potential species, identified nucleotide sequences are clustered using statistical methods (Operating Taxonomic Units—OTUs) and analysed comparatively. None of the methods requires morphological examination of organisms, which is standard practice in the case with classical methods of evolutionary research. This means that—regardless of their size—all organisms and viruses can be identified and studied exclusively on the basis of DNA or RNA fragments (Hebert et al. 2003; Blaxter et al. 2005; Ferri et al. 2009; Chen et al. 2013; Rideout et al. 2014; Merkl 2015).

5.1.4   Special Features of Eukaryotic Cells

Eukaryotic cells are not directly comparable with prokaryotic cells—both because of their structure and because of their functional characteristics. Structurally, key features are the cell nucleus, surrounded by its own membrane, and in the cytoplasm mitochondria and chloroplasts presumably acquired through symbiosis—with their own genomes (Huynen et al. 2013; Allen 2015; Gray 2015; Pittis and Gabaldon 2016). The former—often in an extremely reduced form—ensure the energy supply for metabolism in almost all eukaryotic cells (Tielens et al. 2002; Waller and Jackson 2009). To date, absence of mitochondria and associated adaptation of cell metabolism has been shown only for a group of endobiotic unicellular eukaryotes—(*Monoacercomonoides* sp.) (Karnkowska et al. 2016; Hagström and Andersson 2018). Chloroplasts, on the other hand, are found only in photosynthetically active cells (Fig. 5.5a). Within the Eukaryota, the number of haploid DNA sets is extremely variable, ranging from one (haploid) to a thousand copies (polyploid) in an organism's genome. Even within a species, the number of haploid gene sets can vary between 4 and 40 (Parfrey et al. 2008). Components from archaeans, bacteria and viruses are found in the DNA (Yutin et al. 2008; Cotton and McInerney 2010). The histone assembly and genetic and epigenetic regulatory processes originate from archaeans (Kelman and Kelman 2003). Systemically interpreted, what we see is not a cell but a **symbiotic cell consortium**. Because of the chimerical mixture of features with a very diverse origin, uncertainty persists regarding the origin and phylogenetic status of eukaryotes (Margulis et al. 2000; Gribaldo et al. 2010; McInerney et al. 2014; Rochette et al. 2014; López-Garcia and Moreira 2015).

Still unclear is the eventual classification of an organism from deep-sea sediments of the Pacific Ocean, for the time being called a bacterium (*Parakaryon myojinensis*). The cells, which are about 10 micrometres long and 3 micrometres thick, have a large, membrane-enveloped nucleus and endosymbiotic bacteria, but no mitochondria (Yamaguchi et al. 2012, 2018).

The cell membrane of eukaryotes has a biochemical structure resembling that of bacteria, but it fulfils tasks different from those of prokaryotes. In the latter, the cell membrane separates the cell interior from the outside world and is at the same time the central zone for conversion of energy and substance as well as information processing. The eukaryote cell membrane also separates the cell interior from the environment, but as a selective transfer zone that maintains the chemical conditions in the cytosol—the inner "cell fluid". All essential processes of energy and metabolism, on the other hand, take place on the membranes and inside cell compartments. In addition to the mitochondria already mentioned, all eukaryotic cells contain the membrane-enveloped nucleus, the peroxisomes—functional partners of the mitochondria and the endoplasmic reticulum (ER) (Speijer 2017). The network of the rough endoplasmic reticulum with the attached ribosomes for protein synthesis, the smooth endoplasmic reticulum and the Golgi cisternae with the functionally associated endosomes for the orderly transport of substances, and lysosomes for degradation of toxic compounds (Alberts et al. 2015).

Comparison clearly shows the subdivision of the eukaryotic cell into groups of compartments, contrasting with the—as a rule—undivided cells of prokaryotes. Groups of compartments that are important for energy and metabolism, such as mitochondria and endoplasmic reticulum with the Golgi cisternae, have the largest shares of membrane area with a total of about 97 percent, but take up only about 37 percent of the cell volume. The cell nucleus and the cell organelles, which are, by contrast, predominantly important for transport functions, have only minimal membrane areas (Diekmann and Pereira-Leal 2013). The enveloping cell membrane—dominant in prokaryotes—have a total share of only about 2% of all membrane surfaces in eukaryotes. It should be noted incidentally that in certain prokaryotes (cyanobacteria) photosynthetically active cells also have large inner membrane areas. For photosynthetically active eukaryotic cells, a similar ratio between inner membranes and the cell membrane can be presumed, but with a shift in the proportion of inner membranes in favour of chloroplasts as centres of photosynthesis.

Because of structural differentiation within eukaryotic cells, the framework conditions for physiological processes in cells also change—compared to cells of prokaryotes. The differences can easily be seen in the case of protein synthesis. In prokaryote cells, transfer of genetic information from DNA to messenger RNA (mRNA) is usually coupled directly with subsequent conversion of the information from mRNA into proteins. In eukaryotic cells, however, the first stage—transcription—takes place in the cell nucleus. The mRNA is then transported from the cell nucleus for the second stage—translation—in the cytoplasm. Transcription—transfer of information from DNA to RNA—is regulated in prokaryotes by one RNA polymerase, whereas in eukaryotic cells molecular complexes of four RNA polymerases perform this task. In eukaryotic cells, newly formed mRNA is transported from the cell nucleus through the cytosol to the appropriate ribosomes. In the process, the information content of the mRNA can be modified by alteration of the nucleotide sequences (editing), thus potentially expanding the spectrum of protein synthesis (Graw 2015).

Diverse and variable information transfer from a DNA sequence to protein syntheses is supported by the properties of ribosomes. In greatly simplified terms, the basic structure of all ribosomes consists of one large and one small molecular complex of ribosomal RNA (rRNA) and ribosomal proteins (Frank and Gonzalez 2010; Harish and Caetano-Anollés 2012; Anger et al. 2013; Petrov et al. 2015). The molecular complexes enclose a tunnel in which—under regulation by transfer RNAs (tRNA) and based on information in the mRNA—protein synthesis (translation) takes place. The molecular structures, which are in reality far more complex, presumably developed evolutionarily from different precursor molecules. Here, clues are provided by the distribution of protein components of ribosomes (ribosomal proteins) across the main domains of organisms (archaeans, bacteria and eukaryotes). Across all three domains, about 34 percent of the known forms can be detected. Similarly, about 34 percent are found in archaeans and eukaryotes, but 21 percent exclusively in bacteria 12 percent only in eukaryotes. In eukaryotes, the basic structure of ribosomes is additionally extended by functional protein-RNA and RNA-RNA complexes. Structural differentiation of eukaryotic ribosomes is enabled by significantly more complex formation processes than in prokaryotes. Differences in the regulatory cofactors involved in the formation processes are evident only quantitatively—their number exceeds 200 in eukaryotes compared to around 50 in prokaryotes (Kovacs et al. 2018; Bowman et al. 2020; Londei and Ferreira-Cerca 2021). The formation process of eukaryotic ribosomes begins in a separate part of the cell nucleus, the nucleolus. Induced by cellular signals, the primary forms of the two molecular complexes are formed there separately and then transferred into the plasma of the cell nucleus (nucleoplasm). Subsequently, the primary forms are modified by separation of various molecules, and their molecular structures are finally folded. Only after transfer of the two molecular complexes into the cytoplasm does their "assembly" into functional ribosomes take place. These are not static molecular structures, but highly dynamic molecular systems—only their precise and temporally coordinated structural modifications ensure the synthesis of functional proteins during translation. Ribosomes are not singular cell organelles, but are found in individual cells in large numbers—up to 10 million—and with a wide variety of specializations. Their processes are accordingly hierarchically integrated into cellular control systems in order to form the required proteins in the correct (stoichiometric) ratio to the overall spectrum of all existing proteins—adapted to the relevant requirements. Despite multi-layered systems of control and regulation, errors in the complex processes can lead to abnormal functional developments of cells, which manifest themselves macroscopically in various cancers (Lafontaine and Tollervey 2001; Frank and Gonzalez 2010; Pedersen 2011; Li et al. 2014; Colussi et al. 2015; Melnikov et al. 2012; Brown et al. 2016; Chaker-Margot et al. 2017; Bastide and David 2018; Bock et al. 2018; Lalanne et al. 2018; Baßler and Hurt 2019; Bohnsack and Bohnsack 2019).

5.1.5   Cell Scaffold, Cell Motion and Endogenous Clocks (Zeitgeber)

Spatial functional differentiation of cells, however, is determined not by compartments but by dynamic protein polymers—collectively referred to as the *cytoskeleton*. Unfortunate choice of the term "skeleton", which regrettably triggers misleading associations, was driven by the static observational possibilities available at the time of discovery. Cytoskeletons consist of a wide variety of constantly changing protein polymers that perform a multitude of functions in cells—for example, maintaining structural order between component molecules, transportation of molecules or cell organelles, signal transmission, dynamic shaping of cell shape, cell locomotion and organization of cell division. Functional diversity is based on the combination of molecular sequences in the proteins and the specific features of their morphological structures (coiled coils) (Truebestein and Leonard 2016). Only one of these functions—the structuring of cell shapes—can be interpreted in the broadest sense as analogous to the functions of skeletons known in vertebrates. Development of methods for maintaining archaeans under laboratory conditions permitted demonstration in these organisms (Asgard archaea) of cytoskeleton formation with features resembling those in eukaryotes (Rodrigues-Oliveira et al. 2023). For transportation, the cytoskeleton can either serve as a scaffold for transport proteins (kinesin; in plants, myosin) or itself "pull" or "push" molecules or cell organelles in a directed manner through generation and degradation of the polymers (Vale 2003; Ueda et al. 2010; Peremyslov et al. 2011; Alberts et al. 2015). With the aid of high-resolution microscope technology, direct observation of movements of kinesin molecules proved possible. According to results so far obtained, locomotion along the microtubules takes place by rotating the "stalk" of molecules by 180°, in each case with length increments of around 4 nanometres. In the process, in alternation one of the two anchorage points is temporarily fixed to the surface of the microtubules. In addition, kinesin molecules have been observed to jump between different microtubules (Nanninga 2001; Trachtenberg and Gilad 2001; Pollard 2003; Trachtenberg et al. 2003; Caviston and Holzbaur 2006; Dye and Shapiro 2007; Löwe and Amos 2009; Ross et al. 2008; Vats et al. 2009; Olson and Nordheim 2010; Wang et al. 2010a; Barry and Gitai 2011; van Teeffelen et al. 2011; Wickstead and Gull 2011; Bulmer et al. 2012; Lutkenhaus et al. 2012; Yutin and Koonin 2012; Reimold et al. 2013; Alberts et al. 2015; Deguchi et al. 2023; Wolff et al. 2023). Clues to functional relationships between the cytoskeleton and metabolic processes are provided both by the relationship between proteins and by shared responses to external influencing factors. Hence, proteins of metabolism and cytoskeletons evidently have many common evolutionary roots, and the two systems are linked by shared regulators (Yang and Grégoire 2007; Ingerson-Mahar et al. 2010; Barry and Gitai 2011; Wickstead and Gull 2011; Caino et al. 2013).

The greatest diversity of different protein polymers—at least nine—has now been identified in prokaryotes (Bernandez et al. 2011; Ettema et al. 2011; Evguenieva-Hackenberg et al. 2011; Ingerson-Mahar and Gitai 2011; Pilhofer et al. 2011; Pilhofer and Jensen 2013; Celler et al. 2013; Braun et al. 2015). They significantly influence the shape and functionality of bacterial cells (Margolin 2009). In eukaryotes, three groups are distinguished—actin filaments, tubulin and intermediate filaments, which can be designated as homologues of three protein polymer groups in prokaryotes. Actin filaments are mainly relevant for governing cell shape and cell mobility, tubulin for intracellular transport and order during cell division (mitotic spindles), and intermediate filaments mainly for the mechanical strength of cells (Herrmann et al. 2007; Dawson and Paredez 2013; Jékely 2014; Alberts et al. 2015). Actin also plays an important role in the formation of various cell structures (Chhabra and Higgs 2007). Among cell organelles, chloroplasts also contain their own cytoskeleton resembling that of bacteria (Reski 2002). An indication of pronounced morphological and physiological variability in the cytoskeleton of early eukaryotes is provided by the genus *Naegleria* sp., which can change from an amoeboid to a free-swimming, flagellated life form in about 90 min (Fritz-Laylin et al. 2010).

Translocation of energy conversion from the cell periphery to the mitochondria opened up new degrees of freedom for eukaryotes to achieve metabolic and energetic equilibrium. On the one hand, substances could be absorbed into the cell in an orderly manner at a limited number of locations, and on the other hand, converted energy—predominantly in the form of ATP[1]—could be transported to any location in the cell. For unicellular eukaryotes, this opened up possibilities for focussed food intake through phagocytosis and protection of cells with solid outer shells. It also facilitated the evolution of multicellular organisms and of locomotor systems in animals.

Differences in conversion into mechanical energy can be illustrated most clearly by comparing the drives for flagella in prokaryotes and eukaryotes. For this purpose, prokaryotes have a revolving proton motor in the cell membrane, which can rotate a flagellum both clockwise and counter-clockwise. In gram-negative bacteria, the immobile stator of the motor is anchored in the peptidoglycan wall. Peptidoglycan is very strong as a result of cross-linking of sugar molecules (polysaccharide chains) with short-chain peptides (Fig. 5.8a) (Bardy et al. 2003; Zhao et al. 2014). For locomotion, flagella are rotated at around 125 revolutions per second. The motor is driven by a flow of protons or sodium ions with a power of about 10^{-15} J/s. In relation to the mass of the motors of around $1.3 \cdot 10^{-16}$ g, this is equivalent to the specific power of turboprop engines (Berg 2003, 2008; Sowa and Berry 2008; Watari and Larson 2010; Phillips et al. 2013; Cohen et al. 2017). Because of the specially structured, "denser" cell membrane, the flagellar motors of archaeans are driven by ATP and have a different microstructure (Jarrell et al. 1996; Ng et al. 2008; Albers and Jarrell 2015).

The best-known locomotor pattern of free-swimming prokaryotes consists of irregular forward and backward movements enabled by changing the direction of rotation of the

---

[1] Adenosine triphosphate.

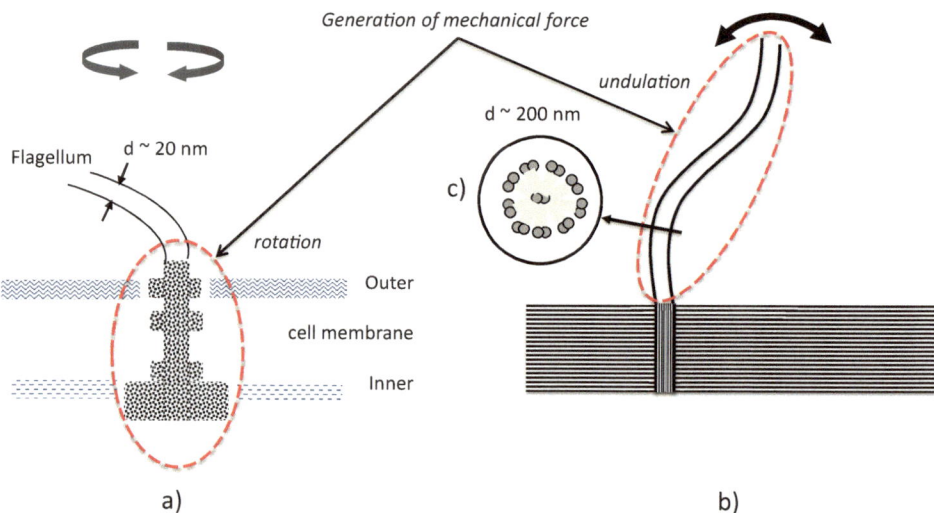

Fig. 5.8 Highly simplified representations of the flagellar drives in (**a**) prokaryotes and (**b**) unicellular eukaryotes; with schematic cross-section of the flagellum and representation of the axoneme with nine peripheral microtubule pairs with radial connections to the two central tubules ($9 \times 2 + 2$ structure) (**c**). In prokaryotes, a flagellum can be rotated in either direction through revolution of the proton motor; in eukaryotes, the flagellum ripples as a result of reciprocal displacement of microtubules, driven by intermediate motor proteins. Sources: Modified from Bardy et al. (2003), Carvalho-Santos et al. (2011), Ounjai et al. (2013)

motor. In this type of locomotion, average speeds are between 3 and 120 micrometres per second (μm/s). More extensive movements over straight distances of up to around 590 μm/s or around 500 body lengths per second have been observed only in archaeans from hot springs (Herzog and Wirth 2012). Similarly high speeds have been observed in chemoautotrophic bacteria of the genus *Thiovulum* (Schulz and Jørgensen 2001). The greatest velocities have so far been measured in the genus *Ovobacter*—with around 400 flagella—at around 1 mm/s (Fenchel and Thar 2004). On surfaces, various prokaryotes can glide over a slime film. In some life forms, movement is achieved by a spiral rotation of the inner cytoskeleton—driven by protons (Hoiczyk and Baumeister 1998; Gilad et al. 2003; Ehlers and Oster 2012). Other life forms, such as myxobacteria, push themselves along with the help of polar cell appendages—type IV pili—or use them—like *Pseudomonas*—to raise themselves up on surfaces (Wall and Kaiser 1999; Skerker and Berg 2001; Wolgemuth et al. 2002; Gibiansky et al. 2010; Zhang et al. 2012; Berry and Pelicic 2015). Remarkably, the most widespread genera of cyanobacteria in marine plankton, *Prochlorococcus* and *Synechococcus*, lack flagella. Nevertheless, they show chemotactic behaviour and can actively seek out higher nutrient concentrations. However, with the loss of the outer cell membrane, the flagellar-less genus Synechococcus sacrificed the capacity to move. The principle underlying its movement through water bodies remains unclear (Willey 1988; Bryant 2003; Seymour et al. 2010; Ehlers and Koiller 2010). Other genera of cyanobacteria such as *Synechocystis*, however, possess flagella (type IV pili) similar to those of bacteria and also employ them for locomotion (Schuergers and Wilde 2015; Wilde and Mullineaux 2017).

These special kinds of locomotion in prokaryotes—and many unicellular eukaryotes—are rendered comprehensible by the physical properties of water at microscopic dimensions. In abstract terms, microscopic organisms move "between water molecules". They barely displace any water during movement, but for them it is extremely viscous—as molasses would be for us. Hydrodynamically, the conditions for locomotion in liquid and gaseous media are described by the ratio of inertial resistance (due to displacement) and viscous resistance by the so-called Reynolds number (Vogel 1994, 2003; Phillips et al. 2013; Denny 2016). Inertial resistance is determined mainly by the size and shape of the organism and viscous resistance mainly by the properties of the medium. For very small Reynolds numbers—for prokaryotes about 10^{-6}—resistance to viscocity alone determines the conditions for locomotion. In such a context, movement with fins, for example, is ineffectual (Purcell 1977; Chattopadhyay and Wu 2009; Watari and Larson 2010; Nguyen et al. 2014). Even with flagellar drives, cells immediately come to a standstill when the motor stops turning.

This is why prokaryotes use free locomotion in water primarily to seek adequate nutrient concentrations or—like the archaeans mentioned above—to escape from life-threatening conditions. They can also generate currents at microscale dimensions—for instance, in a water droplet—through collective coordinated movements, thus ensuring a supply of oxygen or nutrients from the entire water body (Barbara and Mitchell 2003; Cisneros Salerno 2008; Guasto et al. 2012; Stocker 2012, 2015; Petroff and Libchaber 2014). Experimental evidence invalidates—at least for flagellated bacteria—the theoretically asserted impossibility of microorganisms actively foraging and physically influencing their environment. More rapid locomotion also yields advantages for bacteria when foraging in aquatic habitats (Vogel 2003; Son et al. 2016). Depending on the physiological capacities of microorganisms and the particular contexts, distances covered during locomotion range from a few micrometres (µm) to several tenths of a millimetre (Johansen et al. 2002; Barbara and Mitchell 2003; Xie et al. 2011; Stocker and Seymour 2012). Greater distances, on the other hand, can be covered only passively by means of currents—as is also the case with viruses, which are incapable of active locomotion.

Flagellar drives count among the original features of eukaryotic cells. In the course of subsequent evolution, flagellar drives were lost or modified in various ways (Yubuki and Leander 2013). Flagella or motion cilia are driven by motor proteins that bend in different directions through reciprocal displacement of elastic tubes (microtubules) in the polymeric protein scaffold (Fig. 5.8b) (Nicastro et al. 2005; Ginger et al. 2008; Ounjai et al. 2013; Alberts et al. 2015). Unicellular eukaryotes use flagella of various shapes for locomotion in fluid media (Carvalho-Santos et al. 2011). In various life forms, hydrophobic cell surfaces may further reduce flow resistance (Jenkinson and Sun 2014). Depending on environmental conditions, motile cell processes (cilia) can also develop inside organisms (e.g. in bronchi), but options for rotating drives are lost in the process (Sassera et al. 2011).

The basic principle of reciprocal displacement is found in all mobile animal organisms in the sectors for conversion of chemical energy to kinetic energy. For example, the "scaffolds" consist of microtubules in cilia, of actin in muscles, and of dynein, or myosin, in motor proteins (Kull et al. 1998; Mahadevan and Matsudaira 2000; Vale and Milligan 2000; Richards and Cavalier-Smith 2005; Foth et al. 2006; Hwang and Lang 2009; Holzbaur and Goldman 2010; Alberts et al. 2015). In all multicellular animals (Metazoa), slightly modified forms of cilia are of central importance for morphogenesis and intercellular communication processes (see Sect. 6.3).

In fungi, cytoskeletons regulate growth of the hyphae. The tube-like structure specific to fungal hyphae, with open connections between the individual sections, enables the build-up and use of intracellular fluid pressure (turgor) for growth in soils or in organic matter. Growth of hyphae occurs exclusively at their tips through ongoing expansion of cell walls. Specific structures in the cytoskeleton transport membrane vesicles from the Golgi bodies to the substance of the regulating tip. There, the microvesicles are converted into macrovesicles and transported to the cell walls. The hyphae can be elongated by incorporating the vesicle membranes into the cell walls and emptying the vesicle contents externally (exocytosis)—without loss of turgor (Virag and Harris 2006; Harris 2009; Fleißner et al. 2010; Moore et al. 2011; Steinberg et al. 2017). In this way, fungi can penetrate relatively dense inorganic and organic substrates—without active locomotion (Money 2004).

Although the general distribution of cytoskeletons in prokaryotes has been recognized in recent decades as a result of further development of research methods, the *temporal regulators* of cellular processes have not yet been fully explored. As with many physiological processes, early investigations were oriented mainly towards macroscopic phenomena—temporal adaptations of animals and plants to the day-night rhythm. Important aspects were the advantages and disadvantages of the internal—circadian—rhythm of organisms (Johnson et al. 1995; Storch et al. 2002; Bohn et al. 2003; Millar 2004; Dodd et al. 2005; Roenneberg et al. 2005; Reitzel et al. 2010; Tessmar-Raible et al. 2011; Müller et al. 2013). Under the heading "chronobiology", the search for the driving factors of "internal clocks" and their influence subsequently became a core feature of research.

As with many biological phenomena, such abilities were depicted as lacking from the "primitive" prokaryotes. Regulatory feedback loops have been identified between cell nuclei and the cytoplasm involving the genetic transcription and translation system (Transcriptional-Translational Feedback Loop—TTFL) (Glossop and Hardin 2002; Paranjpe and Sharma 2005; Ramsey et al. 2009; Bellet and Sassone-Corsi 2010; Sorrells et al. 2015; Sorrells and Johnson 2015; Kaiser et al. 2016). With the exception of certain beetles and millipedes, these regulatory systems are also found in animals living underground in perpetual darkness (Oda et al. 2000; Cavallari et al. 2011; Friedrich 2013; Chipman et al. 2014; Pasquali and Sbordoni 2014). The temporal cycle of red blood cells, which possess neither a cell nucleus nor mitochondria, at first posed unresolved issues. But this question was answered by discovery of a regulatory system based on an antioxidative protein, independent of genetic control (O'Neill and Reddy 2011). The presence of this regulatory system has also been demonstrated in a eukaryotic alga and in cyanobacteria (Ditty et al. 2003; O'Neill et al. 2011; Loudon 2012). Particularly surprising was the discovery that cyanobacteria also possess the eukaryotic regulatory system and that the two systems are employed both together and separately (Nakajima et al. 2005; Rust et al. 2011; Abe et al. 2015; Cohen and Golden 2015). In the meantime, presence of the antioxidative regulatory cycle has also been demonstrated in many bacteria and archaeans—with the exception of methane-forming forms (Dvornyk et al. 2003; Loudon 2012; Maniscalco et al. 2014; Chaix et al. 2016; Makarova et al. 2017; Tseng et al. 2017; Wang et al. 2017).

Close functional connections with extremely diverse processes of cell metabolism suggest a ubiquitous distribution of molecular endogenous clocks (Zeitgebers) in all organisms, including methanogenic archaeans. However, clarification requires future research (Dunlap 1999; Jeong et al. 2000; Schibler and Sassone-Corsi 2002; Ko and Takahashi 2006; Staels 2006; Yang et al. 2006; Eckel-Mahan and Sassone-Corsi 2009; Bartocci et al. 2010; Bass and Takahashi 2010; Kim et al. 2010; Asher and Schibler 2011; Lu and Thompson 2012; Silver et al. 2012; Panda 2016).

5.2 Biological Regulatory Variables in the Network of Life Forms

5.2.1 Genetic and Epigenetic Processes in Pro- and Eukaryotes

According to the now refuted "central dogma" of genetics, the formation of all cell components should be derived from a standard linear sequence: gene (DNA)—> RNA—> protein (Alberts et al. 1986; Sabin et al. 2013). In this context, transfer of the genetic code from DNA to RNA is referred to as transcription and the subsequent rendering into proteins as translation. According to this dogma, evolution should also proceed through changes in DNA—randomly through mutations, or following Mendelian rules during sexual reproduction—and subsequent external selection of the "fittest" organisms. From this mechanistic-theoretical perspective—already subjected to criticism (Riedl 1975; Dawkins 1986; Noble 2012, 2018)—random factors account for the adaptation of cells and organisms to environmental factors. Despite evidence for external influences under relaxed selection—notably the loss of redundant organs, e.g. eyes in cave fish—information-processing mechanisms supposedly play no role (Blake et al. 2003; Kærn et al. 2005; Raj and van Oudenaarden 2008; Lahti et al. 2009; Balázsi et al. 2011; Cui et al. 2019b). Although random modifications play an essential part in biological processes, arguments based on that are not sufficient to justify adaptation processes. For example, they do not provide an answer to the question of how a cell recognizes whether random changes are beneficial to it or not.

In the same way, a mechanistic view does not take into account the fundamental challenges for perpetuation of evolutionary processes in a constantly and randomly changing environment. Life functions can be maintained over the long term only through a balance between complete rigidity (order) and uncontrollable states (chaos). Seen formally, every heritable genetic change is a wager on future matching environmental conditions.

Because of the complex characteristics of environmental systems (Peak and Frame 1995), all organisms—from prokaryote cells to humans—lack access to requisite information. What is accessible, however, is information about past and present variations in environmental conditions. At the genetic level, a wide range of different strategies can be found for coping with challenges under the framework conditions portrayed. The successful strategy of prokaryotes has been characterized—for billions of years—by the most extensive possible coverage of all manageable environmental conditions with a vast number of very small individuals and rapid exchange of genetic information. At the other end of the spectrum, we find—evolutionarily over only a short timespan—successful strategies of large-bodied, multicellular and long-lived eukaryotes with relatively little genetic flexibility. While organisms with the former strategy can autonomously survive massive changes, the second-mentioned strategy requires not only relatively stable environmental conditions but also complementary underpinning through benefits supplied by other organisms.

Examples of the first strategy can be found in unicellular, photoautotrophic cyanobacteria with an evolutionary history extending over some 2.6 billion years and the greatest number of individuals globally in the marine genus *Prochlorococcus* (Schirrmeister et al. 2016). The autonomously viable cells of this genus, which are only 0.5 to 0.7 µm in size, occur in the oceans worldwide and exhibit great genetic variability (Partensky et al. 1999; Shi and Falkowski 2008; Biller et al. 2014). Similar genetic features are found in the—albeit heterotrophic—marine genus *Pelagibacter* from the—similarly ancient—group of α-proteobacteria (Vergin et al. 2007; Luo et al. 2013), with cells around 0.2 µm in size. Among vertebrates, the shortest lifespan so far discovered is in centimetre-sized dwarf gobies (*Eviota*)—coral reef fish—with a maximum of 60 days. The lifespan of African turquoise killifish (*Nothobranchius*) in temporary freshwater ponds is around 85 days (Depczynski and Bellwood 2005; Blažek et al. 2013, 2016).

Examples of the second strategy can be found among long-lived woody plants—such as a spruce in Sweden that is around 9500 years old or a clone of American trembling poplars that is at least equally old (Öberg and Kullman 2011; Rogers and McAvoy 2018). An extreme age of at least 43,000 years is estimated for the only clone of *Lomatia tasmanica* in Tasmania—consisting of 400–500 stems over an area of about 1.2 km^2 (Arnaud-Haond et al. 2012). As with almost all land plants living under natural conditions, their evolution is highly dependent on functional interactions with microorganisms and mycorrhizal fungi (Berendsen et al. 2012; Strullu-Derrien et al. 2015). Similar strategic characteristics can be seen in the development of the largest living organism in terms of area, the parasitic bulbous honey fungus (*Armillaria bulbosa*) in the US state of Oregon. The clonal fungal tissue covers an area of about 9.6 km^2 and its age is estimated to be between 1900 and about 8600 years (Smith et al. 1992; Schmitt and Tatum 2008). Among animals, the estimated age of bowhead whales (*Balaena mysticus*) at about 200 years or Greenland shark (*Somniosus microcephalus*) of up to 400 years is clearly surpassed by deep-water corals with up to 4000 years (Roark et al. 2009; Dance 2016; Nielsen et al. 2016). As in the case of plants, in a comparison of many life forms no general correlations between individual factors and lifespan can be demonstrated (Healy et al. 2014; Williams and Shattuck 2015; Durkin et al. 2017; McDonald et al. 2017; Montero-Serra et al. 2018). In some groups, however, correlations between lifespan and life form are demonstrable; for example, small mammals that are capable of flight live about three times longer than non-flying mammals of the same size (Austad and Fischer 1991).

In the meantime, an increasing wealth of insights have been gained into the flexibility of genetic regulatory systems. Gene-centred interpretation of evolutionary processes is increasingly being replaced by dynamic interpretations (Fig. 5.9). The genetic structure of DNA by no means forms an unchanging framework for development of the morphological and physiological design of phenotypes, but contains an "archived master plan". Due to the modifications already mentioned, this may well contain ambiguous information. Its interpretation is determined—within the context of the given framework conditions of the environment—only through the complex transcription and translation processes (Wistow 1993; Nadeau and Sankoff 1997; Grenier and Carroll 2000; Dennis et al. 2001; Parra et al. 2006; Young and Badyaev 2007; Conant and Wolfe 2008; Gao and Davidson 2008; Bonduriansky and Day 2009; Malone and Hannon 2009; Jiang et al. 2010; Burke et al. 2011; Lehnert et al. 2014; Alberts et al. 2015; Bier and De Robertis 2015; Philips et al. 2015; Piunti and Shilatifard 2016; Li et al. 2017). In those processes, diverse phenotypic characteristics can arise from individual genes (pleiotropy) (Hodgkin 1998; van Hijum et al. 2009; Wang et al. 2010a, b; Granek et al. 2011; Hill and Zhang 2012), or different phenotypes can develop from identical genes (polygeny) (Ridenhour and Nuismer 2007; Cubillos et al. 2011; Boyle et al. 2017). Flexibility ensures the functional differentiation of cells in tissues of multicellular organisms. Such processes of control and correction can already be observed at the subcellular level, for example in the synthesis of DNA or proteins. The diversity of toxic proteins in venomous snakes can also be attributed to different combinations of gene duplication and gene reduction (Bernales et al. 2006; Lemmon and Schlesinger 2010; Alberts et al. 2015; Giorgianni et al. 2020). Morphological persistence in vertebrate evolution is due to the regulatory effects of pleiotropic genes (Hu et al. 2017). In all vertebrates, the formation of functionally different tissues and body fluids—such as tooth enamel, bone, saliva and breast milk—is based on regulation of a single group of genes. The central regulatory function of the gene group—differentiated through duplications—resides in the formation of secretory, calcium-binding phosphoproteins (Kawasaki and Weiss 2003). Moreover, observation of gene activation in human brain development has revealed clear dynamic adaptations to spatial, tissue-specific and temporal hierarchies (Nowakowski et al. 2017). Let it be noted that pleiotropy and polygeny are problematic for artificial gene manipulation in organisms, as they can give rise to completely unexpected effects.

Epigenetic modification of genetic storage and regulatory systems in the life-cycles of individual organisms (phenotypes) is a feature shared by prokaryotes and eukaryotes (Casadesús and Low 2006; Grewal and Jia 2007; Veening et al. 2008b; Fedoroff 2012;

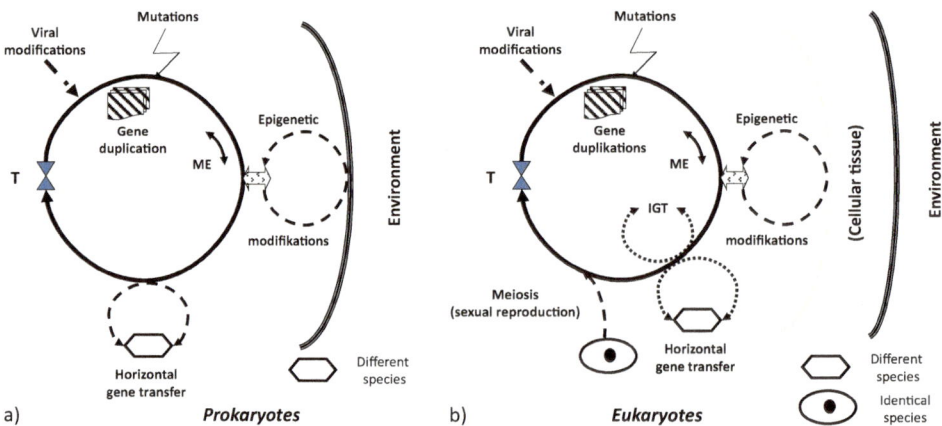

Fig. 5.9 Simplified representation of the continuous modification of genetic storage and regulatory systems (DNA, RNA, proteins) by different factors in prokaryotes and eukaryotes. Abbreviations: T = Transfer of the genetic storage and regulatory system to the next organism (inheritance); IGT = Intracellular gene transfer in eukaryotes between DNA in the cell nucleus and that of mitochondria and chloroplasts; ME = Mobile genome elements (transposons). Compiled from different sources (see text)

Henikoff and Smith 2015; Willbanks et al. 2016; Rajewsky et al. 2017). Modifications can either have a temporary effect or become permanently anchored and heritable. In both prokaryotes and eukaryotes, epigenetic changes can occur through methylation—addition of a methyl group (CH_3) to a nucleobase—adenine or cytosine. In eukaryotes, additional genes can be released or blocked from transcription by acetylation—addition of acetyl groups (C_2H_3O)—and methylation of histones (Fig. 5.7 below). Histones also regulate activation of genes in archaeans and—according to research results obtained so far—also in bacteria (Jaenisch and Bird 2003; Ringrose and Paro 2004; Feil and Fraga 2012; Arrowsmith et al. 2012; Heard and Martiessen 2014; Alberts et al. 2015; Brunk and Martin 2019; Hocher et al. 2023; Ledford 2023).

Results from comparing genomes between marine microplankton species provide evidence for various effects of environmental factors on the genetic make-up of prokaryotes and eukaryotes (Mock and Kirkham 2012). Matching of genes across the genomes of ten selected species in each group was investigated. Despite a common unicellular life form, similar cell sizes and living conditions, there were clear differences between the two groups. Of all genes examined, 2.5 per cent of the genes in prokaryotes and 73.7 percent of the genes in eukaryotes matched in at least nine species. By contrast, 71.2 per cent of the genes in prokaryotes and 0.9 per cent in eukaryotes were found in only a single species (Fig. 5.10a).

If it is assumed that shared genes (Fig. 5.10b) serve primarily to maintain and regulate central cellular processes, the following conclusions can be drawn from those results: In prokaryotes, individual interactions with the environment—especially when involving

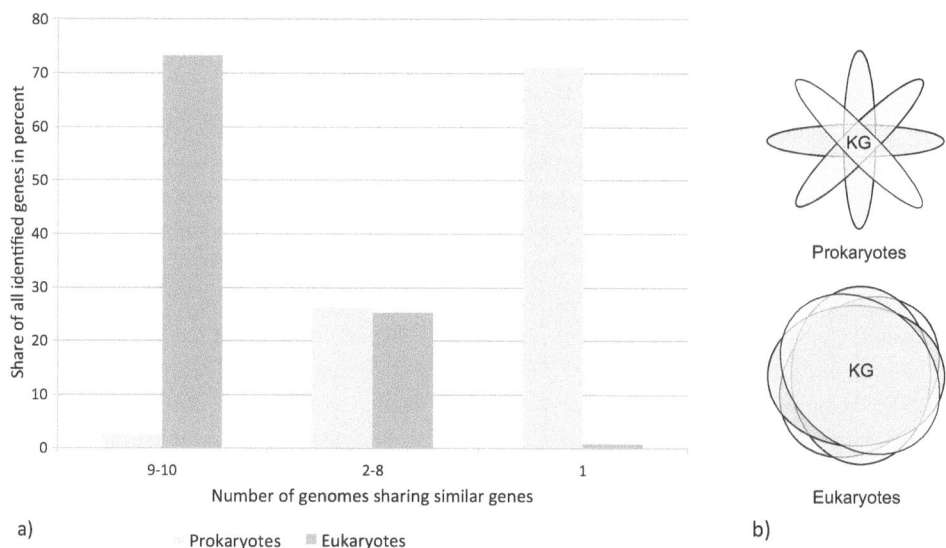

a)

b)

Fig. 5.10 Schematic comparison of matching of genes among different species in prokaryotes and in eukaryotes in marine microplankton. (**a**) Concordance of genes in each of ten species of prokaryotes and eukaryotes studied (Data source: Mock and Kirkham 2012); (**b**) Schematic representation of the proportions of shared genes (KG) in prokaryotes and eukaryotes

energy and substance conversion—are much more firmly embedded in the genes than in eukaryotes (Kunin and Ouzounis 2003; Beiko et al. 2005; Pál et al. 2005; Thomas and Nielsen 2005; Polz et al. 2006; Choi and Kim 2007; Dagan et al. 2008; López-García and Moreira 2008; Lauro et al. 2009; Toft and Andersson 2010; Yooseph et al. 2010; Rhodes et al. 2011; Smilie et al. 2011; Treangen and Rocha 2011; Dini-Andreote et al. 2012; Ryall et al. 2012; Sachs and Hollowell 2012; Williams et al. 2012; Wolf et al. 2012; Bansai et al. 2013; Luo et al. 2013). On the other hand, the organisms' genetic make-up can be flexibly adapted to changes in environmental factors by means of horizontal gene transfer. Eukaryotes are less flexible in this respect (Arnold et al. 2008; Keeling and Palmer 2008; Fitzpatrick 2011; Huang 2013; Husnik and McCutcheon 2018). As a result, they are also dependent on prokaryotes for maintenance of their vital functions—be it through pre-processing of nutrients or as actual food sources.

Ideal conditions for horizontal gene transfer among prokaryotes exist in zones with direct contact between cells—for example in biofilms, in the digestive systems of animals or on particles in water (Daane et al. 1997; Sørensen et al. 2005; Hayes et al. 2010; Aminoc 2011, Caro-Quintero et al. 2011; Metcalf et al. 2014). In soils, horizontal gene transfer is facilitated by migration of bacteria along fungal hyphae—but also along electric cables and water pipes (Berthold et al. 2016). In this context, a decrease in spontaneous mutation rates with increasing population densities of prokaryotes is striking. This points towards yet-to-be-identified intercellular factors that regulate mutations (Krašovec et al. 2017). Under other conditions, however, gene transfer by viruses (phages) is likely to be of greater importance (Angly et al. 2006; Kenzaka et al. 2010; McDaniel et al. 2010; Hurwitz and Sullivan 2013). In eukaryotes, the spread of horizontal gene transfer has not yet been fully investigated, but it has already been demonstrated in various groups of organisms (Gonzalez and Lessions 1999; Bergthorsson et al. 2003; Waller et al. 2006; Keeling and Palmer 2008; Bock 2009; Emiliani et al. 2009; Schmitt and Lumbsch 2009; Liu et al. 2011; Wisecaver et al. 2013; Chang and Stergiopoulos 2015; Debortoli et al. 2016; Shelomi et al. 2016). The extent of horizontal gene transfer in rotifers has been particularly hotly debated (Arnold et al. 2008; Ramulu et al. 2012; Debortoli et al. 2016; Schwander and Oldroyd 2016; Wilson et al. 2018). Intracellular gene exchange (IGT) between a cell's genome—stored in the cell nucleus—and the genomes of symbiotically acquired mitochondria and chloroplasts (cf. Fig. 5.5a) is only partially understood (Martin and Herrmann 1998; Westphal et al. 2003; Kleine et al. 2009; Waller and Jackson 2009; de Paula et al. 2013; Huynen et al. 2013; Allen 2015; Ku et al. 2015). As studies on fungi show, genes can also be regulated by signals from mitochondria (Podholová et al. 2016).

For prokaryotes, there is growing evidence that mobile DNA elements play an essential role in the spread of resistance through horizontal gene transfer (Gillespie et al. 2011; Solheim et al. 2011; Stokes and Gillings 2011; Broaders et al. 2013; Djordjevic et al. 2013; Lee et al. 2015; Brito et al. 2016; El Karkouri et al. 2016). The mobile DNA elements— mobile integrons, also known as genome islands—are gene segments that combine independently with functional DNA segments and can be exchanged between bacteria. They are likely to have evolved during the early stages of bacterial evolution from non-mobile superintegrons, which are also involved in horizontal gene transfer (Mazel 2006; Juhas et al. 2009). In prokaryotes, mobile genes are found mainly in plastids, isolated from the cell's own DNA (Kim and Archibald 2008). In eukaryotes, mobile genes are also significantly involved in horizontal gene transfer. Their functional significance is still largely unclear; it could reside in rapid adaptation to new environmental conditions (Diao et al. 2006; Lowe et al. 2007; Fischer and Suttle 2011; Lynch et al. 2011; Fedoroff 2012; Arkhipova and Rodriguez 2013; Bridier-Nahmias et al. 2015).

The wide distribution of horizontal gene transfer among prokaryotes also renders understandable the difficulties encountered in delimiting species (Blaxter et al. 2005; Gogarten and Townsend 2005; Boto 2010; Rhodes et al. 2011; Treangen and Rocha 2011; Shapiro et al. 2012; Soucy et al. 2015). For example, many bacterial genes are also found in archaeans (Nelson-Sathi et al. 2015). Because of the diversity of different gene configurations, gene distributions of prokaryotes show fuzzy, fractal edges. In summary, all known genes of a bacterial group have been given the name pangenome (Medini et al. 2005; Lapierre and Gogarten 2009; Lefébure et al. 2010; Mira et al. 2010; Collingro et al. 2011; Vernikos et al. 2013; Shin et al. 2016; Moldovan and Gelfand 2018). The close coupling between the genetic apparatus and functional interaction with the environment also explains why prokaryotes could not develop permanent large multicellular organisms. To fulfil the different functions in a multicellular organism, prokaryotic cells would need horizontal gene transfer. However, the necessary conditions for this are present only in one- and two-dimensional cell assemblies (Schooling et al. 2009; Madsen et al. 2012;

Fig. 5.11 Lichen growth on a rock surface illustrating the small-scale diversity of microorganisms

Chimileski et al. 2014; Merod and Wuertz 2014). It is therefore understandable why horizontal gene transfer plays an important role among two-dimensional multicellular organisms, especially in fungi (Richards et al. 2011; Marcet-Houben and Gabaldón 2016; Yin et al. 2016). The relatively stable communities of fungi with algae or cyanobacteria in lichens provide examples of its evolutionary significance (Fig. 5.11) (Schmitt and Lumbsch 2009; Beck et al. 2015). Closed three-dimensional organisms could only develop by means of flexible epigenetic interpretation of a uniform genome. One hypothesis is based on the assumption that the evolution of animals (Metazoa) was significantly influenced by regulatory functions of microRNAs (Peterson et al. 2009).

Here, if not before, an answer is needed for the question—according to classical concepts—of why genomes of prokaryotes lacking sexual reproduction did not suffer extensive degeneration long ago because of harmful mutations. Even in humans, numbers of spontaneous mutations in germ cells increase with age! Originally, this phenomenon was assumed to be due to errors in cell division. In that case, only sperm cells in males would have been affected, since on average their stem cells divide every 16 days until the end of fertility. In principle, studies seemingly confirmed this hypothesis by showing an approximate doubling of mutations in sperm between the ages of 20 to 40. However—over the same age range—an approximately 1.7-fold increase in genetic changes has also been demonstrated for egg cells, although in women the first meiotic cell division (Fig. 5.12) is completed at the embryonic stage. The cause is assumed to be spontaneous mutations during oocyte maturation (Goldmann et al. 2016; Goriely 2016; Dobson et al. 2018). In various arthropod species, basic changes in germ cells caused by parasitic bacteria—ranging from sex change to the killing of males—have been observed. The best-known genus of such bacteria—*Wolbachia*—also shows the largest repertoire of alternative pathways and the broadest range of affected species (Dobson et al. 1999; Cordaux et al. 2011). Completed development of primary stem cells into somatic cells or germ cells can also be secured only with storage in gonads—as in colony-forming protochordates (*Botryllus*) (Stoner and Weissman 1996; Stoner et al. 1999).

For an initial answer to the above question, the quantitative differences between prokaryotes and eukaryotes should be considered. Prokaryotes, with global population sizes of an unimaginable 10^{27} cells and high rates of reproduction, can quickly compensate for even major losses due to adverse mutations (Biller et al. 2014). In addition, bacterial groups can survive severe environmental changes and adapt to new conditions by maintaining subpopulations with different genetic traits (Veening et al. 2008a). Horizontal gene transfer also enables the rapid spread of advantageous genes within cell aggregations. But eukaryotes are subject to very different framework conditions. The significantly smaller population sizes make it difficult to compensate for losses, especially with large-bodied multicellular life forms. As already mentioned,—with the exception of fungi and plants— horizontal gene transfer is far less common. Whereas fungi have consistently used this strategy, with plants it was of crucial importance for the evolution of land plants (Rosewich and Kistler 2000; Keeling and Palmer 2008; Richards et al. 2011; Qiu et al. 2012; Yue et al. 2012; Schönknecht et al. 2013; Soucy et al. 2015). Especially during the initial phase of the transition from algae to mosses, numerous genes for coping with terrestrial living conditions were adopted from bacteria and fungi. In the subsequent evolution of land plants from mosses, other transformation strategies became more important. Various genes were lost, for example for production of motile gametes or development of drought resistance in plants. Instead, genes for signal transmission became more differentiated, drought resistance of seeds improved and the duration of the haploid phase in life cycles was markedly reduced (Rensing et al. 2008; Bowman et al. 2017). This example shows the increased importance of alternative strategies for modification, such as the already mentioned intracellular gene exchange with mitochondria and plastids. In addition, new genes can arise through viral infections, through increases in genome number (polyploidy) or by restructuring the existing genome (Daubin and Ochman 2004; Sorek 2007).

For example, formation of the connective tissue located between maternal and embryonic blood circulations in the placenta of mammals is based on genes of viral origin (Lavialle et al. 2013; Cornelis et al. 2015). The outer boundary layer of the placenta in the maternal tissue—the syncytiotrophoblast—is composed of giant cells with polyploid, 8- to 64-fold gene sets. By contrast, the inner cell layer of the placenta—the cytotrophoblast—is formed by single cells with diploid gene sets. Polyploid cells are also found in the human body in the precursors of blood platelets in bone marrow—the megakaryocytes, in liver cells (hepatocytes), in skeletal muscles and in the osteoclasts of bones. Polyploidy is also suspected to be present in mammary gland cells (Davoli and de Lange 2011). The examples provided indicate that polyploidy is mainly found in cells with specific physiological functions. Augmentation of genomes in individual cells—autopolyploidy—is achieved by changes in the process of mitosis (see Fig. 5.7) (Anisimov 2005; Bergthorsson et al. 2007; Fox and Duronio 2013; Neiman et al. 2013; Frawley and Orr-Weaver 2015). This increases the leeway for interpretation during gene expression—transmission of genetic information to physiological performance. In the simplest case, individual genes contribute only to part of the original functions (subfunctionalization). Modifications of individual gene expression can lead to new physiological functions (neofunctionalization). However, deviations from the regulatory conditions in diploid cells also increase the risk of physiological malfunctions and hence development of cancers (Davoli and de Lange 2011; Gentric and Desdouets 2014; Schoenfelder and Fox 2015). Regardless of the associated disadvantages, polyploidy in glandular cells of animals enables the production of diverse toxic substances from originally identical genetic bases in centipedes (Chilopoda), scorpions, snails and reptiles (Fry et al. 2005, 2008; Kozminsky-Atias et al. 2008; Casewell et al. 2011; Anisimov and Zyumchenko 2012; Roelants et al. 2013; Gorson et al. 2015; Rangel et al. 2015; Undheim et al. 2015).

In contrast to the polyploidy of cells in individual tissues, only a few polyploid species are found among animals, for example rotifers and certain fish or amphibians (Griffiths 1941; Wals and Zhang 1992; Van de Peer et al. 2017). Relatively few polyploid species have so far been found among fungi. In distinction from animals, this is probably due to pronounced genetic variability in combination with adaptation to prevailing environmental conditions. Variability in the genetic make-up of plants applies not only to genome size, but also to exchange of genes between species, which permits polyploidy with different sets of genes—allopolyploidy (Croll et al. 2008; Parfrey et al. 2008; Levin 2013; Cox et al. 2014; Estep et al. 2014; Vanneste et al. 2014; Marcussen et al. 2015; Campbell et al. 2016; Todd et al. 2017; Linder and Barker 2014). Polyploid species spread more rapidly over long distances and predominate among first colonizers in areas that have become vacant—for example due to glacier retreat (Brochmann et al. 2004; Kolář et al. 2013; Vanneste et al. 2014). Polyploidy also occurs more frequently after large-scale destruction of vegetation—for example after the massive changes at the end of the Cretaceous period around 66 million years ago and the end of the Triassic period about 200 million years ago (Kürschner et al. 2013; De Storme and Mason 2014; Freeling 2017; Van de Peer et al. 2017; Wei et al. 2017). The significance of polyploidy for the origin of gymnosperms more than 300 million years ago has not been conclusively resolved (Jiao et al. 2011b; Ruprecht et al. 2017). Polyploidy also enables plants to produce flexibly a wide range of chemical defences (secondary metabolites, see Sect. 6.3.2.1) against harmful influences exerted by other organisms, for example insects (Orians 2000; Mazid et al. 2011). Polyploid plants—for example in the wheat group—also facilitated the development of agriculture (Matsuoka 2011). Geographically, the proportion of polyploid plant species in local vegetation increases with increasing latitude. However, regions with high species diversity, for example the South African Cape province, deviate from this global trend (Oberlander et al. 2016). Indirectly, the contradictory results yielded by single-factor attempts to explain the spread of polyploidy in plants reveal the extremely multi-dimensional nature of interactions between environmental conditions and genome composition (Leitch et al. 2004; Beaulieu et al. 2008; Hodgson et al. 2010; Simonin and Roddy 2018; Forrester et al. 2020; Parshuram et al. 2022).

As with the distribution of polyploid eukaryotic life forms under variable or extreme environmental conditions, polyploid life forms also occur under similar conditions among prokaryotes. Examples of this can be found among archaeans in extremely saline waters and among thermophilic bacteria (Ohtani et al. 2010; Lange et al. 2011; Soppa 2013; Jaakkola et al. 2012). Life forms with polyploidy or mobile genes are found in bacteria that live in eukaryotes. These features are particularly pronounced in bacteria with a wide range of potential host species or with facultative intracellular lifestyles (Bordenstein and Reznikoff 2005; Tobiason and Seifert 2006).

The cited examples of intracellular life forms show how intimately environmental conditions interact with genome evolution. Both bacteria and unicellular eukaryotes show significant reductions in genome size compared to free-living forms. Among intracellular bacteria, the genomes of symbiotic life forms tend to be reduced more than those of parasitic or pathogenic ones. In eukaryotic cells, the functions of mitochondria can be greatly reduced or even completely lost (Karnkowska et al. 2016). The extent of genome reduction also influences further evolutionary possibilities for intracellular life forms. Limited genome reductions permit future switching both between host organisms and between symbiotic and pathogenic relationships. With increasing genome reduction, dependence on the survival of host organisms increases (Moran and Wernegreen 2000; McGrath and Katz 2004; Keeling et al. 2005; Darby et al. 2007; Degnan et al. 2009; Klasson et al. 2009; Casadevall et al. 2011; McCutcheon and Moran 2012). However, given that genome reduction in these life forms can be understood in terms of host organisms taking over physiological functions, a similar change in free-living plants is at first glance surprising. One of the smallest plant genomes is found in the dwarf bladderwort (*Utricularia gibba*), a carnivorous aquatic plant feeding on animal tissue, in which digestion of captured arthropods has led to marked reduction in morphological features and nutrient acquisition (Ibarra-Laclette et al. 2013). The evolutionary development of various cnidarian groups (*Hydra* and *Clytia*) is also based on reduction of various genes and inactivation of transcription factors (Leclère et al. 2019).

Research on globins provides particularly impressive evidence of links between changes in external conditions and genetic/epigenetic adaptations. A few decades ago, only the amino acid sequences of haemo- and myoglobins of vertebrates and haemoglobins of legumes were known. In the meantime, a wide variety of globin variants have been identified. A common feature of this protein family is the integration of an iron-containing organic molecule into upstream and downstream polypeptide chains with various lengths and structures. The evolution and spread of globins was most likely triggered by the increasing concentration of free oxygen during the Precambrian period (see Sect. 8.4.2). The evolution of globins in bacteria enabled them to employ and control oxygen in intracellular metabolic processes. Archaeans and eukaryotes acquired the necessary genes from bacteria through horizontal gene transfer (Vinogradov et al. 2005, 2007). Within the eukaryotes, the functional performance of globins was differentiated by genome duplications and sequence modifications, mainly in adaptation to different environmental conditions (Storz et al. 2011, 2013; Vázquez-Limón et al. 2012; Keppner et al. 2020; Prothmann et al. 2020). On the basis of generalized molecular structural features, three main groups of globins can be distinguished (Vinogradov et al. 2005). Functional differentiation of globins predominantly follows the specific living conditions of individual species rather than their phylogenetic context (Vinogradov and Moens 2008; Gupta et al. 2011; Storz et al. 2011, 2013; Vázquez-Limón et al. 2012; Smith et al. 2018; Keppner et al. 2020; Prothmann et al. 2020). Even within a single organism, globins with specific functional performances are distributed among very diverse organs. For example, in humans haemoglobin is found not only in red blood cells, but also in cells of the lungs, liver, brain, eyes, blood vessels, kidneys, intestinal walls and female reproductive organs (Keppner et al. 2020).

The importance of transposons ("jumping genes") and retrotransposons (RNA-based modifications of gene sets) for maintenance of genetic diversity for non-sexual reproduction is still subject to heated debate. In eukaryotes, they can either change position in the DNA as "jumping" genes (transposons) or—through new generation via RNA—occur as retrotransposons. Although they were originally classified as selfish (Dawkins 2007) or parasitic genes, interpretations are beginning to change as a result of findings from more recent scientific studies. Observations that mobile genes were "switched off" by epigenetic systems were decisive for the original interpretation. In the meantime, it has become apparent that reciprocal functional relationships exist between mobile genes and epigenetic systems. Accordingly, the evolutionary differentiation of proteins and genetic regulation of implantation of fertilized oocytes and pregnancy in mammals can be attributed to transposons (Lynch et al. 2011; Cosby et al. 2021).

The question of the origin and extinction of genes independent of the evolutionary development of phenotypes is still under active discussion. For the origin of new genes, acquisition of coding functions by previously non-coding genes has been discussed (Long et al. 2003; Kaessmann 2010; Tautz and Domazet-Lošo 2011; Carvunis et al. 2012; Neme and Tautz 2013, 2014; McLysaght and Guerzoni 2015; Schlötterer 2015; Moyers and Zhang 2016). For example, in the human genome roughly two thirds of the genes are non-coding (Fedoroff 2012). With increasing understanding of regulatory processes inside cells, their role in the origin of RNA forms that are essential for this purpose is becoming increasingly clear. Non-coding RNA forms regulate both internal cellular processes and their adaptation to the conditions of the cell environment (Amaral et al. 2008; Cech and Steitz 2014).

In eukaryotes, "fragmentation" of genes in the DNA by non-functional sections, so-called introns, is a striking feature. In some cases, sections of other genes can also be embedded in introns. Introns are transferred to messenger RNA (mRNA) during transcription, but are removed before translation. To do this, the intron sections must be excised from the mRNA and the mRNA must then be reconnected—spliced. Four groups of introns are distinguished according to the excision processes and splicing (Graw 2015). The evolutionary origin and significance of introns is still debated (Martin and Koonin 2006; Irimia et al. 2007; Jeffares et al. 2005; Farlow et al. 2011; Shabalina et al. 2010). There is evidence for conversion of introns into exons—i.e. coding sections—and thus into components of new genes (Sorek 2007). In other words, they increase the degrees of freedom in systemic interpretation of the genome. One indication of this is the above-average density of introns in water fleas (*Daphnia*), with their great morphological and physiological adaptability (Li et al. 2009). Triggers for changes are both chemical signals and repeated changes in small-scale currents (Wolf and Mort 1986; Laforsch and Tollrian 2004; Petrusek et al. 2009; Walsh et al. 2014; Weiss et al. 2015).

Epigenetic processes are also of central importance for the evolution of multicellular organisms, which can only safeguard their functional capacity to by reducing individual cell performance to the specific requirements within individual cell groups. For this, the coordinated activation and suppression of gene functions by the epigenetic regulatory systems is decisive. An extreme case is provided by the above-mentioned red blood cells (erythrocytes) of vertebrates. During their formation in bone marrow, they possess all the characteristics of eukaryotic cells. However, before they are released into the bloodstream—i.e. before they become fully functional—they lose their nucleus and shortly afterwards also mitochondria, ribosomes and the endoplasmic reticulum (Fedoroff 2012). Only through these changes can the chemical reactions for oxygen supply to tissue cells and removal of CO_2—of haemoglobin in conjunction with 2,3-bisphosphoglycerate—take place in red blood cells at an adequate level (Amaral et al. 2008; Cech and Steitz 2014). Despite the loss of their genetic apparatus, the cells can remain functional for an average of 120 days. Moreover, their decay is actively and independently initiated in the form of programmed cell death.

Recent research results have increasingly clarified the structuring of epigenetic processes by RNAs, more concretely through recruitment of individual types from the diverse array of RNAs available (Amaral et al. 2008; Alberts et al. 2015). Connections between epigenetic functions and RNA across all organisms become recognizable by virtue of the major differences between the individual domains. Only about 1% of the RNAs compared occur in more than one domain (Hoeppner et al. 2012). The long, non-coding RNAs (lncRNA), which were originally thought to have no function, activate or suppress the transmission of genetic information in the context of prevailing environment conditions (Chu et al. 2011; Lee 2012; Ng et al. 2012; Komienko et al. 2013; Ulitsky and Bartel 2013; Cech and Steitz 2014; Fatica and Bozzoni 2014; Gaiti et al. 2017; Rajewsky et al. 2017). An appropriate example is seen in the functional differentiation of cells in tissues of multicellular organisms, such as during brain development. Small RNAs (sRNA) transmit information in plants, facilitate differentiation between "own" and "foreign" gene information, and secure epigenetic memory—especially for defence against foreign genes in cells of the immune system and in prokaryotes (Jorgensen 2002; Kim 2005; Miska 2005; Wang and Metzlaff 2005; Keene 2007; Kehr and Buhtz 2008; Carthew and Sontheimer 2009; Malone and Hannon 2009; Christodoulou et al. 2010; Saxe and Lin 2011; Shukla et al. 2011; Parent et al. 2012; Shirayama et al. 2012; Durand et al. 2014; Battle et al. 2015; Taylor et al. 2015; Sontheimer and Marraffini 2016; Golden et al. 2017). They can also influence the behaviour of animals (Picao-Osorio et al. 2015).

In multicellular eukaryotes, functional differentiation of individual cells and morphological diversity are achieved by variable combinations in activation of individual genes. Through the combinatorial effects, a significantly higher diversity can be achieved in traits than through direct conversion of individual genes into morphological traits. According to the current state of knowledge, enhancers—specific non-coding gene segments—regulate the combinations of genes in conjunction with specific enhancer RNAs (eRNA). The enhancer system can activate the promoters—essential for the precise transcription of genes—over lengthy DNA segments. Promoters initiate the transcription of a gene into messenger RNA (mRNA) (Graw 2015; Kim and Shiekhattar 2015; Long et al. 2016; Chen et al. 2017a; Liu 2017). Despite great flexibility of the gene combinations concerned, within individual life forms marked robusticity against disturbances in development of organisms is achieved through redundancy in the enhancer systems. Several systems working in parallel—usually five or more—ensure the correct transcription of combined genes in the event of malfunctions of individual systems (Osterwalder et al. 2018). According to gene studies on human embryonic stem cells, the vast majority of enhancers have probably remained stable for around 8 million years. It is therefore not surprising that around 94 per cent of enhancers in humans match those of chimpanzees (Glinsky and Barakat 2019). Although evidence of enhancers was published some thirty years ago, many questions about their function and development remain unanswered because of certain methodological difficulties encountered during their investigation (Collis et al. 1990; Tuan et al. 1992; Chen et al. 2017a, b).

Examples of the impact of environmental influences on the genetic system are provided by convergent evolution of proteins to protect against uncontrolled formation of ice crystals in the tissues of various fish species in polar marine regions, as well as hibernating insects, amphibians and plants (Chen et al. 1997; Zachariassen and Kristiansen 2000; Atici and Nalbantoğlu 2003; Piatigorsky 2007; Deng et al. 2010). Convergent genetic modifications and symbioses enable adaptations of marine unicellular eukaryotes (diatoms) to changes in nutrient conditions in an aquatic environment (Croft et al. 2006; Härnström et al. 2011). Divergent genetic modifications have facilitated the structural and functional diversification of haemoglobins ranging from bacteria to mammals (Hardison 1998; Piatigorsky 2007). During penetration of their hyphae into the tissues of host organisms, parasitic fungi test the defence capacity of individual cells and epigenetically activate specific genes to overcome cell barriers at the appropriate site (Bielska et al. 2014).

Inheritance of epigenetic modifications of the genome is still the subject of active scientific debate (Franklin and Mansuy 2010; Heard and Martiessen 2014; Perfus-Barbeoch et al. 2014; Horsthemke 2018). One key reason for this is failure to consider different pathways of inheritance. For instance, inheritance of epigenetically acquired traits has a higher probability in clonal or parthenogenetic reproduction without fertilization. In sexual reproduction, the probability of inheritance is higher if the gametes are formed from somatic cells. Plants, for example, can inherit epigenetically developed processes for defence against herbivorous (plant-eating) animals (Holeski et al. 2012). Moreover, as a result of differentiation into somatic cells and gametes, the possibility of inheritance in multicellular animals depends on properties of the cells concerned. Changes in the somatic cells mainly affect affected individuals and can be inherited only with low probability. In female honey bees, for example, reproductive capacity is epigenetically regulated exclusively by food composition (Kucharski et al. 2008). On the other hand, changes in germ cells—for example through effects of bacteria of the *Wolbachia* group in arthropods—can be inherited by offspring (Dobson et al. 1999).

The most diverse forms of non-sexual reproduction are found among fungi: haploid and diploid life forms, as well as hyphae with uniform or different cell nuclei. Because of the combination of hyphae with different genomic configurations, different fungi are able to reciprocally reinforce metabolic production (auxotrophy) between their hyphae (Moore et al. 2011; de Jonge et al. 2013; Ropars et al. 2016). Contrary to theoretical expectations, neither signs of consistently persistent gene loss nor damage by competition between different genes are detected in genomes of fungi (Kelkar and Ochman 2011; Vreeburg et al. 2016). Instead, following a connection between hyphae with incompatible genomes, the affected sections are destroyed by closure of partitions and through endogenous regulatory processes (Moore et al. 2011). Primary reasons for this are probably, on the one hand, the above-mentioned horizontal gene transfer and, on the other hand, highly developed—but as yet barely studied—epigenetic regulatory systems (Kuhn et al. 2001; Smith et al. 2012; Aramayo and Selker 2013; Guerriero et al. 2015).

The question raised above regarding long-term preservation of a functional genome can therefore be generally extended to non-sexual reproduction in general. Taking into account the increased probability of inheritance of epigenetic changes in the genome, non-sexual reproduction also enhances adaptability to environmental changes. Under heterogeneous environmental conditions, greater differentiation of evolutionary lineages can also be expected with this form of reproduction than for sexual reproduction. The problem of accumulating mutations with harmful effects is countered at different levels. Defective or harmful changes are corrected over the short term—with varying probability of success—during transcription and translation of genes, and over the medium term by mitotic control processes during cell reproduction. Uncontrollable aberrations inevitably lead to the demise of the evolutionary lineage affected. Additional corrective processes can take effect through facultative sexual reproduction or—as in ciliates—facultative sexual recombination (see Sect. 5.2.2).

The examples provided cast massive doubt on the conventional paradigm of directed biological evolution of organisms from simple to differentiated genetic endowments. Such ideas might appear credible if the evolutionary history of the vast majority of organisms is ignored and biological evolution is seen as an isolated process. In reality, biological evolution—and hence genetic endowment as well—interacts intimately with environmental conditions. The scale ranges of interactions can be seen in the different life forms of organisms. Arguments in favour of this statement are provided by generally observable decreases in genetic differentiation in all groups of organisms studied (Lynch 2006; Kuo and Ochman 2009; Zmasek and Godzik 2011; Morris et al. 2012; Wolf and Koonin 2013). The processes proceed at different rates in the various life forms and are interrupted by relatively short periods in which rapid increases in genetic differentiation occur. A well-known example is provided by rapid differentiation of the genetic make-up and morphological differentiation of animals during the so-called "Cambrian explosion" around 600 million years ago (Rokas et al. 2005). According to studies on marine bony fish (teleosts), environmental factors have a counter-intuitive effect on the dynamics of animal differentiation. Apparently, there are significantly more new life forms in polar marine regions than in tropical regions. Because fewer life forms disappear in the latter regions, however, their biodiversity is significantly higher than in polar waters (Rabosky et al. 2018).

5.2.2   Sexual Reproduction: Disentangling a Paradigmatic Mirror Image

For more than a century, the paradigm of sexual reproduction has been maintained as a basic premise for species diversity and evolution. The theories of F.L.A. Weismann concerning the genetics of sexual reproduction and of R.A. Fisher (1930) regarding natural selection (Fisher 1930; Winther 2001) were—and in many cases still are—foundations for emphatic cultivation of this paradigm. From the perspective of vertebrate evolution, this makes it easy to formulate hypotheses invoking the "fundamental importance of sexual reproduction for evolution". According to this tenet, organisms lacking sexual reproduction should be doomed to extinction. The previous chapter has already presented the predominantly asexual forms of reproduction in prokaryotes—which have been successful for some 3.5 billion years. Although this information may be casually omitted from interpretation of the evolutionary development of "higher organisms", accumulating scientific findings have made perpetuation of the paradigm increasingly questionable even in the case of eukaryotes. Specifically, it has been shown that sexual reproduction does not increase genetic diversity but actually reduces it (Gorelick and Carpinone 2009; Gorelick and Heng 2010)! For example, among rotifers, asexually reproducing groups show comparatively greater genetic bandwidth than sexually reproducing groups (Tang et al. 2014).

Asexual reproduction is found not only among rotifers (Stelzer et al. 2010; Flot et al. 2013; Fontaneto and Barraclough 2015; Bininda-Emonds et al. 2016), tardigrades (Tardigrada) (Jørgensen et al. 2013; Altiero et al. 2015), flatworms (Turbellaria) (Pongratz et al. 2003; Reuter and Kreshchenko 2004), nematodes (Haag 2009; Schwarz 2017), and arthropods (Witzel et al. 2003; Bode et al. 2019; Maniatsi et al. 2011; Schwander et al. 2011; Maccari et al. 2014; Norouzitallab et al. 2015; Nougué et al. 2015; Vogt et al. 2015; Lavanchy et al. 2016; Faddeeva-Vakhrusheva et al. 2017; van der Kooi et al. 2017; Tvedte et al. 2019), but also among vertebrates—such as fish, amphibians and reptiles (Schön et al. 2009; Sites et al. 2011). Asexual reproduction also occurs in plants (Judson and Normark 1996; Holsinger 2000; Johnson et al. 2011; Hojsgaard and Hörandl 2016). Especially in extreme habitats, asexually reproducing lineages can be more geographically widespread than sexual life forms (Maniatsi et al. 2011; Tilquin and Kokko 2016). Asexual life forms are found mainly at the outer limits of the distribution ranges of systematic units, or under particularly variable environmental conditions (Haag and Ebert 2004; Møbjerg et al. 2011; Tilquin and Kokko 2016).

Discussion of genetic changes in asexual reproduction is still influenced by a questionable hypothesis known as "Muller's ratchet". According to this hypothesis, asexually reproducing lineages would be doomed to extinction because of the continuous—and irreversible—accumulation of deleterious mutations (Lynch et al. 1993). Various publications have either confirmed or refuted the hypothesis (Henry et al. 2011; Brandt et al. 2017). The fundamental weakness of the hypothesis resides in the simplistic assumption that cells—for some mysterious reason—can identify and eliminate detrimental genetic changes only during sexual reproduction. Taken alone, the absence of evidence for ageing processes—despite asexual reproduction—in freshwater polyps of the species *Hydra vulgaris* demonstrates that the hypothesis is not universally valid (Martínez and Bridge 2012).

Regardless of that issue, the question arises as to the evolutionary significance of sexual reproduction—which also occurs in unicellular eukaryotes (Dacks and Roger 1999; Struhl 1999; Angert 2005; Robinson and Bell 2005; Schurko et al. 2008; Birky 2009; Amato 2010; Lahr et al. 2011; Levin and King 2013; Umen and Heitman 2013; Ning et al. 2013; Chi et al. 2014; Oliverio and Katz 2014; Tekle et al. 2017). Here, it is necessary to analyse in more detail the core features of the processes involved. One approach is based on the definition of sexual reproduction proposed by Gorelick and Heng, which characterizes it with a combination of meiosis and syngamy (Fig. 5.12) (Gorelick and Heng 2010).

An essential feature of sexual reproduction in eukaryotes is alternation between phases with single gene sets (haploid phase) and double gene sets (diploid phase) in life cycles of the organisms concerned (Lewis 1985). Of central importance for the modification of genomes in eukaryotes is meiosis during the diploid phase. The process begins, as in mitosis (Fig. 5.12), with duplication of the chromosomes (composite structures including DNA and proteins). Subsequently, the four chromosomes are arranged close together with the same orientation. This facilitates the exchange of DNA segments between individual chromosomes and thus the recombination of genes. For this to happen, partial segments of the DNA strands are separated and exchanged in a process known as crossing-over. Depending on the participation of the DNA strands, this results in at least two and at most four DNA sequences with modified gene sets (Alberts et al. 2015). Finally, the four modified

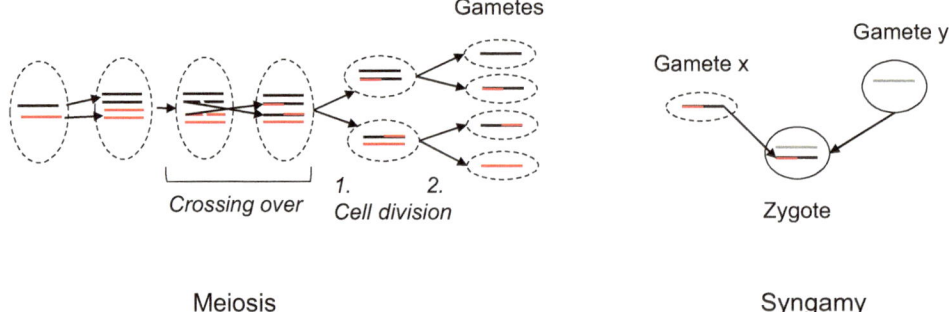

Fig. 5.12 Core features of sexual reproduction—meiosis with modification and reduction of genomes in combination with syngamy, the fusion of two different sex cells (gametes)

chromosomes are separated in two steps and divided into four daughter cells (gametes)—each with halved (haploid) gene sets (Fig. 5.12). The simplified scheme of meiosis presented should not allow us to forget that extremely diverse patterns of meiosis are known, although it is not possible to describe them further here (Archetti 2004, 2010; Stenberg and Saura 2013; Cabral et al. 2014; Heckmann et al. 2014; Niklas et al. 2014; Wang et al. 2015; Lenormand et al. 2016; Lindholm et al. 2016; Zickler and Kleckner 2016). In humans, for example, meiosis produces four sperm cells, but—because of asymmetrical divisions—only one egg cell (Alberts et al. 1986).

Fertilization of an egg by a sperm at first produces a cell with two nuclei, the zygote. in *syngamy*, the second part of sexual reproduction—the challenges of which are easily underestimated—manifold variants exist. A focus on the fusion of nuclear genomes, obscures the fact that autonomous systems with divergent rules and additional subsystems must be united as well—for example, including mitochondria (Hagström and Andersson 2018). Ultimately, a fully functional, complete autonomous system must emerge. Even simplified representations of the various cellular functions provide an idea of the inevitable magnitude of the challenges involved (Maniatis and Reed 2002). This also makes it clear why approximately equally equipped gametes interact only in relatively few cases—all involving unicellular eukaryotes. Observations have revealed that in such cases of—isogamous—sexual reproduction, merging of the cell components is strictly regulated (Bloomfield 2011; Weedall and Hall 2015; Lehtonen et al. 2016). The most diverse forms of sexual reproduction are found among fungi. Haploid forms mitotically produce gametes for sexual reproduction, from which haploid organisms once again arise through meiosis. In combination with meiosis, diploid forms can reproduce either parthenogenetically or sexually (Moore et al. 2011; Billiard et al. 2012).

Interesting examples of differentiated regulatory processes in reproduction can be found in unicellular ciliates, especially well known from the example of the slipper animalcule (*Paramecium*). Individual cells have two nuclei of different sizes, a micronucleus with approximately 0.5–2 percent of the genome volume and a macronucleus with the remaining, dominant share of the genome (Fig. 5.13). All basic information for cell reproduction and maintenance of cell functions is stored in the micronucleus, whereas information for regulating metabolism and interactions with the environment is stored in the macronucleus (Prescott 1994; Gao and Katz 2014; Warren et al. 2017). In general, ciliates reproduce asexually. In this process, both cell nuclei are duplicated mitotically before cell division and each of the pairs of nuclei produced is included in a cell of the next generation. For sexual recombination—which has not yet been observed in all species—two cells combine under conditions that are not fully understood (Dunthorn and Katz 2010). In sexual recombination, only the micronuclei undergo the process of meiosis when one cell comes into contact with a second—equally large—cell. Three of the four micronuclei formed in each cell through meiosis degenerate. The surviving micronucleus duplicates through mitosis. During the subsequent union between two cells, one micronucleus is exchanged in each direction. In both cells, the—different—micronuclei unite to form a cell nucleus (synkaryon). From a newly formed micronucleus, two micronuclei are formed

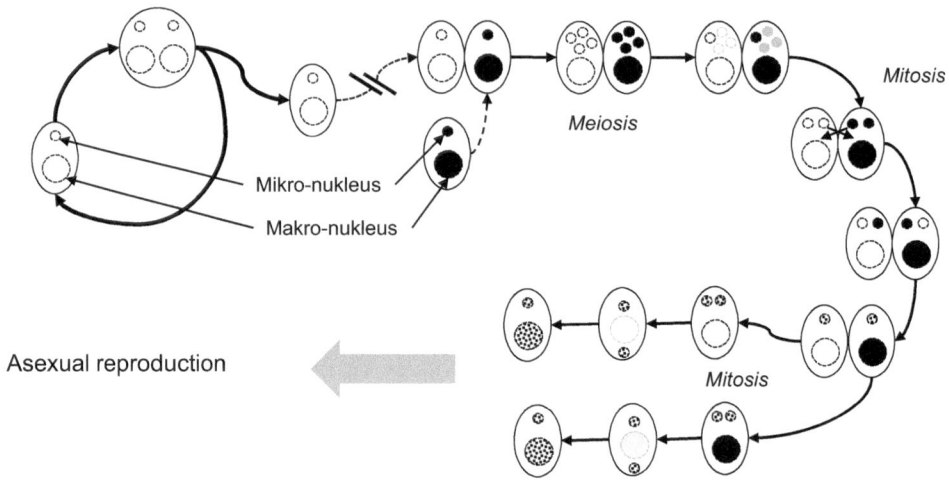

Fig. 5.13 Simplified representation of asexual reproductive cycles and sexual recombination in ciliates. Sources: Modified after Nowacki and Landweber (2009), Orias et al. (2011), Bracht et al. (2013), Yerlici and Landwever (2014). Explanations in the text

mitotically. Under regulation by RNA processes, the original macronuclei are then completely dismantled and the genetic "fragments" are used to expand one micronucleus into a new macronucleus. Since all processes take place within the original cells, there is no cell multiplication; instead far-reaching modifications of the genetic constitution takes place within the cells (Nowacki and Landweber 2009; Nowacki et al. 2011; Orias et al. 2011; Sung et al. 2012; Bracht et al. 2013; Singh et al. 2014; Yerlici and Landwever 2014). Given the global distribution of ciliates—including even anaerobic habitats—and their extensive evolutionary history, probably some 600 million years (Massana and Pedrós-Alió 1994; van Hoek et al. 2000; Schwarz and Frenzel 2005; Foissner et al. 2009; Dunthorn et al. 2015; van Maldegem et al. 2019)—they probably have one of the most successful reproductive strategies among eukaryotes.

Anisogamous sexual reproduction is far more common—indeed almost exclusively present in multicellular eukaryotes—as the mode of propagation (Billiard et al. 2010; Constable and Kokko 2018). One gamete—conventionally designated as female—consists of a fully functional cell. In the second gamete—designated as male—there is usually only a single gene set accompanied by minimal cell components needed for the fertilization process. Because of a much lower coordination effort, the egg cell can absorb the reduced male gamete more. Widespread in animals is the DNA compaction in male gametes (sperms) by replacement of histones with protamines to protect the male DNA during meiosis of the female DNA in the zygote (Sawada et al. 2014; Mignerot and Coelho 2016; Dubruille et al. 2023). In the process, epigenetic information can certainly be preserved. One example is provided by epigenetically regulated imprinting of the genome for regulation of embryonic growth in placental mammals and plants. This has the effect that the male genome stimulates growth, while the female genome antagonistically inhibits growth (Keverne 2014; Pires and Grossniklaus 2014). Among insects, males of the species *Anoplolepis gracilipes* (yellow crazy ants) possess two different haploid gene sets, R and W. When fertilized with the W gene set, eggs develop into males, whereas those fertilized with the R gene set develop into queens (Darras et al. 2023).

Putting it simply, it is easy to conclude from the illustrations that syngamy halves the original functional diversity. This confirms the previous statement that sexual reproduction has a reducing effect. At the same time, it is apparent that cells always aim for optimal regulatory states. This avoids both uncontrollable states and unnecessary regulatory efforts. In an evolutionary context, optimal ranges of regulation are always dependent on the existing internal cellular equipment and prevailing environmental conditions. For example, among the unicellular Alveolata, the free-living dinoflagellate species *Symbiodinium minutum* has a genome about 28 times larger than that of the intracellular malaria pathogen (*Plasmodium falciparum*) (Shoguchi et al. 2013).

In our realm of experience, requirements of syngamy may be roughly compared to a merger between two large corporations at one site. From the original protocols—comparable to DNA—only one can be valid after the merger; regardless of whether one protocol is completely deleted or parts of the original protocols are combined to form a new one. In all realms of responsibility staff duplication must be avoided by "downsizing". Internally, this results in bigger savings and profits, but for society generally the scope for creativity and job diversity is reduced.

In an evolutionary context, the boundaries between asexual and sexual reproduction are far more fluid than comparative presentation of the two forms conveys. The sequences of haploid and diploid phases (phases with single and double chromosome sets), sexual and asexual reproduction, as well as the sex-specific differentiation of haploid cells show an extensive range of variation in fungi, plants and animals. The greatest diversity of genetic make-up and forms of sexual reproduction is found among fungi, ranging from self-fertilization to multiple different forms of reproduction in one species. In parasitic and pathogenic species, correlations between the change from clonal to various forms of sexual reproduction and success in colonizing host organisms can be identified (Lee et al. 2010; de Jonge et al. 2013; Ropars et al. 2016).

The significance of such huge variability in physical reproductive processes, including associated mate choice and gamete selection in the context of ecological processes is, however, incompletely known (Heitman 2006, 2015; Gioti et al. 2012; Pélissié et al. 2014; Sun et al. 2015; van der Kooi and Schwander 2015; Beekman et al. 2016; Hadjivasiliou and Pomiankowski 2016; Haig 2016; Schwander and Oldroyd 2016; Vreeburg et al. 2016). Sexual reproduction can occur either with or without precopulatory selection—for example, through release of gametes in water or pollen in air currents in the latter case or in animals or in flowering plants with insect pollination in the former (Janzen 1967; Babcock et al. 1986; Clifton 1997; Chittka and Thomson 2001; Eibl-Eibesfeldt 2004; Bishop and Pemberton 2006; Maldonaldo and Riesgo 2008; Heitman 2010; Kamel et al. 2010; Boch et al. 2011; Schärer et al. 2011; Barazandeh et al. 2013; Feretzaki and Heitman 2013). Even in animals, in addition to sexual reproduction, various forms of asexual reproduction occur such as cloning—for example in sponges or diverse invertebrate groups in aquatic habitats—or parthenogenesis (reproduction without fertilization) (Ereskovsky and Tokina 2007; Sköld et al. 2009; Martínez and Bridge 2012; Schaible et al. 2015). Parthenogenesis can occur in various combinations with sexual reproduction, for example in alternating phases of life cycles or combined for sex determination in offspring—as in free-living flatworms (Turbellaria) or water fleas (*Daphnia*) (Simon et al. 2003; Decaestecker et al. 2009; D'Souza and Michiels 2010; Neiman et al. 2014). In honeybees, fertilized eggs develop into females and unfertilized eggs into males. Parthenogenesis can arise and disappear again over the course of evolution—as in rotifers, insects or snakes (Schön et al. 2009, Fenwick et al. 2011). The factors that influence geographic parthenogenesis, in which—in addition to populations with sexual reproduction—regional parthenogenetic populations occur, have not yet been definitively clarified. For populations at higher latitudes—for example with insects of the genus *Timema*—this can be explained by dispersal to less favourable habitats (Law and Crespi 2002). For regional parthenogenesis in tropical habitats—for example in scorpions, millipedes (Diplopoda) (Enghoff 1994; Lourenço 2008) or freshwater snails (Vergara et al. 2013), however, this explanation is invalid. Influences of environmental conditions can be seen in sex determination of offspring in individual species (modificatory sex determination). For example, in tuataras, various fish, turtles, crocodiles and lizards, the sex of the maturing embryo is determined by ambient temperature (Barske and Capel 2008; Quinn et al. 2011). Especially in fish, additional factors—for example, external hormones, population density, loss of one of the sexes—can lead to a sex change after embryonic development (sequential hermaphroditism) (Devlin and Nagahama 2002).

Clear correlations between genetic traits and environmental conditions can be observed in the evolution of land plants (Lewis 1985; Qiu et al. 2012). In general, haploid life forms with asexual reproduction (cloning) dominate under conditions of deficiency. However, the extreme longevity of more than 10,000 years in plant clones becomes comprehensible only when epigenetic processes are taken into account (Jenik 1994; van Groenendael et al. 1996; Fritz 2009; de Witte and Stöcklin 2010). Diploid life forms with sexual reproduction, on the other hand, dominate when a sufficient supply of nutrients is available, but there are increased risks of damage from abiotic factors—for example UV radiation—or from other organisms (Qiu et al. 2012). In plants, increased resistance to adverse conditions and increased growth performance are associated with multiple (polyploid) gene sets (Dorken and Eckert 2001; Levy and Feldman 2002; Mable 2004; Amato et al. 2007; Soltis and Soltis 2009; Sattler et al. 2015).

Both the origin and disappearance of genes and their functions—in accordance with the principle "use it or lose it"—can be seen in evolutionary processes. Well-known examples of this are gene reductions or physiological functions in parasitic organisms (Faso and Hehl 2011; Weber et al. 2013). In species of the alga genus *Polytomella*, the entire genome of the non-functional plastids has been lost (Smith and Asmail 2014; Smith and Lee 2014; Figueroa-Martinez et al. 2017).

5.2.3   Microbiomes: Prokaryotic Moderation of Multicellular Evolution

Individual examples of cooperation between bacteria and animals have already been discussed in connection with bioluminescence phenomena. By releasing certain chemical compounds, bacteria of the genus *Vibrio* trigger sexual reproduction in unicellular eukaryotes—choanoflagellates—that are closely related to animals. Other bacteria can also influence the behaviour of larvae of many animals or induce multicellularity in unicellular choanoflagellates or algae (Woznica and King 2018). Drawing on abundant evidence of close symbioses between bacteria and animals, Margaret McFall-Ngai inferred the existence of close connections between animal evolution and bacterial processes (McFall-Ngai 2001, 2002). Results from numerous genetic studies raised the question of the extent to which animal evolution is influenced by bacteria (McFall-Ngai et al. 2013; Flórez et al. 2015). Increasingly powerful methods of genome analysis have permitted detection of bacterial diversity on the body surfaces and in the digestive systems of animals (Moran 2007; Fraune and Bosch 2010). The term microbiome covers not only bacterial communities but also archaeans, viruses and unicellular eukaryotes associated with them (Sharp et al. 1998). Depending on the perspective involved, microbiomes are interpreted as ecosystems (Coyte et al. 2015; Foster et al. 2017) or—together with the multicellular animals or plants they colonize—as hologenomes or holobionts (Zilber-Rosenberg and Rosenberg 2008; Bordenstein and Theis 2015; Theis et al. 2016).

Examples for all interpretations can be found in microbiomes, but none is sufficient for complete characterization. In humans, for example, microbiomes of the skin and of the digestive tract have different functional relationships with bodily processes. The former mainly support immune defence of the skin (Grice and Segre 2011). Functional impacts of microbiomes in the digestive tract are far more diverse. In addition to processing food components into molecules that can be used for human metabolism, microbiomes also influence people's mental states and behaviour (Walter and Ley 2011; Clemente et al. 2012; Cryan and Dinan 2012; Dinan et al. 2015; Archie and Tung 2015; Magnúsdóttir et al. 2015; Fischbach and Segre 2016). On the other hand, clear differences in composition of microbiomes exist between individual regions of the skin and sectors of the digestive system, as well as in connection with a human subject's age (Grice and Segre 2011; Yatsunenko et al. 2012; Donaldson et al. 2016; Tropini et al. 2017). Among people living together—along with pets, for example dogs—similar compositions of their microbiomes are found (Song et al. 2013).

The human microbiome is also important in quantitative terms, regardless of sometimes widely divergent estimates. With an order of magnitude of about 10^{14} genes, the microbial genome is about a hundred times greater than that of humans (Clemente et al.

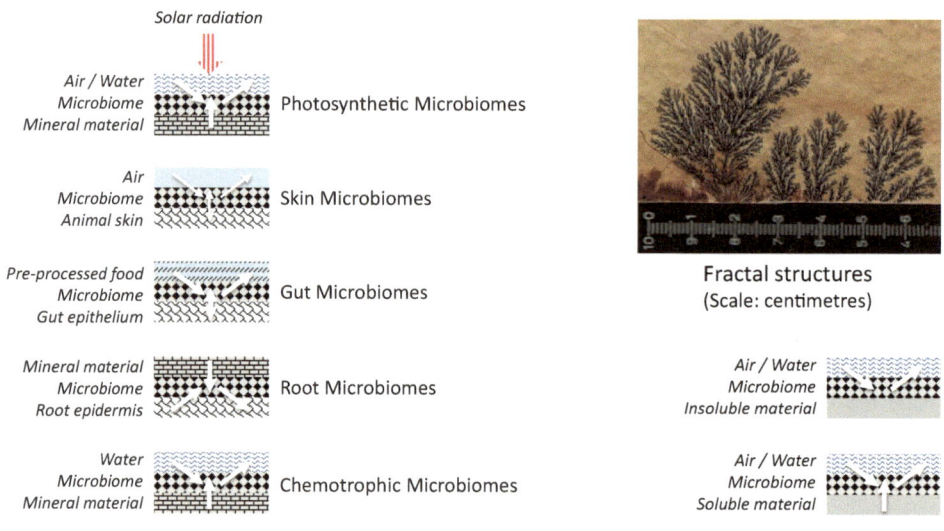

Examples of microbiomes in

natural boundary layers   technical boundary layers

Fig. 5.14 Examples of microbiomes in natural and industrial interfaces; arrows indicate the directions of material transfer. Insert: Example of abiotic fractal structures (dendrite)

2012). However, if cells that move freely in blood are taken into account, the number of cells in the microbiome only slightly exceeds that in the human body (Sender et al. 2016). The biomass of microorganisms is estimated to be about 0.2 to 1 kg, with the microbiome in the large intestine (colon) accounting for a major proportion (Dinan et al. 2015; Sender et al. 2016).

Quantitative data can easily lead us to overlook a core property of microbiomes beyond our understanding: the tiny sizes of the organisms involved. Because of this, microbiomes are able to colonize the tiniest irregularities of surfaces and exert their effects there. As a result, microbiomes colonise an area of around 25 m² on human skin and not 2 m², or an area of around 200 m² in the small intestine and not 0.33 m²—as would be indicated with our habitual methods of measurement. The reason for this is that the fractal properties of natural surfaces (Fig. 5.14, insert), whose extent—in simplified terms—increases disproportionately with decreasing size dimensions (Kunsch 1997; Senesi and Wilkinson 2008; Gallo 2017). *Because of the non-linear properties of fractals and the adaptability of their biochemical transformation processes, microbiomes can—with a similar biomass—influence global metabolic processes to a far greater extent than large-bodied organisms. Unlike multicellular organisms, however, microorganisms cannot mechanically break down or physically modify materials.*

The examples provided show how heavily the existence of multicellular organisms still depends on the activity of microbial systems. Microbiomes facilitate flexible adaptation of multicellular organisms to abiotic and microbial conditions in the environment (Fig. 5.14).

The functions of microbiomes are most clearly apparent at the transitions between plant roots and the mineral environment. At the root surface, interactions with physiological processes of the plant are prevalent, whereas in the outer zones of microbiomes interactions with the mineral environment predominate. Within the microbiomes—referred to as rhizospheres in this context—the systemic processes between microorganisms permit the transfer of mineral nutrients to the plant and the supply of organic nutrients from the plant to microorganisms. Mutual exchange of substances provides the microorganisms with sufficient energetic bases for chemical conversion of mineral nutrients into ionic forms that can be utilized by plants (Kundler et al. 1989; Puente et al. 2004; Ottow 2011; Reinhold-Hurek et al. 2015; Vandenkoornhuyse et al. 2015). In our technological environment, these functions of microorganisms are mostly undesirable, as they accelerate corrosion of metals or decomposition of the surfaces of buildings and works of art. Equally undesirable are changes to surface features of hydrodynamic structures—for example hulls of boats or inner surfaces of pipes—because they adversely affect technical performance values.

Looking at Fig. 5.14, it may be noticed that there are no horizontal arrows to indicate transfers of material within microbiomes. The boundary areas of the physiological activities of microbiomes are hence recognizable. They can indeed protect themselves from external influences through formation of a common extracellular matrix, but—for physico-chemical reasons—are unable to form extensive functional structures. These examples also illustrate the intermediate position of fungi between unicellularity and multicellularity. Fungi complement the functions described above by forming mycorrhizae—associations between plant roots and fungal hyphae (Azcón-Aguilar et al. 2009; Kohler et al. 2015). Because of their multicellular condition, they can transport substances over much greater distances than the microbial systems of rhizospheres. Fungi themselves possess microbiomes, partly in the form of intercellular bacteria, but also on fungal spores or fruiting bodies. Functionally, microbiomes enhance the physiological performance of fungi, improve the germination capacity of fungal spores or determine—for example in truffles—the flavour of fruiting bodies (Roesti et al. 2005; Antony-Babu et al. 2013; Vahdatzadeh et al. 2015; Deveau et al. 2018). Fungi can also occur completely inside plants (as endophytes)—for example grasses—and enhance defence against herbivorous animals (Porras-Alfaro and Bayman 2011). South American leafcutter ants (tribe Attini) carry plant material back to their nests to breed fungi, on which they feed. In doing so, they compete with parasitic micro-fungi that also live on the cultivated fungi. Bacteria (*Pseudonocardia* sp.) in the ants' own microbiome, however, inhibit the growth of the micro-fungi and thus contribute to maintenance of the fungal cultures (Cafaro and Currie 2005). Microbiomes of carrion beetle larvae inhibit rapid decay of a colonized carcase in competition with soil bacteria that promote fast decomposition (Shukla et al. 2018).

As already mentioned in the introduction to this chapter, such interactions—for instance in symbioses as well—are often characterized in static terms. Other interpretations assume a hierarchical control of all systems by one—usually multicellular—organism (Pion et al. 2013; Reinhold-Hurek et al. 2015; Foster et al. 2017). However, genetic studies of microbiomes in various hosts have not revealed intimate relationships between the species of multicellular and unicellular organisms involved. On the other hand, there are clear indications of relationships between the functional performance of microbiomes in the context of small-scale physiological processes of multicellular organisms and prevailing conditions of their environment (Burke et al. 2011; Fan et al. 2012; Coleman-Derr et al. 2016; Verhoeven et al. 2017). Thus, we are dealing with metastable systems that can yield completely different effects if major changes occur in individual factors. Supposedly symbiotic microbiomes can quickly develop into systems that are deleterious for animal or plant organisms (Briggs et al. 2010; Rineau et al. 2013; Scherlach et al. 2013; Netzker et al. 2015; Jones et al. 2017).

Turning to consider the ubiquitous distribution of microbiomes on, or in, plants and animals, the first outlines of their importance for the evolution of multicellular organisms can be recognized (Ley et al. 2008a, b; Hentschel et al. 2012; Bosch 2013; Bragina et al. 2014; Ainsworth et al. 2015; Bringel and Couée 2015; Hacquard et al. 2015; Lee et al. 2018; Lemay et al. 2018). In relation to the core processes of the interactions concerned, microbiomes regulate the evolutionary scope for metabolism of multicellular organisms. As an example, this permitted terrestrial plants to supply themselves with mineral nutrients or carnivorous vertebrates to evolve into herbivores. Illustrations of this are provided by evidence for the evolution of herbivorous beetles through cooperation and ultimately horizontal gene transfer with their microbiomes (Ledón-Rettig et al. 2018; McKenna et al. 2019).

Nonetheless, the question arises as to why multicellular organisms were able to evolve in a world full of microorganisms. Considering only cooperation between microbiomes and living multicellular organisms, extended access to energetic resources is a crucial indicator for their evolution. Multicellular animals can, for instance, gain access to organic material in sediments—inaccessible to microorganisms—because of their relative size and physical constitution. They can also demolish other multicellular organisms and break down the components—together with microorganisms—in their digestive systems for metabolic purposes. Moreover, dead body cells and excreta provide new food sources for microorganisms. If consideration is expanded to cover the entire life cycle of multicellular organisms, additional advantages for microorganisms become apparent. With the death of a multicellular organism—because immune defences are eliminated—the biomass accumulated during its lifetime becomes more easily accessible to microorganisms. In addition, this process can be enhanced by other multicellular organisms, such as scavengers or fungi.

In sum, evolution of multicellular organisms did not create competition, but rather additional developmental potential for microorganisms. This by no means excludes the possibility that various pathogenic viruses, bacteria, fungi—sometimes also in cooperation with multicellular organisms—may destroy individual populations of animals or plants *en masse* to maximize reproduction over the short term. Examples of this are seen in human epidemics, the ongoing dieback of ash trees, the dramatic collapse of amphibian populations caused by fungi, or mass reproduction of bark beetles—in cooperation with fungi (Winkle 1997; Lygis et al. 2005; Briggs et al. 2010; Jordal and Cognato 2012).

5.3   Viruses

Surely no other group highlights the limits of our knowledge of evolutionary processes as clearly as viruses. Just like abiotic factors, viruses have accompanied the evolution of organisms from the outset, yet we are reduced to making assumptions regarding their origin (Webster et al. 1992; Koonin et al. 2006; Dimmock et al. 2007; Duffy et al. 2008; Holmes 2011; Krupovic et al. 2011; Drezen et al. 2014; Forterre and Prangishvilli 2009; Koonin and Dolja 2014; Longdon et al. 2014; Nasir and Caetano-Anollés 2015; Alewsakun and Katzourakis 2017; Hayward 2017; Krupovic and Koonin 2017). Viruses are found in all habitats (Weinbauer et al. 2003; Sano et al. 2004; Weitz 2008; Anderson et al. 2011; Parsons et al. 2012; Holmfeldt et al. 2013). In fact, estimates of their numbers in the oceans and the number of infections they cause yield unimaginably high values of 10^{30} and 10^{23} per second, respectively. Results from studies to date indicate that neither the diversity nor the numbers of viruses are equally distributed on a global scale. For instance, the diversity of marine DNA viruses is lowest in the Antarctic Ocean, higher in the equatorial zone and highest in the Arctic Ocean. For RNA viruses, a study in Chinese coastal waters revealed major differences in infestation of various invertebrate groups (Suttle 1994, 2005, 2007; Wommack and Colwell 2000; Gregory et al. 2019; Zhang et al. 2022). In stark contrast to abiotic factors, however, viruses are completely dependent on appropriate host cells, since they possess no metabolism of their own and must parasitize cells of prokaryotes or eukaryotes in order to reproduce (Dimmock et al. 2007).

5.3.1   Characteristics of Viruses

In general, viruses consist of genetic material (DNA or RNA) and—in many cases—at least one protein for the surrounding capsid. This distinguishes them both from other mobile genetic elements in cells—plasmids for example—and from transmissible proteins such as prions (Krupovic and Bamford 2010; Colby and Prusiner 2011; Manjrekar 2017). The size range alone provides an idea of the diversity of viruses. The smallest viruses—virophages that replicate in other viruses—possess a genome of about 17,000 base pairs with a size of about 35 nanometers. Their potential hosts—nucleocytoplasmic large DNA viruses (NCLDVs), which are some 40 times larger—possess genomes with around 2.5 million base pairs and over 500 proteins (Bekliz et al. 2016; Brandes and Linial 2019; Moniruzzaman et al. 2020).

  We can only speculate about the origin of viruses. Take, for example, the case of polintons—large DNA transposons in the genomes of eukaryotes—where we find evidence of multiple networked co-evolution of viruses, prokaryotes and eukaryotes (Krupovic and Koonin 2015). The diversity of morphological forms of free and enveloped viruses, genetic endowments with single or double-stranded DNA and/or RNA sets, and their interactions with host cells combine to prohibit unequivocal systematic classification. Although viruses do not belong to the category of living organisms, a classification according to the classical

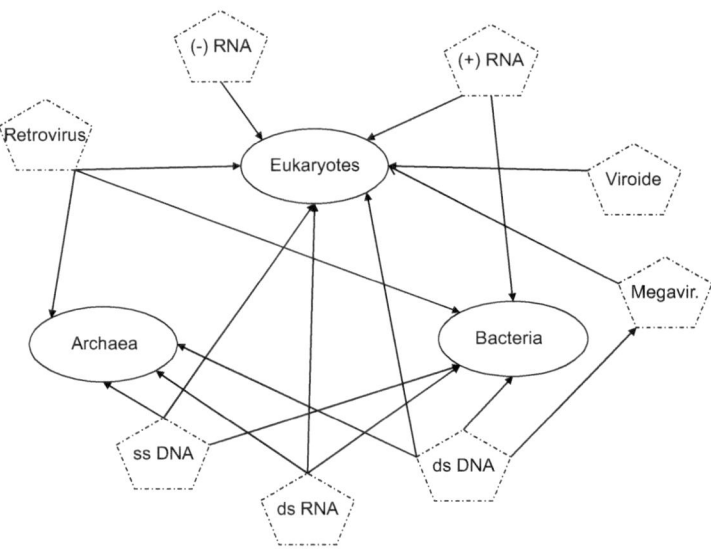

Fig. 5.15 Simplified schematic illustration of the functional network between cellular organisms and functional groups of viruses. Explanation of abbreviations in the text. Sources: supplemented and modified from Koonin and Dolja (2013) and Abergel et al. (2015)

binomial system has been attempted (Adams et al. 2017; Simmonds and Alewsakun 2018). Other classification systems begin with the host organisms or—like the Baltimore system—with genetic characteristics (Dimmock et al. 2007; Breitbart and Rohwer 2005; Lauber and Gorbalenya 2012; Sanjuán 2012; Halary et al. 2016; Sharma et al. 2016). In the meantime, digital databases have been compiled to facilitate combination of different classification systems (Hulo et al. 2011; Masson et al. 2013; Koonin et al. 2020; Nasir et al. 2020; https://talk.ictvonline.org). Viruses can be differentiated according to the following functional characteristic (Fig. 5.15):

- DNA viruses
  - Double-stranded (ds DNA)
  - Single-stranded (ss DNA)
- RNA viruses
  - Double stranded (ds RNA)
  - Single stranded with polar orientation like messenger RNA (mRNA)—[(+) RNA]
  - Single stranded with non-polar orientation [(−) RNA]
  - Retroviruses; the only group of viruses with direct incorporation into cellular DNA during reproduction
- Giant viruses, double-stranded DNA viruses (megaviruses)
- Viroids, small satellite viruses with RNA or DNA that co-occur with other viruses

As the customary notion of viruses links them to disease, references to associated diseases have been deliberately omitted from this brief listing of functional groups. Another reason for this approach is the oft-used distinction between viruses—especially in the context of multicellular eukaryotes—and phages—viruses in the context of prokaryotes (Weinbauer 2004). This historically established distinction makes it difficult to grasp the significance of viruses in an evolutionary context, because current knowledge suggests that only two functional groups—$(-)$ RNA and giant viruses (megaviruses)—are found exclusively in eukaryotes (Fig. 5.15). Of far greater importance is the outlined interconnectedness of viruses with archaeans, bacteria, and eukaryotes.

Direct interactions of viruses with organisms are based exclusively on information; only secondarily do energetic and material processes come into play. When attacking a cell, viruses seek to switch off or subvert its information systems on the surface and inside the cell (Paschos and Allday 2010; Standfuss 2015). The central targets of such attacks are the genetic and epigenetic transformation systems of the cell, which are modified for viral reproduction or replaced by mimicked molecules (Dimmock et al. 2007; Fuller et al. 2007; Sun et al. 2008, 2010; Casjens 2011; Molineux and Panja 2013; Koonin et al. 2020). Extremely simplified, the reproductive processes of viruses engage the same processes as in protein synthesis in organisms. In practice, however, the sequence of processes in individual virus groups deviates from this scheme in a wide variety of ways. Only in viruses are the capsid proteins and the genetic structures of DNA or RNA reproduced in parallel and integrated into the new virus at the end. Because of adaptation processes in the host's immune system, deviations from exact reproduction of the viral genetic code that occur during this process improve future chances of overcoming the immune defences of host organisms, for example in the case of influenza viruses (Farheen and Thattai 2019; Vahey and Fletcher 2019). Reproductive processes are particularly complex in giant viruses with particle diameters of up to 1.5 μm. They possess DNA and RNA, as well as proteins, and can replicate in the cytoplasm of their host cells in self-organized "virus factories" (Fischer 2013, 2016; Gaia et al. 2014; Abergel et al. 2015; Halary et al. 2016; Sharma et al. 2016; Leslie 2017; Abrahão et al. 2018; Koonin et al. 2020; Kee et al. 2022). This allows them to be parasitized themselves by small viruses—virophages (La Scola et al. 2008; Gaia et al. 2014; Raoult 2015).

The material and energy resources of infected cells can be completely exhausted either immediately (lytic cycle) or after an extended residence time in the cell (lysogenic cycle). In some cases, part of the resources can be used for viral reproduction without ultimate destruction of the cell. Which of these alternative paths is taken depends on "decisions" made by the viruses! They employ information about current virus density as well as about physiological conditions of infected cells (Dimmock et al. 2007; Kotzin et al. 2017; Ofir and Sorek 2018; Laganenka et al. 2019; Silpe and Bassler 2019; Brady et al. 2021). Non-enveloped viruses—for example noroviruses—perforate the cell walls of mitochondria in the host cells to release their replicated progeny, thus inducing cell death (Wang et al. 2023). Cells that have been attacked attempt to block viral regulatory systems or—especially in prokaryotes—to paralyze viral genomes with the CRISPR-Cas system.[2] However, there is growing evidence that viruses also possess codes from different CRISPR-Cas systems in their genomes and transfer them into the genomes of infected cells. This presumably blocks multiple infections of a cell by competing viruses (Amital and Sorek 2017; Burstein et al. 2017; Jackson et al. 2017; Kazlauskiene et al. 2017; Al-Shayeb et al. 2022).

5.3.2   Interactions with Organisms

In an evolutionary context, viruses and affected organisms can sustain themselves over the long term only by means of ongoing adaptations of their regulatory capabilities to those of the opponent. Basically, two different options are open to both parties:

(a) improvement of specialized strategies, or
(b) modification of the strategy by altering existing characteristics

Because under natural conditions numbers of interacting cells and viruses are in the range of millions or billions—depending on prevailing conditions—both strategies can be successful. This can also explain the supposed contradiction between specialization and generalization in viruses. The literature contains examples of coevolution of viruses and organisms, as well as cross-domain occurrences of viruses (Hueffer et al. 2003; Arnaud et al. 2007; Duffy et al. 2008; Smith et al. 2009; Streicker et al. 2010; Gray et al. 2011; Krupovic et al. 2011; Gaia et al. 2014; Katzourakis and Aswad 2016).

Over the long term, improvement in the strategy (strategy a) is successful, for example, where extreme environmental conditions—e.g. high temperatures, marked salinity or extreme pH values—impose a high cost for maintaining physiological processes (Prangishvili et al. 2006; DiMaio et al. 2015). In all other cases, however, modification of the strategy (strategy b) is more successful, as the possibilities for specialization are limited. Incidentally, this also shows that interactions with viruses influence the evolution of organisms. In the human genome, for example, there is evidence of long-term interactions with viruses manifested by around 100,000 copies of endogenous retroviruses (Emerman and Malik 2010; Feschotte and Gilbert 2012; Aswad and Katzourakis 2012). The influence of viruses on the evolutionary development of the genomes of all organisms is generally underestimated. This applies not only to modification of defence systems, but also to horizontal transfer of genes and proteins by viruses between different organisms (Chiura 1997; de la Cruz and Davies 2000; Biers et al. 2008; Liu et al. 2010, 2011; Filée 2014).

Just as with abiotic variables, all organisms are confronted with viruses. However, viruses are clearly distinguished from abiotic variables by their ability to specifically intervene in the genetic and epigenetic regulatory systems of cells, as well as to adapt to the

---

[2]Clustered Regularly Interspaced Short Palindromic Repeats Crispr Associated protein.

defence responses of cells. Unlike organisms, they do not use their own energetic or material resources for this purpose (Dimmock et al. 2007; Fuller et al. 2007; Sun et al. 2010; Molineux and Panja 2013). Viruses find ideal conditions for their spread in very large host populations with high spatial densities of individuals. For example, groups of bacteria that are particularly successful in competition for resources often cannot prevail in aquatic ecosystems because they are preferentially attacked and decimated by viruses. This phenomenon is known as *"kill the winner"*(Thingstad 2000).

Anyone who has read Sect. 5.2.1 and critically inspected Fig. 5.11 will ask how horizontal gene transfer can function with such clear demarcations between microbial systems. Current knowledge indicates that viruses are likely to play an important role in this, because in many cases they can also switch between organisms with different genetic constitutions. In general, the probability of host switching decreases with increasing differences in the genetic features of host organisms (Streicker et al. 2010). Genes are transported and incorporated by viruses between different organisms not only through errors in replication, but apparently also in a targeted fashion (Lang et al. 2012). Capacities of viruses to cross genetic boundaries—apparently acquired over the long term—facilitate not only horizontal gene transfer between various organisms, but also co-evolution between viruses and cellular organisms. This applies not just to prokaryotes (Flores et al. 2013; Koskella and Brockhurst 2014; Blanc et al. 2015) but also to eukaryotes. As already mentioned, in placental mammals the formation of the cellular boundary layer—syncytiotrophoblast—between maternal and embryonic tissue in the placenta is founded on retroviral genes (Rawn and Cross 2008; Lavialle et al. 2013; Cornelis et al. 2015). On the other hand, extensive horizontal gene transfer leads to evolutionary changes in viruses, for example from the transfer of a Levi-like RNA virus from bacteria—via fungi and invertebrates—to plants (Dolja and Koonin 2018).

Properties of viruses are also modified by acquisition of genes and proteins from host cells (Roux et al. 2013). Mutations in viral genomes facilitate conquest of cellular defence systems. In turn, their advantages depend on the flexibility of those defence systems (Kamp et al. 2002; Duffy et al. 2008; Sanjuán 2012; Cuevas et al. 2015). However, the extent of the influences involved is not constant and uniform. It depends both on the intensity of other factors—for example, interspecific consumption—and on interactions between organisms. For example, viral infestation of parasites can reduce the burden on the host organism or enhance the toxic effects of bacteria (Brüssow 2007; Roossinck 2011). However, virus infections can also be triggered by intestinal parasites—for example roundworms (*Ascaris*), threadworms (*Strongyloides*) or whipworms (*Trichuris*) (Walson et al. 2010; Osborne et al. 2014). Complex interactions between viruses and cellular organisms are also evident in attempts at formal description with mathematical networks, which have not yet yielded any unambiguous results (Flores et al. 2011; Weitz et al. 2013). Evidence of the influence of abiotic conditions on interactions between organisms and viruses can be found in aquatic habitats. The physico-chemical properties of water and its flow dynamics enable viruses and unicellular organisms to travel over long distances in three-dimensional space. The chaotic dynamics of currents determine the stochasticity of encounters between the groups mentioned. Because of their high rates of reproduction, viruses can respond rapidly to short-term increases in populations of potential hosts, however, and thus also influence entire marine food chains (see Sect. 9.1) (Avrani et al. 2012; Grossart et al. 2020; Zimmerman et al. 2020; Sadeghi et al. 2021; Kolundžija et al. 2022).

Taking into account the general functional principles of viruses (Fuhrman 1999; Weinbauer 2004; Deveau et al. 2008; Fischer and Suttle 2011; Yau et al. 2011; Lang et al. 2012), it thus becomes clear that they establish:

- a general selection network across all groups of organisms,
- and concurrently a transfer network for horizontal gene transfer.

5.3.3   Societal Effects of Viral Infections

The extent to which viruses can affect human societies is aptly illustrated by the SARS-CoV-2 type coronavarius, simplified as Covid-19, which first appeared in Wuhan (China) in 2019. Its designation indicates a close relationship between this type and the SARS-CoV (Severe Acute Respiratory Syndrome Corona Virus) type first detected in Guangdong (China) in 2002, which was transmitted from horseshoe bats (*Rhinolophus*) to humans via masked palm civets (*Paguma larvata*) (Snijder et al. 2003; Zhong et al. 2003; Cui et al. 2019a). Coronaviruses, whose name refers to crown-like structures seen in two-dimensional electron micrographs belong to the category of enveloped positive-strand RNA viruses (+)RNA in Fig. 5.15. They reproduce in the cytoplasm of host cells through autonomously directed protein reactions. A special feature of coronaviruses is the capacity for self-repair of errors in their genome. This also makes them more resistant to chemical therapeutic measures (Ahlquist 2006; Nicholson and White 2014; Cyranoski 2020; Ren et al. 2020; Robson et al. 2020; Gribble et al. 2021). Covid-19 is the seventh lineage of coronaviruses described since 1966 that is pathogenic for humans. Four virus types (NL63, 229E, OC43, HKU1) cause differing, relatively mild symptoms of rhinitis (resembling the "common cold") in humans. Severe to fatal disease progression can occur with infections by SARS-CoV and MERS (Middle East Respiratory Syndrome) viruses transmitted by dromedaries (camels). With the exception of Covid-19, animal transmission routes—mostly bats and rodents—have been demonstrated for all lineages. While the numbers of people infected worldwide by SARS-CoV and MERS-COV remained relatively low at around 8000 and 1600, respectively, for Covid-19 as of 16 March 2023 they had reached around 760 million. Proportions of reported deaths are significantly lower for infections with Covid-19, at around 1.9 per cent, than for SARS-CoV (around 9 per cent) and MERS-CoV (around 36 per cent) (Su et al. 2016; Cui et al. 2019a, b; Zhu et al. 2020; https://covid19.who.int). However, estimates based on global excess mortality—the difference between observed deaths and average expected deaths—in 2020 and 2021 are about 2.8 times higher (Msemburi et al. 2023). The greater infectivity of Covid-19 compared to SARS-CoV is attributed mainly to specific proteins at the docking sites (spikes), which significantly accelerate penetration into host cells (Chu et al. 2020; Hasan et al. 2020; Hoffmann et al. 2020; Wu and Zhao 2021).

Similar docking proteins are found in influenza A viruses, which are members of the family Orthomyxoviridae (included in group (−)RNA in Fig. 5.15), with comparable infectivity but infection processes differing from those of Covid-19 (Taubenberger and Kash 2010; Pflug et al. 2017; Cyranoski 2020). Disease waves with characteristic flu-like symptoms were already described in antiquity. On several occasions, both the acute effects and the fatal consequences of influenza have influenced the course of history—for example, accession of Henry VIII (1491–1547) to the throne after the death of his brother Arthur from a presumed influenza infection, or the outcome of Napoleon's campaign against Russia. Originally, it was assumed that extraneous influences—for example, celestial bodies or particular weather phenomena—caused the waves of illness. In fact, the microbiologist R. Pfeiffer concluded from the discovery of bacteria in the sputum of flu patients during the severe influenza epidemic of 1889–1892 that "bacilli" were the causal pathogens. This misinterpretation had a disastrous effect on treatment of infected patients during the well-known influenza pandemic of 1918–1920. Sick people—especially soldiers—who had no bacteria in their sputum were designated as malingerers and forced to resume their duties. It was not until 1933 that experiments indirectly revealed that viruses were the causal agent in influenza infections (Winkle 1997; Reid et al. 1999; Cunha 2004).

The increasing efficiency of technological investigation methods—ranging from electron microscopes to genome sequencing—has increased public expectations regarding the controllability of infectious diseases. Disease and death from random epidemics are seen as characteristics of "underdeveloped" societies, a condition that is supposedly to be a thing of the past for "developed" societies—as seen in the perception of ceaseless "higher development". Global data for deaths during influenza pandemics seem to support this view, with between 50 and 80 million people dying from influenza in 1918–1920, between 1 and 2 million in 1957–1958, and between 0.5 and 2 million in 1968–1970 (Murray et al. 2006; Morens and Fauci 2007; Saunders-Hastings and Krewski 2016). However, detailed analyses of influenza infection waves reveal neither regular cycles nor stable genetic characteristics of influenza viruses (Morens and Fauci 2007; Nelson and Holmes 2007; Morens and Taubenberger 2011; Bedford et al. 2015; Dunning et al. 2020). Even vaccinations against influenza viruses still resemble wagers on the genome structure of pathogens that are prevalent at a particular time, in spite of the fact that global monitoring systems have been expanded and rates of vaccine adjustment have been augmemted. Such efforts are thwarted, for instance, by rapid increases in air travel and global transportation of live animals (Saunders-Hastings and Krewski 2016; Trovão and Nelson 2020). Even in wealthy countries, the annual recurrence—to varying extents—of influenza waves fails to trigger marked public reponses. Proposals to reduce the risk of infection through public hygiene measures—such as use of protective masks, distance rules, quarantine or and contact tracing—were not planned in advance (Glass et al. 2006; Weber and Stilianakis 2008). A major factor here is the relatively limited number of deaths causally linked to influenza infections, which is primarily attributable to efficient medical care systems. In the USA, for example, the average flu mortality rate is 0.56 per 100,000 inhabitants, although across the entire American continent included the island states—in the under-65 age group alone—it is 2.1 per 100,000 inhabitants (Doshi 2008; Cheng et al. 2015). In the USA, this value was not attained even during the most recent influenza pandemic in the winter of 2009–2010, when the mortality rate was 0.79 per 100,000 inhabitants (Nguyen and Noymer 2013).

The term "pandemic" is not clearly defined and hence provides no information about the health effects of infectious diseases or infection risks. It is therefore used in publications in different ways—especially in a historical context. The term is used by the World Health Organisation (WHO) in the sense of a multi-stage alert system when transmissible pathogens present a threat, to promote preparations at a national level to cope with potential major burdens on health systems (Ferguson et al. 2006; WHO 2009; Qiu et al. 2017; Shearer et al. 2020). This system is designed to avoid bottlenecks in medical care, during which essential treatments are possible only for a proportion of the total number of people who become ill. In this situation, before initiating treatment, medical staff must divide the sick into groups according to their potential chances of survival (triage). Thereafter, only the group with the best chances of survival receives full treatment; other groups receive only emergency or palliative care. In many cases, medical staff involved are confronted with the ethical challenge of having to make such categorizations for other diseases, such as cancer, at the same time (Christian et al. 2006; Downar and Seccareccia 2010; Dietz et al. 2020; Fusi-Schmidhauser et al. 2020).

Key statistics for epidemic processes provide orienting information, but are far less precise than physical key figures from everyday experience, such as commodity weights. For example, the death of flu patients may be associated with various numbers of additional influencing factors, such as simultaneous bacterial infections, prior illnesses or physical stress. Ultimately, it is a matter of judgement which factor is recorded as the determining cause of death. Current mortality rates do not take into account consequential health damage for survivors, which can lead to an increase in other causes of death, such as heart disease (Estabragh and Mamas 2013; Dalen et al. 2014). Quantitative comparisons between infectious diseases with different spreading histories are associated with even greater uncertainties. In such cases, the current impact of a given disease is influenced not only by various ongoing biological processes but also by the differing experience history of medical personnel. Influenza viruses and Covid-19 viruses change continuously through mutations, which are mostly unobservable. However, their long-term spread within a society depends on the random interaction of various—non-viral—influencing factors (Cohen 2000; Chen et al. 2017b; Islam et al. 2020; Kannan et al. 2020; Korber et al. 2020; Malik et al. 2020; Giandhari et al. 2021; Mohammad et al. 2021; Ozono et al. 2021; Tegally et al. 2021; Vilar and Isom 2021). Prior experience with influenza viruses has shown that changes in viral genomes and their effects cannot be predicted even on the basis of long-term observations and computer-assisted data processing (Morens and Taubenberger 2011; Chen et al. 2017a, b; Dunning et al. 2020; Liu et al. 2020).

An initial critical review clearly reveals the discrepancies between the actual trajectory of Covid-19 infections and its social perception (Caduff 2020). Waves of infections caused by emerging forms of viruses are fundamentally unpredictable and unpreventable. First and foremost, sufficient capacities of medical infrastructure and qualified medical personnel are decisive for the real impact of such infection waves in any society. But such a prerequisite is undesirable from a neoliberal viewpoint because, outside times of crisis, it can be presented as economically "inefficient". This discrepancy clearly reveals the general weaknesses of efficiency strategies for coping with—this and other—unforeseeable crises. Driven by numerical fetishism and blind faith in models, the effects of inadequate capacities can trigger massive social over-reactions. This statement is illustrated in Fig. 5.16, which shows a superimposition of standardized cumulative curves for global deaths over confirmed cases of infection, based on data from the Covid-19 WHO dashboard. During the first year of the pandemic (a to a*), there is evidently a more rapid increase in deaths than in cases of infection—as illustrated by the linear extrapolation lines. However, this phenomenon is based not on exact measurements, but on influences exerted by a wide variety of factors. The figures given for deaths are influenced, among other things, by the problems already mentioned facing determination of causes of death, and also by differences in emergency capacities of medical systems in individual countries. Figures for cases of infection may be incomplete at the beginning of a pandemic because of inadequate knowledge and a dearth of methods for recording them. Uncritical adoption of such figures in models inevitably leads to overestimates of expected mortality rates. Dissemination of such information then triggers irrational fears in the societies affected, which can lead to the weakening of democratic structures in favour of authoritarian systems. Because viral pathogens are not recognizable or even perceptible to humans, vague references to "expert knowledge" in society also promote scientific scepticism. Such effects are reinforced with increasing success in the treatment of diseases and increasing immunization in societies, because the numbers for deaths decrease more

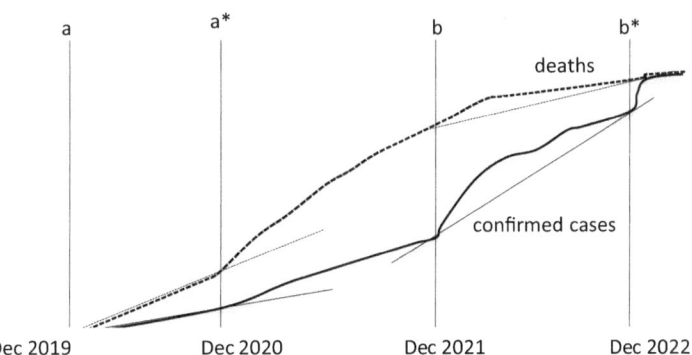

Fig. 5.16 Comparison of standardized global curves for cumulative deaths and confirmed cases of infection based on WHO Dashboard data as of 03/16/2023 (https://covid19.who.int). Explanations in the text

rapidly than those for cases of infection (Fig. 5.16 section b-b*)—and in this way original "scientific" statements are refuted.

5.4 A Change of Perspective

5.4.1 Species: An Outdated Concept?

Before answering this question, it is necessary to clarify what is expected from this word. Does it refer to unambiguous designation of entities or to a concept that is fundamental for the understanding of evolutionary processes? In the first case, the continuing success of the binomial system of nomenclature for plants and animals—originally published by Carl von Linné in 1758 under the title "Systema Naturae"—demonstrates its importance for scientific communication. An independent reference system was defined by specifying designations of genus and species with Latin or Greek names and referring to a preserved voucher specimen for the first description. Principally, this created order in biological collections and comparability between natural history repositories. As a secondary benefit, the concept facilitated description and recording of morphologically distinguishable and visually recognizable organisms in field research—for example, ecological investigations. In this context, difficulties in applying the concept are already apparent. Variation in individual characteristics renders assignment of freshly collected organisms to defined species problematic. Horizontal gene transfer, hybridization, different forms of reproduction, allopolyploidy and epigenetic modifications increase the morphological variability of organisms and increase the practical difficulties encountered in clearly delimiting species (Birky 1996; Koivisto and Braig 2003; Birky et al. 2005; Willis et al. 2006; Soltis and Soltis 2009; Liu et al. 2010, 2011; Albertin and Marullo 2012; Fontaneto et al. 2012; Fontaneto and Barraclough 2015; Gomaa et al. 2015; Tice et al. 2016). This problem has been countered with alternative definitions of species and methods for their delimitation (De Queiroz 2007). Furthermore, a multitude of parallel studies repeatedly leads to numerous descriptions of the same species—a constant challenge for nomenclatural agencies (Adams et al. 2017; Adl et al. 2019).

Approaches using electronic data processing to tackle the problems encountered permit storage and analysis of huge amounts of data. This does not solve the question of the relative—qualitative—significance of different characteristics for the definition of species. This problem was clearly recognizable—and equally clearly addressed—in the context of morphological classification (Riedl 1975). Supported by the success of quantitative analyses of RNA sequences to ascertain phylogenetic relationships (Woese 2000), the methodological focus shifted to quantitative recording of gene sequences. This development was reinforced by a rapid increase in the technical capacities of automated gene sequencing and downstream evaluation procedures (Merkl 2015). Studies in this realm increasingly revealed the inherent ambiguities of life forms and the great evolutionary significance of unicellular organisms (Lake and Rivera 2004; Ciccarelli et al. 2006; Taylor et al. 2006; O'Malley and Koonin 2011; Woyke and Rubin 2014; Hug et al. 2016). Pronounced genetic diversity of morphologically indistinguishable types of organisms (cryptic species) has also become evident (Fernandez et al. 2006; Šlapeta et al. 2006; Amato et al. 2007; Trontelj and Fišer 2009; Poulin 2010; Lumbsch and Leavitt 2011; Funk et al. 2012; Hawksworth 2012; Carstens and Satler 2013). The long-cherished paradigm of the phylogenetic tree (Mayr 1984) is increasingly dissolving into compartments with divergent assignments of genetically similar life forms into hierarchical frameworks with different designations (Roselló-Mora and Amann 2001; Embley and Martin 2006; Kondrashov et al. 2006; Lecointre and Le Guyader 2006; Richter and Roselló-Mlóra 2009; Lauber and Gorbalenya 2012; Groenewald et al. 2012; Ramulu et al. 2012; Nosenko et al. 2013; Kang et al. 2017; Adl et al. 2019). Ultimately, decisions regarding where species are delineated or to which groups they should be allocated always depend on a human perspective. For example, in the scientific literature *Homo sapiens* and *Homo neanderthalensis* have been described as separate species, although frequent interbreeding—also including the Denisovan hominid—has been inferred from genetic analyses of fossil specimens (Abi-Rached et al. 2011; Kuhlwilm et al. 2016; McCoy et al. 2017; Dannemann and Racimo 2018; Slon et al. 2018). There are convincing indications of continual intermingling and differentiation of genomes in the genus *Homo* (Genome Project Consortium 2015; Schlebusch et al. 2017). Due to our comparatively limited observation window—and because we are merely drifting on a present-day sea of change—only current conditions are perceived. If we were to apply standards for species delimitation similar to those we use for marine cyanobacteria of the genus *Prochlorococcus*, we would at least be grouped together with chimpanzees and bonobos—if not also with gorillas and orangutans—in a single species (Wildman et al. 2003; Moldovan and Gelfand 2018).

Discussions of the evolution of organisms largely ignore the phenomenon of *hybridization*. In contrast to horizontal gene transfer, two originally separate developmental lines of multicellular organisms may merge to yield novel life forms. In general, such phenomena are known mainly from animal and plant breeding, during which new life forms with improved, desirable performance characteristics are bred from genetically closely related predecessors—usually referred to as breeds. In fact, however, hybridization occurs as a naturally occurring process of adaptation to changing environmental conditions. An estimated 10–30 per cent of all multicellular organisms regularly produce hybrids. According to these estimates, the probability of hybridization is greater than that of changes due to mutations with estimated magnitudes of 10^{-8}–10^{-9} per base pair and generation (Abbot et al. 2013).

Those with botanical interests are probably familiar with the many hybrid forms of willow trees (*Salix*) (Aichele and Schwegler 1995). Hybridization is probably widespread, especially among flowering plants, and may, for instance, facilitate the colonization of new habitats (Ellstrand and Schierenbeck 2000; Salmon et al. 2005; Paun et al. 2009; Chase et al. 2010; Arnold et al. 2012). Parasitic fungi can adapt more easily to new host organisms through hybridization and hence spread rapidly. For example, the widespread dieback of elm trees (Dutch elm disease) towards the end of the twentieth century was triggered by a hybrid form of fungus (Brasier 2001; Cox et al. 2014; Depotter et al. 2016). Among animals, hybrids are found in almost all groups, including corals, insects, amphibians, lizards, birds and mammals. For example, hybridization with extinct Irish brown bears can be traced through the female line of polar bears (Hotz et al. 1999; Willis et al. 2006; Bi and Bogart 2010; Edwards et al. 2011; Smith and Kronforst 2013; Canestrelli et al. 2016; Jančúchová-Lásková et al. 2015; Olave et al. 2018; Ottenburghs 2019; Leighton et al. 2021). Hybridisation also played an essential role in the evolution of primates, such as guenons, gorillas or humans (Ackermann et al. 2006; Ackermann and Bishop 2009; Roos et al. 2011; Vernot et al. 2016). Taking into account the fact that even eukaryotic cells and their mitochondria show clear characteristics of hybridization, it appears likely that this is an intrinsic feature of evolutionary processes (Williamson 2006; Abbot et al. 2013; McInerney et al. 2014; Gray 2015; Méheust et al. 2015).

This also makes it clear that evolutionary processes do not necessarily take place in a tree-like branching manner, with continual occurrence of new divergences, but must involve networks at different levels. This conclusion is indirectly confirmed by the past fruitless search for "archetypes" of organisms. Despite the avalanche of data from genomes and proteomes (total sets of proteins in individual organisms), uncertainties in determining evolutionary lineages are increasing, not decreasing. An example of this is provided by the comparison of two overviews of classifications of eukaryotes, the first in 2012 and the second in 2019 (Adl et al. 2012, 2019). Whereas in the earlier representation the developmental lines are still depicted as a clearly branched "tree" with a prokaryotic "root", the current representation is reminiscent of a weakly structured bladderwrack frond stranded on a beach. The problem of classificatory uncertainty and phylogenetic lineages (Krell and Cranston 2004; King and Rokas 2017), repeatedly addressed in publications, will be solved only by abandoning the phylogenetic tree paradigm and algorithms based on it, not by larger datasets. An improved understanding of biological evolution might possibly be expected from applying methodological approaches of network analysis (Yamasa and Bork 2009; Huson and Scornavacca 2010; Koonin and Dolja 2014; Sorrells and Johnson 2015).

The main reasons for variability between individual organisms in a given species are their multi-dimensional characteristics and continuous feedback interactions—both biotic and abiotic—with the environment (Fig. 5.17). Evolution is hence not an endogenous process, but is based on interactions between endogenous and exogenous factors. As a result, rates and characteristics of morphological or physiological change are always subject to influences exerted by prevailing environmental conditions. *No organism can exist without interactions with the environment or reproduce itself over the long term without associated feedback. Explicit reference to the individual organism is intentional, to make it clear that—even within a group of similar organisms—different conditions exist for each individual.*

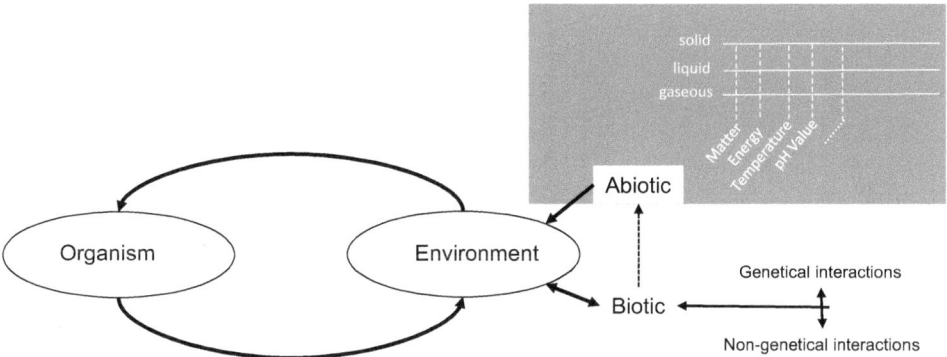

Fig. 5.17 Schematic representation of the central basic principle of the evolution of organisms, with indications of the potential range of different biotic and abiotic framework conditions

5.4.2   Simultaneous Go: A Concept for Evolutionary Processes

Intensity and dynamics of interactions are related, among other things, to adaptability of individual organisms and the probability of inheritance of epigenetically acquired changes during reproduction. Accordingly, prokaryotes can adapt more rapidly to environmental change than multicellular animals. As food for thought, when looking at the front cover image, an attempt should be made to answer the question as to how—without any technical aids or laboratories—organisms can evolve under the general conditions depicted. From the perspective of our technologized living conditions, it is easy to overlook the fact that evolution—especially for organisms with limited capacities for adaptation—depends on the functional networks of a wide variety of life forms. For example, mammals could only evolve in terrestrial ecosystems (see Chap. 10), and they were only able to colonize marine habitats—secondarily—from their original land-based setting. At the same time, it is evident that for a very long period of geological time early mammalian life forms lived in reptile-dominated systems. Mammal-dominated terrestrial systems could only develop after the demise of large-bodied reptiles at the end of the Cretaceous (Stanley 2001).

Such effects of changed framework conditions can be approximately described with reference to a modified version of the game Go. In the classic version of the game, two players take turns placing their stones. In real communities of organisms, however, a far greater number of "players" act simultaneously with different participation rates on constantly changing playing fields—"Simultaneous Go with many players on playing fields with random changes" (Fig. 5.18). In this game, the chances of survival depend not only on the selective influence of major disturbances (Selective Random Events—SREs), but also on the intrinsic evolutionary capacity of the communities of organisms that survive together. To understand biological processes, delineation of "playing fields" by different local and temporary combinations of biotic and abiotic factors is important. For terrestrial systems, the lichen growth shown in Fig. 5.11 provides an initial impression of the two-dimensional, complex spatial structures of the "playing fields". In open water zones of

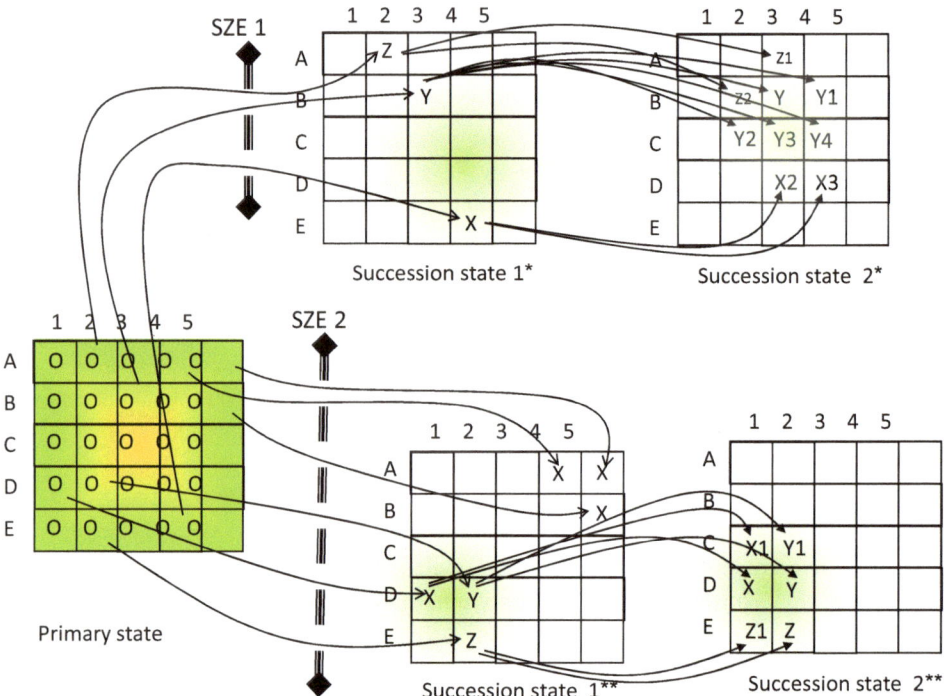

Fig. 5.18 Schematic sketch comparing the effects of selective random events with either high (SRE 1) or low (SRE 2) selectivity on survival rates (succession states 1) of populations of an original community of organisms (initial state). However, longer-term development of individual populations depends on whether suitable conditions—grey markers—exist for them (subsequent state 2′) or not (subsequent state 2″)

aquatic systems, phytoplankton forms similarly structured "playing fields" for relatively short intervals of time (Gallager et al. 1996; Campbell et al. 1997). In reality, however, terrestrial and aquatic systems are always three-dimensional variable structures.

It is highly probable that prokaryotes not only prepare "playing fields", but—when major environmental changes occur—also secure the conditions for new developments following the destruction of previous systems. This interpretation is supported by the marked resistance of prokaryotes to environmental influences, their role in primary successions, and, on the other hand, long-term preservation of regional genetic groups (Grootjans et al. 1997; Belnap and Lange 2003; Dion and Nautiyal 2008). The latter phenomenon has been demonstrated for cyanobacteria of the *Chroococcidiopsis* group inhabiting arid regions (Bahl et al. 2011). Samples collected on a global basis from cold and warm dry regions revealed clear regional differences in ribosomal RNA of these organisms. The results of the phylogenetic analyses indicate that those differences are based on evolutionarily extremely ancient differentiation processes. In fact, differentiation into cold- and warm-adapted life forms probably took place more than 2 billion years ago. In the temporal patterns of geographical differentiation, similarities are virtually absent with respect to the formation dynamics of recent arid regions over the last 100 million years, but are seen instead in the context of far more ancient dynamics of continent formation.

Simple models are typically used to describe organismal interactions, for example the well-known Lotka-Volterra model of predator-prey relationships. Examples of ecosystems that are functionally greatly limited—such as interactions between populations of lemmings and snowy owls—serve to confirm results from models (Schwerdtfeger 1975). Despite their validity for describing interaction processes under the given framework conditions, for several reasons such models lead to incorrect ways of approaching a general understanding of evolutionary processes. One reason resides in their close correspondence with our thought patterns—the model assumptions can be plausibly presented and the results are easily interpretable. A second reason lies in the illusion of a generally valid description of interactions between organisms. But how could peacocks with such conspicuous male plumage evolve under such conditions? Why is it possible to recognise long-term—but spatially different—combinations of organisms in terrestrial communities? One-dimensional and linear approaches do not provide adequate answers to evolutionary questions. Even approximate solutions require—in addition to a genetic perspective—at least consideration of information processes as well as material and energy flows. This also makes it clear that isolated consideration of multicellular organisms fails to provide sufficient insights into evolutionary processes. Apart from abiotic factors, regulation of material and energy flows has been based on the activities of microorganisms—especially prokaryotes (Falkowski et al. 2008; Jiao et al. 2011b; Madsen 2011; Lladó et al. 2017; Noble and Noble 2017)—ever since biological evolution began. Once again, however, the question arises as to why, over the course of the Earth's long geological history, organisms did not all become extinct long ago?

From a constructivist perspective, answers to those diverse questions would have to begin with the properties and characteristics of the "playing fields". However, following the supposition that interested readers are more likely to be familiar with biological topics, the next two chapters deal primarily with the "players".

References

Abbot DS, Voigt A, Li D, Le Hir G, Pierrehumbert RT et al (2013) Robust elements of Snowball Earth atmospheric circulation and oases for life. J Geophys Res Atmos 118:1–11

Abe J, Hiyama TB, Mukaiyama A, Son S, Mori T et al (2015) Atomic-scale origins of slowness in the cyanobacterial circadian clock. Science 349:312–316

Abergel C, Legendre M, Claverie J-M (2015) The rapidly expanding universe of giant viruses: mimivirus, pandoravirus, pithovirus and mollivirus. FEMS Microbiol Rev 39:779–796

Abi-Rached L, Jobin MJ, Kularni S, McWhinnie A, Dalva K et al (2011) The shaping of modern human immune systems by multiregional admixture with archaic humans. Science 334:89–94

Abrahão J, Silva L, Silva LS, Khalil JYB, Rodriguez R et al (2018) Tailed giant Tupanvirus possesses the most complete translational apparatus of the known virosphere. Nat Commun. https://doi.org/10.1038/s41467-018-03168-1

Ackermann RR, Bishop JM (2009) Morphological and molecular evidence reveals recent hybridization between gorilla taxa. Evolution 64(1):271–290

Ackermann RR, Rogers J, Cheverud JM (2006) Identifying the morphological signatures of hybridization in primate and human evolution. J Hum Evol 51(6):632–645

Adams MJ, Lefkowitz EJ, King AMQ, Harrach B, Harrison RL et al (2017) Changes to taxonomy and the International Code of Virus Classification and Nomenclature ratified by the International Committee on Taxonomy of Viruses (2017). Arch Virol 162:2505–2538

Adl SM, Simpson AGB, Lane CE, Lukeš J, Bass D et al (2012) The revised classification of eukaryotes. J Eukaryot Microbiol 59(5):429–493

Adl SA, Bass D, Lane CE, Lukes J, Schoch CL et al (2019) Revisions to the classification, nomenclature, and diversity of eukaryotes. J Eukaryot Microb 66:4–119

Aherfi S, Colson P, La Scola B, Raoult D (2016) Giant viruses of amoebas: an update. Front Microbiol 7:349

Ahlquist P (2006) Parallels among positive-strand RNA viruses, reverse-transcribing viruses and double-stranded RNA viruses. Nat Rev Microbiol 4:371–382

Aichele D, Schwegler H-W (1995) Die Blütenpflanzen Mitteleuropas. Frankh-Kosmos, Stuttgart

Ainsworth T, Krause L, Bridge T, Torda G, Baptise-Raina I et al (2015) The coral core microbiome identifies rare bacterial taxa as ubiquitous endosymbionts. ISME J 9:2261–2274

Albers SV, Jarrell KF (2015) The archaellum: how Archaea swim. Front Microbiol 6:23

Albertin W, Marullo P (2012) Polyploidy in fungi: evolution after whole-genome duplication. Proc Biol Sci 279:2497–2509

Alberts B, Bray D, Lewis J, Raff M, Roberts K, Watson JD (1986) Molekularbiologie der Zelle. VCH, Weinheim

Alberts B, Johnson A, Lewis J, Morgan D, Raff M et al (2015) Molecular biology of the cell. Garland Science, New York

Alewsakun P, Katzourakis A (2017) Marine origin of retroviruses in the early palaeozoic era. Nat Commun. https://doi.org/10.1038/ncomms13954

Allen JF (2015) Why chloroplasts and mitochondria retain their own genomes and genetic systems: colocation for redox regulation of gene expression. PNAS 112(33):10231–10238

Allers T, Mevarech M (2005) Archaeal genetics—the third way. Nat Rev Genet 6:58–73

Al-Shayeb B, Skopintsev P, Soczek KM, Stahl EC, Li Z et al (2022) Diverse virus-encoded CRISPR-Cas systems include streamlined genome editors. Cell 185:4574–4586

Altiero T, Giovannini B, Guidetti R, Rebecchi L (2015) Life history traits and reproductive mode of the tardigrade *Acutuncus antarcticus* under laboratory conditions: strategies to colonize the Antarctic environment. Hydrobiologia. https://doi.org/10.1007/s10750-015-2315-0

Amaral PP, Dinger ME, Mercer TR, Mattick JS (2008) The eukaryotic genome as an RNA machine. Science 319:1787–1789

Amato A (2010) Diatom reproductive biology: living in a crystal cage. Int J Plant Reprod Biol 2(1):1–10

Amato A, Kooistra WHCF, Ghiron JHL, Mann DG, Pröschold T, Montresor M (2007) Reproductive isolation among sympatric cryptic species in marine diatoms. Protist 158:193–207

Aminoc RI (2011) Horizontal gene exchange in environmental microbiota. Front Microbiol 2:158

Amital G, Sorek R (2017) Intracellular signaling in CRISPR-Cas defense. Science 357:550–551

Amodeo AA, Skotheim JM (2016) Cell-size control. Cold Spring Harb Perspect Biol 8:a019083

Anantharaman V, Iyer LK, Aravind L (2007) Comparative genomics of protists: new insights into the evolution of eukaryotic signal transduction and gene regulation. Annu Rev Microbiol 61:453–475

Anderson RE, Brazelton WJ, Baross JA (2011) Is the genetic landscape of the deep subsurface biosphere affected by viruses? Front Microbiol 2:219

Anger AM, Armache J-P, Berninghausen O, Habeck M, Subklewe M et al (2013) Structures of the human and *Drosophila* 80S ribosome. Nature 497:80–85

Angert ER (2005) Alternatives to binary fission in bacteria. Nat Rev Microbiol 3:214–224

Angert ER (2006) The enigmatic cytoarchitecture of *Epulopiscium* spp. Microbiol Monogr. https://doi.org/10.1007/7171_027

Angly FE, Felts B, Breitbart M, Salamon P, Edwards RA et al (2006) The marine viromes of four oceanic regions. PLoS Biol 4(11):e368

Anisimov AP (2005) Endopolyploidy as a morphogenetic factor of development. Cell Biol Int 2009:993–1004

Anisimov AP, Zyumchenko NE (2012) Evolutionary regularities of development of somatic polyploidy in salivary glands of gastropod mollusks: V. Subclasses opisthobranchia and pulmonata. Cell Tissue Biol 6(3):268–279

Antony-Babu S, Deveau A, Van Nostrand JD, Zhou J, Le Tacon F et al (2013) Black truffle-associated bacterial communities during the development and maturation of *Tuber melanosporum* ascocarps and putative functional roles. Environ Microbiol. https://doi.org/10.1111/1462-2920.12294

Aramayo R, Selker EU (2013) *Neurospora crassa*, a model system for epigenetics research. Cold Spring Harb Perspect Biol 5:a017921

Araque A, Carmignoto G, Haydon PG, Oliet SHR, Robitaille R, Volterra A (2014) Gliotransmitters travel in time and space. Neuron 81:728–739

Archetti M (2004) Loss of complementation and the logic of two-step meiosis. J Evol Biol 17:1098–1105

Archetti M (2010) Complementation, genetic conflict, and the evolution of sex and recombination. J Hered 101:S21–S33

Archibald JM (2009) The puzzle of plastid evolution. Curr Biol 19:R81–R88

Archie EE, Tung J (2015) Social behaviour and the microbiome. Curr Opin Behav Sci 6:28–34

Arkhipova IR, Rodriguez F (2013) Genetic and epigenetic changes involving (retro)transposons in animal hybrids and polyploids. Cytogenet Genome Res 140:295–311

Arnaud F, Caporale M, Varela M, Biek R, Chessa B, Alberti A et al (2007) A paradigm for virus-host coevolution: sequential counter-adaptations between endogenous and exogenous retroviruses. PLoS Pathog 3(11):e170

Arnaud-Haond S, Duarte CM, Diaz-Almela E, Marbà N, Sintes T, Serrão EA (2012) Implications of extreme life span in clonal organisms: millenary clones in meadows of the threatened seagrass *Posidonia oceanica*. PLoS One 7(2):e30454

Arnold ML, Sapir Y, Martin NH (2008) Genetic exchange and the origin of adaptations: prokaryotes to primates. Philos Trans R Soc Lond B Biol Sci 363:2813–2820

Arnold ML, Ballerini ES, Brothers AN (2012) Hybrid fitness, adaptation and evolutionary diversification: lessons learned from Louisiana Irises. Heredity 108:159–166

Arrowsmith CH, Bountra C, Fish PV, Lee K, Schapira M (2012) Epigenetic protein families: a new frontier for drug discover. Nat Rev Drug Discov 11:384–400

Asher G, Schibler L (2011) Crosstalk between components of circadian and metabolic cycles in mammals. Cell Metab 13:125–137

Aswad A, Katzourakis A (2012) Paleovirology and virally derived immunity. Trends Ecol Evol 27(11):627–636

Atici Ö, Nalbantoğlu B (2003) Antifreeze proteins in higher plants. Phytochemistry 64:1187–1196

Austad SN, Fischer KE (1991) Mammalian aging, metabolism, and ecology: evidence from the bats and marsupials. J Gerontol 46(2):B47–B53

Avrani S, Schwartz DA, Lindell D (2012) Virus-host swinging party in the oceans. Mob Genet Elements 2(2):88–95

Azcón-Aguilar C, Barea JM, Gianinazzi S, Gianinazzi-Pearson V (eds) (2009) Mycorrhizas. Springer, Berlin

Babcock RC, Bull GD, Harrison PL, Heyward AJ, Oliver JK et al (1986) Synchronous spawning of 105 scleractinian coral species on the Great Barrier Reef. Mar Biol 90:379–394

Bahl J, Lau MCY, Smith GJD, Vijaykrishna D, Cary SC et al (2011) Ancient origins determine global biogeography of hot and cold desert cyanobacteria. Nat Commun. https://doi.org/10.1038/ncomms1167

Balázsi G, van Oudenaarden A, Collins JJ (2011) Cellular decision making and biological noise: from microbes to mammals. Cell 144:910–925

Bansai MS, Banay G, Harlow TJ, Gogarten JP, Shamir R (2013) Systematic inference of highways of horizontal gene transfer in prokaryotes. Bioinformatics 29(5):571–579

Baraquet C, Théraulaz L, Iobbi-Nivol C, Méjean V, Jourlin-Castelli C (2009) Unexpected chemoreceptors mediate energy taxis towards electron acceptors in Shewanella oneidensis. Mol Microbiol 73(2):278–290

Barazandeh M, Davis CS, Neufeld CJ, Coltman DW, Palmer AR (2013) Something Darwin didn't know about barnacles: spermcast mating in a common stalked species. Proc Biol Sci 280:20122919

Barbara GM, Mitchell JG (2003) Bacterial tracking of motile algae. FEMS Microbiol Ecol 44:79–87

Bardy SL, Ng SYM, Jarrell KF (2003) Prokaryotic motility structures. Microbiology 149:295–304

Barry R, Gitai Z (2011) Self-assembling enzymes and the origins of the cytoskeleton. Curr Opin Microbiol 14(6):704–711

Barske LA, Capel B (2008) Blurring the edges in vertebrate sex determination. Curr Opin Genet Dev 18(6):499–505

Bartocci E, Corradini F, Merelli E, Tesei L (2010) Detecting synchronisation of biological oscillators by model checking. Theor Comput Sci 411:1999–2018

Bass J, Takahashi JS (2010) Circadian integration of metabolism and energetics. Science 330:1349–1354

Baßler J, Hurt E (2019) Eukaryotic ribosome assembly. Annu Rev Biochem 88:281–306

Bassler BL, Losick R (2006) Bacterially speaking. Cell 125:237–246

Bastide A, David A (2018) The ribosome, (slow) beating heart of cancer (stem) cell. Oncogenesis. https://doi.org/10.1038/s41389-018-0044-8

Battle A, Khan Z, Wang SH, Mitrano A, Ford MJ et al (2015) Impact of regulatory variation from RNA to protein. Science 347:664–667

Beardall J, Allen D, Bragg J, Finkel ZV, Flynn KJ et al (2009) Allometry and stoichiometry of unicellular, colonial and multicellular phytoplankton. New Phytol 181:295–309

Beaulieu JM, Leitch IJ, Patel S, Pedharkar A, Knight CA (2008) Genome size is a strong predictor of cell size and stomatal density in angiosperms. New Phytol 179:975–986

Beck A, Divakar PK, Zhang N, Molina MC, Struwe L (2015) Evidence of ancient horizontal gene transfer between fungi and the terrestrial alga *Trebouxia*. Org Divers Evol 15:235–248

Beckmann L, Edel KH, Batistič O, Kudla J (2016) A calcium sensor-protein kinase signalling module diversified in plants and is retained in all lineages of Bikonta species. Sci Rep. https://doi.org/10.1038/srep.31645

Bedford T, Riley S, Barr JG, Broor S, Chadha M et al (2015) Global circulation patterns of seasonal influenza viruses vary with antigenic drift. Nature 523:217–220

Beekman M, Nieuwenhuis B, Ortiz-Barrientos D, Evans JP (2016) Sexual selection in hermaphrodites, sperm and broadcast spawners, plants and fungi. Philos Trans R Soc Lond B Biol Sci 371:20150541

Beiko RG, Harlow TJ, Ragan MA (2005) Highways of gene sharing in prokaryotes. PNAS 102(40):14332–14337

Bekliz M, Colson P, La Scola B (2016) The expanding family of virophages. Viruses 8:317

Belkaid Y, Segre JA (2014) Dialogue between skin microbiota and immunity. Science 346:954–959

Bellet MM, Sassone-Corsi P (2010) Mammalian circadian clock and metabolism—the epigenetic link. J Cell Sci 123(22):3837–3848

Belnap J, Lange OL (eds) (2003) Biological soil crusts: structure, function, and management, Ecological studies, vol 150. Springer, Berlin

Berendsen RL, Pieterse CMJ, Bakker PAHM (2012) The rhizosphere microbiome and plant health. Trends Plant Sci 17(8):478–486

Berg HC (2003) The rotary motor of bacterial flagella. Annu Rev Biochem 72:19–54

Berg HC (2008) Bacterial flagellar motor. Curr Biol 18(16):R689–R691

Berg JM, Tymoczko JL, Stryer L (2013) Biochemie. Springer Spektrum, Berlin

Bergthorsson U, Adams KL, Thomason B, Palmer JD (2003) Widespread horizontal transfer of mitochondrial genes in flowering plants. Nature 424:197–201

Bergthorsson U, Andersson DI, Roth JR (2007) Ohno's dilemma: evolution of new genes under continuous selection. PNAS 104(43):17004–17009

Bernales S, Papa FR, Walter P (2006) Intracellular signaling by the unfolded protein response. Annu Res Cell Dev Biol 22:487–408

Bernander R (1998) Archaea and the cell cycle. Mol Microbiol 29(4):955–961

Bernander R, Poplawski A (1997) Cell cycle characteristics of thermophilic archaea. J Bacteriol 179(16):4963–4969

Bernandez R, Lind AE, Ettema TJG (2011) An archaeal origin for the actin cytoskeleton. Mol Microbiol 80:1052–1061

Berry J-L, Pelicic V (2015) Exceptionally widespread nanomachines composed of type IV pilins: the prokaryotic Swiss Army knives. FEMS Microbiol Rev 39:134–154

Berthold T, Centler F, Hübschmann T, Remer R, Thullner M et al (2016) Mycelia as a focal point for horizontal gene transfer among soil bacteria. Sci Rep. https://doi.org/10.1038/srep36390

Bhowmick S, Tripathy S (2014) A tale of effectors; their secretory mechanisms and computational discovery in pathogenic, non-pathogenic and commensal microbes. Mol Biol 3:1

Bi K, Bogart JP (2010) Probing the meiotic mechanism to intergenomic exchanges by genomic in situ hybridization of lambrush chromosomes of unisexual *Ambystoma* (Amphibia: Caudata). Chromosome Res 18:371–382

Bielska E, Higuchi Y, Schuster M, Steinerg N, Kilaru S et al (2014) Long-distance endosome trafficking drives fungal effector during plant infection. Nat Commun. https://doi.org/10.1038/ncomms6097

Bier E, De Robertis EM (2015) BMP gradients: a paradigm for morphogen-mediated developmental patterning. Science 348:aaa5838

Biers EJ, Wang K, Pennington C, Belas R, Chen F, Moran MA (2008) Occurrence and expression of gene transfer agent genes in marine bacterioplankton. Appl Environ Microbiol 74(10):2933–2939

Biller SJ, Berube PM, Lindell D, Chisholm SW (2014) Prochlorococcus: the structure and function of collective diversity. Nat Rev Microbiol 13(1):13–27

Billiard S, López-Villavicencio M, Devier B, Hood ME, Fairhead C, Giraud T (2010) Having sex, yes, but with whom? Inferences from fungi on the evolution of anisogamy and mating types. Biol Rev 86(2):421–442

Billiard S, López-Villavicencio M, Hood ME, Giraud T (2012) Sex, outcrossing and mating types: unsolved questions in fungi and beyond. J Evol Biol 25:1020–1038

Binder BJ, Chisholm SW (1995) Cell cycle regulation in marine Synechococcus sp. strains. Appl Environ Microbiol 61(2):708–717

Bininda-Emonds ORP, Hinz C, Ahlrichs WH (2016) Evidence supporting the uptake and genomic incorporation of environmental DNA in the "ancient asexual" bdelloid Rotifer *Philodina roseola*. Life 6:38

Birky CW (1996) Heterozygosity, heteromorphy, and phylogenetic trees in asexual eukaryotes. Genetics 144:427–437

Birky CW (2009) Giardia sex? Yes, but how and how much? Trends Parasitol 26(2):70–74

Birky CW, Wolf C, Maughan H, Herbertson L, Henry E (2005) Speciation and selection without sex. Hydrobiologia 546:29–45

Bishop JDD, Pemberton AJ (2006) The third way: spermcast mating in sessile marine invertebrates. Integr Comp Biol 46(4):398–406

Blake WJ, Kærn M, Cantor CR, Collins JJ (2003) Noise in eukaryotic gene expression. Nature 422:633–637

Blanc G, Gallot-Lavallée L, Maumus F (2015) Provirophages in the Bigelowiella genome bear testimony to past encounters with giant viruses. PNAS:E5318–E5326

Blaxter M, Mann J, Chapman T, Thomas F, Whitton C et al (2005) Defining operational taxonomic units using DNA barcode data. Philos Trans R Soc Lond B Biol Sci 360:1935–1943

Blažek R, Polačik M, Reichard M (2013) Rapid growth, early maturation and short generation time in African annual fishes. EvoDevo 4:24

Blažek R, Polačik M, Kačer P, Cellerino A, Režucha R et al (2016) Repeated intraspecific divergence in life span and aging of African annual fishes along an aridity gradient. Evolution 71(2):386–402

Bloomfield G (2011) Genetics of sex determination in the social amoeba. Dev Growth Differ 53:608–616

Boch CA, Ananthasubramaniam B, Sweeney AM, Doyle FJ, Morse DE (2011) Effects of light dynamics on coral spawning synchrony. Biol Bull 220:161–173

Bock R (2009) The give-and-take of DNA: horizontal gene transfer in plants. Trends Plant Sci 15(1):11–22

Bock LV, Kolář MH, Grubmüller H (2018) Molecular simulations of the ribosome and associated translation factors. Curr Opin Struct Biol 49:27–36

Bode SNS, Adolfsson S, Lamatsch DK, Martins MJF, Schmit O et al (2019) Exceptional cryptic diversity and multiple origins of parthenogenesis in a freshwater ostracod. Mol Phylogenet Evol 54:542–552

Bohn A, Hinderlich S, Hütt M-T, Kaiser F, Lüttge E (2003) Identification of rhythmic subsystems in the circadian cycle of crassulacean acid metabolism under thermoperiodic perturbations. Biol Chem 384:721–728

Bohnsack KE, Bohnsack MT (2019) Uncovering the assembly pathway of human ribosomes and its emerging links to disease. EMBO J 38:e100278

Bonduriansky R, Day T (2009) Nongenetic inheritance and its evolutionary implications. Annu Rev Ecol Evol Syst 40:103–125

Bordenstein SR, Reznikoff WS (2005) Mobile DNA in obligate intracellular bacteria. Nat Micorbiol 3:688–699

Bordenstein S, Theis KR (2015) Host biology in light of the microbiome: ten principles of holobionts and hologenomes. PLoS Biol 13(8):e1002226

Bosch TCG (2013) Cnidarian-microbe interactions and the origin of innate immunity in metazoa. Annu Rev Microbiol 67:499–518

Bose J, Pottosin II, Shabala SS, Palmgren MG, Shabala S (2011) Calcium efflux systems in stress signalling and adaptation in plants. Front Plant Sci 2:85

Boto L (2010) Horizontal gene transfer in evolution: facts and challenges. Proc Biol Sci 277:819–827

Boughalmi M, Pagnier I, Aherfi S, Colson P, Raoult D, La Scola B (2013) First isolation of a giant virus from wild *Hirudo medicinalis* leech: *Mimiviridae* Isolation in *Hirudo medicinalis*. Viruses 5:2920–2930

Bowman JL, Kohchi T, Yamato KT, Jenkins J, Shu S et al (2017) Insights into land plant evolution garnered from the *Marchantia polymorpha* genome. Cell 171:287–304

Bowman JC, Petrov AS, Frenkel-Pinter M, Penev PI, Williams LD (2020) Root of the tree: the significance, evolution, and origins of the ribosome. Chem Rev 120(11):4848–4878

Boyle EA, Li YI, Pritchard JK (2017) An expanded view of complex traits: from polygenic to omnigenic. Cell 169:1177–1180

Bracht JR, Fang W, Goldman AD, Dolzhenko E, Stein EM, Landweber LF (2013) Genomes in the edge: programmed genome instability in ciliates. Cell 152:406–416

Brady A, Felipe-Ruiz A, Galego del Sol F, Marina A, Quiles-Puchalt N, Penadés JR (2021) Molecular basis of lysis-lysogeny decisions in gram-positive phages. Annu Rev Microbiol 73:26.1–26.19

Bragina A, Oberauner-Wappis L, Zachow C, Halwachs B, Thallinger GG et al (2014) The Sphagnum microbiome supports bog ecosystem functioning under extreme conditions. Mol Ecol 23:4498–4510

Brandes N, Linial M (2019) Giant viruses-big surprises. Viruses 11:404

Brandt A, Schaefer I, Glanz J, Schwander T, Maraun M et al (2017) Effective purifying selection in ancient asexual oribatid mites. Nat Commun. https://doi.org/10.1038/s41467-017-01002-8

Brasier CM (2001) Rapid evolution of introduced plant pathogens via interspecific hybridization. Bioscience 51(2):123–133

Braun T, Orlova A, Valegård K, Lindås A-C, Schröder GF, Egelmann EH (2015) Archaeal actin from a hyperthermophile forms a single-stranded filament. PNAS 112(30):9340–9345

Breitbart M, Rohwer F (2005) Here a virus, there a virus, everywhere the same virus? Trends Microbiol 13(6):278–284

Bresler V, Montgomery WL, Fishelson L, Pollak PE (1998) Gigantism in a bacterium, *Epulopiscium fishelsoni*, correlates with complex patterns in arrangement, quantity, and segregation of DNA. J Bacteriol 180(21):5601–5611

Breuert S, Allers T, Spohn G, Soppa J (2006) Regulated polyploidy in halophilic archaea. PLoS One 1(1):e92

Bridier-Nahmias A, Tchalikian-Cusson A, Baller JA, Menouni R, Fayol H et al (2015) An RNA polymerase III subunit determines sites of retrotransposon integration. Science 348:585–588

Briggs CJ, Knapp RA, Vredenburg VT (2010) Enzootic and epizootoc dynamics of the chytrid fungal pathogen of amphibians. PNAS 107(21):9695–9700

Bringel F, Couée I (2015) Pivotal roles of phyllosphere microorganisms at the interface between plant functioning and atmospheric trace gas dynamics. Front Microbiol 6:486

Brito IL, Yilmaz S, Huang K, Xu L, Jupiter SD et al (2016) Mobile genes in the human microbiome are structured from global to individual scales. Nature 535:435–439

Broaders E, Gahan CGM, Marchesi JR (2013) Mobile genetic elements of the human gastrointestinal tract. Gut Microbes 4(4):271–280

Brochmann C, Brysting AK, Alsos IG, Borgen L, Grundt HH et al (2004) Polyploidy in arctic plants. Biol J Linn Soc 82:521–536

Brocks JJ (2018) The transition from a cyanobacterial to algal world and the emergence of animals. Emerg Top Life Sci. https://doi.org/10.1042/ETLS20180039

Brown L, Wolf JM, Prados-Rosales R, Casadevall A (2015) Through the wall: extracellular vesicles in gram-positive bacteria, mycobacteria and fungi. Nat Rev Microbiol 13(10):620–630

Brown A, Fernández IS, Gordiyenko Y, Ramakrishnan V (2016) Ribosome-dependent activation of stringent control. Nature 534:277–280

Brunk CF, Martin WF (2019) Archaeal histone contributions to the origin of eukaryotes. Trends Microbiol 27(8):703–714

Brüssow H (2007) Bacteria between protists and phages: from antipredation strategies to the evolution of pathogenicity. Mol Microbiol 65(3):583–589

Bryant DA (2003) The beauty in small things revealed. PNAS 100(17):9647–9649

Bulmer DM, Kharraz L, Grant AJ, Dean P, Morgan FJE et al (2012) The bacterial cytoskeleton modulates motility, type 3 secretion, and colonization in *Salmonella*. PLoS Pathog 8(1):e1002500

Burke C, Steinberg P, Rusch D, Kjelleberg S, Thomas T (2011) Bacterial community assembly based on functional genes rather than species. PNAS 108(34):14288–14293

Burki F, Roger AJ, Brown MW, Simpson AGB (2020) The new tree of eukaryotes. Trends Ecol Evol 35(1):43–55

Burstein D, Harrington LB, Strutt SC, Probst AJ, Anantharaman K et al (2017) New CRISPR-Cas systems from uncultivated microbes. Nature 542:237–241

Cabral G, Marques A, Schubert V, Pedrosa-Harand A, Schlögelhofer P (2014) Chiasmatic and achiasmatic inverted meiosis of plants with holocentric chromosomes. Nat Commun. https://doi.org/10.1038/ncomms6070

Caduff C (2020) What went wrong: corona and the world after the full stop. Med Anthropol Q 34(4):467–487

Cafaro MJ, Currie CR (2005) Phylogenetic analysis of mutualistic filamentous bacteria associated with fungus-growing ants. Can J Microbiol 51:441–446

Cai X, Clapham DE (2011) Ancestral Ca^{2+} signaling machinery in early animal and fungal evolution. Mol Biol Evol 29(1):91–100

Cai X, Wang X, Patel S, Clapham DE (2015) Insights into the early evolution of animal calcium signaling machinery: a unicellular point of view. Cell Calcium 57(3):166–173

Caino MC, Chae YC, Vaira V, Ferrero S, Nosotti M et al (2013) Metabolic stress regulates cytoskeletal dynamics and metastasis in cancer cells. J Clin Investig 123(7):2907–2920

Campbell L, Liu H, Nolla HA, Vaulot D (1997) Annual variability of phytoplankton and bacteria in the subtropical North Pacific Ocean at Station ALOHA during the 1991-1994 ENSO event. Deep Sea Res 44(2):167–192

Campbell MA, Ganley ARD, Gabaldón T, Cox MP (2016) The case of the missing ancient fungal polyploids. Am Nat 188(8):602–614

Canestrelli D, Porretta D, Lowe WH, Bisconti R, Carere C, Nascetti G (2016) The tangled evolutionary legacies of range expansion and hybridization. Trends Ecol Evol 31(9):677–688

Caro-Quintero A, Deng J, Auchtung J, Brettar I, Höfle MG et al (2011) Unprecedented levels of horizontal gene transfer among spatially co-occurring *Shewanella* bacteria from the Baltic Sea. ISME J 5:131–140

Carstens BC, Satler JD (2013) The carnivorous plant described as *Sarracenia alata* contains two cryptic species. Biol J Linn Soc 109:737–746

Carthew RW, Sontheimer EJ (2009) Origins and mechanisms of miRNAs and siRNAs. Cell 136:642–655

Carvalho-Santos Z, Azimzadeh J, Pereira-Leal JB, Bettencourt-Dias M (2011) Tracing the origins of centrioles, cilia, and flagella. J Cell Biol 198(2):165–175

Carvunis AR, Rolland T, Wapinski I, Calderwood MA, Yildirim MA et al (2012) Proto-genes and *de novo* gene birth. Nature 487(7404):370–374

Casadesús J, Low D (2006) Epigenetic gene regulation in the bacterial world. Microbiol Mol Biol Rev 70(3):830–856

Casadevall A, Fang FC, Pirofski L (2011) Microbial virulence as an emergent property: consequences and opportunities. PLoS Pathog 7(7):e1002136

Casewell NR, Wagstaff SV, Harrison RA, Renjifo C, Wüster W (2011) Domain loss facilitates accelerated evolution and neofunctionalization of duplicate snake venom metalloproteinase toxin genes. Mol Biol Evol 28(9):2637–2649

Casjens SR (2011) The DNA-packaging nanomotor of tailed bacteriophages. Nature 9:647–657

Cavallari N, Frigato E, Vallone D, Fröhlich N, Lopez-Olmeda JF et al (2011) A blind circadian clock in cavefish reveals that opsins mediate peripheral clock photoreception. PLoS Biol 9(9):e1001142

Caviston JP, Holzbaur ELF (2006) Microtubule motors at the intersection of trafficking and transport. Trends Cell Biol 16(10):530–537

Cech TR, Steitz JA (2014) The noncoding RNA revolution-trashing old rules to forge new ones. Cell 157:77–94

Celler K, Koning RI, Koster AJ, van Wezel GP (2013) Multidimensional view of bacterial cytoskeleton. J Bacteriol 195(8):1627–1636

Chaix A, Zarrinpar A, Panda S (2016) The circadian coordination of cell biology. J Cell Biol 215(1):15–25

Chaker-Margot M, Barandun J, Hunziker M, Klinge S (2017) Architecture of the yeast small subunit processome. Science 355:eaal1880

Chang TC, Stergiopoulos I (2015) Inter- and intra-domain horizontal gene transfer, gain-loss asymmetry and positive selection mark the evolutionary history of the CBM14 family. FEBS J 282:2014–2028

Chase MW, Paun O, Fay MF (2010) Hybridization and speciation in angiosperms: a role for pollinator shifts? BMC Biol 8:45

Chattopadhyay S, Wu X-L (2009) The effect of long-range hydrodynamic interaction on the swimming of a single bacterium. Biophys J 96:2023–2028

Chen L, DeVries AL, Cheng C-H (1997) Convergent evolution of antifreeze glycoproteins in Antarctic notothenoid fish and Arctic cod. PNAS 94:3817–3822

Chen W, Zhang CK, Cheng Y, Zhang S, Zhao H (2013) A comparison of methods for clustering 16S rRNA sequences into OTUs. PLoS One 8(8):e70837

Chen H, Du G, Song X, Li L (2017a) Non-coding transcripts from enhancers: new insights into enhancer activity and gene expression regulation. Genom Proteom Bioinf 15:201–207

Chen W, Yu Q, Zhong Y, Yu H, Shu J et al (2017b) Genetic variation and co-evolutionary relationship of RNA polymerase complex segments in influenza A viruses. Virology 511:193–206

Cheng P-Y, Palekar R, Azziz-Baumgartner E, Iuliano D, Alencar AP et al (2015) Burden of influenza-associated deaths in the Americas, 2002-2008. Influenza Other Respir Viruses 9(S1):13–21

Chhabra ES, Higgs HN (2007) The many faces of actin: matching assembly factors with cellular structures. Nat Cell Biol 9(10):1110–1121

Chi J, Parrow MW, Dunthorn M (2014) Cryptic sex in *Symbiodinium* (Alveolata, Dinoflagellata) is supported by an inventory of meiotic genes. J Eukaryot Microbiol 61:322–327

Chimileski S, Franklin MJ, Papko TT (2014) Biofilms formed by the archaeon Haloferax volcanii exhibit cellular differentiation and social motility, and facilitate horizontal gene transfer. BMC Biol 12:65

Chipman AD, Ferrier DEK, Brena C, Qu J, Hughes DST et al (2014) The first myriapod genome sequence reveals conservative arthropod gene content and genome organisation in the centipede *Strigamia maritima*. PLoS Biol 12(11):e11002005

Chittka L, Thomson JD (eds) (2001) Cognitive ecology of pollination. Cambridge University Press, Cambridge

Chiura HX (1997) Generalized gene transfer by virus-like particles from marine bacteria. Aquat Microbiol Ecol 13:75–83

Cho BC, Azam F (1999) Biogeochemical significance of bacterial biomass in the ocean's euphotic zone. Mar Ecol Prog Ser 63:253–259

Choi I-G, Kim S-H (2007) Global extent of horizontal gene transfer. PNAS 104(11):4489–4494

Christian MD, Hawryluck L, Wax RS, Cook T, Lazar NM et al (2006) Development of a triage protocol for critical care during an influenza pandemic. CMAJ 175(11):1377–1381

Christodoulou F, Raible F, Tomer R, Simakov O, Trachana K et al (2010) Ancient animal microRNAs and the evolution of tissue identity. Nature 25:1084–1088

Chu C, Qu K, Zhong FL, Artandi SE, Chang HY (2011) Genomic maps of long noncoding RNA occupancy reveal principles of RNA-chromatin interactions. Mol Cell 44:667–678

Chu H, Chan JF-W, Yuen TT-T, Shuai N, Yuan S et al (2020) Comparative tropism, replication kinetics, and cell damage profiling of SARS-CoV-2 and SARS-CoV with implications for clinical manifestations, transmissibility, and laboratory studies of COVID-19: an observational study. Lancet Microbe 1:e14–e23

Ciccarelli FD, Doerks T, von Mering C, Creevey CJ, Snel B, Bork P (2006) Toward automatic reconstruction of a highly resolved tree of life. Science 311:1283–1287

Cisneros Salerno L (2008) The organized melee. Dissertation, University of Arizona

Clemente JC, Ursell LK, Wegener Parfrey L, Knight R (2012) The impact of the gut microbiota on human health: an integrative view. Cell 148:1258–1270

Clifton KE (1997) Mass spawning of green algae on coral reefs. Science 275:1116–1118

Cohen ML (2000) Changing patterns of infectious disease. Nature 406:762–767

Cohen SE, Golden SS (2015) Circadian rhythms in cyanobacteria. Microbiol Mol Biol Rev 79(4):373–385

Cohen EJ, Ferreira JL, Ladinsky MS, Beeby M, Hughes KT (2017) Nanoscale-length control of the flagellar driveshaft requires hitting the tethered outer membrane. Science 356(6334):197–200

Cohen D, Melamed S, Millman A, Shulman G, Oppenheimer-Shaanan Y et al (2019) Cyclic GMP-AMP signalling protects bacteria against viral infection. Nature 574:691–695

Colby DW, Prusiner SB (2011) Prions. Cold Spring Harb Perspect Biol 3:a006833

Coleman-Derr D, Desgarennes D, Fonseca-Garcia C, Gross S, Clingenpeel S et al (2016) Plant compartment and biogeography affect microbiome composition in cultivated and native *Agave* species. New Phytol 209:798–811

Collingro A, Tischler P, Weinmaier T, Penz T, Heinz E et al (2011) Unity in variety—the pan-genome of the chlamydiae. Mol Biol Evol 28(12):3253–3270

Collis P, Antoniou M, Grosveld F (1990) Definition of the minimal requirements within the human β-globin gene and the dominant control region for high level expression. EMBO J 9(1):233–240

Colussi TM, Costantino DA, Zhu J, Donohue JP, Korostelev AA et al (2015) Initiation of translation in bacteria by a structured eukaryotic IRES RNA. Nature 519:110–113

Conant GC, Wolfe KH (2008) Turning a hobby into a job: how duplicated genes find new functions. Nature 9:938–950

Constable GWA, Kokko H (2018) The rate of facultative sex governs the number of expected mating types in isogamous species. Nat Ecol Evol. https://doi.org/10.1038/s41559-018-0580-9

Cordaux R, Bouchon D, Gréve P (2011) The impact of endosymbionts on the evolution of host sex-determination mechanisms. Trends Genet 27(8):332–341

Cornelis G, Vernochet C, Carradec Q, Souquere S, Mulot B et al (2015) Retroviral envelope gene captures and syncytyn exaptation for placentation in marsupials. PNAS 112:E487–E496. https://doi.org/10.1073/pnas.1417000112

Cosby RL, Judd J, Zhang R, Zhong A, Garry N et al (2021) Recurrent evolution of vertebrate transcription factors by transposase capture. Science 371:eabc6405

Cotton JA, McInerney JO (2010) Eukaryotic genes of archaebacterial origin are more important than more numerous eubacterial genes, irrespective of function. PNAS 107(40):17252–17255

Cox MP, Dong T, Shen GG, Dalvi Y, Scott DB, Ganley ARD (2014) An interspecific fungal hybrid reveals cross-kingdom rules for allopolyploid gene expression patterns. PLoS Genet 10(3):e1004180

Coyte KZ, Schluter J, Foster KB (2015) The ecology of the microbiome: networks, competition, and stability. Science 350:663–666

Croft MT, Warren MJ, Smith AG (2006) Algae need their vitamins. Eukaryot Cells 5(8):1175–1183

Croll D, Giovannetti M, Koch AM, Sbrana C, Ehinger M et al (2008) Nonself vegetative fusion and genetic exchange in the arbuscular mycorrhizal fungus *Glomus intraradices*. New Phytol 181:924–937

Cryan JF, Dinan TG (2012) Mind-altering microorganisms: the impact of the gut microbiota on brain and behaviour. Nat Neurosci 13:701–712

Cubillos FA, Billi E, Zörgö E, Parts L, Fargier P et al (2011) Assessing the complex architecture of polygenic traits in diverged yeast populations. Mol Ecol 20:1401–1413

Cuevas JM, Geller R, Garijo R, López-Aldeguer J, Sanjuán R (2015) Extremely high mutation rate of HIV-1 in vivo. PLoS Biol. https://doi.org/10.1371/J.pbio.1002251

Cui J, Li F, Shi Z-L (2019a) Origin and evolution of pathogenic coronaviruses. Nat Rev Microbiol 17:181–192

Cui R, Medeiros T, Willemsen D, Iasi LMN, Collier GE et al (2019b) Relaxed selection limits lifespan by increasing mutation load. Cell 178:385–399

Cunha BA (2004) Influenza: historical aspects of epidemics and pandemics. Infect Dis Clin North Am 18:141–155

Curtis BA, Tanifuji G, Burki F, Gruber A, Irimia M et al (2012) Algal genomes reveal evolutionary mosaicism and the fate of nucleomorphs. Nature 492:59–65

Cyranoski D (2020) Profile of a killer virus. Nature 581:22–26

D'Souza TG, Michiels NK (2010) The costs and benefits of occasional sex: theoretical predictions and a case study. J Hered 101(Suppl 1):534–541

Daane LL, Molina JAE, Sadowsky MJ (1997) Plasmid transfer between spatially separated donor and recipient bacteria in earthworm-containing soil microcosms. Appl Environ Microbiol 63(2):679–686

Dacks J, Roger AJ (1999) The first sexual lineage and the relevance of facultative sex. J Mol Evol 48:779–783

Dagan T, Artzy-Randrup Y, Martin W (2008) Modular networks and cumulative impact of lateral transfer in prokaryote genome evolution. PNAS 105(29):10039–10044

Dalen JE, Alpert JS, Goldberg RJ, Weinstein RS (2014) The epidemic of the 20th century: coronary heart disease. Am J Med 127:807–812

Dance A (2016) Life fast, die young. Nature 535:453–455

Dannemann M, Racimo F (2018) Something old, something borrowed: admixture and adaptation in human evolution. Curr Opin Genet Dev 53:1–8

Darby AC, Cho N-H, Fuxelius H-H, Westberg J, Andersson SGE (2007) Intracellular pathogens go extreme: genome evolution in the Rickettsiales. Trends Genet 23(10):511–520

Darras H, Berney C, Hasin S, Drescher J, Feldhaar H, Keller L (2023) Obligate chimerism in male yellow crazy ants. Science 380:55–58

Daubin V, Ochman H (2004) Start-up entities in the origin of new genes. Curr Opin Genet Dev 14:616–619

Davoli T, de Lange T (2011) The causes and consequences of polyploidy in normal development and cancer. Annu Rev Cell Dev Biol 27:585–610

Dawkins R (1986) The blind watchmaker. W. W. Norton & Company, New York

Dawkins R (2007) Das egoistische Gen. Spektrum Verlag, Heidelberg

Dawson SC, Paredez AR (2013) Alternative cytoskeletal landscapes: cytoskeletal novelty and evolution in basal excavate protists. Curr Opin Cell Biol 25(1):134–141

de Duve C (2007) The origin of eukaryotes: a reappraisal. Nat Rev Genet 8:395–403

de Jonge R, Bolton MD, Kombrink A, van den Berg GCM, Yadeta KA, Thomma BPHJ (2013) Extensive chromosomal reshuffling drives evolution of virulence in an asexual pathogen. Genome Res 23:1271–1282

de la Cruz F, Davies J (2000) Horizontal gene transfer and the origin of species: lessons from bacteria. Trends Microbiol 8(3):128–133

de Paula WBM, Agip A-NA, Missirlis F, Ashworth R, Vizcay-Barrena G et al (2013) Female and male gamete mitochondria are distinct and complementary in transcription, structure, and genome function. Genome Biol Evol 5(10):1969–1977

De Queiroz K (2007) Species concepts and species delimitation. Syst Biol 56(6):879–886

De Storme N, Mason A (2014) Plant speciation through chromosome instability and ploidy change: cellular mechanisms, molecular factors and evolutionary relevance. Curr Plant Biol 1:10–33

de Witte LC, Stöcklin J (2010) Longevity of clonal plants: why it matters and how to measure it. Ann Bot 106:859–870

Deatherage BL, Cookson BT (2012) Membrane vesicle release in bacteria, eukaryotes, and archaea: a conserved yet underappreciated aspect of microbial life. Infect Immun 80(6):1946–1957

Debortoli N, Li X, Eyres I, Fontaneto D, Hespeels B et al (2016) Genetic exchange among bdelloid rotifers is more likely due to horizontal gene transfer than to meiotic sex. Curr Biol 26:723–732

Decaestecker E, De Meesters L, Mergeay J (2009) Cyclical parthenogenesis in daphnia: sexual versus asexual reproduction. In: Schön I, Martens K, van Dijk P (eds) Lost sex: the evolutionary biology of parthenogenesis. Springer, Dordrecht, pp 295–316

Degli Esposti M (2014) Bioenergetic evolution in proteobacteria and mitochondria. Genome Biol Evol 6(12):3238–3251

Degli Esposti M, Chouaia B, Comandatore F, Crotti E, Sassera D et al (2014) Evolution of mitochondria reconstructed from the energy metabolism of living bacteria. PLoS One 9(5):e96566

Degnan PH, Yu Y, Sisneros N, Wing RA, Moran NA (2009) *Hamiltonella defensa*, genome evolution of protective bacterial endosymbiont from pathogenic ancestors. PNAS 106(22):9063–9068

Deguchi T, Iwanski MK, Schentarra E-M, Heidebrecht C, Schmidt L et al (2023) Direct observation of motor protein stepping in living cells using MINFLUX. Science 379:1010–1015

Deng C, Cheng CC, Ye H, He X, Cheng L (2010) Evolution of an antifreeze protein by neofunctionalization under escape from adaptive conflict. PNAS 107(50):21593–21598

Dennis PP, Omer A, Lowe T (2001) A guided tour: small RNA function in archaea. Mol Microbiol 40(3):509–519

Denny M (2016) Ecological mechanics. Princeton University Press, Princeton

Depczynski M, Bellwood DR (2005) Shortest recorded vertebrate lifespan found in a coral reef fish. Curr Biol 15(8):R288–R289

Depotter JRL, Seidl MF, Wood TA, Thomma BPHJ (2016) Interspecific hybridization impacts host range and pathogenicity of filamentous microbes. Curr Opin Microbiol 32:7–13

Derelle E, Ferraz C, Rombauts S, Rouzé P, Worden AZ et al (2006) Genome analysis of the smallest free-living eukaryote *Ostreococcus tauri* unveils many unique features. PNAS 103(31):11647–11652

Des Marais D (2003) Biogeochemistry of hypersaline microbial mats illustrates the dynamics of modern microbial ecosystems and early evolution of the biosphere. Biol Bull 204:160–167

Deveau H, Barrangou R, Garneau JE, Labonté J, Fremaux C et al (2008) Phage response to CRISPR-encoded resistance in Streptococcus thermophiles. J Bacteriol 190(4):1390–1400

Deveau A, Bonito G, Uehling J, Paoletti M, Becker M et al (2018) Bacterial-fungal interactions: ecology, mechanisms and challenges. FEMS Microbiol Rev 42:335–352

Devlin RH, Nagahama Y (2002) Sex determination and sex differentiation in fish: an overview of genetic, physiological, and environmental influences. Aquaculture 208:191–364

Diao X, Freeling M, Lisch D (2006) Horizontal transfer of a plant transposon. PLoS Biol 4(1):e5

Diekmann Y, Pereira-Leal JB (2013) Evolution of intracellular compartmentalization. Biochem J 449:319–331

Dietz JR, Moran MS, Isakoff SJ, Kurtzmann SH, Willey SC et al (2020) Recommendations for prioritization, treatment, and triage of breast cancer patients during the COVID-19 pandemic. Breast Cancer Res Treat 181:487–497

DiMaio F, Yu X, Rensen E, Krupovic M, Prangishvili D, Egelman EH (2015) A virus that infects a hyperthermophile encapsidates A-form DNA. Science 6237:914–917

Dimmock NJ, Easton AJ, Leppard KN (2007) Introduction to modern virology. Blackwell, Malden

Dinan TG, Stilling RM, Stanton C, Cryan JF (2015) Collective unconscious: how gut microbes shape human behavior. J Psychiatr Res 63:1–9

Dini-Andreote F, Dini-Andreote F, Araújo WL, Trevors JT, van Elsas JD (2012) Bacterial genomes: habitat specifity and uncharted organisms. Microbiol Evol 64:1–7

Dion P, Nautiyal CS (eds) (2008) Microbiology in extreme soils. Springer, Berlin

Ditty JL, Williams SB, Golden SS (2003) A cyanobacterial circadian timing mechanism. Annu Rev Genet 37:513–543

Djordjevic SP, Stokes HW, Chowdhury PR (2013) Mobile elements, zoonotic pathogens and commensal bacteria: conduits for the delivery of resistance genes into humans, production animals and soil microbiota. Front Microbiol 4:86

Dobson SL, Bourtzis K, Braig HR, Jones BF, Zhou W et al (1999) *Wolbachia* infections are distributed throughout insect somatic and germ line tissues. Insect Biochem Mol Biol 29:153–160

Dobson L, Mészáros B, Tusnády GE (2018) Structural principles governing disease-causing germline mutations. J Mol Biol 430:4955–4970

Dodd AN, Salathia N, Hall A, Kével E, Tóth R et al (2005) Plant circadian clocks increase photosynthesis, growth, survival, and competitive advantage. Science 309:630–633

Dolja VV, Koonin EV (2018) Metagenomics reshapes the concepts of RNA virus evolution by revealing extensive horizontal virus transfer. Virus Res 244:36–52

Dominguez DC, Guragain M, Patrauchan M (2015) Calcium binding proteins and calcium signalling in prokaryotes. Cell Calcium 57(3):1551–1165

Donaldson GP, Lee SM, Mazmanian SK (2016) Gut biogeography of the bacterial microbiota. Nat Rev Microbiol 14:20–32

Doolittle WF (2009a) Eradicating typological thinking in prokaryotic systematics and evolution. Cold Spring Harb Symp Quant Biol 74:197–204

Doolittle WF (2009b) The practice of classification and the theory of evolution, and what the demise of Charles Darwin's tree of life hypothesis means for both of them. Philos Trans R Soc Lond B Biol Sci 364:2221–2228

Dorken ME, Eckert CG (2001) Severely reduced sexual reproduction in northern populations of a clonal plant, *Decodon verticillatus* (Lythraceae). J Ecol 89:339–350

Doshi P (2008) Trends in recorded influenza mortality: United States, 1900-2004. Am J Public Health 98(5):939–945

Douglas S, Zauner S, Fraunholz M, Beaton M, Penny S et al (2001) Highly reduced genome of an enslaved algal nucleus. Nature 410:1091–1096

Downar J, Seccareccia D (2010) Palliating a pandemic: "all patients must be cared for". J Pain Symptom Manag 39(2):291–295

Draper GC, Gober JW (2002) Bacterial chromosome segregation. Ann Rev Microbiol 56:567–597

Drezen J-M, Chevignon G, Louis F, Huguet E (2014) Origin and evolution of symbiotic viruses associated with parasitoid wasps. Curr Opin Insect Sci 2:1–9

Du Q, Kawabe Y, Schilde C, Chen Z-H, Schaap P (2015) The evolution of aggregative multicellularity and cell-cell communication in dictyostelia. J Mol Biol 427:3722–3733

Dubruille R, Herbette M, Revel M, Horard B, Chang C-H, Loppin B (2023) Histone removal in sperm protects paternal chromosomes from premature division at fertilization. Science 382:725–731

Duffy S, Shackelton LA, Holmes EC (2008) Rates of evolutionary change in viruses: patterns and determinants. Nat Rev Genet 9:267–276

Dunlap JC (1999) Molecular bases for circadian clocks. Cell 96:271–290

Dunning J, Thwaites RS, Openshaw PJM (2020) Seasonal and pandemic influenza: 100 years of progress, still much to learn. Mucosal Immunol 13:566–573

Dunthorn M, Katz LA (2010) Secretive ciliates and putative asexuality in microbial eukaryotes. Trends Microbiol 18(5):183–188

Dunthorn M, Lipps JH, Dolan JR, Abboud-Abi Saab M, Aescht E et al (2015) Ciliates—protists with complex morphologies and ambiguous early fossil record. Mar Micropaleontol 119:1–6

Dupraz C, Reid RP, Braissant O, Decho AW, Norman RS, Visscher PT (2009) Processes of carbonate precipitation in modern microbial mats. Earth-Sci Rev 96(3):141–162

Durand E, Méheust R, Soucaze M, Goubet PM, Gallina S et al (2014) Dominance hierarchy arising from the evolution of a complex small RNA regulatory network. Science 346:1200–1208

Durkin A, Fisher CR, Cordes EE (2017) Extreme longevity in a deep-sea vestimentiferan tubeworm and its implications for the evolution of life history strategies. Sci Nat 104:63

Dvornyk V, Vinogradova O, Nevo E (2003) Origin and evolution of circadian clock genes in prokaryotes. PNAS 100(5):2495–2500

Dyall SD, Brown MT, Johnson PJ (2004) Ancient invasions: from endosymbionts to organelles. Science 304:253–257

Dye NA, Shapiro L (2007) The push and pull of the bacterial cytoskeleton. Trends Cell Biol 17(5):239–245

Eckel-Mahan K, Sassone-Corsi P (2009) Metabolism control by the circadian clock and vice versa. Nat Sruct Mol Biol 16(5):462–467

Edwards CJ, Suchard MA, Lerney P, Welch JJ, Barnes I et al (2011) Ancient hybridization and an irish origin for the modern polar bear matriline. Curr Biol 21:1251–1258

Egan ES, Fogel MA, Waldor MK (2005) Divided genomes: negotiating the cell cycle in prokaryotes with multiple chromosomes. Mol Microbiol 56(5):1129–1138

Ehlers KM, Koiller J (2010) Could cell membranes produce acoustic streaming? Making the case for *Synechococcus* self-propulsion. Math Comput Model 53:1489–1504

Ehlers K, Oster G (2012) On the mysterious propulsion of *Synechococcus*. PLoS One 7(5):e36081

Eibl-Eibesfeldt I (2004) Grundriß der vergleichenden Verhaltensforschung. Blank, Vierkirchen-Pasenbach

El Karkouri K, Pontarotti P, Raoult D, Fournier P-E (2016) Origin and evolution of rickettsial plasmids. PLoS One. https://doi.org/10.1371/journal.pone.0147492

Ellison CK, Kan J, Dillard RS, Kysela DT, Ducret A et al (2017) Obstruction of pilus retraction stimulates bacterial surface sensing. Science 358:535–538

Ellstrand NC, Schierenbeck KA (2000) Hybridization as a stimulus for the evolution of invasiveness in plants? PNAS 97(13):7043–7050

Embley TM, Martin W (2006) Eukaryotic evolution, changes and challenges. Nature 440:623–630

Eme L, Spang A, Lombard J, Stairs CW, Ettema TJG (2017) Archaea and the origin of eukaryotes. Nat Rev Microbiol 15:711–723

Emelyanov VV (2003) Mitochondrial connection to the origin of the eukaryotic cell. Eur J Biochem 270:1599–1618

Emerman M, Malik HS (2010) Paleovirology-modern consequences of ancient viruses. PLoS Biol 8(2):e100031

Emiliani G, Fondi M, Fani R, Gribaldo S (2009) A horizontal gene transfer at the origin of phenylpropanoid metabolism: a key adaptation of plants to land. Biol Direct. https://doi.org/10.1186/1745-6150-4-7

Enghoff H (1994) Geographical parthenogenesis in millipedes (Diplopoda). Biogeographica 70(1):25–31

Ereskovsky AV, Tokina DB (2007) Asexual reproduction in homoscleromorph sponges (Porifera: Homoscleromorpha). Mar Biol 151:425–434

Estabragh ZR, Mamas MA (2013) The cardiovascular manifestations of influenza: a systematic review. Int J Cardiol 167:2397–2403

Estep MC, McKain MR, Diaz DV, Zhong J, Hodge JG et al (2014) Allopolyploidy, diversification, and the Miocene grassland expansion. PNAS 111(42):15149–15154

Ettema TJG, Lindås A-C, Bernandez R (2011) An actin-based cytoskeleton in archaea. Mol Microbiol 80(4):1052–1061

Evguenieva-Hackenberg E, Roppelt V, Lassek C, Klug G (2011) Subcellular localization of RNA degrading proteins and protein complexes in prokaryotes. RNA Biol 8(1):49–54

Faddeeva-Vakhrusheva A, Kraaijeveld K, Derks MFL, Anvar SY, Agamennone V et al (2017) Coping with living in the soil: the genome of the parthenogenetic springtail *Folsomia candida*. BMC Genom 18:493

Falk S, Götz M (2017) Glial control of neurogenesis. Curr Opin Neurobiol 47:188–195

Falkowski PG, Fenchel T, Delong EF (2008) The microbial engines that drive earth's biogeochemical cycles. Science 320:1034–1039

Fan L, Reynolds D, Liu M, Stark M, Kjelleberg S et al (2012) Functional equivalence and evolutionary convergence in complex communities of microbial sponge symbionts. PNAS 109:E1878–E1887

Farheen N, Thattai M (2019) Frustration and fidelity in influenza genome assembly. J R Soc Interface 16:20190411

Farlow A, Meduri E, Schlötterer C (2011) DNA double-strand break repair and the evolution of intron density. Trends Genet. https://doi.org/10.1016/j.tig.2010.10.004

Faso C, Hehl AB (2011) Membrane trafficking and organelle biogenesis in *Giardia lamblia*: use it or lose it. Int J Parasitol 41:471–480

Fatica A, Bozzoni I (2014) Long non-coding RNAs: new players in cell differentiation and development. Nat Rev Genet 16:7–21

Fedoroff NV (2012) Transposable elements, epigenetics, and genome evolution. Science 338:758–767

Feil R, Fraga MF (2012) Epigenetics and the environment: emerging patterns and implications. Nat Rev Genet 13:97–109

Fenchel T, Finlay BJ (2006) The diversity of microbes: resurgence of the phenotype. Philos Trans R Soc Lond B Biol Sci 361:1965–1973

Fenchel T, Thar R (2004) "Candidatus Ovobacter propellens": a large conspicuous prokaryote with an unusual motility behaviour. FEMS Microbiol Ecol 48:231–238

Fenwick AM, Greene HW, Parkinson CL (2011) The serpent and the egg: unidirectional evolution of reproductive mode in vipers? J Zool Syst Evol Res. https://doi.org/10.1111/j.1439-0469.2011.00646.x

Feretzaki M, Heitman J (2013) Unisexual reproduction drives evolution of eukaryotic microbial pathogens. PLoS Pathog 9(10):e1003674

Ferguson NM, Cummings DAT, Fraser C, Cajka JC, Cooley PC, Burke DS (2006) Strategies for mitigating an influenza pandemic. Nature 442:448–452

Fernandez C, Lelong B, Via B, Mévy J-P, Rohles C et al (2006) Potential allelopathic effect of *Pinus halepensis* in the secondary succession: an experimental approach. Chemoecology 16:97–105

Ferrell JE, Tsai TY-C, Yang Q (2011) Modeling the cell cycle: why do certain circuits oscillate? Cell 144:874–885

Ferrezuelo F, Colomina N, Palmisano A, Gari E, Gallego C et al (2012) The critical size is set at a single-cell level by growth rate to attain homeostasis and adaptation. Nat Commun 3:1012. https://doi.org/10.1038/ncomms2015

Ferri G, Alù M, Corradini B, Licata M, Beduschi G (2009) Species identification through DNA "barcode". Genet Test Mol Biomarkers. https://doi.org/10.1089/gtmb.2008.0144

Feschotte C, Gilbert C (2012) Endogenous viruses: insights into viral evolution and impact on host biology. Nat Rev Genet 13:283–296

Figueroa-Martinez F, Nedelcu AM, Reyes-Prieto A, Smith DR (2017) The plastid genomes of non-photosynthetic algae are not so small after all. Commun Integr Biol 10(1):e1283080

Filée J (2014) Multiple occurrences of giant virus core genes acquired by eukaryotic genomes: visible part of the iceberg? Virology 466(467):53–59

Finkel ZV, Beardall J, Flynn KJ, Quigg A, Rees TAV, Raven JA (2010) Phytoplankton in a changing world: cell size and elemental stoichiometry. J Plankton Res 32(1):119–137

Fischbach MA, Segre JA (2016) Signaling in host-associated microbial communities. Cell 164:1288–1300

Fischer MG (2013) Wenn Viren Viren infizieren. Biospektrum 06(13):619–621

Fischer MG (2016) Giant viruses come of age. Curr Opin Microbiol 31:50–57

Fischer MG, Suttle CA (2011) A virophage at the origin of large DNA transposons. Science 332:231–234

Fisher RA (1930) The genetical theory of natural selection. Clarendon Press, Oxford

Fitzpatrick DA (2011) Horizontal gene transfer in fungi. FEMS Microbiol Lett 329:1–8

Fleißner A, Simonin AR, Glass NL (2010) Cell fusion in the filamentous fungus, *Neurospora crassa*. In: Chen EH (ed) Cell fusions: overviews and methods. Humana Press, Totowa, pp 21–38

Flindt R (1986) Biologie in Zahlen. Fischer Verlag, Stuttgart

Flood BE, Louw DC, Van der Plaas AK, Bailey JV (2021) Giant sulfur bacteria (Beggiatoaceae) from sediments underlying the Benguela upwelling system host diverse microbiomes. PLoS One 18(11):e0258124

Flores CO, Meyer JR, Valverde S, Farr L, Weitz JS (2011) Statistical structure of host-phage interactions. PNAS 108(28):E288–E297

Flores CO, Valverde S, Weitz JS (2013) Multi-scale structure and geographic drivers of cross-infection within marine bacteria and phages. ISME J 7:520–532

Flórez LV, Biedermann PHW, Engl T, Kaltenpoth M (2015) Defensive symbioses of animals with prokaryotic and eukaryotic microorganisms. Nat Prod Rep 32:904–936

Flot JF, Hespeels B, Li X, Noel B, Arkhipova I et al (2013) Genomic evidence for ameiotic evolution in the Bdelloid rotifer *Adineta vaga*. Nature 500:453–457

Foissner W, Chao A, Katz LA (2009) Diversity and geographic distribution of ciliates (Protista: Ciliophora). In: Foissner W, Hawksworth DL (eds) Protist diversity and geographic distribution. Springer, Dordrecht, pp 111–129

Fontaneto D, Barraclough TG (2015) Do species exist in asexuals? Theory and evidence from bdelloid rotifers. Integr Comp Biol 55(2):253–263

Fontaneto D, Tang CQ, Oberegger U, Leasi F, Barraclough TG (2012) Different diversification rates between sexual and asexual organisms. Evol Biol. https://doi.org/10.1007/s11692-012-9161-z

Forrester NJ, Rebolleda-Gómez M, Sachs JL, Ashman T-L (2020) Polyploid plants obtain greater fitness benefits from a nutrient acquisition mutualism. New Phytol 227:944–954

Forterre P, Prangishvili D (2009) The great billion-year war between ribosome- and capsid-encoding organisms (cells and viruses) as the major source of evolutionary novelties. In: Natural genetic engineering and natural genome editing, vol 1178. Wiley, pp 65–77

Foster KR, Schluter J, Coyte KZ, Nahoum SR (2017) The evolution of the host microbiome as an ecosystem on a leash. Nature 548:43–51

Foth BJ, Goedecke MC, Soldati D (2006) New insights into myosin evolution and classification. PNAS 103(10):3681–3686

Fox DT, Duronio RJ (2013) Endoreplication and polyploidy: insights into development and disease. Development 140:3–12

Frank J, Gonzalez RL Jr (2010) Structure and dynamics of a processive brownian motor: the translating ribosome. Annu Rev Biochem 79:381–412

Franklin TB, Mansuy IM (2010) Epigenetic inheritance in mammals: Evidence for the impact of adverse environmental effects. Neurobiol Dis 39(1):61–65

Fraune S, Bosch TCG (2010) Why bacteria matter in animal development and evolution. Bioessays 32:571–580

Fraústo da Silva JJR, Williams RJP (2001) The biological chemistry of the elements. Oxford University Press, Oxford

Frawley LE, Orr-Weaver TL (2015) Polyploidy. Curr Biol 25:R345–R361

Freeling M (2017) Picking up the ball at the K/Pg boundary: the distribution of ancient polyploidies in the plant phylogenetic tree as a spandrel of asexuality with occasional sex. Plant Cell 29:202–206

Friedrich M (2013) biological clocks and visual systems in cave-adapted animals at the dawn of speleogenomics. Integr Comp Biol 53(1):50–67

Fritz S (2009) Vegetative reproduction and clonal diversity in pleurocarpous mosses (Bryophytina) of mesic habitats. Diss Freie Universität, Berlin

Fritz-Laylin LK, Prochnik SE, Ginger ML, Dacks JB, Carpenter ML et al (2010) The genome of *Naegleria gruberi* illuminates early eukaryotic versatility. Cell 140:631–642

Fry BG, Vidal N, Norman JA, Vonk FJ, Scheib H et al (2005) Early evolution of the venom system in lizards and snakes. Nature 439:584–588

Fry BG, Scheib H, van der Weerd L, Young B, McNaughtan J et al (2008) Evolution of an arsenal. Mol Cell Proteomics 7:215–246

Fuhrman JA (1999) Marine viruses and their biogeochemical and ecological effects. Nature 399:541–548

Fuller DN, Rickgauer JP, Jardine PJ, Grimes S, Anderson DL, Smith DE (2007) Ionic effects on viral DNA packaging and portal motor function in bacteriophage Φ29. PNAS 104(27):11245–11250

Funk WC, Caminer M, Ron SR (2012) High levels of cryptic species diversity uncovered in Amazonian frogs. Proc Biol Sci 279:1806–1814

Fusi-Schmidhauser T, Preston NJ, Keller N, Gamondi C (2020) Conservative management of Covid-19 patients-emergency palliative care in action. J Pain Symptom Manag 60(1):e27–e30

Gage JD, Tyler PA (1999) Deep sea biology. Cambridge University Press, Cambridge

Gaia M, Benamar S, Boughalmi M, Pagnier I, Croce O et al (2014) Zamilon, a novel virophage with *Mimiviridae* host specificity. PLoS One 9(4):e94923

Gaiti F, Calcino AD, Tanurdžić M, Degnan BM (2017) Origin and evolution of the metazoan non-coding regulatory genome. Dev Biol 427:193–202

Gallager SM, Davis CS, Epstein AW, Solow A, Beardsley RC (1996) High-resolution observations of plankton spatial distributions correlated with hydrography in the Great South Channel, Georges Bank. Deep Sea Res 43(7–8):1627–1663

Gallo RL (2017) Human skin is the largest epithelial surface for interactions with microbes. J Investig Dermatol 137:1213–1214

Gao F, Davidson EH (2008) Transfer of a large gene regulatory apparatus to a new developmental address in echinoid evolution. PNAS 105(16):6091–6096

Gao F, Katz LA (2014) Phylogenomic analyses support the bifurcation of ciliates into two major clades that differ in properties of nuclear division. Mol Phylogenet Evol 70:240–243

Genome Project Consortium (2015) A global reference for human genetic variation. Nature 526:68–74

Gentric G, Desdouets C (2014) Polyploidization in liver tissue. Am J Pathol 1842:322–331

Ghosh W, Dam B (2009) Biochemistry and molecular biology of lithotropic sulfur oxidation by taxonomically and ecologically diverse bacteria and archaea. FEMS Microbiol Rev 33:999–1043

Giandhari J, Pillay S, Wilkinson E, Tegally H, Sinayskiy I et al (2021) Early transmission of SARS-CoV-2 in South Africa: an epidemiological and phylogenetic report. Int J Infect Dis 103:234–241

Gibiansky ML, Conrad JC, Jin F, Gordon VD, Motto DA et al (2010) Bacteria use type IV pili to walk upright and detach from surfaces. Science 330:197

Gilad R, Porat A, Trachtenberg S (2003) Motility modes of *Spiroplasma melliferum* BC3: a helical, wall-less bacterium driven by a linear motor. Mol Microbiol 47(3):657–669

Gillespie JJ, Joardar V, Williams KP, Driscoll T, Hostetler JB et al (2011) A *Rickettsia* genome over-run by mobile genetic elements provides insight into the acquisition of gene characteristic of an obligate intracellular lifestyle. J Bacteriol 194(2):376–394

Ginger ML, Portman N, McKean PG (2008) Swimming with protists: perception, motility and flagellum assembly. Nat Microbiol 6:838–850

Giorgianni MW, Dowell NL, Griffin S, Kassner VA, Selegue JE, Carroll SB (2020) The origins and diversification of a novel protein in venomous snakes. PNAS 117(20):10911–10920

Gioti A, Mushegian AA, Strandberg R, Stajich JE, Johanesson H (2012) Unidirectional evolutionary transitions in fungal mating systems and the role of transposable elements. Mol Biol Evol 29(10):3215–3226

Giraud T, Refrégier G, Le Gac M, de Vienne DM, Hood ME (2008) Speciation in fungi. Fungal Genet Biol 45:791–802

Gladfelter AS, Hungerbuehler AK, Philippsen P (2006) Asynchronous nuclear division cycles in multinucleated cells. J Cell Biol 172(3):347–362

Glass RJ, Glass LM, Beyeler WE, Min HJ (2006) Targeted social distancing design for pandemic influenza. Emerg Infect Dis 12(11):1671–1681

Glinsky G, Barakat TS (2019) The evolution of Great Apes has shaped the functional enhancers' landscape in human embryonic stem cells. Stem Cell Res 37:101456

Glossop NRJ, Hardin PE (2002) Central and peripheral circadian oscillator mechanisms in flies and mammals. J Cell Sci 115:3369–3377

Gogarten JP, Townsend JP (2005) Horizontal gene transfer, genome innovations and evolution. Nat Rev Microbiol 3:679–687

Golden RJ, Chen B, Li T, Braun J, Manjunath H et al (2017) An argonaute phosphorylation cycle promotes microRNA-mediated silencing. Nature 542:197–202

Goldmann JM, Wong WSW, Pinelli M, Farrah T, Bodian D et al (2016) Parent-of-origin-specific signatures of de novo mutations. Nat Genet 48:935–939

Gomaa F, Yang J, Mitchell EAD, Zhang W-J, Yu Z et al (2015) Morphological and molecular diversification of Asian endemic *Difflugia tuberspinifera* (Amoebozoa, Arcellinida): a case of fast morphological evolution in protists? Protist 166(1):122–130

Gonzalez P, Lessions HA (1999) Evolution of sea urchin retroviral-like (SURL) elements: evidence from 40 echinoid species. Mol Biol Evol 16(7):938–952

Gooday AJ, Holzmann M, Caulle C, Goineau A, Kamenskaya O et al (2017) Giant protists (Xenophyophores, Foraminifera) are exceptionally diverse in parts of the abyssal eastern Pacific licensed for polymetallic nodule exploration. Biol Conserv 207:106–116

Gorelick R, Carpinone J (2009) Origin and maintenance of sex: the evolutionary joys of self sex. Biol J Linn Soc 98:707–728

Gorelick R, Heng HHQ (2010) Sex reduces genetic variation: a multidisciplinary review. Evolution 65(4):1088–1098

Goriely A (2016) Decoding germline de novo point mutations. Nat Genet 48:823–824

Gorson J, Ramrattan G, Verdes A, Wright EM, Kantor Y et al (2015) Molecular diversity and gene evolution of the venom arsenal of terebridae predatory marine snails. Genome Biol Evol 7(6):1761–1778

Gould SB, Waller RF, McFadden GI (2008) Plastid evolution. Annu Rev Plant Biol 59:491–517

Gould SB, Garg SG, Martin WF (2016) Bacterial vesicle secretion and the evolutionary origin of the eukaryotic endomembrane system. Trends Microbiol 24(7):525–534

Granek JA, Kaikçi Ö, Magwene PM (2011) Pleiotropic signaling pathways orchestrate yeast development. Curr Opin Microbiol 14(6):676–681

Graw J (2015) Genetik. Springer Spektrum, Berlin

Gray MW (2014) The pre-endosymbiont hypothesis: a new perspective on the origin and evolution of mitochondria. Cold Spring Harb Perspect Biol 6:a016097

Gray MW (2015) Mosaic nature of the mitochondrial proteome: implications for the origin and evolution of mitochondria. PNAS 112(33):10133–10136

Gray RR, Parker J, Lerney P, Salemi M, Katzourakis A, Pybus OG (2011) The mode and tempo of hepatitis C virus evolution within and among hosts. BMC Evol Biol 11:131

Gregory TR (2001) Coincidence, coevolution, or causation? DNA content, cell size, and the C-value enigma. Biol Rev 76:65–101

Gregory TE, Nicol JA, Tamm B, Kullman B, Kullman K et al (2007) Eukaryotic genome databases. Nucleic Acid Res. https://doi.org/10.1093/nar/gkl828

Gregory AC, Zayed AA, Conceição-Neto N, Temperton B, Bolduc B et al (2019) Marine DNA viral macro- and microdiversity from pole to pole. Cell 177:1109–1123

Grenier J, Carroll SB (2000) Functional evolution of the ultrabithorax protein. PNAS 97(2):704–709

Grewal SIS, Jia S (2007) Heterochromatin revisited. Nat Rev Genet 8:35–46

Gribaldo S, Poole AM, Daubin V, Forterre P, Brochier-Armanet C (2010) The origin of eukaryotes and their relationship with the archaea: are we at a phylogenomic impasse? Nat Rev Microbiol 8:743–752

Gribble J, Stevens LJ, Agostini ML, Anderson-Daniels J, Chappell JD et al (2021) The coronavirus proofreading exoribonuclease mediates extensive viral recombination. PLoS Pathog 17(1):e1009226

Grice EA, Segre JA (2011) The skin microbiome. Nat Rev Microbiol. 9(4):244–253

Griffiths RB (1941) Triploidy (and haploidy) in the newt, Triturus viridescens, induced by refrigeration of fertilized eggs. Genetics 26:69–88

Groenewald JZ, Nakashima C, Nishikawa J, Shin H-D, Park J-H et al (2012) Species concepts in Cercospora: spotting the weeds among the roses. Stud Mycol 75:115–170

Grootjans AP, van den Ende FP, Walsweer AF (1997) The role of microbial mats during primary succession in calcareous dune slacks: an experimental approach. J Coast Conserv 3:95–102

Grossart H-P, Massana R, McMahon KD, Walsh DA (2020) Linking metagenomics to aquatic microbial ecology and biogeochemical cycles. Limnol Oceanogr 65:S2–S20

Guasto JS, Rusconi R, Stocker R (2012) Fluid mechanics of planktonic microorganisms. Annu Rev Fluid Mech 44:373–400

Guerriero G, Hausman J-F, Strauss J, Ertan H, Siddiqui KS (2015) Destructuring plant biomass: focus on fungal and extremophilic cell wall hydrolases. Plant Sci 234:180–193

Gupta A, Alin D, Bashir R (2004) Single virus particle mass detection using microresonators with nanoscale thickness. Appl Phys Lett 84(11):1976–1978

Gupta KJ, Hebelstrup KH, Mur LAJ, Igamberdiev AU (2011) Plant hemoglobins: important players at the crossroad between oxygen and nitric oxide. FEBS Lett 585:3843–3849

Guy L, Saw JH, Ettema TJG (2014) The archaeal legacy of eukaryotes: a phylogenomic perspective. Cold Spring Harb Perspect Biol 4(6):a016022

Haag ES (2009) Convergent evolution: regulatory lightning strikes twice. Curr Biol 19(21):R977–R979

Haag CR, Ebert D (2004) A new hypothesis to explain geographic parthenogenesis. Ann Zool Fennici 41:539–544

Hacquard S, Garrido-Oter R, González A, Spaepen S, Ackermann G et al (2015) Microbiota and host nutrition across plant and animal kingdoms. Cell Host Microbe 17:603–616

Hadjivasiliou Z, Pomiankowski A (2016) Gamete signalling underlies the evolution of mating types and their number. Philos Trans R Soc Lond B Biol Sci 371:20150531

Hagström E, Andersson SGE (2018) The challenges of integrating two genomes in one cell. Curr Opin Microbiol 41:89–94

Haig D (2016) Living together and living apart: the sexual lives of bryophytes. Philos Trans R Soc Lond B Biol Sci 371:20150535

Haiko J, Westerlund-Wikström B (2013) The role of the bacterial flagellum in adhesion and virulence. Biology 2:1242–1267

Halary S, Temmam S, Raoult D, Desnues C (2016) Viral metagenomics: are we missing the giants? Curr Opin Microbiol 31:34–43

Hallberg KB, Coupland K, Kimura S, Johnson DD (2006) Macroscopic growths in acidic, metal-rich mine waters in North Wales consist of novel and remarkably simple bacterial communities. Appl Environ Microbiol 72(3):2022–2030

Han K, Li Z-F, Peng R, Zhu L-P, Zhou T et al (2013) Extraordinary expansion of a *Sorangium cellulosum* genome from an alkaline milieu. Sci Rep. https://doi.org/10.1038/srep02101

Hardison R (1998) Hemoglobins from bacteria to man: evolution of different patterns of gene expression. J Environ Biol 201:1099–1117

Harish A, Caetano-Anollés G (2012) Ribosomal history reveals origins of modern protein synthesis. PLoS One 7(3):e32776

Härnström K, Ellegaard M, Andersen TJ, Godhe A (2011) Hundred years of genetic structure in a sediment revived diatom population. PNAS 108(10):4252–4257

Harris SD (2009) The Spitzenkörper: a signalling hub or the control of fungal development? Mol Microbiol 73(5):733–736

Hartman H, Fedorov A (2002) The origin of the eukaryotic cell: a genomic investigation. PNAS 99(3):1420–1425

Hasan A, Paray BA, Hussain A, Qadir FA, Attar F et al (2020) A review on the cleavage priming of the spike protein on coronavirus by angiotensin-converting enzyme-2 and furin. J Biomol Struct Dyn. https://doi.org/10.1080/07391102.2020.1754293

Hawksworth DL (2012) Global species numbers of fungi: are tropical studies and molecular approaches contributing to a more robust estimate. Biodivers Conserv 21:2425–2433

Hayes CS, Aoki SK, Low DA (2010) Bacterial contact-dependent delivery systems. Annu Rev Genet 44:71–90

Hayward A (2017) Origin of the retroviruses: when, where, and how? Curr Opin Virol 25:23–27

Healy K, Guillerme T, Finlay S, Kane A, Kelly SBA et al (2014) Ecology and mode-of-life explain lifespan variation in birds and mammals. Proc Biol Sci 281:20140298

Heard E, Martiessen RA (2014) Transgenerational epigenetic inheritance: myths and mechanisms. Cell 157:95–109

Hebert PDN, Cywinska A, Ball SL, deWaard JR (2003) Biological identifications through DNA barcodes. Proc Biol Sci 270:313–321

Heckmann S, Jankowska M, Schubert V, Kumke K, Ma W, Houben A (2014) Alternative meiotic chromatid segregation in the holocentric plant Luzula elegans. Nat Commun. https://doi.org/10.1038/ncomms5979

Heiss AA, Walker G, Simpson AGB (2011) The ultrastructure of *Ancyromonas*, a eukaryote without supergroup affinities. Protist 162:373–393

Heitman J (2006) Sexual reproduction and the evolution of microbial pathogens. Curr Biol 18:R711–R725

Heitman J (2010) Evolution of eukaryotic microbial pathogens via covert sexual reproduction. Cell Host Microbe 8:86–99

Heitman J (2015) Evolution of sexual reproduction: a view from the fungal kingdom supports an evolutionary epoch with sex before sexes. Fungal Biol Rev 29:108–117

Hendrickson H (2009) Order and disorder during Escherichia coli divergence. PLoS Genet 5(1):e1000335

Henikoff S, Smith MM (2015) Histone variants and epigenetics. Cold Spring Harb Perspect Biol 7:a019364

Henry L, Schwander T, Crespi BJ (2011) Deleterious mutation accumulation in asexual *Timema* stick insects. Mol Biol Evol 29(1):401–408

Hentschel U, Piel J, Degnan SM, Taylor MW (2012) Genomic insights into the marine sponge microbiome. Nat Rev Microbiol 10:641–654

Herrmann H, Bär H, Kreplak L, Strelkov SV, Aebi U (2007) Intermediate filaments: from cell architecture to nanomechanics. Nat Rev Mol Cell Biol. https://doi.org/10.1038/nrm2197

Herzog B, Wirth R (2012) Swimming behavior of selected species of archaea. Appl Environ Microbiol 78(6):1670–1674

Hill WG, Zhang X-S (2012) On the pleiotropic structure of the genotype-phenotype map and the evolvability of complex organisms. Genetics 190:1131–1137

Hocher A, Laursen SP, Radford P, Tyson J, Lambert C et al (2023) Histone-organized chromatin in bacteria. bioRxiv. https://doi.org/10.1101/2023.01.26.525422

Hodgkin J (1998) Seven types of pleiotropy. Int J Dev Biol 42:501–505

Hodgson JG, Sharafi M, Jalili A, Díaz S, Montserrat-Marti G et al (2010) Stomatal vs. genome size in angiosperms: the somatic tail wagging the genomic dog? Ann Bot 105:573–584

Hoeppner MP, Gardner PP, Poole AM (2012) Comparative analysis of RNA families reveals distinct repertoires for each domain of life. PLoS Comput Biol 8(11):e1002752

Hoffmann M, Kleine-Weber H, Schroeder S, Krüger N, Herrier T et al (2020) SARS-CoV-2 cell entry depends on ACE2 and TMPRSS2 and is blocked by a clinically proven protease inhibitor. Cell 181:271–280

Hohmann-Marriott MF, Blankenship RE (2011) Evolution of photosynthesis. Annu Rev Plant Biol 62:515–548

Hoiczyk E, Baumeister W (1998) The junctional pore complex, a prokaryotic secretion organelle, is the molecular motor underlying gliding motility in cyanobacteria. Curr Biol 8:1161–1168

Hojsgaard D, Hörandl E (2016) A little bit of sex matters for genome evolution in asexual plants. Front Plant Sci 6:82

Holeski LM, Jander G, Agrawal AA (2012) Transgenerational defense induction and epigenetic inheritance in plants. Trends Ecol Evol 27(11):618–622

Holmes EC (2011) What does virus evolution tell us about virus origin? J Virol 85(11):5247–5251

Holmfeldt K, Solonenko N, Shah M, Corrier K, Riemann L et al (2013) Twelve previously unknown phage genera are ubiquitous in global oceans. PNAS 110(31):12798–12803

Holsinger KE (2000) Reproductive systems and evolution in vascular plants. PNAS 97(13):7037–7042

Holzbaur ELF, Goldman YE (2010) Coordination of molecular motors: from *in vitro* assays to intracellular dynamics. Curr Opin Cell Biol 22(1):4–13

Horner JR (2000) Dinosaur reproduction and parenting. Ann Rev Earth Planet Sci 28:19–45

Horsthemke B (2018) A critical view on transgenerational epigenetic inheritance in humans. Nat Commun. https://doi.org/10.1038/s41467-018-05445-5

Hotz H, Semlitsch RD, Gutmann E, Guex G-D, Beerli P (1999) Spontaneous heterosis in larval life-history traits of hemiclonal frog hybrids. PNAS 96:2171–2176

Hu H, Uesaka M, Guo S, Shimai K, Lu T-M et al (2017) Constrained vertebrate evolution by pleiotropic genes. Nat Ecol Evol. https://doi.org/10.1038/s41559-017-0318-0

Huang J (2013) Horizontal gene transfer in eukaryotes: the weak-link model. Bioessays 35:868–875

Huber H, Hohn MJ, Jahn U, Rachel R (2003) Heiss, klein und "gemein": das neue Phylum "Nanoarchaeota". Biospektrum 4(03):353–357

Huber H, Küper U, Daxer S, Rachel R (2012) The unusual cell biology of the hyperthermophilic Crearchaeon *Ignicoccus hospitalis*. Antonie Van Leeuwenhoek 102:203–219

Hueffer K, Parker JSL, Weichert WS, Geisel RE, Sgro J-Y, Parrish CR (2003) The natural host range shift and subsequent evolution of canine parvovirus resulted from virus-specific binding to the canine transferrin receptor. J Virol 77(3):1718–1726

Hug LA, Baker BJ, Anantharaman K, Brown CT, Probst AJ et al (2016) A new view of the tree of life. Nat Microbiol. https://doi.org/10.1038/nmicrobiol.2016.48

Hulo C, de Castro E, Masson P, Bougueleret L, Bairoch A et al (2011) ViralZone: a knowledge resource to understand virus diversity. Nucleic Acid Res 39:D576–D582

Hurwitz BL, Sullivan MB (2013) The Pacific Ocean Virome (POV): a marine viral metagenomic dataset and associated protein clusters for quantitative viral ecology. PLoS One 8(2):e57355

Husnik F, McCutcheon JP (2018) Functional horizontal gene transfer from bacteria to eukaryotes. Nat Rev Microbiol 16:67–79

Huson DH, Scornavacca C (2010) A survey of combinatorial methods for phylogenetic networks. Genome Biol Evol 3:23–35

Huynen MA, Duarte I, Szklarczyk R (2013) Loss, replacement and gain of proteins at the origin of the mitochondria. Biochim Biophys Acta 1827:224–231

Hwang W, Lang MJ (2009) Mechanical design of translocating motor proteins. Cell Biochem Biophys 54. https://doi.org/10.1007/s12013009-9049-1

Ibarra-Laclette E, Lyons E, Hernández-Guzmán G, Peréz-Torres CA, Carretero-Paulet L et al (2013) Architecture and evolution of a minute plant genome. Nature 498:94–98

Imachi H, Nobu MK, Nakahara N, Morono Y, Ogawara M et al (2020) Isolation of an archaeon at the prokaryote-eukaryote interface. Nature 577:519–525

Ingerson-Mahar M, Gitai Z (2011) A growing family: the expanding universe of the bacterial cytoskeleton. FEMS Microbiol Rev 36:256–266

Ingerson-Mahar M, Briegel A, Werner JN, Jensen GJ, Gitai Z (2010) The metabolic enzyme CTP synthase forms cytoskeletal filaments. Nat Cell Biol 12(8):739–746

Inskeep WP, Ackerman GG, Taylor WP, Kozubal M, Korf S, Macur RE (2005) On the energetics of chemolithotrophy in non equilibrium systems: case studies of geothermal springs in Yellowstone National Park. Geobiology 3:297–317

Ionescu D, Bizic M (2019) Giant bacteria. In: eLS. Wiley. https://doi.org/10.1002/9780470015902.a0020371.pub2

Irimia M, Penny D, Roy SW (2007) Coevolution of genomic intron number and splice sites. Trends Genet. https://doi.org/10.1016/j.tig.2007.04.001

Isayenkov S, Isner JC, Maathuis FJM (2010) Vacuolar ion channels: Roles in plant nutrition and signalling. FEBS Lett 584:1982–1988

Islam MR, Hoque MN, Rahman MS, Ul Alam ASMR, Akther M et al (2020) Genome-wide analysis of SARS-CoV-2 virus strains circulating worldwide implicates heterogeneity. Sci Rep 10:14004

Jaakkola ST, Zerulla K, Guo Q, Liu Y, Ma H et al (2012) Halophilic archaea cultivated from surface sterilized middle-late eocene rock salt are polyployd. PLoS One 9(1):e110533

Jackson SA, MyKenzie RE, Pagerlund RD, Kieper SN, Fineran PC, Brouns SJJ (2017) CRISPR-Cas: adapting to change. Science 356:eaal5056

Jaenisch R, Bird A (2003) Epigenetic regulation of gene expression: how the genome integrates intrinsic and environmental signals. Nat Genet 33:245–254

Jančúchová-Lásková J, Landová E, Frynta D (2015) A genetically distinct lizard species able to hybridize? A review. Curr Zool 61(1):155–180

Janzen DH (1967) Synchronization of sexual reproduction of trees within the dry season in Central America. Evolution 21:620–637

Jarrell KF, Bayley DP, Kostyukova AS (1996) The archaeal flagellum: a unique motility structure. J Bacteriol 178(17):5057–5064

Jarrell KF, Ding Y, Nair DB, Siu S (2013) Surface appendages of archaea: structure, function, genetics and assembly. Life 3:86–117

Jeffares DC, Mourier T, Penny D (2005) The biology of intron gain and loss. Trend Genet. https://doi.org/10.1016/j.tig.2005.10.006

Jékely G (2014) Origin and evolution of the self-organizing cytoskeleton in the networks of eukaryotic organelles. Cold Spring Harb Perspect Biol 6:a016030

Jenik J (1994) Clonal growth in woody plants: a review. Folia Geobot Phytotax 29:291–306

Jenkinson IR, Sun J (2014) Drag increase and drag reduction found in phytoplankton and bacterial cultures in laminar flow: are cell surfaces and EPS producing rheological thickening and a lotus-leaf effect? Deep Sea Res II 101:216–230

Jensen PE, Leister D (2014) Chloroplast evolution, structure and functions. F1000Prime Rep 6:40

Jeong H, Tombor B, Albert R, Oltvai ZH, Barabási A-L (2000) The large-scale organization of metabolic networks. Nature 407:651–654

Jiang H, Guan E, Gu Z (2010) Tinkering evolution of post-transcriptional RNA regulons: Puf3p in fungi as an example. PLoS Genet 6(7):e1001030

Jiao N, Azam F, Sanders S (2011a) The microbial carbon pump in the ocean. The American Association for the Advancement of Science

Jiao Y, Wickett NJ, Ayyampalayam S, Chanderbali AS, Landherr L et al (2011b) Ancestral polyploidy in seed plants and angiosperms. Nature 472:97–100

Johansen JE, Pinhassi J, Blackburn N, Zweifel UL, Hagström Å (2002) Variability in motility characteristics among marine bacteria. Aquat Microb Ecol 28:229–237

Johnson CH, Knight MR, Kondo T, Masson P, Sedbrrok J et al (1995) Circadian oscillations of cytosolic and chloroplastic free calcium in plants. Science 269:1863–1865

Johnson MTJ, FitzJohn RG, Smith SD, Rausher MD, Otto SP (2011) Loss of sexual reproduction and segregation is associated with increased diversification in evening primroses. Evolution 65(11):3230–3240

Jones SE, Ho L, Rees CA, Hill JE, Nodwell JR, Elliot MA (2017) *Streptomyces* exploration is triggered by fungal interactions and volatile signals. eLife 6:e21738

Jordal BH, Cognato AI (2012) Molecular phylogeny of bark and ambrosia beetles reveals multiple origin of fungus farming during periods of global warming. BMC Evol Biol 12:133

Jorgensen RA (2002) RNA traffics information systematically in plants. PNAS 99(18):11561–11563

Jørgensen B (2006) Bacteria and marine biogeochemistry. In: Schulz HD, Zabel M (eds) Marine geochemistry. Springer, Berlin, pp 169–206

Jorgensen P, Rupeš I, Sharom JR, Schneper L, Broach JR, Tyers M (2004) A dynamic transcriptional network communicates growth potential to ribosome synthesis and critical cell size. Genes Dev 18:2491–2505

Jørgensen A, Faurby S, Krog Persson D, Agerlin Halberg K, Møbjerg Kristensen R, Møbjerg N (2013) Genetic diversity in the parthenogenetic reproducing tardigrade *Echiniscus testudo* (Heterotardigrada: Echiniscoidea). J Limnol 72(s1):136–143

Judson OP, Normark BB (1996) Ancient asexual scandals. Trends Ecol Evol 13(2):41–46

Juhas M, van der Meer JR, Gaillard M, Harding RM, Hood DW, Crook DW (2009) Genomic islands: tools of bacterial horizontal gene transfer and evolution. FEMS Microbiol Rev 33:376–393

Kærn M, Elston TC, Blake WJ, Collins JJ (2005) Stochasticity in gene expression: from theories to phenotypes. Nat Rev Genet 6:451–464

Kaessmann H (2010) Origins, evolution, and phenotypic impact of new genes. Genome Res 20:1313–1326

Kaiser TS, Poehn B, Szkiba D, Preussner M, Sedlazeck FJ et al (2016) The genomic basis of circadian and circalunar timing adaptations in a midge. Nature 540:69–73

Kamel SJ, Grosberg RK, Marshall DJ (2010) Family conflicts in the sea. Trends Ecol Evol 25:442–449

Kamennaya NA, Ajo-Franklin CM, Northern T, Jansson C (2012) Cyanobacteria as biocatalysts for carbonate mineralization. Minerals 2:338–364

Kamp C, Wilke CO, Adami C, Bornholdt S (2002) Viral evolution under the pressure of an adaptive immune system - optimal mutation rates for viral escape. arXiv:cond-mat/0209613v1

Kang S, Tice AK, Spiegel FW, Silberman JD, Pánek T et al (2017) Between a pod and a hard test: the deep evolution of amoebae. Mol Biol Evol 34(9):2258–2270

Kannan S, Ali PSS, Sheeza A, Hemalatha K (2020) COVID-19 (novel coronavirus 2019)—recent trends. Eur Rev Med Pharmacol Sci 24:2006–2011

Kappler A, Pasquero C, Konhauser KO, Newman DK (2005) Deposition of banded iron formations by anoxygenic phototrophic Fe(II)-oxidizing bacteria. Geology 33(11):865–868

Karnkowska A, Vacek V, Zubáčová Z, Treitli SC, Petrželková R et al (2016) A eukaryote without a mitochondrial organelle. Curr Biol 26:1274–1284

Katzourakis A, Aswad A (2016) Evolution: endogenous viruses provide shortcuts in antiviral immunity. Curr Biol 26:R427–R429

Kawasaki K, Weiss KM (2003) Mineralized tissue and vertebrate evolution: the secretory calcium-binding phosphoprotein gene cluster. PNAS 100(7):4060–4065

Kazlauskiene M, Kostink G, Venclovas Č, Tamulaitis G, Siksnys V (2017) A cyclic oligonucleotide signaling pathway in type III CRISPR-Cas systems. Science 357:605–609

Kazmierczak J, Kempe S, Kremer B (2013) Calcium in the early evolution of living systems: a biohistorical approach. Curr Org Chem 17:1738–1750

Kee J, Thudium S, Renner DM, Glastad K, Palozola K et al (2022) SARS-CoV-3 disrupts host epigenetic regulation via histone mimicry. Nature 610:381–388

Keeling PJ, Palmer JD (2008) Horizontal gene transfer in eukaryotic evolution. Nat Rev Genet 9:605–618

Keeling PJ, Fast NM, Law JS, Williams BAP, Slamovits CH (2005) Comparative genomics of microsporidia. Folia Parasitol 52:8–14

Keene JD (2007) RNA regulons: coordination of post-transcriptional events. Nat Rev Genet 8:533–543

Kehr J, Buhtz A (2008) Long distance transport and movement of RNA through the phloem. J Exp Bot 58(1):85–92

Kelkar YD, Ochman H (2011) Causes and consequences of genome expansion in fungi. Genome Biol Evol 4(1):13–23

Kelman LM, Kelman Z (2003) Archaea: an archetype for replication initiation studies? Mol Microbiol 48(3):605–615

Kenzaka T, Tani K, Nasu M (2010) High-frequency phage-mediated gene transfer in freshwater environments determined at single-cell level. ISME J 4:648–659

Keppner A, Maric D, Correira M, Koay TW, Orlando IMC et al (2020) Lessons from the post-genomic era: globin diversity beyond oxygen binding and transport. Redox Biol 37:101687

Kerfeld CA, Aussignargues C, Zarzycki J, Cai F, Sutter M (2018) Bacterial microcompartments. Nat Rev Microbiol 16(5):277–290

Kettenmann H, Verkhratsky A (2008) Neuroglia: the 150 years after. Trends Neurosci 31(12):653–659

Keverne EB (2014) Mammalian viviparity: a complex niche in the evolution of genomic imprinting. Heredity 113:138–144

Kim VN (2005) Small RNAs: classification, biogenesis, and function. Mol Cells 19(1):1–15

Kim E, Archibald JM (2008) Diversity and evolution of plastids and their genomes. Plant Cell Monogr. https://doi.org/10.1007/7089_2008_17

Kim BH, Gadd GM (2008) Bacterial physiology and metabolism. Cambridge University Press, Cambridge

Kim T-K, Shiekhattar R (2015) Architectural and functional commonalities between enhancers and promoters. Cell 162:948–959

Kim J-R, Shin D, Jung SH, Heslop-Harrison P, Cho K-H (2010) A design principle underlying the synchronization of aicillations in cellular systems. J Cell Sci 123(4):537–543

King N (2004) The unicellular ancestry of animal development. Dev Cell 7:313–325

King N, Rokas A (2017) Embracing uncertainty in reconstructing early animal evolution. Curr Biol 27:R1081–R1088

Kiørboe T, Jackson GA (2001) Marine snow, organic plumes, and optimal chemosensory behaviour of bacteria. Limnol Oceanogr 46(6):1309–1318

Klasson L, Westberg J, Spountzis P, Näslund K, Lutnaes Y et al (2009) The mosaic genome structure of the *Wolbachia* wRi strain infecting *Drosophila simulans*. PNAS 106(14):5725–5730

Kleine T, Maier UG, Leister D (2009) DNA transfer from organelles to the nucleus: the idiosyncratic genetics of endosymbiosis. Rev Plant Biol 60:115–138

Knoll AH, Bergman KD, Strauss JV (2016) Life: the first two billion years. Philos Trans R Soc B 371:20150493

Ko CH, Takahashi JS (2006) Molecular components of the mammalian circadian clock. Hum Mol Genet 15(2):R271–R277

Kohler A, Kuo A, Nagy LG, Morin E, Barry KW et al (2015) Convergent losses of decay mechanisms and rapid turnover of symbiosis genes in mycorrhizal mutualists. Nat Genet 47(4):410–415

Koivisto RKK, Braig HR (2003) Microorganisms and parthenogenesis. Biol J Linn Soc 79:43–58

Kolář F, Lučanová M, Vít P, Urfus T, Chrtek J et al (2013) Diversity and endemism in deglaciated areas: ploidy, relative genome size and niche differentiation in the *Galium pusillum* complex (Rubiaceae) in Northern and Central Europe. Ann Bot 111:1095–1108

Kolundžija S, Cheng D-Q, Lauro FM (2022) RNA viruses in aquatic ecosystems through the lens of ecological genomics and transcriptomics. Viruses 14:702

Komienko AL, Guenzl PM, Barlow DP, Pauler FM (2013) Gene regulation by the act of long non-coding RNA transcription. BMC Biol 11:59

Kondrashov FA, Koonin EV, Morhunov IG, Finogenova TV, Kondrashova MN (2006) Evolution of glyoxylate cycle enzymes in metazoa: evidence of multiple horizontal transfer events and pseudogene formation. Biol Direct 1:31

Konhauser KO (2007) Introduction to geomicrobiology. Blackwell, Malden

Konhauser KO, Kappler A, Roden EE (2011) Iron in microbial metabolism. Elements 7:89–93

Koonin EV (2015) Energetics and population genetics at the root of eukaryotic cellular and genomic complexity. PNAS 112(52):15777–15778

Koonin EV, Dolja VV (2013) A virocentric perspective on the evolution of life. Curr Opin Virol 3(5):546–557

Koonin EV, Dolja VV (2014) Virus world as an evolutionary network of viruses and capsidless selfish elements. Microbiol Mol Biol Rev 78(2):278–303

Koonin EV, Senkevich TG, Dolja VV (2006) The ancient virus world and evolution of cells. Biol Direct 1:29

Koonin EV, Dolja VV, Krupovic A, Wolf YI, Yutin N et al (2020) Global organization and proposed megataxonomy of the virus world. Microbiol Mol Biol Rev 84:e00061–e00019

Korber B, Fischer WM, Gnanakaran S, Yoon H, Theiler J et al (2020) Tracking changes in SARS-CoV-2 spike: evidence that D614G increases infectivity of the COVID-19 virus. Cell 182:812–827

Koskella B, Brockhurst MA (2014) Bacteria-phage coevolution as a driver of ecological and evolutionary processes in microbial communities. FEMS Microbiol Rev 38:916–931

Kotzin JJ, Mowel WK, Henao-Mejia J (2017) Viruses hijack a host lncRNA to replicate. Science 358:993–994

Kovacs NA, Penev PI, Venapally A, Petrov AS, Williams LD (2018) Circular permutation obscures universality of a ribosomal protein. J Mol Evol. https://doi.org/10.1007/s00239-018-9869-1

Kozminsky-Atias A, Bar-Shalom A, Mishmar D, Zilberberg N (2008) Assembling an arsenal, the scorpion way. BMC Evol Biol. https://doi.org/10.1186/1471-2148/8/333

Krašovec R, Richards H, Gifford DR, Hatcher C, Faulkner KJ et al (2017) Spontaneous mutation rate is a plastic trait associated with population density across domains of life. PLoS Biol 15(8):e2002731

Krell F-T, Cranston PS (2004) Which side of the tree is more basal? Syst Entomol 29:279–281

Krupovic M, Bamford DH (2010) Order to the viral universe. J Virol 84(24):12476–12479

Krupovic M, Koonin EV (2015) Polintons: a hotbed of eukaryotic virus, transposon and plasmid evolution. Nat Rev Microbiol 13:105–115

Krupovic M, Koonin EV (2017) Multiple origins of viral capsid proteins from cellular ancestors. PNAS 114:E2401–E2410

Krupovic M, Prangishvili D, Hendrix RW, Bamford DH (2011) Genomics of bacterial and archaeal viruses: dynamics within the prokaryotic virosphere. Microbiol Mol Biol Rev 75(4):610–635

Ku C, Nelson-Sathi S, Roettger M, Sousa FL, Lockhart PJ et al (2015) Endosymbiotic origin and differential loss of eukaryotic genes. Nature 524:427–432

Kucharski R, Maleszka J, Foret S, Maleszka R (2008) nutritional control of reproductive status in honeybees via DNA methylation. Science 319:1827–1830

Kudla J, Becker D, Grill E, Hedrich R, Hippler M et al (2018) Advances and current challenges in calcium signalling. New Phytol 218:414–431

Kuhlwilm M, Gronau I, Hubisz MJ, de Filippo C, Martinez JP et al (2016) Ancient gene flow from early modern humans into Eastern Neanderthals. Nature 530:429–433

Kuhn G, Hijri M, Sanders IR (2001) Evidence for the evolution of multiple genomes in arbuscular mycorrhizal fungi. Nature 414:745–748

Kull FJ, Vale RD, Fletterick RJ (1998) The case for a common ancestor: kinesin and myosin motor proteins and G proteins. J Muscle Res Cell Motil 19:877–886

Kulp A, Kuehn MJ (2010) Biological functions and biogenesis of secreted bacterial outer membrane vesicles. Annu Rev Microbiol 64:163–184

Kundler P, Steinbrenner K, Smukalski M, Kunze A, Quast J et al (1989) Erhöhung der Bodenfruchtbarkeit. VEB Deutscher Landwirtschaftsverlag, Berlin

Kunin V, Ouzounis CA (2003) The balance of driving forces during genome evolution in prokaryotes. Genome Res 13:1589–1594

Kunsch K (1997) Der Mensch in Zahlen. Gustav Fischer, Stuttgart

Kuo C-H, Ochman H (2009) Deletional bias across the three domains of life. Genome Biol Evol 1:145–152

Kurland CG, Collins LJ, Penny D (2006) Genomics and the irreducible nature of eukaryote cells. Science 312:1011–1014

Kürschner WM, Batenburg SJ, Mander L (2013) Aberrant *Classopollis* pollen reveals evidence for unreduced (2n) pollen in the conifer family Cheirolepidiaceae during the Triassic-Jurassic transition. Proc Biol Sci 280:20131708

La Scola B, Desnues C, Pagnier I, Robert C, Barrassi L et al (2008) The virophage as a unique parasite of the giant mimivirus. Nature 455:100–104

Lafontaine DLJ, Tollervey D (2001) The function and biosynthesis of ribosomes. Nat Rev Mol Cell Biol 2:514–520

Laforsch C, Tollrian R (2004) Extreme helmet formation in *Daphnia cucullata* induced by small-scale turbulence. J Plankton Res 26(1):81–87

Laganenka L, Sander T, Lagonenko A, Chen Y, Link H, Sourjik V (2019) Quorum sensing and metabolic state of the host control lysogeny-lysis switch of bacteriophage T1. mBio 10(5):e01884–e01819

Lahr DJG, Parfrey LW, Mitchell EAD, Katz LA, Lara E (2011) The chastity of amoebae: re-evaluating evidence for sex in amoeboid organisms. Proc Biol Sci. https://doi.org/10.1098/rspb.2011.0289

Lahti DC, Johnson NA, Ajie BC, Otto SP, Hendry AP et al (2009) Relaxed selection in the wild. Trends Ecol Evol 24(9):487–496

Lake JA, Rivera MC (2004) Deriving the genomic tree of life in the presence of horizontal gene transfer: conditioned reconstruction. Mol Biol Evol 21(4):681–690

Lalanne J-B, Taggart JC, Guo MS, Herzel L, Schieler A, Li G-W (2018) Evolutionary convergence of pathway-specific enzyme expression stoichiometry. Cell 173:749–781

Lane CE, Archibald JM (2008) The eukaryotic tree of life: endosymbiosis takes its TOL. Trends Ecol Evol 23(5):268–275

Lane N, Martin W (2010) The energetics of genome complexity. Nature 467:929–934

Lane N, Martin WF (2012) The origin of membrane bioenergetics. Cell 151:1406–1416

Lang AS, Zhaxybayeva O, Beatty JT (2012) Gene transfer agents: phage-like elements of genetic exchange. Nat Rev Microbiol 10(7):472–482

Lange C, Zerulla K, Breuert S, Soppa J (2011) Gene conversion results in the equalization of genome copies in the polyploidy haloarchaeon *Haloferax volcanii*. Mol Microbiol 80(3):666–677

Lapierre P, Gogarten JP (2009) Estimating the size of bacterial pan-genome. Trends Genet 25(3):107–110

Lauber C, Gorbalenya AE (2012) Toward genetics-based virus taxonomy: comparative analysis of a genetics-based classification and the taxonomy of picornaviruses. J Virol 86(7):3905–3915

Lauro FM, McDougald D, Thomas T, Williams TJ, Egan S et al (2009) The genomic basis of trophic strategy in marine bacteria. PNAS 106(37):15527–15533

Lavanchy G, Strehler M, Roman MNL, Lessard-Therrien M, Humbert J-Y et al (2016) Habitat heterogeneity favors axexual reproduction in natural populations of grassthrips. Evolution 70(8):1780–1790

Lavialle C, Cornelis G, Dupressoir A, Esnault C, Heidmann O et al (2013) Paleovirology of 'syncytins', retroviral env genes exapted for a role in placentation. Philos Trans R Soc Lond B Biol Sci 368:20120507

Law JH, Crespi BJ (2002) The evolution of geographic parthenogenesis in *Timema* walking-sticks. Mol Ecol 11:1471–1489

Leclère L, Horin C, Chevalier S, Lapébie P, Dru P et al (2019) The genome of the jellyfish Clytia hemisphaerica and the evolution of the cnidarian life-cycle. Nat Ecol Evol. https://doi.org/10.1038/s41559-019-0833-2

Lecointre G, Le Guyader H (2006) The tree of life. Harvard University Press, Cambridge

Lecroq B, Gooday AJ, Tsuchiya M, Pawlowski J (2009) A new genus of Xenophyophores (Foraminifera) from Japan Trench: morphological description, molecular phylogeny and elemental analysis. Zool J Linn Soc 156:455–464

Ledford H (2023) Dogma-defying bacteria package DNA in unusual ways. Nature 614:401

Ledón-Rettig CC, Moczek AP, Ragsdale EJ (2018) *Diplogastrellus* nematodes are sexually transmitted mutualists that alter the bacterial and fungal communities of their beetle host. PNAS 115(42):10696–10701

Lee JT (2012) Epigenetic regulation by long noncoding RNA. Science 338:1435–1439

Lee SC, Ni M, Li W, Shertz C, Heitman J (2010) The evolution of sex: a perspective from the fungal kingdom. Microbiol Mol Biol Rev 74(2):298–340

Lee C-T, Chen I-T, Yang Y-T, Ko T-P, Huang Y-T et al (2015) The opportunistic marine pathogen *Vibrio parahaemolyticus* becomes virulent by acquiring a plasmid that expresses a deadly toxin. PNAS 112(34):10798–10803

Lee MD, Kling JD, Araya R, Ceh J (2018) Jellyfish life stages shape associated microbial communities, while a core microbiome is maintained across all. Front Microbiol 9:1534

Lefébure T, Bitar PDP, Suzuki H, Stanhope MJ (2010) Evolutionary dynamics of complete *Campylobacter* pan-genomes and the bacterial species concept. Genome Biol Evol 2:646–655

Lehnert EM, Mouchka ME, Burriesci MS, Gallo ND, Schwarz JA, Pringle JR (2014) Extensive differences in gene expression between symbiotic and aposymbiotic cnidarians. G3 (Bethesda) 4:277–295

Lehtonen J, Kokko H, Parker GA (2016) What do isogamous organisms teach us about sex and the two sexes? Philos Trans R Soc Lond B Biol Sci 371:20150532

Leighton GM, Lu L, Holop E, Dobler I, Ligon RA (2021) Sociality and migration predict hybridization across birds. Proc Biol Sci 288:20201946

Leitch AR, Soltis DE, Soltis PS, Leitch IJ, Pires JC (2004) Genome downsizing in polyploid plants. Biol J Linn Soc 82:651–663

Lemay MA, Martone PT, Keeling PJ, Burt JM, Krumhansl KA et al (2018) Sympatric kelp species share a large portion of their surface bacterial communities. Environ Microbiol 20(2):658–670

Lemmon MA, Schlesinger J (2010) Cell signaling by receptor tyrosine kinases. Cell 141:1117–1134

Lenormand T, Engelstädter J, Johnston SE, Wijnker E, Haag CR (2016) Evolutionary mysteries in meiosis. Philos Trans R Soc Lond B Biol Sci 371:20160001

Lenton TM, Daines SJ (2018) The effects of marine eukaryote evolution on phosphorus, carbon and oxygen cycling across the proterozoic-phanerozoic transition. Emerg Top Life Sci 2:267–278

Leslie M (2017) Cell-like giant viruses found. Science 356:15–16

Lever MA, Rogers KL, Lloyd KG, Overmann J, Schink B et al (2015) Life under extreme energy limitation: a synthesis of laboratory- and field-based investigations. FEMS Microbiol Rev 39(5):688–728

Levin DA (2013) The timetable for allopolyploidy in flowering plants. Ann Bot 112:1201–1208

Levin PA, Angert ER (2015) Small but mighty: cell size and bacteria. Cold Spring Harb Perspect Biol 7:a019216

Levin LA, Gooday AJ (1992) Possible roles for xenophyophores in deep-sea carbon cycling. In: Rowe GT, Pariente V (eds) Deep-sea food chains and the global carbon cycle. Kluwer Academic Publishers, Dordrecht, The Netherlands

Levin TC, King N (2013) Evidence for sex and recombination in the choanoflagellate *Salpingoeca rosetta*. Curr Biol 23:2176–2180

Levy A, Feldman M (2002) The impact of polyploidy on grass genome evolution. Plant Physiol 130:1567–1593

Lewalter K, Müller V (2006) Bioenergetics of archaea: ancient energy conserving mechanisms developed in the early history of life. Biochim Biophys Acta 1757:437–445

Lewis WM Jr (1985) Nutrient scarcity as an evolutionary cause of haploidy. Am Nat 125:692–701

Ley RE, Hamady M, Lozupone C, Turnbaugh P, Ramey RR et al (2008a) Evolution of mammals and their gut microbes. Science 320:1647–1651

Ley RE, Lozupone CA, Hamady M, Knight R, Gordon JI (2008b) Worlds within worlds: evolution of the vertebrate gut microbiota. Nat Rev Microbiol 6:776–788

Li W, Tucker AE, Sung W, Thomas WK, Lynch M (2009) Extensive, recent intron gains in daphnia populations. Science 326:1260–1262

Li G-W, Burkhardt D, Gross C, Weissman JS (2014) Quantifying absolute protein synthesis rates reveals principles underlying allocation of cellular resources. Cell 157:624–635

Li X, Kim Y, Tsang EK, Davis JR, Damani FN et al (2017) The impact of rare variation on gene expression across tissues. Nature 550:239–243

Lindås A-C, Karlsson EA, Lindgren MT, Ettema TJG, Bernander R (2008) A unique cell division machinery in the archaea. PNAS 105(48):18942–18946

Linder HP, Barker NP (2014) Does polyploidy facilitate long-distance dispersal? Ann Bot 113:1175–1183

Lindholm AK, Dyer KA, Firman RC, Fishman L, Forstmeier W et al (2016) The ecology and evolutionary dynamics of meiotic drive. Trends Ecol Evol 31(4):315–326

Liu F (2017) Enhancer-derived RNA: a primer. Genomics Proteomics Bioinformatics 15:196–200

Liu H, Fu Y, Jiang D, Li G, Xie J et al (2010) Widespread horizontal gene transfer from double-stranded RNA viruses to eukaryotic nuclear genomes. J Virol 84(22):11876–11887

Liu H, Fu Y, Li B, Yu J, Xie J et al (2011) Widespread horizontal gene transfer from circular single-stranded DNA viruses to eukaryotic genomes. BMC Evol Biol 11:276

Liu Y-C, Kuo R-L, Shih S-R (2020) COVID-19: the first documented coronavirus pandemic in history. Biomed J 43:328–333

Lladó S, López-Mondéjar R, Baldrian P (2017) Forest soil bacteria: diversity, involvement in ecosystem processes, and response to global change. Microbiol Mol Biol Rev 81(2):e00063-16

Lloyd AC (2013) The regulation of cell size. Cell 154:1194–1206

Londei P, Ferreira-Cerca S (2021) Ribosome biogenesis in archaea. Front Microbiol 12:686977

Long M, Betrán E, Thornton K, Wang W (2003) The origin of new genes: glimpses from the young and old. Nat Rev Genet 4:865–875

Long HK, Prescott SL, Wysocka J (2016) Ever-changing landscapes: transcriptional enhancers in development and evolution. Cell 167:1170–1187

Longdon B, Brockhurst MA, Russell CA, Welch JJ, Jiggins FM (2014) The evolution and genetics of virus host shifts. PLoS Pathog 10(11):e1004395

López-Garcia P, Moreira D (1999) Metabolic symbiosis at the origin of eukaryotes. Trends Biochem Sci 24:88–93

López-García P, Moreira D (2008) Tracking microbial biodiversity through molecular and genomic ecology. Res Microbiol 139:67–73

López-Garcia P, Moreira D (2015) Open questions on the origin of eukaryotes. Trends Ecol Evol 30(11):697–708

López-Garcia P, Moreira D (2020) Cultured asgard archaea shed light on eujaryogenesis. Cell 181:232–235

López-Sandoval DC, Rodriguez-Ramos T, Cermeño P, Sobrino C, Marañon E (2014) Photosynthesis and respiration in marine phytoplankton: relationship with cell size, taxonomic affiliation, and growth phase. J Mar Biol Ecol 457:151–159

Loudon ASI (2012) Circadian biology: a 3.5 billion year old clock. Curr Biol 22(14):R570–R571

Lourenço WR (2008) Parthenogenesis in scorpions: some history—new data. J Venom Anim Toxins Incl Trop Dis 14(1):19–44

Lovley DR (2008) Extracellular electron transfer: wires, capacitors, iron lungs, and more. Geobiology 6:225–231

Löwe J, Amos LA (2009) Evolution of cytomotive filaments: the cytoskeleton from prokaryotes to eukaryotes. Int J Biochem Cell Biol 41(2):323–329

Lowe CB, Bejerano G, Haussler D (2007) Thousands of human mobile element fragments undergo strong purifying selection near development genes. PNAS 104(19):8005–8010

Lu C, Thompson CB (2012) Metabolic regulation of epigenetics. Cell Metab 16:9–17

Lumbsch HT, Leavitt SD (2011) Goodbye morphology? A paradigm shift in the delimitation of species in lichenized fungi. Fungal Divers. https://doi.org/10.1007/s13225-011-0123-z

Lundgren M, Bernander R (2007) Genome-wide transcription map of an archaeal cell cycle. PNAS 104(8):2939–2944

Lundgren M, Malandrin L, Eriksson S, Huber H, Bernander R (2008) Cell cycle characteristics of *Crenarchaeota*: unity among diversity. J Bacteriol 190(15):5362–5367

Luo H, Csúros M, Hughes AL, Moran MA (2013) Evolution of divergent life history strategies in marine alphaproteobacteria. mBio 4(4):e00373–e00313

Lutkenhaus J, Pichoff S, Du S (2012) Bacterial cytokinesis: from Z ring to divisome. Cytoskeleton 69(10):778–790

Lygis V, Vasliauskas R, Larsson K-H, Stenlid J (2005) Wood-inhabiting fungi in stems of fraxinus excelsior in declining ash stands of northern Lithuania, with particular reference to *Armillaria cepistipes*. Scand J For Res 20:337–346

Lynch M (2006) Streamlining and simplification of microbial genome architecture. Annu Rev Microbiol 60:327–349

Lynch M, Marinov GK (2015) The bioenergetic costs of a gene. PNAS 112(51):15690–15895

Lynch M, Bürger R, Butcher D, Gabriel W (1993) The mutational meltdown in asexual populations. J Hered 84:339–344

Lynch VJ, Leclerc RD, May G, Wagner GP (2011) Transposon-mediated rewiring of gene regulatory networks contributed to the evolution of pregnancy in mammals. Nat Genet. https://doi.org/10.1038/ng.917

Mable BK (2004) 'Why polyploidy is rarer in animals then in plants': myths and mechanisms. Biol J Linn Soc 82:453–466

Maccari M, Amat F, Hontoria F, Gómez A (2014) Laboratory generation of new parthenogenetic lineages supports contagious parthenogenesis in *Artemia*. PeerJ 2:e439

MacLean LCW, Pray TJ, Onstott TC, Brodie EL, Hazen TC, Southam G (2007) Mineralogical, chemical and biological characterization of an anaerobic biofilm collected from a borehole in a deep gold mine in South Africa. Geomicrobiol J 24:491–504

Madigan MT, Martinko JM, Parker J (2003) Brock biology of microorganisms. Prentice Hall, Upper Saddle River, NJ

Madsen EL (2011) Microorganisms and their roles in fundamental biogeochemical cycles. Curr Opin Biotechnol 22:456–464

Madsen JS, Burmølle M, Hansen LH, Sørensen SJ (2012) The interconnection between biofilm formation and horizontal gene transfer. Fems Immunol Med Microbiol 65:183–195

Magnúsdóttir S, Ravcheev D, de Crécy-Lagard V, Theis I (2015) Systematic genome assessment of B-vitamin biosynthesis suggests co-operation among gut microbes. Front Genet 6:148

Mahadevan L, Matsudaira P (2000) Motility powered by supramolecular springs and ratchets. Science 288:95–99

Makarova KS, Yutin N, Bell SD, Koonin EV (2010) Evolution of diverse cell division and vesicle formation systems in archaea. Nat Rev Microbiol 8(10):731–741

Makarova KS, Galperin MY, Koonin EV (2017) Proposed role for KaiC-like ATPase as major signal transduction hubs in archaea. mBio 8:e01959

Maldonaldo M, Riesgo A (2008) Reproduction in the phylum porifera: a synoptic overview. Biol Reprod 59:29–49

Malik YS, Sircar S, Bhat S, Sharun K, Dhama K et al (2020) Emerging novel coronavirus (2019-nCoV)—current scenario, evolutionary perspective based on genome analysis and recent developments. Vet Q 40(1):68–76

Malone CD, Hannon HJ (2009) Small RNAs as guardians of the genome. Cell 136:656–668

Malvankar NS, Lovley DR (2014) Microbial nanowires for bioenergy application. Curr Opin Biotechnol 27:88–95

Maniatis T, Reed R (2002) An extensive network of coupling among gene expression machines. Nature 416:499–506

Maniatsi S, Baxevanis AD, Kappas I, Deligiannidis P, Triantafyllidis A et al (2011) Is polyploidy a persevering accident or an adaptive evolutionary pattern? The case of the brine shrimp *Artemia*. Mol Phylogenet Evol 58:353–364

Maniscalco M, Nannen J, Sodi V, Silver G, Lowrey PL, Bidle KA (2014) Light-dependent expression of four cryptic archaeal circadian gene homologs. Front Microbiol 5:79

Manjrekar J (2017) Epigenetic inheritance, prions and evolution. J Genet 96:445–456

Marcet-Houben M, Gabaldón T (2016) Horizontal acquisition of toxic alkaloid synthesis in a clade of plant associated fungi. Fungal Genet Biol 86:71–80

Marcussen T, Heier L, Brysting AK, Oxelman B, Jakobsen KS (2015) From gene trees to a dated allopolyploid network: insights from the angiosperm Genus *Viola* (Violaceae). Syst Bot 64(1):84–101

Margolin W (2005) FtsZ and the division of prokaryotic cells and organelles. Nat Rev Cell Biol 6(11):862–871

Margolin W (2009) Sculpting the bacterial cell. Curr Biol 19:R812–R822

Margulis L, Dolan MF, Guerrero R (2000) The chimeric eukaryote: origin of the nucleus from the karyomastigont in amitochondrial protists. PNAS 97(13):6954–6959

Martijn J, Vosseberg J, Guy L, Offre P, Ettema TJG (2018) Deep mitochondrial origin outside the sampled alphaproteobacteria. Nature 557:101–105

Martin W (1999) A briefly argued case that mitochondria and plastids are descendants of endosymbionts, but that the nuclear compartment is not. Proc Biol Sci 266:1387–1395

Martin W (2005) Archaebacteria (Archaea) and the origin of the eukaryotic nucleus. Curr Opin Microbiol 8:630–637

Martin W, Herrmann RG (1998) Gene transfer from organelles to the nucleus: how much, what happens, and why? Plant Physiol 118:9–17

Martin W, Koonin EV (2006) Introns and the origin of nucleus-cytosol compartmentalization. Nature 440:41–45

Martin W, Rotte C, Hoffmeister M, Theissen U, Gelius-Dietrich G et al (2003) Early cell evolution, eukaryotes, anoxia, sulfide, oxygen, fungi first (?), and a tree of genomes revisited. IUBMB Life 55(4–5):193–204

Martin WF, Neukirchen S, Zimorski V, Gould SB, Sousa FL (2016) Energy for two: new archaeal lineages and the origin of mitochondria. Bioessays 38:850–856

Martin WF, Tielens AGM, Mentel M, Garg SG, Gould SB (2017) The physiology of phagocytosis in the context of mitochondrial origin. Microbiol Mol Biol Rev 61(3):e00008–e00017

Martínez DE, Bridge D (2012) *Hydra*, the everlasting embryo, confronts aging. Int J Dev Biol 56:479–487

Massana R, Pedrós-Alió C (1994) Role of anaerobic ciliates in planktonic food webs: abundance, feeding, and impact on bacteria in the field. Appl Environ Microbiol 60(4):1325–1334

Masson P, Hulo C, De Castro E, Bitter H, Gruenbaum L et al (2013) ViralZone: recent updates to the virus knowledge resource. Nucleic Acid Res 41:D579–D583

Matos J, West SC (2014) Holiday junction resolution: regulation in space and time. DNA Repair 19:176–181

Matos J, Blanco MG, Maslen S, Skehel JM, West SC (2011) Regulatory control of the resolution of DNA recombination intermediates during meiosis and mitosis. Cell 147:158–172

Matsuoka Y (2011) Evolution of polyploid *Triticum* wheats under cultivation: the role of domestication, natural hybridization and allopolyploid speciation of their diversification. Plant Cell Physiol 52(5):750–764

Mayer F, Müller V (2014) Adaptations of anaerobic archaea to life under extreme energy limitation. FEMS Microbiol Rev 38:449–472

Mayr E (1984) Die Entwicklung der biologischen Gedankenwelt. Springer, Berlin

Mazel D (2006) Integrons: agents of bacterial evolution. Nat Rev Microbiol 4:608–620

Mazid M, Khan TA, Mohammad F (2011) Role of secondary metabolites in defense mechanisms of plants. Biol Med 3(2):232–249

McCoy RC, Wakefield J, Akey JM (2017) Impacts of neanderthal-introgressed sequences on the landscape of human gene expression. Cell 168:916–927

McCutcheon JP, Moran NA (2012) Extreme genome reduction in symbiotic bacteria. Nat Rev Microbiol 10:13–26

McDaniel LD, Young E, Delaney J, Ruhnau D, Ritchie KB, Paul JH (2010) High frequency of horizontal gene transfer in the oceans. Science 330:50

McDonald JL, Franco M, Townley S, Ezard THG, Jelbert K, Hodgson DJ (2017) Divergent demographic strategies of plants in variable environments. Nat Ecol Evol. https://doi.org/10.1038/s41559-016-0029

McFall-Ngai MJ (2001) Identifying 'prime suspects': symbioses and the evolution of multicellularity. Comp Biochem Physiol B Biochem Mol Biol 129:711–723

McFall-Ngai MJ (2002) Unseen forces: the influence of bacteria on animal development. Dev Biol 242:1–14

McFall-Ngai M, Hadfield MG, Bosch TCG, Carey HV, Domazet-Lošo T et al (2013) Animals in a bacterial world, a new imperative for the life Sci. PNAS 110(9):3229–3236

McGrath CL, Katz LA (2004) Genome diversity in microbial eukaryotes. Trends Ecol Evol 19(1):32–38

McInerney JO, O'Connell MJ, Pisani D (2014) The hybrid nature of the Eukaryota and a consilient view of life on earth. Nat Rev Microbiol 12:449–455

McKenna DD, Shin S, Ahrens D, Balke M, Beza-Beza C et al (2019) The evolution and genomic basis of beetle diversity. PNAS. https://doi.org/10.1073/pnas.1909655116

McLysaght A, Guerzoni D (2015) New genes from non-coding sequence: the role of de novo protein-coding gene in eukaryotic evolutionary innovation. Philos Trans R Soc Lond B Biol Sci 370:20140332

Medini D, Donati C, Tettelin H, Masignani V, Rappuoli R (2005) The microbial pan-genome. Curr Opin Genet Dev 15:589–594

Méheust R, Lopez P, Bapteste E (2015) Metabolic bacterial genes and the construction of high-level composite lineages of life. Trends Ecol Evol 30(3):127–129

Melnikov S, Ben-Shem A, Garreau du Loubresse N, Jenner L, Yusupova G, Yusupov M (2012) One core, two shells: bacterial and eukaryotic ribosomes. Nat Struct Mol Biol 19(6):560–567

Mergaert P, Uchiumi T, Alunni B, Evanno G, Cheron A et al (2006) Eukaryotic control on bacterial cell cycle and differentiation in the *Rhizobium*-legume symbiosis. PNAS 103(13):5230–5235

Merkl R (2015) Bioinformatik. Wiley-VCH, Weinheim

Merod R, Wuertz S (2014) Extracellular polymeric substance architecture influences natural genetics transformation of acinetobacter baylyi in biofilms. Appl Environ Microbiol 80(24):7752–7757

Metcalf JA, Funkhouser-Jones LJ, Brileya K, Reysenbach A-L, Bordenstein SR (2014) Antibacterial gene transfer across the tree of life. eLife 3:e04266

Michiels J, Xi C, Verhaert J, Vanderleyden J (2001) The functions of Ca^{2+} in bacteria: a role for EF-hand proteins? Trends Microbiol 10(2):87–93

Mignerot L, Coelho SM (2016) The origin and evolution of the sexes: novel insights from a distant eukaryotic lineage. C R Biol 339:252–257

Millar AJ (2004) Input signals to the plant circadian clock. J Exp Biol 55(395):277–283

Mills DB, Boyle RA, Daines SJ, Sperling EA, Pisani D et al (2022) Eukaryogenesis and oxygen in Earth history. Nat Ecol Evol 6:520–532

Miot J, Benzerara K, Obst M, Kappler A, Hegler F et al (2009) Extracellular iron biomineralization by photoautotrophic iron-oxidizing bacteria. Appl Environ Microbiol 75(17):5586–5591

Mira A, Martín-Cuadrado AB, D'Auria G, Rodríguez-Valera F (2010) The bacterial pan-genome: a new paradigm in microbiology. Int Microbiol 13:45–57

Miska EA (2005) How microRNAs control cell division, differentiation and death. Curr Opin Genet Dev 15:563–568

Møbjerg N, Halberg KA, Jørgensen A, Persson D, Bjørn M et al (2011) Survival in extreme environments—on the current knowledge of adaptations in tardigrades. Acta Physiol 202:409–420

Mock T, Kirkham A (2012) What can we learn from genomics approaches in marine ecology? From sequences to eco-systems biology! Mar Ecol 33:131–148

Mohammad A, Alshawaf E, Marafie SK, Abu-Farha M, Abubaker J, Al-Mulla F (2021) Higher binding affinity of furin for SARS-CoV-2 spike (S) protein D613G mutant could be associated with higher SARS-CoV-2 infectivity. Int J Infect Dis 193:611–616

Mohr KI (2018) Diversity of myxobacteria—we only see the tip of the iceberg. Microorganisms 6:84

Moissl C, Rachel R, Briegel A, Engelhardt H, Huber R (2005) The unique structure of archaeal 'hami', highly complex cell appendages with nano-grappling hooks. Mol Microbiol 56(2):361–370

Moldovan MA, Gelfand MS (2018) Pangenomic definition of prokaryotic species and the phylogenetic structure of *Prochlorococcus* spp. Front Microbiol 9:428

Molineux IJ, Panja D (2013) Popping the cork: mechanisms of phage genome ejection. Nature 11:194–204

Money NP (2004) The fungal dining habit: a biomechanical perspective. Mycologist 18(2):71–76

Moniruzzaman M, Martinez-Gutierrez CA, Weinheimer AR, Aylward FO (2020) Dynamic genome evolution and complex virocell metabolism of globally-distributed giant viruses. Nat Commun 11:1710

Montero-Serra I, Linares C, Doak DF, Ledoux JB, Garrabou J (2018) Strong linkages between depth, longevity and demographic stability across marine sessile species. Proc Biol Sci 285:20172688

Moore JK, Villareal TA (1996) Size-ascent rate relationships in positively buoyant marine diatoms. Limnol Oceanogr 41(7):1514–1520

Moore D, Robson GD, Trinci APJ (2011) 21st century guidebook to fungi. Cambridge University Press, Cambridge

Moran NA (2007) Symbiosis as an adaptive process and source on phenotypic complexity. PNAS 104:8627–8633

Moran NA, Wernegreen JJ (2000) Lifestyle evolution in symbiotic bacteria: insights from genomics. Trends Ecol Evol 15(8):321–326

Morens DM, Fauci AS (2007) The 1918 influenza pandemic: insights for the 21st century. J Infect Dis 195:1018–1028

Morens DM, Taubenberger JK (2011) Pandemic influenza: certain uncertainties. Rev Med Virol 21(4):262–284

Morono Y, Ito M, Hoshino T, Terada T, Hori T et al (2020) Aerobic microbial life persists in oxic marine sediment as old as 101.5 million years. Nat Commun. https://doi.org/10.1038/s41467-020-17330-1

Morris JJ, Lenski RE, Zinser ER (2012) The black queen hypothesis: evolution of dependencies through adaptive gene loss. mBio 3(2):e00036–e00012

Moyers BA, Zhang J (2016) Evaluating phylostratigraphic evidence for widespread de novo gene birth in genome evolution. Mol Biol Evol 33(5):1245–1256

Msemburi W, Karlinsky A, Knutson V, Aleshin-Guendel S, Chatterji S, Wakefield J (2023) The WHO estimates of excess mortality associated with the COVID.19 pandemic. Nature 613:130–137

Muller F, Brissac T, Le Bris N, Felbeck H, Gros O (2010) First description of giant Archaea (*Thaumarchaeota*) associated with putative bacterial ectosymbionts in a sulfidic marine habitat. Environ Microbiol 12(8):2371–2383

Müller A, Faubert P, Hagen M, Zu Castell W, Polle A et al (2013) Volatile profiles of fungi—chemotyping of species and ecological functions. Fungal Genet Biol 54:25–33

Murat D, Byrne M, Komeili A (2010) Cell biology of prokaryotic organelles. Cold Spring Harb Perspect Biol 2:a000422

Murray CJL, Lopez AD, Chin B, Feehan D, Hill KH (2006) Estimation of potential global pandemic influenza mortality on the basis of vital registry data from the 1918-20 pandemic: a quantitative analysis. Lancet 368:2211–2218

Mußmann M, Hu FZ, Richter M, de Beer D, Preisler A et al (2007) Insights into the genome of large sulfur bacteria revealed by analysis of single filaments. PLoS Biol 5(9):1928–1837

Nadeau JH, Sankoff D (1997) Comparable rates of gene loss and functional divergence after genome duplication early in vertebrate evolution. Genetics 147:1259–1266

Nakajima M, Imai K, Ito H, Nishiwaki T, Murayama Y et al (2005) Reconstitution of circadian oscillation of cyanobacterial kaic phosphorylation in vitro. Science 308:414–415

Nanninga N (2001) Cytokinesis in prokaryotes and eukaryotes: common principles and different solutions. Microbiol Mol Biol Rev 65(2):319–333

Naresh M, Sharma M, Mittla A (2011) Intracellular magneto-spatial organization of magnetic organelles inside intact bacterial cells. J Biomed Nanotechnol 7:572–577

Nasir A, Caetano-Anollés G (2015) A phylogenomic data-driven exploration of viral origins and evolution. Sci Adv 1:e1500527

Nasir A, Romero-Severson E, Claverie J-M (2020) Investigating the concept and origin of viruses. Trends Microbiol 26(12):959–967

Neiman M, Kay AD, Krist AC (2013) Can resource costs of polyploidy provide an advantage to sex? Heredity 110:152–159

Neiman M, Sharbel TF, Schwander T (2014) Genetic causes of transitions from sexual reproduction to asexuality in plants and animals. J Evol Biol 27:1346–1359

Nelson MI, Holmes EC (2007) The evolution of epidemic influenza. Nat Rev Genet 8:196–205

Nelson-Sathi S, Sousa FL, Roettger M, Lozada-Chávez N, Thiergart T et al (2015) Origins of major archaeal clades correspond to gene acquisitions from bacteria. Nature 517:77–80

Neme R, Tautz D (2013) Phylogenetic patterns of emergence of new genes support a model of frequent *de novo* evolution. BMC Genomics 14:117

Neme R, Tautz D (2014) Evolution: dynamics of de novo gene emergence. Curr Biol 24(6):R238–R240

Netzker T, Fischer J, Weber J, Mattern DJ, König CC et al (2015) Microbial communication leading to the activation of silent fungal secondary metabolite gene clusters. Front Microbiol 6:299

Ng SYM, Zolghadr B, Driessen AJM, Albers S-V, Jarrell KF (2008) Cell surface structures of archaea. J Bacteriol 190(18):6039–6047

Ng S-Y, Johnson R, Stanton LW (2012) Human long non-coding RNAs promote pluripotency and neuronal differentiation by association with chromatin modifiers and transcription factors. EMBO J 31:522–533

Nguyen AM, Noymer A (2013) Influenza mortality in the United States, 2009 pandemic: burden, timing and age distribution. PLoS One 8(5):e64198

Nguyen H, Cortez R, Fauci L (2014) Computing flows around microorganisms: slender-body theory and beyond. Am Math Mon. https://doi.org/10.4169/amer.math.monthly.121.09.810

Nicastro D, McIntosh JR, Baumeister W (2005) 3D structure of eukaryotic flagella in a quiescent state revealed by cryo-electron tomography. PNAS 102(44):15889–15894

Nicholson BL, White KA (2014) Functional long-range RNA-RNA interactions in positive-strand RNA viruses. Nat Rev Microbiol 12:493–504

Nielsen J, Hedeholm RB, Heinemeier J, Bushnell PG, Christiansen JS et al (2016) Eye lens radiocarbon reveals centuries of longevity in the Greenland shark (*Somniosus microcephalus*). Science 353:702–704

Niklas KJ, Cobb ED, Kutschera U (2014) Did meiosis evolve before sex and the evolution of eukaryotic life cycles? Bioessays 36:1091–1101

Ning J, Otto TD, Pfander C, Schwach F, Brochet M et al (2013) Comparative genomics in *Chlamydomonas* and *Plasmodium* identifies an ancient nuclear envelope protein family essential for sexual reproduction in protists, fungi, plants and vertebrates. Genes Dev 27:1198–1215

Nisbet EG, Sleep NH (2001) The habitat and nature of early life. Nature 409:1083–1091

Noble D (2012) A theory of biological relativity: no privileged level of causation. Interface Focus 2:55–64

Noble D (2018) Central dogma or central debate? Physiology 33:246–249

Noble R, Noble D (2017) Was the watchmaker blind? Or was she one-eyed? Biology 6:47

Noble R, Tasaki K, Noble PJ, Noble D (2019) Biological relativity requires circular causality but not symmetry of causation: so, where, what and when are rhe boundaries? Front Physiol 10:827

Norouzitallab P, Biswas P, Baruah K, Bossier P (2015) Multigenerational immune priming in an invertebrate parthenogenetic *Artemia* to a pathogenic *Vibrio campbelli*. Fish Shellfish Immunol 42:426–429

Nosenko T, Schreiber F, Adamska M, Adamski M, Eitel M et al (2013) Deep metazoan phylogeny: when different genes tell different stories. Mol Phylogenet Evol 67:223–233

Nougué O, Rode NE, Jabbour-Zahab R, Ségard A, Chevin LM et al (2015) Automixis in artemia: solving a century-old controversy. J Evol Biol 28:2337–2348

Nowack ECM, Melkonian M (2010) Endosymbiotic associations within protists. Philos Trans R Soc Lond B Biol Sci 365:699–712

Nowacki M, Landweber LF (2009) Epigenetic inheritance in ciliates. Curr Opin Microbiol 12(6):638–643

Nowacki M, Shetty K, Landweber LF (2011) RNA-mediated epigenetic programming of genome rearrangements. Annu Rev Genomics Hum Genet 12:367–389

Nowakowski TJ, Bhaduri A, Pollen AA, Alvarado B, Mostajo-Radji MA et al (2017) Spatiotemporal gene expression trajectories reveal developmental hierarchies of the human cortex. Science 358(6308):1318–1323

Nürnberg DJ, Mariscal V, Bornikoel J, Nieves-Morión M, Krauß N et al (2015) Intercellular diffusion of a fluorescent sucrose analog via the septal junctions in a filamentous cyanobacterium. mBio 6(2):e02109–e02114

O'Donnell M, Langston L, Stillman B (2013) Principles and concepts of DNA replication in bacteria, archaea, and eukarya. Cold Spring Harb Perspect Biol 5:a010108

O'Malley MA, Koonin EV (2011) How stands the tree of life a century and a half after the origin? Biol Direct 6:32

O'Neill JS, Reddy AB (2011) Circadian clocks in human red blood cells. Nature 469:498–503

O'Neill JS, van Ooijen G, Dixob LE, Troein C, Corellou F et al (2011) Circadian rhythms persist without transcription in a eukaryote. Nature 469:554–558

Öberg L, Kullman L (2011) Ancient subalpine clonal spruces (*Picea abies*): sources of postglacial vegetation history in the Swedish scandes. Arctic 64(2):183–196

Oberlander KC, Dreyer LL, Goldblatt P, Suda J, Linder HP (2016) Species-rich and polyploidy-poor: Insights into the evolutionary role of whole genome duplication from the Cape flora biodiversity hotspot. Am J Bot 103(7):1336–1347

Oda GA, Caldas IL, Piqueira JRC, Waterhouse JM, Marques MD (2000) Coupled biological oscillators in a cave insect. J Theor Biol 206:515–524

Ofir G, Sorek R (2018) Contemporary phage biology: from classic models to new insights. Cell 172:1260–1270

Ohtani N, Tomita M, Itaya M (2010) An extreme thermophile, *Thermus thermophilus*, is a polyploid bacterium. J Bacteriol 192(20):5499–5505

Olave M, Avila LJ, Sites JW Jr, Morando M (2018) Hybridization could be a common phenomenon within the highly diverse lizard genus *Liolaemus*. J Evol 31:893–903

Oliverio AM, Katz LA (2014) The dynamic nature of genomes across the tree of life. Genome Biol Evol 6(3):482–488

Olson EN, Nordheim A (2010) Linking actin dynamics and gene transcription to drive cellular motile functions. Nat Rev Mol Cell Biol 11(5):353–365

Orians CM (2000) The effects of hybridization in plants on secondary chemistry: implications for the ecology and evolution of plant-herbivore interactions. Am J Bot 87(12):1749–1756

Orias E, Cervantes MD, Hamilton EP (2011) *Tetrahymena thermophila*, a unicellular eukaryote with separate germline and somatic genomes. Rev Microbiol 162(6):578–586

Osborne LC, Monticelli LA, Nice TJ, Sutherland TE, Siracusa MC et al (2014) Virus-helminth coinfection reveals a microbiota-independent mechanism of immunomodulation. Science 345:578–582

Osburn MR, LaRowe DE, Momper LM, Amend JP (2014) Chemolithotrophy in the continental deep subsurface: Sanford Underground Research Facility (SURF), USA. Front Microbiol 5:610

Osterwalder M, Barozzi I, Tissières V, Fukuda-Yuzawa Y, Mannion BJ et al (2018) Enhancer redundancy provides phenotypic robustness in mammalian development. Nature 554:239–243

Ottenburghs J (2019) Multispecies hybridization in birds. Avian Res 10:20

Ottow JCG (2011) Mikrobiologie von Böden. Springer, Berlin

Ounjai P, Kim KD, Liu H, Dong M, Tauscher AN et al (2013) Architectural insights into a ciliary partition. Curr Biol 23:339–344

Owen JA, Osmanovic D, Mirny L (2023) Design principles of 3D epigenetic mempry sysrem. Science 382:eadg3053

Ozono S, Zhang Y, Ode H, Sano K, Tan TS et al (2021) SARS-CoV-2 D614G spike mutation increases entry efficiency with enhanced ACE2-binding affinity. Nat Commun 12:848

Pál C, Papp B, Lercher MJ (2005) Adaptive evolution of bacterial metabolic networks by horizontal gene transfer. Nat Genet 37(12):1372–1375

Panda S (2016) Circadian physiology of metabolism. Science 354:1008–1015

Paranjpe DA, Sharma VK (2005) Evolution of temporal order in living organisms. J Circadian Rhythms 3:7

Parent J-S, de Alba AEM, Vaucheret H (2012) The origin and effect of small RNA signaling in plants. Front Plant Sci 3:179

Parfrey LW, Lahr DJG (2013) Multicellularity arose several times in the evolution of eukaryotes. Bioessays 35:339–347

Parfrey LW, Lahr DJG, Katz LA (2008) The dynamic nature of eukaryotic genome. Mol Biol Evol 25(4):787–794

Parkes RJ, Webster G, Cragg BA, Weightman AJ, Newberry CJ et al (2005) Deep sub-seafloor prokaryotes stimulated at interfaces over geological time. Nature 436:390–394

Parra G, Reymond A, Dabbouseh N, Dermitzakis ET, Castelo R et al (2006) Tandem chimerism as a means to increase protein complexity in the human genome. Genome Res 16:37–44

Parshuram ZA, Harrison TL, Simonsen AK, Stinchcombe JR, Frederickson ME (2022) Nonsymbiotic legumes are more invasive, but only if polyploid. New Phytol 237:758–765

Parsons RJ, Breitbart M, Lomas MW, Carlson CA (2012) Ocean time-series reveals recurring seasonal patterns of virioplankton dynamics in the northwestern Sargasso Sea. ISME J 6:273–284

Partensky F, Hess WR, Vaulot D (1999) *Prochlorococcus*, a marine photosynthetic prokaryote of global significance. Microbiol Mol Biol Rev 63(1):106–127

Paschos K, Allday MJ (2010) Epigenetic reprogramming of host genes in viral and microbial pathogenesis. Trends Microbiol 18(10):439–447

Pasquali V, Sbordoni V (2014) High variability in the expression of circadian rhythm in a cave beetle population. Biol Rhythm Res 45(6):925–939

Paun O, Forest F, Fay MF, Chase MW (2009) Hybrid speciation in angiosperms: parental divergence drives ploidy. New Phytol 182:507–318

Peak D, Frame M (1995) Komplexität—das gezähmte Chaos. Birkhäuser, Basel

Pedersen T (2011) The nucleolus. Cold Spring Harb Perspect Biol 3(a0006):38

Pedrós-Alió C, Calderón-Paz JI, Gasol JM (2000) Comparative analysis shows that bacterivory, not viral lysis, controls the abundance of heterotrophic prokaryotic plankton. FEMS Microbiol Ecol 32:157–165

Pélissié B, Jarne P, Sarda V, Dvid P (2014) Disentangling precopulatory and postcopulatory sexual selection in polyandrous species. Evolution 68:1320–1331

Peremyslov VV, Mocklere TC, Filichkin SA, Fox SE, Jaiswal P et al (2011) Expression, splicing, and evolution of the myosin gene family in plants. Plant Physiol 155:1191–1204

Perfus-Barbeoch L, Castagnone-Sereno P, Reichelt M, Fneich S, Roquis D et al (2014) Elucidating the molecular bases of epigenetic inheritance in non-model invertebrates: the case of the root-knot nematode *Meloidogyne incognita*. Front Physiol 5:211

Pernthaler J (2005) Predation on prokaryotes in the water column and its ecological implications. Nat Rev Microbiol 3:537–546

Perras AK, Wanner G, Klingl A, Mora M, Auerbach AK et al (2014) Grappling archaea: ultrastructural analyses of an uncultivated, cold-loving archaeon, and its biofilm. Front Microbiol 5:397

Peterson KJ, Dietrich MR, McPeek MA (2009) MicroRNAs and metazoan macroevolution: insights into canalization, complexity, and the Cambrian explosion. Bioessays. https://doi.org/10.1002/bies.200900033

Petroff A, Libchaber A (2014) Hydrodynamics and collective behaviour of the tethered bacterium *Thiovulum majus*. PNAS 111(5):E537–E545

Petrov AS, Gulen B, Norris AM, Kovacs NA, Bernier CR et al (2015) History oft he ribosome and the origin of translation. PNAS 112(50):15396–15401

Petrusek A, Tollrian R, Schwenk K, Haas A, Laforsch C (2009) A "crown of thorns" is an inducible defense that protects *Daphnia* against ancient predator. PNAS 106(7):2248–2252

Pflug A, Lukarska M, Resa-Infante P, Reich S, Cusack S (2017) Structural insights into RNA synthesis by influenza virus transcription-replication machine. Virus Res 234:103–117

Philips SJ, Canalizo-Hernandez M, Yildirim I, Schatz GC, Mondragón A, O'Halloran TV (2015) Allosteric transcriptional regulation via changes in the overall topology of the core promoter. Science 349:877–881

Phillips R, Kondev J, Theriot J, Garcia HG, Orme N (2013) Physical biology of the cell. Garland Science, London

Piatigorsky J (2007) Gene sharing and evolution. Harvard University Press, Cambridge, MA

Picao-Osorio J, Johnston J, Landgraf M, Berni J, Alonso CR (2015) MicroRNA-encoded behavior in *Drosophila*. Science 350:815–820

Pilhofer M, Jensen GJ (2013) The bacterial cytoskeleton: more than twisted filaments. Curr Opin Cell Biol 25(1):125–133

Pilhofer M, Ladinsky MS, McDowall AW, Petroni G, Jensen GJ (2011) Microtubules in bacteria: ancient tubulins build a five-protofilament homolog of the eukaryotic cytoskeleton. PLoS Biol 9(12):e1001213

Pion M, Spangenberg JE, Simon A, Bindschedler S, Flury C et al (2013) Bacterial farming by the fungus *Morchella crassipes*. Proc Biol Sci 280:20132242

Pires ND, Grossniklaus U (2014) Different yet similar: evolution of imprinting in flowering plants and mammals. F1000Prime Rep 6:63

Pittis A, Gabaldon T (2016) Late acquisition of mitochondria by a host with chimaeric prokaryotic ancestry. Nature 531:101–104

Piunti A, Shilatifard A (2016) Epigenetic balance of gene expression by Polycomb and COMPASS families. Science 352:aad 9780 1-15

Plattner H (2015) Molecular aspects of calcium signalling at the crossroads of unikont and bikont eukaryote evolution—the ciliated protozoan *Paramecium* in focus. Cell Calcium 57(3):174–185

Podholová K, Plocek V, Rešetárová S, Kučerová H, Hlaváček O et al (2016) Divergent branches of mitochondrial signaling regulate specific genes and the viability of specialized cell types of differentiated yeast colonies. Oncotarget 7(13):15299–15314

Pollard TD (2003) The cytoskeleton, cellular motility and the reductionist agenda. Nature 422:741–745

Polz MF, Hunt DE, Preheim SP, Weinreich DM (2006) Patterns and mechanisms of genetic and phenotypic differentiation in marine microbes. Philos Trans R Soc Lond B Biol Sci 361:2009–2021

Polz MF, Alm EJ, Hanage WP (2013) Horizontal gene transfer and the evolution of bacterial and archaeal population structure. Trends Genet 29(3):170–175

Pongratz N, Stirhas M, Carranza S, Michiels NK (2003) Phylogeography of competing sexual and parthenogenetic forms of a freshwater flatworm: patterns and explanations. BMC Evol Biol 3:23

Porras-Alfaro A, Bayman P (2011) Hidden fungi, emergent properties: endophytes and microbiomes. Annu Rev Phytopathol 49:291–315

Poulin R (2010) Uneven distribution of cryptic diversity among higher taxa of parasitic worms. Biol Lett. https://doi.org/10.1098/rsbl.2010.0640

Prangishvili D, Forterre P, Garrett RA (2006) Viruses of the archaea: a unifying view. Nat Rev Microbiol 4:837–848

Prescott DM (1994) The DNA of ciliated protozoa. Microbiol Rev 58(2):233–267

Prothmann A, Hoffmann FG, Opazo JC, Herbener P, Storz JF et al (2020) The globin gene family in arthropods: evolution and functional diversity. Front Genet 11:858

Puente ME, Bashan Y, Li CY, Lebsky VK (2004) Microbial populations and activities in the rhizoplane of rock-weathering desert plants. I. Root colonization and weathering of igneous rocks. Plant Biol 6:629–642

Purcell EM (1977) Life at low Reynolds number. Am J Phys 45(1):3–11

Qiu Y-L, Taylor AB, McManus HA (2012) Evolution of the life cycle in land plants. J Syst Evol 50(3):171–194

Qiu W, Rutherford S, Mao A, Chu C (2017) The pandemics and its impacts. Health Culture Soc 9–10:3–11

Quinn AE, Sarre SS, Ezaz T, Graves JAM, Georges A (2011) Evolutionary transitions between mechanisms of sex determination in vertebrates. Biol Lett 7:433–448

Rabosky DL, Chang J, Title PO, Cowman PF, Sallan L et al (2018) An inverse latitudinal gradient in speciation rate for marine fishes. Nature 559:392–395

Raj A, van Oudenaarden A (2008) Nature, nurture, or chance: stochastic gene expression and its consequences. Cell 135:216–226

Rajewsky N, Jurga S, Barciszewski J (eds) (2017) Plant epigenetics. Springer, Cham

Ramsey KM, Yoshina J, Brace CS, Abrassart D, Kobayashi Y et al (2009) Circadian clock feedback cycle through NAMPT-mediated NAD⁺ biosynthesis. Science 324:651–654

Ramulu HG, Raoult D, Pontarotti P (2012) The rhizome of life: what about metazoa? Front Cell Infect Microbiol 2:50

Rangel J, Strauss K, Seedorf K, Hjelmen CE, Johnston JS (2015) Endopolyploidy changes with age-related polyethism in the honey bee, Apis mellifera. PLoS One 10(4):e0122208

Raoult D (2015) How the virophage compels the need to readdress the classification of microbes. Virology 477:119–124

Rasmussen T, Jensen RB, Skovgaard O (2007) The two chromosomes of *Vibrio cholerae* are initiated at different time point in the cell cycle. EMBO J 26:3124–3131

Rawn SM, Cross JC (2008) The evolution, regulation, and function of placenta-specific genes. Annu Rev Cell Dev Biol 24:159–181

Raymann K, Brochier-Armanet C, Gribaldo S (2015) The two-domain tree of life is linked to a new root for the archaea. PNAS 112(21):6670–6675

Reardon PN, Mueller KT (2013) Structure of the type IVa major pilin from the electrically conductive bacterial nanowires of *Geobacter sulfurreducens*. J Biol Chem 41(11):29260–29266

Reid AH, Fanning TG, Hultin JV, Taubenberger JK (1999) Origin and evolution of the 1918 "Spanish" influenza virus hemagglutinin gene. PNAS 96:1651–1656

Reimold C, Soufo HJD, Dempwolff F, Graumann PL (2013) Motion of variable-length MreB filaments at the bacterial cell membrane influences cell morphology. Mol Biol Cell 24:2340–2349

Reinhold-Hurek B, Bünger W, Burbano CS, Sabale M, Hurek T (2015) Root shaping their microbiome: global hotspots for microbial activity. Annu Rev Phyropathol 53:403–424

Reitzel AM, Behrendt L, Tarrant AM (2010) Light entrainment rhythmic gene expression in the sea anemone *Nematostella vectensis*: the evolution of the animal circadian clock. PLoS One 5(9):e12805

Ren L-L, Wang Y-M, Wu Z-C, Xiang Z-C, Guo L et al (2020) Identification of a novel coronavirus causing severe pneumonia in human: a descriptive study. Chin Med J 133(9):1015–1024

Rensing SA, Lang D, Zimmer AD, Terry A, Salamoy A et al (2008) The *Physicomitrella* genome reveals evolutionary insights into the conquest of land by plants. Science 319:64–69

Reski R (2002) Rings and networks: the amazing complexity of FtsZ in chloroplasts. Trends Plant Sci 7(3):103–105

Reuter M, Kreshchenko N (2004) Flatworm asexual multiplication implicates stem cells and regeneration. Can J Zool 82:334–356

Reyes-Lamothe R, Nicolas E, Sherratt DJ (2012) Chromosome replication and segregation in bacteria. Annu Rev Genet 46:121–143

Rhodes ME, Spear JR, Oren A, House CH (2011) Differences in lateral gene transfer in hypersaline versus thermal environments. BMC Evol Biol 11:199

Richards TA, Cavalier-Smith T (2005) Myosin domain evolution and the primary divergence of eukaryotes. Nature 436:1113–1118

Richards TA, Leonard G, Soanes DM, Talbot NJ (2011) Gene transfer into the fungi. Fungal Biol Rev 25:98–110

Richter M, Roselló-Mlóra R (2009) Shifting the genomic gold standard for the prokaryotic species definition. PNAS 106(45):19126–19131

Ridenhour BJ, Nuismer SL (2007) Polygenic traits and parasite local adaptation. Evolution 61:368–376

Rideout JR, He Y, Navas-Molina JA, Walters WA, Ursell LK et al (2014) Subsampled open-reference clustering creates consistent, comprehensive OTU definitions and scales to billions of sequences. PeerJ 2:e545

Riedl R (1975) Die Ordnung des Lebendigen. Paul Parey, Hamburg

Rineau F, Shah F, Smits MM, Persson P, Johansson T et al (2013) Carbon availability triggers the decomposition of plant litter and assimilation of nitrogen by an actomycorrhizal fungus. ISME J 7:2010–2022

Ringrose L, Paro R (2004) Epigenetic regulation of cellular memory by the polycomb and trithorax group proteins. Annu Rev Genet 38:413–443

Roark EB, Guilderson TP, Dunbar RB, Fallon SJ, Mucciarone DA (2009) Extreme longevity in proteinaceous deep-sea corals. PNAS 106(13):5204–5208

Robinson NP, Bell SD (2005) Origins of DNA replication in the three domains of life. FEBS J 272:3757–3766

Robson F, Khan KS, Le TK, Paris C, Demirbag S et al (2020) Coronavirus RNA proofreading: molecular basis and therapeutic targeting. Mol Cell 79:710–727

Rochette NC, Brochier-Armanet C, Gouy M (2014) Phylogenomic test of the hypotheses for the evolutionary origin of eukaryotes. Mol Biol Evol 31(4):832–845

Rodrigues-Oliveira T, Wollweber F, Ponce-Toledo RI, Xu J, Rittmann SK-MR et al (2023) Actin cytoskeleton and complex cell architecture in an Asgard archaeon. Nature 613:332–339

Roelants K, Fry BG, Ye L, Stijlemans B, Brys L et al (2013) Origin and functional diversification of an amphibian defense peptide arsenal. PLoS Genet 9(8):e1003662

Roenneberg T, Dragovic Z, Merrow M (2005) Demasking biological oscillators: properties and principles of entrainment exemplified by the *Neurospora* circadian clock. PNAS 102(21):7742–7747

Roesti D, Ineichen K, Braissant O, Redecker D, Wiemken A, Aragno M (2005) Bacteria associated with spores of the arbuscular mycorrhizal fungi glomus geosporum and glomus constrictum. Appl Environ Microbiol 74(11):6673–6679

Rogers PC, McAvoy DJ (2018) Mule deer impede Pando's recovery: implications for aspen resilience from a single-genotype forest. PLoS One 13(10):e0203619

Rokas A, Krüger D, Carroll SB (2005) Animal evolution and the molecular signature of radiations compressed in time. Science 310:1933–1938

Roos C, Zinner D, Kubatko LS, Schwarz C, Yang M et al (2011) Nuclear versus mitochondrial DNA: evidence for hybridization in colobine monkeys. BMC Evol Biol 11:77

Roossinck MJ (2011) The good viruses: viral mutualistic symbioses. Nat Rev Microbiol 9:99–108

Ropars J, Toro KS, Noel J, Pelin A, Charron P et al (2016) Evidence for the sexual origin of heterokaryosis in arbuscular mycorrhizal fungi. Nat Microbiol. https://doi.org/10.1038/NMICROBIOL.2016.33

Roselló-Mora R, Amann R (2001) The species concept for prokaryotes. FEMS Microbiol Rev 25:39–67

Rosewich UL, Kistler HC (2000) Role of horizontal gene transfer in the evolution of fungi. Annu Rev Phytopathol 38:325–363

Ross JL, Ali DM, Warshaw DM (2008) Cargo transport: molecular motors navigate a complex cytoskeleton. Curr Opin Cell Biol 20(1):41–47

Rothschild LJ, Mancinelli RL (2001) Life in extreme environments. Nature 409:1092–1101

Rout MP, Field MC (2017) The evolution of organellar coat complexes and organization of the eukaryotic cell. Annu Rev Biochem 86:637–657

Roux S, Krupovic M, Debroas D, Forterre P, Enault F (2013) Assessment of viral community functional potential from viral metagenomes may be hampered by contamination with cellular sequences. Open Biol 3:130160

Ruggiero MA, Gordon DP, Orrell TM, Bailly N, Bourgoin T et al (2015) A higher level classification of all living organisms. PLoS One 10(4):e0119248

Ruprecht C, Lohaus R, Venneste K, Mutwil M, Nikoloski Z et al (2017) Revisiting ancestral polyploidy in plants. Sci Adv 3:e1603195

Rust MJ, Golden SS, O'Shea EK (2011) Light-driven changes in energy metabolism directly entrain the cyanobacterial circadian oscillator. Science 33:220–223

Ryall B, Eydallin G, Férenci T (2012) Culture history and population heterogeneity as determinants of bacterial adaptation: the adaptomics of a single environmental transition. Microbiol Mol Biol Rev 76(3):597–625

Sabin LR, Delás MJ, Hannon GJ (2013) Dogma derailed: the many influences of RNA on the genome. Mol Cell 49:783–794

Sachs JL, Hollowell AC (2012) The origins of cooperative bacterial communities. mBio 3(3):e00099–e00012

Sadeghi M, Tomaru Y, Ahola T (2021) RNA viruses in aquatic unicellular eukaryotes. Viruses 13:362

Salmon A, Ainouchet ML, Wendel JF (2005) Genetic and epigenetic consequences of recent hybridization and polyploidy in *Spartina* (Poaceae). Mol Biol 14:1163–1175

Sanjuán R (2012) From molecular genetics to phylodynamics: evolutionary relevance of mutation rates across viruses. PLoS Pathog 8(5):e1002685

Sano E, Carlson S, Wegley L, Rohwer F (2004) Movement of viruses between biomes. Appl Environ Microbiol 70:10

Sarmento H (2012) New paradigms in tropical limnology: the importance of the microbial food web. Hydrobiologia 686:1–14

Sassera D, Lo N, Epis S, D'Auria G, Montagna M et al (2011) Phylogenomic evidence for the presence of a flagellum and cbb$_3$ oxidase in the free-living mitochondrial ancestor. Mol Biol Evol 28(12):3285–3296

Sattler MC, Carvalho CR, Clarindo WR (2015) The polyploidy and its key role in plant breeding. Plants. https://doi.org/10.1007/s00425-015-2450.x

Saunders-Hastings PR, Krewski D (2016) Reviewing the history of pandemic influenza: understanding patterns of emergence and transmission. Pathogens 5:66

Sawada H, Inoue N, Iwano M (eds) (2014) Sexual reproduction in animals and plants. Springer, Tokyo

Saxe JP, Lin H (2011) Small noncoding RNAs in the germline. Cold Spring Harb Perspect Biol 3:a002717

Schäfer G, Engelhard M, Müller V (1999) Bioenergetics of the archaea. Microbiol Mol Biol Rev 63(3):570–620

Schaible R, Scheuerlein A, Dańko MJ, Gampe J, Martínez DE, Vaupel JW (2015) Constant mortality and fertility over age in *Hydra*. PNAS 112(51):15701–15706

Schärer L, Littlewood DTJ, Waeschenbach A, Yoshida W, Vizoso DB (2011) Mating behaviour and the evolution of sperm design. PNAS 108(4):1490–1495

Scherlach K, Gaupner K, Hertweck C (2013) Molecular bacteria-fungi interactions: effects on environment, food, and medicine. Annu Rev Microbiol 67:375–397

Schibler U, Sassone-Corsi P (2002) A web of circadian pacemakers. Cell 111:919–922

Schirrmeister BE, Sanchez-Baracaldo P, Wacey D (2016) Cyanobacterial evolution during the Precambrian. Int J Astrobiol 15(3):187–204

Schlebusch CM, Malmström H, Günther T, Sjödin P, Coutinho A et al (2017) Southern African ancient genomes estimate modern human divergence to 350,000 to 260,000 years ago. Science 358:652–655

Schlötterer C (2015) Genes from scratch—the evolutionary fate of *de novo* genes. Trends Genet 31(4):215–219

Schmidt DN, Lazarus D, Young JR, Kucera M (2006) Biogeography and evolution of body size in marine plankton. Earth Sci Rev 78:239–266

Schmitt I, Lumbsch HT (2009) Ancient horizontal gene transfer from bacteria enhances biosynthetic capabilities of fungi. PLoS One 4(2):e4437

Schmitt CL, Tatum ML (2008) The Malheur National Forest; Location of the world's largest living organism. USDA, Pacific Northwest Region

Schneiker S, Perlova P, Kaiser O, Gerth L, Alici A et al (2007) Complete genome sequence of the myxobacterium *Sorangium cellulosum*. Nat Biotechnol 25(11):1281–1289

Schoenfelder KP, Fox DE (2015) The expanding implications of polyploidy. J Cell Biol 209(4):485–491

Schoepp-Cothenet B, van Lis R, Atteia A, Baymann F, Capoviez L et al (2013) On the universal core of bioenergetics. Biochim Biophys Acta 1827:79–93

Schön I, Martens K, van Dijk P (eds) (2009) Lost sex: the evolutionary biology of parthenogenesis. Springer, Dordrecht

Schönknecht G, Weber APM, Lercher MJ (2013) Horizontal gene acquisitions by eukaryotes as drivers of adaptive evolution. Bioessays 36:9–20

Schooling SR, Hubley A, Beveridge TJ (2009) Interactions of DNA with biofilm derived membrane vesicles. J Bacteriol 191(13):4097–4102

Schuergers N, Wilde A (2015) Appendages of the cyanobacterial cell. Life 5:700–715

Schulz H, Jørgensen BB (2001) Big bacteria. Annu Rev Microbiol 55:105–137

Schurko AM, Neiman M, Logsdon JM (2008) Signs of sex: what we know and how we know it. Trends Ecol Evol 24(4):208–217

Schwander T, Oldroyd BP (2016) Androgenesis: where males hijack eggs to clone themselves. Philos Trans R Soc Lond B Biol Sci 371:20150534

Schwander T, Henry L, Crespi BJ (2011) Molecular evidence for ancient asexuality in *Timema* stick insects. Curr Biol 21:1129–1134

Schwarz EM (2017) Evolution: a parthenogenetic nematode shows how animals become sexless. Curr Biol 27:R1060–R1080

Schwarz MV, Frenzel P (2005) Methanogenic symbionts of anaerobic ciliates and their contribution to methanogenesis in an anoxic rice field soil. FEMS Microbiol Ecol 52:93–99

Schwerdtfeger F (1975) Synökologie. Paul Parey, Hamburg

Sender R, Fuchs S, Milo R (2016) Revised estimates for the number of human and bacteria cells in the body. PLoS Biol 14(8):e1002533

Senesi N, Wilkinson KJ (2008) Biophysical chemistry of fractal structures and processes in environmental systems. Wiley, Hoboken, NJ

Seymour JR, Ahmed T, Durham WM, Stocker R (2010) Chemotactic response of marine bacteria to the extracellular products of *Synechococcus* and *Prochlorococcus*. Aquat Microb Ecol 50:161–168

Shabalina SA, Ogurtsov AY, Spiridonov AN, Novichkov PS, Spiridonov NA, Koonin EV (2010) Distinct patterns of expression and evolution of intronless and intron-containing mammalian genes. Mol Biol Evol 27(8):1745–1749

Shapiro BJ, Friedman J, Cordero OX, Preheim SP, Timberlake SC et al (2012) Population genomics of early events in the ecological differentiation of bacteria. Science 336:48–51

Sharma V, Colson P, Pontarotti P, Raoult D (2016) Mimivirus inaugurated in the 21st century the beginning of a reclassification of viruses. Curr Opin Microbiol 31:16–24

Sharp R, Ziemer CJ, Stern MD, Stahl DA (1998) Taxon-specific associations between protozoal and methanogen populations in the rumen and the model rumen system. FEMS Microbiol Ecol 26:71–78

Shearer FM, Moss R, McVernon J, Ross JV, McCaw JM (2020) Infectious disease pandemic planning and response: incorporating decision analysis. PLoS Med 17(1):e1003018

Shelomi M, Danchin EGJ, Heckel D, Wipfler B, Bradler S et al (2016) Horizontal gene transfer of pectinases from bacteria preceded the diversification of stick and leaf insects. Sci Rep 6:26388

Shetty A, Chen S, Tocheva EI, Jensen GJ, Hickey WJ (2011) Nanopods: a new bacterial structure and mechanism for deployment of outer membrane vesicles. PLoS One 6(6):e20725

Shi T, Falkowski PG (2008) Genome evolution in cyanobacteria: the stable core and the variable shell. PNAS 105(7):2510–2515

Shih PM, Matzke NJ (2013) Primary endosymbiosis events date to the later proterozoic with cross-calibrated phylogenetic dating of duplicated ATPase proteins. PNAS 110(30):12355–12380

Shin J, Song Y, Jeong Y, Cho B-K (2016) Analysis of the core genome and pan-genome of autotrophic acetogenic bacteria. Front Microbiol 7:1531

Shirayama M, Seth M, Lee H-C, Gu W, Ishidate T et al (2012) piRNAs initiate an epigenetic memory of nonself RNA in the *C. elegans* germline. Cell 150:65–77

Shoguchi E, Shinzato C, Kawashima T, Gyoja F, Mungpakdee S et al (2013) Draft assembly of the *Symbiodinium minutum* nuclear genome reveals dinoflagellate gene structure. Curr Biol 23:1399–1408

Shukla GC, Singh J, Barik S (2011) MicroRNAs: processing, maturation, target recognition and regulatory functions. Mol Cell Pharmacol 3(3):83–92

Shukla SP, Plata C, Reichelt M, Steiger S, Heckel DG et al (2018) Microbiome-assisted carrion preservation aids larval development in a burying beetle. PNAS 115(44):11274–11279

Silpe JE, Bassler BL (2019) A host-produced quorum-sensing autoinducer controls a phage lysis-lysogenic decision. Cell 176:268–280

Silver AC, Arjona A, Walker WE, Fikrig E (2012) The circadian clock controls toll-like receptor 9-mediated innate and adaptive immunity. Immunity 36:251–261

Simmonds P, Alewsakun P (2018) Virus classification—where do you draw the line? Arch Virol 163:2037–2046

Simon J-C, Delmotte F, Rispe C, Crease T (2003) Phylogenetic relationships between parthenogens and their sexual relatives: the possible routes to parthenogenesis in animals. Biol J Linn Soc 79:151–163

Simonin KA, Roddy AB (2018) Genome downsizing, physiological novelty, and the global dominance of flowering plants. PLoS Biol 16(1):e2003706

Singh DP, Saudemont B, Guglielmi G, Arnaiz O, Goût J-F et al (2014) Genome-defence small RNAs exapted for epigenetic mating-type inheritance. Nature 509:447–452

Sites JW, Reeder TW, Wiens JJ (2011) Phylogenetic insights on evolutionary novelties in lizards and snakes: sex, birth, bodies, niches and venom. Annu Rev Ecol Evol Syst 42:227–244

Skerker JM, Berg HC (2001) Direct observations of extension and retraction of type IV pili. PNAS 98(12):6901–6904

Skerker JM, Prasol MS, Perchuk BS, Biondi EG, Laub MT (2005) Two-component signal transduction pathways regulating growth and cell cycle progression in a bacterium: a system-level analysis. PLoS Biol 3(10):e334

Sköld HN, Obst M, Sköld M, Åkesson B (2009) Stem cells in asexual reproduction of marine invertebrates. In: Rinkevich B, Matranga V (eds) Stem cells in marine organisms. Springer, Dordrecht, pp 105–137

Šlapeta J, López-García P, Moreira D (2006) Global dispersal and ancient cryptic species in the smallest marine eukaryotes. Mol Biol Evol 23(1):23–29

Slater SC, Goldman BS, Goodner B, Setubal JC, Farrand SK et al (2009) Genome sequences of three *agrobacterium* biovars help to elucidate the evolution of multichromosome genomes in bacteria. J Bacteriol 191(8):2501–2511

Slon V, Mafessoni F, Vernot B, de Filippo C, Grote S et al (2018) The genome of the offspring of a Neanderthal mother and a Denisovan father. Nature 561:113–116

Smilie CS, Smith MB, Friedman J, Cordero OX, David LA, Alm EJ (2011) Ecology drives a global network of gene exchange connecting the human microbiome. Nature 480:241–244

Smith DR, Asmail SR (2014) Next-generation sequencing data suggests that certain nonphotosynthetic green plants have lost their plastid genomes. New Phytol 204:7–11

Smith J, Kronforst MR (2013) Do *Heliconius* butterfly species exchange mimicry alleles? Biol Lett 9:20130503

Smith DR, Lee RW (2014) A plastid without a genome: evidence from the nonphotosynthetic green algal genus *Polytomella*. Plant Physiol 164:1812–1819

Smith ML, Bruhn JN, Anderson JB (1992) The fungus *Armillaria bulbosa* is among the largest and oldest living organisms. Nature 356:428–431

Smith GJD, Vijaykrishna D, Bahl J, Lycett SJ, Worobey M et al (2009) Origins and evolutionary genomics of the 2009 swine-origin H1N1 influenza A epidemic. Nature 459:1122–1125

Smith KM, Phatale PA, Bredweg EL, Connolly LR, Pomraning R, Freitag M (2012) Epigenetics of filamentous fungi. In: Meyers RA (ed) Epigenetic regulation and epigenomics. Wiley-VCH, Weinheim, pp 1063–1106

Smith HL, Pavasovic A, Surm JM, Phillips MJ, Prentis PJ (2018) Evidence for a large expansion and subfunctionalization of globin genes in sea anemones. Genome Biol Evol 10(8):1892–1901

Snijder EJ, Bredenbeek PJ, Dobbe JC, Thiel V, Ziebuhr J et al (2003) Unique and conserved features of genome and proteome of SARS-coronavirus, an early split-off from the coronavirus group 2 lineage. J Mol Biol 331:991–1004

Sojo V, Pomiankowski A, Lane N (2014) A bioenergetic basis for membrane divergence in archaea and bacteria. PLoS Biol 12(8):e1001926

Solheim M, Brekke MC, Snipen LG, Willems RIL, Nes IF, Brede DA (2011) Comparative genomic analysis reveals significant enrichment of mobile genetic elements and genes encoding surface structure-proteins in hospital-associated clonal complex 2 *Enterococcus faecalis*. BMC Microbiol 11:3

Soltis PS, Soltis DE (2009) The role of hybridization in plant speciation. Annu Rev Plant Biol 60:561–588

Son K, Menolascina F, Stocker R (2016) Speed-dependent chemotactic precision in marine bacteria. PNAS 113(31):8624–8629

Song SJ, Lauber C, Costello EK, Lozupone CA, Humphrey G et al (2013) Cohabiting family members share microbiota with one another and with their dogs. eLife 2:e00458

Sontheimer EJ, Marraffini LA (2016) CRISPR goes retro. Science 351:920–921

Soppa J (2013) Evolutionary advantages of polyploidy in halophilic archaea. Biochem Soc Trans 41:339–343

Sorek R (2007) The birth of new exons: mechanisms and evolutionary consequences. RNA 13:1603–1608

Sørensen SJ, Bailey M, Hansen LH, Kroer N, Wuertz S (2005) Studying plasmid horizontal transfer *in situ*: a critical review. Nat Rev Microbiol 3:700–701

Sorrells TR, Johnson AD (2015) Making sense of transcription networks. Cell 161:714–723

Sorrells TR, Booth LN, Tuch BB, Johnson AD (2015) Intersecting transcription networks constrain gene regulatory evolution. Nature 523:361–365

Soucy SM, Huang J, Gogarten JP (2015) Horizontal gene transfer: building the web of life. Nat Rev Genet 16:472–482

Sowa Y, Berry RM (2008) Bacterial flagellar motor. Q Rev Biophys 41(2):103–132

Speijer D (2017) Evolution of peroxisomes illustrates symbiogenesis. Bioessays 39(9). https://doi.org/10.1002/bies.201700050

Staels B (2006) When the *Clock* stops ticking, metabolic syndrome explodes. Nat Med 12(1):54–55

Stairs CW, Leger MM, Roger AJ (2015) Diversity and origins of anaerobic metabolism in mitochondria and related organelles. Philos Trans R Soc Lond B Biol Sci 370:20140326

Staley JT, Gosink JJ (1999) Poles apart: biodiversity and biogeography of sea ice bacteria. Annu Rev Microbiol 53:189–215

Standfuss J (2015) Viral chemokine mimicry. Science 347:1071–1072

Stanley SM (2001) Historische geologie. Spektrum Akademischer Verlag, Heidelberg

Stan-Lotter H, Pfaffenhuemer M, Legat A, Busse H-J, Radax C, Gruber C (2002) Halococcos dombrowskii sp. nov., an archaeal isolate from permian alpine salt deposit. Int J Syst Evol Microbiol 52:1807–1814

Steinberg G, Peñalva MA, Riquelme M, Wösten HA, Harris SD (2017) Cell biology of hyphal growth. Microbiol Spectr 5(2). https://doi.org/10.1128/microbiolspec.FUNK-0034-2016

Stelzer C-P, Schmidt J, Wiedlroither A, Riss S (2010) Loss of sexual reproduction and Dwarfing in a small metazoan. PLoS One 5(9):e12854

Stenberg P, Saura A (2013) Meiosis and its deviations in polyploid animals. Cytogenet Genome Res. https://doi.org/10.1159/000351731

Stetter KO (2006) Hyperthermophiles in the history of life. Philos Trans R Soc Lond B Biol Sci 361:1837–1843

Stewart PS (2003) Diffusion in biofilms. J Bacteriol 185(5):1485–1491

Stiller JW, Schreiber J, Yue J, Guo H, Ding Q, Huang J (2014) The evolution of photosynthesis in chromist algae through serial endosymbioses. Nat Commun. https://doi.org/10.1038/ncomms6764

Stocker R (2012) Marine microbes see a sea of gradients. Science 338:628–633

Stocker R (2015) The 100 μm length scale in the microbial ocean. Aquat Microbiol Ecol 76:189–194

Stocker R, Seymour JR (2012) Ecology and physics of bacterial chemotaxis in the ocean. Microbiol Mol Biol Rev 76(4):792–812

Stokes HW, Gillings MR (2011) Gene flow, mobile genetic elements and the recruitment of antibiotic resistance genes into gram-negative pathogens. FEMS Microbiol Rev 35:790–819

Stoner DS, Weissman IL (1996) Somatic germ cell parasitism in a colonial ascidian: possible role for a high polymorphic allorecognition system. PNAS 93:15254–15259

Stoner DS, Rinkevich B, Weissman IL (1999) Heritable germ and somatic cell lineage competitions in chimeric colonial protochordates. PNAS 96:9148–9153

Storch K-F, Lipan O, Leykin I, Viswanathan N, Davis FC et al (2002) Extensive and divergent circadian gene expression in liver and heart. Nature 417:78–83

Storz JF, Opazo JC, Hoffmann FG (2011) Phylogenetic diversification of the globin gene superfamily in chordates. IUBMB Life 63(5):313–322

Storz JF, Opazo JC, Hoffmann FG (2013) Gene duplication, genome duplication, and the functional diversification of vertebrate globins. Mol Phylogenet Evol 66(2):469–478

Streicker DG, Turmelle AS, Vonhof MJ, Kuzmin IV, McCracken GF, Rupprecht CE (2010) Host phylogeny constrains cross-species emergence and establishment of rabies virus in bats. Science 329:676–679

Struhl K (1999) Fundamentally different logic of gene regulation in eukaryotes and prokaryotes. Cell 98:1–4

Strullu-Derrien C, Wawrzyniak Z, Goral T, Kenrick P (2015) Fungal colonization of the rooting system of the early land plant *Asteroxylon mackiei* from the 407-Myr-old Rhynie Chert (Scotland, UK). Bot J Linn Soc 170:201–213

Su S, Wong G, Shi W, Liu J, Lai ACK et al (2016) Epidemiology, genetic recombination, and pathogenesis of coronaviruses. Trends Microbiol 24(6):490–502

Sun S, Kondabagil K, Draper B, Alam TI, Bowman VD et al (2008) The structure of the phage T4 DNA packaging motor suggests a mechanism dependent on electrostatic forces. Cell 135:1251–1262

Sun S, Rao VB, Rossmann MG (2010) Genome packaging in viruses. Curr Opin Struct Biol 20(1):114–120

Sun J, Miller JB, Granqvist E, Wiley-Kalil A, Gobbato E et al (2015) Activation of symbiosis signaling by arbuscular mycorrhizal fungi in legumes and rice. Plant Cell 27:823–838

Sunda WG, Hardison DR (2010) Evolutionary tradeoffs among nutrient acquisition, cell size, and grazing defense in marine phytoplankton promote ecosystem stability. Mar Ecol Prog Ser 401:63–76

Sung W, Tucker AE, Doak TG, Choi E, Thomas WK, Lynch M (2012) Extraordinary genome stability in the ciliate *Paramecium tetraurelia*. PNAS. https://doi.org/10.1073/pnas.1210663109

Suttle CA (1994) The significance of viruses to mortality in aquatic microbial communities. Microb Ecol 28:237–243

Suttle CA (2005) Viruses in the sea. Nature 437:356–361

Suttle CA (2007) Marine viruses—major players in the global ecosystem. Nat Rev Microbiol 5:401–812

Taheri-Araghi S, Bradde S, Sauls JT, Hill NS, Levin PA et al (2015) Cell-size control and homeostasis in bacteria. Curr Biol 25:385–391

Tang CQ, Obertegger U, Fontaneto D, Barraclough TG (2014) Sexual species are separated by larger genetic gaps the asexual species in rotifers. Evolution 68(10):2901–2916

Tasker JG, Oliet SHR, Bains JS, Brown CH, Stern J (2012) Glial regulation of neuronal function: from synapse to systems physiology. J Neuroendocrinol 24(4):566–576

Taubenberger JK, Kash JC (2010) Influenza virus evolution, host adaptation, and pandemic formation. Cell Host Microbe 7:440–451

Tautz D, Domazet-Lošo TD (2011) The evolutionary origin of orphan genes. Nat Rev Genet 12:692–702

Taylor JW, Turner E, Townsend JP, Dettman JR, Jacobson D (2006) Eukaryotic microbes, species recognition and the geographic limits of species: examples from the kingdom Fungi. Philos Trans R Soc Lond B Biol Sci 361:1947–1963

Taylor DW, Zhu Y, Staals RHJ, Kornfeld JE, Shinkai A et al (2015) Structures of the CRISPR-Cmr complex reveal mode of RNA target positioning. Science 348:581–585

Tegally H, Wilkinson E, Lessells RJ, Giandhari J, Pillay S et al (2021) Sixteen novel lineages of SARS-CoV-2 in South Africa. Nat Med. https://doi.org/10.1038/s41591-021-01255-3

Tekle YI, Wood FC, Katz LA, Cerón-Romero MA, Gorfu LA (2017) Amoebozoans are secretly but ancestrally sexual: evidence for sex genes and potential novel crossover pathways in diverse groups of amoebae. Genome Biol Evol 9(2):375–387

Teske A, Reysenbach A-L (2015) Editorial: hydrothermal microbial ecosystems. Front Microbiol 6:884

Tessmar-Raible K, Raible F, Arboleda E (2011) Another place, another timer: marine species and the rhythms of life. Bioessays 33:165–172

Theis KR, Dheilly NM, Klassen JL, Brucker RM, Baines JF et al (2016) Getting the hologenome concept right: an eco-evolutionary framework for hosts and their microbiomes. mSystems 1(2):e00028-16

Thingstad TF (2000) Elements of a theory for the mechanisms controlling abundance, diversity, and biogeochemical role of lytic bacterial viruses in aquatic systems. Limnol Oceanogr 45(6):1320–1329

Thomas CM, Nielsen KM (2005) Mechanisms of, and barriers to, horizontal gene transfer between bacteria. Net Rev Microbiol 3:711–721

Tice AK, Silberman JD, Walthall AC, Le KND, Spiegel FW, Brown MW (2016) *Sorodiplophrys stercorea*: another novel lineage of sorocarpic multicellularity. J Eukaryot Microbiol. https://doi.org/10.1111/jeu.12311

Tielens AGM, Rotte C, van Hellemond JJ, Martin W (2002) Mitochondria as we don't know them. Trends Biochem Sci 27(11):564–572

Tilquin A, Kokko H (2016) What does the geography of parthenogenesis teach us about sex? Philos Trans R Soc Lond B Biol Sci 371:20150538

Tobiason DM, Seifert HS (2006) The obligate human pathogen, *Neisseria gonorrhoeae*, is polyploid. PLoS Biol 4(6):e185

Todd RT, Forche A, Selmecki A (2017) Ploidy variation in fungi—polyploidy, aneuploidy, and genome evolution. Microbiol Spectr. https://doi.org/10.1128/microbiolspec.FUNK-0051-2016

Toft C, Andersson SGE (2010) Evolutionary microbial genomics: insights into bacterial host adaptation. Nat Rev Genet 11:465–475

Trachtenberg S, Gilad R (2001) A bacterial linear motor: cellular and molecular organization of the contractile cytoskeleton of the helical bacterium *Spiroplasma melliferum* BC3. Mol Microbiol 41(4):827–848

Trachtenberg S, Gilad R, Geffen N (2003) The bacterial linear motor of *Spiroplasma melliferum* BC3: from single molecules to swimming cells. Mol Microbiol 47(3):671–687

Treangen TJ, Rocha EPC (2011) Horizontal transfer, not duplication, drives the expansion of protein families in prokaryotes. PLoS Genet. https://doi.org/10.1371/J.pgen.1001284

Treude T, Knittel K, Blumenberg M, Seifert R, Boetius A (2005) Subsurface microbial methanotrophic mats in the black sea. Appl Environ Microbiol 71(10):6375–6378

Trontelj P, Fišer C (2009) Cryptic species diversity should not be trivialised. Syst Biodivers 7(1):1–3

Tropini C, Earle KA, Huang KC, Sonnenburg JL (2017) The gut microbiome: connecting spatial organization to function. Cell Host Microbe 21:433–442

Trovão NS, Nelson MI (2020) When pigs fly: pandemic influenza enters the 21st century. PLoS Pathog 18(3):e1008259

Truebestein L, Leonard TA (2016) Coiled-coils: the long and short of it. Bioessays 38:903–916

Tseng R, Goularte NF, Chavan A, Lou J, Cohen SE et al (2017) Structural basis of the day-night transition in a bacterial circadian clock. Science 355:1174–1180

Tuan D, Kong S, Hu K (1992) Transcription of the hypersensitive site HS2 enhancer in erythroid cells. PNAS 89:11219–11223

Tvedte ES, Logsdon JM, Forbes AA (2019) Sex loss in insects: causes of asexuality and consequences for genomes. Curr Opin Insect Sci 31:77–83

Ueda H, Yokota E, Kutsuna N, Shimada T, Tamura K et al (2010) Myosin-dependent endoplasmic reticulum motility and F-actin organization in plant cells. PNAS 102(15):6894–6899

Ulitsky I, Bartel DP (2013) lincRNAs: genomics, evolution, and mechanisms. Cell 154:28–46

Umen J, Heitman J (2013) Evolution of sex: mating rituals of a pre-metazoan. Curr Biol 23(22):R1006–R1008

Undheim EAB, Hamilton BR, Kurniawan ND, Bowlay G, Cribb BW et al (2015) Production and packaging of a biological arsenal: evolution of centipede venoms under morphological constraint. PNAS 112(13):4026–4031

Vafai SB, Mootha VK (2012) Mitochondrial disorders as windows into an ancient organelle. Nature 491:374–383

Vahdatzadeh M, Deveau A, Splivallo R (2015) The role of the microbiome of truffles in aroma formation: a meta-analysis approach. Appl Environ Microbiol 81:20

Vahey MD, Fletcher DA (2019) Low-fidelity assembly of influenza A virus promotes escape from host cells. Cell 176:281–294

Vale RD (2003) The molecular motor toolbox for intracellular transport. Cell 112:467–480

Vale RD, Milligan RA (2000) The way things move: looking under the hood of molecular motor proteins. Science 288:88–95

Valentine DL (2007) Adaptations to energy stress dictate the ecology and evolution of the Archaea. Nat Rev Microbiol. https://doi.org/10.1038/nrmicro1619

Van de Peer Y, Mizrachi E, Marchal K (2017) The evolutionary significance of polyploidy. Nat Rev Genet. https://doi.org/10.1038/nrg.2017.26

van der Horst MA, Key J, Hellingwerf KJ (2007) Photosensing in chemotrophic, non-phototrophic bacteria: let there be light sensing too. Trends Microbiol 15(12):554–562

van der Kooi CJ, Schwander T (2015) Parthenogenesis: birth of a new lineage or reproductive accident? Curr Biol 25:R654–R676

van der Kooi CJ, Matthey-Doret C, Schwander T (2017) Evolution and comparative ecology of parthenogenesis in haplodiploid arthropods. Evol Lett 1–6:304–316

van Groenendael JM, Klimeš L, Klimešová J (1996) Comparative ecology of clonal plants. Philos Trans R Soc Lond B Biol Sci 351:1331–1339

van Hijum SAFT, Medema MH, Kuipers OP (2009) Mechanisms and evolution of control logic in prokaryotic transcriptional regulation. Microbiol Mol Biol Rev 73(3):481–509

van Hoek AHAM, van Alen TA, Sprakel VSI, Leunissen JAM, Brigge T et al (2000) Multiple acquisition of methanogenic archaeal symbionts by anaerobic ciliates. Mol Biol Evol 17(2):251–258

van Maldegem LM, Sansjofre P, Weijers JWH, Wolkenstein K, Strother PK et al (2019) Bisnorgammacerane traces predatory pressure and the persistent rise of algal ecosystems after snowball earth. Nat Commun. https://doi.org/10.1038/s41467-019-08306-x

van Teeffelen S, Wang S, Furchtgott L, Huang KC, Wingreen NS et al (2011) The bacterial actin MreB rotates, and rotation depends on cell-wall assembly. PNAS 108(38):15822–15827

Vandenkoornhuyse P, Quaiser A, Duhamel M, Le Van A, Dufresne A (2015) The importance of the microbiome of the plant holobiont. New Phytol 206:1196–1206

Vanneste K, Baele G, Maere S, Van de Peer Y (2014) Analysis of 41 plant genomes supports a wave of successful genome duplications in association with the cretaceous-paleogene boundary. Genome Res 24:1334–1347

Vats P, Yu J, Rothfield L (2009) The dynamics nature of the bacterial cytoskeleton. Cell Mol Life Sci 66(20):3353–3361

Vázquez-Limón C, Hoogewijs D, Vinogradov SN, Arrendondo-Peter R (2012) The evolution of land plants hemoglobins. Plant Sci 191-192:71–81

Veening J-W, Smits WK, Kuipers OP (2008a) Bistability, epigenetics, and bet-hedging in bacteria. Annu Rev Microbiol 62:193–210

Veening J-W, Stewart EJ, Berngruber TW, Taddei F, Kuipers OP, Hamoen LW (2008b) Bet-hedging and epigenetic inheritance in bacterial cell development. PNAS 105(11):4393–4398

Vergara D, Lively CM, King KC, Jokela J (2013) The geographic mosaic of sex and infection in lake populations of a New Zealand snail at multiple spatial scales. Am Nat 182(4):484–493

Vergin KL, Tripp HJ, Wilhelm LJ, Denver DR, Rappé MS, Giovannoni SJ (2007) High intraspecific recombination rate in a native population of Candidatus pelagibacter ubique (SAR11). Environ Microbiol 9(10):2430–2440

Verhoeven JTP, Kavanagh AN, Dufour SC (2017) Microbiome analysis shows enrichment for specific bacteria in separate anatomical regions of the deep-sea carnivorous sponge *Chondrocladia grandis*. FEMS Microbiol Ecol 93:fiw 214

Vernikos G, Medini D, Riley DR, Tettelin H (2013) Ten years of pan-genome analyses. Curr Opin Microbiol 23:148–154

Vernot B, Tucci S, Kelso J, Schraiber JG, Wolf AB et al (2016) Excavating Neandertal and Denisovan DNA from the genomes of melanesian individuals. Science 352:235–239

Verret F, Wheeler G, Taylor AR, Farnham G, Brownlee C (2010) Calcium channels in photosynthetic eukaryotes: implications for evolution of calcium-based signalling. New Phytol 187:23–43

Vilar S, Isom DG (2021) One year of SARS-CoV-2: how much has the virus changed? Biology 10:91

Villareal TA, Jospeh L, Brzezinski MA, Shipe RF, Lipschultz F, Altabet MA (1999) Biological and chemical characteristics of giant diatom *Ethmodiscus* (*Bacillariophyceae*) in the central North Pacific gyre. J Physiol 35:896–902

Vinogradov SN, Moens L (2008) Diversity of globin function: enzymatic, transport, storage, and sensing. J Biol Chem 283(14):8773–8777

Vinogradov SN, Hoogewijs D, Bailly X, Arrendondo-Peter R, Guertin M et al (2005) Three globin lineages belonging to two structural classes in genomes from the three kingdoms of life. PNAS 102(32):11385–11389

Vinogradov SN, Hoogewijs D, Bailly X, Mizuguchi K, Dewilde S et al (2007) A model of globin evolution. Gene 198:132–142

Virag A, Harris SD (2006) The Spitzenkörper: a molecular perspective. Mycol Res 110:4–13

Visick KL, Fuqua C (2005) Decoding microbial chatter: cell-cell communication in bacteria. J Bacteriol 187(16):5507–5519

Viswanathan VK (2012) Sizing up microbes. Gut Microbes 3(6):483–484

Vogel S (1994) Life in moving fluids. Princeton University Press, Princeton

Vogel S (2003) Comparative biomechanics. Princeton University Press, Princeton

Vogt G, Falckenhayn C, Schrimpf A, Schmid K, Hanna K et al (2015) The marbled crayfish as a paradigm for saltational speciation by autopolyploidy and parthenogenesis in animals. Biol Open. https://doi.org/10.1241/bio.014241

Vos M, Wolf AB, Jennings SJ, Kowalchuk GA (2013) Micro-scale determinant s of bacterial diversity in soil. FEMS Microbiol Rev 37:936–954

Vreeburg S, Nygren K, Aanen DK (2016) Unholy marriages and eternal triangles: how competition in the mushroom life cycle can lead to genomic conflict. Philos Trans R Soc B Biol Sci 371:20150533

Vreeland RH, Rosenzweig WD, Powers DW (2000) Isolation of a 350 million-year-old halotolerant bacterium from a primary salt crystal. Nature 407:897–900

Wagner JL, Setayeshgar S, Sharon LA, Reilly JP, Brun YV (2006) A nutrient uptake role for bacterial cell envelope extensions. PNAS 103(31):11772–11777

Wall D, Kaiser D (1999) Type IV pili and cell motility. Mol Microbiol 32(1):1–10

Waller RF, Jackson CJ (2009) Dinoflagellate mitochondrial genomes: stretching the rules of molecular biology. BioEssays 31:237–245

Waller RF, Slamovits CH, Keeling PJ (2006) Lateral gene transfer of a multigene region from cyanobacteria to dinoflagellates resulting in a novel plastid-targeted fusion protein. Mol Biol Evol 23(7):1437–1443

Wals EJ, Zhang L (1992) Polyploidy and body size variation in a natural population of the rotifer Euchlanis dilatata. J Evol Biol 5:345–353

Walsh MR, Cooley F IV, Biles K, Munch SB (2014) Predator-induced phenotypic plasticity within- and across-generations: a challenge for theory? Proc Biol Sci 282:20142205

Walson JL, Stewart BT, Sangaré L, Mbogo LW, Otieno PA et al (2010) Prevalence and correlates of Helminth co-infection in Kenyan HIV-1 infected adults. PLoS Negl Trop Dis 4(3):e644

Walter J, Ley R (2011) The human gut microbiome: ecology and recent evolutionary changes. Annu Rev Microbiol 65:411–429

Wang M-B, Metzlaff M (2005) RNA silencing and antiviral defense in plants. Curr Opin Plant Biol 8:216–222

Wang Z, Wu M (2015) An integrated phylogenomic approach toward pinpointing the origin of mitochondria. Sci Rep 5:7949

Wang S, Arellano-Santoyo H, Combs PA, Shaevitz JW (2010a) Actin-like cytoskeleton filaments contribute to cell mechanics in bacteria. PNAS 107(20):9182–9185

Wang Z, Liao B-Y, Zhang J (2010b) Genomic patterns of pleiotropy and the evolution of complexity. PNAS 107(42):18034–18039

Wang S, Zickler D, Kleckner N, Zhang L (2015) Meiotic crossover patterns: obligatory crossover, interference and homeostasis in a single process. Cell Cyle 14(3):305–314

Wang Y, Kuang Z, Yu X, Ruhn KA, Kubo M, Hooper LV (2017) The intestinal microbiota regulates body composition through NFIL3 and the circadian clock. Science 357:912–916

Wang G, Zhang D, Orchard RC, Hancks DC, Reese TA (2023) Norovirus MLKL-like protein inti-
ates cell death to induce viral egress. Nature 616:152–158

Waples RS, Gaggiotti O (2006) What is a population? An empirical evaluation of some genetic
methods for identifying the number of gene pools and their degree of connectivity. Mol Ecol
15:1419–1439

Warren A, Patterson DJ, Dunthorn M, Clamp JC, Achilles-Day UEM et al (2017) Beyond the "code":
a guide to the description and documentation of biodiversity in ciliated protists (Alveolata,
Ciliophora). J Eukaryot Microbiol 64:539–554

Watari N, Larson RG (2010) The hydrodynamics of a run-and-tumble bacterium propelled by polymorphic helical flagella. Biophys J 98:12–17

Weber TP, Stilianakis NI (2008) Inactivation of influenza A viruses in the environment and modes of
transmission: a critical review. J Infect 57:361–373

Weber M, Wey-Fabrizius AR, Podsiadlowski L, Witek A, Schill RO et al (2013) Phylogenetic analyses of endoparasitic acantocephala based on mitochondrial genomes suggest secondary loss of
sensory organs. Mol Phylogenet Evol 66:182–189

Webster RG, Bean WJ, Gorman OT, Chambers TM, Kawaoka Y (1992) Evolution and ecology of
influenza A viruses. Microbiol Rev 56(1):152–179

Weedall GD, Hall N (2015) Sexual reproduction and genetic exchange in parasitic protists.
Parasitology 142:S120–S127

Wei N, Tennessen JA, Liston A, Ashman T-L (2017) Present-day sympatry belies the evolutionary
origin of a high-order polyploid. New Phytol 216:279–290

Weinbauer MG (2004) Ecology of prokaryotic viruses. FEMS Microbiol Rev 28:127–181

Weinbauer MG, Brettar I, Höfle MG (2003) Lysogeny and virus-induced mortality of bacterioplankton in surface, deep, and anoxic marine waters. Limnol Oceanogr 48(4):1457–1465

Weiss LC, Leeese F, Laforsch C, Tollrian R (2015) Dopamine is a key regulator in the signalling
pathway underlying predator induced defences in *Daphnia*. Proc Biol Sci 282:20151440

Weitz JS (2008) Evolutionary ecology of bacterial viruses. Microbe 3(4):171–178

Weitz JS, Poisot T, Meyer JR, Flores CO, Valverde S et al (2013) Phage-bacteria infection networks.
Trends Microbiol 21(2):82–91

Westphal S, Soll J, Vothknecht UC (2003) Evolution of chloroplast vesicle transport. Plant Cell
Physiol 44(2):217–222

White MF (2011) Homologous recombination in the archaea: the means justify the ends. Biochem
Soc Trans 39:15–19

WHO (2009) Whole-of-society pandemic readiness. WHO, Global Influenza Programme

Wickstead B, Gull K (2011) The evolution of the cytoskeleton. J Cell Biol 194(4):513–525

Wilde A, Mullineaux CW (2017) Light-controlled motility in prokaryotes and the problem of directional light perception. FEMS Microbiol Rev 41:900–922

Wildman DE, Uddin M, Liu G, Grossman LI, Goodman M (2003) Implications of natural selection
in shaping 99.4% nonsynonymous DNA identity between humans and chimpanzees: enlarging
genus homo. PNAS 100(12):7181–7188

Willbanks A, Leary M, Greenshields M, Tyminski C, Heerboth S et al (2016) The evolution of epigenetics: from prokaryotes to humans and its biological consequences. Genet Epigenet 8:25–36

Willey JM (1988) Characterization of swimming motility in a marine unicellular cyanobacterium,
Diss MIT

Williams SA, Shattuck MR (2015) Ecology, longevity and naked mole-rats: confounding effects of
sociality? Proc Biol Sci 282:20141664

Williams D, Gogarten P, Papice T (2012) Quantifying homologous replacement of loci between
haloarchaeal species. Genome Biol Evol 4(12):1223–1244

Williamson DI (2006) Hybridization in the evolution of animal form and life-cycle. Zool J Linn Soc 148:585–602

Willis BL, van Oppen MJH, Miller DJ, Vollmer SV, Ayre DJ (2006) The role of hybridization in the evolution of reef corals. Annu Rev Ecol Evol Syst 37:489–517

Wilson CG, Nowell RW, Barraclough TG (2018) Cross-contamination explains "inter and intraspecific horizontal genetic transfers" between asexual bdelloid rotifers. Curr Biol 28:2436–2444

Winkle S (1997) Geißeln der Menschheit: Kulturgeschichte der Seuchen. Artemis & Winkler, Düsseldorf

Winther RG (2001) August Weissmann on germ-plasm variation. J Hist Biol 34:517–555

Wisecaver JH, Brosnahan ML, Hackett JD (2013) Horizontal gene transfer is a significant driver of gene innovation in dinoflagellates. Genome Biol Evol 5(12):2368–2381

Wistow G (1993) Lens crystallins: gene recruitment and evolutionary dynamics. Trends Biochem Sci 18:301–306

Witzel K-P, Zakharov IA, Goryacheva II, Adis J, Golovatch SI (2003) Two parthenogenetic millipede species/lines of the genus *Poratia* Cook & Cook, 1894 (Diplopoda, Polydesmida, Pyrgodesmidae) found free from Wolbachia bacteria. Afr Invertebr 44(1):331–338

Woese CR (2000) Interpreting the universal phylogenetic tree. PNAS 97(15):8392–8396

Woldringh CL, Nanninga N (2006) Structural and physical aspects of bacterial chromosome segregation. J Struct Biol 156:273–283

Wolf YI, Koonin EV (2013) Genome reduction as the dominant mode of evolution. Bioessays 35:829–837

Wolf HG, Mort MA (1986) Inter-specific hybridization underlies phenotypic variability in *Daphnia* populations. Oecologia 68:507–511

Wolf YI, Makarova KS, Yutin N, Koonin EV (2012) Updated clusters of orthologous genes for archaea: a complex ancestor of the archaea and the byways of horizontal gene transfer. Biol Direct 7:46

Wolff JO, Scheiderer L, Engelhardt T, Engelhardt J, Matthias J, Hell SW (2023) MINFLUX dissects the unimpeded walking of kinesin-1. Science 379:1004–1010

Wolgemuth C, Hoiczyk E, Kaiser D, Oster G (2002) How myxobacteria glide. Curr Biol 12:369–377

Wommack KE, Colwell RR (2000) Virioplankton: viruses in the aquatic ecosystems. Microbiol Mol Biol Rev 64(1):69–114

Woyke T, Rubin EM (2014) Searching for new branches on the tree of life. Science 346:698–699

Woznica A, King N (2018) Lessons from simple marine models on the bacterial regulation of eukaryotic development. Curr Opin Microbiol 43:108–116

Wu Y, Zhao S (2021) Furin cleavage sites naturally occur in coronaviruses. Stem Cell Res 50:102115

Wurch L, Giannone RJ, Belisle BS, Swift C, Uttukar S et al (2016) Genomics-informed isolation and characterization of a symbiotic nanoarchaeota system from a terrestrial geothermal environment. Nat Commun 7:12115. https://doi.org/10.1038/ncomms12115

Xie L, Altindal T, Chattopadhyay S, Wu X-L (2011) Bacterial flagellum as a propeller and as a rudder for efficient chemotaxis. PNAS 108(6):2246–2251

Yamaguchi M, Mori Y, Kozuka Y, Okada H, Uematsu K et al (2012) Prokaryote or eukaryote? A unique microorganism from the deep sea. J Electron Microsc 61:423–431

Yamaguchi M, Yamada H, Uematsu K, Honnouchi Y, Chibana H (2018) Electron microscopy and structome analysis of unique amorphous bacteria from the deep sea in Japan. Cytologia 83(3):337–342

Yamasa T, Bork P (2009) Evolution of biomolecular networks-lessons from metabolic and protein interactions. Nat Rev Mol Cell Biol 10:791–803

Yang X-J, Grégoire S (2007) Metabolism, cytoskeleton and cellular signalling in the grip of protein N- and O-acetylation. EMBO Rep 8(6):556–562

Yang X, Downes M, Yu RT, Bookout AL, He W et al (2006) Nuclear receptor expression links the circadian clock to metabolism. Cell 126:801–810

Yatsunenko T, Rey FE, Manary MJ, Trehan I, Dominguez-Bello MG et al (2012) Human gut microbiome viewed across age and geography. Nature 486:222–227

Yau S, Lauro FM, DeMaere MZ, Brown MV, Thomas T et al (2011) Virophage control of antarctic algal host-virus dynamics. PNAS 108(15):6163–6168

Yerlici VT, Landwever LF (2014) Programmed genome rearrangements in the ciliate *Oxytricha*. Microbiol Spectr. https://doi.org/10.1128/microbiolspec.MDNA3-0025-2014

Yin Z, Zhu B, Feng H, Huang L (2016) Horizontal gene transfer drives adaptive colonization of apple trees by the fungal pathogen *Valsa mali*. Sci Rep. https://doi.org/10.1038/srep33129

Yooseph S, Nealson KH, Rusch DB, McCrow JP, Dupont CL et al (2010) Genomic and functional adaptation in surface ocean planktonic prokaryotes. Nature 468:60–66

Young RL, Badyaev AV (2007) Evolution of ontogeny: linking epigenetic remodelling and genetic adaptation in skeletal structures. Integr Comp Biol 47(2):234–244

Yubuki N, Leander BS (2013) Evolution of microtubule organizing centers across the tree of eukaryotes. Plant J 75:230–244

Yue J, Hu X, Sun H, Yang Y, Huang J (2012) Widespread impact of horizontal gene transfer on plant colonization of land. Nat Commun. https://doi.org/10.1038/ncomms2148

Yutin N, Koonin EV (2012) Archaeal origin of tubulin. Biol Direct 7:10

Yutin N, Makarova KS, Mekhedov SL, Wolf YI, Koonin EV (2008) The deep archaeal roots of eukaryotes. Mol Biol Evol 25(8):1619–1630

Zachar I, Szathmáry E (2017) Breath-giving cooperation: critical review of origin of mitochondria hypotheses. Biol Direct 12:19

Zachariassen KE, Kristiansen E (2000) Ice nucleation and antinucleation in nature. Cryobiology 41:257–279

Zhang Y, Ducret A, Shaevitz J, Mignot T (2012) From individual cell motility to collective behaviours: insights from a prokaryote, *Myxococcus xanthus*. FEMS Microbiol Rev 36:149–164

Zhang Y-Y, Chen Y, Wei X, Cui J (2022) Viromes in marine ecosystems reveal remarkable invertebrate RNA virus diversity. Sci China Life Sci 65:426–437

Zhao X, Norris SJ, Liu J (2014) Molecular architecture of the bacterial flagellar motor in cells. Biochemistry 53:4323–4333

Zhong NS, Zheng BJ, Li Y-M, Poon LLM, Xie ZH et al (2003) Epidemiology and cause of severe acute respiratory syndrome (SARS) in Guangdong, People's Republic of China, in February, 2003. Lancet 362:1353–1358

Zhu N, Zhang D, Wang W, Li X, Yang B et al (2020) A novel coronavirus from patients with pneumonia in China, 2019. N Engl J Med 382:727–733

Zickler D, Kleckner N (2016) A few of our favorite things: pairing, the bouquet, crossover interference and evolution of meiosis. Semin Cell Dev Biol 54:135–148

Zilber-Rosenberg I, Rosenberg E (2008) Role of microorganisms in the evolution of animals and plants: the hologenome theory of evolution. FEMS Microbiol Rev 32:723–735

Zimmerman AE, Howard-Varona C, Needham DM, John SG, Worden AZ et al (2020) Metabolic and biogeochemical consequences of viral infection in aquatic ecosystems. Nat Rev Microbiol 18:21–34

Zmasek CM, Godzik A (2011) Strong functional patterns in the evolution of eukaryotic genomes revealed by the reconstruction of ancestral protein domain repertoires. Genome Biol 12:R4

Information Processes

6

6.1   Roots of Biological Information Processes

Surely no other phenomenon in life processes is as riddled with ideological rifts as information. This is where the supposed evidence for human uniqueness is thought to reside. However, when subjected to critical scrutiny none of the arguments put forward stands up free of contradiction. The most common argument favouring the rationality of human decisions is refuted merely by the real-life amount of complete, reliable information that is available. Just consider how often completely recorded and unambiguously certified information has been available for decision-making. In this context, made of the—supposedly—permanent availability of electronic information is frequently mentioned. This justification actually demonstrates the exact opposite—when confronted with a deluge of information nobody can reliably distinguish fakes from facts. Technocrats and populists imply objectivity and security by citing large databases and "algorithms". Inverted commas are deliberately used for that term because in common understanding the algorithm has acquired the mystique of an infallible, omnipotent instrument. *De facto*, algorithms are prescribed (programmed) procedures for obtaining unambiguous answers to relevant questions (Laux 1980). In the context of current applications, the term algorithm refers to self-learning procedures for deriving unambiguous answers from information (MacKay 2003). However, unambiguous answers do not necessarily have any connection with objectivity or even the correctness of decisions. Yet, if they stem from accepted authorities—be they people or computers—people consider unambiguous statements to be true. Here, computers happen to be superior to humans—because their authority cannot be questioned on moral grounds, not because they reach better decisions!

This brings to light several issues that we deem irrational from a social standpoint (Ariely 2010). We believe authorities as long as their moral integrity remains

© The Author(s), under exclusive license to Springer-Verlag GmbH, DE, part of
Springer Nature 2024
M. Knoflacher, *Relativity of Evolution*,
https://doi.org/10.1007/978-3-662-69423-7_6

undoubted—a favoured strategem along the road to seizing power. And we allow belief in authorities to seduce us into all kinds of cruelty, either in experiments (Zimbardo 2007; Milgram 2009) or in real-life dictatorships. From an evolutionary perspective, this reveals how strongly ancient behaviour patterns and emotions govern human decisions (Haidt 2008). Thanks to the development of language and the invention of writing and mathematics, human beings can formulate and communicate abstract notions free of their real-life context (Klix 1993). Because, as a general rule in everyday decisions, "the emotional dog wags the rational tail" (Haidt 2008) we are extremely susceptible to manipulation as a society—due to the effects described above.

Where do the roots of biological information processes lie? Can they—in the sense of Aristotle's *Scala naturae* (Great Chain of Being)—be assigned to a particular stage of evolution (Riedl 1989; Popper and Eccles 1990)? In view of increasing evidence that human behaviour is influenced by intestinal bacteria (Lyte 2013; Farmer et al. 2014), exploring the *Scala naturae* is probably pointless. Interestingly, the Austrian-British philosopher Karl Popper (1902–1994) answered those questions—albeit indirectly—with the title of his compendium *All life is Problem Solving* (Popper 2009). In his work, Popper focussed on phenomena in human society. Indirectly, however, his ideas also make it clear that life is inconceivable without information processes. The crucial connection between life and information cannot be described with greater precision. It is also clear from Popper's book title that biological information is always multidimensional and ambiguous—and not linear, as current societal trends suggest. Biological information is open to unpredictable change and enables solutions to connected problems. However, ambiguity carries the danger of slipping into chaos. Such considerations also reveal similarities to genetic processes. Processes of information and genetics ensure the persistence of evolutionary processes by maintaining a metastable state between chaos and stagnation in absolute order.

This does not answer the question of how information can be interpreted. In the age of digital information processing, technologically inspired interpretations of information processes are rapidly emerging (Farnsworth et al. 2012). With Shannon's information theory (Shannon 1948), we are misled into thinking that we have found the benchmark for all information—yet we actually have no more than the foundations for technological developments. Information can appear in various manifestations—for example, temporal or spatial structures. For life processes, only those that ensure the maintenance of an infinite variety of life functions are important. *A central problem of biological information processes resides in their bases, which are most likely present at the level of quantum processes and are always connected with molecular processes for the preservation of life functions* (Asano et al. 2015). Consequently, information processing is also to be expected in all cellular activities. Isolated consideration of information processes is further complicated by entanglement with processes of metabolism or energy conversion, for example in the photosensitive systems of cyanobacteria or with secondary metabolic products in all groups of organisms (see Sects. 4.1 and 5.2.1) (Chao and Levin 1981; Bryant et al. 1991;

Engel et al. 2002; Bérdy 2005; Keller and Surette 2006; Zeng et al. 2006; Hay 2009; Ma et al. 2012; Duanmu et al. 2013; Enomoto et al. 2015).

The fundamental importance of cellular information and regulation processes for the evolution of all organisms is generally overlooked in discussions. Yet, from the very beginning, the evolution of living cells has been linked to information processing (Lyon 2015). They also regulate interactions between cells at—and between—all structural levels of organic life, from single cells to organically differentiated multicellular organisms (Fig. 6.1a). The same can be assumed for the properties of viruses (Koonin et al. 2006; Brüssow 2009; Ghosh et al. 2009; Villareal and Witzany 2010; Domingo et al. 2012; Koonin and Wolf 2012; Koonin and Dolja 2013; Nasir et al. 2014; Nasir and Caetano-Anollés 2015; Zwart and Elena 2015).

The importance of cellular interactions for evolutionary processes becomes even clearer when their occurrence in interactions between groups of organisms is demonstrated, especially with respect to multicellular organisms (Fig. 6.1b). Because of their extensively

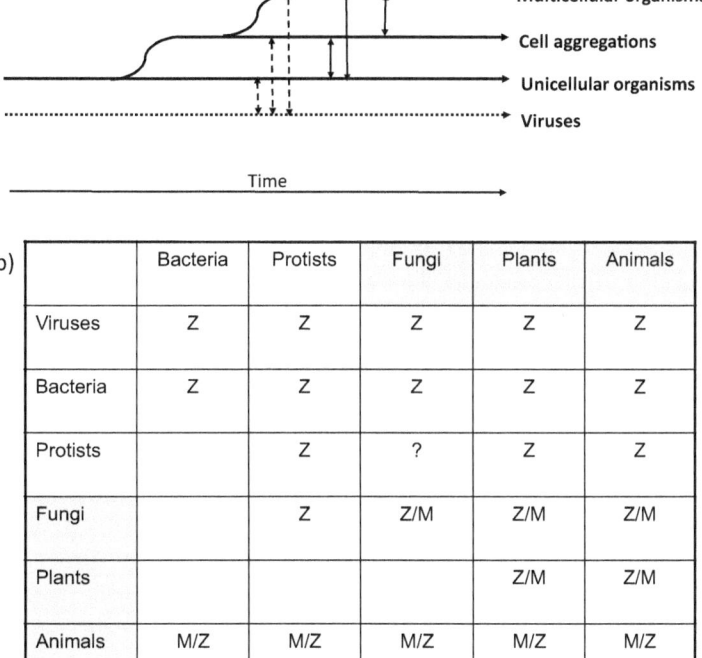

Fig. 6.1 Overview of the evolutionary development of unicellular and multicellular life forms under influences exerted by information and regulatory processes. (**a**) Diagram of the influences of cellular information and regulatory processes on evolutionary development of unicellular to multicellular life forms (double arrows), with suggestion of similar influences in viruses (dashed double arrows). (**b**) Simplified overview of cellular interactions (Z) of viruses, bacteria and eukaryotic organisms; and mechanical interactions (M) between animals and other organisms

documented interactions with multicellular eukaryotes, the only prokaryotes included in the figure are bacteria. Archaeans have been omitted because knowledge of their interactions is still insufficient (Dridi et al. 2006, 2011). Cellular interactions of archaeans with bacteria and viruses are known, notably in the digestive systems of ruminants and termites because of their role in methane formation—although some genus names lead to confusion with bacteria (Madigan et al. 2003; Williams et al. 2009; Leahy et al. 2010; Henderson et al. 2015).

At first glance, the dominance of cellular interactions (Z) compared to mechanical interactions (M) may seem surprising. For various reasons, however, their importance for evolutionary processes has been underestimated. One important reason is the scale of the physiological processes involved, which could only be recorded and observed in the recent past—for example with molecular structural analyses or methods of genetic investigation. By contrast, the phenomena of cellular interactions that are detrimental to humans—for example pestilence—were observed and described very early in human history (Winkle 1997). With the triumph of bacteriological methods in industrialized countries, the primary focus of scientific investigations began to shift to combating adverse impacts. Actions to control viral and bacterial diseases in human and veterinary medicine, as well as in crop breeding, are well known examples. Essential active substances required for such action—for example penicillin—are produced by fungi in cellular interactions with bacteria (Calvo et al. 2002; Madigan et al. 2003; Leveau and Preston 2008; Moore et al. 2011). It is only in the recent past that the increasing spread of various fungal diseases—often with lethal effects—has come to the attention of the disciplines mentioned (Knogge 1996; Kitancharoen et al. 1997; Gindin et al. 2002; Moreira and Barata 2005; Ellis et al. 2006; Howlett 2006; Pfaller et al. 2006; Prado et al. 2009; Frick et al. 2010; Amselem et al. 2011; Cheng et al. 2011; Dean et al. 2012). Socially, this was associated with a subliminal, negative judgement of all cellular interactions as abnormalities and diseases. Unaffected by this general evaluation, the beneficial effects of cellular interactions of bacteria, fungi and plants in the so-called mycorrhizae are being increasingly recognized—especially in forestry (Gadkar et al. 2001; Lévy et al. 2004; Bonfante and Genre 2010; Rinaldi et al. 2008; Rodriguez et al. 2009; van der Heijden and Horton 2009; Maillet et al. 2011; Müller et al. 2013; Fuller et al. 2015).

The evolutionary significance of cellular interactions (Fig. 6.1b) can only be surmised, as only fragmentary evidence of their origins is so far available. However, from the far-reaching regulatory effects of the currently known examples, it can be inferred that cellular interactions play a central role in evolutionary processes. Despite the piecemeal coverage so far achieved, this inference can be supported with a wide variety of examples.

Genetic modifications and transfer of genes across the boundaries of all systematic categories accompanying horizontal gene transfer due to viruses have already been mentioned. Viruses also influence interactions between other organisms at the cellular level. In aquatic habitats, viruses influence the dynamics of nutrient cycles in near-surface zones by dissolving the cells of bacteria they have infected (Wommack and Colwell 2000).

Bacteria—like archaeans—interact with each other and with all other organisms at the cellular level. As is the case for all intercellular processes, the particular effects that occur depend on the relevant context and are therefore ambivalent. The prevailing image of bacteria as causative agents of many diseases obscures a wider view of the broad spectrum of effects of interactions involving bacteria. As explained in the previous chapter, no multicellular organism could have evolved without cellular regulation of interactions with microorganisms in a world packed with bacteria and viruses. Interactions between a multicellular organism and its microbiome—their "second skin" on the outer surface—usually provides protection against effects of harmful organisms.

The interactions of protists—unicellular eukaryotes—like those of bacteria attract public attention primarily as causative agents of diseases, such as the various types of malaria. But people completely overlook the fact that production capacity and interactions between bacteria, protists and viruses have shaped essential framework conditions for the evolution of all organisms in the oceans for at least 1.5 billion years. In this connection, a great diversity of different life forms of protists has also evolved in the oceans, in freshwater environments and ultimately in many terrestrial habitats, covering the entire spectrum from cooperative through competitive to destructive interactions (Visser and Kiørboe 2006).

6.2   General Characteristics of Cellular and Intracellular Information Processes

All biological forms of information handling, along with their evolutionary differentiation, are founded on cellular information processes. Learning processes, storage of information and individual processing of signals can already be observed in archaeans and bacteria. Their functional features are found in all decentralized processes of information processing, be it in multicellular fungi and plants or in the metabolic and immune systems of animals (Lim et al. 2015). On the other hand, neural activities involved in information processing, which we consider to be particularly important, are closely connected with—epigenetically acquired—functional needs and environmental conditions.

In phenomenological terms, biological information processes differ from their technical counterparts predominantly in their multidimensional nature, their contextual variability, emergent generation of new information, and the non-linearity of resulting responses. As far as technical characteristics are concerned, rates of biological processing are significantly slower and storage capacities smaller than with digital processing. A fundamental difference between biological and technical processes of information processing resides in the lack of filtering outside the "receiver". In a biological context, this is an individual cell or individual organism, which is confronted with a vast array of external, fuzzy information (Fig. 6.2). Here, *a priori*, there are no external "channels" to filter the flow of—perceptually tuned—information, as in the technical environment with which we are familiar.

Comparative examination of biological information processes is rendered more difficult by classification according to morphological characteristics (Dayan and Abbott 2001;

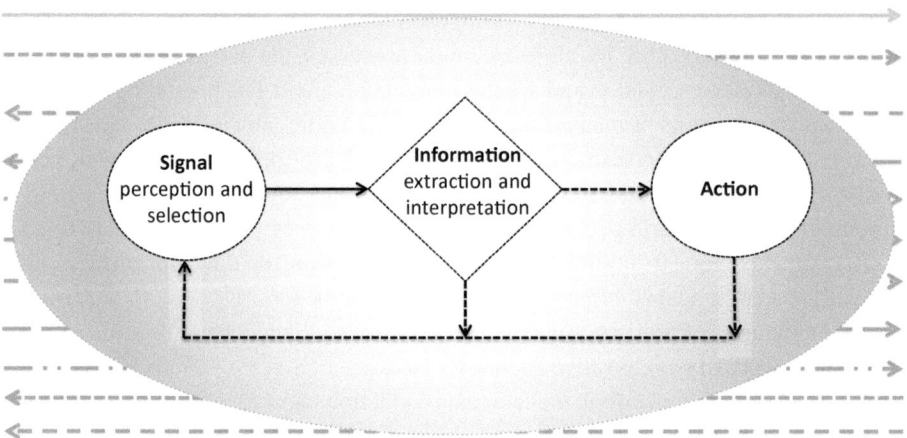

Fig. 6.2 Diagram of the basic cycle of information processing for orientation and maintenance of the capacity of organisms to take action when confronted with a wide spectrum of incoming signals (symbolized by various grey arrows in the background). The process can also be repeated without operative implementation (dashed arrow)

Jorgensen 2002; Madigan et al. 2003; Franklin and Whitelam 2004; Baluška et al. 2006; Wessnitzer and Webb 2006; Jones et al. 2016; Heil and Karban 2009; Zimmermann et al. 2009; Moore et al. 2011; Scherzer et al. 2015), exacerbated by subjective value judgements such as "primitive" or "highly developed". It is easy to overlook the fact that the diversity of information systems and processes seen in organisms is founded on interactions between cellular information processes and environmental conditions. For ease of comparison, here a distinction is drawn only between the following two categories (Fig. 6.3):

 (I) Intracellular information processes
(II) Extracellular information processes

Among other things, intracellular information processes regulate the factors listed as examples in Fig. 6.3, but in each case also process and store the information relevant to the cell. In comparison, the signals involved are more specific and usually less ambiguous than in extracellular information processing. Extracellular information processes ensure transfer of information from the environment that is needed to maintain life functions, as well as transmission of signals to the environment (Fig. 6.3). The range of signals extends from predominantly abiotic variables, though biotic plus abiotic variables—in the case of free-living single cells and loose cellular agglomerations—to purely biotic variables in cell tissues. For energetic reasons, differentiation of extracellular information processes is closely linked to the physiological and morphological characteristics of organisms and prevailing environmental conditions. Reference to Fig. 5.3 provides justification here: Both information processing and synthesis of the biological structures that it necessitates

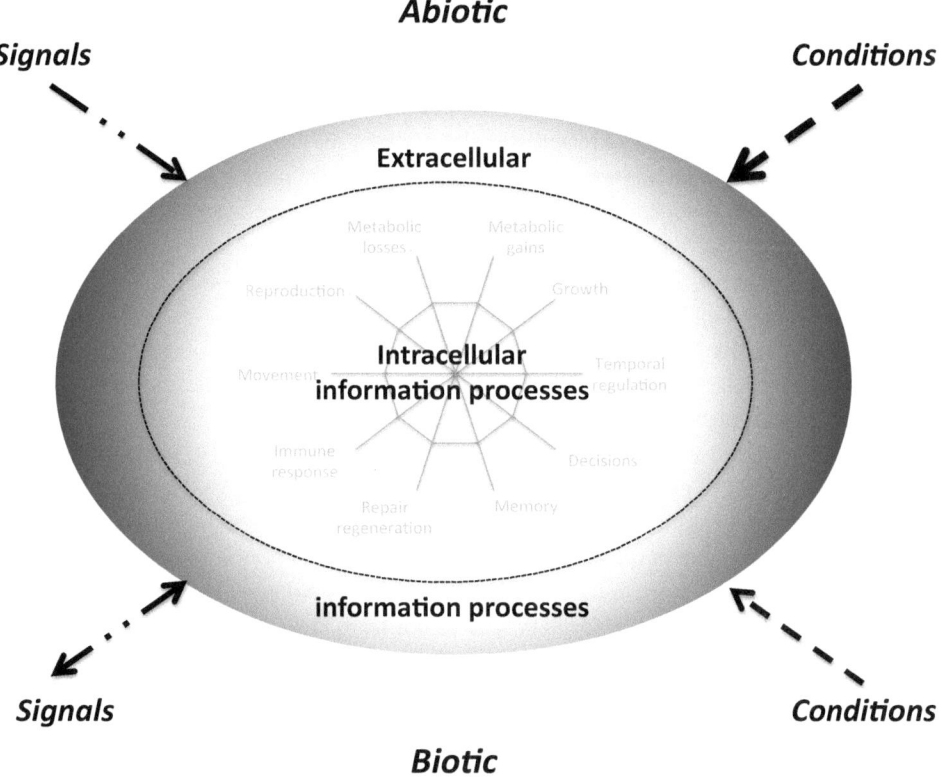

Fig. 6.3 Generalized overview of the basic categories of cellular information processes

require energy and can therefore develop without endangering the survival of the organism only in coordination with other morphological and physiological characteristics. The central nervous system, whose absence would be lethal for vertebrates—but causes no problems for plants—serves as an example of the differing requirements of information processing.

Single-celled organisms fundamentally lack a nervous system. External chemical information is detected by molecular receptors on the cell membrane and selectively conveyed in the form of protons, electrons or ions to specific molecules inside the cell. In eukaryotic cells, the cell organelles—mitochondria and chloroplasts—also participate in the regulation of processes inside the cell (Kleine and Leister 2016). Through changes in the configuration or electrochemical charge of the molecules involved, information is stored or cell responses are triggered in molecular regulatory networks (Holden and Gally 2004; Kollmann et al. 2005; Briegel et al. 2012). In addition, external signals can modify the locations of receptor molecules on the cell membrane (Mauriello et al. 2009). For this to occur, structured coordination of the many processes operating in parallel in a cell—within the context of prevailing conditions of the cell environment—is a prerequisite (see Fig. 6.2)

(Bhattacharyya et al. 2006; Hunter 2007; Schuetz et al. 2012; Suga et al. 2013). To this end, information from the environment must be continuously obtained, interpreted and taken into account in the decision-making process (Perkins and Swain 2009; Balázsi et al. 2011). These capacities have already been demonstrated experimentally by manipulating the temporal frequencies of external signals (Mitchell et al. 2015). Cells of the immune system can also differentially recognize spatial structures of pathogenic microbes (Ugurlar et al. 2018). Context-dependent decision-making processes are also suspected to underly differential behaviour of viruses within the cells they attack—destroying them immediately or settling in for a lengthy residence (Weitz et al. 2008).

In the context of external signals, it is fundamentally essential for all organisms to store essential information and to correct any recognized errors in earlier phases of information processing—in other words: to be capable of learning. Such capacities have been demonstrated in prokaryotes under various experimental configurations (Wolf et al. 2008; John et al. 2009; Vladimirov and Sourjik 2009; Lambert and Kussell 2014; Lan and Tu 2016).

Over the course of evolution, organisms were able to hold their own only if they were able to identify relevant information in a timely fashion, process it and convert it into suitable responses. For this purpose, only their own receptors repertoire and internally stored reference information were and are available to them as "filters". For organisms, the significance and importance of information arise not from abstract rationality but from their own—subjective—context of internal information processing. *Thus, cellular information systems operate in constructivistic and adaptive manner, but not deterministically.*

Those systems are constructivist because properties of the receptors and internally stored information determine the "expectation dimensions" of information reception. On the basis of quantum physical properties, a multi-dimensional network of receptor-related resonance ranges can be presumed for implementation (Fig. 6.4). Responses to external signals depend on the availability of suitable receptors and the excitation states of the relevant signal-processing molecules. The intensity of the excitation states of resonance ranges regulates temporally variable structures and hierarchies of their interaction networks. Changes in the excitation states of individual resonance ranges can be triggered by both external and internal signals. For extraction of information, variable interactions between the resonance ranges of similar activation states are relevant. This permits decentralized information processing with a high degree of flexibility. One indication for the implementation of this principle is so-called "crosstalk" between different signalling molecules in cells (Li and He 2013; Li et al. 2019).

Flexibility of information processing is guaranteed by reversibility of the molecular systems involved. Due to the associated change between different states of order—justified by the second law of thermodynamics—all information processes require energy (Brandizzi et al. 2002; Mehta and Schwab 2012)!

Adaptive cellular information processes stem from the basic need to adapt to variable environmental conditions. Adaptations to different environmental conditions can occur in at least two ways: in the short term by modifying "expected values" in information processing through learning, and over the long term by modifying the provision of receptors.

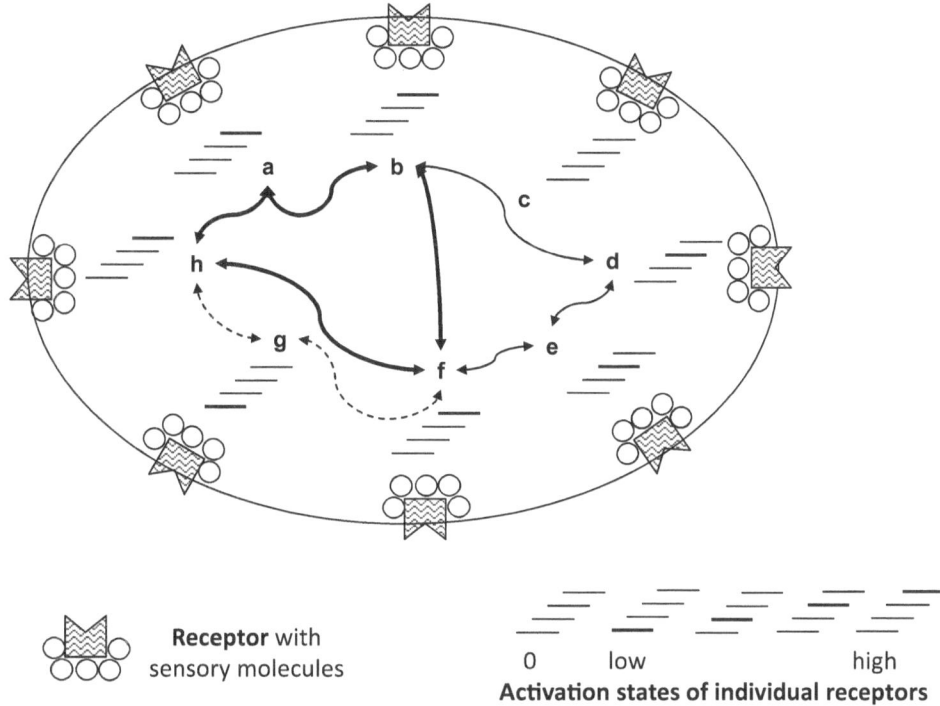

Fig. 6.4 Hypothetical model of cellular information processing through functionally differentiated resonance ranges and their interaction networks. In this example, the resonance ranges a, b, f, and h are excited the most and hence determine interpretation of information. The less excited resonance ranges d, e, and g contribute only to modulation of the information. In this case, resonance range c exerts no influence on information processing

In order to survive, organisms must also be able to make decisions. For example, it is of no value if an organism recognizes two food sources but cannot decide to choose one. It would simply starve. Ideally, a cycle of information processing (Fig. 6.2) should yield a clear decision for operational implementation. This is increasingly true as the time available between information acquisition and the need for action decreases—notably in the case of a lethal threat. For microorganisms, this could be extremely hot currents in the vicinity of submarine hydrothermal vents; for animals it might be the suddenly appearance of a carnivore. Under non-lethal conditions—for example when seeking food—the outcome can be improved if the decision-making processes are repeated several times. Because of the often large number of different influencing factors, better results can be expected if individual cycles of information processing employ different signal combinations. For example, magnetotactic bacteria interpret information from light and the magnetic field in a differentiated manner (Chen et al. 2011). Moreover, individual differences in the operative implementation of information from identical signal combinations can also be observed in microorganisms (Levin et al. 1998). Such correlations between framework conditions and decision-making processes can also be observed in our own field of experience; for example, when we have to leap out of the way of a suddenly approaching vehicle or are planning to furnish a new apartment.

The very deep roots of information processes can be appreciated when pondering the reproduction of viruses—known as phages in the context of prokaryotes. Reproduction is based primarily on manipulations of regulatory processes in infected cells, whereby the viruses themselves regulate the further course of their reproduction (Dimmock et al. 2007; Balázsi et al. 2011). Reproduction can either begin immediately or only after a lengthy period of residence in the infected cell. Infected cells use a multi-level information system to defend themselves and specifically dissect viral DNA using the CRISPR-Cas system already mentioned (Amital and Sorek 2017; Kazlauskiene et al. 2017). This example highlights the capacities of a single cell to identify foreign information and to decide on countermeasures. But even this defence system can be circumvented by certain viruses by substituting a virally designed CRISPR-Cas system (Samson et al. 2013). Various viruses use bacteria as Trojan horses to infect multicellular organisms. Through genetic manipulation, the potential for toxic effects is prepared in the host cells and unleashed through bacterial regulatory processes only after successful infection of the target organism. In some infectious diseases, the associated reproduction of viruses also leads to death of the bacterial host cell, whereas in others the host cell remains functional. Treatments of such infectious diseases with antibiotics can intensify the toxic effects because destruction of the bacterial cells also releases viruses. This correlation was recognized only through investigation of molecular processes accompanying severe infectious diseases—for instance diphtheria or endocarditis (Barksdale 1970; Bensing et al. 2001; Wagner and Waldor 2002; Beier and Gross 2006).

Over recent decades, it has become increasingly clear that intracellular information processing is influenced—if not shaped—by a wide variety of forms of RNA (Amaral et al. 2008; Collin and Schuch 2009; Waters and Storz 2009; Cech and Steitz 2014; Morris and Mattick 2014). Of particular importance are the so-called non-coding RNAs in combination with non-coding DNA segments, previously previously interpreted as useless trash (junk) from former cell functions. For example,—depending on current conditions in the cell environment—RNAs regulate production of proteins by inhibiting or activating the relevant genetic codes by means of the epitranscriptome system (Saletore et al. 2012, 2013). In prokaryotes, regulation is carried out directly by "messenger" RNA (mRNA) in the form of so-called riboswitches. Regulation is triggered by attachment of an emissary molecule to a sensitive section of the mRNA and is accomplished through structural change in the mRNA (Mironov et al. 2002; Mandal and Breaker 2004; Henkin 2008; Breaker 2010; Lemay et al. 2011; Serganov and Nudler 2013; McCown et al. 2014; Nelson et al. 2017). Emissary molecules themselves can be RNAs—especially small or microRNAs—as well as constituents of proteins (Carthew and Sontheimer 2009; Malone and Hannon 2009; Shukla et al. 2011; Montes et al. 2014). However, RNAs can also directly process environmental signals, such as ambient temperature, through mRNA (Johanson 2009). Riboswitches and multi-level cellular regulatory systems have now also been demonstrated in eukaryotes—especially in plants (Croft et al. 2007; Wuichet et al. 2010; Schaller et al. 2011; Vu et al. 2012; Curtis and Liu 2013; Durand et al. 2014; Pekárová et al. 2016). Other forms of RNA—such as long non-coding RNAs (lncRNA)—regulate epigenetic processes in eukaryotes in the context of specific requirements of the respective organ assemblage, including neural processes (Mattick 2011; Shirayama et al. 2012; Komienko et al. 2013; Ulitsky and Bartel 2013; Cech and Steitz 2014; Fatica and Bozzoni 2014; Picao-Osorio et al. 2015).

6.3   Extracellular Information Processes

Intracellular information processes ensure the maintenance of cellular systems, while extracellular information processes by using various signal sources enable target-oriented responses to abiotic signals, the formation of cell assemblages and multicellular organisms, and development of functional networks between organisms (Fig. 6.5). Simplified characterization of abiotic signals (Fig. 6.5b) reveals that, according to the current state of knowledge, external timekeepers (Zeitgebers) for tuning biological processes are predominantly present in near-surface water zones and at the Earth's surface (Tessmar-Raible et al. 2011). Exceptions to this are cyclical changes in gravity due to the moon's orbit, which are used by invertebrates—both in surface zones as well as in the deep sea—as signal generators for reproduction (Richmond and Jokiel 1984; Gaston and Hall 2000; Mercier et al. 2011). In many cases, however, microorganisms are primarily dependent on endogenous, biotic regulatory factors.

Research into the biological effects of abiotic electrical signals at the cellular level has a notable "blind spot". This refers not to the many biological research papers on electrical phenomena in fish or on neural signal transmission. Instead, it alludes to connections between abiotic electrical signals during the build-up to geological processes—for example earthquakes—or atmospheric processes—for example thunderstorms (Johnston 1997; Yoshida et al. 1998; Rycroft et al. 2000; Marshall et al. 2005; Freund 2007a, b; Liu et al. 2009; Harrison et al. 2010; Revil 2013)—and biological responses at the cellular level. In view of the importance of electrical signals for non-neural communication between various cells and metabolic processes, numerous questions arise here (Morris and Gow 1993; Lin et al. 2008; Rabaey and Rozendal 2010; Kiely et al. 2011; Strycharz-Glaven et al. 2011; Kato et al. 2012; Snider et al. 2012; Cortese et al. 2014; Holm and Vikström 2014; Malkin et al. 2014; Malvankar and Lovley 2014; Schauer et al. 2014: Gallé et al. 2015; Prindle et al. 2015; Sutton et al. 2016). For instance, the question of whether animals use variations in electric fields (England et al. 2006) related to geographical longitude for orientation in the open sea has not even been broached.

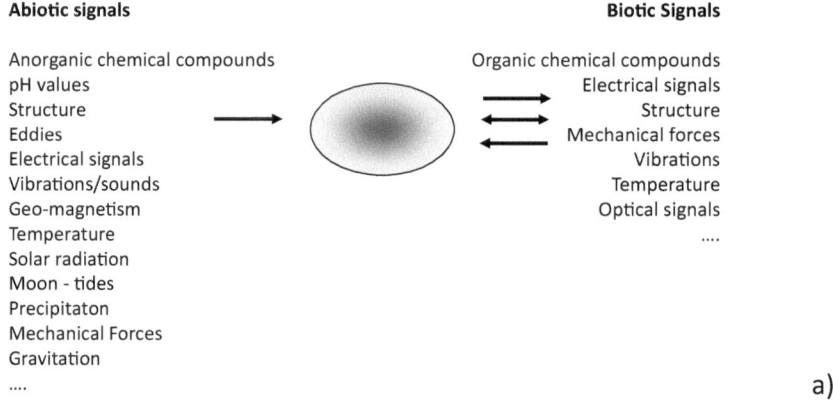

Abiotic signals		Biotic Signals
Anorganic chemical compounds		Organic chemical compounds
pH values		Electrical signals
Structure		Structure
Eddies		Mechanical forces
Electrical signals		Vibrations
Vibrations/sounds		Temperature
Geo-magnetism		Optical signals
Temperature		....
Solar radiation		
Moon - tides		
Precipitaton		
Mechanical Forces		
Gravitation		
....		

a)

Abiotic factor	Micro scale	Macro scale	Dynamics	Biological relevance
Chemical compounds solid/solutions/gaseous	X			Energy, metabolism
pH Value	X			E/S
Structure	X	X		Orientation
Eddies	X	X	irregular	E/S
Electrical potentials	X	X	irregular	?
Vibration/sounds	X	X	irregular	?
Geo-magnetism	X	X	long-term variable	Orientation
Temperature	X	X	variable	E/S
Solar radiation (+earth rotation and revolution)	X	X	cyclical	E/S, coordination, orientation, energy
Moon light/tides	X	X	cyclical	coordination (energy)
Precipitation	X	X	variable	E/S
Gravitation	X	X	permanent	orientation

b)

Fig. 6.5 Simplified overview of extracellular signals and their characteristics. (**a**) Comparative overview of measurable qualities of extracellular abiotic and biotic signals. Cells can obtain only information from abiotic signals, but can also emit biotic signals themselves or use them for communication with other cells; (**b**) Temporal and spatial features, as well as biological significance of abiotic signal sources. Abbreviations: E/S = Energy and metabolism

6.3.1   Abiotic Signals

Obtaining information from abiotic signals is a basic requisite for survival under the dynamic and sometimes hostile conditions of geological and planetary processes. Organisms cannot modify abiotic signals, but they can use them in conjunction with biotic signals. This is shown particularly clearly by the examples of bioluminescence in marine habitats described in Sect. 4.1. In a similar way, cellular adaptations of colours and light intensities at body surfaces are examples of interactions with neural processing of signals and information. The discussion of quantum physical properties of signals provided in

Sect. 4.1 will not be extended here. The main reason for this is the use of classical physical and chemical measurements in the vast majority of documented experiments and publications. As a rule, responses of cells and organisms to signals used are recorded, rather than information processing itself. Depending on the particular organisms studied, observed responses are referred to as taxes—orientation during free locomotion—or nastisms/tropisms—bending or positional adjustment movements in plants. It is common practice to use combinations of signal and response designations, e.g. phototaxis for orientation towards the direction of incoming light.

Basically, it can be assumed that, over the course of evolution, organisms have developed capacities to process abiotic signals that, in the specific context of their habitats, provide sufficiently precise information to permit maintenance of vital functions. Decisive influencing factors for this outcome stem from the abiotic conditions of the habitat and the physiological capacities of organisms themselves. In light-flooded zones of open water, relevant abiotic signals differ from those in deep sea or on land. Temperature changes in surface waters and at the earth's surface are evidently substantial and therefore have a higher information content for organisms than in deep zones of water or soil. Because of the higher density of water—compared to air—gravity influences developmental and locomotor possibilities of organisms to a lesser degree.

Prokaryotes can cope with far more complex challenges than simply orienting themselves along chemical gradients with the familiar process of chemotaxis (Kiørboe and Jackson 2001; Szurmant and Ordal 2004; Stocker et al. 2008). For example, various life forms in temporary communities can coordinate their locomotor activities or complex metabolic processes (Velicer and Yu 2003; Shank and Kolter 2009; Darnton et al. 2010; Handley et al. 2010). Cyanobacteria coordinate their photosynthetic activity with currently available levels of carbon dioxide (CO_2) (Montgomery et al. 2016). Microorganisms can also recognize different properties of solid surfaces or temperature differences and modify their decisions in relation to experience with prior events (Lewus and Ford 1999; Taylor et al. 1999; Galperin 2005; van Duijn et al. 2006; Baker and Stock 2007; John et al. 2009; Karatan and Watnick 2009; Datsenko et al. 2012; Hug et al. 2017; Hughes and Berg 2017). Bacteria can perceive water currents and actively orient themselves in them with flagella. On a macro-scale, animals—such as jellyfish, insects or birds—use other receptors to detect analogous signals (Marcos et al. 2012; Fossette et al. 2015; Chapman et al. 2015). Even supposedly inactive bacterial spores use electrochemical signals to orient themselves to environmental conditions and derive from them their decisions for activating their cell functions (Kikuchi et al. 2022; Lombardino and Burton 2022).

The slime mould *Physarum*, a member of the group Mycetozoa (Amoebozoa) (Fiore-Donno et al. 2010), displays capacities in information processing and optimizing decisions that are barely to be expected for unicellular organisms. Responses of these organisms are easy to observe in experiments, because they extrude a network of cell filaments to search for food and then expand or reduce individual filaments according to local food availability. Because of the associated risk of desiccation, light is a danger signal (Nakagaki et al. 1999). In experiments, slime moulds found optimal solutions in balancing risks and opportunities for complex spatial distributions of food sources of different quality and in zones with different light levels. The error rate was found to be higher for rapid decisions than for lengthier decision processes (Dussutour et al. 2010; Latty and Beekman 2011). In experimental mazes they also found the shortest paths for their cell filaments (Reid et al. 2012). In the meantime, observed decision-making processes—of unicellular organisms—have been formalized in computer programmes and used for applied tasks, for example for navigation in traffic networks (Tero et al. 2006; Zhang et al. 2014).

As shown in Sect. 4.1, organisms can derive a wide variety of kinds of information from light. Observed responses to light signals—both by day and by night—are correspondingly diverse. For example, proteins for processing light signals are widespread among organisms and are even found in cells of nematodes (*Caenorhabditis*) and fungi (Krauss et al. 2009). Photoautotrophic organisms use the information thus obtained to achieve optimal orientation of their photosynthetic apparatus (de Wit et al. 2016). Actively buoyant spores—zoospores—of various fungus species orient themselves to light signals to identify suitable habitats. During the mycelial stage, these signals are used, among other things, to regulate metabolism, to form fruiting bodies or to activate pathogenic factors (Kazama 1972; Cerdá-Olmedo 2001; Deacon 2006; Bowen et al. 2007; Bayram et al. 2010; Idnum et al. 2010; Rodriguez-Romero et al. 2010; Avelar et al. 2014; Fuller et al. 2015; Corrochano et al. 2016). Growth of the fruiting body, on the other hand, is initially oriented to light signals but then in a later phase to gravity (Moore 1991).

Various microorganisms possess photosensory rhodopsin, which in some cyanobacteria has evolved into a molecular "eye" for detecting and processing optical signals (Gärtner and Losi 2003; Richards and Gomes 2015; Nilsson and Colley 2016; Schuergers et al. 2016). Despite the tiny size of the cells, with a diameter of about 3 micrometres (μm), they can detect the direction of incoming light. This is achieved by optically focussing the light with the curved cell wall and detecting the light beam with receptors on the opposite cell wall—a similar principle is used by many unicellular eukaryotes (Schuergers et al. 2016; Ueki et al. 2016; Wilde and Mullineaux 2017). In general, cyanobacteria seek out areas of optimal irradiance and spectral ranges (Purcell and Crosson 2008; Narikawa et al. 2011). Configurations of photosensitive structures in eukaryotic unicellular dinoflagellates are extremely diverse. They range from simple concentrations of photosensitive molecules to eye-like shapes with a molecular lens and optically aligned photomolecules (Gavelis et al. 2015; Hayakawa et al. 2015).

This diversity in photosensitive structures is related to the wide range of life forms among planktonic dinoflagellates. Since they first appeared, presumably in the Mesozoic, both heterotrophic and mixotrophic organisms have evolved from originally photoautotrophic ancestors. In adaption to different lifestyles, some species have lost the original cellulose armour. In addition to the capacity to release toxic substances, cellular cnidoblasts—similar to those found in jellyfish—have evolved in some heterotrophic species (Gavelis 2015; Janouškovec et al. 2017). These examples clearly show that capacities for obtaining information from extracellular signals are determined not by the size of autonomous individuals—be they unicellular or multicellular organisms—but their ecophysiological requirements. Accordingly, attempts to classify the sensory capacities of organisms solely on the basis of body size do not yield meaningful results (Martens et al. 2015; Andersen et al. 2016).

Although only a few studies have succeeded in recording the signals actually used by animals, differences in the detection of light signals can be inferred simply from the different detection perspectives close to the ground or in the air. Evidence of nocturnal orientation to light signals near the ground has been found, for example, in desert lizards (*Uma notata*) (Adler and Phillips 1985), marbled newts (*Triturus marmoratus*) (Diego-Rasilla and Luengo 2002), and insects such as jumping plant lice (*Talitrus saltator*) (Galanti 2009) or African dung beetles *(Scarabaeus zambesianus)* (Dacke et al. 2011, 2013; el Jundi et al. 2015, 2016). The extent to which orientation is influenced by light or sound signals In marbled newts, however, remains a matter for debate (Diego-Rasilla and Luengo 2002). In the case of dung beetles, it is unclear to what extent they orient themselves to polarized moonlight and/or the pattern of the Milky Way. Which light signals or signal patterns of the night sky are used for orientation by flying animals such as insects and birds is also largely unknown (Sotthibandhu and Baker 1979; Able and Able 1996; Åkesson et al. 2001a, b; Mouritsen and Larsen 2001; Muheim et al. 2006). Presumably, combinations of different signal sources are used. As already mentioned in Sect. 4.4, the flexible use of different signals by organisms—for example, in monarch butterflies, sea turtles or eels—makes it difficult to record the signal sources relevant for orientation under experimental conditions (Lohmann and Lohmann 1996; Heinze and Reppert 2011; Naisbett-Jones et al. 2017).

The randomness of many abiotic processes prohibits reliable predictions about their future evolution, thus complicating decision-making by organisms. To augment reliability, organisms often base their decisions on signals from different processes, for example plants employing combinations of changes in daylength and temperature to adjust to seasonal conditions (Tylewicz et al. 2018). Flexibility of information processing is already clearly recognizable in bacteria when, for example, they process complementary signals from the magnetic field and gradients of oxygen concentration in an integrated manner (magneto-aerotaxis), whereas signals from the magnetic field and light irradiation are processed independently (Frankel et al. 1997; Chen et al. 2011). On the other hand, there is no evidence for orientation of bacteria to gravity, whereas signals from that source are perceived by both unicellular and multicellular eukaryotes (Chen et al. 1999; Bräucker et al. 2002; Krause et al. 2010; Azizullah et al. 2013; Hemmersbach et al. 2014). Nevertheless, bacteria in space vessels responded to weightlessness (microgravity) with a significant increase in growth and virulence ("toxicity") (Rosenzweig et al. 2010).

According to the current state of knowledge, changes in gravity due to influences of the moon and the sun affect the formation of melatonin in green macroalgae (*Ulva* sp.) as well as root growth and cell physiology of trees among land plants (Schad 2001; Barlow et al. 2010; Zürcher et al. 2010; Tal et al. 2011; Barlow and Fisahn 2012).

These phenomena reveal the differing importance of abiotic signals for autotrophic and heterotrophic organisms. As already mentioned, animals use signals from the moon's orbit for timing their reproductive cycles. Autotrophic organisms—such as algae in intertidal zones—need these signals to adapt their metabolism to the dry phase conditions at low tide. This requires both protection of cells from intensive radiation and modified exchange of substances with the environment. Here, changes in environmental conditions can be "predicted" relatively accurately by such autotrophic organisms from signals emanating from the lunar orbit.

In contrast, different conditions prevail in the pelagic areas of large bodies of water, where concentrations of important mineral nutrients in particular are determined by random factors. Autotrophic organisms must therefore be able to respond rapidly to both favourable and unfavourable conditions, adjusting their metabolism accordingly. Selective advantages for unicellular organisms associated with this capacity were ultimately decisive for maintaining constancy of unicellular autotrophic life forms in the pelagic areas of large bodies of water throughout the entire course of ascertainable evolutionary history.

6.3.2   Extracellular Biotic Signals

Biotic signals can be received, sent and/or exchanged between cells. Extracellular biotic signals ensure—within the abiotic framework—the evolution of independent information networks between cells. Whereas optical signals are easily perceptible to us, the probably oldest and densest networks of biological interactions of microorganisms in evolutionary history often elude our perception. With their information processing capacities, microorganisms can communicate with one another and coordinate their activities in many ways (Bassler 2002; Vendeville et al. 2005; Bassler and Losick 2006; Fiegna et al. 2006; Keller and Surette 2006; Kato et al. 2007; Williams 2007; Ryan and Dow 2008; Shank and Kolter 2009; Hong et al. 2012; West et al. 2012; Marguet et al. 2013; Prindle et al. 2015). Signals are transmitted selectively between microorganisms in various ways—for example through molecules or electrons—and registered by different receptors on the cell surfaces (Reguera et al. 2005, 2006; Singla and Reiter 2006; Boyer and Wisniewski-Dyé 2009; Collin and Schuch 2009; Wuichet et al. 2010; Kaiser and Warrick 2014; Alberts et al. 2015; Esser et al. 2016). Details of molecular processes involved in the use of light in communication between microorganisms so far remain unknown (Fels 2009; Enomoto et al. 2015).

In contrast to purely material processes, or abiotic signals, in interactions between cells and organisms, the information content of signals is independent of their physical characteristics (Fig. 6.6). All cells—and even viruses—can transmit information content with false signals to one other with no connection to the physical consequences (Ghoul et al. 2013). The evolutionary significance of these facts becomes apparent only in the context of the temporal sequence of prokaryotic and eukaryotic evolution. Because prokaryotes arose considerably earlier, the evolution of eukaryotic organisms was always dependent on their capacities for perceiving prokaryotic signals.

Emitted signals can convey or disguise the real state of the transmitter (cell A). In a theoretical sense, the first case is an unmanipulated encoding of signals, whereas the

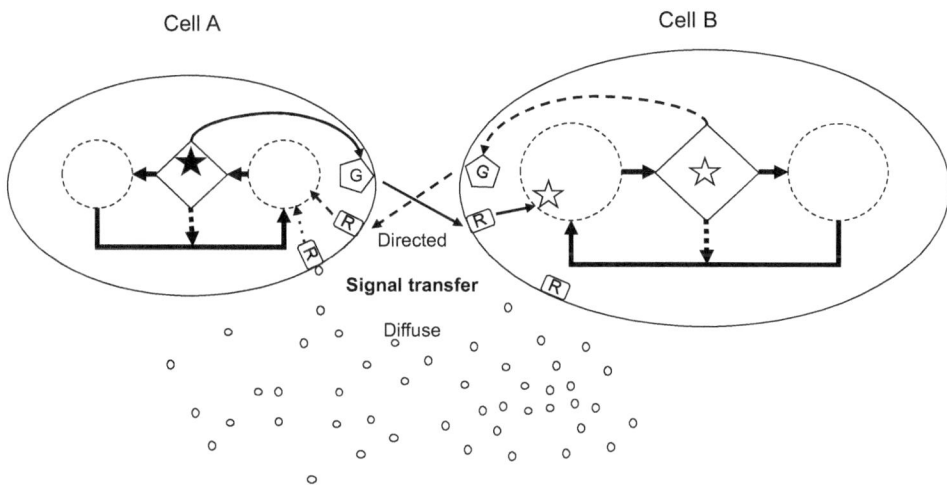

Fig. 6.6 Simplified representation of the relative independence of extracellular biotic signals from their information content. As an example, cell A can specifically generate a false signal (filled star) and send it directly to a receptor of cell B; cell B is misled by the deception if it unwittingly implements the—false—information content (right circle). Risks of misinformation can be reduced with internal feedback using additional information (empty stars), as well as through additional queries—dashed arrows—transmitted to cell A. Explanation of symbols: The contents of unlabelled symbols correspond to Fig. 6.2, with a mirrored arrangement in cell A. G = signaller, R = signal receptor. Further explanations in the text

second is manipulated. Examples for the first case are provided by excretions of metabolic products that are detected by other organisms. The incalculable number of examples for the second case ranges from the viruses already mentioned, through reduction of reproduction effort in various plants by means of mimicked female insect forms of their flowers leading to pollination by deceived male insects (Ellis and Johnson 2010), to falsification in human society. Regardless of this, it is interpretation of signals by the receiver (see Fig. 6.2) that determines the accepted information content. The probability of deviation of accepted information content from the original content is greatest with linear processes of signal transmission and processing. In Fig. 6.6, this would be the path—shown with continuous lines—of a manipulated signal in cell A to the signal transmitter (G) and direct transmission to a receptor (R) in cell B with subsequent unwitting operative implementation. By controlling feedback in cell B (empty stars) and "queries"—represented by dashed lines—transmitted to cell A, differences in the information content can be increasingly neutralized.

No healthy person notices the continuous interactions between bacterial cells and our body cells such as those of the immune or digestive systems. We become aware of these confrontations only when the body's defence responses are boosted—which is recognizable, for example, from a rise in body temperature in a phenomenon we call a fever. On the other hand, we refer to a piece of meat with an unpleasant odour—because of bacterial degradation processes—as rotten. Behind these seemingly different phenomena are the common processes of biochemical signals. In the first case, by manipulating extracellular signals, bacterial cells have managed to outsmart the information processes of the immune system and are able to multiply rapidly in the body. If patients are given an antibiotic to treat a disease, they may receive a substance that—in weak concentrations—microorganisms use for chemical signalling (Linares et al. 2006; Yim et al. 2007; Romero et al. 2011). In the second case, bacteria have won the competition for food by releasing chemical signals that are repulsive for humans and other animals (Burkepile et al. 2006). Moreover, the bacteria thus avoid dealing with the osmotic stress to which they would be exposed during transport through extremely acidic sections of the digestive system if animals were to eat the meat (Sleator and Hill 2001). Flies, on the other hand, find those same signals attractive because they derive from them valuable information for mating and egg-laying. These responses of the flies are in turn exploited by certain fungi—such as the octopus stinkhorn (*Clathrus archeri*)—or plants—for example the carrion plant (*Orbea variegata*) or arum lilies (*Arum* sp.)—to attract flies by releasing chemical substances with a similar smell, to procure transportation of fungal spores or pollination of flowers (Johnson and Jürgens 2010).

6.3.2.1 Examples of Macro-Scale Effects

Abiotic conditions influence communication between free-living single cells most strongly. Under such conditions, the—predominantly biochemical—biotic signals must be clearly distinguishable from abiotic signals. This increases the probability that those signals are also used by other organisms.

An example of the far-reaching effects of chemical signalling substances can be found in the oceans. Single-cell phytoplankton release the sulphur-containing compound DMSP (dimethylsulphoniopropionate $C_5H_{10}O_2S$) (Wolfe 2000) when their cells are destroyed by viruses or ingested by herbivorous organisms such as krill. Zones with high phytoplankton productivity and intensive consumption thereof accordingly have higher DMS concentrations than areas with low productivity. In seawater, the compound decomposes into DMS (dimethyl sulphide CH_3SCH_3), which also escapes into the atmosphere in gaseous form. Some microorganisms are attracted to both compounds, while others are repelled (Seymour et al. 2010). The differing responses are probably due to the fact that some microorganisms use methanethiol (CH_3CH)—also formed during the decay of DMSP—for protein synthesis. Fish orient themselves to concentrations of DMSP and DMS when searching for food in a given locality (DeBose and Nevitt 2007; Foretich et al. 2017). Various marine animals—for example penguins, seals and baleen whales—orient themselves to the signals from DMS concentrations when searching for krill or for fish feeding on them (Cunningham et al. 2008; Lohmann et al. 2008; Thewissen et al. 2011; Kishida et al. 2015a).

It should be noted that baleen whales (Mysticeti) have a much better developed sense of smell than toothed whales (Odontoceti). Unlike toothed whales, baleen whales orient themselves by means of olfaction when searching for food over distances of several hundred kilometers (Kishida and Thewissen 2012; Kishida et al. 2015b; Torres 2017).

However, both whale groups share a weakly developed taste sense (Feng et al. 2014). Comparative analyses between toothed and baleen whales of signal perception reveal the relative importance of individual signal qualities in the context of lifestyle and food acquisition. While baleen whales orient themselves over long distances to DMS signals, toothed whales use acoustic signals to locate potential prey at comparatively short distances of a few hundred meters. Over longer distances, toothed whales prefer to orient themselves to ocean currents and the earth's magnetic field (Torres 2017). This is consistent with the mobility of their prey. "Clouds" of plankton and krill formed by eddy currents are relatively immobile but distributed very irregularly in the oceans (Mackas et al. 1985; Folt and Burns 1999; Murphy et al. 2006; d'Ovidio et al. 2010). The potential prey of toothed whales, on the other hand, are extremely mobile and concentrated in various marine regions with relatively fixed time rhythms.

Atmospheric DMS signals are also used for accurate detection of krill concentrations by bird species inhabiting the high seas and polar waters, such as albatrosses and other species in the order Porcellariiformes (DeBose and Nevitt 2008; Mollo et al. 2014). Extension of the nasal tubes, which give them their common name of "tube-nosed seabirds", is probably a key adaptation for remarkably precise orientation to odour signals, providing aerodynamical support for more precise spatial differentiation of molecule concentrations (Hutchinson and Wenzel 1980; Nevitt 1999, 2008; Nevitt and Bonadonna 2005; Nevitt et al. 2008).

This example clearly reveals the evolutionary influences of a cellular chemical signal for morphological and physiological adaptations of organisms in the context of the energetics of food acquisition, ranging from unicellular organisms to the largest currently living animals (Fig. 6.7). In the context of climate change, it should also be noted that the compensatory effect of cloud formation by DMS emissions from the oceans is under discussion (Ayers and Cainey 2007; Lutterbeck 2012).

On land, microbial activities are distributed discontinuously. As a result, effects of the chemical substances that they release, both as signals and as toxins, cannot be as far-reaching as in the oceans. In their place, on land far-reaching chemical signals of plants predominate. These signals are not so much flower scents, with which we are familiar, but above all a broad range of secondary metabolites—organic carbon compounds such as isoprenes, monoterpenes or sesquiterpenes (Kesselmeier and Staudt 1999; Heil and Karban 2009). These substances are released by plants under stress, be it through infestation by herbivorous insects or exposure to drought (Holopainen and Gershenzon 2010).

Contributions of these compounds—known as VOCs (volatile organic compounds) or NMVOCs (non-methane volatile organic compounds)—to changes in atmospheric chemistry such as the increased formation of ground-level ozone and ultimately to climate change have been discussed in a summary fashion (Lathière et al. 2006; Collins et al. 2010). However, a generalized portrayal of this topic obscures persisting inadequacies in our knowledge of the full spectrum of compounds and relevant triggering factors—including anthropogenic changes in land use.

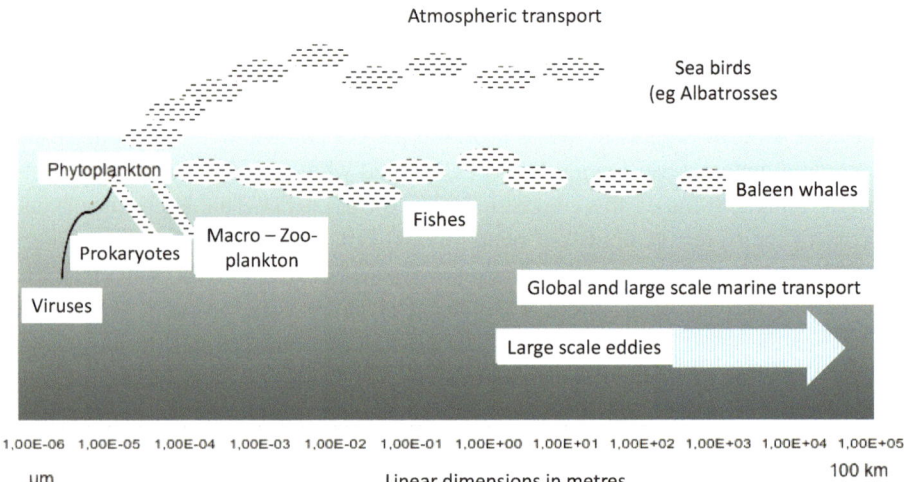

Fig. 6.7 Graphical overview of spatial dimensions of the signalling effect of DMS (dimethyl sulphide)—symbolized by dashed areas—derived from phytoplankton cells in marine regions

In biological terms, long-range volatile organic compounds represent only part of the chemical signalling spectrum of plants, which ranges from large-scale effects down to the level of individual cells. Spatially far-reaching signals are likely to attract mainly parasites and predators of plant-damaging herbivorous insects (Heil and Ton 2008; Dicke et al. 2009). The indirect signalling system is suitable both to reinforce defence against direct infestation by herbivorous insects and to serve as a preventive measure against potential infestation when plant physiology is weakened by other stress factors, such as water shortage. The protective effect can extend to neighbouring plants. However, various volatile organic compounds can also strengthen competitive capacity against neighbouring plants—extending up to complete suppression of their growth (Kost and Heil 2006; Kegge and Pierik 2009).

In the near field of plants, numerous additional compounds are used to ward off herbivorous insects, as well as infections by viruses and bacteria (Unsicker et al. 2009; Helms et al. 2013). Various flowering plants are faced with the conflicting challenge of attracting insects suitable for flower pollination while repelling harmful insects. This goal is achieved by release of differentiated chemical signals in combination with optical signals and selectively acting morphological changes (Holopainen and Blande 2012; Farré-Armengol et al. 2013). Effects of diverse chemical substances used to ward off undesirable organisms range from deterrent odour signals to lethal toxic effects if plant parts are eaten (Bryant et al. 1991; Schaefer et al. 2004). What these effects have in common are their impacts on the signalling and regulatory systems of the target organisms, either on cellular smell and taste receptors or—in the case of plant toxins—on central regulatory processes of cells. Analogous action principles are also found in toxins of bacteria, fungi or animals

(Marquardt and Schäfer 1994). Various insects—such as the monarch butterfly (*Danaus plexippus*)—feed specifically on toxic plants, at least during their juvenile stages, and protect themselves against potential attackers with the ingested toxins (Reddy and Guerrero 2004; Després et al. 2007). Whereas the toxic defences of insects are often signalled by conspicuous colouration in terrestrial habitats, animals in marine areas often signalled by conspicuous colouration in terrestrial habitats, animals in marine areas often employ the chemical defences of algae in combination with optical camouflage. For example, various crabs or sea urchins boost their protection against predator attacks by carrying pieces of algae with them (Hay 2009).

6.3.2.2 Examples of Micro-Scale Effects

At the microscale level, differentiation and intensity of signal exchange are closely linked to distances between neighbouring cells. Free-living single cells, which orient themselves primarily chemotactically, operate at distances above around 10 µm (Stocker and Seymour 2012). In biofilms, individual cells are separated by extracellular polymers. Distances between cells are between about 3 and 10 µm (Drescher et al. 2016) and can vary because individual cells in the biofilm are freely mobile. Signals can be exchanged in different ways in the widely distributed biofilms, for example in the form of electrons through microfilaments or via released chemical substances (Davies et al. 1998; Davey and O'Toole 2000; Boles et al. 2004; Hall-Stoodley et al. 2004; Jefferson 2004; Stanley and Lazazzera 2004; Vuong et al. 2004; De Beer and Stoodley 2006; Battin et al. 2007; Phoenix and Konhauser 2008; Karatan and Watnick 2009; Caldwell and Pagett 2010; Flemming and Wingeneder 2010; Dubey and Ban-Yehuda 2011; Wang and Gerdes 2012; Drescher et al. 2013; Schultz et al. 2013; Dang and Lovett 2016; Flemming et al. 2016). Bacteria are better able to protect themselves against attack from unicellular eukaryotes because of their contributions to formation of intercellular matrix, even if the vast majority participate in biofilm formation. As the number of bacteria not participating in the formation of intercellular matrix increases—to individually benefit from protection while avoiding the energetic cost of extracellular substance formation—the risk of colapse of the entire biofilm increases (Jiricny et al. 2010; Popat et al. 2012). Analogous interactions can be observed in developing cancer cells in multicellular organisms (Li and Neaves 2006).

The signalling system of quorum sensing is widespread among microorganisms. Individual cells release specific signaling substances. Neighbouring bacteria do not automatically respond to such signals, but always do so in the context of the prevailing framework conditions (Diggle et al. 2007; Steindler and Venturi 2007; Mehta et al. 2009; Cornforth et al. 2014). In this way, individual cells receive information about the size of the activated cell populations in their immediate vicinity through concentration of the messenger substances. Experiments have shown that for the diffusion of signaling substances over a—for cells gigantic (Fig. 6.8)—distance of 10 mm takes about 10 h (Dilanji et al. 2012). Above a certain threshold, all cells begin to show the same physiological activity, for example activation of bioluminescence (Anetzberger et al. 2009) or release of toxic substances. In addition, these signals can be used directly by individual cells to

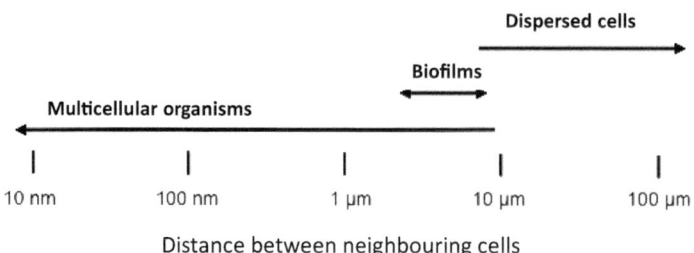

Fig. 6.8 Approximate comparison of the distances between individual cells in biofilms and multi-cellular organisms to those of free-living single cells. Based on data compiled from various sources

coordinate their activities—for instance, biofilm formation. In the reproductive cycle of the genus *Ulva*, a member of the group of green algae, the free-swimming zoospores orient themselves during settlement to particular signals from bacterial quorum sensing (Joint et al. 2007).

In aquatic habitats, unicellular planktonic organisms tune their physiological processes to release toxic substances for defence against competing species or potential consumers (Hay and Kubanek 2002; Wohlrab 2013; Cai et al. 2014; Kurmayer et al. 2016). The diversity of organisms with this capacity ranges from cyanobacteria to unicellular eukaryotes possessing a variety of toxic substances (Janson and Hayes 2006; Bertrand et al. 2007). In the public realm, such phenomena are better known under the euphemistic label "algal blooms". This conceals the far-reaching adverse health consequences that can arise through direct contact with the toxins during water sports, inhalation of marine spray or indirectly through consumption of organisms from aquatic food chains—for example mussels or fish. Depending on the particular toxins ingested and their concentrations, effects range from diarrhoea to death—not only for humans, but also for fish, aquatic birds or seals (Scholin et al. 2000; van Apeldoorn et al. 2001; Fleming et al. 2011). This phenomenon, originally known only from a few marine areas—also known as a "red tide"—can now be observed worldwide, especially at elevated water temperatures. Anthropogenic increases in nutrient inputs from wastewater and fertilizer run-off from agricultural land are considered to be the main factors influencing this development (Glibert et al. 2005; Glibert and Burkholder 2006; Heisler et al. 2008; Paerl et al. 2011; O'Neil et al. 2012).

Striking occurrences of algal blooms make it easy to overlook the variety of chemical signals involved in interactions between different planktonic organisms and their variable, context-dependent effects (Wolfe 2000; Vardi et al. 2002; Legrand et al. 2003; Parkinson et al. 2005; Vos et al. 2006; Pohnert et al. 2007: Van Donk 2007; Hay 2009; Seymour et al. 2010; Roberts et al. 2011; Paul et al. 2012; Poulson-Ellestad et al. 2014; Amin et al. 2015). Two examples are briefly presented as follows:

1. The unicellular flagellate *Prymnesium parvum* augments its competitive power against cyanobacteria and other unicellular plankton organisms by releasing toxins. At high

concentrations of these substances, dinoflagellates of the species *Oxyrrhis marina* can in fact be overwhelmed. However, if concentrations are low, *Prymnesium* itself falls prey to Oxyrrhis (Fisterol et al. 2003; Tillmann 2003).

2. Planktonic radiolarians belonging to the *Acantharia* group have a skeleton composed of strontium sulphate and prey on other planktonic organisms in waters with adequate nutrient supply. In nutrient-poor (oligotrophic) surface waters, however, they form endosymbioses with otherwise also free-living, photosynthetically active microalgae of the genus *Phaeocystis* (Haptophyta). Even as fossils, symbiotic forms of *Acantharia* can be distinguished from other life forms by striking morphological adaptations of their skeletons. This peculiarity made it possible to date the emergence of these symbioses to a period of protracted nutrient deficiency in the oceans around 190 to 100 million years ago (Decelle et al. 2012).

6.3.3 Information Processes in Cell Aggregations

Losses and interference in signal transmission are reduced by aggregations of single cells (Fig. 6.9). This facilitates differentiated responses of cells thus aggregated. On the basis of genetic features of the cells involved, it is possible to distinguish non-clonal aggregations—cells of different lineages—from clonal aggregations—cells with identical genomes.

6.3.3.1 Non-clonal Cell Aggregations
Non-clonal aggregations of cells with complementary properties can coordinate their metabolic processes and physiological functions. Examples include heterospecific biofilms at the interfaces between solid surfaces and water. Embedded in an extracellular matrix,

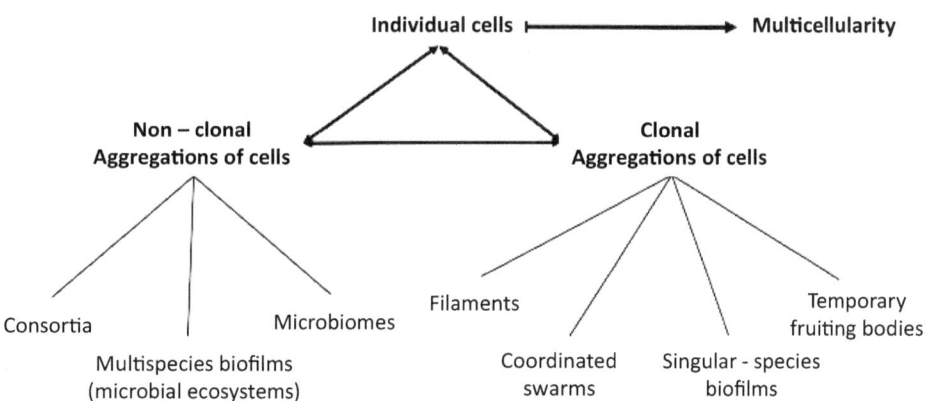

Fig. 6.9 Examples of different forms of cell aggregations, including multicellular organisms. Explanations in the following subchapters

biofilms regulate various aerobic and anaerobic processes of individual cells to optimize abiotic energy conversion.

Groups of different, functionally interacting cells (consortia) exploit inorganic gradients for energy conversion through syntrophic processes (see Sect. 7.3) and/or reduce the energy needed for locomotion. Examples of this can also be found in European lakes in the form of "*Chlorochromatium aggregatum*" (Lake 2009; Overmann 2010; Liu et al. 2013b)—consisting of a central proteobacterium enveloped by green sulphur bacteria—or in the root zones of tropical mangrove forests in the form of a central archaean cell enveloped by sulphur-oxidizing proteobacteria (Muller et al. 2010). In both cases, intensive exchange of substances and energy between the cells has been demonstrated. In the first example, phototactic orientation and locomotion is performed by the central proteobacterium. Different consortia between bacteria and protists are found in the anaerobic habitats of sediments or termite digestive systems (Tamm 1982; Hongoh et al. 2007; Hamann et al. 2016). The flexibility of consortia—often referred to as symbioses—is also illustrated by the example of coral bleaching in the context of nutrient excess and temperature increase, when symbiotic dinoflagellates separate from corals (Bellwood et al. 2004; Silverstein et al. 2012; D'Angelo and Wiedenmann 2014; Roth 2014).

Globally occurring heterospecific biofilms are found mainly at the transitions between different abiotic systems or between abiotic and biotic systems. Important external factors influencing their development are energetic gradients for purely chemotrophic or mixed chemotrophic and phototrophic responses. Complementary responses of the different organisms are regulated both by converted metabolites and by additional chemical or electrical signals (Norris et al. 2002; Decho et al. 2010; Nielsen et al. 2010; Reguera 2011). Since the beginning of biological evolution, interactions between the organisms involved have modulated a great variety of abiotic material cycles (Paerl and Pinckney 1996; Grootjans et al. 1997; Reid et al. 2000; Belnap and Lange 2003; Des Marais 2003; van der Heijden and Sanders 2003; Konhauser 2007; Dion and Nautiyal 2008; Falkowski et al. 2008; Handley et al. 2010; Bahl et al. 2011; Madsen 2011). The functions of heterospecific biofilms most frequently exploited by humans involve transformation and binding of a wide variety of water constituents in soils and geological formations (Paul 2007; Drewniak et al. 2016)—a basic prerequisite for palatability of most spring and ground waters.

All multicellular organisms have intimate functional interactions with microbiomes—non-clonal cell aggregations present on all external and internal surfaces. On the external surfaces of plants and animals, microbiomes protect against attack by pathogenic bacteria or fungi. In the digestive systems of all animals, microbiomes regulate the conversion of ingested substances into physiologically utilizable organic molecules (Belkaid and Segre 2014; Berg et al. 2014; Coleman-Derr et al. 2016; Fischbach and Segre 2016). The composition of microbiomes varies even on and within an individual organism. For example, spatially definable compositions of microorganisms can be distinguished both on human skin and in the intestine (Stearns et al. 2011; Findley et al. 2013). Biofilms can also be used by pathogenic organisms, for example by bacteria or fungi, to shield against defence responses of immune systems (Devaraj et al. 2015; Kernien et al. 2019).

Relative, contextually determined system variability is characteristic of all non-clonal cell aggregations. Changes in individual factors give rise to various effects and compositions, both spatially and temporally. For example, development of the microbiome in the gut of young children influences the risk of autoimmune diseases. Rapid transition from the neonatal to the adult microbiome—for example in Finland—increases the risk, whereas a slow transition—for example in Russia—reduces it (Vatanen et al. 2016). The various effects are caused not only by the appearance of new microorganisms, but also by changes in the physiological functioning of existing microorganisms. Various interactions between bacteria and fungi influence, for instance, health and disease in animals and plants, or production and degradation of food (Frey-Klett et al. 2011; Scherlach et al. 2013).

As examples, bacteria can regulate the synthesis of certain substances by activating genes in fungi, while fungi can parasitize bacteria or algae or cooperate symbiotically with them in the form of lichens (Grube et al. 2015; Netzker et al. 2015).

6.3.3.2 Clonal Cell Aggregations

Clonal cell aggregations usually form for protection against attacks by viruses or other organisms, or to strengthen their own capacities for attacking other organisms. Aggregations of this kind occur in different forms in prokaryotes, for instance as coordinated swarms that spread rapidly over food resources or attack bacteria such as representatives of the genera *Myxococcus* or *Halobacteriovorax* (Velicer and Yu 2003; Berleman and Kirby 2009; Darnton et al. 2010; Dwidar et al. 2011; Zhang et al. 2012; Johnke et al. 2014; Kaiser and Warrick 2014; Keane and Berleman 2016; Welsh et al. 2016). Among eukaryotes, unicellular slime moulds of the genus *Dictyostelium* respond with coordinated swarm formation to food scarcity and high population densities. Subsequently, they form fungus-like structures, the upper parts of which, as temporary fruiting bodies, produce spores for further dissemination, while the underlying cells of the carrier structure die off (Flowers et al. 2010; Schaap 2016; Du et al. 2015; Kawabe et al. 2015; Schaap 2016).

Cyanobacteria form special cells (heterocysts) in filamentous or branched cell assemblies for nitrogen fixation in association with photosynthetic cells (Flores and Herrero 2010; Schirrmeister et al. 2013; Nürnberg et al. 2015). The various physiological needs of two cell types, exchange of their metabolic products and adaptations to different irradiation conditions require a large—so far only partially understood—repertoire of signals to regulate the processes involved (McClung 2006; Zhang et al. 2006; Merino-Puerto et al. 2011; Stucken et al. 2012; Herrero et al. 2016). Various magnetotactic bacteria form permanent, radially symmetrically arranged aggregates of single cells with identically aligned magnetosomes. In contrast to isolated single cells, the aggregates are buoyant because of the formation of flagella in individual cells (Keim et al. 2004; Abreu et al. 2006, 2014; Winklhofer et al. 2007; Martins et al. 2009; Zhou et al. 2012; Chen et al. 2014). Presumably, the qualities of magnetic orientation are improved by signal exchange between cells and flagellar movements are also coordinated—showing similarity to the cell aggregates of the eukaryotic green alga *Volvox* (Arakaki et al. 2013).

In many cases, coordination of clonal cell aggregates is based on quorum sensing processes. In this context, cellular regulation of information processing by various non-coding RNAs plays an important role (Tu et al. 2010; Feng et al. 2015). Differentiated information—for example for activation of processes for defence against antibiotics—can be transmitted by means of additionally released substances. In this way, bacteria can drastically multiply the small effect of a single cell and thus overcome the immune systems of multicellular organisms. This is why many bacterial infectious diseases become virulent in the affected organisms only when the bacteria are able to form biofilms. Hyphae of moulds (*Rhizopus*) are decomposed by bacteria (*Serratia*) only when a sufficient number of bacteria are distributed along the hyphae and have formed local biofilms (Hover et al. 2016).

Similar processes can also be observed in the infection of multicellular eukaryotes by unicellular pathogenic fungi (Hogan 2006; Wongsuk et al. 2016). Of particular medical importance is the spread of resistance to antibiotics through horizontal gene transfer in biofilms. With the increasing resistance of bacteria to antibiotics, hopes for disease control are pinned on prevention of biofilm formation and quorum sensing. However, medically relevant effects of quorum sensing are only one aspect of its utilization by bacteria. For instance, bacteria use quorum sensing to coordinate defence against viral infections or to establish symbioses with eukaryotes (Lyon and Muir 2003; Horswill et al. 2007; Williams 2007; Anetzberger et al. 2009; Atkinson and Williams 2009; Uroz et al. 2009; Popat et al. 2012; Yong and Zhong 2013; Høyland-Kroghsbo et al. 2013; Solano et al. 2014; Lee and Zhang 2015; Papenfort and Bassler 2016).

Viral attacks on clonal cell aggregates can be contained only by interrupting infection pathways, ideally through timely destruction of infected cells. The process is initiated by affected cells themselves through programmed cell death, in which the cell destroys itself by means of a complex endogenous regulatory process (Alberts et al. 2015). Similarly, infected cells of protists destroy themselves and thus block viral reproduction. In conjunction with self-destruction of a cell, signalling substances are released that stimulate unaffected cells to activate physiological defence processes against viral attack (Fig. 6.10) (Bidle and Falkowski 2004; Toth et al. 2004; Mutschler et al. 2011; Huysmans et al. 2017). Through these processes, populations of unicellular plankton organisms can maintain themselves in spite of the continuous presence of viruses Thyrhaug et al. 2003).

Such relationships between different organisms are by no means static, but are subject to ongoing dynamic changes. It is known, at least from studies on plants, that pathogenic bacteria and fungi can block the regulatory processes of programmed cell death by manipulating cellular signal transmission—and thus successfully continue the infection (Mukhtar et al. 2016). Programmed cell death can also be observed in various bacterial populations experiencing deficiency. The self-destruction of some of the cells releases nutrients they contain and thus enables the survival of the remaining cells (Engelberg-Kulka et al. 2006; Bayles 2014; Peeters and de Jonge 2018). In all probability, programmed cell death can be counted among the evolutionarily oldest defence methods against viral infections. The philosophical question of whether development of the defence strategy has altruistic roots must remain unanswered here (Ameisen 2002; Nedelcu et al. 2010). It may seem

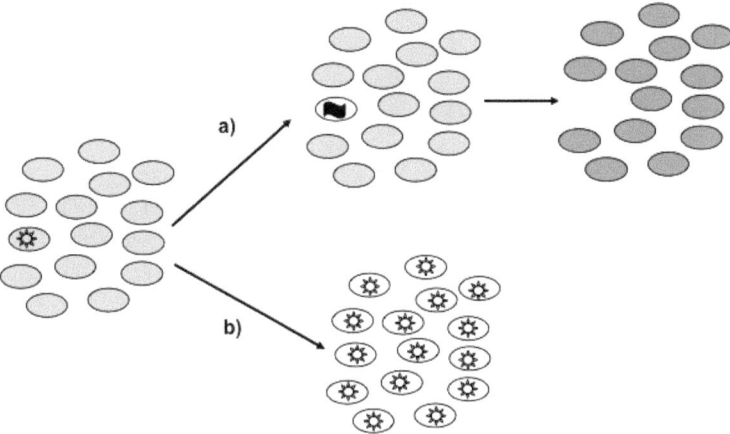

Fig. 6.10 Simplified comparison of the effects of programmed cell death of a cell originally infected with viruses. (**a**) Dissolution of the infected cell and persistence of the remaining population; (**b**) without such a response, the originally affected cell infects the entire population

paradoxical, but programmed cell death contributes significantly to ensuring the further evolutionary development of organic life.

Connections between the accumulation of identical cells and the effort required to defend against attacks by heterotrophic organisms are most clearly observed in actinobacteria of the genus *Streptomyces*. If resources are scarce, their own stored resources are used to form spore carriers above the existing substrate. This process is accompanied by a significantly increase in antibiotic production to protect the own cells (McCormick and Flärdh 2012; Claessen et al. 2014).

Due to the predominantly identical metabolic demands of all cells, clonal cell aggregations are also exposed to greater risks of resource exhaustion and/or accumulation of critical metabolic waste. Because of the paucity of options for metabolic change, colonies are faced with the decision of sacrificing at least some cells or perishing altogether. Bacteria and unicellular eukaryotes meet such challenges with various regulatory strategies. (Kelemen et al. 2001; Bonner 2003; Justice et al. 2008; Herron et al. 2009, 2013; Lefèvre et al. 2010; McCormick and Flärdh 2012; Fisher et al. 2013; Parfrey and Lahr 2013; Ratcliff et al. 2013; Wilking et al. 2013; Claessen et al. 2014; Hammerschmidt et al. 2014; Lyons and Kolter 2015; Tecon and Leveau 2016; Hillmann et al. 2018; Kapsetaki et al. 2016) In the simplest case, cell colonies can grow at the periphery, while cells in the interior the die due to lack of resources. Signals indicating expected attacks on the cell colony lead to a behavioural change in which nutrients are also cyclically supplied to the inner cells. This increases the probability of survival of the cell colony, since it is mainly peripheral cells that are destroyed during attacks (Liu et al. 2015).

Bacteria of the genus *Myxococcus* and various unicellular eukaryotes form "fruiting bodies" on elevated, often stalk-like structures. Cells in the "fruiting body" are transformed into spores and transported into the environment by abiotic flow (Fig. 6.11). The

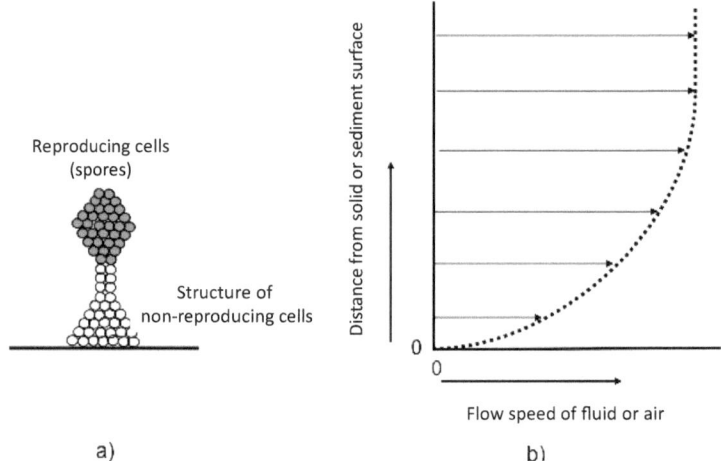

Fig. 6.11 Schematic representation of the use of abiotic flow for reproduction of unicellular organisms through structure formation. (**a**) Use of abiotic flow forces for reproduction in various prokaryotes and unicellular eukaryotes through formation of three-dimensional structures; (**b**) Diagram showing the increase in flow velocity from approximately zero at the lowest cell layer up to the level of cells capable of reproduction

remaining cells dissolve completely or partially through programmed cell death. In the latter case, surviving cells feed on their remains (Nowell and Jumars 1984; Shimkets 1990; Zusman et al. 2007; Du et al. 2015; Kawabe et al. 2015; Persat et al. 2015). In this way, part of the energy loss can be compensated by formation of non-reproductive morphological structures.

It has been shown for the eukaryotic slime mould *Dictyostelium* that swarm formation is triggered by quorum sensing. However, regulation of subsequent stages—from cell aggregation to the formation of spore bodies—is based on direct intercellular communication processes (Loomis 2014; Schaap 2016). From a systemic perspective, the processes can be described as **competent swarm intelligence**, in which single cells contribute their individual capacities in a complementary and constructive way to ensure reproductive capacity. The process clearly differs from the strictly coordinated behaviour of large groups—for example with fish or humans—under certain external conditions, which is called swarm intelligence, especially in an economic or technological context (Korinets et al. 2008; Parpinelli and Lopes 2011).

A non-cooperative strategy to overcome bottlenecks has been observed in the bacterial genus *Paenibacillus*. In this case, colonies of genetically identical cells mutually inhibit cell proliferation by means of toxic proteins until the smaller colonies perish (Be'er et al. 2009; Be'er et al. 2010).

6.4   Information Processes of Multicellular Organisms

6.4.1   General Conditions and Characteristics of Multicellularity

Classical theories of evolution refer to multicellular organisms for a number of reasons. This may be simply because of an initial lack of possibilities for observing microscopic organisms, or because of the paradigmatic perspective of "higher" development over the course of evolution. For example, there has been discussion about evolution of individuality or multicellularity through the interplay between genetic modification and external selection (Okasha 2004; Rainey and Kerr 2010; Ispolatov et al. 2012; Goodnight 2015). While the first aspect cannot be resolved conclusively, the many examples of different transitions between unicellular and multicellular life forms—already discussed in previous chapters—provide illumination. However, it should also be evident that unicellular organisms have been evolutionarily successful over the long term and are not "primitive" forms of life (Grosberg and Strathmann 1998, 2007; Donoghue and Antcliffe 2010; Suga and Ruiz-Trillo 2013; Brown et al. 2012; Parfrey and Lahr 2013; Schirrmeister et al. 2013; Claessen et al. 2014).

On the one hand, evolution of permanent multicellularity can resolve the challenges of coordinating predominantly identical cell demands through functional differentiation. On the other hand, this is associated with permanently increased energy expenditure and—because of the concentration of many, biochemically largely identical cells—greater risks of attack by heterotrophic organisms. Compared to unicellular life forms, the challenges of coordinated functional differentiation of cells and regulation of the exchange of substances between them are novel. Functional differentiation also increases requirements for differentiation between own and foreign cells, both in the establishment of symbioses and for defence against potentially pathogenic organisms and viruses. Regardless of the size and complexity of multicellular organisms, these interactions must always occur at the cellular level. As a result, each individual multicellular organism is confronted with a large number of different microbial interaction systems. So far, these interactions have been incompletely researched only for individual multicellular organisms. As a result, there is a scarcity of adequate facts for comparative observations (Sandmeier and Tracy 2014; Netea et al. 2019).

Knowledge regarding endogenous mechanisms of defence against pathogenic or parasitic organisms by multicellular life forms—their *immune systems*—is somewhat more extensive. But here, too, coverage extends to a maximum of 5 per cent of all known multicellular organisms. In principle, immune systems have been found in all multicellular organisms studied to date—fungi, macroalgae, plants and animals (Rodríguez et al. 2012; Thomas et al. 2014; Ipcho et al. 2016; Uehling et al. 2017; Han 2019; DeFalco and Zipfel 2021; Ngou et al. 2022a, b). From the data so far available, major shared features can be identified among multicellular organisms, especially for—cellular—innate immune systems. Pointers to these common features are provided by the evolutionary characteristics of intracellular proteins for identifying invading pathogens, for example the leucine-rich repeat (NLR) superfamily. Differentiated proteins of this superfamily identify and inhibit pathogens in plant and animal cells (Duxbury et al. 2016; Jones et al. 2016; Uehling et al. 2017; Białas et al. 2021). Proteins for perforating the membrane of foreign cells are common to all prokaryotic and eukaryotic organisms. The diverse proteins of the originally separately described groups—membrane attack complex/perforin (MACPF) and cholesterol dependent cytolysins (CDC)—are derived evolutionarily from the same protein superfamily MCPF/CDC (Anderluh and Gilbert 2014). In general, it can be inferred for both plants and animals that differentiation of immune systems in multicellular organisms increases along with that of cell functions. In animals, this trend is further enhanced by greater mobility and higher energy turnover. This affects both the complement system and the adaptive immune system. Stated simply, the former is used for "labelling" (opsonization) of pathogens, the latter for flexible adaptation of molecules of the immune system—for example immunoglobulins—through recombination of genes (Rodríguez et al. 2012; Neely and Flajnik 2016; Martin et al. 2020).

In multicellular organisms, functional differentiation of cells is not limited to the period between first cell division and complete morphological development, but extends over the entire lifespan. In long-lived multicellular organisms, only a few cells attain full age. These relationships can be observed most clearly in the dynamic changes of fungal mycelia. Even in supposedly long-lived trees, the largest proportion consists of dead cells, and only a small part is composed of comparatively young, living cells. In temperate latitudes—in deciduous trees—the renewal and demise of plant cells can be directly observed in the annual rhythms of budding and leaf fall. Similar temporal rhythms—but observable only through investigation—occur in the fine roots of trees (McClaugherty et al. 1982). In animals, cells are permanently preserved only in a few tissues. Differences from plants are most clearly illustrated by the example of supporting tissues. Trees can adapt their mechanical properties to changes in loading only through newly formed tissues, but they can also change their outer shape in the process. Thanks to cellular dynamics, the bones of vertebrates can be continuously adapted to changes in load by building up and breaking down. The entire evolutionary lineage of cartilaginous fish (Chondrychthyes)—best known by sharks and rays—has no skeletal bones (Zhu 2014). Apart from pathological effects, their outer form remains largely unchanged. Even in our bodies, some cell types—such as certain white blood cells—reach an age of only a few hours. Others accompany us throughout our entire lives—for example nerve cells (Kunsch 1997). Moreover, in these cells the mitochondria—crucial for energy conversion—are constantly renewed (Westermann 2010; Devireddy et al. 2015).

Potentials for the evolution of multicellular life forms are found in the genomes of many unicellular organisms with optional multicellular life phases (Sebé-Pedrós et al. 2013, 2016; Suga and Ruiz-Trillo 2013; Torruella et al. 2015). Genetic analyses show that the "building set" for the morphological differentiation of eukaryotes was already present from the outset of their evolution (Carroll 2000; Hedges et al. 2006; Erwin 2009; David and Alm 2011; de Mendoza et al. 2013). Morphological changes in multicellular life forms over the course of evolution are predominantly based on epigenetic regulation of the genetic toolkit. Micro-RNAs (miRNAs) play an important role in epigenetic networks (Kettler et al. 2007; Peterson et al. 2009; Krol et al. 2010; Sato et al. 2011; Shea et al. 2011; Hansen et al. 2013; Liu and Pan 2015; Sorrells and Johnson 2015; Sorrells et al. 2015). Their importance can be seen above all in their increasing diversity during the evolution of multicellular animals (Metazoa) (Erwin et al. 2011; Ameres and Zamore 2013), as well as in their role in the regulation of immune systems or nervous systems (Baltimore et al. 2008; Szulwach et al. 2010).

In terms of set theory, the total potential for information processing in all multicellular organisms is determined by the genetic and epigenetic capacity of the cell of origin. Subsets of the total potential are required to varying degrees for functional regulation in the organism (Fig. 6.12a). In general, a distinction can be drawn between regulation still needed by the individual cells, overall regulation of morphology and physiology, and the information potential required for immune defence. Accordingly, in all multicellular organisms only a part of the total potential for processing environmental information is available to the organism as a whole. Because of the sensory-physiological apparatus—specific for any life form—capacities for perceiving signals in spatial and temporal resolution are also limited. According to the context, an organism can only modulate its perceptual space (Fig. 6.12b) within these—evolutionarily developed—framework conditions.

This type of presentation becomes more easily understandable when the biological framework conditions for evolution of multicellular organisms are taken into account. In a world full of microorganisms, their evolution could only be successful if this resulted in advantages for both forms of life. The prerequisite for this is continuous information at all functional levels of multicellular organisms. Characteristics and effects of these information processes remain largely unexplored, but should always be considered in studies of multicellular organisms. According to current knowledge, the bandwidths of the effects of the microbiomes already mentioned (Sect. 5.2.3) range from symbionts of herbivorous insects to the influence of intestinal bacteria on neural processes in humans (Frago et al. 2012; Lyte 2013; Farmer et al. 2014). In parallel, the small intestine of mice has been shown to have its own sensory cells for detecting sugar, the signals from which are transmitted directly to the brain along the vagus nerve (Kaelberer et al. 2018). In interactions between multicellular organisms, relationships between the individual information categories also vary, for example in the different strategies of plants to defend themselves against herbivores (Hanley et al. 2007).

In this context, comment on a special feature of the evolution of multicellular animals is needed. Whereas—according to the current state of knowledge—the evolution of multicellular fungi and plants took place predominantly under terrestrial conditions, for animals two different combinations of factors can be distinguished. The more ancient evolution lineage proceeded within the framework conditions of marine habitats, which predominantly encompassed unicellular life forms and determined interactions between animal organisms at that stage. By contrast, framework conditions for the more

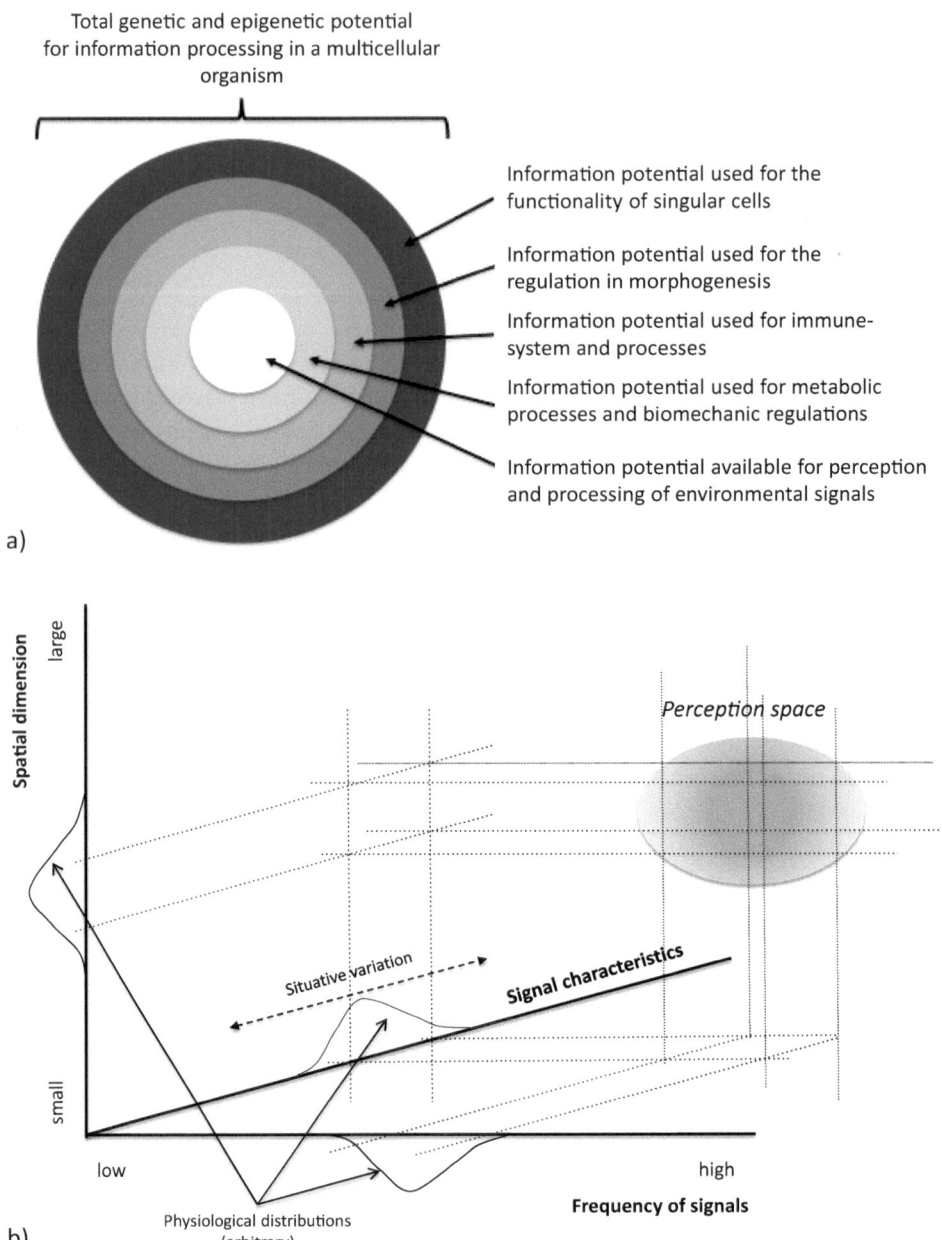

Fig. 6.12 (**a**) Representation using set theory of the distribution of the genetically and epigeneti-cally determined potential for information processing in multicellular organisms across different functional domains; (**b**) Simplified representation of variables influencing the perceptual space of organisms

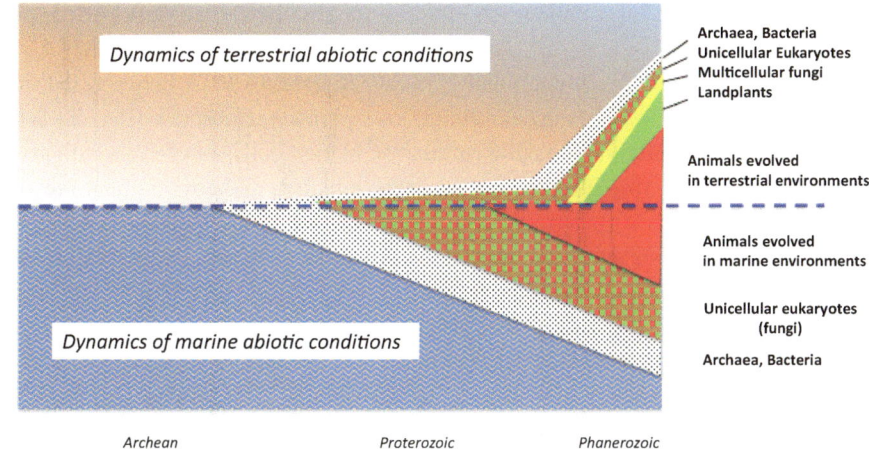

Dynamics of terrestrial abiotic conditions

Archaea, Bacteria
Unicellular Eukaryotes
Multicellular fungi
Landplants

Animals evolved
in terrestrial environments

Animals evolved
in marine environments

Unicellular eukaryotes
(fungi)

Archaea, Bacteria

Dynamics of marine abiotic conditions

Hadean   Archean   Proterozoic   Phanerozoic

Fig. 6.13 Schematic representation of the different frameworks for multicellularity of fungi and plants compared to animals. The patterns for unicellular organisms and prokaryotes symbolize their diverse possibilities for energy conversion—ranging from autotrophic to heterotrophic. Further explanation in the text

recent—terrestrial—evolutionary lineage included from the very beginning the additional presence of multicellular fungi and plants (Fig. 6.13).

Multiple examples of the representations in Fig. 6.12 can be seen when looking at an aquarium containing organisms from tropical coral reefs. The scene teems with diverse life forms that often resemble plants. In fact, they are always animals, such as various corals, sponges, sea pens or tube worms. Although many can move freely through the water only as juvenile stages, such life forms have colonized marine habitats for over 500 million years. Despite having nervous systems that are greatly reduced or even absent, these animals are able to filter their food—mostly microorganisms—from the surrounding water and avoid attacks by other animals. This lifestyle is permitted not only by hard tissue structures and defence toxins, but also by morphology and movements that are barely noticeable or active. At the same time, structures of sessile individuals provide habitats for various mobile organisms, the best-known example being provided by mutual relationships between clownfish and sea anemones (Marcionetti et al. 2019), made popular by the film *Nemo*. Figure 6.13 is designed to show that in marine habitats all macroscopic life forms of animals have evolved exclusively through interactions between animals. During the early phases of these interactions—at the beginning of the Phanerozoic—in addition to sessile creatures, mobile life forms with powerful optical sensory organs are already identifiable—for example trilobites, a long extinct group of arthropods (Clarkson et al. 2006; Schoenemann and Clarkson 2013; Scholtz et al. 2019; Plotnick et al. 2023). Perhaps this makes it somewhat clearer that the diverse sessile animal life forms found in the oceans are not in fact primitive. They are, instead, examples of successful organisms overcoming environmental challenges with various distributions of information potentials (Fig. 6.12).

6.4.2   Relationships Between Life Form and Information Processing

Both morphologically and functionally, major differences exist between the three major groups of multicellular organisms—fungi, land plants and animals (Fig. 6.14). Processes of morphological and functional differentiation in their cells are correspondingly different. Because of the differing interactions between organism and environment (see Fig. 5.18), attempts to establish a uniform evolutionary scheme exclusively on a genetic foundation contribute little to our understanding of the processes involved (Domazet-Lošo and Tautz 2010; Kalinka et al. 2010; Kalinka and Tomancak 2012; Quint et al. 2012; Irie and Kuratani 2014; Buchholz 2015; Cheng et al. 2015; Drost et al. 2015, 2016; Cridge et al. 2016).

One of the most important aspects of multicellularity is engagement with mechanical-physical variables, such as currents or gravity (Vogel 1994; Denny 2016). To a marked degree, determination by mechanical-physical dimensions influences possibilities for morphological development of multicellular organisms—especially animals and plants (Thompson 1942). Prevailing scientific and social conceptions of evolution are shaped by these phenomena—because macrofossils remain at the centre of scientific discussions and serve as illustrations. Theories concerning this topic are correspondingly diverse and basically deal with a single aspect—the emergence of order (Thompson 1942; Kauffman 1993; Riedl 2000; Mayr 2003; Newman and Bhat 2008; Hernández-Hernández et al. 2012; Newman 2012; Newman and Linde-Medina 2013).

The appearances of present organisms make it easy to forget that evolution of the three groups of multicellular organisms—fungi, animals and plants—followed different trajectories. Fungi are confronted with the physical variables mentioned only on a temporary basis—during the formation of fruiting bodies. Their permanent multicellular structures consist of networks in soils and organic matter, whose development is primarily oriented

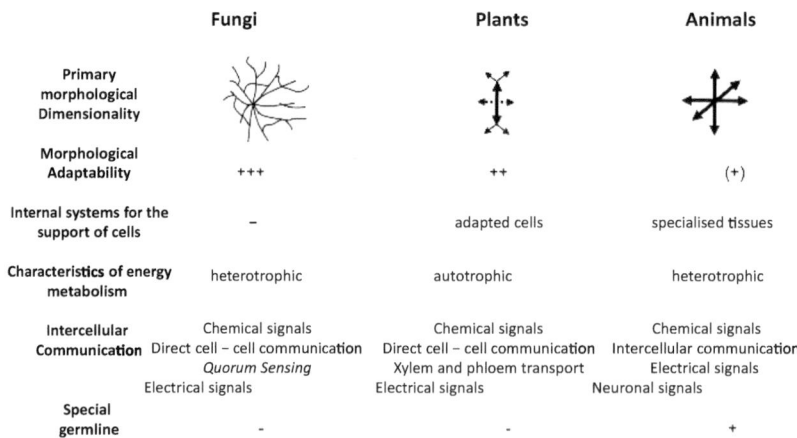

	Fungi	Plants	Animals
Primary morphological Dimensionality			
Morphological Adaptability	+++	++	(+)
Internal systems for the support of cells	−	adapted cells	specialised tissues
Characteristics of energy metabolism	heterotrophic	autotrophic	heterotrophic
Intercellular Communication	Chemical signals Direct cell – cell communication Quorum Sensing Electrical signals	Chemical signals Direct cell – cell communication Xylem and phloem transport Electrical signals	Chemical signals Intercellular communication Electrical signals Neuronal signals
Special germline	-	-	+

Fig. 6.14 Simplified comparison of the life forms of multicellular organisms

towards the spatial distribution of material resources. Multicellular animals evolved in the sea around 150 million years before the first multicellular terrestrial plants evolved (Becker and Marin 2009; Steemans et al. 2009; Cunningham et al. 2017). This means that the inter-relationships—known from terrestrial ecosystems—are only a late evolutionary phenomenon.

6.4.2.1 Multicellular Fungi

Multicellular fungi are widely known from their fruiting bodies, which are temporarily formed from reticulate, vegetative parts of organisms—mycelia—for dissemination of fungal spores (Nowrousian 2014). The less conspicuous vegetative mycelia colonize a wide variety of habitats and interact with other organisms in many different ways (Moore et al. 2011). Various species can also develop under anaerobic conditions (Kurakov et al. 2008). Fungi feed osmotrophically by selective exchange of substances across their cell walls (Darby et al. 1968; Konhauser 2007; Naranjo-Ortiz and Gabaldón 2019). Their classical description as purely heterotrophic organisms—feeding on organic materials—was questioned some while ago (Wainwright 1988). Despite the capacities of fungi for selective accumulation of mineral substances (Tzeferis et al. 1994; Valix et al. 2001), which has already been exploited for technological purposes, evidence has only recently emerged for purely autolithotrophic metabolism (energy conversion from inorganic compounds) in fungi (Hornick 2017; Xu et al. 2018). Multiple reports of observations of electrical signals in fungi and orientation of fungal hyphae to proton fluxes during formation of mycorrhiza indicate multimodal functions of osmotic processes (Gow 1984; Ramos et al. 2008; Adamatzky 2022). With this evidence, it is somewhat easier to understand the diverse—over 450 million years old—symbioses of fungi with plants in terrestrial regions (Divakar et al. 2015; Schneider et al. 2016; Brundrett and Tedersoo 2018). In symbioses, fungi not only extract minerals from dead materials but also employ their hyphae to tap new mineral resources from inorganic materials—for example rock (Martin et al. 2012; Danger and Chauvet 2013). However, this also makes it clear that our knowledge regarding the real information and regulation functions of fungi can be no more than fragmentary.

6.4.2.2 Land Plants

The evolution of multicellularity in land plants is constrained by a vertical separation in the availability—for autotrophic energy conversion—of sunlight in the air and of water and mineral nutrients in soil. This is associated with novel requirements for adaptation of cell structures to mechanical stresses, as well as intercellular transport of water, nutrients and signals from roots to leaves. Changes in mechanical and material requirements, because of plant growth and the limited local availability of mineral nutrients, necessitate continual adaptation of root morphology to subsoil structure conditions. Despite these multidimensional requirements, no independent organs for transmission or processing of information are found in plants. Even the conducting vessels for transport of aqueous solutions along a longitudinal axis consist of coupled individual cells that have transverse walls with different perforation patterns. Mechanical structures for water transport are absent, although the lifting height often clearly exceeds the range of capillary transfer height (Denffer et al. 1971). In terrestrial plants, multicellularity and morphological differentiation also open up greater potential for infestation by pathogenic and parasitic organisms than in aquatic plants possessing just one or a few cells. Compared to the latter, terrestrial plants also possess a significantly larger repertoire of molecules in their immune systems for recognition and defence against various microbial pathogens (Han 2019).

The nanotubes of cellulose fibres from lignified plants can be used in experiments as various building blocks for devices transmitting electronic information (Zhu et al. 2016; Huang et al. 2019; Tran et al. 2023). It is still unclear to what extent living plants also physiologically utilize potentials for electrical signal transmission.

6.4.2.3 Animals

In the early stages of multicellular development in the aqueous environment of the egg cell, animals are free of any external physiological or mechanical constraints (Newman 2011). This potentially allows cells to differentiate in all three dimensions of space. This is especially true under the conditions for primary evolution of multicellular life forms in marine habitats (Fig. 6.14). Here, diverse three-dimensional juvenile stages of a wide variety of systematic groups can still be found, but also radially symmetrical adult life forms— for example among jellyfish and their relatives (Cnidaria) and comb jellies (Ctenophora). This is not valid for echinoderms (sea stars, sea urchins, sea cucumbers, sea lilies) with pentaradial body plans but derived from bilaterian origins (Emlet 1991; Rouse 2000; Raff 2008; Nielsen 2009; Jenner 2014; Formery et al. 2023). Differentiation of body axes was probably brought about by additional environmental factors, such as contact with a solid substrate or interactions with other organisms. Clear influences of substrate and feeding form can be seen in the disc-shaped *Dickinsonia* of the Ediacaran period, in placozoans and in sponges. Because of the clear dominance of a longitudinal body axis, with bilaterally symmetrical arrangement of morphology on the transverse axis, organisms could both penetrate loose solid substrates and prey on other animals. Examples of such life forms are worms and fish, both from the morphological group Bilateria (bilaterally symmetrical animals) (Telford et al. 2015; Hoekzema et al. 2017; Bobrovskiy et al. 2018; Varoqueaux et al. 2018). The recent evolutionary lineage of terrestrial animals was established exclusively by organisms in this group. Morphological differentiation is regulated by interactions between genes and epigenetic processes in cells and cell tissue (Gilmour et al. 2017). After processes of growth and differentiation have been completed, basic morphological patterns remain largely unchanged. Even in sponges—according to the classical definition of species—local biodiversity can be estimated with adequate precision from morphological characteristics alone (Bell and Barnes 2001). Consequently, relevant connections cannot be expected for morphologically similar but genetically distinct—cryptic—species (Bode et al. 2011; Funk et al. 2012). Moving away from the standard organisms in textbooks, morphological studies reveal extreme deviations from developmental processes described for models (Lemaire and Marcellini 2003). In this respect, genetic and epigenetic analyses of evolutionary processes differ markedly, yielding indications of similar developmental processes in morphologically different life forms (Lecuit and Le Goff 2007; Lemaire 2011; Ryan et al. 2013; Hejnol and Pang 2016; Loh et al. 2016; Gold et al. 2019). The results of such analyses also raise questions about classical systematic categories such as Bilateria or the classificatory distinction between Protostomia (organisms with a single mouth) and Deuterostomia (organisms with a second mouth) (Hejnol and Martindale 2008; Martindale and Hejnol 2009; Simakov et al. 2015). Viable organisms do not always require use of a complete gene repertoire for development. According to the spectrum of genes that are activated, water bears (Tardigrada)—over their entire life cycle—larvae of acorn worms (Hemichordata) or adult echinoderms, for example, could be aptly described as "living heads". In acorn worms, genes for formation of the trunk are activated during transformation to adult life forms (Smith et al. 2016; Gonzalez et al. 2017; Formery et al. 2023). There are also major differences between adult organisms in the capacity to regenerate, ranging from complete reconstruction of the body from tissue fragments—for example in flatworms (planarians)—to minimal regeneration of individual organs—for instance in mammals (Galis et al. 2003; Slack 2017). As the regenerative capacity of life forms increases, the separation between somatic cells and gametes (sex cells) shifts from embryogenesis to the adult stage. This means that, even in animals, smooth transitions exist between differentiation of the so-called germ line and differentiation of somatic cells in the individual organism (Stoner et al. 1999; Extavour and Akam 2003; Juliano and Wessel 2010; Johnson et al. 2011).

In animals, heterotrophic nutrition through phagocytosis or mechanical uptake of food requires complete development at an early stage of all morphological structures and physiological functions that are decisive for food acquisition. The spectrum ranges from direct phagocytosis—for instance in sponges—to differentiated digestive systems—for example in vertebrates. Not only sponges but also jellyfish and their relatives (Cnidaria) lack blood vessel systems. This condition is also found in various small-bodied or approximately two-dimensional organisms—for instance flatworms. Enclosed blood vessel systems are found in vertebrates and cephalopods (Cephalopoda) (Monahan-Earley et al. 2013). Nervous systems are lacking only in Placozoa (literally "flat animals") and sponges (Dunn et al. 2014; Liebeskind et al. 2016; Eitel et al. 2018).

In contrast to fungi and plants, morphological evolution of animals is greatly influenced by the dynamic demands of their lifestyles. Morphological changes (metamorphosis) during development of organisms—from embryo to adult—can also be associated with a shift between different habitats. This occurs, for example, in insects or amphibians with aquatic immature stages (Kukalova-Peck 1978; Laudet 2011; Pechenik 1999; Truman and Riddiford 1999; Van Buskirk and Saxer 2001; Müller and Leitz 2002; Jackson et al. 2002; Hodin 2006; Degnan and Degnan 2010; Belles 2011; Grasso et al. 2011; Laudet 2011; Bonett et al. 2013; Huang et al. 2013; Gonzalez et al. 2017).

6.5   Information Exchange in Multicellular Organisms Between Cells and the Environment

In a multicellular organism, each individual cell is in permanent functional interaction with genetically identical cells. In the simplest case—as in fungi—each cell is connected with two neighbouring cells in a "chain" (hyphae). Accordingly, each cell is also exposed to direct influences from its environment. Within a consortium of hyphae—often consisting of a widely branching network—cells exchange nutrients and information. Tip cells of hyphae are very important for exploiting new resources and developing associated sensory functions. Fungal life forms allow extremely flexible adaptation to a great diversity of environmental conditions. Various species—especially parasites—can change from unicellular to multicellular as well as from multicellular to unicellular life forms, notably when infecting their hosts (Deacon 2006). Like bacteria, various fungi can form biofilms in the bodies of infected hosts to provide protection against defensive responses (Ramage et al. 2012). Large three-dimensional fruiting bodies—widely called mushrooms—are formed only temporarily during reproduction and dispersal of fungal spores.

In contrast to fungi, cells of plants and animals are permanently integrated into two- or three-dimensional tissues. This is connected with increasing functional differentiation of individual cells. All cells develop from genetically identical, functionally undifferentiated stem cells that are concentrated in specific tissue zones—stem cell niches (Heidstra and Sabatini 2014).

In plants, the distribution of stem cell tissues—meristems—across all growth-related areas facilitates ongoing adaptation of organs to changing environmental conditions and physiological needs. Visually evident examples of this are annual cycles of foliage and the morphological development of branches and trunks of trees. By contrast, complementary development of root systems is not usually visible. In concert, albeit in different ways, leaves and roots contribute to the overall metabolism of plants. In addition to water and minerals, energetic conversion of solar radiation in photosynthetically active cells of leaves requires sufficient irradiation and facilitation of gas exchange with the environment. The first two conditions are regulated by structural shapes and orientation of leaves to sunlight and by currents in surrounding media (Niinemets et al. 2015; Keenan and Niinemets 2016). The latter two conditions, on the other hand, are regulated by the performance of the root system. The roots also ensure the static anchoring of plants as well as serving the function of nutrient storage in various life forms to bridge periods of deficiency. Incidentally, let it be noted here that the morphological and physiological differentiation outlined has evolved under terrestrial habitat conditions only over about 450 million years (Steemans et al. 2009; Sørensen et al. 2014; Delwiche et al. 2017). In the evolutionarily much older aquatic lineages of photoautotrophic organisms—cyanobacteria and algae—unicellular life forms with diverse types of differentiation of their physiological functions are still dominant.

In contrast to plants, the greatest diversity of animal life forms has evolved in marine habitats. In addition to vertebrates, examples include extremely diverse forms of sponges, virtually two-dimensional placozoans, the tree-shaped and radial body forms of jellyfish and their relatives, the diverse life forms of spiny-skinned sea-urchins and related species (Echinodermata), and mollusks. Only some of the basic life forms have evolved further on land—in intimate interactions with vegetation. These include, among others, insects, mammals, reptiles and birds, which are usually the primary focus of evolutionary investigations. Definitive morphological and physiological differentiation during early life phases is common to almost all animals. In the later stages of life, stem cells activity is reduced to functional maintenance of organs by replacing dead cells with new ones. In contrast to plants, newly formed cells actively migrate into the appropriate tissue and undergo functional differentiation after arriving.

Neither the preceding explanations nor the sight of a healthy living organism can convey an adequate impression of the complexity of underlying systemic regulatory functions. Unlike individual cells, stem cells—which are potentially capable of unlimited division—may only reproduce in coordination with the relevant needs of individual tissues. In spite of differences in the physiological performance of functionally differentiated cells, which are often extreme, deviations must be recognized and corrected by means of signals, or deactivation of the cells concerned must be inhibited. The spread of potentially harmful viruses, prokaryotes or eukaryotes in host organisms should be detected, if possible, prevented in a timely fashion and. To cope with manifold requirements, organisms use intercellular information pathways (Fig. 6.15), which can only be outlined in the following subchapters.

Intercellular information processes enable plants not only to tap into new resources, but also to spread vegetatively. For example, members of the grass family—bamboo, for instance—can invade neighbouring plant systems through horizontal shoots—rhizomes— and displace them with rapidly growing new plants (van Groenendael et al. 1996; Song et al. 2016), which can decouple from the original plant and then act completely autonomously. Similar phenomena can also be observed within individual plants. Here, individual sectors decouple and form units that are functionally completely autonomous. Physiological disturbances in an autonomous sector are limited to the affected region alone and do not affect the whole plant (Orians et al. 2004). On the other hand, various plants can exchange water and nutrients through the mycorrhizal systems (Bücking et al. 2016). The intercellular signalling systems also enable responses to various environmental influences, for example defence against herbivores or competing organisms using chemical compounds (secondary metabolites). Development of competing plants can also be inhibited by targeted shading or by depriving their root zones of nutrients.

Like plants, animals lacking nervous systems—for example sponges—respond to attacks by other organisms and use chemical compounds for defence. Their intercellular information systems also permit coordinated cell contractions, for example to remove disruptive particles from their canal systems (Thacker et al. 1998; Elliott and Leys 2007; Pawlik 2011). The basic requirements for transmission and processing of information are found in all living cells. This is indicated by the general distribution of calcium ion-based signalling systems in the cells of all organisms, ranging from unicellular prokaryotes to

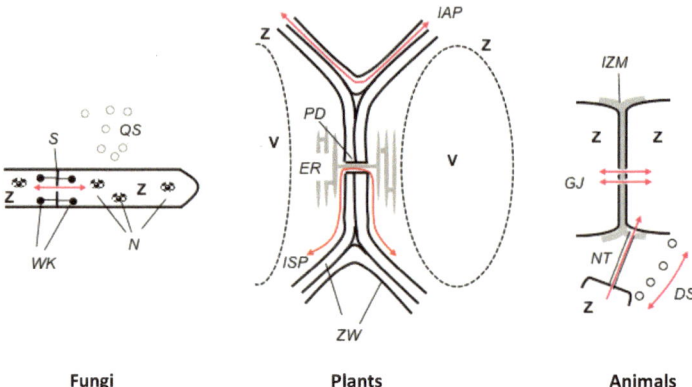

Fig. 6.15 Simplified comparison of direct intercellular communication pathways in plants, fungi and animals. In plants and animals, only transitions between individual cells (Z) are shown. Abbreviations: Fungi: N nuclei; QS quorum sensing; S septum; WK Woronin body. Plants: ER endoplasmic reticulum; IAP apoplastic information pathways; ISP symplastic information pathways; PD plasmodesmata; V vacuole; ZW cell walls. Animals: DS solutes; GJ gap junctions; IZM intercellular matrix; NT nanotubules. Explanations in the text

multicellular eukaryotes (Dodd et al. 2010; Verkhratsky and Parpura 2014; Beckmann et al. 2016; Choi et al. 2016; Marchader et al. 2016; Edel et al. 2017). Cells of animal nervous systems are no exception, using a rich array of calcium ion-based signalling systems (Rosenberg and Spitzer 2011; Brini et al. 2014; Puhl III et al. 2014; Bazargani and Attwell 2016; Umpierre et al. 2020). The protein equipment for encoding and decoding the signals differs (Edel et al. 2017). The Information processes in fungi and plants are as yet largely unexplored, although the diversity and contextual appropriateness of their responses to different environmental conditions indicate their presence (Baluška et al. 2006; Johnson and Gilbert 2015; Huber and Bauerle 2016; Schmid 2016; Adamatzky 2018; Canales et al. 2018; Deshibi and Adamatzk 2020). Indications of commonalities in cellular signal transmission are also provided by blocking of physiological processes in plants with medical anaesthetics (Yokawa et al. 2018).

Examples of mechanical movements outside of metazoans are the catapult-like movements of the spore capsules in some fungi, the leaf movements of mimosas among plants, or movements of the seizing apparatus of various carnivorous life forms (Trail 2007; Yafetto et al. 2008; Stolze-Rybczynski et al. 2009; Volkov et al. 2010; Böhm et al. 2016). What makes them different from animals are their predominantly sessile life forms and predominantly very slow mechanical responses. Whereas movements of the spore capsules of fungi are triggered by a sufficiently high humidity level, in plants the stimuli are mechanical or electrical. In the case of both mimosas and Venus-flytraps (*Dionaea*), selective responses to different stimuli can be detected with electrophysiological methods (Volkov et al. 2010; Böhm et al. 2016). Such evidence has is still lacking for the suction traps of bladderworts (*Utricularia*). Static charge differences between the inner and outer walls of the chambers can be detected on traps that are ready to be sprung. Following digestion of the prey—mostly aquatic invertebrates—the negative pressure in the traps can be restored by osmotic transportation of part of the water contained therein (Vincent et al. 2011; Masi et al. 2016; Poppinga et al. 2017). As the trap flaps open inwards and extremely rapidly, the closing force may be based not on mechanical but on electrostatic forces.

The principle of extremely rapid movements in plants and animals is similar to that of a mousetrap. Kinetic energy is released—as with a metal spring—by relaxation of elastic material. In the suction chambers of bladderworts, elastic walls are involved, whereas animals use elastic outer skeletons or tendons. Tension is generated in bladderworts by a hydrostatic vacuum in the chambers, but muscle power is the agency in animals (Reifenrath et al. 2006; Astley and Roberts 2012; Patek 2015; Westermeier et al. 2017; Larabee et al. 2018).

For the vast majority of all animal movements, coordinated, well-regulated processes are necessary. Only on this basis—assisted by appropriate sensory organs—can they move freely in space, acquire food or respond appropriately to attacks by other organisms. Morphological and physiological differentiation of nervous systems are the prerequisites and results of meeting these requirements (Moroz et al. 2014; Castelfranco and Hartline 2016; Schmidt-Rhaesa et al. 2016; Wang and Clandinin 2016; Liebeskind et al. 2017). Through them—independently of intermediary tissues—signals can be transmitted between receptive and motor sectors and information can be processed in variously centralized nerve networks. Hence, nervous systems are prerequisites not for information processing but for the motor capacities of animals.

Anyone who attempts to rationalize the special status of human beings—for instance by invoking the extremely low probability that a chimpanzee could compose the works of Shakespeare (Anderson 2011)—should also test the argument on other organisms. There are enough examples to do this. For example, how likely is it that plant flowers would become adapted to imitate female insect forms, or that complex animal traps would evolve in plants (Merbach et al. 2002; Bauer et al. 2007; Schaefer and Ruxton 2008; Davies et al. 2012; Papadopulos et al. 2013; Phillips et al. 2014)? We are confronted here with open questions, but we do know that the number of evolutionary steps available for this is relatively small. Much larger—but also finite—are the numbers of possible attempts by viruses to mimic proteins (Standfuss 2015). In any case, the number of steps is never infinitely large, as would be required for probabilistic arguments. What all phenomena have in common is their emergence in the context of a perceivable environment—be it an individual infected cell or the Elizabethan Era in England. But differences also exist in pertinent spatial and temporal dimensions. In spatial terms, billionths of a millimetre are involved for viruses, whereas in Shakespeare's case the range lies between decimetres and several hundred kilometres. Temporal dimensions of the human behaviour exquisitely described by Shakespeare occupy orders of magnitude between thousands and tens of thousands of years. Temporal dimensions of modes of action in viruses evolved over a timespan of billions of years.

6.5.1 Fungi

Multicellularity in fungi is found in separate systematic groups and is hence not a useful feature for classifying them (Hibbett et al. 2007). The following remarks refer primarily to multicellular forms in terrestrial habitats. Compared with the enormous diversity of life forms in fungi, scientific knowledge regarding their information processes is as yet pitifully sparse (Deacon 2006; Blackwell 2011). This may come as a surprise because there is an overabundance of scientific literature on economically or medically relevant groups of fungi (Moore et al. 2011). But we must not overlook the fact that, among eukaryotes, fungi show the greatest adaptability to an extensive range of environmental conditions (Muszewska et al. 2017). They live under both anaerobic and aerobic conditions. Examples of extreme habitats are underground rock crevices, extremely dry and hot rock surfaces (Hoffland et al. 2003; Zakharova et al. 2013; Drake et al. 2017), or even fuel tanks of fighter planes (Thomas and Hill 1976). These examples suggest similarities to the lifestyles of prokaryotes—characteristics that are also reflected in a diversity of interactions with other organisms. Symbioses with cyanobacteria or algae in the form of lichens, or with the roots of land plants in the form of various mycorrhizae, are surely widely known (van der Heijden and Sanders 2003; Finlay 2004; Nash III 2008; Azcón-Aguilar et al. 2009; Chagnon et al. 2016). Adverse interactions with plants and animals in the form of various fungal diseases are of medical and economic importance (Gundermann et al. 1991; Walsh et al. 2004; Pfaller et al. 2006; Gauthier and Klein 2008; Briggs et al. 2010; Scheele et al. 2019). Less well known are plants that parasitize fungi—for example certain orchids (Leake 2005; Selosse and Roy 2008). This should suffice to show that our knowledge of the evolutionary significance of fungi is still rudimentary and that the following examples of information processing connected with their life forms can, at best, be no more than anecdotal.

Compared to plants, the mycelia of multicellular fungi show extremely flexible differentiation of cellular units. Cells can contain different numbers of nuclei (N) and can also be subsequently subdivided by formation of partition walls—septa (S) (Fig. 6.15). In various life forms, openings in those partition walls can, if necessary, be closed by their own structures, the Woronin bodies (WK) (Soundararajan et al. 2004; Moore et al. 2011; Beck 2013). Marked flexibility in formation and resorption of new cell tissue, together with communication between individual fungal hyphae by means of quorum sensing (QS) (Albuquerque and Casadevall 2012; Steinberg et al. 2017), shows great similarities to the dynamics of prokaryotes. The tubular hyphal shape is maintained by hydrostatic internal pressure, which can reach up to 8 megapascals (MPa)—about 40 times the pressure in a car tyre (Geitmann 2006). Exchange of substances for intracellular energy conversion takes place osmotrophically through the cell membrane.

Morphological development is based primarily on linear growth of fungal hyphae and secondarily on cross-linking between them (Glass et al. 2004; Read 2011; Riquelme 2013). Externally, cell membranes of hyphae may be enclosed by chitin-containing cell walls (Latgé 2007; James and Berbee 2011; Moore et al. 2011). Variations in hyphal networks are closely related to conditions in the relevant substrate—such as soils, living or dead plants—and losses due to fungivorous animals (Bebber et al. 2007; Heaton et al. 2012; Takeshita 2016). Changes in growth direction result on the one hand from mechanical resistance of substrates and on the other from orientation towards signals indicating potential food sources, or neighbouring fungal hyphae. To take one example, the effort required to maintain large mycelial networks and their signalling power is related to prevailing states of the interacting plants. Under experimental conditions, originally coherent mycelial networks in dead plant material dissolved into separate subnetworks. In the context of living plants, it was possible to demonstrate material transportation and signal transmission in mycorrhizal networks (Barto et al. 2012; Babikova et al. 2013; Klein et al. 2016). Moreover, selective preference for chemically and structurally optimally decomposable mineral particles by fungal hyphae—in coordination with prevailing nutrient requirements of the plant—has been demonstrated experimentally (Leake et al. 2008). Hyphal walls contain receptors for perception of almost all signal qualities (Bahn et al. 2007). Fungi are thus able to cooperate both with one another and—in mycorrhizae—with plants (Hahn and Mendgen 2001; Hogan 2006; Albuquerque and Casadevall 2012; Bucher et al. 2014; Carbonnel and Gutjahr 2014; Wongsuk et al. 2016). In combination with differentiation of particular morphological structures—such as loops or harpoons—fungi can employ their sensory functions to actively catch and resorb small soil animals—notably nematodes (Yang et al. 2011; Liu et al. 2012a; Hsueh et al. 2013; Soares et al. 2018) Knowledge of intracellular information processes is, however, only rudimentary (Pereira-Leal 2008; Lai et al. 2012; Takeshita et al. 2014).

In connection with sexual reproduction, many species temporarily form multicellular fruiting bodies—colloquially known as mushrooms—on surfaces of the appropriate habitat (Moore et al. 2011). For successful reproduction, the relatively short-lived fruiting bodies must be formed under suitable climatic conditions, such that height and orientation of spore carriers permit optimal dispersal of the fungal spores. Destruction of fruiting bodies by other organisms should also have as little impact as possible on spore dispersal. The complex challenges of timing and spatial orientation of growth, according to the concentration of metabolic products, are managed intracellularly by a differentiated information system (Hasenstein 1999; Palmer and Horton 2006; Fischer et al. 2010; Rodriguez-Romero et al. 2010; Halbwachs et al. 2016). Under favourable atmospheric flow conditions, fungal spores—especially when small—can be transported well beyond a thousand kilometres (Hovmøller et al. 2002).

In view of the very large surface area of mycelia—compared to their volume—the question arises as to how fungi could have evolved at all under habitat conditions preferred by prokaryotes. Free surfaces offer ideal possibilities for attack by unicellular organisms, which multicellular organisms are barely able to resist. Theoretically, fungi with mycelia should have become extinct long ago. Indications of this are provided by mycophagous bacteria that use fungal hyphae as a food source along various pathways (Leveau and Preston 2008). Figures 6.13 and 6.14 provide clues to resolving this issue—fungi with mycelia evolved over more than four hundred million years and could only have done so in cooperation with unicellular organisms. Cooperation requires interspecific communication, which is also found in different forms in fungi. Evidence for this is provided by the experimentally observed signal exchange with volatile organic compounds (VOCs) between fungi and bacteria (Schmidt et al. 2016). Endosymbioses with bacteria enable the synthesis of toxic substances for defence or permit fungi to attack plants and animals (Partida-Martines and Hertweck 2005; Wargo and Hogan 2006; Kobayashi and Crouck 2009). Details of these communication processes are as little understood as those of symbioses between fungi and cyanobacteria, or with algae in lichens. Experiments with lichens provided a second indication of important prerequisites for the successful development of symbioses (Kranner et al. 2005). In isolation, neither the fungus nor the alga could survive the extreme stresses of solar radiation and oxygen radicals to which they are commonly exposed when living in symbiosis. Fungi and algae form lichens in other habitats, each with different species, and under favourable conditions some algae can also live without symbioses (Bubrick et al. 1984; Kroken and Taylor 2000; Piercey-Normore 2006).

6.5.2 Land Plants

The regulatory effort required for growth and metabolism in land plants shows much greater diversity. Cellular growth in plants can be triggered by sexual reproduction or, in many cases, by external signals to tissue regions. In general, a fully functional embryo develops along either pathway with bipolar orientation of cell growth at the shoot and root apex (Leljak-Levanić et al. 2015). Whereas cell volume decreases continuously during the initial cell divisions, in subsequent divisions cell sizes vary according to their functional differentiation (West and Harada 1993; Yoshida et al. 2014; Horst et al. 2016). In photosynthetically active parts of a plant—in order to optimize utilization of light—cell tissues are typically formed in a geometric arrangement. Similar patterns are also found in flowers and seed-bearing structures (Thompson 1942; Kappraff 2004). As photoautotrophic organisms, they are far more dependent on abiotic factors than fungi. Their development is influenced by at least five groups of abiotic factors—solar radiation, carbon dioxide, mineral nutrients, water and temperature. Ideal combinations of factors are found, for example—in stagnant nutrient-rich, tropical aquatic habitats—in the central distribution areas of water lilies (Nymphaeales), notably including the well-known giant water lily (*Victoria amazonica*) (Borsch et al. 2008; Löhne et al. 2008). In reality, conditions deviate in various ways from hypothetical optimal conditions because of both abiotic and biotic influences. In a geographical context, it is hence possible to define different vegetation communities that are influenced by the dynamics of abiotic and biotic processes in an evolutionary context (Walter and Breckle 1983; Willis and McElwain 2002).

As plants grow taller, spatial separation of the leaves and roots poses a special challenge for metabolism. Unlike animals—with their centralized internal digestive systems (Fig. 6.16)—land plants convert energy and substances in structural regions that are often spatially widely separated. An additional challenge for plants arises from the different

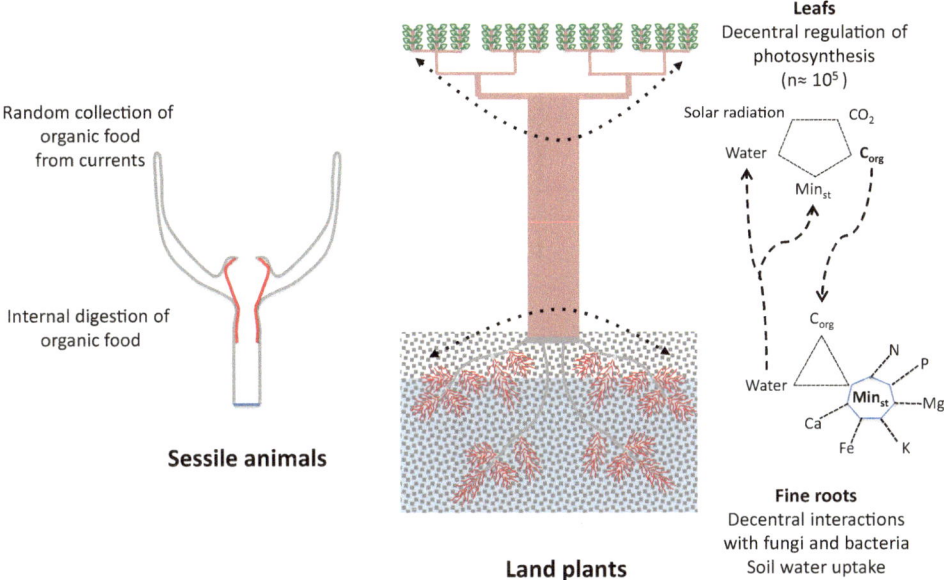

Fig. 6.16 Schematic comparison of various requirements for energy regulation and conversion of matter in animals and land plants. Abbreviations: C_{org} = organic carbon compounds; Min_{st} = sum of mineral elements in the required stoichiometric ratio. Explanation of symbols: curved dashed double arrows symbolize exchange of substances within the crown and root zones. Further explanations in the text

kinds of "food intake". Animals ingest exclusively organic food with extensively coordinated chemical compositions, which is further broken down by the microbiomes in their digestive systems. Land plants must meet their mineral requirements from differently distributed resources in their environments and selectively take them up in ratios that are stoichiometrically suitable for their metabolic machinery (Min_{st} in Fig. 6.16). Roots have no *a priori* information about potentially available mineral nutrients in their environment and must therefore search for suitable mineral nutrients in a trial-and-error process with a large number of fine roots. Potentially, this strategy can be supported through cooperative arrangements with microorganisms and fungi (mycorrhiza) (Johnson 2010). While microorganisms mainly contribute to breakdown of minerals in various chemical compounds, fungal hyphae can increase the spatial density and range of exploitation. Roots need organic carbon compounds (C_{org} in Fig. 6.16) from photosynthetic processes in leaves for such cooperation. By means of decentralized, cellular information processing, the individual fine roots can regulate their interactions with microorganisms and fungi directly and in coordination with local conditions. In addition to the local processes shown in individual fine roots, substances and water can also be exchanged between roots (dashed double arrow in Fig. 6.16) (Tester and Leigh 2001; Gorska et al. 2008; Raven 2014; Wu et al. 2016). However, water is needed mainly for photosynthesis processes in leaves and must therefore be transported—together with minerals—from the root zone to the tree crown. Although physiological processes and environmental conditions in roots and tree crowns have completely different dynamic characteristics, exchange of solutions with mineral and organic contents—to attain functional equilibrium (Brouwer 1983)—takes place solely on the basis of cellular information processing. This is also supported by the fact that phototropism of plants depends on micro-air channels between hypocotyl to enable cellular information processing. Cellular regulation in conjunction with physiological buffer processes also makes it possible to use the energy—obtained from hundreds of thousands of dynamic conversion systems—in a decentralized manner for the organism as a whole (Meng et al. 2018; Nawkar et al. 2023). These are functions that can, at best, be provided in technological systems by large-scale digital arrays!

The common and familiar life form of land plants with roots, stem and leaves makes it easy to forget the challenging adaptation processes underlying their evolution. Parasitic or carnivorous species reveal just how much the framework conditions shown in Fig. 6.17 influence the evolution of plant life forms. Common to both of those groups is mixotrophy, which is already widespread among unicellular planktonic algae (Selosse et al. 2017). Phototrophic organisms also use minerals and/or carbon compounds from other organisms for their own metabolism. The spectrum ranges from the support of photosynthetic metabolism to its extensive replacement by heterotrophic metabolism. Parasitic plants obtain organic nutrients directly from the metabolism of their host plants or indirectly via the fungal hyphae of mycorrhizae, as is the case with many orchids (Woltz et al. 1994; Bidartondo et al. 2003; Feild and Brodribb 2005; Westwood et al. 2010; Gebauer et al. 2016). To achieve a direct supply, parasitic plants penetrate the pathways of their host plants with tissue strands—haustoria—from their roots or stems (Yoshida et al. 2016). Obligate parasites do not carry out photosynthesis and cover their energy needs completely with synthetic products derived from their host plants. Examples of this are many summerwort species (*Orobanche*) or the corpse flower (*Rafflesia arnoldii*), which has the largest flowers in the world—up to 1 m in diameter (Westwood et al. 2010; Nikolov et al. 2014). Studies of parasitic plants that simultaneously infest several host plants have provided evidence of signal flows between the organisms involved. For dodders (*Cuscuta*), for example, transport of alarm substances between plants with and without insect infestation has been demonstrated (Aly et al. 2011; Hettenhausen et al. 2017).

If the generalized term "parasite" may obscure the diversity of interrelationships between the organisms concerned, this is all the more true of the collective term mycorrhizae. Quantitative features of the two groups themselves provide indications of this: Only about 1% of all flowering plants are parasitic, but about 94% interact with fungi in mycorrhizae. Moreover, with the exception of deciduous mosses (Bryophyta), mycorrhizae are found in almost all groups of land plants (Johnson et al. 1997; Westwood et al. 2010; Wang and Qiu 2006; Brundrett and Tedersoo 2018; Hoysted et al. 2018). Mycorrhizae are far from static symbioses between plant roots and fungi, but instead opportunistic associations whose existence and development are influenced by a wide variety of biotic and abiotic factors (Johnson et al. 1997; van der Heijden and Sanders 2003; Gehring et al.

2006; Kauserud et al. 2008; Gutjahr and Parniske 2017). The considerable flexibility of the associations involved is based on the capacities of all organisms involved for communication by means of various signals, both locally in the zone of directly interacting cells and throughout the entire extent of the consortium of organisms (Johnson and Gilbert 2015; Sun et al. 2015). Communication between fine roots and bacteria also counts among the prerequisites for associations to develop, needed to supply plants with minerals in biologically usable forms. To this end, plants create morphological preconditions, for instance by enlarging root surfaces with brush-like accumulations of fine roots—cluster roots—or by binding microbiomes in thickened roots. The latter are characteristic for associations involving nitrogen-fixing bacteria (Lamont 2003; Shane and Lambers 2005; Kinkema et al. 2006; Shane et al. 2006; Sprent and James 2007; Downie 2014; Lamont et al. 2014; Sasse et al. 2018; Hansen et al. 2020). By contrast, development of coarse root structures—especially in trees—is influenced to a far greater extent by physical and chemical soil conditions (Hinsinger et al. 2005; Grossman and Rice 2012; Wang et al. 2019).

A general feature of such associations is overall higher energetic turnover in the plants involved, although they give up a proportion of the assimilated matter to microorganisms and fungi in exchange for a substantial supply of mineral nutrients (Katschuk et al. 2010). Opposite effects occur when the signalling systems of plants are overwhelmed and manipulated by a broad array of parasitic life forms. One example of this is provided by nematodes (*Meloidogyne*) that invade plant roots, inducing growth of feeder cells and development of an enveloping layer of tuberous protective tissue (Caillaud et al. 2008). For reasons that have yet to be deciphered, roots with mycorrhizae are better protected against infestation by nematodes than non-mycorrhizal roots (Schouteden et al. 2015).

In contrast to fungi, no nematode traps have yet been found in plant roots. However, underground traps for nematodes consisting of sticky leaves are known for a Brazilian plant genus (*Philcoxa*) (Pereira et al. 2012). As with all so-called carnivorous plants, minerals from the animal carcasses are used to supplement or replace the supply through the roots (Adlassnig et al. 2005). Among carnivorous flowering plants, a wide variety of trapping devices have evolved—mainly from leaves—ranging from sticky leaf surfaces, through cage traps to ambush traps employing either a mechanical flap or suction (Ellison and Gotelli 2001; Król et al. 2012; Adamec 2013).

Leaves can also take up nutrients from the atmosphere via transformation activities by bacteria and fungi on their surfaces. Such ways of supplying minerals and water can be demonstrated in all vascular plants. Extreme examples are epiphytic bromeliads with funnel-shaped leaves to collect and store rainwater and deposited matter. These "cisterns" provide habitats for many different organisms and also spawning sites for specialized amphibians (Sabagh et al. 2011; Leroy et al. 2016; Males 2016; Louca et al. 2017; Males and Griffiths 2018; Fernandez et al. 2020). Deciduous mosses are completely adapted to being supplied through rainfall and substance deposition. Their nutrient supply is supported by species-rich epiphytic bacterial and fungal communities. Most species of these mosses lack vascular systems; instead, nutrients and photosynthetic products are transported directly from the cells. Dense aggregations of these plants in moss cushions completely absorb—within the species-specific physiological bandwidths—atmospheric water and deposited substances. Rudimentary roots (rhizoids) serve predominantly for attachment to the substrate. This enables deciduous mosses to develop on a wide variety of materials and also to hold their own against vascular plants (Ligrone and Duckett 1994; Cornelissen et al. 2007; Crandall-Stotler and Bartholomew-Began 2007; Davey et al. 2009; Pouliot et al. 2009; Pressel et al. 2010; Lindo et al. 2013; Liu et al. 2013a; Sorg et al. 2016; Roos et al. 2019).

Differing locations of growth zones in leaves and roots are also closely linked to prevailing environmental conditions. In leaves, cell division take places over the entire surface during the initial growth phase; in subsequent phases, the growth zone remains at the base of the leaf until cell growth is completed. This allows grasses, for example, to grow continuously even when heavily grazed by ungulates (Nelissen et al. 2016). In contrast, roots grow at their tips. Through signal processing and cell division in the tip zone, roots can orient their direction of growth to prevailing environmental conditions. Longitudinal growth takes place only in the root region behind the tip, through axial stretching of cells (Beemster et al. 2003; Grossman and Rice 2012). Stems of plants, like roots, grow at their tips. In many plant groups, the stability of stems and roots is continuously adapted to changes in loading by means of secondary growth in thickness (Rowe and Speck 2005; Spicer and Groover 2010; Speck and Burgert 2011). The great diversity of morphological form in plants—and associated processes for distribution of material resources—is enabled by ongoing signal exchange between the cells (Geitmann and Ortega 2009; Ruts et al. 2012; Wolf et al. 2012; Sablowski 2016). The fact that decision-making processes for the organism as a whole take place on this basis is demonstrated by root formation in trees on sites with extremely different local conditions (Hodge 2009). On rocky outcrops, for instance, roots of trees on the valley side can grow uphill—against gravity—to reach more easily accessible soil on the mountain side. This requires a change of about 180 degrees in relation to the original direction of horizontal growth[1]!

Although plant cells—in contrast to those of fungi and animals—are usually enclosed by a solid framework of cell walls (Lo et al. 2000; Plotnikov et al. 2012), they have different information pathways at their disposal (Fig. 6.15). It should not be forgotten that the formation of cell walls is itself based on cellular processes (Cosgrove 2005, 2016; Popper and Tuohy 2010). Depending on the constituents of cell walls, plant tissue structures can remain long after physiological death of the individual cells. Symplastic (ISP) and apoplastic (IAP) information pathways can be distinguished in the tissues. Information pathways in vascular plants through the conduits for water and mineral nutrients—the xylem—or assimilates—the phloem—are not shown in the figure (Diaz-Espejo and Hernandez-Santana 2017). All information pathways are connected by membrane bridges to form communication networks. For the regulation of all cellular and trans-cellular metabolic and transport processes in plants and interactions with microorganisms, a central role

---

[1] Personal observation of a Swiss pine rootstock in the Turracher Höhe Pass (Austria).

is played by diverse aquaporins—molecular channels in all internal and peripheral membranes of plant cells (Maurel et al. 2015; Wang et al. 2018).

The symplastic information pathway is made possible by direct connection of the endoplasmic reticulum between individual cells through openings in their walls (plasmodesmata). Information can hence be transmitted directly from cell to cell via the symplasm—the interconnected cytoplasms (ISP). At the same time, by regulating the plasmodesmata and dividing the cell space into the cytoplasm and a vacuole—separated by a membrane (tonoplast)—plant cells also have relative local regulatory autonomy. By regulating the plasmodesmata, information can be transmitted and the spread of viruses or bacteria in the symplasm can be inhibited (Lee and Lu 2011; Lee 2015). Subdivision of the cell space enables local regulation of cell metabolism through the exchange of substances between cytoplasm and vacuole (Ebine et al. 2014; Eisenach et al. 2015). Under the framework conditions of the cell association, this enables the individual cell to realize all of the vital functions of a unicellular organism. On this basis, substances can be stored in the vacuole and recovered when needed. Toxic substances can be removed from the metabolic machinery by permanent storage in the vacuole. By dissolving the tonoplast, attacks by viruses, bacteria and fungi can be warded off either by initiating programmed cell death or by activating toxic responses (Hara-Nishimura and Hatsugai 2011). For the benefit of maintaining the "overall system" of the plant and its capacity to reproduce, framework conditions for the autonomous regulation of the individual cells are constantly changing. Physiologically active cells are, for example, subject to diurnal and latitude-dependent annual cycles of solar radiation, irregular changes in atmospheric conditions and water supply (Fig. 6.17). In connection with their own life cycles and the physiological cycles of the plant as a whole, they can—during growth—both receive nutrients from other cells and—during the senescent phase—make them available to other cells (Chapin III et al. 1990; Loescher et al. 1990; Eckstein et al. 1999; Ainsworth and Bush 2011).

Because of their location between the double cell walls, the apoplastic information pathways are largely independent of signal modifications by individual cells. Results of physiological studies suggest that the apoplastic information pathway plays a central role in coordination of physiological processes throughout the entire plant. By means of this pathway, information relating to stress conditions in the leaf and root zones is exchanged on the basis of ions, protons or electrical signals, and defence responses are triggered in the event of plant injury (Zimmermann et al. 2009; Pechanova et al. 2010; Maathuis et al. 2014; Choi et al. 2017; Geilfus 2017; Qi et al. 2017; Sukhov et al. 2018). Apoplastic processes in flowering plants also regulate the exchange of substances between cells and assimilates in the phloem pathways and aqueous solutions in the xylem pathways from the soil (Wilkinson and Davies 1997; Hartung et al. 2002; Shatil-Cohen and Moshelion 2012; Regmi et al. 2016). In addition to transportation of sugar molecules, signal transmission by RNA molecules and ions has also been detected in the phloem (De Schepper et al. 2013; Fromm et al. 2013; Pallas and Gómez 2013; Ham and Lucas 2017).

The apoplastic information system is also the first barrier confronting viruses, bacteria and fungi invading mechanically undamaged plants. Regulation at this level determines

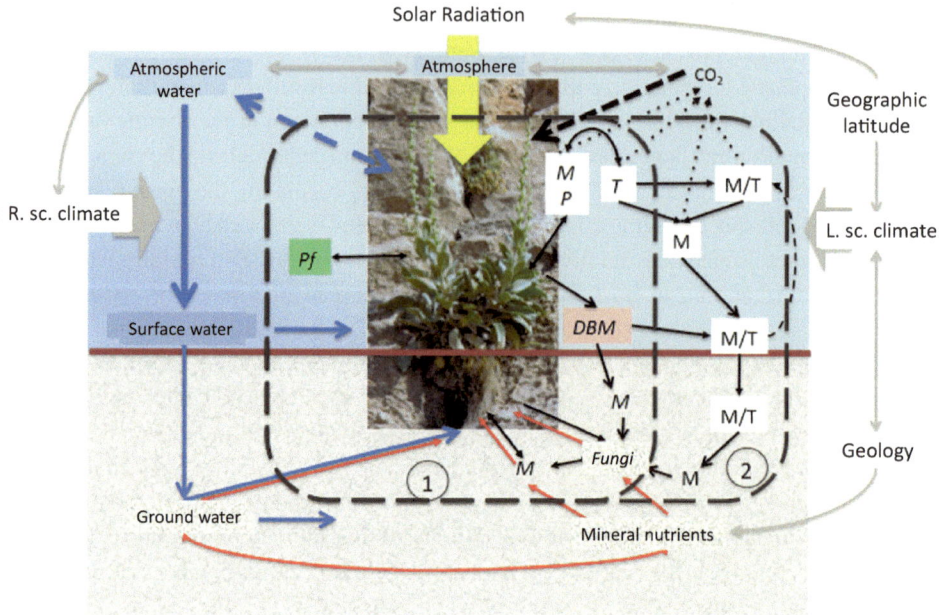

Fig. 6.17 Simplified diagram of requirements for plants to process information and regulate processes in the context of relevant environmental conditions. Abbreviations: 1 = immediate environment of the plant; 2 = extended environment of the plant; DBM = dead plant material; M = microorganisms; P = fungi; Pf = interactions with other plants; T = animals; L. sc. climate = large scale climate; R. sc. climate = small scale climate

the further development of symbioses or plant diseases. Essential symbioses with microorganisms and fungi for the supply of nitrogen or phosphorus are established and also regulated through this pathway. In the simplest case, atmospheric nitrogen is converted by symbiotic cyanobacteria in the intercellular space. Among legumes, plants regulate the nitrogen conversion of bacteria (Rhizobiaceae) in complex tissues in the root area (Oono et al. 2010a, b; Santi et al. 2013; Cannon et al. 2014). Between plant roots and fungi, in ectomycorrhizae, morphological connections are restricted to the intercellular space alone. The fungal hyphae of arbuscular mycorrhizae are additionally connected to the intracellular space of root cells. Through the intensive exchange of signals between plants and fungi, mycorrhizae form an interspecific regulatory system for the extended acquisition of information and nutrients (Schaarschmidt et al. 2007; Takeda et al. 2009; Barto et al. 2012; Bárzana et al. 2012; Luginbuehl and Oldroyd 2017; Wang et al. 2017; Brundrett and Tedersoo 2018). Gradually, the focus of action shifts from the exploitation of abiotic mineral resources of ectomycorrhizae to enhancement of biological material cycles and regulatory processes in arbuscular mycorrhizae (van Schöll et al. 2008; Bonfante and Anca 2009; Brundrett 2009; McGuire et al. 2010; Snaddon et al. 2012). The information processes regulating symbioses and their transition to pathogenic effects of endophytic organisms (organisms living in plants) are as yet largely obscure (Eaton et al. 2011; Saikkonen et al. 2015).

Processes for information transfer between plants and animals are extremely complex and still virtually unexplored. Images of plants lushly growing for the benefit of placid herbivores and of industrious bees working to ensure fruit production obscure the enormous potentials of plants to process and implement information. Given the extensive risk assessment capacities of unicellular organisms—as evident in slime moulds (Sect. 6.3.1)—observations of similar capacities in plants for regulating root growth are hardly surprising (Dener et al. 2016, Schmid 2016). Research to date has focussed mainly on the study of chemical substances and related genetic or physiological processes (Erb and Reymond 2019). Reflections on the actual mechanisms that plants use to process information must therefore remain speculative in many cases (Baluška and Manusco 2007, 2009; Baluška 2009). Examples of interactions between plants and animals described below are hence intended only to provide some idea of their diversity.

Mechanical damage to cell structures by animals can trigger various responses in plants. The most widespread are releases of toxic compounds and signalling substances (Will and van Bel 2006; Maffei et al. 2007; Walling 2008; Heil 2009; Bellandi et al. 2022). In mosses, the chemical substances concerned are still stored in oil bodies of individual cells. In vascular plants, various secretory cell structures, developed over the course of evolution, can release different secretions, for example resins or latex, both when surfaces are touched and when plant tissue is damaged (Langenheim 2003; Lange 2015; Ponce de León et al. 2015). In addition, plants can be supported by endosymbionts, for example bacteria in sequoias (*Sequoia* and *Sequoiadendron*), toxic bacteria in South African plants or endophytic fungi in grasses (Saikkonen et al. 2004, 2013; Verstraete et al. 2011; Carrell and Frank 2015). In turn, herbivorous insects also use symbioses with bacteria to overcome chemical defences of plants or—as in bark beetles (Scolytinae)—with fungi to break down resistant plant tissue (Jordal and Cognato 2012; Six 2012; Chung et al. 2013).

In the simplest case, plants seal off damaged tissue is with rapid, disorderly cell division (callus formation) (Ikeuchi et al. 2013). Certain insects can specifically control plants responses by releasing plant hormones and—sometimes in symbiosis with fungi—stimulate formation of organized tissues (galls) on leaves. The insects concerned use galls for protection and nutrition of larvae growing inside them (Raman 2012; Joy 2013).

Insects—especially ants—often protect plants against herbivores, but also from competition from other plants. Various plant groups therefore provide shelter for ants in specially developed cavities of leaves or stems (domatia) and sometimes food as well by means of specially produced tissues or extrafloral nectar on leaf bases (Janzen 1969; González-Teuber and Heil 2015). Individual ant species deliberately destroy flowers, thus castrating the plants to promote leaf growth. Affected plants may evade this problem by dissolving the domatia before the onset of the bloom period (Izzo and Vaconcelos 2002).

Interactions between animals and plants have been influenced not only by avoidance of losses, but also by exploitation of animal characteristics by plants. Using appropriate signals, plants use animals both for their own reproduction and for spatial dispersal. For reproduction, animals transfer pollen or sperm cells to the ovules of plants. With respect to dispersal, animals ensure spatial distribution of plant seeds. Probably the best-known example of exploiting animals for reproduction is provided by bees fertilizing flowering plants. Basic features of this mode of reproduction can already be seen in mosses, with microarthropods transporting sperm to the female archegonia (Cronberg et al. 2006; Rosenstiel et al. 2012). Information processes that result in morphological and olfactory imitation of insect forms and scents by plants are far more complex and have yet to be clarified. Biological assessment of the phenomenon generally referred to as mimicry is purely classificatory in nature and does not make any further contribution to our understanding of underlying evolutionary processes (Danchin et al. 2008). In this context, various plants exploit the sex drive of male insects, which—without any additional reward through nectar—visit the flowers far more intensively than females and also take up more pollen because of their copulatory motions (Ellis and Johnson 2010; Bohman et al. 2014; Xu et al. 2017). Various arum plants (Araceae) achieve the same outcome by temporarily "trapping" insects attracted to their flowers (Chartier et al. 2013). In sum, evidence from various studies suggests that plants interact dynamically with their environment and can optimize both morphological characteristics and physiological performances in the context of several different factors (Marino et al. 2009; Kessler et al. 2010; Labandeira 2013; Schiestl and Johnson 2013; Gianoli and Carrasco-Urra 2014; Raguso 2016; Smith 2016; Peris et al. 2017; Zhou et al. 2017).

6.5.3   Animals: Cellular Processes

In a world of microorganisms, animals cannot survive without direct interactions between individual cells. Just like fungi or plants, they are directly confronted with microorganisms specifically during food transformation. As with plants, immune systems in the digestive tracts of animals must not only distinguish their own cells from foreign cells, but also recognize pathogenic forms among the latter. To do so, immune system cells must continuously "learn" the characteristics of pathogenic cells via multi-stage processes and inactivate them in a targeted fashion (Hooper et al. 2012; Flajnik and Kasahara 2010; Flajnik 2014; Akagbosu et al. 2022). In many cases, this is achieved through endocellular decomposition in specialized mobile cells (macrophages). As in the case of food acquisition by unicellular eukaryotes and many invertebrates, an infected cell is enveloped by a macrophage and decomposed in the internal vacuole thus formed (Rosales and Uribe-Querol 2017; Hartenstein and Martinez 2019; Steinmetz 2019). Like viruses, various bacteria and unicellular parasites have also developed strategies to prevent or evade decomposition, for example the tuberculosis pathogen (*Mycobacterium tuberculosis*) or the causative agents of leishmaniasis (*Leishmania* sp.) (Uribe-Querol and Rosales 2017).

The multicellular state in animals ranges from the virtually flat, discoid placozoans consisting of just a few cell layers all the way to complex three-dimensional organisms such as arthropods or vertebrates. Developmental processes proceeding from the egg cell to the adult organism are correspondingly variable (Lemaire and Marcellini 2003; Solnica-Krezel 2005; Heisenberg and Solnica-Krezel 2008). Through sequential cell divisions, extremely generalized, hollow spheres (blastulae) are formed. During embryogenesis, morphological structures with functionally specialized organs differentiate by means of folding of initial cell assemblies, migration of cells within and between them, as well as cell formation and cell death (Tyler 2003; Bryant and Mostov 2008; Heisenberg and Solnica-Krezel 2008; Loh et al. 2016; Yasuoka et al. 2016; Ogura and Sasakura 2017). Potential connections with organisms of the *Rangeomorpha* group from the Ediacaran Period around 560 to 575 million years ago have yet to be clarified. The body structures of these organisms are modular, and at least some individual life forms exhibited fractal patterns during reproduction (Narbonne 2004; Mitchell et al. 2015).

For various reasons, apart from mere geometrical similarities, those processes have nothing to do with the development of spherical cell assemblies in green algae belonging to the *Volvox* group. In this group, multicellular life forms emerged only about 200 million years ago—much later than land plants or multicellular animals. As a phototrophic organism, *Volvox* interacts with the environment in a manner that is significantly different from that of heterotrophic animals. *Volvox* is one of many different examples of independent evolution of multicellular life forms. It is therefore simply inadmissible to explain the development of multicellular life forms on this comparative foundation alone (Herron et al. 2009; Arakaki et al. 2013). Coordinated light responses can also be found in multicellular assemblies of choanoflagellates—unicellular organisms phylogenetically related to animals (Brunet et al. 2019). Proteins for formation of differentiated epithelial cells are already present in organisms that are evolutionarily far older, for example in social amoebae (*Dictyostelium*) (Dickinson et al. 2012; Fairclough et al. 2013; Murray and Zaidel-Bar 2014; Belahbib et al. 2018). Cell barriers—Casparian strips—resembling those in animal epithelia are otherwise found only in the roots of terrestrial plants, but they evolved completely independently (Geldner 2013; Kania et al. 2014). Proteins involved in formation and repair of animal epithelia have, however, been detected in fungi (Paré et al. 2012).

In the forms of their cell connections, animals differ substantially from plants (Sugimoto et al. 2011; Oda and Takeichi 2011). In fact, this is associated with the differing lifespan of animal cells, which in human tissues, for example, can range from about a day in bone marrow to the full lifespan of the entire body in nerve cells (Kunsch 1997). Comparisons with the short-lived formation of multicellular life forms from unicellular organisms provide some idea of the diversity of processes involved in metazoan evolution. Epithelial stages of cells at the edges of tissues are of prime importance for morphological and physiological differentiation and maintenance of functional capacity in the different organs. The label "epithelial cells" was deliberately avoided here because cells and cell groups can switch between epithelial and mesenchymal tissues. Alternation between different functional properties and mobility of cells ensures the differentiation of organs during embryogenesis and their functional maintenance throughout the life cycles of animals (Pérez-Pomares and Muñoz-Chápuli 2002; Hay 2005; Kalluri 2009; Thiery et al. 2009; Sheng 2015). Because of functional flexibility of epithelia, over the course of their evolution animals have also attained far greater morphological diversity than plants. This is particularly evident in terrestrial vertebrates, whose organs "dangle" relatively loosely from an internal, rigid skeleton and must become stabilized (Tyler 2003; Engler et al. 2006; Keller 2006; Adams et al. 2010; Leys and Riesgo 2012; Gattazzo et al. 2014; Jodoin et al. 2015). The only exceptions to this requirement are the brain and spinal cord. Compared to invertebrates alone, the variety of different proteins for formation of epithelial tissues in vertebrates is more than twice as great. Yet this is founded on combinatorial effects of genes that originated predominantly—more than 70 per cent—from unicellular organisms (Murray and Zaidel-Bar 2014)!

An important role in continual renewal of cells is played by stem cells in the mesenchyme—for example in bone marrow, but also in other organ systems (Yoshimura et al. 2007; Maurer 2011; Park et al. 2012; Caplan and Correa 2011). The differing requirements of stability and dynamics in the various cell assemblies are managed by relatively uniform principles of communication and cooperation between individual cells. The need for such cooperation can be seen in the development of tumours, which are induced by "uncoupling" of individual cells from integrative communication networks. Because animal cells can migrate relatively easily through tissues and form new colonies (metastases) in other regions of the body, the consequences of tumours are generally more threatening than in plants (Scheel and Weinberg 2011).

Epithelial stages of cells regulate the exchange of substances and information between the regions they enclose and the environment. Enclosed areas can be mesenchymal tissue but also cavities—for example blood vessels. For this purpose, cell clusters must be formed such that they are as tightly enclosed as possible, with individual cells having different properties on their "outer" and "inner" sides. In many cases, a dynamically modifiable extracellular matrix (basement membrane) is attached to the tissue sides of cells. Firm cohesion between epithelial cells is ensured by intercellular protein connections, which are also connected to internal organelles of adjacent cells. Several groups of connections—tight junctions, adherens junctions, septate junctions, desmosomes and gap junctions—have been distinguished according to their molecular structures. Combinations of connections differ both between animal groups and between the epithelia of individual organs (Tyler 2003; Cereijido et al. 2004; Leys and Riesgo 2012; Nekrasova and Green 2013; Ganot et al. 2014). Contrary to earlier notions, these connections are not "cell adhesives" but dynamic structures that interact with cell tissue through information processes both during organ formation and during continual renewal of cells (Matter and Balda 2003; Blanpain and Fuchs 2009; Citi et al. 2014; Rübsam et al. 2018; Vasileva and Citi 2018). In parallel, the differing requirements for exchange of substances between the respective tissues and their environment must be managed, along with prevention of penetration by viruses, parasites or other harmful agents. To take one example, skin epithelia of terrestrial animals are confronted with challenges from solar radiation, mechanical impacts and material stresses. They are also supported by locally differing microbiomes. The question of whether animals can survive without microbiomes under certain conditions is also under discussion. During digestion of foods with different chemical compositions, epithelial cells of the digestive organs interact with a host of different of microorganisms (Magalhaes et al. 2007; Findley et al. 2013; Jablonski and Chaplin 2013; Naik et al. 2015; Elias and Williams 2016; Hammer et al. 2019; Lin et al. 2019; Bergstrom et al. 2020).

It has been demonstrated for zebrafish that ciliary movements of the zygote regulate the correct pairing of chromosomes and hence establish conditions for an orderly course of morphogenesis (Mytlis et al. 2022). Morphogenesis is influenced by epithelial tissues not only through forces and processes at the cellular level, but also through formation of cilia in neighbouring cavities of the embryo's body. In this process, signals from the cavity fluids can be detected by sensory (primary) cilia, while directed flow in fluids can be generated by motile cilia (Singla and Reiter 2006; Eggenschwiler and Anderson 2007; Satir and Christensen 2007; Goetz and Anderson 2010; Drummond 2012; Durdu et al. 2014; Ludeman et al. 2014; Avidor-Reiss and Leroux 2015; Malicki and Johnson 2017). Formation and orientation of motile cilia, primarily controlled by molecular regulatory processes, is ultimately responsible for hydrodynamical induction of chiral—left/right-handed—macromorphological differentiation of many animal organs (Nonaka et al. 2005; Blum et al. 2009; Guirao et al. 2010; Chien et al. 2015; Ferreira and Vermot 2017; Minegishi et al. 2017; Wu and Mlodzik 2017; Smith et al. 2019).

In tissues of the organs, directly adjacent cells can actively regulate communication by establishing and degrading protein channels—gap junctions (GJ Fig. 6.15)—through epigenetic processes (Oyamada et al. 2013; Bargiello et al. 2018). Like plasmodesmata in plants, they allow direct contact between the cytoplasms of individual cells (Meşe et al. 2007; Nielsen et al. 2012). However, in animal tissues individual compounds usually have a lifespan of only a few hours, after which the structure-forming proteins are dissolved once more (Matsuuchi and Naus 2013; Falk et al. 2014; Solan and Lampe 2018). In vertebrates, the group of structure-forming proteins consists of connexins and pannexins; in invertebrates, only pannexins (originally called innexins) occur (Panchin 2005; Moroz and Kohn 2016). Proteins of the pannexin group mainly form structures for exchange of signals between cells and their surroundings (Meşe et al. 2007). Gap junctions can be regarded as cellularly controlled parts of the dynamic regulatory networks present in tissues. They transmit both electrical signals and various molecules (Pereda et al. 2013). Research on the best-studied connexins suggests that gap junctions are involved in the differentiation of tissues—including the brain—and the maintenance of dynamic equilibria in all cases (Elias et al. 2007; Potolicchio et al. 2012; Hervé 2013). The differing synapses of nerve cells evolved from gap junctions as well (Moroz and Kohn 2016). Functional disturbances can trigger a wide variety of disease conditions in affected organs, for example in skin or in nervous systems (Abrams and Scherer 2012; Scott et al. 2013). In combination with pannexins, connexins influence the extent of tissue damage in heart attacks (Rodríguez-Sinovas et al. 2012). In the non-structural form, connexins influence the migration of cells in tissues as well as formation and death of cells (Kameritsch et al. 2012; Vinken et al. 2012). By influencing the exchange of information between cells and their environments, pannexins also promote macroscopic changes, for example in differentiation of nervous systems or in regulation of blood pressure (Nualart-Marti et al. 2013; Penuela et al. 2013).

Gap junctions enable signal transmission in nerveless epithelial tissues of siphonophores, hydromedusae and sponges (Mackie and Passano 1968; Mackie 1976; Josephson and Schwab 1979; Leys and Meech 2006; Meech 2015). In individual glass sponges—in a similar manner to plants—the cytoplasms of neighbouring cells are permanently connected. However, speeds of signal transmission are ten to a hundred times slower than in plants, whose transmission speeds are approached only by comb jellies (Ctenophora) (Leys 2015).

Non-adjacent cells in animal tissues can contact each other via membrane tubes—tunneling nanotubes (NT Fig. 6.15)—and transfer information or cell organelles (Rustom et al. 2004). Distances between cells connected in this way can amount to about ten times the cell length. At the morphological level, tunneling nanotubes may release various molecules in the vicinity of a target cell (DS Fig. 6.15), extend only as far as the membrane of the contacted cell, or connect the cytoplasms of both cells. Associated functions are correspondingly diverse, for instance regulation of cell structure during tissue formation, communication between pigment cells in the retina of a vertebrate eye or identification and dissolution of foreign cells in the course of immune defence (Gurke et al. 2008; Chauveau et al. 2010; Abounit and Zurzolo 2012; Wittig et al. 2012; Gerdes et al. 2013; Austefjord et al. 2014). However, tunneling nanotubes can also serve as a conduit to transfer viruses and prions between cells or to amplify tumour formation (Belting and Wittrup 2008; Ariazi et al. 2017; Nawaz and Fatima 2017).

Research over recent decades has increasingly yielded insights into the diversity of intercellular communication processes in animal tissues and their putative evolutionary pathways. The number of communication processes has seemingly decreased in current publications compared to earlier publications, reflecting the clustering of functional groups. The main reason for this is the use of different features in earlier classifications. In reality, the number of known signalling substances and receptors has exploded, for example in the realm of signal transduction through extracellular vesicles (Colombo et al. 2014). In addition, increasingly more variants and combinations of communication processes are being detected. Clear differences in the interactions of communication processes can be observed between embryogenesis and the fully developed organism. During the developmental phase, ongoing changes in processes culminate in functional differentiation of the organism as a whole. During the adult phase, only processes for maintenance and regeneration of organ functions are still required, whereas "relapses" in embryonic processes can lead to lethal damage—as with cancer. The seven currently distinguished processes of intercellular communication and regulation probably developed from intracellular regulation systems in unicellular organisms, undergoing further differentiation (Gerhart 1999; Guo and Wang 2009; Richards and Degnan 2009; Zhang 2009; Sebé-Pedros et al. 2012; Babonis and Martindale 2016; Barberán et al. 2016; Loh et al. 2016; Schenkelaars et al. 2017).

Without extracellular processes of regulation and synthesis, animals would at best have been limited to largely unstructured assemblages of cells. The first indications of the importance of extracellular processes are provided by the—often seemingly abstract—diversity of forms in the exoskeletons of unicellular organisms (Thompson 1942; Gussone et al. 2006; de Nooijer et al. 2014; Haeckel 2014). In multicellular animals, functional and morphological differentiation of organs is achieved through the diversity of proteins engaged and their combinations in the extracellular matrix (Aouacheria et al. 2006; Yurchenco 2011; Hynes and Naba 2012; Daley and Yamada 2013; Freedman et al. 2015). Through interactions between intracellular and extracellular regulatory processes, the extracellular matrix is dynamically involved in processes of embryonic development, for example, spatial orientation of organs or formation of alveoli (Czirók et al. 2004; Rozario and DeSimone 2010; Pulina et al. 2011; Harunaga et al. 2014). Biochemical modifications of extracellular substances are of central importance for formation of the musculoskeletal system in animals, for instance involving muscle, tendon, cartilage or bone tissue (Leucht et al. 2008; Exposito et al. 2010; Ramirez and Sakai 2010; Batra et al. 2012; Freedman et al. 2015). In the evolutionary history of vertebrates alone, a multitude of different materials has been involved, which is further extended by components of body covering—such as scales, feathers or hair—and the dentition (Cotton and Page 2002; Narayanan et al. 2003; Helfman et al. 2009; Johanson et al. 2010; Mahamid et al. 2010; Janis et al. 2012; Hirasawa and Kuratani 2015; Keating and Donoghue 2016). In non-vertebrates, the variety of materials and resulting structures is even greater. Examples are provided by different support structures of sponges, increase in body stability through hydrostatic forces in worm-like organisms, combination of endoskeletons and exoskeletons in cnidarians, the exoskeletons of arthropods, or extremely hard beaks in squids (Gutmann 1981; Miserez et al. 2008; Nikolov et al. 2011; Mendoza-Becerril et al. 2016; Michels et al. 2016; Dohrmann et al. 2017). Also to be added are various combinations of inorganic elements—biominerals—present in all animal groups, which have left geologically significant vestiges, for example from corals and molluscs (Weiner and Dove 2003; Anderson et al. 2005; Lloyd et al. 2008; Furuhashi et al. 2009; Gorzelak et al. 2011; Falini et al. 2015; Takeuchi et al. 2016). Macroscopically conspicuous accumulations of material easily obscure the diverse cellular regulatory processes that lead to formation of biominerals—for example in bacteria, fungi, echinoderms, sponges, algae and unicellular life forms such as coccolithophores or foraminiferans (Joubert et al. 2010; Das et al. 2012; Görgen et al. 2021; Gilbert et al. 2022).

However, the examples of supporting and locomotor structures of animals also reveal the varied dimensions of factors that influence their formation. Their fundamental features are already determined in embryogenesis by an evolutionary heritage of genetic and epigenetic regulatory factors. In addition, currently prevailing environment factors—for example temperature, nutrient supply or toxic substances—influence developmental processes. Over the course of later development (juvenile and adult phases), fundamental characteristics cannot be altered but only fine-tuned—for example, strength of bones as opposed to their specific shape (Cubo et al. 2008).

6.5.4   Animals, Coordination of Movement and Information: Neural Systems

Compared to prokaryotic systems, those of animals differ mainly in their mobility at macroscale dimensions and their capacities for mechanically fragmenting food. This innovation created new challenges for information processing in the organisms concerned. At macroscale dimensions, coordination and precision of motor functions significantly influence the survival capacity of organisms. To this end, neural systems must not only coordinate locomotor sequences as precisely as possible, but also be able to derive unambiguous decisions from the plentitude of environmental information.

In addition to cross-cellular transportation of signalling substances through vascular systems, in most animal groups nervous systems have evolved for transmission and processing of information. Signals—received by internal and external receptors—are interpreted in the context of the whole organism and translated into coordinated actions. Among all animal groups, only sponges (Porifera) and placozoans have been found to lack nerve cells. All other groups have nerve networks or nerve tracts, often with recognizable local assemblies of nerve cells (brains). Nerve cells can arise from different embryonic cells and—during the life cycle of an organism—adapt their functional performance to a variety of requirements (Hejnol and Rentzsch 2015). Possible indications of different evolutionary developments are provided by the interconnected multicellular (syncytial) neural networks without synapses found in comb jellies (Burkhardt et al. 2023).

In the overall context of cellular information processing in all organisms, the question arises as to the functional significance of the—energetically costly—neural systems of animals. From an evolutionary point of view, it must be taken into account that animals differentiated almost exclusively in their own heterotrophic food webs over at least 100 million years of their early evolution. The entire development from the late Ediacaran to the Ordovician was dependent on the energetic framework conditions of primary production in plankton (Mángano and Buatois 2014; Budd and Jackson 2016; Paulin and Cahill-Lane 2021). Long-term evolution of the individual life forms was energetically dependent on their ability to overcome other animals without being consumed as food themselves. Unlike fungi or plants, the morphological evolution of predominantly mobile life forms allowed new degrees of freedom in interactions with other organisms and abiotic factors.

The new degrees of freedom became available when the total effort for foraging and ingestion of food and potential losses incurred for individuals through other organisms were smaller than the magnitude of convertible energy gained. The range of life forms suited to this balancing act extends from sessile filter feeders to highly mobile carnivores—trilobites, for example. Even among ecologically/functionally similar life forms, the challenges have been met in different ways over hundreds of millions of years. Most sponges regulate the ciliary movements of their filter organs by means of cellular information processes, corals have nerve networks, and sessile feather duster worms (phylum Annelida, family Sabellidae) have eyes and fast neurons. Sponges defend by chemical compounds, corals through a combination of chemical agents and symbioses with other organisms, sessile tube worms can retreat into their dwelling-tubes at lightning speed if danger threatens (Gochfeld 2004; Chen et al. 2008; Haber et al. 2011; McKeon et al. 2012; Leys 2015; Bok et al. 2016).

The examples mentioned are representative of three completely different—but successful over the long term—developmental strategies of animals. Individual sponge species can reach body heights of up to three metres or—at much smaller sizes—also feed carnivorously on animals (Wang et al. 2011; Hestetun et al. 2016; Wagner and Kelley 2017). Sponges are not only the oldest metazoan group, but also extremely diverse, with about 8500 different species (Van Soest et al. 2012). Corals—despite their simple nervous system—have been no less successful in evolutionary terms and are also creators of the largest structures built by organisms: coral reefs. Unlike sponges, they cooperate not only with microorganisms but also with a great variety of algae and animals (Hughes 1989; Stolarski et al. 2011; Liu et al. 2012b; Graham and Nash 2013).

Tube worms provide an example of strategies employed for immediate confrontations between carnivores. In this context, rapid evolution of efficient sensory organs and neural systems in combination with appropriate morphological adaptation are crucial for survival. In abbreviated anthropocentric form, this evolutionary phenomenon is known by the labels "arms race" or "Red Queen hypothesis".[2] Evidence for this is provided by—the variety and presumably late evolutionary emergence of eyes—secondarily—developed on tentacles. The original eyes on a tube worm's body are hidden inside the dwelling-tube and have undergone extensive degeneration (Bok et al. 2016). Indications of the high-performance nervous systems of annelids (segmented worms)—which include tube worms—can be detected in the fossil record as early as the Cambrian period (Parry and Caron 2019). The annelids group includes the most diverse life forms in almost all marine, freshwater and terrestrial habitats. Their survival strategies and associated sensory and neurophysiological equipment are correspondingly diverse. Current knowledge indicates that a "ladder-like" nervous system, once considered characteristic of annelids, is in fact a special case rather than a defining feature (Schmidt-Rhaesa et al. 2016).

The latter example shows that—energetically costly—sensory and nervous systems are developed only in the context of particular living conditions. Well-known examples of the modification of sensory organs are kiwis in New Zealand or the regression of eyes in cave fish (Niven 2007; Moore et al. 2017). Far more differentiated morphological adaptations are found in the diverse nervous systems of individual groups of organisms (Kandel et al. 2013; Schmidt-Rhaesa et al. 2016). For example, the number of nerve cells can vary even in morphologically small organisms such as nematodes (Han et al. 2016; Schafer 2016). Among marine tunicates (sea squirts and their relatives), the relatively differentiated nervous systems of the short-lived planktonic salps (Thaliacea) are preserved throughout the life cycle, whereas in sea squirts themselves (Ascidia) the nervous system becomes extremely reduced during metamorphosis from the mobile juvenile stage to the sessile adult phase (Manni et al. 2004; Mackie et al. 2006; Manni and Burighel 2006).

---

[2] Named after the Red Queen in *Alice in Wonderland*.

Of the two basic patterns of nerve cells (Fig. 6.18), the unipolar type is found most commonly in invertebrates. Due to the separate connection of the conducting cell components to the cell body, nerves can supply extremely thin body organs in tiny organisms, for example in parasitic wasps (*Megaphragma*; Hymenoptera) only 170 μm in size (Polilov 2012). Bipolar neurons, with a variety of morphological variants, predominate in vertebrates. For example, the number of dendrites can range from a few thousand on cells of the spinal cord to a few billion in brain regions (Kandel et al. 2013). To maintain functionality, all nerve cells depend on an adequate exchange of substances with neighbouring tissue. In the central nervous systems of vertebrates, for example, this is ensured through a connection with blood vessels by means of a special cell type—astrocytes (Laughlin et al. 1998; DiNuzzo et al. 2012; Kandel et al. 2013).

Transmission rates and velocities within the nervous systems are closely connected with the biological significance of particular signals. Above all, high transmission velocities are needed for coordination of escape movements and, in carnivorous species, also for capturing prey.

The requisite performance level is achieved by increasing diameters of nerve fibres (axons) or by encasing them in myelin sheaths (Fig. 6.18). Strikingly thickened nerve fibres (giant axons) are known mainly from squids, but are also found in nerve networks of jellyfish medusae, where they coordinate rapid pumping movements of the mantle during escape from attackers (Mackie 2004; Hartline and Colman 2007). Disadvantages of non-sheathed axons—which are widespread among invertebrates—are the relatively high energy requirement for signal transmission and the space requirement for high

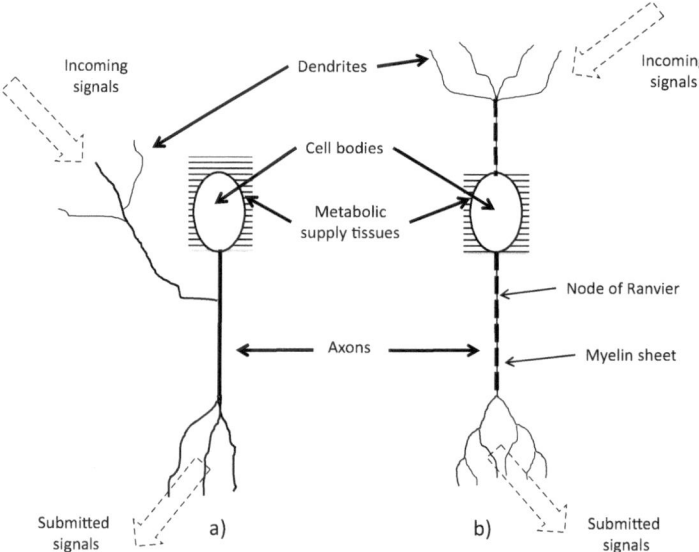

Fig. 6.18 Schematic diagram of (**a**) unipolar and (**b**) polar nerve cells and their directions of signal transmission. Source: Modified after Kandel et al. (2013)

transmission speeds. Diameters of axons range from 0.5 to a few micrometres (µm), while giant axons have diameters between around 40 and 500 µm. Sheathed axons require less energy for signal transmission because of their insulation from neighbouring tissue and the significantly smaller number of "relay stations" per unit length needed to maintain signal strength. Because of the significantly larger distances between "relay stations"—required for regulation of ion concentrations—at interruptions in the myelin sheath (nodes of Ranvier), significantly higher transmission speeds are achieved for a given axon diameter (Figs. 6.18 and 6.19) (Castelfranco and Hartline 2016).

The distribution of differing types of nerve fibres bears no clear relationship to animal classification and can differ even within systematic groups (Nishikawa 2002; Katz 2016a). An example of this is provided by a group of marine animals known as copepods, which contains some species with sheathed axons (Myelinata) and others lacking them (Amyelinata). Myelinata dominate in the upper levels (epipelagic zones) of seas at all latitudes and in subtropical estuaries (wide river mouths), where they are exposed to substantial predation risks. By contrast, Amyelinata rise to the surface zones of the open oceans only at night and migrate back down to depths of several hundred metres at dawn. Amyelinata thus reduce the risk of loss through predation while avoiding the higher energy costs for formating and maintaining myelin sheaths (Lenz 2012). This reveals further advantages and disadvantages of sheathed *versus* non-sheathed axons. The former allow faster and more precise escape movements that can be repeated more often, but require additional energy expenditure for formation and maintenance of their myelin sheaths. Because of the insulating effects of the membrane, exchange of substances between axons is limited to the nodes of Ranvier. Non-enveloped axons avoid the additional energy

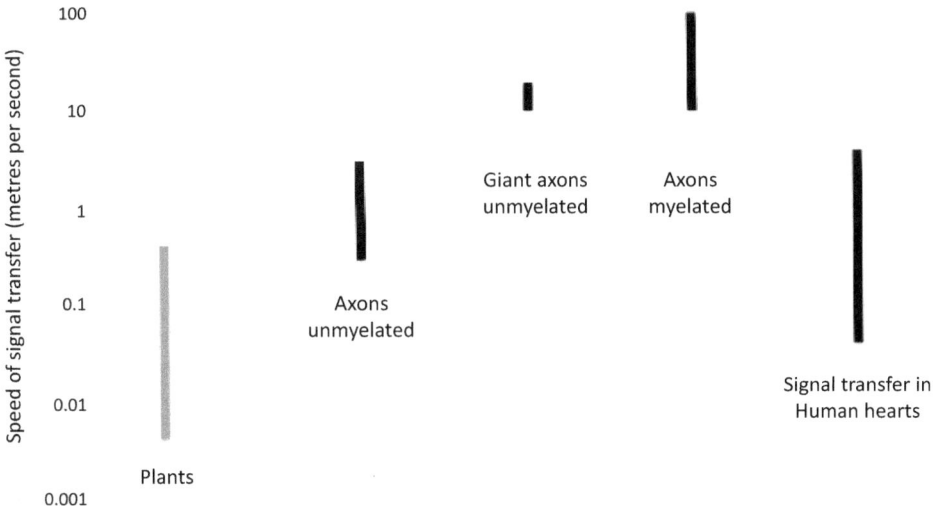

Fig. 6.19 Comparison of signal transmission speeds in plants, various animal nerve pathways and impulse conduction in the human heart. Data source: Castelfranco and Hartline (2016)

expenditure for formation of myelin sheaths, but permit slower responses with fewer repetitions—because of the higher energy costs for signal transmission (Lenz et al. 2000; Buskey et al. 2017).

In the hypothetically simplest case, a nerve cell is connected to a sensory cell and a motor cell through two transition sites—synapses. In synapses, transmission of signals can take place across gap junctions (electrical synapses) or by means of vesicles with chemical messengers (chemical synapses). Complex cellular processes in synapses and the energy they require limit maximum frequencies of information processing to around 1 kHz (Attwell and Gibb 2005; Watanabe et al. 2013a, b). In principle, with respect to effects on signal transmission, two kinds of synapses—reinforcing and inhibiting—can be distinguished. Because nerve cells connect dynamically through synapses in different ways, each nerve network can modify combinations of signals and hence information as well (Larkum 2013). In central nervous systems, these effects are further enhanced by networks in the extracellular matrix (perineural networks) and changes in the insulation of sheathed axons brought about by glial cells (Dityatev and Rusakov 2011; Ye and Miao 2013; Araque et al. 2014; Sorg et al. 2016; Reichelt et al. 2019). Distributions of the two tissue types are extremely variable and are closely related to the physiological life cycles of individual organisms, as well as their interactions with the environment (Bercury and Macklin 2015; Salzer and Zalc 2016; Kaller et al. 2017). This directly influences the associative and combinatorial capacities of information processing and the short-term memory of organisms. In addition, regular repetitions of certain interactions—training and learning—influence the dynamics of synaptic patterns in neural systems (Ryan and Grant 2009; Sengupta et al. 2010; Kandel et al. 2013; Baraban et al. 2016; Dutta et al. 2018; Wang et al. 2020). Long-term memory also involves molecular processes within cells, including epigenetic and genetic modifications (Kandel et al. 2013, 2014; Marathe et al. 2014; Ryan et al. 2015; Constantinescu et al. 2016; Kronschläger et al. 2016; Chatterjee et al. 2022). Damage to synapses—for example by viruses—can severely impair the performance of neural networks (Vasek et al. 2016).

At this point, viewed from the perspective of evolutionary development, interfaces with morphological features of organisms and environmental conditions become recognizable. Morphological features determine the scope for motor implementation. For example because of its anatomical structure, a bottlenose dolphin—despite its considerable mental creativity—cannot paint pictures. But such capacities have been experimentally demonstrated in primates with a different anatomical configuration (Eibl-Eibesfeldt 2004; van Schalk et al. 2016). Differences in morphological degrees of freedom related to the locomotor system are even more evident in comparisons between molluscs and vertebrates. Molluscs typically lack a skeleton. In various groups—for example bivalve shellfish or snails—hard shells provide protection against attacks by certain carnivores. At the same time, however, shells also restrict the range of possible movements of the organisms concerned, culminating in the sessile life forms of many bivalves. Among cephalopods, a completely different development has led to evolution of the greatest possible degree of motor freedom in octopuses. Their arms can be moved flexibly in three-dimensional space and magnitudes of body measurements can be modified over an extensive range (Hooper 2015; Levy et al. 2015). Degrees of freedom in locomotion are generally limited by the internal skeleton among vertebrates, but by the external skeleton in arthropods. Adaptations to different habitats or food properties are possible only within the scope of this "construction kit". In morphological comparisons of vertebrates, many homologies are inevitably found between differentially modified body parts—for example, the anterior limbs of birds, terrestrial animals or marine mammals. Hydrostatic structures—as found in molluscs and other animal groups—are found only in individual organs, for example tongues or elephant trunks. Among hominids, degrees of motor freedom of the tongue—yielding the potential for acoustic articulation—contributed significantly to the development of differentiated language and hence to the special status of human beings (Kier and Smith 1985; Wilson et al. 1991; Takemoto 2001; Kier 2012; Stone et al. 2016).

With the advent of nervous systems, degrees of freedom for spatially and qualitatively differentiated sensory perception were augmented in animals. Because of the additional energy costs incurred, development and diversity of sensory organs and their performance capacities are closely related to the morphology and lifestyle of individual species. Even the free-swimming larval stages of sessile sponges can orient themselves to light stimuli. According to current knowledge, signals can be transmitted only by means of cellular connections—as is also the case with adult animals (Leys 2015). To control water currents, sessile adult sponges possess sensitive cilia in their chamber systems (Maldonaldo et al. 2003; Ludeman et al. 2014).

All echinoderms (starfish and their relatives) have photosensitive nerve cells distributed over their bodies to perceive light stimuli. In a similar way to starfish (Asteroidea), individual species of sea urchins (Echinoidea) can orient themselves optically to contours. In sea urchins, such abilities are achieved by neural interconnections of specific groups of sensory cells distributed on the suckers. Starfish have different optical sensory organs, ranging up to simple compound eyes in some species, especially at the tips of their arms (Ullrich-Lüter et al. 2011; Garm and Nilsson 2014; Petie et al. 2016). A correlation between endowment with sensory organs and associated energy expenditure becomes increasingly evident as their efficiency increases. Mantis shrimps, for example, possess extremely potent—although not very many—optical receptors (see Fig. 4.3). Like primates with their large number of much simpler sensory cells in eyes with a focussing lens, by means of active scanning they gather similar colour information from their environment (Zaidi et al. 2014).

This example shows the close interrelationships between physiological and neurophysiological mechanisms in animals. Isolated analyses of the evolution of sensory organs contribute little to our understanding of the variable and highly adaptive interactions between environmental conditions and life forms with their morphological and physiological functional characteristics (Fernald 2006; Buschbeck and Friedrich 2008; Gabbott et al. 2016; Müller et al. 2018). The most impressive examples of sensory and neurophysiological adaptations arising solely through interactions between organisms are found in the deep sea. From a purely physical viewpoint, a significant decrease in photosensitive organisms is to be expected with increasing depth due to light absorption of water, comparable to that in subterranean waters (Niven 2015). In fact, specifically among deep-sea fish morphologically and physiologically very differently equipped optical sense organs exist, ranging from greatly reduced forms—via two-dimensional receptor fields—to extremely sensitive and adaptable eyes. In this deep-sea habitat, similarly diverse optical sensory capacities can also be found among invertebrates (Munk 1966; Warrant and Locket 2004; Wagner et al. 2009; Davis and Fielitz 2010; Landgren et al. 2014; de Busserolles and Marshall 2017). Such diversity becomes partially understandable if account is taken of complementary sensory organs, for example lateral line organs for perception of local water currents, chemoreceptors or electroreceptors in cartilaginous fish (Baker et al. 2013; Bellono et al. 2018; Marranzino and Webb 2018). However, the great variability of visual sensory performance in deep-sea organisms only becomes fully comprehensible when consideration is given to the diversity of optical signals in different organisms attributable to bioluminescence (see Sect. 4.3).

Examples from marine habitats also reveal the influence of information processes on colonization of habitats. Seabirds primarily orient themselves visually in marine photic zones when hunting. Accordingly, deep-diving species such as king penguins (*Aptenodytes patagonicus*) feed mainly on bioluminescent fish. Similar hunting behaviour has been documented in elephant seals (*Mirounga leonina*), which also make deep dives (Kooyman et al. 1992; Vacquié-Garcia et al. 2012).

Among cetaceans living permanently in the oceans, different systems of signal processing—novelties for marine fauna—have evolved. Toothed whales successfully use sound signals to locate and catch their prey (Miller et al. 2004; Berta et al. 2014). At least one group of their preferred prey—squids—has improved visual capacity as a primary counter-strategy, but they are unable to perceive the acoustic signals emitted by whales (Wilson et al. 2007; Nilsson et al. 2012). By contrast, in terrestrial habitats insects have acquired not only the capacity to perceive acoustic signals of the bats hunting them but also the capacity to interfere with them (Barber and Kawahara 2013). Only in a few vertebrate species that have secondarily migrated to aquatic habitats have sensory organs developed for perception of electrical signals, for example in hammerhead sharks (Sphyrnidae), platypuses (*Ornithorhynchus*) and river dolphins (*Sotalia*) (Mello 2009; Czech-Damal et al. 2013; Crampton 2019).

When hunting free-swimming marine organisms—for example fish or krill—marine mammals also use novel behavioural patterns that have proved to be evolutionarily successful over the long term. When threatened, potential prey group together in dense shoals, which make it difficult for individually hunting carnivores to orient themselves towards a particular target animal. Humpback whales (*Megaptera*) and also killer whales (*Orcinus*) reinforce the swarm effect by creating circular visual walls of air bubbles around schools of fish. Within these bubble walls humpback whales catch their prey directly with a large-volume filtering apparatus, while killer whales catch their prey indirectly after stunning them with blows from their tail flukes (Jourdain and Vongraven 2017; Cade et al. 2020).

In terrestrial habitats, the scope for evolution of biological information systems has been substantially expanded by terrestrial vegetation. Because of "noise", spatial structures—depending on their size—render visual perception of prey more difficult. At the same time, they offer potential prey improved camouflage possibilities by means of adaption of their body shapes to match plant structures. In closed tropical and subtropical forests, the diversity of visual and chemical signals of flowers and fruits of flowering plants (angiosperms), combined with their morphological mutability, offers far more options for developing and modifying interspecific information relationships. Plants are by no means passively involved in this, but influence the mutual information interactions with animals by imitating visual and chemical signals (mimicry). This promotes development of animal capacities for information processing, especially of visual and olfactory signals. Indeed, this phenomenon involves further development not only of physical performance of sensory organs, but also of neural capacities for selective identification of relevant signals and transmission to central nervous systems (Lucas 2013; Vroman et al. 2013; Clark and Demb 2016; Franke et al. 2017; Sonoda et al. 2020).

In concrete terms, conflicting requirements for information processing exist. On the one hand, the potentially relevant information must be isolated from a multitude of different signals. On the other hand, it must be assumed that only incomplete information is available. As a result, an animal can reach decisions about further action only on the basis of internal concepts regarding the target object and in conjunction with other—for example olfactory or acoustic—signals. Inevitably, the less those internal concepts have been tested by multiple applications under different environmental conditions, the more such processes are prone to error (Kostaki and Vatakis 2016; Persike and Meinhardt 2016). In general, the greater the neural capacity to store information and the longer an organism lives, the more opportunities exist for testing and adjusting internal concepts. Capacities for storing integrated information are directly related to the neural dimensions of central nervous systems. Inadequate testing of internal concepts is inevitable in inexperienced, "naïve" organisms, but they can also occur in experienced organisms because of significant changes in environmental conditions.

6.5.5   Centralization of Nervous Systems

Many discussions about evolutionary processes focus on central nervous systems—to be more specific, brains—and on their associated functions (Schoenemann 2006; Balanoff et al. 2013; Smaers et al. 2017). However, functional connections and interactions with all other organs in the whole organism, in the context of the relevant specific environmental conditions, are completely ignored. For instance, an evolutionary connection between the development of the cerebral cortex in mammals and the original olfactory sensory organs has been demonstrated (Rowe and Shepherd 2016).

Aggregations of nerve cells in spatial proximity to motor organs are found in almost all animals with nervous systems. In the simplest case, autonomous local systems are in this way coordinated with the rhythms of the organism as a whole. In conjunction with different sensory organs, increasingly complex aggregations within nervous systems develop near the mouthparts (Carr et al. 1987; Mead et al. 1999; Grillner et al. 2005; Packard 2006; Büschges et al. 2011; Büschges and Borgmann 2013; Smarandache-Wellmann et al. 2014; Arendt et al. 2015; Strausfeld et al. 2016). Differentiation and structures of the remaining nervous systems in organisms are closely linked to morphological and functional differentiation of the species concerned (Northcutt 2012; Oisi et al. 2013; Martin and Mayer 2014; Hochner and Glanzman 2016; Schmidt-Rhaesa et al. 2016; Arendt 2018; Martin-Durán et al. 2018).

Examples of bilaterians that are often invoked to explain progressive increases in brain size are untenable for several reasons. One reason is the manner of presentation, following the paradigm of the "*Scala naturae*", which fails to take into account the parallel evolution seen in "lower" and "higher" animals, already mentioned several times (Emery and Clayton 2005). Another reason are examples of centralized nervous systems existing outside the Bilateria. Among cnidarians, box jellyfish (Cubomedusae) possess—instead of a nerve network—a differentiated nervous system with four sensory centres (rhopalia). Each rhopalium is equipped with a total of six eyes—involving four different structural types—together with a statolith for perception of spatial position. Moreover, each rhopalium is freely connected to the umbrella by a stalk. Box jellyfish can orient themselves visually and—like fish—hunt actively. Instead of using teeth—which they lack—their prey is immobilized and subsequently digested by highly toxic substances released from their exploding cnidocytes (Nilsson et al. 2005; Garm et al. 2006, 2016; Garm and Mori 2009; Katsuki and Greenspan 2013; Kondo et al. 2018). Unlike cnidarians that have neural networks, box jellyfish are able not only to flee from danger, but also to orient themselves in their environment and actively recognize and pursue their prey. To do this, they need to have concepts regarding their surrounding stored in sense organs—concerning the current environmental situation. Referring to the functional principle of organisms processing external information (Fig. 6.2), decisions for further activities can once again be derived. From an energetic perspective, box jellyfish are successful because the effort expended to synthesize toxic compounds and to process information facilitates exploitation of energy-rich food sources.

Centralization of nervous systems connects and regulates different functional areas, and their differentiation takes place within the context of the morphological and environmental conditions of the organisms concerned:

– Within the organism, coordination of physiological processes in individual organs is ensured. Transmission of signals can take place via nervous systems as well as through messenger substances (hormones) in the bloodstream. An example of this is the autonomic nervous system in humans, which coordinates the performance of the internal organs through antagonistic effects of the parasympathetic and sympathetic nervous systems (Espinosa-Medina et al. 2016). Reproductive processes are also triggered by these systems when they stimulate external processes. Both endoparasites (Flegr 2013;

Lafferty and Shaw 2013) and microbiomes of digestive organs (Ronald and Beutler 2010; Cordaux et al. 2011; Heintz and Mair 2014; Colston and Jackson 2016) influence neurophysiological processes by means of these systems.

- Already at the interfaces between internal and external conditions, peripheral nervous systems coordinate the processes of motor action. Functionally, the nervous apparatus of an individual limb can autonomously regulate its motion. In the simplest case, coordination of movements of all limbs only needs induction by a central pacemaker (Lillvis and Katz 2013; Fiore et al. 2015; Katz 2016b).
- Through sensory organs—of the organisms concerned—relevant information from the outside world is selectively received, converted into chemo-electrical signals and passed on to the central nervous system. Selection of external information is highly variable and can be influenced by external signals as well as by the central nervous system itself (Kandel 2012; Xiao et al. 2023).

Comparisons between molluscs and vertebrates also yield insights into the influence exerted by environmental conditions on evolution of morphological and neurophysiological characteristics. Evolution of morphological diversity ranging from molluscs to the largest invertebrates—for example, squids of the genus *Architeuthis* with a total length of around 17 m (McClain et al. 2015; da Fonseca et al. 2020)—was feasible only under the physical conditions of aquatic habitats. The high density of water allows development of—theoretically unlimited—large body size of organisms without fixed support structures. Further examples of this are the sometimes extremely large life forms of jellyfish (Cnidaria) (Costello et al. 2008). However, food availability and the risk of loss due to carnivores exert a limiting effect. Within the potential range, animals can develop iteratively in accordance with their particular morphological and physiological features. Among molluscs, predation is associated with substantial neural capacities, connected either to high mobility—as in squids—or to great adaptability—as in octopuses. For example, octopuses and squids have chemotactic sensors in the suckers of their arms for parallel detection of chemical and tactile stimuli (Allard et al. 2023; Kang et al. 2023). In both groups, several species have higher relative brain weights than either bony fish or amphibians (Packard 1972; Hochner and Glanzman 2016). What differs, however, is the distribution of neurons throughout the nervous system. In octopuses, 60 per cent of the approximately 500 million neurons are distributed among the eight arms, 8 per cent in their neural systems and 32 per cent in the central nervous system. Indeed, the central nervous system proper accounts for 10 per cent and the optic lobes for 22 per cent (Fig. 6.20) (Hochner 2012). Despite this "decentralized centralization" of their nervous system, octopuses exhibit a particularly wide range of adaptive change to the prevailing environmental context in their body markings and shapes, as well as astonishing learning capacities under experimental conditions (Shomrat et al. 2008; Hochner 2012; Zarrella et al. 2015; Levy and Hochner 2017).

The variety of situationally adapted behavioural patterns and the learning capacity shown by octopuses attests to the efficiency of neural networks when they are sufficiently

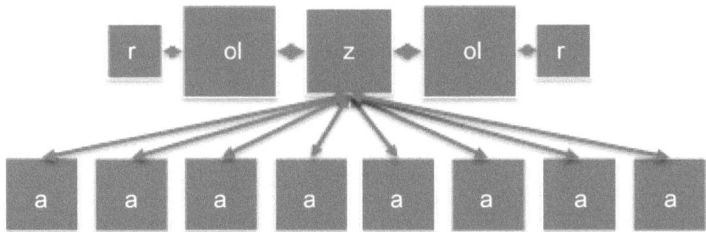

Fig. 6.20 Schematic diagram of the distribution of neurons in the nervous system of an octopus. Area sizes correspond approximately to relative proportions of the total number of neurons. Abbreviations: a arm; ol optical lobe; r retina; z central nervous system. Data source: Hochner (2012)

large. Octopuses—like chameleons—can not only change structural features and colour patterns of their skin surface, but can also adapt their body shape to the prevailing environmental context or to mimic other animals (Norman et al. 2001; Finn et al. 2009; Zarrella et al. 2015). Dynamic mimicry requires not only a genetic foundation—as with the static mimicry of various plants and animals—but also internal concepts for situationally adapted responses of the organism as a whole. Capacities for learning and solving complex tasks provide evidence for the combinatorial performance abilities of octopuses. Only about one tenth of all neurons are available centrally for this purpose; all other neurons are organized as decentralized groups (Fig. 6.20).

Like vertebrates, representatives of one particular group of molluscs—gastropods—migrated to terrestrial habitats during the Phanerozoic, but without profound modifications of their neural processing capacities (Zaitseva 1999; Hochner and Glanzman 2016; Kiss 2017). By comparison, after moving to new habitats, other groups of invertebrates have shown adaptations that are far more distinct. Among crustaceans (Crustacea), predominantly terrestrial fiddler crabs (members of Brachyura) have both enhanced neural capacities for visual perception and conspicuously asymmetrical morphology in the pincers of males. The central nervous system of the largest land-dwelling crab—the coconut crab (*Birgus*)—shows a pronounced enhancement of olfactory capacities. Pan-tropical cleaner shrimps (*Stenopus hispidus*) have particularly well differentiated brains (Zeil and Hemmi 2006; Krieger et al. 2010, 2020), adapted to the diverse demands of communication, along with sensory and motor performance. Despite their extensive morphological adaptation to terrestrial conditions and major differentiation to become the most species-rich arthropod group, the central nervous systems of insects show many similarities to those of phylogenetically related crustaceans (Malacostraca) (Schmidt-Rhaesa et al. 2016).

Among vertebrates, the greatest functional diversity of life forms evolved in ray-finned fishes (Helfman et al. 2009). Despite their neural capacities for processing diverse types of information, quantitative relationships between brain weight and body weight in individual species (brain:body indices) are somewhat beneath those in cartilaginous fish (Northcutt 2002). A possible reason for this could be the specialized adaptation of individual sensory brain areas to specific conditions of their habitats and food acquisition (Kotrschal et al.

1998; Iglesias et al. 2015; Tsuboi et al. 2016). Brain:body relationships similar to those in ray-finned fishes are found in amphibians—the basal group of all land-living vertebrates (terrestrial tetrapods) (Northcutt 2002; Helfman et al. 2009). Their evolution began around 390 million years ago in the Devonian and continued progressively during the Carboniferous (Clack 2012; Schoch 2014). There is a striking temporal overlap between colonization of land by vertebrates and the onset of modification of stream structures by solid roots of terrestrial vegetation (Fig. 9.10) (Tomescu and Rothwell 2006; Algeo and Scheckler 2010; Davies and Gibling 2010; Gibling and Davies 2012; Gibling et al. 2014; Hetherington and Dolan 2017). Brain:body relationships also remain largely unchanged in the reptiles that subsequently evolved from amphibians. Among reptiles, various life forms migrated into aquatic habitats at several different times—including such well-known groups as ichthyosaurs, sea turtles or sea snakes.

A distinct increase in relative brain sizes is seen in bird and mammal species that evolved under framework conditions dominated by flowering plants (angiosperm flora) (Jarvis et al. 2005). Among dolphins and whales (cetaceans)—phylogenetically related to even-toed hoofed mammals—adaptation to aquatic habitats, extending over about 50 million years, was associated with divergent evolution of brain:body relationships. In general, relative brain sizes among baleen whales increased only slowly. Moreover, in very large-bodied cetacean species—including toothed whales—relative brain weights even decreased over the past few millions of years. The main reason for this was more rapid increase in body mass. By contrast, a brain-body relationship similar to that seen in humans evolved in dolphins (*Delphinus*) and porpoises (*Tursiops*) (Marino 2007; Gatesy et al. 2013; Montgomery et al. 2013; Berta et al. 2016). Despite the same long expanse of evolutionary time—and increasing food availability due to an increase in more productive groups of algae (coccoliths, dinoflagellates and diatoms)—no comparable evolution of brain capacities has so far been found among fish.

General comparisons of brain:body relationships (Northcutt 2002) reveal higher values in predominantly endothermic vertebrates—birds and mammals—but do not provide any evidence of an evolutionary trend towards ever greater capacities of the central nervous system. Even among fishes, there are individual examples of comparatively greater brain:body indices in individual groups, such as various freshwater elephantfish (Mormyridae) among ray-finned fishes or manta rays (Mobulidae) among cartilaginous fishes (Ari 2011; Sukhum et al. 2019). Among mammals, however, aquatic manatees and sea-cows (Sirenia) have the smallest brain:body indices (Boddy et al. 2012).

Even tiny arthropods with about one millionth of the brain mass of an octopus, such as web-building spiders, can achieve complicated feats. Unlike octopuses, they can adapt to changing environmental conditions only to a limited extent (Eberhard 2007; Eberhard and Weislo 2011). Social insects—for example termites, ants and bees—can compensate for this handicap through cooperative sharing of tasks. To achieve this, eventual morphological and neural performance spectra of individuals are epigenetically determined during reproduction or through selective feeding. Any individual animal can perform only a part of the total repertoire of tasks. Their contribution to overall performance is ensured by direct or indirect information from other animals or through local structural information (Robinson 1992; Choe and Crespi 1997; Tschinkel 1999; Foster and Ratnieks 2001; Wenseleers et al. 2004; Oldroyd and Fewell 2007; Maleszka 2008). Examples of this are the waggle dances of bees, odour signals in ants or local wall structure in termite burrows (Root-Bernstein 2010; Czaczkes et al. 2011; Calovi et al. 2019). On this basis, insects can respond flexibly to changes in food supply without permanently losing their information-processing capacities. It is possible that the flexible properties of the "social brain", in combination with the deterministic establishment of individual information-processing capacities have been key factors in the long-term successful evolution of social communities of insects. Bees are thought to have evolved their social structures around 20 million years ago (Hölldobler and Wilson 1990; Grimaldi and Engel 2005; Lihoreau et al. 2012; Barden and Grimaldi 2016; Engel et al. 2016).

As a basic prerequisite, large central nervous systems require an adequate energy supply (Laughlin 2001; Isler and van Schalk 2006, 2009; Niven and Laughlin 2008; Sukhum et al. 2016). Because the total energy turnover of animals increases with body size, a regular increase in the performance of central nervous systems is to be expected. However, the scope for differentiated development of neural systems depends on the life forms concerned and on environmental factors. Stimulating factors for the growth of the central nervous system are, for example, highly structured habitats, diverse and variable food, heavy competitive pressure, complex spatial and temporal distribution patterns of food, or increased requirements for interpreting complex signals. Inhibiting factors are, for instance, predominance of plant food, limited foraging challenges, living conditions with persistent requirements for rapid responses or locomotion, or monotonous habitats (Rehkämper et al. 1991; Fish and Lockwood 2003; Kruska 2003; Farris and Roberts 2005; Boddy et al. 2012; Dicke and Roth 2016; Farris 2015; Karten 2015; Bear et al. 2016; Eriksson et al. 2019).

Due to combined effects of various influencing factors, relationships between brain and body weights do not increase steadily, but remain fairly stable above a certain value. According to data for primates (Fish and Lockwood 2003), the range of maximum index values (between 0.5 and 0.65)[3] is already reached at body weights of about 2 kg. Brain:body weight relationships among the generally heavier toothed whales (Marino 2007) lie within the range mentioned (Fig. 6.21), but show a slight trend towards decrease. That notwithstanding, summarized data analyses indicate tendencies, as exemplified by the trend lines for primates and the combined data in Fig. 6.21. Independently of formal data analyses, distributions of individual values for primates and toothed whales show a clustering of maximum values between body weights of about 50 and 300 kg. Perhaps this is an optimal range for the evolution of large central nervous systems in mammals? In this context, it is remarkable that the brain volumes of currently living humans (*Homo sapiens sapiens*) are smaller than those of Neanderthals (*Homo sapiens neandertalensis*) (Roth and Dicke 2005; Falk et al. 2009).

---

[3] Value = logarithm of brain weight (g)/logarithm of body weight (g).

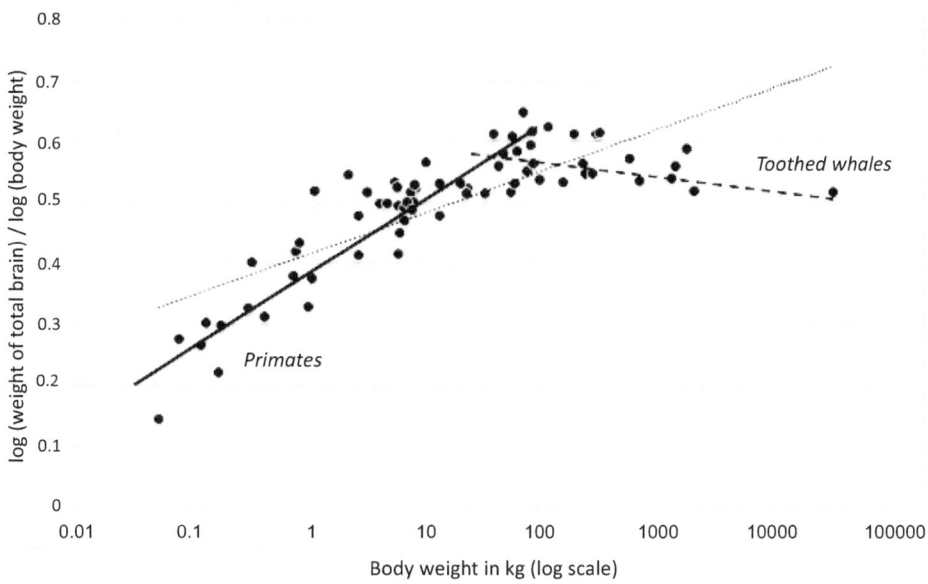

Fig. 6.21 Comparison of different interpretations of brain:body weight relationships taking the examples of primates and toothed whales. Trend lines are provided for the combined data set (dotted) and separately for primates (solid) and toothed whales (dashed). Data sources: Fish and Lockwood (2003), Marino (2007)

The widely discussed hypothesis of fundamental links between social organization and increased size of central nervous systems does not stand up to closer scrutiny. Core weaknesses of the purported evidence are selective choice of underlying data sets, use of inappropriate indicators and fuzzy definitions of social organization in different animal groups (Holmes et al. 2007; Finarelli and Flynn 2009; Farris and Schulmeister 2011; Shultz and Dunbar 2010; O'Connell and Hofmann 2011, 2012; Muscedere et al. 2014). Especially in connection with hominid evolution, effects of various confounding variables and self-reinforcing processes—closely related to tool use—during the later phases of evolution should be taken into account (Knoflacher 2017).

Disproportionate growth of central nervous systems—as with all organs—is not associated exclusively with advantages, but also with numerous disadvantages. Apart from the energetic problems already mentioned, for physical reasons alone an increase in number of neurons and synapses leads to longer processing and response times. Such effects were lethal under the conditions of early evolution and are at least undesirable in present human societies. Increased capacities of central nervous systems do not lead to better perception of the environment, but rather expand the scope for developing internal concepts of our environment. An expanded memory yields a broader repertoire for comparisons of temporally and spatially different perceptions. However, because of the fundamentally random nature and complexity of environmental processes, enhanced performance capacities do not permit improved estimation of future developments!

As yet, mass-based comparisons of central nervous systems have not yielded any reliable indication of functional correlations with the complete nervous systems of animals. Across all mammalian species studied, however, comparative analyses of fractional volumes of anatomically distinguishable brain regions reveal—independently of body size—the greatest consistency, with an average proportion of 13 per cent. This suggests that the cerebellum is the central regulatory region for all neural functions in mammals (Clark et al. 2001; Barton 2012). This conclusion is further supported by comparisons of the relative proportions of different features of the cerebrum and the cerebellum, as well as other parts of human brains. In relation to a total human brain weight averaging about 1500 g, the cerebrum (cerebral cortex) accounts for 82 per cent, the cerebellum for 10 per cent, and the rest of the brain for 8 per cent. Of the brain's approximately 86 billion nerve cells, however, 19 per cent are located in the cerebrum, 80 per cent in the cerebellum and 1 per cent in the rest of the brain. Moreover, out of approximately 85 billion non-neural cells—for example glial cells—72 per cent are in the cerebrum, 19 per cent in the cerebellum and 9 per cent in the rest of the brain (Fig. 6.22) (Azevedo et al. 2009; Herculano-Houzel 2009).

Comparison of Fig. 6.20 with 6.22b reveals, in quantitative terms, similar principles underlying functional divisions of nerve cells in human and octopus brains. But, other than substantial differences in total numbers of nerve cells, the relative proportions of nerve cells in the central area of the human brain are about 70 per cent higher than in octopus brains.

Genetic and biochemical studies have yielded a new perspective on the significance of the six histologically distinguishable layers of the cortex (pallium) in the mammalian brain. Cells with identical functions but with a distinctly different histological structure have also been identified in bird brains. Available evidence clearly indicates that the structure of the "neocortex" was determined by evolution from olfactory cortex. To take one

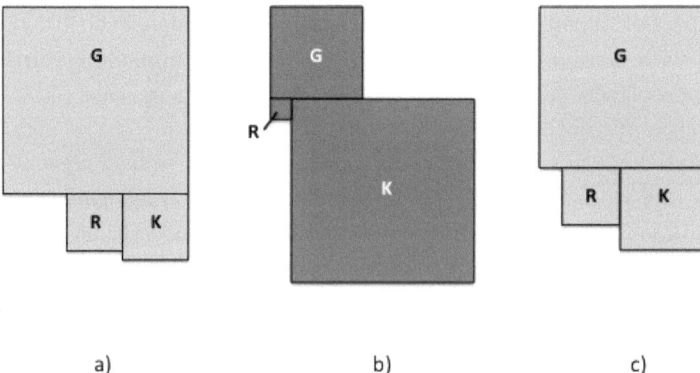

a)   b)   c)

Fig. 6.22 Comparison of relative proportions of cerebrum (G), cerebellum (K) and remaining regions (R) of an average human brain. (**a**) Of total weight, (**b**) of nerve cells and (**c**) of non-neural cells. Sizes of brain regions correspond approximately to their respective proportions. Data source: Azevedo et al. (2009)

example, it has emerged that development of bird song and human language are based on fundamentally similar neural circuits, the origins of which are suspected to have been already present in fish (Aboitiz and Montiel 2003; Grillner et al. 2008; Tessmar-Raible et al. 2007; Shepherd 2011; Bass and Chagnaud 2012; Dugas-Ford et al. 2012; Karten 2012; Samu et al. 2014; Karten 2015; Rowe and Shepherd 2016; Tosches and Laurent 2019).

6.6 Overview of Information

– Biological information processes always take place in a dynamic environment with a variable and multi-layered flow of information. Organisms can survive only if they can obtain the information relevant to them and make sufficiently rapid decisions. For all organisms, evolutionarily developed morphological and physiological features limit the specific potentials of information processing. Endogenous states of expectation and stored information modulate the prevailing perceptual space of an individual organism.
– Information processes take place in all cells; discrete nervous systems are not prerequisites for information processing.
– The diversity of abiotic signals and structures yields parameters for orientation that organisms use in different combinations.
– Cellular information processes—such as those occurring under the influence of viruses—are of central importance for biological evolution. Each individual cell must—in coordination with prevailing dynamic environmental conditions—be able to adequately regulate its internal processes and meet from them its needs for external variables. Each cell itself sends out information and thus influences other single- or multicellular organisms in different ways—ranging from cooperation to manipulation.
– Information flows between cells and organisms accordingly establish—under the influence of additional framework conditions—metastable states for their interactions. As a result, different cell societies of can form—at least temporarily. All multicellular organisms are—at the cellular level—integrated into the network of unicellular organisms, for example with specific microbiomes. Because of the metastable properties of interactions, this can result in effects that are either beneficial or detrimental for multicellular organisms.
– Similarly, the multicellular condition in organisms—plants, fungi and animals—can only be achieved and maintained through information exchange between cells. Depending on morphological differentiation, the exchange of information can take place at various structural levels and also over long distances. Even in the largest-bodied organisms, cellular information systems are hence able to regulate decentralized, multimodal energetic and material processes.
– Channelling of the flow of information in nervous systems ensures that mobile animals harmonize their movements in coordination with signals selectively perceived from the environment in combination with endogenous physiological signals. To meet these requirements, information in nervous systems must be processed deterministically and

causally. Environmental conditions and local organisms influence the endowment of animals with sensory organs and the differentiation of centralized nervous systems. Reciprocal interactions occur between structures in centralized nervous systems and learning capacities, and also between long-term information storage and endocellular processes.

References

Able K, Able MA (1996) The flexible migratory orientation system of the Savannah sparrow (*Passerculus sandwichensis*). J Exp Biol 199:3–8

Aboitiz F, Montiel J (2003) One hundred million years of interhemispheric communication: the history of the corpus callosum. Braz J Med Biol Res 36(4):409–420

Abounit S, Zurzolo C (2012) Wiring through tunnelling nanotubes—from electrical signals to organelle transfer. J Cell Sci 125:1–10

Abrams CE, Scherer SS (2012) Gap junctions in inherited human disorders of the central nervous system. Biochim Biophys Acta 1818:2030–2047

Abreu F, Silva KT, Martins JL, Lins U (2006) Cell viability in magnetotactic multicellular prokaryotes. Int Microbiol 9:267–272

Abreu F, Morillo V, Nascimento FF, Werneck C, Cantão ME et al (2014) Deciphering unusual uncultured magnetotactic multicellular prokaryotes through genomics. ISME J 8:1055–1068

Adamatzky A (2018) On spiking behaviour of oyster fungi Pleurotus djamor. Sci Rep 8:7873

Adamatzky A (2022) Language of fungi derived from their electrical spiking activity. R Soc Open Sci 9:211926

Adamec L (2013) Foliar mineral nutrient uptake in carnivorous plants: what do we know and what should we know? Front Plant Sci 4:10

Adams EDM, Goss GG, Leys SP (2010) Freshwater sponges have functional, sealing epithelia with high transepithelial resistance and negative transepithelial potential. PLoS One 5(11):e15040

Adlassnig W, Peroutka M, Lambers H, Lichtscheidl IK (2005) The roots of carnivorous plants. Plant Soil 274:127–140

Adler K, Phillips JB (1985) Orientation in a desert lizard (*Uma notata*): time-compensated compass movement and polarotaxis. J Comp Physiol A 156:547–552

Ainsworth EA, Bush DE (2011) Carbohydrate export from the leaf: a highly regulated process and target to enhance photosynthesis and productivity. Plant Physiol 155:64–69

Akagbosu B, Tayyebi Z, Shibu G, Iza GAP, Deep D et al (2022) Novel antigen-presenting cell imparts Tr_{eg}-dependent tolerance to gut microbiota. Nature 610:752–760

Åkesson S, Morin J, Muheim R, Ottosson U (2001a) Avian orientation at steep angles of inclination: experiments with migratory white-crowned sparrows at the magnetic North Pole. Proc Biol Sci 268:1907–1913

Åkesson S, Walinder G, Karlsson L, Ehnbom S (2001b) Reed warbler orientation: initiation of nocturnal migratory flights in relation to visibility of celestial cues at dusk. Anim Behav 61:181–189

Alberts B, Johnson A, Lewis J, Morgan D, Raff M et al (2015) Molecular biology of the cell. Garland Science, New York

Albuquerque P, Casadevall A (2012) Quorum sensing in fungi—a review. Med Mycol 50:317–345

Algeo TJ, Scheckler SE (2010) Land plant evolution and weathering rate changes in the Devonian. J Earth Sci 21:75–78

Allard CAG, Kang G, Kim JJ, Valencia-Montoya WA, Hibbs RA, Bellono NW (2023) Structural basis of sensory receptor evolution in octopus. Nature 616:373–377

Aly R, Hamamouch N, Abu-Nassar J, Wolf S, Joel DM et al (2011) Movement of protein and macromolecules between host plants and the parasitic weed *Phelipanche aegyptiaca* Pers. Plant Cell Rep 30:2233–2241

Amaral PP, Dinger ME, Mercer TR, Mattick JS (2008) The eukaryotic genome as an RNA machine. Science 319:1787–1789

Ameisen JC (2002) On the origin, evolution, and nature of programmed cell death: a timeline of four billion years. Cell Death Differ 9:367–393

Ameres SL, Zamore PD (2013) Diversifying microRNA sequence and function. Nat Rev Mol Cell Biol 14:475–488

Amin SA, Hmelo LR, van Tol HM, Durham BP, Carlson LT et al (2015) Interaction and signalling between a cosmopolitan phytoplankton and associated bacteria. Nature 522:98–191

Amital G, Sorek R (2017) Intracellular signaling in CRISPR-Cas defense. Science 357:550–551

Amselem J, Cuomo CA, van Kan JAL, Viaud M, Benito EP et al (2011) Genomic analysis of the necrotrophic fungal pathogens *Sclerotinia sclerotiorum* and *Botrytis cinerea*. PLoS Genet 7(8):e 1002230

Anderluh G, Gilbert R (2014) MACPF/CDC proteins—agents of defence, attack and invasion. Springer, Dordrecht

Andersen KH, Berge T, Gonçalves RJ, Hartvig M, Heuschele J et al (2016) Characteristic sizes of life in the oceans, from bacteria to whales. Annu Rev Mar Sci 8:217–241

Anderson J (2011, December) A million monkeys and Shakespeare. Significance 8(4):190–192

Anderson HC, Garimella R, Tague SE (2005) The role of matrix vesicles in growth plate development and biomineralization. Front Biosci 10:822–837

Anetzberger C, Pirch T, Jung K (2009) Heterogeneity in quorum sensing-regulated bioluminescence of *Vibrio harveyi*. Mol Microbiol 73(2):267–277

Aouacheria A, Geourjon C, Aghajari N, Navratil V, Deléage G et al (2006) Insights into early extracellular matrix evolution: spongin short chain collagen-related proteins are homologous to basement membrane Type IV collagens and form a novel family widely distributed in invertebrates. Mol Biol Evol 23(12):2288–2302

Arakaki Y, Kawai-Toyooka H, Hamamura Y, Higashiyama T, Noga A et al (2013) The simplest integrated multicellular organism unveiled. PLoS One 8(12):e81641

Araque A, Carmignoto G, Haydon PG, Oliet SHR, Robitaille R, Volterra A (2014) Gliotransmitter travel in time and space. Neuron 81:728–739

Arendt D (2018) Animal evolution: convergent nerve cords? Curr Biol 28:R225–R227

Arendt D, Benito-Gutierrez E, Brunet T, Marlow H (2015) Gastric pouches and the mucociliary sole: setting the stage for nervous system evolution. Philos Trans R Soc Lond B Biol Sci 370:20150286

Ari C (2011) Encephalization and brain organization of mobulid rays (myliobatiformes, elasmobranchii) with ecological perspectives. Open Anat J 3:1–13

Ariazi J, Benowitz A, De Biasi V, Den Boer ML, Cherqui S et al (2017) Tunneling nanotubes and gap junctions—their role in long-range intercellular communication during development, health, and disease conditions. Front Mol Neurosci 10:333

Ariely D (2010) Predictability irrational. Harper Perennial, New York

Asano M, Khrennikov A, Ohya M, Tanaka Y, Yamato I (2015) Quantum adaptivity in biology: from genetics to cognition. Springer, Dordrecht

Astley HC, Roberts TJ (2012) Evidence for a vertebrate catapult: elastic energy storage in the plantaris tendon during frog jumping. Biol Lett 8:386–389

Atkinson S, Williams P (2009) Quorum sensing and social networking in the microbial world. J R Soc Interface. https://doi.org/10.1098/rsif.2009.0203

Attwell D, Gibb A (2005) Neuroenergetics and the kinetic design of excitatory synapses. Nat Rev Neurosci 6:841–849

Austefjord MW, Gerdes H-H, Wang X (2014) Tunneling nanotubes. Commun Integr Biol 7(1):e27934

Avelar GM, Schumacher RI, Zaini PA, Leonard G, Richards TA, Gomes SL (2014) A rhodopsin-guanylyl cyclase gene fusion functions in visual perception in a fungus. Curr Biol 24:1234–1240

Avidor-Reiss T, Leroux MR (2015) Shared and distinct mechanisms of compartmentalized and cytosolic ciliogenesis. Curr Biol 25:R1143–R1150

Ayers GP, Cainey JM (2007) The CLAW hypothesis: a review of the major developments. Environ Chem 4:366–374

Azcón-Aguilar C, Barea JM, Gianinazzi S, Gianinazzi-Pearson V (eds) (2009) Mycorrhizas. Springer, Berlin

Azevedo FAC, Carvalho LRB, Grinberg LT, Farfel JM, Ferretti REL et al (2009) Equal numbers of neuronal and nonneuronal cells make the human brain an isometrically scaled-up primate brain. J Comp Neurol 513:532–541

Azizullah A, Murad W, Adnan M, Ullah W, Häder D-P (2013) Gravitactic orientation of *Euglena gracilis*—a sensitive endpoint for ecotoxicological assessment of water pollutants. Front Environ Sci 1:4

Babikova Z, Gilbert L, Bruce TJA, Birkett M, Caulfield JC et al (2013) Underground signals varied through common mycelial networks warn neighbouring plants of aphid attack. Ecol Lett 16:835–843

Babonis LS, Martindale MQ (2016) Phylogenetic evidence for the modular evolution of metazoan signalling pathways. Philos Trans R Soc Lond B Biol Sci 372:20150477

Bahl J, Lau MCY, Smith GJD, Vijaykrishna D, Cary SC et al (2011) Ancient origins determine global biogeography of hot and cold desert cyanobacteria. Nat Commun. https://doi.org/10.1038/ncomms1167

Bahn Y-S, Xue C, Idnurm A, Rutherford JC, Heitman J, Cardenas ME (2007) Sensing the environment: lessons from fungi. Nat Rev Microbiol 5:57–69

Baker MD, Stock JB (2007) Signal transduction: networks and integrated circuits in bacterial cognition. Curr Biol 17(23):R1021–R1024

Baker CVH, Modrell MS, Gillis JA (2013) The evolution and development of vertebrate lateral line electroreceptors. J Exp Biol 281:2515–2522

Balanoff AM, Bever GS, Rowe TB, Norell MA (2013) Evolutionary origins of the avian brain. Nature 501:93–96

Balázsi G, van Oudenaarden A, Collins JJ (2011) Cellular decision making and biological noise: from microbes to mammals. Cell 144:910–925

Baltimore D, Boldin MP, O'Connell RM, Rao DS, Taganov K (2008) MicroRNAs: new regulators of immune cell development and function. Nature Immunol 9(8):839–844

Baluška F (ed) (2009) Plant-environment interactions. Springer, Berlin

Baluška F, Manusco S (2007) Plant neurobiology as a paradigm shift not only in the plant science. Plant Signal Behav 2(4):205–207

Baluška F, Manusco S (eds) (2009) Signaling in plants. Springer, Berlin

Baluška F, Mancuso S, Volkmann D (eds) (2006) Communication in plants. Springer, Berlin

Baraban M, Mensch S, Lyons DA (2016) Adaptive myelination from fish to man. Brain Res 1641:149–161

Barber JR, Kawahara AY (2013) Hawkmoths produce anti-bat ultrasound. Biol Lett 9:20130161

Barberán S, Martin-Durán JM, Cebrià F (2016) Evolution of the EGFR pathway in metazoa and its diversification in the planarian *Schmidtea mediterranea*. Sci Rep. https://doi.org/10.1038/srep28071

Barden B, Grimaldi DA (2016) Adaptive radiation in socially advanced stem-group ants from the cretaceous. Curr Biol 26:515–521

Bargiello TA, Oh S, Tang Q, Bargiello NK, Dowd TL, Kwon T (2018) Gating of connexin channels by transjunctional-voltage: conformations and models of open and closed stated. Biochim Biophys Acta 1860:22–39

Barksdale L (1970) *Corynebacterium diphteriae* and its relatives. Bacteriol Rev 34(4):378–422

Barlow PW, Fisahn J (2012) Lunisolar tidal force and the growth of plant roots, and some other of its effects on plant movements. Ann Bot 110:301–318

Barlow PW, Mikulecký M, Střeštík J (2010) Tree-stem diameter fluctuates with the lunar tides and perhaps with geomagnetic activity. Protoplasma 247:25–43

Barto EK, Weidenhamer JD, Cipollini D, Rillig MC (2012) Fungal superhighways: do common mycorrhizal networks enhance below ground communication? Trends Plant Sci 17(11):633–637

Barton RA (2012) Embodied cognitive evolution and the cerebellum. Philos Trans R Soc Lond B Biol Sci 367:2097–2107

Bárzana G, Aroca R, Paz JA, Chaumont F, Martinez-Ballesta MC et al (2012) Arbuscular mycorrhizal symbiosis increases relative apoplastic water flow in roots of the host plant under both well-watered and drought stress conditions. Ann Bot 109:1009–1017

Bass AH, Chagnaud BP (2012) Shared developmental and evolutionary origins for neural basis of vocal-acoustic and pectoral-gestural signalling. PNAS 109(1):10677–10684

Bassler BL (2002) Small talk: cell-to-cell communication in bacteria. Cell 109:421–424

Bassler BL, Losick R (2006) Bacterially speaking. Cell 125:237–246

Batra N, Kar R, Jiang JX (2012) Gap junctions and hemichannels in signal transmission, function and development of bone. Biochim Biophys Acta 1818:1909–1918

Battin TJ, Sloan WT, Kjelleberg S, Daims H, Head IM et al (2007) Microbial landscapes: new paths to biofilm research. Nat Rev Microbiol 5:76–81

Bauer U, Bohn HF, Federle W (2007) Harmless nectar source or deadly trap: *Nepenthes* pitchers are activated by rain, condensation and nectar. Proc Biol Sci. https://doi.org/10.1098/rspb.2007.1402

Bayles KW (2014) Bacterial programmed cell death: making sense of a paradox. Nat Rev Microbiol 12(1):63–69

Bayram Ö, Braus GH, Fischer R, Rodriguez-Romero J (2010) Spotlight on *Aspergillus nidulans* photosensory systems. Fungal Genet Biol 47:900–908

Bazargani N, Attwell D (2016) Astrocyte calcium signaling: the third wave. Nat Neurosci 19:182–189

Be'er A, Zhang HP, Florin E-L, Payne SM, Ben-Jacob E, Swinney HL (2009) Deadly competition between sibling bacterial colonies. PNAS 106(7):428–433

Be'er A, Ariel G, Kalisman O, Helman Y, Sirota-Madi A et al (2010) Lethal protein produced in response to competition between sibling bacterial colonies. PNAS. https://doi.org/10.1073/pnas.1001062107

Bear DM, Lassance J-M, Hoekstra HE, Datta SR (2016) The evolving neural and genetic architecture of vertebrate olfaction. Curr Biol 26:R1039–R1049

Bebber DP, Hynes J, Darrah PR, Boddy L, Fricker MD (2007) Biological solutions to transport network design. Proc Biol Sci 274:2307–2315

Beck J (2013) Woronin Körper von *Aspergillus fumigatus*, ihre Verankerung am Hyphenseptum und Bedeutung für die Stressresistenz und Virulenz dieses pathogenen Schimmelpilz. Dissertation Ludwig-Maximilians-Universität München

Becker B, Marin B (2009) Streptophyte algae and the origin of embryophytes. Ann Bot 103:999–1004

Beckmann L, Edel KH, Batistič O, Kudla J (2016) A calcium sensor-protein kinase signalling module diversified in plants and is retained in all lineages of Bikonta species. Sci Rep. https://doi.org/10.1038/srep.31645

Beemster GTS, Fiorani F, Inzé D (2003) Cell cycle: the key to plant growth control? Trends Plant Sci 8(4):154–158

Beier D, Gross R (2006) Regulation of bacterial virulence by two-component systems. Curr Opin Microbiol 9:143–152

Belahbib H, Renard E, Santini S, Jourda C, Claverie M et al (2018) New genomic data and analyses challenge the traditional vision of animal epithelium evolution. BMC Genomics 19:393

Belkaid Y, Segre JA (2014) Dialogue between skin microbiota and immunity. Science 346:954–959

Bell JJ, Barnes DKA (2001) Sponge morphological diversity: a qualitative predictor of species diversity? Aquat Conserv Mar Freshw Ecosyst 11:109–121

Bellandi A, Papp D, Breakspear A, Joyce J, Johnston MG et al (2022) Diffusion and bulk flow of amino acids mediate calcium waves in plants. Sci Adv 8:eabo6693

Belles X (2011) Origin and evolution of insect metamorphosis. In: Encyclopedia of life sciences. Wiley, Chichester

Bellono NW, Leitch B, Julius D (2018) Molecular tuning of electroreception in sharks and skates. Nature 558:122–126

Bellwood DR, Hughes TP, Folke C, Nyström M (2004) Confronting the coral reef crisis. Nature 429:827–833

Belnap J, Lange OL (eds) (2003) Biological soil crusts: structure, function, and management, Ecological studies 150. Springer, Berlin

Belting M, Wittrup A (2008) Nanotubes, exosomes, and nucleic acid-binding peptides provide novel mechanisms of intercellular communication in eukaryotic cells: implications in health and disease. J Cell Biol 183(7):1187–1191

Bensing BA, Siboo IR, Sullam PM (2001) Proteins PBIA and PBIB of *Streptococcus mitis*, which promote binding to human platelets, are encoded within a lysogenic bacteriophage. Infect Immun 69(10):6186–6192

Bercury KK, Macklin WB (2015) Dynamics and mechanisms of CNS myelination. Dev Cell 32:447–458

Bérdy J (2005) Bioactive microbial metabolites. J Antibiot 58(1):1–26

Berg G, Grube M, Schloter M, Smalla K (2014) Unraveling the plant microbiome: looking back and future perspectives. Front Microbiol 5:148

Bergstrom K, Shan X, Casero D, Batushansky A, Lagishetty V et al (2020) Proximal colon-derived O-glycosylated mucus encapsulates and modulates the microbiota. Science 370:467–472

Berleman JE, Kirby JR (2009) Deciphering the hunting strategy of a bacterial wolfpack. FEMS Microbiol Rev 33:942–957

Berta A, Ekdale EG, Cranford TW (2014) Review of the cetacean nose: form, function, and evolution. Anat Rec 297:2205–2215

Berta A, Lanzetti A, Ekdale EG, Deméré TA (2016) From teeth to baleen and raptorial to bulk filter feeding in mysticete cetaceans: the role of paleontological, genetic, and geochemical data in feeding evolution and ecology. Integr Comp Biol 56(6):1271–1284

Bertrand N, Cathala D, Delahaie S (2007) Les toxines marines sur le littoral français, état des connaissances. ENSP, Rennes

Bhattacharyya RP, Remény A, Yeh B-J, Lim WA (2006) Domains, motifs, and scaffolds: the role of modular interactions in the evolution and wiring of cell signaling circuits. Annu Rev Biochem 75:655–680

Białas A, Langner T, Harant A, Contreras MP, Stevenson CEM et al (2021) Two NLR immune receptors acquired high-affinity binding to a fungal effector through convergent evolution of their integrated domain. eLife 10:e66961

Bidartondo MI, Bruns TD, Weiß M, Sérgio C, Read DJ (2003) Specialized cheating of the ectomycorrhizal symbiosis by an epiparasitic liverwort. Proc Biol Sci 270:835–842

Bidle KD, Falkowski PG (2004) Cell death in planktonic photosynthetic microorganisms. Nat Rev Microbiol 2:643–655

Blackwell M (2011) The fungi: 1, 2, 3 … 5.1 million species? Am J Bot 98(3):426–438

Blanpain C, Fuchs E (2009) Epidermal homeostasis: a balancing act of stem cells in the skin. Nat Rev Mol Cell Biol 10(3):207–217

Blum M, Weber T, Beyer T, Vick P (2009) Evolution of leftward flow. Semin Cell Dev Biol 20:464–471

Bobrovskiy I, Hope JM, Ivantsov A, Nettersheim BJ, Hallmann C, Brocks JJ (2018) Ancient steroids establish the Ediacaran fossil Dickinsonia as one of the earliest animals. Science 361:1246–1249

Boddy AM, McGowen MR, Sherwood CC, Grossman LI, Goodman M, Wildman DE (2012) Comparative analysis of encephalization in mammals reveals relaxed constraints on anthropoid primates and cetacean brain scaling. J Evol Biol 25:981–994

Bode SNS, Adolfson S, Lamatsch DK, Martins MJF, Schmit O et al (2011) Exceptional cryptic diversity and multiple origins of parthenogenesis in a freshwater ostracod. Mol Phylogenet Evol 54:542–552

Böhm J, Scherzer S, Krol E, Kreuzer I, von Meyer K et al (2016) The venus flytrap dionaea muscipula counts prey-induced action potentials to induce sodium uptake. Curr Biol 26:1–10

Bohman B, Phillips RD, Menz MHM, Berntsson BW, Flematti GR et al (2014) Discovery of pyrazines as pollinator sex pheromones and orchid semiochemicals: implications for the evolution of sexual deception. New Phytol 203:939–957

Bok MJ, Capa M, Nilsson D-E (2016) Here, there and everywhere: the radiolar eyes of fan worms (Annelida, Sabellidae). Integr Comp Biol 56(5):784–796

Boles BR, Thoendel M, Singh PK (2004) Self-generated diversity produces "insurance effects" in biofilm communities. PNAS 101(47):16630–16635

Bonett RM, Steffen MA, Lambert SM, Wiens JJ, Chippindale PT (2013) Evolution of paedomorphosis in Plethodontid salamanders: ecological correlates and re-evolution of metamorphosis. Evolution 68(2):466–482

Bonfante P, Anca I-A (2009) Plants, mycorrhizal fungi, and bacteria: a network of interactions. Annu Rev Microbiol 63:363–383

Bonfante P, Genre A (2010) Mechanisms underlying beneficial plant-fungus interactions in mycorrhizal symbiosis. Nat Commun. https://doi.org/10.1038/ncomms1046

Bonner JT (2003) On the origin of differentiation. J Biosci 28(4):523–528

Borsch T, Löhne C, Wiersma J (2008) Phylogeny and evolutionary patterns in Nymphaeales: integrating genes, genomes and morphology. Taxon 57(4):1052–1081

Bowen AD, Davidson FA, Keatch R, Gadd GM (2007) Induction of contour sensing in *Aspergillus niger* by stress and its relevance to fungal growth mechanics and hyphal tip structure. Fungal Genet Biol 44:484–491

Boyer M, Wisniewski-Dyé F (2009) Cell-cell signalling in bacteria: not simply a matter of quorum. FEMS Microbiol Ecol 70:1–19

Brandizzi F, Snapp EL, Roberts AG, Lippincott-Schwartz J, Hawes C (2002) Membrane protein transport between the endoplasmic reticulum and the golgi in tobacco leaves is energy dependent but cytoskeleton independent: evidence from selective photobleaching. Plant Cell 14:1293–1309

Bräucker R, Cogoli A, Hemmersbach R (2002) Gravireception and graviresponse at the cellular level. In: Horneck G, Baumstark-Khan C (eds) Astrobiology. Springer, Berlin, pp 287–295

Breaker RR (2010) Riboswitches and the RNA world. Cold Spring Harb Perspect Biol. https://doi.org/10.1101/cshperspect.a003566

Briegel A, Li X, Bilwes AM, Hughes KT, Jensen GJ, Crane BR (2012) Bacterial chemoreceptor arrays are hexagonally packed trimers of receptor dimmers networked by rings of kinase and coupling proteins. PNAS 109(10):3766–3771

Briggs CJ, Knapp RA, Vredenburg VT (2010) Enzootic and epizootic dynamics of the chytrid fungal pathogen of amphibians. PNAS 107(21):9695–9700

Brini M, Cali T, Ottolini D, Carafoli E (2014) Neuronal calcium signalling: function and dysfunction. Cell Mol Life Sci. https://doi.org/10.1007/s0018-013-1550-7

Brouwer R (1983) Functional equilibrium: sense or nonsense? Neth J Agric Sci 31:335–348

Brown MW, Kolisko M, Silberman JD, Roger AJ (2012) Aggregative multicellularity evolved independently in the eukaryotic supergroup rhizaria. Curr Biol 22:1123–1127

Brundrett MC (2009) Mycorrhizal associations and other means of nutrition of vascular plants: understanding the global diversity of host plants by resolving conflicting information and developing reliable means of diagnosis. Plant Soil 320:37–77

Brundrett MC, Tedersoo L (2018) Evolutionary history of mycorrhizal symbioses and global host plant diversity. New Phytol. https://doi.org/10.1111/nph.14967

Brunet T, Larson BT, Linden TA, Vermeij MJA, McDonald K, King N (2019) Light-regulated collective contractility in a multicellular choanoflagellate. Science 366:326–334

Brüssow H (2009) The not so universal tree of life or the place of viruses in the living world. Philos Trans R Soc Lond B Biol Sci 364:2263–2274

Bryant DM, Mostov KE (2008) From cells to organs: building polarized tissue. Nat Rev Mol Cell Biol 9:887–901

Bryant JP, Provenza FD, Pastor J, Reichardt PB, Clausen TP, du Toit JT (1991) Interactions between woody plants and browsing mammals mediated by secondary metabolites. Annu Rev Ecol 22:431–446

Bubrick P, Galun M, Frensdorff A (1984) Observations on free-living *Trebouxia de puymaly* and *Pseudotrebouxia archibald*, and evidence that both symbionts from *Xanthoria parietina* (L.) Th. Fr. can be found free-living in nature. New Physiol 97:455–462

Bucher M, Hause B, Krajinski F, Kuster H (2014) Through the doors of perception to function in arbuscular mycorrhizal symbioses. New Phytol 204:833–840

Buchholz DR (2015) More similar than you think: frog metamorphosis as a model of human perinatal endocrinology. Dev Biol 408:188–195

Bücking H, Mensah JA, Fellbaum CR (2016) Common mycorrhizal networks and their effect on the bargaining power of the fungal partner in the arbuscular mycorrhizal symbiosis. Commun Integr Biol 9(1):e1107684

Budd GE, Jackson ISC (2016) Ecological innovations in the Cambrian and the origin of the crown group phyla. Philos Trans R Soc Lond B Biol Sci 371:20150287

Burkepile DE, Parker JD, Woodson CB, Mills HJ, Kubanek J et al (2006) Chemically mediated competition between microbes and animals: microbes as consumers in food webs. Ecology 87(11):2821–2831

Burkhardt P, Colgren J, Medhus A, Digel L, Naumann B et al (2023) Syncytial nerve net in a ctenophore adds insights on the evolution of nervous systems. Science 380:293–297

Buschbeck EK, Friedrich M (2008) Evolution of insect eyes: tales of ancient heritage, deconstruction, reconstruction, remodeling, and recycling. Evol Edu Outreach 1:448–462

Büschges A, Borgmann A (2013) Network modularity: back to the future of motor control. Curr Biol 23(20):R936–R938

Büschges A, Scholz H, El Manira A (2011) New moves in motor control. Curr Biol 21:R513–R524

Buskey EJ, Strickler JR, Bradley CJ, Hartline DK, Lenz PH (2017) Escapes in copepods: comparison between myelinate and amyelinate species. J Exp Biol 220:754–758

Cade DE, Carey N, Domenici P, Potvin J, Goldbogen JA (2020) Predator-informed looming stimulus experiments reveal how large filter feeding whales capture highly maneuverable forage fish. PNAS 117(1):472–478

Cai Y, Hua H, Schiffbauer JD, Sun B, Yuan X (2014) Tube growth patterns and microbial mat-related lifestyles in the Ediacaran fossil *Cloudina*, Gaojiashan Lagerstätte, South China. Gondwana Res 25:1008–1018

Caillaud M-C, Dubreuil G, Quentin M, Perfus-Barbeoch L, Lecomte P et al (2008) Root-knot nematodes manipulate plant cell functions during a compatible interaction. J Plant Physiol 165:104–113

Caldwell GS, Pagett HE (2010) Marine glycobiology: current status and future perspectives. Mar Biotechnol. https://doi.org/10.1007/s10126-010-9263-5

Calovi DS, Bardunias P, Carey N, Turner JS, Nagpal R, Werfel J (2019) Surface curvature guides early construction activity in mound-building termites. Philos Trans R Soc Lond B Biol Sci 374:20180374

Calvo AM, Wilson RA, Bok JW, Keller NP (2002) Relationship between secondary metabolism and fungal development. Microbiol Mol Biol Rev 66(3):447–459

Canales J, Henriquez-Velencia C, Brauchi S (2018) The integration of electrical signals originating in the root of vascular plants. Front Plant Sci 8:2173

Cannon SB, McKain MR, Harkess A, Nelson MN, Dash S et al (2014) Multiple polyploidy events in the early radiation of nodulating and nonnodulating legumes. Mol Biol Evol 32(1):193–210

Caplan AI, Correa D (2011) The MSC: an injury drugstore. Cell Stem Cell 9:11–15

Carbonnel S, Gutjahr C (2014) Control of arbuscular mycorrhizal development by nutrient signals. Front Plant Sci 5:462

Carr WES, Ache BW, Gleeson RA (1987) Chemoreceptors of crustaceans: similarities to receptors for neuroactive substances in internal tissues. Environ Health Perspect 71:31–46

Carrell AA, Frank AC (2015) Bacterial endophyte communities in the foliage of coast redwood and giant sequoia. Front Microbiol 6:1008

Carroll SB (2000) Endless forms: the evolution of gene regulation and morphological diversity. Cell 101:577–580

Carthew RW, Sontheimer EJ (2009) Origins and mechanisms of miRNAs and siRNAs. Cell 136:642–655

Castelfranco AM, Hartline DK (2016) Evolution of rapid nerve condition. Brain Res 1641:11–23

Cech TR, Steitz JA (2014) The noncoding RNA revolution-trashing old rules to forge new ones. Cell 157:77–94

Cerdá-Olmedo E (2001) *Phycomyces* and the biology of light and color. FEMS Microbiol Rev 25:503–512

Cereijido M, Contreras RG, Shoshani L (2004) Cell adhesion, polarity, and epithelia in the dawn of metazoans. Physiol Rev 84:1229–1262

Chagnon P-L, Rineau F, Kaiser C (2016) Mycorrhizas across scales: a journey between genomics, global patterns of biodiversity and biogeochemistry. New Phytol 209:913–916

Chao L, Levin BR (1981) Structured habitats and the evolution of anticompetitor toxins in bacteria. PNAS 78(10):6324–6328

Chapin FS III, Schulze E-D, Mooney HA (1990) The ecology and economics of storage in plants. Annu Rev Ecol Syst 21:423–447

Chapman JW, Nilsson C, Lim KS, Bäckman J, Reynolds DR et al (2015) Detection of flow direction in high-flying insect and songbird migrants. Curr Biol 25:R733–R752

Chartier M, Gibernau M, Renner SS (2013) The evolution of pollinator-plant interaction types in the Araceae. Evolution 68(5):1533–1543

Chatterjee S, Bahl E, Mukherjee U, Walsh EN, Shetty MS et al (2022) Endoplasmatic reticulum chaperone genes encodes effectors of long-term memory. Sci Adv 8:eabm6063

Chauveau A, Aucher A, Eissmann P, Vivier E, Davis DM (2010) Membrane nanotubes facilitate long-distance interactions between natural killer cells and target cells. PNAS 107(12):5545–5550

Chen R, Rosen E, Masson PH (1999) Gravitropism in higher plants. Plant Physiol 120:343–350

Chen E, Stiefel KM, Sejnowski TJ, Bullock TH (2008) Model of travelling waves in a coral nerve network. J Comp Physiol A 194:195–200

Chen C, Ma Q, Jiang W, Song T (2011) Phototaxis in the magnetotactic bacterium *Magnetospirillum magneticum* strain AMB-1 is independent of magnetic fields. Appl Microbiol Biotechnol 90:269–275

Chen Y-R, Zhang R, Du H-J, Pan H-M, Zhang W-Y et al (2014) A novel species of ellipsoidal multicellular magnetotactic prokaryotes from Lake Yuehu in China. Environ Microbiol. https://doi.org/10.1111/1462-2920.12480

Cheng TL, Rovito SM, Wake DB, Vredenburg VT (2011) Coincident mass extirpation of neotropical amphibians with the emergence of the infectious fungal pathogen *Batrachochytrium dendrobatidis*. PNAS 108(23):9502–9507

Cheng X, Hui JHL, Lee YY, Law PTW, Kwan HS (2015) A "developmental hourglass" in fungi. Mol Biol Evol 32(6):1556–1566

Chien Y-H, Keller R, Kintner C, Shook DR (2015) Mechanical strain determines the axis of polar polarity in ciliated epithelia. Curr Biol 25:2774–2784

Choe JC, Crespi BJ (1997) Social behavior in insects and arachnids. Cambridge University Press, Cambridge

Choi W-G, Hilleary R, Swanson SJ, Kim S-H, Gilroy S (2016) Rapid, long-distance electrical and calcium signaling in plants. Annu Rev Plant Biol 67:28.1–28.21

Choi W-G, Miller G, Wallace I, Harper J, Mittler R, Gilroy S (2017) Orchestrating rapid long-distance signalling in plants with Ca^{2+}, ROS and electrical signals. Plant J 90:698–707

Chung SH, Rosa C, Scully ED, Peiffer M, Tooker JF et al (2013) Herbivore exploits orally secreted bacteria to suppress plant defenses. PNAS 110(39):15728–15733

Citi S, Guerrea D, Spadaro D, Shah J (2014) Epithelial junctions and Rho family GTPases: the zonular signalosome. Small GTPases 5(4):e973760

Clack JA (2012) Gaining ground—the origin and evolution of tetrapods. Indiana University Press, Bloomington

Claessen D, Rozen DE, Kuipers OP, Søgaard-Andersen L, van Wezel GP (2014) Bacterial solutions to multicellularity: a tale of biofilms, filaments and fruiting bodies. Nat Rev Microbiol 12:115–124

Clark DA, Demb JB (2016) Parallel computations in insect and mammalian visual motion processing. Curr Biol 26:R1062–R1072

Clark DA, Nitra PP, Wang SS-H (2001) Scalable architecture in mammalian brains. Nature 411:189–193

Clarkson E, Levi-Setti R, Horváth G (2006) The eyes of trilobites: the oldest preserved visual system. Arthropod Struct Dev 35:247–259

Coleman-Derr D, Desgarennes D, Fonseca-Garcia C, Gross S, Clingenpeel S et al (2016) Plant compartment and biogeography affect microbiome composition in cultivated and native *Agave* species. New Phytol 209:798–811

Collin M, Schuch R (eds) (2009) Bacterial sensing and signaling. Karger, Basel

Collins WJ, Sitch S, Boucher O (2010) How vegetation impacts affect climate metrics for ozone precursors. J Geophys Res 115. https://doi.org/10.1029/2010JD014187

Colombo M, Raposo G, Théry C (2014) Biogenesis, secretion, and intercellular interactions of exosomes and other extracellular vesicles. Annu Rev Cell Dev Biol 30:255–289

Colston TJ, Jackson CE (2016) Microbiome evolution along divergent branches of the vertebrate tree of life: what is known and unknown. Mol Ecol. https://doi.org/10.1111/mec.13730

Constantinescu AO, O'Reilly JX, Behrens TEJ (2016) Organizing conceptual knowledge in humans with a gridlike code. Science 352:1464–1468

Cordaux R, Bouchon D, Gréve P (2011) The impact of endosymbionts on the evolution of host sex-determination mechanisms. Trends Genet 27(8):332–341

Cornelissen JHC, Lang SI, Soudzilovskaia NA, During HJ (2007) Comparative cryptogam ecology: a review of bryophyte and lichen traits that drive biogeochemistry. Ann Bot 99:987–1001

Cornforth DM, Popat R, McNally L, Gurney J, Scott-Phillips TC et al (2014) Combinatorial quorum sensing allows bacteria to resolve the social and physical environment. PNAS 111(11):4280–4284

Corrochano LM, Kuo A, Marcet-Houben M, Polaino S, Salamov A et al (2016) Expansion of signal transduction pathways in fungi by extensive genome duplication. Curr Biol 26:1577–1584

Cortese B, Palamà IE, D'Amone S, Gigli G (2014) Influence of electrotaxis on cell behaviour. Integr Biol 6(9):817–830

Cosgrove DJ (2005) Growth of the plant cell wall. Nat Rev Mol Cell Biol 6:850–861

Cosgrove DJ (2016) Plant cell wall extensibility: connecting plant cell growth with cell wall structure, mechanics, and the action of wall-modifying enzymes. J Exp Biol 67(2):463–476

Costello JH, Colin SP, Dabiri JO (2008) Medusan morphospace: phylogenetic constraints, biomechanical solutions, and ecological consequences. Invertebr Biol 127(3):265–290

Cotton JA, Page RDM (2002) Going nuclear: gene family evolution and vertebrate phylogeny reconciled. Proc Biol Sci 269:1555–1561

Crampton WGR (2019) Electroreception, electrogenesis and electric signal evolution. Fish Biol 95:92–124

Crandall-Stotler BJ, Bartholomew-Began SE (2007) Morphology of mosses (phylum bryophyta). In: Flora of North America, vol 27. Oxford University Press, pp 3–13

Cridge AG, Dearden PK, Brownfield LR (2016) Convergent occurrence of the developmental hourglass in plant and animal embryogenesis? Ann Bot 117:833–843

Croft MT, Moulin M, Webb ME, Smith AG (2007) Thiamine biosynthesis in alga is regulated by riboswitches. PNAS 104(52):20770–20775

Cronberg N, Natcheva R, Hedlund K (2006) Microarthropods mediate sperm transfer in mosses. Science 313:1255

Cubo J, Legendre P, de Ricqlès A, Montes L, de Margerie E et al (2008) Phylogenetic, functional, and structural components of variation in bone growth rate of amniotes. Evol Dev 10(2):217–227

Cunningham GB, Strauss V, Ryan PG (2008) African penguins (Spheniscus demersus) can detect dimethyl sulphide, a prey-related odour. J Exp Biol 211:3123–3127

Cunningham JA, Liu AG, Bengtson S, Donoghue PCJ (2017) The origin of animals: can molecular clocks and the fossil record be reconciled? Bioessays 39(1):1–12

Curtis EA, Liu DR (2013) Discovery of widespread GTP-binding motifs in genomic DNA and RNA. Chem Biol 20:521–532

Czaczkes TJ, Grüter C, Jones SM, Ratnieks FLW (2011) Synergy between social and private information increases foraging efficiency in ants. Biol Lett 7:521–524

Czech-Damal NU, Dehnhardt G, Manger P, Hanke W (2013) Passive electroreception in aquatic mammals. J Comp Physiol A 199:555–561

Czirók A, Rongish BJ, Little CD (2004) Extracellular matrix dynamics during vertebrate axis formation. Dev Biol 268:111–122

D'Angelo C, Wiedenmann J (2014) Impacts of nutrient enrichment on coral reefs: new perspectives and implications for coastal management and reef survival. Curr Opin Environ Sustain 7:82–93

d'Ovidio F, De Monte S, Alvain S, Dandonneau Y, Lévy M (2010) Fluid dynamical niches of phytoplankton types. PNAS 107(43):18366–18370

da Fonseca RR, Couto A, Machado AM, Brejova BB, Albertin CB et al (2020) A draft genome sequence of the elusive giant squid, Architeuthis dux. Giga Sci 9:1–12

Dacke M, Byrne MJ, Baird E, Scholtz CH, Warrant EJ (2011) How dim is dim? Precision of the celestial compass in moonlight and sunlight. Philos Trans R Soc Lond B Biol Sci 366:697–702

Dacke M, Baird E, Byrne M, Scholtz CH, Warrant EJ (2013) Dung beetles use the milky way for orientation. Curr Biol 23:298–300

Daley WP, Yamada KM (2013) ECM-modulated cellular dynamics as a driving force for tissue morphogenesis. Curr Opin Genet Dev 23(4):408–414

Danchin È, Giraldeau L-A, Cézilly F (eds) (2008) Behavioural ecology. Oxford University Press, Oxford

Dang H, Lovett CR (2016) Microbial surface colonization and biofilm development in marine environments. Microbiol Mol Biol Rev 80(1):91–138

Danger M, Chauvet E (2013) Elemental composition and degree of homeostasis of fungi are aquatic hyphomycetes more like metazoans, bacteria or plants? Fungal Ecol 6(5):453–457

Darby RT, Simmons EG, Wiley BJ (1968) A survey of fungi in a military aircraft fuel supply system. Int Biodeterior Biodegr 4(1):39–41

Darnton NC, Turner L, Rojevsky S, Berg HC (2010) Dynamics of bacterial swarming. Biophys J 98:2082–2090

Das SK, Liang J, Schmidt M, Laffir F, Marsili E (2012) Biomineralization mechanism of gold by zygomycete fungi rhizopous oryzae. Acs Nano 6(7):6165–6173

Datsenko KA, Pougach J, Tikhonov A, Wanner BL, Severinov K, Semenova E (2012) Molecular memory of prior infections activates the CRISPR/Cas adaptive bacterial immunity system. Nat Commun 3:945. https://doi.org/10.1038/ncomms1937

Davey ME, O'Toole G (2000) Microbial biofilms: from ecology to molecular genetics. Microbiol Mol Biol Rev 64(4):847–867

Davey ML, Nybakken L, Hauserud H, Ohlson M (2009) Fungal biomass associated with the phyllosphere of bryophytes and vascular plants. Mycol Res 113:1254–1260

David LA, Alm EJ (2011) Rapid evolutionary innovation during an archaean genetic expansion. Nature 469:93–96

Davies NS, Gibling MR (2010) Cambrian to Devonian evolution of alluvial systems: the sedimentological impact of the earliest land plants. Earth Sci Rev 98:171–200

Davies DG, Parsek MR, Pearson JP, Iglewski BH, Costerton JW, Greenberg EP (1998) The involvement of cell-to-cell signals in the development of a bacterial biofilm. Science 280:295–298

Davies KM, Albert NW, Schwinn KE (2012) From landing lights to mimicry: the molecular regulation of flower colouration and mechanisms for pigmentation patterning. Funct Plant Biol 39(8):619–638

Davis MP, Fielitz C (2010) Estimating divergence times of lizardfishes and their allies (Euteleostei: Aulopiformes) and the timing of deep-sea adaptations. Mol Phylogenet Evol 57:1194–1208

Dayan P, Abbott LF (2001) Theoretical neurobiology. MIT Press, Cambridge

De Beer D, Stoodley P (2006) Microbial biofilms. In: Dworkin M, Falkow S, Rosenberg E, Schleifer K-H, Stackebrandt E (eds) The prokaryotes: Vol. 1: symbiotic associations, biotechnology, applied microbiology. Springer, New York, pp 903–937

de Busserolles F, Marshall NJ (2017) Seeing the deep-sea: visual adaptations in lanternfishes. Philos Trans R Soc Lond B Biol Sci 372:20160070

de Mendoza A, Sebé-Pedrós A, Šestak MS, Matejčić M, Torruella G et al (2013) Transcription factors evolution in eukaryotes and the assembly of the regulatory toolkit in multicellular lineages. PNAS:E4858–E4866

de Nooijer LJ, Spero HJ, Erez J, Bijma J, Reichart GJ (2014) Biomineralization in perforate foraminifera. Earth Sci Rev 135:48–58

De Schepper V, De Swaef T, Bauwaeraerts I, Steppe K (2013) Phloem transport: A review of mechanisms and controls. Exp Bot 64(16):4839–4850

de Wit M, Keuskamp DH, Bongers FJ, Hornitschek P, Gommers CMM et al (2016) Integration of phytochrome and cryptochrome signals determines plant growth during competition for light. Curr Biol 26:3320–3326

Deacon J (2006) Fungal biology. Blackwell, Malden

Dean R, van Kan JAL, Pretorius ZA, Hammond-Kosack KE, di Pietro A et al (2012) The top 10 fungal pathogens in molecular plant pathology. Mol Plant Pathol 13(4):414–430

DeBose JL, Nevitt GA (2007) Investigating the association between pelagic fish and dimethylsulfoniopropionate in a natural coral reef system. Mar Freshw Res 58:720–724

DeBose JL, Nevitt GA (2008) The use of odors at different spatial scales: comparing birds with fish. J Chem Ecol. https://doi.org/10.1007/s10886-008-9493-4

Decelle J, Probert J, Bittner L, Desdevises Y, Colin S et al (2012) An original mode of symbiosis in open ocean plankton. PNAS 109(44):18000–18005

Decho AW, Norman RS, Visscher PT (2010) Quorum sensing in natural environments: emerging views from microbial mats. Trends Microbiol 18(2):73–80

DeFalco TA, Zipfel C (2021) Molecular mechanisms of early plant pattern-triggered immune signaling. Mol Cell 81:3449–3467

Degnan SM, Degnan BM (2010) The initiation of metamorphosis as an ancient polyphonic trait and its role in metazoan life-cycle evolution. Philos Trans R Soc Lond B Biol Sci 365:641–651

Delwiche CF, Goodman CA, Chang C (2017) Land plant model systems branch out. Cell 171:265–266

Dener E, Kacelnik A, Shemesh H (2016) Pea plants show risk sensitivity. Curr Biol 26:1763–1767

Denffer DV, Schumacher W, Mägdefrau K, Ehrendorfer F (eds) (1971) Lehrbuch der Botanik. Fischer, Stuttgart

Denny M (2016) Ecological mechanics. Princeton University Press, Princeton

Des Marais D (2003) Biogeochemistry of hypersaline microbial mats illustrates the dynamics of modern microbial ecosystems and early evolution of the biosphere. Biol Bull 204:160–167

Deshibi MM, Adamatzk YA (2020) Electrical activity of fungi: Spikes detection and complexity analysis. arXiv:2008.10276v1

Després L, David J-P, Gallet C (2007) The evolutionary ecology of insect resistance to plant chemicals. Trends Ecol Evol 22(6):298–307

Devaraj A, Justice SS, Bakaletz LO, Goodman SD (2015) DNABII proteins play a central role in UPEC biofilm structure. Mol Microbiol 96(6):1119–1135

Devireddy S, Liu A, Lampe T, Hollenbeck PJ (2015) The organization of mitochondrial quality control and life cycle in the nervous system in vivo in the absence of PINK1. J Neurosci 35(25):9391–9401

Diaz-Espejo A, Hernandez-Santana V (2017) The phloem-xylem consortium: until death do them apart. Tree Physiol 37:847–850

Dicke U, Roth G (2016) Neuronal factors determining high intelligence. Philos Trans R Soc Lond B Biol Sci 371:20150180

Dicke M, van Loon JJA, Soler R (2009) Chemical complexity of volatiles from plants induced by multiple attack. Nat Chem Biol 5(5):317–324

Dickinson DJ, Nelson WJ, Weis WI (2012) An epithelial tissue in *Dictyostelium* challenges the traditional origin of metazoan multicellularity. Bioessays 34:833–840

Diego-Rasilla FJ, Luengo RM (2002) Celestial orientation in the marbled newt (*Triturus marmoratus*). J Ethol 20:137–141

Diggle SP, Griffin AS, Campbell GS, West SA (2007) Cooperation and conflict in quorum-sensing bacterial population. Nature 450:411–414

Dilanji GE, Langebrake JB, De Leenheer P, Hagen SJ (2012) Quorum activation at a distance: spatiotemporal patterns of gene regulation from diffusion of an autoinducer signal. J Am Chem Soc 134:5618–5626

Dimmock NJ, Easton AJ, Leppard KN (2007) Introduction to modern virology. Blackwell Publishing, Malden

DiNuzzo M, Mangia S, Maraviglia B, Giove F (2012) The role of astrocytic glycogen in supporting the energetics of neuronal activity. Neurochem Res 37(11):2432–2438

Dion P, Nautiyal CS (eds) (2008) Microbiology in extreme soils. Springer, Berlin

Dityatev A, Rusakov DA (2011) Molecular signals of plasticity at the tetrapartite synapse. Curr Opin Neurobiol 21(2):353–359

Divakar PK, Crespo A, Wedin M, Leavitt SD, Hawksworth DL et al (2015) Evolution of complex symbiotic relationships in a morphologically derived family of lichen-forming fungi. New Phytol 208:1217–1226

Dodd AN, Kudla J, Sanders D (2010) The language of calcium signaling. Annu Rev Plant Biol 61:593–620

Dohrmann M, Kelley C, Kelly M, Pisera A, Hooper JNA, Reiswig HM (2017) An integrative systematic framework helps to reconstruct skeletal evolution of glass sponges (Porifera, Hexactinellida). Front Zool 14:18

Domazet-Lošo T, Tautz D (2010) A phylogenetically based transcriptome age index mirrors ontogenetic divergence patterns. Nature 468:815–818

Domingo E, Sheldon J, Perales C (2012) Viral quasispecies evolution. Microbiol Mol Biol Rev 76(2):159–216

Donoghue PCJ, Antcliffe JB (2010) Origins of multicellularity. Nature 466:41–42

Downie JA (2014) Legume nodulation. Curr Biol 24(5):R184–R190

Drake H, Ivarsson M, Bengtson S, Heim C, Siljeström S et al (2017) Anaerobic consortia of fungi and sulphate reducing bacteria in deep granite fractures. Nat Commun. https://doi.org/10.1038/s41467-017-00094-6

Drescher K, Shen Y, Bassler BL, Stone HA (2013) Biofilm streamers cause catastrophic disruption of flow with consequences for environmental and medical systems. PNAS 110(11):4345–4350

Drescher K, Dunkel J, Nadell CD, van Teeffelen S, Grnja I et al (2016) Architectural transitions in *Vibrio cholerae* biofilms at single-cell resolution. PNAS:E2066–E2072

Drewniak L, Krawczyk PS, Mielnicki S, Adamska D, Sobczak A et al (2016) Physiological and metagenomic analyses of microbial mats involved in self-purification of mine waters contaminated with heavy metals. Front Microbiol 7:1252

Dridi B, Henry M, Khéchine AE, Raoult F, Drancourt M (2006) High Prevalence of *Methanobrevibacter smithii* and *Methanospaera stadtmanae* detected in the human gut using an improved DNA detection protocol. PLoS One 4(9):e7063

Dridi B, Raoult F, Drancourt M (2011) Archaea as emerging organisms in complex human microbiomes. Anaerobe 17:56–63

Drost H-G, Gabel A, Grosse I, Quint M (2015) Evidence for active maintenance of phylotranscriptomic hourglass patterns in animal and plant embryogenesis. Mol Biol Evol 32(5):1221–1231

Drost H-G, Bellstädt J, Ó'Maolédigh DS, Silva AT, Gabel A et al (2016) Post-embryonic hourglass patterns mark ontogenetic transitions in plant development. Mol Biol Evol 33(5):1158–1163

Drummond IA (2012) Cilia functions in development. Curr Opin Cell Biol 24(1):24–30

Du Q, Kawabe Y, Schilde C, Chen Z-H, Schaap P (2015) The evolution of aggregative multicellularity and cell-cell communication in dictyostelia. J Mol Biol 427:3722–3733

Duanmu D, Casero D, Dent RM, Gallaher S, Yang W et al (2013) Retrograde bilin signaling enables *Chlamydomonas* greening and phototrophic survival. PNAS 110(9):3621–3626

Dubey GP, Ban-Yehuda S (2011) Intercellular nanotubes mediate bacterial communication. Cell 144:590–600

Dugas-Ford J, Rowell JJ, Ragsdale CW (2012) Cell-type homologies and the origins of the neocortex. PNAS 109(42):16974–16979

Dunn CW, Giribet G, Edgecombe GD, Hejnol A (2014) Animal phylogeny and its evolutionary implications. Annu Rev Ecol Evol Syst 45:371–395

Durand E, Méheust R, Soucaze M, Goubet PM, Gallina S et al (2014) Dominance hierarchy arising from the evolution of a complex small RNA regulatory network. Science 346:1200–1208

Durdu S, Iskar M, Revenu C, Schieber N, Kunze A et al (2014) Luminal signalling links cell communication to tissue architecture during organogenesis. Nature 515:120–124

Dussutour A, Latty T, Beekman M, Simpson SJ (2010) Amoeboid organism solves complex nutritional challenges. PNAS 107(10):4606–4611

Dutta DJ, Woo DH, Lee PR, Pajevic S, Bukalo O et al (2018) Regulation of myelin structure and conduction velocity by perinodal astrocytes. PNAS 115(48):11932–11837

Duxbury Z, Ma Y, Furzer OJ, Huh SU, Cevik V et al (2016) Pathogen perception by NLRs in plants and animals: parallel worlds. Bioessays 38:769–781

Dwidar M, Monnappa AK, Mitchell RJ (2011) The dual probiotic and antibiotic nature of Bdellovibrio bacteriovorus. BMB Rep 45:71–78

Eaton CJ, Cox MP, Scott B (2011) What triggers grass endophytes to switch from mutation to pathogenism? Plant Sci 180:190–198

Eberhard WG (2007) Miniaturized orb-weaving spiders: behavioural precision is not limited by small size. Proc Biol Sci 274:2203–2209

Eberhard WG, Weislo WT (2011) Grade changes in brain-body allometry: morphological and behavioural correlates of brain size in miniature spiders, insects and other invertebrates. In: Casas J (ed) Advances in insect physiology, vol 60. Academic Press, Burlington, pp 155–214

Ebine K, Inoue T, Ito J, Ito E, Uemura T et al (2014) Plant vacuolar trafficking occurs through distinctly regulated pathways. Curr Biol 24:1375–1382

Eckstein RL, Karlsson PS, Weih M (1999) Leaf life span and nutrient resorption as determinants of plant nutrient conservation in temperate-arctic regions. New Phytol 143:177–189

Edel KH, Marchadler E, Brownlee C, Kudla J, Hetherington AM (2017) The evolution of calcium-based signalling in plants. Curr Biol 27:R667–R679

Eggenschwiler JT, Anderson KV (2007) Cilia and developmental signalling. Annu Rev Cell Dev Biol 23:345–373

Eibl-Eibesfeldt I (2004) Grundriss der vergleichenden Verhaltensforschung. BuchVertrieb Blank, Vierkirchen-Pasenbach

Eisenach C, Francisco R, Martinoia E (2015) Plant vacuoles. Curr Biol 25(4):136–137

Eitel M, Francis WR, Varoqueaux F, Daraspe J, Osigus H-J et al (2018) Comparative genomics and the nature of placozoan species. PLoS Biol 18(7):e2005359

el Jundi B, Warrant EJ, Byrne MJ, Khaldy L, Baird E et al (2015) Neural coding underlying the cue preference for celestial orientation. PNAS 112(36):11395–11400

el Jundi B, Foster JJ, Khaldy L, Byrne MJ, Dacke M, Baird E (2016) A snapshot-based mechanism for celestial orientation. Curr Biol 26:1456–1462

Elias PM, Williams ML (2016) Basis for the gain and subsequent dilution of epidermal pigmentation during human evolution: the barrier and metabolic conservation hypotheses revisited. Am J Phys Anthropol 161(2):189–207

Elias LAB, Wang DD, Kriegstein AR (2007) Gap junction adhesion is necessary for radial migration in the neocortex. Nature 448:901–907

Elliott GRD, Leys SP (2007) Coordinated concentrations effectively expel water from the aquiferous system of a freshwater sponge. J Exp Biol 210:3736–3748

Ellis AG, Johnson SD (2010) Floral mimicry enhances pollen export: the evolution of pollination by sexual deceit outside the orchidaceae. Am Nat 176(5):E143–E151

Ellis J, Catanzariti A-M, Dodds P (2006) The problem of how fungal and oomycete avirulence proteins enter plant cells. Trends Plant Sci 11(2):61–63

Ellison AM, Gotelli NJ (2001) Evolutionary ecology of carnivorous plants. Trends Ecol Evol 16(11):623–629

Emery NJ, Clayton NS (2005) Evolution of the avian brain and intelligence. Curr Biol 15(23):R946–R950

Emlet RB (1991) Functional constraints on the evolution of larval forms of marine invertebrates: experimental and comparative evidence. Am Zool 31:707–725

Engel S, Jensen PR, Fenical W (2002) Chemical ecology of marine microbial defense. J Chem Ecol 28(10):1971–1985

Engel MS, Barden P, Riccio ML, Grimaldi DA (2016) Morphologically specialized termite castes and advanced sociality in the early cretaceous. Curr Biol 26:522–530

Engelberg-Kulka H, Amital S, Kolodkin-Gal I, Hazan R (2006) Bacterial programmed cell death and multicellular behavior in bacteria. PLoS Genet 2(10):e135

England SL, Maus S, Immel TJ, Mende SB (2006) Longitudinal variation of the E-region electric fields cause by atmospheric tides. Geophys Res Lett 33:L21105

Engler AJ, Sen S, Sweeney HL, Discher DE (2006) Matrix elasticity directs stem cell lineage specification. Cell 126:677–689

Enomoto G, Win N-N, Narikawa R, Ikeuchi M (2015) Three cyanobacteriochromes work together to form a light color-sensitive input system for c-di-GMP signaling of cell aggregation. PNAS 112(26):8082–8087

Erb M, Reymond P (2019) Molecular interactions between plants and insect herbivores. Annu Rev Plant Biol 70:527–557

Eriksson M, Nylin S, Carlsson MA (2019) Insect brain plasticity: effects of olfactory input on neuropil size. R Soc Open Sci 6:190875

Erwin DH (2009) Early origin of the bilaterian development toolkit. Philos Trans R Soc Lond B Biol Sci 364:2253–2261

Erwin DH, Laflamme M, Tweedt SM, Sperling EA, Pisani D, Peterson KJ (2011) The Cambrian Conundrum: early divergence and later ecological success in the early history of animals. Science 334:1091–1097

Espinosa-Medina T, Saha O, Boismoreau F, Chettouh Z, Rossi F et al (2016) The sacral autonomic outflow is sympathetic. Science 354:893–897

Esser D, Hoffmann L, Pham TK, Bräsen C, Qiu W et al (2016) Protein phosphorylation and its role in archaeal signal transduction. FEMS Microbiol Rev 40:625–647

Exposito J-Y, Valcourt U, Cluzel C, Lethias C (2010) The fibrillar collagen family. Int J Mol Sci 11:407–426

Extavour CG, Akam M (2003) Mechanisms of germ cell specification across the metazoans: epigenesist and preformation. Development 130:5869–5864

Fairclough SR, Chen Z, Kramer E, Zeng Q, Young S et al (2013) Premetazoan genome evolution and the regulation of cell differentiation in the choanoflagellate *Salpingoeca rosetta*. Genome Biol 14(2):R15

Falini G, Fermani S, Goffredo S (2015) Coral biomineralization: a focus on intra-skeletal organic matrix and calcification. Semin Cell Dev Biol 46:17–28

Falk D, Hildebolt C, Smith K, Morwood MJ, Sutkina T et al (2009) LB1's virtual endocast, microcephaly, and hominin brain evolution. J Hum Evol 57(5):597–607

Falk MM, Kells RM, Berthoud VM (2014) Degradation of connexions and gap junctions. FEBS Lett 588:1221–1229

Falkowski PG, Fenchel T, Delong EF (2008) The microbial engines that drive earth's biogeochemical cycles. Science 320:1034–1039

Farmer AD, Randall HA, Aziz Q (2014) It's a gut feeling: how the gut microbiota affects the state of mind. J Physiol 592(14):2981–2988

Farnsworth KD, Nelson J, Gershenson C (2012) Living is information processing: from molecules to global systems. arXiv:1210.5908v1

Farré-Armengol G, Filella I, Llusia J, Peñuelas J (2013) Floral volatile organic compounds: between attraction and deterrence of visitors under global change. Perspect Plant Ecol Evol Syst 15:56–67

Farris SM (2015) Evolution of brain elaboration. Philos Trans R Soc Lond B Biol Sci 370:20150054

Farris SM, Roberts NS (2005) Coevolution of generalist feeding ecologies and gyrencephalic mushroom bodies in insects. PNAS 102(48):17394–17399

Farris SM, Schulmeister S (2011) Parasitoidism, not sociality, is associated with the evolution of elaborate mushroom bodies in the brains of hymenopteran insects. Proc Biol Sci 278:940–951

Fatica A, Bozzoni I (2014) Long non-coding RNAs: new players in cell differentiation and development. Nat Rev Genet 16:7–21

Feild TS, Brodribb TJ (2005) A unique mode of parasitism in the conifer coral tree *Parasitaxus ustus* (Podocarpaceae). Plant Cell Environ 28:1316–1325

Fels D (2009) Cellular communication through light. PLoS One 4(4):e5086

Feng P, Zheng J, Rositter SJ, Wang D, Zhan H (2014) Massive losses of taste receptor genes in toothed and baleen whales. Genome Biol Evol 6(6):1254–1265

Feng L, Rutherford ST, Papenfort K, Bagert JD, van Kessel JC et al (2015) A Qrr noncoding RNA deploys four different regulatory mechanisms to optimize quorum-sensing dynamics. Cell 160:228–240

Fernald RD (2006) Casting a genetic ligth on the evolution of eyes. Science 313:1914–1918

Fernandez V, Gil-Peregrin E, Eichert T (2020) Foliar water and solute absorption: an update. Plant J. https://doi.org/10.1111/tpj.15090

Ferreira RR, Vermot J (2017) The balancing roles of mechanical forces during left-right patterning and asymmetric morphogenesis. Mech Dev 144:71–80

Fiegna F, Yu Y-TN, Kadam SV, Velicer GJ (2006) Evolution of an obligate social cheater to a superior cooperator. Nature 441:310–314

Finarelli JA, Flynn JJ (2009) Brain-size evolution and sociality in Carnivora. PNAS 106(23):9345–9349

Findley K, Oh J, Yang J, Conlan S, Deming C et al (2013) Topographic diversity of fungal and bacterial communities in human skin. Nature 498:367–370

Finlay RD (2004) Mycorrhizal fungi and their multifunctional roles. Mycologist 18(2):91–96

Finn JK, Tregenza T, Norman MD (2009) Defensive tool use in a coconut-carrying octopus. Curr Biol 19(23):R1069–R1070

Fiore VG, Dolan RJ, Strausfeld NJ, Hirth F (2015) Evolutionarily conserved mechanisms for the selection and maintenance of behavioural activity. Philos Trans R Soc Lond B Biol Sci 370:20150053

Fiore-Donno AM, Nikolaev SI, Nelson M, Pawlowski J, Cavalier-Smith T, Baldauf SL (2010) Deep phylogeny and evolution of slime moulds (mycetozoa). Protist 161:55–70

Fischbach MA, Segre JA (2016) Signaling in host-associated microbial communities. Cell 164:1288–1300

Fischer MWF, Stolze-Rybczynski JL, Cui Y, Money NP (2010) How far and how fast can mushroom spores fly? Physical limits on ballistospore size and discharge distance in the Basidiomycota. Fungal Biol 114(8):669–675

Fish JL, Lockwood CA (2003) Dietary constraints on encephalization in primates. Am J Phys Anthropol 120:171–181

Fisher RM, Cornwallis CK, West SA (2013) Group formation, relatedness, and the evolution of multicellularity. Curr Biol 23:1120–1125

Fisterol GO, Legrand C, Granéli E (2003) Allelopathic effect of *Prymnesium parvum* on a natural plankton community. Mar Ecol Prog Ser 255:115–123

Flajnik MF (2014) Re-evaluation of the immunological big bang. Curr Biol 24:R1060–R1065

Flajnik MF, Kasahara M (2010) Origin and evolution of the adaptive immune system: genetic events and selective pressure. Nat Rev Genet 11(1):47–59

Flegr J (2013) How and why *Toxoplasma* makes us crazy. Trends Parasitol 29(4):156–163

Fleming LE, Kirkpatrick B, Backer LC, Walsh CJ, Nierenberg K et al (2011) Review of Florida red tides and human health effects. Harmful Algae 10(2):224–233

Flemming H-C, Wingeneder J (2010) The biofilm matrix. Nat Rev Microbiol 8:623–633

Flemming H-C, Wingeneder J, Szewzyk U, Steinberg P, Rice SA, Kjelleberg S (2016) Biofilms: an emergent form of bacterial life. Nat Rev Microbiol 14:563–575

Flores E, Herrero A (2010) Compartmentalized function through cell differentiation in filamentous cyanobacteria. Nat Rev Microbiol 8:39–50

Flowers JM, Li SI, Stathos A, Saxer G, Ostrowski EA et al (2010) Variation, sex, and social cooperation: molecular population genetics of the social amoeba *Dictyostelium discoideum*. PLoS Genet 6(7):e1001013

Folt CL, Burns CW (1999) Biological drivers of zooplankton patchiness. Tree 14(8):300–305

Foretich MA, Paris CB, Grosell M, Stieglitz JD, Benetti DD (2017) Dimethyl sulfide is a chemical attractant for reef fish larvae. Sci Rep. https://doi.org/10.1030/s41598-017-02675-3

Formery L, Peluso P, Kohnle I, Malnick J, Thompson JR et al (2023) Molecular evidence of anterio-posterior patterning in adult echinoderms. Nature 623:555–561

Fossette S, Gleiss AC, Chatumeau J, Bastian T, Armstrong CD et al (2015) Current-oriented swimmung by jellyfish and its role in bloom maintenance. Curr Biol 25:342–347

Foster KR, Ratnieks FLW (2001) Convergent evolution of worker policing by egg eating in the honeybee and common wasp. Proc Biol Sci 268:169–174

Frago E, Dicke M, Godfray HCJ (2012) Insect symbionts as hidden players in insect-plant interactions. Trends Ecol Evol 27(12):705–711

Franke K, Berens P, Schubert T, Bethge M, Euler T, Baden T (2017) Inhibition decorrelates visual feature representations in the inner retina. Nature 542:439–444

Frankel RB, Bazylinski DA, Johnson MS, Taylor BL (1997) Magneto-aerotaxis in marine coccoid bacteria. Biophys J 73:994–1000

Franklin KA, Whitelam GC (2004) Light signals, phytochromes and cross-talk with other environmental cues. J Exp Bot 55(395):271–276

Freedman BR, Bade ND, Riggin CN, Zhang S, Haines PG et al (2015) The (dys)functional extracellular matrix. Biochim Biophys Acta 1853:3153–3164

Freund ET (2007a) Pre-earthquake signals—Part I: deviatoric stresses turn rocks into a source of electric currents. Nat Hazards Earth Syst Sci 7:535–541

Freund ET (2007b) Pre-earthquake signals—Part II: flow of battery currents in the crust. Nat Hazards Earth Syst Sci 7:543–548

Frey-Klett P, Burlinson P, Deveau A, Barret M, Tarkka M, Sarniguet A (2011) Bacterial-fungal interactions: hyphens between agricultural, clinical, environmental, and food microbiologists. Microbiol Mol Biol Rev 75(4):583–609

Frick WF, Pollock JF, Hicks AC, Langwig KE, Reynolds DS et al (2010) An emerging disease causes regional population collapse of a common North American bat species. Science 329:679–692

Fromm J, Hajirezael M-R, Becker VK, Lautner S (2013) Electrical signalling along the phloem and its physiological responses in the maize leaf. Front Plant Sci 4:239

Fuller KK, Loros JJ, Dunlap JC (2015) Fungal photobiology: visible light as a signal for stress, space and time. Curr Genet 61(3):275–288

Funk WC, Caminer M, Ron SR (2012) High levels of cryptic species diversity uncovered in Amazonian frogs. Proc Biol Sci 279:1806–1814

Furuhashi T, Schwarzinger C, Miksik I, Smrz M, Beran A (2009) Molluscan shell evolution with review of shell calcification hypothesis. Comp Biochem Physiol B Biochem Mol Biol 154:351–371

Gabbott SE, Donoghue PCJ, Sansom RS, Vinther J, Dolocan A, Purnell MA (2016) Pigmented anatomy in Carboniferous cyclostomes and the evolution of the vertebrate eye. Proc Biol Sci 283:20161151

Gadkar V, David-Schwartz R, Kunik T, Kapulnik Y (2001) Arbuscular mycorrhizal fungal colonization. Factors involved in host recognition. Plant Physiol 127:1493–1499

Galanti G (2009) Percezione e riconoscimento di fattori orientanti astronomici e celesti in *Talitrus saltator*. Dissertation University of Firenze

Galis F, Wagner GP, Jockusch EL (2003) Why is limb regeneration possible in amphibians but not in reptiles, birds, and mammals? Evol Dev 5(2):208–220

Gallé A, Lautner S, Flexas J, Fromm J (2015) Environmental stimuli and physiological responses: the current view on electrical signalling. Environ Exp Bot 134:15–21

Galperin MY (2005) A census of membrane-bound and intracellular signal transduction proteins in bacteria: bacterial IQ, extroverts and introverts. BMC Microbiol 5:35

Ganot P, Zoccola D, Tambutté E, Voolstra CR, Aranda M et al (2014) Structural molecular components of septate junctions in cnidarians point to the origin of epithelial junctions in eukaryotes. Mol Biol Evol 32(1):44–62

Garm A, Mori S (2009) Multiple photoreceptor systems control the swim pacemaker activity in box jellyfish. J Exp Biol 212:3951–3960

Garm A, Nilsson D-E (2014) Visual navigation in starfish: first evidence for the use of vision and eyes in starfish. Proc Biol Sci 281:20133011

Garm A, Ekström P, Boudes M, Nilsson D-E (2006) Rhopalia are integrated parts of the central nervous system in box jellyfish. Cell Tissue Res 325:333–343

Garm A, Bielecki J, Petie R, Nilsson D-E (2016) Hunting in bioluminescent light: vision in the nocturnal box jellyfish *Copula sivickisi*. Front Physiol 7:99

Gärtner W, Losi A (2003) Crossing the borders: archaeal rhodopsins go bacterial. Trends Microbiol 11(9):405–407

Gaston GR, Hall J (2000) Lunar periodicity and bioluminescence of swarming *Odontosyllis luminosa* (Polychaeta: Syllidae) in Belize. Gulf Carrib Res 12:47–51

Gatesy J, Geisler JH, Chang J, Buell C, Berta A et al (2013) A phylogenetic blueprint for a modern whale. Mol Phylogenet Evol 66(2):479–506

Gattazzo F, Urciuolo A, Bonaldo P (2014) Extracellular matrix: a dynamic microenvironment for stem cell niche. Biochim Biophys Acta 1840:2506–2519

Gauthier G, Klein BS (2008) Insights into fungal morphogenesis and immune evasion. Microbe Wash DC 3(9):416–423

Gavelis GS (2015) Evolution of complex organelles in dinoflagellates. Dissertations, University of British Columbia

Gavelis GS, Hayakawa S, White RA III, Gojobori T, Suttle CA et al (2015) Eye-like ocelloids are built from different endosymbiotically acquired components. Nature 523:204–207

Gebauer G, Preiss K, Gebauer AC (2016) Partial mycoheterotrophy is more widespread among orchids than previously assumed. New Phytol 211(1):11–15

Gehring CA, Mueller RC, Whitham TG (2006) Environmental and genetic effects on the formation of ectomycorrhizal and arbuscular mycorrhizal associations in cottonwoods. Oecologia 149:158–164

Geilfus C-M (2017) The pH of the apoplast: dynamic factor with functional impact under stress. Mol Plant 10:1371–1388

Geitmann A (2006) Plant and fungi cytomechanics: quantifying and modelling cellular architecture. Can J Bot 84(4):581–593

Geitmann A, Ortega JKE (2009) Mechanics and modeling of plant cell growth. Trends Plant Sci 14(9):467–478

Geldner N (2013) Casparian strips. Curr Biol 23(23):1025–1026

Gerdes H-H, Rustom A, Wang X (2013) Tunneling nanotubes, an emerging intercellular communication route in development. Mech Dev 130:381–387

Gerhart J (1999) 1998 Warkany lecture signaling pathways in development. Teratology 60:226–239

Ghosh Z, Mallick B, Chakrabarti J (2009) Cellular versus viral microRNAs in host-virus interaction. Nucleic Acid Res 37(4):1035–1045

Ghoul M, Griffin AS, West SA (2013) Toward and evolutionary definition of cheating. Evolution 68(2):318–331

Gianoli E, Carrasco-Urra F (2014) Leaf mimicry in a climbing plant protects against herbivory. Curr Biol 24:984–987

Gibling MR, Davies NS (2012) Palaeozoic landscapes shaped by plant evolution. Nat Geosci 5:99–105

Gibling MR, Davies NS, Falcon-Lang HJ, Bashforth AR, DiMichele WA et al (2014) Paleozoic co-evolution of rivers and vegetation: a synthesis of current knowledge. Proc Geol Assoc 125:524–533

Gilbert PUPA, Bergmann KD, Boekelheide N, Tambutté S, Mass T et al (2022) Biomineralization: integrating mechanism and evolutionary history. Sci Adv 8:eabl9653

Gilmour D, Rembold M, Leptin M (2017) From morphogen to morphogenesis and back. Nature 541:311–320

Gindin G, Samish M, Zangi G, Mishoutchenko A, Glazer I (2002) The susceptibility of different species and stages of ticks to entomopathogenic fungi. Exp Appl Acarol 28:283–288

Glass NL, Rasmussen C, Roca MG, Read ND (2004) Hyphal homing, fusion and mycelial interconnectedness. Trends Microbiol 12(3):135–141

Glibert PM, Burkholder JM (2006) The complex relationships between increases in fertilization of the earth, coastal eutrophication and proliferation of harmful algal blooms. In: Granéli E, Turner JT (eds) Ecology of harmful algae. Springer, Berlin, pp 341–354

Glibert PM, Anderson DM, Gentien P, Granéli E, Sellner KG (2005) The global, complex phenomena of harmful algal blooms. Oceanography 18(2):136–147

Gochfeld DJ (2004) Predation-induced morphological and behavioural defenses in a hard coral: implications for foraging behaviour of coral-feeding butterflyfishes. Mar Ecol Prog Ser 267:145–156

Goetz SC, Anderson KV (2010) The primary cilium: a signaling center during vertebrate development. Nat Rev Genet 11(5):331–344

Gold DA, Katsuki T, Li Y, Yan X, Regulski M et al (2019) The genome of the jellyfish Aurelia and the evolution of animal complexity. Nat Ecol Evol 3:96–104

Gonzalez P, Uhlinger KR, Lowe CJ (2017) The adult body plan of indirect developing hemichordates develops by adding a hox-patterned trunk to an anterior larval territory. Curr Biol 27:87–95

González-Teuber M, Heil M (2015) Comparative anatomy and physiology of myrmecophytes: ecological and evolutionary perspectives. Res Rep Biodivers Stud 4:21–32

Goodnight CJ (2015) Multilevel selection theory and evidence: a critique of Gardner, 2015. J Evol Biol 28:1734–1746

Görgen S, Benzerara K, Skouri-Panet F, Gugger M, Chauvat F, Cassier-Chauvat C (2021) The diversity of molecular mechanisms of carbonate biomineralization of bacteria. Discov Mater 1:2

Gorska A, Ye Q, Holbrook M, Zwieniecki MA (2008) Nitrate control of root hydraulic properties in plants: translating local information to whole plant response. Plant Phys 148:1159–1167

Gorzelak P, Stolarski J, Dubois P, Kopp C, Meibom A (2011) ^{26}Mg labelling of the sea urchin regenerating spine: insights into echinoderm biomineralization process. J Struct Biol 176:119–126

Gow NAR (1984) Transhyphal electrical currents in fungi. J Gen Microbiol 130:3313–3318

Graham NAJ, Nash KL (2013) The importance of structural complexity in coral reef ecosystems. Coral Reefs 32:315–326

Grasso LC, Negri AP, Fôret S, Saint R, Hayward DC et al (2011) The biology of coral metamorphosis: molecular responses of larvae to inducers of settlement and metamorphosis. Dev Biol 353:411–419

Grillner S, Markram H, De Schutter E, Silberberg G, LeBeau FEN (2005) Microcircuits in action—from CPGs to neocortex. Trends Neurosci 28(10):525–533

Grillner S, Wallén P, Saitoh K, Kozlov A, Robertson B (2008) Neural bases of goal-directed locomotion in vertebrates—an overview. Brain Res Rev 57:2–12

Grimaldi D, Engel MS (2005) Evolution of the insects. Cambridge University Press, Cambridge

Grootjans AP, van den Ende FP, Walsweer AF (1997) The role of microbial mats during primary succession in calcareous dune slacks: an experimental approach. J Coast Conserv 3:95–102

Grosberg RK, Strathmann RR (1998) One cell, two cell, red cell, blue cell: the persistence of a unicellular stage in multicellular life histories. Trends Ecol Evol 13(3):112–116

Grosberg RK, Strathmann RR (2007) The evolution of multicellularity: a minor major transition? Annu Rev Ecol Evol Syst 38:621–654

Grossman JD, Rice JJ (2012) Evolution of root plasticity responses to variation in soil nutrient distribution and concentration. Evol Appl 5:850–857

Grube M, Cernava T, Soh J, Fuchs S, Aschenbrenner I et al (2015) Exploring functional contexts of symbiotic sustain within lichen-associated bacteria by comparative omics. ISME J 9:412–424

Guirao B, Meunier A, Mortaud S, Aguilar A, Corsi J-M et al (2010) Coupling between hydrodynamic forces and planar cell polarity orients mammalian motile cilia. Nat Cell Biol 12(4):341–350

Gundermann KO, Rüden H, Sonntag H-G (eds) (1991) Lehrbuch der Umwelthygiene. Fischer, Stuttgart

Guo X, Wang X-F (2009) Signaling cross-talk between TGF-β/BMP and other pathways. Cell Res 19:71–88

Gurke S, Barroso JFV, Gerdes H-H (2008) The art of cellular communication: tunneling nanotubes bridge the divide. Histochem Cell Biol 129:539–550

Gussone N, Langer G, Thoms S, Nehrke G, Eisenbauer A et al (2006) Cellular calcium pathways and isotope fractionation in *Emiliana huxleyi*. Geology 34(8):625–628

Gutjahr C, Parniske M (2017) Cell biology: control of partner lifetime in a plant–fungus relationship. Curr Biol 27:R420–R423

Gutmann WF (1981) Relationships between invertebrate phyla based on functional-mechanical analysis of the hydrostatic skeleton. Am Zool 21:63–81

Haber M, Carbone M, Mollo R, Gavagnin M, Ilan M (2011) Chemical defense against predators and bacterial fouling in the Mediterranean sponges *Axinella polyploides* and *A. verrucosa*. Mar Ecol Prog Ser 422:113–122

Haeckel E (2014) Kunstformen der Natur - Kunstformen aus dem Meer. Neudrucke der Originalausgaben. Prestel Verlag, Munich

Hahn M, Mendgen K (2001) Signal and nutrient exchange at biotrophic plant-fungus interfaces. Curr Opin Plant Biol 4:322–327

Haidt J (2008) The emotional dog and its rational tail: a social intuitionist approach to moral judgement. In: Adler JE, Rips LJ (eds) Reasoning: studies of human inference and its foundations. Cambridge University Press, Cambridge

Halbwachs H, Simmel J, Bässler C (2016) Tales and mysteries of fungal fruiting: how morphological and physiological traits affect a pileate lifestyle. Fungal Biol Rev 30:36–61

Hall-Stoodley L, Costerton JW, Stoodley P (2004) Bacterial biofilms: from the natural environment to infectious diseases. Nat Rev Microbiol 2:95–108

Ham B-K, Lucas WJ (2017) Phloem-mobile RNAs as systemic signaling agents. Annu Rev Plant Biol 68:173–195

Hamann E, Gruber-Vodicka H, Kleiner M, Tegetmeyer HE, Riedel D et al (2016) Environmental breviatea harbour mutualistic *Arcobacter* epibionts. Nature 534:254–258

Hammer TC, Sanders JG, Fierer N (2019) Not all animals need a microbiome. FEMS Microbiol Lett 366:fnz117

Hammerschmidt K, Rose CJ, Kerr B, Rainey PB (2014) Life cycles, fitness decoupling and the evolution of multicellularity. Nature 515:75–79

Han G-Z (2019) Origin and evolution of the plant immune system. New Phytol 222:70–83

Han Z, Boas S, Schroeder NE (2016) Unexpected variation in neuroanatomy among diverse nematode species. Front Neuroanat 9:162

Handley KM, Boothman C, Mills RA, Pancost RD, Lloyd JR (2010) Functional diversity of bacteria in a ferruginous hydrothermal sediment. ISME J. https://doi.org/10.1038/ismej.2010.38

Hanley ME, Lamont BB, Fairbanks MM, Rafferty CM (2007) Plant structural traits and their role in anti-herbivore defence. Perspect Plant Ecol Evol Syst 8:157–178

Hansen TB, Jensen TI, Clausen BH, Bramsen IB, Pinsen B et al (2013) Natural RNA circles function as efficient microRNA sponges. Nature 495:384–388

Hansen BL, Pessotti RDC, Fischer MS, Collins A, El-Hifnawi L et al (2020) Cooperation, competition, and specialized metabolism in a simplified root nodule microbiome. mBio 11:e01917–e01920

Hara-Nishimura I, Hatsugai N (2011) The role of vacuole in plant cell death. Cell Death Differ 18:1298–1304

Harrison RG, Aplin KL, Rycroft MJ (2010) Atmospheric electricity coupling between earthquake regions and the ionosphere. J Atmos Sol Terr Phys 72:376–2010

Hartenstein V, Martinez P (2019) Phagocytosis in cellular defense and nutrition: a food centered approach to the evolution of macrophages. Cell Tissue Res. https://doi.org/10.1007/s00441-019-03096-6

Hartline DK, Colman DR (2007) Rapid conduction and the evolution of giant axons and myelinated fibers. Curr Biol 17:R29–R35

Hartung W, Sauter A, Hose E (2002) Abscisic acid in the xylem: where does it come from, where does it go to? J Exp Bot 53(366):27–32

Harunaga JS, Doyle AD, Yamada KM (2014) Local and global dynamics of the basement membrane during branching morphogenesis require protease activity and actomyosin contractibility. Dev Biol 394:197–205

Hasenstein KH (1999) Gravisensing in plants and fungi. Adv Space Res 24(6):677–685

Hay ED (2005) The mesenchymal cell, its role in the embryo and the remarkable signaling mechanisms that create it. Dev Dyn 233:708–720

Hay ME (2009) Marine chemical ecology: chemical signals and cues structure marine populations, communities, and ecosystems. Ann Rev Mar Sci 1:193–212

Hay M, Kubanek J (2002) Community and ecosystem level consequences of chemical cues in the plankton. J Chem Ecol 28(10):2001–2016

Hayakawa S, Takaku Y, Hwang JS, Horiguchi T, Suga H et al (2015) Function and evolutionary origin of unicellular camera-type eye structure. PLoS One 10(3):e 0118415

Heaton L, Obara B, Grau V, Jones N, Nakagaki T et al (2012) Analysis of fungal networks. Fungal Biol Rev 26(1):12–29

Hedges SB, Battistuzzi FU, Blair JE (2006) Molecular timescale of evolution in the proterozoic. In: Xiao S, Kaufman AJ (eds) Neoproterozoic geobiology and paleobiology. Springer, New York, pp 199–229

Heidstra R, Sabatini S (2014) Plant and animals stem cells: similar yet different. Nat Rev Mol Cell Biol 15:301–312

Heil M (2009) Damaged-self recognition in plant herbivore defence. Trends Ecol Evol 14(7):356–363

Heil M, Karban R (2009) Explaining evolution of plant communication by airborne signals. Trends Ecol Evol 25(3):137–144

Heil M, Ton J (2008) Long-distance signalling in plant defence. Trends Plant Sci 13(6):264–272

Heintz C, Mair W (2014) You are what you host: microbiome modulation of the aging process. Cell 156:408–411

Heinze S, Reppert SM (2011) Sun compass integration of skylight cues in migratory monarch butterflies. Neuron 69:345–358

Heisenberg C-P, Solnica-Krezel L (2008) Back and forth between cell fate specification and movement during vertebrate gastrulation. Curr Opin Genet Dev 18:311–316

Heisler J, Glibert PM, Burkholder JM, Anderson DM, Cochlan W et al (2008) Eutrophication and harmful algal blooms: a scientific consensus. Harmful Algae 8:1–13

Hejnol A, Martindale MQ (2008) Acoel development indicates the independent evolution of the bilaterian mouth and anus. Nature 456:382–386

Hejnol A, Pang K (2016) Xenacoelomorpha's significance for understanding bilaterian evolution. Curr Opin Genet Dev 39:48–54

Hejnol A, Rentzsch F (2015) Neural nets. Curr Biol 25:R782–R786

Helfman GS, Collette BB, Facey DE, Bowen BW (2009) The diversity of fishes. Wiley-Blackwell, Chichester

Helms AM, De Moraes CM, Tooker JF, Mescher MC (2013) Exposure of *Solidago altissima* plants to volatile emissions of an insect antagonist (*Eurosta solidaginis*) deters subsequent herbivory. PNAS 110(1):199–204

Hemmersbach R, Simon A, Waßer K, Hauslage J, Christiansen PCM et al (2014) Impact of a high magnetic field on the orientation of gravitactic unicellular organisms—a critical consideration about the application of magnetic fields to mimic functional weightlessness. Astrobiology 14(3):205–215

Henderson G, Cox F, Ganesh S, Jonker A, Young W et al (2015) Rumen microbial community composition varies with diet and host, but a core microbiome is found across a wide geographical range. Sci Rep. https://doi.org/10.1038/srep14567

Henkin TM (2008) Riboswitch RNAs: using RNA to sense cellular metabolism. Genes Dev 22:3383–3390

Herculano-Houzel S (2009) The human brain in numbers: a linearily scaled-up primate brain. Front Hum Neurosci 3:31

Hernández-Hernández V, Niklas KJ, Newman SA, Benítez M (2012) Dynamical patterning modules in plant development and evolution. Int J Dev Biol 56:661–674

Herrero A, Stavans J, Flores E (2016) The multicellular nature of filamentous heterocyst-forming cyanobacteria. FEMS Microbiol Rev 40(6):831–854

Herron MD, Hackett JD, Aylward FO, Michod RE (2009) Triassic origin and early radiation of multicellular volvocine algae. PNAS 106(9):3254–3258

Herron MD, Rashidi A, Shelton DE, Driscoll WW (2013) Cellular differentiation and individuality in the "minor" multicellular taxa. Biol Rev Camb Philos Soc 88(4):844–861

Hervé JH (2013) The communicating junctions, roles and dysfunctions. Biochim Biophys Acta 1828:1–3

Hestetun JT, Vacelet J, Boury-Esnault N, Borchiellini C, Kelly M et al (2016) The systematics of carnivorous sponges. Mol Phylogenet Evol 94:327–345

Hetherington AJ, Dolan L (2017) Bilaterally symmetric axes with rhizoids composed the rooting structure of the common ancestor of vascular plants. Philos Trans R Soc Lond B Biol Sci 373:20170042

Hettenhausen C, Li J, Zhuang H, Sun H, Xu Y et al (2017) Stem parasitic plant *Cuscuta australis* (dodder) transfers herbivory-induced signals among plants. PNAS 114(32):E6703–E6709

Hibbett DS, Binder M, Bischoff JF, Blackwell M, Cannon PF et al (2007) A higher-level phylogenetic classification of the fungi. Mycol Res 111:509–547

Hillmann D, Forbes G, Novohradská S, Ferling I, Riege K et al (2018) Multiple roots to fruiting body formation in amoebozoa. Genome Biol Evol 10(2):591–606

Hinsinger P, Gobran GR, Gregory PJ, Wenzel WW (2005) Rhizosphere geometry and heterogeneity arising from root-mediated physical and chemical processes. New Phytol 168:293–303

Hirasawa T, Kuratani S (2015) Evolution of the vertebrate skeleton: morphology, embryology, and development. Zool Lett 1:2

Hochner B (2012) An embodied view of octopus neurobiology. Curr Biol 22:R887–R892

Hochner B, Glanzman DL (2016) Evolution of highly diverse forms of behaviour in molluscs. Curr Biol 26:R965–R971

Hodge A (2009) Root decisions. Plant Cell Environ 32:628–640

Hodin J (2006) Expanding networks: signaling components in and a hypothesis for the evolution of metamorphosis. Integr Comp Biol 46(6):719–742

Hoekzema RS, Brasier MD, Dunn FS, Liu AG (2017) Quantitative study of developmental biology confirms Dickinsonia as a metazoan. Proc Biol Sci 284:20171348

Hoffland E, Giesler R, Jongmans AG, van Breemen N (2003) Feldspar tunneling by fungi along natural productivity gradients. Ecosystems 6:739–746

Hogan DA (2006) Talking to themselves: autoregulation and quorum sensing in fungi. Eukaryot Cell 5(4):613–619

Holden NJ, Gally DL (2004) Switches, cross-talk and memory in Escherichia coli adherence. J Mol Microbiol 53:585–593

Hölldobler B, Wilson EO (1990) The ants. Belknap, Harvard University Press, Cambridge

Holm A, Vikström E (2014) Quorum sensing communication between bacteria and human cells: signals, targets, and functions. Front Plant Sci 5:309

Holmes MM, Rosen GJ, Jordan CL, de Vries GJ, Goldman BD, Forger NG (2007) Social control of brain morphology in a eusocial mammal. PNAS 104(25):10548–10552

Holopainen JK, Blande JD (2012) Molecular plant volatile communication. In: López-Larrea C (ed) Sensing in nature. Landes Bioscience, New York, pp 17–31

Holopainen JK, Gershenzon J (2010) Multiple stress factors and the emission of plant VOCs. Trends Plant Sci 15(3):176–184

Hong K-W, Koh C-L, Sam C-K, Yin W-F, Chan K-G (2012) Quorum quenching revisited-from signal decays to signalling confusion. Sensors 12:4661–4696

Hongoh Y, Sato T, Dolan MF, Noda S, Ui S et al (2007) The motility symbiont of the termite gut flagellate caduceia versatilis is a member of the "synergistes" group. Appl Environ Microbiol 73(19):6270–6276

Hooper SL (2015) Octopus movement: push right, go left. Curr Biol 25:R366–R368

Hooper LA, Littman DR, Macpherson AJ (2012) Interactions between the microbiota and the immune system. Science 336:1268–1273

Hornick BL (2017) Fungal iron oxidation in Brazilian iron caves. University of Akron, Honors Res Project, p 432

Horst NA, Katz A, Pereman I, Decker EL, Ohad N, Reski R (2016) A single homeobox gene triggers phase transition, embryogenesis and asexual reproduction. Nat Plants. https://doi.org/10.1038/NPLANTS.2015.209

Horswill AR, Stoodley P, Stewart PS, Parsek MR (2007) The effect of the chemical, biological, and physical environment on quorum sensing in structured microbial communities. Anal Bioanal Chem 387:371–380

Hover T, Maya T, Ron S, Sandovsky H, Shadkchan Y et al (2016) Mechanisms of bacterial (*Serratia marcescens*) attachment to, migration along, and killing of fungal hyphae. Appl Environ Microbiol 82(9):2585–2594

Hovmøller MS, Justesen AF, Brown JKM (2002) Clonality and long-distance migration of *Puccinia striiformis f.sp. tritici* in north-west Europe. Plant Pathol 51:24–32

Howlett BJ (2006) Secondary metabolite toxins and nutrition of plant pathogenic fungi. Curr Opin Plant Biol 9:371–375

Høyland-Kroghsbo NM, Mærkedahl RB, Svenningsen SL (2013) A quorum-sensing-induced bacteriophage defense mechanism. mBio 4(1):e00362

Hoysted GA, Kowal J, Jacob A, Rimington WR, Duckett JG et al (2018) A mycorrhizal revolution. Curr Opin Plant Biol 44:1–6

Hsueh Y-P, Mahanti P, Schroeder FC, Sternberg PW (2013) Nematode-trapping fungi eavesdrop on nematode pheromones. Curr Biol 23:83–86

Huang J-H, Lozano J, Belles X (2013) Broad-complex functions in postembryonic development of the cockroach *Blatella germanica* shed new light on the evolution of insect metamorphosis. Biochim Biophys Acta 1830:2178–2187

Huang J, Zhao B, Liu T, Mou J, Jiang Z et al (2019) Wood-derived material for advanced electrochemical energy storage devices. Adv Funct Mat 29:1902255

Huber AE, Bauerle TL (2016) Long-distance plant signalling pathways in response to multiple stressors: the gap in knowledge. J Exp Biol 67(7):2063–2079

Hug I, Deshpande S, Sprecher KS, Pfohl T, Jenal U (2017) Second messenger-mediated tactile response by a bacterial rotary motor. Science 358:531–534

Hughes TP (1989) Community structure and diversity of coral reefs: the role of history. Ecology 70(1):275–279

Hughes KT, Berg HC (2017) The bacterium has landed. Science 358:446–447

Hunter T (2007) The age of crosstalk: phosphorylation, ubiquination, and beyond. Mol Cell 28:730–738

Hutchinson LV, Wenzel BM (1980) Olfactory guidance in foraging by procellariiforms. Condor 82:314–319

Huysmans M, Saul Lema A, Coll NS, Nowack MK (2017) Dying two deaths—programmed cell death regulation in development and disease. Curr Opin Plant Biol 35:37–55

Hynes RO, Naba A (2012) Overview of the matrisome—an inventory of extracellular matrix constituents and functions. Cold Spring Harb Perspect Biol 4:a004903

Idnum A, Verma S, Corrochano LM (2010) A glimpse into the basis of vision in the kingdom *Mycota*. Fungal Genet Biol 47(11):881–892

Iglesias TL, Dornburg A, Brandley MC, Alfaro ME, Warren DL (2015) Life in the unthinking depths: energetic constraints on encephalization in marine fishes. J Evol Biol 28:1080–1090

Ikeuchi M, Sugimoto K, Iwase A (2013) Plant callus: mechanisms of induction and repression. Plant Cell 25:3159–3173

Ipcho S, Sundelin T, Erbs G, Kistler HC, Newman M-A, Olsson S (2016) Fungal innate immunity induced by bacterial microbe-associated molecular patterns (MAMPs). G3 6:1585–1595

Irie N, Kuratani S (2014) The developmental hourglass model: a predictor of the basic body plan? Development 141:4649–4655

Isler K, van Schalk C (2006) Costs of encephalization: the energy trade-off hypothesis tested on birds. J Hum Evol 51:228–243

Isler K, van Schalk CP (2009) The expensive brain: a framework for explaining evolutionary changes in brain size. J Hum Evol 57:392–400

Ispolatov I, Ackermann M, Doebeli M (2012) Division of labour and the evolution of multicellularity. Proc Biol Sci 279:1768–17776

Izzo TJ, Vaconcelos HL (2002) Cheating the cheater: domatia loss minimizes the effects of ant castration in an Amazonian ant-plant. Oecologia 133:200–205

Jablonski NG, Chaplin G (2013) Epidermal pigmentation in the human lineage is an adaptation to ultraviolet radiation. J Hum Evol 65:671–675

Jackson D, Leys SP, Hinman VF, Woods R, Lavin MF, Degnan BM (2002) Ecological regulation of development: induction of marine invertebrate metamorphosis. Int J Dev Biol 46:679–686

James TY, Berbee ML (2011) No jacket required—new fungal lineage defies dress code. Bioessays 34:94–102

Janis CM, Devlin K, Warren DE, Witzmann F (2012) Dermal bone in early tetrapods: a palaeophysiological hypothesis of adaptation for terrestrial acidosis. Proc R Soc B. https://doi.org/10.1098/rspb.2012.0558

Janouškovec J, Gavelis GS, Burki F, Dinh D, Bachvaroff TR et al (2017) Major transitions in dinoflagellate unveiled by phylotranscriptomics. PNAS 114:E171–E180

Janson S, Hayes PK (2006) Molecular taxonomy of harmful algae. In: Granéli E, Turner JT (eds) Ecology of harmful algae. Springer, Berlin, pp 9–21

Janzen DH (1969) Allelopathy by myrmecophytes: the ant azteca as an allelopathic agent of cecropia. Ecology 50(1):147–153

Jarvis ED, Güntürkün O, Bruce L, Csillag A, Karten H et al (2005) Avian brains and a new understanding of vertebrate brain. Nat Rev Neurosci 6(2):151–159

Jefferson KK (2004) What drives bacteria to produce a biofilm? FEMS Microbiol Lett 236:163–173

Jenner RA (2014) Macroevolution of animal body plans: is there science after the tree? BioScience 64:653–664

Jiricny N, Diggle SP, West SA, Evans BA, Ballantynes G et al (2010) Fitness correlates with the extent of cheating in a bacterium. J Evol Biol 23:738–747

Jodoin JN, Coravos JS, Chanet S, Vasquez CG, Tworoger M et al (2015) Stable force balance between epithelial cells arises from F-actin turnover. Dev Cell 35:685–697

Johanson J (2009) RNA thermosensors in bacterial pathogens. In: Collin M, Schuch R (eds) Bacterial sensing and signaling. Karger, Basel, pp 150–160

Johanson Z, Kearsley A, den Blaauwen J, Newman M, Smith MM (2010) No bones about it: an enigmatic devonian fossil reveals a new skeletal framework—a potential role of loss of gene regulation. Semin Cell Dev Biol 21:414–423

John M, Rubick R, Schmitz RPH, Rakoczy J, Schubert T, Diekert G (2009) Retentive memory of bacteria: long-term regulation of dehalospiration in *Sulfurospirillum multivorans*. J Bacteriol 191(5):1650–1655

Johnke J, Cohen Y, de Leeuw M, Kushmaro A, Jurkevitch E, Chatzinotas A (2014) Multiple micro-predators controlling bacterial communities in the environment. Curr Opin Biotechnol 27:185–190

Johnson NC (2010) Resource stoichiometry elucidates the structure and function of arbuscular mycorrhizas across scales. New Phytol 185:631–647

Johnson D, Gilbert L (2015) Interplant signalling through hyphal networks. New Phytol 205:1448–1453

Johnson SD, Jürgens A (2010) Convergent evolution of carrion and faecal scent mimicry in fly-pollinated angiosperm flowers and a stinkhorn fungus. S Afr J Bot 76:796–807

Johnson NC, Graham JH, Smith FA (1997) Functioning of mycorrhizal associations along mutualism-parasitism continuum. New Phytol 135:575–585

Johnson AD, Richardson E, Bachvarova RF, Crother BJ (2011) Evolution of the germ line-soma relationship in vertebrate embryos. Reproduction 141:291–300

Johnston MJS (1997) Review of electric and magnetic fields accompanying seismic and volcanic activity. Surv Geophys 18:441–475

Joint I, Talt K, Wheeler G (2007) Cross-kingdom signalling: exploitation of bacterial quorum sensing molecules by the green seaweed *Ulva*. Philos Trans R Soc Lond B Biol Sci 362:1223–1233

Jones JDG, Vance RE, Dangl JL (2016) Intracellular innate immune surveillance devices in plants and animals. Science 354:aaf6395

Jordal BH, Cognato AI (2012) Molecular phylogeny of bark and ambrosia beetles reveals multiple origin of fungus farming during periods of global warming. BMC Evol Biol 12:133

Jorgensen RA (2002) RNA traffics information systematically in plants. PNAS 99(18):11561–11563

Josephson RK, Schwab WE (1979) Electrical properties of an excitable epithelium. J Gen Physiol 74:213–236

Joubert C, Piquemal D, Marie B, Manchon L, Pierrat F et al (2010) Transcriptome and proteome analysis of *Pinctada margaritifera* calcifying mantle and shell: focus on biomineralization. BMC Genomics 11:613

Jourdain E, Vongraven D (2017) Humpback whale (*Megaptera novaeangliae*) and killer whale (*Orcinus orca*) feeding aggregations for foraging in herring (*Clupea harengus*) in Northern Norway. Mamm Biol 86:27–32

Joy JB (2013) Symbiosis catalyses niche expansion and diversification. Proc Biol Sci 280:20122820

Juliano C, Wessel G (2010) Versatile germline genes. Science 329:640–641

Justice SS, Hunstad DA, Cegelski L, Hultgren SJ (2008) Morphological plasticity as a bacterial survival strategy. Nat Rev Microbiol 6:162–168

Kaelberer MM, Buchanan KL, Klein ME, Barth BB, Montoya MM et al (2018) A gut-brain neural circuit for nutrient sensory transduction. Science 361:eaat5236

Kaiser D, Warrick H (2014) Transmission of a signal that synchronizes cell movements in swarms of *Myxococcus xanthus*. PNAS 111(36):13105–13110

Kalinka AT, Tomancak P (2012) The evolution of early animal embryos: conservation or divergence? Trends Evol Ecol 27(7):385–393

Kalinka AT, Varga KM, Gerrard DT, Preibisch S, Corcoran DL et al (2010) Gene expression divergence recapitulates the developmental hourglass model. Nature 468:811–814

Kaller MS, Lazari A, Blanco-Duque C, Sampaio-Baptista S, Johansen-Berg H (2017) Myelin plasticity and behaviour—connecting the dots. Curr Opin Neurobiol 47:86–92

Kalluri R (2009) EMT: when epithelial cells decide to become mesenchymal-like cells. J Clin Investig 119:1417–1419

Kameritsch P, Pogoda K, Pohl U (2012) Channel-independent influence of connexion 43 on cell migration. Biochim Biophys Acta 1818:1993–2001

Kandel E (2012) Das Zeitalter der Erkenntnis. Siedler, Munich

Kandel ER, Schwartz JH, Jessell TM, Siegelbaum SA, Hudspeth AJ (eds) (2013) Principles of neural science. McGraw Hill Medical, New York

Kandel ER, Dudai Y, Mayford MR (2014) The molecular and systems biology of memory. Cell 157:163–186

Kang G, Allard CAG, Valencia-Montoya WA, van Giessen L, Kim JJ et al (2023) Sensory specializations drive octopus and squid behaviour. Nature 616:378–383

Kania U, Fendrych M, Friml J (2014) Polar delivery in plants: commonalities and differences to animal epithelial cells. Open Biol 4:140017

Kappraff J (2004) Growth in plants: a study in number. Forma 19:335–354

Kapsetaki SE, Fisher RM, West SA (2016) Predation and the formation of multicellular group in algae. Evol Ecol Res 17:651–669

Karatan E, Watnick P (2009) Signals, regulatory networks, and materials that build and break bacterial biofilms. Microbiol Mol Biol Rev 73(2):310–347

Karten HR (2012) Neocortical evolution: neuronal circuits arise independently of lamination. Curr Biol 23(1):R12–R15

Karten HJ (2015) Vertebrate brains and evolutionary connectomics: on the origins of the mammalian 'neocortex'. Philos Trans R Soc Lond B Biol Sci 370:20150060

Kato A, Mitrophanov AY, Groisman EA (2007) A connector of two-component regulatory systems promotes signal amplification and persistence of expression. PNAS 104(29):12063–12068

Kato S, Hashimoto K, Watanabe K (2012) Microbial interspecies electron transfer via electric currents through conductive minerals. PNAS 109(25):10042–10046

Katschuk G, Leffelaar PA, Giller KE, Alberton O, Hungria M, Kuyper TW (2010) Responses of legumes to rhizobia and arbuscular mycorrhizal fungi: a meta-analysis of potential photosynthate limitation of symbioses. Soil Biol Biochem 42:125–127

Katsuki T, Greenspan RJ (2013) Jellyfish nervous systems. Curr Biol 23(14):R592–R594

Katz PS (2016a) Evolution of central pattern generators and rhythmic behaviours. Philos Trans R Soc Lond B Biol Sci 371:20150057

Katz PS (2016b) Phylogenetic plasticity in the evolution of molluscan neural circuits. Curr Opin Neurobiol 41:8–16

Kauffman SA (1993) The origins of order. Oxford University Press, Oxford

Kauserud H, Mathiesen C, Ohlson M (2008) High diversity of fungi associated with living parts of boreal forest bryophytes. Botany 86:1326–1333

Kawabe Y, Schilde C, Chen Z-H, Du Q, Lawal H, Schaap P (2015) The evolution of developmental signalling in dictyostelia from an amoebozoan stress response. In: Ruiz-Trillo I, Bedelcu AM (eds) Evolutionary transitions to multicellular life. Springer, Dordrecht, pp 451–467

Kazama FY (1972) Ultrastructure and phototaxis of the zoospores of *Phlyctochytrium* sp., an estuarine chytrid. J Gen Microbiol 71:555–566

Kazlauskiene M, Kostink G, Venclovas Č, Tamulaitis G, Siksnys V (2017) A cyclic oligonucleotide signaling pathway in type III CRISPR-Cas systems. Science 357:605–609

Keane R, Berleman J (2016) The predatory life cycle of *Myxococcus xanthus*. Microbiology 162:1–11

Keating JN, Donoghue PCJ (2016) Histology and affinity of anaspids, and the early evolution of the vertebrate dermal skeleton. Proc Biol Sci 283:20152917

Keenan TF, Niinemets Ü (2016) Global leaf trait estimates biased due to plasticity in the shade. Nat Plants 3:16201

Kegge W, Pierik R (2009) Biogenic volatile organic compounds and plant competition. Trends Plant Sci 15(3):126–132

Keim CN, Martins JL, Abreu F, Rosado AS, de Barros HL et al (2004) Multicellular life cycle of magnetotactic prokaryotes. FEMS Microbiol Lett 240:203–208

Kelemen GH, Viollier PH, Tenor JL, Marri L, Buttner MJ, Thompson CJ (2001) A connection between stress and development in the multicellular *Streptomyces coelicolor* A3(2). Mol Microbiol 40(4):804–814

Keller R (2006) Mechanisms of elongation in embryogenesis. Development 133:2291–2302

Keller L, Surette MG (2006) Communication in bacteria: an ecological and evolutionary perspective. Nat Microbiol 4:249–258

Kernien JF, Snarr BD, Sheppard DC, Nett JE (2019) The interface between fungal biofilms and innate immunity. Front Immunol 8:1968

Kesselmeier J, Staudt M (1999) Biogenic volatile organic compounds (VOC): an overview on emission, physiology and ecology. J Atmos Chem 33:23–88

Kessler D, Diezel C, Baldwin IT (2010) Changing pollinators as a means of escaping herbivores. Curr Biol 20:237–242

Kettler GC, Martiny AC, Huang K, Zucker J, Coleman ML et al (2007) Patterns and implications of gene gain and loss in the evolution of prochlorococcus. PLoS Genet 3(12):e231

Kiely PD, Regan JM, Logan BE (2011) The electric picnic: synergistic requirements for exoelectro-genic microbial communities. Curr Opin Biotechnol 22:378–385

Kier WM (2012) The diversity of hydrostatic skeletons. J Exp Biol 215:1247–1257

Kier WM, Smith KK (1985) Tongues, tentacles and trunks: the biomechanics of movement in muscular-hydrostats. Zool J Linn Soc 83:307–324

Kikuchi K, Galera-Laporta L, Weatherwax C, Lam JY, Moon EC et al (2022) Electrochemical potential enables dormant spores to integrate environmental signals. Science 378:43–49

Kinkema M, Scott PT, Gresshoff PM (2006) Legume nodulation: successful symbiosis through short- and long-distance signalling. Funct Plant Biol 33:707–721

Kiørboe T, Jackson GA (2001) Marine snow, organic plumes, and optimal chemosensory behaviour of bacteria. Limnol Oceanogr 46(6):1309–1318

Kishida T, Thewissen JGM (2012) Evolutionary changes of the importance of olfaction in cetaceans based on the olfactory marker protein gene. Gene 492(2):349–353

Kishida T, Thewissen JGM, Hayakawa T, Imai H, Agata K (2015a) Aquatic adaptation and the evolution of smell and taste in whales. Zool Lett 1:9

Kishida T, Thewissen JGM, Usip S, Suydam RS, George JC (2015b) Organization and distribution of glomeruli in the bowhead whale olfactory bulb. PeerJ 3:e897

Kiss T (2017) Do terrestrial gastropods use olfactory cues to locate and select food actively? Invert Neurosci 17:9

Kitancharoen N, Yamamoto A, Hatai K (1997) Fungicidal effect of hydrogen peroxide on fungal infection of rainbow trout eggs. Mycoscience 38:375–378

Klein T, Siegwolf RTW, Körner C (2016) Belowground carbon trade among tall trees in a temperate forest. Science 352:342–344

Kleine T, Leister D (2016) Retrograde signaling: organelles go networking. Biochim Biophys Acta 1857:1313–1325

Klix F (1993) Erwachendes Denken. Spektrum, Heidelberg

Knoflacher M (ed) (2017) Herausforderungen der evolutionären Komplexität. LIT, Wien

Knogge W (1996) Fungal infections of plants. Plant Cell 8:1711–1722

Kobayashi DY, Crouck JA (2009) Bacterial/fungal interactions: from pathogens to mutualistic endo-symbionts. Annu Rev Phytopathol 47:63–82

Kollmann M, Løvdok L, Bartholomé K, Timmer J, Sourjik V (2005) Design principles of a bacterial signalling network. Nature 438:504–507

Komienko AL, Guenzl PM, Barlow DP, Pauler FM (2013) Gene regulation by the act of long non-coding RNA transcription. BMC Biol 11:59

Kondo Y, Okada S, Ohtsika S, Hirabayashi T, Adachi A et al (2018) Piscivory of the Japanese giant box jellyfish *Morbakka virulenta*. Plankton Benthos Res 13(2):66–74

Konhauser KO (2007) Introduction to geomicrobiology. Blackwell, Malden

Koonin EV, Dolja VV (2013) A virocentric perspective on the evolution of life. Curr Opin Virol 3(5):546–557

Koonin EV, Wolf YJ (2012) Evolution of microbes and viruses: a paradigm shift in evolutionary biology? Cell Infect Biol 2:110

Koonin EV, Senkevich TG, Dolja VV (2006) The ancient virus world and evolution of cells. Biol Direct 1:29

Kooyman GL, Cherel Y, Le Maho Y, Croxall JP, Thorson PH et al (1992) Diving behaviour and energetics during foraging cycles in King Penguins. Ecol Monogr 62(1):143–163

Korinets RV, Hemetsberger A, Schau HJ (2008) The wisdom of consumer crowds. J Macromarket 28(4):339–354

Kost C, Heil M (2006) Herbivore-induced plant volatiles induce an indirect defence in neighbouring plants. J Ecol 94:619–628

Kostaki M, Vatakis A (2016) Crossmodal binding rivalry: a "race" for integration between unequal sensory inputs. Vision Res 127:165–176

Kotrschal K, Van Staaden MJ, Huber R (1998) Fish brains: evolution and environmental relationships. Rev Fish Biol Fish 8:373–408

Kranner I, Cram WJ, Zorn M, Wornik S, Yoshimura I et al (2005) Antioxidants and photoprotection in a lichen as compared with its isolated symbiotic partners. PNAS 102(8):3141–3146

Krause M, Bräucker R, Hemmersbach R (2010) Gravikinesis in *Stylonychia mytilus* is based on membrane potential changes. J Exp Biol 213:161–171

Krauss U, Minh BQ, Losi A, Gärtner W, Eggert T et al (2009) Distribution and phylogeny of light-oxygen-voltage-blue light-signaling proteins in the three kingdoms of life. J Bacteriol 191(23):7234–7242

Krieger J, Sandeman RE, Sandeman DC, Hansson BS, Harzsch S (2010) Brain architecture of the largest living land arthropod, the Giant Robber Crab *Birgus latro* (Crustacea, Anomura, Coenobitidae): evidence for a prominent central olfactory pathway? Front Zool 7:25

Krieger J, Hörnig MK, Sandeman RE, Sandeman DC, Harzsch S (2020) Masters of communication: the brain of the banded cleaner shrimp *Stenopus hispidus* (Olivier, 1811) with an emphasis on sensory processing areas. J Comp Neurol 28:1561–1587

Kroken S, Taylor JW (2000) Phylogenetic species, reproductive mode, and specificity of the green alga *Trebouxia* forming lichens with the fungal genus letharia. Bryologist 103(4):645–660

Krol J, Loedige I, Filipowicz W (2010) The widespread regulation of microRNA biogenesis, function and decay. Nat Rev Genet 11:597–610

Król E, Plachno BJ, Adamec L, Stolarz M, Dziubińska H, Trębacz K (2012) Quite a few reasons for calling carnivores 'the most wonderful plants in the world'. Ann Bot 109:47–64

Kronschläger MT, Drdla-Schutting R, Gassner M, Honsek SD, Teuchmann HL, Sandkühler J (2016) Gliogenic LTP spreads widely in nociceptive pathways. Science 354:1144–1148

Kruska DCT (2003) On the evolutionary significance of encephalization in some eutherian mammals: effects of adaptive radiation, domestication, and fertilization. Brain Behav Evol 65:73–108

Kukalova-Peck J (1978) Origin and evolution of insect wings and their relation to metamorphosis, as documented by the fossil record. J Morphol 156:53–126

Kunsch K (1997) Der Mensch in Zahlen. Gustav Fischer, Stuttgart

Kurakov AV, Lavrent'ev RB, Nechitailo TY, Golyshin PN, Zvyagintsev DG (2008) Diversity of facultatively anaerobic microscopic mycelial fungi in soils. Mikrobiologiia 77(1):90–98

Kurmayer R, Deng L, Entfellner E (2016) Role of toxic and bioactive secondary metabolites in colonization and bloom formation by filamentous cyanobacteria Planktothrix. Harmful Algae 54:69–86

Labandeira CC (2013) A paleobiologic perspective on plant-insect interactions. Curr Opin Plant Biol 16:414–421

Lafferty KD, Shaw JC (2013) Comparing mechanisms of host manipulation across host and parasite taxa. J Exp Biol 216:56–66

Lai J, Koh CH, Tjota M, Pleuchot L, Raman V et al (2012) Intrinsically disordered proteins aggregate at fungal cell-to-cell channels and regulate intercellular connectivity. PNAS 109(39):15781–15786

Lake JA (2009) Evidence for an early prokaryotic endosymbiosis. Nature 460:967–971

Lambert G, Kussell E (2014) Memory and fitness optimization of bacteria under fluctuating environments. PLoS Genet 10(9):e1004556

Lamont BB (2003) Structure, ecology and physiology of root clusters—a review. Plant Soil 248:1–19

Lamont BB, Pérez-Fernández M, Rodríguez-Sánchez J (2014) Soil bacteria hold the key to root cluster formation. New Phytol 206:1156–1162

Lan G, Tu Y (2016) Information processing in bacteria: memory, computation, and statistical physics: a key issue review. Rep Prog Phys 79(5):052601

Landgren E, Fritsches K, Brill R, Warrant E (2014) The visual ecology of a deep-sea fish, the escolar *Lepidocybium flavobrunneum* (Smith, 1984). Philos Trans R Soc Lond B Biol Sci 369:20130039

Lange BM (2015) The evolution of plant secretory structures and emergence of terpenoid chemical diversity. Annu Rev Plant Biol 66:139–159

Langenheim JH (2003) Plant resins. Timber Press, Portland

Larabee FJ, Smith AA, Suarez AV (2018) Snap-jaw morphology is specialized for high-speed power amplification in the Dracula ant, Mystrium camillae. R Soc Open Sci 5:181447

Larkum M (2013) A cellular mechanism for cortical associations: an organizing principle for the cerebral cortex. Trends Neurosci 36:141–151

Latgé J-P (2007) The cell wall: a carbohydrate armour for the fungal cell. Mol Microbiol 66(2):279–290

Lathière J, Hauglustaine DA, Friend AD, De Noblet-Ducoudré N, Viovy N, Folberth GA (2006) Impact of climate variability and land use changes on global biogenic volatile organic compound emissions. Atmos Chem Phys 6:2129–2146

Latty T, Beekman M (2011) Speed-accuracy trade-offs during foraging decisions in the acellular slime mould *Physarum polycephalum*. Proc Biol Sci 278:539–545

Laudet V (2011) The origins and evolution of vertebrate metamorphosis. Curr Biol 21:R726–R737

Laughlin SB (2001) Energy as a constraint on the coding and processing of sensory information. Curr Opin Neurobiol 11:475–480

Laughlin SB, de Ruyter van Steveninck RR, Anderson JC (1998) The metabolic cost of neural information. Nature Neurosci 1(1):36–41

Laux G (1980) Kybernetik. Akademie Verlag, Berlin

Leahy SC, Kelly WJ, Altermann E, Ronimus RS, Yeoman CJ et al (2010) The genome sequence of the rumen methanogen *Methanobrevibacter ruminantium* reveals new possibilities for controlling ruminant methane emissions. PLoS One 3(1):e8926

Leake JR (2005) Plants parasitic on fungi: unearthing the fungi in myco-heterotrophs and debunking the 'saprophytic' plant myth. Mycologist 19(3):113–122

Leake JR, Duran AL, Hardy KE, Johnson I, Beerling DJ et al (2008) Biological weathering in soil: the role of symbiotic root-associated fungi biosensing minerals and directing photosynthate-energy into grain-scale mineral weathering. Mineral Mag 72(1):85–89

Lecuit T, Le Goff L (2007) Orchestrating size and shape during morphogenesis. Nature 450:189–192

Lee J-Y (2015) Plasmodesmata: a signalling hub at the cellular boundary. Curr Opin Plant Biol 27:133–140

Lee J-Y, Lu H (2011) Plasmodesmata: the battleground against intruders. Trends Plant Sci 18(4):201–210

Lee J, Zhang L (2015) The hierarchy quorum sensing network in *Pseudomonas aeruginosa*. Protein Cell 6(1):26–41

Lefèvre CT, Abreu F, Lins U, Bazylinski DA (2010) Nonmagnetotactic multicellular prokaryotes from low-saline nonmarine aquatic environments and their unusual negative phototactic behavior. Appl Environ Microbiol 76(10):3220–3227

Legrand C, Rengefors K, Fistarol GO, Granéli E (2003) Allelopathy in phytoplankton—biochemical, ecological and evolutionary aspects. Phycology 42(4):406–419

Leljak-Levanić D, Mihaljević S, Bauer N (2015) Somatic and zygotic embryos share common developmental features at the onset of plant embryogenesis. Acta Physiol Plant 37:127

Lemaire P (2011) Evolutionary crossroads in developmental biology: the tunicates. Development 138:2143–2152

Lemaire P, Marcellini S (2003) Early animal embryogenesis: why so much variability? Biologist 50(3):136–140

Lemay J-F, Desnoyers G, Blouin S, Heppell B, Bastet L et al (2011) Comparative study between transcriptionally- and translationally-acting adenine riboswitches reveals key differences in ribo-switch regulatory mechanisms. PLoS Genet 7(1):e1001278

Lenz PH (2012) The biogeography and ecology of myelin in marine copepods. J Plant Res 34(7):575–589

Lenz PH, Hartline DK, Davis AD (2000) The need for speed. I. Fast reactions and myelinated axons in copepods. J Comp Physiol A 186:337–345

Leroy C, Carrias J-F, Céréghino R, Corbara B (2016) The contribution of microorganisms and meta-zoans to mineral nutrition in bromeliads. J Plant Ecol 9(3):241–255

Leucht P, Kim J-B, Amasha R, James AW, Girod S, Helms JA (2008) Embryonic origin and Hox status determine progenitor cell fate during adult bone regeneration. Development 135:2845–2854

Leveau JH, Preston GM (2008) Bacterial mycophagy: definition and diagnosis of a unique bacterial-fungal interaction. New Phytol 177:859–876

Levin MD, Morton-Firth CJ, Abouhamad WN, Bourret RB, Bray D (1998) Origins of individual swimming behavior in bacteria. Biophys J 74:175–181

Levy G, Hochner B (2017) Embodied organization of *Octopus vulgaris* morphology, vision, and locomotion. Front Physiol 8:164

Lévy J, Bres C, Geurts R, Chalhoub B, Kulikova O et al (2004) A putative Ca^{2+} and calmodulin-dependent protein kinase required for bacterial and fungal symbioses. Science 303:1361–1364

Levy G, Flash T, Hochner B (2015) Arm coordination in octopus crawling involves unique motor control strategies. Curr Biol 25:1195–1200

Lewus P, Ford RM (1999) Temperature-sensitive motility of *Sulfolobus acidocaldarius* influences population distribution in extreme environments. J Bacteriol 181(13):4020–4025

Leys SP (2015) Elements of a 'nervous system' in sponges. J Exp Biol 218:581–591

Leys SP, Meech RW (2006) Physiology of coordination in sponges. Can J Zool 84:288–306

Leys SP, Riesgo A (2012) Epithelia, an evolutionary novelty of metazoans. J Exp Zool 318(6):438–447

Li Q-F, He J-X (2013) Mechanisms of signalling crosstalk between brassinosteroids and gibberel-lins. Plant Signal Behav 8(7):e24686

Li L, Neaves WB (2006) Normal stem cells and cancer stem cells: the niche matters. Cancer Res 66(9):4553–4557

Li N, Han X, Feng D, Yuan D, Huang L-J (2019) Signaling crosstalk between salicylic acid and ethylene/jasmonate in plant defense: do we understand what they are whispering? Int J Mol Sci 20:671

Liebeskind BJ, Hillis DM, Zakon HH, Hofmann HA (2016) Complex homology and the evolution of nervous systems. Trends Ecol Evol 31(2):127–135

Liebeskind BJ, Hofmann HA, Hillis DM, Zakon HH (2017) Evolution of animal neuronal systems. Annu Rev Ecol Evol Syst 48:377–398

Ligrone R, Duckett JG (1994) Cytoplasmic polarity and endoplasmic microtubules associated with the nucleus and organelles are ubiquitous features of food-conducting cells in bryoid mosses (bryophyta). New Phytol 127:601–614

Lihoreau M, Latty T, Chittka L (2012) An exploration of the social brain hypothesis in insects. Front Physiol 3:442

Lillvis JL, Katz PS (2013) Parallel evolution of serotonergic neuromodulation underlies independent evolution of rhythmic motor behavior. J Neurosci 33(6):2709–2717

Lim W, Mayer B, Pawson T (2015) Cell signaling. Garland Science, New York

Lin F, Baldessari F, Gyenge CC, Sato T, Chambers RD et al (2008) Lymphocyte electrotaxis in vitro and in vivo. J Immunol 181:2465–2471

Lin T-K, Man M-Q, Abuabara K, Wakefield JS, Sheu H-M et al (2019) By protecting against cuta-neous inflammation, epidermal pigmentation provided an additional advantage for ancestral humans. Evol Appl 12:1960–1970

Linares JF, Gustafsson I, Baquero F, Martinez JL (2006) Antibiotics as intermicrobial signaling agents instead of weapons. PNAS 103(51):19484–19489

Lindo Z, Nilsson M-C, Gundale MJ (2013) Bryophyte-cyanobacteria associations as regulators of the northern latitude carbon balance in response to global change. Glob Chang Biol. https://doi.org/10.1111/geb.12175

Liu N, Pan T (2015) RNA epigenetics. Transl Res 165(1):28–35

Liu C-M, Liu L-G, Pirjola R, Wang Z-Z (2009) Calculation of geomagnetically induced currents in mid- to low-latitude power grids based on the plane wave method: a preliminary case study. Space Weather 7:S04005

Liu K, Tian J, Xiang M, Liu X (2012a) How carnivorous fungi use three-celled constricting rings to trap nematodes. Protein Cell 3(5):325–328

Liu M, Fan L, Zhong L, Kjelleberg S, Thomas T (2012b) Metaproteogenomic analysis of a community of sponge symbionts. ISME J 6:1515–1525

Liu X-Y, Koba K, Makabe A, Li X-D, Yoh M, Liu C-Q (2013a) Ammonium first: natural mosses prefer atmospheric ammonium but vary utilization of dissolved organic nitrogen depending on habitat and nitrogen deposition. New Phytol 199:407–419

Liu Z, Müller J, Li T, Alvey RM, Vogl K et al (2013b) Genomic analysis reveals key aspects of prokaryotic symbiosis in the phototrophic consortium '*Chlorochromatium aggregatum*'. Genome Biol 14:R127

Liu J, Prindle A, Humphries J, Gabalda-Sagarra M, Asally M et al (2015) Metabolic co-dependence gives rise to collective oscillations within biofilms. Nature 523:550–554

Lloyd JR, Pearce CI, Coker VS, Pattrick RAD, Van der Laan G et al (2008) Biomineralization: linking the fossil record to the production of high value functional materials. Geobiology 6:285–297

Lo C-M, Wang H-B, Dembo M, Wang Y-L (2000) Cell movement is guided by the rigidity of the substrate. Biophys J 79:144–152

Loescher WH, McCamant T, Keller JD (1990) Carbohydrate reserves, translocation, and storage in woody plant roots. Hort Sci 25(3):274–281

Loh KM, van Amerongen R, Nusse R (2016) Generating cellular diversity and spatial form: wnt signaling and the evolution of multicellular animals. Dev Cell 38:643–655

Lohmann KJ, Lohmann CMF (1996) Detection of magnetic field intensity by sea turtles. Nature 380:59–61

Lohmann KJ, Lohmann CMF, Endres CS (2008) The sensory ecology of ocean navigation. J Exp Biol 211:1719–1728

Löhne C, Yoo M-J, Borsch T, Wiersma J, Wilde V et al (2008) Biogeography of nymphaeales: extant patterns and historical events. Taxon 57(4):1123–1146

Lombardino J, Burton BM (2022) An electric alarm clock for spores. Science 378:25–26

Loomis WF (2014) Cell signalling during development of *Dictyostelium*. Dev Biol 391:1–16

Louca S, Jacques SMS, Pires APF, Leal JS, González AL et al (2017) Functional structure of the bromeliad tank microbiome is strongly shaped by local geochemical conditions. Environ Microbiol 19(8):3132–3151

Lucas RJ (2013) Mammalian inner retinal photoreception. Curr Biol 23:R125–R133

Ludeman DA, Farrar N, Riesgo A, Paps J, Leys SP (2014) Evolutionary origins of sensation in metazoans: functional evidence for a new sensory organ in sponges. BMC Evol Biol 14:3

Luginbuehl LH, Oldroyd GED (2017) Understanding the arbuscule at the heart of endomycorrhizal symbioses in plants. Curr Biol 27:R952–R963

Lutterbeck H (2012) Die Verteilung von DMS/DMSP/DMSO während des SOPRAN Mesokosmen Experiments 2011 in Bergen (Norwegen). DA Christian-Albrechts-Universität zu Kiel

Lyon P (2015) The cognitive cell: bacterial behavior reconsidered. Front Microbiol 6:264

Lyon GJ, Muir TW (2003) Chemical signaling among bacteria and its inhibition. Chem Biol 10:1007–1021

Lyons NA, Kolter R (2015) On the evolution of bacterial multicellularity. Curr Opin Microbiol 24:21–28

Lyte M (2013) Microbial endocrinology in the microbiome-gut-brain axis: how bacterial production and utilization of neurochemicals influence behavior. PLoS Pathog 9(11):e1003726

Ma Q, Hua H-H, Chen Y, Liu B-B, Krämer AL et al (2012) A rising tide of blue absorbing biliprotein photoreceptors—characterization of seven such bilin-binding GAF domains in *Nostoc* sp. PCC7120. FEBS J 279:4095–4108

Maathuis FJM, Ahmad I, Patishtan J (2014) Regulation of Na^+ fluxes in plants. Front Plant Sci 5:467

Mackas DL, Denman KL, Abbott MR (1985) Plankton patchiness: biology in the physical vernacular. Bull Mar Sci 37(2):652–674

MacKay DJC (2003) Information theory, inference, and learning algorithms. Cambridge University Press, Cambridge

Mackie GO (1976) Propagated spikes and secretion in a coelenterate glandular epithelium. J Gen Physiol 68:313–325

Mackie GO (2004) Central neural circuitry in the jellyfish *Aglantha*. Neurosignals 13:5–19

Mackie GO, Passano LM (1968) Epithelial conduction in hydromedusae. J Gen Physiol 52:600–621.

Mackie GO, Burighel P, Caicci F, Manni L (2006) Innervation of ascidian siphons and their responses to stimulation. Can J Zool 84:1146–1162

Madigan MT, Martinko JM, Parker J (2003) Brock biology of microorganisms. Prentice Hall, Upper Saddle River

Madsen EL (2011) Microorganisms and their roles in fundamental biogeochemical cycles. Curr Opin Biotechnol 22:456–464

Maffei ME, Mithöfer A, Boland W (2007) Insects feeding on plants: Rapid signals and responses preceding the induction of phytochemical release. Phytochemistry 68:2946–2959

Magalhaes JG, Tattoli I, Girardin SE (2007) The intestinal epithelial barrier: how to distinguish between the microbial flora and pathogens. Semin Immunol 19:106–115

Mahamid J, Aichmayer B, Shimoni E, Ziblat R, Li C et al (2010) Mapping amorphous calcium phosphate transformation into crystalline mineral from the cell to the bone in zebrafish fin rays. PNAS 107(14):6316–6321

Maillet F, Poinsot V, André O, Puech-Pagès V, Haouy A et al (2011) Fungal lipochitooligosaccharide symbiotic signals in arbuscular mycorrhiza. Nature 469:58–63

Maldonaldo M, Durfort M, McCarthy Young CM (2003) The cellular basis of photobehavior in the tufted parenchymella larva of demosponges. Mar Biol 143:427–441

Males J (2016) Think tank: water relations of Bromeliaceae in their evolutionary context. Bot J Linn Soc 181:415–440

Males J, Griffiths H (2018) Economic and hydraulic divergences underpin ecological differentiation in the Bromeliaceae. Plant Cell Environ 41:64–78

Maleszka R (2008) Epigenetic integration of environmental and genomic signals in honey bees. Epigenetics 3(4):188–192

Malicki JJ, Johnson CA (2017) The cilium: cellular antenna and central processing unit. Trends Cell Biol 27(2):126–140

Malkin SY, Rao AMF, Seitaj D, Vasquez-Cardenas D, Zetsche E-M et al (2014) Natural occurrence of microbial sulphur oxidation by long-range electron transport in the seafloor. ISME J 8:1843–1854

Malone CD, Hannon HJ (2009) Small RNAs as guardians of the genome. Cell 136:656–668

Malvankar NS, Lovley DR (2014) Microbial nanowires for bioenergy application. Curr Opin Biotechnol 27:88–95

Mandal M, Breaker RR (2004) Gene regulation by riboswitches. Nat Rev Mol Cell Biol 5:451–463

Mángano MG, Buatois LA (2014) Decoupling of body-plan diversification and ecological structuring during the Ediacaran-Cambrian transition: evolutionary and geobiological feedbacks. Proc Biol Sci 281:20140018

Manni L, Burighel P (2006) Common and divergent pathways in alternative developmental processes of ascidians. Bioessays 28(9):902–912

Manni L, Lane NJ, Joly J-S, Gasparini F, Tiozzo S et al (2004) Neurogenic and non-neurogenic placodes in ascidians. J Exp Zool B Mol Dev Evol 302(5):483–504

Marathe R, Meel C, Schmidt NC, Dewenter L, Kurre R et al (2014) Bacterial twitching motility is coordinated by a two-dimensional tug-of-war with directional memory. Nat Commun. https://doi.org/10.1038/ncomms4759

Marchader E, Oates ME, Fang H, Donoghue CJ, Hetherington AM, Gough J (2016) Evolution of the calcium-based intracellular signaling system. Genome Biol Evol 8(7):2118–2132

Marcionetti A, Rossier V, Roux N, Salis P, Laudet V, Salamin N (2019) Insights into the genomics of clownfish adaptive radiation: genetic basis of the mutualism with sea anemones. Genome Biol Evol 11(3):869–882

Marcos FHC, Powers TR, Stocker R (2012) Bacterial rheotaxis. PNAS. https://doi.org/10.1073/pnas.1120955109

Marguet E, Gaudin M, Gauliard E, Fourquaux I, du Ploy SB et al (2013) Membrane vesicles, nanopods and/or nanotubes produced by hyperthermophilic archaea of the genus *Thermococcus*. Biochem Soc Trans 41:436–442

Marino L (2007) Cetacean brains: how aquatic are they? Anat Rec 290:694–700

Marino P, Raguso R, Goffinet B (2009) The ecology and evolution of fly dispersed dung mosses (Family Splachnaceae): manipulating insect behavior through odour and visual cues. Symbiosis 47:61–76

Marquardt H, Schäfer SG (eds) (1994) Lehrbuch der Toxikologie. BI Wissenschaftsverlag, Mannheim

Marranzino AN, Webb JF (2018) Flow sensing in the deep sea: the lateral line system of stomiiform fishes. Zool J Linn Soc 183:945–965

Marshall TC, Stolzenbrug M, Maggio CR, Coleman M, Krehbiel PR et al (2005) Observed electric fields associated with lightning initiation. Geophys Res Lett 32:L03813

Martens EA, Wadhwa N, Jacobsen NS, Lindemann C, Andersen KH, Visser A (2015) Size structures sensory hierarchy in ocean life. Proc Biol Sci 282:20151346

Martin C, Mayer G (2014) Neuronal tracing of oral nerves in a velvet worm—implications for the evolution of the ecdysozoan brain. Front Neuroanat 8:7

Martin G, Guggiari M, Bravo D, Zopfl J, Cailleau G et al (2012) Fungi, bacteria and soil pH: the oxalate-carbonate pathway as a model for metabolic interactions. Environ Microbiol 14(11):2960–2970

Martin EC, Vicari C, Tsakou-Ngouafo L, Pontarotti P, Petrescu AJ, Schatz DG (2020) Identification of RAG-like transposons in protostomes suggests their ancient bilaterian origin. Mobile DNA 11:17

Martindale MQ, Hejnol A (2009) A developmental perspective: changes in the position of the blastopore during bilaterian evolution. Dev Cell 17:162–174

Martin-Durán JM, Pang K, Børve A, Lê HS, Furu A et al (2018) Convergent evolution of bilaterian nerve cords. Nature 553:45–50

Martins JL, Silveira TS, Silva KT, Lins U (2009) Salinity dependence of the distribution of multicellular magnetotactic prokaryotes in a hypersaline lagoon. Int Microbiol 12:193–201

Masi E, Ciszak M, Colzi I, Adamer L, Mancuso S (2016) Resting electrical network activity in traps of the aquatic carnivorous plants of the genera *Aldrovanda* and *Utricularia*. Sci Rep 6:24989

Matsuuchi L, Naus CC (2013) Gap junction proteins on the move: connexins, the cytoskeleton and migration. Biochim Biophys Acta 1828:94–108

Matter K, Balda MS (2003) Signalling to and from tight junctions. Nat Rev Mol Cell Biol 4:225–236

Mattick JS (2011) The central role of RNA in human development and cognition. FEBS Lett 585:1600–1616

Maurel C, Boursiac Y, Luu D-T, Santoni V, Shahzad Z, Verdoucq L (2015) Aquaporins in plants. Physiol Rev 95:1321–1358

Maurer MH (2011) Proteomic definitions of mesenchymal stem cells. Stem Cells Int 2011:704256

Mauriello EMF, Astling DP, Sliusarenko O, Zusman DR (2009) Localization of a bacterial cytoplasmic receptor is dynamic and changes with cell-cell contacts. PNAS 106(12):4852–4857

Mayr E (2003) Das ist Evolution. Bertelsmann, Munich

McClain CR, Balk MA, Benfield MC, Branch TA, Chen C et al (2015) Sizing ocean giants: patterns of intraspecific size variation in marine megafauna. PeerJ. https://doi.org/10.7717/peerj.715

McClaugherty CA, Aber JD, Melillo JM (1982) The role of fine roots in the organic matter and nitrogen budgets of two forested ecosystems. Ecology 63(5):1481–1490

McClung CR (2006) Two component signaling provides the major output from the cyanobacterial circadian clock. PNAS 103(32):11819–11820

McCormick JR, Flärdh K (2012) Signals and regulators that govern *Streptomyces* development. FEMS Microbiol Rev 36:206–231

McCown PJ, Liang JJ, Weinberg Z, Breaker RR (2014) Structural, functional, and taxonomic diversity of three PreQ$_1$ riboswitch classes. Chem Biol 21:880–889

McGuire KL, Zak DR, Edwards IP, Blackwood CB, Upchurch R (2010) Slowed decomposition is biotically mediated in an ectomycorrhizal, tropical rain forest. Oecologia 164:785–795

McKeon CS, Stier AC, McIlroy SE, Bolker BM (2012) Multiple defender effects: synergistic coral defense by mutualist crustaceans. Oecologia 169:1095–1103

Mead KS, Koehl MAR, O'Donnell MJ (1999) Stomatopod sniffing: the scaling of chemosensory sensillae and flicking behaviour with body size. J Exp Mar Biol Ecol 241:235–261

Meech RW (2015) Electrogenesis in the lower Metazoa and implications for neuronal integration. J Exp Biol 218:537–550

Mehta P, Schwab DJ (2012) Energetic costs of cellular computation. PNAS 109(44):17978–17982

Mehta P, Goyal S, Long T, Bassler BL, Wingreen NS (2009) Information processing and signal integration in bacterial quorum sensing. Mol Syst Biol 5:325

Mello W (2009) The electrosensorial pore system of the cephalofoil in the four most common species of hammerhead shark (Elasmobranchii: Sphyrnidae) from the Southwestern Atlantic. C R Biol 332:404–412

Mendoza-Becerril MA, Maronna MM, Pacheco MLAF, Simões MG, Leme JM et al (2016) An evolutionary comparative analysis of the medusozoan (Cnidaria) exoskeleton. Zool J Linn Soc 178:206–225

Meng S, Jia Q, Zhou G, Zhou H, Liu Q, Yu J (2018) Fine root biomass and its relationship with aboveground traits of *Larix gmelinii* trees in Northeastern China. Forests 9:35

Merbach MA, Merbach DJ, Maschwitz U, Booth WE, Fiala B, Ziska G (2002) Mass march of termites into the deadly trap. Nature 415:36–37

Mercier A, Sun Z, Baillon S, Hamel J-F (2011) Lunar rhythms in the deep sea: evidence from reproductive periodicity of several marine invertebrates. J Biol Rhythms 26(1):82–86

Merino-Puerto V, Schwarz H, Maldener I, Mariscal V, Mullineaux CW et al (2011) FraC/FraD-dependent intercellular molecular exchange in the filaments of a heterocyst-forming cyanobacterium Anabaena sp. Mol Microbiol 82(1):87–98

Meşe G, Richard G, White TW (2007) Gap junctions: basic structure and function. J Integr Dermatol 127:2516–2524

Michels J, Appel E, Gorb SN (2016) Functional diversity of resilin in Arthropoda. Beilstein J Nanotechnol 7:1241–1259

Milgram S (2009) Das Milgram experiment. Rowohlt, Reinbeck

Miller PJO, Johnson MP, Tyack PL (2004) Sperm whale behaviour indicates the use of echolocation click buzzes 'creaks' in prey capture. Proc Biol Sci 271:2239–2247

Minegishi K, Hashimoto M, Ajima R, Takaoka K, Shinohara K et al (2017) A Wnt5 activity asymmetry and intracellular signaling via PCP proteins polarize node cells for left-right symmetry breaking. Dev Cell 40:439–452

Mironov AS, Gusarov I, Rafikov R, Lopez LE, Shatalin K et al (2002) Sensing small molecules by nascent RNA: a mechanism to control transcription in bacteria. Cell 111:747–756

Miserez A, Schneberk T, Son C, Zok FW, Waite JH (2008) The transition from stiff to compliant materials in squid beaks. Science 319:1816–1819

Mitchell EG, Kenchington CG, Liu AG, Matthews JJ, Butterfield NJ (2015) Reconstructing the reproductive mode of an Ediacaran macro-organism. Nature 524:343–346

Mollo E, Fontana A, Roussis V, Polese G, Amodeo P, Ghiselin MT (2014) Sensing marine biomolecules: Smell, taste, and the evolutionary transition from aquatic to terrestrial life. Front Chem 2:92

Monahan-Earley R, Dvorak AM, Aird WC (2013) Evolutionary origins of the blood vascular system and endothelium. J Thromb Homeost 11:46–66

Montes RAC, Rosas-Cárdens FF, De Paoli E, Accerbi M, Rymarquis LA et al (2014) Sample sequencing of vascular plants demonstrates widespread conservation and divergence of microRNAs. Nat Commun. https://doi.org/10.1038/ncomms4722

Montgomery SH, Geisler JH, McGowen MR, Fox C, Marino L, Gatesy J (2013) The evolutionary history of cetacean brain and body size. Evolution 67(11):3339–3353

Montgomery BL, Lechno-Yossef S, Kerfeld CA (2016) Interrelated modules in cyanobacterial photosynthesis: the carbon-concentrating mechanism, photorespiration, and light perception. J Exp Bot 67(10):2931–2940

Moore D (1991) Perception and response to gravity in higher fungi—a critical appraisal. New Phytol 117:3–23

Moore D, Robson GD, Trinci APJ (2011) 21st century guidebook to fungi. Cambridge University Press, Cambridge

Moore BA, Paul-Murphy JR, Tennyson AJD, Murphy CJ (2017) Blind free-living kiwi offer a unique window into the ecology and evolution of vertebrate vision. BMC Biol 15:85

Moreira PL, Barata M (2005) Egg mortality and early embryo hatching caused by fungal infection of Iberian rock lizard (*Lacerta monticola*) clutches. Herpetol J 15:265–272

Moroz LL, Kohn AB (2016) Independent origins of neurons and synapses: insights from ctenophores. Philos Trans R Soc Lond B Biol Sci 371:20150041

Moroz LL, Kocot KM, Citarella MR, Dosung S, Norekian TP et al (2014) The ctenophore genome and the evolutionary origins of neural systems. Nature 510:109–114

Morris BM, Gow NAR (1993) Mechanism of electrotaxis of zoospores of phytopathogenic fungi. Phytopathology 83:877–882

Morris KV, Mattick JS (2014) The rise of regulatory RNA. Nat Rev Genet 15(6):423–437

Mouritsen H, Larsen ON (2001) Migrating songbirds tested in computer-controlled Emlen funnels using stellar cues for a time-independent compass. J Exp Biol 204:3855–3865

Muheim R, Moore FR, Phillip JB (2006) Calibration of magnetic and celestial compass cues in migratory birds—a review of cue-conflict experiments. J Exp Biol 209:2–17

Mukhtar MS, McCormack ME, Argueso CT, Pajerowska-Mukhtar KM (2016) Pathogen tactics to manipulate plant cell death. Curr Biol 26:R608–R619

Müller WA, Leitz T (2002) Metamorphosis in the Cnidaria. Can J Zool 80:1755–1771

Muller F, Brissac T, Le Bris N, Felbeck H, Gros O (2010) First description of giant Archaea (*Thaumarchaeota*) associated with putative bacterial ectosymbionts in a sulfidic marine habitat. Environ Microbiol 12(8):2371–2383

Müller A, Faubert P, Hagen M, Castell WZ, Polle A et al (2013) Volatile profiles of fungi—chemotyping of species and ecological functions. Fungal Genet Biol 54:25–33

Müller J, Bickelmann C, Sobral G (2018) The evolution and fossil history of sensory perception in amniote vertebrates. Annu Rev Earth Planet Sci 46:495–519

Munk O (1966) Ocular degeneration in deep-sea fishes. Galathea Rep 8:21–31

Murphy EJ, Watkins JL, Trathan PN, Reid K, Meredith MP et al (2006) Spatial and temporal operation of the Scotia Sea ecosystem: a review of large-scale links in a krill centred food web. Philos Trans R Soc Lond B Biol Sci 362:113–148

Murray PS, Zaidel-Bar R (2014) Pre-metazoan origins and evolution of the cadherin adhesome. Biol Open 3:1183–1195

Muscedere ML, Gronenberg W, Moreau CS, Traniello JFA (2014) Investment in higher order central processing regions is not constrained by brain size in social insects. Proc Biol Sci 281:20140217

Muszewska A, Stepniewska-Dziubinska MM, Steczkiewicz K, Pawlowska J, Dziedzic A, Ginalski K (2017) Fungal lifestyle reflected in serine protease repertoire. Sci Rep. https://doi.org/10.1038/s41598-017-09644-w

Mutschler H, Gebhardt M, Shoeman RL, Meinhart A (2011) A novel mechanism of programmed cell death in bacteria by toxin-antitoxin systems corrupts peptidoglycan synthesis. PLoS Biol 9(3):e1001033

Mytlis A, Kumar V, Qiu T, Deis R, Hart N et al (2022) Control of meiotic chromosomal bouquet and germ cell morphogenesis by the zygotene cilium. Science 376:eabh3104

Naik S, Bouladoux N, Linehan JL, Han SJ, Harrison OJ et al (2015) Commensal-dendritic-cell interaction specifies a unique protective skin immune signature. Nature 520:104–108

Naisbett-Jones LC, Putman NF, Stephenson JF, Ladak S, Young KA (2017) A magnetic map leads juvenile European eels to the gulf stream. Curr Biol 27:1–5

Nakagaki T, Yamada H, Ueda T (1999) Modulation of cellular rhythm and photoavoidance by oscillatory irradiation in the *Physarum* plasmodium. Biophys Chem 83:23–28

Naranjo-Ortiz MA, Gabaldón T (2019) Fungal evolution: major ecological adaptations and evolutionary transitions. Biol Rev 94:1443–1476

Narayanan K, Ramachandran A, Hao J, He G, Park KW et al (2003) Dual functional roles of dentin matrix protein 1. J Biol Chem 278(19):17500–17508

Narbonne GM (2004) Modular construction of early ediacaran complex life forms. Science 305:1141–1144

Narikawa R, Suzuki F, Yoshihara S, Higashi S-I, Watanabe M, Ikeuchi M (2011) Novel photosensory two-component system (PixA-NixB-NixC) involved in the regulation of positive and negative phototaxis of cyanobacterium *Synechocystis* sp. PCC 6803. Plant Cell Physiol 52(12):2214–2224

Nash TH III (ed) (2008) Lichen biology. Cambridge University Press, Cambridge

Nasir A, Caetano-Anollés G (2015) A phylogenomic data-driven exploration of viral origins and evolution. Sci Adv 1:e1500527

Nasir A, Forterre P, Kim KM, Caetano-Anollés G (2014) The distribution and impact of viral lineages in domains of life. Front Microbiol 5:194

Nawaz M, Fatima F (2017) Extracellular vesicles, tunneling nanotubes, and cellular interplay: synergies and missing links. Front Mol Biosci 4:50

Nawkar GM, Legris M, Goyal A, Schmid-Siegert E, Fleury J et al (2023) Air channels create a directional light signal to regulate hypocotyl phototropism. Science 382:935–940

Nedelcu AM, Driscoll WW, Durand PM, Herron MD, Rashidi A (2010) On the paradigm of altruistic suicide in the unicellular world. Evolution 65(1):3–20

Neely HR, Flajnik MF (2016) Emergence and evolution of secondary lymphoid organs. Annu Rev Cell Dev Biol 32:693–711

Nekrasova O, Green KJ (2013) Desmosome assembly and dynamics. Trends Cell Biol 23(11):537–546

Nelissen H, Gonzalez N, Inzé D (2016) Leaf growth in dicots and monocots: so different yet so alike. Curr Opin Plant Biol 33:72–76

Nelson JW, Atilho RM, Sherlock ME, Stockbridge RB, Breaker RR (2017) Metabolism of free guanidine in bacteria is regulated by a widespread riboswitch class. Mol Cell 65:220–230

Netea MG, Schlitzer A, Placek K, Joosten LAB, Schultze JL (2019) Innate and adaptive immune memory: an evolutionary continuum in the host's response to pathogens. Cell Host Microbe 9:13–26

Netzker T, Fischer J, Weber J, Mattern DJ, König CC et al (2015) Microbial communication leading to the activation of silent fungal secondary metabolite gene clusters. Front Microbiol 6:299

Nevitt G (1999) Olfactory foraging in Antarctic seabirds: a species-specific attraction to krill odors. Mar Ecol Prog Ser 177:235–241

Nevitt GA (2008) Sensory ecology on the high seas: the odor world of the procellariiform seabirds. J Exp Biol 211:1706–1713

Nevitt GA, Bonadonna F (2005) Sensitivity to dimethyl sulphide suggests a mechanism for olfactory navigation by seabirds. Biol Lett 1:303–305

Nevitt GA, Losekoot M, Weimerskirch H (2008) Evidence for olfactory search in wandering albatross, *Diomedea exulans*. PNAS 105(12):4576–4581

Newman SA (2011) Animal egg as evolutionary innovation: a solution to the "embryonic hourglass" puzzle. J Exp Zool 316:467–483

Newman SA (2012) Physico-genetic determinants in the evolution of development. Science 338:217–219

Newman SA, Bhat R (2008) Dynamical patterning modules: physic-genetic determinants of morphological development and evolution. Phys Biol 5:015008

Newman SA, Linde-Medina M (2013) Physical determinants in the emergence and inheritance of multicellular form. Biol Theory 8:274–285

Ngou BPM, Ding P, Jones JDG (2022a) Thirty years of resistance: zig-zag through the plant immune system. Plant Cell 34:1447–1478

Ngou BPM, Jones JDG, Ding P (2022b) Plant immune networks. Trends Plant Sci 27(3):255–273

Nielsen C (2009) How did indirect development with planktotrophic larvae evolve? Biol Bull 216:203–215

Nielsen LP, Risgaard-Petersen N, Fossing H, Christensen PB, Sayama M (2010) Electric currents couple spatially separated biogeochemical processes in marine sediment. Nature 463:1071–1074

Nielsen MS, Axelsen LN, Sorgen PL, Verma V, Delmar M, Holstein-Rathlou N-H (2012) Gap junctions. Compr Physiol. https://doi.org/10.1002/cphy.c110051

Niinemets Ü, Keenan TF, Hallik L (2015) A worldwide analysis of within-canopy variations in leaf structural, chemical and physiological traits across plant functional types. New Phytol 205:973–993

Nikolov S, Fabritius H, Petrov M, Friák M, Lymperakis L et al (2011) Robustness and optimal use of design principles of exoskeleton studied by ab initio-based multiscale simulations. J Mech Behav Biomed Mater 4:129–145

Nikolov LA, Tomlison PB, Manickam S, Endress PK, Kramer EM, Davis CC (2014) Holoparasitic Rafflesiaceae possess the most reduced endophytes and yet give rise to the world's largest flowers. Ann Bot 114:233–242

Nilsson D-E, Colley NJ (2016) Comparative vision: can bacteria really see? Curr Biol 36:R369–R371

Nilsson D-E, Gislén L, Coates MM, Skogh C, Garm A (2005) Advanced optics in a jellyfish eye. Nature 435:201–205

Nilsson D-E, Warrant EJ, Johnsen S, Hanlon R, Shaahar N (2012) A unique advantage for giant eyes in giant squid. Curr Biol:683–688

Nishikawa KC (2002) Evolutionary convergence in nervous systems: insights from comparative phylogenetic studies. Brain Behav Evol 59:240–249

Niven JW (2007) Evolution: convergent eye losses in fishy circumstances. Curr Biol 18(1):R27–R29

Niven JE (2015) Neural evolution: costing the benefits of eye loss. Curr Biol 25:R840–R841

Niven JE, Laughlin SB (2008) Energy limitation as a selective pressure on the evolution of sensory systems. J Exp Biol 211:1792–1804

Nonaka S, Yoshiba S, Watanabe D, Ikeuchi S, Goto T et al (2005) De novo formation of left-right asymmetry by posterior tilt of nodal cilia. PLoS Biol 3(8):e268

Norman MD, Finn J, Trgenza T (2001) Dynamic mimicry in an Indo-Malayan octopus. Proc Biol Sci 268:1755–1758

Norris TB, McDermott TR, Castenholz RW (2002) The long-term effects of UV exclusion on the microbial composition and photosynthetic competence of bacteria in hot-spring microbial mats. FEMS Microbiol Ecol 39:193–209

Northcutt RG (2002) Understanding vertebrate brain evolution. Integr Comp Biol 42:743–756

Northcutt RG (2012) Evolution of centralized nervous systems: two schools of evolutionary thought. PNAS 109:10626–19633

Nowell ARM, Jumars PA (1984) Flow environments of aquatic benthos. Ann Res Ecol Syst 15:303–328

Nowrousian M (2014) Genomics and transcriptomics to analyze fruiting body development. In: Nowrousian M (ed) Fungal genomics. Springer, Heidelberg

Nualart-Marti A, Solsona C, Fields RD (2013) Gap junction communication in myelinating glia. Biochim Biophys Acta 1828:69–78

Nürnberg DJ, Mariscal V, Bornikoel J, Nieves-Morión M, Krauß N et al (2015) Intercellular diffusion of a fluorescent sucrose analog via the septal junctions in a filamentous cyanobacterium. mBio 6(2):e02109–e02114

O'Connell LA, Hofmann HA (2011) The vertebrate mesolimbic reward system and social behavior network: a comparative synthesis. J Comp Neurol 519:3599–3639

O'Connell LA, Hofmann HA (2012) Evolution of a vertebrate social decision-making network. Science 336:1154–1157

O'Neil JM, Davis TW, Burford MA, Gobler CJ (2012) The rise of harmful cyanobacteria blooms: the potential roles of Eutrophication and climate change. Harmful Algae 14:313–334

Oda H, Takeichi M (2011) Structural and functional diversity of cadherin at the adherens junction. J Cell Biol 193(7):1137–1146

Ogura Y, Sasakura Y (2017) Emerging mechanisms regulating mitotic synchrony during animal embryogenesis. Dev Growth Differ 59:565–579

Oisi V, Ota KG, Kuraku S, Fujimoto S, Kuratani S (2013) Cranofacial development of hagfishes and the evolution of vertebrates. Nature 493:175–180

Okasha S (2004) Multi-level selection, covariance and contextual analysis. Br J Philos Sci 55:481–504

Oldroyd BP, Fewell JH (2007) Genetic diversity promotes homeostasis in insect colonies. Trends Ecol Evol 22(8):408–413

Oono R, Anderson CG, Denison RF (2010a) Failure to fix nitrogen by non-reproductive symbiotic rhizobia triggers host sanctions that reduce fitness of their reproductive clonemates. Proc Biol Sci. https://doi.org/10.1098/rspb.2010.2193

Oono R, Schmitt I, Sprent JI, Denison RF (2010b) Multiple evolutionary origins of legume traits leading to extreme rhizobial differentiation. New Phytol 187:508–520

Orians CM, van Vuuren MMI, Harris NL, Babst BA, Ellmore GS (2004) Differential sectoriality in long-distance transport in temperate tree species: evidence from dye flow, ^{15}N transport, and vessel element pitting. Trees 18:501–509

Overmann J (2010) The phototrophic consortium "chlorochromatium aggregatum" - a model for bacterial heterologous multicellularity. In: Hallenbeck PC (ed) Recent advances in phototrophic prokaryotes. Springer, New York, pp 15–29

Oyamada M, Takebe K, Oyamada Y (2013) Regulation of connexion expression by transcription factors and epigenetic mechanisms. Biochim Biophys Acta 1828:118–133

Packard A (1972) Cephalopods and fish: the limits of convergence. Biol Rev 47:241–307

Packard A (2006) Contribution to the whole (H). Can squids show us anything that we did not know already? Biol Philos 21:189–211

Paerl HW, Pinckney JL (1996) A mini-review of microbial consortia: their roles in aquatic production and biogeochemical cycling. Microb Ecol 31:225–247

Paerl HW, Hall NS, Calandrino ES (2011) Controlling harmful cyanobacterial blooms in a world experiencing anthropogenic and climatic-induced change. Sci Total Environ 409:1739–1745

Pallas V, Gómez G (2013) Phloem RNA-binding proteins as potential components of the long-distance RNA transport system. Front Plant Sci 4:130

Palmer GE, Horton JS (2006) Mushrooms by magic: making connections between signal transduction and fruiting body development in the basidiomycete fungus *Schizophyllum commune*. FEMS Microbiol Lett 262:1–8

Panchin YV (2005) Evolution of gap junction proteins—the pannexin alternative. J Exp Biol 208:1415–1419

Papadopulos AST, Powell MP, Pupulin F, Warner J, Hawkins JA et al (2013) Convergent evolution of floral signals underlies the success of neotropical orchids. Proc R Soc B 280:20130960

Papenfort K, Bassler B (2016) Quorum-sensing signal-response systems in gram-negative bacteria. Nat Rev Microbiol 14(9):576–588

Paré A, Kim M, Juarez MT, Brody S, McGinnis W (2012) The functions of grainy head-like proteins in animals and fungi and the evolution of apical extracellular barriers. PLoS One 7(5):e36254

Parfrey LW, Lahr DJG (2013) Multicellularity arose several times in the evolution of eukaryotes. Bioessays 35:339–347

Park D, Spencer JA, Koh BI, Kobayashi T, Fujisaki J et al (2012) Endogenous bone marrow MSCs are dynamic, fate-restricted participants in bone maintenance and regeneration. Cell Stem Cell 10:259–272

Parkinson JS, Ames P, Studdert CA (2005) Collaborative signaling by bacterial chemoreceptors. Curr Opin Microbiol 8:116–121

Parpinelli RS, Lopes HS (2011) New inspirations in swarm intelligence: a survey. Int J Bio-Inspired Comput 3(1):1–16

Parry L, Caron J-B (2019) Canadia spinosa and the early evolution of the annelid nervous system. Sci Adv 5:eaax5858

Partida-Martines LP, Hertweck C (2005) Pathogenic fungus harbours endosymbiotic bacteria for toxin production. Nature 437:884–888

Patek SN (2015) The most powerful movements in biology. Am Sci 103:330–337

Paul EA (ed) (2007) Soil microbiology, ecology, and biochemistry. Academic Press, Amsterdam

Paul C, Mausz MA, Pohnert G (2012) A co-culturing/metabolomics approach to investigate chemically mediated interactions of planktonic organisms reveals influence of bacteria on diatom metabolism. Metabolomics. https://doi.org/10.1007/s11306-012-0453-1

Paulin MG, Cahill-Lane J (2021) Events in early nervous system evolution. Top Cogn Sci 13:25–44

Pawlik JR (2011) The chemical ecology of sponges on caribbean reefs: natural products shape natural systems. BioScience 61:888–898

Pechanova O, Hsu C-Y, Adams JP, Pechan T, Vandervelde L et al (2010) Apoplast proteome reveals that extracellular contributes to multistress response in poplar. BMC Genomics 11:674

Pechenik JA (1999) On the advantages and disadvantages of larval stages in benthic marine invertebrate life cycles. Mar Ecol Prog Ser 177:269–297

Peeters SH, de Jonge M (2018) For the greater good: programmed cell death in bacterial communities. Microbiol Res 207:161–169

Pekárová B, Szmitkowska A, Dopitová R, Degtjank O, Židek L, Hejátko J (2016) Structural aspects of multistep phosphorelay-mediated signaling in plants. Mol Plant 9:71–85

Penuela S, Gehi R, Laird DW (2013) The biochemistry and function of pannexin channels. Biochim Biophys Acta 1828:15–22

Pereda AE, Curti S, Hoge G, Cachope R, Flores CE, Rash JE (2013) Gap junction-mediated electrical transmission: regulatory mechanisms and plasticity. Biochim Biophys Acta 1828:134–146

Pereira CG, Almenara DP, Winter CE, Fritsch PW, Lambers H, Oliveira RS (2012) Underground leaves of *Philcoxia* trap and digest nematodes. PNAS 109(4):1154–1158

Pereira-Leal JB (2008) The Ypt/Rab Family and the evolution of trafficking in fungi. Traffic 9:27–38

Pérez-Pomares JM, Muñoz-Chápuli R (2002) Epithelial-mesenchymal transitions: a mesodermal cell strategy for evolutive innovation in metazoans. Anat Rec 268:343–351

Peris D, Pérez-de la Fuente R, Peñalver E, Delciòs X, Barrón E, Labandeira CC (2017) False blister beetles and the expansion of gymnosperm-insect pollination modes before angiosperm dominance. Curr Biol 27:1–8

Perkins TJ, Swain PS (2009) Strategies for cellular decision-making. Mol Syst Biol 5:326

Persat A, Nadell CD, Kim MK, Ingremeau F, Siryapom A et al (2015) The mechanical world of bacteria. Cell 161:988–997

Persike M, Meinhardt G (2016) Contour integration with corners. Vision Res 127:132–140

Peterson KJ, Dietrich MR, McPeek MA (2009) MicroRNAs and metazoan macroevolution: insights into canalization, complexity, and the Cambrian explosion. Bioessays. https://doi.org/10.1002/bies.200900033

Petie R, Garm A, Hall MR (2016) Crown-of-thorns starfish have true image forming vision. Front Zool 13:41

Pfaller MA, Pappas PG, Wingard JR (2006) Invasive fungal pathogens: current epidemiological trends. Clin Infect Dis 43:S3–S14

Phillips RD, Scaccabarozzi D, Retter BA, Hayes C, Brown GR et al (2014) Caught in the act: pollination of sexually deceptive trap-flowers by fungus gnats in *Pterostylis* (Orchidadeae). Ann Bot 113:629–641

Phoenix VR, Konhauser KO (2008) Benefits of bacterial biomineralization. Geobiology 6:303–308

Picao-Osorio J, Johnston J, Landgraf M, Berni J, Alonso CR (2015) MicroRNA-encoded behavior in *Drosophila*. Science 350:815–820

Piercey-Normore MD (2006) The lichen-forming ascomycete *Evernia mesomorpha* associates with multiple genotypes of *Trebouxia jamesii*. New Phytol 169:331–344

Plotnick RE, Young GA, Hagadorn JW (2023) An abundant sea anemone from the carboniferous Mazon Creek Lagerstätte, USA. In: Papers in palaeontology. Wiley. https://doi.org/10.1002/spp2.1479

Plotnikov SV, Pasapera AM, Sabass B, Waterman CM (2012) Force fluctuations within focal adhesions mediate ECM-rigidity sensing to guide directed cell migration. Cell 151:1513–1527

Pohnert G, Steinke M, Tollrian R (2007) Chemical cues, defence metabolites and the shaping of pelagic interspecific interactions. Trends Ecol Evol 22(4):198–204

Polilov AA (2012) The smallest insects evolve anucleate neurons. Arthropod Struct Dev 41:29–34

Ponce de León I, Hamberg M, Castresana C (2015) Oxylipins in moss development and defense. Front Plant Sci 6:483

Popat R, Crusz SA, Messina M, Williams P, West SA, Diggle SP (2012) Quorum-sensing and cheating in bacterial biofilms. Proc Biol Sci 279:4765–4771

Popper KR (2009) Alles Leben ist Problemlösen. Piper

Popper KR, Eccles JC (1990) Das Ich und sein Gehirn. Piper, Munich

Popper ZA, Tuohy MG (2010) Beyond the green: understanding the evolutionary puzzle of plant and algal cell walls. Plant Physiol 153:373–383

Poppinga S, Daber LE, Westermeier AS, Kruppert S, Horstmann M et al (2017) Biomechanical analysis of prey capture in the carnivorous Southern bladderwort (*Utricularia australis*). Sci Rep 7:1776

Potolicchio B, Cigliola V, Velazquez-Garcia S, Klee P, Valjevac A et al (2012) Connexin-dependent signalling in neuro-hormonal systems. Biochim Biophys Acta 1818:1919–1936

Pouliot R, Rochefort L, Gauthier G (2009) Moss carpets constrain the fertilizing effects of herbivores on graminoid plants in arctic polygon fens. Botany 87:1209–1222

Poulson-Ellestad KL, Jones CM, Roy J, Viant MR, Fernandez FM et al (2014) Metabolomics and proteomics reveal impact of chemically mediated competition on marine phytoplankton. PNAS. https://doi.org/10.1073/pnas.1402130111

Prado M, da Silva MB, Laurenti R, Travassos LR, Taborda CP (2009) Mortality due to systemic mycoses as a primary cause of death or in association with AIDS in Brazil: a review from 1996 to 2006. Mem Inst Oswaldo Cruz 104(3):513–521

Pressel S, Bidartondo MI, Ligrone R, Duckett JG (2010) Fungal symbioses in bryophytes: new insights in the twenty first century. Phytotaxa 9:238–253

Prindle A, Liu J, Asally M, Ly S, Garcia-Ojalvo J, Süel GM (2015) Ion channels enable electrical communication in bacterial communities. Nature 527:59–63

Puhl HL III, Lu VB, Won Y-J, Sasson Y, Hirsch JA et al (2014) Ancient origins of RGK protein function: modulation of voltage-gated calcium channels preceded the protostome and deuterostome split. PLoS One 9(7):e100694

Pulina MV, Hou S-Y, Mittal A, Julich D, Whittaker CA et al (2011) Essential roles of fibronectin in the development of the left-right embryonic body plan. Dev Biol 354:208–220

Purcell EE, Crosson S (2008) Photoregulation in prokaryotes. Curr Opin Microbiol 11:168–178

Qi J, Wang J, Gong Z, Zhou J-M (2017) Apoplastic ROS signalling in plant community. Curr Opin Plant Biol 38:92–100

Quint M, Drost H-G, Gabel A, Ullrich KK, Bönn M, Grosse I (2012) A transcriptomic hourglass in plant embryogenesis. Nature 490:98–101

Rabaey K, Rozendal RA (2010) Microbial electrosynthesis—revisiting the electrical route for microbial production. Nat Rev Microbiol 8:706–716

Raff RA (2008) Origins of the other metazoan body plans: the evolution of larval forms. Philos Trans R Soc Lond B Biol Sci 363:1473–1479

Raguso RA (2016) Plant evolution: repeated loss of floral scent—a path of least resistance? Curr Biol 26:R1272–R1296

Rainey PB, Kerr B (2010) Cheats as first propagules: a new hypothesis for the evolution of individuality during the transition from single cells to multicellularity. Bioessays 32(10):872–880

Ramage G, Rajendran R, Sherry L, Williams C (2012) Fungal biofilm resistance. Int J Microbiol. https://doi.org/10.1155/2012/528521

Raman A (2012) Gall induction by hemipteroid insects. J Plant Interact 7(1):29–44

Ramirez F, Sakai LY (2010) Biogenesis and function of fibrillin assemblies. Cell Tissue Res 339(1):71–82

Ramos AC, Façanha AR, Feijó JA (2008) Proton (H+) flux signature for the presymbiotic development of the arbuscular mycorrhizal fungi. New Phytol 178:177–188

Ratcliff WC, Herron MD, Howell K, Pentz J, Rosenzweig F, Travisano M (2013) Experimental evolution of an alternating uni- and multicellular life cycle in *Chlamydomonas reinhardtii*. Nat Commun. https://doi.org/10.1038/ncomms3742

Raven JA (2014) Speedy small stomata? J Exp Bot 65(6):1415–1424

Read ND (2011) Exocytosis and growth do not occur only at hyphal tips. Mol Microbiol 81(1):6–7

Reddy GVP, Guerrero A (2004) Interactions of insect pheromones and plant semiochemicals. Trends Plant Sci 9(5):253–261

Regmi KC, Zhang S, Gaxiola RA (2016) Apoplasmic loading in the rice phloem supported by the presence of sucrose synthases and plasma membrane-localized proton pyrophosphatase. Ann Bot 117:247–268

Reguera G (2011) When microbial conversations get physical. Trends Microbiol 19(3):105–113

Reguera G, McCarthy KD, Mehta T, Nicoll JS, Tuominen MT, Lovley DR (2005) Extracellular electron transfer via microbial nanowires. Nature 435:1098–1101

Reguera G, Nevin KP, Nicoll JS, Covalia SF, Woodard TL, Lovley DR (2006) Biofilm and nanowire production leads to increased current in *Geobacter sulfurreducens* fuel cells. Appl Environ Microbiol 71(11):7345–7348

Rehkämper G, Schuchmann K-L, Schleicher A, Ziller K (1991) Encephalization in hummingbirds (Trochilidae). Brain Behav Evol 37:85–91

Reichelt AC, Hare DJ, Bussey TJ, Saksida LM (2019) Perineuronal nets: plasticity, protection and therapeutic potential. Trends Neurosci 42(7):458–470

Reid RP, Visscher PT, Decho AW, Stolz JF, Bebout BM et al (2000) The role of microbes in accretion lamination and early lithification of modern marine microbialites. Nature 406:989–992

Reid CR, Latty T, Dussutour A, Beekman M (2012) Slime mold uses an externalized spatial "memory" to navigate in complex environments. PNAS 109(43):17490–17494

Reifenrath K, Theisen I, Schnitzler J, Porembski S, Barthlott W (2006) Trap architecture in carnivorous *Utricularia* (Lentibulariaceae). Flora 201:597–605

Revil A (2013) On charge accumulation in heterogeneous porous rocks under the influence of an external electric field. Geophysics 78(4):D271–D291

Richards GS, Degnan BM (2009) The dawn of developmental signaling in the metazoa. Cold Spring Harb Symp Quant Biol LXXIV. https://doi.org/10.1101/sqb.2009.74.028

Richards TA, Gomes S (2015) How to build a microbial eye. Nature. https://doi.org/10.1038/nature14630

Richmond RH, Jokiel PL (1984) Lunar periodicity in larva release in the reef coral *Pocillopora damicornis* at Enewetak and Hawaii. Bull Mar Sci 34(2):280–287

Riedl R (1989) Die Strategie der Genesis. Piper, Munich

Riedl R (2000) Strukturen der Komplexität. Springer, Berlin

Rinaldi AC, Comandini O, Kuyper TW (2008) Extomycorrhizal fungal diversity: separating the wheat from the chaff. Fungal Divers 33:1–45

Riquelme M (2013) Tip growth in filamentous fungi: a road trip to the apex. Annu Rev Microbiol 67:587–609

Roberts EC, Legrand C, Steinke M, Wootton EC (2011) Mechanisms underlying chemical interactions between predatory planktonic protists and their prey. J Plankton Res 33(6):833–841

Robinson GE (1992) Regulation of division of labor in insect societies. Annu Rev Entomol 37:637–665

Rodriguez RJ, White JF, Arnold AE, Redman RS (2009) Fungal endophytes: diversity and functional roles. New Phytol 182:314–330

Rodríguez RM, López-Vázquez A, López-Larrea C (2012) Immune systems evolution. In: López-Larrea C (ed) Sensing in nature. Springer, New York, pp 237–251

Rodriguez-Romero J, Hedtke M, Kastner C, Müller S, Fischer R (2010) Fungi, hidden in soil or up in the air: light makes a difference. Annu Rev Microbiol 64:585–610

Rodríguez-Sinovas A, Sánchez JA, Fernandez-Sanz C, Ruiz-Meana M, Garcia-Dorado D (2012) Connexin and pannexin as modulators of myocardial injury. Biochim Biophys Acta 1818:1962–1970

Romero D, Traxler MF, López D, Kolter R (2011) Antibiotics as signal molecules. Chem Rev 111(9):5492–5505

Ronald PC, Beutler B (2010) Plant and animal sensors of conserved microbial signatures. Science 330:1061–1064

Roos RE, van Zuijlen K, Birkemoe T, Klanderud K, Lang SI et al (2019) Contrasting drivers of community-level trait variation for vascular plants, lichens and bryophytes across an elevational gradient. Funct Ecol 33:2430–2446

Root-Bernstein M (2010) Displacement activities during the honeybee transition from waggle dance to foraging. Anim Behav 79:935–938

Rosales C, Uribe-Querol E (2017) Phagocytosis: a fundamental process in immunity. Biomed Res Int. https://doi.org/10.1155/2017/9042851

Rosenberg SS, Spitzer NC (2011) Calcium signaling in neuronal development. Cold Spring Harb Perspect Biol 3:a004259

Rosenstiel TN, Shortlidge EE, Melnychenko AN, Pankow JF, Eppley SM (2012) Sex-specific volatile compounds influence microarthropod-mediated fertilization of moss. Nature 489:431–433

Rosenzweig JA, Abogunde O, Thomas K, Lawal A, Nguyen YU et al (2010) Spaceflight and modelled microgravity effects on microbial growth and virulence. Appl Microbiol Biotechnol 85:885–891

Roth MS (2014) The engine of the reef: photobiology of the coral-algal symbiosis. Front Microbiol 5:422

Roth G, Dicke U (2005) Evolution of the brain and intelligence. Trends Cogn Sci 9(5):250–257

Rouse GW (2000) Polychaetes have evolved feeding larvae numerous times. Bull Mar Sci 67(1):391–409

Rowe TB, Shepherd GM (2016) The role of ortho-retronasal olfaction in mammalian cortical evolution. J Comp Neurol 524(3):471–495

Rowe N, Speck T (2005) Plant growth forms: an ecological and evolutionary perspective. New Phytol 166:61–72

Rozario T, DeSimone DW (2010) The extracellular matrix in development and morphogenesis: a dynamic view. Dev Biol 341:126–140

Rübsam M, Broussard JA, Wickström SA, Nekrasova O, Green KJ, Niessen CM (2018) Adherens junctions and desmosomes coordinate mechanics and signaling to orchestrate tissue morphogenesis and function: an evolutionary perspective. Cold Spring Harb Perspect Biol 10:a029207

Rustom A, Saffrich R, Markovic I, Walther P, Gerdes H-H (2004) Nanotubular highways for intercellular organelle transport. Science 303:1007–1010

Ruts T, Matsubara S, Wiese-Klinkenberg A, Walter A (2012) Diel patterns of leaf and root growth: endogenous rhythmicity or environmental response? J Exp Bot 63(9):3339–3351

Ryan RP, Dow JM (2008) Diffusible signals and interspecies communication in bacteria. Microbiology 154:1845–1858

Ryan TJ, Grant SGN (2009) The origin and evolution of synapses. Nat Rev Neurosci 10:701–712

Ryan JF, Pang K, Schnitzler CE, Nguyen A-D, Moreland T et al (2013) The genome of the ctenophore *Mnemiopsis leidyi* and its implications for cell type evolution. Science 342:1242592

Ryan TJ, Roy DS, Pignatelli M, Arons A, Tonegawa S (2015) Engram cells retain memory under retrograde amnesia. Science 348:1007–1013

Rycroft MJ, Israelsson S, Price C (2000) The global atmospheric electric circuit, solar activity and climate change. J Atmos Sol Terr Phys 62:1563–1576

Sabagh LT, RjP D, Branco CWC, Rocha CFD (2011) News records of phoresy and hyperphoresy among treefrogs, ostracods, and ciliates in bromeliad of Atlantic forest. Biodivers Conserv 20:1837–1841

Sablowski R (2016) Coordination of plant cell growth and division: collective control of mutual agreement. Curr Opin Plant Biol 34:54–60

Saikkonen K, Wäli P, Helander M, Faeth SH (2004) Evolution of endophyte-plant symbioses. Trends Plant Sci 9(6):275–280

Saikkonen K, Gundel PE, Helander M (2013) Chemical ecology mediated by fungal endophytes in grasses. J Chem Ecol 39:962–968

Saikkonen K, Mikola J, Helander M (2015) Endophytic phyllosphere fungi and nutrient cycling in terrestrial ecosystems. Curr Sci 109(1):121–126

Saletore Y, Meyer K, Korlach J, Vilfan ID, Jaffrey S, Mason CE (2012) The birth of the epitranscriptome: deciphering the function of RNA modifications. Genome Biol 13:175

Saletore Y, Chen-Kiang S, Mason CE (2013) Novel RNA regulatory mechanisms revealed in the epitranscriptome. RNA Biol 10(3):342–346

Salzer JL, Zalc B (2016) Myelination. Curr Biol 26:R971–R975

Samson JE, Magadán AH, Sabri M, Moineau S (2013) Revenge of the phages: defeating bacterial defences. Nat Rev Microbiol 11:675–687

Samu D, Seth AK, Nowotny T (2014) Influence of wiring cost on the large-scale architecture of human cortical connectivity. PLoS Comput Biol 10(4):e1003557

Sandmeier FC, Tracy RC (2014) The metabolic pace-of-life model: incorporating ectothermic organisms into the theory of vertebrate ecoimmunology. Integr Comp Biol 53(3):387–395

Santi C, Bogusz D, Franche C (2013) Biological nitrogen fixation in non-legume plants. Ann Bot 111:743–767

Sasse J, Martinoia E, Northen T (2018) Feed your friends: do plant exudates shape the root microbiome? Trends Plant Sci 23(1):25–41

Satir P, Christensen ST (2007) Overview of structure and function of mammalian cilia. Annu Rev Physiol 69:377–400

Sato F, Tsuchiya S, Meitzer SJ, Shimizu K (2011) MicroRNAs and epigenetics: FEBS J 278:1598–1609

Schaap P (2016) Evolution of developmental signalling in Dictyostelid social amoebas. Curr Opin Genet Dev 39:29–34

Schaarschmidt S, Kopka J, Ludwig-Müller J, Hause B (2007) Regulation of arbuscular mycorrhization by apoplastic invertases: enhanced invertase activity in the leaf apoplast affects the symbiotic interaction. Plant J 51:390–405

Schad W (2001) Lunar influence on plants. Earth Moon Planets 85–86:405–409

Schaefer HM, Ruxton GD (2008) Fatal attraction: carnivorous plants roll out the red carpet to lure insects. Biol Lett 4:153–155

Schaefer HM, Schaefer V, Levey DJ (2004) How plant-animal interactions signal new insights in communications. Trends Ecol Evol 19(11):577–584

Schafer W (2016) Nematode nervous systems. Curr Biol 26:R955–R959

Schaller E, Shiu S-H, Armitage JP (2011) Two-component systems and their co-option review for eukaryotic signal transduction. Curr Biol 21:R320–R330

Schauer R, Risgaard-Petersen N, Kjeldsen KU, Bjerg JJT, Jørgensen BB et al (2014) Succession of a cable bacteria and electric currents in marine sediment. ISME J 8:1314–1322

Scheel C, Weinberg RA (2011) Phenotypic plasticity and epithelial mesenchymal transitions in cancer and normal stem cells. Int J Cancer 129:2310–2314

Scheele BC, Pasmans F, Skerratt LF, Berger L, Martel A et al (2019) Amphibian fungal panzootic causes catastrophic and ongoing loss of biodiversity. Science 363:1459–1463

Schenkelaars Q, Pratlong M, Kodjabachian L, Fierro-Constain L, Vacelet J et al (2017) Animal multicellularity and polarity without Wnt signaling. Sci Rep. https://doi.org/10.1038/s41598-017-15557-5

Scherlach K, Gaupner K, Hertweck C (2013) Molecular bacteria-fungi interactions: effects on environment, food, and medicine. Annu Rev Microbiol 67:375–397

Scherzer S, Böhm J, Krol E, Shabala L, Kreuzer I et al (2015) Calcium sensor kinase activates potassium uptake systems in gland cells of Venus flytrap. PNAS 112(23):7309–7314

Schiestl FP, Johnson SD (2013) Pollinator-mediated evolution of floral signals. Trends Ecol Evol 28(5):307–315

Schirrmeister BE, de Vos JM, Antonelli A, Bagheri HC (2013) Evolution of multicellularity coincided with increased diversification of cyanobacteria and the great oxidation event. PNAS 110(5):1791–1796

Schmid B (2016) Decision-making: are plants more rational than animals? Curr Biol 26:R667–R668

Schmidt R, Etalo DW, de Jager V, Gerards S, Zweers H et al (2016) Microbial small talk: volatiles in fungal-bacterial interactions. Front Microbiol 6:1496

Schmidt-Rhaesa A, Harzsch S, Purschke G (2016) Structure and evolution of invertebrate nervous systems. Oxford University Press, Oxford

Schneider K, Resl P, Spribille T (2016) Escape from the cryptic species trap: lichen evolution on both sides of a cynobacterial acquisition event. Mol Ecol 25:3453–3468

Schoch RR (2014) Amphibian evolution. Wiley Blackwell, Oxford

Schoenemann PT (2006) Evolution of the size and functional areas of the human brain. Annu Rev Anthropol 35:379–406

Schoenemann B, Clarkson ENK (2013) Discovery of some 400 million year-old sensory structures in the compound eyes of trilobites. Sci Rep 3:1429

Scholin CA, Gulland F, Doucette GJ, Benson S, Busman M et al (2000) Mortality of sea lions along the central California coast linked to toxic diatom bloom. Nature 403:80–84

Scholtz G, Staude A, Dunlop JA (2019) Trilobite compound eyes with crystalline cones and rhabdoms show mandibulate affinities. Nat Commun 10:2503

Schouteden N, De Waele D, Panis B, Vos CM (2015) Arbuscular mycorrhizal fungi for the biocontrol of plant-parasitic nematodes: a review oft the mechanisms involved. Front Microbiol 6:1280

Schuergers N, Lenn T, Kampmann R, Meissner MV, Esteves T et al (2016) Cyanobacteria use micro-optics to sense light direction. eLife. https://doi.org/10.7554/eLife.12620

Schuetz R, Zamboni N, Zampieri M, Heinemann M, Sauer U (2012) Multidimensional optimality of microbial metabolism. Science 336:601–604

Schultz D, Lu M, Stavropoulos T, Onuchic J, Ben-Jacob E (2013) Turning oscillations into opportunities: lessons from a bacterial decision gate. Sci Rep. https://doi.org/10.1038/srep01668

Scott CA, Tattersall D, O'Toole EA, Kelsell D-P (2013) Connexins in epidermal homeostasis and skin disease. Biochim Biophys Acta 1828:1952–1961

Sebé-Pedros A, Zheng Y, Ruiz-Trillo I, Pan D (2012) Premetazoan origin of the hippo signaling pathway. Cell Rep 1:13–20

Sebé-Pedrós A, Irimia M, del Campo J, Parra-Acero H, Russ C et al (2013) Regulated aggregative multicellularity in a close unicellular relative of metazoa. eLife 2:e01287

Sebé-Pedrós A, Peña MI, Capella-Gutiérrez S, Antó M, Gabaldón T et al (2016) High-throughput proteomics reveals the unicellular roots of animal phosphosignaling and cell differentiation. Dev Cell 39:186–197

Selosse M-A, Roy M (2008) Green plants that feed on fungi: facts and questions about mixotrophy. Trends Plant Sci 14(2):64–70

Selosse M-A, Charpin M, Not F (2017) Mixotrophy everywhere on land and in water: the grand écart hypothesis. Ecol Lett 20:246–263

Sengupta B, Stemmler M, Laughlin SB, Niven JE (2010) Action potential energy efficiency varies among neuron types in vertebrates and invertebrates. PLoS Comput Biol 6(7):e1000840

Serganov A, Nudler E (2013) A decade of riboswitches. Cell 152:17–24

Seymour JR, Simó R, Ahmed T, Stocker R (2010) Chemoattraction to dimethylsulfoniopropionate throughout the marine microbial food web. Science 329:342–345

Shane MW, Lambers H (2005) Cluster roots: a curiosity in context. Plant Soil 274:101–125

Shane MW, Cawthray GR, Cramer MD, Kuo J, Lambers H (2006) Specialized 'dauciform' roots of Cyperaceae are structurally distinct, but functionally analogous with 'cluster' roots. Plant Cell Environ 29:1989–1999

Shank EA, Kolter R (2009) New developments in microbial interspecies signaling. Curr Opin Microbiol 12(2):205–214

Shannon CE (1948) A mathematical theory of communication. Bell Syst Tech J 27:379–423

Shatil-Cohen A, Moshelion M (2012) Smart pipes. Plant Signal Behav 7(9):1085–1091

Shea N, Pen I, Uller T (2011) Three epigenetic information channels and their different roles in evolution. J Evol Biol 24:1178–1187

Sheng G (2015) Epiblast morphogenesis before gastrulation. Dev Biol 401:17–24

Shepherd GM (2011) The microcircuit concept applied to cortical evolutions: from three-Layer to six-layer cortex. Front Neuroanat 5:30

Shimkets LJ (1990) Social and developmental biology of the myxobacteria. Microbiol Rev 54(4):473–501

Shirayama M, Seth M, Lee H-C, Gu W, Ishidate T et al (2012) piRNAs initiate an epigenetic memory of non-self RNA in the C. elegans germline. Cell 150:65–77

Shomrat T, Zarrella I, Fiorito G, Hochner B (2008) The octopus vertical lobe modulates short-term learning rate and uses LTP to acquire long-term memory. Curr Biol 18:337–342

Shukla GC, Singh J, Barik S (2011) MicroRNAs: processing, maturation, target recognition and regulatory functions. Mol Cell Pharmacol 3(3):83–92

Shultz S, Dunbar R (2010) Encephalization is not a universal macroevolutionary phenomenon in mammals but is associated with sociality. PNAS 107(50):21582–21586

Silverstein RN, Correa AMS, Baker AC (2012) Specificity is rarely absolute in coral-algal symbiosis: implications for coral response to climate change. Proc Biol Sci 279:2609–2618

Simakov O, Kawashima T, Marlétaz F, Jenkins J, Koyanagi R et al (2015) Hemichordate genomes and deuterostomes origins. Nature 527:459–465

Singla V, Reiter JF (2006) The primary cilium as the cell's antenna: signaling at a sensory organelle. Science 313:629–633

Six DL (2012) Ecological and evolutionary determinants of Bark Beetle-fungus symbioses. Insects 3:339–366

Slack JMW (2017) Animal regeneration: ancestral character of evolutionary novelty? EMBO Rep 18(9):1497–1508

Sleator RD, Hill C (2001) Bacterial osmoadaptation: the role of osmolytes in bacterial stress and virulence. FEMS Microbiol Rev 26:49–71

Smaers JB, Gómez-Robles A, Parks AN, Sherwood CC (2017) Exceptional evolutionary expansion of prefrontal cortex in great apes and humans. Curr Biol 27:714–720

Smarandache-Wellmann CR, Weller C, Mulloney B (2014) Mechanisms of coordination in distributed neural circuits: decoding and integration of coordinating information. J Neurosci 34(3):793–803

Smith SD (2016) Pleiotropy and the evolution of floral integration. New Phytol 209:80–85

Smith FW, Boothby TC, Giovannini I, Rebecchi L, Jockusch EL, Goldstein B (2016) The compact body plan of tardigrades evolved by the loss of a large body region. Curr Biol 26:224–229

Smith DJ, Montenegro-Johnson TD, Lopes SS (2019) Symmetry-breaking cilia-driven flow in embryogenesis. Annu Rev Fluid Mech 51:105–128

Snaddon JL, Turner EC, Fayle TM, Khen CV, Eggleton P, Foster WA (2012) Biodiversity hanging by a thread: the importance of fungal litter-trapping systems in tropical rainforests. Biol Lett 8:397–400

Snider RM, Strycharz-Glaven SM, Tsoi SD, Erickson JS, Tender LM (2012) Long-range electron transport in *Geobacter sulfurreducens* biofilms is redox gradient-driven. PNAS. https://doi.org/10.1073/pnas.1209829109

Soares DEF, Sufiate BL, de Queiroz JH (2018) Nematophagous fungi: far beyond the endoparasite, predator and ovicidal groups. Agric Nat Resourc 52:1–8

Solan JL, Lampe PD (2018) Spatio-temporal regulation of connexin43 physphorylation and gap junction dynamics. Biochim Biophys Acta 1860:83–90

Solano C, Echeverz M, Lasa I (2014) Biofilm dispersion and quorum sensing. Curr Opin Microbiol 18:96–104

Solnica-Krezel L (2005) Conserved patterns of cell movements during vertebrate gastrulation. Curr Biol 15:R213–R228

Song X, Peng C, Zhou G, Gu H, Li Q, Zhang C (2016) Dynamic allocation and transfer of non-structural carbohydrates, a possible mechanism for the explosive growth of Moso bamboo (*Phyllostachys heterocycla*). Sci Rep. https://doi.org/10.1038/srep25908

Sonoda T, Li JY, Hayes NW, Chan JC, Okane Y et al (2020) A noncanonical inhibitory circuit dampens behavioural sensitivity to light. Science 368:527–531

Sørensen I, Fei Z, Andreas A, Willats WGT, Domozych DS, Rose JKC (2014) Stable transformation and reverse genetic analysis of *Penium margaritaceum*: a platform for studies of charophyte green algae, the immediate ancestors of land plants. Plant J 77:339–351

Sorg BA, Berretta S, Blacktop JM, Fawcett JW, Kitagawa H et al (2016) Casting a wide net: role of perineuronal nets in neural plasticity. J Neurosci 36(45):11459–11468

Sorrells TR, Johnson AD (2015) Making sense of transcription networks. Cell 161:714–723

Sorrells TR, Booth LN, Tuch BB, Johnson AD (2015) Intersecting transcription networks constrain gene regulatory evolution. Nature 523:361–365

Sotthibandhu S, Baker RR (1979) Celestial orientation by the large yellow underwing moth, *Noctua pronuba* L. Anim Behav 27:786–800

Soundararajan S, Jedd G, Li X, Ramos-Pamplofa M, Chua NH, Naqvi NI (2004) Woronin body function in *Magnaporthe grisea* is essential for efficient pathogenesis and for survival during nitrogen starvation stress. Plant Cell 16:1564–1574

Speck T, Burgert I (2011) Plant stems: functional design and mechanics. Annu Rev Mater Res 41:169–191

Spicer R, Groover A (2010) Evolution of development of vascular cambia and secondary growth. New Phytol 186:577–592

Sprent JI, James EK (2007) Legume evolution: where do nodules and mycorrhizas fit in? Plant Physiol 144:575–581

Standfuss J (2015) Viral chemokine mimicry. Science 347:1071–1072

Stanley NR, Lazazzera BA (2004) Environmental signals and regulatory pathways that influence biofilm formation. Mol Microbiol 52(4):917–924

Stearns JC, Lynch MDJ, Senadheera DB, Tenenbaum HC, Goldberg MB et al (2011) Bacterial bio-geography of the human digestive tract. Sci Rep 1:170

Steemans P, Le Hérissé A, Melvin J, Miller MA, Paris F et al (2009) Origin and radiation of the earliest vascular land plants. Science 324:353

Steinberg G, Peñalva MA, Riquelme M, Wösten HA, Harris SD (2017) Cell biology of hyphal growth. Microbiol Spectr 5(2):FUNK-0034-2016

Steindler L, Venturi V (2007) Detection of quorum-sensing N-acyl homoserine lactone signal molecules by bacterial biosensors. FEMS Microbiol Lett 266:1–9

Steinmetz PRH (2019) A non-bilaterian perspective on the development and evolution of animal digestive systems. Cell Tissue Res 377:321–339

Stocker R, Seymour JR (2012) Ecology and physics of bacterial chemotaxis in the ocean. Microbiol Mol Biol Rev 76(4):792–812

Stocker R, Seymour JR, Samadani A, Hunt DE, Polz MF (2008) Rapid chemotactic response enables marine bacteria to exploit ephemeral microscale nutrient patches. PNAS 108(11):4209–4214

Stolarski J, Kitahara MV, Miller DJ, Cairns SD, Mazur M, Meibom A (2011) The ancient evolutionary origins of Scleractinia revealed by azooxanthellate corals. Evol Biol 11:316

Stolze-Rybczynski JL, Cui Y, Stevens HH, Davis DJ, Fischer MWF, Money NP (2009) Adaptation of the spore discharge mechanism in the basidiomycota. PLoS One 4(1):e4163

Stone M, Woo J, Lee J, Poole T, Seagraves A et al (2016) Structure and variability in human tongue muscle anatomy. Comput Methods Biomech Biomed Eng Imaging Vis. https://doi.org/10.108 0/21681163.2016.1162752

Stoner DS, Rinkevich B, Weissman IL (1999) Heritable germ and somatic cell lineage competitions in chimeric colonial protochordates. PNAS 96:9148–9153

Strausfeld NJ, Ma X, Edgecombe GD (2016) Fossils and the evolution of the arthropod brain. Curr Biol 26:R989–R1000

Strycharz-Glaven SM, Snider RM, Guiseppi-Elie A, Tender LM (2011) On the electrical conductivity of microbial nanowires and biofilms. Energy Environ Sci. https://doi.org/10.1039/c1ee01753e

Stucken K, Ilhan J, Roettger M, Dagan T, Martin WF (2012) Transformation and conjugal transfer of foreign genes into the filamentous multicellular cyanobacteria (Subsection V) fischerella and chlorogloeopsis. Curr Microbiol 65:552–560

Suga H, Ruiz-Trillo I (2013) Development of ichthyosporeans sheds light on the origin of metazoan multicellularity. Dev Biol 377:284–292

Suga H, Chen Z, de Mendoza A, Sebé-Pedrós A, Brown MW et al (2013) The Capsaspora genome reveals a complex unicellular prehistory of animals. Nat Commun. https://doi.org/10.1038/ ncomms3325

Sugimoto K, Gordon SP, Meyerowitz EM (2011) Regeneration in plants and animals: dedifferentiation, transdifferentiation, or just differentiation? Trends Cell Biol 21(4):212–218

Sukhov V, Sukhova E, Vodeneev V (2018) Long-distance electrical signals as a link between the local action of stressors and the systemic physiological responses in higher plants. Prog Biophys Mol Biol. https://doi.org/10.1016/j.pbiomolbio.2018.11.009

Sukhum KV, Freiler MK, Wang R, Carlson BA (2016) The costs of a big brain: extreme Encephalization results in higher energetic demand and reduced hypoxia tolerance in weakly electric African fishes. Proc Biol Sci 283:20162157

Sukhum KV, Freller MK, Carlson BA (2019) Intraspecific energetic trade-offs and costs of encephalization vary from interspecific relationships in three species of mormyrid electric fishes. Brain Behav Evol 93:196–205

Sun J, Miller JB, Granqvist E, Wiley-Kalil A, Gobbato E et al (2015) Activation of symbiosis signaling by arbuscular mycorrhizal fungi in legumes and rice. Plant Cell 27:823–838

Sutton GP, Clarke D, Morley EL, Robert D (2016) Mechanosensory hairs in bumblebees (*Bombus terrestris*) detect weak electric fields. PNAS 113(26):7261–7265

Szulwach KE, Li X, Smrt RD, Li Y, Luo Y et al (2010) Cross talk between microRNA and epigenetic regulation in adult neurogenesis. J Cell Biol 1:127–141

Szurmant H, Ordal GW (2004) Diversity in chemotaxis mechanisms among the bacteria and archaea. Microbiol Mol Biol Rev 68(2):301–319

Takeda N, Sato S, Asamizu E, Tabata S, Parniske M (2009) Apoplastic plant subtilases support arbuscular mycorrhiza development in *Lotus japonicus*. Plant J 58:766–777

Takemoto H (2001) Morphological analyses of the human tongue musculature for three-dimensional modeling. J Speech Lang Hear Res 44:195–107

Takeshita N (2016) Coordinated process of polarized growth in filamentous fungi. Biosci Biotechnol Biochem 80(9):1693–1699

Takeshita N, Manck R, Grün N, de Vega SH, Fischer R (2014) Interdependence of the actin and the microtubule cytoskeleton during fungal growth. Curr Opin Microbiol 20:34–41

Takeuchi T, Yamada L, Shinzato C, Sawada H, Satoh N (2016) Stepwise evolution of coral biomineralization revealed with genome-wide proteomics and transcriptomics. PLoS One 11(6):e0156424

Tal O, Haim A, Harel O, Gerchman Y (2011) Melatonin as an antioxidant and its semi-lunar rhythm in green macroalga *Ulva* sp. J Exp Bot 62(6):1903–1910

Tamm SL (1982) Flagellated ectosymbiotic bacteria propel a eucaryotic cell. J Cell Biol 94:697–709

Taylor BL, Zhulin IB, Johnson MS (1999) Aerotaxis and other energy-sensing behavior in bacteria. Annu Rev Microbiol 53:103–128

Tecon R, Leveau JHJ (2016) Symplasmata are a clonal, conditional, and reversible type of bacterial multicellularity. Sci Rep. https://doi.org/10.1038/srep31914

Telford MJ, Budd GE, Philippe H (2015) Phylogenomic insights into animal evolution. Curr Biol 25:R876–R887

Tero A, Kobayashi R, Nakagaki T (2006) Physarum solver: a biologically inspired method of road-network navigation. Phys A Stat Mech Appl 363(1):115–119

Tessmar-Raible K, Raible F, Christodoulou F, Guy K, Rembold M et al (2007) Conserved sensory-neurosecretory cell types in annelid and fish forebrain: insights into hypothalamus evolution. Cell 129:1389–1400

Tessmar-Raible K, Raible F, Arboleda E (2011) Another place, another timer: marine species and the rhythms of life. Bioessays 33:165–172

Tester M, Leigh RA (2001) Partitioning of nutrient transport processes in roots. J Exp Bot 52:445–457

Thacker RW, Becerro MA, Lumbang WA, Paul VJ (1998) Allelopathic interactions between sponges on a tropical reef. Ecology 79(5):1740–1750

Thewissen JGM, George J, Rosa C, Kishida T (2011) Olfaction and brain size in the bowhead whale (*Balaena mysticetus*). Mar Mammal Sci 27(2):282–294

Thiery JP, Acloque H, Huang RYJ, Nieto MA (2009) Epithelial-mesenchymal transitions in development and disease. Cell 139:871–890

Thomas AR, Hill EC (1976) Aspergillus fumigatus and supersonic aviation. 2. Corrosion. Int Biodeterior Bull 12(4):116–119

Thomas F, Cosse A, Le Panse S, Kloareg B, Potin P, Lebland C (2014) Kelps feature systemic defense responses: insights into the evolution of innate immunity in multicellular eukaryotes. New Phytol 204:567–576

Thompson DW (1942) On growth and form. Republication 1992, Dover, New York

Thyrhaug R, Larsen A, Thingstad TF, Bratbak G (2003) Stable coexistence in marine algal host-virus systems. Mar Ecol Prog Ser 254:27–35

Tillmann U (2003) Kill and eat your predator: a winning strategy of the planktonic flagellate *Prymnesium parvum*. Aquat Microbial Ecol 32:73–84

Tomescu AMF, Rothwell GW (2006) Wetlands before tracheophytes: thalloid terrestrial communities of the Early Silurian Passage Creek biota (Virginia). Geol Soc Am Spec Pap 399:41–56

Torres LG (2017) A sense of scale: foraging cetaceans' use of scale-dependent multimodal sensory systems. Mar Mammal Sci 33(4):1170–1193

Torruella G, de Mendoza A, Grau-Bové X, Antó M, Chaplin MA et al (2015) Phylogenomics reveals convergent evolution of lifestyles in close relatives of animals and fungi. Curr Biol 25:2404–2410

Tosches MA, Laurent G (2019) Evolution of neuronal identity in the cerebral cortex. Curr Opin Neurobiol 56:199–208

Toth GB, Norén F, Selander E, Pavia H (2004) Marine dinoflagellates show induced life-history shifts to escape parasite infection in response to water-borne signals. Proc Biol Sci 271:733–738

Trail F (2007) Fungal cannons: explosive spore discharge in the Ascomycota. FEMS Microbiol Lett 276:12–18

Tran VC, Mastantuoni GG, Zabihipour M, Li L, Berglund L et al (2023) Electrical current modulation in wood electrochemical transistor. PNAS 120(18):e2218380120

Truman JW, Riddiford LM (1999) The origins of insect metamorphosis. Nature 401:447–452

Tschinkel WR (1999) Sociometry and sociogenesis of colonies of the harvester ant, *Pogonomyrmex badius*: distribution of workers, brood and seeds within the nest in relation to colony size and season. Ecol Entomol 24:222–237

Tsuboi M, Kotrschal A, Hayward A, Buechel SD, Zidar J et al (2016) Evolution of brain-body allometry in Lake Tanganyika cichlids. Evolution 70(7):1559–1568

Tu KC, Long T, Svenningsen SL, Wingreen NS, Bassler BL (2010) Negative feedback loops involving small regulatory RNAs precisely control the *Vibrio harvey* quorum-sensing response. Mol Cell 37:567–579

Tyler S (2003) Epithelium—the primary building block for metazoan complexity. Integr Comp Biol 43:55–63

Tylewicz S, Petterle A, Martilla S, Miskolczi P, Azeez A et al (2018) Photoperiodic control of seasonal growth is mediated by ABA acting on cell-cell communication. Science 360:212–215

Tzeferis PG, Agatzini S, Nerantzis ET (1994) Mineral leaching of non-sulphide nickel ores using heterotrophic micro-organisms. Lett Appl Microbiol 18:209–213

Uehling J, Deveau A, Paoletti M (2017) Do fungi have an innate immune response? An NLR-based comparison to plant and animal immune system. PLoS Pathog 13(10):e1006578

Ueki N, Ide T, Mochiji S, Kobayashi Y, Tokutsu R et al (2016) Eyespot-dependent determination of phototactic sign in *Chlamydomonas reinhardtii*. PNAS 113(19):5299–5304

Ugurlar D, Howes SS, de Kreuk B-J, Koning RI, de Jong RN et al (2018) Structures of C1-IgG1 provides insights into how danger pattern recognition activates complement. Science 359:794–797

Ulitsky I, Bartel DP (2013) lincRNAs: genomics, evolution, and mechanisms. Cell 154:28–46

Ullrich-Lüter EM, Dupont S, Arboleda E, Hausen H, Arnone MI (2011) Unique system of photoreceptors in sea urchin tube feet. PNAS 108(20):8367–8372

Umpierre AD, Bystrom LL, Ying Y, Liu YU, Worrell G, Wu L-J (2020) Microglial calcium signalling is attuned to neuronal activity in awake mice. eLife 9:e56502

Unsicker SB, Kunert G, Gershenzon J (2009) Protective perfumes: the role of vegetative volatiles in plant defense against herbivores. Curr Opin Plant Biol 12:479–485

Uribe-Querol E, Rosales C (2017) Control of phagocytosis by microbial pathogens. Front Immunol 8:1368

Uroz S, Dessaux Y, Oger P (2009) Quorum sensing and quorum quenching: the Yin and Yang of bacterial communication. Chembiochem 10:205–216

Vacquié-Garcia J, Royer F, Dragon A-C, Viviant M, Bailleul F, Guinet C (2012) Foraging in the darkness of the southern ocean: influence of bioluminescence on a deep diving predator. PLoS One 7(8):e43565

Valix M, Usai F, Malik R (2001) Fungal bio-leaching of low grade laterite ores. Miner Eng 14(2):197–203

van Apeldoorn ME, van Egmond HP, Speijers GJA (2001) Neurotoxic shellfish poisoning: a review. RIVM, Bilthoven

Van Buskirk J, Saxer G (2001) Delayed costs of an induced defense in tadpoles? Morphology, hopping, and development rate at metamorphosis. Evolution 66(4):821–829

van der Heijden MGA, Horton TR (2009) Socialism in soil? The importance of mycorrhizal fungal networks for facilitation in natural ecosystems. J Ecol 97:1139–1150

van der Heijden MGA, Sanders IR (eds) (2003) Mycorrhizal ecology. Springer, Berlin

Van Donk E (2007) Chemical information transfer in freshwater plankton. Ecol Inform 2:112–120

van Duijn M, Keijzer F, Franken D (2006) Principles of minimal cognition: casting cognition as sensorimotor coordination. Adapt Behav 14(2):157–170

van Groenendael JM, Klimeš L, Klimešová J (1996) Comparative ecology of clonal plants. Philos Trans R Soc Lond B Biol Sci 351:1331–1339

van Schalk CP, Burkart J, Damerius L, Forss SIF, Koops K et al (2016) The reluctant innovator: orangutans and the phylogeny of creativity. Philos Trans R Soc Lond B Biol Sci 371:20150183

van Schöll L, Kuyper TW, Smits MM, Landweert R, Hoffland E, van Breemen N (2008) Rock-eating mycorrhizas: their role in plant nutrition and biogeochemical cycles. Plant Soil 303:35–47

Van Soest RWM, Boury-Esnault N, Vacelet J, Dohrmann M, Erpenbeck D et al (2012) Global diversity of sponges (Porifera). PLoS One 7(4):e35105

Vardi A, Schatz D, Beeri K, Motro U, Sukenik A et al (2002) Dinoflagellate-cyanobacterium communication may determine the composition of phytoplankton assemblage in a mesotrophic lake. Curr Biol 12:1787–1772

Varoqueaux F, Williams EA, Grandemange S, Truscello L, Kamm K et al (2018) High cell diversity and complex peptidergic signaling underlie placozoan behavior. Curr Biol 28:3495–3501

Vasek MJ, Garber C, Dorsey D, Durrant DM, Bollman B et al (2016) A complement-microglial axis drives synapse loss during virus-induced memory impairment. Nature 534:538–543

Vasileva E, Citi S (2018) The role of microtubules in the regulation of epithelial junctions. Tissue Barriers 6(3):e1539596

Vatanen T, Kostic AD, d'Hennezel E, Siljander H, Franzosa EA et al (2016) Variation in microbiome LPS immunogenicity contributes to autoimmunity in humans. Cell 165:842–853

Velicer GJ, Yu Y-TN (2003) Evolution of novel cooperative swarming in the bacterium *Myxococcus xanthus*. Nature 425:75–78

Vendeville A, Winzer K, Heurlier K, Tang CM, Hardie KR (2005) Making 'sense' of metabolism: autoinducer-2, LuxS and pathogenic bacteria. Nat Microbiol 3:383–396

Verkhratsky A, Parpura V (2014) Calcium signalling and calcium channels: evolution and general principles. Eur J Pharmacol 739:1–3

Verstraete B, Van Elst D, Steyn H, Van Wyk B, Lamaire B et al (2011) Endophytic bacteria in toxic South African plants: identification, phylogeny and possible involvement in Gousiekte. PLoS One 6(4):e19265

Villareal LP, Witzany G (2010) Viruses are essential agents within the roots and stem of the tree of life. J Theor Biol 262(4):698–710

Vincent O, Weißkopf C, Poppinga S, Masselter T, Speck T et al (2011) Ultra-fast underwater suction traps. Proc Biol Sci 278:2909–2914

Vinken M, Decrock E, Leybaert L, Bultynck G, Himpens B et al (2012) Non-channel function of connexions in cell growth and cell death. Biochim Biophys Acta 1818:2002–2008

Visser AW, Kiørboe T (2006) Plankton motility patterns and encounter rates. Oecologia 148:538–546

Vladimirov N, Sourjik V (2009) Chemotaxis: how bacteria use memory. Biol Chem 390:1097–1104

Vogel S (1994) Life in moving fluids. Princeton University Press, Princeton

Volkov A, Foster JC, Ashby TA, Walker RK, Johnson JA, Markin VS (2010) *Mimosa pudica*: electrical and mechanical stimulation of plant movements. Plant Cell Environ 33:163–173

Vos M, Vet LEM, Wäckers FL, Middleburg JJ, van der Putten WH et al (2006) Infochemicals structure marine, terrestrial and freshwater food webs: implications for ecological informatics. Ecol Inform 1:23–32

Vroman R, Klaassen LJ, Kamermans M (2013) Ephaptic communication in the vertebrate retina. Front Hum Neurosci 7:612

Vu MMK, Jameson NE, Masuda SJ, Lin D, Larraide-Ridaura R, Lupták A (2012) Convergent evolution of adenosine aptamers spanning bacterial, human, and random sequences revealed by structure-based bioinformatics and genomic SELEX. Chem Biol 19:1247–1254

Vuong C, Kocianova S, Voyich JM, Yao Y, Fischerl ER et al (2004) A crucial role for exopolysaccharide modification in bacterial biofilm formation, immune evasion, and virulence. J Biol Chem 279(52):54881–54886

Wagner D, Kelley CD (2017) The largest sponge in the world? Mar Biodivers 47:367–368

Wagner PL, Waldor MK (2002) Bacteriophage control of bacterial virulence. Infect Immun 70(8):3985–3993

Wagner H-J, Douglas RH, Frank TM, Roberts NW, Partridge JC (2009) A novel vertebrate eye using both refractive and reflective optics. Curr Biol 19:108–114

Wainwright M (1988) Metabolic diversity of fungi in relation to growth and mineral cycling in soil—a review. Trans Br Mycol Soc 90(2):159–170

Walling LL (2008) Avoiding effective defenses: strategies employed by phloem-feeding insects. Plant Physiol 146:859–866

Walsh TJ, Groll A, Hiemenez J, Fleming R, Roilides E, Anaissie E (2004) Infections due to emerging and uncommon medically important fungal pathogens. Clin Microbiol Infect 10(Suppl 1):48–66

Walter H, Breckle S-W (1983) Ökologie der Erde Band 1, Ökologische Grundlagen in globaler Sicht. Fischer, Stuttgart

Wang IE, Clandinin TR (2016) The influence of wiring economy on nervous system evolution. Curr Biol 26:R1101–R1108

Wang X, Gerdes H-H (2012) Long-distance electrical coupling via tunneling nanotubes. Biochim Biophys Acta 1818:2082–2086

Wang B, Qiu Y-L (2006) Phylogenetic distribution and evolution of mycorrhizas in land plants. Mycorrhiza 16:299–363

Wang X, Gan L, Jochum KP, Schröder HC, Müller WEG (2011) The largest bio-silica structure on earth: the giant basal spicule from the deep-sea glass sponge *Monorhaphis chuni*. Evid Based Complement Alternat Med 540:987

Wang W, Shi J, Xie Q, Jiang Y, Yu N, Wang E (2017) Nutrient exchange and regulation in arbuscular mycorrhizal symbiosis. Mol Plant 10:1147–1158

Wang R, Wang M, Chen K, Wang S, Mur LAJ, Guo S (2018) Exploring the roles of aquaporins in plant-microbe interactions. Cells 7:267

Wang D, Qin M, Liu L, Liu L, Zhou Y et al (2019) The most extensive devonian fossil forest with small lycopsid trees bearing the earliest stigmarian roots. Curr Biol 29:1–12

Wang C, Yue H, Hu Z, Shen Y, Ma J et al (2020) Microglia mediate forgetting via complement-dependent synaptic elimination. Science 367:688–694

Wargo MJ, Hogan DA (2006) Fungal-bacterial interactions: a mixed bag of mingling microbes. Curr Opin Microbiol 9:359–364

Warrant EJ, Locket NA (2004) Vision in the deep sea. Biol Rev 79:671–712

Watanabe S, Liu Q, Davis MW, Hollopeter G, Thomas N et al (2013a) Ultrafast endocytosis at *Caenorhabditis elegans* neuromuscular junctions. eLife 2:e00723

Watanabe S, Rost BR, Camacho-Pérez M, Davis MW, Kielczynski BS et al (2013b) Ultrafast endocytosis at mouse hippocampal synapses. Nature 504:242–247

Waters LS, Storz G (2009) Regulatory RNAs in bacteria. Cell 136:615–628

Weiner S, Dove PM (2003) An overview of biomineralization processes and the problem of the vital effect. Rev Mineral Geochem 54(1):1–29

Weitz JS, Mileyko Y, Joh RI, Voit EO (2008) Collective decision making in bacterial viruses. Biophys J 95:2673–2680

Welsh RM, Zaneveld JR, Rosales SM, Payet JP, Burkepile DE, Thurber RV (2016) Bacterial predation in a marine host-associated microbiome. ISME J 19:1540–1544

Wenseleers T, Helanterä H, Hart A, Ratnieks FLW (2004) Worker reproduction and policing in insect societies: an ESS analysis. J Evol Biol 17:1035–1047

Wessnitzer J, Webb B (2006) Multimodal sensory integration in insects—towards insect brain control architectures. Bioinspir Biomim 1(3):63–75

West MAL, Harada JJ (1993) Embryogenesis in higher plants: an overview. Plant Cell 5:1361–1369

West SA, Winzer K, Gardner A, Diggle SP (2012) Quorum sensing and the confusion about diffusion. Trends Microbiol 20(12):586–594

Westermann B (2010) Mitochondrial fusion and fission in cell life and death. Nat Rev Mol Cell Biol 11:872–884

Westermeier AS, Fleischmann A, Müller K, Schäferhoff B, Rubach C et al (2017) Trap diversity and character evolution in carnivorous bladderworts (*Utricularia*, Lentibulariaceae). Sci Rep 7:12052

Westwood JH, Yoder JI, Timko MP, dePamphilis CW (2010) The evolution of parasitism in plants. Trends Plant Sci 15(4):227–235

Wilde A, Mullineaux CW (2017) Light-controlled motility in prokaryotes and the problem of directional light perception. FEMS Microbiol Rev 41:900–922

Wilking JN, Zaburdaev V, De Volder M, Losick R, Brenner MP, Weitz DA (2013) Liquid transport facilitated by channels in *Bacillus subtilis* biofilms. PNAS 110(3):848–852

Wilkinson S, Davies WJ (1997) Xylem Sap pH increase: a drought signal received at the apoplastic face of the guard cell that involves the suppression of saturable abscisic acid uptake by the epidermal symplast. Plant Physiol 113:559–573

Will T, van Bel AJE (2006) Physical and chemical interactions between aphids and plants. J Exp Bot 47(4):729–737

Williams P (2007) Quorum sensing, communication and cross-kingdom signalling in the bacterial world. Microbiology 153:3923–3938

Williams YJ, Popovski S, Rea SM, Skillman LC, Toovey AF et al (2009) A vaccine against rumen methanogens can alter the composition of archaeal populations. Appl Environ Microbiol 75(7):1860–1866

Willis KJ, McElwain JC (2002) The evolution of plants. Oxford University Press, Oxford

Wilson JF, Mahajan U, Wainwright SA, Croner LJ (1991) A continuum model of elephant trunks. J Biomech Eng 113:79–84

Wilson M, Hanlon RT, Tyack PL, Madsen PT (2007) Intense ultrasonic clicks from echolocating toothed whales do not elicit anti-predator responses or debilitate the squid *Loligo pealeii*. Biol Lett 3:225–227

Winkle S (1997) Geißeln der Menschheit; Kulturgeschichte der Seuchen. Artemis & Winkler, Düsseldorf

Winklhofer M, Abraçado LG, Davila AF, Keim CN, de Barros HGPL (2007) Magnetic optimization in a multicellular magnetotactic organism. Biol J 92:661–670

Wittig D, Wang X, Walter C, Gerdes H-H, Funk RHW, Roehlecke C (2012) Multi-level communication of human retinal pigment epithelial cells via tunneling nanotubes. PLoS One 7(3):e33195

Wohlrab S (2013) Characterization of grazer-induced responses in the marine dinoflagellate *Alexandrinum tamarense*. Diss Univ Bremen

Wolf DM, Fontaine-Bodin L, Bischofs O, Price G, Keasling J, Arkin AP (2008) Memory in microbes: quantifying history-dependent behavior in a bacterium. PLoS One 3(2):e1700

Wolf S, Hématy K, Höfte H (2012) Growth control and cell wall signalling in plants. Annu Rev Plant Biol 63:381–407

Wolfe GV (2000) The chemical defense ecology of marine unicellular plankton: constraints, mechanisms, and impacts. Biol Bull 198:225–244

Woltz P, Stockey RA, Gondran M, Cherrier J-F (1994) Interspecific parasitism in the gymnosperms: unpublished data on two endemic new caledonian podocarpaceae using scanning electron microscopy. Acta Bot Gallica 141(6–7):731–746

Wommack KE, Colwell RR (2000) Virioplankton: viruses in the aquatic ecosystems. Microbiol Mol Biol Rev 64(1):69–114

Wongsuk T, Pumeesat P, Lupertlop N (2016) Fungal quorum sensing molecules: role in fungal morphogenesis and pathogenicity. J Basic Microbiol 56:440–447

Wu J, Mlodzik M (2017) Wnt/PCP instructions for cilia in left-right asymmetry. Dev Cell 40:423–424

Wu Q-S, Cao M-Q, Zou Y-N, Wu C, He X-H (2016) Mycorrhizal colonization represents functional equilibrium on root morphology and carbon distribution of trifoliate orange grown in a split-root system. Sci Hortic 199:95–102

Wuichet K, Cantwell BJ, Zhulin IB (2010) Evolution and phyletic distribution of two-component signal transduction systems. Curr Opin Microbiol 13:1–7

Xiao N, Xu S, Li Z-K, Tang M, Mao R et al (2023) A single photoreceptor splits perception and entrainment by cotransmission. Nature 623:562–570

Xu H, Bohman B, Wong DCJ, Rodriguez-Delgado C, Scaffidi A et al (2017) Complex sexual deception in an orchid is achieved by co-opting two independent biosynthetic pathways for pollinator attraction. Curr Biol 27:1857–1877

Xu H-B, Tsukuda M, Takahara Y, Sato T, Gu J-D, Katayama Y (2018) Lithoautotrophical oxidation of elemental sulphur by fungi including *Fusarium solani* isolated from sandstone Angkor temples. Int Biodeter Biodegrad 16:95–102

Yafetto L, Carroll L, Cui Y, Davis DJ, Fischer MWF et al (2008) The fastest flights in nature: high-speed spore discharge mechanisms among fungi. PLoS One 3(9):e3237

Yang J, Wang L, Ji X, Feng LX et al (2011) Genomic and proteomic analyses of the fungus *Arthrobotrys oligospora* provide insights into nematode-trap formation. PLoS Pathog 7(9):e1002179

Yasuoka Y, Shinzato C, Satoh N (2016) The mesoderm-forming gene *brachyury* regulates ectoderm-endoderm demarcation in the coral *Acropora digitifera*. Curr Biol 26:2885–2892

Ye Q, Miao Q-L (2013) Experience-dependent development of perineuronal nets and chondroitin sulphate proteoglycan receptors in mouse visual cortex. Matrix Biol 32:352–363

Yim G, Wang HH, Davies J (2007) Antibiotics as signalling molecules. Philos Trans R Soc Lond B Biol Sci 362:1195–1200

Yokawa K, Kagenishi T, Pavlovič A, Gall S, Weiland M et al (2018) Anaesthetics stop diverse plant organ movements, affect endocytic vesicle recycling and ROS homeostasis, and block action potentials in Venus flytraps. Ann Bot 122:747–756

Yong Y-C, Zhong J-J (2013) Impacts of quorum sensing on microbial metabolism and human health. Adv Biochem Eng Biotechnol 131:25–61

Yoshida S, Clint OC, Sammonds PR (1998) Electric potential changes prior to shear fracture in dry and saturated rocks. Geophys Res Lett 25(10):1577–1590

Yoshida S, de Reuille PB, Lane B, Bassel GW, Prusinkiewicz P et al (2014) Genetic control of plant development by overriding a geometric division rule. Dev Cell 29:75–87

Yoshida S, Cui S, Ichihashi Y, Shirasu K (2016) The haustorium, a specialized invasive organ in parasitic plants. Annu Rev Plant Biol 67:643–667

Yoshimura H, Muneta T, Nimura A, Yokoyama A, Koga H, Sekiya I (2007) Comparison of rat mesenchymal stem cells derived from bone marrow, synovium, periosteum, adipose tissue, and muscle. Cell Tissue Res 327:449–462

Yurchenco PD (2011) Basement membranes: cell scaffoldings and signaling platforms. Cold Spring Harb Perspect Biol 3:a004911

Zaidi Q, Marshall J, Thoen H, Conway BR (2014) Evolution of neural computations: Mantis shrimp and human color decoding. i-Perception 5:492–496

Zaitseva OV (1999) Principles of the structural organization of the chemosensory systems of freshwater gastropod mollusks. Neurosci Behav Phys 29(5):581–593

Zakharova K, Tesei D, Marzban G, Dijksterhuis J, Wyatt T, Sterflinger K (2013) Microcolonial fungi in rocks: a life in constant drought? Mycopathologie 175:537–547

Zarrella I, Ponte G, Baldascino E, Fiorito G (2015) Learning and memory in *Octopus vulgaris*: a case of biological plasticity. Curr Opin Neurobiol 35:74–79

Zeil J, Hemmi JM (2006) The visual ecology of fiddler crabs. J Comp Physiol A 192:1–25

Zeng L, Jacobs MW, Swalla BJ (2006) Coloniality has evolved once in Stolidobranch Ascidians. Integr Comp Biol 46(3):255–268

Zhang YE (2009) Non-Smad pathways in TGF-β signalling. Cell Res 19:128–139

Zhang C-C, Laurent S, Sakr S, Peng L, Bédu S (2006) Heterocyst differentiation and pattern formation in cyanobacteria: a chorus of signals. Mol Microbiol 59(2):367–375

Zhang Y, Ducret A, Shaevitz J, Mignot T (2012) From individual cell motility to collective behaviours: insights from a prokaryote, *Myxococcus xanthus*. FEMS Microbiol Rev 36:149–164

Zhang X, Zhang Y, Zhang Z, Mahadevan S, Adamatzky A, Deng Y (2014) Rapid physarum algorithm for shortest path problem. Appl Soft Comput 23:19–25

Zhou K, Zhang W-Y, Pan H-M, Li J-H, Yue H-D et al (2012) Adaptation of spherical multicellular magnetotactic prokaryotes to the geochemically variable habitat of an intertidal zone. Environ Microbiol. https://doi.org/10.1111/1462-2920.12057

Zhou W, Kügler A, McGale E, Haverkamp A, Knaden M et al (2017) Tissue-specific emission of (E)-α-bergamotene helps resolve the dilemma when pollinators are also herbivores. Curr Biol 27:1336–1341

Zhu M (2014) Bone gain and loss: insights from genome and fossils. Nat Sci Rev 1:490–492

Zhu H, Luo W, Ciesielski PN, Fang Z, Zhu JY et al (2016) Wood-derived materials for green electronics, biological devices, and energy applications. Chem Rev 116:9305–9374

Zimbardo P (2007) The Lucifer effect. Rider, London

Zimmermann MR, Maischak H, Mithöfer A, Boland W, Felle HH (2009) System potentials, a novel electrical long-distance apoplastic signal in plants, induced by wounding. Plant Physiol 149:1593–1600

Zürcher E, Schlaepfer R, Conedera M, Giudici F (2010) Looking for differences in wood properties as a function of the felling date: lunar phase-correlated variations in the drying behavior of Norway Spruce (*Picea abies* Karst.) and Sweet Chestnut (*Castanea sativa* Mill.). Trees 24:31–41

Zusman DR, Scott AE, Yang Z, Kirby JR (2007) Chemosensory pathways, motility and development in *Myxococcus xanthus*. Nat Rev Microbiol 5:862–872

Zwart MP, Elena SS (2015) Matters of size: genetic bottlenecks in virus infection and their potential impact on evolution. Annu Rev Virol 2:161–179

Biological Energy Transformation 7

7.1 The Basic Pattern

Based on phenomena observable in the macroscopic world, processes of biological energy conversion are usually depicted in the form of food chains. The starting-point is conversion of sunlight by green plants into (plant) biomass, which is in turn converted by animals into (animal) biomass and eventually everything is broken down by detritivores (waste recyclers), yielding basic building blocks once more. The diversity of energetic conversion processes in microbes cannot be inferred from this illustration—but all of them are based on a central basic principle:

> *Every biological energy conversion is based on balancing energy potentials between protons and electrons, or ions connected with them. The basic prerequisite for cellular life functions is maintenance of a sufficient intracellular charge potential, i.e. an energetic imbalance.*

This basic principle can be interpreted using both quantum-physical approaches and electrochemical approaches. The former facilitate understanding of conversion processes from radiant energy and the non-conventional energy balance through electron bifurcation, tunneling effects or functional variability of bacteriorhodopsins (see Sect. 4.5) (Page et al. 1999; May and Kühn 2011; Buckel and Thauer 2013; Spudich et al. 2014; Wolf et al. 2014; Metcalf 2016). The latter are particularly helpful for assessing convertible energy from chemical reactions and voltage potentials at biological membranes (Berg et al. 2013; Phillips et al. 2013; Alberts et al. 2015; Sojo-Martínez 2016).

Biologically educated readers may shake their heads at the word "disequilibrium", since according to widespread notions everything in nature is in perfect balance. However, the sentence also contains the word "potential", and in this light the presumed contradiction can be clarified with an example taken from everyday life: A mobile phone only works if the battery has sufficient voltage potential. If the voltage potential is lost because of frequent use of the device, the battery is actually in energetic balance—but it is nevertheless impossible to make a phone call. By charging the battery from an external source, imbalance can be restored in the battery and the device is ready to function once again. When it comes to maintaining the viability of organisms, a similar situation exists. Furthermore, during reproduction, some viruses take a small charge potential from the infected cell in order to employ it in a new infection (Madigan et al. 2003; Sun and Wirtz 2006; Dimmock et al. 2007).

To simplify matters, Fig. 7.1 has omitted the distinction between energy conversion for assembling cell substances (catabolic energy conversion) or for maintaining cellular processes (anabolic energy conversion), along with the distinction between conversion in an oxygen-containing (aerobic) environment and one that is oxygen-free (anaerobic). The various by-products of chemoautotrophic energy conversion—for example nitrogen compounds—are usually subsumed under the term "material cycles".

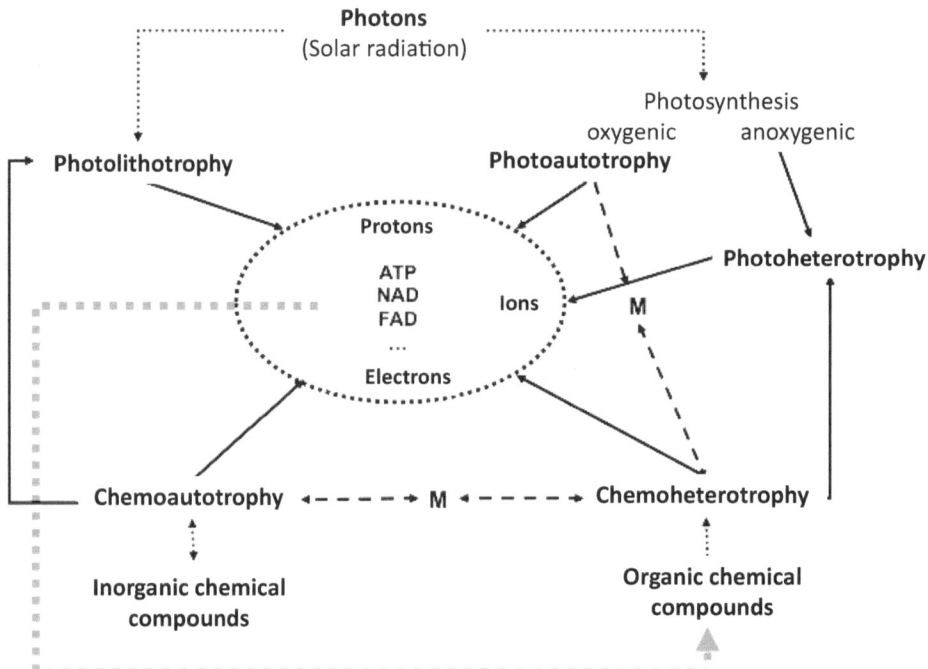

Fig. 7.1 Highly simplified diagram showing potential biological pathways for conversion from extracellular to biologically usable energy. Biological energy carriers are symbolized by adenosine triphosphate (ATP), nicotinamide adenine dinucleotide (NAD), flavin adenine dinucleotide (FAD). Certain organisms can switch between different conversion pathways—mixotrophy (M). While the supply of radiant energy and of inorganic compounds is primarily determined by abiotic conditions, the supply of organic compounds depends solely on biological production (symbolized by the thick, dashed grey arrow)

The great variety of combinations of transformation processes in prokaryotic and eukaryotic microorganisms is scarcely indicated by reference to mixotrophy (M). Organisms can cover their energy requirements—adapted to prevailing environmental conditions—through both autotrophic and heterotrophic transformation (Jansson et al. 1996; Eiler 2006; Muehe et al. 2009). The latter process can also be associated with direct extraction of mineral nutrients from biological materials. This mode of nutrition hence differs from all forms of exchange of substances between different organisms, for example syntrophy. Mixotrophy should in no way be confused with omnivory—the ability to convert energy heterotrophically from both animal and plant matter. The manifold ways of converting radiant energy are also striking, ranging from the widely known mechanism of photosynthesis—associated with synthesis of new organic molecules—to energetic support of conversion processes from inorganic or organic compounds (photoautotrophy *versus* photoheterotrophy). The latter process is based on a conversion of radiation energy without release of oxygen (anoxygenically).

In general, microbial processes of energy conversion constitute the qualitative outer limits of energy intake for all organisms. They include not only processes of synthesizing organic compounds but also some providing inorganic and organic compounds for all metabolic processes!

The food chains mentioned at the outset would be completely covered by conventional notions of photosynthesis and chemoheterotrophy (Fig. 7.1). In an evolutionary context, however, organisms would not have been able to develop at all in the absence of chemoautotrophy under the conditions existing during the Archean Eon (see Sect. 8.3), because, among the energy sources mentioned, only reaction potentials of inorganic chemical compounds were available (Wächtershäuser 1990, 2006; Russell et al. 2003; Mulkidjanian et al. 2008; Russell et al. 2010; Mulkidjanian 2011; Poehlein et al. 2012; Schoepp-Cothenet et al. 2012; Nitschke et al. 2013; Braakman and Smith 2014; Martin et al. 2014; Sojo et al. 2014; Schönheit et al. 2016; Sousa et al. 2016). In the early phases, radiation energy could be used only to a limited extent because of the presence of biologically harmful high energetic short-wave components (X-rays and UV) (Cockell and Horneck 2001; Cnossen et al. 2007).

By contrast, older hypotheses assume that organic compounds were utilized during the early phases of evolution (Follmann and Brownson 2009). Solely from a thermodynamic perspective, this is unlikely because it presupposes a perpetual motion machine in which more convertible energy would be available at the end than was converted at the beginning. Moreover, it must be taken into account that all known heterotrophic life forms possess only a limited repertoire of enzymes derived from autotrophic life forms. Their material and energetic transformation potentials are therefore limited to components of other cells. From this, it inevitably follows that the occurrence of organic compounds of any kind would be insufficient for heterotrophic life forms to emerge (Bada and Lazcano 2002; Orgel 2008; Schönheit et al. 2016). This limitation also clashes with hypotheses that postulate that life originated from organic compounds borne on meteorites (Sephton 2002).

Various microorganisms can utilize energy from electron transfer in reactions involving inorganic compounds. Electron exchange between different organisms—for example archaeans and bacteria—can also be used to increase the energetic conversion potential under anaerobic conditions (Amend et al. 2003; Christner et al. 2014; Osburn et al. 2014; Yu and Leadbetter 2020). Such reactions should be familiar from chemistry lessons as redox reactions, in which one substance (reducing agent) donates electrons that are taken up by a second substance (oxidizing agent). The propensity to give or take up individual substances is determined experimentally under standard conditions with a hydrogen electrode in a galvanic cell as the electrical potential. The achievable reaction energy can be calculated from the potential difference of the respective reaction partners.

Differences in electrochemical potentials under standard conditions can be determined from the electrochemical voltage ranges (Fig. 7.2 a). It should be noted that values differ between general chemical and biological voltage ranges because different standard conditions apply. General chemical voltage ranges refer to a pH value of 0, biochemical voltage ranges to a pH value of 7, along with a particular temperature (+25 °C) and a specified ambient pressure (101.3 kPa). Quantity ratios for initial and end products under standard conditions are, however, the same. In formulas, the differences are recognizable from the symbols used for standard potentials (Berg et al. 2013), with E^o in a general chemical context and $E^{o'}$ under biochemical conditions.

Energy representations for microorganisms easily give rise to confusion because different orders of magnitude are often used. As a rule, the achievable free energy of reactions—also called Gibb's enthalpy ($\Delta G^{o'}$)—is given in kilojoules per mole (kJ M^{-1}). In the examples shown in Fig. 7.2 a, these are—from the largest to the smallest difference[1]—around -239 kJ M^{-1}, -158 kJ M^{-1}, and -27 kJ M^{-1}. Defined simply, a mole is the molecular weight in grams.[2] The body mass of microorganisms is in the range of 10^{-12} to 10^{-14} grams and their metabolic turnover rates lie between 10^{-10} and 10^{-17} W.[3] Millions (10^8) to billions (10^9) or even more microorganisms can live in one thousandth of a litre of water. In general comparisons of transformation processes—for the sake of clarity—only differences in redox potentials are shown. However, this should not be taken to derive direct conclusions about the energy turnover of a single bacterium.

7.2   "Eating Stones": Chemoautotrophy

Viewed from the perspective of present-day habitats, the barrenness of living conditions in the Archean is difficult to imagine. Energy sources available to organisms were mainly inorganic chemical compounds, the potential of which—due to a lack of free oxygen—could be exploited only to a limited extent (Fig. 7.2 b). Despite this limitation, an immense

---

[1] By convention, the negative sign indicates a reaction with the release of energy.

[2] In scientific terms calculated from the Avogadro constant = $6.022140857 \times 10^{23}$ particles per mole.

[3] 1 watt = 1 J per second

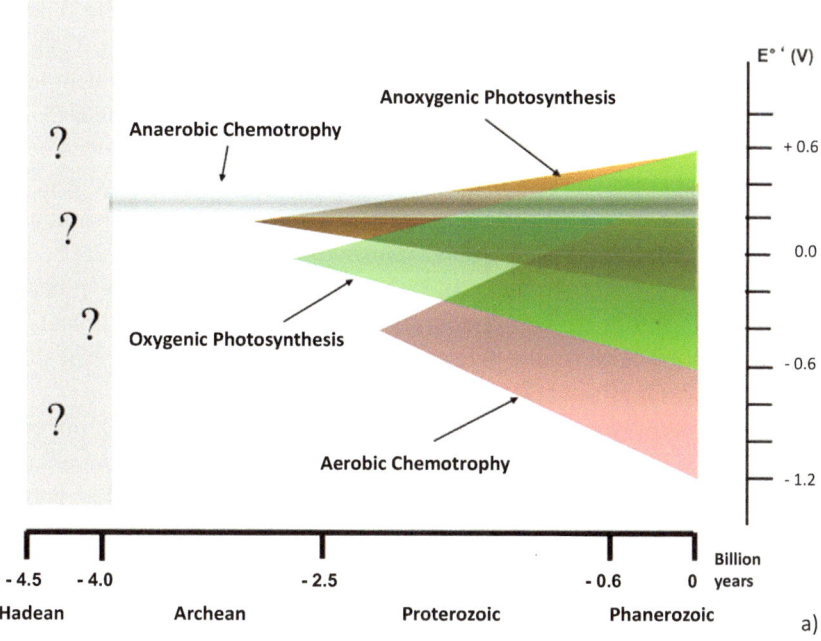

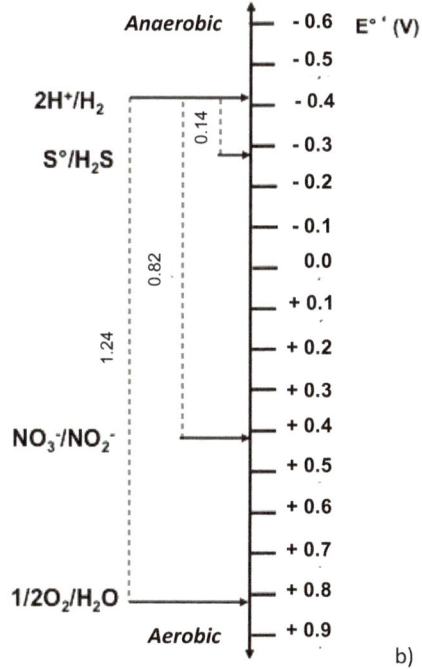

Fig. 7.2 Simplified diagram showing the potentials involved and hypothetical evolution of autotrophic energy conversion. (**a**) Simplified representation of electrochemical voltage potentials with examples of individual redox reactions. (**b**) Hypothetical representation of the evolution of autotrophic energetic conversion potentials. Data source: Schoepp-Cothenet et al. (2013)

variety of energy conversion processes developed for maintenance of life functions and reproduction of organisms. Examples of electron donors are hydrogen (H_2), methane (CH_4), hydrogen sulphide (H_2S), sulphur (S0), divalent iron (Fe(II)), divalent manganese (Mn(II)) and metal compounds. Examples of electron acceptors are sulphur (S), sulphate (SO_4^{2-}), trivalent iron (Fe(III)), tetravalent manganese (Mn(IV)), or nitrogen compounds with various oxidation states. With the advent of release of molecular oxygen, the range was further extended by aerobic reactions with higher energy yields (Amend and Shock 2001; Brzezinski 2004; King 2007; Konhauser 2007; Herrmann et al. 2008; Kelly and Wood 2013; Schoepp-Cothenet et al. 2013). Many prokaryotes can adjust their metabolism to utilization of different energy sources either by possessing functionally versatile gene sets or by means of horizontal gene transfer (Kolstø 1997; Madigan et al. 2003; Reva and Tümmler 2008; Ghosh and Dam 2009; Strnad et al. 2010).

In prokaryotes, redox reactions are usually catalyzed outside the cell, directly on the membrane surface, through connections with electron-conducting filamentous cell attachments (pili), or with the aid of shuttle molecules (Sapra et al. 2003; Horisberger 2004; Gralnick and Newman 2007; Kellosalo et al. 2012; Schoepp-Cothenet et al. 2013; McGlynn et al. 2015; Wegener et al. 2015; Shi et al. 2016). Electron energy thus generated is used by proteins to transport protons or sodium ions out of the cell (proton pump). The resulting proton or ion gradient between outside and inside ultimately brings about energy transfer to the proton motor (Fig. 4.12) of ATP synthase (Fig. 7.3 a). The effect of proton exportation on ATP synthesis increases with decreasing distance between the proton pump and ATP synthase (Sjöholm et al. 2017).

A remarkable exception to the rule is the bacterium *Thiomargarita namibiensis*—found in shelf sediments off the coast of Namibia. Its huge cells, up to 750 µm in diameter, contain an internal membrane-enclosed vacuole, which—depending on prevailing environmental conditions—serves as a store for sulphide or nitrate. This provides cells over a period of several years with a reserve for catalyzing redox reactions in sulphite or nitrate zones in sediments. The physiological peculiarities involved permit adaptation to irregular upheavals in the sediments—mostly triggered by gas leaks—and the associated passive transportation of these bacteria to other sediment zones (Schulz et al. 1999; Schulz 2002).

In marine sediments formed in shallow water zones, sulphur-oxidizing bacteria of the family Desulfobulbaceae follow a different pathway to exploiting concentration gradients. The bacterial cells—living in the anaerobic sediment zone—produce filaments up to 30 millimetres long (!) that extend up to the oxygen-rich surface of the sediment. These "cables" permit the requisite transfer of electrons between the reaction partners, oxygen and hydrogen sulphide (Pfeffer et al. 2012; Snider et al. 2012; Malkin et al. 2014; Schauer et al. 2014).

Bacteriorhodopsins—complementing the redox processes—opened up the possibility of directly using photon energy from solar radiation to transfer protons and sodium ions through cell membranes, permitting phototrophy (Fig. 7.3 b). It is currently still unclear when this form of energy conversion emerged. On the one hand, in the early history of Earth unfiltered ultraviolet radiation probably excluded direct exposure of organisms to

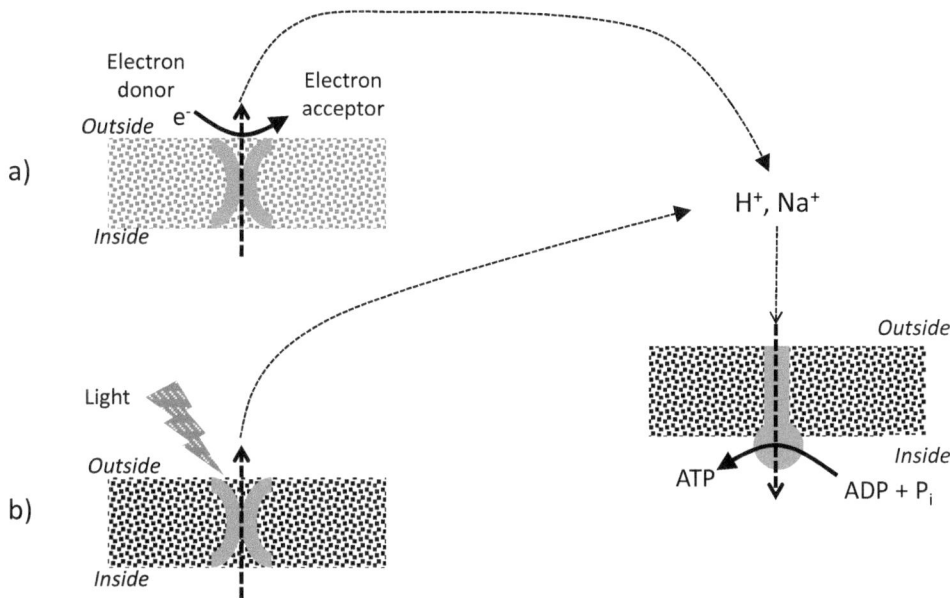

Fig. 7.3 Schematic comparison of differences in energy generation and similarities chemo- and phototrophic energy conversion in ATP. (**a**) Simplified scheme of chemoautotrophic energy generation; (**b**) Simplified scheme of—evolutionarily younger—phototrophic energy generation. Abbreviations: ATP = adenosine triphosphate, ADP = adenosine diphosphate, Pi = orthophosphate. Sources: Modified from Schoepp-Cothenet et al. (2013), incorporating other sources (see text)

light; on the other hand, the extent to which microhabitats and pigmentation of organisms (Wynn-Williams et al. 2002) enabled energetic light conversion at least to some degree remains open. Proteins related to bacteriorhodopsins play an energetic role in halobacteria and a sensory role in photoreceptors of animals.

Despite combination of different pathways of primary energy conversion, the global amount of biomass generated was significantly below the current level. Under the anaerobic conditions prevailing over 2.5 billion years ago, compared with current total global production, only about 3 per cent of organically bound carbon was produced along chemolithotrophic pathways (Raven 2009).

Such low biomass production probably impeded any significant evolution of predatory life forms, which emerged at a late stage in the phylogeny of prokaryotes (Yair et al. 2003). It did, however, permit the evolution and proliferation of viruses (Bamford et al. 2005; Nasir et al. 2012; Koonin and Dolja 2014; Nasir and Caetano-Anollés 2015; Krupovic and Koonin 2017). This interpretation is strengthened by the occurrence of viruses in archaeans and bacteria found in extreme habitats (Rice et al. 2001; Le Romancer et al. 2007; Williamson et al. 2008; Krupovic et al. 2011; DiMaio et al. 2015). However, there are also correlations between the diversity of viruses and the energetic conditions of their hosts, since rates of viral reproduction are also influenced by their total biomass (Erez et al.

2017). Accordingly, only about 5 per cent of viruses in bacteria and 2.7 per cent of viruses in eukaryotes have been detected in archaeans, which live predominantly under extreme environmental conditions (Nasir et al. 2014). The question as to how far viruses have also influenced biogeochemical processes by destroying their host cells during reproduction has yet to be clearly answered (Desnues et al. 2008; Pacton et al. 2014).

Living conditions of bacteria and archaeans exhibiting chemolithotrophy (energy conversion from inorganic compounds) usually deviate greatly from the standard conditions for the defined electrochemical potentials (Fig. 7.4). Not only individual parameters such as temperature, pH values and pressure, but also their combinations can diverge strongly from values defined for standard conditions (Rothschild and Mancinelli 2001; Cox and Battista 2005; Singh and Gabani 2011). Moreover, chemical reaction partners are only rarely constantly available in ideal—stoichiometric—concentration ratios. This alone results in different values for the free chemical energy of the reactions concerned (Amend and Shock 2001; Inskeep et al. 2005; King 2007; Osburn et al. 2014). It is notable that two-dimensional representations and logarithmic quantities scarcely reveal the diversity and variability of the actual living conditions of microorganisms. An inexperienced eye will not register the fact that on a logarithmic scale the value 4 is not twice as big as the

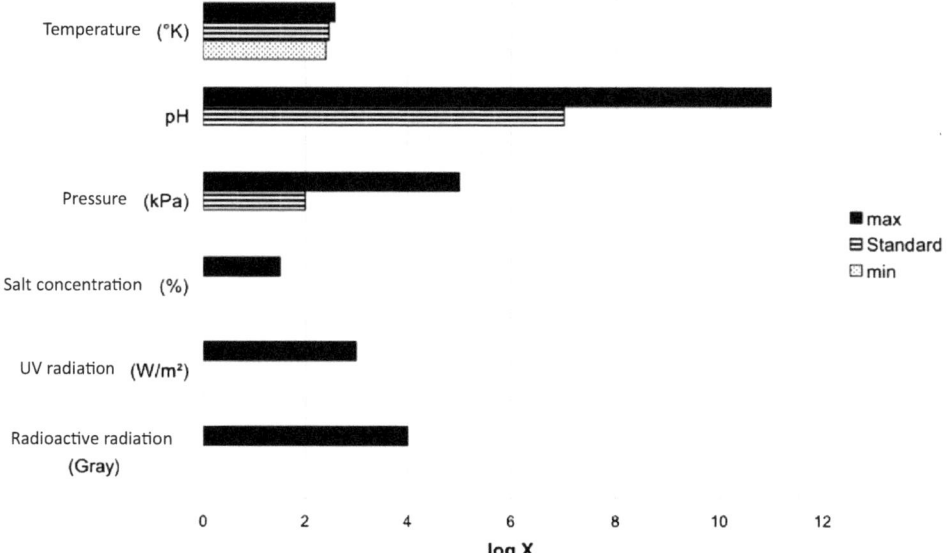

Fig. 7.4 Representation of bandwidths of individual environmental parameters in habitats of bacteria and archaeans, as well as standard values for calculating free energy of chemical reactions. All values on a logarithmic scale. Data sources: Rothschild and Mancinelli (2001), Cox and Battista (2005), Singh and Gabani (2011)

value 2, but a hundred-fold bigger.[4] Furthermore, the living conditions that microorganisms overcome extend far beyond our own experience. For instance, the marine anaerobic (oxygen-free) zones—e.g. in the Black Sea, the Baltic Sea or the Gulf of Mexico—are often referred to as "dead zones", despite the fact that they are populated by microorganisms (Rabalais et al. 2002; Kuypers et al. 2003; Helly and Levin 2004; Diaz and Rosenberg 2008; Grote et al. 2008; Stramma et al. 2008). The same is true of sediments in fresh- and saltwater zones, underground rock, waste-disposal sites or fermenting vats of winemakers (Stevens 1997; Madigan et al. 2003; Konhauser 2007; Osburn et al. 2014). Microorganisms can survive combinations of high salt concentrations with temperatures around −20 °C (253° Kelvin (K)) in polar sea ice as well as temperatures around +60 °C in hot salt springs. They also endure immense periods of desiccation—over millions of years—mostly in the form of resting stages (spores) (Potts 1994; Priscu et al. 1998; Vincent et al. 2000; Rothschild and Mancinelli 2001; Deming 2002; Thomas and Dieckmann 2002; Alpert 2005; Miteva and Brenchley 2005; Mock and Thomas 2005; Alpert 2006; Jiang et al. 2006; Pikuta et al. 2007; Oarga 2009; Koh et al. 2010; Leggett et al. 2012). Various microorganisms also tolerate high doses of short-wave ultraviolet radiation over extensive periods of time, or—as with bacteria of the genus *Deinococcus* sp.—exposure to radioactive radiation of up to 1000 Gray units,[5] a dose that is lethal for human beings (Cox and Battista 2005; Rainey et al. 2005; Cockell et al. 2008; Wassmann et al. 2010; Singh and Gabani 2011).

In extreme arid regions of Antarctica communities of soil bacteria maintain themselves chemotrophically by utilizing atmospheric trace gases—carbon dioxide, hydrogen and methane (Greening et al. 2015; Ji et al. 2017). This allows those bacterial communities to avoid direct solar radiation and hence reduce additional stresses from thermal and short-wave (ultraviolet) radiation. Results from a relatively new study reinforce earlier assumptions that colonization of land by terrestrial life forms of bacteria (Terrabacteria) began around three billion years ago (Battistuzzi and Hedges 2009).

Anaerobic-chemolithotrophic microorganisms often face the challenge of having to withstand high energy losses at extremely low energy conversion potentials—for example, due to high temperatures or extreme pH values. Morphologically and physiologically, these challenges are met by adaptations that sometimes have contradictory effects. For example, increasingly impermeable cell membranes impede exchange of molecules with the environment, but at the same time also reduce losses of energetically important protons and ions. Cell membranes of archaeans, for example, are up to 120 times less permeable to various compounds and up to 17 times less permeable to protons than membranes of bacteria (van de Vossenberg et al. 1998; Mathai et al. 2001; Macalady et al. 2004; Lane and Martin 2012). Low energy potentials are exploited to the limits of thermodynamic possibility by electron bifurcation and proton tunneling (Page et al. 1999; Biegel et al. 2011; Poehlein et al. 2012; Buckel and Thauer 2013; Schoepp-Cothenet et al. 2013;

---

[4] $10^2 = 100$, $10^4 = 10,000$

[5] 1Gray = 1 J/kg

Buckel 2015), as well as by modifications of ATP synthases (Valentine 2007; Walker 2013; Grüber et al. 2014; Mayer and Müller 2014; Lever et al. 2015). Additional challenges arise with extraction of carbon from CO_2 or HCO_3^- (hydrogen carbonate ions) to build organic cell substance. Among the various forms of anabolic carbon fixation in archaeans, compared to the Calvin-Benson cycle the Wood-Ljungdahl cycle reduces the specific effort from 7 to 1 ATP equivalents (Berg et al. 2010; Wrighton et al. 2016). Presumably, the Wood-Ljungdahl cycle is one of the earliest forms of carbon fixation (Braakman and Smith 2012; Weiss et al. 2016).

An even more efficient means of carbon fixation was found in an autotrophic, thermophilic bacterium. In the newly discovered—but presumably widespread—roTCA (reversed oxidative tricarboxylic acid) cycle, the specific effort for carbon fixation is around 0.5 ATP equivalents Nunoura et al. 2018; Ragsdale 2018). The apparent contradiction between "widespread" and "newly discovered" is explained by the lack of specific signatures in the genes, which is why this type of metabolism cannot be detected solely with currently employed metagenomic analyses (Mall et al. 2018).

Members of the group Thaumarchaeota—belonging to the archaean assemblage—also use an energy-efficient anabolic cycle (hydroxypropionate/hydroxybutyrate) for carbon fixation under aerobic conditions. By reducing the specific energy input by about 25 per cent compared to the Calvin-Benson cycle—at low ammonia concentrations— Thaumarchaeota outperform competing ammonia-oxidizing bacteria (AOB) (Könneke et al. 2014).

Because of interactions between all modifications, synthesis of biochemical molecules in anaerobic microorganisms requires only about 8 percent of the comparable energy input needed by aerobic microorganisms (Lever et al. 2015). In combination with the usually low biomass, this makes it possible to cope with the relatively large repair effort for damaged molecules and production of additional molecules (chaperones) to ensure correct protein folding under extreme environmental conditions (Archibald et al. 1999; Yoshida et al. 2002; Laksanalamai et al. 2008; Santra et al. 2017). These factors also affect growth of prokaryotes. For organisms living in extreme environments, the ratio between growth and the resting state for specific energy turnover is between about 3 and 5, whereas under favourable conditions it is between about 8 and 80 (Makarieva et al. 2005)! Under extreme conditions, prokaryote cells are not only smaller—for the reasons stated above—but can also multiply only slowly (DeLong et al. 2010). Doubling of their population size extends over intervals of months or years and not—as under laboratory conditions—just 20 min. Nevertheless, cell size should by no means be used to infer the living conditions of these microorganisms, as extremely small cells are often found in pathogenic or parasitic forms. In such cases, cell size is determined not by precarious nutritional conditions, but by the defence systems of infected organisms and osmotic uptake of food through the cell membrane (Yus et al. 2009).

During starvation phases, microorganisms constantly living at the thermodynamic limit of energy conversion reduce their carbohydrates by up to 80 percent, their proteins by up to 50 percent and their RNA by up to 75 percent. The DNA content, on the other hand, remains constant or increases by up to 50 per cent in individual life forms. This shows a clear difference from microorganisms living in energy-rich environments—for example parasites—which reduce their DNA content by up to 75 per cent as well during starvation phases (Lever et al. 2015, 2016).

Chemolithotrophic organisms expanded the foundations for life during the early phases of Earth's history and still maintain essential material frameworks for the evolution of all other organisms (Schlesinger 2005; Zerkle et al. 2005; Williams and Fraústo da Silva 2006; Falkowski et al. 2008). This functionally modifies long-term abiotic material transformation cycles with average intervals between 200 and 600 million years (Veizer 1973). These cycles are constituted by various geodynamic processes—such as subduction (submergence of lithospheric plates) and volcanism—and by abiotic chemical processes in liquid and gaseous states. Biological modifications are also multidimensional. However, their cycle times are much shorter and the metabolic variability of microorganisms permits processes to be adapted to prevailing abiotic conditions (Fig. 7.5). Effects of the interactions between abiotic and biotic processes are generally recognized under the term Global Material Cycles, for instance of carbon, oxygen, nitrogen, phosphorus, sulphur or iron (Schlesinger 2005; Weber et al. 2006; Stein and Klotz 2016).

A good example of a Global Material Cycle is provided by biological nitrogen fixation, which—according to current knowledge—has accompanied the evolution of organic life for around 3.2 billion years (Thomazo and Papineau 2013; Stüeken et al. 2015). Widespread dependence of the processes concerned on completely oxygen-free (anaerobic) conditions provides evidence for early development of nitrogen fixation in archaeans and bacteria (Madigan et al. 2003; Raymond et al. 2004). Presumably, the complex processes for fixing atmospheric nitrogen evolved in several stages (Leigh 2000; Canfield et al. 2010; Boyd and Peters 2013; Hoffman et al. 2014). The question remains open as to what role horizontal gene transfer played in this process. Further developments of the processes enabled photosynthetically active cyanobacteria in addition to fix nitrogen in the same cell under minimal oxygen concentrations. The associated challenges are overcome by either temporal (day/night) or spatial separation of the processes. The latter requires formation of

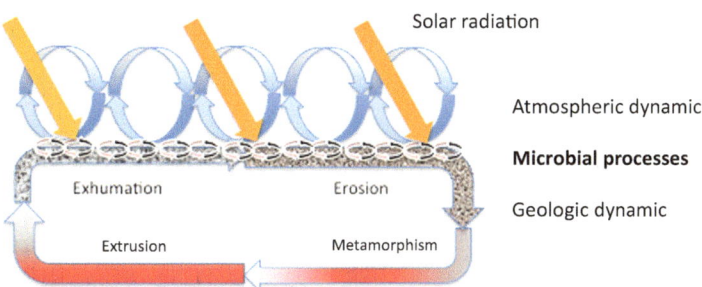

Fig. 7.5 Schematic representation of modification of material environmental conditions by chemolithotrophic microorganisms under the dynamic framework conditions of geological and atmospheric processes as well as solar radiation

multicellular associations with specialized cells (heterocysts) without photosynthesis. Heterocysts fix nitrogen and supply other cells in the association, which in turn provide the requisite metabolic products from photosynthesis (Berman-Frank et al. 2003; Bothe et al. 2010).

The suggestive power of the term "material cycle" easily disguises our still very incomplete knowledge of the many—often significant—sub-processes, as well as their sensitivity to supposedly minimal changes. Examples of the diversity of biological processes in nitrogen cycles include discovery of anaerobic oxidation of ammonium by so-called anammox bacteria (Mulder et al. 1995; Kuenen 2008; Kartal et al. 2011), anaerobic oxidation of methane in marine sediments (Boetius et al. 2000; Treude 2003), energetic utilization of methane (methanotrophy) for nitrogen fixation (Dedysh et al. 2004; Trotsenko and Khmelenina 2005; Dekas et al. 2009; Khadem et al. 2010) or conversion of ammonium into nitrate in a single bacterial cell (van Kessel et al. 2015). Diverse symbioses of nitrogen-fixing microorganisms with a wide variety of eukaryotes respond in unpredictable ways to the supply of mineral nitrogen compounds (Rengel and Marschner 2005; Kneip et al. 2007; Frey-Klett et al. 2011; Cooper and Smith 2015). Responses to different tree species can also be observed in methanotrophic bacteria in forest soils, which oxidize significantly less methane under conifers than under deciduous trees (Degelmann 2010). Important factors for the variability of material processes are the great metabolic adaptability of individual microorganisms and adaptability of microbial consortia to prevailing framework conditions.

The construct of global material cycles complicates access to understanding of evolutionary processes for various reasons:

- The first reason resides in dimensions of the processes under consideration, which cannot be anticipated physiologically or cognitively by any single organism.
- The second reason lies in the underlying assumption of harmoniously coordinated and long-term stable interactions between all components considered in alternative models—most clearly recognizable in the Gaia hypothesis (Lovelock 1992). The reasons mentioned above are enough to refute this fiction.
- The third reason is failure to consider the integrative processes of metabolic processes in organisms. These phenomena are usually described with the term biological or ecological stoichiometry (Sterner and Elser 2002). Descriptions are usually highly simplified because in most studies only ratios of carbon to nitrogen and to phosphorus are taken into account. In reality, the number of elements converted in parallel in biological processes is much greater. In order to understand evolutionary developments, it is also necessary to distinguish which substances are enriched in the biological systems over the long term and which substances, for various reasons, are continuously eliminated from biological processes. Examples of the latter are the formation of biominerals to protect against ultraviolet radiation in bacteria (Phoenix et al. 2001), the formation of reefs by a wide variety of organisms such as bacteria, corals, mussels or sponges (Riding et al. 1991; Stanley and Hardie 1998; Banfield et al. 1999; Nauhaus 2003;

Treude 2003; Grotzinger et al. 2005; Stanley and Hardie 1998; Adachi and Ezaki 2007; Chu and Leys 2010; Wienberg et al. 2018). Here, too, no sharp boundaries can be drawn because, for example, iron compounds can be deposited by bacteria both externally and internally—for orientation in the magnetic field (Scheffel et al. 2008; Miot et al. 2009); calcium compounds are found not only in reefs but also in the bones of many vertebrates or in shells of moss animals (bryozoans) or molluscs (Jacobs et al. 2000; Addadi et al. 2006; Donoghue et al. 2006; Taylor et al. 2015).

– This reveals a fourth reason—the flexible, but rather limited, adaptability of organisms and organism communities to changes in framework conditions. Directly effective functional relationships are always relevant for the adaptation of organisms. As a result, even among organisms of the same phylogenetic group, different demands on the supply of substances can develop—so-called "stoichiometric phenotypes (Geider and La Roche 2002; Makino et al. 2003; Carrillo et al. 2006; Johnson and Jürgens 2010; Weber and Deutsch 2010; Abbas et al. 2013; Mooshammer et al. 2014; Leal et al. 2017). Different accumulations of substances in the local biological substance cycles also exert a decisive influence on evolutionary processes (Martin et al. 2008; Kiessling et al. 2010). This is in clear contrast to the mechanistic understanding of material cycles, according to which each component contributes deterministically and indefinitely to maintaining functional capacity.

A typical example of misconception is provided by current widespread use of nanoparticles, which can modify a wide variety of biological reactions for physical reasons alone (Navrotsky 2004; Mahendra et al. 2008; Radha et al. 2010). This affects not only microorganisms, but the metabolism of all organisms. Nanoparticles can enter the brains, placenta or gonads of vertebrates or be transferred to infants with their mother's milk (Melnik et al. 2013; Wang et al. 2014; Li et al. 2015; Teleanu et al. 2018; Bongaerts et al. 2020). It is precisely these properties that make nanoparticles so attractive, seen from a laboratory perspective, for a wide range of technological applications (see Fig. 3.1) (Calderón-Jiménez et al. 2017). In a global context, however, on top of hundreds of thousands of new chemical compounds and changes in climatic and structural factors, an additional array of anthropogenic interventions is being released with unforeseeable consequences (Ferreira et al. 2016; Wang et al. 2020).

Evolutionary accounts largely ignore the influence exerted by microorganisms on transportation of minerals from the continents to the oceans. Extremely simplified assumptions in geochemical models are buttressed with questionable hypotheses to reconcile results with geologically recorded data (Boucot and Gray 2001). For example, changes in erosion rates are attributed to propagation of the first land plants (Algeo et al. 1995; Algeo and Scheckler 1998; Berner 1998; Masuda and Ezaki 2009). This completely ignores the fact that such changes brought about by microorganisms began at least two billion years earlier. Because of the evolution of photosynthesis—described in the next chapter—cyanobacteria, at least, were able to proliferate in terrestrial zones. Existing capacities for dissolving mineral compounds enabled them to migrate into rock layers near the surface, with the advantages of protection from ultraviolet radiation, water storage and access to mineral nutrients. These processes still take place today on free rock surfaces and are estimated to accelerate chemical weathering hundreds to thousands of times (Schwartzman and Volk 1989; Büdel et al. 2004). Higher values are mainly achieved by symbioses of phototrophic organisms with fungi, as well as archaeans and bacteria.

Fungi are, however, found in almost all extreme habitats together with archaeans and bacteria, for instance in sediments, on basalts or around active volcanoes in the deep sea, as well as in deep rock formations (Connell et al. 2009; Ivarsson et al. 2015; Purkamo et al. 2018; Rojas-Jimenez et al. 2020). This is remarkable because archaeans and bacteria convert their energy at the outer cell membranes, whereas eukaryotic fungi do so at intracellular membranes. What they have in common is their osmotrophic mode of nutrition and the wide range of different substances used for energy conversion (Wainwright 1988; Bindschedler et al. 2016; Burghelea et al. 2018; Drake and Ivarsson 2018). Globally, instead of the minimal proportions originally estimated, almost half of biologically induced rock weathering is probably due to cellular processes alone (Li et al. 2016). In a technological context, the wide metabolic bandwidths of fungi are exploited for a wide variety of processes, including removal of heavy metals and toxic compounds from mine dumps and soils (Banfield et al. 1999; Arvieu et al. 2003; Hoffland et al. 2004; Gadd 2007; Gadd et al. 2007; Finlay et al. 2009; Smits et al. 2009; Harms et al. 2011; Cordier et al. 2012; Hadibarata et al. 2012; Morel et al. 2013). Despite many indications (Wainwright 1988)—mentioned only in abbreviated form here—potential nutrition of fungi from inorganic compounds, which long remained unheeded, was eventually confirmed by investigations of samples from the sandstone of Angkor Wat (Xu et al. 2018).

Lichens—the earliest form of terrestrial symbioses—can also colonize extreme habitats and initiate soil formation (Belnap and Lange 2003; Nash 2008; Su et al. 2011; Elbert et al. 2012; Porada et al. 2014). Because of these properties, lichens may have prepared the way for the evolution of terrestrial ecosystems more than two billion years ago (Retallack and Mindszenty 1994; Watanabe et al. 2000; Battistuzzi et al. 2004). In connection with increasing mountain formation, transportation of liberated nutrients by wind and run-off water also facilitated higher productivity in marine algae (Bindeman et al. 2018). Almost one and a half billion years later, the principle of such symbioses in the form of mycorrhizae—connections between fungi, prokaryotes and root zones of plants—enabled the spread of multicellular plants across the continents (van der Heijden and Sanders 2003). In terrestrial soils, higher (alkaline) pH values and temperatures favour growth of bacteria, while contrary conditions favour growth of fungi (Pietikäinen et al. 2005; Rousk et al. 2010).

Geological and climatic processes are also presumably major triggers for changes in nutrient inputs into the oceans, up until the end of the last ice age. However, their extent is influenced by accumulation of biologically modified weathering products. For example, over millions of years erosion materials from mountain ranges far from the coast can be deposited in alluvial plains on coastlines or over geologically short periods of time—induced by uplift of mountains—washed into the oceans. Over long periods of time, however, erosional materials can also be continuously carried from the mainland into the oceans or onto ice surfaces by strong winds or through large river systems. Extensive areas of the tropical Atlantic zone off the coasts of South America, for instance, are regularly fed by nutrient inputs from the Amazon (Subramaniam et al. 2008). On glaciers and sea ice, dust particles deposited in so-called cryoconite holes enable a relatively high level of microbial productivity (Born and Böcher 2001; Anesio et al. 2009; Kaczmarek et al. 2015).

7.3   Syntrophy

A topic often discussed in publications concerns the stage at which cooperation between organisms first occurred over the course of evolution. Here, as a rule, the diverse interactions between microorganisms are completely ignored. Chemolithotrophic life forms undoubtedly provide the oldest forms of cooperation between organisms. The term *syntrophy* is used to summarize processes of mutual assistance between organisms for energy conversion (Fig. 7.6).

Prevailing environmental conditions exert a significant influence on combinations of conversion processes. For example, under conditions present in submarine hydrothermal

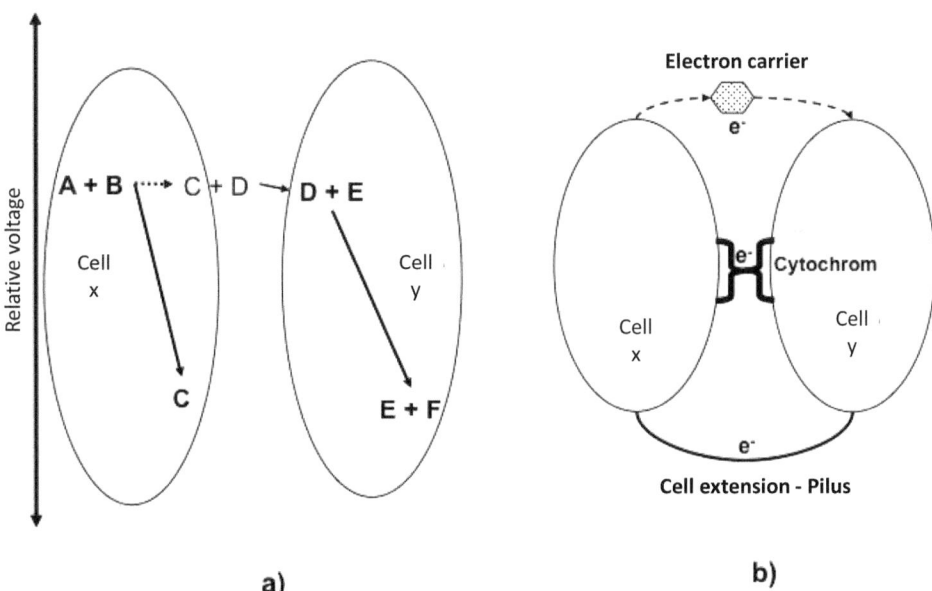

Fig. 7.6 Schematic representation of the principles of action in syntrophic processes. In example (**a**), an energetically sufficient reaction potential is achieved through ongoing consumption of a reaction product (D) by cell y, leading to a response by cell x. In example (**b**), three different possibilities for obtaining energetically favourable reaction conditions in both cells are shown: Transportation with an electron carrier or direct transfer via either a cytochrome or a cell appendage (pilus)

vents at around 100 °C, reactions of methanogenesis, methane oxidation, and reduction of sulphate and elemental sulphur are energetically significantly more favourable than at around 25 °C (Shock and Holland 2004). Moreover, physiologically different organisms can modify concentrations of reaction products and thus modify the energy yield compared to standard conditions (Schink 1997; Boetius et al. 2000; Jackson and McInerney 2002; Raghoebarsing et al. 2006; Lam et al. 2007; Treude et al. 2007; Pernthaler et al. 2008; Kato and Watanabe 2010). This is made possible by the variety of redox reactions in which novel energetically usable potentials can be rendered accessible through combination with other reaction partners (Fig. 7.6 a). In the minimum case, this involves exchange of electrons between different organisms (Fig. 7.6 b) (Lovley 2008; Summers et al. 2010; Kato et al. 2012; Nagarajan et al., 2013; Kouzuma et al. 2015; McGlynn et al. 2015; Wegener et al. 2015; Shen et al. 2016). In microbial biofilms, for example, cyanobacteria transfer electrons to other microorganisms when there is a surplus of light. In this way, the cyanobacteria can avoid energy-related damage, while other organisms can use the photosynthetically converted energy (Lea-Smith et al. 2016). As a result of the degrees of freedom of the interactions involved, autotrophic microbial cooperation can change or even dissipate completely when environmental conditions change (Müller et al. 2009; Wintermute and Silver 2010a; Werner et al. 2014). Inevitably, changes in compositions of chemical end products from the biological processes are associated with this.

Just how far syntrophic communities can extend is shown by the example of between of two bacterial species ("*Chlorochromatium aggregatum*") occurring in lakes. A large, autotrophic and mobile bacterium is completely enveloped by a larger number of smaller phototrophic green sulphur bacteria. The outer cell membranes fuse at various places, facilitating direct exchange of vitamins and various other substances. To optimize metabolic processes, the cell consortium orients itself to light gradients and sulphur concentration of in the surrounding water (Wanner et al. 2008; Müller and Overmann 2011; Liu et al. 2013; Cerqueda-Garcia et al. 2014).

In the examples provided, either use of chemical energy is rendered possible only through syntrophy or the total energy yield is augmented. In addition, chemical reactions in the organisms concerned can also be enabled by electron bifurcation (Fig. 4.13). Examples of syntrophic communities are seen in combinations of nitrogen fixation with oxidation of methane or sulphur in deep-sea habitats (Boetius et al. 2000; Raghoebarsing et al. 2006; Pernthaler et al. 2008; Dekas et al. 2009; Plugge et al. 2011; Lau et al. 2016).

Viewed from a systemic perspective, syntrophy is just a category of reciprocal benefits between different organisms to reduce losses—in physical terms decreasing entropy—in communities of organisms (Fig. 7.7). The special facet of syntrophic interactions, however, resides in synergistic expansion of biological energy conversion potential beyond the sum of isolated conversion potentials of all individual cells. Without the basic principle of synergy, organic life would have remained at best a negligible marginal phenomenon of abiotic processes and would never have reached the evolutionary dimensions observed to date. The levels of order achieved in each case in turn determine options for the evolution of individual levels of order among organisms (Fig. 7.7). Thus, in each case the long-term

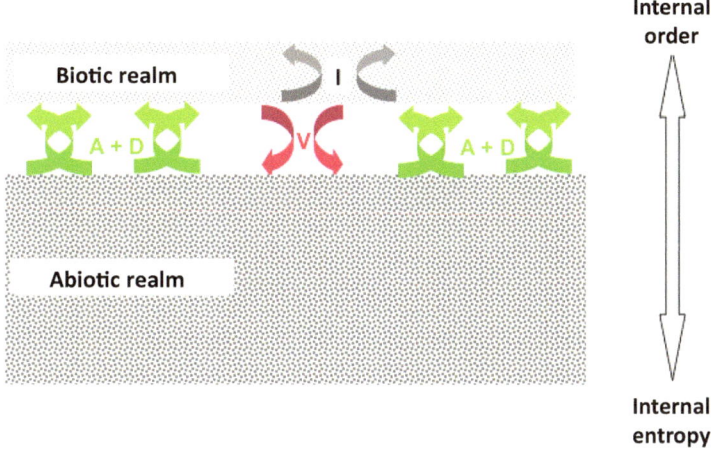

Fig. 7.7 Schematic representation of effects of the Basic Principle of Synergy for the evolution of organisms. Abbreviations: A + D = autotrophs and D-heterotrophs; I = potential for individual evolution of organisms; V = V-heterotrophs

evolution of each individual species depends crucially on its contributions to maintaining the requisite level of order. The question of the importance of heterotrophic organisms for evolutionary processes will be considered in detail in subsequent chapters.

In the scientific literature, common features of the principle are lost in the categorization of individual phenomena. Accordingly, the term auxotrophy is used for exchange of synthesized biomolecules. Examples of such biomolecules are vitamins or amino acids. Microorganisms play a central role in these phenomena—regardless of whether they involve mutual exchange with other microorganisms (Emmett and Kloos 1979; Rollefson et al. 2009; Tripp et al. 2009; Findon et al. 2010; Singh et al. 2010; Wintermute and Silver 2010b; Zelezniak et al. 2010), plants (Droop 1957; Prell et al. 2010; Helliwell et al. 2011; Bertrand and Allen 2012; Wheeler et al. 2015) or animals (Robinson and Cavanaugh 1995; Johnson et al. 2002; Torrallardona et al. 2003; Ereskovsky et al. 2005; Newton et al. 2007; Dattagupta et al. 2009; Douglas 2009; Sczesnak et al. 2011; Price et al. 2014; Brown et al. 2015; Murray 2016; Malcicka et al. 2018). They also reduce energetic losses. The effects of these phenomena reach deep into genetic regulatory systems. On the one hand, a regular supply of synthesized biomolecules renders superfluous any maintenance of an organism's own genes needed for that purpose; on the other hand, those phenomena can also lead to morphological modifications (Chapman and Margulis 1998).

Under the dynamic framework conditions of abiotic processes, duration of the interactions described is quite short. Currents separate cells or drive them together, and as a result of dilution large distances between cells permit the effects of molecules to decay rapidly (Fig. 7.8 above). For the reasons outlined in Sect. 5.1 alone, long-term maintenance of functional cell communities fails to occur. Other framework conditions evolved in biofilms—formed by organisms—on solid surfaces (De Beer and Stoodley 2006; McDougald

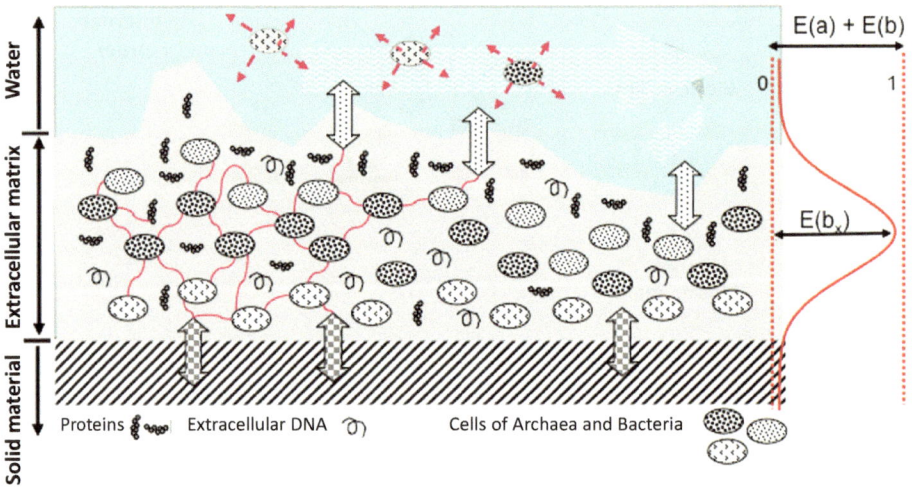

Fig. 7.8 Schematic representation of a biofilm on the surface of a solid substrate. The extracellular matrix with embedded cells, extracellular DNA and extracellular proteins are all shown, along with examples of interactions between cells (thin red lines) and material flows in transition zones. For comparison, possibilities for discontinuous interaction of freely floating cells in water currents are indicated (short arrows). The diagram on the right schematically illustrates the increase in proportion of biotic energy conversion in the biofilm ($E(b_x)$) compared with the total for potential abiotic and biotic energy conversion ($E(a) + E(b)$)

et al. 2011; Madsen et al. 2012; Flemming et al. 2016; Guerrero and Berlanga 2016). In principle, a biofilm consists of aggregations of cells enclosed by a collectively formed extracellular matrix (Fig. 7.8).

The extracellular matrix is by no means homogeneous, but spatially structured with polymers and contains, among other things, extracellular DNA (eDNA), proteins and lipids. Structured spatial distributions permit chemical signals to be transmitted more reliably within functionally interacting cells than in free-floating life forms (Gantner et al. 2006). Cells are not tightly confined in the matrix and can redistribute themselves in accordance with prevailing conditions (Bebout and Garcia-Pichel 1995; Barbara and Mitchell 1996). This is rendered possible by temporary changes in consistency of the matrix due to rotational movements of cell flagella. On the one hand, the extracellular matrix offers mechanical protection against circulation or loss of substances through flow; on the other hand, internal efficiency gradients are built up through selective filtering of external physical variables and substances (Allen et al. 2009; Flemming and Wingender 2010). This also renders biofilms important in a medical context, as they can lead to infections even under the sterile conditions of operating theatres—for example in association with implants. Because of effects of the extracellular matrix, bacteria in biofilms are particularly resistant to responses of immune systems and treatment with antibiotics (Devaraj et al. 2015; Khatoon et al. 2018).

Because of the ubiquitous occurrence of biofilms under modern living conditions and their associated undesirable consequences—for instance in medicine, materials technology or boat hulls—it is easy to forget that, in terms of evolutionary history, they are probably the oldest forms of cross-cellular biological structures (Hall-Stoodley et al. 2004).

Because of the dominance of biogenic energy conversion within biofilms ($E(b_x)$ in Fig. 7.8), biological processes achieve the highest degrees of freedom for self-organization. However, in each case dimensions of the total energetic conversion potential ($E(a) + E(b)$ in Fig. 7.8) are determined by external energy gradients that are present. In the biofilms considered here, under early evolutionary conditions, gradients exclusively involve abiotic energy. In later phases of evolution, external energy gradients are usually composed of both abiotic and biotic factors, or—as with medically important biofilms—are composed exclusively of biotic factors.

The principle of syntrophic biofilms probably also facilitated the conquest of dry land areas by lichens—communities of fungi with cyanobacteria or algae (Kenny and Knauth 2001; Nash 2008). Up to the present day, adaptation of these life forms to a wide diversity of environmental conditions has ensured their dispersal to the most extremely arid locations. This is due to the great flexibility of combinations of different partners and modifications of properties of their cell physiology.

The decisive factor for effects of biofilms—based on converted energy—is development of complementary functions in the microorganisms involved. This can be seen most clearly in microbial mats occupying current shallow water zones, where photosynthetic processes, over distances of a few millimetres, interact with anaerobic processes of sulphur conversion (van Gemerden 1993; Laval et al. 2000; Reid et al. 2000; Dupraz and Visscher 2005). Marine biological studies in the Bahamas have shown that, under certain conditions, microbial mats also control the population size and composition of multicellular organisms living within them (Tarhan et al. 2013). Interactions with water currents lead to hydrodynamic adaptations of the macroscopic structures of microbial mats (Suosaari et al. 2016). By shaping microscopic structures on their surfaces, microbial mats optimize—through feedback with metabolic processes—the exchange of matter and energy with the environment (Petroff et al. 2010).

This highlights the diversity of different evolutionary conditions for biofilms. Differences exist not only between potential sources of external energy, but also among all relevant dimensions such as size, temporal duration and development patterns. Above all else, from a systemic perspective the descriptor of "evolutionary incubators" can be attributed to biofilms that have been able to persist over the long term under relatively constant energy gradients. Examples of this during early evolutionary history can be found in biofilms in the vicinity of submarine hydrothermal vents (Brazelton and Baros 2009; Ishibashi et al. 2015) and in near-surface zones. Such intimate and long-term interactions between prokaryotes provided numerous opportunities for adaptation through horizontal gene transfer and selective testing within a restrictive autotrophic energetic framework. Low energetic transformation potentials and inherent interdependencies probably forced the emergence of new cooperative life forms during the earliest phases of evolution, while at the same time restricting the development of non-cooperative life forms.

This inference is based on widespread distribution of biofilms around vents of deep-sea hydrothermal sources and the very early occurrence of microbial mats in fossil deposits, the age of which has been estimated at around 3.4 billion years (Tice and Lowe 2004; Westall et al. 2006). Studies of biofilms from vents in the "Lost City"on the Mid-Atlantic Ridge reveal multidimensional adaptations of microorganisms to potentials and dynamics of abiotic processes. From the biofilms of a single vent, it proved possible to reconstruct changes in microbial communities over a period of about a thousand years. Over the course of time, various life forms disappeared from the originally diverse biocoenosis. The originally predominant archaeans were increasingly edged out by bacteria, although they still remained in the biofilms (Fig. 7.9 a) (Brazelton et al. 2010). In a biofilm colonized by only one archaean species, however, functional differentiation of cells extending up to morphological specialization—supported by horizontal gene transfer—was increasingly observed (Brazelton et al. 2011). The associated emergence of syntrophic processes

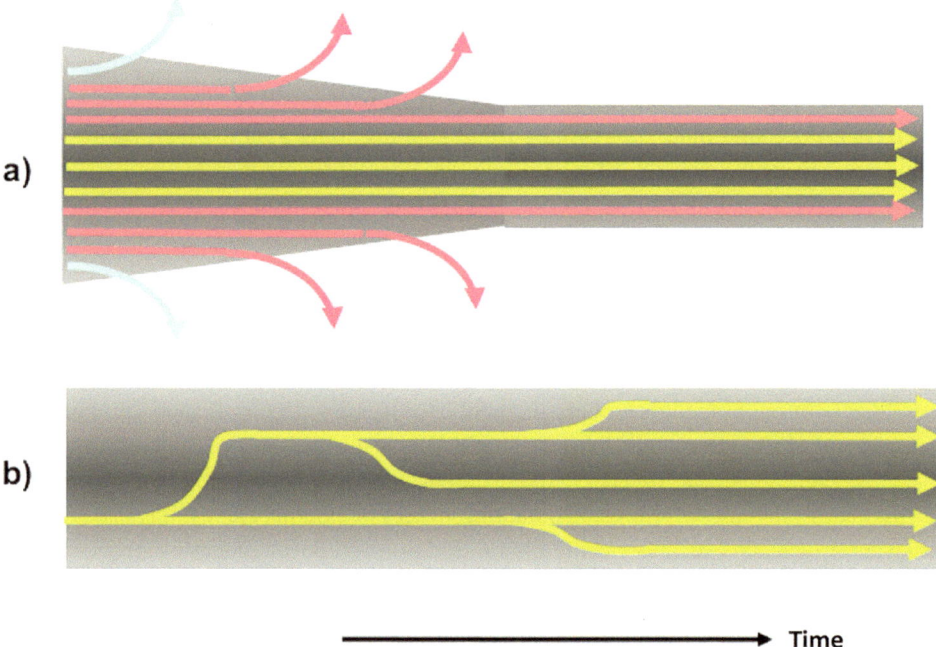

Fig. 7.9 Generalized diagrams of biological changes observed in biofilms from vents of deep-sea hydrothermal sources. Potential abiotic development options are represented schematically by grey bands with transitions from optimal (dark grey) to suboptimal (light grey). (**a**) In biofilms originally colonized by different organisms, as a result of changes in abiotic conditions, individual groups of organisms disappear over time. A shift in dominance ratios occurs among the organisms remaining. (**b**) In biofilms originally colonized by just a single life form, the entire abiotic transformation potential is utilized over time by means of functional differentiation

permitted increasing exploitation of the available inorganic transformation potential (Fig. 7.9 b).

From a geological-palaeontological perspective, fossilized microbial mats (microbialites) in sedimentation zones are of focal interest, as their traces are relatively easy to recognize even in rock that is billions of years old. Mineral material can be washed in by water currents or originate from metabolic processes of the microorganisms themselves (Nisbet and Fowler 1999; Riding 2006; Noffke 2008; Allwood et al. 2009; Kamennaya et al. 2012). A corresponding nomenclature reflects distinctions between differing fine structures (Riding 2008, 2011). Among the best-known forms are multi-layered stromatolites from shallow water zones. Less well known are branched forms such as thrombolites or dendrolites.

For the development of organic life during the Precambrian (earlier than 635 million years ago), fossils of stromatolites provide valuable outline information. This mainly concerns structures influenced by organisms, but in rare cases also indicates the type of organisms involved. In many cases, compositions of present-day stromatolites are used to infer that the organisms that shaped them were cyanobacteria—an approach that recent studies have called into question. As already mentioned, biofilms—and hence stromatolites as well—can be formed by a wide variety of organisms. Detailed microbial investigations have revealed that cyanobacteria probably originated over 3.5 billion years ago, but as fossils can be traced only as far back as around 2.5 billion years (Schirrmeister et al. 2015, 2016).

The earliest stromatolites were probably formed by chemoautotrophic, anaerobic communities. Limited energetic conversion potentials would account for the low diversity of stromatolites and the rarity of their occurrence over three billion years ago (Schopf 2006; Schopf et al. 2007). It must be borne in mind that organisms can form stromatolites only under suitable abiotic conditions.

Single-celled organisms can, however, evolve under a wide variety of conditions. Indications of this are provided by comparative physiological studies of optimal growth temperatures. According to results obtained from such investigations, increasing thermotolerance is likely to have developed during an initial evolutionary phase—in archaeans and bacteria. During a subsequent evolutionary phase—including the emergence of eukaryotes—augmented thermotolerance was secondarily lost (Boussau et al. 2008). Another indication of survivability was provided by experiments with bacterial spores that were exposed to environmental conditions resembling those accompanying a direct asteroid impact. Despite a low survival rate of about one in 10,000 (10^{-4})—because bacterial spores numbered between 10^6 and 10^9—this did not lead to any lasting constraints on their further evolution.

It is therefore unsurprising that early—albeit debated—traces of life can already be found for the period between 3.95 and 3.5 billion years ago (Mojzsis et al. 1996; Nisbet and Sleep 2001; Westall et al. 2006, 2011; Wacey et al. 2011; Tashiro et al. 2017). The biological origin of single-cell fossils washed up on a seashore about 3.43 billion years ago has been convincingly confirmed (Brasier et al. 2015). Evidence for biological influences can also be found in changes in geochemical processes, for example methane formation around 3.5 billion years ago (Ueno et al. 2006) or nitrogen fixation around 3.2 billion years ago (Stüeken et al. 2015). An increase in sulphur input into the oceans around 2.8 billion years ago is attributed to terrestrial oxidation by microorganisms. Bacterial chemolithotrophic sulphur oxidation in rocks (endolithic) or photolithotrophic sulphur oxidation have been discussed as potential triggers for these processes (Walker and Pace 2007; Buick 2012; Stüeken et al. 2012; Planavsky et al. 2014; Lalonde and Konhauser 2015). This would require propagation of the first microbial biofilms to have taken place in terrestrial habitats, and initial indications for this are provided by remains of microbial mats from arid regions dated at around 2 billion years old (Simpson et al. 2013). In contrast, traces of marine anaerobic microbial processes (photoferrotrophy) are clearly preserved in deposits of ribbon iron ore, which were formed as early as 3.2 billion years ago and reached their greatest extent between 2.7 and 2.4 billion years ago. Subsequent iron deposits, on the other hand, are based on combinations of biological and inorganic oxidative processes (Li et al. 2011; Posth et al. 2013).

7.4   Photosynthesis: An Initial Liberation from Abiotic Contingencies

The diversity and ubiquity of phototrophic organisms such as plants and the atmosphere's current composition make it easy to forget the challenges associated with the emergence of photosynthesis. And the diversity of different functions performed by photosynthetic organisms in aquatic and terrestrial habitats is often overlooked as well. Ignorance of evolutionary connections becomes particularly evident when the word photosynthesis is replaced by the term "biological pump" (Passow and Carlson 2012; Robinson et al. 2014). Behind this term—which has become increasingly popular in connection with climate change—is a linear technocratic approach, according to which plants are seen as exclusively serving to bind excess carbon dioxide. Inevitably, this leads to different notions about how to increase the "pumping capacity"—from "replacing" huge forests with "efficient" plantations to "fertilizing" oceans. Carried away by enthusiasm for these novel ideas, in connection with the latter example nobody notices the nonsensical use of such an expression—in physical terms alone. What the term in fact means is passive sinking of organic material and not active transportation of carbon by organisms.

As shown in Sect. 4.2, solar radiation reaching the Earth's surface is comprised of a broad spectrum of wavelengths. Only parts of that range are suitable for energy conversion by organisms. Short-wave radiation components can destroy biomolecules, while intensive long-wave radiation can lead to thermal destruction of cells (Cockell 2000; Raven et al. 2013). Damage to individual cell components—for example DNA—increases the energy needed for damage repair. Moreover, phototrophic organisms cannot directly influence the intensity of usable parts of solar radiation. Globally, irradiation intensity is determined primarily by solar activity and secondarily by the geographical latitude of the site receiving the radiation. In addition, other factors—such as composition of the atmosphere, geomorphology or cloud formation—influence radiation conditions for the organisms concerned (White et al. 1992; Dickey and Falkowaki 2002; Farquhar and Wing 2003; Klausmeier et al. 2004; Laskar et al. 2004; Hessen 2008; Ueno et al. 2009; Nisbet and Fowler 2011; Feulner 2012; Tan et al. 2016). In near-surface zones of aquatic habitats, turbulence and suspended matter further increase the range of variation and dynamics of solar irradiation (MacIntyre et al. 2000). Apart from these influencing factors, solar radiation has advantages because of recurring cycles in its intensity and spectral composition. Well-known examples of this are the characteristic, latitude-specific day-night cycles and their pattern of variation across the year. By contrast, energy sources for chemolithotrophic organisms are discontinuously distributed in spatial terms and subject to unpredictable temporal changes.

It is hard to estimate the quantitative increase in global energy turnover that accompanied the transition from an anoxygenic, chemolithotrophic context to the present oxygenic, phototrophic living environment in which geothermal energy flow is ultimately exceeded by a factor of about three (Rosing et al. 2006). In addition, the release of oxygen permitted a massive, not yet quantitatively estimated, expansion of the energetic potential of chemolithotrophic energy conversion. Qualitatively, the number of mineral compounds increased—through reactions with oxygen—to about 4500 (Hazen et al. 2014).

Organisms find the greatest degrees of freedom for large-scale selection of different irradiation conditions in oceanic water bodies. Water depth, dissolved substances and suspended matter give rise to extensive zones with differing irradiation intensity and composition of the wavelength spectrum (Dunne and Brown 1996; Six et al. 2007; Stomp et al. 2007; Kirk 2011). Degrees of freedom are constrained by the particular profile of dissolved inorganic substances in the water, including nitrogen and phosphorus, as well as carbon—especially CO_2 and HCO_3^-—for building organic cell substance (Badger et al. 2002; Sterner and Elser 2002; Williams and Fraústo da Silva 2006).

Purple bacteria and green sulphur bacteria, for example, use reduced sulphur compounds, nitrite, arsenic compounds or hydrogen as electron donors. Bacteriochlorophylls have an identical core structure—four pyrrole rings surrounding a central magnesium atom. Depending on different peripheral compounds, nine bacteriochlorophylls can be distinguished, each with a specific light-absorption spectrum (Fuller et al. 1985; Madigan et al. 2003; Bryant and Frigaard 2006; Ohashi et al. 2010; Vogl et al. 2012; Orf et al. 2013). From studies of carbonate deposits in South Africa, it has been inferred that microorganisms with anoxygenic photosynthesis influenced sedimentation in marine shallow water zones already about 3.4 billion years ago (Beukes 2004; Tice and Lowe 2006). Purple bacteria and green sulphur bacteria may have been present in large numbers in the anoxic (oxygen-free) oceans rich in iron compounds and may have contributed significantly to large-scale deposition of ribbon iron ore as a result of their anoxic phototrophy (photoferrotrophy) (Camacho et al. 2017). Because of their wide physiological ranges, anoxygenic microorganisms can colonize a wide variety of aquatic habitats, such as salt lakes, microbial mats in shallow water zones, sediments, and free water bodies up to the limits of oceanic solar irradiation (Garlick et al. 1977; Ollivier et al. 1994; Yurkov and Beatty 1998; Borrego and Garcia-Gil 1995; Madigan et al. 2003; Visscher and Stolz 2005; Griffin et al. 2007; Trouwborst et al. 2007; Kulp et al. 2008; Koblížek 2015). Discovery of an obligate photosynthetic life form of green sulphur bacteria in deep-sea hydrothermal vents raises the—as yet unresolved—question of infrared radiation conversion capacities of these organisms (Beatty et al. 2005). Despite their one-step, and hence theoretically limited, photosynthetic energy conversion (Fig. 7.11 a), under suitable living conditions anoxygenic photosynthetic life forms have presumably survived for more than 3 billion years. This is due not only to differentiation of photosynthetic systems, such as the extremely conversion-efficient antenna complexes (chlorosomes) of green sulphur bacteria, or extensive folding of the inner cell membrane in purple bacteria (Bird et al. 2011; Orf and Blankenship 2013). Appropriately adapted metabolic processes for adequate production of organic matter—because of engagement of different chemical reactions in energy conversion—are also important (Madigan et al. 2003).

Theoretical comparisons between energetic conversion potentials alone do not suffice to explain evolutionary developments. For instance, for physiological reasons, real values can deviate quite widely from theoretically determinations. Various microorganisms can still switch between the two forms of photosynthesis (Oren et al. 1977; Madigan et al. 2003). Far more fundamental is the contextual relationship of evolutionary change. In contrast to laboratory conditions, in real-life habitats evolutionary developments can only ever take place on the basis of existing conditions. Inevitably—because of the multiple interactions involved—changes in a broad variety of factors are associated with different evolutionary outcomes, with far-reaching consequences for all organisms concerned. The undertaking of searching for decisive factors driving the evolution of oxygenic from anoxygenic photosynthesis is correspondingly difficult.

In the absence of sufficient evidence, the issue of the origin of oxygenic photosynthesis remains unresolved. Speculation regarding origination ranges from green non-sulphur bacteria (Chloroflexi), through horizontal gene transfer from purple bacteria and green sulphur bacteria, and on to cyanobacteria (Dismuskes et al. 2001; Olson and Blankenship 2004; Allen 2005; Mulkidjanian et al. 2006; Allen and Martin 2007; Blankenship 2010; Hohmann-Marriott and Blankenship 2011; Sousa et al. 2012; Sánchez-Baracaldo 2015; Blankenship 2017; Soo et al. 2017). Current hypotheses are based on the inference that an original reaction center differentiated through gene duplication and horizontal gene transfer and subsequently divided into two distinct reaction centers. Both lineages have separately undergone further differentiation in anaerobic habitats. In cyanobacteria, oxidative photosynthesis presumably evolved through combination of the two systems (Fig. 7.10). Details of the evolutionary processes involved are still unclear, given that phylogenetic analyses of individual molecular components have led to contradictory results (Fischer 2016; Martin et al. 2018; Cardona 2019; Chernomor et al. 2021; Ward and Shih 2021).

In this context, the relatively undemanding physiology of the possible "main culprits"—cyanobacteria—is striking. They probably originated as thermophilic forms more than 3 billion years ago and were widespread in oceans around 2.8 billion years ago (Dvořák

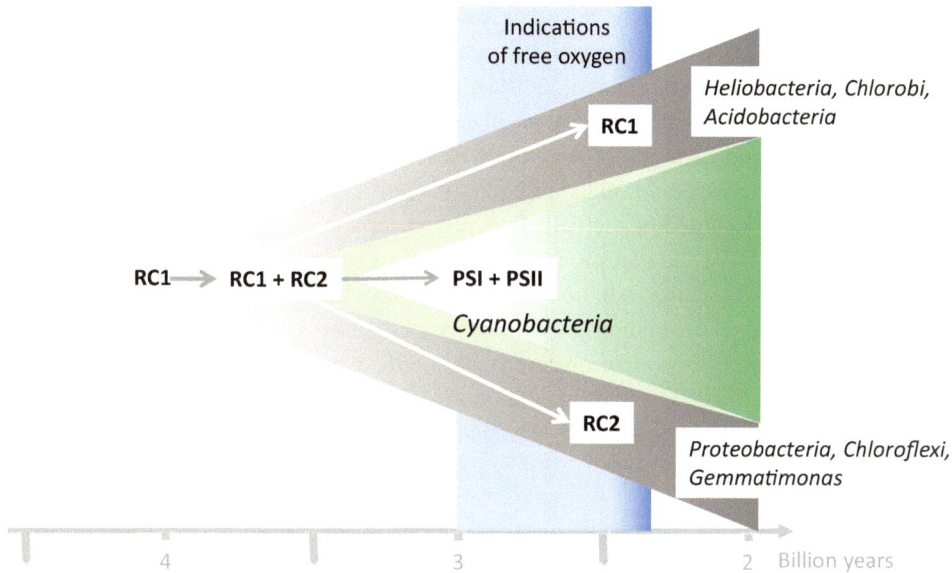

Fig. 7.10 Simplified diagram of the early evolution of phototrophic systems as currently envisaged. According to hypotheses by Martin et al. (2018), and Cardona (2019)

et al. 2014). Many members of the group do not require vitamins, and they are able to absorb nitrogen in the form of nitrate or ammonium, convert gaseous nitrogen and assimilate simple organic substances (Madigan et al. 2003). Life forms of the group *Synechococcus*—even when the supply of iron is inadequate—can maintain photosynthesis by forming a chlorophyll-protein complex (Boekema et al. 2001). The wavelength of the physiological optimum for representatives of the group *Prochlorococcus* permits photosynthesis at the maximum depths of aquatic solar light irradiation (Stomp et al. 2007). This facilitates access to sufficient concentrations of mineral nutrients in relatively deep-water bodies (Karl 2002; Mann and Lazier 2006; Kolber 2007). These indications reinforce the suspicion that oxygenic photosynthesis originated in cyanobacteria. Because of the microscopically small size of the organisms concerned that has already been mentioned, the necessary physiological abilities could have evolved much earlier than the first geologically detectable traces of the existence of free oxygen. The issue of the timespan needed for evolution of the relevant physiological changes remains unanswered, as does the question of the origin of cyanobacteria. This lack of explanation also applies to the structurally different proteins involved in conversion of light energy, the chlorophylls a, b, c, d and f (Larkum et al. 1994; Madigan et al. 2003; Kühl et al. 2005; Zhang et al. 2007; Behrendt et al. 2011; Tomo et al. 2014). Experimental studies have shown that cyanobacteria can synthesize the more energetically effective form of chlorophyll relatively rapidly (Mielke et al. 2011; Airs et al. 2014).

What is well documented, on the other hand, is the enormous adaptability of cyanobacteria to a wide range of conditions in both aquatic and terrestrial habitats (Ting et al. 2002; Madigan et al. 2003; Tomitani et al. 2006; Coleman and Chisholm 2007). A prerequisite for this was the evolution of suitable regulatory processes to ensure optimal use of light energy and to avoid damage from non-convertible radiant energy. Optimal use of light energy requires production of matching quantities of ATP and NADPH to meet the existing physiological demand of the cell. In cyanobacteria, the allocation of light energy to production of NADPH and of protons (H+ in Fig. 7.11 b) is regulated within a few minutes by distribution of incoming radiation to the two photosystems (I and II). This is achieved by means of the upstream antenna proteins (phycobilisomes), thereby controlling the level of ATP production. Excess electrons are used to reduce protons via "electron valves" and are released from the cell as hydrogen atoms. Within minutes, surplus radiant energy can be guided past both photosystems by orange carotenoids and converted to heat. Slower to

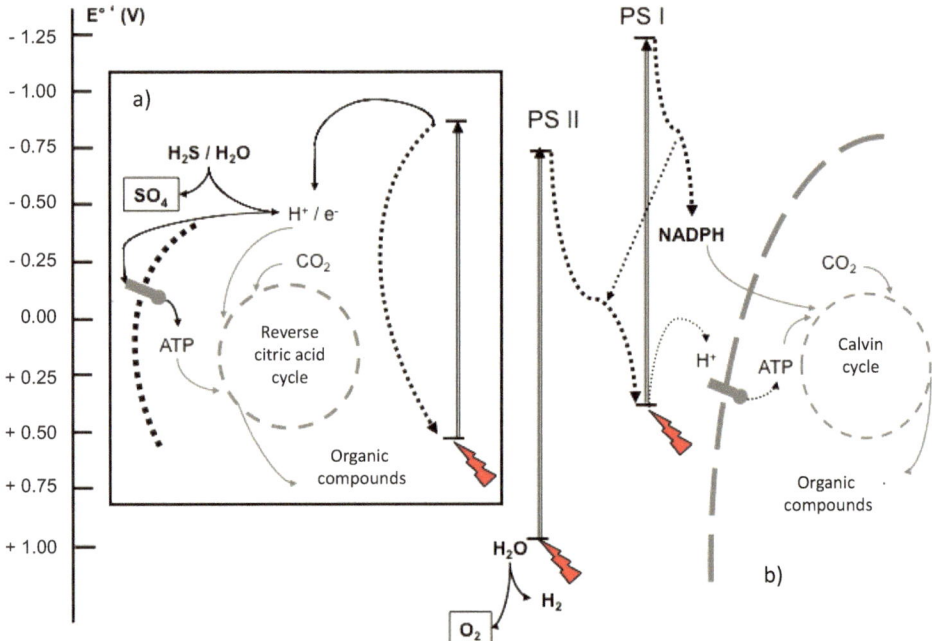

Fig. 7.11 Greatly simplified comparison between the basic principles of (**a**) anoxygenic and (**b**) oxygenic photosynthesis. Portrayed (approximately to scale compared to electrochemical dimensions) are the single light reaction in A (symbolized by jagged red arrows) and the double light reaction in the twin photosystems (PS I and PS II) in **b**). The chemical processes—quantified but not shown in detail—operate in a manner complementary to the light reaction for energy conversion in a); in contrast, in **b**) energy conversion is primarily borne by the two light reactions. Also of considerable importance with respect to overall evolutionary impact are the various "waste products" from the processes (marked by boxes). Sources: Madigan et al. (2003), Konhauser (2007), Heldt and Piechulla (2008), Nobel (2009), Berg et al. (2013)

react are various flavoproteins that divert surplus electrons from the photosystems, for example for the reaction of oxygen. Even longer reaction times are needed for translocation of proteins in the thylakoid membranes to decouple electrons from photosynthetic processes. The slowest regulatory processes of all are those involving changes in gene expression (Zhang et al. 2009; Mullineaux 2014; Allahverdiyeva et al. 2015a).

Current knowledge indicates that eukaryotes acquired the ability to photosynthesize more than 1.5 billion years ago through endosymbiosis with cyanobacteria (Fig. 7.12) (Hohmann-Marriot and Blankenship 2011). Evidence from fossils (*Bangiomorpha pubescens*) dates the first appearance of photosynthesis in eukaryotes to about 1.05 billion years ago (Gibson et al. 2018). From simple endosymbiosis, chloroplasts evolved in the cells of green algae and subsequently in all land plants. In aquatic habitats, the diversity of different algal forms evolved through repeated endosymbioses. Over the course of evolution, various unicellular forms developed the capacity to perform photosynthesis and evolved into heterotrophic organisms with predatory or parasitic lifestyles—for example the malaria pathogen *Plasmodium falciparum* (Obornik et al. 2009; Gould et al. 2008; Archibald 2009; Jensen and Leister 2014; Stiller et al. 2014).

Even in habitat dimensions as small as a few tenths of a millimeter, as is relevant for bacteria, initial oxygen release must have had fatal consequences for many microorganisms. For instance, uptake of carbon from CO_2 or HCO_3- was impeded because free oxygen was a competing binding partner for the catalytic enzyme ribulose biphosphate carboxylase/oxygenase (RuBisCO). The resulting photorespiration inevitably reduced the

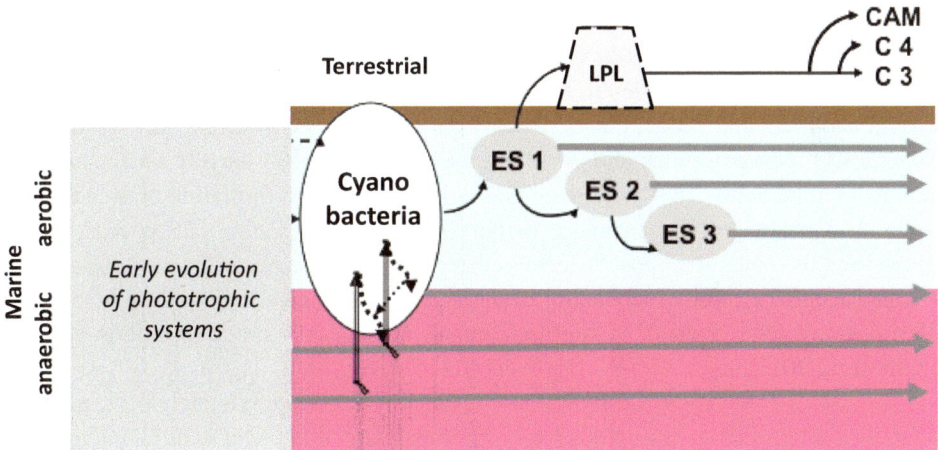

Fig. 7.12 Schematic overview of the presumed evolutionary shift in photosynthesis from anoxygenic forms of Green Sulphur Bacteria (GSB) and Purple Bacteria (PB) to oxygenic forms of Cyanobacteria and their endosymbioses (ES) with eukaryotes. Indicated are the multiple endosymbioses (ES 2-ES 3) in algae and functional adaptation of photosynthetic cells in land plants (LPFL; C_3, C_4, CAM). Horizontal arrows indicate temporal continuity of the respective functional principles; white ovals symbolize prokaryotes and gray ovals indicate eukaryotes

efficiency of photosynthesis. Anatomically and physiologically, this problem was solved in cyanobacteria by increasing the concentration of carbon compounds in cells using new selective transport systems and concentrating RuBisCO in protein-enveloped cell organelles (carboxysomes) (Badger et al. 2002; Badger and Price 2003). At the same time, a closed membrane system (thylakoid) evolved to secure the complex photosynthetic reaction processes structurally inside cells (Nelson and Ben-Shem 2004; Eberhard et al. 2008; Nickelsen et al. 2010; Rast et al. 2015).

Eukaryotic algae—especially green algae (Raven et al. 2012)—accumulate carbon in different ways in the form of hydrogen carbonates (HCO_3-) in the cell plasma or at chloroplast membranes. This allows a sufficiently high concentration of carbon dioxide to be achieved during reconversion inside the chloroplasts (Moroney and Somachi 1999; Giordano et al. 2005). Over the course of evolution, modifications of the RuBisCO proteins towards lower reactivity with oxygen reduced the need for an elevated concentration of carbon dioxide in many plant forms (Tabita et al. 2008).

Various analyses have revealed that these basic physiological patterns of CO_2 accumulation in terrestrial plants had already been adopted in the Carboniferous by lycopods (*Selaginella*) (Green 2010; Chi et al. 2014). More extensive physiological adaptations evolved in flowering plants following the decline in atmospheric carbon dioxide levels about 30 million years ago and subsequent global decreases in precipitation and temperature (Edwards et al. 2010; Beerling and Royer 2011; Sage et al. 2012; Griffiths et al. 2013; Lundgren et al. 2014; Schlüter and Weber 2016). In response to the decrease in carbon dioxide concentrations, cell structures (Kranz anatomy) and physiological processes evolved to increase carbon dioxide concentrations in the photosynthetic apparatus in some groups of angiosperms—especially in grasses. In this process, atmospheric carbon dioxide is concentrated into a molecule with 4 carbon atoms (malate) in one type of cell (mesophyll cell) and then transferred to the photosynthetically active cells (bundle sheath cells) (Heldt and Piechulla 2008). Plants with C4 photosynthesis can hence grow at low atmospheric CO_2 concentrations and—within physiological limits—largely independently of ambient temperature. They require slightly less water for the formation of organic molecules than plants without enhanced carbon dioxide concentration (C3 photosynthesis). However, this comes at the cost of decreased desiccation resistance in C4 plants (Ghannoum 2009; Sage 2017). The combination of these ecophysiological traits apparently enabled the spread of C4 grasslands in warmer climates of the Earth about 10 million years ago (Strömberg 2011).

The risk of desiccation was reduced—presumably over approximately the same period of time—in various angiosperm groups through the evolution of a different metabolic process (Edwards and Ogburn 2012; Bouchenak-Khelladi et al. 2014; Edwards 2019). Crassulacean acid metabolism (CAM) permits uptake and storage of carbon dioxide during the night. For photosynthesis, the carbon dioxide is released again within the same cell. This allows the stomata to remain closed during the day and significantly reduces water evaporation (Heldt and Piechulla 2008). Apart from a wide variety of physiologically transitional forms, the current proportions of CAM and C4 species among all flowering plants are estimated to be about 6 percent and 3 percent, respectively (Silvera et al. 2010; Sage et al. 2012). In terms of global vegetation area, the proportion of C4 plants is estimated to be about 18 percent (Still et al. 2003). Meaningful figures are prohibited for CAM plants because many of them—orchids, for example—live epiphytically on other plants (Silvera et al. 2010).

Different patterns are clearly evident in the evolution of photosynthetic systems on land and in water. Multiple endosymbioses (ES)—and hence increasing differentiation of photosynthetic machinery by means of multiple membranes (Gould et al. 2008)—are found only in aquatic and predominantly unicellular eukaryotes. Secondary endosymbioses are found, for instance, in Cryptomonadales, Haptophyta, Stramenopiles, Ciliata, Apicomplexa and Dinoflagellata. Tertiary endosymbioses with various photosynthetically active eukaryotes occur in various genera among Dinoflagellata (Delwiche 1999; Reyes-Prieto et al. 2007; Archibald 2009; Elias and Archibald 2009; Hohmann-Marriott and Blankenship 2011). Causes and functional significance of multiple endosymbioses (Fig. 7.12) are still not definitively understood (Hehenberger et al. 2014; Park et al. 2014). One of the reasons why it is difficult to resolve this issue is that many aquatic unicellular eukaryotes can switch between phototrophy and heterotrophy—that is, they are *mixotrophs* (Sanders 1991; Hansen 2011; Burkholder et al. 2008; Moore 2013). Failure to consider this fact in marine biological and limnological studies continues to impede any expansion of knowledge regarding evolutionary and ecological relationships in aquatic systems. It should be noted here, incidentally, that mixotrophy is also found in chemolithotrophic archaeans (Qin et al. 2014). Evidence of algae from marine benthic habitats in high mountain streams of the Alps provides an indication of extreme adaptive capabilities in photosynthetic organisms (Rott et al. 2006). These examples should also stimulate a critical examination of traditional notions of evolutionary change.

Primary endosymbioses are found in the Archaeplastida (Glaucophyta, Rhodophyceae and Chloroplastida) (Hohmann-Marriott and Blankenship 2011; Jackson et al. 2015). From the latter group—also known as green algae—multicellular land plants developed over the course of evolution—along with the physiological adaptations already mentioned (Stiller 2007). Far less frequently, various forms of mixotrophy are found in land plants, for example for the extraction of additional carbon compounds from fungi as well as for additional extraction of nitrogen and phosphorus compounds from animal tissue in so-called "carnivorous" plants (Tedersoo et al. 2007; Selosse and Roy 2008; Ellison and Gotelli 2009; Merckx et al. 2009; Volkov et al. 2011; Yagame et al. 2012; Fukushima et al. 2017).

This overview does not take into account the many adaptations of the photosynthetic antenna systems and chloroplast morphology to different habitat conditions (Delwiche 1999; Westphal et al. 2003; Mizutani and Ohta 2010; Solymosi 2012; Solymosi and Keresztes 2012; Hori et al. 2014; Kunugi et al. 2016), nor does it consider the differentiation of regulatory processes to avoid damage by radiation (Horton et al. 2005; Pollastri and Tattini 2011; Gerotto et al. 2012; Jahns and Holzwarth 2012; Agati et al. 2013; Allahverdiyeva et al. 2015b; Derks et al. 2015).

7.5   Heterotrophy: Reduction and Increase of Entropy in Biological Systems

In regard to Sect. 7.2 categorically thinking readers may ask themselves why the two forms of chemotrophic energy conversion were not treated together. After all, they function according to the same basic principles; the only difference is their definition according to the origin of converted carbon compounds. If more than 50 per cent of converted carbon comes from inorganic compounds—for example CO_2—the conversion process is called autotrophic. The term heterotrophic, on the other hand, refers to all processes in which at least 50 per cent of carbon originates from cellular components (Schönheit et al. 2016). This itself shows that classification of the conversion processes refers to extreme examples, between which in reality a wide variety of mixed forms can be found—as indicated in Fig. 7.1.

The motive underlying the chosen classification can be justified from an evolutionary/ ecosystem perspective. Autotrophic processes convert energy and substances from abiotic sources and thus create the foundation for the evolution of heterotrophic life forms. *Accordingly, the existence of all heterotrophic life forms depends on the output of autotrophic life forms.*

The definition provided in the first paragraph provides clues to the possible evolution of heterotrophy. By avoiding the more general term "organic compound", a clear functional connection is established. Use of the term "cellular components" expresses the fact that in heterotrophic energy conversions various transformation steps—essential for autotrophs—can be eliminated. Presumed evolution of a heterotrophic lifestyle from fermentation of cellular components (Schönheit et al. 2016) hence seems plausible for several reasons:

- First, locally accumulated material from dead cells offered new possibilities for utilization to generate energy or to build up an individual's own cell material.
- Second, the most likely anaerobic environment permitted only fermentative processes.
- Third, the switch from anaerobic autotrophic to anaerobic heterotrophic processes did not necessitate too many changes in metabolic processes.

Use of stoichiometrically optimized combinations of substances from dead cell material or metabolic waste resulted in reduced transformation costs for this form of *D-heterotrophy* (detritivore heterotrophy). At the same time, in an ecosystem context, utilization of primarily converted energy improved (Fig. 7.13). If heterotrophic transformation processes also produce substances that are directly usable by autotrophic organisms (3 in Fig. 7.13), this results in a positive feedback loop with a further reduction in the primary transformation costs. Examples of such systems are communities of fungi with algae in the form of lichens, or of fungi with plant roots (mycorrhizae) or bacteria (Frey-Klett et al. 2011; Moore et al. 2011). Effects of D-heterotrophy become particularly evident in connection with the mechanical digestion capacities of animals. As a result, for instance, from 60 to

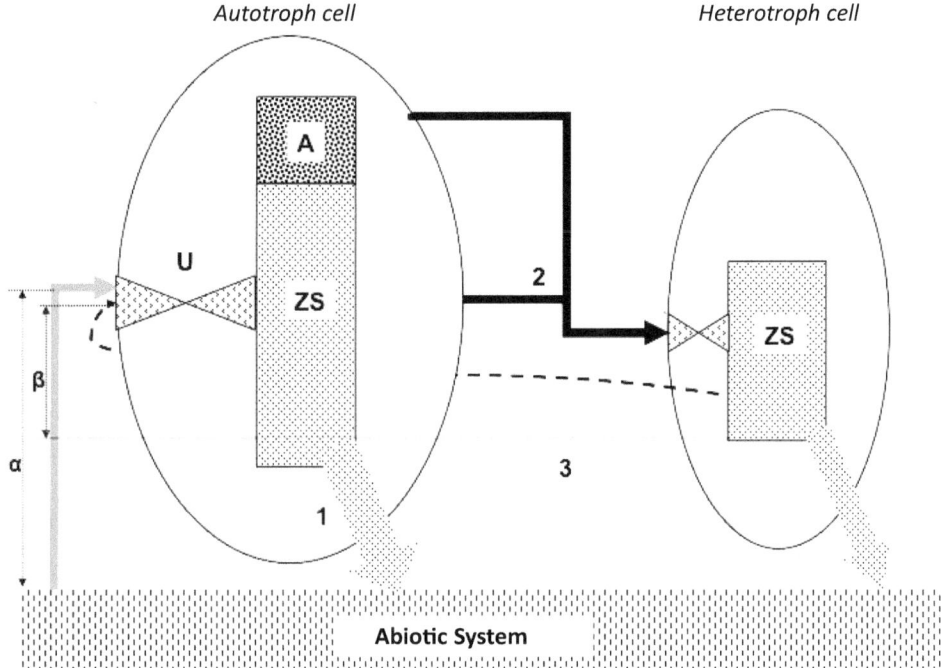

Fig. 7.13 Schematic diagram of changes in biological energy fluxes due to the occurrence of D-heterotrophs (with exclusive utilization of excreta or dead cell substance from autotrophic organisms). Without heterotrophic organisms, all organic material from metabolic processes (A) and cell substance (ZS) of autotrophic organisms would be lost to the abiotic system (1). Through heterotrophic organisms (2), organic material accumulating from excretion (A) or dead cells can be maintained in the biological system longer, while energy gained during primary transformation (U) can be better utilized. A further improvement in utilization results when substances—directly usable by autotrophs—are generated once again during heterotrophic processes (3). The diagram is simplified by comparing energy input between conversion of substances from the abiotic system (α) and recovery from biotic processes (β)

over 90 percent of the mineral nutrient requirements of plants in temperate forests are recovered from dead plant material through integrated actions of D-heterotrophs (Lambers et al. 2006). In this way, *D-heterotrophs* support *a reduction in total entropy of biological systems*.

This does not take into account stochastic characteristics of abiotic processes in the context of the multidimensional requirements of autotrophic organisms. For energy conversion, autotrophs require aqueous solutions with various chemical elements in specific (stoichiometric) proportions and states of energetic activation. Due to the inherently chaotic dynamics of abiotic processes, such prerequisites are met only for limited periods that are for unpredictable organisms (Doebeli and Ispolatov 2014). Purely autotrophic organisms can convert energy into biomass and maintain it only during these periods. Outside them, biomass is continuously lost—due to the absence of preconditions for new formation. Accordingly, evolution of purely autotrophic communities of organisms is obliged to follow the randomness of abiotic processes and does not permit any further differentiation of life forms. Through heterotrophy, however, energy—converted into organic compounds by autotrophs—becomes available for formation of new life forms. This opens up an energetic relay race with framework conditions that are determined by abiotic factors—relevant for autotrophic organisms. In general, greater regularity in inputs from optimal abiotic factor constellations promotes an increase in functional connections in the relevant biotic systems. By contrast, decreasing probabilities of occurrence lead to more open and flexible connections in biotic systems.

Characteristic examples of the effects of abiotic contingencies are provided by collapses in populations of fish, seabirds and marine mammals on the west coast of South America due to variations in atmospheric currents during so-called El Niño episodes. Examples of optimal factor constellations are provided by tropical rainforests, while deserts exemplify irregular conditions (Trillmich and Limberger 1985).

A priori other cause-effect relationships result from destruction of living autotrophic cells by *V-heterotrophs*[6]—for example various genera of "predatory" bacteria (Jurkevitch 2007; Dashiff et al. 2010; Fenton et al. 2010; Kadouri et al. 2013: Shanks et al. 2013; Keane and Berleman 2016; Korp et al. 2016). As a consequence, there is an *increase in total entropy of biological systems*. This lifestyle is nevertheless advantageous for organisms because it is associated with lower transformation costs than autotrophic life. This results in effects in the same direction as for abiotic processes. Those effect are diminished by contributions of V-heterotrophs to recovery of substances from biotic processes (3 in Fig. 7.13).

For an initial answer to the question of why V-heterotrophs were able to develop successfully in evolutionary terms, it is necessary to consider aspects of population dynamics. In general, all populations of organisms tend to show exponential growth (pop 0 in Fig. 7.14). Deviations observed in real life are due to influences exerted by various external factors and corresponding adaptations of the organisms concerned. Abiotic and biotic limiting factors can be distinguished for populations of autotrophic organisms. Abiotically, population growth is limited by physical factors and availability of material resources. Resources can limit population growth as non-renewable quantities—static (RS)—or with a particular order of magnitude as a continuously renewable flow—dynamic (RD). In the first case, populations inevitably collapse as resources are depleted. In the second case, with appropriate adaptation, population size can be maintained relatively constant over the long term (Fig. 7.14 a).

As a guide to developing a general understanding of limiting effects of biological factors, the earliest form of destruction of prokaryotic cells by viruses—in this specific context also synonymously referred to as phages—is appropriate. These interactions can be described in simplified terms by relationships for rates of reproduction and adaptive responses in the prokaryotic cells affected (Fig. 7.14 b). Growth of a prokaryotic

---

[6]V for Vita (life)

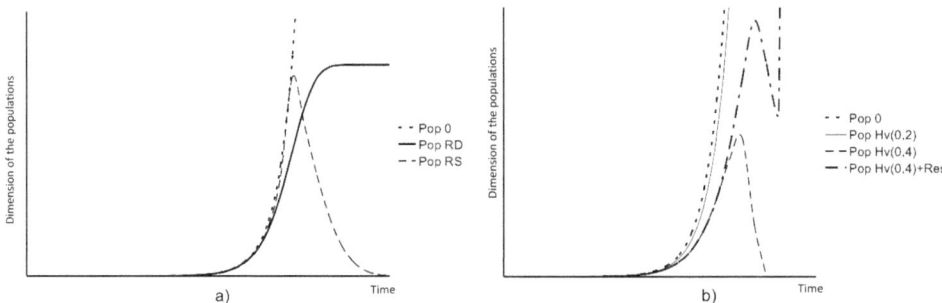

Fig. 7.14 Simplified representation of effects of growth-limiting factors on populations of autotrophic organisms. (**a**) Limitation by non-regenerable (Pop RS) and regenerable (Pop RD) resources. (**b**) Limitation by heterotrophic organisms with about 20% higher growth rate (Pop Hv(0,2)), about 40% higher growth rate (Pop Hv(0,4)) and about 40% higher growth rate, but accompanied by increasing development of defence strategies in the autotrophic population. The last two examples illustrate the comparison between failure and overcoming a competitive barrier. Pop 0 = unlimited population growth

population always begins at the intersection of the axes in the diagram, but a detectable increase is seen only after about ten reproductive cycles. In the first example (Pop Hv(0,2)), growth of the virus population—which is about 20 per cent faster—begins after five reproductive cycles, but does not lead to any significant deviations in growth of the prokaryote population. In the second example (Pop Hv(0,4)), the starting point for the virus population is the same, but its growth is about 40 percent faster. After a short interval, this reduces growth of the prokaryotic population and ultimately leads to its decline. In the third example (Pop Hv(0,4) + Res), increasingly resistant forms begin to develop in the prokaryotic population around twelve cycles after the virus attack, until eventually around 80 percent of prokaryotes are resistant. This leads to a short-term decline in the growth of the prokaryotic population, but not to collapse of the population. Examples of such elimination of a **competitive barrier** can be found both in the dynamics of marine bacterioplankton and in human population dynamics after introduction of vaccinations against various pathogens responsible for epidemics (Winkle 1997; Middelboe et al. 2001; Brüssow 2007).

The impact factors of growth limitations portrayed represent the basic patterns of all interactions. Competition between organisms manifests itself, for example, through changes in access to resources. Biogenic losses in populations can occur, for example, through lethal effects of other organisms or viruses. In real systems, the processes presented separately here overlap in many ways. In addition, interactions between the organisms involved change due to adaptive changes in their genetic, sensory, chemical and physical properties. Depending on prevailing contextual conditions, microorganisms in particular can display a wide variety of metabolic activities. All organisms—including V-heterotrophs—can contribute to an overall decrease in entropy in organismal communities, as long as their effects are limited by functional interactions between organisms (Fig. 7.15).

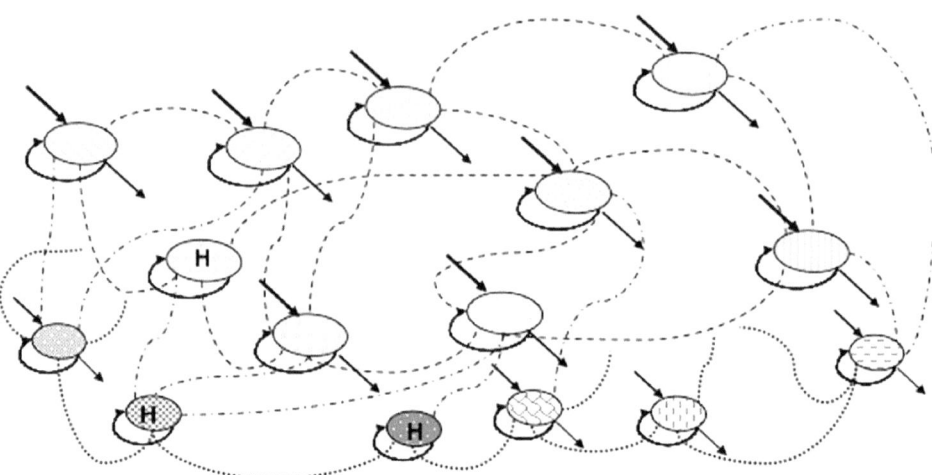

Fig. 7.15 Generalized diagram of possible interactions between populations of autotrophic organisms (unlabelled ovals with straight arrows indicating material flow) and heterotrophic organisms (labelled ovals) at two levels of impact (grey *versus* light patterns in the ovals). Possible limitations of population growth—arising from competition for resources (right sector)—or effects of heterotrophic organisms (left sector)—are symbolized by curved arrows in the ovals

A distinctive feature of many V heterotrophs is the potential ability to autoregulate population sizes in the event of food scarcity or overpopulation. Common to all phenomena is predation on and consumption of living individuals of their own species. The spectrum ranges from an individual eating its own clutches to killing and eating individuals of all ages. The anthropomorphic term cannibalism also includes a wide variety of types of consumption of components of dead conspecifics. Such phenomena have been documented in numerous animal species—including humans—and are particularly common in industrial factory farms (Shankman 1969; Polis 1981; Glass et al. 1985; Takahata 1985; Hecht and Pienaar 1993; Ebensperger 1998; Tablante et al. 2000; Claessen et al. 2003; Saltzman et al. 2006; Richardson et al. 2010; Saladié et al. 2012; Abriel and Jais 2013; Lukas and Huchard 2014; Lietzenmayer et al. 2022; Rosenheim und Schreiber 2022; Chang et al. 2023). The behaviour is always triggered by individual attitudes of animal actors in connection with perception of particular situational conditions. Since descriptions of such phenomena are usually limited to directly observable circumstances, their population-dynamic significance becomes apparent only under special conditions (Wise 2006). Examples are provided by isolated populations of Arctic charr (*Salvelinus alpinus*) in lakes where large individuals can feed only on smaller conspecifics cannibalistically. To maintain the population, sexually mature individuals must reproduce regularly, as only juveniles and small specimens can feed on plankton (Berg et al. 2010; Borgstrøm et al. 2015).

As interactions become denser and more diverse in biotic communities, extreme developments occur less frequently. An indication of potential diversity can be provided by a comparative view of physiological performance (Fig. 5.4), the multiplicity of different orders of magnitude (Fig. 5.5 b) and possible interactions (Fig. 7.15). Although—because of simplification—Fig. 7.14 b may appear to indicate only one direction of effect for the various factors concerned, that interpretation is unjustified. Interrelationships between different organisms work not only through superior speed but also through combinations of a variety of factors—even including a slower pace!

Studies of occurrences that humans find undesirable, such as so-called "algal blooms"in coastal waters, show how difficult it is to identify triggering factors. In fact, these are caused by mass reproduction of different planktonic species of cyanobacteria or eukaryotic protozoans, which often release toxic substances. Possible factors driving the increasing global incidence of such events include over-fertilization (eutrophication) of oceans by agricultural runoff, overfishing or climatic factors (Anderson et al. 2002; Heisler et al. 2008; Roy et al. 2013). With reference to Fig. 7.14 b, for mass propagation of cyanobacteria to occur, at least two competitive barriers—the first in the relationship with viruses and the second in the relationship with protists—had to be overcome. However, it remains challenging to identify underlying changes in the complex regulatory networks of marine plankton communities (Ouverney and Fuhrman 2000; Pedrós-Alió et al. 2000; Larsen et al. 2001; Cloern and Dufford 2005; Kent et al. 2007; Teeling et al. 2012).

The diversity of animal life forms evolved within the context of marine habitats during the late Ediacaran and Cambrian—around 600 million years ago. Because of insufficient fossil finds, times for the origin of animal evolution are difficult to determine. Physiologically, however, there are many indications that the first developments had already begun under anaerobic conditions. For example, the energy needed for synthesis of all cellular macromolecules under anaerobic conditions is about 13 times lower than under aerobic conditions (Lever et al. 2015). If all processes of biological synthesis in cells were aerobic, animals would have only minor advantages over purely anaerobic organisms—despite the approximately 16-fold higher energy yield in production of ATP (Alberts et al. 2015).

Accordingly, many molecular cellular processes continue to take place under anaerobic conditions. In order to be able to exploit the energy advantages of anaerobic and aerobic processes, in bacteria molecular protective mechanisms have already become established to control and regulate the oxygen balance (Barth et al. 2018). In all animal cells, gene transcription is protected against oxygen by processes based on HIF (Hypoxia-Inducible Factor). Disturbance of these processes triggers many pathological changes in humans, for example in the immune system, in the heart, and in inflammations or tumours (Eckle et al. 2008; Palazon et al. 2014; Pereira et al. 2014; Masoud and Li 2015). HIF processes also influence morphological development, for example in regression of acid-sensitive neurons in nematodes (Nematoda) in environments with low oxygen concentrations (Chang and Bargmann 2008). Various unicellular and multicellular eukaryotes avoid high oxygen loads by means of strictly or facultatively anaerobic lifestyles (Branicky and Schafer 2008; Hampl et al. 2011; Müller et al. 2012; Maguire and Richards 2014; Stairs et al. 2014).

Molecular protection against oxygen damage inevitably leads to energy losses from overall cell metabolism, the extent of which can be estimated only indirectly. According to comparative studies of anaerobic and aerobic exercise in human athletes, summative aerobic energy yield is about four to seven times higher than the anaerobic yield (Spanghero et al. 2018). Despite this difference, fish predominantly use anaerobic energy conversion for locomotion—around 80 to 90 percent (Goolish 1991).

Even in our own bodies, glucose is rapidly converted anaerobically into adenosine triphosphate (ATP) under short-term pronounced stress—for example weightlifting or short-distance running (Spanghero et al. 2018). In aerobic energy production from organic macromolecules, the higher—but slower—energy yield is also achieved through production of additional ATP molecules from reactions of oxygen with the transport molecules in the form of NADH (reduced form of nicotinamide adenine dinucleotide) (Alberts et al. 2015; Schut et al. 2016). It can be recognized that:

– the central processes of energy conversion evolved under anaerobic conditions, and
– oxygen reactions in cells can take place only under strictly controlled conditions.

Based on the chemical equilibrium constants for oxygen in animal mitochondria, their cellular energy conversion systems are thought to have evolved under oxygen concentrations ranging from 0.04 per cent to 4 per cent of the current level (Zimorski et al. 2019). Included within this range—at around 0.25 per cent—is the oxygen concentration at which marine sponges can survive. This evidence indicates that evolution of the first animal life forms might have begun much earlier—between 850 and 1500 million years ago (Mills et al. 2014, 2018; Zimorski et al. 2019). Because of the lack of fossil evidence, such assumptions cannot be evaluated at present (dos Reis et al. 2015). From an energetic perspective, sufficient convertible organic material—available over the long term—was needed for the evolution of multicellular animals to take places. Despite this, no sharp line can be drawn, as both sizes and physiological demands of animals vary over wide ranges.

It is fairly well established that the evolution of all animal life forms was based on microbial primary production under marine conditions (Mills and Canfield 2016; Bobrovskiy et al. 2018; Cohen and Kodner 2022). Morphological and sensory-physiological differentiation developed subsequently primarily, through animal interactions under marine habitat conditions. Because of the general absence of feedbacks to autotrophic organisms (3 in Fig. 7.13), in energetic terms the evolutionary dynamics were predominantly determined by abiotic processes in near-surface oceanic zones (Fig. 7.14 a, Pop RD). Trace fossils so far known indicate that the energetic basis was already adequate for evolution of burrowing life forms in marine sediments near the end of the Ediacaran—about 550 million years ago (Seilacher et al. 2005; Cai et al. 2014).

Each new life form opened up further developmental possibilities for animal organisms living on it. An example of this is the morphological diversity of Cambrian trilobites, members of the Arthropoda group (Hughes 2007; Webster 2007). While carnivorous trilobites or arrow worms (Chaetognatha) often grew to only a few decimetres, plankton-consuming species—for example Anomalocarididae among Ordovician arthropods—attained body sizes of up to two metres (Van Roy et al. 2015; Briggs and Caron 2017). Food acquisition, habitat type and locomotor behaviour significantly influenced morphological and sensory-physiological differentiation of the organisms concerned. Independently of this, differentiation of animals opened up new scope for the development of parasitic life forms (Waloszek et al. 2006). Population dynamic processes (Fig. 7.14 b)—because of interrelationships that were functionally largely uniform—may have significantly determined the emergence and demise of specific life forms.

Whereas gas exchange in unicellular organisms can still be regulated at the cellular level alone, the need for cross-cellular supply and regulation systems grows with increasing body size and cell number. Organisms with only a few cell layers—for example sponges or cnidarians—can still manage the exchange of metabolic products at the cellular level alone. With increasingly three-dimensional expansion of tissues, the supply of individual cells can be ensured only by transportation fluids. On a molecular level, the regulated transport of oxygen was achieved by adopting the iron-containing haem molecule—a key component of haemoglobin—from prokaryotes (Hsia et al. 2013). A surprising degree of morphological and physiological diversity is evident in the supply systems—collectively referred to as blood vessel systems (Monahan-Earley et al. 2013). In smaller organisms, gas exchange tends to be regulated by open vascular systems. In aqueous environments, gas exchange can take place directly through the skin, as the oxygen content per unit volume in water is only about 3 per cent compared to air (Hsia et al. 2013). This means there is less risk of cell damage due to excessive oxygen concentrations. Larger organisms are generally equipped with closed blood vessel systems that permit concentrated gas exchange with the environment by means of specialized organs—for example gills or lungs. Morphological structures and physiological regulation mechanisms ensure a sufficient supply of oxygen to the organism as a whole without damaging individual cells through excessive oxygen loads (O'Regan and Majcherczyk 1982). In the multi-stage transportation processes between uptake and delivery to the individual cell, the oxygen content is reduced to concentrations suitable for cell physiology (Hsia et al. 2013). This is associated with the converse process of removing carbon dioxide to the environment. Examples of morphological adaptations in marine areas are specific gill sizes (cm^2/g body weight) of organisms in oxygen-poor hydrothermal waters, which are about four times larger than those of organisms in coastal areas (Hourdez and Lallier 2007).

Whereas aquatic animals can take up oxygen dissolved directly in water, in the case of air respiration the organism must also manage the dissolution process. In connection with colonization of terrestrial habitats, different respiratory systems developed in individual animal groups, for example tracheae in invertebrates or lungs in vertebrates. What they all have in common—during the first stage of the respiratory process—is dissolution of oxygen in fluid within the animal body (Burggren 1982). These adaptations can be seen most clearly in modifications of the respiratory systems of land-dwelling crustaceans. Whereas crabs or isopods, which have regular contact with water, have only simple morphological modifications to protect their gills from desiccation, in permanently terrestrial species lung-like organs have evolved (Burggren 1992; Paoli et al. 2002). Many invertebrates with tracheal systems—insects, for example—regulate oxygen concentrations in the air they breathe by means of discontinuous respiration. The concentration of oxygen in the trachea is lowered to a physiologically acceptable level by temporary closure during pauses in breathing (Contreras and Bradley 2009). In vertebrates, respiratory systems developed in close association with morphological changes of the skull and trunk (Schoch 2012). Differentiation of respiratory systems in the context of adaptations of the blood circulation and physiological specializations of muscle cells enabled the development of a wide variety of vertebrate life forms in terrestrial habitats (Peck and Chapelle 2003; Wittenberg and Wittenberg 2003; Perry and Sander 2004; Kamga et al. 2012; Schoch 2012; Hsia et al. 2013).

7.6 Interactions with Mechanical and Thermal Factors

As they increase in size, multicellular organisms are subject to influences exerted by mechanical forces. This affects both interactions with the environment and the evolutionary potential of their inherent morphology. In general, in all organisms morphological phenomena are based on cellular processes. This also applies to movements of cell assemblies of fungi and plants. Even movements of animals are based on cellular processes, although they are coordinated in superordinate morphological units. Less conspicuous—but very relevant to the organisms themselves—are interactions of fungi with their environment (see Sect. 6.5.1). This concerns both penetration of substrates with fungal hyphae and temporary formation of fruiting bodies. Far more striking are the three-dimensional structures of plants, especially of tall trees. Among animals, the massive endoskeletons of large-bodied life forms or the flying skills of colourful butterflies with external skeletons are immediately eye-catching. It is easy to forget that many animal "blueprints" pass muster without fixed "frameworks". What almost all animal blueprints have in common is that their functional capacities are ensured only through intensive interactions with sensory organs and nervous systems. Exceptions to this are sponges—which often have skeletal elements composed of lime or silicate compounds (Uriz et al. 2003; Dohrmann and Wörheide 2017)—and placozoans.

As with neuronal systems, questions arise—from an energetic point of view—in connection with multicellularity about factors influencing its evolution. Multicellularity is always associated with formation of supporting tissue in immotile life forms and additional motor tissue linked to tissue for its control and coordination. Inevitably, depending on size and degree of functional differentiation, the need for conversion of energy and substances is higher than in unicellular organisms (Fig. 7.16). In evolutionary terms, therefore, multicellular organisms could evolve only where sufficient food resources were available and could be tapped more rapidly than by unicellular organisms. From observable correlations between current spatial distributions of unicellular and multicellular life forms of plants or fungi with large mycelia, it can be seen that such conditions do not exist

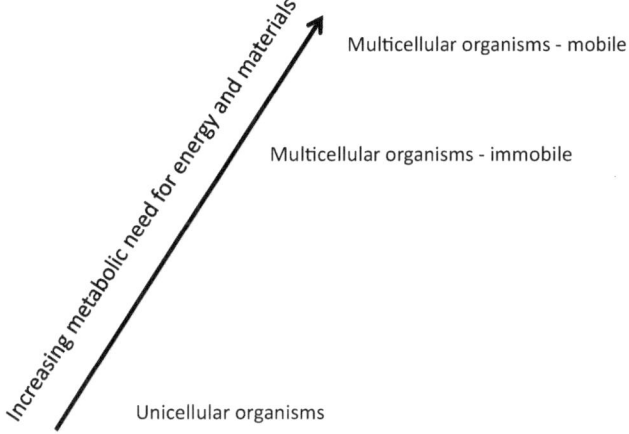

Fig. 7.16 Simplified representation of the increasing demand for conversion of energy and substances, from unicellular to motile multicellular life forms

over extensive areas. Areas far from the shores of large lakes and oceans (Hamm et al. 1999; Geider and La Roche 2002; Badger et al. 2006; Beardall et al. 2009; Smetacek 2012; Zubkov 2014; Biard et al. 2016), as well as terrestrial areas with extreme environmental conditions (Belnap and Lange 2003; Chan et al. 2012; Pointing and Belnap 2012),—e.g. deserts—are still dominated by unicellular organisms. In extremely nutrient-poor areas, photosynthesis relies exclusively on various species of cyanobacteria. In retrospect, such conditions can hence also be excluded as evolutionary zones for multicellular plants. Among animals, the diversity of morphological "blueprints" is far greater in oceans than on land (Nielsen 2001).

The decisive factor for evolution of multicellular organisms is hence the availability of sufficient convertible energy and convertible substances to meet the additional demand.

Therefore, release of oxygen through oxygenic photosynthesis cannot have been the sole triggering factor for the evolution of multicellular organisms. This interpretation is supported both by persistence of anaerobic metabolism or survivability under low oxygen concentrations in different eukaryotes and by the sensitivity of eukaryotic metabolism to reactive oxygen compounds (Kennedy et al. 1992; Aguirre et al. 2005; Jackson and Colmer 2005; Abe et al. 2007; Hourdez and Lallier 2007; Stoimenova et al. 2007; Mentel and Martin 2008; Hug et al. 2010; Payne et al. 2010; Borgonie et al. 2011; Müller et al. 2012; Atteia et al. 2013; Knoll 2014; Mills et al. 2014; Stairs et al. 2014; Sousa et al. 2016; Mills et al. 2018; Zimorski et al. 2019). Because of the higher energy gain of aerobic energy conversion (Alberts et al. 2015), and the faster growth of multicellular organisms that this makes possible, oxygen can be described in the broadest sense as an accelerator of evolutionary processes. During the relatively rapid evolution of different life forms, an important role was played by the tendency for oxygen-rich marine water to extend into ocean depths (Fike et al. 2006; Wang et al. 2012; Chen et al. 2015). Discontinuities in the

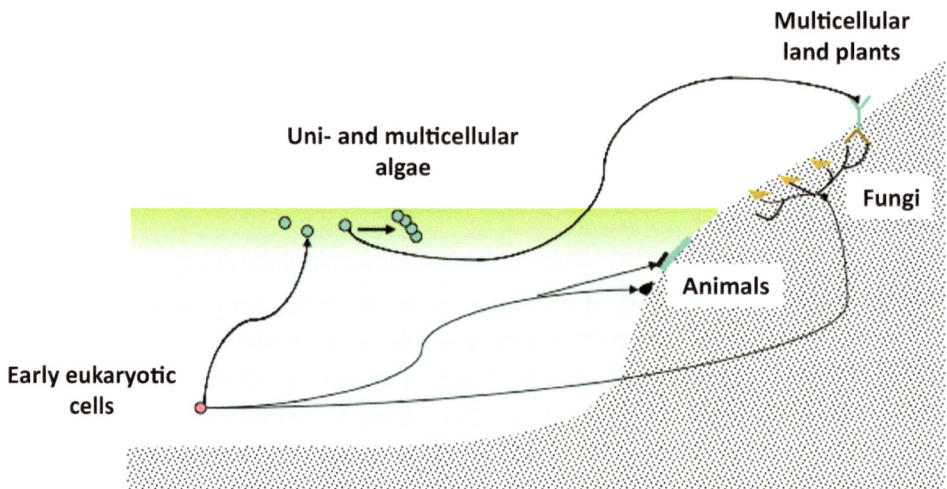

Fig. 7.17 Scenarios showing potential zones for evolution of multicellular life forms in fungi, animals and plants. Explanations in the text

process, recognizable from intermittent expansions of the anoxic water body, led to recurrent collapses of the biotic communities that were affected (McFadden et al. 2008; Darroch et al. 2015; Kurzweil et al. 2015; Sahoo et al. 2016).

Favourable conditions for the evolution of multicellular organisms can be inferred for zones with enriched food resources (Fig. 7.17). For aquatic environments, biofilms and realms enriched with suspended organic matter may be considered, whereas local deposits of organic material would be relevant on land. According to this scenario, multicellularity would have to have evolved first in heterotrophic organisms. In aquatic regions, the first animals (Metazoa) with filtering—such as sponges (Porifera)—and active absorption of food—as in placozoans—would be potential candidates (Misevic et al. 2007; Renard et al. 2009; Schierwater et al. 2009; Eitel et al. 2013; Renard et al. 2013; Smith et al. 2014; Dunn et al. 2015). Both life forms show a low degree of cell differentiation—derivable from the condition in unicellular and multicellular choanoflagellates—and largely decentralized regulation of multicellularity (Maldonaldo 2004; Hoffmeyer and Burkhardt 2016). It should be noted here that this scenario is one of one of many possibilities—because absence of nervous systems, for instance, has also been discussed as a potential secondary development from originally more complex life forms (Jékely et al. 2015; Ryan and Chiodin 2015; Wray 2015). On the other hand, molecular analyses have revealed that nervous systems have evolved several times independently and that their performance is adapted to fit particular ecophysiological requirements—especially spatial movement (Arendt 2008; Nickel 2010; Steinmetz et al. 2012; Moroz et al. 2014).

On land, on the other hand, accumulations of organic material presented evolutionary possibilities for multicellular life forms in fungi. With their hyphae, fungi can quickly expand into new zones and utilize located resources for all of their cells (Moore et al.

2011). The high proportion of root symbioses (mycorrhizae) with plants revealed by fossil discoveries as well as present today suggests that proliferation of multicellular fungi was a crucial prerequisite for the evolution of multicellular land plants (Remy et al. 1994; Selosse and Le Tacon 1998; Heckman et al. 2001; Carafa et al. 2003; Read and Perez-Moreno 2003; Dotzler et al. 2006; Winther and Friedman 2007; Brundrett 2009; Pressel et al. 2010; Philippot et al. 2013; Strullu-Derrien et al. 2014, 2015; Brundrett and Tedersoo 2018; Hoysted et al. 2018).

The many examples of multicellular life forms that still exist—and are dominant in many habitats—show that multicellularity can by no means be regarded as the result of directed evolution. On the other hand, permanently multicellular organisms were only able to evolve under environmental conditions in which they experienced sufficiently large competitive advantages compared to unicellular life forms. For such considerations, it must be taken into account that evolutionary processes are always associated with very large numbers. Changes in life forms can therefore develop rapidly through positive feedback and disappear just as quickly in cases where negative feedback occurs. Examples of this can be found in fungi and plants in connection with the evolution of permanently multicellular life forms. In both cases, life forms in pelagic zones of the oceans were multicellular at best, whereas on land a great diversity of permanently multicellular organisms evolved.

7.6.1   Plants: At the Limits of Human Imagination

7.6.1.1 Diverse Kinds of Appearance and Scientific Myths

Among plants, multicellular life forms dominate on land; in open waters, on the other hand, unicellular algae predominate. With the exception of the pelagic brown alga genus *Sargassum*, multicellular life forms occur predominantly in coastal and inland waters that are sufficiently clear—for example in kelp forests (Prince and O'Neal 1979; Prince 1980; Steneck et al. 2002; Stiger et al. 2003; Silberfeld et al. 2010; Niklas and Kutschera 2010; Leliaert et al. 2011; De Clerck et al. 2012; Umen 2014). This raises the question as to why a significantly greater diversity of multicellular plants has developed on land than in oceans (Falkowski et al. 2004). To understand the correlations involved, the sum formula of photosynthesis (Heldt and Piechulla 2008) needs to be recollected here:

$$CO_2 + 2H_2O \xrightarrow{Light} [CH_2O] + H_2O + O_2$$

Plants require the compounds carbon dioxide and water to convert the photon energy of light. Depending on acidity, carbon dioxide reacts with water to form carbonic acid. Free carbon dioxide is found mainly in acidic water at low pH values. In the biologically favourable neutral range (pH 7), carbonic acid decomposes in favour of hydrogen carbonate (HCO_3-), and at the same time the concentration of carbon dioxide falls significantly (Wetzel 1975; Lautenschläger et al. 2001). Accordingly, in aquatic habitats, free carbon dioxide is present only to a limited extent (Reinfelder 2011). Here—because of their more favourable surface-to-volume ratios—unicellular life forms are at an advantage. But, in order to survive in these habitats, even cells as small as those of the photosynthetic cyano-bacteria must chemically concentrate carbon from carbon dioxide (Montgomery et al. 2016). Eukaryotic phytoplankton—with significantly larger cell dimensions—often supplement their carbon acquisition mixotrophically through phagocytosis of other organisms (Worden et al. 2015).

According to current knowledge, land plants evolved from freshwater green algae (Charophyta). However, rather than originating from the morphologically differentiated stoneworts (Charophyceae), they presumably evolved from conjugating green algae (Zygnematophyceae), which are often unicellular (McCourt et al. 2004; Wickett et al. 2014; de Vries and Archibald 2018; Del-Bem 2018).

In terrestrial habitats, carbon dioxide has been available in sufficient quantities for plants over the past 450 million years—albeit with marked variations in concentration (Berner 1998; Franks et al. 2014). The great challenge was to manage the water supply by developing suitable morphological structures in conjunction with the requisite regulatory processes. Outside of permanent water bodies, water is available on Earth's surfaces only temporarily, for the longest timespan in the microscopic gaps between mineral particles. These are conditions that have been globally exploited by cyanobacteria and algae—mostly together with fungi—in the form of microbial soil crusts and lichens for at least 415 million years (Honegger 1993; Belnap and Lange 2003; Cardon et al. 2008; Honegger et al. 2013). Core prerequisites for the long-term existence of such ecosystems are capacities of the organisms concerned to survive long periods of desiccation, intense solar radiation and pronounced temperature fluctuations. Although—according to current estimates—around 7 per cent of global photosynthesis and about half of biological nitrogen fixation are carried out by these systems (Elbert et al. 2012), they have only a minor impact on water evaporation. In arid regions, only about 1 to 4 per cent of rainfall evaporates whereas in tropical rainforests up to 35 per cent is evaporated (Schlesinger and Jasechko 2014).

Where sufficient rainfall or near-surface water resources permit higher rates of photosynthesis, multicellular plants are superior to microbial soil crusts. In physiological terms, they are confronted with novel physical challenges. Physiologically, water is needed not only for transportation of mineral nutrients from the roots to the leaves and, in the opposite direction, for transportation of carbon compounds for cellular energy conversion or storage. According to classical ideas, dissolved minerals in vascular plants are transported to the leaves with water from the roots through the rigidified xylem cells. Dissolved products of photosynthesis—various sugars, signalling substances and other compounds—are transported up and down through the cell system of the phloem to all actively growing areas—for example roots, branches or trunk (Ruiz-Medrano et al. 2001; van Bel 2003; Aoki et al. 2005; Haywood et al. 2005; Banerjee et al. 2006; Kehr and Buhtz 2008; Nobel 2009; Mencuccini and Hölttä 2009; Melnyk et al. 2011; Kehr and Kragler 2018). In addition—in plants—water is exchanged between the two systems (van Bel 1990; Jeschke and

Pate 1995; Pfautsch et al. 2015). In reality, plants also transport atmospheric water from the photosynthetic organs to the roots, which distribute it through the soil, or transported at night from deeper zones to upper soil horizons and carried up from there with dissolved minerals at the onset of photosynthesis (Jobbágy and Jackson 2001; Jobbágy and Jackson 2004; González et al. 2011; Prieto et al. 2012; Neumann and Cardon 2012; Goldsmith et al. 2013; Ishii et al. 2014). In persistently or permanently recurrently flooded locations—for instance in swamps or mangrove forests—roots supply themselves with oxygen from photosynthetically active plant tissues. In dry areas, deep-reaching roots supply water to nutrient-rich zones near the surface to optimize provision for the plant.

Mechanical work is hence performed in land plants, but in the absence of muscles. In addition, various substances—according to the physiological needs of individual cells—must be distributed throughout the plant body. For example, combinations of substances needed for formation of new cells differ from those required to maintain photosynthesis. In scientific studies, separate consideration of cellular processes and transportation between roots and leaves has led to established paradigms that are sometimes scientifically untenable (Caldwell and Richards 1989; Armstrong et al. 2000; Herrera et al. 2008; Hogarth 2010; Cardon et al. 2013). Since plants do not have muscles, the question arises as to how they can lift water over greater differences in height in a network of dead plant cells. Explanations for this are sought mainly in water evaporation on leaf surfaces (Canadell et al. 1996; Koch et al. 2004; Burgess et al. 2006; Moles et al. 2009; Nobel 2009; Lefsky 2010; Venturas et al. 2017). Without considering the transport heights involved, these approaches seem plausible because about 98 percent of the water taken up evaporates directly and only 2 percent is needed for photosynthesis (Jensen et al. 2016). This also explains the water supply in small plants, because in largely airless capillaries—due to atmospheric air pressure—water can rise to almost 10 m at sea level. However, no water can be sucked higher than this, as a "negative vacuum" would be required. Physically, such states are not achievable, since by definition not a single atom should ever be in an absolute vacuum. Therefore, theoretical discussions about cohesion of water molecules are idle (Tyree 1997). The search for a "negative vacuum" therefore has a relation to reality similar to the question of how many individuals have to get out of an empty vehicle so that three less people continue their journey. In fact, a negative vacuum has never been measured; all data are based on inadmissible transformations of values from pressure measurements. For this purpose, cut branches are clamped upside down in pressure chambers up to the interface. In the chambers, the pressure is increased with inert gases until liquid emerges from the interface and the measured pressure value is interpreted as a vacuum value. Comparative series of measurements show that the greatest values are found not with the highest tree species, but with species from arid zones mangrove stands (Stroock et al. 2014).

7.6.1.2 Osmosis: A Multi-Faceted Phenomenon

Osmotic processes occur in all organisms, but because of their special importance for plants, only a brief overview of their diverse functions will be provided here.

A basic principle of biological-osmotic processes is based on molecular interactions between water and substances contained therein, usually in ionic or—when integrated into organic molecules—chelated form. Water molecules are electrically polar due to their asymmetrical structure and charge distribution. In comparison with other fluids, water thus possesses a variety of physical and chemical peculiarities, which also influence, for example, the diffusion (distribution) of molecular features with polar or non-polar characteristics (Nelson 2008; Collins 2012). Structure-forming reactions between ions and water molecules also influence processes in biological molecules. The effects of different ions on proteins had already been described in rudimentary form in 1888 by Franz Hofmeister (1850–1922), in the Hofmeister series that is named after him. As a result of further development of analytical methods and theoretical findings, the broad significance of different ions for biological processes is becoming increasingly apparent (Kiriukhin and Collins 2002; Collins et al. 2007; Okur et al. 2017). Another basic principle is that, when solutions with different concentrations are brought together, they seek to attain an equilibrium state and release energy in the process (Nelson 2008). If the two solutions come into contact through a semi-permeable membrane, the effect is developed as a—dynamic—osmotic force (Fig. 7.18). In simplified terms, the semi-permeable membrane is permeable to only one substance—for instance water. This increases the volume in the compartment with the—initially—more highly concentrated solution until opposing forces terminate the

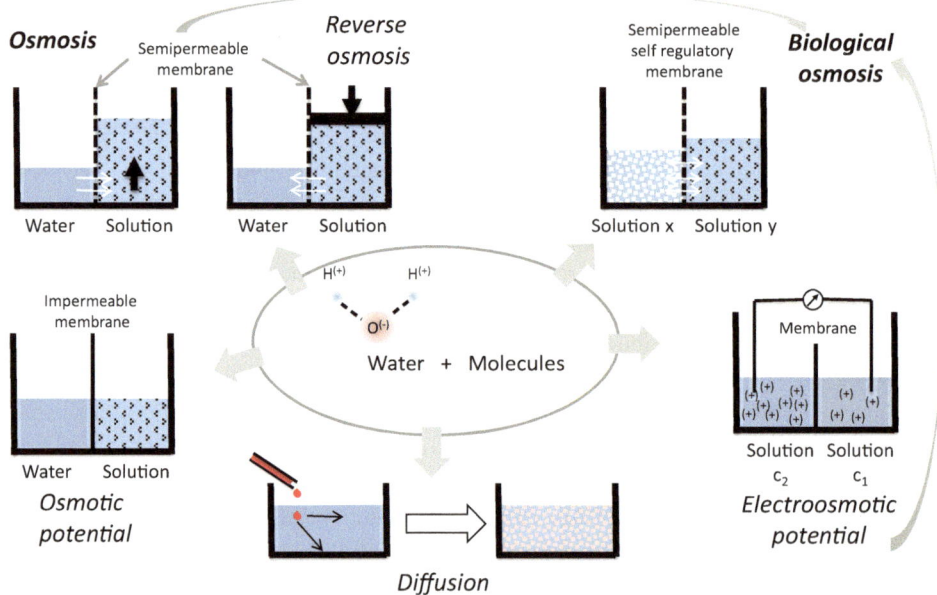

Fig. 7.18 The multiple effects of aqueous solutions are scientifically described and studied in different ways according to phenomena observed in any individual case. In biological processes, most effects proceed superimposed and in diverse interactions. In order to understand their significance, it is important to realize that—in the case of osmosis too—processes are always dynamic

process. In closed systems, instead of changing the volume of the solution, a pressure increase occurs. Simultaneously, a negative pressure is created in the compartment containing the water. This occurs, for instance, in the root tissues of grasses (Cao et al. 2012). By applying external pressure to the more concentrated solution, the processes can be driven in the opposite direction as reverse osmosis. If an impermeable membrane separates the two compartments, only a—static—osmotic potential exists (Fig. 7.18).

Processes In biological cells are far more complex not only because there are different mixtures of substances in the solutions, but also because membrane properties are constantly changing. Membranes play a central role in adaptive regulation and energetic boosting of transportation of water and substances. As a result, reverse osmosis requires no external force. Functionally, these processes are mainly furnished by aquaporins—proteins that regulate transportation of water through cell membranes (Tournaire-Roux et al. 2003; Maurel et al. 2008; Heinen et al. 2009; Wudick et al. 2009; Chaumont and Tyerman 2014; Osakabe et al. 2014; Maurel et al. 2015). In photosynthetically active zones of plants, transpiration and gas exchange with the environment through stomata is regulated by intracellular water pressure—between 1 and 4 megapascals (MPa)—in the opening/closing guard cells. Additionally, light-induced modifications of starch metabolism can strengthen the regulatory processes of the guard cells (Franks et al. 2001; Horrer et al. 2016). Various cells use water pressure to move leaves, other plant parts or the special prey-catching mechanisms of carnivorous species (Dumais and Forterre 2012). Dynamic variation of ion concentrations in biological tissues of multicellular organisms also changes electro-osmotic potentials (Fig. 7.18), which can in fact be observed macroscopically with suitable measuring equipment (Gow 1984; Fromm et al. 2013; Adamatzky 2022).

7.6.1.3 Complex Challenges and Variable Solutions

To maintain vital functions—according to specific cellular requirements—multicellular plants need to distribute water, mineral nutrients and photosynthetic products between their tissues. In terrestrial habitats, plants face different multidimensional challenges. In vegetation-free zones, a large proportion of precipitation runs off rapidly over the surface, collects in geological depressions, or flows into the oceans. Mineral substances deposited on the surface are flushed away, deposited in standing waters or transported into the oceans. The remaining proportion of precipitation seeps into geologically deeper layers, where it is stored as groundwater or continues to run off across the surface. Plants can use a—hitherto unmentioned—sub-process of the water cycle for their own water supply: evaporation through solar radiation and atmospheric processes. To do so, plants must absorb water and transfer it to photosynthetically active zones before allowing evaporation in a controlled fashion. For this, the same forces are used as for evaporation from open areas. Scientifically, however, evaporation from vegetation-free land surfaces is designated as evaporation and that from plants as transpiration (Baird and Wilby 1999).

In order to control their own water balance, plants need to overcome two different challenges. At the surfaces of standing water, high humidity renders transpiration difficult. Plants therefore exploit additional radiant energy by forming large leaves—for example, in water lilies—or by growing in height, thus shifting their evaporative organs to zones with lesser humidity and stronger air exchange. On arid land areas, plants can sustain themselves only if they are able to compensate for irregular and short-term supply of water from precipitation through storage or by extending plant organs into groundwater zones.

In simplified terms, three strategies for coping with terrestrial environmental conditions can be distinguished among the life forms of multicellular land plants:

– Mosses lack structures for internal water transport, but they can store water up to about 14 times their own dry weight (Pypker et al. 2006; Michel et al. 2013). Many species survive extended periods of desiccation of the entire plant (Hedderson and Longton 1996; Crawford et al. 2009).
– Ferns have structures for internal water transport and adjustable stomata for transpiration (Ruszala et al. 2011). In the alternation of generations between the diploid sporophytes usually called fern plants and the inconspicuous haploid gametophytes, the latter can survive extended periods of drought. Fern spores are also drought-resistant (Watkins et al. 2007; Pittermann et al. 2013; Haufler et al. 2016; Ballesteros et al. 2017).
– Seed plants primarily have structures for internal water transport and variously adjustable stomata for transpiration (Haworth et al. 2011; Ruszala et al. 2011). Only in a few life forms is the plant itself resistant to desiccation—for example the well-known resurrection plant ("Rose of Jericho") (Oliver et al. 2000; Hegazy et al. 2006). Many plant seeds are desiccation-resistant (Alpert 2000; Tweddle et al. 2003).

Contrary to classical accounts of the evolution of terrestrial plants, it can be assumed that all three strategic lineages developed in parallel and under reciprocal influences. Thus, among the 407-million-year-old fossils of the Scottish Rhynie chert flora, original features of both bryophytes and vascular plants can be found (Edwards et al. 2017; Kerp 2017). Species of all three strategic lineages are found in dry areas (Rothfels et al. 2008; Farrant and Moore 2011; Sigel et al. 2011; Shortlidge et al. 2012; Stark et al. 2013). With the exception of gymnosperms, various species in all lineages have secondarily re-established themselves in freshwater habitats (Chambers et al. 2008; Wanke 2011). During recent Earth history, the functional diversity of mosses and ferns has markedly increased (Laenen et al. 2014; Testo and Sundue 2016). Similarly, species diversity among palm ferns—usually associated with dinosaurs—has increased significantly over the past 10 million years (Nagalingum et al. 2011). Geographical expansion of flowering plants (angiosperms) did not displace the above-mentioned plant groups (Willis and McElwain 2002), but instead promoted their differentiation.

The decisive factor here is probably the greater physiological flexibility of angiosperms and the resulting augmented feedback effects on environmental factors. Through their diverse capacities for regulating physiological processes, angiosperms influence environmental conditions at differing scales. Through their capacity for differentially regulating transpiration, angiosperms facilitated the evolution of tropical rainforests in the form known to us (Brodribb and Holbrook 2004; Brodribb et al. 2009; Berry et al. 2010; Boyce et al. 2010; de Boer et al. 2012). In addition, various angiosperms can actively regulate their water supply from underground sources. These phenomena have so far been documented for woody plants in steppe regions, for example the umbrella thorn acacia (*Vachellia* [*"Acacia"*] *tortilis*), velvet mesquite (*Prosopis velutina*) or big sagebrush (*Artemisia tridentata*). The root systems of these plants irrigate shallow soil zones with groundwater from deeper layers of the soil (Caldwell and Richards 1989; Horton and Hart 1998; Ludwig et al. 2003; Cardon et al. 2013). Detailed measurements on Australian silky oaks (*Grevillea*) and eucalypts (*Eucalyptus*) have revealed that they can flexibly distribute water up and down between moist and dry soil layers (Burgess et al. 1998).

In coordination with photosynthetic processes, water—enriched with minerals—is transported from near-surface soil zones into leaves (Horton and Hart 1998; Ludwig et al. 2003; Scott et al. 2008; Cardon et al. 2013). In combination with the leaves, the extensively ramified branches create a variety of climatically differing microhabitats. This opens up new potentials not only for the above-mentioned differentiation of mosses and ferns, but also for evolution of extremely diverse life forms among angiosperms. The plants concerned can grow directly on stems or branches (epiphytic) or on leaves as well (epiphyllous) (Hietz and Hietz-Seifert 1995; Andrade and Nobel 1997; Lücking and Bernecker-Lücking 2000; Affeld 2008; Hennequin et al. 2008; Dong et al. 2012; Almeida et al. 2013; Kraichak 2013). Basically, both lifestyles are not evolutionarily novel developments. There are already indications of epiphytes on predominantly unbranched stems of gymnosperms in fossils from the Carboniferous period, around 310 million years ago (Pšenička and Opluštil 2013). Epiphyllous mosses have so far been found on angiosperm leaves only as far back as about 95 million years (Barclay et al. 2013).

The transition in developmental processes from water-dependent life forms—such as mosses—to extremely drought-resistant cacti or bromeliads can be traced at all levels—from cellular processes to the morphological features of plants (Qiu and Palmer 1999; Willis and McElwan 2002; Lambers et al. 2006; Stein et al. 2007; Laity 2008; Taylor et al. 2009; Doyle 2013). Early terrestrial life forms such as hornworts, liverworts or mosses have at most simple tissue structures for water transportation (Raven 2003; Sperry 2003). Because their transpiration can hardly be regulated physiologically, their distribution is usually confined to sites with an adequate water supply or low solar radiation (Price and Whittington 2010; Lindo and Gonzalez 2010). Many species survive desiccation of the plant body, and specialized moss species (*Syntrichia* sp.) occur permanently in arid areas and obtain water from dew and mist droplets (Alpert 2000; Li et al. 2010; Pan et al. 2016). Cacti and bromeliads physiologically reduce their water losses following the same principle. In both groups, many species regulate gas exchange for photosynthesis through CAM processes (see Sect. 7.4). During dry periods, gas exchange is restricted to the night hours, which allows the stomata to remain closed in the daytime, hence avoiding water loss. In many cases, the species concerned can switch back to C3 photosynthetic metabolism with open stomata during the day in periods with sufficient water availability. Habitats with irregular, generally extended dry periods, but—averaged over the year—an adequate water supply characterize the distribution of both groups (cacti and bromeliads).

Ecophysiologically and morphologically, however, the associated challenges are met in different ways.

The cacti—relatives of purslanes (Portulaceae)—are plant life forms that tend to completely lack leaves, generally possessing sparsely branched, photosynthetically active stems. For protection against herbivores, the surfaces of cactuses commonly bear a dense covering of spines. The stem tissues also serve to store water. Elastic longitudinal folds on the stem surfaces permit their volume to be variably adapted to any given water content. Even during short bouts of precipitation, shallow root systems allow rapid absorption of water (Nobel 2002).

Comparatively, bromeliads (Bromeliaceae)—relatives of grasses (Poaceae)—show a greater diversity of life forms. In addition to certain terrestrial representatives—such as the pineapple plant (*Ananas*)—epiphytic life forms predominate with a range of different life forms extending up to highly specialized species of air plants (*Tillandsia*), which can even grow on power lines (Wester and Zotz 2010; Givnish et al. 2011; Givnish et al. 2014). Unlike cactuses, epiphytic bromeliads store rainwater in funnel-shaped "tanks"(phytotelmata) formed by leaves. Because leaves grow from the base (Kerns et al. 1936), young leaf tissues are better protected from direct sunlight and well supplied with water and nutrients contained therein. Many air plants store atmospheric water in "micro-tanks"scattered over the leaf surface. These are covered on the outside by a multilayer of cells directly connected to adjacent squamous cells (trichomes). When the plant lacks water, the squamous cells are straightened perpendicularly to the leaf surface through cell tensions, thus allowing water droplets with their contents to condense from surrounding mist. The molecular structure of the lid cells ensures that water is quickly absorbed, filling the micro-tanks. Simultaneously, the lid cells bulge outwards, thus forming small cavities in the micro-tanks, and the squamous cells flatten down to the leaf surface. Because of the approximately 5800-fold higher resistance to vaporization, water loss is significantly reduced during dry periods (Benzing et al. 1985; Pierce 2007; Raux et al. 2020).

The extreme diversity of life forms among bromeliads offers impressive examples of the effects of interactions between species on evolution of organisms. While photosynthetic C4 metabolism dominates in grasses, the evolutionarily related bromeliads use C3 and CAM metabolism in various combinations. Certain air plants have developed ecophysiological and morphological characteristics analogous to those of mosses, which are in fact reflected in common names such as "Spanish moss"or "ball moss". Such developments are possible only by means of interactions—also endophytic in nature—with microorganisms and fungi for the uptake of nutrients, which are borne in either from the atmosphere or by animals. The latter use bromeliads in many ways as habitats and/or as food sources. At the same time—particularly in dense forests—animals accelerate the dispersion and differentiation of bromeliads through pollination and seed transportation (Kessler and Krömer 2000; Schmid et al. 2011; Leroy et al. 2012; Angelini and Silliman 2014; Crayn et al. 2015; Wolowski and Freitas 2015; Leroy et al. 2016; Queiroz et al. 2016; Males and Griffiths 2017; Males and Griffiths 2018; Borst et al. 2019; Hermida-Carrera et al. 2020; Tellez et al. 2020). These examples provide an indication of the multiple material, energetic, spatial and temporal dimensions on which organisms interact in a flexible manner. They also reveal that the concept of the holobiont (see Sect. 5.2.3) should be interpreted primarily in functional terms and not at all in a classificatory sense (Douglas and Werren 2016; Haag 2018; Roughgarden et al. 2018; Flores-Núñez et al. 2020).

The evolutionary history of tree-shaped plants clearly shows the challenges—at a cellular level—for the self-organized development of functional morphological structures (Raven and Edwards 2001; Willis and McElwain 2002; Langdale 2008; Taylor et al. 2009; Geitmann 2010; Spicer and Groover 2010; Popper et al. 2011). It is intercellular regulatory processes that enable the height growth of hundred-metre-high trunks with sufficient strength and, at the same time, the coordination of photosynthetic processes in the crowns with water uptake in the roots (Sperry 2003; Irvine et al. 2005; Geitmann and Ortega 2009; Pittermann et al. 2011; Bidhendi and Geitmann 2018)! Morphologically, tasks were "solved" by formation of roots for the uptake of water and mineral nutrients in connection with the differentiation of transpiration organs to leaves, as well as connecting transportation pathways (Sperry 2003; Langdale 2008; Jones and Dolan 2012; Doyle 2013; Kenrick and Strullu-Derrein 2014). The morphological structures were and are always shaped by the cellular implementation of signals in the context of the respective cell association. In terrestrial habitats, photosynthetically active cells must not only be oriented to the incidence of light, but also protected from excessive doses of radiation or thermal stresses by cellular regulation (Baas et al. 2004; Lau and Deng 2010; Legris et al. 2017; Yin and Ulm 2017). Under the heterogeneous conditions of natural geological formations or soils, root cells orient their growth through integrated processing of various—for instance mechanical, electrical, chemical or optical—cues (Chiatante et al. 2003; Masi et al. 2009; Baluška et al. 2010; Kutschera and Briggs 2012; Osakabe et al. 2013). Even photosynthesis can be taken over by roots, as in leafless epiphytic orchids of the group Vandeae (Carlsward et al. 2006).

As the size of a plant increases, the root system must ensure not only a supply of water and mineral nutrients, but also—with dynamic adaptation—mechanical stability of the plant as a whole (Comas et al. 2002; Mickovski 2002; Genet et al. 2011). Interactions between mechanical abiotic forces and plants have significantly increased since sufficiently deep roots evolved in the Devonian (Driese et al. 1997; Leuschner et al. 2007; Algeo and Scheckler 2010; Kenrick and Strullu-Derrien 2014). Mechanical strength of roots not only ensures the physical stability of plants, but also modifies the nature of the geological substrate. The range of changes involved extends from loosening of solid rock to stabilization of unconsolidated sediments in sloping terrain or on the banks of water courses (Carrara and Carroll 1979; Abernethy and Rutherfurd 2000; De Baets et al. 2008; Gibling and Davies 2012). Tree species with horizontally extending buttress roots can develop adequate mechanical stability and closed forest structures even on muddy subsoil or peat soils, for example in mangrove forests or in tropical rainforests of Southeast Asia (Crook et al. 1997; Tang et al. 2011; Méndez-Alonzo et al. 2015). People even exploit the properties of roots to construct living bridges, for example in the Indian state of Meghalya (Ludwig et al. 2019).

With respect to the examples given, it should not be forgotten that plant roots not only ensure mechanical stability, but also fulfil a large number of different functions. In doing so, they interact closely with other organisms, including prokaryotes. Examples are dissolution of mineral compounds from rocks or dead cells, conversion of atmospheric nitrogen into ammonium compounds, and conversion of food particles into energetically usable molecules in the visceral cavities of all animals (Emmett and Kloos 1979; Benson and Silvester 1993; Hansen and Olafsen 1999; Johnson et al. 2002; Torrallardona et al. 2003; Ereskovsky er al. 2005; Dattagupta et al. 2009; Douglas 2009; Bright and Bulgheresi 2010; Prell et al. 2010; Helliwell et al. 2011; Sachs et al. 2011; Sczesnak et al. 2011; Di Mauro et al. 2013; McFall-Ngai et al. 2013; Sachs et al. 2014; Stilling et al. 2014; Belkaid and Segre 2014; Mayer et al. 2014; Lebeis et al. 2015; Sampson and Mazmanian 2015; Wheeler et al. 2015; Malcicka et al. 2018).

Unlike roots, above-ground plant parts interact with currents of air or water. Intensities and frequencies of the currents constitute the framework conditions for the evolution of mechanical properties of the tissue and the pattern of growth. Whereas weak water currents permit the evolution of differing tissue structures and growth patterns, in the presence of strong currents only plants with flexible but tensile tissues can evolve (Holbrook et al. 1991; Vogel 2003). Strong currents facilitate exchange of substances between photosynthetic tissues and water and hence induce the growth of leaves with smooth surfaces, for example in giant kelp. In the presence of weak currents, the same plants form corrugated leaves, which induce micro-turbulence that ensures the exchange of substances on their surfaces (Koehl et al. 2008).

In contrast with conditions in water, low density and much higher flow velocities of air permit evolution of tall, elastic plant bodies. The elastic deformability of stems and branches increases resistance of plants to air currents by reducing dimensions of geometric engagement surfaces. Effect relationships are non-linear, as under plant-specific flow conditions dynamic oscillations of the plant body can be amplified until the tissue breaks (de Langre 2008; Gardiner et al. 2016). Vegetation structures in turn alter air flows towards increased turbulence. In addition to thermal effects, this also assists transfer of substances to photosynthetically active organs. In dense vegetation near ground level, increased feedback from air currents promotes formation of local climatic conditions (Birnbaum 2001; Gardiner et al. 2016; Santana et al. 2018). Finally, during reproduction various plants exploit air currents to transport pollen and seeds (Niklas 1982; Greene and Johnson 1993; Friedman and Barrett 2009).

7.6.1.4 Seeking Unknown Forces

In light of the explanations provided in the preceding subchapter, it should be clear that transportation of water and substances in plants (Fig. 6.16) is managed in various ways by cellular processes. This is also indicated by the finding that no evolutionary trends can be inferred from morphological characteristics of cells in the transport systems of plants (Liesche et al. 2017). Further indicators are provided by the different strategies of growth in height of woody plants. Palm trees or grasses such as bamboos have already developed the full stem diameter by the onset of increase in height. By contrast, conifers and deciduous trees increase their stem diameters in parallel with height growth. This strategy requires continuous regeneration of cell systems for transportation of water and matter in narrow, circular growth zones between the inner—woody—stem zone and the protective outer—bark—zone. In both zones, cells are dead, but they differ in their physical properties and contain different stored substances (Ellenberg et al. 1986). This growth strategy has been documented in fossils, starting in the Cretaceous period (Spicer and Groover 2010; Wheeler and Baas 2019). For this reason, it is not expected that the forces underlying the strategy can be identified by studying individual model plant species or employing questionable physical methods (see 7.6.1.1).

In the search for answers, all hypotheses that contradict the laws of physics should be excluded (Zimmermann et al. 2004). In-depth investigations of osmotic processes seem promising, as plants can only develop mechanical forces through these pathways. This idea was already proposed in 1930 by the German botanist Ernst Münch (1876–1946), but was for a long time ignored (Zimmermann et al. 2004; Stroock et al. 2014). In the foreground was the paradigm of continuous conduction pathways, which focussed primarily on water transportation in the dead cells of the xylem (Zwieniecki et al. 2001; Nardini et al. 2011). The persisting cell structure of the system and its cross-connections via membrane pores (pits) with physiologically active accompanying cells—cells of the parenchyma and of the phloem—were largely ignored. Here, too, in connection with different types of stem growth, it can be assumed that specific exchange processes are to be expected. In many bamboo species, for instance, it has been possible to demonstrate water supply to the trunks by means of osmotic pressure in the roots (Wang et al. 2011). In the meantime, exchange of substances and water between cells in leaves and roots—as postulated by Münch—has been confirmed. Depending on the plant species concerned, 4 to 8 cells are involved in the exchange processes. In root tips, dissolved substances are released continuously, whereas proteins are released in successive packages (batch unloading). In energetic terms, the selectively regulated transfer processes are based on combinations of osmosis and reverse osmosis (Hu et al. 2011; Cabrita et al. 2013; Stroock et al. 2014; Schulz 2015; Ross-Elliott et al. 2017). Various indications support the hypothesis that all transportation of water and substances in trunks is also regulated by cellular processes, for example higher mineral concentrations in the phloem solution, detection of RNA in regulatory processes of the phloem, transfer of injected dyes from the phloem into the xylem or—compared to angiosperms—the higher flow resistances with greater distances between the sieve plates in gymnosperms (Jeschke and Pate 1995; Pate et al. 1998; Jensen et al. 2012; van Bel et al. 2013; Pfautsch et al. 2015; Jensen et al. 2016; Knoblauch et al. 2016; Ham and Lucas 2017; Liesche et al. 2017; Jensen 2018). In view of the multimodality of osmotic processes in cells (see 7.6.1.2), it is also understandable that it has not yet proved possible to obtain causal evidence for central underlying forces through experiments.

7.6.1.5 Significance for the Evolution of Animals

For at least 400 million years, through their physiological and structural functions, plants have influenced the evolution of animals on land. In comparison to marine habitats, special pathways of morphological development are associated with this connection—especially among arthropods and vertebrates—and with allied functional and structural changes in food webs. The complexity of interactions involved is difficult to comprehend even for present-day systems, so analyses of those interactions over extended periods of time are virtually impossible. *However, comparisons of evolutionary trajectories of marine and terrestrial metazoans* (see Chaps. 9 and 10) *suggest that plants have significantly influenced the evolution of animals on land—which is the customary focus of evolutionary considerations.*

7.6.2　Animals

Because of the heterotrophic nutrition of animals, their morphological evolution involves an interplay with abiotic factors and food properties. Accordingly, oceans *a priori* permit a greater diversity of "blueprints" than terrestrial habitats. This is due not only to physical differences between water and air, but also to the influence of gravity. The relationships concerned are directly recognizable from comparisons of morphological features betweeen marine and terrestrial animal groups.

Indirectly, gravity influences the distribution of organic materials and nutrients. Basically, substances are continuously exported from land surfaces to the oceans through effects of gravity. A higher concentration of organic particles permits the evolution of diverse sessile filter feeders in the oceans. Passive filter feeders—such as rangeomorphs—can already be detected among fossils from the Ediacaran period some 750 million years ago (Fig. 7.19) (Sperling et al. 2011; Cuthill and Morris 2014). Passive filter feeders, for example cnidarian sea pens (Pennatulacea) and active filter feeders such as sponges, continue to colonize aquatic habitats (Gaino et al. 1999; Antcliffe and Brassier 2007; Wörheide et al. 2012; Dolan et al. 2013). Passive filter feeders do not expend energy to move water, but they are only able to colonize habitats with suitable flow conditions. Active filter feeders lack such a restriction on their distribution, but require a higher energy input for the flagellar movements needed to propel water.

Among mobile animals, only flat-bodied placozoans move by means of flagella (cilia) (Eitel et al. 2013); all others have a locomotor apparatus—functioning according to the same basic principle. To understand that basic principle, it is only necessary to consider "internal drives". Every bodily movement—from the writhing of a worm to the flapping of a bird's wings—is based on antagonistic effects of pressure and traction forces in the body. In all animals, tensile forces are triggered in various types of muscle tissue by neural signals. Depending on the relevant morphological functions, body parts can be fixed or moved by reducing lengths of muscles.

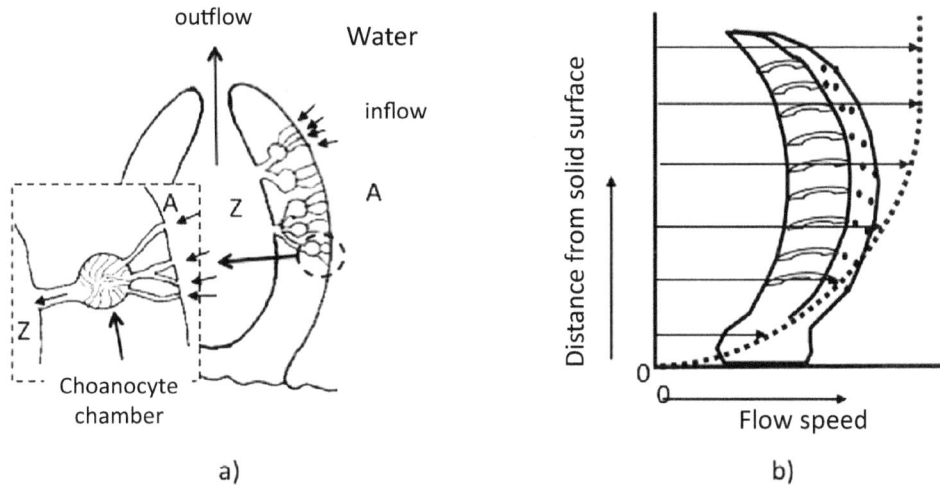

Fig. 7.19 Comparative illustration of the functional principles of active and passive filtering. (**a**) Example of a sponge with active transportation of water from the outside (A) via flagellar chambers into the central space (Z), but with passive outflow. (**b**) Schematic illustration of a sea pen as an example of the passive use of flowing water

Organs for absorbing compressive forces are more diverse. According to the materials involved, a simplified distinction can be made between hydraulic and rigid skeletons. Hydraulic skeletons are based on the generally incompressible nature of water. If its walls are sufficiently flexible, a body can be enlarged solely by a supply of fluid, for example the suckers of echinoderms or penises of various animal species. The body is deformed and moved by the pull of internal muscles—for example in earthworms or nematodes. In the simplest case, the alternating pull of parallel longitudinal muscles (l_1 and l_2 in Fig. 7.20) can bend the body in either direction. Tension in annular muscles (r) can constrict the body at particular points and expand it in adjacent regions. Muscles (sr) running obliquely to the longitudinal axis and combinations of different muscle positions can be used to perform increasingly differentiated movements—for example in earthworms, snails or cephalopods. Earthworms can, for instance, penetrate soil matter in this way (Ruiz et al. 2015). Among vertebrates, cartilaginous fish—for example sharks—use this principle for locomotion. Unlike bony fish, their locomotor muscles are connected not to the skeleton but to the skin. Sharks have a lower specific energy expenditure for locomotion compared to bony fish (Helfman et al. 2009). In tuna fish, hydrodynamically optimal positions of the dorsal fins are hydraulically controlled by a specialized lymphatic system (Pavlov et al. 2017). Arachnids use the hydraulic principle to extend their legs. When people lift heavy loads, they employ this principle by tensing the abdominal muscles to reduce forces acting on the spine (Quillin 1999; Vogel 2003; Kier 2012; Cohen and Sanders 2014).

Exoskeletons capable of movement are especially common among arthropods. Molecularly complex exoskeletons—incorporating various chitin compounds (Fabritius

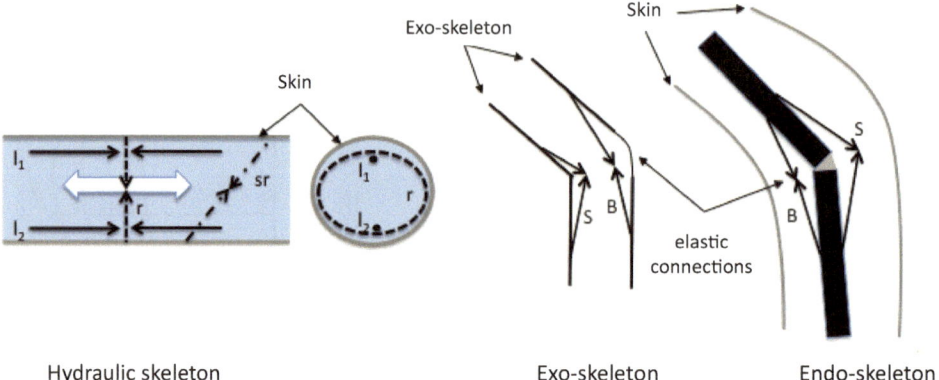

Fig. 7.20 Schematic representation of the differing locomotor systems of metazoans. Abbreviations: B - flexors, S - extensors; l_1 and l_2 - longitudinal muscles; r - radial muscle, sr - oblique radial muscle. The white arrow in the hydraulic skeleton symbolizes compressive forces in the body of fluid

et al. 2009)—fulfil the functions of a skin as well as those of support and locomotion. Hardness of the material can be significantly augmented by inclusion of metal ions, for example in the mouthparts of spiders (Politi et al. 2012). Advantages of increased protection against attacks are offset by disadvantages in growth requirements, among other things. To grow, an animal is obliged to discard its old exoskeleton and then wait, relatively unprotected, until the new—larger—skeleton hardens. An exoskeleton also complicates rapid adaptation of muscle masses to different loads (Fig. 7.20) and respiration. One particular kind of adaptation is represented by asynchronous flight muscles in many insect species. In simplified terms, this phenomenon—which has been known for some time—is characterized by the absence of synchronization between impulses in motor neurons and the frequencies of wingbeats. Recent studies have been able to demonstrate that the frequency of wing beats—based on direct cellular signal transmissions—is controlled by small intercellular networks between motor neurons (Josephson et al. 2000; Deora et al. 2021; Hürkey et al. 2023). In juvenile stages, various insect groups avoid the disadvantages mentioned by having particular life forms possessing hydrostatic skeletons, for example butterfly caterpillars (van Griethuijsen and Trimmer 2014).

Elastic properties of exoskeletons enable some ant species, termites and mantis shrimps to make the fastest movements so far observed in the animal kingdom. Used for defence or to catch prey, mouthparts are stretched—rather like bows—and expeditiously released. The mandibles of ants accelerate at around 10^6 m/s^2 and reach speeds of around 300 km/h. Mantis shrimps, for example, prey on fish with their mouthparts striking at speeds of over 100 km/h in water (Zack et al. 2009; Larabee and Suarez 2014; Patek 2015; Patek and Summers 2017; Larabee et al. 2018). On the other hand, the rigidity of the exoskeleton enables large crabs, such as the coconut crab (*Birgus*), to achieve biting forces that no vertebrate of the same size is able to match (Oka et al. 2016).

Endoskeletons of vertebrates allow both iterative growth of the entire organism and flexible adaptation of muscle masses to different loads. Muscles are attached to bones with elastic material directly or indirectly by tendons. Mechanically, tendons function as elastic connections between muscles. At the same time, they usually contain sensory cells (Golgi sensory cells) for detection of current muscle tension. Neurologically, Golgi sensory cells are the sensory elements in fine motor control circuits for muscle movement (Jami 1992; Alexander 2002; Gregory et al. 2002; Roberts 2002; Mileusnic and Loeb 2006). This also applies to turtles, whose shell—despite the visual impression formed—originated from the endoskeleton (Hirasawa et al. 2013). Early jawless "bony-armoured" fish"(Osteostraca) possessed exoskeletons. In currently living vertebrates, fish scales, teeth, clavicle and skullcap are ectodermal[7] in origin (Hirasawa and Kuritani 2015). Physiologically, bones and muscles communicate with each other as well as with other organs through endocrine systems (Boyle et al. 2003; Wolf 2008; Houssaye 2009; Houssaye et al. 2016; Karsenty and Olson 2016).

Cartilage and tendons serve different functions in vertebrate locomotion. Cartilage regions—elastic elements in rigid bones—increase mobility of the spine, for instance, or dampen mechanical shocks during locomotion on land. Accordingly, large dinosaurs may have had significantly thicker layers of cartilage in their joints than seen in mammals (Bonnan et al. 2013).

As with exoskeletons, mobility is limited by largely inflexible skeletal elements. Range of motion is determined by the distribution of joints, their mechanical characteristics and elastic intercalations between bones (Fig. 7.20). Tiny threadsnakes (Leptotyphlopidae) have for example additional joints in their mandibles to ingest ant larva with a mechanics like a swinging door (Full and Tu 1990; Romer 1971; Kley and Brainerd 1999; Polly 2007; Chen et al. 2008; Hasenfuss 2008). The elastic properties of tendons—resembling those of exoskeletons—permit faster movement sequences than would be with muscle movements alone. Tendons also serve as dynamic energy stores. Through them, portions of the kinetic energy of muscles as well as the impact energy when limbs contact the ground are re-used for subsequent movement sequences (Alexander 2002; Roberts 2002; Cavagna and Legramandi 2009; Astley and Roberts 2011). Actualization of such potentials is related to overall morphology of organisms and their ecological framework conditions (Gleiss et al. 2011). Identifying general correlations between body dimensions and energetic transformation can hence only provide guidelines for more penetrating investigation (Martin et al. 2005).

The range of bone structure realized is shown by a comparison between bird bones with high material density but—because of their construction—low weight and the massive, heavy bones of aquatic tetrapods (Houssaye 2009; Dumont 2010). Evolution of hollow bones occurred long before early bird-like life forms appeared in pterosaurs (Pterosauria) (Benson et al. 2012). Studies of diving waterfowl show that the multidimensional

---

[7] The term refers to the embryonic cell layers (germ layers): entoderm (inner cell layer), mesoderm (intermediate layer) and ectoderm (outer layer).

requirements of flight and of locomotion in water can be managed only approximately in individual life forms. Among tetrapods capable of flight, species such as cormorants (*Phalacrocorax*) or guillemots (*Uria*) have the greatest specific energy consumption for flight. Underwater, cormorants have a higher specific energy expenditure for locomotion than guillemots. However, energy costs in the latter still exceed values for penguins, which are unable to fly (Elliott et al. 2013). Extreme long-distance oceanic flyers exploit either differences in air currents above the waves (dynamic soaring)—for example albatrosses—or, for thermal soaring flight, thermal currents caused by clouds—for example frigatebirds (Weimerskirch et al. 2016; Bousquet et al. 2017; Kempton et al. 2022). Swifts (*Apus apus*) and Alpine swifts (*Tachymarptis melba*) can remain airborne continuously for over 200 days (Liechti et al. 2013; Hedenström et al. 2016). Among exclusively terrestrial birds, the most species-rich group containing songbirds (Passeriformes) has the highest average specific energy consumption for flight. This is a result of morphological adaptation of the wings and the motor system for flight movements within dense vegetation (Gavrilov 2011). In hovering flight, leaf-nosed bats (Phyllostomidae) require less energy per unit body weight than humming-birds (Trochilidae), while the latter need less than insects such as hawk moths (Sphingidae) (Voigt and Winter 1999).

Various examples of modular, iterative adaptations of morphology to changed requirements can also be identified through comparative analyses of fossils. For example, differentiation of jaw morphology began only with the evolution of the amniotes (see Sect. 6.4.1) at the end of the Carboniferous—about 80 million years after the first evidence for vertebrate migration into terrestrial habitats (Anderson et al. 2013). Fossils of herbivorous dinosaurs (sauropods) in Niger attest to the morphological diversity that resulted. In addition to sensationally gigantic species, relatively small life forms also evolved around 110 million years ago. Evolution of differences in body size was probably linked to physiological adaptations in metabolism (mesothermy), which can also be found in leatherback turtles, for instance (Sereno et al. 1999; Grady et al. 2014). Questions regarding the conditions that led to extensive reduction of the caudal spine in individual groups of dinosaurs around 140 million years ago, thus rendering possible the evolution of birds capable of flight, remain unanswered (Rashid et al. 2014). However, here it must also be taken into account that over the course of bird evolution the capacity for flight was often secondarily lost under appropriate environmental conditions—for example on isolated islands (McNab 1994; Wood et al. 2012; Worthy et al. 2017; Stervander et al. 2019).

Surface friction between solid materials, in combination with gravity, permits the well-known patterns of locomotion on land, from legless snakes to running or jumping insects, spiders or vertebrates (Hu et al. 2009; Astley et al. 2015; Denny 2016). Impressive examples of the interdependence between morphological and neurological characteristics (Fig. 6.12a) with the consequences for eco-physiological variabilities are provided by squamata (lizards and snakes). The loss of limbs and the shift to locomote on their bellies have snakes freed from morphological constraints of tetrapod skeletons. Consequently, snakes could evolve faster adaptations to spectacular diversities of locomotion, feeding and sensory processing than lizards (Simões et al. 2016; Da Silva et al. 2018; Macrì et al. 2019; Capano 2016; Tingle et al. 2024; Title et al. 2024). Land snails, on the other hand, move on a self-generated slime film—independent of any particular substrate. The dynamic viscoelastic properties of the film allow both traction through adhesion and forward movement of snails on surfaces with a wide range of inclinations. In a similar way, various freshwater snail species can progress while suspended from the underside of the water surface, grazing on the organic materials present (Vogel 2003; Lee et al. 2008; Lai et al. 2010).

Various insect and spider species can walk on water surfaces without sinking. Water-repellent—hydrophobic—body surfaces and suitably adapted leg morphology permit exploitation of the surface tension of water. Dynamic deformation of the water surface is used for locomotion instead of friction between solid bodies, which is lacking. The best-known representatives of these life forms are water striders (Gerridae), which belong to the true water bug group of insects. Water striders occur in a wide variety of inland waters, as well as on coasts and in warmer oceanic high seas (Hu et al. 2003; Bush and Hu 2006; Ikawa et al. 2012; Zheng et al. 2015). Larger land animals with heavier bodies, such as bipedal basilisk lizards—members of the iguana group—run across water, but pierce its surface. Locomotion is based on rapid water displacement by piston-like movements of the legs and the air that is dragged along in the process (Hsieh and Lauder 2004; Bush and Hu 2006).

The example of water striders in high sea zones illustrates the differing framework conditions in aquatic and terrestrial habitats. In the mesoscale spatial dimensions of the water strider's direct habitat, surface tension serves as a barrier between aquatic and terrestrial habitats. Planktonic organisms that are transported across the boundary layer by turbulence are no longer able to immerse themselves in water. Current knowledge indicates that they also provide the food base—for example in the form of fish larvae or planktonic crabs—for water striders on open oceans (Cheng 1976). From an energetic viewpoint, maintenance of terrestrial locomotion is disadvantageous for water striders because—in comparison to land—they have to move their legs about 2.5 times faster for a given speed of locomotion (Zheng et al. 2015). Globally, in addition to the temperature, the dynamics of large-scale water currents at ocean surfaces determine the general conditions for distribution of water striders. For example, in the Pacific Ocean for at least 20,000 years the equatorial current has divided the species *Halobates sericeus* into two genetically distinct populations (Ikawa et al. 2012; Leo et al. 2012).

In sum, morphological developments in various combinations with neurological and physiological functions expand the scope of energetic transformations in food webs. Without multicellular animals, neither crushing of large plant or animal tissues, nor energetic conversion of organic materials in soils and sediments would be possible. Indirectly, animals also accelerate the recovery of mineral nutrients for plants living on land (Koehl and Strickler 1981; Blackledge et al. 2009; Dorgan 2015; Sakamoto et al. 2019). In parallel, photosynthetic primary production defines the context for differentiation of animals.

7.7   Biological Energy Conversion at a Glance

– All biological energy conversion is based on exploitation of imbalances in distributions of electrons and protons in an aqueous environment. Biological membranes—ranging from cell membranes to intracellular partitions—provide sites for energy conversion and regulation.
– Primary sources of biological energy production (autotrophy) are inorganic electrochemical voltage potentials and photon energy from a sector of the electromagnetic radiation spectrum (see Chap. 4).
– Biological energy production from inorganic materials is fundamentally limited by geochemical potentials. In connection with this form of energy production, organisms modify global inorganic processes in many ways. By utilizing solar irradiation for energy, organisms amplify these effects. Increasing primary biological energy conversion (Fig. 7.2) potentially paves the way for evolution of differentiated, large-bodied life forms.
– Prokaryotes extend the limits of the use of abiotic energy potentials of a single cell by forming dynamically interacting consortia of cells with complementary physiological endowments (syntrophy). To protect against substance loss and to maintain interactions between cells, they are usually encased in an extracellular matrix (biofilm). Microorganisms are involved in many energy conversion processes in a wide variety of forms.
– Only multicellular plants can form permanent three-dimensional structures for primary energy conversion. Essential internal nutrient transportation is implemented and regulated by osmotic processes. Furthermore. because of associated storage and evaporation processes, land plants additionally modify the dynamics of global water cycles.
– Organic material can be biochemically utilized by microorganisms extracellularly or—at microscopic dimensions—used intracellularly for energy acqustion (heterotrophy) following enclosure and ingestion (phagotrophy). Various microorganisms can convert energy both autotrophically and heterotrophically (mixotrophy).
– Evolutionary possibilities for heterotrophic animals depend on the organic material available in any particular case. Depending on morphological endowments of the animals concerned, organic matter can be tapped in a variety of ways, for example by comminuting plant or animal tissues or by directly ingesting liquid or filtered organic material. As a result of the spatial discontinuity of potential food sources, many animal life forms can actively displace, crush or circulate materials in various ways. The resulting feedback effects in overall biological systems, which depend on the functional interactions that are present, cannot be determined *a priori*.

References

Abbas M, Ebeling A, Oelmann Y, Ptacnik R, Roscher C et al (2013) Biodiversity effects on plant stoichiometry. PLoS One 8(3):e58179

Abe T, Hoshino T, Nakamura A, Takaya N (2007) Anaerobic elemental sulfur reduction by fungus *Fusarium oxysporum*. Biosci Biotechnol Biochem 71(10):2402–2407

Abernethy B, Rutherfurd ID (2000) The effect of riparian tree roots on the mass-stability of riverbanks. Earth Surf Process Landf 25:921–932

Abriel M, Jais C (2013) Influence of housing conditions on the appearance of cannibalism in weaning piglets. Landtechnik 68(6):389–393

Adachi N, Ezaki Y (2007) Microbial impacts on the genesis of lower Devonian reefal limestones, eastern Australia. Palaeoworld 16:301–310

Adamatzky A (2022) Language of fungi derived from their electrical spiking activity. R Soc Open Sci 9:211926

Addadi L, Joester D, Nudelman F, Weiner S (2006) Mollusk Shell formation: a source of new concepts for understanding biomineralization processes. Dent Concepts 12:980–987

Affeld K (2008) Spatial complexity and microclimatic responses of epiphyte communities and their invertebrate fauna in the canopy of northern rata (*Metrosideros robusta* A. Cunn.: Myrtaceae) on the west coast of the South Island, New Zealand. Dis Univ Lincoln

Agati G, Brunetti C, Di Fernando M, Ferrini F, Pollastri S, Tattini M (2013) Functional roles of flavonoids in photorespiration: new evidence, lessons from the past. Plant Physiol Biochem 72:35–45

Aguirre J, Rios-Momberg M, Hewitt D, Hansberg W (2005) Reactive oxygen species and development in microbial eukaryotes. Trends Microbiol 13(3):111–118

Airs RL, Temperton B, Sambles C, Farnham G, Skill SC, Llewellyn CA (2014) Chlorophyll f and chlorophyll d are produced in the cyanobacterium *Chlorogloepsis fritschii* when cultured under natural light and near-infrared radiation. FEBS Lett 588:3770–3777

Alberts B, Johnson A, Lewis J, Morgan D, Raff M et al (2015) Molecular biology of the cell. Garland Science, New York

Alexander RMCN (2002) Tendon elasticity and muscle function. Comp Biochem Physiol A 133:1001–1011

Algeo TJ, Scheckler SE (1998) Terrestrial-marine teleconnections in the Devonian: links between the evolution of land plants, weathering processes, and marine anoxic events. Phil Trans R Soc Lond B 353:113–130

Algeo TJ, Scheckler SE (2010) Land plant evolution and weathering rate changes in the Devonian. J Earth Sci 21:75–78

Algeo TJ, Berner RA, Maynard JB, Scheckler SE (1995) Late Devonian oceanic anoxic events and biotic crises: "Rooted" in the evolution of vascular land plants? GSA Today 5/3:45–66

Allahverdiyeva Y, Suorsa M, Tikkanen M, Aro E-M (2015a) Photoprotection of photosystems in fluctuating light intensities. J Exp Bot 66(9):2427–2435

Allahverdiyeva Y, Isojärvi J, Zhang P, Aro E-M (2015b) Cyanobacterial oxygenic photosynthesis is protected by Flavodiiron proteins. Life 5:716–743

Allen JF (2005) A redox switch hypothesis for the origin of two light reactions in photosynthesis. FEBS Lett 579:963–968

Allen JF, Martin W (2007) Out of thin air. Nature 445:610–612

Allen MA, Goh F, Burns BP, Neilan BA (2009) Bacterial, archaeal and eukaryotic diversity of smooth and pustular microbial mat communities in the hypersaline lagoon of Shark Bay. Geobiology 7:82–96

Allwood AC, Grotzinger JP, Knoll AH, Burch IW, Anderson MS et al (2009) Controls on development and diversity of early Archean stromatolites. PNAS 106(24):9548–9555

Almeida OJG, Paoli LA, Souza LA, Cota-Sánchez JH (2013) Seedling morphology and development in the epiphytic cactus *Epiphyllum phyllanthus* (L.) Haw. (Cactaceae: Hyloceraceae). J Torrey Bot Soc 140(2):196–214

Alpert P (2000) The discovery, slope, and puzzle of desiccation tolerance in plants. Plant Ecol 151:5–17

Alpert P (2005) The limits and Frontiers of desiccation-tolerant life. Integr Comp Biol 45:685–695

Alpert P (2006) Constraints of tolerance; why are desiccation-tolerant organisms so small or rare? J Exp Biol 209:1575–1584

Amend JP, Shock EL (2001) Energetics of overall metabolic reactions of thermophilic and hyperthermophilic Archaea and Bacteria. FEMS Microb Rev 25:175–243

Amend JP, Rogers KL, Shock EL, Gurrieri S, Inguaggiato S (2003) Energetics of chemolithoautotrophy in the hydrothermal system of Vulcano Island, southern Italy. Geobiology 1:37–58

Anderson DM, Gilbert PM, Burkholder JM (2002) Harmful algal blooms and eutrophication: nutrient sources, composition, and consequences. Estuaries 25(4b):704–726

Anderson PSL, Friedman M, Ruta M (2013) Late on the table: diversification of tetrapod mandibular biomechanics lagged behind the evolution of terrestriality. Integr Comp Biol 53:197–208. https://doi.org/10.1093/ich/ict006

Andrade JL, Nobel PS (1997) Microhabitats and water relations of epiphytic cacti and ferns in a lowland neotropical forest. Biotropica 29(3):261–270

Anesio AM, Hodson AJ, Fritz A, Psenner R, Sattler B (2009) High microbial activity on glaciers: importance of the global carbon cycle. Glob Chang Biol. https://doi.org/10.1111/j.1365-2486.2008.01758.x

Angelini C, Silliman BR (2014) Secondary foundation species as drivers of trophic and functional diversity: evidence from a tree-epiphyte system. Ecology 95(1):185–196

Antcliffe JB, Brasier MD (2007) *Charnia* and sea pens are poles apart. J Geol Soc 164:49–51

Aoki K, Suzuki N, Fujimaki S, Dohmae N, Yonekura-Sakakibara K et al (2005) Destination-selective long-distance movement of phloem proteins. Plant Cell 17:1801–1814

Archibald JM (2009) The puzzle of plastid evolution. Curr Biol 19:R81–R88

Archibald JM, Logsdon JM, Doolittle WF (1999) Recurrent paralogy in the evolution of archaeal chaperonins. Curr Biol 9:1053–1056

Arendt D (2008) The evolution of cell types in animals: emerging principles from molecular studies. Nat Rev Genet 9:868–882

Armstrong W, Cousins D, Armstrong J, Turner DW, Beckett PM (2000) Oxygen distribution in wetland plant roots and permeability barriers to gas-exchange with the rhizosphere: a microelectrode and modelling study with *Phragmites australis*. Ann Bot 86:687–703

Arvieu J-C, Leprince F, Plassard C (2003) Release of oxalate and protons by exctomycorrhizal fungi in response to P-deficiency and carbonate in nutrient solution. Ann For Sci 60:815–821

Astley HC, Roberts TJ (2011) Evidence for a vertebrate catapult: elastic energy storage in the plantaris tendon during frog jumping. Biol Lett 8:386–389

Astley HC, Gong C, Dai J, Travers M, Serrano MM et al (2015) Modulation of orthogonal body waves enables high maneuverability in sidewinding locomotion. PNAS 112(19):6200–6205

Atteia A, van Lis R, Tielens AGM, Martin WF (2013) Anaerobic energy metabolism in unicellular photosynthetic eukaryotes. Biochim Biophys Acta 1827:210–223

Baas P, Ewers FW, Davis SD, Wheeler EA (2004) Evolution of xylem physiology. In: Hensley AR, Poole I (eds) The evolution of plant physiology. Elsevier, Amsterdam, pp 273–295

Bada JL, Lazcano A (2002) Some like it hot, but not the first biomolecules. Science 296:1982–1983

Badger MR, Price GD (2003) CO_2 concentrating mechanisms in cyanobacteria: molecular components, their diversity and evolution. J Exp Bot 54(383):609–622

Badger MR, Hanson D, Price GD (2002) Evolution and diversity of CO_2 concentrating mechanisms in cyanobacteria. Funct Plant Biol 29:161–173

Badger ME, Price GD, Long BM, Woodger FJ (2006) The environmental plasticity and ecological genomics of the cyanobacterial CO_2 concentrating mechanism. J Exp Bot 57(2):249–265

Baird AJ, Wilby RL (1999) Eco-hydrology. Routledge, London

Ballesteros D, Hill LM, Walters C (2017) Variation of desiccation tolerance and longevity in ferns. J Plant Physiol 211:53–62

Baluška F, Mancuso S, Volkmann D, Barlow PW (2010) Root apex transition zone: a signalling-response nexus in the root. Trends Plant Sci 15(7):402–408

Bamford DH, Grimes JM, Stuart DI (2005) What does structure tell us about virus evolution? Curr Opin Struct Biol 15:655–663

Banerjee AK, Chatterjee M, Yu Y, Suh S-G, Miller WA, Hannapel DJ (2006) Dynamics of a mobile RNA of potato involved in a Long-distance signaling pathway. Plant Cell 18:3443–3457

Banfield JF, Barker WW, Welch SA, Taunton A (1999) Biological impact on mineral dissolution: application of the lichen model to understanding mineral weathering in the rhizosphere. PNAS 96:3404–3411

Barbara GM, Mitchell JG (1996) Formation of 30- to 40-micrometer-thick laminations by high-speed marine bacteria in microbial mats. Applied and Environmental Microb 62(11):3985–3990

Barclay RS, McElwain JC, Duckett JG, van Es MH, Mostaert AS et al (2013) New methods reveal oldest known fossil epiphyllous moss: *Bryiidites utahensis* GEN. ET SP. NOV. (Bryidae). Am J Bot 100(12):2450–2457

Barth C, Weiss MC, Roettger M, Martin WF, Unden G (2018) Origin and phylogenetic relationships of (4Fe-4S)-containing O_2-sensors in bacteria. Environ Microbiol 20(12):4567–4586

Battistuzzi FU, Hedges SB (2009) A major clade of prokaryotes with ancient adaptations to life on ôLand. Mol Biol Evol 26(2):335–343

Battistuzzi FU, Feijao A, Hedges SB (2004) A genomic timescale of prokaryote evolution: insights into the origin of methanogenesis, phototrophy, and the colonization of land. BMC Evol Biol 4/44

Beardall J, Allen D, Bragg J, Finkel ZV, Flynn KJ et al (2009) Allometry and stoichiometry of unicellular, colonial and multicellular phytoplankton. New Phytol 181:295–309

Beatty JT, Overmann J, Lince MT, Manske AK, Lang AS et al (2005) A obligately photosynthetic bacterial anaerobe from a deep-sea hydrothermal vent. PNAS 102(26):9306–9310

Bebout BM, Garcia-Pichl F (1995) UV B-induced vertical migrations of cyanobacteria in a microbial mat. Appl Environ Microb 61(12):4215–4222

Beerling DJ, Royer DL (2011) Convergent cenozoic CO_2 history. Nat Geosci 4:418–420

Behrendt L, Larkum AWD, Norman A, Qvortrup K, Chen M et al (2011) Endolithic chlorophyll *d*-containing phototrophs. The ISME J 5:1072–1076

Belkaid Y, Segre JA (2014) Dialogue between skin microbiota and immunity. Science 346:954–959

Belnap J, Lange OL (eds) (2003) Biological soil crusts: structure, function, and management. Ecological studies, vol 150, Springer

Benson DR, Silvester WB (1993) Biology of *Frankia* strains, actinomycete symbiont of actinorhizal plants. Microbiol Rev 57(2):293–319

Benson RBJ, Butler RJ, Carrano MT, O'Connor PM (2012) Air-filled postcranial bones in theropod dinosaurs: physiological implications and the 'reptile'-bird transition. Biol Rev 87:168–193

Benzing DH, Givnish TJ, Bermudes D (1985) Absorptive trichomes in *Brocchinia reducta* (Bromeliaceae) and their evolutionary and systematic significance. Syst Bot 10(1):81–91

Berg OK, Finstad AG, Olsen PH, Arnekleiv JV, Nilssen K (2010) Dwarfs and cannibals in the Arctic: production of Arctic char (*Salvelinus alpinus* (L.)) at two trophic levels. Hydrobiologie 652:337–347

Berg JM, Tymoczko JL, Stryer L (2013) Biochemie. Springer Spektrum, Berlin

Berman-Frank I, Lundgren P, Falkowski P (2003) Nitrogen fixation and photosynthetic oxygen evolution in cyanobacteria. Res Microbiol 154:157–164

Berner RA (1998) The carbon cycle and CO_2 over Phanerozoic time: the role of land plants. Phil Trans R Soc Lond B 353:75–82

Berry JA, Beerling DJ, Franks PJ (2010) Stomata: key players in the earth system, past and present. Curr Opin Plant Biol 13:233–240

Bertrand EM, Allen AE (2012) Influence of vitamin B auxotrophy on nitrogen metabolism in eukaryotic phytoplankton. Front Microbiol 3:375

Beukes N (2004) Early options in photosynthesis. Nature 431:522–523

Biard T, Stemmann L, Picheral M, Mayot N, Vandromme P et al (2016) In situ imaging reveals the biomass of giant protists in the global ocean. Nature 532:504–507

Bidhendi AJ, Geitmann A (2018) Finite element modelling of shape changes in plant cells. Plant Physiol 176:41–56

Biegel E, Schmidt S, González JM, Müller V (2011) Biochemistry, evolution and physiological function of the Rnf complex, a novel ion-motive electron transport complex in prokaryotes. Cell Mol Life Sci 68:613–634

Bindeman IN, Zakharov DO, Palandri I, Greber ND, Dauphas N et al (2018) Rapid emergence of subaerial landmasses and onset of a modern hydrological cycle 2.5 billion years ago. Nature 557:545–548

Bindschedler S, Cailleau G, Verrecchia E (2016) Role of fungi in the biomineralization of calcite. Fortschr Mineral 6/41

Bird LJ, Bonnefoy V, Newman DK (2011) Bioenergetics challenges of microbial iron metabolism. Trends Microbiol 19(7):330–340

Birnbaum P (2001) Canopy surface topography in a French Guiana forest and the folded forest theory. Plant Ecol 153:293–300

Blackledge TA, Scharff N, Coddington JA, Szüts T, Wenzel JW et al (2009) Reconstructing web evolution and spider diversification in the molecular era. PNAS 106(13):5229–5234

Blankenship RE (2010) Early evolution of photosynthesis. Plant Physiol 154:434–438

Blankenship RE (2017) How cyanobacteria went green. Science 355:1372–1373

Bobrovskiy I, Hope JM, Ivantsov A, Nettersheim BJ, Hallmann C, Brocks JJ (2018) Ancient steroids establish the Ediacaran fossil Dickinsonia as one of the earliest animals. Science 361:1246–1249

Boekema EJ, Hifney A, Yakushevska AE, Piotrowsli M, Keegstra W et al (2001) A giant chlorophyll-protein complex induced by iron deficiency in cyanobacteria. Nature 412:745–748

Boetius A, Ravenschlag K, Schubert CJ, Rickert D, Widdel F et al (2000) A marine microbial consortium apparently mediating anaerobic oxidation of methane. Nature 407:623–626

Bongaerts E, Nawrot TS, Van Pee T, Ameloot M, Bové H (2020) Translocation of (ultra)fine particles and nanoparticles across the placenta; a systematic review on the evidence of in vitro, ex vivo, and in vivo studies. Part Fibre Toxicol 17/56

Bonnan MF, Wilhite DR, Masters SL, Yates AM, Gardner CK, Agular A (2013) What lies beneath: sub-articular Long bone shape scaling in eutherian mammals and saurischian dinosaurs suggests different locomotor adaptations for gigantism. PLoS One 8(10):e75216

Borgonie G, Garcia-Moyano A, Lithauer D, Bert W, Bester A et al (2011) Nematoda from the terrestrial subsurface of South Africa. Nature 477:79–82

Borgstrøm R, Isdahi T, Svenning M-A (2015) Population structure, biomass, and diet of landlocked Arctic charr (*Salvelinus alpinus*) in a small, shallow high arctic lake. Polar Biol 38:309–317

Born EW, Böcher J (2001) The ecology of Greenland. Atuakkiorfik Education, Nuuk

Borrego CM, Garcia-Gil LJ (1995) Rearrangement of light harvesting bacteriochlorophyll homologues as a response of green sulfur bacteria to low light intensities. Photosynth Res 45:21–30

Borst AC, Angelini C, ten Berge A, Lamers L, Derksen-Hooijberg M, van der Heide T (2019) Food or furniture: separating trophic and non-trophic effects of Spanish moss to explain its high invertebrate diversity. Ecosphere 10(9):e02846

Bothe H, Tripp HJ, Zehr JP (2010) Unicellular cyanobacteria with a new mode of life: the lack of photosynthetic oxygen evolution allows nitrogen fixation to proceed. Arch Microbiol 192:783–790

Bouchenak-Khelladi Y, Muasya AM, Linder HP (2014) A revised evolutionary history of Poales: origins and diversification. Bot J Linn Soc 175(1):4–16

Boucot AJ, Gray J (2001) A critique of phanerozoic climatic models involving changes in the CO_2 content of the atmosphere. Earth-Sci Rev 56:1–159

Bousquet GD, Triantafyllou MS, Slotine J-JE (2017) Optimal dynamic soaring consists of successive shallow arcs. J R Soc Interface 14:20170496

Boussau B, Blanquart S, Necsulea A, Lartillot N, Gony M (2008) Parallel adaptations to high temperatures in the Archean eon. Nature 456:942–945

Boyce CK, Lee J-E, Field TS, Brodribb TJ, Zwieniecki MA (2010) Angiosperms helped put the rain in the rainforests: the impact of plant physiological evolution of tropical biodiversity. Ann Missouri Bot Gard 97:527–540

Boyd ES, Peters JW (2013) New insights into evolutionary history of biological nitrogen fixation. Front Microbiol 4:201

Boyle WJ, Simonet WS, Lacey DL (2003) Osteoclast differentiation and activation. Nature 423:337–342

Braakman R, Smith E (2012) The emergence and early evolution of biological carbon-fixation. PLoS Comput Biol 8(4):e1002455

Braakman R, Smith E (2014) Metabolic evolution of a deep-branching hyperthermophilic chemoautotrophic bacterium. PLoS One 9(2):e87950

Branicky RS, Schafer WR (2008) Oxygen homeostasis: how the worm adapts to variable oxygen levels. Curr Biol 15(13):R559–R560

Brasier MD, Antcliffe J, Saunders M, Wacey D (2015) Changing the picture of Earth's earliest fossils (3.5-1.9 Ga) with new approaches and new discoveries. PNAS 112(16):4859–4864

Brazelton WJ, Baross JA (2009) Abundant transposases encoded by the metagenome of a hydrothermal chimney biofilm. ISME J 3:1420–1424

Brazelton WJ, Ludwig KA, Sogin ML, Andreishcheva EN, Kelley DS et al (2010) Archaea and bacteria with surprising microdiversity show shifts in dominance over 1,000-year time scales in hydrothermal chimneys. PNAS 107(4):1612–1617

Brazelton WJ, Mehta MP, Kelley DS, Baross JA (2011) Physiological differentiation within a single-species biofilm fueled by serpentinization. MBio 2(4):e00127–e00111

Briggs DEG, Caron J-B (2017) A large Cambrian Chaetognath with supernumerary grasping spines. Curr Biol 27:2536–2543

Bright M, Bulgheresi S (2010) A complex journey: transmission of microbial symbionts. Nat Rev Microbiol 8(3):218–230

Brodribb TJ, Holbrook NM (2004) Stomatal protection against hydraulic failure: a comparison of coexisting ferns and angiosperms. New Phytol 162:663–670

Brodribb TJ, McAdam SAM, Jordan GJ, Field TS (2009) Evolution of stomatal responsiveness to CO_2 and optimization of water-use efficiency among land plants. New Phytol 183:839–847

Brown AMV, Howe DK, Wasala SK, Peetz AB, Zasada IA, Denver DR (2015) Comparative genomics of a plant-parasitic nematode endosymbiont suggest a role in nutritional symbiosis. Genome Biol Evol 7(9):2727–2746

Brundrett MC (2009) Mycorrhizal associations and other means of nutrition of vascular plants: understanding the global diversity of host plants by resolving conflicting information and developing reliable means of diagnosis. Plant and Soil 320:37–77

Brundrett MC, Tedersoo L (2018) Evolutionary history of mycorrhizal symbioses and global host plant diversity. New Phytol. https://doi.org/10.1111/nph.14967

Brüssow H (2007) Bacteria between protists and phages: from antipredation strategies to the evolution of pathogenicity. Molecular Microb 65(3):583–589

Bryant DA, Frigaard N-U (2006) Prokaryotic photosynthesis and phototrophy illuminated. Trends Microbiol 14(11):488–496

Brzezinski P (2004) Redox-driven membrane-bound proton pumps. Trends Biochem Sci 29(7):380–387

Buckel W (2015) Wie anaerobe Bakterien und Archaeen Energie konservieren MPG Jahrbuch 2014/2015

Buckel W, Thauer RK (2013) Energy conservation via electron bifurcating ferredoxin reduction and proton/Na$^+$ translocating ferredoxin oxidation. Biochim Biophys Acta 1827:94–113

Büdel B, Weber B, Kühl M, Pfanz H, Sültemeyer D, Wessels D (2004) Reshaping of sandstone surfaces by cryptoendolithic cyanobacteria: bioalkalization causes chemical weathering in arid landscapes. Geobiology 2:261–268

Buick R (2012) Geobiology of the Archean eon. In: Knoll AH, Canfield DE, Konhauser KO (eds) Fundamentals of geobiology. Wiley-Blackwell, Oxford, pp 351–370

Burgess SSO, Adams MA, Turner NC, Ong CK (1998) The redistribution of soil water by tree root systems. Oecologia 115:306–311

Burgess SSO, Pittermann JA, Dawson TE (2006) Hydraulic efficiency and safety of branch xylem increases with height in *Sequoia sempervirens* (D. Don) crowns. Plant Cell Environ 29:229–239

Burggren WW (1982) Pulmonary blood plasma filtration in reptiles: a "Wet" vertebrate lung? Science 215:77–78

Burggren WW (1992) Respiration and circulation in land crabs: novel variations on the marine design. Am Zool 32:417–427

Burghelea CI, Dontsova K, Zaharescu DG, Maier RM, Huxman T et al (2018) Trace element mobilization during incipient bioweathering of four rock types. Geochim Cosmochim Acta 234:98–114

Burkholder JM, Glibert PM, Skelton HM (2008) Mixotrophy, a major mode of nutrition for harmful algal species in eutrophic waters. Harmful Algae 8:77–93

Bush JWM, Hu DL (2006) Walking on water: biolocomotion at the interface. Annu Rev Fluid Mech 38:339–369

Cabrita P, Thorpe M, Huber G (2013) Hydrodynamics of steady state phloem transport with radial leakage of solute. Front Plant Sci 4:531

Cai Z, Zhu H, Duan S (2014) Allelopathic interactions between the red-tide causative dinoflagellate *Prorocentrum donghaiense* and the diatom *Phaeodactylum tricornutum*. Oceanologia 56(3):639–650

Calderón-Jiménez B, Johnson ME, Bustos ARM, Murphy KE, Winchester MR, Baudrit JRV (2017) Silver nanoparticles: technological advances, societal impacts, and metrological challenges. Front Chem 5:6

Caldwell MM, Richards JH (1989) Hydraulic lift: water efflux from upper roots improves effectiveness of water uptake by deep roots. Oecologia 79:1–5

Camacho A, Walter XA, Picazo A, Zopfi J (2017) Photoferrotrophy: remains of an ancient photosynthesis in modern environments. Front Microbiol 8:323

Canadell J, Jackson RB, Ehleringer JR, Mooney HA, Sala OE, Schulze E-D (1996) Maximum rooting depth of vegetation types at the global scale. Oecologia 108:583–595

Canfield DE, Glazer AN, Falkowski PG (2010) The evolution and future of earth's nitrogen cycle. Science 330:192–196

Cao K-F, Yang S-J, Zhang Y-J, Brodribb TJ (2012) The maximum height of grasses is determined by roots. Ecol Lett. https://doi.org/10.1111/j.1461-0248.2012.01783.x

Capano JG (2016) Reaction forces and rib function during locomotion in snakes. Integr Comp Biol 60(1):215–231

Carafa A, Duckett JG, Ligrone R (2003) Subterranean gametophytic axes in the primitive liverwort *Haplomitrium* harbour a unique tape of endophytic association with aseptate fungi. New Phytol 160:185–197

Cardon ZG, Gray DW, Lewis LA (2008) The green algal underground: evolutionary secrets of desert cells. BioSci 58(2):114–122

Cardon ZG, Stark JM, Herron PM, Rasmussen JA (2013) Sagebrush carrying out hydraulic lift enhances surface soil nitrogen cycling and nitrogen uptake into inflorescences. PNAS 110(47):18988–18993

Cardona T (2019) Thinking twice about the evolution of photosynthesis. Open Biol 9:180426

Carlsward BS, Stern WS, Bytebier B (2006) Comparative vegetative anatomy and systematics of the Angraecoids (*Vandeae, Orchidaceae*) with an emphasis on the leafless habit. Faculty Res and Creative Activity 259, Eastern Illinois University

Carrara PE, Carroll TR (1979) The determination of erosion rates from exposed tree roots in the Piceance Basin, Colorado. Earth Surf Process 4:307–317

Carrillo P, Medina-Sánchez JM, Villar-Argaiz M, Delgado-Molina JA, Bullejos FJ (2006) Complex interactions in microbial food webs: stoichiometric and functional approaches. Limnetica 25(1–2):189–204

Cavagna GA, Legramandi MA (2009) The bounce of the body in hopping, running and trotting: different machines with the same motor. Proc R Soc B 276:4279–4285

Cerqueda-García D, Martínez-Castilla L, Falcón LI, Delaye L (2014) Metabolic analysis of *Chlorobium chlorochromatii* CaD3 reveals clues of symbiosis in '*Chlorochromatium aggregatum*'. The ISME J 8:991–938

Chambers PA, Lacoul P, Murphy KJ, Thomaz SM (2008) Global diversity of aquatic macrophytes in freshwater. Hydrobiologia 595:9–26

Chan Y, Lacap DC, Lau MCY, Ha KY, Warren-Rhodes KA et al (2012) Hypolithic microbial communities: between a rock and a hard place. Environ Microb 14(9):2272–2282

Chang AJ, Bargmann CI (2008) Hypoxia and the HIF-1 transcriptional pathway reorganize a neuronal circuit for oxygen-dependent behaviour in *Caenorhabditis elegans*. PNAS 105(20):7321–7326

Chang H, Cassau S, Krieger J, Guo X, Knaden M et al (2023) A chemical defense deters cannibalism in migratory locusts. Science 380:537–543

Chapman MJ, Margulis L (1998) Morphogenesis by symbiogenesis. Int Microbiol 1:319–326

Chaumont F, Tyerman SD (2014) Aquaporins: highly regulated channels controlling plant water relations. Plant Physiol 164:1600–1618

Chen P-Y, Lin AY-M, McKittrick J, Meyers MA (2008) Structure and mechanical properties of crab exoskeletons. Acta Biomater 4:587–596

Chen X, Ling H-F, Vance D, Shields-Zhou GA, Zhu M et al (2015) Rise to modern levels of ocean oxygenation coincided with the Cambrian radiation of animals. Nat Commun. https://doi.org/10.1038/ncomms8142

Cheng L (1976) Marine insects. North Holland, Amsterdam

Chernomor O, Peters L, Schneidewind J, Loeschcke A, Knieps-Grünhagen E et al (2021) Complex evolution of light-dependent protochlorophyllide oxidoreductases in aerobic anoxygenic phototrophs: origin, phylogeny, and function. Mol Biol Evol 38(3):819–837

Chi S, Wu S, Yu J, Wang X, Tang X, Liu T (2014) Phylogeny of C_4-photosynthesis enzymes based on algal transcriptomic and genomic data supports an archaeal/proteobacterial origin and multiple duplication for most C_4-related genes. PLoS One 9(19):e1101154

Chiatante D, Scippa SG, Di Iorio A, Sarnataro M (2003) The influence of steep slopes on root system development. J Plant Growth Regul 21:247–260

Christner BC, Priscu JC, Achberger AM, Barbante C, Carter SP et al (2014) A microbial ecosystem beneath the West Antarctic ice sheet. Nature 512:310–313

Chu JWF, Leys SP (2010) High resolution mapping of community structure in three glass sponge reefs (Porifera, Hexactinellida). Mar Ecol Prog Ser 417:97–113

Claessen D, de Roos AM, Persson L (2003) Population dynamic theory of size-dependent cannibalism. Proc R Soc Lond B 271:333–340

Cloern JE, Dufford R (2005) Phytoplankton community ecology: principles applied in San Francisco Bay. Mar Ecol Prog Ser 285:11–28

Cnossen I, Sanz-Forcada J, Favata F, Witasse O, Zegers T, Arnold NF (2007) Habitat of early life: solar X-ray and UV radiation at Earth's surface. J Geophys Res 112:E02008. https://doi.org/10.1029/2006JE002784

Cockell CS (2000) Ultraviolet radiation and the photobiology of earth's early oceans. Orig Life Evol Biosph 30:467–499

Cockell CS, Horneck G (2001) The history of the UV radiation climate of the earth—theoretical and space-based observations. Photochem Photobiol 73(4):447–451

Cockell CS, McKay CP, Warren-Rhodes K, Horneck G (2008) Ultraviolet radiation-induced limitation to epilithic microbial growth in arid deserts—Dosimetric experiments in the hyperarid core of the Atacama Desert. J Photochem Photobiol B Biol 90:79–87

Cohen PA, Kodner RB (2022) The earliest history of eukaryotic life: uncovering an evolutionary story through the integration of biological and geological data. Trends Ecol Evol 37(3):246–256

Cohen N, Sanders T (2014) Nematode locomotion: dissecting the neuronal-environmental loop. Curr Opin Neurobiol 25:99–106

Coleman ML, Chisholm SW (2007) Code and context: *Prochlorococcus* as a model for cross-scale biology. Trends Microbiol 15(9):398–407

Collins KD (2012) Why continuum electrostatics theories cannot explain biological structure, polyelectronic or ionic strength effects in ion-protein interactions. Biophys Chem 167:43–59

Collins KD, Neilson GW, Enderby JE (2007) Ions in water: characterizing the forces that control chemical processes and biological structure. Biophys Chem 128:95–104

Comas LH, Bouma TJ, Eissenstat DM (2002) Linking root traits to potential growth rate in six temperate tree species. Oecologia 132:34–43

Connell L, Barrett A, Templeton A, Staudigel H (2009) Fungal diversity associated with an active deep sea volcano: Vailulu'u seamount, Samoa. Geomicrobiol J 26:597–605

Contreras HL, Bradley TJ (2009) Metabolic rate controls respiratory pattern in insects. The J of Experimental Biology 212:424–428

Cooper MB, Smith AG (2015) Exploring mutualistic interactions between microalgae and bacteria in the omics age. Curr Opin Plant Biol 26:147–153

Cordier T, Robin C, Capdevielle X, Desprez-Loustau M-L, Vacher C (2012) Spatial variability of phyllosphere fungal assemblages: genetic distance predominates over geographic distance in a European beech stand (*Fagus sylvatica*). Fungal Ecol 5(5):509–520

Cox MM, Battista JR (2005) *Deinococcus radiodurans* the consummate survivor. Nature 3:882–892

Crawford M, Jesson LK, Garnock-Jones PJ (2009) Correlated evolution of sexual system and life-history traits in mosses. Evolution 63(5):1129–1142

Crayn DM, Winter C, Schulte K, Smith JAC (2015) Photosynthetic pathways in Bromeliaceae: phylogenetic and ecological significance of CAM and C_3 based on carbon isotope ratios for 1893 species. Botanical J of the Linnean Society 178:169–221

Crook MJ, Ennos AR, Banks JR (1997) The function of buttress roots: a comparative study of the anchorage systems of buttressed (*Aglaia* and *Nephelium ramboutan* species) and non-buttressed (*Mallotus wrayi*) tropical trees. J Exp Biol 48(314):1703–1716

Cuthill JFH, Morris SC (2014) Fractal branching organizations of Ediacaran rangeomorph fronds reveal a lost Proterozoic body plan. PNAS 111(36):13122–13126

Da Silva FO, Fabre A-C, Savriama Y, Ollonen J, Mahlow K et al (2018) The ecological origins of snakes as revealed by skull evolution. Nat Commun 9/376

Darroch SAF, Sperling EA, Boag TH, Racicot RA, Mason SJ et al (2015) Biotic replacement and mass extinction of the Ediacara biota. Proc R Soc B 282:20151003

Dashiff A, Junka RA, Libera M, Kadouri DE (2010) Predation of human pathogens by the predatory bacteria *Micavibrio aeruginosavorus* and *Bdellovibrio bacteriovorus*. J Appl Microb 110:431–444

Dattagupta S, Schaperdoth I, Montanari A, Mariani S, Kita N et al (2009) A novel symbiosis between chemoautotrophic bacteria and a freshwater cave amphipod. The ISME J 3:935–943

De Baets S, Poesen J, Reubens B, Wemans K, De Baerdemacker J, Muys B (2008) Root tensile strength and root distribution of typical Mediterranean plant species and their contribution to soil shear strength. Plant and Soil 305:207–226

De Beer D, Stoodley P (2006) Microbial biofilms. In: Dworkin M, Falkow S, Rosenberg E, Schleifer K-H, Stackebrandt E (eds) The prokaryotes, Symbiotic associations, biotechnology, applied microbiology, vol 1. Springer, New York, pp 903–937

de Boer HJ, Eppinga MB, Wassen MJ, Dekker SC (2012) A critical transition in leaf evolution facilitated the cretaceous angiosperm revolution. Nat Commun. https://doi.org/10.1038/ncomms2217

De Clerck O, Bogaert KA, Leliaert F (2012) Diversity and evolution of algae: primary endosymbiosis. Adv Bot Res 64:55–86

de Langre E (2008) Effects of wind on plants. Annu Rev Fluid Mech 40:141–168

de Vries J, Archibald JM (2018) Plant evolution: landmarks on the path to terrestrial life. New Physiol 217:1428–1434

Dedysh SN, Ricke P, Liesack W (2004) NifH and NifD phylogenies: an evolutionary basis for understanding nitrogen fixation capabilities of methanotrophic bacteria. Microbiology 150:1301–1313

Degelmann DM (2010) Bedeutung der Baumart für die Aktivität, Diversität und Abundanz methanoxidierender Bakterien in temperaten Waldböden. Diss Univ Bayreuth

Dekas AE, Poretsky RS, Orphan VJ (2009) Deep-sea archaea fix and share nitrogen in methane-consuming microbial consortia. Science 326:422–426

Del-Bem L-E (2018) Xyloglucan evolution and the terrestrialization of green plants. New Phytol 291:1150–1153

DeLong JP, Okie JG, Moses ME, Sibly RM, Brown JH (2010) Shifts in metabolic scaling, production, and efficiency across major evolutionary transitions of life. PNAS 107(29):12941–12945

Delwiche CF (1999) Tracing the thread of plastid diversity through the tapestry of life. Am Nat 154:S164–S177

Deming JW (2002) Psychrophiles and polar regions. Curr Opin Microb 5:301–309

Denny M (2016) Ecological mechanics. Princeton University Press, Princeton

Deora T, Sane SS, Sane SP (2021) Wings and halters act as coupled dual oscillators in flies. Elife 10:e53824

Derks A, Schaven K, Bruce D (2015) Diverse mechanisms for photoprotection in photosynthesis. Dynamic regulation of photosystem II excitation in response to rapid environmental change. Biochim Biophys Acta 1847:468–485

Desnues C, Rodriguez-Brito B, Rayhawk S, Kelley S, Tran T et al (2008) Biodiversity and biogeography of phages in modern stromatolites and thrombolites. Nature. https://doi.org/10.1038/nature06735

Devaraj A, Justice SS, Bakaletz LO, Goodman SD (2015) DNABii proteins play a central role in UPEC biofilm structure. Mol Microbiol 96(6):1119–1135

Di Mauro A, Neu J, Riezzo G, Raimondi F, Martinelli D et al (2013) Gastrointestinal function development and microbiota. Ital J Pediatr 39:15

Diaz RJ, Rosenberg R (2008) Spreading dead zones and consequences for marine ecosystems. Science:926–929

Dickey TD, Falkowski PG (2002) Solar energy and its biological-physical interactions in the sea. In: Robinson AR, McCarthy JJ, Rothschild BJ (eds) The sea. Wiley, New York, pp 401–440

DiMaio F, Yu X, Rensen E, Krupovic M, Prangishvili D, Egelman EH (2015) A virus that infects a hyperthermophile encapsidates A-form DNA. Science 6237:914–917

Dimmock NJ, Easton AJ, Leppard KN (2007) Introduction to modern virology. Blackwell Publishing, Malden

Dismuskes GC, Klimov VV, Baranov SV, Kozlov YN, DasGupta J, Tyryshkin A (2001) The origin of atmospheric oxygen on earth: the innovation of oxygenic photosynthesis. PNAS 98(5):2170–2175

Doebeli M, Ispolatov I (2014) Chaos and unpredictability in evolution. Evolution 68(5):1365–1373

Dohrmann M, Wörheide G (2017) Dating early animal evolution using phylogenomic data. Sci Rep 7:3599

Dolan E, Tyler PA, Yesson C, Rogers AD (2013) Phylogeny and systematics of deep-sea sea pens (Anthozoa: Octocorallia: Pennatulacea). Mol Phylogenet Evol 69:610–618

Dong S, Schäfer-Verwimp A, Meinecke P, Feldberg K, Bombosch A et al (2012) Tramps, narrow endemics and morphologically cryptic species in the epiphyllous liverwort *Diplasiolejeunea*. Mol Phylogenet Evol 65:582–594

Donoghue PCJ, Sansom IJ, Downs JP (2006) Early evolution of vertebrate skeletal tissues and cellular interactions, and the canalization of skeletal development. J Exp Zool B Mol Dev Evol. https://doi.org/10.1002/jez.b

Dorgan KM (2015) The biomechanics of burrowing and boring. J Exp Biol 218:176–183

dos Reis M, Thawomwattana Y, Angelis K, Telford MJ, Donoghue PCJ, Yang Z (2015) Uncertainty in the timing of origin of animals and the limits of precision in molecular timescales. Curr Biol 25:2939–2950

Dotzler N, Krings M, Taylor TN, Agerer R (2006) Germination shields in *Scutellospora* (Glomeromycota: Diversisporales, Gigasporaceae) from the 400 million-year-old Rhynie chert. Mycol Progress. https://doi.org/10.1007/s11557-006-0511-z

Douglas AE (2009) The microbial dimension in insect nutritional ecology. Funct Ecol 23:38–47

Douglas AE, Werren JH (2016) Holes in the hologenome: why host-microbe symbioses are not holobionts. MBio 7(2):e02099–e02015

Doyle JA (2013) Phylogenetic analyses and morphological innovations in land plants. Annu Plant Rev 45:1–50

Drake H, Ivarsson M (2018) The role of anaerobic fungi in fundamental geochemical cycles in the deep biosphere. Fungal Biol Rev 32(1):20–25

Driese SG, Mora CI, Elick JM (1997) Morphology and taphonomy of root and stump casts of earliest trees (middle to late Devonian), Pennsylvania and New York, U.S.A. Palaios, vol 12, pp 524–537

Droop MR (1957) Auxotrophy and organic compounds in the nutrition of marine phytoplankton. J Gen Microbiol 16:286–298

Dumais J, Forterr Y (2012) "Vegetable Dynamicks": the role of water in plant movements. Annu Rev Fluid Mech 44:453–478

Dumont ER (2010) Bone density and the lightweight skeletons of birds. Proc R Soc B 277:2193–2198

Dunn CW, Leys SP, Haddock SHD (2015) The hidden biology of sponges and ctenophores. Trends Ecol Evol 30(5):282–291

Dunne R, Brown BE (1996) Penetration of solar UVB radiation in shallow tropical waters and its potential biological effects on coral reefs; results from Central Indian Ocean and the Andaman Sea. Mar Ecol Prog Ser 114:109–118

Dupraz C, Visscher PT (2005) Microbial lithification in marine stromatolites and hypersaline mats. Trends Microbiol 13(9):429–438

Dvořák P, Casamatta DA, Poulíčková A, Hašler P, Ondřej V, Sanges R (2014) *Synechococcus*: 3 billion years of global dominance. Mol Ecol 23:5538–5551

Ebensperger LA (1998) Strategies and counterstrategies to infanticide in mammals. Biol Rev 73:321–346

Eberhard S, Finazzi G, Wollman F-A (2008) The dynamics of photosynthesis. Annu Rev Genet 42:463–515

Eckle T, Köhler D, Lehmann R, El Kasmi KC, Eltzschig HK (2008) Hypoxia-inducible factor-1 is central to cardioprotection. Circulation 118:166–175

Edwards EJ (2019) Evolutionary trajectories, accessibility and other metaphors: the case of the C_4 and CAM photosynthesis. New Phytol 223:1742–1755

Edwards EJ, Ogburn RM (2012) Angiosperm responses to a low-CO_2 world: CAM and C_4 photosynthesis as parallel evolutionary trajectories. Int J Plant Sci 173(6):724–733

Edwards EJ, Osborne CP, Strömberg CAE, Smith SA (2010) The origins of C_4 grasslands: integrating evolutionary and ecosystem science. Science 328:587–591

Edwards D, Kenrick P, Dolan L (2017) History and contemporary significance of the Rhynie cherts—our earliest preserved terrestrial ecosystem. Philos Trans R Soc B 373:20160489

Eiler A (2006) Evidence for the ubiquity of Mixotrophic bacteria in the Upper Ocean: implications and consequences. Applied and Environmental Microb 72(12):7431–7437

Eitel M, Osigus H-J, DeSalle R, Schierwater B (2013) Global diversity of the Placozoa. PLoS One 8(4):e57131

Elbert W, Weber B, Burrows S, Steinkamp J, Büdel B, Andreae MO, Pöschl U (2012) Contribution of cryptogamic covers to the global cycles of carbon and nitrogen. Nature GeoSci 5:459–462

Elias M, Archibald JM (2009) Sizing up the genomic footprint of endosymbiosis. Bioessays 31:1273–1279

Ellenberg H, Mayer R, Schermann J (eds) (1986) Ökosystemforschung Ergebnisse des Sollingprojekts 1966–1986. Ulmer, Stuttgart

Elliott KH, Ricklefs RE, Gaston AJ, Hatch SA, Speakman JR, Davoren GK (2013) High flight costs, but low dive costs, in auk support the biomechanical hypothesis for flightlessness in penguins. PNAS 110(23):9380–9384

Ellison AM, Gotelli NJ (2009) Energetics and the evolution of carnivorous plants—Darwin's 'most wonderful plants in the world'. J Exp Bot 60(1):19–42

Emmett M, Kloos WE (1979) The nature of arginine auxotrophy in cutaneous populations of staphylococci. J Gen Microbiol 110:305–314

Ereskovsky AV, Gonobobleva E, Vishnyakov A (2005) Morphological evidence for vertical transmissions of symbiotic bacteria in the viviparous sponge *Halisarca dujardini* Johnston (Porifera, Demospongiae, Halisarcida). Mar Biol 146:869–875

Erez Z, Steinberger-Levy I, Shamir M, Doron S, Stokar-Avihail A et al (2017) Communications between viruses guides lysis-lysogeny decisions. Nature 541:488–493

Fabritius H-O, Sachs C, Triguero PR, Raabe D (2009) Influence of structural principles on the mechanics of a biological fiber-based composite material with hierarchical organization: the exoskeleton of the lobster *Homarus amricanus*. Adv Mater 21:391–400

Falkowski PG, Katz ME, Knoll AH, Quigg A, Raven JA et al (2004) The evolution of modern eukaryotic phytoplankton. Science 305:354–360

Falkowski PG, Fenchel T, Delong EF (2008) The microbial engines that drive earth's biogeochemical cycles. Science 320:1034–1039

Farquhar J, Wing BA (2003) Multiple sulfur isotopes and the evolution of the atmosphere. Earth Planet Sci Lett 6707:1–13

Farrant JM, Moore JP (2011) Programming desiccation-tolerance: from plants to seeds to resurrection plants. Curr Opin Plant Biol 14:340–345

Fenton AK, Kanna M, Woods RD, Aizawa S-I, Sockett RE (2010) Shadowing the actions of a predator: backlit fluorescent microscopy reveals synchronous nonbinary septation of predatory *Bdellovibrio* inside prey and exit through discrete bdelloplast pores. J Bacteriol 192(24):6329–6335

Ferreira P, Fonte E, Soares ME, Carvalho F, Guilhermino L (2016) Effects of multi-stressors on juveniles of the marine fish *Pomastachius microps*: gold nanoparticles, microplastics and temperature. Aquat Toxicol 170:89–103

Feulner G (2012) The faint young sun problem. arXiv:1204.4449v1

Fike DA, Grotzinger JP, Pratt LM, Summons RE (2006) Oxidation of the Ediacaran Ocean. Nature 444:744–747

Findon H, Calcagno-Pizarelli A-M, Martínez JL, Spielvogel A, Markina-Iñarraegui A et al (2010) Analysis of a novel calcium auxotrophy in *Aspergillus nidulans*. Fungal Genet Biol 47:647–655

Finlay R, Wallander H, Smits M, Holstrom S, Van Hees P et al (2009) The role of fungi in biogenic weathering in boreal forest soils. Fungal Biol Rev 23:101–106

Fischer WW (2016) Breathing room for early animals. PNAS 113(7):1686–1688

Flemming H-C, Wingeneder J (2010) The biofilm matrix. Nat Rev Microbiol 8:623–633

Flemming H-C, Wingeneder J, Szewzyk U, Steinberg P, Rice SA, Kjelleberg S (2016) Biofilms: an emergent form of bacterial life. Nat Rev Microbiol 14:563–575

Flores-Núñez VM, Fonseca-Garcia C, Desgarennes D, Eloe-Fadrosh E, Woyke T, Partida-Martinez LP (2020) Functional signatures of the epiphytic prokaryotic microbiome of agaves and cacti. Front Microbiol 10:3044

Follmann H, Brownson C (2009) Darwin's warm little pond revisited: from molecules to the origin of life. Naturwissenschaften. https://doi.org/10.1007/s00114-009-0602-1

Franks PJ, Buckley TN, Shope JC, Mott KA (2001) Guard cell volume and pressure measured concurrently by confocal microscopy and the cell pressure probe. Plant Physiol 125:1577–1584

Franks PJ, Royer DL, Beerling DJ, Van de Water PK, Cantrill DJ et al (2014) New constraints on atmospheric CO_2 concentration for the phanerozoic. Geophys Res Lett 41:4685–4694

Frey-Klett P, Burlinson P, Deveau A, Barret M, Tarkka M, Sarniguet A (2011) Bacterial-fungal interactions: hyphens between agricultural, clinical, environmental, and food microbiologists. Microb and Molecular Biology Rev 75(4):583–609

Friedman J, Barrett SCH (2009) Wind of change: new insights on the ecology and evolution of pollination and mating in wind-pollinated plants. Ann Bot 103:1515–1527

Fromm J, Hajirezael M-R, Becker VK, Lautner S (2013) Electrical signalling along the phloem and its physiological responses in the maize leaf. Front Plant Sci 4:239

Fukushima K, Fang X, Alvarez-Ponce D, Cai H, Carretero-Paulet L et al (2017) Genome of the pitcher plant *Cephalotus* reveals genetic changes associated with carnivory. Nat Ecol Evol 1:0059

Full RJ, Tu MS (1990) Mechanics of six-legged runners. J Exp Biol 148:129–146

Fuller RC, Sprague SG, Gest H, Blankenship RE (1985) A unique photosynthetic reaction center from *Heliobacterium chlorum*. Febs 182(2):345–349

Gadd GM (2007) Geomycology: biogeochemical transformations of rocks, minerals, metals and radionuclides by fungi, bioweathering and bioremediation. Mycol Res 111:3–49

Gadd GM, Watkinson SC, Dyer PS (eds) (2007) Fungi in the environment. Cambridge University Press, Cambridge

Gaino E, Bavestrello G, Magnino G (1999) Self/non-self recognition in sponges. Ital J Zool 66:299–315

Gantner S, Schmid M, Dürr C, Schuhegger R, Steidle A et al (2006) *In situ* quantitation of the spatial scale of calling distance and population density-independent N-acylhomoserine lactone-mediated communication by rhizobacteria colonized on plant roots. FEMS Microbiol Ecol 56:188–194

Gardiner B, Berry P, Moulia B (2016) Wind impacts on plant growth, mechanics and damage. Plant Sci 245:94–118

Garlick S, Oren A, Padan E (1977) Occurrence of facultative anoxygenic photosynthesis among filamentous and unicellular cyanobacteria. J Bacteriol 129(2):623–629

Gavrilov VM (2011) Energy expenditure for flight, aerodynamic quality, and colonization of forest habitats by birds. Biol Bull 38(8):779–788

Geider RJ, La Roche J (2002) Redfield revisited: variability of C:N:P in marine microalgae and its biochemical basis. Eur J Phycol 37:1–17

Geitmann A (2010) Mechanical modelling and structural analysis of the primary plant cell wall. Curr Opin Plant Biol 13:693–699

Geitmann A, Ortega JKE (2009) Mechanics and modeling of plant cell growth. Trends Plant Sci 14(9):467–478

Genet M, Li M, Luo T, Fourcaud T, Clément-Vidal A, Stokes A (2011) Linking carbon supply to root cell-wall chemistry and mechanics at high altitudes in *Abies georgei*. Ann Bot 107:311–320

Gerotto C, Alboresi A, Giacometti GM, Bassi R, Morosinotto T (2012) Coexistence of plant and algal energy dissipation mechanisms in the moss *Physcomitrella patens*. New Phytol 196:763–773

Ghannoum O (2009) C_4 photosynthesis and water stress. Ann Bot 103:635–644

Ghosh W, Dam B (2009) Biochemistry and molecular biology of lithotropic sulfur oxidation by taxonomically and ecologically diverse bacteria and archaea. FEMS Microbiol Rev 33:999–1043

Gibling MR, Davies NS (2012) Palaeozoic landscapes shaped by plant evolution. Nat Geosci 5:99–105

Gibson TM, Shih PM, Cumming VM, Fischer WW, Crockford PW et al (2018) Precise age of *Bangiomorpha pubescens* dates the origin of eukaryotic photosynthesis. Geology 46:135–138

Giordano M, Beardall J, Raven JA (2005) CO_2 concentrating mechanisms in algae: mechanisms, environmental modulation and evolution. Annu Rev Plant Biol 56:99–131

Givnish TJ, Barfuss MHJ, Van Ee B, Riina R, Schulte K et al (2011) Phylogeny, adaptive radiation, and historical biogeography in Bromeliaceae: insights from an eight-locus plastid phylogeny. Am J Bot 98(5):872–895

Givnish TJ, Barfuss MHJ, Van Ee B, Riina R, Schulte K et al (2014) Adaptive radiation, correlated and contingent evolution, and net species diversification in Bromeliaceae. Mol Phylogenet Evol 71:55–78

Glass GE, Holt RD, Slade NA (1985) Infanticide as an evolutionary stable strategy. Anim Behav 33:384–391

Gleiss AC, Jorgensen SJ, Liebsch N, Sala JE, Norman B et al (2011) Convergent evolution in loco-motory patterns of flying and swimming animals. Nat Commun. https://doi.org/10.1038/ncomms1350

Goldsmith GR, Matzke NJ, Dawson TE (2013) The incidence and implications of clouds for cloud forest plant water relations. Ecol Lett 16:307–314

González AL, Fariña JM, Pinto R, Pérez C, Weathers KC et al (2011) Bromeliad growth and stoichi-ometry: responses to atmospheric nutrient supply in fog-dependent ecosystems of the hyper-arid Atacama Desert, Chile. Oecologia 167:835–845

Goolish EM (1991) Aerobic and anaerobic scaling in fish. Biol Res 66:33–56

Gould SB, Waller RF, McFadden GI (2008) Plastid evolution. Annu Rev Plant Biol 59:491–517

Gow NAR (1984) Transhyphal electrical currents in fungi. J Gen Microbiol 130:3313–3318

Grady JM, Enquist BJ, Dettweiler-Robinson E, Wright NA, Smith FA (2014) Evidence for meso-thermy in dinosaurs. Science 344:1268–1272

Gralnick JA, Newman DK (2007) Extracellular respiration. Molecular Microb 65(1):1–11

Green WA (2010) The function of the aerenchyma in arborescent lycopsids: evidence of an unfamil-iar metabolic strategy. Proc R Soc B 277:2257–2267

Greene DF, Johnson EA (1993) Seed mass and dispersal capacity in wind-dispersed diaspores. Oikos 67:69–74

Greening C, Carere CR, Rushton-Green R, Harold LK, Hards K et al (2015) Persistence of the domi-nant soil phylum *Acidobacteria* by trace gas scavenging. PNAS 112(33):10497–10502

Gregory JE, Brockert CL, Morgan DL, Whitehead NP, Proske U (2002) Effect of eccentric muscle contractions on Golgi tendon organ responses to passive and active tension in the cat. J Physiol 538(1):209–218

Griffin BM, Schott J, Schink B (2007) Nitrite, an electron donor for anoxygenic photosynthesis. Science 316:1870

Griffiths H, Weller G, Toy LFM, Dennis RJ (2013) You're so vein: bundle sheath physiology, phy-logeny and evolution in C3 and C4 plants. Plant Cell Environ 36:249–261

Grote J, Jost G, Labrenz M, Herndl GJ, Jürgens K (2008) *Epsilonproteobacteria* represent the major portion of chemoautotrophic bacteria in sulfidic waters of pelagic redoxclines of the baltic and black seas. Applied and Experimental Microb 74(24):7456–7551

Grotzinger J, Adams EE, Schröder S (2005) Microbial-metazoan reefs of the terminal Proterozoic Nama Group (c. 550-543 Ma). Namibia Geolog Mag 142(5):499–517

Grüber G, Manimekalai MSS, Mayer F, Müller V (2014) ATP synthases from archaea: the beauty of a molecular motor. Biochim Biophys Acta 1837:940–952

Guerrero R, Berlanga M (2016) From the cell to the ecosystem: the physiological evolution of sym-biosis. Evol Biol 43(4):543–552

Haag KL (2018) Holobionts and their hologenomes: evolution with mixed modes of inheritance. Genet Mol Biol 41(1):189–197

Hadibarata T, Yusoff ARM, Aris A, Kristanti RA (2012) Identification of naphthalene metabolism by white rot fungus *Armillaria* dp. F022. J Environ Sci 24(4):728–732

Hall-Stoodley L, Costerton JW, Stoodley P (2004) Bacterial biofilms: from the natural environment to infectious diseases. Nat Rev Microbiol 2:95–108

Ham B-K, Lucas WJ (2017) Phloem-mobile RNAs as systemic signaling agents. Annu Rev Plant Biol 68:173–195

Hamm CE, Simson DA, Merkel R, Smetacek V (1999) Colonies of *Phaeocystis globosa* are pro-tected by a thin but tough skin. Mar Ecol Prog Ser 187:101–111

Hampl V, Stairs CW, Roger AJ (2011) The tangled past of eukaryotic enzymes involved in aerobic metabolism. Mob Genet Elem 1(1):71–74

Hansen PJ (2011) The role of photosynthesis and food uptake for the growth of marine mixotrophic dinoflagellates. J Eukaryot Microb 58(3):203–214

Hansen GH, Olafsen JA (1999) Bacterial interactions in early life stages of marine cold water fish. Microb Ecol 38:1–26

Harms H, Schlosser D, Wick LY (2011) Untapped potential: exploiting fungi in bioremediation of hazardous chemicals. Nat Rev Microbiol 9:177–192

Hasenfuss I (2008) The evolutionary pathway to insect flight—a tentative reconstruction. Arthropod Syst Phylo 66(1):19–35

Haufler CH, Pryer KM, Schuettpelz E, Sessa EB, Farrar DR et al (2016) Sex and the single game-tophyte: revising the homosporous vascular plant life cycle in light of contemporary research. BioSci 66(11):928

Haworth M, Elliott-Kingston C, McElwain JC (2011) Stomatal control as a driver of plant evolution. J Exp Bot 62(8):2419–2423

Haywood V, Yu T-S, Huang N-C, Lucas WJ (2005) Phloem long-distance trafficking of GIBBERELLIC ACID-INSENSITIVE RNA regulates leaf development. Plant J 42:49–68

Hazen RM, Liu X-M, Downs RT, Golden J, Pires AJ et al (2014) Mineral evolution: episodic metallogenesis, the supercontinent cycle, and the coevolving geosphere and biosphere. Soc Econ Geol Spec Publ 18:1–15

Hecht T, Pienaar AG (1993) A review of cannibalism and its implications in fish Larviculture. J World Aqucult Soc 24(2):246–261

Heckman DS, Geiser DM, Eidell BR, Stauffer RL, Kardos NL, Hedges SB (2001) Molecular evidence for the early colonization of land by fungi and plants. Science 293:1129–1133

Hedderson TA, Longton RE (1996) Life history variation in mosses: water relations, size and phylogeny. Oikos 77:31–43

Hedenström A, Norelik G, Warfvinge K, Andersson A, Bäckman J, Åkesson S (2016) Annual 10-month aerial life phase in the common swift *Apus apus*. Curr Biol 26:3066–3070

Hegazy AK, Barakat HN, Kabiel HF (2006) Anatomical significance of the hygrochastic movement *in Anastatica hierochuntica*. Ann Bot 97:47–55

Hehenberger E, Imanian B, Burki F, Keeling PJ (2014) Evidence for the retention of two evolutionary distinct plastids in dinoflagellates with diatom endosymbionts. Genome Biol 6(9):2321–2334

Heinen RB, Ye Q, Chaumont F (2009) Role of aquaporins in leaf physiology. J Exp Bot 60(11):2971–2985

Heisler J, Glibert PM, Burkholder JM, Anderson DM, Cochlan W et al (2008) Eutrophication and harmful algal blooms: a scientific consensus. Harmful Algae 8:1–13

Heldt HW, Piechulla B (2008) Pflanzenbiochemie. Spektrum, Heidelberg

Helfman GS, Collette BB, Facey DE, Bowen BW (2009) The diversity of fishes. Wiley-Blackwell, Chichester

Helliwell KE, Wheeler GL, Leptos KC, Goldstein RE, Smith AG (2011) Insights into the evolution of vitamin B_{12} auxotrophy from sequenced algal genomes. Mol Biol Evol 28(10):2921–2933

Helly JJ, Levin LA (2004) Global distribution of naturally occurring marine hypoxia on continental margins. Deep-Sea Res I 51:1159–1168

Hennequin S, Schuettpelz E, Pryer KM, Ebihara A, Dubuisson J-Y (2008) Divergence times and the evolution of epiphytism in filmy ferns (*Hymenophyllaceae*) revisited. Int J Plant Sci 169(9):1278–1287

Hermida-Carrera C, Fares MA, Font-Canascosa M, Kapralov MV, Koch MA et al (2020) Exploring molecular evolution of Rubisco in C_3 and CAM Orchidaceae and Bromeliaceae. BMC Evol Biol 20/11

Herrera A, Tezara W, Marin O, Rengifo E (2008) Stomatal and non-stomatal limitations of photosynthesis in trees of a tropical seasonally flooded forest. Physiol Plant 134:41–48

Herrmann G, Jayamani E, Mai G, Buckel W (2008) Energy conservation via electron-transferring flavoprotein in anaerobic bacteria. J Bacteriol 190(3):784–791

Hessen DO (2008) Solar radiation and the evolution of life. In: Bjertness E (ed) Solar radiation and human health. The Norwegian Academy of Sci and Letters, Oslo, pp 123–136

Hietz P, Hietz-Seifert U (1995) Composition and ecology of vascular epiphyte communities along an altitudinal gradient in Central Veracruz, Mexico. J Veg Sci 6:487–498

Hirasawa T, Kuratani S (2015) Evolution of the vertebrate skeleton: morphology, embryology, and development. Zool Lett 1/2

Hirasawa I, Nagashima H, Juratani S (2013) The endoskeletal origin of the turtle carapace. Nat Commun. https://doi.org/10.1038/ncomms3107

Hoffland E, Kuyper TW, Wallander H, Plassard C, Gorbushina AA et al (2004) The role of fungi in weathering. Front Ecol Environ 2(5):258–264

Hoffman BM, Lukoyanov D, Yang Z-Y, Dean DR, Seefelt LC (2014) Mechanism of nitrogen fixation by nitrogenase: the next stage. Chem Rev 114:4041–4062

Hoffmeyer TT, Burkhardt P (2016) Choanoflagellate models—*Monosiga brevicollis* and *Salpingoeca rosetta*. Curr Opin Genet Dev 39:42–47

Hogarth PJ (2010) The biology of mangroves and seagrasses. Oxford University Press, Oxford

Hohmann-Marriott MF, Blankenship RE (2011) Evolution of photosynthesis. Annu Rev Plant Biol 62:515–548

Holbrook NM, Denny MW, Koehl MAR (1991) Intertidal "trees": consequences of aggregation on the mechanical and photosynthetic properties of sea-palms *Postelis palmaeformis* Ruprecht. J Exp Mar Biol Ecol 146:39–67

Honegger R (1993) Developmental biology of lichens. New Phytol 125:659–677

Honegger R, Edwards D, Axe L (2013) The earliest records of internally stratified cyanobacterial and algal lichens from the Lower Devonian of the Welsh Borderland. New Phytol 197:264–275

Hori K, Maruyama F, Fujisawa T, Togashi T, Yamamoto N et al (2014) *Klebsormidium flaccidum* genome reveals primary factors for plant terrestrial adaptation. Nat Commun 5:3978. https://doi.org/10.1038/ncomms4978

Horisberger JD (2004) Recent insights into structure and mechanism of the sodium pump. Physiology 19:377–387

Horrer D, Flütsch S, Pazmino D, Matthews JSA, Thalmann M et al (2016) Blue light induces a distinct starch degradation pathway in guard cells for stomatal opening. Curr Biol 26:362–370

Horton JL, Hart SC (1998) Hydraulic lift: a potentially important ecosystem process. TREE 13(6):232–235

Horton P, Wentworth M, Ruban A (2005) Control of the light harvesting function of chloroplast membranes: the LHCH-aggregation model for non-photochemical quenching. FEBS Lett 579:4201–4206

Hourdez S, Lallier FH (2007) Adaptations to hypoxia in hydrothermal-vent and cold-seep invertebrates. Re Environ Sci Biotechnol 6:143–159

Houssaye A (2009) "Pachyostosis" in aquatic amniotes: a review. Integr Zool 4:325–340

Houssaye A, Waskow K, Hayashi S, Cornette R, Lee AH, Hutchinson JR (2016) Biomechanical evolution of solid bones in large animals: a microanatomical investigation. Biol J Linn Soc 117(2):350–371

Hoysted GA, Kowal J, Jacob A, Rimington WR, Duckett JG et al (2018) A mycorrhizal revolution. Curr Opin Plant Biol 44:1–6

Hsia CCW, Schmitz A, Lamberts M, Perry SF, Maina JN (2013) Evolution of air breathing: oxygen hoemostasis and the transition from water to land and sky. Compr Physiol 3(2):849–915

Hsieh ST, Lauder GV (2004) Running on water: three-dimensional force generation by basilisk lizards. PNAS 101(48):16784–16788

Hu DL, Chan B, Bush JWM (2003) The hydrodynamics of water strider locomotion. Nature 424:663–666

Hu DL, Nirody J, Scott T, Shelley MJ (2009) The mechanics of slithering locomotion. PNAS 106(25):10081–10085

Hu L, Sun H, Li R, Zhang L, Wang S et al (2011) Phloem unloading follows an extensive apoplasmic pathway in cucumber (Cucumis sativus L.) fruit from anthesis to marketable matunring stage. Plant Cell Environ 34:1835–1848

Hug LA, Stechmann A, Roger AJ (2010) Phylogenetic distributions and histories of proteins involved in anaerobic pyruvate metabolism in eukaryotes. Mol Biol Evol 27(2):311–324

Hughes NC (2007) The evolution of trilobite body patterning. Annu Rev Earth Planet 35:402–434

Hürkey S, Niemeyer N, Schleimer J-H, Ryglewski S, Schreiber S, Duch C (2023) Gap junctions desynchronize a neural circuit to stabilize insect flight. Nature 618:118–125

Ikawa T, Okabe H, Cheng L (2012) Skaters of the seas—comparative ecology of nearshore and pelagic *Halobates* species (Hemiptera: Gerridae), with special reference to Japanese species. Mar Biol Res 8:915–936

Inskeep WP, Ackerman GG, Taylor WP, Kozubal M, Korf S, Macur RE (2005) On the energetics of chemolithotrophy in non equilibrium systems: case studies of geothermal springs in Yellowstone National Park. Geobiology 3:297–317

Irvine J, Law BE, Kurpius MR (2005) Coupling of canopy gas exchange with root and rhizosphere respiration in a semi-arid forest. Biogeochemistry 73:271–282

Ishibashi J-I, Okino K, Sunamura M (eds) (2015) Subseafloor biosphere linked to hydrothermal systems. Spinger, Tokyo

Ishii HR, Azuma W, Kuroda K, Sillett SC (2014) Pushing the limits to tree height: could foliar water storage compensate for hydraulic constraints in *Sequoia sempervirens*? Funct Ecol 28:1087–1093

Ivarsson M, Bengtson S, Skogby H, Lazor P, Broman C et al (2015) A fungal-prokaryotic consortium at the basalt-zeolite interface in subseafloor igneous crust. PLoS One 10(10):e0140106

Jackson MB, Colmer TD (2005) Response and adaptation by plants to flooding stress. Ann Bot 96:501–505

Jackson BE, McInerney MJ (2002) Anaerobic microbial metabolism can proceed close to thermodynamic limits. Nature 415:454–456

Jackson C, Clayden S, Reyes-Prieto A (2015) The Glaucophyta: the blue-green plants in a nutshell. Acta Soc Bot Pol 84(2):149–165

Jacobs DK, Wray CG, Wedeen CJ, Kostriken R, DeSalle R et al (2000) Molluscan engrailed expression, serial organization, and shell evolution. Evol Dev 26:340–347

Jahns P, Holzwarth AR (2012) The role of the xanthophyll cycle and of lutein in photoprotection of photosystem II. Biochim Biophys Acta 1817:182–193

Jami L (1992) Golgi tendon organs in mammalian skeletal muscle: functional properties and central actions. Physiol Rev 72(3):623–666

Jansson M, Blomqvist P, Jonsson A, Bergström A-K (1996) Nutrient limitation of bacterioplankton, autotrophic and mixotrophic phytoplankton, and heterotrophic nanoflagellates in Lake Örträsket. Limnol Oceanogr 41(7):1552–1559

Jékely G, Keijzer F, Godfrey-Smith P (2015) An option space for early neural evolution. Philos Trans R Soc B 370:20150181

Jensen KH (2018) Phloem physics: mechanisms, constraints, and perspectives. Curr Opin Plant Biol 43:96–100

Jensen PE, Leister D (2014) Chloroplast evolution, structure and functions. F1000Prime Rep 6:40

Jensen KH, Liesche J, Bohr T, Schulz A (2012) Universality of phloem transport in seed plants. Plant Cell Environ 35:1065–1076

Jensen KH, Berg-Sørensen K, Bruus H, Holbrook NM, Liesche J et al (2016) Sap flow and sugar transport in plants. Rev Mod Phys 88(3):035007

Jeschke WD, Pate JS (1995) Mineral nutrition and transport in xylem and phloem of *Banksia prionotes* (Proteaceae), a tree with dimorphic root morphology. J Exp Bot 46(289):895–905

Ji M, Greening C, Vanwonterghem I, Carere CR, Bay SK et al (2017) Atmospheric trace gases support primary production in Antarctic desert surface soil. Nature 552:400–403

Jiang H, Dong H, Zhang G, Yu B, Chapman LR, Fields MW (2006) Microbial diversity in water and sediment of Lake Chaka, an Athalassohaline Lake in northwestern China. Appl Exp Microb 72(6):3832–3845

Jobbágy EG, Jackson RB (2001) The distribution of soil nutrients with depth: global patterns and the imprint of plants. Biogeochemistry 53:51–77

Jobbágy EG, Jackson RB (2004) The uplift of soil nutrients by plants: biogeochemical consequences across scales. Ecology 85(9):2380–2389

Johnson SD, Jürgens A (2010) Convergent evolution of carrion and faecal scent mimicry in fly-pollinated angiosperm flowers and a stinkhorn fungus. S Afr J Bot 76:796–807

Johnson MA, Fernandez C, Pergent G (2002) The ecological importance of an invertebrate chemo-autotrophic symbiosis to phanerogam seagrass beds. Bull Mar Sci 71(3):1343–1351

Jones VAS, Dolan L (2012) The evolution of root hairs and rhizoids. Ann Bot 110:205–212

Josephson RK, Malamud JG, Stokes DR (2000) Asynchronous muscle: a primer. J Exp Biol 203:2713–2722

Jurkevitch E (2007) Predatory behaviors in bacteria—diversity and transitions. Microbe 2(2):67–73

Kaczmarek Ł, Jakubowska N, Celewicz-Gołdyn S, Zawierucha K (2015) The microorganisms of cryoconite holes (algae, Archaea, bacteria, cyanobacteria, fungi, and Protista): a review. Polar Record. https://doi.org/10.1017/S0032247415000637

Kadouri DE, To K, Shanks RM, Doi Y (2013) Predatory bacteria: a potential ally against multidrug-resistant gram-negative pathogens. eLife One 8(5):e63397

Kamennaya NA, Ajo-Franklin CM, Northern T, Jansson C (2012) Cyanobacteria as biocatalysts for carbonate mineralization. Fortschr Mineral 2:338–364

Kamga C, Krishnamurthy S, Shiva S (2012) Myoglobin and mitochondria: a relationship bound by oxygen and nitric oxide. Nitric Oxide 26(4):251–258

Karl DM (2002) Hidden in a sea of microbes. Nature 415:590–591

Karsenty G, Olson EN (2016) Bone and muscle endocrine functions: unexpected paradigms of inter-organ communication. Cell 164:1248–1256

Kartal B, Maalcke WJ, de Almeida NM, Cirpus I, Gloerich J et al (2011) Molecular mechanism of anaerobic ammonium oxidation. Nature 479:127–130

Kato S, Watanabe K (2010) Ecological and evolutionary interactions in syntrophic methanogenic consortia. Microbes Environ 25(3):145–151

Kato S, Hashimoto K, Watanabe K (2012) Microbial interspecies electron transfer via electric currents through conductive minerals. PNAS 109(25):10042–10046

Keane R, Berleman J (2016) The predatory life cycle of *Myxococcus xanthus*. Microbiology 162:1–11

Kehr J, Buhtz A (2008) Long distance transport and movement of RNA through the phloem. J Exp Bot 58(1):85–92

Kehr J, Kragler F (2018) Long distance RNA movement. New Phytol 218:29–40

Kellosalo J, Kajander T, Kogan K, Pokharel K, Goldman A (2012) The structure and catalytic cycle of a sodium-pumping pyrophosphatase. Science 337:473–476

Kelly DP, Wood AP (2013) The chemolithotrophic prokaryotes. In: Rosenberg E, DeLong EF, Stackebrandt E, Lory S, Thompson F (eds) The prokaryotes-prokaryotic communities and eco-physiology. Springer, Berlin, pp 275–287

Kempton JA, Wynn J, Bond S, Evry J, Fayet AL et al (2022) Optimization of dynamic soaring in flap-gliding seabirds affects its large-scale distribution at sea. Sci Adv 8:eabo0200

Kennedy RA, Rumpho ME, Fox TC (1992) Anaerobic metabolism in plants. Plant Physiol 100:1–6

Kenny R, Knauth LP (2001) Stable isotope variations in the neoptoroterozoic Beck Spring Dolomite and Mesoproterozoic Mescal Limestone paleokarst: implications for life on land in the Precambrian. GSA Bull 113(5):650–658

Kenrick P, Strullu-Derrien C (2014) The origin and early evolution of roots. Plant Physiol 166:570–580

Kent AD, Yannarell AC, Rusak JA, Triplett EW, McMahon KD (2007) Synchrony in aquatic microbial community dynamics. ISME J 1:38–47

Kerns KR, Collins JL, Kim H (1936) Developmental studies of the pine-apple *Ananas comosus* (L.) Merr. New Phytol 35(4):305–317

Kerp H (2017) Organs and tissues of Rhynie chert plants. Philos Trans R Soc B 373:20160495

Kessler M, Krömer T (2000) Patterns and ecological correlates of pollination modes among bromeliad communities of Andean forests in Bolivia. Plant Biol 2:659–669

Khadem AF, Pol A, Jetten MSM, Opdencamp HJM (2010) Nitrogen fixations by the verrucomicrobial methanotroph '*Methylacidiphilum fumariolicum*' SoLV. Microbiology 156:1052–1059

Khatoon Z, McTiernan CD, Suuronen EJ, Mah T-F, Alarcon EI (2018) Bacterial biofilm formation on implantable devices and approaches to its treatment and prevention. Heliyon 4:e01067

Kier WM (2012) The diversity of hydrostatic skeletons. J Exp Biol 215:1247–1257

Kiessling W, Simpson C, Foote M (2010) Reefs as cradles of evolution and sources of biodiversity in the phanerozoic. Science 327:196–198

King GM (2007) Chemolithotrophic bacteria: distributions, functions and significance in volcanic environments. Microbes Environ 22(4):309–319

Kiriukhin MY, Collins KD (2002) Dynamic hydration numbers for biologically important ions. Biophys Chem 99:155–168

Kirk JTO (2011) Light and photosynthesis in aquatic ecosystems. Cambridge University Press, Cambridge

Klausmeier CA, Litchmann E, Daufresne T, Levin SA (2004) Optimal nitrogen-to-phosphorus stoichiometry of phytoplankton. Nature 429:171–174

Kley NJ, Brainerd EL (1999) Feeding by mandibular raking in a snake. Nature 402:369–370

Kneip C, Lockhart P, Voß C, Maier U-G (2007) Nitrogen fixation in eukaryotes—new models for symbiosis. BMC Evol Biol. https://doi.org/10.1186/1471-2148-7-55

Knoblauch M, Knoblauch J, Mullendore DL, Savage JA, Babst BA et al (2016) Testing the Münch hypothesis of long distance phloem transport. Elife 5:e15341

Knoll AH (2014) Paleobiological perspectives on early eukaryotic evolution. Cold Spring Harb Perspect Biol 6:a016121

Koblížek M (2015) Ecology of aerobic anoxygene phototrophs in aquatic environments. FEMS Microbiol Rev 39:854–870

Koch GW, Sillett SC, Jennings GM, Davis SD (2004) The limits to tree height. Nature 428:851–854

Koehl MAR, Strickler JR (1981) Copepod feeding currents: food capture at low Reynolds number. Limnol Oceanogr 26(6):1062–1073

Koehl MAR, Silk WK, Liang H, Mahadevan L (2008) How kelp produce blade shapes suited to different flow regimes: a new wrinkle. Integr Comp Biol 48(6):834–851

Koh EY, Atamna-Ismaeel N, Martin A, Cowie ROM, Beja O et al (2010) Proteorhodopsin-bearing bacteria in Antarctic Sea ice. Applied and Experimental Microb 76(17):5918–5925

Kolber Z (2007) Energy cycles in the ocean: powering the microbial world. Oceanography 20(2):79–88

Kolstø AB (1997) Dynamic bacterial genome organization. Molecular Microb 24(2):241–248

Konhauser KO (2007) Introduction to geomicrobiology. Blackwell, Malden

Könneke M, Schubert DM, Brown PC, Hügler M, Standfest S et al (2014) Ammonia-oxidizing archaea use the most energy-efficient aerobic pathway for CO_2 fixation. PNAS 111(22):8239–8244

Koonin EV, Dolja VV (2014) Virus world as an evolutionary network of viruses and Capsidless Selfish elements. Microb and Molecular Biology Rev 78(2):278–303

Korp J, Gurovic MSV, Nett M (2016) Antibiotics from predatory bacteria. Beilstein J Org Chem 12:594–607

Kouzuma A, Kato S, Watanabe K (2015) Microbial interspecies interactions: recent findings in syntrophic consortia. Front Microbiol 6:477

Kraichak E (2013) Adaptive traits and community assembly of epiphyllous bryophytes. Dissertations, University of California, Berkeley

Krupovic M, Koonin EV (2017) Multiple origins of viral capsid proteins from cellular ancestors. PNAS, pp E2401–E2410

Krupovic M, Prangishvili D, Hendrix RW, Bamford DH (2011) Genomics of bacterial and archaeal viruses: dynamics within the prokaryotic virosphere. Microb and Molecular Biology Rev 75(4):610–635

Kuenen JG (2008) Anammox bacteria: from discovery to application. Nat Rev Microbiol 6:320–326

Kühl M, Chen M, Ralph PJ, Schreiber U, Larkum WD (2005) A niche for cyanobacteria containing chlorophyll d. Nature 433:820

Kulp TR, Hoeft SE, Asao M, Madigan MT, Hollibaugh JT et al (2008) Arsenic(III) fuels anoxygenic photosynthesis in hot spring biofilms from Mono Lake, California. Science 321:967–970

Kunugi M, Satoh S, Ihara K, Shibata K, Yamagishi Y (2016) Evolution of green plants accompanied changes in light-harvesting systems. Plant Cell Physiol 57(6):1231–1243

Kurzweil F, Drost K, Pašava J, Wille M, Taubald H et al (2015) Coupled sulfur, iron and molybdenum isotope data from black shales of the Teplá-Barrandian unit argue against deep ocean oxygenation during the Ediacaran. Geochim Cosmochim Acta 171:121–142

Kutschera U, Briggs WR (2012) Root phototropism: from dogma to the mechanism of blue light perception. Planta. https://doi.org/10.1007/s00425-012-1597-y

Kuypers MMM, Sllekers AO, Lavik G, Schmid M, Jørgensen BB et al (2003) Anaerobic ammonium oxidation by anammox bacteria in the Black Sea. Nature 422:608–611

Laenen B, Shaw B, Schneider H, Goffinet B, Paradis E et al (2014) Extant diversity of bryophytes emerged from successive post-mesozoic diversification bursts. Nat Commun. https://doi.org/10.1038/ncomms6134

Lai JH, del Alamo JC, Rodríguez-Rodríguez J, Lasheras JC (2010) The mechanics of the adhesive locomotion of terrestrial gastropods. J Exp Biol 213:3920–3933

Laity J (2008) Deserts and desert environments. Wiley-Blackwell, Chichester

Laksanalamai P, Narayan S, Luo H, Robb FT (2008) Chaperone action of a versatile small heat shock protein from *Methanococcoides burtonii*, a cold adapted chaperon. Proteins 75:275–281

Lalonde SV, Konhauser KO (2015) Benthic perspective on earth's oldest evidence for oxygenic photosynthesis. PNAS 112(4):995–1000

Lam P, Jensen MM, Lavik G, McGinnis DF, Müller B et al (2007) Linking crenarchaeal and bacterial nitrification to anammox in the Black Sea. PNAS 104(17):7104–7109

Lambers H, Chapin FS III, Pons TL (2006) Plant physiological ecology. Springer, New York

Lane N, Martin WF (2012) The origin of membrane bioenergetics. Cell 151:1406–1416

Langdale JA (2008) Evolution of developmental mechanisms in plants. Curr Opin Genet Dev 18:368–373

Larabee FJ, Suarez AV (2014) The evolution and functional morphology of trap-jaw ants (Hymenoptera: Formicidae). Mymercol News 20:25–36

Larabee FJ, Smith AA, Suarez AV (2018) Snap-jaw morphology is specialized for high-speed power amplification in the Dracula ant, Mystrium camillae. R Soc Open Sci 5:181447

Larkum AWD, Scaramuzzi C, Cox GC, Hiller RG, Turner AG (1994) Light-harvesting chlorophyll c-like pigment in *Prochloron*. PNAS 91:679–683

Larsen A, Castberg T, Sandaa RA, Brussaard CPD, Egge J et al (2001) Population dynamics and diversity of phytoplankton, bacteria and viruses in a seawater enclosure. Mar Ecol Process Ser 221:47–57

Laskar J, Robutel P, Joutel F, Gastineau M, Correia ACM, Levrard B (2004) A long term numerical solution for the insolation quantities of the earth. Astron Astrophys 428:281–285

Lau OS, Deng XW (2010) Plant hormone signalling lightens up: integrators of light and hormones. Curr Opin Plant Biol 13:571–577

Lau MCY, Kleft TL, Kuloyo O, Linage-Alvarez B, van Heerden E et al (2016) An oligotrophic deep-subsurface community dependent on syntrophy is dominated by sulfur-driven autotrophic denitrifiers. PNAS:E7927–E7936

Lautenschläger K-H, Schröter W, Teschner J, Bibrack H (2001) Taschenbuch der Chemie. Deutsch, Frankfurt am Main

Laval B, Gady SL, Pollack JC, McKay CP, Bird JS et al (2000) Modern freshwater microbialite analogues for ancient dendritic reef structures. Nature 407:626–629

Le Romancer M, Gaillard M, Geslin C, Prieur D (2007) Viruses in extreme environments. Rev Environ Sci Biotechnol 6(1–3):17–31

Leal MC, Seehausen O, Matthews B (2017) The ecology and evolution of stoichiometric phenotypes. Trend Ecol Evol 32(2):108–117

Lea-Smith DJ, Bombelli PB, Vasudevan R, Howe CJ (2016) Photosynthetic, respiratory and extracellular electron transport pathways in cyanobacteria. Biochim Biophys Acta 1857:247–255

Lebeis SL, Paredes SH, Lundberg DS, Breakfield N, Gehring J et al (2015) Salicylic acid modulates colonization of the root microbiome by specific bacterial taxa. Science 349:860–864

Lee S, Bush JWM, Hosoi AE, Lauga E (2008) Crawling beneath the free surface: water snail locomotion. arXiv:0806.3651v1

Lefsky MA (2010) A global forest canopy height map from the moderate resolution imaging spectroradiometer and the geoscience laser altimeter system. Geophys Res Lett 37:L15401

Leggett MJ, McDonnell G, Denyer SP, Setlow P, Maillard J-Y (2012) Bacterial spore structures and their protective role in biocide resistance. Applied Microb 113:485–498

Legris M, Nieto C, Sellaro R, Prat S, Casal JJ (2017) Perception and signalling of light and temperature cues in plants. Plant J 90:683–697

Leigh JA (2000) Nitrogen fixation in methanogens: the archaeal perspective. Curr Issues Mol Biol 2(4):125–131

Leliaert F, Verbruggen H, Zechman FW (2011) Into the deep: new discoveries at the base of the green plant phylogeny. Bioessays 33:683–692

Leo SST, Cheng L, Sperling FAH (2012) Genetically separate populations of the ocean-skater *Halobates sericeus* (Heteroptera: Gerridae) have been maintained since the late Pleistocene. Biol J Linn Soc 105:797–805

Leroy C, Corbara B, Pélozuclo L, Carrias J-F, Dejean A, Céréghino R (2012) Ants species identity mediates reproductive traits and allocation in an ant-garden bromeliad. Ann Bot 109:145–152

Leroy C, Carrias J-F, Céréghino R, Corbara B (2016) The contribution of microorganisms and metazoans to mineral nutrition in bromeliads. J Plant Ecol 9(3):241–255

Leuschner C, Moser G, Bertsch C, Röderstein M, Hertel D (2007) Large altitudinal increase in tree root/shoot ration in tropical mountain forests of Ecuador. Basic Appl Ecol 8:219–230

Lever MA, Rogers KL, Lloyd KG, Overmann J, Schink B et al (2015) Life under extreme energy limitation: a synthesis of laboratory- and field-based investigations. FEMS Microb Rev 39(5):688–728

Levy AT, Lee KH, Hanson TE (2016) *Chlorobaculum tepidum* modulates amino acid composition in response to energy availability, as revealed by a systemic exploration of the energy landscape of phototrophic sulfur oxidation. Applied and Environmental Microb 82(21):6431–6439

Li Y, Wang Z, Xu T, Tu W, Liu C et al (2010) Reorganization of photosystem II is involved in the rapid photosynthetic recovery of desert moss *Syntrichia caninervis* upon rehydration. J Plant Physiol 167:1390–1397

Li Y-L, Konhauser KO, Cole DR, Phelps TJ (2011) Mineral ecophysiological data provide growing evidence for microbial activity in banded-iron formations. Geology 39(8):707–710

Li J, Ying G-G, Jones KC, Martin FL (2015) Real-world carbon nanoparticle exposures induces brain and gonadal alterations in zebrafish (*Danio rerio*) as determined by biospectroscopy techniques. Analyst 140:2687–2695

Li C, Zhu M, Chu X (2016) Atmospheric and oceanic oxygenation and evolution of early life on earth: new contributions from China. J Earth Sci 27(2):167–169

Liechti F, Witvliet W, Weber R, Bächler E (2013) First evidence of a 200-day non-stop flight in a bird. Nat Commun 4:2554

Liesche J, Pace MR, Xu Q, Li Y, Chen S (2017) Height-related scaling of phloem anatomy and the evolution of sieve element end wall types in woody plants. New Phytol 214:245–256

Lietzenmayer LB, Goldstein LM, Pasche JM, Taylor LA (2022) Extreme natural size variation in both sexes of a sexually cannibalistic mantidfly. R Soc Open Sci 9:220544

Lindo Z, Gonzalez A (2010) The bryosphere: an integral and influential component of the earth's biosphere. Ecosystems 13:612–627

Liu Z, Müller J, Li T, Alvey RM, Vogl K et al (2013) Genomic analysis reveals key aspects of prokaryotic symbiosis in the phototrophic consortium '*Chlorochromatium aggregatum*'. Genome Biol 14:R127

Lovelock J (1992) Gaia, die Erde ist ein Lebewesen. Fischer-Scherz, Frankfurt

Lovley DR (2008) Extracellular electron transfer: wires, capacitors, iron lungs, and more. Geobiology 6:225–231

Lücking R, Bernecker-Lücking A (2000) Lichen feeders and lichenicolous fungi: do they affect dispersal and diversity in tropical foliicolous lichen communities? Ecotropica 6:23–41

Ludwig F, Dawson TE, Kroon H, Berendse F, Prins HHT (2003) Hydraulic lift in *Acacia tortilis* trees on an east African savanna. Oecologia 134:293–300

Ludwig F, Middleton W, Gallenmüller F, Rogers P, Speck T (2019) Living bridges using aerial roots of *ficus elastica*—an interdisciplinary perspective. Sci Rep. https://doi.org/10.1038/s41598-019-48652-w

Lukas D, Huchard E (2014) The evolution of infanticide by males in mammalian society. Science 346:841–844

Lundgren MR, Osborne CP, Christin P-A (2014) Deconstructing Kranz anatomy to understand C_4 evolution. J Exp Bot 65(13):3357–3369

Macalady JL, Vestling MM, Baumler D, Bockelheide N, Kaspar CW, Banfield JF (2004) Tetraether-linked membrane monolayers in *Ferroplasma* spp: a key to survived in acid. Extremophiles 8:411–419

MacIntyre HL, Kans TM, Gelder RJ (2000) The effect of water motion on short-term rates of photosynthesis by marine phytoplankton. Trends Plant Sci 5(1):12–17

Macrì S, Savriama Y, Khan I, Di-Poï N (2019) Comparative analysis of squamate brains unveils multi-level variation in cerebellar architecture associated with locomotor specialization. Nat Commun 10:5560

Madigan MT, Martinko JM, Parker J (2003) Brock biology of microorganisms. Prentice Hall, Upper Saddle River

Madsen JS, Burmølle M, Hansen LH, Sørensen SJ (2012) The interconnection between biofilm formation and horizontal gene transfer. FEMS Immunol Med Microbiol 65:183–195

Maguire F, Richards TA (2014) Organelle evolution: a mosaic of 'Mitochondrial' Functions'. Curr Biol 24(11):R518–R520

Mahendra S, Zhu H, Colvin VL, Alvarez PJ (2008) Quantum Dot weathering results in microbial toxicity. Environ Sci Technol 42:9424–9430

Makarieva AM, Gorshkov VG, Li B-L (2005) Energetics of the smallest: do bacteria breathe at the same rate as whales? Proc R Soc B 272:2219–2224

Makino W, Cotner JB, Sterner RW, Elser JJ (2003) Are bacteria more like plants or animals? Growth rate and resource dependence of bacterial C:N:P stoichiometry. Funct Ecol 17:121–130

Malcicka M, Visser B, Ellers J (2018) An evolutionary perspective on linolic acid synthesis in animals. Evol Biol 45:15–26

Maldonaldo M (2004) Choanoflagellates, choanocytes, and animal multicellularity. Invertebr Biol 123(1):1–22

Males J, Griffiths H (2017) Specialized stomatal humidity responses underpin ecological diversity in C_3 bromeliads. Plant Cell Environ 40:2931–2945

Males J, Griffiths H (2018) Economic and hydraulic divergences underpin ecological differentiation in the Bromeliaceae. Plant Cell Environ 41:64–78

Malkin SY, Rao AMF, Seitaj D, Vasquez-Cardenas D, Zetsche E-M et al (2014) Natural occurrence of microbial sulphur oxidation by long-range electron transport in the seafloor. ISME J 8:1843–1854

Mall A, Sobotta J, Huber C, Tschirner C, Kowarschik S et al (2018) Reversibility of citrate synthase allows autotrophic growth of a thermophilic bacterium. Science 359:563–567

Mann KH, Lazier JRN (2006) Dynamics of marine ecosystems. Blackwell, Malden

Martin RD, Genoud M, Hemelrijk CK (2005) Problems of allometric scaling analysis: examples from mammalian reproductive biology. J Exp Biol 208:1731–1747

Martin RE, Quigg A, Podkovyrov V (2008) Marine biodiversification in response to evolving phytoplankton stoichiometry. Palaeogeogr Palaeoclimatol Paleoecol 258:277–291

Martin WF, Sousa FL, Lane N (2014) Energy at life's origin. Science 344(6188):1092–1093

Martin WF, Bryant DA, Beatty JT (2018) A physiological perspective on the origin and evolution of photosynthesis. FEMS Microb Rev 42:205–231

Masi E, Ciszak M, Stefano G, Renna L, Azzarello E et al (2009) Spatiotemporal dynamics of the electrical network activity in the root apex. PNAS 106(10):4048–4053

Masoud GN, Li W (2015) HIF-1α pathway: role, regulation and intervention for cancer therapy. Acta Pharm Sin B 5(5):378–389

Masuda F, Ezaki Y (2009) A great evolution of the earth-surface environment: linking the bioinvasion onto the land and the Ordovician radiation of marine organisms. Paleontol Res 13(1):3–8

Mathai JC, Sprott GD, Zeidel ML (2001) Molecular mechanisms of water and solute transport across archaebacterial lipid membranes. J Biol Chem 276(29):27266–27271

Maurel C, Verdoucq L, Luu D-T, Santoni V (2008) Plant aquaporins: membrane channels with multiple integrated functions. Annu Rev Plant Biol 59:595–624

Maurel C, Boursiac Y, Luu D-T, Santoni V, Shahzad Z, Verdoucq L (2015) Aquaporins in plants. Physiol Rev 95:1321–1358

May V, Kühn O (2011) Charge and energy transfer dynamics in molecular systems. Wiley-VCH, Darmstadt

Mayer F, Müller V (2014) Adaptations of anaerobic archaea to life under extreme energy limitation. FEMS Microb Rev 38:449–472

Mayer EA, Knight R, Mazmanian SK, Cryan JF, Tillisch K (2014) Gut microbes and the brain: paradigm shift in neuroscience. J Neurosci 34(46):15490–15496

McCourt RM, Delwiche CF, Karol KG (2004) Charophyte algae and land plant origins. Trends Ecol Evol 19(12):660–666

McDougald D, Rice SA, Barraud N, Steinberg PD, Kjelleberg S (2011) Should we stay or should we go: mechanisms and ecological consequences for biofilm dispersal. Nat Rev Microbiol 10:39–50

McFadden KA, Huang J, Chu X, Jiang G, Kaufman AJ et al (2008) Pulsed oxidation and biological evolution in the Ediacaran Doushantuo formation. PNAS 105(9):3197–3202

McFall-Ngai M, Hadfield MG, Bosch TCG, Carey HV, Domazet-Lošo T et al (2013) Animals in a bacterial world, a new imperative for the life Sci. PNAS 110(9):3229–3236

McGlynn SE, Chadwick GL, Kempes CP, Orphan VJ (2015) Single cell activity reveals direct electron transfer in methanotrophic consortia. Nature 526:531–534

McNab BK (1994) Energy conservation and the evolution of flightlessness in birds. Am Nat 144(4):628–642

Melnik EA, Buzulukov YP, Demin VF, Demin VA, Gmoshinski IV et al (2013) Transfer of silver nanoparticles through the placenta and breast Milk during *in vivo* experiments on rats. Acta Nat 5(3):107–115

Melnyk CW, Molnar A, Baulcombe DC (2011) Intercellular and systemic movement of RNA silencing signals. EMBO J 30:3553–3563

Mencuccini M, Höltta T (2009) The significance of phloem transport for the speed with which canopy photosynthesis and belowground respiration are linked. New Phytol 185:189–203

Méndez-Alonzo R, Montezuma C, Ordoñez VR, Angeles G, Marínez AJ, López-Portillo J (2015) Root biomechanics in *Rhizophora mangle*: anatomy, morphology and ecology of mangrove's flying buttresses. Ann Bot 115:833–840

Mentel M, Martin W (2008) Energy metabolism among eukaryotic anaerobes in light of Proterozoic Ocean chemistry. Philos Trans R Soc B 363:2717–2729

Merckx V, Bidartondo MI, Hynson NA (2009) Myco-heterotrophy: when fungi host plant. Ann Bot 104:1255–1261

Metcalf WW (2016) Classic spotlight: electron bifurcation, a unifying concept for energy conservation in anaerobes. J Bacteriol 198(9):1358

Michel P, Payton IJ, Lee WG, During HJ (2013) Impact of disturbance on above-ground water storage capacity of bryophytes in New Zealand indigenous tussock grassland ecosystems. New Zealand J Ecol 37(1):114–126

Mickovski SB (2002) Anchorage mechanics of different root systems. Dissertations, University of Manchester

Middelboe M, Hagström A, Blackburn N, Sinn B, Fischer U et al (2001) Effects of bacteriophages on the population dynamics of four strains of pelagic marine bacteria. Microb Ecol 42:395–406

Mielke SP, Kiang NY, Blankenship RE, Gunner MR, Mauzerall D (2011) Efficiency of photosynthesis in a Chl d-utilizing cyanobacterium is comparable to or higher than that in Chl a-utilizing oxygenic species. Biochim Biophys Acta 1807:1231–1236

Mileusnic MP, Loeb GE (2006) Mathematical models of Proprioreceptors. II. Structure and function of the Golgi Tendon Orga. J Neurophysiol 96:1789–1802

Mills DB, Canfield DE (2016) A trophic framework for animal origins. Geobiology:1–14

Mills B, Lenton TM, Watson AJ (2014) Proterozoic oxygen rise linked to shifting balance between seafloor and terrestrial weathering. PNAS 111(25):9073–9078

Mills DB, Francis WR, Vargas S, Larsen M, Elemans CPH et al (2018) The last common ancestor of animals lacked the HIF pathway and respired in low-oxygen environments. Elife (7):e31176

Miot J, Benzerara K, Obst M, Kappler A, Hegler F et al (2009) Extracellular iron biomineralization by photoautotrophic iron-oxidizing bacteria. Appl Environ Microb 75(17):5586–5591

Misevic GN, Ripoll C, Norris J, Norris V, Guerardel Y et al (2007) Evolution of multicellularity in Porifera via self-assembly of glycoconectin carbohydrates. Biodiversity, Innovation and Sustainability, Porifera Res, pp 79–88

Miteva VI, Brenchley JE (2005) Detection and isolation of ultrasmall microorganisms from a 120,000-year-old Greenland glacier ice core. Applied and Experimental Microb 71(12):7806–7818

Mizutani M, Ohta D (2010) Diversification of P450 genes during land plant evolution. Annu Rev Plant Biol 61:291–315

Mock T, Thomas DN (2005) Recent advances in sea-ice microbiology. Environmental Microb 7(5):605–619

Mojzsis SJ, Arrhenius G, Mc Keegan KD, Harrison TM, Nutman AP, Friend CRL (1996) Evidence for life on earth before 3,800 million years ago. Nature 384:55–59

Moles AT, Warton DI, Warman L, Swenson NG, Laffan SW et al (2009) Global patterns in plant height. J Ecol 97:923–932

Monahan-Earley R, Dvorak AM, Aird WC (2013) Evolutionary origins of the blood vascular system and endothelium. J Thromb Homeost 11:46–66

Montgomery BL, Lechno-Yossef S, Kerfeld CA (2016) Interrelated modules in cyanobacterial photosynthesis: the carbon-concentrating mechanism, photorespiration, and light perception. J Exp Bot 67(10):2931–2940

Moore LM (2013) More mixotrophy in the marine microbial mix. PNAS 110(21):8323–8324

Moore D, Robson GD, Trinci APJ (2011) 21st century guidebook to fungi. Cambridge University Press, Cambridge

Mooshammer M, Wanek W, Zechmeister-Boltenstern S, Richter A (2014) Stoichiometric imbalances between terrestrial decomposer communities and their resources: mechanisms and implications of microbial adaptations to their resources. Front Microbiol 5:22

Morel M, Meux E, Mathieu Y, Thuillier A, Chibani K et al (2013) Xenomic networks variability and adaptation traits in wood decaying fungi. J Microbial Biotechnol 6(3):248–263

Moroney JV, Somanchi A (1999) How do algae concentrate CO_2 to increase the efficiency of photosynthetic carbon fixation? Plant Physiol 119:9–16

Moroz LL, Kocot KM, Citarella MR, Dosung S, Norekian TP et al (2014) The ctenophore genome and the evolutionary origins of neural systems. Nature 510:109–114

Muehe EM, Gerhardt S, Schink B, Kappler A (2009) Ecophysiology and the energetic benefit of mixotrophic FE(II) oxidation by various strains of nitrate-reducing bacteria. FEMS Microbiol Ecol 70:335–343

Mulder A, van der Graaf AA, Robertson LA, Kuenen JG (1995) Anaerobic ammonium oxidation discovered in a denitrifying fluidized bed reactor. FEMS Microb Ecol 16:177–184

Mulkidjanian AY (2011) Energetics of the first life. In: Egel R, Lankenau D-H, Mulkidjanian AY (eds) Origins of life: the primal self-organization. Springer, Heidelberg, pp 3–33

Mulkidjanian AY, Koonin EV, Makarova KS, Mekhedov SL, Sorokin A et al (2006) The cyanobacterial genome core and the origin of photosynthesis. PNAS 103(35):13126–13131

Mulkidjanian AY, Galperin MY, Makarova KS, Wolf YI, Koonin EV (2008) Evolutionary primacy of bioenergetics. Biol Direct 3:13. https://doi.org/10.1186/1745-6150-3-13

Müller J, Overmann J (2011) Close interspecies interactions between prokaryotes from sulfureous environments. Front Microbiol 2:146

Müller S, Vogt C, Laube M, Harms H, Kleinsteuber S (2009) Community dynamics within a bacterial consortium during growth on toluene under sulfate-reducing conditions. FEMS Microbial Ecol 70:586–596

Müller M, Mentel M, van Hellemond JJ, Henze K, Woehle C et al (2012) Biochemistry and evolution of anaerobic energy metabolism in eukaryotes. Microb Mol Biol 76(2):444–495

Mullineaux CW (2014) Electron transport and light-harvesting switches in cyanobacteria. Front Plant Sci 5:7

Murray PJ (2016) Amino acid auxotrophy as immunological control nodes. Nat Immunol 17(2):132–139

Nagalingum NS, Marshall CR, Quental TB, Rai HS, Little DP, Mathews S (2011) Recent synchronous radiation of a living fossil. Science 334:796–799

Nagarajan H, Embree M, Rotaru A-E, Shrestha PM, Feist AM et al (2013) Characterization and modelling of interspecies electron transfer mechanisms and microbial community dynamics of a syntrophic association. Nat Commun 4/208. https://doi.org/10.1038/ncomms3809

Nardini A, Salleo S, Jansen S (2011) More than just a vulnerable pipeline: xylem physiology in the light of ion-mediated regulation of plant water transport. J Exp Bot 62(14):4701–4718

Nash TH III (ed) (2008) Lichen biology. Cambridge University Press, Cambridge

Nasir A, Caetano-Anollés G (2015) A phylogenomic data-driven exploration of viral origins and evolution. Sci Adv 1:e1500527

Nasir A, Kim KM, Caetano-Anollés G (2012) Viral evolution. Mob Genet Elem 2(5):247–252

Nasir A, Forterre P, Kim KM, Caetano-Anollés G (2014) The distribution and impact of viral lineages in domains of life. Front Microbiol 5:194

Nauhaus K (2003) Mikrobiologische Studien zur anaeroben Oxidation von Methan (AOM). Dissertations, University of Bremen

Navrotsky A (2004) Energetic clues to pathways to biomineralization: precursors, clusters, and nanoparticles. PNAS 101(33):12096–12101

Nelson B (2008) Biological physics. Freeman and Company, New York

Nelson N, Ben-Shem A (2004) The complex architecture of oxygenic photosynthesis. Nat Rev Mol Cell Biol 5

Neumann RB, Cardon ZG (2012) The magnitude of hydraulic redistribution by plant roots: a review and synthesis of empirical and modelling studies. New Phytol 194:337–352

Newton ILG, Woyke T, Auchtung TA, Dilly GF, Dutton RJ et al (2007) A window into hydrothermal vent endosymbioses: the *Calyptogena magnifica* chemoautotrophic symbiont genome. Science 315:998–1000

Nickel M (2010) Evolutionary emergence of synaptic nervous systems: what can we learn from th non-synaptic, nerveless Porifera? Invertebr Biol 129(1):1–16

Nickelsen J, Rengstl B, Stengel A, Schottkowski M, Sill J, Ankele E (2010) Biogenesis of the cyanobacterial thylakoid membrane system—an update. FEMS Microbiol Lett 315:1–5

Nielsen C (2001) Animal evolution. Oxford University Press, Oxford

Niklas KJ (1982) Simulated and empiric wind pollination patterns of conifer ovulate cones. PNAS 79:510–514

Niklas KJ, Kutschera U (2010) The evolution of the land plant life cycle. New Phytol 185:27–41

Nisbet EG, Fowler CMR (1999) Archaean metabolic evolution of microbial mats. Proc R Soc Lond B 266:2375–2382

Nisbet E, Fowler CMR (2011) The evolution of the atmosphere in the Archaean and early Proterozoic. Chin Sci Bull 56:4–13

Nisbet EG, Sleep NH (2001) The habitat and nature of early life. Nature 409:1083–1091

Nitschke W, McGlynn SE, Milner-White EJ, Russell MJ (2013) On the antiquity of metalloenzymes and their substrates in bioenergetics. Biochim Biophys Acta 1827:871–881

Nobel PS (2002) Cacti. University of California Press, Berkeley

Nobel PS (2009) Physiochemical and environmental plant physiology. Academic Press, Oxford

Noffke N (2008) Turbulent lifestyle: microbial mats on earth's sandy beaches-today and 3 billion years ago. GSA Today 18(10):4–9

Nunoura T, Chikaraishi Y, Izaki R, Suwa T, Sato T et al (2018) A primordial and reversible TCA cycle in a facultatively chemolithoautotrophic thermophile. Science 359:559–563

O'Regan RG, Majcherczyk S (1982) Role of peripheral chemoreceptors and central chemosensitivity in the regulation of respiration and circulation. J Exp Biol 100:23–40

Oarga A (2009) Life in extreme habitats. Revista de Biologia Ciéncias da Terrain 9:1

Obornik M, Janouškovec J, Chrudimský T, Lukeš J (2009) Evolution of the apicoplast and its hosts: from heterotrophy to autotrophy and back again. Int J Parasitol 39:1–12

Ohashi S, Iemura T, Okada S, Itoh S, Furukawa H et al (2010) An overview on chlorophylls and quinines in the photosystem I-type reaction centers. Photosynth Rev 104:305–319

Oka S-I, Tomita T, Miyamoto K (2016) A mighty claw: pinching force of the coconut crab, the largest terrestrial crustacean. PLoS One 11(11):e0166108

Okur HI, Hladílková J, Rembert KB, Cho Y, Heyda J et al (2017) Beyond the hofmeister series: ion-specific effects on proteins and their biological functions. J Phys Chem B 121:1997–2014

Oliver M, Tuba Z, Mishler BD (2000) The evolution of vegetative desiccation tolerance in land plants. Plant Ecol 151:85–100

Ollivier B, Caumette P, Garcia J-L, Mah RA (1994) Anaerobic bacteria from hypersaline environments. Microbiol Rev 58(1):27–38

Olson JM, Blankenship RE (2004) Thinking about the evolution of photosynthesis. Photosynth Res 80:373–386

Oren A, Padan E, Avron M (1977) Quantum yields for oxygenic and anoxygenic photosynthesis in the cyanobacterium *Oscillatoria limnetica*. PNAS 74/r:2152–2156

Orf GS, Blankenship RE (2013) Chlorosome antenna complexes from green photosynthetic bacteria. Photosynth Res 116:314–331

Orf GS, Tank M, Vogl K, Niedzwiedzki DM, Bryant DA, Blankenship RE (2013) Spectroscopic insights into the decreased efficiency of chlorosomes containing bacteriochlorophyll f. Biochim Biophys Acta 1827:493–501

Orgel LE (2008) The implausibility of metabolic cycles on the prebiotic earth. PLoS Biol 6(1):e18

Osakabe Y, Yamaguchi-Shinozaki K, Shinozaki K, Tran L-SP (2013) Sensing the environment: key roles of membrane-localized kinases in plant perception and response to abiotic stress. J Exp Bot 64(2):445–458

Osakabe Y, Osakabe K, Shinozaki K, Tran L-SP (2014) Response of plants to water stress. Front Plant Sci 5:86

Osburn MR, LaRowe DE, Momper LM, Amend JP (2014) Chemolithotrophy in the continental deep subsurface: Sanford Underground Res Facility (SURF), USA. Front Microbiol 5:610

Ouverney CC, Fuhrman JA (2000) Marine planktonic archaea take up amino acids. Applied and Environmental Microb 66(11):4829–4833

Pacton M, Wacey D, Corinaldesi C, Tangherlini M, Kilburn MR et al (2014) Viruses as new agents of organomineralization in the geological record. Nat Commun. https://doi.org/10.1038/ncomms5298

Page CP, Moser CC, Chen X, Dutton L (1999) Natural engineering principles of electron tunnelling in biological oxidation-reduction. Nature 402:47–52

Palazon A, Goldrath AW, Nizet V, Johnson RS (2014) HIF transcription factors, inflammation, and immunity. Immunity 41:518–528

Pan Z, Pitt WG, Zhang Y, Wu N, Tao Y, Truscott TT (2016) The upside-down water collection system of *Syntrichia caninvervis*. Nat Plants. https://doi.org/10.1038/NPLANTS.2016.76

Paoli P, Ferrara F, Taiti S (2002) Morphology and evolution of the respiratory apparatus in the family Eubelidae (Crustacea, Isopoda, Oniscidea). J Morphol 253:272–289

Park MG, Kim M, Kim S (2014) The acquisition of plastids/phototrophy in heterotrophic dinoflagellates. Acta Protozool 53:39–50

Passow U, Carlson CA (2012) The biological pump in a high CO_2 world. Mar Ecol Prog Ser 470:240–271

Pate J, Shedley E, Arthur D, Adams M (1998) Spatial and temporal variations in phloem sap composition of plantation-grown *Eucalyptus globulus*. Oecologia 117:312–322

Patek SN (2015) The most powerful movements in biology. Am Sci 103:330–337

Patek SN, Summers AP (2017) Invertebrate biomechanics. Curr Biol 27:R371–R375

Pavlov V, Rosental B, Hansen NF, Beers JM, Parish G et al (2017) Hydraulic control of tuna fins: a role for the lymphatic system in vertebrate locomotion. Science 357:310–314

Payne JL, McClain CR, Boyer AG, Brown JH, Finnegan S et al (2010) The evolutionary consequences of oxygenic photosynthesis: a body size perspective. Photosynth Res. https://doi.org/10.1007/s11120-010-959-1

Peck LS, Chapelle G (2003) Reduced oxygen at high altitude limits maximum size. Proc R Soc Lond B (Suppl) 270:A166–S167

Pedrós-Alió C, Calderón-Paz JI, Gasol JM (2000) Comparative analysis shows that bacterivory, not viral lysis, controls the abundance of heterotrophic prokaryotic plankton. FEMS Microb Ecol 32:157–165

Pereira E, Frudd K, Awad W, Hendershot LM (2014) Endoplasmatic reticulum (ER) stress and hypoxia response pathways interact to potentiate hypoxia-inducible factor 1 (HIF-1) transcriptional activity of targets like vascular endothelial growth factor (VEGF). J Biol Chem 289(6):3352–3364

Pernthaler A, Dekas AE, Brown CT, Goffredi SK, Embaye T, Orphan VJ (2008) Diverse syntrophic partnerships from deep-sea methane vents revealed by direct cell capture and metagenomics. PNAS 105(19):7052–7097

Perry SF, Sander M (2004) Reconstructing the evolution of the respiratory apparatus in tetrapods. Respir Physiol Neurobiol 144(2–3):125–139

Petroff AP, Sin MS, Maslov A, Krupenin M, Rothman DH, Bosak T (2010) Biophysical basis for the geometry of colonial stromatolites. PNAS 107(22):9956–9961

Pfautsch S, Renard J, Tjoelker MG, Salih A (2015) Phloem as capacitor: radial transfer of water into xylem of tree stems occurs via symplastic transport in ray parenchyma. Plant Physiol 167:963–971

Pfeffer C, Larsen S, Song J, Dong M, Besenbacher F et al (2012) Filamentous bacteria transport electrons over centimetre distances. Nature 491:218–221

Philippot L, Raaijmakers JM, Lemanceau P, van der Putten WH (2013) Going back to the roots: the microbial ecology of the rhizosphere. Nat Rev Microbiol 11:789–799

Phillips R, Kondev J, Theriot J, Garcia HG, Orme N (2013) Physical biology of the cell. Garland Science, London

Phoenix VR, Konhauser KO, Adams DG, Bottrell SH (2001) Role of biomineralization as an ultraviolet shield: implications for Archean life. Geology 29(9):823–826

Pierce S (2007) The jewelled armor of *Tillandsia* multifaceted of elongated trichomes provide photoprotection. Aliso 23:44–52

Pietikäinen J, Pettersson M, Bååth E (2005) Comparison of temperature effects on soil respiration and bacterial and fungal growth rates. FEMS Microb Ecol 52:49–58

Pikuta EV, Hoover RB, Tang J (2007) Microbial extremophiles at the limits of life. Crit Rev Microb 33:183–209

Pittermann J, Limm E, Rico C, Christman MA (2011) Structure-function constraints of tracheid-based xylem: a comparison of conifers and ferns. New Phytol 192:449–461

Pittermann J, Brodersen C, Watkins JE (2013) The physiological resilience of fern sporophytes and gametophytes: advances in water relations offer new insights into an old lineage. Front Plant Sci 4:285

Planavsky NJ, Asael D, Hofmann A, Reinhard CT, Lalonde SV et al (2014) Evidence for oxygenic photosynthesis half a billion years before the great oxidation event. Nat Geosci 7:283–286

Plugge CM, Zhang W, Scholten JCM, Stams AJM (2011) Metabolic flexibility of sulfate-reducing bacteria. Front Microbiol 2:81

Poehlein A, Schmidt S, Kaster A-K, Goenrich M, Vollmers J et al (2012) An ancient pathway combining carbon dioxide fixation with the generation and utilization of a sodium ion gradient for ATP synthesis. PLoS One 7(3):e33439. https://doi.org/10.1371/journal.pone.0033439

Pointing SB, Belnap J (2012) Microbial colonization and controls in dryland systems. Nat Rev Microbiol 10:551–562

Polis GA (1981) The evolution and dynamics of intraspecific predation. Annu Rev Ecol Syst 12:225–251

Politi Y, Priewasser M, Pippel E, Zaslansky P, Hartmann J et al (2012) A Spider's Fang: how to design an injection needle using chitin-based composite material. Adv Funct Mater. https://doi.org/10.1002/adfm.201200063

Pollastri S, Tattini M (2011) Flavonols: old compounds for old roles. Ann Bot 108:1225–1233

Polly PD (2007) Limbs in mammalian evolution. In: Hall BK (ed) Fins into limbs: evolution, development, and transformation. University of Chikago Press, Chikago, pp 245–268

Popper ZA, Michel G, Hervé C, Domozych DS, Willats WGT et al (2011) Evolution and diversity of plant cell walls: from algae to flowering plants. Annu Rev Plant Biol 62:567–590

Porada P, Weber W, Elbert W, Pöschl U, Kleidon A (2014) Estimating impacts of lichens and bryophytes on global biogeochemical cycles. Global Biogeochem Cycles 28:71–85

Posth NR, Konhauser KO, Kappler A (2013) Microbiological processes in banded iron formation deposition. Sedimentology 60:1733–1754

Potts M (1994) Desiccation tolerance of prokaryotes. Microbiol Rev 58(4):755–805

Prell J, Bourdés A, Kumar S, Ludwig E, Hosie A et al (2010) Role of symbiotic auxotrophy in the *Rhizobium*-legume symbioses. PLoS One 5(11):e13933

Pressel S, Bidartondo MI, Ligrone R, Duckett JG (2010) Fungal symbioses in bryophytes: new insights in the twenty first century. Phytotaxa 9:238–253

Price JS, Whittington PN (2010) Water flow in Sphagnum hummocks: Mesocosm measurements and modelling. J Hydrol 381:333–340

Price CTD, Richards AM, Von Dwingelo JE, Samara HA, Kwaik YA (2014) Amoeba host-*Legionella* synchronization of amino acid auxotrophy and its role in bacterial adaptation and pathogenic evolution. Environ Microbiol 16(2):350–358

Prieto I, Armas C, Pugnaire FI (2012) Water release through plant roots: new insights into its consequences at the plant and ecosystem level. New Phytol 193:830–841

Prince JS (1980) The ecology of *Sargassum pteropleuron* Grunow (Phaeophycaeae, Fucales) in the waters off South Florida. II. Seasonal photosynthesis and respiration of *S. pteropleuron* and comparison of its phenology with that of *S. polyceratium* Montagne. Phycologia 19(3):190–193

Prince JS, O'Neal SW (1979) The ecology of *Sargassum pteropleuron* Grunow (Phaeophycaeae, Fucales) in the waters off South Florida. I. Growth, reproduction and population structures. Phycologia 18(2):109–114

Priscu JC, Fritsen CH, Adams EE, Giovannoni SJ, Paerl HW et al (1998) Perennial Arctic Lake ice: an oasis for life in a Polar Desert. Science 280:2095–2098

Pšenička J, Opluštil S (2013) The epiphytic plants in the fossil record and its example from *in situ* tuff from Pennsylvanian of Radnice Basin (Czech Republic). Bull GeoSci 88(2):401–416

Purkamo L, Kietäväinen R, Miettinen H, Sohlberg E, Kukkonen I et al (2018) Diversity and functionality of archaeal, bacterial and fungal communities in deep Archaean bedrock groundwater. FEMS microb. Ecology 94:fiy116

Pypker TG, Unsworth MH, Bond BJ (2006) The role of epiphytes in rainfall interception by forests in the Pacific northwest. I. Laboratory measurements of water storage. Can J For Res 36:809–818

Qin W, Amin SA, Martens-Habbena W, Walker CB, Urakawa H et al (2014) Marine ammonia-oxidizing archaeal isolates display obligate mixotrophy and wide ecotypic variation. PNAS 111(34):12504–12509

Qiu Y-L, Palmer JD (1999) Phylogeny of early land plants: insights from genes and genomes. Trends Plant Sci 4(1):26–30

Queiroz JA, Quirino ZGM, Lopes AV, Machado LC (2016) Vertebrate mixed pollination system in *Encholirium spectabile*: a bromeliad pollinated by bats, opossum and hummungbirds in a tropical dry forest. J Arid Environ 125:21–30

Quillin KJ (1999) Kinematic scaling of locomotion by hydrostatic animals: ontogeny of peristaltic crawling by the earthworm *Lumbricus terrestris*. J Exp Biol 202:661–674

Rabalais NN, Turner RE, Wiseman WJ (2002) Gulf of Mexico hypoxia, A.K.A. "The Dead Zone". Annu Rev Ecol Syst 33:235–263

Radha AV, Forbes TZ, Killian CE, Gilbert PUPA, Navrostky A (2010) Transformation and crystallization energetics of synthetic and biogenic amorphous calcium carbonate. PNAS 107(38):16438–16443

Raghoebarsing AA, Pol A, van de Pas-Schoonen KT, Smolders AJP, Ettwig KF et al (2006) A microbial consortium couples anaerobic methane oxidation to denitrification. Nature 440:918–921

Ragsdale SW (2018) Stealth reactions driving carbon fixation. Science 359:517–518

Rainey FE, Ray K, Ferreira M, Gatz BZ, Nobre MF et al (2005) Extensive diversity of ionizing-radiation-resistant bacteria recovered from Sonoran Desert soil and description of nine new species of the genus *Deinococcus* obtained from a single soil sample. Applied and Environmental Microb 71(9):5225–5235

Rashid DJ, Chapman SC, Larsson HCE, Organ CL, Bebin A-G et al (2014) From dinosaurs to birds: a tail of evolution. EvoDevo 5/25

Rast A, Heinz S, Nickelsen J (2015) Biogenesis of thylakoid membranes. Biochim Biophys Act 1847:821–830

Raux PS, Gravelle S, Dumais J (2020) Design of a unidirectional water valve in *Tillandsia*. Nat Commun 11:396

Raven JA (2003) Long-distance transport in non-vascular plants. Plant Cell Environ 26:73–85

Raven JA (2009) Contributions of anoxygenic and oxygenic phototrophy and chemolithotrophy to carbon and oxygen fluxes in aquatic environments. Aquat Microb Ecol 56:177–192

Raven JA, Edwards D (2001) Roots: evolutionary origins and biogeochemical significance. J Exp Bot 52:381–401

Raven JA, Giordano M, Beardall J, Maberly SC (2012) Algal evolution in relation to atmospheric CO2: carboxylases, carbon-concentrating mechanisms and carbon oxidation cycles. Philos Trans R Soc B 367:493–507

Raven JA, Beardall J, Larkum AWD, Sánchez-Baracaldo P (2013) Interactions of photosynthesis with genome size and function. Philos Trans R Soc B 368:20120264

Raymond J, Siefert JL, Staples CR, Blankenship RE (2004) The natural history of nitrogen fixation. Mol Biol Evol 21(3):541–554

Read DJ, Perez-Moreno J (2003) Mycorrhizas and nutrient cycling in ecosystems—a journey towards relevance? New Phytol 157:475–492

Reid RP, Visscher PT, Decho AW, Stolz JF, Bebout BM et al (2000) The role of microbes in accretion lamination and early lithification of modern marine microbialites. Nature 406:989–992

Reinfelder JR (2011) Carbon concentrating mechanisms in eukaryotic marine phytoplankton. Ann Rev Mar Sci 3:291–315

Remy W, Taylor TN, Hass H, Kerp H (1994) Four hundred-million-year-old vesicular arbuscular mycorhizae. PNAS 91:11841–11843

Renard E, Vacelet J, Gazave E, Lapébie P, Borchiellini C, Ereskovsky AV (2009) Origin of the neuro-sensory system: new and expected insights from sponges. Integrative Zoology 4:294–308

Renard E, Gazave E, Fierro-Constain L, Schenkelaars Q, Ereskovsky A et al (2013) Porifera (sponges): recent knowledge and new perspectives. eLS. https://doi.org/10.1002/970015902. a0001562.pub2

Rengel Z, Marschner P (2005) Nutrient availability and management in the rhizosphere exploiting genotypic differences. New Phytol 168:305–312

Retallack GJ, Mindszenty A (1994) Well preserved Precambrian paleosols from Northwest Scotland. J Sediment Res A64(2):264–281

Reva O, Tümmler B (2008) Think big-giant genes in bacteria. Environ Microb. https://doi.org/10.1111/j.1462-29200.01500.xx

Reyes-Prieto A, Weber APM, Bhattacharya D (2007) The origin and establishment of the plastid in algae and plants. Annu Rev Genet 41:147–168

Rice G, Stedman K, Snyder J, Wiedenheft B, Willits D et al (2001) Viruses from extreme thermal environments. PNAS 98(23):13341–13345

Richardson ML, Mitchell RF, Reagel PF, Hanks LM (2010) Causes and consequences of cannibalism in noncarnivorous insects. Annu Rev Entomol 55:39–53

Riding R (2006) Microbial carbonate abundance compared with fluctuations in metazoan diversity over geological time. Sediment Geol 185:229–238

Riding R (2008) Abiogenic, microbial and hybrid authigenic carbonate crusts: components of Precambrian stromatolites. Geologia Croatia 61(2–3):73–103

Riding R (2011) Microbialite, stromatolites, and thrombolites. In: Reitner J, Thiel V (eds) Encyclopedia of geobiology. Springer, Heidelberg, pp 635–654

Riding R, Awramik SM, Winsborough BM, Griffin KM, Dill RF (1991) Bahamian giant stromatolites: microbial composition of surface mats. Geol Mag 128(3):227–234

Roberts TJ (2002) The integrated function of muscles and tendons during locomotion. Comp Biochem Physiol A 133:1087–1099

Robinson JJ, Cavanaugh CM (1995) Expression of form I and form II Rubisco in chemoautotrophic symbioses: implications for the interpretation of stable carbon isotope values. Limnol Oceanogr 40(8):1496–1502

Robinson J, Popova EE, Yool A, Srokosz M, Lampit RS, Blundell RJ (2014) How deep is deep enough? Ocean iron fertilization and carbon sequestration in the Southern Ocean. Geophys Res Lett 41:2489–2495

Rojas-Jimenez K, Grossart H-P, Cordes E, Cortés J (2020) Fungal communities in sediments along a depth gradient in the eastern tropical pacific. Front Microbiol 11:575207

Rollefson JB, Levar CE, Bond DE (2009) Identification of genes involved in biofilm formation and respiration via mini-*Himar* transposon mutagenesis of geobacter sulfurreducens. J Bacteriol 191(13):4207–4217

Romer AS (1971) Vergleichende Anatomie der Wirbeltiere Paul Parey, Hamburg

Rosenheim JA, Schreiber SJ (2022) Pathways to the density-dependent expression of cannibalism, and consequences for regulated population dynamics. Ecology 103:e3785

Rosing MT, Bird DK, Sleep NH, Glassley W, Albarede F (2006) The rise of continents—an essay on the geologic consequences of photosynthesis. Palaeogeo Palaeoclim Palaeoecol 232:99–113

Ross-Elliott TJ, Jensen KH, Haaning KS, Wager BM, Knoblauch J et al (2017) Phloem unloading in Arabidopsis roots is convective and regulated by the phloem-pole pericycle. Elife 6:e24125

Rothfels CJ, Windham MD, Grusz AI, Gastony GJ, Pryer KM (2008) Toward a monophyletic *Notholaena* (Pteridaceae): resolving patterns of evolutionary convergence in xerix-adapted ferns. Taxon 57(3):712–724

Rothschild LJ, Mancinelli RL (2001) Life in extreme environments. Nature 409:1092–1101

Rott E, Cantonati M, Füreder L, Pfister P (2006) Benthic algae in high altitude streams of the Alps—a neglected component of the aquatic biota. Hydrobiologia 562:195–216

Roughgarden J, Gilbert SF, Rosenberg E, Zilber-Rosenberg I, Lloyd EA (2018) Holobionts as units of selection and a model of their population dynamics and evolution. Biol Theory 13:44–65

Rousk J, Brookes PC, Bååth E (2010) Investigating the mechanisms for the opposing pH relationships of fungal and bacterial growth in soil. Soil Biol Biochem 42:926–934

Roy S, Pospelova V, Montresor M, Cembella A (eds) (2013) Global ecology and oceanography of harmful algal blooms, GEOHAB core res project: HABS in fjords and coastal embayments. IOC, Paris

Ruiz S, Or D, Schymanski SJ (2015) Soil penetration by earthworms and plant roots-mechanical energetics of bioturbation of compacted soils. PLoS One 10(6):e0128914

Ruiz-Medrano R, Xoconostle-Cázres B, Lucas WJ (2001) The phloem as a conduit for inter-organ communication. Curr Opin Plant Biol 4:202–209

Russell MJ, Hall AJ, Mellersh AR (2003) On the dissipation of thermal and chemical energies on the early earth: the onsets of hydrothermal convection, chemiosmosis, genetically regulated metabolism and oxygenic photosynthesis. In: Ikan R (ed) Natural and laboratory-simulated thermal geochemical processes. Kluwer Academic Publishers, Dordrecht, pp 325–388

Russell MJ, Hall AJ, Martin W (2010) Serpentinization as a source of energy at the origin of life. Geobiology 8:355–371

Ruszala EM, Beerling DJ, Franks PJ, Chater C, Casson SA et al (2011) Land plants acquired active stomatal control early in their evolutionary history. Curr Biol 21:1030–1035

Ryan JF, Chiodin M (2015) Where is my mind? How sponges and placozoans may have lost neural cell types. Philos Trans R Soc B 370:20150059

Sachs JL, Skophammer RG, Regus JU (2011) Evolutionary transitions in bacterial symbiosis. PNAS 108:10800–10807

Sachs JL, Skophammer RG, Bansal N, Stajich JE (2014) Evolutionary origins and diversification of proteobacterial mutualists. Proc R Soc B 281:20132146

Sage RF (2017) A portrait of the C_4 photosynthetic family on the 50th anniversary of its discovery: species number, evolutionary lineages, and Hall of Fame. J Exp Bot 68(2):e11–e28

Sage RF, Sage TL, Kocacinar F (2012) Photorespiration and the evolution of C_4 photosynthesis. Annu Rev Plant Biol 63:19–47

Sahoo SK, Planavsky NJ, Jiang G, Kendal lB, Owens JD et al (2016) Oceanic oxygenation events in the anoxic Ediacaran Ocean. Geobiology. https://doi.org/10.1111/gbi.12182

Sakamoto M, Ruta M, Venditti C (2019) Extreme and rapid bursts of functional adaptations shape bite force in amniotes. Proc R Soc B 286:20181932

Saladié P, Huguet R, Rodríguez-Hidalgo A, Cáceres I, Esteban-Nadal M et al (2012) Intergroup cannibalism in the European early Pleistocene: the range expansion and imbalance of power hypotheses. J Hum Evol 63(5):682–695

Saltzman W, Ahmed S, Fahimi A, Wittwer DJ, Wegner FH (2006) Social suppression of female reproductive maturation and infanticidal behaviour in cooperatively breeding Mongolian gerbils. Horm Behav 49:527–537

Sampson TR, Mazmanian SK (2015) Control of brain development, function, and behavior by the microbiome. Cell Host Microbe 17:565–576

Sánchez-Baracaldo P (2015) Origin of marine planktonic cyanobacteria. Sci Rep 5:17418. https://doi.org/10.1039/srep17418

Sanders RW (1991) Mixotrophic protists in marine and freshwater ecosystems. J Protozool 38(1):76–81

Santana RA, Dias-Júnior CQ, Tóta da Silva J, Fuentes JD, Souza do Vale R et al (2018) Air turbulence characteristics at multiple sites in and above the Amazon rainforest canopy. Agric For Meteorol 260–261:41–54

Santra M, Farrell DW, Dill KA (2017) Bacterial proteostasis balances energy and chaperons utilization efficiency. PNAS:E2654–E2661

Sapra R, Bagramyan K, Adams MWW (2003) A simple energy-conserving system: proton reduction coupled to proton translocation. PNAS 100(13):7545–7550

Schauer R, Risgaard-Petersen N, Kjeldsen KU, Bjerg JJT, Jørgensen BB et al (2014) Succession of a cable bacteria and electric currents in marine sediment. ISME J 8:1314–1322

Scheffel A, Gärdes A, Grünberg K, Wanner G, Schüler D (2008) The major magnetosome proteins MamGFDC are not essential for magnetite biomineralization in *Magnetospirillum gryphiswaldense* but regulate the size of magnetosome crystals. J Bacteriol 190(1):377–386

Schierwater B, Eitel M, Jakob W, Osigus H-J, Hadrys H et al (2009) Concatenated analysis sheds light on early metazoan evolution and fuels a modern "Urmetazoan" hypothesis. PLoS Biol 7(1):e1000020

Schink B (1997) Energetics of syntrophic cooperation in methanogenic degradation. Microb Mol Biol Rev 61(2):262–280

Schirrmeister BE, Gugger M, Donoghue PCJ (2015) Cyanobacteria and the great oxidation event: evidence from genes and fossils. Paleontology 58(5):769–785

Schirrmeister BE, Sanchez-Baracaldo P, Wacey D (2016) Cyanobacterial evolution during the Precambrian. Int J Astrobiol 15(3):187–204

Schlesinger WH (2005) Biogeochemistry. Elsevier, Amsterdam

Schlesinger WH, Jasechko S (2014) Transpiration in the global water cycle. Agric For Meteorol 189–190:115–117

Schlüter U, Weber APM (2016) The road to C_4 photosynthesis: evolution of a complex trait via intermediary states. Plant Cell Physiol 57(5):881–889

Schmid S, Schmid VS, Zillikens A, Harter Marques B, Steiner J (2011) Bimodal pollination system of the bromeliad *Aechmea nudicaulis* involving hummingbirds and bees. Plant Biol 13:41–40

Schoch RR (2012) Character distribution and phylogeny of the dissorophid temnospondyls. Foss Res 15(2):121–137

Schoepp-Cothenet B, van Lis R, Philippot P, Magalon A, Russell MJ, Nitschke E (2012) The ineluctable requirement for the trans-iron elements molybdenum and/or tungsten in the origin of life. Sci Rep 2:263

Schoepp-Cothenet B, van Lis R, Atteia A, Baymann F, Capoviez L et al (2013) On the universal core of bioenergetics. Biochim Biophys Acta 1827:79–93

Schönheit P, Buckel W, Martin WF (2016) On the origin of heterotrophy. Trends Microbiol 24(1):12–25

Schopf JW (2006) Fossil evidence of Archaean life. Phil Trans E Soc 361:869–885

Schopf JW, Kudryavtsev AB, Czaja AD, Tripathi AB (2007) Evidence of Archean life: stromatolites and microfossils. Precambrian Res 158:141–155

Schulz HN (2002) *Thiomargarita namibiensis*: Giant microbe holding its breath. ASM News 68(3):122–127

Schulz A (2015) Diffusion or bulk flow: how plasmodemata facilitate pre-phloem transport of assimilates. J Plant Ras 128:49–61

Schulz HN, Brinkhoff T, Ferdelman TG, Hernández Meriné M, Teske A, Jørgensen BB (1999) Dense populations of a giant sulfur bacterium in Namibian shelf sediments. Science 284(5413):493–495

Schut GJ, Zadvornyy O, Wu C-H, Peters JW, Boyd ES et al (2016) The role of geochemistry and energetics in the evolution of modern respiratory complexes from a proton-reducing ancestor. Biochim Biophys Acta 1857:958–970

Schwartzman DW, Volk T (1989) Biotic enhancement of weathering and the habitability of earth. Science 340:457–460

Scott RL, Cable WL, Hultine KR (2008) The ecohydrologic significance of hydraulic redistribution in a semiarid savannah. Water Resour Res 44:W02440

Sczesnak A, Segata N, Qin X, Gevers D, Petrosino JF et al (2011) The genome of Th17 cell-inducing segmented filamentous bacteria reveals extensive auxotrophy and adaptations to the intestinal environment. Cell Host Microbe 10:260–272

Seilacher A, Buatois LA, Mángano MG (2005) Trace fossils in the Ediacaran Cambrian transition: behavioral diversification, ecological turnover and environmental shift. Palaeogeo Palaeoclim Palaeoecol 227:323–356

Selosse MA, Le Tacon F (1998) The land flora: a phototroph-fungus partnership? TREE 13(1):15–20

Selosse M-A, Roy M (2008) Green plants that feed on fungi: facts and questions about mixotrophy. Trends Plant Sci 14(2):64–70

Sephton MA (2002) Organic compounds in carbonaceous meteorites. Nat Prod Rep 19:292–311

Sereno PC, Beck AL, Duthell DB, Larsson HCE, Lyon GH et al (1999) Cretaceous sauropods from the Sahara and the Uneven rate of skeletal evolution among dinosaurs. Science 286:1342–1347

Shankman P (1969) Le Rôti et le Builli: Lévi-Strauss' theory of cannibalism. Amer Anthrop 71:54–69

Shanks RMQ, Davra VR, Romanowski EG, Brothers KM, Stella NA et al (2013) An eye to a kill: using predatory bacteria to control gram-negative pathogens associated with ocular infections. PLoS One 8(6):e66723

Shen L, Zhao Q, Wu X, Li X, Li Q, Wang Y (2016) Interspecies electron transfer in syntrophic methanogenic consortia: from cultures to bioreactors. Renew Sustain Energy Rev 54:1358–1357

Shi L, Dong H, Reguera G, Beyenal H, Lu A et al (2016) Extracellular electron transfer mechanisms between microorganisms and minerals. Nat Rev Microbiol 14:651–662

Shock EL, Holland ME (2004) Geochemical energy sources that support the subsurface biosphere. The subseafloor biosphere at mid-ocean ridges, geophys monograph series, vol 144, pp 153–165

Shortlidge EE, Rosenstiel TN, Eppley SM (2012) Tolerance to environmental desiccation in moss sperm. New Phytol 194:741–750

Sigel EM, Windham MD, Huiet L, Yatskievych G, Pryer KM (2011) Species relationships and Farina evolution in the Cheilanthoid Fern Genus *Argyrochosma* (Pteridaceae). Syst Bot 36(3):554–564

Silberfeld T, Leigh JW, Verbruggen H, Cruaud C, de Reviers B, Rousseau F (2010) A multi-locus time-calibrated phylogeny of the brown algae (Heterokonta, Ochrophyta. Phaeophyceae): investigating the evolutionary nature of the "brown algal crown radiation". Mol Phylogenet Evol 56:659–674

Silvera K, Neubig KM, Whitten WM, Williams NH, Winter K, Cushman JC (2010) Evolution along the crassulacean acid metabolism continuum. Funct Plant Biol 37:995–1010

Simões BF, Sampaio F, Douglas RH, Kodandaramaiah U, Casewell NR et al (2016) Visual pigments, ocular filters and the evolution of Snake vision. Mol Biol Evol 33(10):2483–2495

Simpson EL, Heness E, Bumby A, Eriksson PG, Eriksson KA et al (2013) Evidence for 2.0 Ga continental microbial mats in a paleodessert setting. Precambrian Res 237:36–50

Singh OV, Gabani P (2011) Extremophiles: radiation resistance microbial reserves and therapeutic implications. Appl Microb 110:851–861

Singh R, Ray P, Das A, Sharma M (2010) Enhanced production of exopolysaccharide matrix and biofilm by a menadione-auxotrophic *Staphylococcus aureus* small-colony variant. J Med Microb 59:521–527

Six C, Finkel ZV, Irwin AJ, Campbell DA (2007) Light variability illuminates niche-partitioning among marine picocyanobacteria. PLoS One 12:e1341

Sjöholm J, Bergstrand J, Nilsson T, Sachi R, von Ballmoos C et al (2017) The lateral distance between a proton pump and ATP synthase determines the ATP-synthesis rate. Sci Rep 7:2926

Smetacek V (2012) Making sense of ocean biota: how evolution and biodiversity of land organisms differ from that of the plankton. J Biosci 37(4):589–607

Smith CL, Varoqueaux F, Kittelmann M, Azzam RN, Cooper B et al (2014) Novel cell types, neuro-secretory cells, and body plan of the early-diverging metazoan *Trichoplax adhaerens*. Curr Biol 24:1565–1572

Smits MM, Herrmann AM, Duane M, Duckworth OW, Bonneville S et al (2009) The fungal-mineral interface: challenges and considerations of micro-analytical developments. Fungal Biol Rev 23:122–131

Snider RM, Strycharz-Glaven SM, Tsoi SD, Erickson JS, Tender LM (2012) Long-range electron transport in *Geobacter sulfurreducens* biofilms is redox gradient-driven. PNAS. https://doi.org/10.1073/pnas.1209829109

Sojo V, Pomiankowski A, Lane N (2014) A bioenergetic basis for membrane divergence in archaea and bacteria. PLoS Biol 12(8):e1001926

Sojo-Martínez V (2016) Membrane bioenergetics at major transitions in evolution. Dissertations, University College London

Solymosi K (2012) Plastid structure, diversification and interconversions I. Algae. Curr Chemical Biology 6:167–186

Solymosi K, Keresztes Á (2012) Plastid structure, diversification and interconversions II. Land plants. Curr Chem Biol 6:187–204

Soo RM, Hemp J, Parks DH, Fischer WW, Hugenholtz P (2017) On the origins of oxygenic photo-synthesis and aerobic respiration in cyanobacteria. Science 355:1436–1440

Sousa F, Shavit-Grievink L, Allen JF, Martin WF (2012) Chlorophyll biosynthesis gene evolution indicates photosystem gene duplication, not photosystem merger, at the origin of oxygenic pho-tosynthesis. Genome Biol 5(1):200–216

Sousa FL, Nelson-Sathi S, Martin WF (2016) One step beyond a ribosome: the ancient anaerobic core. Biochim Biophys Acta 1857:1027–1038

Spanghero GM, Albuquerque C, Fernandes TL, Hernandez AJ, Mady CEK (2018) Exergy analysis of the musculoskeletal system efficiency during aerobic and anaerobic activities. Entropy 20:119

Sperling EA, Peterson KJ, Laflamme M (2011) Rangeomorphs, *Thectardis* (Profera?) and dissolved organic carbon in the Ediacaran oceans. Geobiology 9:24–33

Sperry JS (2003) Evolution of water transport and xylem structure. Int J Plant Sci 164:S115–S127

Spicer R, Groover A (2010) Evolution of development of vascular cambia and secondary growth. New Phytol 186:577–592

Spudich JL, Sineshchekov OA, Govorunova EG (2014) Mechanism divergence in microbial rhodop-sins. Biochim Biophys Acta 1837:546–552

Stairs CW, Eme L, Brown MW, Mutsaers C, Susko E et al (2014) A SUF Fe-S cluster biogen-esis system in the mitochondrion-related organelles of the anaerobic protist *Pygsula*. Curr Biol 24:1176–1186

Stanley SM, Hardie LA (1998) Secular oscillations in the carbonate mineralogy of reef-building and sediment-producing organisms driven by tectonically forced shifts in seawater chemistry. Paleogeogr Paleoclimatol Paleoecol 144:3–19

Stark LR, Greenwood JL, Brinda JC, Olivier MJ (2013) The desert moss *Pterygoneurum lamellatum* (Pottiaceae) exhibits and inducible ecological strategy of desiccation tolerance: effects of rate of drying on shoot damage and regeneration. Am J Bot 100(8):1522–1531

Stein LV, Klotz MG (2016) The nitrogen cycle. Curr Biol 26:R83–R101

Stein WE, Mannolini F, Hernick LVA, Landing E, Berry CM (2007) Giant cladoxylopsid trees resolve the enigma of the Earth's earliest forest stumps at Gilboa. Nature 446:904–907

Steinmetz PRH, Kraus JEM, Larroux C, Hammel JU, Amon A et al (2012) Independent evolution of striated muscles in cnidarians and bilaterians. Nature 487:231–234

Steneck RS, Graham MH, Bourque BJ, Corbett D, Erlandson JM et al (2002) Kelp forest ecosys-tems: biodiversity, resilience and future. Environ Conserv 29(4):436–459

Sterner RW, Elser JJ (2002) Ecological stoichiometry. Princeton University Press, Princeton

Stervander M, Ryan PG, Melo M, Hansson B (2019) The origin of the world's smallest flight-less bird, the Inaccessible Island rail *Atlantisia rogersi* (Aves: Rallidae). Mol Phylogenet Evol 130:92–98

Stevens T (1997) Lithoautotrophy in the subsurface. FEMS Microbiol Rev 20:327–337

Stiger V, Horiguchi T, Yoshida T, Coleman AW, Masuda M (2003) Phylogenetic relationships within the genus *Sargassum* (Fucales, Phaeophyceae), inferred from ITS-2 nrDNA, with an emphasis on the taxonomic subdivision of the genus. Phycol Res 51:1–10

Still CJ, Berry JA, Collatz GJ, DeFries RS (2003) Global distribution of C_3 and C_4 vegetation: carbon cycle implications. Global Biochem Cycles 17(1):1006

Stiller JW (2007) Plastid endosymbiosis, genome evolution and the origin of green plants. Trends Plant Sci 12(9):391–396

Stiller JW, Schreiber J, Yue J, Guo H, Ding Q, Huang J (2014) The evolution of photosynthesis in chromist algae through serial endosymbioses. Nat Commun. https://doi.org/10.1038/ncomms6764

Stilling RM, Bordenstein SR, Dinan TG, Cryan JF (2014) Friends with social benefits: host-microbe interactions as a driver of brain evolution and development? Front Cell Infect Microbiol 4:147

Stoimenova M, Igamberdiev AU, Gupta KJ, Hill RD (2007) Nitrite-driven anaerobic ATP synthesis in barley and rice root mitochondria. Planta 226:465–474

Stomp M, Huisman J, Stal LJ, Matthijs HCP (2007) Colorful niches of phototrophic microorganisms shaped by vibrations of the water molecule. ISME J 1:271–282

Stramma L, Johnson GC, Sprintall J, Mohrholz V (2008) Expanding oxygen-minimum zones in the tropical oceans. Science 320:655–658

Strnad H, Lapidus A, Paces J, Ulbrich P, Vlcek C et al (2010) Complete genome sequence of the photosynthetic purple nonsulfur bacterium *Rhodobacter capsulatus* SB 1003. J Bacteriol 192(13):3545–3546

Strömberg CAE (2011) Evolution of grasses and grassland ecosystems. Annu Rev Earth Planet Sci 39:517–544

Stroock AD, Pagay VV, Zwieniecki MA, Holbrook NM (2014) The physicochemical hydrodynamics of vascular plants. Annu Rev Fluid Mech 46:615–642

Strullu-Derrien C, Kenrick P, Pressel S, Duckett JG, Rioult J-P, Strullu D-G (2014) Fungal associations in *Horneophyton ligneri* from the Rhynie Chert (c. 407 million year old) closely resemble those in extant lower land plants: novel insights into ancestral plant-fungus symbioses. New Phytol 203:964–979

Strullu-Derrien C, Wawrzyniak Z, Goral T, Kenrick P (2015) Fungal colonization of the rooting system of the early land plant *Asteroxylon machiei* from the 407-Myr-old Rhynie Chert (Scotland, UK). Bot J Linn Soc 170:201–213

Stüeken EE, Catling DC, Buick R (2012) Contributions to late Archaean sulphur cycling by life on land. Nat GeoSci. https://doi.org/10.1038/NGEO1585

Stüeken EE, Buick R, Guy BM, Koehler MC (2015) Isotopic evidence for biological nitrogen fixation by molybdenum-nitrogenase from 3.2 Gyr. Nature 520:666–669

Su Y-G, Zhao X, Li A-X, Li X-R, Huang G (2011) Nitrogen fixation in biological soil crusts from the Tengger desert, northern China. Eur J Soil Biol 47:182–187

Subramaniam A, Yager PL, Carpenter EJ, Mahaffey C, Björkman K et al (2008) Amazon River enhances diazotrophy and carbon sequestration in the tropical North Atlantic Ocean. PNAS 105:10460–10465

Summers ZM, Fogarty HE, Leang C, Franks AE, Malvankar NS, Lovley DR (2010) Direct exchange of electrons within aggregates of an evolved syntrophic coculture of anaerobic bacteria. Science 330:1413–1415

Sun SX, Wirtz D (2006) Mechanics of enveloped virus entry into host cells. Biophys J Biophys Lett. https://doi.org/10.1529/biophysj.105.074203

Suosaari EP, Reid RP, Playford PE, Foster JS, Stolz JF et al (2016) New multi-scale perspectives on the stromatolites of Shark Bay, Western Australia. Sci Rep 6:20557

Tabita FR, Satagopan S, Hanson TE, Kreel NE, Scott SS (2008) Distinct form I, II, III, and IV Rubisco proteins from the three kingdoms of life provide clues about Rubisco evolution and structure/function relationships. J Exp Bot 59(7):1515–1524

Tablante NL, Vaillancourt J-P, Martin SW, Shoukri M, Estevez I (2000) Spatial distribution of cannibalism mortalities in commercial laying hens. Poult Sci 79:705–708

Takahata Y (1985) Adult male chimpanzees kill and eat a male newborn infant: newly observed intragroup infanticide and cannibalism in Mahale National Park, Tanzania. Folia Primatol 44:161–170

Tan I, Storelvmo T, Zelinka MD (2016) Observational constraints on mixed-phase clouds imply higher climate sensitivity. Science 352:224–227

Tang Y, Yang X, Cao M, Baskin CC, Baskin JM (2011) Buttress trees elevate soil heterogeneity and regulate seedling diversity in a tropical rainforest. Plant and Soil 338:301–309

Tarhan LG, Planavsky NJ, Laumer CE, Stolz JF, Reid RP (2013) Microbial mat controls on infaunal abundance and diversity in modern microbialites. Geobiology 11:485–497

Tashiro T, Ishida A, Hori M, Igisu M, Koike M et al (2017) Early trace of life from 3.95 Ga sedimentary rocks in Labrador, Canada. Nature 549:516–518

Taylor TW, Taylor EL, Krings M (2009) Paleobotany. Academic Press, Burlington

Taylor PD, Lombardi C, Cocito S (2015) Biomineralization in bryozoans: present, past and future. Biol Rev 90(4):1118–1150

Tedersoo L, Pellet P, Kõljalg U, Selosse M-A (2007) Parallel evolutionary paths to mycoheterotrophy in understorey Ericaceae and Orchidaceae: ecological evidence for mixotrophy in Pyroleae. Oecologia 151:206–217

Teeling H, Fuchs BM, Becher D, Klockow C, Gardebrecht A et al (2012) Substrate-controlled succession of marine bacterioplankton populations induced by a phytoplankton bloom. Science 336:608–611

Teleanu DM, Chircov C, Grumezescu AM, Volceanov A, Teleanu RI (2018) Impact of nanoparticles on brain health: an up to date overview. J Clin Med 7:490

Tellez PH, Woods CL, Formel S, Van Bael SA (2020) Relationships between Foliar Fungal Endophyte Communities and Ecophysiological Traits of CAM and C_3 epiphytic bromeliads in a neotropical rainforest. Diversity 12:378

Testo W, Sundue M (2016) A 4000-species dataset provides new insight into the evolution of ferns. Mol Phylogenet Evol 105:200–211

Thomas DN, Dieckmann GS (2002) Antarctic Sea ice—a habitat for extremophiles. Science 295:641–644

Thomazo C, Papineau D (2013) Biogeochemical cycling of nitrogen on the early earth. Elements 9:345–351

Tice MM, Lowe DR (2004) Photosynthetic microbial mats in the 3,416-Myr-Old Ocean. Nature 431:549–552

Tice MM, Lowe DR (2006) Hydrogen-based carbon fixation in the earliest known photosynthetic organisms. Geology 34(1):37–40

Ting CS, Rocap G, King J, Chisholm SW (2002) Cyanobacterial photosynthesis in the oceans: the origins and significance of divergent light-harvesting strategies. Trends Microbiol 10(3):134–142

Tingle JL, Garner KL, Astley HV (2024) Functional diversity of snake locomotor behaviors: a review of the biological literature for bioinspiration. Ann N Y Acad Sci. https://doi.org/10.1111/nyas.15109

Title PO, Singhal S, Grundler MC, Costa GC, Paron RA et al (2024) The macroevolutionary singularity of snakes. Science 383:918–923

Tomitani A, Knoll AH, Cavanaugh CM, Ohno T (2006) The evolutionary diversification of cyanobacteria: molecular-phylogenetic and paleontological perspectives. PNAS 103(14):5442–5447

Tomo T, Shinoda T, Chen M, Allakhverdiev SI, Akimoto S (2014) Energy transfer processes in chlorophyll f-containing cyanobacteria using time-resolved fluorescence spectroscopy on intact cells. Biochim Biophys Acta 1837:1484–1489

Torrallardona D, Harris CI, Fuller MF (2003) Pigs' gastrointestinal microflora provide them with essential amino acids. J Nutr 133:1127–1131

Tournaire-Roux C, Sutka M, Javot H, Gout E, Gerbeau P et al (2003) Cytosolic pH regulates root water transport during anoxic stress through gating of aquaporins. Nature 425:393–397

Treude T (2003) Anaerobic oxidation of methane in marine sediments. Dissertations, University of Bremen

Treude T, Orphan V, Knittel K, Gieseke A, House CH, Boetius A (2007) Consumption of methane and CO_2 by methanotrophic microbial Mats from gas seeps of the Anoxic Black Sea. Appl Environ Microbiol 73(7):2271–2283

Trillmich F, Limberger D (1985) Drastic effects of El Niño on Galapagos pinnipeds. Oecologia 67:19–22

Tripp HJ, Schwalbach MS, Meyer MM, Kitner JB, Breaker R, Giovannoni SJ (2009) Unique glycine-activated riboswitch linked to glycine-serine auxotrophy in SAR11. Environ Microbiol 11(1):230–238

Trotsenko YA, Khemelina VN (2005) Aerobic methanotrophic bacteria of cold ecosystems. FEMS Microb Ecology 53:15–26

Trouwborst RE, Johnston A, Koch G, Luther GW, Pierson BK (2007) Biogeochemistry of Fe(II) oxidation in a photosynthetic microbial mat: implications for Precambrian Fe(II) oxidation. Geochim Cosmochim Acta 71:4629–4643

Tweddle JC, Dickie JB, Baskin CC, Baskin JM (2003) Ecological aspects of seed desiccation sensitivity. J Ecol 91:294–304

Tyree MT (1997) The cohesion-tension theory of sap ascent: current controversies. J Exp Biol 48(315):1753–1765

Ueno Y, Yamada K, Yoshida N, Maruyama S, Isozaki Y (2006) Evidence from fluid inclusions for microbial methanogenesis in the early Archaean era. Nature 440(516):519

Ueno Y, Johnson MS, Danielache SO, Eskebjerg C, Pandey A, Yoshida N (2009) Geological sulfur isotopes indicate elevated OCS in the Archean atmosphere, solving faint young sun paradox. PNAS 106(35):14784–14789

Umen JG (2014) Green algae and the origins of multicellularity in the plant kingdom. Cold Spring Harb Perspect Biol 6:a016170

Uriz MJ, Turon X, Becerro MA, Agell G (2003) Siliceous spicules and skeleton frameworks in sponges: origin, diversity, ultrastructural patterns, and biological functions. Microsc Res Tech 62:279–299

Valentine DL (2007) Adaptations to energy stress dictate the ecology and evolution of the Archaea. Nat Rev Microbiol. https://doi.org/10.1038/nrmicro1619

van Bel AJE (1990) Xylem-phloem exchange via the rays: the undervalued route of transport. J Exp Bot 41(227):631–644

van Bel AJE (2003) The phloem, a miracle of ingenuity. Plant Cell Environ 26:125–149

van Bel AJE, Helariutta Y, Thompson GA, Ton JA, Dinant S et al (2013) Phloem: the integrative avenue for resource distribution signalling, and defense. Front Plant Sci 4:471

van de Vossenberg JLCM, Driessen AJM, Konings WN (1998) The essence of being extremophilic: the role of the unique archaeal membrane lipids. Extremophiles 2:163–170 van Gemerden H (1993) Microbial mats: a joint venture. Mar Geol 113:3–25

Van Griethuijsen LL, Trimmer BA (2014) Locomotion in caterpillars. Biol Rev 89(3):656–670

van Kessel MAHJ, Speth DR, Allertsen M, Nielsen PH, Op den Camp HJM et al (2015) Complete nitrification by a single microorganism. Nature 528:555–559

Van Roy P, Daley AC, Briggs DEG (2015) Anomalocaridid trunk limb homology revealed by a giant filter-feeder with paired flaps. Nature 522:77–82

Veizer J (1973) Sedimentation in geological history: recycling vs. evolution or recycling with evolution. Contrib Mineral Petrol 38:261–278

Venturas MD, Sperry JS, Hacke UG (2017) Plant xylem hydraulics: what we understand, current research, and future challenges. J Integr Plant Biol 59(6):356–389

Vincent WF, Gibson JAE, Pienitz R, Villeneuve V (2000) Ice shelf microbial ecosystems in the high Arctic and implications for life on snowball earth. Naturwissenschaften 87:137–141

Visscher PT, Stolz JF (2005) Microbial mats as bioreactors: populations, processes, and products. Palaeogeo Palaeoclim Palaeoecol 219:87–100

Vogel S (2003) Comparative biomechanics. Princeton University Press, Princeton

Vogl K, Tank M, Orf GS, Blankenship RE, Bryant DA (2012) Bacteriochlorophyll f: properties of chlorosomes containing the "forbidden chlorophyll". Front Microbiol 3:298

Voigt CC, Winter Y (1999) Energetic cost of hovering flight in nectar-feeding bats (Phyllostomidae: Glossophaginae) and its scaling in moths, birds and bats. J Comp Physiol B 169:38–38

Volkov AG, Pinnock M-R, Lowe DC, Gay MS, Markin VS (2011) Complete hunting cycle of *Dionaea muscipula*: consecutive steps and their electrical properties. J Plant Physiol 168:109–120

Wacey D, Kilburn MR, Saunders M, Cliff J, Brasier MD (2011) Microfossils of Sulphur-metabolizing cells in 3,4-billion-year-old rocks of Western Australia. Nature GeoSci. https://doi.org/10.1038/NGEO1238

Wächtershäuser G (1990) Evolution of the first metabolic cycles. PNAS 87:200–204

Wächtershäuser G (2006) From volcanic origins of chemoautotrophic life of bacteria, archaea and Eukarya. Phil Trans R Soc 361:1787–1808

Wainwright M (1988) Metabolic diversity of fungi in relation to growth and mineral cycling in soil—a review. Trans Br Mycol Soc 90(2):159–170

Walker JE (2013) The ATP synthase: the understood, the uncertain and the unknown. Biochem Soc Trans 41:1–16

Walker JJ, Pace NR (2007) Phylogenetic composition of Rocky Mountain endolithic microbial ecosystems. Appl Environ Microbiol 73(11):3497–3504

Waloszek D, Repetski JE, Maas A (2006) A new late Cambrian pentastomid and a review of the relationships of this parasitic group. Transcriptions of the Royal Society of Edinburgh. Earth Sci 96:163–176

Wang F, Tian X, Ding Y, Wan X, Tyree MT (2011) A survey of root pressure in 53 Asian species of bamboo. Ann For Sci 68:783–791

Wang J, Chen D, Yan D, Wei H, Xiang L (2012) Evolution from an anoxic to oxic deep ocean during the Ediacaran-Cambrian transition and implications for bioradiation. Chem Geol 306–307:129–138

Wang H, Ho KT, Scheckel KG, Wu F, Cantwell MG et al (2014) Toxicity, bioaccumulation, and biotransformation of Silver nanoparticles in marine organisms. Environ Sci Technol 48:13711–13717

Wang Z, Walker GW, Muir DCG, Nagatani-Yoshida K (2020) Toward a global understanding of chemical pollution: a first comprehensive analysis of national and regional chemical inventories. Environ Sci Technol 54:2575–2584

Wanke D (2011) The ABA-mediated switch between submersed and emersed life-style in aquatic macrophytes. J Plant Res 124:467–475

Wanner G, Vogl K, Overmann J (2008) Ultrastructural characterization of the prokaryotic Symbiosis in '*Chlorochromatium aggregatum*'. J Bacteriol 190(10):3721–3730

Ward LM, Shih PM (2021) Granick revisited: synthesizing evolutionary and ecological evidence for the late origin of bacteriochlorophyll via ghost lineages and horizontal gene transfer. PLoS One 16(1):e0239248

Wassmann M, Moeller R, Reitz G, Rettberg P (2010) Adaptation of *Bacillus subtilis* cells to Archean-like UV climate: relevant hints of microbial evolution to remarkably increased radiation resistance. Astrobiology 10(6):605–615

Watanabe Y, Martini JEJ, Ohmoto H (2000) Geochemical evidence for terrestrial ecosystems 2.6 billion years ago. Nature 408:574–578

Watkins JE Jr, Mack MC, Sinclair TR, Mulkey SS (2007) Ecological and evolutionary consequences of desiccation tolerance in tropical fern gametophytes. New Phytol 176:708–717

Weber TS, Deutsch C (2010) Ocean nutrient ratios governed by plankton biogeography. Nature 467:550–554

Weber KA, Achenbach LA, Coates JD (2006) Microorganisms pumping iron: anaerobic microbial iron oxidation and reduction. Nat Rev Microbiol 4:752–764

Webster M (2007) A Cambrian peak in morphological variation within trilobite species. Science 317:499–502

Wegener G, Krukenberg V, Riedel D, Tegetmeyer HE, Boetius A (2015) Intercellular wiring enables electron transfer between methanotrophic archaea and bacteria. Nature 526:587–590

Weimerskirch H, Bishop C, Jeanniard-du-Dot T, Prudor A, Sachs G (2016) Frigate birds track atmospheric conditions over month-long transoceanic flights. Science 353:74–78

Weiss MC, Sousa FL, Mrnjavac N, Neukirchen S, Roettger M et al (2016) The physiology and habitat of the last common ancestor. Nat Microb. https://doi.org/10.1038/NMICROBIOL.2016.116

Werner JJ, Garcia ML, Perkins SD, Yarasheki KE, Smith SR et al (2014) Microbial community dynamics and stability during an ammonia-induced shift to syntrophic acetate oxidation. Appl Environ Microbiol 80(11):3375–3383

Westall F, de Ronde CEJ, Southam G, Grassineau N, Colas M et al (2006) Implications of a 3.472-3.333 Gyr-old subaerial microbial mat from the Barberton greenstone belt, South Africa for the UV environmental conditions on the early earth. Phil Trans R Soc 361:1857–1875

Westall F, Cavalazzi B, Lemelle L, Marrocchi Y, Rouzaud J-N et al (2011) Earth Planet Sci Lett 310(3–4):468–479

Wester S, Zotz G (2010) Growth and survival of *Tillandsia flexuosa* on electrical cables in Panama. J Trop Ecol 26:123–126

Westphal S, Soll J, Vothknecht UC (2003) Evolution of chloroplast vesicle transport. Plant Cell Physiol 44(2):217–222

Wetzel RG (1975) Limnology. Saunders, Philadelphia

Wheeler EA, Baas P (2019) Wood evolution: Baileyan trends and functional traits in the fossil record. IAWA J 40(3):488–529

Wheeler G, Ishikawa T, Pornsaksit V, Smirnoff N (2015) Evolution of alternative biosynthetic pathways for vitamin C following plastid acquisition in photosynthetic eukaryotes. Elife. https://doi.org/10.7554/eLife.06369

White ID, Mottershead DN, Harrison SJ (1992) Environmental systems. Chapman and Hall, London

Wickett NJ, Mirarab S, Nguyen N, Warnow T, Carpenter E et al (2014) Phylotranscriptomic analysis of the origin and early diversification of land plants. PNA:E4859–E4868

Wienberg C, Titschak J, Freiwald A, Frank N, Lundalv T et al (2018) The giant Mauritanian cold-water coral mound province: oxygen control on coral mound formation. Quat Sci Rev 185:135–152

Williams RJP, Fraústo da Silva JJR (2006) The chemistry of evolution. Elsevier, Amsterdam

Williamson SJ, Cary SC, Williamson KE, Helton RR, Bench SR et al (2008) Lysogenic virus-host interactions predominate at deep-sea diffuse-flow hydrothermal vents. The ISME J 2:1112–1121

Willis KJ, McElwain JC (2002) The evolution of plants. Oxford University Press, Oxford

Winkle S (1997) Geißeln der Menschheit; Kulturgeschichte der Seuchen. Artemis and Winkler, Düsseldorf

Wintermute EH, Silver PA (2010a) Dynamics in the mixed microbial concourse. Genes Dev 24:2603–2614

Wintermute EH, Silver PA (2010b) Emergent cooperation in microbial metabolism. Mol Syst Biol 6:407

Winther JL, Friedman WE (2007) Arbuscular mycorrhizal associations in Lycopodiaceae. New Phytol 177:790–801

Wise DH (2006) Cannibalism, food limitation, intraspecific competition, and the regulation of spider populations. Annu Rev Entomol 51:441–465

Wittenberg JB, Wittenberg BA (2003) Myoglobin function reassessed. The J of Experimental Biology 206:2011–2020

Wolf G (2008) Energy regulation by the skeleton. Nutr Rev 66(4):229–233

Wolf S, Freier E, Gerwert K (2014) A delocalized proton-binding site within a membrane protein. Biophys J 107:174–184

Wolowski M, Freitas L (2015) An overview on pollination of the Neotropical Poales. Rodriguésia 66(2):329–336

Wood JR, Wilmhurst JM, Worthy TH, Holzapfel AS, Cooper A (2012) A lost link between a flightless parrot and a parasitic plant and the potential role of coprolites in conservation paleobiology. Conserv Biol 26(6):1091–1099

Worden AZ, Follows MJ, Giovannoni SJ, Wilken S, Zimmerman AE, Keeling PJ (2015) Rethinking the marine carbon cycle: factoring in the multifarious lifestyles of microbes. Science 347:1257594

Wörheide G, Dohrmann M, Erpenbeck D, Larroux C, Maldonaldo M et al (2012) Deep phylogeny and evolution of sponges (Phylium Porifera). In: Becerro MA, Uriz MJ, Maldonaldo M, Turon X (eds) Advances in marine biology, vol 61. Academic Press, Amsterdam, pp 1–78

Worthy TH, Degrange FJ, Handley WD, Lee MSY (2017) The evolution of giant flightless birds and novel phylogenetic relationships for extinct fowl (Aves, Galloanseres). R Soc Open Sci 4:170975

Wray GA (2015) Molecular clocks and the early evolution of metazoan nervous systems. Philos Trans R Soc B 370:20150046

Wrighton KC, Castelle CJ, Varaljay VA, Satagopan S, Brown CT et al (2016) RubisCO of a nucleoside pathway known from archaea is found in diverse uncultivated phyla in bacteria. ISME J 10:2702–2714

Wudick MM, Luu D-T, Maurel C (2009) A look inside: localization patterns and functions of intracellular plant aquaporins. New Phytol 184:289–302

Wynn-Williams DD, Edwards HGM, Newton EM, Holder JM (2002) Pigmentation as a survival strategy for ancient and modern photosynthetic microbes under high ultraviolet stress on planetary surfaces. Int J Astrobiol 1(1):34–39

Xu H-B, Tsukuda M, Takahara Y, Sato T, Gu J-D, Katayama Y (2018) Lithoautotrophical oxidation of elemental sulphur by fungi including *Fusarium solani* isolated from sandstone Angkor temples. Int Biodeterior Biodegeneration 16:95–102

Yagame T, Orihara T, Selosse M-A, Yamato M, Iwase K (2012) Mixotrophy of *Platanthera minor*, an orchid associated with ectomycorrhiza-forming Ceratonasidiaceae fungi. New Phytol 193:178–187

Yair S, Yaacov D, Susan K, Jurkevitch E (2003) Small eats big: ecology and diversity of *Bdellovibrio* and like organisms, and their dynamics in predator-prey interactions. Agronomic 23:433–439

Yin R, Ulm R (2017) How plants cope with UV-B: from perception to response. Curr Opin Plant Biol 37:42–48

Yoshida T, Ideno A, Suzuki R, Yohda M, Maruyama T (2002) Two kinds of archaeal group II chaperonin subunits with different thermostability in *Thermococcus* strain KS-1. Mol Microb 44(3):761–769

Yu H, Leadbetter JR (2020) Bacterial chemolithoautotrophy via manganese oxidation. Nature 583:453–458

Yurkov VV, Beatty JT (1998) Aerobic anoxygenic phototrophic bacteria. Microb Mol Biol Rev 62(3):695–724

Yus E, Maier T, Michalodimitrakis K, van Noort V, Yamada T et al (2009) Impact of genome reduction on bacterial metabolism and its regulation. Science 326:1263–1268

Zack TI, Claverie T, Patek SN (2009) Elastic storage in the mantis shrimp's fast predatory strike. J Exp Biol 212:4002–4009

Zelezniak A, Andrejev S, Ponomarova O, Mende DR, Bork P, Patil KR (2010) Metabolic dependencies drive species co-occurrence in diverse microbial communities. PNAS 112(20):6449–6454

Zerkle AL, House CH, Brantle SL (2005) Biogeochemical signatures through time as interfered from whole microbial genomes. Am J Sci 305:467–502

Zhang Y, Chen M, Zhaou BB, Jermiin LS, Larkum AWD (2007) Evolution of the inner light-harvesting antenna protein family of cyanobacteria, algae, and plants. J Mol Evol 64:321–331

Zhang P, Allahverdiyeva Y, Eisenhut M, Aro E-M (2009) Flavodiiron proteins in oxygenic photosynthetic organisms: photoprotection of photosystem II by Flv2 and Flv4 in *Synechocystis* sp. PCC 6803. PLoS One 4(4):e5331

Zheng J, Yu K, Zhang J, Wang J, Li C (2015) Modeling of the propulsion hydrodynamics for the water strider on water surface. Procedia Eng 126:280–284

Zimmermann U, Schneider H, Wegner LH, Haase A (2004) Water ascent in tall trees: does evolution of land plants rely on a high metastable state? New Phytol 162:575–615

Zimorski V, Mentel M, Tielens AGM, Martin WF (2019) Energy metabolism in anaerobic eukaryotes and Earth's late oxygenation. Free Radic Biol Med 140:279–294

Zubkov MV (2014) Faster growth of the major prokaryotic versus eukaryotic CO_2 fixers in the oligotrophic ocean. Nat Commun. https://doi.org/10.1038/ncomms4776

Zwieniecki MA, Melcher PJ, Holbrook NM (2001) Hydrogel control of xylem hydraulic resistance in plants. Science 291:1059–1062

Imbalances: Sources of All Change

8.1   General Relationships

Among scientific interpretations of complex systems (Eigen 1971; Nicolis and Prigogine 1987; Peitgen et al. 1992; Peak and Frame 1995; Thurner et al. 2018), the example of bifurcation—represented by a sphere rolling from an inclined plane towards a divergence between two valleys—is probably the most widely known. The slope symbolizes the acting force (gravity), while the branching valleys represent increasing structural self-organization. The drawback in this representation is mediated independence between the sphere and the symbolic "terrain". In real terms, in self-organizing systems movement of the "sphere" also influences the structure-forming framework conditions of the "terrain". Self-change in systems becomes easier to understand using the sandpile analogy for the concept of self-organized criticality (Bak 1996; Solé et al. 1999). Here, a flow of energy—for instance through a child's hand holding a shovel—continuously piles up sand in a heap. The weight of the newly added sand triggers both small and large cascades. Although the general shape of the sand pile remains unaltered, its internal structural conditions are constantly changing. This concept is used, for example, to describe the atmospheric formation of clouds and precipitation (Yano et al. 2012). Because of their static information content, I have deliberately omitted illustrations for the examples provided. Self-organizing structures can be observed both in flowing water and—under windy conditions—in stagnant water. Experimental observation of the dynamics of a sand pile can be conducted anywhere—given a sufficient amount of sand.

On the foundation of concepts of self-organization, a question arises regarding the extent to which similar conditions can be expected for biological evolution in water and on land. Physically, a clear difference is immediately recognizable for the density of water, which is around a thousand times greater than that of air. In systemic terms, however, that

M. Knoflacher, *Relativity of Evolution*, https://doi.org/10.1007/978-3-662-69423-7_8

seemingly simple question cannot be answered clearly. Because of atmospheric events, processes taking place on land are closely interwoven with those occurring in oceans. In terrestrial regions, water can be found in different forms and quantitative dimensions: above ground extending from the smallest dewdrop through temporary channels and streams to large lakes; below ground ranging from soil water through groundwater to water bound in crystal structures. Outside bodies of water above ground—depending on climatic conditions—water can cover land areas in liquid form for a brief period of time or as ice for up to hundreds of thousands of years.

Seen in a long-term evolutionary context, key characteristics can be identified most clearly by comparing oceans and land. Over the long term, expanses of land and oceans are connected through atmospheric transportation of water and particles, as well as by continuous removal of minerals from land through erosion and subsequent deposition in oceans. This influences the chemical composition of mineral substances dissolved in the water. The latter can be found not only in the hydrochemistry of current inland waters but also in the geological history of sediments in early oceans (see also Fig. 8.1) (Löfgren and Boström 1989; Druschel et al. 2004; Morley et al. 2005). Accumulation of substances in oceans is clearly recognizable from the salinity levels caused by evaporation. Even in oceans, concentrations and composition of dissolved substances do not remain constant over the long term, but change under the influence of numerous additional factors (Lowenstein et al. 2001; Stanley 2001).

As regards morphodynamics, geological structures—across all spatial dimensions, from grains of sand to sea basins—form the framework conditions for inland waters and

Fig. 8.1 Metaphor for self-organization in biological evolution, which always moves "uphill" in its environment and is exposed to opposing—unpredictable— abiotic processes

Abiotic direction of processes (entropic)

Biological direction of evolution (negentropic)

oceans. Within these framework conditions, the self-similar structures outlined in Fig. 8.20 are repeated in water bodies. In the context of concentrations of mineral constituents and level of solar radiation—especially with deep bodies of water—this results in selective framework conditions influencing the size of organisms in the "phytoplankton". Analogous selective effects exist for chemoautotrophic organisms in the vicinity of hydrothermal vents. As a result of the cumulative properties of oceans, organic materials are not lost but deposited on the seabed in varying distributions. Depending on renewal rates of the organic material and the conditions of deep waters as part of their ecophysiological capacity, this provides a potential food source that heterotrophic organisms can utilize.

Compared to oceans, the abiotic conditions of land areas offer a far less favourable basis for the evolution of organisms. The latter must cope not only with potential loss of substances through erosion, but also with availability of water, which is often highly variable. Additional stress factors are large bandwidths and rapid changes in ambient temperature, as well as a higher intensity of short-wave radiation from sunlight.

Explanations of biological evolutionary processes using the aforementioned "rolling sphere model" are radically wrong (Balázsi et al. 2011). In terms of energy, evolutionary processes generally run in the opposite direction—uphill. The processes involved are more appropriately compared to climbers on a mountainside attempting to push themselves to the limits of their abilities—against the abiotic forces at work (Fig. 8.1). The basic processes of biological self-organization have often been subsumed under the thermodynamical term "negentropy" (Schrödinger 1943; Riedl 1989), which refers to an increase in internal order or reduction in entropy in biological systems. Quite aside from discussions about the meaning of this term, it does not go far enough, as organisms must not only organize themselves; they are also constantly exposed to the entropy of abiotic processes (see also Fig. 4.16).

Metaphorically speaking—due to their low weight, low energy consumption and extremely high adaptability—the smallest organisms can travel the furthest. Because of their potentially large numbers, losses due to wrong turns play no part. The successful paths that they "discover" can be followed by larger organisms with a more limited tolerance for losses. Technically, this statement does not apply to all life forms. As explained in more detail in Chap. 7, some heterotrophic life forms act energetically in the same direction as abiotic processes and thus increase entropy in biological systems. Collapses—metaphorical "landslides"—of ecological systems can therefore be triggered both by abiotic processes and by effects that organisms exert. Such collapses have recurred several times during Earth's history and have brought about the development of new biotic communities (see also Fig. 5.18) (Brook et al. 2008; Barash 2013; Hoffman 2013).

This in turn allows conclusions to be drawn about the significance of organism size in an evolutionary context. With equally large amounts of free energy and substances, more small organisms capable of reproduction can form than large ones. The lowest limit of a viable unit is defined as a cell. Hence, for energetic reasons, single-celled organisms can form the largest populations. Because of the generally very small dimensions of these organisms—in the range of thousandths of a millimetre—they can also colonize very small-scale habitats. In marine sediments, for example, tens to hundreds of thousands of prokaryotic cells can be found on a single grain of sand. Their total global quantity in marine sediments is estimated at around 1.7×10^{28} (Probandt et al. 2017). Under the heterogeneous conditions of natural geological formations, this further increases the probability of survival in the event of large-scale destructive events such as asteroid impacts or major volcanic eruptions.

The ability to exchange and pass on information is crucial for the persistence of biological evolutionary processes (Fig. 8.2). This applies to both intracellular—for example genetic and epigenetic (see Sect. 5.2)—and extracellular processes (see Chap. 6). However, organisms remain subject to the physical and chemical principles of abiotic processes. Although they are therefore unable to escape those conditions, they can develop independent solutions to maintain their capacity to reproduce.

8.2   Coping with Complexity

8.2.1   Self-Organization in the Trajectory of Biological Evolution

Theoretical treatises on biological evolutionary processes focus on two main topics. One addresses the question of the origin of living organisms, the other addresses evolutionary lineages of life forms and species. However, there is almost no discussion of the question as to why entire organismal systems were able to re-evolve following dramatic collapses (see Fig. 5.18)—usually constituted by other life forms.

In light to the explanations provided in Chaps. 5 to 7, the search for answers cannot be expected to yield adequate results if only multicellular organisms are taken into account. An initial indication of the features of possible processes is provided by Fig. 5.11. Here it can be recognized macroscopically that a rock surface is covered by irregular lichens of different sizes. Lichens are not individual organisms, but communities of different

Fig. 8.2 Core features of biological evolutionary processes

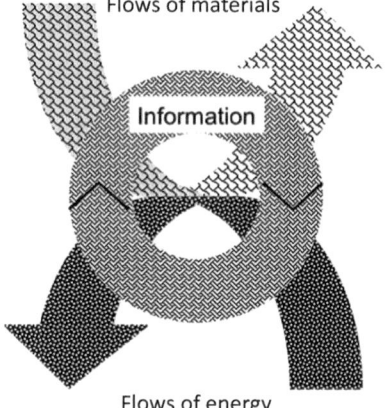

organisms that are perceived macroscopically as a single unit. In each individual community, functionally different groups interact according to similar principles. Under certain conditions, different communities of organisms can disintegrate and re-form in other configurations—in other words, they are metastable.

8.2.2   Integrating Factors in Biological Evolution

At a microscopical level, similar system characteristics can be found in almost all microbiomes (Fig. 5.14) that are able to develop on solid surfaces. In contrast to the constant dynamic-chaotic changes of flow in open water, solid surfaces offer—at least provisionally—conditions for continuously repeated interactions between the same organisms. A central *attractor* for the formation of chemoautotrophic organism communities is the development of additional energetic transformation potential through syntrophy (see Sect. 7.3); *co-operation is energetically determined*. The overall energetic potential of the attractor is determined solely by prevailing environmental conditions and is independent of the physiological characteristics of the organisms involved. However, their composition and physiological characteristics determine *the relevant potential for biological energy conversion.*

In systemic terms, however, major challenges must be overcome. Under the particular conditions of the evolutionary environment, prokaryotic cells are faced with a variety of potential abiotic reaction partners. However, a single cell's own energetic conversion potential is sufficient to utilize the potential of only a few reaction partners. Energetic conversion potential can generally be increased by syntrophic connections of the reaction processes with cells that are complementarily equipped. These effects can be further enhanced by cooperation with phototrophic unicellular organisms. Since both the abiotic conditions and the potential reaction partners present in each case are influenced by the random nature of prior processes, only pragmatic solutions are to be expected in the formation of new syntrophic systems (Fig. 8.3a). Important prerequisites for this are adequate physiological complementarity and overlapping perceptual spaces (Fig. 6.12) of the cells involved. Give the pragmatic nature of formation processes, syntrophic systems can utilize the prevailing overall energetic potentials only incompletely. Unutilized components trigger evolutionary developments both in the organisms and in the composition of organismal communities (see Figs. 5.17 and 7.9).

Integration of D-heterotrophs (see Sect. 7.5) augments recovery of mineral nutrients from dead organic material in organism communities. The associated partial decoupling of fluxes of geochemical material increases the latitude for evolution of phototrophic organisms. Following massive disturbance events, the enormous spatiotemporal diversity of combinations of external influencing factors and functional organism communities enables rapid restoration of biological functionality (Fig. 5.18). The range and magnitude of the changed framework conditions can induce the development of new organismal communities. However, elevated, small-scale diversity increases the probability that less adapted

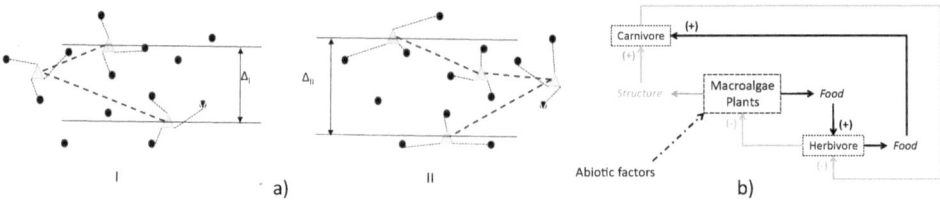

Fig. 8.3 Extremely simplified representations of functional relationships between energy fluxes and interactions in autotrophic and heterotrophic organism communities. (**a**) Energetically induced cooperation of chemo-autotrophic organisms; expansion of the utilization of abiotic energy potentials (symbolized by circles) through increasing cooperation between synthrophic organisms (symbolized by triangles) from state I to II. To illustrate this, differences between the conversion potentials achieved in each case (ΔI and ΔII) are shown; see Fig. 7.11. (**b**) Basic scheme of indirect regulation of energy fluxes (black lines) in auto-/heterotrophic food chains through indirect feedback effects (grey lines). Symbols: (+) promoting, (−) inhibiting effects

communities will be constrained, but not completely displaced. Moreover, high spatial diversity hinders the rapid spread of viruses and thus offers structural protection against epidemic disruptive events.

Despite their hugely diverse capacities for physiological transformation, unicellular organisms can syntrophically exploit the potential for two-dimensional development only on solid surfaces. In aquatic habitats, sediment surfaces in particular are available to them for this purpose. Superimposition of freshly added sediments leads to a loss of transformation potential and ultimately to the demise of the organisms concerned, as can be recognized, for instance, from the layered sequences of stromatolites (see Sect. 7.3).

The potential for transformation in three-dimensional space can be realized only by multicellular organisms, albeit through different systemic relationships. In marine habitats, turnover of sediments by animals (bioturbation) increases production rates of microorganisms and hence in turn availability of food resources for animal species living on them. Bioturbation is intensified by species that retreat into sediments to conceal themselves from predators or potential prey. The increase in microbial conversion rates in deeper waters has no effect on photosynthetic production in surface zones. Indirectly, however, mineral nutrients can be transported into these upper zones by water currents, thus stimulating photosynthetic transformation processes.

Conditions existing on land are different. Here, photosynthetic processes are directly linked to microbial mineral transformation carried out by plants. In many cases, the spatial range of interactions in the subsurface zone is greatly increased by integrated effects of widely distributed fungal hyphae (mycorrhizae). As with microbiomes, D-heterotrophs can reduce loss of mineral nutrients by means of adaptive recycling and hence more effectively decouple photosynthetic energy conversion from geochemical processes. By integrating land plants into microbial processes, terrestrial ecosystems follow the same principles as in microbiome systems. Macroscopically, impacts in vegetation can be recognized both in the varying compositions of plant communities and in trends evident in their modification over time (successions). Scale-independent validity of integrative principles for the development of organism communities—also known as holarchy (Günther and Folke 1993)—can be traced back to the energetic attractors of chemoautotrophic processes.

8.2.3 Evolutionary Processes Betwixt Competition and Cooperation

Phototrophic energy conversion takes place intracellularly and—exclusively for energetic reasons—without requiring co-operation between organisms. Examples of this are provided by unicellular organisms in aquatic habitats. For the reasons mentioned above, land plants are dependent on co-operation with other organisms in the root zone, but photosynthetically active tissues are in competition with other plants. The resulting contradictory requirements generate various different strategies that optimize individual requirements for energy conversion (see Sect. 7.6.1). Accordingly, plant populations are also in metastable states, which manifest themselves in different ways under the influence of abiotic and biotic factors (e.g. climate and fungi or animals, respectively).

In energetic terms, as heterotrophic organisms, animals are entirely dependent on availability of organic biomass. Theoretically, if they all fed directly on autotrophic organisms, animals would have the greatest potential for energetic conversion. In evolutionary terms, the question hence arises as to how such diverse life forms and diets could develop among animals in largely unstructured ocean habitats (Fig. 6.13). Purely mechanistic modelling assumptions (Schwerdtfeger 1975) indicate that populations of a few species could have remained in a dynamically fluctuating state of equilibrium. In reality, however, all basic animal life forms differentiated over a relatively short period of term—in geological terms. The responsible factors are complex, so they can be outlined only in greatly simplified form. One group of factors can be derived from dimensions, light conditions and dynamics of marine habitats. These conditions alone preclude complete dynamic feedback between simple populations of food organisms and animals. This results in numerous and diverse energetic possibilities for the evolution of new life forms. As explained in Chap. 6, for energetic reasons, each organism can only capture a limited range of information (perceptual space) regarding its environment—and available capacities are closely related to its morphological and physiological make-up (Fig. 6.12). For each life form, this influences the scope for strategies of food acquisition and reduction of predation risk. Differing life forms of cnidarians and vertebrates, along with incompletely known variation in bioluminescence (see Sect. 4.3), serve to illustrate possible variations. Structure-forming, sessile animal groups—such as sponges or corals—offer further, small-scale options for diversification of animal life forms.

In many marine shallow water zones, the energetic advantages of phototrophy enable macroalgae (kelp) in particular to form large biogenic structures. Energetically, they are a very suitable potential food for herbivorous animals such as sea urchins. In a purely thermodynamic context, such potential would be rapidly converted into animal biomass and kelp stocks would be eliminated. Such an outcome is prevented by an additional animal species with specific behaviour patterns—carnivorous sea otters. Systemically, the effect comes about through regulation of energy flow (black lines in Fig. 8.3b) by means pf non-energetic factors (grey lines). As explained in Sect. 9.1.6, the metastability of such simple interactions can quickly be lost due to disruptive factors and lead to the above-mentioned collapse of kelp stocks. *The key to understanding the far more complex interactions in terrestrial ecosystems is the central importance of non-energetic factors—such as biogenic structures and functional diversity of life forms—for achieving and maintaining dynamic stability.*

8.3   Evidence for Interactions Between Abiotic Factors and Organisms in the Archaean Eon

Preservation of independent systems is by no means self-evident If allowance is made not only for the obvious imbalance in energy fluxes but also for the often irregular dynamics of abiotic processes. Over the approximately 3.5 billion years that have elapsed since the earliest trace evidence for life (Nutman et al. 2016; Dodd et al. 2017; Rasmussen and Muhling 2023), the shape of Earth's surface, physical and chemical conditions of the atmosphere and oceans, as well as influential extraterrestrial factors, have changed repeatedly. For example, the intensity of solar radiation has increased by around 30 per cent over this period, while geothermal energy flow has decreased by around 70 per cent (Rosing et al. 2006; Stacey and Davis 2008). Cyclical processes, such as changes in Earth's orbit around the sun or oscillation of Earth's axis of rotation, influence solar radiation and climate. Large-scale glaciers, the distribution of continental and oceanic regions and anthropogenic interventions in the global water balance also influence the position of Earth's axis of rotation. Despite the slow but steady increase in daylength due to the tides—which are influenced by the moon's orbit—both the cyclical changes in seasonal solar radiation and tides have become important time indicators (*Zeitgeber*) for organisms (Schad 2001; Zürcher 2001; Boden and Kennaway 2006; Foster and Roenneberg 2008; Mercier et al. 2011; Tessmar-Raible et al. 2011; Müller et al. 2013; Zantke et al. 2013; Seo et al. 2023). Effects of events with irregular time cycles—e.g. changes in radiation intensity or reversals of the terrestrial magnetic field—remain largely obscure (Gradstein et al. 2012; Hanslmeier 2014).

It has been inferred that huge supercontinents on Earth formed and subdivided again four times, accompanied by mountain formation, changes in sea level by several hundred metres, gigantic volcanic eruptions and asteroid impacts (Fig. 8.4) (Stanley 2001; Eyles 2008; Heaman 2008; Keller 2008, Reddy and Evans 2009; Bradley 2011; Nance et al. 2014; Condie et al. 2015; Pirajno and Santosh 2015; Vérard et al. 2015; Tang et al. 2016). The literature provides differing opinions regarding timing and dimensions. The main reason for this is exponentially increasing uncertainty with increasing timespans. For example, an average error of 45° has been estimated for the geographical latitudes of lithospheric

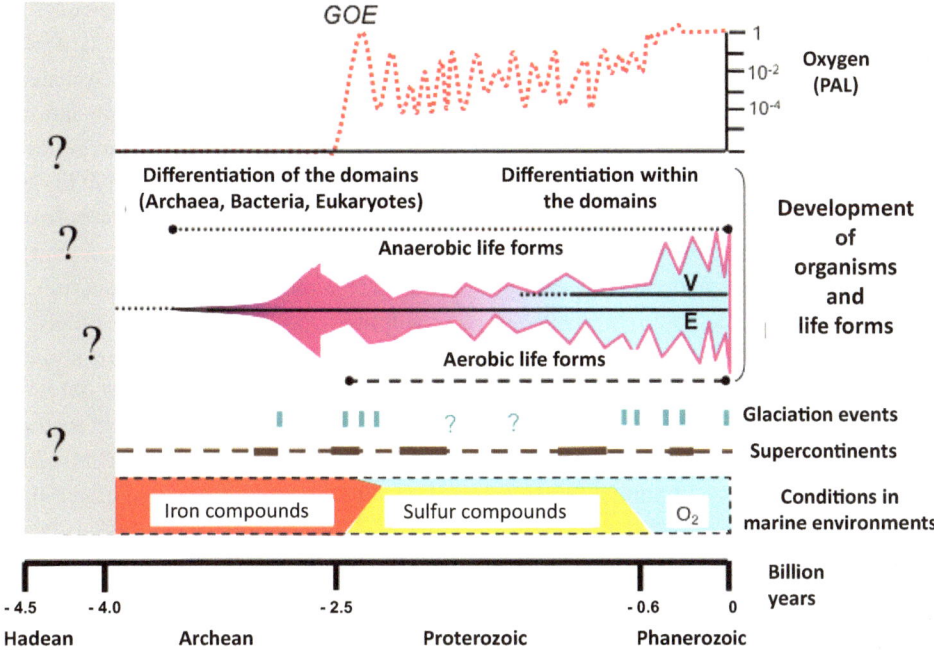

Fig. 8.4 Simplified overview of changes in individual abiotic conditions in interaction with organisms over the course of Earth's history. Explanations: Question marks (?) indicate unsubstantiated developments; E = unicellular life forms; V = multicellular life forms; GOE = Great Oxidation Event; PAL = presumed oxygen concentrations in relation to current size with logarithmic scale; O$_2$ = oxygen in the oceans—presumably only in near-surface areas during the Proterozoic. Presumed timespans for supercontinents are marked by thicker, solid lines. In the band representing organisms, predominantly anaerobic life forms are symbolized in purple and aerobic forms in blue. Curves for development of oxygen concentration and organisms are intended simply to indicate fluctuations and do not represent measured values. Compiled from various sources, see text

plates around 600 million years ago (Scotese 2004). This also explains why geographical positions of the ancient land masses of Gondwana and Avalon are variably specified in published accounts (Landing and McGabhann 2010). Particularly crucial are statements and representations—obtained using powerful computer modelling—based on meagre data. One example of this is the representation of the Paratethys Sea in the region between present-day Europe and the Indian Ocean. In a comparison between model-generated "time series" of the geographical distribution of the mainland and oceans and differentiated palaeontological-geological findings. The former indicates that the Paratethys had already disappeared 15 million years ago, whereas the latter yield evidence that it was still present until the middle Miocene—around 13 million years ago (Harzhauser and Piller 2007; Scotese 2021). This example once again reveals the limits of model-based processing of large datasets (big data) to investigate processes in non-linear systems.

Even climatic developments in the Earth's recent history can be reconstructed only with great uncertainty. In view of the currently melting ice masses in Antarctica and Greenland, the question arises as to when they were formed. The range of estimates is astonishingly wide. For Arctic mainland areas—including Greenland—the beginning of glaciations is estimated to be between 3.5 and 45 million years before present; for the Antarctic around 30 to 35 million years ago (Shackleton et al. 1984; Coxall et al. 2005; Moran et al. 2006; Eldrett et al. 2007; DeConto et al. 2008; Lunt et al. 2008). The great uncertainties become understandable when the limited possibilities for recording the temporal and spatial development of ice sheets and their substantial dynamics are taken into account. As a rule, data are available from indirect indications, such as sediment deposits or certain isotope compositions. Using different modelling assumptions, conclusions can be drawn about the possible extent of the glaciations. This inevitably results in simplifications, as only a few of the large number of different influencing factors can be taken into account in the calculations (Pollard and DeConto 2005; Abe-Ouchi et al. 2013; Gasson et al. 2016). Comparative analyses of publications about the most recent glaciation cycle in geological history indicate globally different temporal patterns and a lack of synchrony in the progression of glaciations (Hughes et al. 2013). Evidence of extreme ice formation—for example, the thickness of the Arctic sea ice of presumably 1000 m around 150,000 years ago—must therefore be seen in the context of regional climatic developments. Isotope analyses suggest that freshwater lakes also formed in the ice cap over the long term (Jakobsson et al. 2016; Geibert et al. 2021).

In a best-case scenario, for periods further back in time sedimentary deposits and chemical traces provide indications of environmental conditions and organic life (Hart 1978; Pavlov et al. 2000; Watanabe et al. 2000; Cockell and Horneck 2001; Mojzsis et al. 2001; Nisbet and Sleep 2001; Wilde et al. 2001; Miyakawa et al. 2002; Brock et al. 2003; Kaufman and Xiao 2003; Xiao 2004; Papineau et al. 2005, 2007; Tian et al. 2005; Valley et al. 2005; Ohmoto et al. 2006; Canfield et al. 2007; Martin et al. 2007; Battistuzzi and Hedges 2009; Rasmussen et al. 2008; Scott et al. 2008; Papineau 2010; Gradstein et al. 2012; Mulkidjanian et al. 2012; Beraldi-Campesi 2013; Reinhard et al. 2013; Schirrmeister et al. 2013; Wordsworth and Pierrehumbert 2013; Xiao 2013; Butterfield 2014; Grosch and McLoughlin 2014; Retallack 2014; Bell et al. 2015; Sperling et al. 2015; Fischer et al. 2016; Knoll et al. 2016; Zhu et al. 2016; Bengtson et al. 2017; Knoll and Nowak 2017). Radioactive isotopes can be used to determine the age of individual minerals on the basis of their characteristic decay rate (half-life) or through traces of radioactive decay in minerals (fission track method) (Bahlburg and Breitkreuz 2012). Zircon grains are predominantly used to determine age in the early phases of Earth's history—due to their global distribution and relatively precise time determination from isotopes of the uranium decay series that they contain. Nevertheless, even age determinations with radioactive decay series show relatively large uncertainties due to varying influences in individual time phases. In geology, different methods are therefore combined to determine the age of individual sites (Gradstein et al. 2012).

Simply stated, biological activities can be determined from fossils, traces of biological activities (ichnofossils), biomarkers (hydrocarbons) or the effects of metabolism via ratios of characteristic isotopes (Ziegler 1992; Brocks 2005; Michener and Lajtha 2007). It should be borne in mind that combinations of life traces found in each case are influenced by a wide variety of factors—raising issues that are addressed by taphonomy as a subdiscipline of palaeontology. Organisms can die for a variety of reasons, for instance because of illness, old age or external influences. After death, other influences can modify the organism and its position, for example partial consumption by other organisms, decomposition processes or transportation by water currents. The soft parts of organisms particularly affected by these agents can lose important morphological features and lead to misinterpretation of fossils (Sansom et al. 2010). Simplifying to the extreme, there is a greater probability of preservation of dead organisms in oceans than on mountain ridges. Inevitably, terrestrial traces of life—especially from early periods of Earth's history—are found much more rarely than marine remains. The imbalance in fossil finds is exacerbated when large-scale glaciations additionally destroy terrestrial traces that were once present. The exact opposite is true for the probability of finding preserved traces of life. In terrestrial areas, the closer the traces are to the surface, the greater the probability of discovery. Ideally, high rock faces yield extended time series of fossil deposits. After initial deposition, microbial, geochemical or geomechanical processes can further alter composition and form (Behrensmeyer et al. 2000; Konhauser et al. 2003; Brasier et al. 2006, 2011; Konhauser 2007; Brasier 2014). Recognizing, recording and interpreting all these factors requires both forensic instinct and careful methodological procedures, making it all the more difficult to record the dynamic living conditions of the organisms prior to their death.

In rare cases, entire ecosystems are destroyed in concert by major events—for example volcanic eruptions, landslides or submarine mudslides—leaving behind tiny "still images" of the composition and lifestyles of organisms (Allmon and Bottjer 2001; Selden and Nudds 2012). Far more frequently, individual "puzzle pieces" of information from various individual finds have to be assembled to form images that are often somewhat blurred (Knoll et al. 2012). In extreme cases, depictions of important events—such as the occurrence of asteroid impacts in the early phase of Earth's history—are based solely on analogies and modelling assumptions (Sleep et al. 1989; Ryder et al. 2000; Abramov and Mojzsis 2009).

Just how uncertain such statements can be is revealed by examination of evidence for eukaryotes from around 2.7 billion years ago based on chemical compounds (biomarkers) from a site in Australia. In fact, a new investigation —carefully avoiding contamination during drilling of the test horizons and precisely analyzing geothermal conditions in the area of the deposits —was unable to confirm the original findings (French et al. 2015). The sensitivity of biomarkers to external influences was also determined when their validity was tested using time-dated fossil sponges from museum collections (Gold et al. 2016). Based on the findings of both studies, the author of the last-mentioned publication re-evaluated the chronological classification of data derived from biomarkers (Gold et al. 2017).

In view of the often sparse fossil evidence and associated uncertainties, the idea of a molecular clock—independent of environmental influences—which emerged along with advances in molecular biology was received enthusiastically (Zuckerkandl and Pauling 1965; Kimura 1991). Technological improvements in molecular biological methods and the resulting surge in decoding of molecular structures and the genetic code of various organisms led to accumulation of vast amounts of data. In the scientific field of bioinformatics, increasingly sophisticated methods of structural analyses have been developed to manage and evaluate the data amassed (Merkl 2015). Based on general assumptions of evolutionary change, this facilitated rapid identification of relationships between organisms. Increasing knowledge led to reorganization of organisms into the three domains of archaeans, bacteria and eukaryotes (Fig. 2.1). Exchange of genes between different species (horizontal gene transfer) and adaptability of the cellular regulatory system to different environmental conditions also became detectable (Osawa et al. 1992, Hobert 2008; Wagner 2008; Huang 2011). This increasingly raised questions about the classical picture of a phylogenetic tree of organisms and rigid delimitation of species (Mann 1999; Taylor et al. 2000; Dagan and Martin 2006; Konstantinidis et al. 2006; Naor et al. 2012; Ryan et al. 2012). However, contradictions in chronological inferences based on different molecular clocks also emerged, leading to ongoing discussions regarding evolution of the three organismal domains—especially during the early periods of Earth's history (Bonen and Vogel 2001; Lynch and Conery 2003; Huang 2004; Kumar 2005; Peterson and Butterfield 2005; Ho and Larson 2006; Pulquério and Nichols 2007; Lee et al. 2009a, b; Baele et al. 2012; Wolf and Koonin 2013). For this reason, the current state of discussion about the origins of the three main biological domains during the early history of Earth is not reviewed here. Instead, in the overview portrayed in Fig. 8.4, only a simplified distinction is drawn between unicellular (E) and multicellular organisms (V).

The simple fact is that we are reaching the limits of our capacity to recognize and record developments in the temporal and spatial dimensions of evolutionary processes. As the limits loom closer, emerging images become increasingly blurred because low probabilities of preservation and detection rapidly decreases the number of clues discovered (Fig. 8.5). It should also be borne in mind that very short-term events can often lead to lasting changes in morphological characteristics of animals. Changes in beak shape of Darwin's finches on the Galapagos island of Daphne Major provide an impressive example of this. Evolutionary change was triggered by changes in vegetation as a result of a dry period lasting almost 2 years associated with an El Niño event (Grant and Grant 2014; Lamichhaney et al. 2015). We must accept this factual evidence and proceed with caution when interpreting individual results.

The factors that led to multiple morphological changes in biotic communities throughout Earth's history remain subject to heated debate in scientific circles. Originally used primarily for demarcation between geological epochs, temporal changes in the fossil record—following the publication of data analyses by J.J. Sepkoski in 1984 (White 2002)—have been increasingly analyzed and discussed in search of possible causes. Depending on the authors concerned, the term "mass extinction" has been used to label

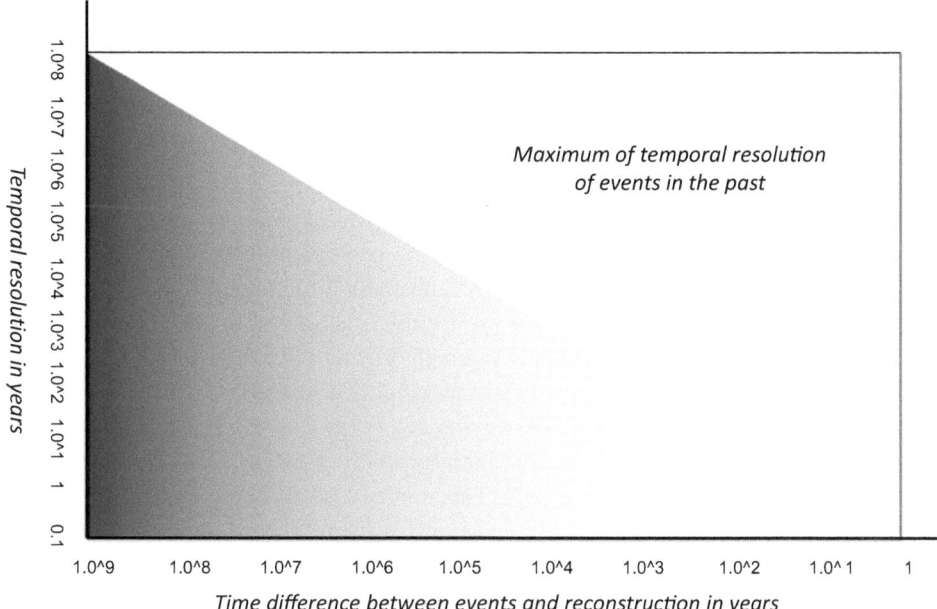

Temporal resolution in years

Maximum of temporal resolution of events in the past

Time difference between events and reconstruction in years

Fig. 8.5 Simplified schematic representation of the boundaries for temporal differentiation in geological time periods. For recent human history, the boundaries have vanished because of the development of written and scientific records

between five and eleven individual events (Benton 1995; Stanley 2001; Bambach et al. 2004; Glikson 2005; Keller 2005; Twitchett 2006; McElwain and Punyasena 2007; Hull et al. 2011; McGhee 2013; Roopnarine and Angielczyk 2015; Hochuli et al. 2016). Loss of species at the end of the Permian period around 251 million years ago is scientifically categorized as a particularly severe event. In the realm of public discussion, on the other hand, the most recent event to date—at the end of the Cretaceous period around 65 million years ago, when most dinosaur species became extinct—has become the centre of interest. Despite numerous publications reviewing those two extinction events, none of the various causal theories—notably invoking an asteroid impact or massive volcanic eruption—has been able to yield conclusive evidence that a single trigger was responsible for the major loss of species that occurred (D'Hondt et al. 1998; Arens and West 2008; Berner 2002; Peters and Foote 2002; White 2002; Wilf et al. 2003; D'Hondt 2005; Coxall et al. 2006; Knoll et al. 2007; Sahney and Benton 2008; Friedman 2009; Payne et al. 2010; Hull et al. 2011; Longrich et al. 2011; Sobolev et al. 2011; Alegret et al. 2012; Mitchell et al. 2012; Fröbisch 2013; Clarkson et al. 2015; Renne et al. 2015; Roopnarine and Angielczyk 2015; Schoene et al. 2015; Petersen et al. 2016). The main reasons for this—seen globally—are sporadic availability of fossil evidence difficulties arising with temporal assignment of potential trigger events (Signor and Lipps 1982; Smith et al. 2001). Regardless of this, the question arises as to why organisms were able to survive over the long term.

8.4   Bundling of Biological Functions

To answer the foregoing question, we must once again refer back to Fig. 4.16 and the explanations presented in Sect. 8.2. In view of the extreme imbalance between abiotic and biotic energy fluxes, the phenomenon of ecosystems could not have developed without coordinated alignment of physiological functions among organisms.

In principle, it can be assumed that any species can evolve only if it interacts with a sufficient diversity of other organisms according to ecological-functional principles.

As is revealed by explanations of energy conversion processes (Chap. 7), this general requirement is fulfilled only by certain life forms. From a purely energetic perspective, heterotrophic organisms—which use living cells as fuel for energy conversion—do not fulfil the requirement *a priori*. Their fundamental effects augment entropy in ecosystems and can hence be viewed in the same way as abiotic processes (Fig. 8.1). This demonstrates that ecosystems can be destroyed both by abiotic factors ("from the outside") and by biological factors ("from the inside"). This makes it all the more fascinating to search for answers to the question regarding processes that permit the origin and long-term revitalization of ecosystems.

8.4.1   Origin and Integration of New Life Forms

As already mentioned, oxygenic photosynthesis in cyanobacteria had already evolved prior to oxygen enrichment of the atmosphere (Fig. 8.6: GOE—Great Oxidation Event), although the time taken for that evolutionary development remains to be determined. Anaerobic microorganisms were already confronted by free oxygen—at least in microscopic realms—and were obliged to adapt physiologically or perish. The extent and intensity of the associated challenges are incomparable with any other event in evolutionary history (see Sect. 8.4.2). In addition to oxygenic photosynthesis, cyanobacteria are characterized by their capacity to bind atmospheric nitrogen while remaining independent of other organisms. Combination of these properties was an important prerequisite for migration into new, near-surface water zones.

In this habitat, microorganisms faced the challenge of finding areas in which sufficient radiation energy was available for photosynthesis, while ensuring that the dose of short-wave (ultraviolet) radiation received did not destroy their biomolecules. In addition, requisite mineral nutrients—for instance phosphorus, nitrate or iron—were usually available only in small quantities in the surface zones of oceans. Spatial and temporal increases in concentration depend on large-scale currents, which microorganisms are obliged to endure. In an initial phase, migration into this zone enabled cyanobacteria to develop a wide variety of life forms without pressure from predatory heterotrophic organisms—yet under the constant influence of viruses. The first forms living in open water were probably tiny unicellular picoplanktonic species (cells smaller than 4 µm in diameter), with

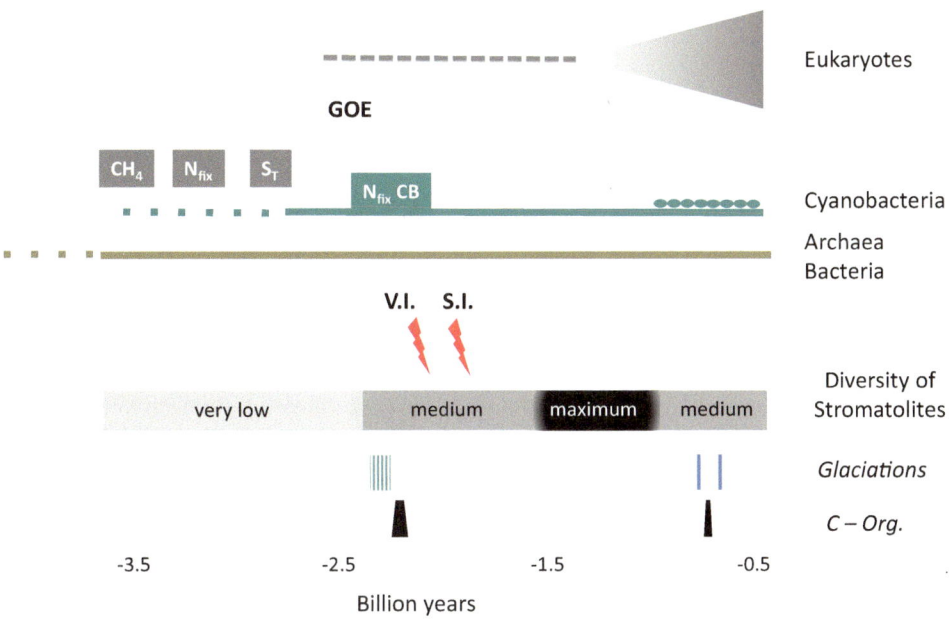

Fig. 8.6 Development of stromatolite diversity in the Precambrian in the context of environmental factors. Atmospheric oxygen content, presumed ice ages, accumulation of organic carbon in marine sediments (C-Org.) and asteroid impacts with global effects (V.I. = Vredefort impact; S.I. = Sudbury impact) are all shown schematically

cyclically alternating photosynthesis and nitrogen fixation (Stockner 1988; Urbach et al. 1998; Ernst et al. 2003; Schirrmeister et al. 2013). Geochemical evidence for aerobic nitrogen fixation in marine sediments also fits within this time interval (Zerkle et al. 2017). Initial indications of multicellularity are found during the period in which the first increase in atmospheric oxygen content (GOE) occurred. Morphological differentiation into photosynthetic and nitrogen-fixing cells has been dated to the period between 1.5 and 1 billion years ago. Subsequently, cyanobacteria underwent further differentiation, taking the form of branching in the originally linear cell clusters (Schirrmeister et al. 2016).

Explanations: Letters in grey boxes indicate the earliest periods of presumed biological influences on methane formation (CH_4), nitrogen fixation (N_{fix}) and sulphur inputs from land surfaces (S_T), the green box (N_{fix} CB) indicates presumed development of the capacity of cyanobacteria to fix nitrogen. Compiled from various sources (see text).

Because of the change in chemical reaction conditions, highly specialized chemoautotrophic organisms abruptly had no suitable sources for energy conversion and were therefore unable to maintain their vital functions. With increasing abundance of free oxygen, highly efficient anaerobic syntrophic communities lost most of their original habitats or had to adapt to the new conditions by incorporating new life forms. The most favourable conditions for this were once again provided by biofilms or microbial mats in which the oxygen concentration drops to zero beyond a few millimetres beneath the surface—which remains within the dimensions of microbial interactions. At the same time, the concentration of substances that are important for anaerobic processes—such as sulphur—rapidly increased towards the lower boundary of biofilms (Des Marais 2003). It is therefore not surprising that the diversity of stromatolites increased sharply at the onset of the first major oxygen enrichment (Fig. 8.6).

It is, however, striking that the evolution of stromatolite diversity remained stagnant over almost a billion years, in combination with one or more decreases in atmospheric oxygen content. At least two groups of factors are thought to be responsible for this (Yamaguchi 2002; Le Guerroue 2006; Cohen et al. 2009; Grotzinger et al. 2011; Och and Shields-Zhou 2012; Young 2013; Kump 2014; Retallack 2014; Lee et al. 2015; Brocks et al. 2017; Cui et al. 2017). One group of factors is caused by photosynthetic organisms themselves: Because of the rapid increase in photosynthesis, carbon dioxide from the atmosphere was increasingly converted into biomass. As heterotrophic life forms were still largely lacking, conversion of organic carbon back into carbon dioxide was absent. As a consequence of loss of carbon dioxide from the atmosphere, its thermal influence (greenhouse effect) increasingly declined—resulting in multiple large-scale glaciations (first symbol for ice ages in Fig. 8.6) (Kopp et al. 2005). Most of the organic material was deposited in marine sediments, in the so-called Lomagundi isotope anomaly (C-Org., first symbol in Fig. 8.6) (Condie 2005; Blank and Sánchez-Baracaldo 2010; Planavsky et al. 2012; Schirrmeister et al. 2013; Sánchez-Baracaldo et al. 2014). However, habitats of the organisms responsible remain unknown. As a result of the development of autonomous nitrogen fixation (N_{fix} CB in Fig. 8.6), cyanobacteria might have already colonized open oceans. However, the observed biomass could also have been formed in nutrient-rich shallow seawater zones and in freshwater areas lying inland.

Independently of this, two massive asteroid impacts probably wiped out most organisms. The crater of the first impact—the Vredefort impact—was located in present-day South Africa, while the crater of the second—the Sudbury impact—was located in Canada (Young 2013). Relatively rapid succession of these two major events—first impact around 2.02 billion years ago; second impact about 1.85 billion years ago—could have contributed significantly to the extensive hiatus in diversification of stromatolites.

In connection with these events, the issue of eukaryote evolution arises. Our view of the evolutionary history of eukaryotes is often obscured by focusing on the evolution of animals. The enormous diversity of the domain of eukaryotes is easily forgotten (Lane and Archibald 2008). Our perspective is greatly constrained by focusing on aerobic life forms without considering the diversity of anaerobic life forms among eukaryotes (Müller et al. 2012). With such a presupposition, the period under investigation is inevitably limited to approximately the last 0.7 to 1 billion years of Earth's history (Stanley 2001; Douzery et al. 2004; Berney and Pawlowski 2006; Budd 2008; Parfrey et al. 2011; Dohrmann and Wöhrheide 2017). A different picture emerges when existing fossil evidence for eukaryotic characteristics is analysed without *a priori* restrictions. In an initial approach, characteristics of the cell nucleus and mitochondria are of importance, whereas an expanded approach can include uncertain chemical biomarkers. This approach extends the period under consideration to at least 2.5 billion years, back to the first chemical traces of possible eukaryotic processes (Anbar and Knoll 2002; Knoll et al. 2006; Sergeev et al. 2007, 2010; Xiao 2013; Dacks et al. 2016; Porter 2020). From around 2.1 billion years ago, evidence begins to accumulate for early—still debated—eukaryotic life forms, such as "gabonionts": *Grypania*, *Valvimorpha* or *Tappania* (Butterfield 2000; Rasmussen et al. 2002; Yoon et al. 2004; Porter 2006; El Albani et al. 2010; Moczydłowska et al. 2011; De Clerck et al. 2012; Xiao 2013; Butterfield 2014; Planavsky et al. 2014; Zhu et al. 2016). Fossils of multicellular algae from Chinese sites provide evidence of increasing morphological differentiation between the Palaeoproterozoic around 1.6 billion years ago and the Cambrian around 500 million years ago (indicated by the superimposed triangles in Fig. 8.6) (Xiao and Dong 2006; Shi et al. 2014; Ye et al. 2015; Wang et al. 2017).

In view of the functional relationships already described, a question arises as to the plausibility of an early emergence of the first eukaryotic cells. In a systemic context, this question concerns not only the origin of life forms, but also their survival in a world dominated by prokaryotes. The associated scenario is based on confrontation of anaerobic chemoautotrophic communities with the appearance of free oxygen due to the onset of photosynthesis. Under the resulting energetic constraints, at least one life form of α-proteobacteria must have successfully switched to aerobic metabolism. The first eukaryotic life forms with mitochondria could have developed subsequently under the conditions of syntrophic communities in biofilms. In an extension of the scenario, these proto-eukaryotes would have gained an advantage if they were able to obtain raw materials for other cells in the community from the remains of doubtless numerous dead cells (Brasier et al. 2010).

The first fossil evidence of eukaryotic cells has been found in deposits around 1.7 billion years old (Knoll et al. 2006; Sergeev et al. 2010; Cohen and Kodner 2022). In accordance with the classification used here, they must have been D-heterotrophs—osmotrophic fungus-like organisms in the broadest sense. This scenario is not entirely implausible if the very broad ecophysiological range of fungi—exceeding that of all other eukaryotic life forms—is taken into account. Fungi are found in all areas of life—from deep-sea hydrothermal vents to extremely arid or polar habitats—in various combinations with other organisms. Their life forms range from anaerobic unicellular endosymbionts or parasites to the largest known aerobic organisms (Carlile et al. 2001; Naranjo-Ortiz and Gabaldón 2019). Although there is no fossil or genetic evidence to support this assumption, various authors have already surmised that fungi were the original eukaryotes (Martin et al. 2003). Given that fungi are not known to ingest large food particles directly (phagocytosis), evolution into V-heterotrophs must have taken place at a later stage. For successful further development, this mode of nutrition requires sufficiently large biomasses of autotrophic organisms—a prerequisite that became energetically feasible only with the spread of photosynthesis.

However, an alternative scenario attributes the evolution of mitochondria to confrontations with V-heterotrophic bacteria (predatory bacteria). Different bacteria use various methods to utilize other bacteria as energy sources facultatively or continuously (Jurkevitch

2007; Nandy et al. 2007; Dori-Bachash et al. 2008; Berleman and Kirby 2009; Fenton et al. 2010; Buchholz et al. 2012; Keane and Berleman 2016; Korp et al. 2016; Welsh et al. 2016). In one method, the predatory bacterium nests between the double membranes of its victim and exploits its components for its own reproduction and growth of daughter cells. In this scenario, a similar process—albeit with a surviving host cell—can be inferred for the evolution of mitochondria. Supposition of a relatively early emergence of predatory life forms in bacteria does not seem implausible in view of the presumably substantial biomass of anaerobic photoferrotrophic bacteria and cyanobacteria (Guerrero et al. 1986; Rashidan and Bird 2001; Jurkevitch and Davidov 2007; Davidov and Jurkevitch 2009). However, a diametrically opposed evolutionary sequence is also conceivable, in which colonization by parasitic bacteria ultimately leads to ectosymbiosis (Smetacek 2012). In principle, however, the question of the origin of the first eukaryotes has yet to be answered (Embley and Martin 2006).

Phagocytosis also facilitates the evolution of chloroplasts through incorporation of cyanobacteria and hence the origin of photosynthetic eukaryotes (Raven et al. 2009). As mentioned in Sect. 7.4, incorporation of photosynthetic organisms has taken place repeatedly. Moreover, many examples of endosymbiosis of heterotrophic eukaryotes with both photoautotrophic and chemoautotrophic organisms also be found today. In photic marine zones, to expand their energy supply, various heterotrophic eukaryotes such as marine gastropods—for example the sap-sucking slug *Elysia viridis*—either take up chloroplasts from algae or—for instance like the unicellular slipper animalcule *Paramecium*—entire algae (Gorelova et al. 2009; Maeda et al. 2012; Cruz et al. 2013; Hill 2014; Baumgartner et al. 2015; Lowe et al. 2016). In submarine hydrothermal vents, chemoautotrophic bacteria are found in ectosymbioses—for example involving the Yeti crab *Kiwa*—and in endosymbioses with far-reaching morphological adaptations—for example the tubeworm *Riftia* (Girguis et al. 2002; Minic and Hervé 2004; Dubilier et al. 2008; Goffredi 2010).

Emergence of the first symbiosis with cyanobacteria is thought to have occurred around 1.6 billion years ago (Yoon et al. 2004), i.e. after biological systems had recovered from the disruptive effects by ice ages and asteroid impacts (Fig. 8.6). Recovery of biological systems was possibly also accompanied by differentiation and spread of V-heterotrophic life forms, but this has not been clearly shown. Possible influences of heterotrophic organisms on the decline in diversity of stromatolites and the resulting accumulation of phosphorus in marine sediments, which began around a billion years ago, have also been discussed (Bengtson 2002; Riding 2006; Reinhard et al. 2017). It is still unclear whether morphological changes in cyanobacteria were triggered by the increasing feeding pressure of protists (Pernthaler 2005; Schirrmeister et al. 2015).

8.4.2   The "Good" Catastrophe

The above lists fail to consider what is probably the greatest catastrophe in evolutionary history, which—by all accounts—was triggered by organisms. The trivializing English term "Great Oxidation Event" (GOE) barely gives any hint to how close organisms affected by it came to complete extinction (Fig. 8.4). The primary trigger for the event, in an originally anaerobic world, was the release of gaseous oxygen (O_2) by abiotic processes and cyanobacteria. Through the interaction of biological processes, the greatest chemical imbalance—in thermodynamic terms—in the entire solar system developed in Earth's ocean-atmosphere system. In an initial phase about 2.5 billion years ago, the atmospheric oxygen content rose to values between 0.001 and 1 per cent of the current concentration (Kump 2008; Farquhar et al. 2010; Krissansen-Totton et al. 2016; Zhang et al. 2016; Zerkle et al. 2017; Ward and Shih 2021). According to geochemical analyses—using cerium content as an indicator—the increase was not linear, but involved multiple rises and falls in oxygen concentrations. Around 550 million years ago—during the Cryogenian, after the end of the Gaskiers glaciation period—the oxygen concentration at first decreased and then rose steeply for around 20 million years. Concentrations then fell again to reach values similar to those existing before the Cryogenian. Only towards the end of the Ordovician—with the initial differentiation of terrestrial vegetation—did the atmospheric oxygen concentration rise to attain values close to those of the present day. In parallel, anoxic zones in the oceans are also likely to have declined significantly (Holland 2002; Bekker et al. 2004; Canfield 2005; Catling and Claire 2005; Kopp et al. 2005; Anbar et al. 2007; Kump and Barley 2007; Lyons and Reinhard 2009; Crowe et al. 2013; Lyons et al. 2014; Mills et al. 2014; Wallace et al. 2017; Tostevin and Mills 2020).

Those outlined trends in global oxygen concentrations are founded on model-based hypotheses. The quantitative basis for the hypotheses usually consists of a small number of measured chemical indicators, from which conclusions can be drawn—with limited accuracy —only about relevant oxygen concentrations (Boyce et al. 2023).

A focus on methodological aspects makes it easy to forget the global systemic changes brought about by the evolution of oxidative photosynthesis, which continue to the present day. With their origination, the release of highly reactive oxygen molecules—energetically dependent on solar radiation—commenced. This fundamentally changed chemical reaction conditions (Chap. 7) throughout oxygen-enriched aquatic and terrestrial zones. It is accordingly difficult to analyze the overall effects on chemical processes. Attribution of mineral genesis solely to triggering formation factors leads to different outcomes, depending on the categorization principles selected. For instance, one particular working group initially estimated the proportion of biogenically formed minerals at around 75 per cent, but reduced this to around 35 per cent in a more recent publication (Hazen and Papineau 2012; Kendall et al. 2012; Hazen et al. 2014; Hazen and Morrison 2022).

The problems outlined above for reconstructing both oxygen concentrations and mineral formation are caused by complex systemic interactions that can be outlined only in extremely simplified form (Fig. 8.7). Even under anoxic conditions, chemical substances interact in non-linear spatially heterogeneous processes, primarily as a result of geological and atmospheric dynamics. Furthermore, global sea level fluctuations due to alternating exposure and submersion of continental areas influence the availability of chemical reactants (Fig. 8.7a).

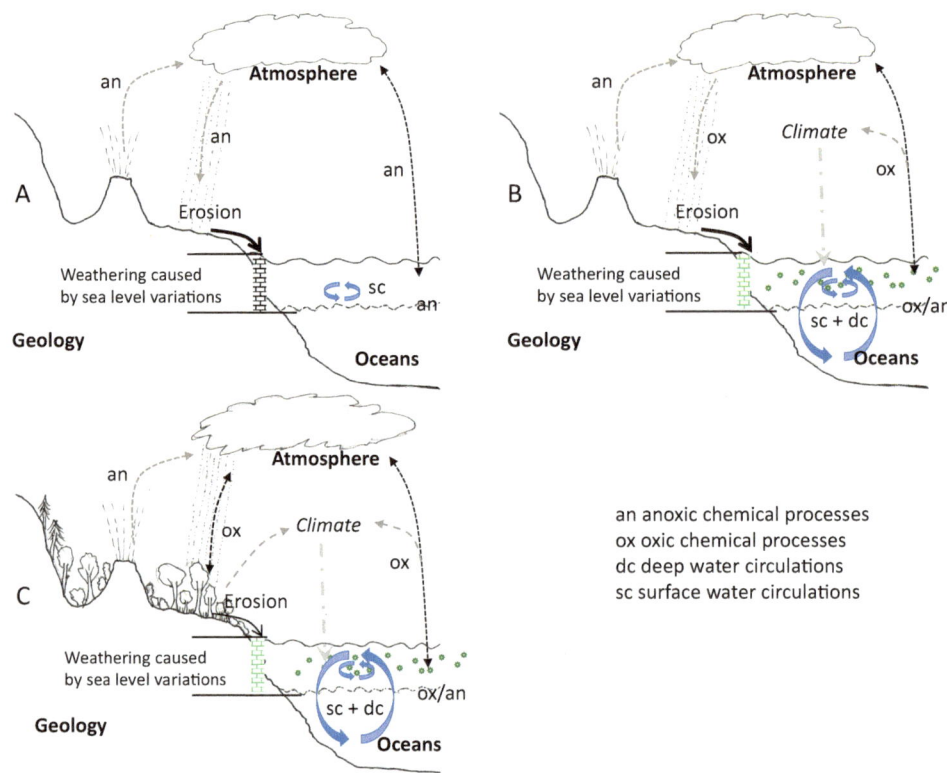

Fig. 8.7 Simplified overview of the effects of oxic photosynthesis on originally anoxic global chemical environmental processes (**a**), following the evolution of aquatic cyanobacteria (**b**), and multicellular land plants (**c**). Further explanations in the text

With the advent of oxidative photosynthesis, a new reagent with largely independent release dynamics is propagated in these systems. Energetic dynamics and global spatial distribution of photosynthesis are primarily dominated by planetary factors and solar radiation, but moderated by terrestrial abiotic factors. Aquatic cyanobacteria can enrich the oxygen content in surface zones of water bodies and in the atmosphere, but not in deeper water zones. It is possible that the juvenile planktonic life forms of many marine animal groups are an outcome of limited oxygen availability at substantial ocean depths during the Cambrian (Strathmann 2020). Theoretically, ocean currents in the oxygen-rich surface zones might have made it possible to overcome the anoxic barriers when spreading into new habitats. Climate fluctuations induced by changes in atmospheric-chemical conditions can trigger deep currents in oceans and thus contribute to oxygen enrichment in all deep zones. Biofilms with photosynthetically active microorganisms ultimately also modify chemical processes associated with sea level fluctuations (Fig. 8.7b).

With the development of multicellular land plants, photosynthetic organisms expanded into new orders of magnitude (Fig. 8.7c). This is associated with various functional changes:

- Minerals from terrestrial weathering processes are incorporated directly on site by plants and—during the existence of the plant population—enter oceans to a lesser extent through erosion;
- Transpiration of plants—which is linked to photosynthesis—creates new dynamic components in the global water cycle in terrestrial regions
- During the initial phases, the magnitude of oxygen release exceeds the potential of microbial respiration, thereby intensifying the effects on atmospheric chemical processes and climate dynamics mentioned in connection with Fig. 8.7b. These effects are attenuated only by amplification of microbial processes by heterotrophic multicellular organisms, fungi and animals acting effectively at these scales.

The deeper zones of marine water bodies remained oxygen-free for almost two billion years because of an apparent lack of intermixing. Presumably, oxygenic and anoxygenic photosynthesis processes could take place in parallel in the transition zone between chemically different water bodies (chemocline) (Johnston et al. 2009; Kendall et al. 2010; Hamilton et al. 2016). In the present day, examples of this can be found in the Black Sea and the Cariaco Basin off Venezuela (Manske et al. 2005; Lin et al. 2006), as well as in lakes that are incompletely mixed (meromictic) (Boehrer et al. 2009; Halm et al. 2009; Pouliot et al. 2009; Walter et al. 2014). However, it would be incorrect to draw direct conclusions about communities and processes more than 2.5 billion years ago from observations of present conditions. Anaerobic phototrophic organisms first had to adapt to limited light conditions in deeper water bodies and presumably to changed chemical conditions in transition zones. Organisms had to overcome a wide variety of challenges in areas with limited—but nevertheless present—oxygen enrichment in the surface zones of the oceans and on land. A core internal challenge was the control of toxic reactive oxygen compounds in metabolic machinery (see Sect. 7.4) (Aguirre et al. 2005; Saul and Schwartz 2007; Thannickal 2009; Shaughnessy et al. 2014). This was overcome by development of globins in bacteria and their spread through all groups of organisms by means of horizontal gene transfer and subsequent genetic modifications (see Sect. 5.2.1) (Vinogradov et al. 2005; Vinogradov et al. 2007; Vinogradov and Moens 2008; Storz et al. 2013; Keppner et al. 2020). At the same time, various environmental factors changed—also with the cooperation of organisms. One of those factors was loss of easily convertible nitrogen compounds in the form of ammonium (NH_4^+), replaced by gaseous nitrogen (N_2), which can be converted only by a few prokaryotes (Berman-Frank, et al. 2003; Zerkle et al. 2017).

In connection with the initial increase in atmospheric oxygen content (GOE), multiple large-scale glaciations—summarized under the term Huronian Ice Age—occurred around 2.3 billion years ago (Hoffman and Schrag 2002; Bekker and Kaufman 2007; Hoffman 2013; Tang and Chen 2013; Bekker 2014). Subsequently, increasing biological productivity accompanied by changes in geobiochemical cycles may have led to exceptionally large deposits of organic carbon (Lomagundi anomaly) in the period between 2.3 and 2 billion years ago (Maheshwari et al. 2010; Planavsky et al. 2012; Canfield et al. 2013; Bachan and Kump 2015). Anoxic conditions persisted in deep marine water bodies—with transient fluctuations in their extent (Scott et al. 2014)—for almost two billion years. Near-surface marine zones and terrestrial areas were exposed to at least two major asteroid impacts during this period (Fig. 8.6). Moreover, atmospheric oxygen content and availability of biologically important trace elements progressively declined (Large et al. 2014).

Around a billion years ago, dramatic changes in living conditions began, for which the triggering factors are still subject to debate. Despite the decreasing oxygen content of the atmosphere, the Cryogenian age began around 717 million years ago with the large-scale Sturtian glaciation—popularized by the name "Snowball Earth" (Hoffman et al. 1998; Kirschvink et al. 2000; Donnadieu et al. 2004; Boyle et al. 2007; Eyles 2008; Maruyama and Santosh 2008; Gaucher et al. 2009; Macdonald et al. 2010; Tziperman et al. 2011; Pu et al. 2022; Wei-Haas 2023). In contrast to popular images of an Earth frozen under ice, that epoch—which lasted around 60 million years—is likely to have been characterized by regionally varying and repeatedly occurring glaciations with intervening thaw periods (Peltier et al. 2007; Eyles 2008; Le Heron 2012). After only 20 million years, the Marinoan glaciation, lasting around 5 million years, began (Hoffman and Li 2009; Rooney et al. 2015). By definition, this was followed by the Ediacaran—the final geological period of the Neoproterozoic.

It should be noted here that abiotic processes do not occur in isolation, but instead overlap one another with differing rates of change. For example, before that period, daylength varied between about 16 and 22 h. Daylength increased from around 19 h to the current 24 h only over the last 1.5 billion years or so, under the influence of superimposed forces. The best known among these are the—decelerating—tidal forces brought about by the moon's orbit, the magnitude of which is influenced by the global quantity of fluid water. Without the effects of atmospheric thermal tides—which act in the opposite direction due to solar radiation—daylength would now be 65 h (Williams 2000; Wu et al. 2023). This also makes it easier to understand that many drastic changes continued into the Cambrian (Gaucher et al. 2009) and that irreversible abiotic processes will ultimately exert a significant influence on the overall status of Earth until the end of the solar system.

Various factors have been discussed as potential triggers for large-scale glaciations, such as a decline in the climate-relevant methane content of the atmosphere, changes in proportions of land and sea brought about by dynamics of the lithospheric plates, shielding from solar radiation, volcanic eruptions, changes in Earth's magnetic field, or the extraterrestrial cloud of a supernova (Meert and van der Voo 1994; Donnadieu et al. 2004; Liang et al. 2006; Eyles 2008; Maruyama and Santosh 2008; Stern et al. 2008; Ueno et al. 2009; Passchier and Erikanure 2010; Tziperman et al. 2011; Abbott et al. 2013; Kataoka et al. 2014; Lenton et al. 2014; McKenzie et al. 2014; Macdonald and Wordsworth 2017). Triggers for smaller glaciations in the Ediacaran—for example the Gaskiers glaciation

(Hoffman and Li 2009; Hebert et al. 2010; Retallack 2013; Pu et al. 2016)—have also been a topic for discussion. Fossilized littoral deposits provide evidence of large ice-free marine zones for this period (Qi et al. 2023)—at least at low latitudes. One of the triggers for increasing intermixture of oceanic waters was probably drifting of lithospheric plates connected with disintegration of the supercontinent Rodinia, which started around 850 million years ago (Canfield et al. 2008; Stern 2008; McKenzie et al. 2014; Kendall et al. 2015). Climate changes are also possible sources of energy for deep ocean currents. For energetic reasons alone, biological driving forces—for example, the pumping power of marine sponges (Tatzel et al. 2017)—can be ruled out.

Barely discussed, but far more important for the entire course of subsequent biological evolution in aquatic habitats, was the spread of an oxygen-rich zone in the uppermost layer of water bodies. Regardless of the size and extent of anoxic zones in deeper zones, in oceans this created a globally encompassing arena for potential dispersal of animals. Simultaneously, surface currents—generated mainly by winds—provided opportunities for transportation, albeit stochastically, for unicellular and microscopically small multicellular organisms along with eggs and immature stages of large animals.

The subsequent Cambrian period is characterized by both high concentrations of atmospheric CO_2 and elevated temperatures (Wallmann 2004; Royer 2006; Bao et al. 2008). Nonetheless, traces of small, localized glaciations can also be found during this period (Averbuch et al. 2005; Eyles 2008; Ghienne et al. 2014; Goddéris et al. 2014). Broadly coinciding with an increase in the atmospheric oxygen content to over 30 per cent—presumably due to the evolution of terrestrial vegetation—the last major global epoch of glaciations began between 350 and 250 million years ago (Crowell 1983; Berner 1998; Isbell et al. 2003; Ghienne et al. 2014).

On the other hand, the most recent glaciations in Earth's history affected the northern or southern hemisphere and are attributed primarily to cyclical changes in Earth's orbit and oscillation of Earth's axis (Milanković cycles) (Hays et al. 1976; Berger 1977; DeConto et al. 2008). When glaciations peaked, the sea level was around 100 m lower than at present because of huge ice masses bound to the mainland, while the ice layer in the Arctic Ocean was around 1000 m thick (Jakobsson et al. 2016; Hasenclever et al. 2017). As the glaciations receded, the rise in sea level was not steady but at times abrupt. As a result, tropical reefs were rapidly flooded and repeatedly died off (Blanchon and Shaw 1995).

For biological evolutionary processes, diverse regional changes brought both opportunities and risks of complete annihilation (Maloof et al. 2010; Swanson-Hysell et al. 2010; Konhauser et al. 2011; Kump 2012; Retallack 2016). Possible indications of the demise of numerous organisms are the—from a geological time perspective—short but quantitatively vast accumulations of organic carbon in the Shuram Formation (Oman) about 600 million years ago (Le Guerroue 2006; Grotzinger et al. 2011; Retallack et al. 2014; Lee et al. 2015; Cui et al. 2017).

For various reasons, the dramatic nature of upheavals in the late Neoproterozoic has been underestimated. On the one hand, because of the intensity of surface glaciations and changes in Earth's crust, few fossil traces have been left. In many cases, only chemical traces in minerals provide evidence for the influence of biological processes (Konhauser 2007; Knoll et al. 2012). This means that the lost world of organisms remains largely invisible to the layman; there are no menacing or emotion-arousing life forms—such as dinosaurs—to trigger arousal. From our current perspective, the upheavals might certainly be seen as a "good catastrophe", as it was the increase in oxygen content of air and water that created conditions in which evolution of multicellular organisms could occur.

A secondary effect of atmospheric oxygen enrichment became apparent only with increasing growth of plants on land—the emergence of fires. Prerequisites for this are adequate oxygen concentrations in the atmosphere and presence of combustible organic material. In fact, the term "fire window" is used in the palaeontological literature to refer to the physical and chemical properties of plant material. Characteristic atmospheric oxygen contents of 15 to 35 per cent relate to ignition of completely dry plant material at lower values and spontaneous propagation of fires through terrestrial vegetation—regardless of its water content—in the upper range. Lightning strikes are thought to be the most important natural triggers of fires (Glasspool und Scott 2010).

Self-ignition of accumulated organic material through various combinations of anaerobic microbial and chemical reactions is rarely mentioned. In contrast to direct aerobic combustion, these processes produce combustible carbon compounds through self-heating, which ignite on contact with oxygen. Well-known present-day examples of spontaneous combustion occur in compost, rubbish dumps, coal beds and landfill sites (Buggeln and Rynk 2002; Hogland and Marques 2003; Moqbel 2009; Fabiańska et al. 2013; Karabyn et al. 2018). As such conditions for origination of fires are by no means novel, they can also be considered as potential triggers for fires during early stages of terrestrial evolution. Regardless of anthropogenic influences, accumulation of large quantities of dead plant material may well have occurred, or geological processes may have exposed or even ignited coal deposit (Sokol and Volkova 2007; Grasby et al. 2015).

Traces of fires can already be found in Silurian deposits that are around 420 million years old. During the Carboniferous and Permian periods around 350 to 250 million years ago, there is increasing fossil evidence for fires, associated with an increase in atmospheric oxygen levels (Wildman et al. 2004; Scott and Glasspool 2006; Jasper et al. 2013; Glasspool et al. 2016; Wallace et al. 2017). But morphological and physiological adaptations of plants—such as conifers—suggest that fires were actually a persistent feature of the evolution of terrestrial vegetation (Bowman et al. 2009; He et al. 2012, 2016). During the Miocene around 10 to 5 million years ago, the readily flammable nature of aboveground biomass of dead C4-grasses also facilitated expansion of savannah landscapes at the expense of forests (Keeley and Rundel 2005; Strömberg 2011; Hoetzel et al. 2013; Maurin et al. 2014; Karp et al. 2018). Non-C4-grasses were able to develop in steppe areas—under pressure from frequently recurring fires—only by shifting their biomass into the soil and adapting the growth dynamics of their above-ground structures (Simon et al. 2009a, b; Scheiter et al. 2012; Maurin et al. 2014). During the period mentioned above, there was also a notable increase in numbers of different plant species showing above-ground adaptations of seed formation and reproductive cycles to counter fire stress (Crisp et al. 2011). The capacity to rapidly form new leaves from woody plant parts after fires is widespread among correspondingly adapted plant species (Whelan 1995; Schwilk and Ackerly 2001; Bond and Scott 2010; Bradshaw et al. 2011; Keeley et al. 2011; Pausas and Schwilk 2012; Pausas and Keeley 2017). Adaptations to frequently occurring fires are also found in various animal species, such as insects—including European representatives (Wikars 1997; Pausas and Parr 2018). Observations suggest that the construction of conspicuous mating sites of Australian bowerbirds also serves to prevent damage to their bowers during fires (Mikami et al. 2010).

Current knowledge indicates that morphological characteristics of primates differentiated in such ecosystems, ultimately leading to evolution of the genus *Homo* (Laden and Wrangham 2005; Jones et al. 2009; Wood and Baker 2011). Learning to use fire around a million years ago opened up new possibilities for human ancestors to influence their environment and also facilitated food acquisition (Jones et al. 2009; Archibald et al. 2012; Berna et al. 2012). Even today, fires are deliberately set to raze natural ecosystems in order to expand farmland or settlement areas, for example in Indonesia or Brazil (Harrison et al. 2010; Pivello 2011). Although we now ignite far more fires every day in technical installations than occur spontaneously out in the open, we are still unable to prevent the latter. Reasons for this reside not in a lack of knowledge of the individual factors involved, but in an insufficient capacity to control the inherent randomness of their interactions (White et al. 2006; Power et al. 2008; Field et al. 2009; Pechony and Shindell 2010; Busch et al. 2015; Phillips and Nogrady 2020).

8.4.3 Colonization of New Habitats

Quantitative decoupling of energy flows from material flows opened up new freedoms for organisms in their choice of habitats—provided that essential inorganic elements for the formation and maintenance of cell substance were available in suitable proportions. Particularly important here was the availability of inorganic carbon, which was present at far higher concentrations in the atmosphere than in seawater (Ohmoto et al. 2004). Because of their dependence on water, during an initial phase unicellular organisms were only able to colonize terrestrial freshwater habitats. This inference is based on various lines of evidence. Among cyanobacteria, genetic analyses provide clear evidence of early evolutionary lineages inhabiting freshwater zones (Blank and Sánchez-Baracaldo 2010; Sánchez-Baracaldo 2015). Particularly exciting and revealing are clear indications of emergence of the first phototrophic eukaryotes in freshwater zones—i.e. the first endosymbioses of eukaryotes with cyanobacteria—around 1.6 billion years ago (Javaux et al. 2001; Strother et al. 2011; Blank 2013). From the resulting groups of red or green algae, the latter still colonize fresh waters almost exclusively (Delwiche and Cooper 2015). Cyanobacteria and green algae are also currently the photosynthetic partners for symbioses with fungi in lichens—life forms with the capacity to colonize extreme terrestrial habitats (Belnap and Lange 2003; Nash 2008). Geochemical data suggest that a dispersal of lichens to non-aquatic land areas occurred around a billion years ago (Heckman et al. 2001; Kump 2014). In addition to supporting the above interpretation, this evidence suggests that the heterogeneous conditions of terrestrial habitats played a far greater role in the evolution of organic diversity than indicated by the fossil record so far available.

In order to understand the further course of evolutionary processes, however, it is necessary to compare the differing conditions for photosynthesis on land and in the oceans. In this context, the term "land" refers not only to dry land areas but also to inland waters and coastal shallow water zones of oceans. For the sake of simplicity, no influences from viruses or animal organisms are taken into account in the proposed scenario (Fig. 8.8). All organisms considered—prokaryotes, fungi and photosynthetic organisms (PO: cyanobacteria and algae)—can therefore be dispersed only by water currents or winds. This is an entirely realistic assumption, as such dispersal processes have been demonstrated for all organisms, at least at individual life stages (Lutz et al. 2015; Tesson et al. 2016). Only zones of potential autotrophic production are taken into account. In oceans, a vertical distinction can be drawn between the light-bathed, near-surface epipelagic zone with photosynthesis and the deep zone of the sea floor (benthic) with chemoautotrophic energy conversion. The central argument in favour of this distinction is the biologically functional coupling between lithotrophic and phototrophic processes, which are always based on symbioses between different organisms. The presence or absence of coupling between these processes has a decisive influence on the evolutionary possibilities for phototrophic organisms and thus also on the ecosystems that depend upon them. If there is no direct coupling—as in the epipelagic zones of open seas—photosynthetic processes are significantly influenced by the randomness of abiotic processes involved in the supply of mineral nutrients. According to current knowledge, insufficient availability of phosphorus in the Proterozoic restricted primary production of plankton up to the Ediacaran period (Reinhard et al. 2017; Laakso and Schrag 2018).

Examples of abiotic transportation processes are water enriched with nutrients rising from deep levels or input of nutrients resulting from storms coming from dry land areas. The effects of such random changes are best known from the so-called El Niño phenomenon (Cane 2005; Chatterjee et al. 2017). At irregular intervals during these events, nutrient-rich deep waters fail to reach surface zones of the east Pacific coast. This immediately triggers a dramatic decline in plankton production and populations of all organisms that depend upon that food source in affected regions as far away as the Galapagos Islands (Barber and Chavez 1983; Trillmich and Limberger 1985). Many people are familiar with the irregularly recurring reports of Saharan dust pollution in Central Europe, but have little notion of resulting changes in plankton production in marine areas that are also affected (Griffin and Kellogg 2004; Jickells et al. 2005; Duarte et al. 2006).

The random nature of the supply of mineral nutrients is exacerbated by ongoing loss thereof associated with dead organisms sinking to the sea floor. Only a small fraction can be recovered for phototrophic production by bacteria in the near-surface zone. The processes involved are dependent on influences exerted by viruses and heterotrophic

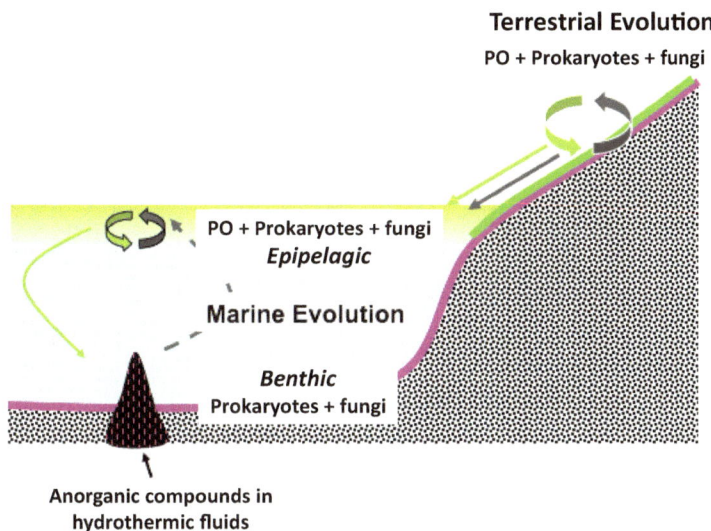

Fig. 8.8 Summary diagram portraying framework conditions for the evolution of autotrophic production in oceans and on land, including shallow water zones. It does not take into account influences from viruses and animals, which do not become functionally effective until later. Functional links between phototrophic zones and lithotrophic zones on the boundary to the lithosphere are decisive for further evolution of phototrophic organisms. In open seas, the two areas are connected only through abiotic processes. Also shown are locally occurring chemolithotrophic communities in geologically active zones around hydrothermal vents. PO = Phototrophic Organisms

organisms (Eppley and Peterson 1979; Miki and Jaquet 2008; Zonneveld et al. 2010; Ramanan et al. 2016). The sinking mineral and organic substances—aside from active hydrothermal vents—facilitate life of prokaryotes and fungi on sea floor sediments (benthic zone). The processes concerned are also extensively characterized by abiotic conditions—such as ocean currents—and cannot be actively influenced by organisms.

In both near-surface and deep ocean zones, framework conditions for autotrophic organisms are determined by spatial and temporal randomness in the supply of mineral nutrients. As a result of spatial separation between the two zones and energetic dimensions of the mediating water currents, randomness of the abiotic influences on organisms cannot be diminished.

Close spatial overlap of phototrophic lithotrophic zones enables direct functional feedback of phototrophic organisms with chemotrophic fungi and prokaryotes on land (Cane 2005; Chatterjee et al. 2017). With the exchange of photosynthetically generated carbohydrates for minerals, more organic matter can be formed and mineral nutrients can be released more rapidly from the geological subsurface (Barber and Chavez 1983; Trillmich and Limberger 1985). *At least at the surface, communities of organisms—with their energetic turnover capacity, which is around three times higher on average* (Griffin and Kellogg 2004; Jickells et al. 2005; Duarte et al. 2006)—*are hence able to modify geological processes structurally over large areas. On this basis, multicellular plants were able to spread on land over the course of evolution, whereas in oceans this was possible only in shallow water zones and in a few deep-sea areas with weak water currents and an adequate supply of minerals—such as the Sargasso Sea* (Eppley and Peterson 1979; Miki and Jaquet 2008; Zonneveld et al. 2010; Ramanan et al. 2016).

Depending on the chemical properties of the geological subsoil, mineral nutrients can be continuously dissolved on land—by using the energy obtained through photosynthesis—and employed by organisms for growth and compensation of losses. Outside of shallow water zones, the randomness of abiotic processes on land influences the growth of symbioses through precipitation and by overcoming gravity with air currents for spatial dispersal.

This also makes it easier to understand the long time lag between evolutionary developments in oceans and subsequently on land. For the evolution of significant phototrophic production to take place away from water, symbioses had to develop between organisms from at least three different groups, whose encounters at a common location were determined solely by coincidences in anaerobic transportation processes. Furthermore, evolutionary possibilities for symbioses depended on local characteristics of the geological subsoil and climatic conditions. Too much or too little sunlight could prevent the evolution of complete symbioses just as much as over-long periods of drought or intense precipitation—influencing factors to which marine organisms have never been exposed with such intensity.

8.5   Transition from the Proterozoic to the Phanerozoic: No "Explosion"

In both popular and scientific literature, the rapid burgeoning of animal life forms at the beginning of the Phanerozoic is referred to as the "Cambrian Explosion". Following on from the purportedly hostile conditions of "Snowball Earth" in the Cryogenian and the life forms of the Ediacaran at the end of the Proterozoic, the rapid evolution of animal life forms known to us—if viewed uncritically—may appear puzzling. In particular, this has stimulated proliferation of an almost overwhelming array of hypotheses about possible causes and triggers (von Bloh et al. 2003; Ginsburg and Jablonka 2010; Zhang et al. 2014; Briggs 2015). Discussions are amplified by numerous attempts to identify the triggers and their temporal occurrence using model-based genome analyses. Well-designed graphic overviews readily convey an impression of identification of the "roots" of the "family tree of life". According to results obtained, the main steps in genetic differentiation could have taken place approximately between 1.1 and 0.6 billion years ago—i.e. well before the "Cambrian explosion" for the oldest date but also after the Cambrian—in the Ordovician—for the most recent (Douzery et al. 2004; Labandeira 2005; Peterson et al. 2005, 2008; Cartwright and Collins 2007; Qun et al. 2007; Erwin 2009, 2015; Neuweiler et al. 2009; Wheeler et al. 2009; Erwin et al. 2011; Sperling et al. 2013; Knoll 2014; Dohrmann and

Wörheide 2017; del Rey et al. 2022). For several reasons, readers might find these results thought-provoking.

As already described in previous chapters, fossil traces of photosynthetically active eukaryotic cells can be traced back to around 1.6 billion years ago (Han and Runnegar 1992; Teyssèdre 2006; Agić et al. 2015). Current knowledge indicates that the capacity for photosynthesis originally evolved through symbiosis between a non-photosynthetic eukaryotic cell and a cyanobacterium. Division of evolutionary pathways between plants, fungi and animals must therefore have taken place prior to this development—around a billion years earlier than the estimates from genome analyses. Otherwise, animals and fungi would necessarily have evolved from plants. However, individual fossil finds also suggest that non-photosynthetically-active eukaryotes were present for at least 1.4 billion years (Porter 2004; Knoll et al. 2006).

Discussion is further complicated by the predominant "terrestrial" perspective on evolutionary processes, which is characterized primarily by a dichotomy between plants and animals. This also obscures our view of the evolution in the aquatic realm of unicellular eukaryotes (protists), with their notable diversity of life forms. The distinction between phytoplankton and zooplankton continues to dominate the scientific literature, although many species have been shown to possess different combinations of phototrophic and V-heterotrophic energy conversion—summarized under the term mixotrophy (Chap. 7) (Burkholder et al. 2008; Feng et al. 2010; Moore 2013; Nakajima et al. 2014; Mitra et al. 2016).

This also makes it difficult to interpret the functional characteristics of fossilized unicellular eukaryotes—which are grouped under the collective name acritarchs. Only in a few cases can evidence of morphological relationships between unicellular fossils and later protists, as well as unicellular fungi, be found (Teyssèdre 2006; Bosak et al. 2011, 2012; Dunthorn et al. 2015; Retallack 2015). Temporal development of the diversity of forms among acritarchs yields a very generalized notion of the evolution of plankton communities from the late Proterozoic to the end of the Triassic, when data for this group largely dry up (Knoll 1994; Katz et al. 2004; Huntley et al. 2006; van Soelen and Kürschner 2018).

It is possible that functional complexity of plankton communities in the late Proterozoic —including participation of fungi—was far greater than the fossil record suggests. Fungi in the early plankton communities might have been chytrids (Chytidiomycota) because of their original life-cycle with flagellated, freely motile spores and their wide distribution in marine plankton communities (Muehlstein et al. 1988; Richards et al. 2012; Kagami et al. 2014).

Representatives of chytrid fungi with varying properties are found in all habitats, ranging from marine epipelagic zones and soils to parasitism in amphibians and symbiosis in the stomachs of ruminants. The extent to which their genetic systems are influenced by particular environments is highlighted by an example from exploration of their evolutionary occurrence using molecular biological methods (Berbee and Taylor 2001). Because a representative of the symbionts in ruminant stomachs was used for investigation, for interpretation of the results their evolutionary age was estimated to be only around 200 million years. This clearly conflicts with their primordial characteristics, which would suggest that they appeared at the beginning of fungal evolution. Instead, the period determined could be more easily explained by emergence of this life form in the digestive systems of herbivorous dinosaurs. This example clearly shows the caution required when interpreting molecular biological determinations of evolutionary processes in the absence of adequate fossil evidence. In many cases, only organisms that can be maintained in laboratories under controlled conditions are used for study, hence representing no more than a special case in the evolutionary history of a large group of systematically related organisms.

The scarcity of available fossil evidence is due not only to the vast periods of time between the emergence of certain organisms and the present, but also to the massive geological and climatic upheavals that occurred around one billion to 0.5 billion years ago. The supercontinent Rodinia began to break up around one billion years ago—accompanied by volcanic eruptions with far-reaching effects on oceans and the atmosphere (Fig. 8.4). Details and extent of the processes are still subject to debate—as is the incipient merging of individual lithospheric plates at the beginning of the Phanerozoic (Eyles 2008; Bradley 2011; Roberts 2012; Young 2013; Nance et al. 2014; Oriolo et al. 2017). During the long-lasting glaciation episodes of the Cryogenian, most fossil remnants were probably destroyed by glacial movements on land and by icebergs and sea ice in shallow water zones.

By contrast, chemolithotrophic primary production, and the associated release of minerals, by syntrophic communities of microorganisms on rocks beneath the glaciers was not suppressed. Similarly, photosynthetic primary production may have been present on ice surfaces in small sedimentation cavities (cryoconite holes), in and under the sea ice, as well as in current-induced ice-free areas of seas (polynyas). Moreover, because of the rising sea level, marine habitats expanded. This modified the chemical composition of seawater and the species spectrum of planktonic eukaryotic organisms (Nagy et al. 2009; Hood et al. 2011; Margesin and Collins 2019). Evidence of life on the sea floor is provided by isolated fossilized feeding traces left by heterotrophic organisms and chemical traces of sponges, for which occurrence before the beginning of the Cryogenian has been proposed (Brocks and Butterfield 2009; Love et al. 2009; Brain et al. 2012; Porter 2016). Sponges actively filter food particles from seawater pumped through their tube systems. This is not in serious contradiction to the reported low oxygen content in the oceans, as sponges—at least if small in size—can survive over the long term at oxygen concentrations of 0.5 to 4 per cent of the current level (Payne et al. 2010; Knoll and Sperling 2014; Mills et al. 2014). Dates given in scientific publications for fossil evidence of sponges—usually small fragments of the original organisms (microfossils)—vary between about 630 and 540 million years (Bowyer et al. 2017; Slater and Bohlin 2022).

Isotope analyses of carbon deposits also indicate biological productivity on land—possibly by lichens (Knauth and Kennedy 2009; Kump 2014; Delwiche and Cooper 2015; Brasier et al. 2016). However, rates of photosynthesis by lichens are too low to explain the increase in oxygen concentration that occurred around 800 million years ago. The increase in the photosynthetic performance of algae in nutrient-rich wetlands, with their 100 to 1000-fold higher production rates per unit area—compared to lichens—seems more plausible (Lieth and Whittaker 1975). Traces of salt lakes with independent organism communities give some idea of the diversity of different forms of wetlands (Schintele and Brocks 2017). Contemporary conditions in the Ediacaran and Cambrian oceans—relevant for marine primary production (see Chap. 9)—are completely unknown. Overall, the conditions opened up possibilities for the evolution and differentiation of new life forms of V-heterotrophs (Porter 2011), which may also have contributed to the decline of stromatolites.

It is clear that many organisms were wiped out by the ice ages and that traces of their existence—especially on land and in shallow-water zones—were eradicated. As a result, there is little evidence of life forms in these areas. By contrast, fossils and traces of life from marine areas are better preserved (Corsetti et al. 2003). The associated imbalances in scientific reports inevitably lead to a distorted view of living environments in the Precambrian. This problem is unlikely to be solved in the foreseeable future by an adequate number of new finds, but it should not be forgotten when interpreting organism communities in subsequent periods of Earth's history.

Ice ages by no means wiped out all life, but they exerted a very selective influence on the evolutionary development of organisms. Small organisms—including multicellular forms—were able to survive to a limited degree both on the ice surface and in and under sea ice. Water areas left open by ice movements (polynyas) offered additional options for evolution of local plankton and dependent benthic communities. Long-term isolation of local communities in open-water areas promoted genetic differentiation of originally relatively homogeneous populations. Once the restrictive environmental conditions were removed, the surviving organisms provided the starting point for new, genetically and ultimately also functionally differentiated evolutionary lineages. The effects of such processes have also been demonstrated by genetic studies of European mammals that lived before and after the last ice age (Hofreiter et al. 2004).

From this perspective, the Cryogenian can be seen as a key period for all subsequent evolutionary processes for several reasons:

- Large-scale and long-term glaciations gave marine life forms a significant evolutionary advantage by inhibiting, or even eliminating, evolutionary processes on land.
- Fragmentation of relatively homogeneous communities of organisms into small-scale, isolated units provided fundamental impulses for new genetically and functionally differentiated evolutionary lineages.
- Glaciation and thaw phases, in conjunction with ice movements, significantly accelerated the transportation of weathering products into oceans and—in conjunction with geological processes—led to alteration of chemical and physical conditions in seas.

The complex effects of these factors can be outlined only in broad terms, as various random processes influence their interaction. In an initial stage, the survival probability of organisms or organismal communities depends on characteristics and randomness in the spatial distribution of intensities and in the temporal occurrence of abiotic selective events. In the manageable dimensions of our own perception, for example, a single falling stone might kill one person in a randomly assembled group of people, while all others remain unharmed. For many, a far greater degree of randomness was observed with effects of the tsunami that occurred along the coasts of Thailand on 26 December 2004. What we cannot directly perceive are influences of organism size and resistance on survival probability under various impacts (see e.g. Fig. 7.4).

Following a selective event, under modified environmental conditions, surviving organisms must develop functional interactions with other organisms in newly composed communities in a temporally randomized sequence. Over the long term, only those communities of organisms can survive in which conditions for supply of energy and material are adequately fulfilled for each population. Conditions are "sufficiently" fulfilled if—despite the fact that population sizes and ratios are constantly changing—no population collapses completely. The spectrum between randomly formed communities of organisms and communities that are stable over the long term is influenced by a substantial number of biotic and abiotic factors. Dimensions of heterotrophically convertible primary biological production and the scope of abiotic factors that can be utilized by organisms are important influencing factors. The first issue can be illustrated by differences between primary production and plant biomass on land and in oceans. Rough estimates indicate that annual totals of primary production on land and in oceans are similar, but whereas about 99.5 per cent of global plant biomass is stored on land, only around 0.5 per cent is stored in oceans (Smetacek 2012). Accordingly, with similar primary production, a significantly higher proportion can be converted into heterotrophic biomass in oceans than on land. The second issue can be illustrated using the example of altitudinal distribution of biomass production on mountains (Körner 1999).

In abstract terms, each population concerned adapts to existing framework conditions, but in doing so it modifies future framework conditions for all other populations and ultimately finds itself adapting to conditions that have been modified by other populations.

As regards developments from the Proterozoic to the Phanerozoic, effects of the Cryogenian can be understood in terms of the sequences in the sequence from state 1″ to state 2″ in Fig. 5.18, during which only marine organisms made the transition to the subsequent Ediacaran and Cambrian periods. On the other hand, for marine organisms, changes in abiotic conditions created new potential zones for dynamic development of their life forms. Documented developments in form diversity among acritarchs indicate that these effects were further intensified by the relatively short Gaskiers glaciation epoch during the Ediacaran. Additional abiotic factors must have delayed a renewed increase in diversity of acritarch forms until the Ordovician. Possible influencing factors include a massive asteroid impact during the Gaskier glaciation around 580 million years ago and a second glaciation epoch in the Ediacaran (Hill et al. 2007; Gostin et al. 2010; Vernhet et al. 2012).

8.6   Abiotic Framework Conditions During the Phanerozoic

8.6.1   Influences of Planetary Factors on Dynamics of Terrestrial Systems

In everyday life, it is easy to forget the framework conditions on which the lives of all organisms—including humans—depend and are influenced. This is because we simply do not recognize many influencing factors, while taking others—because of their presumed consistency—for granted as immutable. Seen from our—in evolutionary time dimensions—extremely brief perspective, Earth is stable and unchanging. With regular rhythmic timing, sun and moon rise in the east and set in the west. While earthquakes, volcanic eruptions and asteroid impacts trigger fear and uncertainty, the global input of cosmic material—estimated at 5000 tonnes per year from micrometeorites (Rojas et al. 2021)—is barely noticed. Most people also scarcely perceive seasonal changes in daylight and darkness because their life rhythms are governed by a socially prescribed timescale. This perception is reinforced by the widespread availability of artificial light sources. Accordingly, it is easier to understand why many people also believe that control of the global climate is possible.

Here, it is necessary to remember that we live on a planet—with unique characteristics—in the solar system. At the same time, we need to realize that this does not guarantee that living conditions will always remain stable. Even a very simplified illustration reveals the variety of different—and interacting—planetary factors that influence long-term living conditions on Earth (Fig. 8.9). For example, Earth's axis of rotation oscillates at an angle to its orbit around the sun (obliquity) and also rotates—with a different

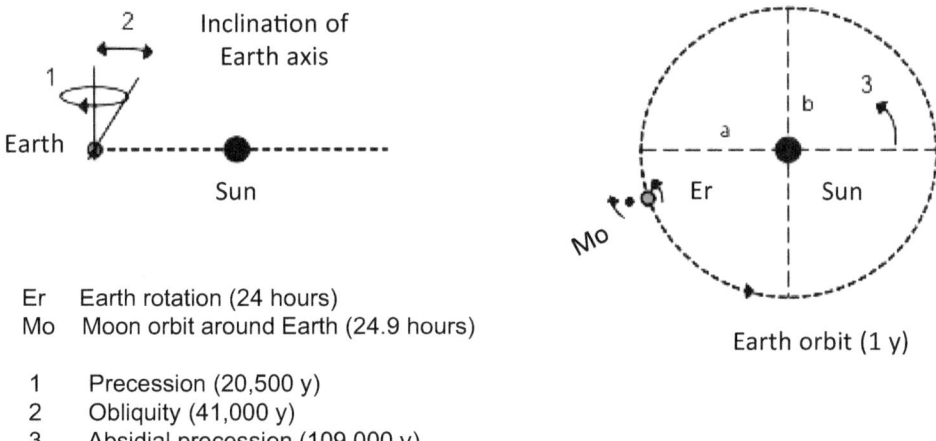

Er   Earth rotation (24 hours)
Mo   Moon orbit around Earth (24.9 hours)

1   Precession (20,500 y)
2   Obliquity (41,000 y)
3   Absidial precession (109,000 y)
4   Eccentricity $(r_a - r_b)/(r_a - r_b)$ (109,000 and 413,000 y)

Fig. 8.9 Greatly simplified overview of different dynamic processes involved in terrestrial planetary motion

periodicity—around the virtual perpendicular to its orbit (precession). The virtual connecting line (apsidal line) between the point closest to the sun (perihelion) and the point furthest from the sun (aphelion) of the slightly oval Earth orbit also rotates around the sun about every 109,000 years. Earth's orbit alternates between a circular and elliptical geometry (eccentricity) over a similar period (Clauser 2014; Hanslmeier 2014). As a result of the combined effect of all factors mentioned, geographical locations of direct solar radiation on Earth change continuously. At the equator, the virtual point of solar irradiation covers a distance of 1669 km/h. Because of Earth's spherical shape, at different latitudes the intensity of solar radiation deviates from its global average—340 watts per square metre (Stephens et al. 2012). During summer periods of Earth's hemispheres, irradiation values —especially at subtropical latitudes—are significantly higher than the global average. During winter periods, irradiation values decline polewards from subtropical latitudes down to zero (Roedel 1994). In conjunction with movement in Earth's orbit, this continuously generates dynamic effects—particularly in atmospheric processes. Comparative modelling calculations indicate that the dynamics of Pacific cold water currents—known, for example, in connection with the El Niño phenomenon—are also influenced by changes in planetary conditions (Fig. 8.9) (Chiang et al. 2022).

The order of magnitude of solar energy flow can be surmised most easily by observing the tidal range—for example on individual stretches of the Atlantic coast. Although underlying energy flow is only 150,000th of that of solar irradiation, it can elevate vast quantities of water by several metres. As the moon orbits Earth in the opposite direction to our planet's rotation, these forces also reduce our planet's rotational velocity. Due to the mutual attraction between Earth and the moon, the latter rotates around their common centre of mass and not precisely around its own axis. Therefore, Earth also moves in a spiral trajectory along its orbit around the sun. According to more recent analyses, the delay caused by additional influences is not uniform, but varies around an average value of 1.8 ms per century. This value is in fact less than the calculated delay of 2.3 ms per century due to the influence of tidal forces alone (Stephenson et al. 2016).

Even if we are unable to observe the above-mentioned dynamic changes directly because of their lengthy periodic duration, we are at least familiar with the terrestrial phenomena of short-term dynamics. We are familiar with the duration of an orbital journey around the sun as years, or the Earth's own rotation as days. We can perceive the effects of gravitational force on falling objects or as a cause of effort when climbing a steep slope. Although Earth as a whole is subject to the dynamic influences of gravitational forces of the sun and moon, we perceive their terrestrial effects primarily as tides on ocean beaches. Combined effects of the moon and the sun—moderated by the coastline—are recognizable from the magnitude of the tidal range.

Although they remain largely unnoticed, two effects of Earth's rotation influence all movement processes on our planet (White et al. 1992):

- The Coriolis force describes summated effects on the movement of a body on a rotating disc. Anti-clockwise rotations of the disc—analogous to a view of Earth's rotation from the North Pole—lead all movements to deviate to the right. Clockwise rotations of the disc —analogous to viewing Earth's rotation from the South Pole—lead all movements to deviate to the left. As the orientation of Earth's surface corresponds to the position of the hypothetical disc only at the poles, that is where the Coriolis force is maximal. Following a sine function with latitude, the influence of the Coriolis force decreases to zero at the equator.

- For all bodies on Earth, the approximately spherical shape of the planet results in different distances from its axis, depending on latitude and—because of Earth's rotation—differing centrifugal accelerations. In contrast to the Coriolis force, these are greatest at the equator and decrease to zero at the poles, following a cosine function according to latitude. The superimposed influences of the Coriolis force and centrifugal acceleration—within the relevant geomorphological framework conditions—significantly influence all large-scale flow in water and in the atmosphere (Talley et al. 2011).

The sun itself does not have a uniform intrinsic speed of rotation. Velocity of rotation is lower in the inner—presumably solid—core than in the enveloping—gaseous—region. Since the solar system formed, average radiation intensity of the sun has increased by around 25 to 30 per cent (Rosing et al. 2006). Over the short term, the values vary over an 11-year cycle and probably over a 90-year cycle as well. Indicators of changes in radiation intensity are so-called sunspots, which were in fact directly observable from Earth before the advent of satellite-based measurements (Hanslmeier 2014). In addition, short-term eruptions occur at irregular intervals, sometimes with intense effects on the terrestrial magnetic field (de Jager 2005; Emslie et al. 2012; Baker et al. 2013). Even Earth's magnetic field exhibits long-term chaotic patterns with change in the geographical positions of magnetic poles and repeated polarity reversals over the course of Earth's history (Van der Voo and Torsvik 2001; Stacey and Davis 2008; Russell et al. 2017; Witze 2019). The earlier dynamics of solar activity over the longer term can be reconstructed by analyzing concentrations of beryllium isotopes (^{10}Be) in deposits. This has made it possible to find correlations between long-term phases of low solar activity and cold phases—such as the Maunder (1675) and Dalton (1810) minima—during the past 1000 years. In general, the intensity of solar radiation exhibits chaotic characteristics over the long term (de Jager 2005; Mann et al. 2005; Goosse and Renssen 2006; Lee et al. 2009a, b; Meehl et al. 2009; Ineson et al. 2011; Swingedouw et al. 2011; Goosse et al. 2012; Thiéblemont et al. 2015; Zeebe and Lourens 2019).

The marked regularities of planetary processes—as perceived by humans—presumably inspired precise observations and eventually calculations at an early stage in human history (Pearson et al. 2007; Maul 2013; Ossendrijver 2016; Sweatman and Tsikritsis 2017). Associated expectations of improved prediction of terrestrial events continue to stimulate a wide variety of investigations into connections between extraterrestrial and terrestrial processes. Meanwhile, technological possibilities for storing and analyzing huge quantities of data are fuelling hopes for continual improvements in long-term predictions. However, error ranges for direct and indirect fluxes in models of global atmospheric energy balance reveal just how quickly even sophisticated models hit their limits (Fig. 8.10) (Stephens et al. 2012).

While direct correlations—for example between lunar cycles and biological rhythms (Barlow et al. 2010; Zürcher et al. 2010; Mercier et al. 2011)—are easy detectable, studies of multifactorial correlations rapidly reach the limits of reliable demonstrability. Well-known examples of this are studies of relationships between dynamics of the solar system and climatic developments. Theoretically, a cycle duration of 100,000 years can be derived from the interaction of parameters 1–4 (Fig. 8.9). Empirically, these statements can be confirmed only for recent Earth history (Milankovitch 1941; Abe-Ouchi et al. 2013). Formation and duration of ice ages are determined primarily by superposition of different cycles in the tilt and rotation of Earth's axis—obliquity and precession in Fig. 8.9 (Bajo et al. 2020). In geological deposits, a cycle of around 405,000 years extending over a period of around 250 million years has been recognized. This has been attributed to a temporal influence on Earth's orbit exerted by special alignments of Jupiter and Venus (Laskar et al. 2004; Hinnov and Hilgen 2012).

The main reason for deviations in terrestrial phenomena—such as climate—from modelled relationships between planetary movements and solar activity resides in terrestrial modulation of energy flow. On its way through the atmosphere, around 45 per cent of radiation is converted, for example into kinetic or thermal energy. The remaining radiation that reaches land and water surfaces, averaging 180–190 watts per square metre—known

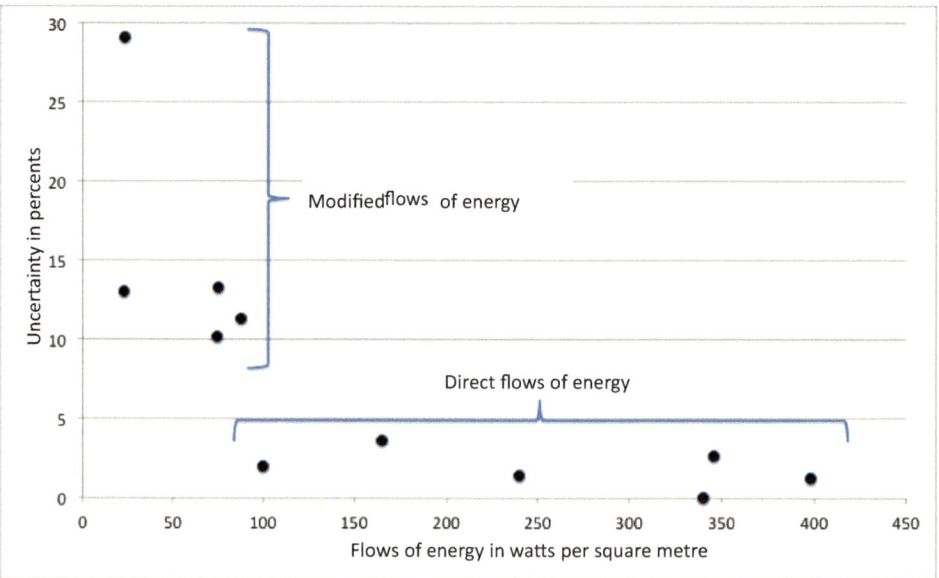

Fig. 8.10 Indications of the effects of complex terrestrial processes derived from differences in reported error ranges for direct and modified energy fluxes in balance models. Data source: Stephens et al. (2012)

as global radiation—is in turn converted in various ways. Some of it is reflected directly (albedo). Depending on irradiation conditions and surface properties, the proportion can be between 4 per cent (with vertical irradiation in water) and 90 per cent (with tangential irradiation of snow or water in polar regions) (Roedel 1994; Häckel 2021). The remaining energy flow is mainly converted into thermal energy or—if water is present—into evaporation energy. Thermodynamically, this results in fundamental differences between the lithosphere and hydrosphere, as well as from the atmosphere. Land and water are heated by solar radiation only at their surfaces. They can temporarily store thermal energy but cannot convert it directly into kinetic energy. About 100 times more solar heat energy per unit area can be stored in oceans than in the lithosphere (Stewart 2008).

Heating and thermal storage in the hydrosphere and lithosphere at the lower atmospheric boundary induce conversion of solar energy into kinetic energy in air currents—in analogy with the principle of a heat engine. The same principle drives hydrothermal currents in near-surface zones of the lithosphere—albeit on the basis of geothermal energy. Heated water can emerge at the surface both on land and in the hydrosphere in the form of springs with differing elevated temperatures (Scheirer et al. 2006; Staudigel and Clague 2010; Kiryukhin et al. 2012; Clauser 2014). Comparatively small amounts of heat from hydrothermal vents in oceans are only enough to induce small-scale currents over a few hundred metres, not large-scale currents (Lupton et al. 1985; Thomson et al. 1995, 2009; Pruis and Johnson 2004). Under the influence of gravity, cooling effects due to heat radiation and resulting changes in density trigger downward currents in both the atmosphere and the hydrosphere.

In reality, fluxes are always three-dimensional, with energetically dominant main trajectories and intensive coupling between atmosphere and hydrosphere (Zachos et al. 2001; Small et al. 2008; Lagerloef et al. 2010; Ma et al. 2015, 2016; Delplace et al. 2017; Su et al. 2018; Abram et al. 2020). As every flow causes new energetic imbalances in its environment, flow paths constantly induce eddy currents of different sizes and directions of movement. In two dimensions, similarities in flow patterns on surfaces of water bodies— whether they be in rivers or in the Gulf Stream in the Atlantic—are directly observable (Knighton 1998; Talley et al. 2011). The self-similarity of structural patterns of turbulent systems across all known spatial dimensions was formally described by A.N. Kolmogorov in 1941 (Frisch 1991; Gibson 1991). Two-dimensional representations of fractals in publications (Mandelbrot 1991) or on the Internet provide a first impression of self-similar structures. However, those often striking graphical representations make it easy to forget that real-life processes in the atmosphere and hydrosphere take place dynamically in three dimensions and can be greatly modified, for example, by volcanic eruptions (Du et al. 2022). Due to their fundamentally chaotic properties, future developments can at best be estimated only for short periods of time—never over the long term. These fundamental limits to long-term predictability in the chaotic systems of the atmosphere and hydrosphere cannot be overcome by using any of the methods currently available to us, such as computer-based modelling (Arnol'd 1991).

Transformation of global radiation on Earth's surface is influenced to a great extent by the geographical distribution of water bodies and land areas, as well as their geomorphological structures. Historically, these factors are also subject to ongoing changes due to lithospheric dynamics. In terms of energy, this is presumably driven by cyclical processes in Earth's interior (Stacey and Davis 2008). The geodynamic processes involved are comparatively slow and long-lasting. Lithospheric plates shift horizontally by a few centimetres per year. Vertically, mountains can rise at rates of between 0.1 and 10 millimetres per year (White et al. 1992; Slaymaker et al. 2009; Brune et al. 2016). As a result, the structure of Earth's surface has been continuously and irreversibly altered over the entire recordable timespan of the planet's history (Hay 1996; Stanley 2001; Slaymaker et al. 2009). This affects both sizes and structures, as well as the geographical positions of continental areas and vertical structuring of the zones of transition to the atmosphere and to the hydrosphere. Enlargement and modifications in shallow water zones along mainland coastlines are particularly important for marine biological evolution. Lithospheric processes can be observed at the surface, especially along the plate margins, either as compression processes in collision and subduction zones or as expansion processes in spreading zones (Clauser 2014). Characteristic side-effects of plate tectonics are earthquakes and volcanic eruptions with varying intensities and consequences (Kanamori and Brodsky 2001; Sulpizio et al. 2006; Staudigel and Clague 2010; Clauser 2014; Obara and Kato 2016). Regardless of this, volcanoes can also develop within lithospheric plates (Staudigel and Clague 2010; Yang and Faccenda 2020). There is also evidence for connections between the water cycle in geological formations and volcanic processes, for example in the Lesser Antilles (Cooper et al. 2020).

In social terms, after events occur, only short-term effects on human lives and the economy are foreseen, within a relatively narrow time-frame. One example is the tsunami wave in December 2004—triggered by a submarine earthquake off Sumatra—whose wave fronts even reached the coasts of Antarctica, Africa and Australia (Ishii et al. 2005; Titov et al. 2005; Galetzka et al. 2015). This event remains in public memory primarily because of the large number of fatalities in various countries. On the other hand, eruption of the Eyjafjallajökull volcano in Iceland in 2010 (Sigmundsson et al. 2010), whose impact was primarily economic, has remained in social memory because its name is so challenging for non-Icelanders. Events that occur repeatedly but at longer intervals, such as eruptions of Mount Vesuvius or earthquakes in California, keep scientists and insurance companies busy, but do not restrain expansion of major settlements within the threatened regions (Guatteri et al. 2005; Sulpizio et al. 2006).

Effects of geological processes have a far more enduring impact on the climatic and geological conditions of biological evolution (Gaillard et al. 2011; Costa et al. 2012, 2014; McKenzie et al. 2016; Malavelle et al. 2017). The best-known scientific example of this is the largest global mass extinction of organisms—with estimated species losses of up to 90 per cent—at the end of the Permian period. Emissions from magma eruptions in the Siberian Plate (Siberian Traps) and further volcanic eruptions at the end of the Permian period around 252 million years ago probably led to changes in global climate that triggered biological repercussions as a result of chain reactions (Heydari et al. 2008; Shen et al. 2012; Burgess and Bowring 2015; Clarkson et al. 2016; Benton 2018). In addition to these impacts and the influx of sulphur compounds, the collapse of terrestrial vegetation is likely to have led to drastic augmentation of erosion and hence to the spread of anoxic zones in marine ecosystems (Saunders and Reichow 2009; Sidor et al. 2013; Song et al. 2014; Cascales-Miñana et al. 2015).

The extent to which monocausal—usually model-based—claims can be misleading is shown by the example of—supposed—intensification of erosion processes by terrestrial vegetation. This assertion, made in many publications, can be challenged most effectively by analysing watercourses in the Amazon basin, which has vegetation cover extending over an immense area. Sources of individual watercourses are located in geologically differing regions, such as the Amazon in the Andes, the Rio Negro in the Precambrian Guiana Shield and the Tapajós in the archaic Central Brazilian Shield. The total sediment load—estimated at around 900 million tonnes per year—emanates almost exclusively from the Andes region. In fact, because of colouration caused by sediments, the upper reaches of the Amazon (Solimões) are also known as white-water rivers. Black-water rivers such as the Rio Negro mainly transport humic products from plant material, while clear-water rivers such as the Tapajós contain only small quantities of mineral or humic substances. Comparisons with river systems under other climatic and geomorphological conditions render the diversity of different influencing factors even more evident. For example, the Nile—with a slightly greater overall length but a catchment area only half the size—transports around 100 million tonnes of sediment per year. Under Arctic climate conditions, however, around 180 million tonnes of sediment per year are transported into the Arctic Ocean by the Lena River—with a lesser gradient and a catchment area that is around 25 percent smaller (Filizola et al. 2011; Forbes 2011; Biswas and Tortajada 2012; Ríos-Villamizar et al. 2020).

Because all organismic communities live in open systems, monocausal explanations generally do not suffice to explain changes in ecosystems. Even for emissions from fissure eruptions, which are quantitatively large and long-lasting, no systematic correlations with changes in communities of organisms have been shown for the entire duration of the Phanerozoic (Prokoph et al. 2013; Ernst et al. 2021).

The mass extinction that occurred around 66 million years ago—at the end of the Cretaceous period—is firmly anchored in public consciousness because of the extinction of numerous dinosaur species. However, during this event—which was apparently triggered by an asteroid impact—only around half as many species became extinct as was the case at the end of the Permian period (Bond and Grasby 2017). Asteroid impacts have led to species losses throughout Earth's history, not only at the end of the Cretaceous period. However, the particular consequences of an impact do not depend solely on forces released but also on other factors—for example, the global distribution of biotic communities in conjunction with the locations of the collision site. For the last 250 million years alone, at least two—possibly four—additional asteroid impacts can also be linked to lesser-known mass extinctions (Glikson 2005; Wittke et al. 2013; Rampino and Caldeira 2017).

Although we can apply scientific methods for approximate reconstruction of changes that have occurred in lithospheric plates over several hundred million years, our understanding of processes in the hydrosphere and in the atmosphere is woefully incomplete. An example comparing our direct perception to scientific interpretation provides support for this argument:

- In our everyday perception, we interpret propagation of water waves as horizontal movements of water. In fact, what we observe is propagation of the energetic shock wave, while the water particles move up and down whilst locally unchanged.
- Textbooks and publications are packed with descriptions of so-called material cycles, often combined with references to the equilibrium of natural processes. In reality, all processes—usually shown separately—are based on dynamic equalization processes influencing energetic imbalances in extraterrestrial and terrestrial energy fluxes through gases and liquids (Figs. 8.11 and 8.12). This is what creates the conditions for the transportation of different substances in interaction with biotic and abiotic processes for binding and release.

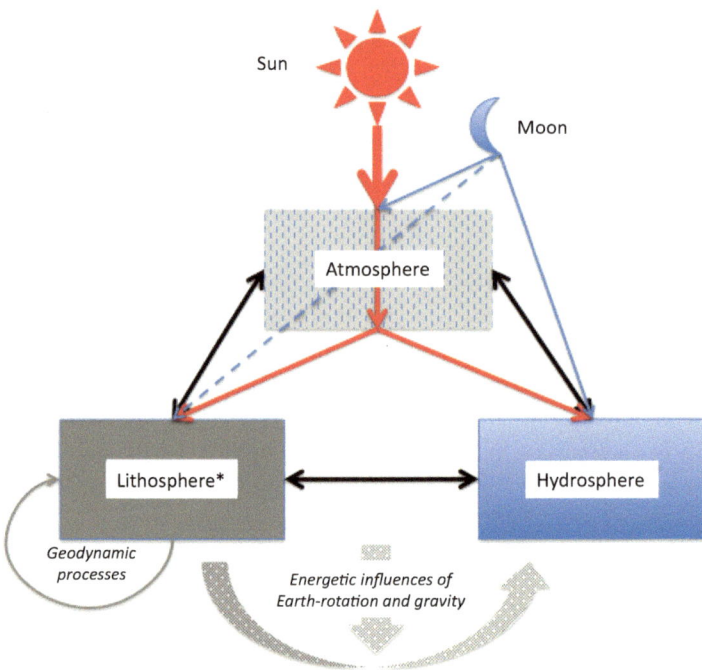

Fig. 8.11 Overview of interactions between atmosphere, hydrosphere and lithosphere (grey arrows) under the energetic influences of solar radiation (black arrows), the gravitational force of the moon (dashed arrows), together with geodynamic processes, Earth's rotation and gravity. The asterisk (*) indicates the non-cyclic dynamics of lithospheric surface structures

In simplified terms, the atmosphere and hydrosphere can be characterized as self-organizing systems that are constantly being pushed out of equilibrium by the energetic influences of solar radiation, extraterrestrial gravitational forces and Earth's rotation. Their dynamic processes are caused mainly by ongoing dissipation of solar energy and only to a small extent through radiation of geothermal energy. They are modulated by influences of gravity, the Coriolis force, angular momentum and prevailing structural characteristics of the lithosphere. Over the course of Earth's history, structures, extents and geographical locations of continental and marine areas have continuously changed (Scotese 2021). Accordingly, the framework conditions for atmospheric and hydrospheric processes and their interactions have also changed (Ghil 1994; Chang et al. 1995; Masoller et al. 1995; Ferguson and Messier 1996; Koshel and Prants 2006; Li and Ding 2011; Kirtman et al. 2017; Timmermann et al. 2018). Comparative figures provide a global idea of the diversity of lithospheric structures. For example, the current mean sea level of land areas is 743 m and the maximum height is 8848 m, while the mean sea depth is 3734 m and the maximum is 11,034 m (Talley et al. 2011). At least 8500 elevations—mainly on geologically younger plate zones—protrude from deep-sea floors more than a thousand metres into open water of oceans. Many islands in the Pacific—and in some cases also in the Atlantic and Indian Oceans—are "summit regions" of deep-sea mountains at least 5000 m in height (Kim and Wessel 2011).

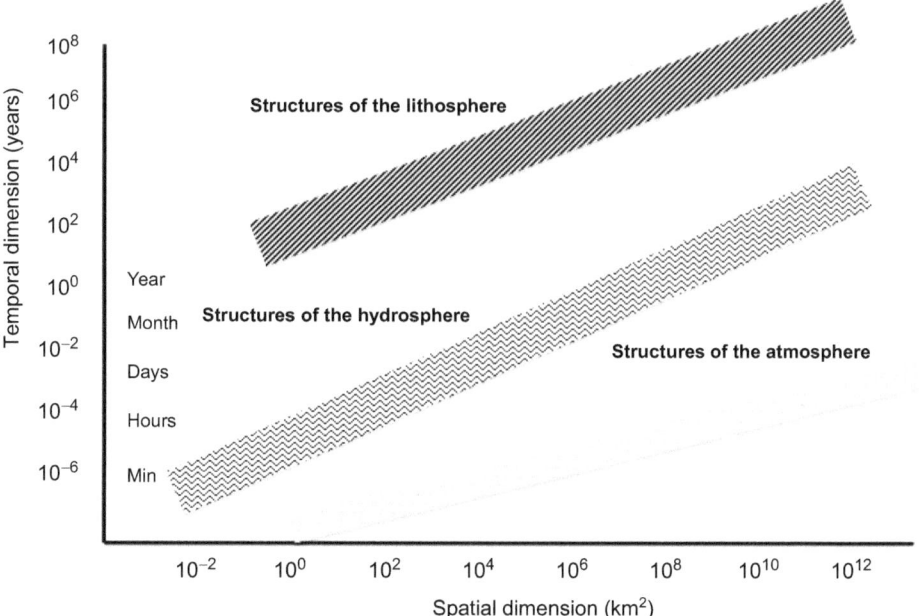

Fig. 8.12 Simplified comparison of spatio-temporal orders of magnitude of structural changes in the lithosphere, hydrosphere and atmosphere. Source: According to data in Dickey (1991) and Delcourt and Delcourt (1992), as well as the author's own estimates

The intensity of interactions between the individual spheres can be characterized to a first approximation by the differences in specific densities of air, water and rock at 1.2, 1000 and 2600 kg/m^3. The differences become even clearer when comparing estimated total masses. In relation to the total mass of Earth, the proportion of hydrosphere and atmosphere combined is around 0.2 per thousand. If only Earth's crust, with an average depth of around 17 km, is taken for the comparison, the fractional mass of the hydrosphere increases to around 5.6 per cent and that of the atmosphere to about 0.02 per cent (Mason and Moore 1985).

8.6.2   The Lithosphere

In terms of dynamic change, the lithosphere shows the slowest rate (Fig. 8.12). The cycle time between the formation of two supercontinents(Wilson cycle) is estimated at around 300 to 500 million years. Over the past 500 million years, centres of mass of the continental plates have moved from almost 80° south latitude originally to around 40° north latitude now. Differences in thermal energy flow from the Earth's mantle through continental and marine plates are assumed to be the driving forces behind lithospheric movements (Versteegh 2005; Coltice et al. 2007, 2009; Grotzinger et al. 2008; Landuyt and Bercovici 2009; Gavin et al. 2011; Rolf et al. 2012; Sorrel et al. 2012; Torsvik et al. 2012; Abe-Ouchi et al. 2013; Hanslmeier 2014; Bercovici et al. 2015). Earth's magnetic polarity shifts—for reasons that remain obscure—at irregular intervals of between 10^5 and 10^7 years (Ogg 2012). In the hydrosphere, the duration of different cycles ranges from around 1 week for atmospheric water to around two million years in Antarctic ice (Wetzel 1975; Oki and Kanae 2006; Yan et al. 2019). In the atmosphere, particles usually last only a number of days, and gas compounds up to hundreds of years (Roedel 1994). In addition to mechanical properties, these ratios are particularly relevant for physical interactions between the individual spheres.

Because of its significantly higher density, the lithosphere determines the structural framework conditions for dynamic processes in the hydrosphere and atmosphere (Fig. 8.13). Its chemical composition has influenced material conditions for all abiotic and biotic processes at all scales—from microscopic to global dimensions—ever since Earth was formed (White 2013). Energetically, elevations on land surfaces—for example mountains—create the kinetic potential for outflow of water to the oceans. In water bodies, geological structures primarily modify flow dynamics (Wetzel 1975; Garrison and Ellis 2016). At the same time, at its interfaces the lithosphere is subject to ongoing changes due to influences of dynamic processes in the hydrosphere and atmosphere (Molnar and England 1990; Spotila et al. 2004; Hornung and Brandner 2005; Berger et al. 2008; Sinclair 2017). Examples of this are erosion by wind, precipitation. Water runoff over land or erosion by glaciers (Tardy et al. 1989; Harrison 1994; Willenbring and von Blanckenburg 2010; Koppes et al. 2015; Kim et al. 2016). Associated with this are chemical and physical weathering processes of rock substance (White 2013). Interaction between water runoff and solid surfaces results in organization of self-similar—fractal—structures with continuous and predictable directions of water flow (Mandelbrot 1991; Rodríguez-Iturbe and Rinaldo 2001). Over the long term, mineral substances are continuously transported in solid or dissolved form from land surfaces into the oceans.

The lithosphere consists of mixtures of minerals with different strengths, packing densities and chemical compositions (Grotzinger et al. 2008). Material properties and relatively low transformation rates permit the evolution of sessile life forms of plants and animals. Sessile animals are found mainly at the boundaries of the hydrosphere, ranging from inland waters—e.g. freshwater sponges—to deep-sea marine floors—e.g. deep-sea corals. Fixed life forms of plants form biological structures between aquatic shallow water zones and high mountains. Fixed animals and algae "adhere" to mineral surfaces by means of their own organic compounds.

Vascular plants anchor themselves with roots in crevices or in the fabric of the mineral subsoil. Their retention force is achieved—utilizing the surface friction of solid materials—through radial compressive forces of the roots and morphological differentiation of the root system (Schwarz et al. 2010, 2011). Distribution of the holding forces over multiple branched roots with varying dimensions not only ensures a supply of water and mineral nutrients, but also minimizes the risk of total loss if individual roots are overloaded. Impressive examples of multifactorial adaptations of roots can be found in mangrove forests with different substrates in intertidal zones of tropical and subtropical waters (Méndez-Alonzo et al. 2015; Srikanth et al. 2016). For similar reasons, mussels also attach themselves with a large number of individual fibres (byssus threads), rather than with a continuous surface of rigid attachments. The principle of a high degree of redundancy is also found in morphological structures on the foot soles of insects or geckos. Independently of the influence of gravity, for mobile animals multiple microscopic "hairs" ensure a

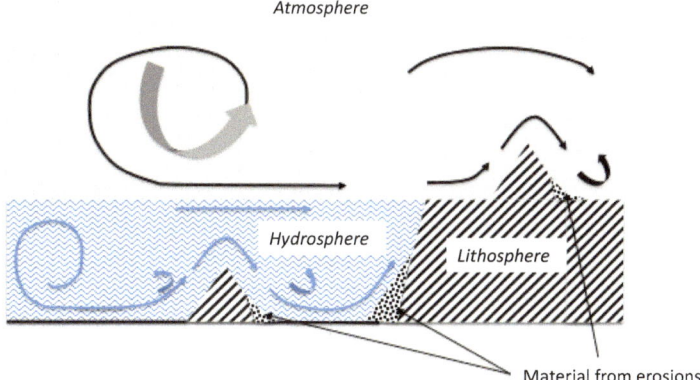

Fig. 8.13 Schematic diagram of dynamic interactions between the spheres. Explanations in the text

dynamic grip on solid bodies with adhesive forces (Vogel 2003; Autumn et al. 2006; Federle 2006; Lee et al. 2007).

Increasing strength and surface friction of mineral substances increases the stability of plants, but impedes root growth and activities of soil animals such as earthworms (Ruiz 2018; Ruiz and Or 2018). Whereas animal activity in mineral substrates on land is limited to loose sediments, the spectrum in oceans ranges from occupation of sediments to boring into solid rock—for example, by paddocks (lithophagous clams) (Dorgan et al. 2006; Suter et al. 2011; Nederlof and Müller 2012; Dorgan 2015).

8.6.3   The Hydrosphere

Categorization into spheres make it easy to forget that the hydrosphere is not composed exclusively of oceans. Seen as a whole, the three-dimensional magnitude of potential habitats ranges from the smallest dewdrop to the global water body of an ocean. Thanks to the surface tension of water, even the smallest droplet can retain its shape under suitable conditions and afford specific living conditions within. Even in solid form—as ice—water provides habitats for microorganisms (Born and Böcher 2001; Christner et al. 2003; Simon et al. 2009a, b; Tebo et al. 2015).

Surface tension of water is based on its polymer-like internal structuring. Within every water body, each molecule is connected by hydrogen bonds to four neighbouring water molecules. At the boundary with air, asymmetrical force ratios exist because suitable bonding partners are available only on the water side. Water molecules at the boundary with air are accordingly more strongly attracted to neighboring water molecules than to air molecules. Various factors, such as temperature, chemical substances or salt content, influence surface tension of the water (Wetzel 1975).

In bodies of water, in contrast to terrestrial habitats, sunlight—in the blue wavelength range—can penetrate to a depth of up to about 200 m. Other wavelengths—for example in the infrared range—are already converted into heat at lesser depths. Photosynthesis is hence also possible in water bodies—if free of suspended matter—far below the surface, up to the limits stated. The photosynthetically utilizable depth of sunlight in various bodies of water is influenced by numerous factors, such as geographical latitude, frequency of cloud cover, light transmission of any existing ice cover, or presence of suspended matter (Littler et al. 1986; Kirk 2011).

Because of the solubility of many substances in water, water bodies differ in their physical and, above all, chemical properties. For oceans, variability is most easily illustrated by changes over the course of Earth's history (Fig. 8.4). Currently, the diversity of chemical substances in terrestrial waters testifies to the range of possible constituents, for example in volcanic waters, springs, deep waters or salt lakes (Wetzel 1975; Druschel et al. 2004; Humphreys 2006; Stetter 2006; Mesbah et al. 2007; Weltzer 2011; White 2013). Compared to the estimated total volume of the hydrosphere of around 1.38×10^9 km^3, present-day exchange of water between oceans and land, at around 40,103 km^3 per year, is relatively insignificant in quantitative terms. In oceans, it is exceeded more than 300-fold by surface currents alone, driven by winds (Trenberth et al. 2011; Garrison and Ellis 2016). These ratios are by no means constant over geological timespans. Model-based estimates assume up to 60 per cent higher exchange rates between oceans and land for the early Phanerozoic—around 505 to 400 million years ago (Tardy et al. 1989).

Dynamics of ice formation on land also influence the relationship between the atmospheric transportation of water on dry land and the outflow of water into oceans. The seemingly static status of ice formation—currently around 2 per cent of global water reserves are stored in glaciers—has undergone highly dynamic changes over the last 100,000 years. During this period alone, extensive glaciations on the mainland have led to multiple fluctuations in global sea level of up to 90 m or more (Wetzel 1975; Trenberth et al. 2011; Hughes et al. 2013). The geographical distribution of glaciations has a significant influence on global thermal exchange processes in the atmosphere and—with the associated near-surface winds—the surface currents (Fig. 8.13) of large bodies of water. Sea ice—in conjunction with the effects of solar radiation on the density and salinity of water at low latitudes—amplifies the influences of wind currents through so-called thermohaline effects (Fig. 8.14).

Thermohaline effects are integrated changes in the specific density of water under the influence of salinity, temperature and pressure. Due to the great variety of possible combinations of factors, some research groups reject that label, preferring to use the term "mass effects". Ultimately, neither term is self-explanatory and those labels should therefore

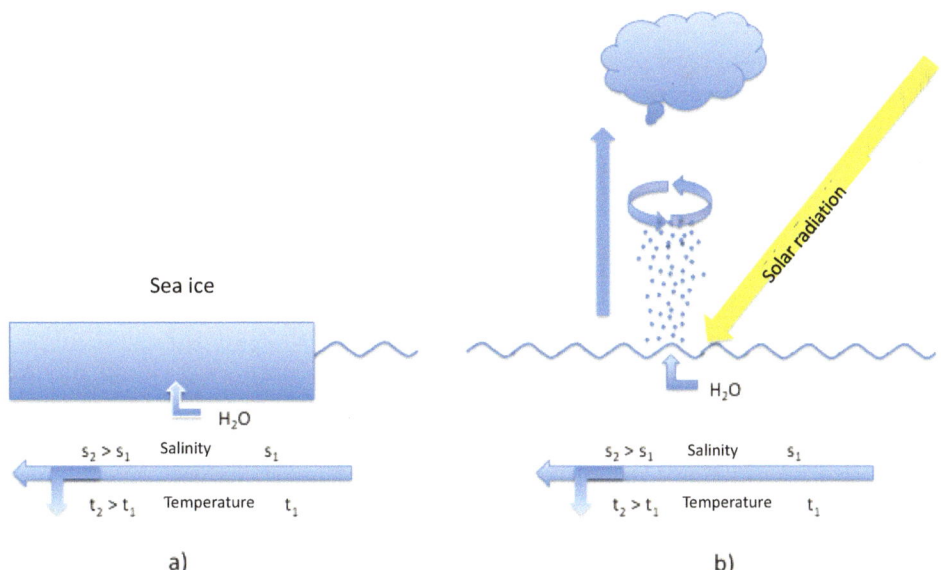

Fig. 8.14 Schematic representation of factors influencing the development of thermohaline circulation: (**a**) by cooling, (**b**) by warming

always be used in conjunction with clear definitions of the influencing factors considered in each case (Wunsch 2002; Hátún et al. 2005; Stouffer et al. 2006; Stewart 2008; Toggweiler and Russell 2008).

In the absence of external influences, layers with increasing densities form from the surface to the deepest zones in stagnant water bodies. In contrast to density differences at the surface—between water and air—differences within the water body are usually much smaller. With decreasing differences in density, the probability of vertical water currents due to external influences increases. Salt-free water reaches its highest specific gravity at around +4 °C, whereas density falls with increasing or decreasing temperatures. At 0 °C, when ice forms, the specific density decreases abruptly by around 8 per cent. Because of its lower density, ice floats on the water surface. Physical processes in fresh water are therefore primarily dominated by thermal effects and—at greater water depths—also by the influence of water pressure. As the salt content of the water increases, its specific density increases and its physical properties change in tandem. Up to a salinity of 24.7 parts per thousand, falling reference temperatures approach the freezing point and maximum density. At higher salinities, maximum density and the freezing point are reached at the same temperature, in seawater with an average salinity of around −2 °C. Because of additional influences of the salt content, physical processes exhibit greater variability in seawater than in fresh water. For example, due to its higher temperature in the Atlantic Ocean, high-salinity Mediterranean water flowing out through the Strait of Gibraltar cannot sink as deeply as the lower-salinity but cooler Atlantic water (Talley et al. 2011; Garrison and Ellis 2016). Even in fresh water, temperature-related density differences alone can bring about the exchange between deep and surface water through local currents (Pannard et al. 2011; Troitskaya et al. 2015).

Sea ice is formed predominantly in polar regions through cooling of open seawater at the edge of ice-free areas—polynyas—by cold air currents (Alexandrov et al. 2000; Dimitrenko et al. 2009; Bauer et al. 2013). The crystal lattice of the ice is formed primarily by water molecules. The residual brine that results partially sinks into the water beneath, increasing its salt content. In highly compressed ice floes, the seeping brine is increasingly replaced by air in the cavities (Nakawo 1983; Hellmer and Olbers 1989; Thomas and Dieckmann 2002). Both in the internal cavity structures and on its underside, sea ice provides habitats for a wide variety of microorganisms (Gosselin et al. 1997; Staley and Gosink 1999; Lizotte 2001; Mock and Thomas 2005; Stecher 2015).

Other effects are associated with formation of ice from fresh water—especially from rainfall. Depending on the processes of formation and storage, the ice can be completely free of air pockets or contain variously distributed cavities of with trapped air compressed up to around 4 bars (Lipenkov et al. 1997). On land, large-scale ice accumulations in the form of glaciers provide habitats for various biotic communities. Chemolithotrophic and syntrophic communities develop in the gaps between the underside of the ice and rock beneath (Boyd et al. 2014). Due to the mineral dust input, photosynthetic communities—often concentrated in small sedimentation cavities (cryoconite holes)—develop on surfaces, as is also the case on sea ice. The mineral subsoil is eroded by ice movements and transported away with meltwater (glacial milk) (Fig. 8.15). With the increase in solar radiation in polar waters, the mineral nutrient input in coastal areas permits rapid development of marine primary production. Through the pumping effect of rising freshwater—by recycling sinking nutrients—fragmenting icebergs reinforce the nutrient cycle in their surroundings (Vincent et al. 2000; Born and Böcher 2001; Arrigo et al. 2002; Mock and Kroon 2002; Aescht 2005; Bodiselitsch et al. 2005; Pollard and Kasting 2005; Elie et al. 2007; Stern et al. 2008; Anesio et al. 2009; Macdonald et al. 2010; Koh et al. 2011, 2012; Boetius et al. 2013; Lutz et al. 2015; Rooney et al. 2015; Stecher 2015; Chen et al. 2016; Shields-Zhou et al. 2016).

Opposing effects have been observed with very large icebergs in the Antarctic. Under unfavourable drift conditions, in comparison with annual average values, an iceberg measuring around 10,000 km^2 reduced biomass production of phytoplankton in the Ross Sea by around 40 per cent (Arrigo et al. 2002; Convey et al. 2009).

An impressive present-day example of the combinatorial effects of thermohaline factors and wind currents is provided by the global current cycle at the surface and in the depths of the world's oceans, with a circulation time of around a thousand years. The amount of kinetic energy converted annually is estimated at around 41,018 joules, which

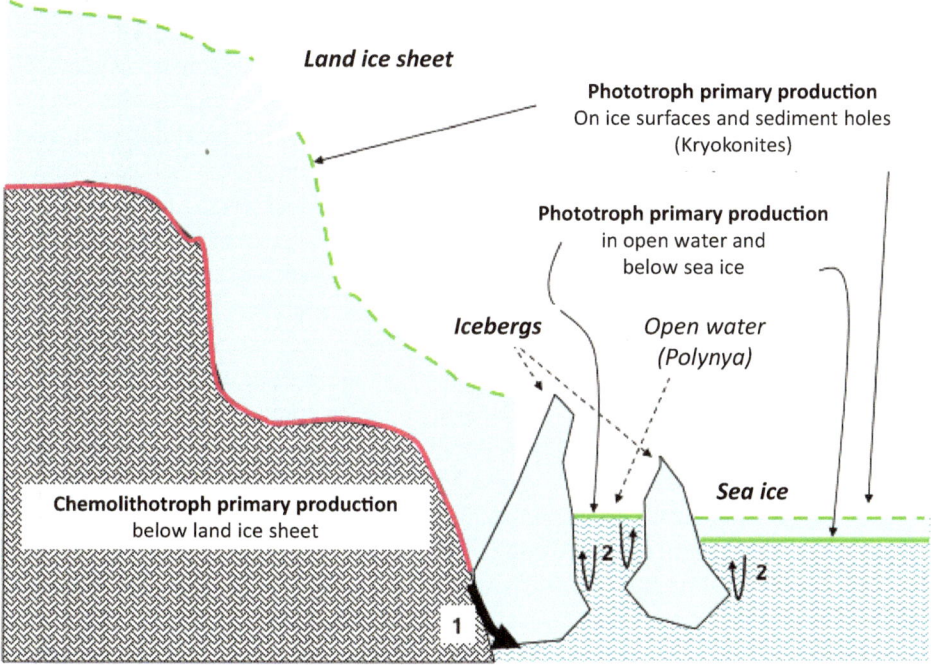

Fig. 8.15 Summarized overview diagram of factors influencing primary production processes during large-scale glaciations. Processes: 1 = transportation of inorganic and organic materials from the mainland to the oceans by ice movements; 2 = transportation of nutrients to the sea surface by rising meltwater on the flanks of icebergs. According to various source, see text

would correspond to around one per cent of global anthropogenic energy conversion in 2004 (Kaltschmitt et al. 2006; Lumpkin and Johnson 2013). The global current receives its greatest impetus from the inflow of the Gulf Stream, with its relatively high salt content, into the Arctic Ocean, and associated cooling of the water. The Gulf Stream is hence part of the global current, surface sections of which are geographically subdivided with numerous regional designations. Due to the shallow depth of the Bering Strait, deep water can flow back into the Atlantic only through the Fram Strait east of Greenland. This generates a deep current—essential for the oxygen supply of benthic habitats—through the Atlantic, which is re-intensified under the sea ice of the Antarctic and channelled into the Pacific. Mixing of deep-sea layers with surface water ensures distribution of oxygen to deep-sea trenches (Gage and Tyler 1999; Jamieson et al. 2010). Due to continuous warming and decreasing salinity, part of the current in the Pacific rises and is transported back to the Atlantic via the Indian Ocean by surface winds. In this way, the Atlantic Ocean receives more freshwater from the South-East Asian region of the Pacific than from all the rivers on the Atlantic coasts (Lohmann 2003; Talley 2008). Because of substantial evaporation in the subtropical zones, salt content in the current increases on the way to the North Atlantic. Hence, in a global comparison of the world's oceans, specific conditions prevail in the Atlantic, such as continuous heat transport from the southern hemisphere across the equator to the Arctic, along with the high salt content in the North Atlantic (Rahmstorf 2002; Talley et al. 2011).

On the ocean surfaces, the global, meandering "current band" is connected to numerous large-scale currents, for Instance: the Antarctic Circumpolar Current, the Pacific equatorial current system and many extensive eddy currents in subtropical and polar regions (Talley et al. 2011). At relatively regular intervals, a comparatively small eddy current in the Arabian Sea transports large quantities of nutrient-rich deep water to the surface. The resulting high biological productivity even allows the existence of a local population of humpback whales (Mann and Lazier 2006; Fleming and Jackson 2011).

Lack of mixing leads to the development of anoxic deep zones even in water bodies with very shallow water depths—for instance the Black Sea or various inland lakes (Wetzel 1975; Arthur and Dean 1998). Supercritical concentrations of gaseous compounds—such as methane or carbon dioxide—can be released in sudden eruptions and exert lethal effects on the surface. One example of this is the abrupt release of large quantities of carbon dioxide of geological origin from Africa's Lake Nyos in August 1986 (Cotel 1999).

Even in the geologically short periods to date of scientific observation of ocean currents—under the influence of the Coriolis force and tidal forces—continuous changes in current patterns have been observed (Krueger and Winston 1974; Fraedrich et al. 1992; Sparrow et al. 2002; Talley et al. 2011). Analysis of satellite images from recent decades has revealed a long-term shift of large-scale oceanic eddy currents to higher latitudes (Yang et al. 2020). In many cases, the direction of the currents can also change at irregular intervals, for example in the Arctic Ocean (Proshutinsky and Johnson 1997; Willmes et al. 2011; Gramling 2015; Proshutinsky et al. 2015). Over geological timespans, processes of plate tectonics and orbital changes in terrestrial solar radiation determine the intensity and spatial course of large-scale ocean currents (Müller et al. 2008). The opening of the Fram Strait around 18 million years ago and ultimately the geological modification of the Bering Strait around 4.5 million years ago were prerequisites for the above-mentioned flow through the Arctic Ocean (Jakobsson et al. 2007; De Schepper et al. 2015; Gasson et al. 2016).

The latter developments coincide with other dramatic changes, such as the temporary, extensive drying out of the Mediterranean around 5.5 million years ago, or interruption of the Central American connection between the Pacific and Atlantic due to uplift of the Isthmus of Panama about 4 million years ago (Haug and Tiedemann 1998; Haug et al. 2001; Warny et al. 2003; O'Dea et al. 2016). Opening of the Drake Strait—the southernmost connection between the Atlantic and the Pacific—began much earlier, around 50 million years ago (Livermore et al. 2007). In conjunction with the subsequent deepening of the oceanic connection between Tasmania and Antarctica around 35 to 30 million years ago, a geological framework was created for development of the circumpolar—eastward— Antarctic current (Stickley et al. 2004; Barker et al. 2007; Bijl et al. 2013; Scher et al. 2015). At about the same time—for reasons that are still debated—the Antarctic Ocean began to freeze over (Coxall et al. 2005; Merico et al. 2008; Pagani et al. 2011; Feakins et al. 2014). In contrast to the Arctic Ocean, this generally isolated marine habitats along the Antarctic coast. As a result, many endemic organisms—for example benthic communities with life forms similar to those in the Cambrian—as well as specific plankton communities were able to evolve (Gili et al. 2006; Houben et al. 2013).

We can only make assumptions about the large-scale course of ocean currents over extensive geological periods. Nevertheless, comparisons between documentation of climate changes by means of fossil finds and on the basis of continental distributions indicate that extensive oceanic areas or continuous ocean connections at lower latitudes are associated with more balanced global climate conditions (Boucot et al. 2013). One indication of this is the polar distribution of dinosaurs in the late Cretaceous period in the context of a continuous marine connection that traversed Europe and the American continent (Benton 1991; Godefroit et al. 2009). Open marine connections to tropical marine regions (Paratethys) were also associated with subtropical conditions in Europe until the late Miocene around 12 million years ago (Rögl 1999; Latal et al. 2006; Harzhauser and Piller 2007; Harzhauser et al. 2011).

Development of thermohaline processes and their effects on exchange between upper and deep water bodies in the oceans can be determined only sporadically. Different research groups have therefore reached divergent conclusions about the extent and interdependencies of anoxic conditions in the oceans (Takashima et al. 2005; Jenkyns 2010). Interpretations based on individual indicators—such as isotope ratios—lead to incorrect conclusions even for recent geological periods (Piller and Harzhauser 2005). Minimal changes in climatic conditions can rapidly alter large-scale current systems and, for example, result in the spread of oxygen-free zones in the depths of seas (Schulz et al. 1998; Erbacher et al. 2001; Beauchamp and Baud 2002; Clark et al. 2002; Rahmstorf et al. 2005; Crucifix 2012; Talley et al. 2016). Reliable reconstruction of events and assessment of biological effects are rendered challenging by the great number of potential influencing factors and their interactions. Large-scale conditions for marine life during early Earth history can be inferred only approximately from congruent results of analyses for different sites (Ye et al. 2015; Knoll and Follows 2016). Detailed analyses of individual sites indicate, for example, repeated expansions of oxygen-free zones at various depths for the period around 253 to 248 million years ago—at the end of the Permian (Clarkson et al. 2016). Due to a lack of adequate evidence, farther-reaching scenarios for living conditions—for example in the Cryogenian—must remain speculative (Trindade and Macouin 2007; Le Heron 2012; Hoffman et al. 2017; Yang et al. 2017).

Turbulent currents can also be triggered in deep sea zones by waves with a very long wavelength—the distance between two adjacent wave crests. In simplified terms, wave movements extend to a depth of half the wavelength. Characteristics and effects of waves depend on whether their lower limit is in open water (deep water wave) or in contact with a solid subsurface (shallow water wave). Waves induced by wind reach maximum wavelengths of around 600 meters, whereas seismic waves—such as tsunamis—attain around 200 km and tidal waves half the circumference of the earth. Apart from coastal areas, wind-induced waves in oceans are always deep-water waves, whereas seismic waves or tidal waves are shallow-water waves. The velocity of deep-water waves is related to their wavelength, while that of shallow-water waves is related to prevailing water depth. The former can reach maximum speeds of around 110 km/h, the latter about 790 km/h at a water depth of 5 km (Garrison and Ellis 2016). If a shallow water wave hits one of the above-mentioned underwater mountains or an island, rapid changes in velocity create turbulence with powerful vertical water movement (D'Asaro et al. 2011; Rosso et al. 2015; Alford et al. 2016). Tidal waves induce water movement over all irregularities of sea basins in regular time rhythms, but seismic waves do so only sporadically in connection with geological events (Fahnenstiel et al. 1988; Holloway and Merrifield 1999; Niwa and Hibiya 2001; Johnston and Merrifield 2003; Garrett and Kunze 2007; Wang et al. 2007; Helo et al. 2011; Gudmundsson 2012; Muacho et al. 2013; McGillicuddy 2016). In the vicinity of "underwater mountains", water currents not only influence dispersal of the specialized fauna of hydrothermal vents, but by means of vertical exchange of nutrients also permit a high diversity of different marine organisms (Cannon and Pashinski 1997; Bower et al. 2002; White and Mohn 2002; Thomson et al. 2003; Tyler et al. 2003; Pitcher et al. 2007; Sutton et al. 2008; Adams et al. 2011; Sievert and Vetriani 2012; Jantzen et al. 2013; Priede et al. 2013; Zhang et al. 2014).

In surface zones of water bodies, wave action exerts a significant influence on the development of phototrophic communities. Key factors are the simultaneous availability of mineral nutrients, sufficient solar radiation for primary production by photosynthetically active organisms, and mechanical forces of wave action. In shallow coastal water areas—which are supplied via flowing water with mineral nutrients from erosion of land surfaces—it is primarily mechanical forces of wave action that influence organismal evolution.

Completely different conditions prevail in zones over deep waters far from the coast. In calm water bodies, photosynthetic organisms rapidly exhaust mineral nutrients borne in by wind transportation and sink rapidly to deeper zones along with the organic remains of dead organisms. Under such conditions, recovery of mineral nutrients through microbial decomposition chains is possible only to a limited extent.

In reality, open water surfaces are exposed to a wide variety of wind currents with specific effects on conditions for organisms in near-surface water zones. Due to surface tension and internal friction (viscosity) of water, even low wind speeds cause the closed water surface to vibrate slightly, which is visually recognizable as small waves (capillary waves). As wind speed increases, stationary cyclical currents develop, which are recognizable on the water surface as waves at right angles to the wind direction (Fig. 8.16). Two-dimensional representations make it easy to forget that pressure differences in liquid and gaseous media always lead to three-dimensional equalization processes. At the interface between water and air, wind and Coriolis forces act on the viscous inertial force of the water in a simplified way (Fig. 8.17).

Depending on the duration, spatial extent and intensity of air currents, as well as the particular framework conditions, differing processes develop in water bodies. On open bodies of water, persistent wind forces of more than 3 to 5 m/s form counter-rotating circulation currents running parallel to the wind direction (Langmuir circulation). Their dynamic structure suggests self-organization processes in which wind-induced "primary circulations" form weaker and counter-rotating "secondary circulations" (Fig. 8.18). The spiraling circulation currents delay sinking of any included plankton along with its nutrients (Wetzel 1975; Noh et al. 2004; Thorpe 2004; Polton and Belcher 2007; Talley et al. 2011; Belcher et al. 2012; Garrison and Ellis 2016). Surface water is also mixed by upward and downward currents between the circulation currents. In shallow water zones, the powerful downward currents also detach and destroy algal growth (Dierssen et al. 2009).

Under large-scale conditions, for instance when the direction of action transitions at the equator or along continental coasts, the effects of the Coriolis force can trigger deeper

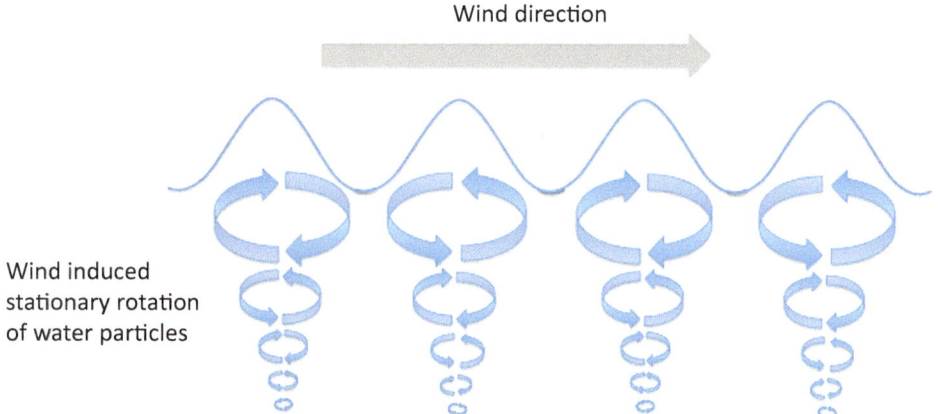

Fig. 8.16 Schematic representation of wind-induced cyclic water flows beneath water surfaces at low wind speeds

Fig. 8.17 Simplified force diagram for wind effects on water surfaces

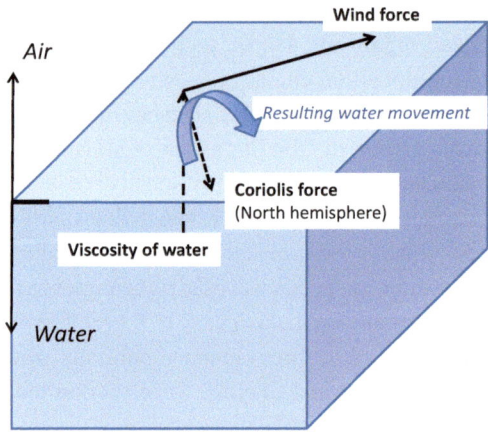

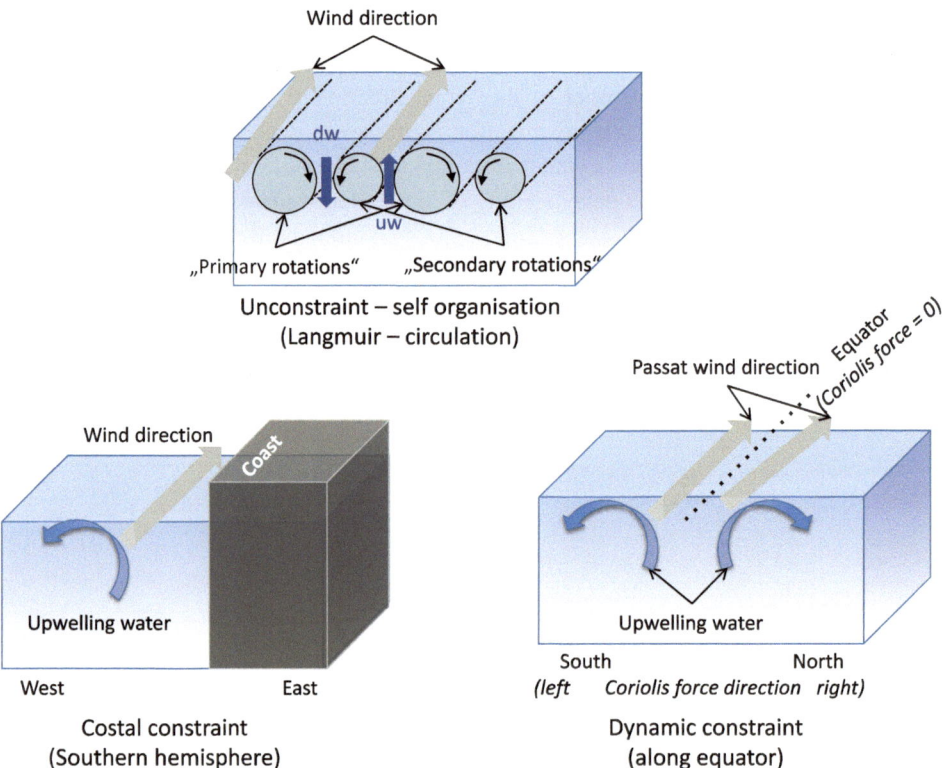

Fig. 8.18 Schematic representation of the different effects of large-scale wind effects on open water surfaces, on coastlines and along the equator. Abbreviations: dw = downward flow, uw = upward flow

uplift processes (Fig. 8.18). The overall effects are based on a complex interplay of dis-placements of surface water, resulting changes in pressure distribution in the water body and the associated uplift of deeper water layers. Such conditions induce a rapid increase in the production of marine algae and all marine organisms that depend on them in certain coastal regions, for example on the Peruvian Pacific coast. Changes in air currents—known as "El Niño"or ENSO (El Niño-Southern Oscillation)—trigger dramatic collapses of the entire food chain (Barber and Chavez 1983; Trillmich and Limberger 1985).

Anomalous warming of tropical oceans at irregular intervals of 2 to 7 years is not a present-day phenomenon. Over around 9000 to 7000 years in the recent geological past, its effects can be detected indirectly in sediment deposits or fossil corals (Conroy et al. 2008; Cobb et al. 2013). Triggered by complex interactions between marine and atmo-spheric processes (Bjerkness feedback), the associated phenomena occur in various zones of the tropical Pacific Ocean. The ecological effects mentioned above are associated with ongoing changes on the eastern Pacific coast (Eastern Pacific El Niño). Changes in the area of the International Date Line (Central Pacific El Niño) primarily affect weather pat-terns in Southeast Asia. In global terms, the first phenomenon is associated with more profound weather anomalies than the second (Walker 1925; Bjerknes 1969; Capotondi et al. 2015; Timmermann et al. 2018; Cai et al. 2019, 2020; Silva et al. 2021).

In addition to the upwelling zone of the Humboldt Current, there are currently three other large global areas with similar functional relationships: the Benguela Current off the coast of Namibia in the southern hemisphere, the California Current off the coast of North America in the northern hemisphere, and the Canary Current off the coasts of Portugal and North-West Africa (Wang et al. 2015). Given the contrary influence of the Coriolis force in the northern hemisphere, the latter two current systems are driven by southward-directed coastal winds.

Large-scale Ekman effects on vertical displacement of water layers under the influence of trade winds permit high-level biological productivity within a band, a few hundred kilometres in width, on either side of the Equator (Fig. 8.18) (Mann and Lazier 2006). The significance of internal ocean currents for the nutrient supply of planktonic organisms is usually greatly underestimated. For example, around 98 per cent of the silicon incorpo-rated into diatoms or radiolarians comes from internal recycling in the oceans. On average, each silicon atom in oceans is recycled between 23 and 53 times by organisms (Tréguer et al. 1995). As a result, linear approaches to explaining the evolution of planktonic organ-isms solely through short-term changes in nutrients imported from terrestrial ecosystems usually do not yield meaningful results (Boyce and Lee 2011; Cermeño et al. 2015).

Due to the influence of the Coriolis force, Ekman effects also occur in eddy currents, persisting independently for several months to years, that are generated by large-scale cur-rents. In addition to major eddy currents—which extend over thousands of kilometres—smaller eddies with diameters of up to about 100 km become detached from the global currents (Fig. 8.19, left-hand image). Within zones of influence connected with eddies, temperature stratification in the oceans can be displaced vertically by several hundred metres (Talley et al. 2011; Callies et al. 2015). Moreover, the eddies in their entirety drift

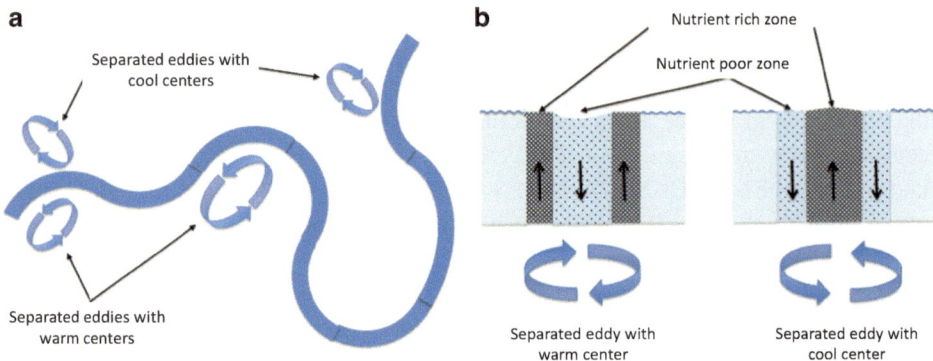

Fig. 8.19 Highly schematic representation of formation of eddy currents in connection with large-scale surface currents (left-hand image), as well as resulting framework conditions for development of biological food chains in zones with eddies (right-hand image). Legend: Blue arrows = directions of rotation of eddies (for the northern hemisphere); black arrows = directions of displacement of water bodies in zones influenced by eddies. Source: considerably modified after Mann and Lazier (2006)

through the oceans in random patterns. Under the influence of the Coriolis force, depending on the direction of rotation, deeper water layers are elevated either in the core (eddy with a cold centre) or in the rotating outer cylinder (eddy with a warm centre) (Fig. 8.19, right-hand image). The eddies rotate anti-clockwise in the northern hemisphere and anti-clockwise in the southern hemisphere. Due to the associated uplift of nutrient-rich water from deeper layers, eddies in nutrient-poor ocean surface zones function like oases, with multiple differentiated food chains. However, overall the effects of eddies depend on the randomness of their drifting paths. In particular, transportation of eddies with warm centres to unsuitable ocean areas can lead to a die-off of organisms living in those areas—for example, individual generations of juvenile fish (Garzoli et al. 1999; Mann and Lazier 2006).

As a result of turbulent currents in water bodies, planktonic organisms are constantly confronted with new, unpredictable challenges from abiotic and biotic factors, following the principle illustrated in Fig. 8.20. Due to fractal characteristics of turbulent currents—which fan out into ever smaller flow patterns—there is a high probability of dynamic formation of "local" populations among microscopically small organisms (Sinclair et al. 1981; Lande and Yentsch 1988; Ragueneau et al. 1996; Peters and Marrasé 2000; Alcaraz et al. 2002; Sarmiento et al. 2003; Winder and Hunter 2008; Lévy et al. 2012; Durham et al. 2013).

Additional challenges for planktonic organisms arise in regions with persistent unfavourable conditions. Examples of this are coastal upwelling zones or polar waters. Under such conditions, organisms must be able to bridge long periods spent under unfavourable conditions and to respond rapidly to favourable conditions. In contrast to terrestrial

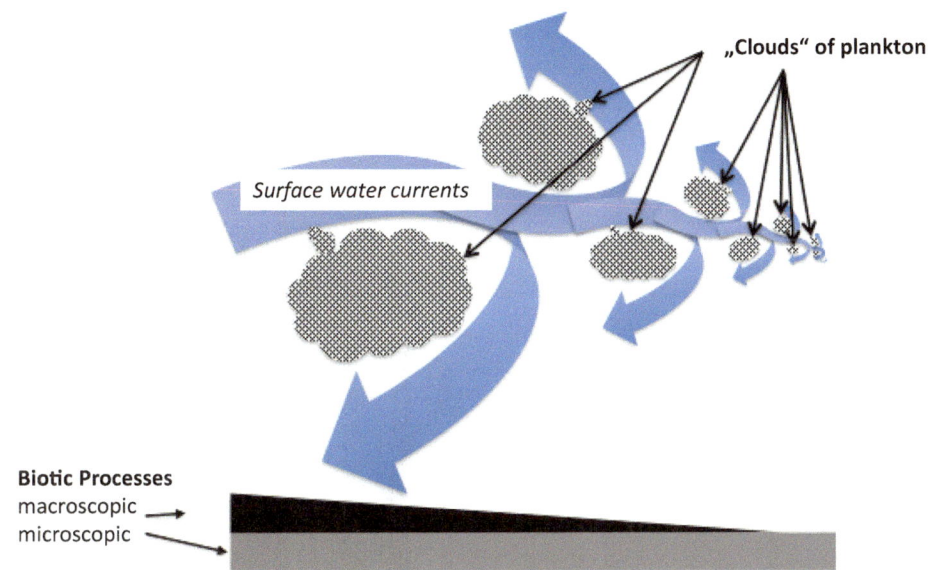

Fig. 8.20 Simplified illustration of dynamic fractal separation of planktonic communities by turbulent water currents. Explanations in the text

habitats, the spatial location of sites for development of favourable conditions is unpredictable.

Comparison of globally prevailing marine currents with the seasonal distribution of chlorophyll concentrations (Fig. 8.21) summarizes the importance of different influencing factors for primary production of plankton. Under present-day conditions, high chlorophyll concentrations are clearly recognizable in the relevant summer periods of the polar seas. In coastal regions, high chlorophyll concentrations are brought about by nutrient inputs from land areas or by atmospherically induced upwelling zones—for example off Peru and Chile or in the Benguela Current. In open seas, large zones with very low chlorophyll concentrations (dark areas) prevail. Only when trade winds travel over sufficiently long distances—for instance in the Pacific Ocean—is the converted energy adequate to yield a more-or-less continuous supply of nutrients to plankton in the equatorial region. However, the nutrient supply for photosynthetic planktonic organisms can also be complemented by a wind-driven input of solid matter from land areas, for example in the Atlantic Ocean from the Sahara region or in the Indian Ocean between the Red Sea and South-East Asia (Westberry et al. 2023). When applied to changes occurring in the Phanerozoic, these examples illustrate the diverse consequences for planktonic organisms associated with dynamics of the lithosphere. Geographical location, size of mainland areas, coastal structures and neighbouring shallow water zones (Zaffos et al. 2017) all influenced global climate and nutrient conditions for the evolution of planktonic organisms during individual timespans through retroactive effects on flow processes in the atmosphere and in water surface zones.

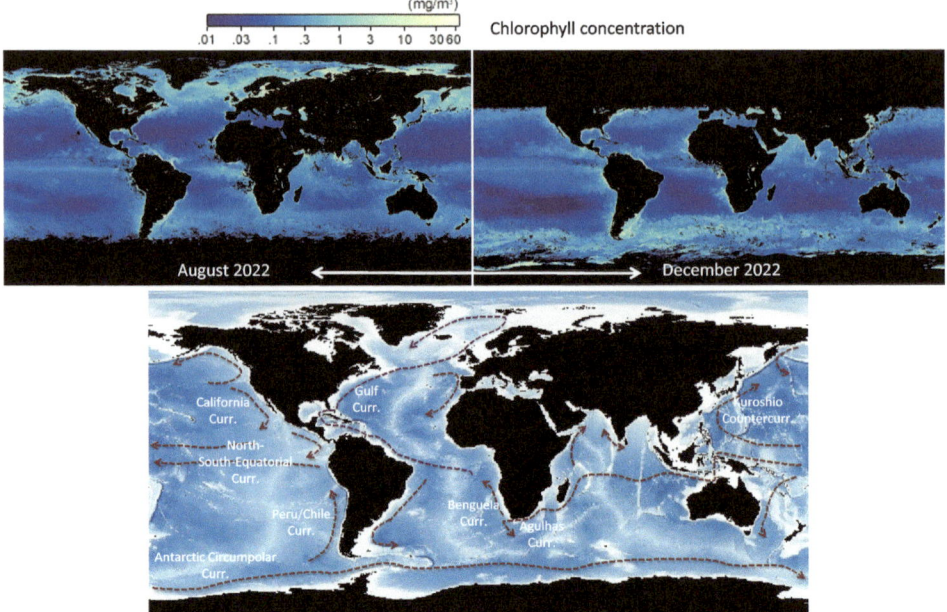

Fig. 8.21 Comparison of distributions of marine chlorophyll concentrations in August and December 2022 with global surface currents. Seasonally varying concentrations in polar regions and relatively uniformly high concentrations in equatorial currents in the Pacific and Atlantic Oceans can be recognized. Further explanations in the text. Credits: Chlorophyll Maps and Ocean Contour Map: NASA Earth Observations (NEO); Ocean Surface Currents: modified from a map provided by R. Lumpkin, NOAA/Atlantic Oceanographic and Meteorological Laboratory

8.6.4   Atmosphere

As Earth's enveloping boundary layer, the atmosphere, with dynamics driven by the solar energy flow, enables transportation of substances against gravity over large distances between land and water surfaces. This alone, in combination with geomorphological conditions, permits exchange of water between oceans and landmasses—the so-called water cycle. Turbulent currents ensure—despite decreasing air pressure—that the molecular conditions of the gas mixture are maintained up to altitudes of around 80 km. With stable air stratification and a temperature of 0 °C, oxygen, for example, would occur only up to an altitude of around 7.5 km (Roedel 1994). Even a simple comparison of quantitative dimensions suggests that biological evolution is confronted with different abiotic conditions in the oceans and on land.

On land, evolutionary conditions for organisms are closely connected to slow local changes in geochemical conditions and rapid events close to the ground associated with fundamentally three-dimensional atmospheric processes (Geiger 1961; Schachtschabel et al. 1989; Claussen et al. 2001). On a global scale, atmospheric flow patterns are generated by thermally triggered differences in air density, while their trajectories are influenced by the Coriolis force and by angular momentum.

Under the influence of solar radiation, heat from condensation induces high-rising air currents on both sides of the equator, which trigger the frequently recurring trade winds in the form of self-similar circulation systems—so-called Hadley cells—near the ground (Fig. 8.22). With respect to geography, trade winds move symmetrically only over large ocean areas in two sets of currents north and south of the equator, usually in a westerly direction. Over continents, both the geographical trajectory and the direction of flow can change significantly over the course of the year (Roedel 1994). Under present-day climatic conditions, circulation cells of varying extent form in the polar regions, whose near-surface currents are deflected westwards by the Coriolis force. The big temperature differences between the polar currents (polar cells) and the subtropical Hadley cells generate three-dimensional currents at mid-latitudes, predominantly in easterly directions (Ferrell cells). Seasonally—influenced by the geographical distribution of water and land areas, along with mountains—chaotic, self-similar flow patterns develop in these cells (Hoinka 1985; Busse 1989; Roedel 1994; Schuster 1994; Röttger 2000; Volkert et al. 2003; Schneider and Walker 2006; Fasullo and Trenberth 2008a; Fasullo and Trenberth 2008b; Jiang and Doyle 2008; Heaviside and Czaja 2013; Horton et al. 2015; Rusotto and Ackerman 2018). Accordingly, geological processes such as mountain uplift lead to long-term changes in the spatial distribution of precipitation. For example: intensification, as currently happens in monsoon regions; attenuation, as in the Taklamakan Desert area in the Tarim Basin due to the uplift of the Himalayas; or occurrence of opposing effects on mountain flanks with different orientations, as in the Andes. However, it should also be noted here that geomorphological conditions influence climatic conditions only in interaction with other factors. Monsoon phenomena, for example, are based on large-scale shifts in trade wind currents. In the northern summer, they are deflected far to the south by a large area of high pressure over Mongolia, whereas during the monsoon season deflection far to the north is caused by low pressure areas over India (Zhisheng et al. 2001; DeCelles et al. 2004; Sun and Liu 2006; Antonelli et al. 2009; Häckel 2021). Air currents also transport all gaseous substances, water vapour and small solid particles. Their spatial displacement is influenced by other factors such as flow velocities, air pressure, concentration of individual components, or interactions between them (Rosenfeld et al. 2008; Dacre et al. 2015; Rädel et al. 2015; Maher et al. 2010). Hence, regional weather developments can be predicted only for very short periods of time. Timespans for estimating locally expected conditions are even shorter (Rind 1999; Peters et al. 2002; Peters and Neelin 2006; Schneider and Walker 2006; Yano et al. 2012).

On land surfaces, terrestrial vegetation has an impact on atmospheric processes within the large-scale induced framework of trends in precipitation and temperature (Walter and Breckle 1983; Zeng et al. 1999; Wang 2003). For example, conversion of solar radiation, near-surface flow conditions of air, or the terrestrial water balance may change (Geiger 1961; van Eimern and Häckel 1979; Keddy 2007; Schlesinger and Jasechko 2014; Wang et al. 2014; Evaristo et al. 2015). Associated changes in water storage durations also

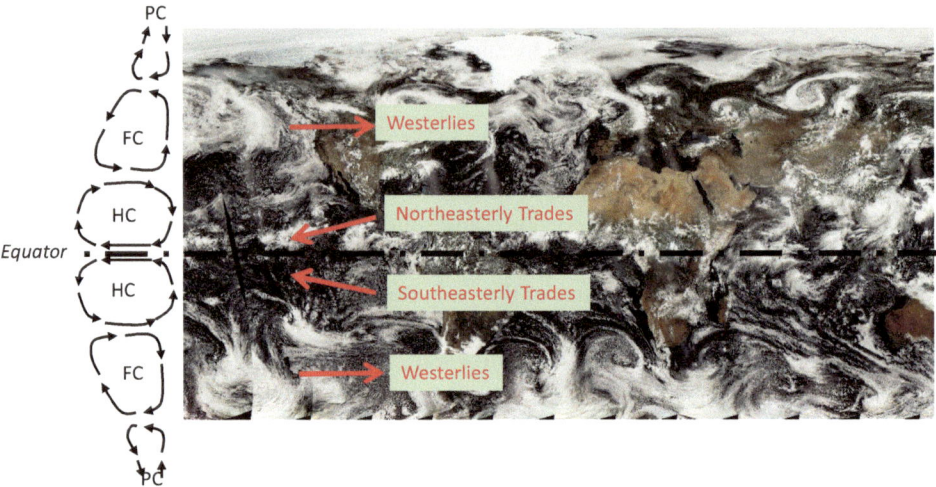

Fig. 8.22 Schematic representation of vertical and horizontal atmospheric flow patterns against the background of global cloud formations on 2 July 2022 (southern polar zone not visible due to the polar night). Abbreviations: FC = Ferrel Cells; HC = Hadley Cells; PC = Polar Cells. Map Credits: NASA Earth Observations (NEO)

intensify development of terrestrial cycles of evaporation and precipitation (Trenberth 1999; Trenberth et al. 2011). Also functionally linked to this are reductions in surface erosion and—in co-action with microorganisms—modifications to geochemical processes (Schachtschabel et al. 1989; Morgan 1999; Keddy 2007; Paul 2007). These feedback effects of vegetation in turn depend on stochastic variability of abiotic processes. Changes in climatic conditions or extreme events—such as heavy rainfall or volcanic eruptions— can transform or destroy plant populations, such that the macroscopic interactions are lost (Zhang et al. 2003; Borelli et al. 2013; Kim et al. 2016). Detailed analyses of approximately 42,000-year-old tree fossils in New Zealand show just how far-reaching those interactions can be. The polarity of Earth's magnetic field was reversed during the lifetime of kauri trees (*Agathis australis*). Associated phases of attenuation in the magnetic field presumably triggered global climate changes with far-reaching consequences. During this period, for example, the Laurentide ice sheet in North America expanded and the Intertropical Convergence Zone moved further south. Accompanying changes may also have contributed to the extinction of the Neanderthals (Cooper et al. 2021). The spectrum of interactions between organisms and abiotic conditions on land was—and is—much broader than conveyed by one-dimensional depictions in the current climate debate (Jung et al. 2011; Knoflacher 2017).

While plant growth leads to self-reinforcing developments in terrestrial ecosystems, dynamics of the atmosphere promote the input of mineral nutrients from vegetation-poor land areas into the oceans. Examples of this are the large-scale atmospheric input of dust from the Sahara into the Atlantic Ocean or clear indications in sediments of the Pacific

Ocean of similar processes connected with temporarily exposed areas resulting from large-scale melting of glaciers after ice ages (Rea 1990; Walsh and Steidinger 2001; Okin et al. 2011; Zielhofer et al. 2017).

Waves and water currents with various effects are generated on water surfaces by the transfer of atmospheric kinetic energy (Wetzel 1975). Rotational movements of waves bring about stationary vertical mixing of water, while currents transport it horizontally (Fig. 8.18). Flow direction and transportation distance are influenced primarily by consistency of the air currents and structures of adjacent coasts, as well as by impulses from Earth's rotation (Coriolis force). While air movements mix water bodies only up to a depth of about 150 m, different water densities generate currents throughout the entire water body. Because such currents were first observed in oceans, they are also referred to as thermohaline currents—referring to the main influencing factors (Urakawa and Hasumi 2009). However, such currents can also develop in large freshwater lakes, for example under a closed covering of ice in Lake Baikal (Kouraev et al. 2019). As a result, oxygen-rich water can be transported down to the floor of water bodies and nutrient-rich water up to the surface from deep zones. Examples of this are the present-day currents in the Atlantic Ocean, which are driven mainly by water that has been cooled by Arctic ice sinking down from the Gulf Stream (Böhm et al. 2015). Over the long term, these currents are just as unstable as wind-induced currents (Schmittner 2016; van Westen et al. 2024).

References

Abbot DS, Voigt A, Li D, Le Hir G, Pierrehumbert RT et al (2013) Robust elements of snowball earth atmospheric circulation and oases for life. J Geophys Res Atm 118:1–11

Abe-Ouchi A, Saito F, Kawamura K, Raymo ME, Okuno J et al (2013) Insolation driven 100,000-year glacial cycles and hysteresis of ice-sheet volume. Nature 500:190–193

Abram NJ, Wright NM, Ellis B, Dixon BC, Wurtzel JB et al (2020) Coupling of Indo-Pacific climate variability over the last millennium. Nature 579:385–393

Abramov O, Mojzsis SJ (2009) Microbial habitability of the Hadean Earth during the late heavy bombardment. Nature 459:419–422

Adams DK, McGillicuddy DJ, Zamudio L, Thurnherr AM, Liang X et al (2011) Surface-generated mesoscale eddies transport Deep-Sea products from hydrothermal vents. Science 332:580–583

Aescht E (2005) Ciliaten (Protozoa) im Eisstaub (Kryokonit) zweier Gletscher der Ötztaler Alpen (Tirol, Österreich). Ber Nat Med Verein Innsbruck 92:89–93

Agić H, Moczydłowska M, Yin L-M (2015) Affinity, life cycle, and intracellular complexity of organic-walled microfossils from Mesoproterozoic of Shanxi, China. J Paleontol 89(1):28–50

Aguirre J, Rios-Momberg M, Hewitt D, Hansberg W (2005) Reactive oxygen species and development in microbial eukaryotes. Trends Microbiol 13(3):111–118

Alcaraz M, Marrasé C, Peters F, Arin L, Malits A (2002) Effects of turbulence conditions on the balance between production and respiration in marine communities. Mar Ecol Prog Ser 242:63–71

Alegret L, Thomas E, Lohmann KC (2012) End-cretaceous marine mass extinction not caused by productivity collapse. PNAS 109(3):728–732

Alexandrov VY, Martin T, Kolatschek J, Eicken H, Kreyscher M, Makshtas AP (2000) Sea ice circulation in the Laptev Sea and ice export to the Arctic Ocean: results from satellite remote sensing and numerical modelling. J Geophys Res 105(C7):17143–17159

Alford MH, MacKinnin JA, Simmons HL, Nash JD (2016) Near-inertial internal gravity waves in the ocean. Ann Rev Mar Sci 8:95–123

Allmon WD, Bottjer DJ (2001) Evolutionary paleoecology. Columbia University Press, Chichester

Anbar AD, Knoll AH (2002) Proterozoic ocean chemistry and evolution: a bioinorganic bridge? Science 297:1137–1142

Anbar AD, Duan Y, Lyons TW, Arnold GL, Kendall B et al (2007) A whiff of oxygen before the great oxidation event? Science 317:1903–1906

Anesio AM, Hodson AJ, Fritz A, Psenner R, Sattler B (2009) High microbial activity on glaciers: importance of the global carbon cycle. Glob Chang Biol. https://doi.org/10.1111/j.1365-2486.2008.01758.x

Antonelli A, Nylander JAA, Persson C, Sanmartin I (2009) Tracing the impact of the Andean uplift on Neotropical plant evolution. PNAS 106(24):9749–9754

Archibald S, Staver AC, Levin SA (2012) Evolution of human-driven fire regimes in Africa. PNAS 109(3):847–852

Arens NC, West ID (2008) Press-pulse: a general theory of mass extinction? Paleobiology 34(4):456–471

Arnol'd VI (1991) Kolmogorov's hydrodynamic attractors. Proc R Soc Lond A 434:19–22

Arrigo KR, van Dijken GL, Ainley DG, Fahnestock MA, Thorsten M (2002) Ecological impact of a large Antarctic iceberg. Geophys Res Lett 29(7):1104

Arthur MA, Dean WE (1998) Organic-matter production and preservation and evolution of anoxia in the Holocene Black Sea. Paleoceanography 15(4):395–411

Autumn K, Dittmore A, Santos D, Spenko M, Cutkosky M (2006) Frictional adhesion: a new angle on gecko attachment. J Exp Biol 209:3569–3579

Averbuch O, Tribovillard N, Devleeschouwer X, Riquier L, Mistiaen B, van Vliet-Lanoe B (2005) Mountain building-enhanced continental weathering and organic carbon burial as major causes for climate cooling at the Frasnian-Famennian boundary (c. 376 Ma)? Terra Nova 17:25–34

Bachan A, Kump LR (2015) The rise of oxygen and siderite oxidation during the Lomagundi event. PNAS 112(21):6562–6567

Baele G, Lerney P, Bedford T, Rambaut A, Suchard MA, Alekseyenko AV (2012) Improving the accuracy of demographic and molecular clock model comparison while accommodating phylogenetic uncertainty. Mol Biol Evol 29(9):2157–2167

Bahlburg H, Breitkreuz C (2012) Grundlagen der Geologie. Springer Spektrum, Berlin

Bajo P, Drysdale RN, Woodhead JD, Hellstrom JC, Hodell D et al (2020) Persistent influence of obliquity on ice age terminations since the middle Pleistocene transition. Science 367:1235–1239

Bak P (1996) How nature works. Copernicus, New York

Baker DN, Li X, Pulkkinen A, Ngwira CM, Mays ML et al (2013) A major solar eruptive event in July 2012: defining extreme space weather scenarios. Space Weather 11:585–591

Balázsi G, van Oudenaarden A, Collins JJ (2011) Cellular decision making and biological noise: from microbes to mammals. Cell 144:910–925

Bambach RK, Knoll AH, Wang SC (2004) Origination, extinction, and mass depletions of marine diversity. Paleobiology 30(4):522–542

Bao H, Lyons JR, Zhou C (2008) Triple oxygen isotope evidence for elevated CO_2 levels after a Neoproterozoic glaciation. Nature 453:504–506

Barash MS (2013) Interaction of the reasons for the mass biota extinctions in the phanerozoic. Okeanologiya 53(6):825–837

Barber RT, Chavez FP (1983) Biological consequences of El Niño. Science 222:1203–1210

Barker PF, Filipelli GM, Florindo F, Martin EE, Scher HD (2007) Onset and role of the antarctic circumpolar current. Deep-Sea Res II 54:2388–2398

Barlow PW, Mikulecký M Sr, Střeštík J (2010) Tree-stem diameter fluctuates with the lunar tides and perhaps with geomagnetic activity. Protoplasma 247:25–43

Battistuzzi FU, Hedges SB (2009) A major clade of prokaryotes with ancient adaptations to life on ôLand. Mol Biol Evol 26(2):335–343

Bauer M, Schröder D, Heinemann G, Willmes S, Ebner L (2013) Quantifying polynya ice production in the Laptev Sea with the COSMO model. Polar Res 32:20922

Baumgartner FA, Pavia H, Toth GB (2015) Acquired phototrophy through retention of functional chloroplasts increases growth efficiency of the sea slug *Elysia viridis*. PLoS One 10(4):e0120974

Beauchamp B, Baud A (2002) Growth and demise of Permian biogenic chert along northwest Pangea: evidence for end-Permian collapse of thermohaline circulation. Palaeogeo Palaeoclim Palaeoecol 184(1–2):37–63

Behrensmeyer AK, Kidwell SM, Gastaldo RA (2000) Taphonomy and paleobiology. Paleobiology 26(4):103–147

Bekker A (2014) Huronian glaciation. In: Encyclopedia of astrobiology. Springer, Berlin. https://doi.org/10.1007/978-3-642-27833-4-742-4

Bekker A, Kaufmann AJ (2007) Oxidative forcing of global climate change: a biogeochemical record across the oldest Paleoproterozoic ice age in North America. Earth Planet Sci Lett 258:486–499

Bekker A, Holland HD, Wang PL, Rumble D, Stein HJ et al (2004) Dating the rise of atmospheric oxygen. Nature 427:117–120

Belcher SE, Grant ALM, Hanley KE, Fox-Kemper B, Van Roekel L et al (2012) A global perspective on Langmuir turbulence in the ocean surface boundary layer. Geophys Res Lett 39:L18605

Bell EA, Boehnke P, Harrison TM, Mao WL (2015) Potentially biogenic carbon preserved in a 4.1 billion-year-old zircon. PNAS 112(47):14518–14521

Belnap J, Lange OL (eds) (2003) Biological soil crusts: structure, function, and management. Ecological studies, vol 150, Springer, Ther Ber

Bengtson S (2002) Origins and early evolution of predation. Paleontological Society Papers, vol 8, pp 289–318

Bengtson S, Sallstedt T, Belivanova V, Whitehouse M (2017) Three-dimensional preservation of cellular and subcellular structures suggests a1.6 billion-year-old crown-group red algae. PLoS Biol. https://doi.org/10.1371/J.pbio.2000735

Benton MJ (1991) Polar dinosaurs and ancient climate. TREE 6(1):28–30

Benton MJ (1995) Diversification and extinction in the history of life. Science 268:52–58

Benton MJ (2018) Hyperthermal-driven mass extinctions: killing models during the Permian-Triassic mass extinction. Phil Trans R Soc A 376:20170076

Beraldi-Campesi H (2013) Early life on land and the first terrestrial ecosystems. Ecol Process 2(1)

Berbee ML, Taylor JW (2001) Fungal molecular evolution: gene trees and geologic time. In: McLaughlin DJ, McLaughlin EG, Lemke PA (eds) The Mycota VII part B. Springer, Berlin, pp 220–245

Bercovici D, Tackley PJ, Ricard Y (2015) The generation of plate tectonics from mantle dynamics. In: Schubert G (ed) Treatise on geophysics, Mantle dynamics, vol 7. Elsevier, Amsterdam, pp 271–318

Berger A (1977) Long-term variations of the earth's orbital elements. Celest Mech 15:53–74

Berger AL, Gulick SPS, Spotila JA, Upton P, Jaeger JM et al (2008) Quaternary tectonic response to intensified glacial erosion in an orogenic wedge. Nature GeoSci 1:793–799

Berleman JE, Kirby JR (2009) Deciphering the hunting strategy of a bacterial wolfpack. FEMS Microbiol Rev 33:942–957

Berman-Frank I, Lundgren P, Falkowski P (2003) Nitrogen fixation and photosynthetic oxygen evolution in cyanobacteria. Res Microbiol 154:157–164

Berna F, Goldberg P, Horwitz LK, Brink J, Holt S et al (2012) Microstratigraphic evidence of in situ fire in the Acheulean strata of Wonderwerk Cave, Northern Cape province, South Africa. PNAS:e1215–e1220

Berner RA (1998) The carbon cycle and CO_2 over Phanerozoic time: the role of land plants. Phil Trans R Soc Lond B 353:75–82

Berner RA (2002) Examination of hypotheses for the Permo-Triassic boundary extinction by carbon cycle modeling. PNAS 99(7):4172–4177

Berney C, Pawlowski J (2006) A molecular time-scale for eukaryote evolution recalibrated with the continuous microfossil record. Proc R Soc B 273:1867–1872

Bijl PH, Bendle JAP, Bohaty SM, Pross J, Schouter S et al (2013) Eocene cooling linked to early flow across the Tasman Gateway. PNAS 110(24):9645–9650

Biswas AK, Tortajada C (2012) Impacts of the Hich Aswan Dam. In: Tortajada C, Altinbilek D, Biswas AK (eds) Impacts of large dams a global assessment. Springer, Berlin, pp 329–395

Bjerknes J (1969) Atmospheric teleconnections from the equatorial pacific. Mon Weather Rev 97(3):163–172

Blanchon P, Shaw J (1995) Reef drowning during the last deglaciation: evidence for catastrophic sea-level rise and ice-sheet collapse. Geology 23(1):4–8

Blank CE (2013) Origin and early evolution of photosynthetic eukaryotes in freshwater environments: reinterpreting proterozoic paleobiology and biogeochemical processes in light of trait evolution. J Phycol 49(6):1040–1055

Blank CE, Sánchez-Baracaldo P (2010) Timing of morphological and ecological innovations in the cyanobacteria—a key to understanding the rise in atmospheric oxygen. Geobiology 8:1–23

Boden MJ, Kennaway DJ (2006) Circadian rhythms and reproduction. Reproduction 132:379–392

Bodiselitsch B, Koeberl C, Master S, Reimold WU (2005) Estimating duration and intensity of neoproterozoic snowball glaciations from Ir anomalies. Science 308:239–242

Boehrer B, Dietz S, von Rohden C, Kiwel U, Jöhnk KD et al (2009) Double-diffusive deep water circulation in an iron-meromictic lake. Geochem Geophys Geosyst 10(6). https://doi.org/10.1029/2009GC002389

Boetius A, Albrecht S, Bakker K, Bienhold C, Felden J et al (2013) Export of algal biomass from the melting Arctic Sea ice. Science 339:1430–1432

Böhm E, Lippold J, Gutjahr M, Frank M, Blaser P et al (2015) Strong and deep Atlantic meridional overturning circulation during the last glacial cycle. Nature 517:73–76

Bond DPG, Grasby SE (2017) On the causes of mass extinctions. Palaeogr Palaeoclimatol Palaeoecol 478:3–29

Bond WJ, Scott AC (2010) Fire and the spread of flowering plants in Cretaceous. New Phytol 188:1137–1150

Bonen L, Vogel J (2001) The ins and outs of group II introns. Trends Genet 17(6):322–331

Borelli P, Robinson DA, Fleischer LR, Lugato E, Ballabio C et al (2013) An assessment of the global impact of 21st century land use change on soil erosion. Nat Commun 8. https://doi.org/10.1038/s41467-017-02142-7

Born EW, Böcher J (2001) The ecology of Greenland. Atuakkiorfik Education, Nuuk

Bosak T, Macdonald F, Lahr D, Matys E (2011) Putative Cryogenian ciliates from Mongolia. Geology 39(121):1123–1126

Bosak T, Lahr DJG, Pruss SB, Macdonald FA, Gooday AJ et al (2012) Possible early foraminiferans in post-Sturtian (716-636) cap carbonates. Geology 49(1):67–70

Boucot A, Xu C, Scotese CR, Morley RJ (2013) Phanerozoic paleoclimate: an atlas of lithologic indicators of climate. SEPM, Tulsa

Bower AS, Le Cann B, Rossby T, Zenk W, Gould J et al (2002) Directly measured mid-depth circulation in the northeastern North Atlantic Ocean. Nature 419:603–607

Bowman DMJS, Balch JK, Artazo P, Bond WJ, Carlson JM et al (2009) Fire in the earth system. Science 324:481–484

Bowyer F, Wood RA, Poulton SW (2017) Controls on the evolution of Ediacaran metazoan ecosystems: a redox perspective. Geobiology 15:516–551

Boyce CK, Lee J-E (2011) Could land plant evolution have fed the marine evolution? Paleontol Res 15(2):100–105

Boyce CK, Ibarra DE, D'Antonio MPD (2023) What we talk about when we talk about the long term carbon cycle. New Phytol 237:1550–1557

Boyd ES, Hamilton TL, Havig JR, Skidmore ML, Shock EL (2014) Chemolithotrophic primary production in a subglacial ecosystem. Applied and Environmental Microb 80(19):6146–6153

Boyle RA, Lenton TM, Williams HTP (2007) Neoproterozoic, snowball earth glaciations and the evolution of altruism. Geobiology. https://doi.org/10.1111/j.1472-4669

Bradley DC (2011) Secular trends in the geologic record and the supercontinent cycle. Earth Sci Rev 108:16–33

Bradshaw SD, Dixon KW, Hopper SD, Lambers H, Turner SR (2011) Little evidence for fire-adapted plant traits in Mediterranean climate regions. Trends Plant Sci 16(2):69–76

Brain CK, Prave AR, Hoffmann K-H, Fallick AE, Botha A et al (2012) The first animals: ca. 760-million-year-old sponge-like fossils from Namibia. S Afr J Sci 108(1–2) Art #658

Brasier AT (2014) Archean soils, lakes and springs: looking for signs of life. In: Dilek Y, Furnes H, Muehlenbachs K (eds) Links between geological processes, microbial activities and evolution of life. Springer Science+Business Media, pp 367–384

Brasier M, McLoughlin N, Green O, Wacey D (2006) A fresh look at the fossil evidence for early Archaean cellular life. Philos Trans R Soc B 361:887–902

Brasier MD, Callow RHT, Menon LR, Liu AG (2010) Osmotrophic biofilms: from modern to ancient. In: Seckbach J, Oren A (eds) Microbial mats. Cellular origin, life in extreme habitats and astrobiology, vol 14. Springer, Dordrecht, pp 131–148

Brasier MD, Antcliffe JB, Callow RHT (2011) Evolutionary trends in remarkable fossil preservation across Ediacaran-Cambrian transition and the impact of metazoan mixing. In: Allison PA, Bottjer DJ (eds) Taphonomy: processes and bias through time. Springer Science+Business Media, pp 519–567

Brasier AT, Culwick T, Battison L, Callow RHT, Brasier MD (2016) Evaluating evidence from the Torridonian Supergroup (Scotland, UK) for eukaryotic life on land in the proterozoic. Geol Soc Lond Spec Publ 448:121–144

Briggs DEG (2015) The Cambrian explosion. Curr Biol 25:R845–R875

Brock JJ, Buick R, Summons RE, Logan GA (2003) A reconstruction of Archean biological diversity based on molecular fossils from the 2.78 to 2.43 billion-year-old Mount Bruce Supergroup, Hamersley Basin, Western Australia. Geochim Cosmochim Acta 67(22):4321–4335

Brocks JJ (2005) Sedimentary hydrocarbons, biomarkers for early life. In: Schlesinger WH (ed) Biogeochemistry. Elsevier, Amsterdam

Brocks JJ, Butterfield NJ (2009) Early animals out of the cold. Nature 457:672–673

Brocks JJ, Jarrett AJM, Sirantoine E, Hallmann C, Hoshino Y, Liyanage T (2017) The rise of algae in Cryogenian oceans and the emergence of animals. Nature 548:578–581

Brook BW, Sodhi NS, Bradshaw CJA (2008) Synergies among extinction drivers under global change. Trends Ecol Evol 23(8):453–460

Brune S, Williams S, Butterworth SP, Müller RD (2016) Abrupt plate accelerations shape rifted continental margins. Nature 536:201–204

Buchholz F, Lerchner J, Mariana F, Kuhlicke U, Neu TR et al (2012) Chip-calorimetry provides real time insights into the innovations of biofilms by predatory bacteria. Biofouling 28(3):351–362

Budd GE (2008) The earliest fossil record of the animals and its significance. Philos Trans R Soc B 363:1425–1434

Buggeln R, Rynk R (2002) Self-heating in yard trimmings: conditions leading to spontaneous combustion. Compost Sci Util 10(2):162–182

Burgess SD, Bowring SA (2015) High-precision geochronology confirms voluminous magmatism before, during and after earth's most severe extinction. Sci Adv:e1500470

Burkholder JM, Glibert PM, Skelton HM (2008) Mixotrophy, a major mode of nutrition for harmful algal species in eutrophic waters. Harmful Algae 8:77–93

Busch J, Ferretti-Gallon K, Engelmann J, Wright M, Austin KG et al (2015) Reductions in emissions from deforestation from Indonesia's moratorium on new oil palm, timber, and logging concessions. PNAS 112(5):1328–1333

Busse F (1989) Dynamische Strukturbildung in Flüssigkeiten und Planetenatmosphären. In: Gerok W (ed) Ordnung und Chaos. Hirzel, Stuttgart, pp 283–296

Butterfield NJ (2000) *Bangiomorpha pubescens* n. gen., n. sp.; implications for the evolution of sex, multicellularity, and the mesoproterozoic/neoproterozoic radiation of eukaryotes. Paleobiology 26(3):386–404

Butterfield NJ (2014) Early evolution of the eukaryota. Palaeontology:1–13

Cai W, Wu L, Lengaigne M, Li T, McGregor S et al (2019) Pantropical climate interactions. Science 363:eaav4236

Cai W, Ng B, Geng T, Wu L, Santoso A, McPhaden MJ (2020) Butterfly effects and self-modulating El Niño response to global warming. Nature 585:68–73

Callies J, Ferrari R, Klymak JM, Gula J (2015) Seasonality in submesoscale turbulence. Nat Commun. https://doi.org/10.1038/ncomms7862

Cane MA (2005) The evolution of El Niño, past and future. Earth Planet Sci Lett 230:227–240

Canfield DE (2005) The early history of atmospheric oxygen. Annu Rev Earth Planet Sci 33:1–36

Canfield DE, Poulton SW, Narbonne GM (2007) Late-neoproterozoic deep-ocean oxygenation and the rise of animal life. Science 315:92–95

Canfield DE, Poulton SW, Knoll AH, Narbonne GM, Ross G et al (2008) Ferruginous conditions dominated later neoproterozoic deep-water chemistry. Science 321:949–952

Canfield DE, Ngombi-Pemba L, Hammarlund EU, Bengtson S, Chaussidon M et al (2013) Oxygen dynamics in the aftermath of the great oxidation of earth's atmosphere. PNAS 110(42):16736–16741

Cannon GA, Pashinski DJ (1997) Variations in mean currents affecting hydrothermal plumes on the Juan de Fuca Ridge. J Geophys Res 102(C11):24965–24976

Capotondi A, Wittenberg AT, Newman M, Di Lorenzo E, Yu J-Y et al (2015) Understanding ENSO diversity. BAMS. https://doi.org/10.1175/BAMS-D-13-00117.1

Carlile MJ, Watkinson SC, Gooday GW (2001) The fungi. Elsevier, Amsterdam

Cartwright P, Collins A (2007) Fossils and phylogenies: integrating multiple lines of evidence to investigate the origin of early major metazoan lineages. Integr Comp Biol 47(5):744–751

Cascales-Miñana B, Diez JB, Gerrienne P, Cleal CJ (2015) A palaeobotanical perspective on the great end-Permian biotic crisis. Hist Biol. https://doi.org/10.1080/08912963.2015.1103237

Catling DC, Claire MW (2005) How earth's atmosphere evolved to an oxic state: a status report. Earth Planet Sci Lett 237:1–20

Cermeño P, Falkowski PG, Romero OE, Schaller MF, Vallina SM (2015) Continental erosion and the Cenozoic rise of marine diatoms. PNAS 112(14):4239–4244

Chang P, Ji L, Wang B, Li T (1995) Interactions between the seasonal cycle and El Niño-Southern Ocean oscillation in an intermediate coupled ocean-atmosphere model. American Meteorological Society, pp 2353–2372

Chatterjee A, Gierach MM, Sutton AJ, Feely RA, Crisp D et al (2017) Influence of El Niño on atmospheric CO_2 over the tropical Pacific Ocean: findings from NASA's OCO-2 mission. Science 358:eamm5776

Chen T, Robinson LF, Beasley MP, Claxton LM, Andersen MB et al (2016) Ocean mixing and ice-sheet control of seawater $^{234}U/^{238}U$ during the last deglaciation. Science 354:626–629

Chiang JCH, Atwood AR, Vimont DJ, Nicknish PA, Roberts WHG et al (2022) Two annual cycles of the Pacific cold tongue under orbital precession. Nature 611:295–300

Christner BC, Mosley-Thompson E, Thompson LG, Reeve JN (2003) Bacterial recovery from ancient glacial ice. Environmental Microb 5(5):433–436

Clark PU, Pisias PU, Stocker TF, Weaver AJ (2002) The role of the thermohaline circulation in abrupt climate change. Nature 415:863–869

Clarkson MO, Kasemann SA, Wood RA, Lenton TM, Daines SJ et al (2015) Ocean acidification and the Permo-Triassic mass extinction. Science 348(6283):229–232

Clarkson MO, Wood RA, Poulton SW, Richoz S, Newton RJ et al (2016) Dynamic anoxic ferruginous conditions during the end-Permian mass extinction and recovery. Nat Commun. https://doi.org/10.1038/ncomms12236

Clauser C (2014) Einführung in die Geophysik. Springer Spektrum, Berlin

Claussen M, Brovkin V, Ganopolski A (2001) Biogeophysical versus biogeochemical feedbacks of large-scale land cover change. Geophys Res Lett 28(6):1011–1014

Cobb KM, Westphal N, Sayani HR, Watson JT, Di Lorenzo E et al (2013) Highly variable El Niño-southern oscillation throughout the holocene. Science 339:67–70

Cockell CS, Horneck G (2001) The history of the UV radiation climate of the earth—theoretical and space-based observations. Photochem Photobiol 73(4):447–451

Cohen PA, Kodner RB (2022) The earliest history of eukaryotic life: uncovering an evolutionary story through the integration of biological and geological data. Trends Ecol Evol 37(3):246–256

Cohen PA, Knoll AH, Kodner RB (2009) Large spinose microfossils in Ediacaran rocks as resting stages of early animals. PNAS 106(16):6519–6524

Coltice N, Phillips BR, Bertrand H, Ricard Y, Rey P (2007) Global warming of the mantle at the origin of flood basalts over supercontinents. Geology 35(5):391–394

Coltice N, Bertrand H, Rey P, Jourdan F, Phillips BR, Ricard Y (2009) Global warming of the mantle beneath continents back to the Archaean. Gondw Res 15:254–266

Condie KC (2005) Earth as an evolving planetary system. Elsevier, Amsterdam

Condie K, Pisarevsky SA, Korenaga J, Gardoll S (2015) Is the rate of supercontinent assembly changing with time? Precambrian Res 259:278–289

Conroy JL, Overpeck JT, Cole JE, Shanahan TM, Steinitz-Kannan M (2008) Holocene changes in eastern tropical Pacific climate inferred from a Galápagos lake sediment record. Quaternary Sci Rev 27:1166–1180

Convey P, Bindschadler R, Di Prisco G, Fahrbach E, Gutt J et al (2009) Antarctic climate change and the environment. Antarct Sci 21(6):541–563

Cooper GF, Macpherson CG, Blundy JD, Maunder B, Allen RW et al (2020) Variable water input controls evolution of the Lesser Antilles volcanic arc. Nature 582:525–529

Cooper A, Turney CSM, Palmer J, Hogg A, McGlone M et al (2021) A global environmental crisis 42,000 years ago. Science 371:811–818

Corsetti FA, Awramik SM, Pierce D (2003) A complex microbiota from snowball earth times: microfossils from the neoproterozoic kingston peak formation, Death Valley, USA. PNAS 199(8):4399–4404

Costa A, Folch A, Macedonio G, Giaccio B, Isaia R, Smith VC (2012) Quantifying volcanic ash dispersal and impact of the Campanian Ignimbrite super-eruption. Geophys Res Lett 39:L10310

Costa A, Smith VC, Macedonio G, Matthews NE (2014) The magnitude and impact of the youngest Toba tuff. Front Earth Sci 2(16)

Cotel AJ (1999) A trigger mechanism for the Lake Nyos disaster. J Volcanol Geotherm Res 88:343–347

Coxall HK, Wilson PA, Pälike H, Lear CH, Backman J (2005) Rapid stepwise onset of Antarctic glaciation and deeper calcite compensation in the Pacific Ocean. Nature 433:53–57

Coxall HK, D'Hondt S, Zachos JC (2006) Pelagic evolution and environmental recovery after the Cretaceeous-Paleogene mass extinction. Geology 34(4):297–300

Crisp MD, Burrows GE, Cook LG, Thornhill AH, Bowman DMJS (2011) Flammable biomes dominated by eucalypts originated at the Cretaceous-Palaeogene boundary. Nat Commun. https://doi.org/10.1038/ncomms1191

Crowe SA, Døssing LN, Beukes NJ, Bau M, Kruger SJ et al (2013) Atmospheric oxygenation three billion years ago. Nature 501:535–539

Crowell LC (1983) Ice ages recorded on Gondwana continents. Trans Geol Soc South Afr 86:237–262

Crucifix M (2012) Oscillators and relaxation phenomena in Pleistocene climate theory. Phil Trans R Soc A 379:1140–1165

Cruz S, Calado R, Serôdio J, Cartaxana P (2013) Crawling leaves: photosynthesis in sacoglossan sea slugs. J Exp Bot 64(13):3999–4009

Cui H, Kaufman AJ, Xiao S, Zhou C, Liu X-M (2017) Was the Ediacaran Shuram excursion a globally synchronized early diagenetic event? Insights from methane-derived authigenic carbonates in the uppermost Doushantuo formation, South China. Chem Geol 450:59–80

D'Asaro E, Lee C, Rainville L, Harcourt R, Thomas L (2011) Enhanced turbulence and energy dissipation at ocean fronts. Science 332:318–322

D'Hondt S (2005) Consequences of the cretaceous/paleogene mass extinction for marine ecosystems. Annu Rev Ecol Evol Syst 36:295–317

D'Hondt S, Donaghay P, Zachos JC, Luttenberg D, Lindinger M (1998) Organic carbon fluxes and ecological recovery from the cretaceous-tertiary mass extinction. Science 282:276–279

Dacks JB, Field MC, Buick R, Eme L, Gribaldo S et al (2016) The changing view of eukaryogenesis—fossils, cell, lineages and how they all come together. J Cell Sci 129:3695–3703

Dacre HF, Clark PA, Martinez-Alvarado O, Stringer MA, Lavers DA (2015) How do atmospheric rivers form? BAMS:1243–1255

Dagan T, Martin W (2006) The tree of one percent. Genome Biol 7(10):118

Davidov Y, Jurkevitch E (2009) Predation between prokaryotes and the origin of eukaryotes. Bioessays 9999:1–10

De Clerck O, Bogaert KA, Leliaert F (2012) Diversity and evolution of algae: primary endosymbiosis. Adv Bot Res 64:55–86

De Jager C (2005) Solar forcing ot climate. 1: solar variability. Space Sci Rev 120:197–241

De Schepper S, Schreck M, Beck KM, Matthiessen J, Fahl K, Mangerud G (2015) Early Pliocene onset of modern Nordic Seas circulation related to ocean gateway changes. Nat Commun 6. https://doi.org/10.1038/ncomms9659

DeCelles PG, Gehrels GE, Najman Y, Martin AJ, Carter A, Garzanti E (2004) Detrital geochronology and geochemistry of cretaceous-early Miocene strata of Nepal: implications for timing and diachroneity of initial Himalayan orogenesis. Earth Planet Sci Lett 227:313–330

DeConto RM, Pollard D, Wilson PA, Pälike H, Lear CH, Pagani M (2008) Thresholds for Cenozoic bipolar glaciation. Nature 455:652–656

del Rey Á, Rasmussen CMØ, Calner M, Wu R, Asael D, Dahl TW (2022) Stable Ocean redox during the main phase of the great ordovician biodiversification event. Commun Earth Environ 3:220

Delcourt PA, Delcourt HR (1992) Ecotone dynamics in space and time. In: Hansen AJ, di Castri F (eds) Landscape boundaries. Springer, Berlin, pp 19–54

Delplace P, Marston JB, Venaille A (2017) Topological origin of equatorial waves. Science 358:1075–1077

Delwiche CF, Cooper ED (2015) The evolutionary origin of a terrestrial flora. Curr Biol 25:R899–R910

Des Marais D (2003) Biogeochemistry of hypersaline microbial mats illustrates the dynamics of modern microbial ecosystems and early evolution of the biosphere. Biol Bull 204:160–167

Dickey T (1991) Concurrent high resolution physical and bio-optical measurements in the upper ocean and their applications. Rev Geophys 29:383–413

Dierssen HM, Zimmermann RC, Drake LA, Burdige DJ (2009) Potential export of unattached benthic macroalgae to the deep sea through wind-driven Langmuir circulation. Geophys Res Lett 36:L04602

Dimitrenko IA, Kirillov SA, Tremblay LB, Bauch D, Willmes (2009) Sea-ice production over the Laptev Sea shelf inferred from historical summer-to-winter hydrographic observations 1960s-1990s. Geophys Res Lett 36:L13605

Dodd MS, Papineau D, Grenne T, Slack JF, Rittner M et al (2017) Evidence for early life in Earth's oldest hydrothermal vent precipitates. Nature 543:60–64

Dohrmann M, Wörheide G (2017) Dating early animal evolution using phylogenomic data. Sci Rep 7:3599

Donnadieu Y, Goddéris Y, Ramstein G, Nédélec A, Meert J (2004) A 'snowball earth' climate triggered by continental break-up through changes in runoff. Nature 428:303–306

Dorgan KM (2015) The biomechanics of burrowing and boring. J Exp Biol 218:176–183

Dorgan KM, Jumars PA, Johnson BD, Boudreau BP (2006) Macrofaunal burrowing: the medium is the message. Oceanogr Mar Biol 44:85–121

Dori-Bachash M, Dassa B, Pietrokovsky S, Jurkevich E (2008) Proteome-based comparative analyses of growth stages reveal new cell cycle-dependent functions in the predatory bacterium *Bdellovibrio bacteriovorus*. Applied and Environmental Microb 74(23):7152–7162

Douzery EJP, Snell EA, Bapteste E, Delsuc F, Philippe H (2004) The timing of eukaryotic evolution: does a relaxed molecular clock reconcile proteins and fossils? PNAS 101(43):15386–15391

Druschel GK, Bakes BJ, Gihring TM, Banfield JF (2004) Acid mine drainage biogeochemistry at Iron Mountain, California. Geochem Trans 5(2):13–32

Du J, Mix AC, Haley BA, Belanger CL, Sharon (2022) Volcanic trigger of ocean deoxygenation during Cordilleran ice sheet retreat. Nature 611:74–80

Duarte CM, Dachs J, Llabres M, Alonso-Laita P, Gasol JM et al (2006) Aerosol inputs enhance new production in the subtropical northeast Atlantic. J Geophys Res 111:G04006

Dubilier N, Bergin C, Lott C (2008) Symbiotic diversity in marine animals: the art of harnessing chemosynthesis. Nat Rev Microbiol 6:725–740

Dunthorn M, Lipps JH, Dolan JR, Abboud-Abi Saab M, Aescht E et al (2015) Ciliates—Protists with complex morphologies and ambiguous early fossil record. Mar Micropaleontol 119:1–6

Durham WM, Climent E, Barry M, De Lillo F, Boffetta G et al (2013) Turbulence drives microscale patches of motile phytoplankton. Nat Commun. https://doi.org/10.1038/ncomms3148

Eigen M (1971) Selforganization of matter and the evolution of biological macromolecules. Naturwissenschaften 58(10):465–523

El Albani A, Bengtson S, Canfield DE, Bekker A, Macchiarelli R et al (2010) Large colonial organisms with coordinated growth in oxygenated environment 2.1 Gyr ago. Nature 466:100–104

Eldrett JS, Harding IC, Wilson PA, Butler E, Roberts AP (2007) Continental ice in greenland during the eocene and oligocene. Nature 446:176–179

Elie M, Nogueira ACR, Nédélec A, Trindade RIF, Kenig F (2007) A red algal bloom in the aftermath of the Marinoan Snowball Earth. Terra Nova 19:303–308

Embley TM, Martin W (2006) Eukaryotic evolution, changes and challenges. Nature 440:623–630

Emslie AG, Dennis BR, Shih AY, Chamberlin PC, Mewaldt RA et al (2012) Global energetics of thirty-eight large solar eruption events. Astrophys J 759:71

Eppley RW, Peterson BJ (1979) Particulate organic matter flux and planktonic new production in the deep ocean. Nature 282:677–680

Erbacher J, Huber BT, Norris RD, Markey M (2001) Increased thermohaline stratification as a possible cause for an ocean anoxic event in the cretaceous period. Nature 309:325–327

Ernst A, Becker S, Wollenzien UIA, Postius C (2003) Ecosystem-dependent adaptive radiations of picocyanobacteria inferred from 16S rRNA and IST-1 sequence analysis. Microbiology 149:217–228

Ernst RE, Dickson AJ, Bekker A (2021) Large igneous provinces: a driver of global environmental and biotic changes. Geophys monograph, vol 255, American Geophysical Union

Erwin DH (2009) Early origin of the bilaterian development toolkit. Phil Trans R Soc 364:2253–2261

Erwin DH (2015) Early metazoan life: divergence, environment and ecology. Philos Trans R Soc B 370:20150036

Erwin DH, Laflamme M, Tweedt SM, Sperling EA, Pisani D, Peterson KJ (2011) The Cambrian conundrum: early divergence and later ecological success in the early history of animals. Science 334:1091–1097

Evaristo I, Jasechko S, McDonnell JJ (2015) Global separation of plant transpiration from groundwater and streamflow. Nature 525:91–94

Eyles N (2008) Glacio-epochs and the supercontinent cycle after ~3.0 Ga: tectonic boundary conditions for glaciation. Palaeogeo Palaeoclim Palaeoecol 238:89–129

Fabiańska MJ, Ciesielczuk J, Kruszewski L, Misz-Kennan M, Blake DR et al (2013) Gaseous compounds and efflorescences generated in self-heating coal-waste dumps—a case study from upper and lower Silesian coal basins (Poland). Int J Coal Geol 116–117:247–261

Fahnenstiel GL, Scavia D, Lang GA, Saylor JH, Miller GS, Schwab DJ (1988) Impact of inertial period internal waves on fixed-depth primary production estimates. J Plankton Res 10(1):77–87

Farquhar J, Zerkle AL, Bekker A (2010) Geological constraints on the origin of oxygenic photosynthesis. PhotosynthRes. https://doi.org/10.1007/s11120-010-9594-0

Fasullo JZ, Trenberth KE (2008a) The annual cycle of the energy budget. Part I: global mean and Land-Ocean exchanges. J Climate 21:2297–2312

Fasullo JZ, Trenberth KE (2008b) The annual cycle of the energy budget. Part II: meridional structures and poleward transports. J Climate 21:2213–2325

Feakins SJ, Warny S, DeConto RM (2014) Snapshot of cooling and drying before onset of antarctic glaciation. Earth Planet Sci Lett 404:154–166

Federle W (2006) Why are so many adhesive pads hairy? J Exp Biol 209:2611–2621

Feng X, Tang K-H, Blankenship RE, Tang YJ (2010) Metabolic flux analysis of the mixotrophic metabolisms in the green sulfur bacterium Chlorobaculum tepidum. J Biol Chem 285(50):39544–39550

Fenton AK, Kanna M, Woods RD, Aizawa S-I, Sockett RE (2010) Shadowing the actions of a predator: backlit fluorescent microscopy reveals synchronous nonbinary septation of predatory *Bdellovibrio* inside prey and exit through discrete Bdelloplast pores. J Bacteriol 192(24):6329–6335

Ferguson SH, Messier F (1996) Ecological implications of a latitudinal gradient in inter-annual climatic variability: a test using fractal and chaos theories. Ecography 19:382–392

Field RD, van der Werf GR, Shen SSP (2009) Human amplification of drought-induced biomass burning in Indonesia since 1960. Nat Geosci 2:185–188

Filizola N, Guyot J-L, Wittmann H, Martinez J-M, de Oliveira E (2011) The significance of suspended sediment transport determination on the Amazonian hydrological scenario. In: Manning A (ed) Sediment transport in aquatic environments. InTech, Rijeka, pp 45–64

Fischer WW, Hemp J, Johnson JE (2016) Evolution of oxygenic photosynthesis. Annu Rev Earth Planet Sci 44:467–483

Fleming A, Jackson J (2011) Global review of humpback whales. NOAA-TM-NMFS-SWFSC-474

Forbes DL (ed) (2011) State of the Arctic coast 2010—scientific review and outlook. Int Arct Comm, Helmholtz-Zentrum Geesthacht

Foster RG, Roenneberg T (2008) Human responses to the geophysical daily, annual and lunar cycles. Curr Biol 18:R784–R794

Fraedrich K, Müller K, Kuglin R (1992) Northern hemisphere circulation regimes during the extremes of the El Niño/southern oscillation. Tellus 44A:33–40

French KL, Hallmann C, Hope JM, Schoon PL, Zumberge JA et al (2015) Reappraisal of hydrocarbon biomarkers in Archean rocks. Proc Natl Acad Sci U S A 112(19):5915–5920

Friedman M (2009) Ecomorphological selectivity among marine teleost fishes during the end-cretaceous extinction. PNAS 106(13):5218–5223

Frisch U (1991) From global scaling, à la Kolmogorov, to local multifractal scaling in fully developed turbulence. Proc R Soc Lond A 434:89–99

Fröbisch J (2013) Vertebrate diversity across the end-Permian mass extinction—separating biological and geological signals. Palaeogeo Palaeoclim Palaeoecol 372:50–61

Gage JD, Tyler PA (1999) Deep sea biology. Cambridge University Press, Cambridge

Gaillard F, Scaillet B, Arndt NT (2011) Atmospheric oxygenation caused by a change in volcanic degassing pressure. Nature 478:229–232

Galetzka J, Melgar D, Genrich JF, Geng J, Owen S et al (2015) Slip pulse and resonance of the Kathmandu basin during the 2015 Gorkha earthquake, Nepal. Science 349:1091–1095

Garrett C, Kunze E (2007) Internal tide generation in the Deep Ocean. Annu Rev Fluid Mech 30:57–87

Garrison T, Ellis R (2016) Oceanography. Cengage Learning, Boston

Garzoli SL, Richardson PL, Rae CMD, Fratantoni DM, Goñi GJ, Roubicek AJ (1999) Three Agulhas rings observed during the Benguela current experiment. J Geophys Res 104(9):20971–20985

Gasson E, DeConto R, Pollard D, Levy RH (2016) Dynamic Antarctic ice sheet during early to mid-miocene. PNAS 113(13):3459–3464

Gaucher C, Sial AN, Halverson GP, Frimmel HE (2009) The Neoproterozoic and Cambrian: a time of upheavals, extremes and innovations. Dev Precambrian Geol 16:3–11

Gavin DG, Henderson ACG, Westover KS, Fritz SC, Walker IR et al (2011) Abrupt Holocene climate change and potential response to solar forcing on western Canada. Quat Sci Rev 30:1243–1255

Geibert W, Matthiessen J, Stimac I, Wollenburg J, Stein R (2021) Glacial episodes of a freshwater Arctic Ocean covered by a thick ice shelf. Nature 490:97–102

Geiger R (1961) Das Klima der bodennahen Luftschichten. Vieweg, Braunschweig

Ghienne J-F, Desrochers A, Vandenbroucke TRA, Achab A, Asselin E et al (2014) A Cenozoic-style scenario for the end-Ordovician glaciation. Nat Commun. https://doi.org/10.1038/ncomms5485

Ghil M (1994) Cryothermodynamics: the chaotic dynamics of paleoclimate. Physica D 77:130–159

Gibson CH (1991) Kolmogorov similarity hypotheses for scalar fields: sampling intermittent turbulent mixing in the ocean and galaxy. Proc R Soc Lond A 434:149–164

Gili J-M, Arntz WE, Palanques A, Orejas C, Clarke A et al (2006) A unique assemblage of epibenthic sessile suspension feeders with archaic features in the high Antarctic. Deep-Sea Res II 53:1029–1052

Ginsburg S, Jablonka E (2010) The evolution of associative learning: a factor in the Cambrian explosion. J Theor Biol 266:11–20

Girguis PR, Childress JJ, Freytag JK, Klose K, Stuber R (2002) Effects of metabolite uptake on proton-equivalent elimination by two species of deep-sea vestimentiferan tubeworm, *Riftia pachyptila* and *Lamellibrachia* cf *luymesi*: proton elimination is a necessary adaptation to sulfide-oxidizing chemoautotrophic symbionts. J Exp Biol 205:3055–3066

Glasspool IJ, Scott AC (2010) Phanerozoic concentrations of atmospheric oxygen reconstructed from sedimentary charcoal. Nat Geosci 3:627–630

Glasspool IJ, Scott AC, Waltham D, Pronina N, Shao L (2016) The impact of fire on the late paleozoic earth system. Front Plant Sci 6:756

Glikson A (2005) Asteroid/comet impact clusters, flood basalts and mass extinctions: significance of isotopic age overlaps. Earth Planet Sci Lett 236:933–937

Goddéris Y, Donnadieu Y, Le Hir G, Lefebvre V, Nardin E (2014) The role of palaeogeography in the phanerozoic history of atmosphere CO_2 and climate. Earth Sci Rev 128:122–138

Godefroit P, Goloneva L, Shchepetov S, Garcia G, Alekseev P (2009) The last polar dinosaurs: high diversity of latest cretaceous arctic dinosaurs in Russia. Naturwissenschaften 96:495–501

Goffredi SK (2010) Indigenous ectosymbiotic bacteria associated with diverse hydrothermal vent invertebrates. Environmental Microb Rep. https://doi.org/10.1111/j.1758-2229.2010.00136.x

Gold DA, O'Reilly SS, Luo G, Briggs DEG, Summons RE (2016) Prospects for sterane preservation in sponge fossils from museum collections and the utility of sponge biomarkers for molecular clocks. Bull Peabody Mus Nat Hist 57(2):181–189

Gold DA, Caron A, Fournier GP, Summons RE (2017) Paleoproterozoic sterol biosynthesis and the rise of oxygen. Nature 543:420–423

Goosse H, Renssen H (2006) Regional response of the climate system to solar forcing: the role of the ocean. Spce Sci Rev 125:227–235

Goosse H, Guiot J, Mann ME, Dubinkina S, Sallaz-Damaz Y (2012) The medieval climate anomaly in Europe: comparison of the summer and annual mean signals in two reconstructions and in simulations with data assimilation. Global Planet Change 84–85:35–47

Gorelova OA, Kosevich IA, Baulina OI, Fedorenko TA, Torshkhoeva AZ, Lobakova ES (2009) Associations between the White Sea invertebrates and oxygen-evolving phototrophic microorganisms. Mosc Univ Biol Sci Bull 64(1):16–22

Gosselin M, Levasseur M, Wheeler PA, Horner RA, Booth BC (1997) New measurements of phytoplankton and ice al.gal. production in the Arctic Ocean. Deep-Sea Res II 44(8):1623–1644

Gostin VA, McKirdy DM, Webster LJ, Williams GE (2010) Ediacaran ice-rafting and coeval. asteroid impact, South Austral.ia: insights into the terminal. Proterozoic environment. Australian J of Earth Sci 57:859–869

Gradstein FM, Ogg JG, Schmitz MD, Ogg GM (eds) (2012) The geological time scale 2012, vol 1+2. Elsevier, Amsterdam

Gramling C (2015) Arctic impact. Science 347:818–821

Grant PR, Grant BR (2014) 40 years of evolution. Princeton University Press, Princeton

Grasby SE, Sanei H, Beauchamp B (2015) Latest Permian chars may derive from wildfires, not coal combustion. Geol Forum:e358

Griffin DW, Kellogg CA (2004) Dust storms and their impact on ocean and human health: dust in earth's atmosphere. Ecohealth 1:284–295

Grosch EG, McLoughlin N (2014) Reassessing the biogenicity of earth's oldest trace fossil with implications for biosignatures in the search for early life. PNAS 111(23):8380–8385

Grotzinger J, Jordan TH, Press F, Siever R (2008) Allgemeine geologie. Springer, Berlin

Grotzinger JP, Fike DA, Fischer WW (2011) Enigmatic origin of the largest-known carbon isotope excursion in Earth's history. Nature GeoSci 4:285–292

Guatteri M, Berlogg M, Castaldi A (2005) A shake in insurance history the 1906 San Francisco earthquake. Swiss Reinsurance Company, Zurich

Gudmundsson A (2012) Strengths and strain energies of volcanic edifices: implications for eruptions, collapse calderas, and landslides. Nat Hazards Earth Syst 12:2241–2258

Guerrero R, Pedrós-Alió C, Esteve I, Mas J, Chase D, Margulis L (1986) Predatory prokaryotes: predation and primary consumption evolved in bacteria. PNAS 83:2138–2142

Günther F, Folke C (1993) Characteristics of nested living systems. J Biol Syst 1(3):257–274

Häckel H (2021) Meteorologie. Eugen Ulmer, Stuttgart

Halm H, Musat N, Lam P, Langlois R, Musat F et al (2009) Co-occurrence of denitrification and nitrogen fixations in a meromictic lake, Lake Cadagno (Switzerland). Environ Microb. https://doi.org/10.1111/J.1462-2920.2009.01917.x

Hamilton TL, Bryant DA, Macalady JL (2016) The role of biology in planetary evolution: cyanobacterial primary production in low-oxygen proterozoic oceans. Environ Microb 18(2):325–340

Han T-M, Runnegar B (1992) Megascopic eukaryotic algae from the 2.1-billion-year-old Negaunee iron-formation, Michigan. Science 257:232–235

Hanslmeier A (2014) Einführung in Astronomie und Astrophysik. Springer Spektrum, Berlin

Harrison CGA (1994) Rates of continental erosion and mountain building. Geol Rundsch 83:431–447

Harrison SP, Marlon JR, Bartlein PJ (2010) Fire in the earth system. In: Dodson J (ed) Changing climate, earth systems and society. Springer Science+Business Media, pp 21–48

Hart MH (1978) The evolution of the atmosphere of the earth. Icarus 33:23–39

Harzhauser M, Piller WE (2007) Benchmark data of a changing sea—palaeogeography, palaeogeography and events in the central parathetys during the miocene. Palaeogeo Palaeoclim Palaeoecol 253:8–31

Harzhauser M, Piller WE, Müllegger S, Grunert P, Micheels A (2011) Changing seasonality in Central Europe from miocene climate optimum to miocene climate transition deduced from the *Crassostrea* isotope archive. Global Planet Change 76:77–84

Hasenclever J, Knorr G, Rüpke LH, Köhler P, Morgan J et al (2017) Sea level fall during glaciation stabilized atmospheric CO_2 by enhanced volcanic degassing. Nat Commun. https://doi.org/10.1038/ncomms15867

Hátún H, Sandø AB, Drange H, Hansen B, Valdimarsson H (2005) Influence of the Atlantic subpolar gyre on the thermohaline circulation. Science 309:1841–1844

Haug GH, Tiedemann R (1998) Effect of the formation of the isthmus of Panama on the Atlantic Ocean thermohaline circulation. Nature 393:673–676

Haug GH, Tiedemann R, Zahn R, Ravelo AC (2001) Role of Panama uplift on oceanic freshwater balance. Geology 29(3):207–210

Hay WW (1996) Tectonics and climate. Geol Rundsch 85:409–437

Hays JD, Imbrie J, Shackleton NJ (1976) Variations in the earth's orbit: pacemaker of the ice ages. Science 194:1121–1132

Hazen RM, Morrison SM (2022) On the paragenetic modes of minerals: a mineral evolution perspective. Am Mineral 107(7):1262–1287

Hazen RM, Papineau D (2012) Mineralogical co-evolution of the geosphere and biosphere. In: Knoll AH, Canfield DE, Konhauser KO (eds) Fundamentals of geobiology. Wiley-Blackwell, Oxford, pp 333–350

Hazen RM, Liu X-M, Downs RT, Golden J, Pires AJ et al (2014) Mineral evolution: episodic metallogenesis, the supercontinent cycle, and the coevolving geosphere and biosphere. Soc Econ Geol Spec Publ 18:1–15

He T, Pausas JG, Belcher CM, Schwilk DW, Lamont BB (2012) Fire-adapted traits of *Pinus* arose in the Cretaceous. New Phytol 194:751–759

He T, Belcher CM, Lamont BB, Lim SL (2016) A 350-million-year legacy of fire adaptation among conifers. J Ecol 104:352–363

Heaman LM (2008) Precambrian large igneous provinces: an overview of geochronology, origins and impact of earth evolution. J Geol Soc India 59:15–34

Heaviside C, Czaja A (2013) Deconstructing the Hadley cell heat transport. Q J Roy Meteorol Soc 139:2181–2189

Hebert CL, Kaufman AJ, Penniston-Dorland SC, Martin AJ (2010) Radiometric and stratigraphic constraints on terminal Ediacaran (post-Gaskiers) glaciation and metazoan evolution. Precambrian Res 182:402–412

Heckman DS, Geiser DM, Eidell BR, Stauffer RL, Kardos NL, Hedges SB (2001) Molecular evidence for the early colonization of land by fungi and plants. Science 293:1129–1133

Hellmer HH, Olbers DJ (1989) A two-dimensional model for the thermohaline circulation under an ice shelf. Antarct Sci 1(4):325–336

Helo C, Longpré M-A, Shimizu N, Clague DA, Stix J (2011) Explosive eruptions at mid-ocean ridges driven by CO_2-rich magmas. Nature GeoSci 4:260–263

Heydari E, Arzani N, Hassanzadeh J (2008) Mantle plume: the invisible serial killer—application to the Permian-Triassic boundary mass extinction. Palaeogeo Palaeoclim Palaeoecol 264:147–162

Hill MS (2014) Production possibility frontiers in phototroph:heterotroph symbioses: trade-offs in allocating fixed carbon pools and the challenges these alternatives present for understanding the acquisition of intracellular habitats. Front Microbiol 5:357

Hill AC, Haines PW, Grey K, Willman S (2007) New records of Ediacaran Acraman ejecta in drillholes from the Stuart Shelf and Officer Basin, South Australia. Meteorit Planet Sci 42(11):1883–1891

Hinnov LA, Hilgen FJ (2012) Cyclostratigraphy and astrochronology. In: Gradstein FM, Ogg JG, Schmitz MD, Ogg GM (eds) The geological time scale 2012, vol 1+2. Elsevier, Amsterdam, pp 63–83

Ho SYW, Larson G (2006) Molecular clocks: when times are a-changin. Trends Genet 22(2):79–83

Hobert O (2008) Gene regulation by transcription factors and MicroRNAs. Science 391:1785–1786

Hochuli PA, Sanson-Barrera A, Schneeneli-Hermann E, Bucher H (2016) Severest crisis overlooked-Worst disruption of terrestrial environments postdates the Permian-Triassic mass extinction. Sci Rep 6/28372

Hoetzel S, Dupont L, Schefuß E, Rommerskirchen F, Wefer G (2013) The role of fire in miocene to pliocene C_4 grassland and ecosystem evolution. Nature GeoSci 6:1027–1030

Hoffman PF (2013) The great oxidation and a Siderian snowball earth: MIF-S based correlation of paleoproterozoic glacial epoch. Chem Geol 362:143–156

Hoffman PF, Li Z-X (2009) A palaeogeographic context for neoproterozoic glaciation. Paleogeogr Paleoclimatol Paleoecol 277(3–4):158–172

Hoffman PF, Schrag DP (2002) The snowball earth hypothesis: testing the limits of global change. Terra Nova 14:129–155

Hoffman PF, Kaufman AJ, Halverson GP, Schrag DP (1998) A neoproterozoic snowball earth. Science 281:1342–1346

Hoffman PF, Abbot DS, Ashkenazy Y, Benn DI, Brocks JJ et al (2017) Snowball earth climate dynamics and Cryogenian geology-geobiology. Sci Adv 3:e1600983

Hofreiter M, Serre D, Rohland N, Rabeder G, Nagel D et al (2004) Lack of phylogeography in European mammals before the last glaciation. PNAS 101(35):12963–12968

Hogland W, Marques M (2003) Physical, biological and chemical processes during storage and spontaneous combustion of waste fuel. Resour Conserv Recycl 40:53–69

Hoinka KP (1985) Observation of the airflow over the Alps during a foehn event. Q J Roy Meteorol Soc 111:199–224

Holland HD (2002) Volcanic gases, black smokers, and the great oxidation event. Geochim Cosmochim Acta 66(21):3811–3826

Holloway PE, Merrifield MA (1999) Internal tide generation by seamounts, ridges, and islands. J Geophys Res 104(C11):29937–29951

Hood AVS, Wallace MW, Drysdale RN (2011) Neoproterozoic aragonite-dolomite seas? Widespread marine dolomite precipitation in Cryogenian reef complexes. Geology 39(9):871–874

Hornung T, Brandner R (2005) Biochronostratigraphy of the Reingraben Turnover (Hallstatt Facies Belt): local black shale events controlled by regional tectonics, climatic change and plate tectonics. Facies 51:460–479

Horton DE, Johnson NC, Singh D, Swain DL, Rajaratnam B, Diffenbaugh NS (2015) Contribution of changes in atmospheric circulation patterns to extreme temperature trends. Nature 522:465–469

Houben AJP, Bijl PK, Pross J, Bohaty SM, Passchier S et al (2013) Reorganization of Southern Ocean Plankton Ecosystem at the onset of antarctic glaciation. Science 340:341–344

Huang S (2004) Back to the biology in systems biology: what can we learn from biomolecular networks? Brief Funct Genomic Proteomic 2(4):279–297

Huang S (2011) The molecular and mathematical basis of Waddington's epigenetic landscape: a framework for post-Darwinian biology? Bioessays. https://doi.org/10.1002/bies.201100031

Hughes PD, Gibbard PL, Ehlers J (2013) Timing of glaciation during the last glacial cycle: evaluating the concept of a global 'Last Glacial Maximum' (LGM). Earth Sci Rev 125:171–198

Hull PM, Norris RD, Bralower TJ, Schueth JD (2011) A role for chance in marine recovery from the end-cretaceous extinction. Nat Geosci 4:856–860

Humphreys WF (2006) Aquifers: the ultimate groundwater-dependent ecosystems. Aust J Bot 54:115–132

Huntley JW, Xiao S, Kowalewski M (2006) On the morphological history of proterozoic and cambrian acritarchs. In: Xia S, Kaufman AJ (eds) Neoproterozoic geobiology and paleobiology. Springer, Dordrecht, pp 23–56

Ineson S, Scaife AA, Knight JR, Manners JC, Dunstone NJ et al (2011) Solar forcing of winter climate variability in the northern hemisphere. Nat Geosci 4:753–757

Isbell JL, Miller MF, Wolfe KL, Lenaker PA (2003) Timing of late Paleozoic glaciation in Gondwana: was glaciation responsible for the development of northern hemisphere cyclothems? In: Chan MA, Archer AW (eds) Extreme depositional environments: mega end members in geological time, Geological Society of America Special Paper, vol 370, pp 5–24

Ishii M, Shearer PM, Houston H, Vidale JE (2005) Extent, duration and speed of the 2004 Sumatra-Andaman earthquake imaged by the hi-net array. Nature 435:933–936

Jakobsson M, Backman J, Rudels B, Nycander J, Frank M et al (2007) The early Miocene onset of a ventilated circulation regime in the Arctic Ocean. Nature 447:986–990

Jakobsson M, Nilsson J, Anderson L, Backman J, Björk G et al (2016) Evidence for an ice shelf covering the Central Arctic Ocean during the penultimate glaciation. Nat Commun. https://doi.org/10.1038/ncomms10365

Jamieson AJ, Fujii T, Mayor DJ, Solan M, Priede IG (2010) Hadal trenches: the ecology of the deepest places on earth. TREE 25(3):190–197

Jantzen C, Schmidt GM, Wild C, Roder C, Khokiattiwong S, Richter C (2013) Benthic reef primary production in respones to large amplitude internal waves at the Similan Islands (Andaman Sea, Thailand). PLoS One 8(11):e81834

Jasper A, Guerra-Sommer M, Abu Hamad AMB, Bamford M, Bernardes-de-Oliveira MEC et al (2013) The burning of Gondwana: Permian fires on the southern continent—a palaeobotanical approach. Gondw Res 24(1):148–160

Javaux EJ, Knoll AH, Walter MR (2001) Morphological and ecological complexity in early eukaryotic ecosystems. Nature 412:66–69

Jenkyns HC (2010) Geochemistry of oceanic events. Geochem Geophys Geosyst 11:Q03004

Jiang Q, Doyle JD (2008) On the diurnal variation of mountain waves. J Atmos Sci 65:1360–1377

Jickells TD, An ZS, Andersen KK, Baker AR, Bergametti G et al (2005) Global iron connections between desert dust, ocean biogeochemistry, and climate. Science 308:67–71

Johnston TMS, Merrifield MA (2003) Internal tide scattering at seamounts, ridges, and islands. J Geophys Res 108(C6):3180

Johnston DT, Wolfe-Simon F, Pearson A, Knoll AH (2009) Anoxygenic photosynthesis modulated Proterozoic oxygen and sustained Earth's middle age. PNAS 106(40):16925–16929

Jones S, Martin R, Pilbeam D (eds) (2009) Human evolution. Cambridge University Press, Cambridge

Jung M, Reichstein M, Margolis HA, Cescatti A, Richardson AD et al (2011) Global patterns of land-atmosphere fluxes of carbon dioxide, latent heat, and sensible heat derived from eddy covariance, satellite, and meteorological observations. J Geophys Res 16:G00307

Jurkevitch E (2007) Predatory behaviors in bacteria—diversity and transitions. Microbiology 2(2):67–73

Jurkevitch E, Davidov Y (2007) Phylogenetic diversity and evolution of predatory prokaryotes. In: Jurkevitch E (ed) Predatory prokaryotes. Springer, Berlin, pp 11–56

Kagami M, Miki T, Takimoto G (2014) Mycoloop: chytrids in aquatic food webs. Front Microbiol 6(166)

Kaltschmitt M, Streicher W, Wiese A (eds) (2006) Erneuerbare Energien. Springer, Berlin

Kanamori H, Brodsky EE (2001) The physics of earthquakes. Phys Today 54(6):34–40

Karabyn V, Shtain B, Popovych V (2018) Thermal regimes of spontaneous firing coal washing waste sides. Ser Geol Tech Sci Kazakhstan 3(429):64–79

Karp AT, Behrensmeyer A, Freeman KH (2018) Grassland fire ecology has roots in the late miocene. PNAS 115(48):12130–12135

Kataoka R, Ebisuzaki T, Miyahara H, Nimura T, Tomida T et al (2014) The nebula winter: the united view of the snowball earth, mass extinctions, and explosive evolution in the late Neoproterozoic and Cambrian periods. Gondw Res 25(3):1153–1163

Katz ME, Finkel ZV, Grzebyk D, Knoll AH, Falkowski PG (2004) Evolutionary trajectories and biogeochemical impacts of marine eukaryotic phytoplankton. Annu Rev Ecol Evol Syst 35:523–556

Kaufman AJ, Xiao S (2003) High CO_2 levels in the Proterozoic atmosphere estimated from analysis of individual microfossils. Nature 425:279–282

Keane R, Berleman J (2016) The predatory life cycle of *Myxococcus xanthus*. Microbiology 162:1–11

Keddy PA (2007) Plants and vegetation. Cambridge University Press, Cambridge

Keeley JE, Rundel PW (2005) Fire and the Miocene expansion of C_4 grasslands. Ecol Lett 8:683–690

Keeley JE, Pausas JG, Rundel PW, Bond WJ, Bradstock RA (2011) Fire as an evolutionary pressure shaping plant traits. Trends Plant Sci 16(8):406–411

Keller G (2005) Impacts, volcanism and mass extinction: random coincidence or cause and effect? Aust J Earth Sci 52:725–757

Keller G (2008) Cretaceous climate volcanism, impacts, and biotic effects. Cretac Res 29:754–771

Kendall B, Reinhard CT, Lyons TW, Kaufman AJ, Poulton SW, Anbar AD (2010) Pervasive oxygenation along late Archaean Ocean margins. Nature GeoSci 3:647–652

Kendall B, Anbar AD, Kappler A, Konhauser KO (2012) The global iron cycle. In: Knoll AH, Canfield DE, Konhauser KO (eds) Fundamentals of geobiology. Wiley-Blackwell, Oxford, pp 65–92

Kendall B, Komiya T, Lyons TW, Bates SM, Gordon GW et al (2015) Uranium and molybdenum isotope evidence for an episode of widespread ocean oxygenation during the late Ediacaran period. Geochim Cosmochim Acta 156:173–193

Keppner A, Maric D, Correira M, Koay TW, Orlando IMC et al (2020) Lessons from the postgenomic era: globin diversity beyond oxygen binding and transport. Redox Biol 37:101687

Kim S-S, Wessel P (2011) New global seamount census from altimetry-derived gravity data. Geophys J Int 186:615–631

Kim J, Ivanov VY, Fatichi S (2016) Environmental stochasticity controls soil erosion variability. Sci Rep. https://doi.org/10.1038/srep22065

Kimura M (1991) The neutral theory of molecular evolution: a review of recent evidence. Jpn J Genet 66:367–386

Kirk JTO (2011) Light and photosynthesis in aquatic ecosystems. Cambridge University Press, Cambridge

Kirschvink JL, Gaidos EJ, Bertani LE, Beukes NJ, Gutzmer J et al (2000) Paleoproterozoic snowball earth: extreme climatic and geochemical change and its biological consequences. PNAS 97(4):1400–1405

Kirtman BP, Perlin N, Siqueira L (2017) Ocean eddies and climate predictability. Chaos 27:126902

Kiryukhin AV, Rychkova TV, Dubrovskaya IK (2012) Formation of the hydrothermal system in Geysers Valley (Kronotsky nature reserve, Kamchatka) and triggers of the Giant landslide. Appl Geochem 27:1753–1766

Knauth LP, Kennedy MJ (2009) The late Precambrian greening of the earth. Nature 460:728–732

Knighton D (1998) Fluvial forms and processes. Arnold, London

Knoflacher M (Hg.) (2017) Herausforderungen der evolutionären Komplexität. LIT, Wien

Knoll AH (1994) Proterozoic and early Cambrian protists: evidence for accelerating evolutionary tempo. PNAS 91:6743–6750

Knoll AH (2014) Paleobiological perspectives on early eukaryotic evolution. Cold Spring Harb Perspect Biol 6:a016121

Knoll AH, Follows MJ (2016) A bottom-up perspective on ecosystem change in Mesozoic oceans. Proc R Soc B 283:20161/ETLS20170153

Knoll AH, Nowak MA (2017) The timetable of evolution. Sci Adv 3:e1603076

Knoll AH, Sperling EA (2014) Oxygen and animals in earth history. PNAS 111(11):3907–3908

Knoll AH, Javaux EJ, Hewitt D, Cohen P (2006) Eukaryotic organisms in proterozoic oceans. Philos Trans R Soc B 361:1023–1038

Knoll AH, Bambach RK, Payne JL, Pruss S, Fischer WW (2007) Paleophysiology and end-Permian mass extinction. Earth Planet Sci Lett 256:295–313

Knoll AH, Canfield DE, Konhauser KO (eds) (2012) Fundamentals of geobiology. Wiley Blackwell, Chichester

Knoll AH, Bergman KD, Strauss JV (2016) Life: the first two billion years. Philos Trans R Soc B 371:20150493

Koh EY, Phua W, Ryan KG (2011) Aerobic anoxygenic phototrophic bacteria in Antarctic Sea ice and seawater. Environ Microb Rep. https://doi.org/10.1111/j.1758-2229.2011.00286.x

Koh EY, Martin AR, McMinn A, Ryan KG (2012) Recent advances and future perspectives in microbial phototrophy in antarctic Sea ice. Biology 1:542–556

Konhauser KO (2007) Introduction to geomicrobiology. Blackwell, Malden

Konhauser KO, Jones B, Reysenbach A-L, Renaut RW (2003) Hot spring sinters: keys to understanding earth's earliest life forms. Can J Earth Sci 40:1713–1724

Konhauser KO, Lalonde SV, Planavsky NJ, Pecoits E, Lyons TW et al (2011) Aerobic bacterial pyrite oxidation and acid rock drainage during the great oxidation event. Nature 478:369–373

Konstantinidis KT, Ramette A, TIedje JM (2006) The bacterial species definition in the genomic era. Philos Trans R Soc B 361:1929–1940

Kopp RE, Kirschvink JL, Hilburn IA, Nash CZ (2005) The Paleoproterozoic snowball earth: a climate disaster triggered by the evolution of oxygenic photosynthesis. PNAS 102(32):11131–11136

Koppes M, Hallet B, Rignot E, Mouginot J, Smith Wellner J, Boldt K (2015) Observed latitudinal variations in erosion as a function of glacier dynamics. Nature 526:100–103

Körner C (1999) Alpine Plant Life. Springer, Berlin

Korp J, Gurovic MSV, Nett M (2016) Antibiotics from predatory bacteria. Beilstein J Org Chem 12:594–607

Koshel KV, Prants SV (2006) Chaotic advection in the ocean. Physics-Uspekhi 49(11):1151–1178

Kouraev AV, Zakharova EA, Rémy F, Kostianoy AG, Shimaraev MN et al (2019) Giant ice rings on lakes and field observations of lens-like eddies in the Middle Baikal (2016-2017). Limnol Oceanogr 64:2738–2754

Krissansen-Totton J, Bergsman DS, Catling DC (2016) On detecting biospheres from chemical thermodynamic disequilibrium in planetary atmospheres. Astrobiology 16(1):39–67

Krueger AF, Winston JS (1974) A comparison of the flow over the tropics during two contrasting circulation regimes. J Atmos Sci 31:358–370

Kumar S (2005) Molecular clocks: four decades of evolution. Nature 6:654–662

Kump LR (2008) The rise of atmospheric oxygen. Nature 451:277–278

Kump LR (2012) Sulfur isotopes and the stepwise oxygenation of the biosphere. Elements 8:410–411

Kump LR (2014) Hypothesized link between neoproterozoic greening of the land surface and the establishment of an oxygen-rich atmosphere. PNAS 111(39):14062–14065

Kump LR, Barley ME (2007) Increased subaerial volcanism and the rise of atmospheric oxygen 2.5 billion years ago. Nature 448:1033–1036

Laakso TA, Schrag DP (2018) Limitations on limitation. Global Biogeochem Cycles 32:486–496

Labandeira CC (2005) Invasion of the continents: cyanobacterial crusts to tree-inhabiting arthropods. Trends Ecol Evol 20(5):253–262

Laden G, Wrangham R (2005) The rise of the hominids as an adaptive shift in fallback foods: plant underground storage organs (USOs) and australopith origins. J Hum Evol 49(4):482–498

Lagerloef G, Schmitt R, Schanze J, Kao H-Y (2010) The ocean and the global water cycle. Oceanography 23(4):82–93

Lamichhaney S, Berglund J, Almén MS, Maqhool K, Grabherr M et al (2015) Evolution of Darwin's finches and their beaks revealed by genome sequencing. Nature 518:371–375

Lande R, Yentsch CS (1988) Internal waves, primary production and the compensation depth of marine phytoplankton. J Plankton Res 10(3):565–571

Landing E, McGabhann BA (2010) First evidence for Cambrian glaciation provided by sections in Avalonian New Brunswick and Ireland: additional data for Avalon-Gondwana separation by the earliest Palaeozoic. Palaeogeo Palaeoclim Palaeoecol 285:174–185

Landuyt W, Bercovici D (2009) Variations in planetary convection via the effect of climate on damage. Earth Planet Sci Lett 277:29–37

Lane CE, Archibald JM (2008) The eukaryotic tree of life: endosymbiosis takes its TOL. Trends Ecol Evol 23(5):268–275

Large RR, Halpin JA, Danyushevsky LD, Maslennikov VV, Bull SW et al (2014) Trace element content of sedimentary pyrite as a new proxy for deep-time ocean-atmosphere evolution. Earth Planet Sci Lett 389:209–220

Laskar J, Robutel P, Joutel F, Gastineau M, Correia ACM, Levrard B (2004) A long term numerical solution for the insolation quantities of the earth. Astron Astrophys 428:281–285

Latal C, Piller WE, Harzhauser M (2006) Shifts in oxygen and carbon isotope signals in marine molluscs from the central Paratethys (Europe) around the lower/middle Miocene transition. Palaeogeo Palaeoclim Palaeoecol 231(3–4):347–360

Le Guerroue E (2006) Sedimentology and chemostratigraphy of the Ediacaran Shuram Formation, Nafun Group, Oman. Diss ETH Zürich

Le Heron DP (2012) The location and styles of ice-free "Oases" during neoproterozoic glaciations with evolutionary implications. Geosciences 2:90–108

Lee H, Lee BP, Messersmith PB (2007) A reversible wet/dry adhesive inspired by mussels and geckos. Nature 448:338–341

Lee JN, Shindell DT, Hameed S (2009a) The influence of solar forcing on tropical circulation. J Climate 22:5870–2885

Lee MSY, Oliver PM, Hutchinson MN (2009b) Phylogenetic uncertainty and molecular clock calibrations: a case study of legless lizards (Pygopodidae, Gekkos). Mol Phylogenet Evol 50:661–666

Lee C, Love GD, Fischer WW, Grotzinger JP, Halverson GP (2015) Marine organic matter cycling during the Ediadaran Shuram excursion. Geology 43(12):1103–1106

Lenton TM, Boyle RA, Poulton SW, Shields-Zhou GA, Butterfield NJ (2014) Co-evolution of eukaryotes and ocean oxygenation in the neoproterozoic. Nat Geosci 7:257–265

Lévy M, Ferrari R, Franks PJS, Martin AP, Rivière P (2012) Bringing physics to life at the submesoscale. Geophys Res Lett 19:L14602

Li J, Ding R (2011) Temporal-spatial distribution of atmospheric predictability limit by local dynamical analogs. American Meteorological Society, pp 3265–3283

Liang M-C, Hartman H, Kopp RE, Kirschvink JL, Yung YL (2006) Production of hydrogen peroxide in the atmosphere of a snowball earth and the origin of oxygenic photosynthesis. PNAS 103(50):18896–18899

Lieth H, Whittaker RH (eds) (1975) Primary productivity of the biosphere. Springer, Berlin

Lin X, Wakeham SG, Putnam IF, Astor YM, Scranton MI et al (2006) Comparison of vertical distributions of prokaryotic assemblages in the anoxic Cariaco Basin and Black Sea by use of fluorescence in situ hybridization. Appl Environ Microb 72(4):2679–2690

Lipenkov VY, Salamatin AN, Duval P (1997) Bubbly-ice densification in ice sheets: II. Applications. J Glaciol 43(145):397–407

Littler MM, Littler DS, Blair SM, Norris JN (1986) Deep-water plant communities from an uncharted seamount off San Salvador Island, Bahamas: distribution, abundance, and primary productivity. Deep-Sea Res 33(7):881–892

Livermore R, Hillenbrand C-D, Meredith M, Eagles G (2007) Drake passage and cenozoic climate: an open and shut case? Geochem Geophys Geosyst 8:Q01005

Lizotte MP (2001) The contributions of sea ice algae to antarctic marine primary production. Am Zool 41:57–73

Löfgren S, Boström B (1989) Interstitial water concentrations of phosphorus, iron and manganese in a shallow, eutrophic Swedish lake—implications for phosphorus cycling. Water Res 23(9):1115–1125

Lohmann G (2003) Atmospheric and oceanic freshwater transport during weak Atlantic overturning circulation. Tellus 55A:438–449

Longrich NR, Tokaryk T, Field DJ (2011) Mass extinction of birds at the cretaceous-paleogene (K-PG) boundary. PNAS 108(37):15253–15257

Love GD, Grosjean E, Stalvies C, Fike DA, Grotzinger JP, Bradley AS et al (2009) Fossil steroids record the appearance of Demospongiae during the Cryogenian period. Nature 457:718–721

Lowe CD, Minter EJ, Cameron DD, Brockhurst MA (2016) Shining a light on exploitative host control in a photosynthetic endosymbiosis. Curr Biol 26:207–211

Lowenstein TK, Timofeeff MN, Brennan ST, Hardie LA, Demicco RV (2001) Oscillations in phanerozoic seawater chemistry: evidence from fluid inclusions. Science 294:1086–1088

Lumpkin R, Johnson GC (2013) Global Ocean surface velocities from drifters: mean, variance, El Niño-southern oscillation response, and seasonal cycle. J of Geophys Res Oceans 118:2992–3006

Lunt DJ, Foster GL, Haywood AM, Stone EJ (2008) Late pliocene greenland glaciation controlled by a decline in atmospheric CO_2 levels. Nature 454:1102–1105

Lupton JE, Delaney JR, Johnson HP, Tivey MK (1985) Entrainment and vertical transport of deep-ocean water by buoyant hydrothermal plumes. Nature 316:621–623

Lutz S, Anesio AM, Edwards A, Benning LG (2015) Microbial diversity on Icelandic glaciers and ice caps. Front Microbiol 6:307

Lynch M, Conery JS (2003) The origins of genome complexity. Science 302:1401–1404

Lyons TW, Reinhard CT (2009) Oxygen for heavy-metal fans. Nature 461:179–181

Lyons TW, Reinhard CT, Planavsky NJ (2014) The rise of oxygen in earth's early ocean and atmosphere. Nature 506:302–315

Ma X, Chang P, Saravanan R, Montuoro R, Hsieh J-S et al (2015) Distant influence of Kuroshio Eddies on North Pacific weather patterns? Sci Rep. https://doi.org/10.1038/srep17785

Ma X, Jing Z, Chang P, Liu X, Montuoro R et al (2016) Western boundary currents regulated by interaction between ocean eddies and the atmosphere. Nature 535:533–537

Macdonald FA, Wordsworth R (2017) Initiation of snowball earth with volcanic sulfur aerosol emissions. Geophys Res Lett 44. https://doi.org/10.1002/2016GL072335

Macdonald FA, Schmitz MD, Crowley JL, Roots CF, Jones DS et al (2010) Calibrating the cryogenian. Science 327:1241–1243

Maeda T, Hirose E, Chikaraishi Y, Kawato M, Takishita K et al (2012) Algivore or phototroph? *Plakobranchus ocellatus* (Gastropoda) continuously acquires Kleptoplasts and nutrition from multiple algal species in nature. PLoS One 7(7):e42024

Maher BA, Prospero JM, Mackie D, Gaiero D, Hesse PP, Balkanski Y (2010) Global connections between Aeolian dust, climate and ocean biogeochemistry at the present day and at the last glacial maximum. Earth Sci Rev 99:61–97

Maheshwari A, Sial AN, Gaucher C, Bossi J, Bekker A et al (2010) Global nature of the Paleoproterozoic Lomagundi carbon isotope excursion: a review of occurrences in Brazil, India, and Uruguay. Precambrian Res 182:274–299

Malavelle FF, Haywood JM, Jones A, Gettelman A, Clarisso L et al (2017) Strong constraints on aerosol-cloud interactions from volcanic eruptions. Nature 546:485–491

Maloof AC, Rose CV, Beach R, Samuels BM, Calmet CC et al (2010) Possible animal body fossils in pre-Marinoan limestones from South Australia. Nature GeoSci. https://doi.org/10.1038/NGEO934

Mandelbrot BB (1991) Die fraktale Geometrie der Natur. Birkhäuser, Basel

Mann DG (1999) The species concept in diatoms. Phycologia 38(6):437–494

Mann KH, Lazier JRN (2006) Dynamics of marine ecosystems. Blackwell, Malden

Mann ME, Cane MA, Zebiak SE, Clement A (2005) Volcanic and solar forcing of the tropical pacific over the past 1000 years. J Climate 18:447–456

Manske AK, Glaeser J, Kuypers MMM, Overmann J (2005) Physiology and phylogeny of green sulfur bacteria forming a monospecific phototrophic assemblage at a depth of 100 meters in the Black Sea. Applied and Environmental Microb 71(12):8049–8060

Margesin R, Collins T (2019) Microbial ecology of the cryosphere (glacial and permafrost habitats): current knowledge. Appl Microbiol Biotechnol 103:2537–2549

Martin W, Rotte C, Hoffmeister M, Theissen U, Gelius-Dietrich G et al (2003) Early cell evolution, eukaryotes, anoxia, sulfide, oxygen, fungi first (?), and a tree of genomes revisited. Life 55(4–5):193–204

Martin RS, Mather TA, Pyle DM (2007) Volcanic emissions and the early Earth atmosphere. Geochim Cosmochim Acta 71:3673–3685

Maruyama S, Santosh M (2008) Models of snowball earth and Cambrian explosion: a synopsis. Gondw Res 14:22–32

Masoller C, Schifino ACS, Romanelli L (1995) Characterization of strange attractors of Lorenz model of general circulation of the atmosphere. Chaos Solitons Fractals 6:357–366

Mason B, Moore CB (1985) Grundzüge der geochemie. Enke, Stuttgart

Maul SS (2013) Die Wahrsagekunst im Alten Orient. Beck, München

Maurin O, Davies TJ, Burrows JF, Daru BH, Yessoufou K et al (2014) Savanna fire and the origins of the 'underground forests' of Africa. New Phytol 204:201–214

McElwain JC, Punyasena SW (2007) Mass extinction events and the plant fossil record. Trends Ecol Evol 22(10):548–567

McGhee GR (2013) When the invasion on land failed. Columbia University Press, New York

McGillicuddy DJ (2016) Mechanisms of physical-biological-biochemical interactions at the oceanic mesoscale. Ann Rev Mar Sci 8:125–159

McKenzie NR, Hughes NC, Gill BC, Myrow PM (2014) Plate tectonic influences on neoproterozoic-early paleozoic climate and animal evolution. Geology 42(2):127–130

McKenzie NR, Horton BK, Loomis SE, Stockli DF, Planavsky NJ, Lee C-TA (2016) Continental arc volcanism as the principal driver of icehouse-greenhouse variability. Science 352:444–447

Meehl GA, Arblaster JM, Matthes K, Sassi F, van Loon H (2009) Amplifying the pacific climate system response to a small 11-year solar cycle forcing. Science 325:1114–1118

Meert JG, van der Voo R (1994) The neoproterozoic (1000-540 Ma) glacial intervals: no more snowball earth? Earth Planet Sci Lett 123:1–13

Méndez-Alonzo R, Montezuma C, Ordoñez VR, Angeles G, Marínez AJ, López-Portillo J (2015) Root biomechanics in *Rhizophora mangle*: anatomy, morphology and ecology of mangrove's flying buttresses. Ann Bot 115:833–840

Mercier A, Sun Z, Baillon S, Hamel J-F (2011) Lunar rhythms in the Deep Sea: evidence from reproductive periodicity of several marine invertebrates. J Biol Rhythms 26(1):82–86

Merico A, Tyrrell T, Wilson PA (2008) Eocene/oligocene ocean de-acidification linked to antarctic glaciation by sea-level fall. Nature 452:979–982

Merkl R (2015) Bioinformatik. Wiley-VCH, Weinheim

Mesbah NM, Abou-El-Ela SH, Wiegel J (2007) Novel and unexpected prokaryotic diversity in water and sediments of the alkaline, hypersaline lakes of the Wadi An Natrun, Egypt. Microb Ecol 54:598–617

Michener R, Lajtha K (2007) Stable isotopes in ecology and environmental science. Blackwell, Malden

Mikami OK, Katsuno Y, Yamashita DM, Noske R, Eguchi K (2010) Bowers of the great bowerbird (Chlamydera nuchalis) remains unburned after fire: is it an adaptation to fire? J Ethol 28:15–20

Miki T, Jacquet S (2008) Complex interactions in the microbial world: underexplored key links between viruses, bacteria and protozoan grazers in aquatic environments. Aquat Microb Ecol 51:195–208

Milankovitch M (1941) Kanon der Erdbestrahlung. Königlich Serbische Akademie Ed Spec Tome 33

Mills DB, Ward LM, Jones C, Sweeten B, Forth M et al (2014) Oxygen requirements of the earliest animals. PNAS 111(11):4164–4172

Minic Z, Hervé G (2004) Biochemical and enzymological aspects of the endosymbiosis between the deep-sea tubeworm *Riftia pachyptila* and its bacterial endosymbiont. Eur J Biochem 271:3093–3102

Mitchell JS, Roopnarine PD, Angielczyk KD (2012) Late cretaceous restructuring of terrestrial communities facilitated the end-cretaceous mass extinction in North America. PNAS 109(46):18857–18861

Mitra A, Flynn KJ, Tillmann U, Raven JA, Caron D et al (2016) Defining planktonic protist functional groups on mechanisms for energy and nutrient acquisition: incorporation of diverse myxotrophic strategies. Protist 167:106–120

Miyakawa S, Yamanashi H, Kobayashi K, Cleaves HJ, Miller SL (2002) Prebiotic synthesis from CO atmospheres: implications for the origins of life. PNAS 99(23):14628–14631

Mock T, Kroon BMA (2002) Photosynthetic energy conversion under extreme conditions II: the significance of lipids under light limited growth in Antarctic sea ice diatoms. Phytochemistry 61:53–60

Mock T, Thomas DN (2005) Recent advances in sea-ice microbiology. Environ Microb 7(5):605–619

Moczydłowska M, Landing E, Zang W, Palacios T (2011) Proterozoic phytoplankton and the timing of chlorophyte algae origins. Paleontology 54(4):721–733

Mojzsis SJ, Harrison TM, Pidgeon RT (2001) Oxygen-isotope evidence from ancient zircons for liquid water at the earth's surface 4,300 Myr ago. Nature 409:178–181

Molnar P, England P (1990) Late cenozoic uplift of mountain ranges and global climate change: chicken or egg? Nature 346:29–34

Moore LM (2013) More mixotrophy in the marine microbial mix. PNAS 110(21):8323–8324

Moqbel SY (2009) Characterizing spontaneous fires in landfills. Dissertations, University of Central Florida

Moran K, Backman J, Brinkhuis H, Clemens SC, Cronin T et al (2006) The cenozoic palaeoenvironment of the arctic ocean. Nature 441:601–605

Morgan RPC (1999) Bodenerosion und Bodenerhaltung. Enke, Stuttgart

Morley DW, Leng MJ, Mackay AW, Sloane HJ (2005) Late glacial and Holocene environmental change in the Lake Baikal region documented by oxygen isotopes from diatom silica. Global Planet Change 46:221–233

Muacho S, da Silva JCB, Brotas V, Oliveira PB (2013) Effect of internal waves on near-surface chlorophyll concentration and primary production in the Nazaré Canyon (west of the Iberian Peninsula). Deep-Sea Res I 81:89–96

Muehlstein LK, Amon JP, Leffler DL (1988) Chemotaxis in the marine fungus *Rhizophydium littoreum*. Appl Environ Microbiol 54(7):1668–1672

Mulkidjanian AV, Bychkov AY, Dibrova DV, Galperin MY, Koonin EV (2012) Origin of first cells at terrestrial, anoxic geothermal fields. PNAS. https://doi.org/10.1073/pnas.1117774109

Müller RD, Sdrolias M, Gaina C, Steinberger B, Heine C (2008) Long-term sea-level fluctuations driven by ocean basin dynamics. Science 319:1357–1362

Müller M, Mentel M, van Hellemond JJ, Henze K, Woehle C et al (2012) Biochemistry and evolution of anaerobic energy metabolism in eukaryotes. Microb Mol Biol 76(2):444–495

Müller WEG, Schröder HC, Pisignano D, Markl JS, Wang X (2013) Metazoan circadian rhythm: towards an understanding of a light-base Zeitgeber in sponges. Integr Comp Biol. https://doi.org/10.1093/iicb/ict001

Nagy RM, Porter SM, Dehler CM, Shen Y (2009) Biotic turnover driven by eutrophication before the Sturtian low-latitude glaciation. Nat GeoSci. https://doi.org/10.1038/NGEO525

Nakajima T, Kajihata S, Yoshikawa K, Matsuda F, Furusawa C et al (2014) Integrated metabolic flux and omics analysis of Synechocystis sp. PCC 6803 under mixotrophic and photoheterotrophic conditions. Plant Cell Physiol 55(9):1605–1612

Nakawo M (1983) Measurements on air porosity of sea ice. Ann Glaciol 4:204–208

Nance RD, Murphy JB, Santosh M (2014) The supercontinent cycle: a retrospective essay. Gondw Res 25(1):4–29

Nandy SK, Bapat PM, Venkatesh KV (2007) Sporulating bacteria prefers predation to cannibalism in mixed cultures. FEBS Lett 581:151–156

Naor A, Lapierre P, Mevarech M, Papke RT, Gophna U (2012) Low species barriers in halophilic archaea and the formation of recombinant hybrids. Curr Biol 22:1444–1448

Naranjo-Ortiz MA, Gabaldón T (2019) Fungal evolution: major ecological adaptations and evolutionary transitions. Biol Rev 4:1443–1476

Nash TH III (ed) (2008) Lichen biology. Cambridge University Press, Cambridge

Nederlof R, Müller M (2012) A biomechanical model of rock drilling in the paddock *Barnea candida* (Bivalvia; Mollusca). J R Soc Interface 9:2947–2958

Neuweiler F, Turner EC, Burdige DJ (2009) Early Neoproterozoic origin of the metazoan clade recorded in carbonate rock texture. Geology 37(5):475–478

Nicolis G, Prigogine I (1987) Die Erforschung des Komplexen. Piper, München

Nisbet EG, Sleep NH (2001) The habitat and nature of early life. Nature 409:1083–1091

Niwa Y, Hibiya T (2001) Numerical study of the spatial distribution of the M_2 internal tide in the pacific ocean. J Geophys Res 106(C10):22441–22449

Noh Y, Min HS, Raasch S (2004) Large Eddy simulation of the ocean mixed layer: the effects of wave breaking and Langmuir circulation. J Phys Oceanogr 34:720–735

Nutman AP, Bennett VC, Friend CRL, Kranendonk V, Chivas AR (2016) Rapid emergence of life shown by discovery of 3,700-million-year-old microbial structures. Nature 537:535–538

O'Dea A, Lessios HA, Coates AG, Eytan RI, Restrepo-Moreno SA et al (2016) Formation of the Isthmus of Panama. Sci Adv 2:e1600883

Obara K, Kato A (2016) Connecting slow earthquakes to huge earthquakes. Science 353:253–257

Och LM, Shields-Zhou GA (2012) The neoproterozoic oxygenation event: environmental perturbations and biogeochemical cycling. Earth Sci Rev 110:26–57

Ogg JG (2012) Geomagnetic polarity time scale. In: Gradstein FM, Ogg JG, Schmitz MD, Ogg GM (eds) The geological time scale 2012, vol 1+2. Elsevier, Amsterdam, pp 82–113

Ohmoto H, Watanabe Y, Kumazawa K (2004) Evidence from massive siderite beds for a CO2-rich atmosphere before ~1.8 billion years ago. Nature 429:395–399

Ohmoto H, Watanabe Y, Ikemi H, Poulson SB, Taylor BE (2006) Sulphur isotope evidence for an oxic Archaean atmosphere. Nature 442:908–911

Oki T, Kanae S (2006) Global hydrological cycles and world water resources. Science 313:1068–1072

Okin GS, Baker AR, Tegen I, Mahowald NM, Dentener FJ et al (2011) Impacts of atmospheric nutrient deposition on marine productivity: rules of nitrogen, phosphorus, and iron. Global Biogeochem Cycles 25:GB2022

Oriolo S, Oyhantçabal P, Wemmer K, Siegesmund S (2017) Contemporaneous assembly of Western Gondwana and final Rodinia break-up: implications for the supercontinent-cycle. Geosci Front 8:1431–1445

Osawa S, Jukes TH, Watanabe K, Muto A (1992) Recent evidence for evolution of the genetic code. Microbiol Rev 56(1):229–264

Ossendrijver M (2016) Ancient Babylonian astronomers calculated Jupiter's position from the area under a time-velocity graph. Science 331:482–484

Pagani M, Huber M, Liu Z, Bohaty SM, Henderiks J et al (2011) The role of carbon dioxide during the onset of antarctic glaciation. Science 334:1261–1264

Pannard A, Beisner BE, Bird DF, Braun J, Planas D, Bormans M (2011) Recurrent internal waves in a small lake: potential ecological consequences for metalimnetic phytoplankton populations. Limnol Oceanogr 1:91–109

Papineau D (2010) Global biogeochemical changes at both ends of the proterozoic: insights from phosphorites. Astrobiology 10(2):165–181

Papineau D, Mojzsis SJ, Coath CD, Karhu JA, McKeegan KD (2005) Multiple sulfur isotopes of sulfides from sediments in the aftermath of Paleoproterozoic glaciations. Geochim Cosmochim Acta 69(21):5033–5060

Papineau D, Mojzsis SJ, Schmitt AK (2007) Multiple sulfur isotopes from Paleoproterozoic Huronian interglacial sediments and the rise of atmospheric oxygen. Earth Planet Sci Lett 255:188–212

Parfrey LW, Lahr DJG, Knoll AH, Katz LA (2011) Estimating the timing of early eukaryotic diversification with multigene molecular clocks. PNAS 108(33):13624–13629

Passchier S, Erukanure E (2010) Palaeoenvironments and weathering regime of the neoproterozoic squantum 'Tillite' Boston basin: no evidence of a snowball earth. Sedimentology 57:1526–1544

Paul EA (ed) (2007) Soil microbiology, ecology, and biochemistry. Academic Press, Amsterdam

Pausas JG, Keeley JE (2017) Epicormic resprouting in fire-prone ecosystems. Trends Plant Sci 22(12):1008–1015

Pausas JG, Parr CL (2018) Towards an understanding of the evolutionary role of fire in animals. Evol Ecol 32:113–125

Pausas JG, Schwilk D (2012) Fire and plant evolution. New Phytol 193:301–303

Pavlov AA, Kasting JF, Brown LL, Rages KA, Freedman R (2000) Greenhouse warming by CH_4 in the atmosphere of early earth. J Geophys Res 105(E5):11981–11990

Payne JL, Turchyn AV, Paytan A, De Paolo DJ, Lehrmann DJ et al (2010) Calcium isotope constraints on the end-Permian mass extinction. PNAS 107(19):8543–8548

Peak D, Frame M (1995) Komplexität—das gezähmte Chaos. Birkhäuser, Basel

Pearson MP, Cleal R, Marshall P, Needham S, Pollard J et al (2007) The age of stonehenge. Antiquity 81:617–639

Pechony O, Shindell DT (2010) Driving forces of global wildfires over the past millennium and the forthcoming century. PNAS 107(45):19167–19170

Peitgen H-O, Jürgens H, Saupe D (1992) Chaos and Fractals. Springer, New York

Peltier WR, Liu Y, Crowley JW (2007) Snowball Earth prevention by dissolved organic carbon rich remineralization. Nature 450:813–818

Pernthaler J (2005) Predation on prokaryotes in the water column and its ecological implications. Nat Rev Microbiol 3:537–546

Peters SE, Foote M (2002) Determinants of extinction in the fossil record. Nature 316:420–424

Peters F, Marrasé C (2000) Effects of turbulence on plankton: an overview of experimental evidence and some theoretical considerations. Mar Ecol Prog Ser 205:291–306

Peters O, Neelin JD (2006) Critical phenomena in atmospheric precipitation. Nat Phys 2:393–396

Peters O, Hertlein C, Christensen K (2002) A complexity view of rainfall. Phys Rev Lett 88(1):018701

Petersen SV, Dutton A, Lohmann KC (2016) End-cretaceous extinction in Antarctica linked to both Deccan volcanism and meteorite impact via climate change. Nat Commun. https://doi.org/10.1038/ncomms12079

Peterson KJ, Butterfield NJ (2005) Origin of the Eumetazoa: testing ecological predictions of molecular clocks against the Proterozoic fossil record. PNAS 102(27):9547–9552

Peterson KJ, McPeek MA, Evans DAD (2005) Tempo and mode of early animal evolution: inferences from rocks, Hox, and molecular clocks. Paleobiology 31(2):36–55

Peterson KJ, Cotton JA, Gehling JG, Pisani D (2008) The Ediacaran emergence of bilaterians: congruence between the genetic and the geological fossil records. Philos Trans R Soc B 363:1435–1443

Phillips N, Nogrady B (2020) The climate link to Australasia's fires. Nature 577:610–612

Piller WE, Harzhauser M (2005) The myth of the brackish Sarmatian Sea. Terra Nova 17:450–455

Pirajno F, Santosh M (2015) Mantle plumes, supercontinents, intracontinental rifting and mineral systems. Precambrian Res 259:243–261

Pitcher TJ, Morato T, Hart PJB, Clark MR, Haggan N, Santos RS (2007) Seamounts: ecology, fisheries and conservation. Blackwell Publishing, Oxford

Pivello VR (2011) The use of fire in the Cerrado and Amazonian rainforests of Brazil: past and present. Fire Ecol 7(1):24–39

Planavsky NJ, Bekker A, Hofmann A, Owens JD, Lyons TW (2012) Sulfur record of rising and falling marine oxygen and sulfate levels during the Lomagundi event. PNAS 109(45):18300–18305

Planavsky NJ, Reinhard CT, Wang X, Thomson D, McGoldrick P et al (2014) Low mid-proterozoic atmospheric oxygen levels and the delayed rise of animals. Science 346:635–638

Pollard D, DeConto RM (2005) Hysteresis in cenozoic antarctic ice-sheet variations. Global Planet Change 45:9–21

Pollard D, Kasting JF (2005) Snowball earth: a thin-ice solution with flowing sea glaciers. J Geophys Res 110:C07010

Polton JA, Belcher SE (2007) Langmuir turbulence and deeply penetrating jets in an unstratified mixed layer. J Geophys Res 112:C09020

Porter SM (2004) The fossil record of early eukaryotic diversification. Paleontological Society Papers, vol 10, pp 35–50

Porter SM (2006) The proterozoic fossil record of heterotrophic eukaryotes. In: Xiao S, Kaufman AJ (eds) Neoproterozoic geobiology and paleobiology. Springer, Dordrecht, pp 1–21

Porter SM (2011) The rise of predators. Geology 39(6):607–608

Porter SM (2016) Tiny vampires in ancient seas: evidence for predation via perforation in fossils from the 780-740 million-year-old Chuar Group, Grand Canyon, USA. Proc R Soc B 283:20160221

Porter SM (2020) Insights into eukaryogenesis from the fossil record. Interface Focus 10:20190105

Pouliot J, Galand PE, Lovejoy C, Vincent WF (2009) Vertical structure of archaeal communities and the distribution of ammonia monooxygenase A gene variants in two meromictic High Arctic lakes. Environmental Microb 11(3):687–699

Power MJ, Marlon J, Ortiz N, Bartlein PJ, Harrison SP et al (2008) Changes in fire activity since the last glacial maximum: an assessment based on a global synthesis and analysis of charcoal data. Climate Dynam 30(7–8):887–907

Priede IG, Bergstad OA, Miller PI, Vecchione M, Gebruk A et al (2013) Does presence of a mid-ocean ridge enhance biomass and biodiversity? PLoS One 8(5):e61550

Probandt D, Eickhorst T, Ellrott A, Amann R, Knittel K (2017) Microbial life on a sand grain: from bulk sediment to single grain. ISME J. https://doi.org/10.1038/ismej.2017.197

Prokoph A, El Bilali H, Ernst R (2013) Periodicities in the emplacement of large igneous provinces through the phanerozoic: relations to ocean chemistry and marine biodiversity evolution. Geoscience Front 4:263–276

Proshutinsky AY, Johnson MA (1997) Two circulation regimes of the wind-driven arctic ocean. J Geophys Res 102(C6):12493–12514

Proshutinsky A, Dukhovskoy D, Timmermans ML, Krishfield R, Bamber JL (2015) Arctic circulation regimes. Phil Trans R Soc A 373:20140160

Pruis MJ, Johnson HP (2004) Tapping into the sub-seafloor: examining diffuse flow and temperature from an active seamount on the Juan de Fuca Ridge. Earth Planet Sci Lett 217:379–388

Pu JP, Bowring SA, Ramezani J, Myrow P, Raub TD et al (2016) Dodging snowballs: geochronology of the Gaskiers glaciation and the first appearance of the Ediacaran biota. Geolog Soc Am. https://doi.org/10.1130/G38248.1

Pu JP, Macdonald FA, Schmitz MD, Rainbird RH, Bleeker W et al (2022) Emplacement of the Franklin large igneous province and initiation of the Sturtian Snowball Earth. Sci Adv 8(eadc):9430

Pulquério MJF, Nichols RA (2007) Dates from the molecular clock: how wrong can we be? Trend Ecol Evol 22(4):180–184

Qi L, Hou M, Gawood PA, Lang X, Zhu S, Zhang M (2023) Neoptroterozoic storm deposits in western Yangtze: implications for the sea conditions during the middle Sturtian glaciation. Precambrian Res 384:106945

Qun Y, Junye M, Xiaoyan S, Peiyun C (2007) Phylochronology of early metazoans: combined evidence from molecular and fossil data. Geol J 42:281–295

Rädel G, Shine KP, Ptashnik IV (2015) Global radiative and climate effect of the water vapour continuum at visible and near-infrared wavelengths. Q J Roy Meteorol Soc 141:727–738

Ragueneau O, Quéguiner B, Tréguer P (1996) Contrast in the biological responses to tidally-induced vertical mixing for two macrotidal ecosystems of Western Europe. Estuar Coast Shelf Sci 42:645–665

Rahmstorf S (2002) Ocean circulation and climate during the past 120,000 years. Nature 419:207–214

Rahmstorf S, Crucifix M, Ganopolski A, Goosse H, Kamenkovich I et al (2005) Thermohaline circulation hysteresis: a model intercomparison. Geophys Res Lett 32:L23605

Ramanan R, Kim B-H, Cho D-H, Oh H-M, Kim H-S (2016) Algae-bacteria interactions: evolution, ecology and emerging applications. Biotechnol Adv 14:14–29

Rampino MR, Caldeira K (2017) Correlation of the largest craters, strigraphic impact signatures, and extinction events over the past 250 Myr. Geoscience Front 8:1241–1245

Rashidan KK, Bird DF (2001) Role of predatory bacteria in the termination of a cyanobacterial bloom. Microb Ecol 41:97–105

Rasmussen B, Muhling JR (2023) Organic carbon generation in 3.5-billion-year-old basalt-hosted seafloor hydrothermal vent systems. Sci Adv 9:eadd7925

Rasmussen B, Bengtson S, Fletcher IR, McNaughton NJ (2002) Discoidal impressions and trace-like fossils more than 1200 million years old. Science 296:1112–1115

Rasmussen B, Fletcher IR, Brocks JJ, Kilburn MR (2008) Reassessing the first appearance of eukaryotes and cyanobacteria. Nature 455:1101–1104

Raven JA, Beardall J, Flynn KJ, Maberly SC (2009) Phagotrophy in the origins of photosynthesis in eukaryotes and as a complementary mode of nutrition in phototrophs: relation to Darwin's insectivorous plants. J Exp Bot 60(14):3975–3987

Rea DK (1990) Aspects of atmospheric circulation: the Late Pleistocene (0-959,000 yr) record of eolian deposition in the Pacific Ocean. Palaeogeogr Palaeoclimatol Palaeoecol 78:217–222

Reddy SM, Evans DAD (2009) Palaeoproterozoic supercontinents and global evolution: correlations from core to the atmosphere. In: Reddy SM, Mazumder R, Evans DAD, Collins AS (eds) Palaeoproterozoic supercontinents and global evolution, vol 323, Geological Society, Special Publications, London, pp 1–26

Reinhard CT, Planavsky NJ, Robbins LJ, Partin CA, Gill BC et al (2013) Proterozoic ocean redox and biogeochemical stasis. PNAS 110(14):5357–5362

Reinhard C, Planavsky NJ, Gill BC, Ozaki K, Robbins L et al (2017) Evolution of the global phosphorus cycle. Nature 541:386–389

Renne PR, Sprain CJ, Richards MA, Self S, Vanderkhysen L, Pande K (2015) State shift in Deccan volcanism at the cretaceous-paleogene boundary, possibly induced by impact. Science 350:76–78

Retallack GJ (2013) Ediacaran Gaskiers Glaciation of newfoundland reconsidered. J Geol Soc Lond 170:19–36

Retallack GJ (2014) Precambrian life on land. Palaeobotanist 63:1–15

Retallack GJ (2015) Acritarch evidence for an Ediacaran adaptive radiation of fungi. Bot Pac 4(2):19–33

Retallack GJ (2016) Astropedology: palaeosols and the origin of life. Geol Today 32(5):172–178

Retallack GJ, Marconato A, Osterhout JT, Watts K, Bindeman IN (2014) Revised Wonoka isotopic anomaly in South Australia and late Ediacaran mass extinction. J of the Geol Soc London 171:709–722

Richards TA, Jones MDM, Leonard G, Bass D (2012) Marine fungi: their ecology and molecular diversity. Ann Rev Mar Sci 4:495–522

Riding R (2006) Microbial carbonate abundance compared with fluctuations in metazoan diversity over geological time. Sediment Geol 185:229–238

Riedl R (1989) Die Strategie der Genesis. Piper, München

Rind D (1999) Complexity and climate. Science 284:105–107

Rios-Villamizar EA, Adeney JM, Fernandez Piedade MT, Junk WJ (2020) Hydrochemical classification of Amazonian rivers: a systematic review and meta-analysis. Caminhos de Geografia 21(78):211–226

Roberts NMW (2012) Increased loss of continental crust during supercontinent amalgamation. Gondw Res 21:994–1000

Rodríguez-Iturbe I, Rinaldo A (2001) Fractal river basins. Cambridge University Press, Cambridge

Roedel W (1994) Physik unserer Umwelt: Die Atmosphäre. Springer, Berlin

Rögl F (1999) Mediterranean and Paratethys. Facts and hypotheses of an oligocene to miocene paleogrography (short overview). Geol Carpath 30(4):339–349

Rojas J, Duprat J, Engrand C, Dartois E, Delauche L et al (2021) The micrometeorite flux at Dome C (Antarctica), monitoring the accretion of extraterrestrial dust on earth. Earth Planet Sci Lett 560:116794

Rolf T, Coltice N, Tackley PJ (2012) Linking continental drift, plate tectonics, and the thermal state of the earth's mantle. Earth Planet Sci Lett 351–352:134–146

Rooney AD, Strauss JV, Brandon AD, Macdonald FA (2015) A cryogenian chronology: two long-lasting synchronous neoproterozoic glaciations. Geology 43(5):459–462

Roopnarine PD, Angielczyk KD (2015) Community stability and selective extinction during the Permian-Triassic mass extinction. Science 350:90–93

Rosenfeld D, Lohmann U, Raga GB, O'Dowd CD, Kulmala M et al (2008) Flood or drought: how do aerosols affect precipitation? Science 321:1309–1313

Rosing MT, Bird DK, Sleep NH, Glassley W, Albarede F (2006) The rise of continents—an essay on the geologic consequences of photosynthesis. Palaeogeo Palaeoclim Palaeoecol 232:99–113

Rosso I, Hogg AM, Kiss AE, Gayen B (2015) Topographic influence on submesoscale dynamics in the Southern Ocean. Geophys Res Lett 42:1139–1147

Röttger J (2000) ST radar observations of atmospheric waves over mountainous areas: a review. Ann Geophys 18:750–765

Royer DL (2006) CO_2-forced climate thresholds during the phanerozoic. Geochim Cosmochim Acta 70:5665–5675

Ruiz S (2018) Mechanics and energetics of soil bioturbation by earthworms and growing plant roots. Dissertations, ETH Zürich

Ruiz S, Or D (2018) Biomechanical limits to soil penetration by earthworms: direct measurements of hydroskeletal pressures and peristaltic motions. J R Soc Interface 15:20180127

Rusotto RD, Ackerman TP (2018) Energy transport, polar amplification, and ITCZ shifts in the GeoMIP G1 ensemble. Atmos Chem Phys 18:2287–2305

Russell CT, Strangeway RJ, Zhao C, Anderson BJ, Baumjohann W et al (2017) Structure, force balance, and topology of earth's magnetopause. Science 356:960–963

Ryan CJ, Roguev A, Patrick K, Xu J, Jahari H et al (2012) Hierarchical modularity and the evolution of genetic interactomes across species. Mol Cell 46:691–704

Ryder G, Koeberl C, Mojzsis SJ (2000) Heavy bombardment on the earth at ~3.85 Ga: the search for petrographic and geochemical evidence. In: Canup R, Righter K (eds) Origin of earth and moon. University of Arizona Press, Tucson, pp 475–492

Sahney S, Benton MJ (2008) Recovery from the most profound mass extinction of all time. Proc R Soc B 275:759–765

Sánchez-Baracaldo P (2015) Origin of marine planktonic cyanobacteria. Sci Rep 5:17418. https://doi.org/10.1039/srep17418

Sánchez-Baracaldo P, Ridgwell A, Raven JA (2014) A neoproterozoic transition in the marine nitrogen cycle. Curr Biol 24:652–657

Sansom RS, Gabbott SE, Purnell MA (2010) Non-random decay of chordate characters causes bias in fossil interpretation. Nature 463:797–800

Sarmiento JL, Gruber H, Brzezinski MA, Dunne JP (2003) High-latitude controls of thermocline nutrients and low latitude biological productivity. Nature 427:56–60

Saul JM, Schwartz L (2007) Cancer as a consequence of the rising level of oxygen in the Late Precambrian. Lethaia 40:211–220

Saunders A, Reichow M (2009) The Siberian traps and the end-Permian mass extinction: a critical review. Chin Sci Bull 54(1):20–37

Schachtschabel P, Blume H-P, Brümmer G, Hartge K-H, Schwertmann U (1989) Lehrbuch der Bodenkunde. Enke, Stuttgart

Schad W (2001) Lunar influence on plants. Earth Moon Planet 85–86:405–409

Scheirer DS, Shank TM, Fornari DJ (2006) Temperature variations at diffuse and focused flow hydrothermal vent sites along the northern east pacific rise. Geochem Geophys Geosyst 7:3

Scheiter S, Higgins SI, Osborne CP, Bradshaw BS, Lunt D et al (2012) Fire and fire-adapted vegetation promoted C_4 expansion in the late Miocene. New Phytol 195:653–666

Scher HD, Whittaker JM, Williams SE, Latimer JC, Kordesch WEC, Delaney ML (2015) Onset of antarctic circumpolar current 30 million years ago as Tasman gateway aligned with westerlies. Nature 523:580–583

Schintele R, Brocks JJ (2017) Paleoecology of Neoproterozoic hypersaline environments: biomarker evidence for haloarchaea, methanogens, and cyanobacteria. Geobiology 2017:1–23

Schirrmeister BE, de Vos JM, Antonelli A, Bagheri HC (2013) Evolution of multicellularity coincided with increased diversification of cyanobacteria and the great oxidation event. PNAS 110(5):1791–1796

Schirrmeister BE, Gugger M, Donoghue PCJ (2015) Cyanobacteria and the great oxidation event: evidence from genes and fossils. Paleontology 58(5):769–785

Schirrmeister BE, Sanchez-Baracaldo P, Wacey D (2016) Cyanobacterial evolution during the Precambrian. Int J Astrobiol 15(3):187–204

Schlesinger WH, Jasechko S (2014) Transpiration in the global water cycle. Agric For Meteorol 189–190:115–117

Schmittner A (2016) The smoking gun for Atlantic circulation changes. Science 353:445–446

Schneider T, Walker CC (2006) Self-organization of atmospheric macroturbulence into critical states of weak nonlinear Eddy-Eddy interactions. J Atm Soc 63:1569–1586

Schoene B, Samperton KM, Eddy MP, Keller G, Adatte T et al (2015) U-Pb geochronology of the Deccan traps and relation to the end-cretaceous mass extinction. Science 347:182–184

Schrödinger E (1943) What is life? Trinity College, Dublin. Deutsche Übersetzung (1989) "Was ist Leben". Piper, München

Schulz H, von Rad U, Erlenkeuser H (1998) Correlation between Arabian Sea and Greenland climate oscillations of the past 110,000 years. Nature 393:54–57

Schuster HG (1994) Deterministisches Chaos. VCH, Weinheim

Schwarz M, Cohen D, Or D (2010) Root-soil mechanical interactions during pullout and failure of root bundles. J Geophys Res 115:F04035

Schwarz M, Cohen D, Or D (2011) Pullout tests of root analogs and natural root bundles in soil: experiments and modelling. J Geophys Res 116:F02007

Schwerdtfeger F (1975) Synökologie. Paul Parey, Hamburg

Schwilk DW, Ackerly DD (2001) Flammability and serotiny as strategies: correlated evolution in pines. Oikos 94:326–336

Scotese CR (2004) The continental drift flipbook. J Geol 112:729–741

Scotese CR (2021) An atlas on phanerozoic paleogeographic maps: the seas come in and the seas go out. Annu Rev Earth Planet Sci 49:669–718

Scott AC, Glasspool IJ (2006) The diversification of Paleozoic fire systems and fluctuations in atmospheric oxygen concentration. PNAS 103(29):10861–10865

Scott C, Lyons TW, Bekker A, Shen Y, Poulton SW et al (2008) Tracing the stepwise oxygenation of the proterozoic ocean. Nature 452:456–459

Scott C, Wing BA, Bekker A, Planavsky NJ, Medvedev P et al (2014) Pyrite multiple-sulfur isotope for rapid expansion and contraction of the early paleoproterozoic seawater sulfate reservoir. Earth Planet Sci Lett 389:95–104

Selden P, Nudds J (2012) Evolution of fossil ecosystems. Manson Publishing, London

Seo K-W, Ryu D, Eom J, Jeon T, Kim J-S et al (2023) Drift of earth's pole confirms groundwater depletion as a significant contributor to Global Sea level rise 1993-2010. Geophys Res Lett 50:e2023GL103509

Sergeev VN, Semikhatov MA, Fedonkin MA, Veis AF, Vorob'eva NG (2007) Principal stages in the evolution of Precambrian organic world: communication 1. Archean and early proterozoic. Stratigr Geol Correl 15(2):141–160

Sergeev VN, Semikhatov MA, Fedonkin MA, Veis AF, Vorob'eva NG (2010) Principal stages in the evolution of Precambrian organic world: communication 2. The late proterozoic. Stratigr Geol Correl 18(6):561–592

Shackleton NJ, Backman J, Zimmerman H, Kent DV, Hall MA et al (1984) Oxygen isotope calibration on the onset of ice-rafting and history of glaciation in the North Atlantic region. Nature 307:620–623

Shaughnessy DT, McAllister K, Worth L, Haugen AC, Meyer JN, Domann FE et al (2014) Mitochondria, energetics, epigenetics, and cellular responses to stress. Environ Health Perspect 122(12):1271–1278

Shen J, Algeo TJ, Zhou L, Feng Q, Yu J, Ellwood B (2012) Volcanic perturbations of the marine environment in South China preceding the latest Permian mass extinction and their biotic effects. Geobiology 10:82–103

Shi M, Feng Q, Zhu S (2014) Biotic evolution and its relation with geological events in the Proterozoic Yanshan Basin, North China. Science 57(5):903–918

Shields-Zhou GA, Porter S, Halverson GP (2016) A new rock-based definition for the Cryogenian period (circa 720-635 Ma). Episodes 39(1):3–8

Sidor CA, Vilhena DA, Angielczyk KD, Huttenlocker AK, Nesbitt SJ et al (2013) Provincialization of terrestrial faunas following the end-Permian mass extinction. PNAS 110(20):8129–8133

Sievert SM, Vetriani C (2012) Chemoautotrophy at deep-sea vents: past, present, and future. Oceanography 25(1):218–233

Sigmundsson F, Hreinsdóttir S, Houper A, Pedersen R, Roberts MJ et al (2010) Intrusion triggering of the 2010 Eyjafjallajökull explosive eruption. Nature 468:426–430

Signor PW, Lipps JH (1982) Sampling bias, gradual extinction patterns and catastrophes in the fossil record. Geol Soc Am Spec Paper 190:291–296

Silva P, Wainer I, Khodri M (2021) Changes in the equatorial mode of the tropical Atlantic in terms of the Bjerknes feedback index. Climate Dynam 56:3005–3024

Simon C, Wiezer A, Strittmatter AW, Daniel R (2009a) Phylogenetic diversity and metabolic potential revealed in a glacier ice metagenome. Appl Environ MIcrobiol 75(23):7519–7526

Simon MF, Grether R, de Queiroz LP, Skema C, Pennington RT, Hughes CE (2009b) Recent assembly of the Cerrado, a neotropical plant diversity hotspot, by in situ evolution of adaptations to fire. PNAS 106(48):20359–20364

Sinclair H (2017) Making a mountain out of a plateau. Nature 542:41–42

Sinclair M, Rao DVS, Couture R (1981) Phytoplankton temporal distributions in estuaries. Oceanol Acta 4(2):239–246

Slater BJ, Bohlin MS (2022) Animal origins: the record from organic microfossils. Earth Sci Rev 232:104107

Slaymaker O, Spencer T, Embleton-Hamann C (eds) (2009) Geomorphology and global environmental change. Cambridge University Press, Cambridge

Sleep NH, Zahnle KJ, Kasting JF, Morowitz HJ (1989) Annihilation of ecosystems by large asteroid impacts on the early earth. Nature 342:139–142

Small RJ, deSzoeke SP, Xie SP, O'Neill L, Seo H et al (2008) Air-sea interaction over ocean fronts and eddies. Dyn Atmos Oceans 45:274–319

Smetacek V (2012) Making sense of ocean biota: how evolution and biodiversity of land organisms differ from that of the plankton. J Biosci 37(4):589–607

Smith AB, Gale AS, Monks NEA (2001) Sea-level change and rock-record bias in the cretaceous: a problem for extinction and biodiversity studies. Paleobiology 27(2):241–253

Sobolev SV, Sobolev AV, Kuzmin DV, Krivolutskaya NA, Petrunin AG et al (2011) Linking mantle plumes, large igneous provinces and environmental catastrophes. Nature 477:312–316

Sokol EV, Volkova NJ (2007) Combustion metamorphic events resulting from natural coal fires. In: Stracher GB (ed) Geology of coal fires: case studies from around the world. Geol Soc Am doi: https://doi.org/10.1130/2007.4118(07)

Solé RV, Manrubia SC, Benton M, Kauffman S, Bak P (1999) Criticality and scaling in evolutionary ecology. TREE 14(4):156–160

Song H, Wignall PB, Chu D, Tong J, Sun Y et al (2014) Anoxia/high temperature double whammy during the Permian-Triassic marine crisis and its aftermath. Sci Rep. https://doi.org/10.1038/srep04132

Sorrel P, Debret M, Billeaud I, Jaccard SL, McManus JF, Tessier B (2012) Persistent non-solar forcing of holocene storm dynamics in coastal sedimentary archives. Nat Geosci. https://doi.org/10.1038/NGEO1619

Sparrow M, Boebel O, Zervakis V, Zenk W, Cantos-Figuerola A, Gould W (2002) Two circulation regimes of the Mediterranean outflow revealed by Lagrangian measurements. J Phys Oceanogr 32:1322–1330

Sperling EA, Galen PH, Knoll AH, Macdonald FA, Johnston DT (2013) A basin redox transect at the dawn of animal life. Earth Planet Sci Lett 371–372:143–155

Sperling EA, Wolock CJ, Morgan AS, Gill BC, Kunzmann M et al (2015) Statistical analysis of iron geochemical data suggests limited late proterozoic oxygenation. Nature 523:451–454

Spotila JA, Buscher JT, Meigs AJ, Reiners PW (2004) Long-term glacial erosion of active mountain belts: example of the Chugach-St. Elias Range, Alaska. Geology 32(6):501–504

Srikanth S, Lum SKY, Chen Z (2016) Mangrove root: adaptations and ecological importance. Trees 30:451–465

Stacey FD, Davis PM (2008) Physics of the earth. Cambridge University Press, Cambridge

Staley JT, Gosink JJ (1999) Poles apart: biodiversity and biogeography of sea ice bacteria. Annu Rev Microbiol 53:189–215

Stanley SM (2001) Historische Geologie. Spektrum Akademischer Verlag, Heidelberg

Staudigel H, Clague DA (2010) The geological history of deep-sea volcanoes. Oceanography 23(1):58–71

Stecher A (2015) Functional biodiversity of sea ice-associated protists in the Central Arctic Ocean. Dissertations, University of Konstanz

Stephens GL, Li J, Wild M, Clayson CA, Loeb N et al (2012) An update on earth's energy balance in light of the latest global observations. Nat Geosci 5:691–696

Stephenson FR, Mossison LV, Hohenkerk CY (2016) Measurement of earth's rotation: 720 BC to AD 2015. Proc R Soc A 472:20160404

Stern RJ (2008) Neoproterozoic crustal growth: the solid earth system during a critical episode of earth history. Gondw Res 14(1–2):33–50

Stern RJ, Avigad D, Miller N, Beyth M (2008) From volcanic winter to snowball earth: an alternative explanation for neoproterozoic biosphere stress. In: Dilek Y, Furnes H, Muehlenbachs K (eds) Links between geological processes, microbial activities and evolution of life. Springer Science+Business Media, pp 313–317

Stetter KO (2006) Hyperthermophiles in the history of life. Philos Trans R Soc B 361:1837–1843

Stewart RH (2008) Introduction to physical oceanography. Texas University, Department of Oceanography

Stickley CE, Brinkhuis H, Schellenberg SA, Sluijs A, Röhl U et al (2004) Timing and nature of the deepening of the Tasman gateway. Paleoceanography 19:PA4027

Stockner JG (1988) Phototrophic picoplankton: an overview from marine and freshwater ecosystems. Limnol Oceanogr 33(4):765–775

Storz JF, Opazo JC, Hoffmann FG (2013) Gene duplication, genome duplication, and the functional diversification of vertebrate globins. Mol Phylogenet Evol 66(2):469–478

Stouffer RJ, Yin J, Gregory JM, Dixon KW, Spelman MJ et al (2006) Investigating the causes of the response of the thermohaline circulation to past and future climate changes. J Climate 19:1365–1387

Strathmann RR (2020) Multiple origins of feeding head larvae by the early Cambrian. Can J Zool 98(12):761–776

Strömberg CAE (2011) Evolution of grasses and grassland ecosystems. Annu Rev Earth Planet Sci 39:517–544

Strother PK, Battison L, Brasier MD, Wellman CH (2011) Earth's earliest non-marine eukaryotes. Nature 473:505–509

Su Z, Wang J, Klein P, Thompson AF, Menemenlis D (2018) Ocean submesoscales as a key component of the global heat budget. Nat Commun. https://doi.org/10.1038/s41467-018-02983-w

Sulpizio R, Zanchetta G, Demi F, Di Vito MA, Pareschi MT, Santacruse R (2006) The Holocene syneruptive volcaniclastic debris flows in the Vesuvian area: geological data as a guide for hazard assessment. In: Siebe C, Macias JL, Aguirre-Diaz GJ (eds) Neogene-quaternary continental margin volcanism: a perspective from Mexico, Geol Soc Am Spec Pap, vol 402, pp 203–221

Sun J, Liu T (2006) The age of the Taklimakan Desert. Science 312:1621

Suter RB, Stratton GE, Miller PR (2011) Mechanics and energetics of excavation by burrowing wolf spiders, *Geolycosa* spp. J Insect Sci 11:22

Sutton TT, Porteiro FM, Heino M, Byrkjedal I, Langhelle G et al (2008) Vertical structure, biomass and topographic associations of deep-pelagic fishes in relation to a mid-ocean ridge system. Deep-Sea Res II 55:161–184

Swanson-Hysell NL, Rose CV, Calmet CC, Halverson GP, Hurtgen MT, Maloof AC (2010) Cryogenian glaciation and the onset of carbon-isotope decoupling. Science 328:608–6011

Sweatman MB, Tsikritsis D (2017) Decoding Göbekli Tepe with archaeoastronomy: what does the fox say? Mediter Archaeol Archaeom 17(1):233–250

Swingedouw D, Terray L, Cassouc C, Voldoire A, Salas-Mélia D, Servonnat J (2011) Natural forcing of climate during the last millennium: fingerprint of solar variability. Climate Dynam 36:1349–1364

Takashima R, Nishi H, Huber BT, Leckie RM (2005) Greenhouse world and the Mesozoic Ocean. Oceanography 19(4):82–93

Talley LD (2008) Freshwater transport estimates and the global overturning circulation: shallow, deep and throughflow components. Prog Oceanogr 78:257–303

Talley LD, Pickard GL, Emery WJ, Swift JH (2011) Descriptive physical oceanography. Elsevier, Amsterdam

Talley LD, Feely RA, Sloyan BM, Wanninkhof R, Baringer MO et al (2016) Changes in ocean heat, carbon content, and ventilation: a review of the first decade of GO-SHIP global repeat hydrography. Ann Rev Mar Sci 8:19.1–19.31

Tang H, Chen Y (2013) Global glaciations and atmospheric change at 2.3 Ga. Geoscience Front 4(5):583–596

Tang M, Chen K, Rudnick RL (2016) Archean upper crust transition from mafic to felsic marks the onset of plate tectonics. Science 351:372–375

Tardy Y, N'Kounkou R, Probst J-L (1989) The global water cycle and continental erosion during phanerozoic time (570 my). Am J Sci 289:455–483

Tatzel M, von Blanckenburg F, Oelze M, Bouchez J, Hippler D (2017) Late neoproterozoic seawater oxygenation by siliceous sponges. Nat Commun. https://doi.org/10.1038/s41467-017-00586-5

Taylor JW, Jacobson DJ, Kroken S, Kasuga T, Geiser DM et al (2000) Phylogenetic species recognition and species concepts in fungi. Fungal Genet Biol 31:21–32

Tebo BM, Davis RE, Anitori RP, Connell LB, Schiffman P, Staudigl H (2015) Microbial communities in dark oligotrophic volcanic ice cave ecosystems of Mt. Erebus, Antarctica. Front Microbiol 6:179

Tessmar-Raible K, Raible F, Arboleda E (2011) Another place, another timer: marine species and the rhythms of life. Bioessays 33:165–172

Tesson SVM, Skjøth CA, Šanti-Temkiv T, Löndahl J (2016) Airborne microalgae: insights, opportunities, and challenges. Appl Environ Microbiol 82(7):1978–1991

Teyssèdre B (2006) Are the green algae (phylum Viridiplantae) two billion years old? Carnets de Géologie 2006/03

Thannickal VJ (2009) Oxygen in the evolution of complex life and the price we pay. Am J Respir Cell Mol Biol 40:507–510

Thiéblemont R, Matthes K, Omrani N-E, Kodera K, Hansen F (2015) Solar forcing synchronizes decadal North Atlantic climate variability. Nat Commun. https://doi.org/10.1038/ncomms9268

Thomas DN, Dieckmann GS (2002) Antarctic Sea ice—a habitat for extremophiles. Science 295:641–644

Thomson RE, Davis EE, Bard BJ (1995) Hydrothermal venting and geothermal heating in Cascadia Basin. J Geophys Res 100(B4):6121–6141

Thomson RE, Mihály SF, Rabinovich AB, McDuff RE, Veirs SR, Stahr FR (2003) Constrained circulation at Endeavour ridge facilitates colonization by vent larvae. Nature 424:545–548

Thomson RE, Subbotina MM, Anisimov MV (2009) Numerical simulation of mean currents and water property anomalies at Endeavour ridge: hydrothermal versus topographic forcing. J Geophys Res 114:C09020

Thorpe SA (2004) Langmuir circulation. Annu Rev Fluid Mech 36:55–79

Thurner S, Hanel R, Klimek P (2018) Introduction into the theory of complex systems. Oxford University Press, Oxford

Tian F, Toon OB, Pavlov AA, De Sterck H (2005) A hydrogen-rich early earth atmosphere. Science 308:1014–1017

Timmermann A, An SI, Kug S, Jin FF, Cai E et al (2018) El Niño-Sothern Oscillation complexity. Nature 559:535–545

Titov V, Rabinovich AB, Mofjeld HO, Thomson RE, González FI (2005) The global reach of the 26 December 2004 Sumatra tsunami. Science 309:2045–2048

Toggweiler JR, Russell J (2008) Ocean circulation in a warming climate. Nature 451:286–288

Torsvik TH, Van der Voo R, Preeden U, Mac Niocaill C, Steinberger B et al (2012) Phanerozoic polar wander, palaeogeography and dynamics. Earth Sci Rev 114:325–368

Tostevin R, Mills BJW (2020) Reconciling proxy records and models of earth's oxygenation during the neoproterozoic and palaeozoic. Interface Focus 10:20190137

Tréguer P, Nelson DM, Van Bennekom AJ, DeMaster D, Leynaert A, Quéguiner B (1995) The silica balance in the World Ocean: a Reestimate. Science 268:375–379

Trenberth KE (1999) Atmospheric moisture recycling: role of advections and local evaporation. J Climate 12:1368–1381

Trenberth KE, Fasullo JT, Mackaro J (2011) Atmospheric moisture transports from ocean and global energy flows in Reanalyses. J Climate 24:4907–4924

Trillmich F, Limberger D (1985) Drastic effects of El Niño on Galapagos pinnipeds. Oecologia 67:19–22

Trindade RIF, Macouin M (2007) Palaeolatitude of glacial deposits and palaeogeography of neoproterozoic ice ages. C R Geosci 339:200–211

Troitskaya E, Blinov V, Ivanov V, Zhdanov A, Gnatovsky R et al (2015) Cyclonic circulation and upwelling in Lake Baikal, Aquatic Sciences, vol 77, pp 171–182

Twitchett RJ (2006) The palaeoclimatology, palaeoecology and palaeoenvironmental analysis of mass extinction events. Palaeogeo Palaeoclim Palaeoecol 232:190–213

Tyler PA, German CR, Ramirez-Llodra E, Van Dover CL (2003) Understanding the biogeography of chemosynthetic ecosystems. Oceanol Acta 25:227–241

Tziperman E, Halevy I, Johnston DT, Knoll AH, Schrag DP (2011) Biologically induced initiation of neoproterozoic snowball-earth events. PNAS. https://doi.org/10.1073/pnas.1016361108

Ueno Y, Johnson MS, Danielache SO, Eskebjerg C, Pandey A, Yoshida N (2009) Geological sulfur isotopes indicate elevated OCS in the Archean atmosphere, solving faint young sun paradox. PNAS 106(35):14784–14789

Urakawa LS, Hasumi H (2009) The energetics of global thermohaline circulation and its wind enhancement. J Phys Oceanogr:1715–1728

Urbach E, Scanlan DJ, Distel DL, Waterbury JB, Chisholm SW (1998) Rapid diversification of marine picophytoplankton with dissimilar light-harvesting structures inferred from sequences of *Prochlorococcus* and *Synechococcus* (Cyanobacteria). J Mol Evol 46:188–201

Valley JW, Lackey JS, Cavossie AJ, Clechenko CC, Spicuzza MJ et al (2005) 4.4 billion years of crustal maturation: oxygen isotope ratios of magmatic zircon. Contrib Mineral Petrol 150:561–580

Van der Voo R, Torsvik TH (2001) Evidence for late paleozoic and mesozoic non-dipole fields provides an explanation for the Pangea reconstruction problems. Earth Planet Sci Lett 187:71–81

van Eimern J, Häckel H (1979) Wetter- und Klimakunde. Ulmer, Stuttgart

van Soelen EE, Kürschner WM (2018) Late Permian to early Triassic changes in acritarch assemblages and morphology in the boreal Arctic: new data from the Finnmark platform. Palaeogeo Palaeoclim Palaeoecol 505:120–127

van Westen RM, Kliphuis M, Dijkstra HA (2024) Physics-based early warning signal shows that AMOC is on tipping course. Sci Adv 10:eadk1189

Vérard C, Hochard C, Baumgartner PO, Stampfi GM (2015) Geodynamik evolution of the earth over the Phanerozoic: plate tectonic activity and plalaeoclimatic indicators. J Palaeogeogr 4(2):167–188

Vernhet E, Youbi N, Chellai EH, Villeneuve M, El Archi A (2012) The Bou-Azzer glaciation: evidence for an Ediacaran glaciation on the West African Craton (Anti-Atlas, Morocoo). Precambrian Res 196–197:106–112

Versteegh GM (2005) Solar forcing of climate. 2: evidence from the past. Space Sci Rev 120:243–286

Vincent WF, Gibson JAE, Pienitz R, Villeneuve V (2000) Ice shelf microbial ecosystems in the high Arctic and implications for life on snowball earth. Naturwissenschaften 87:137–141

Vinogradov SN, Moens L (2008) Diversity of globin function: enzymatic, transport, storage, and sensing. J Biol Chem 283(14):8773–8777

Vinogradov SN, Hoogewijs D, Bailly X, Arrendondo-Peter R, Guertin M et al (2005) Three globin lineages belonging to two structural classes in genomes from the three kingdoms of life. PNAS 102(32):11385–11389

Vinogradov SN, Hoogewijs D, Bailly X, Mizuguchi K, Dewilde S et al (2007) A model of globin evolution. Gene 198:132–142

Vogel S (2003) Comparative biomechanics. Princeton University Press, Princeton

Volkert H, Keil C, Kiemle C, Poberaj G, Chaboureau J-P, Richard E (2003) Gravity waves over the eastern Alps: a synopsis of the 25 October 1999 event (IOP 10) combining in situ and remote-sensing meadurements with a high-resolution simulation. Q J Roy Meteorol Soc 129:777–797 von Bloh W, Bounama C, Franck S (2003) Cambrian explosion triggered by geosphere-biosphere feedbacks. Geophys Res Lett 30:18

Wagner A (2008) Robustness and evolvability: a paradox resolved. Proc R Soc B 275:91–100

Walker GT (1925) Correlation in seasonal variations of weather—a further study world weather. Mon Weather Rev 1925:252–254

Wallace MW, AvS H, Shuster A, Greig A, Planavsky NJ, Reed CP (2017) Oxygenation history of the Neoproterozoic to early Phanerozoic and the rise of land plants. Earth Planet Sci Lett 466:12–19

Wallmann K (2004) Impact of atmospheric CO_2 and galactic cosmic radiation on phanerozoic climate change and the marine $\delta^{18}O$ record. Geochem Geophys Geosyst 5. https://doi.org/10.1029/2003GC000683

Walsh JJ, Steidinger KA (2001) Saharan dust and Florida red tides: the cyanophyte connection. J Geophys Res 106(C6):11597–11612

Walter H, Breckle S-W (1983) Ökologie der Erde Band 1, Ökologische Grundlagen in globaler Sicht. Fischer, Stuttgart

Walter XA, Picazo A, Miracle MR, Vicente E, Camacho A et al (2014) Phototrophic Fe(II)-oxidation in the chemocline of a ferruginous meromictic lake. Front Microbiol 5:713

Wang G (2003) Reassessing the impact of North Atlantic oscillation on the sub-Saharan vegetation productivity. Glob Chang Biol 9:493–499

Wang Y-H, Dai C-F, Chen YY (2007) Physical and ecological processes of internal waves on an isolated reef ecosystem in the South China Sea. Geophys Res Lett 34:L18609

Wang L, Good SP, Caylor KK (2014) Global synthesis of vegetation control on evapotranspiration partitioning. Geophys Sci Lett 41:6753–6757

Wang D, Gouhier TC, Menge BA, Ganguly AR (2015) Intensification and spatial homogenization of coastal upwelling under climate change. Nature 518:390–394

Wang Y, Wang Y, Du W (2017) A rare disc-like holdfast of the Ediacaran macroalga from South China. J Paleo. https://doi.org/10.1017/jpa.2017.43

Ward LM, Shih PM (2021) Granick revisited: synthesizing evolutionary and ecological evidence for the late origin of bacteriochlorophyll via ghost lineages and horizontal gene transfer. PLoS One 16(1):e0239248

Warny SA, Bart PJ, Suc J-P (2003) Timing and progression of climatic tectonic and glacioeustatic influences on the Messinian Salinity Crisis. Palaeogeo Palaeoclim Palaeoecol 202(1–2):59–66

Watanabe Y, Martini JEJ, Ohmoto H (2000) Geochemical evidence for terrestrial ecosystems 2.6 billion years ago. Nature 408:574–578

Wei-Haas M (2023) Lava outburst may have led to snowball earth. Science 381:120

Welsh RM, Zaneveld JR, Rosales SM, Payet JP, Burkepile DE, Thurber RV (2016) Bacterial predation in a marine host-associated microbiome. ISME J 19:1540–1544

Weltzer ML (2011) Microbial community assembly and diversification of the genus Chloroflexus along an alkaline hot spring gradient. Dissertations, University of Montana

Westberry TK, Behrenfed MJ, Shi YR, Yu H, Remer LA, Bian H (2023) Atmospheric nourishment of global ocean ecosystems. Science 380:515–519

Wetzel RG (1975) Limnology. Saunders, Philadelphia

Wheeler BM, Heimberg AM, Moy VN, Sperling EA, Holstein TW et al (2009) The deep evolution of metazoan microRNAs. Evol Dev 11(1):50–68

Whelan RJ (1995) The ecology of fire. Cambridge University Press, Cambridge

White RV (2002) Earth's biggest 'whodunnit': unravelling the clues in the case of the end-Permian mass extinction. Phil Trans R Soc Lond A 360:2963–2985

White WM (2013) Geochemistry. Wiley-Blachwell, Chichester

White M, Mohn C (2002) Seamounts: a review of physical processes and their influence on the seamount ecosystems. OASIS report, University of Hamburg

White ID, Mottershead DN, Harrison SJ (1992) Environmental systems. Chapman and Hall, London

White I, Wade A, Worthy M, Mueller N, Danieli T, Wasson R (2006) The vulnerability of water supply catchments to bushfires: impacts of the January 2003 wildfires on the Australian Capital Territory. Aust J Water Ressour 10(2):179–194

Wikars L-O (1997) Effects of forest fire and the ecology of fire-adapted insects. Dissertations, University of Uppsala

Wilde SA, Valley JW, Peck WH, Graham CM (2001) Evidence from detrital zircons for the existence of continental crust and oceans on the earth 4.4 Gyr ago. Nature 409:175–178

Wildman RA, Hickey LJ, Dickinson MB, Berner RA, Robinson JM (2004) Burning on forest materials under late paleozoic high atmospheric oxygen level. Geology 32(5):457–460

Wilf P, Johnston KR, Huber BT (2003) Correlated terrestrial and marine evidence for global climate changes before mass extinction at the cretaceous-Paleogene boundary. PNAS 100(2):599–604

Willenbring JK, von Blanckenburg F (2010) Long-term stability of global erosion rates and weathering during late-Cenozoic cooling. Nature 465:211–214

Williams GE (2000) Geological constraints on the Precambrian history of earth's rotation and the Moon's orbit. Rev Geophys 38(1):37–59

Willmes S, Adams S, Schröder D, Heinemann G (2011) Spatio-temporal variability of polynya dynamics and ice production in the Laptev Sea between the winters of 1979/80 and 2007/08. Polar Res 30:5971

Winder M, Hunter DA (2008) Temporal organization of phytoplankton communities linked to physical forcing. Oecologia 156:179–192

Wittke JH, Weaver JC, Bunch TE, Kennett JP, Kennett DJ et al (2013) Evidence for deposition of 10 million tonnes of impact spherules across four continents 12,800 y ago. PNAS:E2088–E2097

Witze A (2019) Earth's magnetic field is acting up. Nature 565:143–144

Wolf YI, Koonin EV (2013) Genome reduction as the dominant mode of evolution. Bioessays 35:829–837

Wood B, Baker J (2011) Evolution in the genus homo. Annu Rev Ecol Evol Syst 42:47–69

Wordsworth R, Pierrehumbert R (2013) Hydrogen-nitrogen greenhouse warming in earth's early atmosphere. Science 339:64–67

Wu H, Murray N, Menou K, Lee C, Leconte J (2023) Why the day is 24 hours long: the history of earth's atmospheric thermal tide, composition, and mean temperature. Sci Adv 9:eadd2499

Wunsch C (2002) What is the thermohaline circulation? Science 298:1179–1191

Xiao S (2004) Neoproterozoic glaciations and the fossil record. The extreme proterozoic: geology, geochemistry, and climate geophys monograph series, vol 146, pp 199–214

Xiao S (2013) Written in stone: the fossil record of early eukaryotes. In: Trueba G, Montúfar C (eds) Evolution from the Galapagos, social and ecological interactions in the Galapagos Islands 2. Spinger, Berlin, pp 107–124

Xiao S, Dong L (2006) On the morphological and ecological history of proterozoic macroalgae. In: Xiao S, Kaufman AJ (eds) Neoproterozoic geobiology and paleobiology. Springer, Dordrecht, pp 57–90

Yamaguchi K (2002) Geochemistry of Archean-Paleoproterozoic black shales: the early evolution of the atmosphere, oceans, and biosphere. Dissertations, Pennsylvania State University

Yan Y, Bender MI, Brook EJ, Clifford HM, Kemeny PC et al (2019) Two-million-year-old snapshots of atmospheric gases from Antarctic ice. Nature 574:663–668

Yang J, Faccenda M (2020) Intraplate volcanism originating from upwelling hydrous mantle transition zone. Nature 579:88–91

Yang J, Jansen MF, Macdonald FA, Abbot DS (2017) Persistence of a freshwater surface ocean after a snowball earth. Geology 45(7):615–618

Yang H, Lohmann G, Krebs-Kanzow U, Ionita M, Shi X et al (2020) Poleward shift of the major ocean gyres detected in a warming climate. Geophys Res Lett 47:5

Yano J-I, Liu C, Moncrieff MW (2012) Self-organized criticality and homeostasis in atmospheric convective organization. J Atm Soc 69:3449–3462

Ye Q, Tong J, Xiao S, Zhu S, An Z et al (2015) The survival of benthic macroscopic phototrophs on a neoproterozoic snowball earth. Geology 43(6):507–510

Yoon HS, Hackett JD, Cinigla C, Pinto G, Bhattacharya D (2004) A molecular timeline for the origin of photosynthetic eukaryotes. Mol Biol Evol 21(5):809–818

Young GM (2013) Precambrian supercontinents, glaciations, atmospheric oxygenation, metazoan evolution and an impact that may have changed the second half of earth history. Geoscience Front 4:247–261

Zachos J, Pagani M, Sloan L, Thomas E, Billups K (2001) Trends, rhythms, and aberrations in global climate 65 Ma to present. Science 292:686–693

Zaffos A, Finnegan S, Peters SE (2017) Plate tectonic regulation of global marine animal diversity. PNAS 114(22):5653–5658

Zantke J, Ishikawa-Fujiwara T, Arboleda E, Lohs C, Schipany K et al (2013) Circadian and circalunar clock interactions in a marine annelid. Cell Rep 5:99–113

Zeebe RE, Lourens LJ (2019) Solar system chaos and the Paleocene-Eocene boundary age constrained by geology and astronomy. Science 365:926–929

Zeng N, Neelin JD, Lau K-M, Tucker CJ (1999) Enhancement of Interdecadal climate variability in the Sahel by vegetation interactions. Science 286:1537–1540

Zerkle AL, Poulton SW, Newton RJ, Mettam C, Claire MW et al (2017) Onset of aerobic nitrogen cycle during the great oxidation event. Nature 542:465–467

Zhang Y, Liu B-Y, Zhang Q-C, Xie Y (2003) Effect of different vegetation types on soil erosion in water. Acta Bot SInica 45(10):1204–1209

Zhang Y, Liu Z, Zhao Y, Wang W, Li J, Xu J (2014) Mesoscale eddies transport deep-sea sediments. Sci Rep. https://doi.org/10.1038/srep05937

Zhang S, Wang X, Wang H, Bjøerrum CJ, Hammerlund EU et al (2016) Sufficient oxygen for animal respiration 1,400 million year ago. PNAS 113(7):1731–1736

Zhisheng A, Kutzbach JE, Prell WL, Porter SC (2001) Evolution of Asian monsoons and phased uplift of the Himalaya-Tibetan plateau since late Miocene times. Nature 411:62–66

Zhu S, Zhu M, Knoll AH, Yin Z, Zhao F et al (2016) Decimetre-scale multicellular eukaryotes from the 1.56-billion-year-old Gaoyuzhuang formation in North China. Nat Commun. https://doi.org/10.1038/ncomms11500

Ziegler B (1992) Allgemeine Paläontologie. E. Schweizerbart'sche Verlagsbuchhandlung, Stuttgart

Zielhofer C, von Suchodoletz H, Fletcher WJ, Schneider B, Dietze E et al (2017) Millenial-scale fluctuations in Saharan dust supply across the decline of the African humid period. Quaternary Sci Rev 171:119–135

Zonneveld KAF, Versteegh GJM, Eglinton TI, Emeis K-C, Huguet C et al (2010) Selective preservation of organic matter in marine environments; processes and impact on the sedimentary record. Biogeosciences 7:483–511

Zuckerkandl E, Pauling L (1965) Molecules as documents of evolutionary history. J Theoret Biol 8:357–366

Zürcher E (2001) Lunar rhythms in forestry traditions—lunar-correlated phenomena in tree biology and wood protection. Earth Moon Planet 85–86:463–478

Zürcher E, Schlaepfer R, Conedera M, Giudici F (2010) Looking for differences in wood properties as a function of the felling date: lunar phase-corrected variations in the drying behavior of Norway Spruce (Picea abies Karst.) and Sweet Chestnut (Castanea sativa Mill.). Trees 24:31–41

Biological Evolutionary Lineages in Marine Habitats of the Phanerozoic

9

9.1 Fortuity and Self-Similarity in Near-Surface Zones

9.1.1 Abiotic Factors

As explained in the previous chapter, abiotic processes in aquatic systems predominantly exhibit stochastic, self-similar characteristics. Challenges arising from this can be seen most clearly in biotic communities in epipelagic zones—reached by sunlight—of water bodies that are sufficiently deep and extensive. As on land, all new biomass is produced by photosynthetic organisms. Among the abiotic requirements needed for this (Fig. 6.16), water is the only resource that is always available in unlimited quantities. However, various spatial and temporal distributions of saline conditions—for example in intertidal zones—can present major challenges for the osmotic regulation capacity of marine organisms. Even over the long term, average marine salinity in the Phanerozoic varied between around 4.4 per cent in the Carboniferous and about 3.3 per cent at the end of the Cretaceous (Warren 2010). However, spatial combinations of solar radiation and mineral nutrients are primarily determined by dynamics of flow processes (Dickes 1991; Horne and Platt 1984; Peters and Marrasé 2000; Friedrichs and Hofmann 2001; Cheriton et al. 2009; Hoecker-Martínez and Smyth 2012). On a regional basis, favourable combinations of adequate availability of mineral nutrients and solar radiation usually last only a few days or weeks. There are many reasons for this. For example, currents can carry nutrients or organisms out of the epipelagic zone. Moreover, organisms with fast growth rates can absorb most available nutrients, thus inhibiting further growth. Last but not least, because of suspended matter, the intensity of solar radiation can vary continuously (MacIntyre et al. 2000; Verheye 2000; Lovejoy et al. 2001; Seuront et al. 2001; Arrigo and van Dijken 2003;

© The Author(s), under exclusive license to Springer-Verlag GmbH, DE, part of
Springer Nature 2024
M. Knoflacher, *Relativity of Evolution*,
https://doi.org/10.1007/978-3-662-69423-7_9

Kestener and Arneodo 2004; Huisman et al. 2006; Mann and Lazier 2006; Evans et al. 2008; Seuront 2008).

Organisms in epipelagic habitats are exposed to the greatest variety of random environmental factors with respect to convertible energy potentials and abiotic framework conditions. For photosynthetic organisms, locations, times and intensities of favourable combinations of factors are all unpredictable. By comparison, in coral reefs and shallow water zones photosynthetic organisms are firmly established, but their development is strongly influenced by dynamics of mineral nutrient inputs. Over the long term, the existence of their habitats depends on physical processes in the hydrosphere and geosphere. For chemoautotrophic organisms, quite regular nutrient flows remain available over the geological duration of hydrothermal vents (Fig. 9.1).

Rates of change of individual factors in epipelagic zones significantly exceed the average lifespan of many aquatic organisms. Variability in chemoautotrophic habitats near hydrothermal vents—with relatively long-term fluxes of convertible energy potentials that are primarily determined by dynamics of hydrogeological processes—is far more limited. Comparable conditions exist in benthic habitats, which are primarily exposed to randomness in the occurrence of mechanical forces—such as water currents and sediment

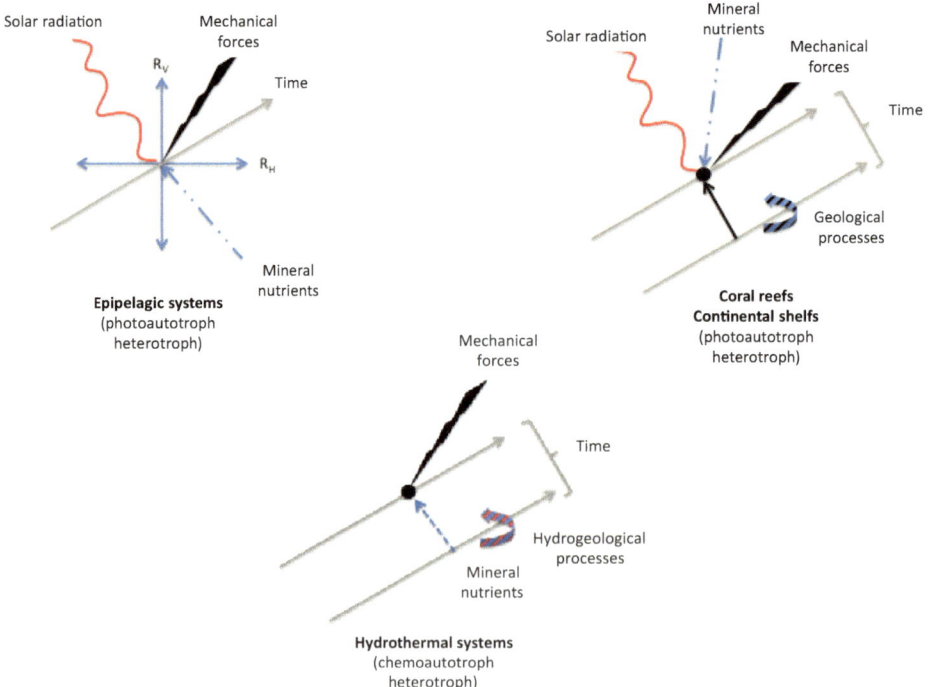

Fig. 9.1 Comparison between different combinations of energetic and abiotic conditions in autotrophic zones of aquatic habitats. Explanations in the text

displacement—as well as by the supply of organic material (Cathles et al. 1997; Danovaro et al. 2014, 2017).

In evolutionary terms, only photosynthetic organisms that can both cope with ongoing, unpredictable changes and attain reproductive capacity in a short period of time can evolve under the conditions in surface zones of oceans (Fig. 9.2). As an initial stage, this suggests clear competitive advantages for small organisms with high rates of reproduction. This alone does not resolve the "paradox of the plankton"(Bracco et al. 2000). According to this concept, a single species would be expected to prevail in the—supposedly—homogeneous habitat of near-surface water zones. If this concept is complemented with the observed variability and short-term nature of availability of mineral nutrients, that species would have to be small and fast-growing. The consequences would be mass proliferation of this species over the short term leading to exhaustion of available nutrient supplies and subsequent complete population collapse. Such an outcome is in fact observable in water bodies with little flow and—with increasing frequency—as a result of human influences (Abelmann et al. 2006; Paerl et al. 2011; Purcell 2012; Behrenfeld and Boss 2014; Daniels et al. 2015; Takahashi et al. 2015).

The size spectra of phototrophic plankton groups can be described most simply according to combinations of the factors nutrient availability and solar radiation, along with the dynamics of occurrence of those factor combinations. In simplified terms, very small photosynthetically active organisms have competitive advantages in habitats—constantly changing at random—with a low level of solar radiation and unpredictable changes in nutrient concentrations. Under contrasting conditions, larger-bodied photosynthetic

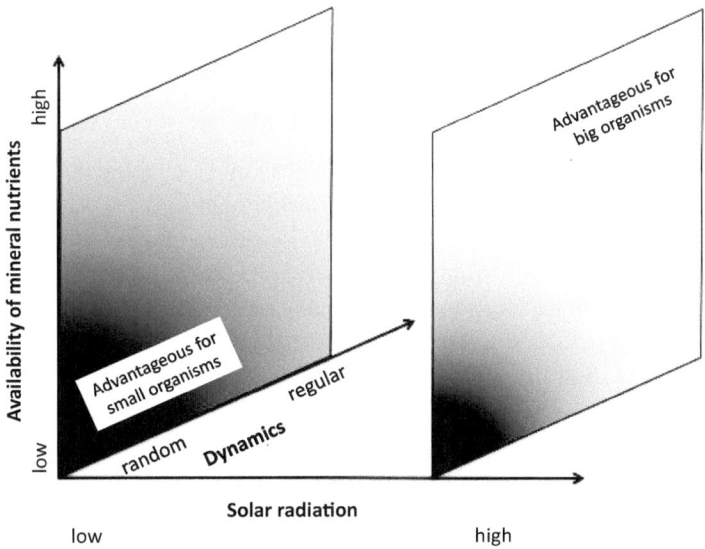

Fig. 9.2 Highly simplified illustration of the influences of abiotic conditions on evolutionary possibilities for photosynthetically active organisms in aquatic habitats. Explanations in the text

organisms have a competitive advantage (Fig. 9.2). Any number of variations in combinations of factors between the two conditions is possible. Influences exerted by dynamic characteristics are of central importance, not only for an understanding of the overall trajectory of biological evolution, but also for practical comprehension of processes within ecosystems. However, in aquatic habitats observations of photosynthetic processes are constrained because on-site measurements are possible only sporadically and global measurements using satellites can capture only near-surface photosynthetic processes (Garrison and Ellis 2016). As a result, uncertainties still exist in estimations of biological primary production in water at great depths and hence of aquatic carbon balances (Grob et al. 2007; Worden et al. 2015).

In reality, photosynthetic organisms in plankton exhibit variety over linear dimensions between around 0.5 micrometres and a few centimetres (Fenchel and Finlay 2006; Biller et al. 2014). Reflecting mesh sizes of fishing nets, the classic classification of plankton is in fact based on linear dimensions in successive steps of one power of ten, ranging from 0.2–2 micrometres (picophytoplankton) up to the category including anything over 2 millimetres (macrophytoplankton) (Chisholm 1992; Reynolds 2006; Finkel et al. 2010; Edwards et al. 2012). In general, in nutrient-poor zones of oceans and inland waters picophytoplankton dominates. Where a greater supply of nutrients is regionally available, the proportion of larger-bodied life forms in phytoplankton can increase (Marañon et al. 2001; Callieri and Stockner 2002; Callieri 2008; Bec et al. 2011; Acevedo-Trejos et al. 2013). Sample analyses based on this classification alone provide little information about biological processes, as identical species may also occur in very different size categories. The widespread marine foam alga (*Phaeocystis* sp.) can occur both as individual cells with a size of around 2 to 8 micrometres and as closed colonies with sizes between 25 micrometres and 3 centimetres (Schoemann et al. 2005). Nevertheless, the question arises as to why phototrophic communities have been able to exist for billions of years in surface zones of oceans.

An initial answer to this question is provided by *hydrodynamic differentiation* in aquatic habitats. Small organisms—including bacteria—can move over the microscale range (see Sect. 5.1.5) (Jiménez 1997; Svendsen 1997; Blackburn et al. 1998; Barbara and Mitchell 2003; Currie and Roff 2006; Ginger et al. 2008; Locsei and Pedley 2008). The usual concept of plankton as passively drifting organisms applies only to a limited extent to currents with large spatial dimensions and high flow velocities (Carrick et al. 1993; Seuront et al. 1996; Lovejoy et al. 2001; Rodriguez et al. 2001; Malits et al. 2004; Iversen et al. 2010; Doubell et al. 2014). For this reason, short-term changes in currents can lead to the demise of entire populations, even under the relatively favourable conditions of equatorial upwelling zones. On the other hand, the diversity of different current patterns contributes to functional and physiological differentiation of phytoplankton (Fig. 8.20) (Bracco et al. 2000; Alvain et al. 2008; d'Ovidio et al. 2010; Rusconi and Stocker 2016; Basterretxea et al. 2020). Observations on phytoplankton show that risks can be reduced by diversifying response patterns. To this end, subgroups of the original population attempt to float in different directions out of the crucial current (Sengupta et al. 2017). Hydrodynamic properties—such as individual buoyancy—can be modified either by formation of cell clusters or physiologically (Beardall et al. 2009). While cluster formation is relatively easy to observe, physiological processes can usually be studied only in living organisms. Because just a few species can be kept alive under laboratory conditions, many kinds of adaptation, collectively described by the label "vital factor", remain unknown (Moore and Villareal 1996; Naselli-Flores and Barone 2011). As a rule, such adaptations have a multi-factorial relationship with physiological and biotic factors—for example, increased losses to herbivorous organisms (Schoemann et al. 2005; Sunda and Hardison 2010; Peperzak and Gäbler Schwarz 2012; Herrero et al. 2016; Kapsetaki et al. 2016; Lovecchio et al. 2019). For photosynthesis, phytoplankton in deeper, nutrient-rich water bodies utilize internal waves for vertical uplift into the photic zone (Fahnenstiel et al. 1988; Huisman et al. 2006; Evans et al. 2008).

Diversity and variability of life cycles of different plankton organisms are partly related to this. Many organisms can reproduce both sexually and asexually (Nehring 1993; Figueroa and Rengefors 2006; Meier et al. 2007; Kremp 2013; Jue et al. 2016). To bridge unfavourable living conditions, many species form abundant permanent stages (cysts) with minimal metabolic turnover (Nehring 1995; Belmonte et al. 1997; Marcus and Boero 1998; Jarnagin et al. 2004; McQuoid and Godhe 2004; Rengefors et al. 2004; Figueroa et al. 2006, 2008; Kremp and Parrow 2006). Sinking to sediments on the sea floor facilitates coordination of their reactivation with the onset of favourable conditions, because they are flushed into surface zones by rising water currents at the same time as nutrients. In analogy to sediment transportation in flood waves (Knoflacher et al. 2003), it can be assumed that the highest nutrient concentrations occur at the leading edge of rising water masses and thus provide optimal opportunities for phytoplankton to develop. Cysts can remain viable in sediments for decades, up to about one hundred years (Ribeiro et al. 2011; Ellegaard et al. 2016). This provides an indication of the development of substantial capacities for temporal resilience over the course of evolution.

This also indicates the magnitude of the potential influence of currents on population dynamics of phytoplankton. Low-intensity upwelling currents loosen only the uppermost sediment layers and thus activate cysts only from the most recent generation. More strongly acting—usually rarer—currents can also exert an impact on deeper sediment layers and thus activate cysts belonging to different generations. Such occurrences permit gene exchange between different generations during reproduction. As a result, with regularly recurring patterns in currents—for example in fjords—genetically stable regional plankton populations can arise (Härnström et al. 2011). For example, cyanobacteria could have evolved from originally benthic life forms into photoautotrophic organisms through transportation by currents into the photic zones of oceans (Sánchez-Baracaldo 2015).

Similar patterns of life cycles alternating between long-term benthic phases (polyps) and temporary pelagic phases (medusae—better known as jellyfish) have also developed in many cnidarians (Phylum Cnidaria) and comb jellies (Phylum Ctenophora) (Boero et al. 2008).

A second answer can be found in the *physiological differentiation* of phototrophic planktonic organisms. Phototrophic organisms living in the hydrosphere are confronted with randomly varying factors in solar radiation. Photosynthetically utilizable habitats depend on suspended matter content and can range in depth between a few centimetres and about 200 metres. Just below the water surface, the intensity of solar radiation changes rapidly over large bandwidths because of wave movements and suspended matter (MacIntyre et al. 2000; Stomp et al. 2007). Among phototrophic planktonic organisms— especially prokaryotes—almost all physiological variants of energetic conversion of different spectral ranges of sunlight can be found (see Sects. 4.1 and 7.4) (Béjà et al. 2000; Ting et al. 2002; Schwalbach and Fuhrman 2005; Gómez-Consarnau et al. 2010; López-Sandoval et al. 2014). Symbioses—facultative or permanent—broaden both the physiological range of phototrophic organisms and direct utilization of phototrophy for many heterotrophic plankton organisms (Caron et al. 1995; Norris 1996; Shaked and de Vargas 2006; Malfatti and Azam 2009; Stoecker et al. 2009; Morris et al. 2011; Walker et al. 2011; Pittera et al. 2014; Takagi et al. 2019).

In oceans, large-scale and long-term distributions of the concentrations of mineral nutrients, their relative ratios (stoichiometry) and concentrations are determined by geological conditions and the random nature of water currents (Sarmiento et al. 2003; Holzer and Primeau 2013). Distributions of nitrogen, which can be biologically fixed by various cyanobacteria, deviate from the general pattern (Zubkov et al. 2003; Latysheva et al. 2012; Ascani et al. 2013). Under the influence of different combinations of factors, an almost unbridled variety of photosynthetically active life forms has developed (Margalef 1978; Estrada and Berdalet 1997; Peters and Marrasé 2000; Alcaraz et al. 2002; Ciotti et al. 2002; Irwin et al. 2006; Cermeño and Falkowski 2009; Key et al. 2010; Ghiglione et al. 2012; Lima-Mendez et al. 2015; Walsh et al. 2015). What they have in common is adaptation of their physiological cycles to the temporal course of solar radiation. This is associated with general synchronization of many processes in aquatic ecosystems (Beaufort et al. 1997; Hewson et al. 2006; Zinser et al. 2009; Waldbauer et al. 2012; Ottesen et al. 2014; Fuhrman et al. 2015; Ribalet et al. 2015; Bunse and Pinhassi 2017; Raina et al. 2022).

With further developments in research methods, it is becoming increasingly clear that prokaryotes—archaeans and bacteria—themselves modify these conditions, at least in oceanic water bodies. In many cases, auxotrophic co-operation between different groups of organisms exists—for example in the supply of vitamin B12 to eukaryotic microalgae by cyanobacteria (Martens et al. 2002; Bonnet et al. 2010; Bertrand and Allen 2012; Grossman 2016; Helliwell et al. 2016). The wide range of physiological adaptation to available chemical elements—for example metal ions—and the physiological capacities of microorganisms present are as yet incompletely known (Keshtacher-Liebson et al. 1995; Butler 1998). The diversity of functional co-operation between microorganisms tends to decrease with increasing concentrations of carbon, nitrogen and phosphorus compounds (Biddanda et al. 2001; Rier and Stevenson 2002; Medina-Sánchez et al. 2004; Forsström et al. 2013). Traditionally developed notions of clearly definable functional characteristics—such as with phyto- and zooplankton—as well as stable ratios of nutrient requirements—for example the well-known Redfield ratio—can be confirmed only for a few realms of life. Natural water bodies are heterogenous dynamic systems and not homogeneous solutions. For example, the compositions of prokaryote communities—and consequently the bio-chemical processes—in marine water bodies are highly variable (Mestre et al. 2017; Xue et al. 2020).

With more recent findings on diversity in the metabolic performance of prokaryotes, it is becoming increasingly clear that they regulate dynamics of nitrogen and phosphorus compounds in oceans to a far greater extent than previously assumed (Fig. 9.3) (Lv et al. 2008; Martiny et al. 2009; Van Mooy et al. 2009, 2015; Villareal-Chiu et al. 2012; Yu et al. 2013; Zehr and Capone 2020). Prokaryotes form the evolutionary attractor of aquatic primary production, from which—in conjunction with the associated supply of mineral nutrients—diverse phototrophic life forms evolve among eukaryotes (Finkel et al. 2010; Quigg et al. 2011; Martin and Quigg 2012). The influence exerted by parasitic aquatic fungi (chytrid fungi) on the dynamics of nutrient supply (Fig. 9.3) is as yet incompletely understood (Frenken et al. 2020; Klawonn et al. 2021a, b).

Investigations conducted in connection with the lava eruption of the Hawaiian volcano Kilauea in August 2018 have revealed how quickly the composition of the phototrophic

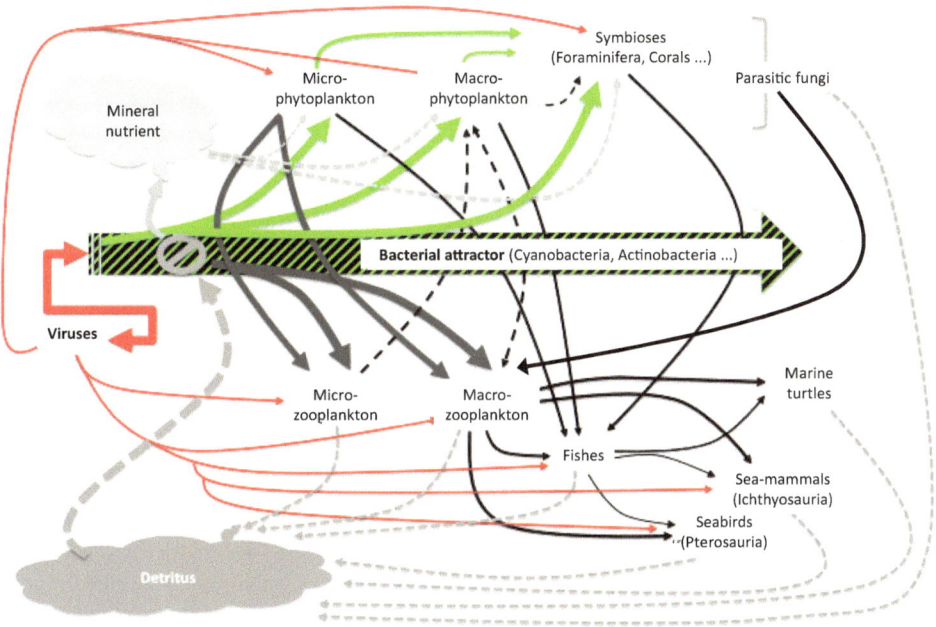

Fig. 9.3 Simplified overview illustrating the central importance of bacteria in epipelagic ecosystems. Explanation of symbols: light grey solid lines—temporary or permanent symbioses of cyanobacteria with other organisms; black dashed lines—processes influenced by viruses; black lines—heterotrophic and mixotrophic food fluxes; light grey dashed lines—fluxes of dead organic material

plankton can change, even over the short term. Within a few days, nutrient input from the lava shifted dominance in the species composition towards eukaryotic algae. After the unusual nutrient input ceased, dominance by cyanobacteria returned within a similarly short timespan (Wilson et al. 2019).

9.1.2   Interactions with Viruses

Interactions with **viruses** are of central importance for continuous renewal of physiological flexibility in marine prokaryotes (Gobler et al. 2004; Suttle 2005, 2007; Rodriguez-Valera et al. 2009; Breitbart 2012; Roux et al. 2016). Through their influence on horizontal gene transfer, they enable prokaryotes to acquire new physiological capacities in the context of changing environmental conditions. In estuaries, for instance, effects of viruses are significantly weakened at low tide, but intensified at high tide (Zhaxybayeva et al. 2009; Avrani et al. 2011; Moniruzzaman et al. 2017; Chen et al. 2019). The—better known—destructive effect of viruses has consequences not only for ongoing evolutionary differentiation of prokaryotes, but also for the ecological dynamics of plankton systems as a whole. In general, viral infection leads to cell death only in part of the overall population. Large sectors of the population develop various kinds of resistance, in tandem with development of new physiological features.

Under the influence of viruses, functionally and geographically differentiated subgroups, often with complementary properties, have developed in cyanobacteria belonging to the groups of *Synechococcus* and *Prochlorococcus*. Simplified ecophysiological classifications of *Synechococcus* as a near-surface group and *Prochlorococcus* as a group inhabiting deeper photic zones reflect their functional diversity only partially. Among *Prochlorococcus*, for example, two large subgroups can be distinguished with adaptations to high (HL) and low (NL) irradiation conditions and correspondingly preferred depth distributions. Genetically, however, the NL group of *Prochlorococcus* has greater similarities with *Synechococcus* (Johnson et al. 2006; Zwirglmaier et al. 2008; Zhaxybayeva et al. 2009; Flombaum et al. 2013; Biller et al. 2014).

Accordingly, it is not surprising that a far greater number of different life forms have now been identified in this "forgotten unicellular world of evolution" (Fig. 5.6) than in the "multicellular world of evolutionary stars" (Adl et al. 2012, 2019). Large-bodied organisms—for example marine mammals or seabirds—can feed only on plankton concentrations that are sufficiently large (Fig. 8.20: large plankton cloud) or on communities that depend upon them. Prerequisites for this are a high degree of intrinsic mobility and the capacity for spatial localization of large plankton concentrations (see Fig. 6.7). The rare—from a microscopic perspective—and widely distributed biotic communities can generally be consumed in their entirety without detrimental effects for microorganisms. However, excessive losses of mesoscale organisms in food chains, such as sardines or krill, are decisive Cury et al. 2000; Amarasekare 2008).

Destructive effects of viruses lead not only to mass mortality among marine vertebrates—such as fish and seals—but also to ongoing devastation of cells of planktonic
organisms. Because of the high population density of viruses, which is about 15 times
greater than that of marine prokaryotes in photic zones of oceans, they are also able to
terminate mass reproduction of unicellular organisms prematurely, corresponding to the
principle of "killing the winner"(Suttle 2007; Rohwer and Thurber 2009; Brum et al.
2016). In deep-sea benthic ecosystems, around 80 per cent of prokaryotes are decomposed
by viruses (Danovaro et al. 2008). Exceptions are found in the vicinity of hydrothermal
vents, which have significantly smaller numbers of viruses compared to prokaryotes
(Yoshida-Takashima et al. 2011). Viruses—especially giant viruses that parasitize eukaryotes—can themselves also be destroyed by viruses (virophages) (Derelle et al. 2008; Yau
et al. 2011; Santini et al. 2013). In an evolutionary context, the negative effects of viruses—
from an anthropocentric viewpoint—take on a completely different meaning. Only premature destruction of cells—under the dynamic conditions of the phytoplankton world—can
preclude dominant development of the most successful species and rapid loss of all
absorbed nutrients in the photic zone (Fig. 9.4). Self-organized agglomeration of individual particles into floccules several micrometres across also renders organic materials
attractive as food for larger protists (Kerner et al. 2003). The interplay of viral lysis and
molecular self-organization not only fractures the bacterial recycling loop, but also turns
bacteria into food for protists (Sibbald and Albright 1988; Landry and Calbet 2004;

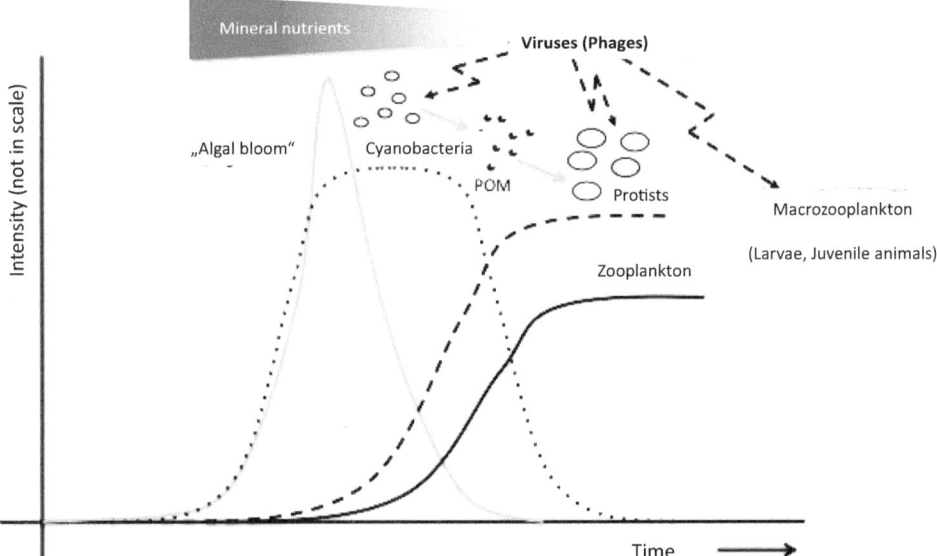

Fig. 9.4 Schematic outline of the importance of viruses for development of planktonic food chains
for multicellular organisms. Viral decomposition of bacteria permits more slowly reproducing, mixotrophic protists to take up mineral nutrients from organic particles (POM). Potential beneficiaries
are competing multicellular eukaryotes in the macrozooplankton. Further explanations in the text

Buchan et al. 2014). This ultimately creates conditions for transfer of organic material—as an energy source—towards the macroscopic world, along with the minerals it contains, for formation of cellular structures (Evans et al. 2009).

9.1.3   Wolves in Sheep's Clothing: Mixotrophy

Such preconditions would, however, be ineffective if plankton organisms were to behave strictly according to the frequently used categorization into phyto- and zooplankton based on their energy conversion. This schematic categorization makes it easy to forget the underlying biological diversity of organisms ranging from prokaryotes (cyanobacteria) to eukaryotes, with different kinds of symbioses (Beardall et al. 2009). Functionally, the term "phytoplankton" suggests a physiologically standardized form of energy conversion for all organisms grouped together under that name. In reality—due to the unpredictability of their living conditions—many organisms can convert energy both photoautotrophically and heterotrophically (Flynn et al. 2013). The spectrum of so-called mixotrophy ranges from photoautotrophic organisms—which obtain mineral nutrients by the heterotrophic route—to heterotrophic organisms, which also feed photoautotrophically when food is scarce. In the latter situation, organisms usually engage in symbioses of varying duration with photoautotrophic cyanobacteria or eukaryotes (Hammer and Pitchford 2005; Adolf et al. 2006; Zubkov and Tarran 2008; Stoecker et al. 2009, 2017; Feng et al. 2010; Jeong et al. 2010; Kang et al. 2011; Hartmann et al. 2013; Moore 2013; Hehenberger et al. 2014; Mitra et al. 2014, 2016; Moreira and López-Garcia 2014; Ward and Follows 2016; Selosse et al. 2017).

Mixotrophy is by no means a speciality of eukaryotes, but a widespread strategy—also found among archaeans and bacteria—for reducing the influence of random abiotic factors (Eiler 2006; Muehe et al. 2009; Czypionka et al. 2011). In addition to the effects in eukaryotes already mentioned, the diversity of variants in the conversion of solar radiation energy can lead to unexpected phenomena. In humus-rich inland waters, for example, bacterial secondary production can be higher than the primary production of phytoplankton (Jansson et al. 1996). A seeming contradiction with the second law of thermodynamics is explained by non-photosynthetic utilization of solar radiation by bacteria for conversion of washed-in (allochthonous) organic materials.

9.1.4   Oceans Are Neither Meadows Nor Farmlands

The wide distribution of mixotrophy under the extremely random and short-lived conditions of planktonic communities indicates that formation of biomass cannot be a primary goal of biological energy conversion. This argument is illustrated by estimates of the transfer of aquatic and terrestrial primary production into fossil energy carriers, calculated using classical approaches. In the first case—conversion of phytoplankton into crude oil—around 0.1 per cent is stored in fossil fuels; in the second—conversion of terrestrial plants into peat and coal—the estimate is around 13 per cent (Smil 2017).

Failure to consider such aspects is evident in many theoretical approaches, for example when "fertilization" of phytoplankton is proposed or when concepts from terrestrial systems are applied to aquatic habitats (Hoffmann et al. 2006; Holt 2009; Harfoot et al. 2014).

As current estimates show, ratios of biomass bear no relation to dimensions of photosynthetic energy conversion in the oceans compared to land. In the oceans, around 48,109 tonnes of carbon are converted annually, whereas only about 1109 tonnes are permanently bound in plant biomass—predominantly in picophytoplankton. On land, at around 56,109 tonnes, somewhat more is converted, but significantly more carbon is sequestered in plant biomass over the long term, amounting to 450,109 tonnes. On the other hand, the ratio of sequestered carbon for heterotrophic *versus* phototrophic organisms in the oceans is around 5, compared to about 0.04 on land (Garcia-Pichel et al. 2003; Geider et al. 2001; Buitenhuis et al. 2012; Bar-On et al. 2018). The seeming contradiction with thermodynamic principles begins to dissolve once the amount of carbon transferred between phototrophic and heterotrophic organisms is considered. It is approximately two thirds lower in oceans than on land.

I. A major reason for this resides in the irregular dynamics of pelagic primary production in oceans (Woods and Onken 1982; Fennel 2001; Cloern and Dufford 2005; Henson and Thomas 2007a, b; Widdicombe et al. 2019; Clayton et al. 2014). Instead of continuous processes, relatively short phases with spatially self-similar distribution patterns dominate marine primary production. Due to time delays in the subsequent population development of heterotrophic organisms, large parts of the plant biomass escape consumption by eukaryotes, especially at the beginning of production phases. Due to the time lag and higher rates of energetic conversion, bacteria can therefore form independent food webs, which are in fact detectable in investigations (Gilbert et al. 2012).

II. Another reason lies in the enormous differences in body size between phototrophic and heterotrophic organisms. For example, the body mass of a krill crustacean (*Euphausia* sp.) is around 7109 times larger than that of a cell in the preferred food—phytoplankton (Kils 1983). A hypothetical comparison with terrestrial ecosystems—based on the biomass of a blade of grass—would yield a weight of around 40 tonnes for the consumer, i.e. with roughly the dimensions of a *Brachiosaurus* (Sander et al. 2011). Whereas size ratios between marine and terrestrial food chains have been hitherto comparable, the next stage of the marine food chain exceeds all terrestrial dimensions. Based on the body mass (Woodward et al. 2006) of the largest krill consumer—the blue whale (*Balaenoptera musculus*)—by terrestrial standards this would result in a gigantic predator with a hypothetical body weight of 2 billion tonnes. In comparison with a body weight of approximately 8 tonnes for *Tyrannosaurus* (Ruxton and Houston 2003), such a monster would be virtually unimaginable, even for film producers. Baleen whales avoid problems by filtering their food out of seawater, but nevertheless live in confined energy niches with a high risk of extinction if environmental conditions change. To meet their energy requirements, food organisms must be present in sufficient abundance and high concentration (Platt and Methven 1992). Such conditions are much rarer (see Fig. 8.20) than smaller aggregations of food organisms. If baleen whales feed on larger organisms—such as fish—they increase the length of their food chains. This means that instead of around 5 per cent—as in the case of feeding on krill—only around 1 per cent of the photosynthetically bound biomass is available to them (Fig. 9.5 (a) and (b)) (Kaiser et al. 2011). Carnivorous fish—such as tuna (*Thunnus*)—are also inevitably subject to the constraining conditions of decreasing potential for convertible energy within food chains.

III. Another reason is simply widespread mixotrophy in pelagic ecosystems, which does not allow a sufficiently precise differentiation between autotrophic and heterotrophic energy transformations with currently available methods.

Heterotrophic organisms of all sizes meet challenges I and II in plankton with various different strategies (de Vargas et al. 2015). With relatively small size differences between heterotrophic organisms and their food, direct capture and a wide variety of forms of

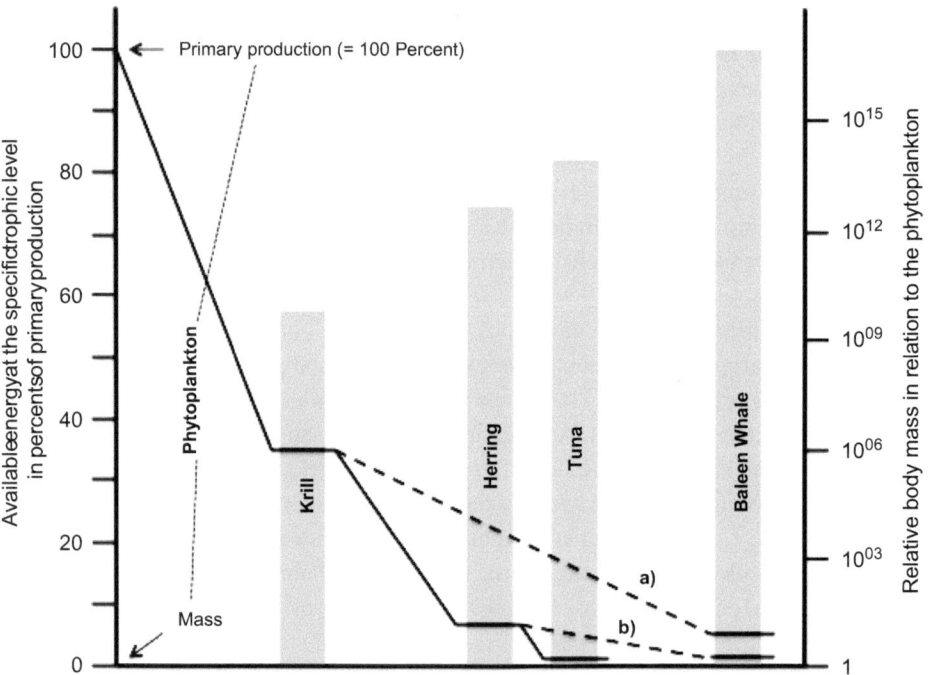

Fig. 9.5 Simplified overview of mass ratios (vertical bars) and proportions of convertible energy in marine pelagic food chains (lines). For baleen whales: (**a**) direct diet from krill, (**b**) indirect diet from fish. Right scale logarithmic. Further explanations in the text

ingestion—for example uncrushed (with phagotrophy) or crushed—dominate (Boenigk and Arndt 2000; Sherr and Sherr 2002; Kiørboe 2011; Boyce et al. 2015. A particularly wide spectrum of capture and ingestion methods can be found among dinoflagellates. In analogy with the stinging capsules of cnidarians, these unicellular organisms use "harpoons" to capture food and protruding "snouts" for extracellular digestion. Many species can open their cell walls at various sites for phagotrophic ingestion of prey (Jacobson and Anderson 1996; Jacobson 1999; Gavelis et al. 2017).

9.1.5   Physiological and Morphological Adaptations

Sensory-physiological prerequisites for detection of food organisms have so far been researched only incompletely, but have been demonstrated in isolated cases ranging from bacteria to eukaryotic protists (Stocker and Seymour 2012; Taylor and Stocker 2012). The photosynthetic "plant-animal" *Euglena* was long regarded as a curiosity because of its "eye spot". With increasing knowledge about mixotrophic life forms, it is becoming increasingly clear just how widely organs for processing optical signals—extending from pigment spots to adaptive lens eyes—are distributed in supposed phytoplankton (Jékely 2009; Cusick and Widder 2014; Colley and Nilsson 2016; Gavelis et al. 2017). The diversity of sensory-physiological functions clearly reveals dynamic influences of interactions between microscopic organisms. That diversity is accompanied by a rich variety of morphological structures and material compounds with differing toxic effects. Thanks to further advances in the development of methods of electron microscopy, the relationships concerned are detectable down to organisms of the smallest size (Leander and Keeling 2003; Leander et al. 2007; Lukeš et al. 2009).

Morphological diversity among protists is due not only to interactions between eutrophic organisms, but also to the influence of viruses. For instance, formation of the heavy, silica-based armour of diatoms is attributed not only to protection against predators, but also to protection against viruses. This interpretation is founded on rapid sinking of damaged cells from the photic zone and associated elimination of infected cells from healthy cell populations (Raven and Waite 2004). Clear connections between the influence of viruses and particular life forms can be found in calcareous flagellates (Coccolithophorida). Fossilized deposits of these organisms—forming chalk cliffs—are well known. In nutrient-poor waters with low virus densities, the diploid life form, densely armoured with calcareous platelets, dominates. In nutrient-rich waters with high virus densities, on the other hand, the barely armoured—but virus-resistant—haploid life form dominates (Iglesias-Rodríguez et al. 2002; Read et al. 2013; Taylor et al. 2017).

As already mentioned in connection with the example of baleen whales, in pelagic habitats multicellular organisms are confronted with the problem of length of food chains and associated energy losses. Free-floating eggs and larvae permit many invertebrates and fish to feed on short—and hence energetically favourable—food chains, at least during their first life stage (Markle and Frost 1985; Lazzaro 1987; Johnson and Shanks 2003;

Villanueva and Norman 2008; Shanks 2009). However, due to the dynamics of marine planktonic ecosystems described above, they are confronted with a high risk of loss. According to one estimate, on average only 0.1 per cent of fish larvae survive in marine plankton, and around 5.3 per cent in the plankton of inland waters (Houde 1994). In part, high reproductive losses of marine fish are offset by large egg numbers and small egg volumes. In some cases, the larvae of different fish reduce their losses through morphological similarities (mimicry) to toxic or hard-to-digest planktonic organisms (Greer et al. 2016).

Depending on development of the gill rakers—filter structures in the gullet region of the gills—the diet of bony fish can range between purely planktivorous and fully carnivorous. The morphological structures concerned are not evolutionarily invariable, but adapt relatively rapidly to the available food spectrum (Lazzaro 1987; Palkovacs et al. 2008; Friedman et al. 2010; Friedman 2012; Liston 2013; Collard et al. 2017). Around eight million years ago, for example, several planktivorous species of Pacific salmon (*Oncorhynchus*) appeared (Eiting and Smith 2007). Among present-day cartilaginous fish, several shark and ray species—including the largest extant representatives, the whale shark (*Rhincodon*), basking shark (*Cetorhinus*) and giant manta ray (*Mobula*)—feed on planktonic organisms (Helfman et al. 2009).

Because of heavy losses and competitive pressure in the planktonic phase, overexploited fish stocks—such as cod—recover far more slowly than theoretically expected. This is mainly due to changes in planktonic food systems caused by rapid increases in populations of other fish species, such as herring (Hutchings 2000; Finn et al. 2002; Frank et al. 2005).

The main causes of such developments are technological fishing methods, linear principles of social regulation and incomplete modelling assumptions. Huge quantities of fish are caught with nets whose mesh sizes determine minimum sizes of fish captured. Economic and population-dynamical effects are taken into account when determining permissible mesh sizes for fishing nets. In economic terms, the aim is to maximize profits. As regards population dynamics, the aim is to safeguard stocks by maintaining age categories that are capable of reproduction. Optimization calculations define the size of the youngest age category still capable of reproducing and thus the basis for minimum mesh dimensions. However, this drastically reduces the overall reproduction rates of the fish stock because younger individuals produce fewer and smaller eggs than larger and older fish.

In analogy with the diet of baleen whales, planktivorous species up to two metres in length evolved among radiodonts as early as the Cambrian (Lerosey-Aubril and Pates 2018). Among extant animals, chordates (Tunicata) with body sizes of a few centimetres feed on planktonic organisms (Fuchs and Franks 2010). Dimensions and composition of the food spectrum are determined both by physical characteristics of filtering organs and by physiological processes of food intake (Bone et al. 2003; Katechakis et al. (2004); Henschke et al. 2016; Conley et al. 2017; Conley et al. 2018). With their temporary filter structures, larvaceans (Appendicularia) can capture extremely small particles—the size of viruses—and use them as food (Nakamura et al. 1997; Gorsky et al. 1999; Sato et al. 2001;

Sutherland et al. 2010; Lawrence et al. 2018). If necessary—for example if filters become blocked—the structures can be shed and rebuilt. Transparent shells—consisting of gelatinous material—of some deep-sea species can reach dimensions of up to 1 metre in diameter (Katija et al. 2020). Free-floating sea pickles (*Pyrosoma*, Tunicata) presumably attract potential prey with bioluminescence, as significantly more food is to be found in them at night than during daytime (Perissinotto et al. 2007).

Taking into account the explanations given above, it also becomes clearer why planktonic systems fail to achieve stability even under experimental conditions, showing clearly chaotic reaction patterns instead (Scheffer et al. 2003; Benincà et al. 2008). In terms of population dynamics, viruses and bacteria can respond very rapidly to changes in environmental conditions by multiplying *en masse*. Response and development speeds of inhibiting factors, such as viruses, are decisive for the actual outcome—in relation to the dynamics of the bacterial population concerned (see Fig. 7.14). Accordingly, different combinations of promoting and inhibiting factors can lead repeatedly to excessive population increases—so-called algal blooms (Fig. 9.4). In principle, strong impulses at the starting-points of planktonic food chains are propagated through all subsequent transformation stages. Under such chaotic abiotic framework conditions, only biotic systems consisting of organisms with pronounced functional adaptability to variable interactions with other organisms can survive over the long term (Huisman and Weissing 2001; Brown et al. 2002; Jessup et al. 2004; Sommer and Sommer 2006; Armbrust 2009; Flores and Holmgren 2021; Amin et al. 2012; Våge and Thingstad 2015). As with many watercourses, only biotic communities with a high degree of flexibility in their overall functions and the lowest possible degree of specialization in interspecific interactions can develop (Hasting 2004; Doebeli and Ispolatov 2014). In contrast to theoretical models, convertible energy can reach larger organisms through different food chains—for example via cnidarians (Brodeur et al. 2011; Hays et al. 2018). Animal species that specialize on this food source can attain remarkable sizes, such as moonfish (*Mola*), which can reach up to four metres in height and weigh over two tonnes, or leatherback turtles (*Dermochelys*), with weights of about 500 kilograms (Helfman et al. 2009; Bailey et al. 2012; Heaslip et al. 2012; Nakamura et al. 2015).

Because of differing anthropogenic evaluations of the phenomena concerned, the general pattern is easily overlooked. Mass proliferation of protists—for example toxic dinoflagellates, tunicates or jellyfish—is rated negatively, while planktonic crustaceans—for example krill—or fish—are ranked positively. Which phenomena in the "relay race" of pelagic food chains ultimately occur is not exclusively dependent on the variety of promoting and inhibiting factors in individual stages. The number and readiness of the "relay runners" in individual stages also determine whether individual organism groups can participate or not. The example of cod stocks already mentioned should be borne in mind here.

Nevertheless, the validity of this statement must always be viewed in the context of characteristics and dimensions of the systems considered. Relatively regular seasonal patterns of mass reproduction—for example in polar waters—have enabled mobile heterotrophic organisms to develop extremely large-bodied, long-lived life forms, such as baleen whales (Sims 1999; Jackson et al. 2009; Grady et al. 2019). Bear in mind, however, that the estimated lifespan of around 200 years for bowhead whales (*Balaena mysticus*) years is almost doubled in Greenland sharks (*Somniosus microcephalus*) (Keane et al. 2015; Nielsen et al. 2016).

Morphology and physiology of marine vertebrates are influenced to a great extent by physical properties of water and properties of their food sources (Kelley and Motani 2015; Kelley and Pyenson 2015). Feeding on sessile or drifting organisms has allowed vertebrates greater morphological independence from hydrodynamic conditions—for example in sea turtles, manatees or moonfish (Velez-Juarbe et al. 2012; Scott et al. 2014; Nakamura et al. 2015). Mobile or spatially widely distributed food leads to adaptations of body shapes and physiology to hydrodynamic conditions—clearly recognizable in the convergent evolution of tuna, deep-sea sharks, dolphins and penguins (Fish and Hui 1991; Bernal et al. 2001; Kelley and Motani 2015). Baleen whales share the feature of high energy expenditure—due to flow resistance with open jaws—when feeding (Ridgway 1972; Goldbogen et al. 2007, 2008, 2011).

Contrasting with vertebrates, cnidarians (phylum Cnidaria) and comb jellies (phylum Ctenophora) have been employing completely different strategies to feed on planktonic organisms for around 800 million years (Cartwright et al. 2007; Haddock 2007; Park et al. 2012). Rather than incurring a high energy input for seeking and capturing food, energy expenditure is minimized mainly by the size of the capture organs, which are adapted to potential prey (Bernal et al. 2001; Colin et al. 2003; Ball et al. 2004; Robison 2004; Dunn 2005; Dunn et al. 2005; Dunn and Wagner 2006; Sims et al. 2006; Fautin 2009; Gemmell et al. 2013; Mapstone 2014). However, linear sizes of the organisms concerned vary widely, from a few centimetres to around 40 metres (Van Iten et al. 2014; Kim et al. 2019). Active hunters with differentiated optical sensory perception are found mainly among box jellyfish (Cubozoa) living close to the coast (Liegertová et al. 2015). Owing to the similarly minimized effort invested to form the—usually transparent—biomass, their biology can be investigated only by employing underwater robots, because fishing nets destroy their bodies (Robison 2004).

The importance of the physiological make-up of organisms for stability in the face of disturbances can be recognized in the context of the dynamic phenomena of mass reproduction and mass mortality. Groups of organisms at the ends of food chains with high energy conversion rates (Fig. 9.3) are far more sensitive to disturbances than groups with low, flexible energy conversion rates at early stages of food chains.

Perhaps this explanation makes it easier to understand why causal interpretations of the extremely long evolutionary history of planktonic systems are impossible. However, the multiplicity of different influencing factors and their interactions opens up considerable scope for hypotheses and speculative interpretations (Kidder and Erwin 2001; Pawlowsky et al. 2003; Kaminski et al. 2010; Simon et al. 2009; Falkowski et al. 2004; Knoll and Follows 2016; Schmidt et al. 2006; Servais et al. 2016).

9.1.6 Sedentary Existence and Structures in Coral Reefs and Shallow Water Zones

Because of the lack of spatial variability, the number of variable influencing factors in relation to the lifespans of organisms is lower on coral reefs and in shallow water zones (Fig. 9.1). Over the long term, however, framework conditions for these habitats will in fact change due to geological and climatic processes (Leprieur et al. 2016). Effects of regular tidal dynamics on water levels and currents are more pronounced (Hatcher 1990; Green et al. 2018; Juva et al. 2020). Accordingly, sessile organisms can differentiate themselves morphologically to a greater extent than planktonic organisms—whether it be to utilize water currents to obtain food or to exchange metabolic products. For many animals, structural differentiation of habitats by primary producers—such as seaweed—reduces the risk of discovery and predation by carnivores (Vanderklift et al. 2009). At the same time, for sessile primary producers, the risk of destruction by herbivorous organisms increases. Strong currents and carnivorous organisms offset these effects. For example, expansion of kelp forests on the North American Pacific coast is closely linked to populations of sea otters (*Enhydra lutris*). Sea otters feed mainly on sea urchins—the dominant herbivores of kelp populations. For this reason, reductions in sea otter populations due to human hunting or killer whales (*Orcinus orca*) lead to population decline in kelp forests. The effect is amplified in connection with losses caused by killer whales, because structural protection for sea otters is also lost when kelp forests decline (Estes et al. 1998, 2004; Vanderklift et al. 2009; Poore et al. 2012).

In tropical and subtropical regions, intensity and dynamics of nutrient fluxes particularly influence the composition and functional interactions of communities. Above all, nutrient deficiency promotes photosynthesis-based, diverse communities of organisms with close functional interactions—for example coral reefs. Such communities, for instance the colourful diversity of present-day corals and coral reef fish, can develop during geologically relatively short periods of a few million years (Wood 1998; Riegl et al. 2009; Goldberg 2013; Bellwood et al. 2017). Excess nutrients can reduce diversity to a small number of heterotrophic organisms—for example, with inflows of untreated wastewater from residential areas or intensively farmed areas (Romanuk et al. 2006; Worm and Lotze 2006; Keatley et al. 2011).

As in plankton communities, functional relationships between organisms are not rigidly fixed in any of the examples mentioned. Coral reefs have been found at several times during the Phanerozoic period—but only for relatively short geological periods. Over much longer periods, there are only a few indications of reefs, mostly with other reef-building organisms such as mussels (Kiessling 2009). Various species of the dinoflagellate genus *Symbiodinium*—as photosynthetically active zooxanthellae—also occur in endosymbiosis with coral organisms in present-day reefs. Specific cell types regulate endosymbiosis in corals (Hu et al. 2020). Under nutrient-poor—oligotrophic—conditions, symbiosis is advantageous for both partners. Ciliary motions of the corals increase the exchange of substances in the physical boundary layer at the surface of the organisms by around 400 per cent (Carpenter et al. 1991; Shapiro et al. 2014; Stocking et al. 2016). *Symbiodinium* utilizes the coral's CO_2 and nitrogenous excretions and in return supplies the coral with around 90 percent of the organic nutrients it needs. In addition, the algae also reinforce the formation of calcareous skeletons. Under less restrictive conditions, many algae of the genus *Symbiodinium* can also survive without corals, while various corals can also feed directly on bacteria (Bak et al. 1998; Baker 2003; Baker et al. 2004; Stanley 2006; Decelle et al. 2018; LaJeunesse et al. 2018). Abot 50 per cent of known stony coral species do not form symbioses with photosynthetic organisms; symbioses with chemoautotrophic prokaryotes have been detected in individual species. Around 75 per cent of these species form benthic reefs at depths of up to around 6000 metres (Freiwald et al. 2004; Kayal et al. 2013; Middelburg et al. 2015; Zapata et al. 2015; Wienberg et al. 2018).

9.2 Deep Seas: Exclusive Largest Habitat of Animals

9.2.1 Evolution of Intrinsic Rules in Exclusively Heterotrophic Systems

We still have extremely little knowledge of biological processes beneath the photic zones of oceans—the largest habitat on Earth (Helfman et al. 2009). For example, it was not until 2009 that it was demonstrated that the three species originally described, each allocated to a different fish family, are larvae and the two adult sexes of a single species—belonging to the whalefish family (Cetomimidae) (Johnson et al. 2009). Here, estimates of biomass production reach methodological limits as photosynthetic primary production in eddy currents of nutrient-poor marine regions can be concentrated at the lowest limit of solar radiation (Mann and Lazier 2006). Below these zones, heterotrophic interactions between organisms moving freely horizontally (RH) and vertically (RV) through—in principle—structureless space predominate (Fig. 9.6).

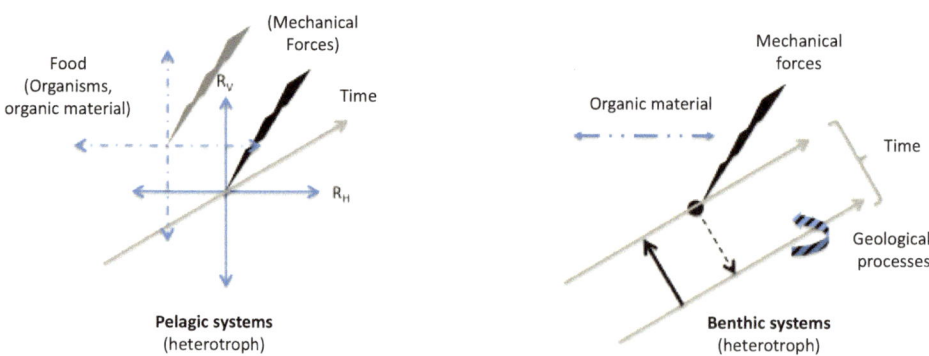

Fig. 9.6 Comparison of differing combinations of energetic and abiotic conditions in heterotrophic zones of aquatic habitats. Explanations in the text

Bizarre life forms of fish with huge jaws and oversized—in some cases transparent—teeth and comparatively long, thin bodies (Stomiiformes) are well known, as are deep-sea anglerfish (Ceratioidei) with mobile lures and often tiny dwarf males (Pietsch and Orr 2007; Helfman et al. 2009; Velasco-Hogan et al. 2019) firmly attached to females. Less well known are the diverse sensory-physiological adaptations to the special habitat conditions, for example in the eyes and lateral line organs. Unlike cave fish (McGaugh et al. 2014), which live largely isolated in the dark, deep-sea fish have a wide variety of morphological and sensory eye types. The spectrum ranges from reduced, flat eyes lacking lenses on the roof of the skull (*Ipnops*) to morphologically or photoreceptively adapted eyes (*Dolichopteryx, Maurolicus*) (Wagner et al. 2009; Davis and Fielitz 2010; de Busserolles et al. 2017). In many species, perception of water movements—and hence of potential prey—is significantly enhanced by a dense distribution of lateral line organs in the vicinity of the mouth (Marranzino and Webb 2018).

However, in many regions a sufficient basis for differentiated food webs must be provided by vertically migrating organisms and sinking organic material. This inference is supported by the diversity of life forms and physiological characteristics of deep-sea organisms (Robison 2004; Helfman et al. 2009). Diving depths of several hundred metres to over two kilometres observed with penguins and marine mammals seeking food also provide evidence of a rich food supply (Wilson et al. 1993; Kooyman and Ponganis 1998; Watwood et al. 2006; Schorr et al. 2014).

It should be noted that, in the North Atlantic alone, before commercial whaling began whale populations were probably up to twenty times larger than they are today (Roman and Palumbi 2003).

In relation to current catch results, estimates of fish stocks based on echo-soundings down to depths of 1000 metres indicate higher population densities (Irigolen et al. 2014). Optical counts with submersibles off the Californian coast revealed particularly high densities of organisms at depths ranging around 2000 metres. These were, however, predominantly species of cnidarians (phylum Cnidaria) with only a few fish species (Robison et al. 2017). In fact, this makes it easier to understand why many species of deep-sea fish swarm to form a circle when threatened, imitating the life forms of inedible cnidarians (Robison 2004). The widespread occurrence of bioluminescence among all organisms—from bacteria to vertebrates—suggests that optical information is of great importance at such depths (Widder et al. 1983; Robison et al. 2003; Haddock et al. 2010; Craig et al. 2011; Davis et al. 2014; Straube et al. 2015; de Busserolles et al. 2017). Morphological findings of huge eyes with a diameter of around 25 centimetres in deep-sea squid and ichthyosaurs from the Jurassic and Cretaceous periods, combined with results of theoretical analyses, support this inference (Humphries and Ruxton 2002; Nilsson et al. 2012, 2014; Davis et al. 2016). Southern elephant seals (*Mirounga leonina*) apparently orient themselves to light signals emitted by their prey when foraging—at depths of up to 1500 metres—while sperm whales (*Physeter macrocephalus*)—like many other toothed whale species—employ acoustic methods of localization (Miller et al. 2004; Vacquié-Garcia et al. 2012; Mouriam and Orliac 2017).

9.2.2 The Benthic Zone: Life in the Transitional Realm

If open water bodies (pelagic zone) constitute the largest three-dimensional habitat on Earth, the sea floor (benthic zone) (Fig. 9.6) is the largest realm of transition between two spheres. General conditions in this habitat are determined by dynamics of the water currents, associated distributions of oxygen concentrations and properties of solid material (McCave et al. 1995). Depending on current velocities, composition of the solid material varies from predominantly mineral to predominantly organic components. Impressive documentation of the dynamics of this habitat in Earth's history are provided by the many mountain ranges formed from organic deposits, often thousands of meters high (Veizer 1973; Stanley 2001).

Global dimensions of the overall network are easily overlooked because of the different scientific and methodological approaches used to investigate individual areas of the aquatic-geological transition zone. In functional terms, however, there are diverse relationships between open, standing or flowing water bodies, groundwater and fissure water bodies. The range of flow velocities is extremely wide, extending from velocities like those at the surface down to zero. Considering only the differing geometric dimensions—from underwater caves to the finest crevices hundreds of meters below the seabed (Riedl 1966; Kostylev et al. 2005; Parkes et al. 2005; Lomstein et al. 2012; Tetu et al. 2013)—provides an initial indication of the fractal structures of the overall system.

Prokaryotes colonize all dimensions of the overall system, and their metabolism significantly influences processes of chemical exchange between the hydrosphere and lithosphere (Humphreys 2006; Taniguchi et al. 2006; Kato et al. 2009; Fang et al. 2010; Orcutt et al. 2011; Teske 2013; Gorelick and Zheng 2015; de Graaf et al. 2019). Microbial biofilms are composed of small-scale differentiated—but highly adaptive—communities that can maintain their functionality even in isolated systems over millions of years (Bethke et al. 2008; Inagaki et al. 2015). Dispersal possibilities for eukaryotes are limited by the dimensions of cavities as well as by chemical and thermal environmental factors. Through their mechanical activities (bioturbation), multicellular eukaryotes modify microbial processes in sediments and mineral substrates—for instance by digging or drilling tunnels—as well as energetic conversion of voluminous organic materials—for example by crushing plant parts or animal carcasses (Fig. 9.7) (Yonge 1955; Østergaard et al. 1991; Laverock et al. 2010; Mermillod-Blondin 2011). Differentiation of marine heterotrophic food webs began around 550 million years ago with the advent of various kinds of mouthparts in animal life forms (Klug et al. 2017; Moysiuk and Caron 2019). Over the course of the Phanerozoic, the processes concerned were subject to multiple regional and global changes due to abiotic and biotic factors (McGowan and Smith 2008), although these led to only limited changes in functional life forms in benthic habitats.

From a macrobiological perspective, the habitation area can be simplified by referring to immediate transition zones in open bodies of water. Functionally, a distinction can be drawn between epifauna—organisms living upon the substrate—and infauna—organisms living within the substrate (Fig. 9.7). Flow velocities significantly influence the

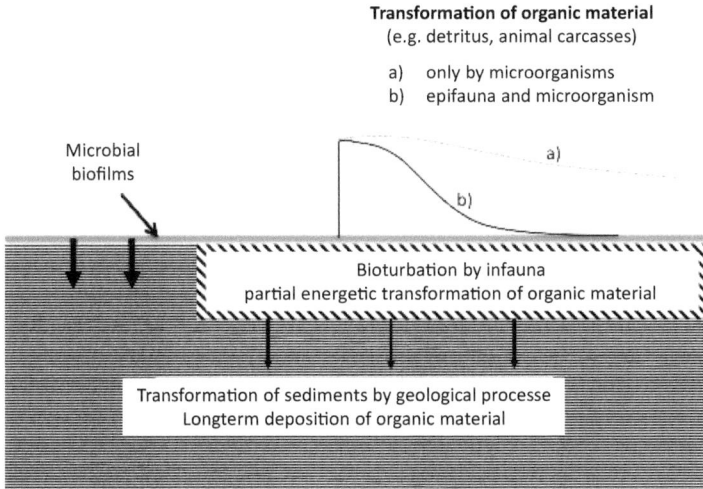

Fig. 9.7 Schematic representation of effects of activities of animal organisms on and in sediments (right side of figure) compared to microbiological processes alone (left side of figure)

distribution of mobile and sessile life forms within the epifauna. Many sessile life forms actively select suitable local conditions for colonization during the mobile larval stages of their life cycles (Butman 1987; Abelson et al. 1993; Pawlik and Butman 1993). For feeding, various organisms exploit particle transportation by water currents. Functionally, the spectrum ranges from morphological developments exerting hydrodynamic effects to motor regulation of local currents by cilia or muscles. In various cases, the degree of hydrodynamic influence is enhanced by adapted collective spatial arrangements of groups of organisms—for example sand dollars (*Dendraster*), tunicates (*Botryllus*) or bryozoans (*Membranipora*) (Vogel 1994; Riisgård and Larsen 2010).

As a result of their active process of colonization, sessile life forms are constantly exposed to the danger of overgrowth by other organisms. Accordingly, such organisms use various strategies to defend themselves against this threat—ranging from formation of defensive substances in their own metabolic processes to symbioses to keep their body surfaces clear (Bowman 2007; Marhaeni et al. 2011; Stowe et al. 2011; Piazza et al. 2014). In technological facilities or on artificial surfaces in contact with natural water bodies—for example pipes or hulls of boats—overgrowth is also known as "biofouling"and is often countered with extremely toxic substances. Use of substitute substances with lower toxicity is now at least being discussed (Dafforn et al. 2011; Guardiola et al. 2012; Amini et al. 2017).

Organisms that burrow in sediments—representatives of the infauna—reinforce the latter effect by transporting mineral materials to the sediment surface and strengthening microbial processes through their own metabolic activity. For energetic reasons, the extent of sediment turning and mixing is closely related to functional characteristics of the life forms involved and the organic content of the sediments. For life forms that use sediments solely for protection, depth of penetration is primarily related to the particular body size and the risk of loss to carnivory. The extent of sediment overturning by organisms feeding directly on sediment constituents is primarily related to concentrations and qualities of organic components present. In shallow waters—as a food source—the infauna triggers sediment overturning by large vertebrates such as gray whales (*Eschrichtius*), bottlenose dolphins (*Tursiops*) or walruses (*Odobenus*) (Kvitek 1986; Mann et al. 2008; Clark et al. 2019).

9.2.3   An Experimental Arena for Morphological Differentiation of Animals

Mobile life forms of epifauna range from protists (*Gromia*) to fish, and feed on all kinds of energetically convertible organic materials (Gage and Tyler 1999; Turner 2002; Witte et al. 2003a; Matz et al. 2008). Physical characteristics of the substrate influence the morphological differentiation of the infauna, as well as the interstitial fauna living in gaps within the sediments or the mineral subsoil (Duplisea and Drgas 1999; Mermillod-Blondin 2011). Depending on dimensions of cavities in the mineral substrate, a wide range of different life forms can be found therein—from miniature versions of the epifauna to highly specialized organisms (Danielopol 1976; Rundell and Leander 2010). In general, the mechanical activities of animals in substrates expand the microbiological latitude for microorganisms—especially archaeans and bacteria—and hence also microbiological turnover rates (Fig. 9.8) (Dobbs and Guckert 1988; Gilbert et al. 2003; Kogure and Wada 2005; Pischedda et al. 2008; Nogaro et al. 2009). In addition to providing easier access to deeper sediment layers, the modified oxygen distribution and organic excretions in the

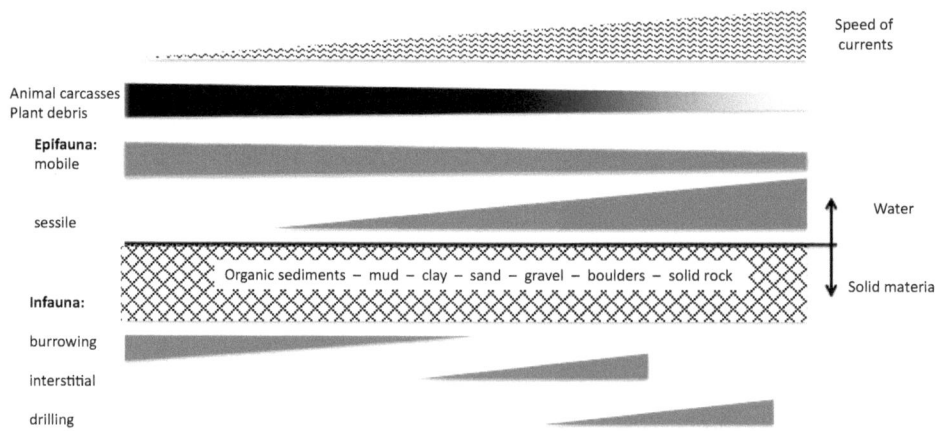

Fig. 9.8 Schematic representation of functional relationships between biotic (left text column) and abiotic (right text column) factors in benthic systems

vicinity of animal burrows also make a contribution. Animals utilize the increased micro-biological turnover rates to varying extents. Life forms that live in sediments primarily for protection against attack, but feed on organic sediments on the surface—for example spoon worms (Echiurida)—derive less benefit from this than bacterivorous or symbiotic life forms—for example sea cucumbers (Holothuria) (Gage and Tyler 1999; Lee et al. 2018).

Physical and chemical diversity of habitats in transition zones between water bodies and solid materials, ranging from deep-sea floors to intertidal zones, has contributed significantly to evolutionary differentiation of animal life forms. For quantitative simplification, corresponding relationships can be assessed on the basis of systematically defined animal phyla. Of 33 defined animal phyla, 17 are found exclusively in seas and 15 in both salt and fresh water. Of the latter group, 11 are also found in terrestrial habitats. One animal phylum—Onychophora—lives only in terrestrial habitats, but probably originates from Cambrian marine Lobopodia (Stella et al. 2011; Zhang et al. 2016). Further differentiated quantitative comparisons across all organisms—for example at the species level—cannot yield meaningful information for a number of reasons. On the one hand, knowledge of the diversity of life forms based on morphological features alone for individual groups of organisms and different habitats is extremely skewed (Mora et al. 2011). Inclusion of prokaryotes and viruses in so-called "species lists" is particularly problematic in this context (Costello and Chaudhary 2017). On the other hand, even—hypothetically—complete lists of species do not provide any information about functional relationships between the different species in their various habitats (Smetacek 2012).

Marine benthic systems are supplied with convertible energy mainly by means of organic materials and—in the vicinity of hydrothermal vents or gas seeps—via inorganic chemical compounds. The principle of chemoautotrophic energy conversion can be assumed to have been constant since the Proterozoic, and symbioses of bacteria and macro-organisms that depend directly on that source for survival change in close association with the evolution of marine animals (Little and Vrijenhoek 2003; Vrijenhoek 2010). However, the composition and structure of organic materials carried in reflects ongoing changes in organisms and their activities in oceans and on land. In other words, the evolution of benthic organisms was essentially promoted by development of photosynthetic energy conversion in the epipelagic zone of oceans, with subsequent consequences on land.

Two clearly different responses to external evolutionary dynamics can be recognized in the epifauna. Evolutionary lineages of mobile life forms with direct interactions between different animal species— for example predator-prey relationships—are generally known. Indeed, they also partly promoted animal evolution on land. The relative constancy of life forms among sessile organisms, which feed mainly on small organisms and organic particles, is less evident. Their morphological differentiation is often based on formation of colonies of individual organisms with varying degrees of functional modification, for example sponges (Porifera), cnidarians, bryozoans or tunicates (Cartwright 2003; Hughes 2005; Carver et al. 2006; Barbeitos et al. 2010; Winston 2010). Colonial life forms may have evolved as early as 1.5 billion years ago, although the systematic classification of many fossil forms is still subject to discussion (Fedonkin and Michelson 2002; Yang et al. 2023).

Among the animal groups mentioned, the oldest fossil evidence for sponges is currently estimated to be around 585 million years old (Nettersheim et al. 2019). The origin of sessile cnidarians and bryozoans is assumed to be in the Cambrian period, but no definite evidence has so far been obtained for sessile tunicates (Taylor and Waeschenbach 2015; Mendoza-Becerril et al. 2016; Zhang et al. 2021). What is certain, however, is their persistent global distribution on solid substrates ranging from the benthic zone—given sufficient flow—to the intertidal zones of oceans (Riedl 1966; Sebens 1986; Gabriele et al. 1999; Kuklinski 2009). Long-term evolution of sessile colonial organisms is characterized by recurrent collapses, dynamic shifts to solitary life forms and adaptations to changes in chemical and physical environmental conditions (Coma et al. 1998; Barbeitos et al. 2010; Stolarski et al. 2011; Taylor et al. 2015; Taylor and Waeschenbach 2015; Mendoza-Becerril et al. 2016). Over the course of their evolution, sponges, cnidarians and bryozoans have successfully migrated into freshwater habitats (Dumont 1994; Manconi and Pronzato 2008; Annenkova et al. 2011; Erpenbeck et al. 2011; Müller et al. 2013; Koletić et al. 2014). This makes it easy to recognize that these life forms follow an independent, successful evolutionary trajectory and are not "forgotten" phylogenetic relics. Persistence of unicellular organisms in aquatic primary production and energetic advantages of the sessile, cooperative lifestyle are probably decisive for their sustainable evolution.

In the benthic realm, organisms are often massive in comparison to other members of their respective overall groups. Examples of this among unicellular eukaryotes are Xenophyophores, sponges, cnidarians or amphipods, whose size can be measured in decimetres (Gage and Tyler 1999; Wang et al. 2011; Kamenskaya et al. 2013; Wagner and Kelley 2017). This phenomenon is often explained solely on the basis of low metabolic rates and consequent greater scope for formation of body substance. From ecological analyses it can be inferred that combinatorial effects of different factors promote increase in size in particular cases (Brey and Clarke 1993; Verberk and Atkinson 2013). In addition to physiological factors, systemic pathways of biogenic energy fluxes must also be taken into account. On the one hand, low food input densities permit only slow growth; on the other hand, this is associated with a low risk of being covered by sedimentary deposits. Low food input densities combined with chaotic temporal distributions also render problematic the evolution of long food chains including large carnivorous populations. The chances of survival for slow-growing and long-lived organisms in the overall system are thus increased.

9.2.4 Utilization of Larger Habitation Zones: Mobile Life Forms

Evolution of mobile benthic organisms on the sea floor attracts far greater attention. But in this case, too, the diversity of life forms that inhabit benthic habitats—despite the restrictive energetic conditions—is easily overlooked (Gage and Tyler 1999; McClain et al.

2012). Composition of the mobile fauna is much more strongly influenced by epipelagic and terrestrial conditions than that of the sessile fauna, as it can energetically convert organic material regardless of its size—from bacterioplankton to whale carcasses. Apart from generalists, various species—in symbioses with prokaryotes—have specialized in energy conversion of organic materials that are difficult to decompose. For example, mussels on wood and other plant remains (Wolff 1979; Distel et al. 2002) or snails and bone-worms (*Osedax*) on whale bones (Goffredi et al. 2007; Vrijenhoek et al. 2009; Katz et al. 2011; Huusgaard et al. 2012; Aronson et al. 2017). The example of the genus *Osedax* shows how closely the evolution of specialized deep-sea fauna is linked to the particular food supply of whale bones, because no traces thereof have yet been found on the bones of ichthyosaurs (Danise et al. 2014).

The spatial distribution of deposited organic material is—in simplified terms—influenced by the original size of the organisms concerned and their distribution in the upper marine zones, as well as by benthic currents and geomorphological structures (Thistle et al. 1991; Zhang et al. 2014; Simon-Lledó et al. 2019). Microscopic organic material is to be expected—at varying densities—in all benthic zones. Burrowing or sediment-collecting organisms can hence remain in one place for a long time if their own energy requirements can be met by the average production of organic material (Gage and Tyler 1999; D'Hondt et al. 2002; Kogure and Wada 2005; Teal et al. 2008). Large quantities of organic material—such as plant remains or animal carcasses—are less likely to accumulate on the seabed in areas far removed from the coast, and their spatial density is lower. The resulting "food oases"initiate a dynamic sequence of differing organismic communities (succession) extending from initial decomposition of macroscopic material to consumption of remaining bacterial films by specialized sea cucumbers (Lopez 1988; Roberts et al. 2000; Rosenberg 2001; Witte et al. 2003b; Lundsten et al. 2010; Jamieson et al. 2011; Smith et al. 2014, 2015).

Observations of experimentally deposited fish morsels at a depth of around 7.7 kilometres in the Japanese Trench of the Pacific Ocean revealed the rapidity with which sinking material is discovered by deep-sea organisms. Within an hour the first fish of the genus *Pseudoliparis* arrived at the bait and their numbers increased exponentially during the observation period. Their target was not the bait itself, but amphipods (*Eurythenes*) feeding upon it (Fuji et al. 2010). On one hand, this example demonstrates the capacity of deep-sea organisms to rapidly locate potential food sources, and on the other hand it reveals needs for locomotor capacity of carnivorous organisms in present-day marine benthic ecosystems with adequately oxic conditions.

In view of the huge dimensions of the above-discussed habitat, a general decrease in living biomass with increasing depth, but not a homogeneous horizontal distribution, may be expected. The general vertical trend can also be found in the diversity of depth distribution in bony fishes, some of which occur at maximal depths. Among cartilaginous fish, some—such as sharks and rays—are found at depths of up to around 4000 meters, while cyclostomes—such as hagfish (class Myxini)—occur only at depths up to around 3000 meters (Fuji et al. 2010; Priede and Froese 2013; Linley et al. 2016).

Density and diversity of food also exert an effect on diversification and specialization of invertebrates. At depths of up to 200 metres, the fauna of hydrothermal vents is barely distinguishable from that of the surrounding area; at greater depths, the fauna consists mainly of specialized species (Tarasov et al. 2005; Bernardino et al. 2012). Sharks and rays use hydrothermal vents as "incubators" for their eggs in the relatively cold deep waters (Treude et al. 2011; Salinas-de-León et al. 2018).

For fish, clear deviations from generalized depth distributions have been observed along the Mid-Atlantic Ridge. The greatest density of fish biomass was found in the zone up to 200 metres above the sea floor, followed by the zone between 1500 and 2330 metres in depth. In terms of numbers of individuals, the greatest densities were in the zones from 0 to 200 metres in depth and up to 200 metres above the seabed. In simple terms, juvenile fish dominate in near-surface zones and adult fish in deeper sea zones (Sutton et al. 2008). The vertical distribution of fish along the Mid-Atlantic Ridge—because of the high proportion of juvenile fish at depths of up to 750 metres—is probably not attributable solely to geological and current-related factors. It is reasonable to assume that the results of study are also influenced by the intensity of fishing in these regions.

Global estimates indicate that most bony fish species—about 45.4 per cent—inhabit areas with reefs and littoral zones, while around 41.2 per cent live in inland waters (Fig. 9.9). In deep-sea regions, the distribution of species is estimated at 1.2 per cent in the epipelagic zone, 5 per cent in the pelagic sector and 6.5 per cent in the benthic zone. The proportion of species migrating between oceans and inland waters (diadromous) is

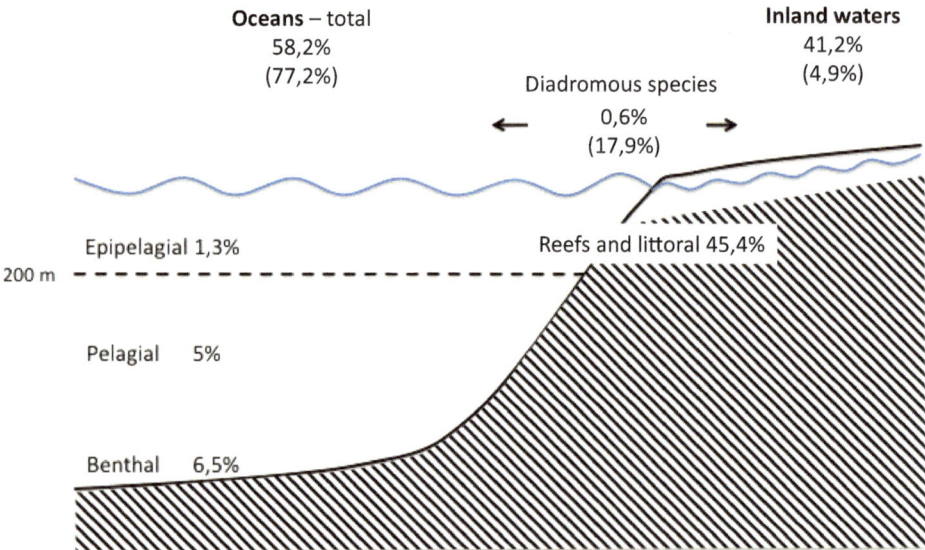

Fig. 9.9 Global distribution of species of cartilaginous fish and of bony fish in various marine habitats and inland waters. Values for bony fishes without brackets, values for cartilaginous fishes in brackets. Data basis: Helfman et al. (2009)

estimated at 0.6 per cent of the total of around 16,000 species. The approximately 950 species of cartilaginous fish live predominantly in marine habitats (77.2 per cent), while around 17.9 per cent can at least temporarily migrate into inland waters (Helfman et al. 2009).

Markedly structured habitats permit a high diversity of species and thus increase the stability of biotic communities when confronted with biological disturbances. Examples for this are coral reefs or coastal zones. The decisive factor is the associated diversity of physiological life forms and small structural spaces, which for energetic reasons both inhibit the spread of parasites and preclude the development of disproportionately large life forms. For example, carnivores that currently feed on reef organisms reach maximum lengths of about three metres and devour a wide range of different prey (Kiessling 2005; Rocha et al. 2005; Frisch et al. 2016).

9.3 Early Differentiation and Ongoing Adaptations

Because fossil evidence of early organisms—presumably small and consisting solely of soft cell tissues—is lacking, estimates of evolutionary timespans for multicellular organisms are subject to many uncertainties. As a result, estimates based on molecular or genetic methods cannot be verified with fossils (Cunningham et al. 2016). Whereas older publications inferred earlier periods of origination, current estimates suggest that the primary origin of Metazoa occurred between 835 and 650 million years ago. According to known finds of microfossils, however, differentiation could have taken place as early as the Tonian—between 1000 and 720 million years ago (Wang et al. 1999; dos Reis et al. 2015; Xiao and Tang 2018). According to these estimates and fossil finds, differentiation into most animal life forms may have already taken place in the Ediacaran period (635 to 541 million years ago)—i.e. prior to the Cambrian (Shore et al. 2021). A significant increase in multicellular animal organisms can be observed in the fossil record around 630 million years ago, at the beginning of the Ediacaran. Structures in fossils from the 609-million-year-old deposits of the Doushantuo Formation in China (*Caveasphaera*) are interpreted as the first indications of embryonic development (Vannier et al. 2010; Mángano and Buatois 2017; Budd and Jackson 2016; Yin et al. 2019; Shore et al. 2021). Many early life forms of the Ediacaran period were sessile, feeding on plankton and organic material by passively filtering it out of water currents—resembling the morphologically analogous sea pens (Pennatulacea) in deep zones of present-day oceans (Fig. 7.19). There is also evidence for the presence of sponges (Brasier et al. 1997; Li et al. 1998). Around 570 million years ago, there is increasing evidence of mobile life forms in body fossils—for example the flat, oval *Kimberella*—and feeding traces in microbial mats.

The question of whether sufficient oxygen concentrations in oceans existed for the evolution of multicellular animals during these periods is often discussed. As a counterargument, it is highly probable that zones with differing oxygen concentrations were present in the oceans (Canfield et al. 2007; Li et al. 2016a, b; Sahoo et al. 2016). Moreover, representatives of early animal groups have been shown to have a high tolerance for oxygen deficiency and increased sulphur levels. For example, individual sponge species survive O_2 concentrations of around 0.25% of the current level (Mills et al. 2014, 2018). Functional properties of the mitochondria of bristle worms (Polychaeta), bivalve molluscs or brush heads (Loricifera) permit metabolism under anaerobic conditions, at least intermittently (Völkel and Grieshaber 1997; Doeller et al. 1999; Danovaro et al. 2010; Mentel and Martin 2010; Stefano et al. 2015). In view of the low probability of fossil finds, it is also not surprising that pre-Cambrian habitat types (microbial mats) are found sporadically in Cambrian deposits (Buatois et al. 2014). Evidence for possible colonization of land areas by fungi and algae during these periods and the subsequent Cambrian period is still largely inconclusive (Yang et al. 2004; Knauth and Kennedy 2009; Rota-Stabelli et al. 2013; Berbee et al. 2017). Sparsity of evidence is mainly a result of large-scale glaciations and the associated destruction of terrestrial traces of life (Landing and MacGabhann 2010; Pu et al. 2016).

Relative distributions of extant species in oceans and inland waters clearly reveal the differing evolutionary trajectories of cartilaginous and bony fishes. Over the 420 million years of their evolution, cartilaginous fish have developed in comparatively structureless habitats (Long et al. 2019). Morphological and physiological characteristics of the three main recent groups of cartilaginous fish—sharks (Selachii), rays (Batoidea) and ratfish (Holocephali)—clearly display the influence of feeding habits and hydrodynamic conditions. In these groups, only minor changes in body shape can be recognized over hundreds of millions of years. However, adaptations and major variations in their diets can be found in the morphology of their dentition and the shapes of their regenerable teeth. Energetically, they show clear adaptations to the restrictive conditions of marine habitats. At rest, sharks require only about a third of the energy needed by bony fish. Due to their specific morphological and physiological make-up, the energy metabolism of sharks increases around threefold during active movement, whereas in bony fish the level is ten times higher than the basal metabolic rate (Helfman et al. 2009).

The two main groups of bony fish—ray-finned fish (Actinopterygii) and lobe-finned fish (Sarcopterygii)—also reveal the influence of structural conditions on morphological evolution. Among ray-finned fishes, diverse life forms—often with extreme morphological adaptations—can be found in all aquatic habitats. Among lobe-finned fish, only two groups have remained permanently in aquatic habitats—coelacanths (Coelacanthomorpha) in oceans and lungfish (Dipnoi) in inland waters. All terrestrial vertebrates evolved from the third large group (Tetrapodomorpha), which is more closely related to lungfishes and gave rise to individual groups that secondarily migrated back into aquatic habitats (Janvier 2007; Pough and Janis 2019).

Extinct life forms trigger associations with qualifiers such as primitive or modern. In the context of biological evolution, such terms reinforce the baseless idea of an intrinsically directed trajectory. To avoid such misunderstanding, the oft-used distinction between temporally assigned faunal groups has been omitted from Fig. 9.10 (Sepkoski 1998). The reason for this resides in nuanced assessment of the evolution of higher-level systematic

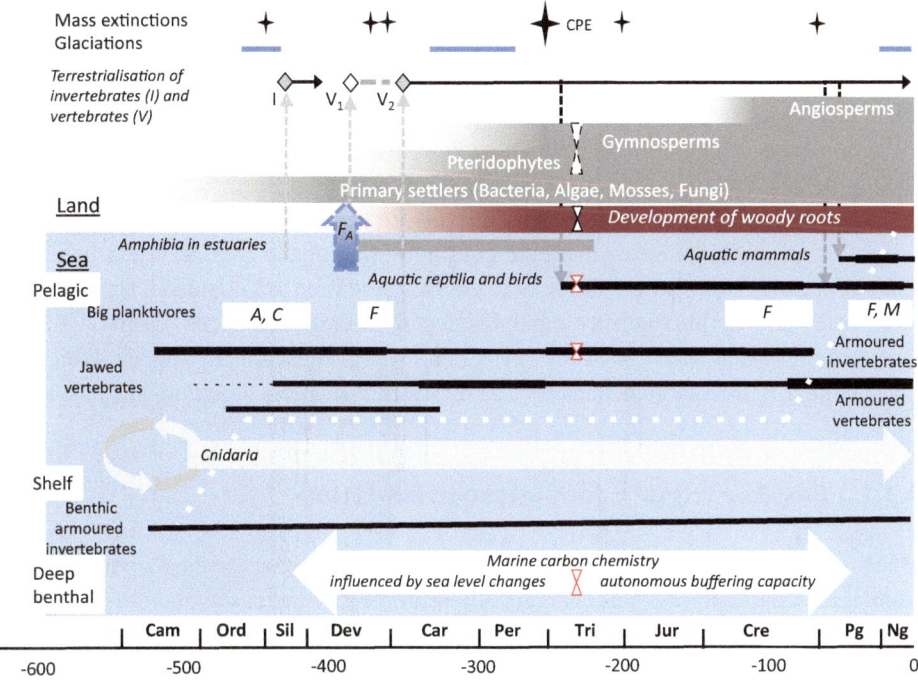

Fig. 9.10 Simplified illustration of the differing evolutionary processes in pelagic and benthic marine habitats. In the pelagic zone, competition among bilaterians for primary food sources has promoted the evolution of life forms with increasing efficiency in food acquisition, combined with an elevated metabolism. Return migrations by terrestrial vertebrates—reptiles, birds and mammals—indicate that marine evolutionary processes could not bring about exploitation of all food niches. Throughout the Phanerozoic, limited food resources in deeper benthic areas were utilized mainly by life forms with a scaled-down metabolism. Benefiting from this strategy and certain hydrodynamic adaptations, cnidarians were also able to spread into pelagic habitats. In connection with the evolution of land plants, terrestrial habitats could only be colonized by invertebrates (I) and vertebrates (V). Abbreviations: *M* Mass extinction, *A* Arthropoda, *C* Cephalopoda, *F* Fishes, *FA* Immigration of ray-finned fishes (Actinopterygii) into freshwater habitats, *M* Mammals, *CPE* Carnian Pluvial Event, with reference to changes following the "hourglass principle". White dotted line: abstracted trend in Phanerozoic diversity of marine invertebrates. Ages: *Dev* Devonian; Jurassic; Cambrian; Cretaceous; *Ng* Neogene; Ordovician; *Per* Permian; *Pg* Paleogene; Triassic. Literature and data sources: Willis and McElwain (2002); Alroy (2010); Laurin and Soler-Gijón (2010); Fortuny et al. (2011); Kröger et al. (2011); Kenrick et al. (2012); Donoghue and Keating (2014); Pyenson et al. (2014); Schoch (2014); Friedman (2015); Porada et al. (2016); Yao et al. (2016); Bond and Grasby (2020); Dal Corso et al. (2020); Andreev et al. (2022); Bault et al. (2022); Benton and Wu (2022)

groups of marine organisms (Sepkoski 1981; Alroy 2010). According to our current state of knowledge, 97 per cent of groups originated in the Palaeozoic. Around half of them disappeared in the Palaeozoic, while around a quarter have survived to the present day without any major differentiation. Morphological differentiation has increased up to the present in only 17 per cent of the original groups.

Because of major uncertainties, presentation of dynamized reconstructions of environmental conditions and changes in diversity during the Phanerozoic has also been omitted (Katz et al. 2004; Kiessling 2005; Miller et al. 2005; Algeo and Scheckler 1998; Alroy 2010; Bradley 2011; Hannisdal and Peters 2011; Vérard et al. 2015; Mills et al. 2017). The question of the extent to which animal evolution was influenced by the differing conditions in pelagic and benthic habitats, as described in chapter Sects. 9.1 and 9.2, was investigated instead. It should be noted here that living conditions in the deep sea were also subject to ongoing changes. Given the enormous size of the habitat, it can be assumed that—especially during periods of marked fragmentation of mainland areas—changes took place with varying intensity and dynamics in different subspaces (Zaffos et al. 2017). Contrary to popular belief, the deep sea is not a static "museum" of archaic living conditions, but a dynamic habitat characterized by specific energetic conditions.

9.3.1   Possible Triggers for Metazoan Evolution

Trace fossils and body fossils in marine sediments suggest that the development of multicellular animal organisms was energetically promoted by increased phototrophic primary production. Minimal requirements for increased phototrophic production were sufficiently large inputs of mineral nutrients from land areas. Among the essential—non-energetic—conditions for the evolution of animal organisms, enrichment of oxygen in the atmosphere and water bodies is directly linked to phototrophic production. The development of animal life in benthic habitats also required the exchange between near-surface and deep waters brought about by ocean currents. The interplay of the above-mentioned factors may have driven the evolution of morphological diversity in mobile animal life forms during the transitional phase from the Ediacaran to the Cambrian (Shields-Zhou and Zhu 2013; Buatois et al. 2014, 2018; Erwin 2015; Yin et al. 2015). In the absence of adequate exchange between surface and deep waters, anoxic zones spread rapidly through the depths (Lenton and Daines 2017; Liu et al. 2018), which are colonized only by specialized microorganisms. Animals quickly perish in anoxic zones, as is evidenced by the many fossilized remains of mobile arthropods (Collette et al. 2010).

Let it be noted here that logically nonsensical terms—such as "agricultural revolution" (Mángano and Buatois 2017)—have been used in the literature in connection with these early evolutionary changes. Such usage is based on the false notion that bioturbation—in analogy with cultivation of agricultural soils—directly increased marine productivity. In functional terms, the differentiation of animal life forms was solely dependent on the conditions mentioned above. Increasing mechanical overturning of sedimentary material by animals testifies to extended possibilities for animals to exploit energetic resources. This applies both to the capacity to transform organic material in sediments and to inclusion of carnivorous diets in heterotrophic food webs (Klug et al. 2017; Mángano and Buatois 2017).

According to findings so far available, quantities and durations of raised marine primary production sufficed for the evolution of large planktivorous organisms during only a few time periods (Fig. 9.10). The body length of Cambrian planktivorous arthropods (*Anomalocaris*), at about one metre, was approximately doubled in the genus *Aegirocassis* during the Ordovician. The latter shared the same food source with cephalopods of the genus *Endoceris*, whose straight shells were up to 6 metres in length. Judging by the dimensions of their filter openings, the species concerned presumably fed mainly on animals around one millimetre in diameter. Evidence of the first large planktivorous vertebrates has so far been found in the Late Devonian (Famenian). Jaw anatomy of *Titanichthys* (Arthrodira)—presumed to have been over 5 metres long—indicates a planktivorous diet. Species in the Arthrodira group—belonging to the class Placodermi—are characterized by an additional joint at the base of the skull, which allowed them to fold up the upper jaws when catching prey (Paterson et al. 2011; Daley et al. 2013; Klug et al. 2014; Vinther et al. 2014; Boyle and Ryan 2017; Lerosey-Aubril and Pates 2018; Coatham et al. 2020; Mironenko 2020).

During the Jurassic, the largest planktivorous life form to date evolved among bony fish of the genus *Leedsichthys* (family Pachycormiformes), with a body length of around 9 metres.[1] Medium-sized genera (*Ohmedia*) in this family, with a body length of around two metres, inhabited the oceans for over 100 million years—up to the end of the Cretaceous period. Contrary to theoretical assumptions, some marine reptiles and pterosaurs probably also fed on small creatures by filter-feeding (Collin and Janis 1997; Helfman et al. 2009; Friedman et al. 2010; Motani et al. 2015; Martill et al. 2023).

During the Cenozoic, four groups of large planktivorous life forms—basking shark (Cetorhinidae), megamouth shark (Megachasmidae), manta ray (Mobulidae), and whale shark (Rhincodontidae)—evolved from benthic cartilaginous fish (Chondrichthyes). According to currently prevailing interpretations, evolution of those adaptations occurred at around the time of the Paleocene-Eocene Thermal Maximum (PETM) about 55 million years ago. Significantly more recently, around 30 million years ago—with the advent of the Antarctic Circumpolar Current—differentiation of baleen whales began among marine mammals, beginning with the development of suction-based feeding behaviour. Their largest life forms—such as blue whales—evolved only in connection with recurrent glaciations and associated changes in ocean currents around 3.6 million years ago (Zachos et al. 2001; Marx et al. 2016; Slater et al. 2017; Bianucci et al. 2019; Pimiento et al. 2019; Ekdale and Deméré 2022).

9.3.2 Factors Influencing the Evolution of Metazoans

Rapid differentiation of all basic morphological animal life forms in the Cambrian can —hypothetically—be interpreted as the iterative development and occupation of potential niches in heterotrophic food webs. In turn, the latitude for systemic self-organization depends on the genetic and epigenetic potentials of the organisms concerned.

---

[1] Earlier estimates of 15 metres body length are doubtful.

However, degrees of differentiation achieved in each case depend on the randomness of phototrophic production and abiotic processes. Extremely abstracted, three timespans in the Phanerozoic with differing dynamic features can be distinguished from summarized data for the diversity of invertebrates (Fig. 9.10, dotted white line) (Alroy 2010):

- an early differentiation phase from the Cambrian to the Ordovician,
- a subsequent phase of multiple upheavals lasting until the Late Cretaceous, and
- a subsequent late differentiation phase extending up to the Neogene.

In simplified terms, the initial period for pelagic habitats can be characterized by predominance of invertebrates, the second by multiple changes in dominance of invertebrates and vertebrates, and the third by dominance of vertebrates. This does not take into account the incompletely documented—and still ongoing (Fig. 9.10)—evolution of cnidarians with their diverse pelagic life forms (medusae), whose life cycles are to varying degrees associated with sessile life phases (polyps). Ecophysiologically, many pelagic cnidarians are better adapted to marine conditions than fish or crustaceans, as their metabolism is not diminished by low water temperatures. For this reason, fish feeding on medusae—for example the ocean sunfish (*Mola mola*)—must regularly visit warmer surface waters to adjust their body temperatures (Thuesen and Childress 1994; Park et al. 2012; Van Iten et al. 2014; Nakamura et al. 2015; Song et al. 2021). Other deep-diving fish—such as the tropical scalloped hammerhead shark (*Sphryna lewini*)—lower their respiratory rate in cold water zones to stabilize body temperature or maintain it through endothermic metabolism, which is more energy-intensive, for instance the moonfish (*Lampris guttatus*) (Wegner et al. 2015; Royer et al. 2023).

The decisive factor for survival or demise of individual species is maintenance of their ability to reproduce under the general conditions constituted by abiotic and biotic factors. In this respect, because of their three-dimensional expanse, marine habitats differ significantly from terrestrial habitats. Because—in simplified two-dimensional terms—phototrophic primary production can take place only within a narrow surface zone, interactions in biotic communities are determined exclusively by heterotrophic organisms in most of the three-dimensional habitat. As explained, biological feedback between the two habitats at greater water depths is only weak. Instead, random, abiotic water currents dominate interactions between the two systems. Moreover, conditions in water bodies are influenced by inputs from land areas, which are also dominated by random, abiotic processes. Inputs of mineral nutrients through the atmosphere or from watercourses stimulate the dynamics of phototrophic organisms. Inputs of biological materials directly influence heterotrophic organisms, but only indirectly affect phototrophic life forms.

Functional relationships between phototrophic and heterotrophic organisms similar to those on land can be found only in shallow water areas (neritic zones). The main abiotic factors influencing shallow water zones are sizes of mainland areas, their coastline structures and prevailing sea level. Biotically formed shallow water zones—reefs—can be created by a wide variety of communities of organism. Whereas reefs formed by microorganisms are still grouped under the collective term stromatolites, metazoan reefs are named after the structure-forming animal group involved—for example sponges or coral reefs. This makes it easy to forget that the oldest and largest structures ever formed by organisms are always the outcome of close functional cooperation between different groups of organisms (Kiessling 2009).

In benthic zones during the Cambrian, around 520 million years ago—approximately 20 million years after the appearance of the genus *Anomalocaris*—rapid evolution of carnivorous arthropods (trilobites) ensued. Morphological differentiation and a sometime extreme optical capacity of their eyes point to highly competitive food webs. Conversion of various trilobite species to intrasedimentary lifestyles in the Ordovician—associated with reduction of the eyes—provides evidence of this (Clarkson et al. 2006; Hughes 2007; Friedman and Sallan 2012; Hopkins 2013). Similar developments in other groups of invertebrates—for instance molluscs (Nautiloidea), brachiopods or cnidariad corals—and simultaneously high rates of new formation and loss of species indicate multiple changes in environmental conditions (Alroy 2008; Kröger et al. 2011; Buatois et al. 2016; Harper et al. 2023).

The associated dynamics of oxygen-free zones because of temporarily high deposition rates of organic material—in addition to abiotic influences—may also be a result of incompletely developed food webs in photic zones (Lehmann et al. 2007; Babcock et al. 2015; Liu et al. 2018). This is indicated by the fact that planktivorous life forms appear in different groups of organisms only during the subsequent Ordovician period, for example the evolution of planktonic larval forms in molluscs or of planktivorous fish (Nützel et al. 2006; Helfman et al. 2009; Servais et al. 2016). A prerequisite for the development of planktivorous fish was sufficiently increased productivity of the phytoplankton, which was presumably promoted by various factors and their interactions: The increasing spread of the first terrestrial vegetation led to decreases in the carbon dioxide content of the atmosphere, combined with an increase in oxygen content, and thus triggered climate changes with far-reaching effects on deep-reaching ocean currents. Increased nutrient concentrations, occurring in upwelling zones and generated by inputs from land areas, are to be expected from this complex of factors—at least in cyclical successive time periods. Increased nutrient inputs due to volcanic activity are also suspected (Parnell and Foster 2012; Huff et al. 2014; Morris et al. 2018; Macdonald 2020; Scotese et al. 2021; Gurung et al. 2022).

Among other things, these developments were favoured by relatively high sea levels and subdivision of the major landmasses into several continents (Bradley 2011; Hannisdal and Peters 2011), as prerequisites for the spread of shallow water zones. Fossil plant spores and traces of millipedes indicate colonization of land areas by the first plants and invertebrates in the Ordovician (Labandeira 2005; Rubinstein et al. 2010; Kenrick et al. 2012; Wellman et al. 2013; Xiong et al. 2013; Porada et al. 2016). Increasing reduction of atmospheric carbon dioxide due to fixation of carbon in land plants probably not only led to an increase in atmospheric and aquatic oxygen levels, but also favoured large-scale glaciations (Masuda and Ezaki 2009; Lenton et al. 2012, 2016; Quirk et al. 2015).

9.3.3   Development of Basic Construction Plans in Vertebrates

Current knowledge indicates that vertebrates were already represented in the Cambrian by the group of—predominantly parasitic—conodonts. This animal group is best known for the great variety in shapes of their teeth, which are barely a millimetre in size but extremely hard and sharp (Murdock et al. 2013, 2014; Martinez-Pérez et al. 2016). Because body fossils have rarely been found, phylogenetic affinities of the eel-like animals, which are only a few centimetres in length, remain unclear. Isotope analyses of the teeth suggest that this group was present in all marine habitats for over 400 million years—from the end of the Proterozoic eon to the Triassic (Ziegler 1998; Lai et al. 2001; Sobstel et al. 2006; Dzik 2007; Helfman et al. 2009; Rigo and Joachimski 2010; Purnell and Jones 2012). In some publications, conodonts are systematically classified as vertebrates. Regardless of that, the basic principle of the diet of this life form has been preserved to the present day in hagfishes and lampreys (Goudemand et al. 2011; Iannicelli 2017).

Systematic classification of some 520-million-year-old Cambrian fish-like fossils (*Haikouichthys, Myllokunmingia, Metaspriggina*) (Shu et al. 1999; Zhao et al. 2012; Conway Morris and Caron 2014) has also not been definitively clarified. Consensus already exists for classification of jawless armoured fish, fossils of which have been found in Ordovician deposits about 430 million years old. They are thought to have evolved in coastal marine zones, arising across a variety of marine habitats, including estuaries (Blieck 1985; Groh 2014; Ferrón and Botella 2017; Sallan et al. 2018). This group, originally clustered together as Ostracodermi, is now divided into six subgroups, most of which possess planktivorous feeding characteristics (Helfman et al. 2009).

Among the originally jawless armoured fishes, species with functionally differentiated jaw skeletons evolved as early as 420 million years ago—in the Silurian (Zhu et al. 2013, 2016, 2019a; Brazeau and Friedman 2015). They already possessed all the bones of the facial skull of vertebrates, including humans (Long 2016; Gai et al. 2017; Zhu et al. 2019b). The first evidence of the formation of tiny teeth was found in fossils (Radotina) of the early Devonian (Vaškaninova et al. 2020). The development of internal fertilization in armoured fish (Placodermi) is also documented by fossils (Long et al. 2009).

9.3.4   Collapse and Regeneration

Potential effects of changes in abiotic factors on biotic communities have already been discussed in Sect. 8.5. In view of the vast period of over 500 million years involved, it should be borne in mind that the five major mass extinctions usually cited offer only limited insights into overall processes during the Phanerozoic. Beyond statistical data, dynamics of the events and the duration of after-effects—which lasted for thousands to millions of years—are usually barely mentioned. As a result, effects on biological communities exerted by changes in different combinations of factors have been covered only incompletely.

An example for this is the Hangenberg Crisis during the transition from the Devonian to the Carboniferous (Fig. 9.10), the causes of which are still subject to debate. Although the Hangenberg Crisis is not included among the great mass extinctions, it led to the demise of almost all trilobite groups as well as the extinction of jawless fish (Luppold et al. 1994; Hughes 2007; Friedman and Sallan 2012; Kaiser et al. 2015; Sansom et al. 2015; Yao et al. 2016). During the subsequent period, an initial significant increase in diversity among bony fishes is noticeable, lasting up to the end of the Permian. During that same period, many original groups of shell-bearing cephalopods disappeared. The abiotic disturbances at the end of the Permian led to a trend reversal. The diversity of bony fishes declined dramatically and continued to fall until the Cretaceous. It was not until the Cenozoic that a second significant increase in the diversity of bony fishes occurred. Evolutionary change among shell-bearing cephalopods (ammonites) shows a completely different pattern. Their diversity increased significantly from the Jurassic onwards and abruptly collapsed at the end of the Cretaceous. The evolution of a group of marine reptiles (Omphalosauridae) (McGowan and Smith 2007; Keupp et al. 2016; Puttick et al. 2017; Qiao et al. 2022) that specialized on this food source also provides evidence of the great biomass of ammonites. In present-day marine habitats, only a few species of shell-bearing cephalopods are represented, belonging to the orders Sepiida (cuttlefish), Spirulida (ram's horn) and Nautiloidea (nautilus) (Kröger et al. 2011).

As explained in Sect. 9.1.6, most biotic structures in marine habitats—such as coral reefs—are in fact formed by animals. Hence, deleterious impacts on animals affect individuals not only directly, but also indirectly through the collapse of large biotic structures. Accordingly, large-scale impacts also result in habitat loss for smaller species. This results in extinction rates that are largely independent of the sizes of organisms, as has been demonstrated, for instance, with respect to the mass extinction at the end of the Permian (Puttick et al. 2017).

The last-mentioned processes reveal the influence of climatic and terrestrial factors on marine living conditions. For example, the glaciations at the end of the Ordovician and in the Carboniferous (Fig. 9.10) were probably caused at least in part by expansion of land plants and associated removal of carbon dioxide from the atmosphere (Mills et al. 2017). Degradation of terrestrial communities can lead secondarily to massive changes in marine living conditions, for example as a result of significantly increased volcanic activity at the ends of the Devonian, Permian and Triassic (Chen and Tucker 2003; Whiteside et al. 2007; Algeo et al. 2011; Stigall 2012; Cascales-Miñana et al. 2015; Capriolo et al. 2020; Marshall et al. 2020). Organisms cannot forestall the effects of such events. However, adaptability of surviving organisms to changed environmental conditions is crucial for the long-term evolution of organic life (see Fig. 5.18) (Ezard et al. 2011).

Fossil evidence provides only incomplete information regarding functional changes in benthic communities. Archaeans and bacteria are of central importance for energetic conversion of organic materials. However, for mechanical exploitation of food resources they depend on co-operation with animals (Metazoa). In many cases, the organisms concerned are relatively small and soft-bodied—for example representatives of segmented worms

(Annelida) or nematodes (Nematoda)—for which only sparse fossil evidence is available (Gambi et al. 2003; Baliński et al. 2013; Parry et al. 2014; Purschke et al. 2014; Chen et al. 2020). In marine habitats—together with other benthic organisms—they increase secondary biomass production not only through energetic conversion of organic material, but also by releasing the minerals it contains.

In order to understand the herewith associated significance for biological evolution in marine habitats, it is necessary to recall the functional relationships of primary production in plankton and to compare developments on land and in oceans. As already explained, the minerals needed for biomass production are obtained by marine phototrophic via different pathways. From land, mineral particles can be blown in directly by air currents or carried, in either dissolved or solid form, through precipitation, watercourses and—in polar regions—movements of glaciers (Fig. 8.15). Ideal conditions for this are provided by vegetation-free land areas with adequate quantities of weathered minerals, for example following the retreat of glaciers or arid zones.

Independently of this, internal transportation—of recycled nutrients from sediments—by upwelling water currents (Figs. 8.18 and 8.19) also contributes significantly to production of phototrophic plankton. However, the amount of mobilizable nutrients in sediments depends on the intensity of bioturbation by the infauna (Fig. 9.8) (de Lucas Pardo et al. 2013). For energetic reasons, activities of the infauna are closely linked to the proportions of organic materials in sediments (Rosenberg 2001; Smith and Rabouille 2002). Enrichment of sediments with organic components through terrestrial inputs also increases the intensity of bioturbation and hence recycling of bound mineral nutrients.

Here, the coverage is simplified because additional factors influence the interactions involved. For example, organisms can convert organic material qualitatively only with certain compositions and quantitatively only up to certain amounts per unit of space and time. In sedimentation zones, the oxygen content of the water must permit metazoans to survive—an important framework condition for which is adequate exchange between near-surface and deep water. If these conditions are not met, bioturbation collapses and energetic transformation is reduced to microbial processes alone. Macroscopically, such changes can be recognized by the formation of anaerobic sludge and ultimately anaerobic zones in water bodies.

Over the course of the Phanerozoic, conditions for the input of organic materials into oceans were generated exclusively by the evolution of terrestrial vegetation (Fig. 9.10). From the Ediacaran to the Ordovician, bioturbation was solely dependent on organic matter sinking in the water body and was thus inevitably limited.

As already explained in Sect. 8.6.1, there are no direct functional links between natural terrestrial vegetation and the erosion of land areas. However, increased leaching of mineral nutrients into aquatic systems can occur due to natural or anthropogenic disturbance of vegetation (Stark and Jordan 1978; Kopáček et al. 1995; Likens and Bormann 1995; Mander et al. 1998). Nevertheless, the associated dynamics of quantities and chemical compositions differ significantly from natural mineral inputs from erosion of vegetation-free areas by glaciers (Fig. 8.15), or abiotic and microbial processes (Born and Böcher

2001; Mladenov et al. 2011; Kammerlander et al. 2015). As a result of increased binding of minerals in the biomass of expanding terrestrial vegetation, direct input from vegetation-free land areas into oceans has therefore inevitably decreased. On the other hand, the input of organically bound minerals in dead plant parts—which cannot be utilized by photo-trophic plankton—is increasing. This is indicated by increasing organically enriched sediments from the Devonian to the Carboniferous (Willis and McElwain 2002; McMahon and Davies 2018). Increasing deposition of organic materials during the late Carboniferous—according to fossil evidence—may have induced increased bioturbation and thus increased nutrient recycling in the sediments (Ausich and Bottjer 2003; Kelley and Hansen 2003). The time delay prior to intensive bioturbation is possibly attributable to insufficient flow through deeper marine zones.

Increased bioturbation was succeeded by a significant increase in marine diversity during the Permian. However, this collapsed abruptly at the end of the epoch due to global environmental changes (Cao et al. 2009). A considerable loss of species in the oceans was probably triggered by expansion of oxygen-free, hydrogen-sulphide-enriched zones in the deep water and a marked warming of surface water (Song et al. 2014). Several million years before the final collapse, anoxic zones probably expanded in irregular cycles owing to factors that have not been conclusively clarified. Sessile organisms in deeper sea zones were particularly affected (Schneebeli-Hermann 2012; Song et al. 2014; Clarkson et al. 2016; Benton 2018). Ongoing basal volcanic eruptions in the Siberian plate (forming the Siberian Traps) played a significant role in this. The question of whether associated gaseous emissions alone triggered global changes or whether additional factors—such as additional emissions from ignited coal layers— played a role is under discussion (Heydari et al. 2008; Cao et al. 2009; Saunders and Reichow 2009; Nielsen et al. 2010; Shen et al. 2012; Burgess and Bowring 2015; He et al. 2015).

The biological effects of events and developments during the regeneration phase, which lasted around five million years, show that no deterministic regenerative processes are identifiable in marine benthic communities (Knoll et al. 2007; Scheyer et al. 2014; Brayard et al. 2017). Evidence does, however, exist for active regeneration of chemical conditions by prokaryotes (Li et al. 2013). Apart from the problem of inadequate fossil evidence (Alroy 2008), several other aspects must to be taken into consideration. The vast size of the habitat alone means that the effects outlined cannot have occurred everywhere in the same way. This points to the likely continued existence of local ecosystems. This inference is supported by rich fossil finds from deposits in Idaho (USA) formed about 1.3 million years after the event. Organisms represented include both Palaeozoic and Mesozoic species. These include a sponge species from the Cambrian that had supposedly died out 200 million years ago and cephalopods whose appearance is clearly documented only 50 million years later (Brayard et al. 2017).

Such refugia have presumably permitted the survival of various species of bony fishes and amphibians (Stereospondyli) with almost no impact (Ruta and Benton 2008; Schoch 2009; Smithwick and Stubbs 2018). Moreover, local ecosystems are ideal bases for dispersal of various organisms into vacated habitats after the disturbances have ceased (Powell et al. 2015). As already mentioned, because of the dominance of unicellular organisms, phototrophic primary production can develop very rapidly under suitable regional conditions. Mixotrophy among plankton organisms and the dominance of carnivorous life forms among marine metazoans allow for great variability in pelagic food webs.

In benthic communities, abiotic factors significantly influence the great spatiotemporal variability and randomness of food input. The evolution of benthic organisms is determined primarily by these framework conditions and only to a small extent by competition between species (Antell et al. 2020). As a result, a wide variety of life forms such as filter-feeding sponges and omnivorous sea anemones (Cnidaria)—which can dissolve and digest plant and animal food up to the size of entire bird carcasses—were able to evolve throughout the Phanerozoic (Tsurpalo and Kostina 2003; Guy et al. 2014a, b).

This makes it clear that the evolution of marine benthic communities is determined not by systematic groups, but primarily by abiotic conditions. Multidimensional interactions have led to differing assessments of species losses in scientific publications. Depending on the organism groups examined and classifications based on them, faunal changes can be assessed as relatively minor or as full-blown mass extinctions (Sepkoski 1998; Droser et al. 2000; Klug et al. 2010; McGhee et al. 2013). With increasing numbers of metazoan groups considered and new fossil finds, up to the end of the Permian only gradual changes can be recognized within the diversity of life forms that had already evolved (Alroy 2010; Friedman and Sallan 2012; Donoghue and Keating 2014; Sansom et al. 2015; Whalen and Briggs 2018; Otoo et al. 2019). Even the evolution and differentiation of jaw-bearing vertebrates did not trigger any profound faunal changes in marine habitats during this period, although there are clear indications of increased carnivory in fossils (Anderson et al. 2011; Dupret et al. 2014; Syverson et al. 2018; Randle and Sansom 2019). Prior to the end of the Permian, feather stars (Crinoidea)—members of the echinoderm group (Echinodermata), first recorded in the Ordovician—developed predominantly sessile benthic life forms. Only from the Triassic onwards did the proportion of mobile life forms increase (Baumiller and Messing 2007; Baumiller 2008).

The evolutionary processes involved were influenced not only by the widely known mass extinctions, but also by—geologically—relatively short-term changes in environmental conditions. One example of this is the Carnian Pluvial Event in the Triassic (PCE in Fig. 9.10), which occurred around 20 million years after the end of the Permian. According to current knowledge, around 232 million years ago increased volcanic activity in the large igneous provinces of Wrangellia triggered climate changes with massively increased precipitation (megamonsoons). Over a putative period of two million years, this led to large-scale erosion of land areas and extreme sediment input into oceans. One indication of this is the largest alluvial fan ever recorded, with an area of around 1.6 million square kilometres, from the Triassic in the Arctic Ocean off the current coast of Norway (Klausen et al. 2019; Dal Corso et al. 2020; Benton and Wu 2022).

Those relatively brief events—in geological terms—permitted propagation of planktonic organisms with calcareous shells and the evolution of large coral reefs in the seas. Through accumulation of sunken calcareous shells on every sea floor, sufficient capacity developed in the oceans to compensate autonomously for fluctuations in global carbon dioxide concentrations (Fig. 9.10) (Falkowski et al. 2004; Ridgwell and Zeebe 2005; Suchéras-Marx et al. 2019).

Biological effects of events in the oceans have been only partially researched to date. A clear decline in species diversity can be observed in bony fish, but the overall impact cannot yet be estimated due to incompleteness of the data available from the Early Triassic (Romano et al. 2016; Romano 2021). Effects on the reptile fauna—which migrated secondarily into oceans—are somewhat more recognizable. Over the 20 million years—especially in marine shallow water zones—a great diversity of life forms evolved, ranging from relatively small herbivores to megacarnivores with a body length of around five metres. Of these, only a few groups—such as Ichthyosauria and Sauropterygia—survived the PCE. In subsequent time periods, along with newly arriving groups, they were able to occupy vacated feeding niches with more mobile and energetically more demanding life forms (Motani 2009; Pyenson et al. 2014; Kelley and Pyenson 2015; Chun et al. 2016; Jiang et al. 2020; Spiekman et al. 2021).

9.3.5 Emigration and Multiple Returns of Tetrapods

Interest in the evolution of (mostly) four-legged land animals from fish overshadows the diversity of evolutionary adaptations of different animal groups to conditions in the boundary zone between marine and terrestrial habitats. In addition to the evolution of invertebrate animal groups that are fully adapted to terrestrial environmental conditions, a wide variety of life forms have also adapted to terrestrial-aquatic conditions in freshwater habitats. This is a development that first occurred among vertebrate fish in the Silurian (Fig. 9.10). Although various fish species are temporarily active or undergo embryonic development outside of water—for example intertidal mudskippers (Oxudercinae)—complete development of terrestrial life forms was limited to just a few species (Von Reumont et al. 2012; Martin and Carter 2013; Capobianco and Friedman 2019; van Straalen 2021; Schnetz et al. 2022; Aiello et al. 2023).

Colonization of terrestrial habitats by vertebrates in the Devonian and Carboniferous periods (Fig. 9.10) opened up new possibilities for evolution and differentiation. The diversity of the resulting life forms continues to stimulate discussions about categorization and systematic classification of fossils known to date (Ruta and Benton 2008; Tsuji and Müller 2009; Modesto et al. 2015; Laurin and Piñeiro 2017; MacDougall et al. 2018; Marjanović and Laurin 2019). However, it is relatively certain that the first quadrupedal—tetrapod—life forms, known as amphibians, moved away from water only temporarily (Clack 2012; Schoch 2014). After colonizing terrestrial habitats, representatives of tetrapod groups have repeatedly migrated back to aquatic habitats. Depending on conditions of the colonized habitat and the method of food acquisition, this involved varying degrees of morphological and physiological adaptation. Morphological differences between hippopotamuses, dolphins and baleen whales alone serve to illustrate this. Physiologically, sea snakes (e.g. *Laticauda*, *Pelamis*) are not fully adapted to marine living conditions. Because of the insufficiency of their osmotic regulation capacities, they can cover their fluid requirements only with freshwater, for example from freshwater lenses on the sea surface following precipitation (Lillywhite et al. 2019; Motani and Vermeij 2021).

Species classified as reptiles (Mesosauria) probably migrated into marine habitats only during the Permian, but then became extinct again (Modesto 2010; Piñeiro et al. 2012; Kelley and Pyenson 2015; Villamil et al. 2016; da Silva and Sedor 2017; Silva et al. 2017; Nuñez Demarco et al. 2018). It can be concluded from rapid diversification and increase in size of marine reptiles—following the major changes at the end of the Permian—that there were sufficiently abundant, largely unutilized food resources in the oceans. Although many species disappeared at the latest in association with the environmental changes at the end of the Triassic, some lineages of plesiosaurs and their relatives (Sauropterygia) continued evolving—over some 140 million years—until the end of the Cretaceous, while the ichthyosaurs (Ichthyosauromorpha) did so until the Middle Cretaceous (Benson et al. 2012; Diedrich 2013a; Pyenson et al. 2014; Kelley and Pyenson 2015; Fischer et al. 2017; Motani et al. 2017; Vermeij and Motani 2018). Until the end of the Cretaceous period about 66 million years ago, morphologically similar mosasaurs took over many ecological niches of the ichthyosaurs (Lindgren et al. 2010; Polcyn et al. 2014). Among turtles, which emerged around 250 million years ago, the first groups migrated into marine habitats in the Jurassic. Evolution of the sea turtles still found today began only around 100 million years ago, during the Cretaceous period (Bever et al. 2015; Kelley and Pyenson 2015).

Following the extinction of marine reptile groups at the end of the Cretaceous, various groups of birds and mammals migrated into the ecological niches that had thus become vacant (Fordyce and Barnes 1994; Ksepka et al. 2006; Lindberg and Pyenson 2007; Geisler et al. 2011; Pyenson et al. 2014; Kelley and Pyenson 2015; Slater et al. 2017). Morphologically, they evolved life forms analogous to those of marine reptiles, adapting to the specific conditions of their habitats (Williams 1999; Houssaye 2009; Gleiss et al. 2011; Motani et al. 2015). Energetically and ecophysiologically, new biological conditions thus developed for the persisting aquatic fauna. Among marine reptiles, at least, ichthyosaurs were able to surpass the sensory-physiological and motor performance of the aquatic fauna by means of higher energetic turnover. With the differentiation of generally homothermic aquatic birds and mammals, such effects were further intensified—especially in zones with low water temperatures. The evolution of endothermy in some species of pelagic fish, such as swordfish and tuna, is also a striking development (Cairns et al. 2008; Steeman et al. 2009; Motani 2010; Estes et al. 2016; Grady et al. 2019; Wu et al. 2021).

When average sea level fell around three million years ago, approximately 50 per cent of marine mammal species that had existed up to that point became extinct, along with about 35 per cent of seabirds and around 40 per cent of sea turtles. Species living in coastal zones were particularly affected (Pimiento et al. 2017). The giant shark known as megalodon (*Otodus megalodon*)—up to 18 metres in length— also became extinct during the same period, whereas the great white shark (*Carcharodon*) survived to the present day.

Both genera evolved in parallel with marine birds and mammals. In the genus *Otodus*, the temporal pattern of tooth size development suggests specialization in feeding. The giant shark may therefore have become extinct due to the loss of a suitable food source, whereas loss of species among marine mammals and birds cannot be attributed to a single cause (Naafs et al. 2010; Diedrich 2013b; Pimiento and Balk 2015; O'Dea et al. 2016; Boessenecker et al. 2019; Shimada 2019).

9.3.6 Smooth Transitions

Despite multiple collapses in marine biodiversity, many Cambrian life forms survived into the early Mesozoic. The last conodonts, for example, became extinct only during the Triassic (Stanley 2009). Only the massive disturbance events at the end of the Permian and at the end of the Triassic induced renewed evolution of diversity of forms among cartilaginous fish (Elasmobranchia). According to model calculations based on molecular analyses, ray-finned fishes (Actinopterygii) could have evolved as early as the beginning of the Carboniferous (Giles et al. 2017). Fossil evidence indicates a much later diversification during the Cretaceous period (Near et al. 2012; Guinot and Cavin 2015). This is probably linked to morphological changes in shell structures of marine gastropods, which indicate a significant increase in carnivorous organisms (Vermeij 1977).

In deep seas, too, repeated expansions of anoxic zones led to multiple collapses of biotic communities (Smith and Stockley 2005; Jenkyns 2010; Thuy et al. 2012; Priede and Froese 2013; Straube et al. 2015). Because of restrictive abiotic conditions—such as water pressure, low temperatures and darkness—and limited nutrient availability, many life forms remained intact over the long term, despite some biological evolution in individual groups of organisms. One example of this is the low evolutionary variability of coelacanths (order Actinista), including the genus *Latimeria*, which has existed for some 100 million years. Since the earliest evidence of this group in the Permian, only a few morphological features have changed to the present day. Far more numerous changes can be observed in bony fish (Actinopterygii) and tetrapods over the same period (Cavin and Guinot 2014). In both of those groups, considerable morphological differentiation is associated, among other things with highly structured and dynamic habitats. In bony fish, for instance, this involved differentiation of reef-building corals, the spread of seagrass beds or dynamic freshwater habitats (Teske and Beheregaray 2009; Bellwood et al. 2017; Salzburger 2018; Miller and Román-Palacios 2021). Factors influencing the differentiation of tetrapods are discussed in the next chapter.

The falling sea level and associated loss of shallow water zones must inevitably have affected all organisms in these habitats. Interestingly, published analyses of fish evolution do not show any decline in diversity during the last few millions of years of Earth's history (Guinot and Cavin 2015). One reason for this could be use of higher systematic categories that do not record changes at the species level. This means that it is not possible to state whether fish-eating species were also affected by shifts in the food spectrum. Local studies on fish and invertebrates show very different effects. While certain individual species are observed to be disappearing, in other groups species diversity is increasing due to structural changes in habitats or immigration (Kirby and Jackson 2004; Smith and Roy 2006; Murray et al. 2009; Shen et al. 2011; Avila et al. 2020). Species diversity of various pelagic fish families—such as tuna—actually increased during this period (Miya et al. 2013).

References

Abelmann A, Gersonde R, Cortese G, Kuhn G, Smetacek V (2006) Extensive phytoplankton blooms in the Atlantic sector of the glacial Southern Ocean. Paléo 21:PA1013

Abelson A, Miloh T, Loya Y (1993) Flow patterns induced by substrata and body morphologies of benthic organisms, and their roles in determining availability of food particles. Limnol Oceanogr 18(6):1116–1124

Acevedo-Trejos E, Brandt G, Merico A, Smith SL (2013) Biogeographical patterns of phytoplankton community size structure in the oceans. Glob Ecol Biogeogr 22:1060–1070

Adl SM, Simpson AGB, Lane CE, Lukeš J, Bass D et al (2012) The revised classification of eukaryotes. J Eukaryotic Microb 59(5):429–493

Adl SA, Bass D, Lane CE, Lukes J, Schoch CL et al (2019) Revisions to the classification, nomenclature, and diversity of eukaryotes. Eukaryotic Microb 66:4–119

Adolf JE, Stoecker DK, Harding LW (2006) The balance of autotrophy and heterotrophy during mixotrophic growth in Kardolinium micrum (Dinophyceae). J Plankton Res 28(8):737–751

Aiello BR, Bhamia MS, Gau J, Morris JGL, Bomar K et al (2023) The origin of blinking in both mudskippers and tetrapods is linked to life on land. PNAS 120(18):e2220404120

Alcaraz M, Marrasé C, Peters F, Arin L, Malits A (2002) Effects of turbulence conditions on the balance between production and respiration in marine communities. Mar Ecol Prog Ser 242:63–71

Algeo TJ, Scheckler SE (1998) Terrestrial-marine teleconnections in the Devonian: links between the evolution of land plants, weathering processes, and marine anoxic events. Philos Trans R Soc Lond B 353:113–130

Algeo TJ, Chen ZQ, Fraiser ML, Twitchett RJ (2011) Terrestrial-marine teleconnections in the collapse and rebuilding of early Triassic marine ecosystems. Palaeogeogr Palaeoclimatol Palaeoecol 308:1–11

Alroy J (2008) Dynamics of origination and extinction in the marine fossil record. PNAS 105(1):11536–11542

Alroy J (2010) The shifting balance of diversity among major marine animal groups. Science 329:1191–1194

Alvain S, Moulin C, Dandonneau Y, Loisel H (2008) Seasonal distribution and succession of dominant phytoplankton groups in the global ocean: a satellite view. Glob Biogeochem Cycles 22:GB3001

Amarasekare P (2008) Spatial dynamics of foodwebs. Annu Rev Ecol Evol Syst 39:479–500

Amin SA, Parker MS, Armbrust EV (2012) Interactions between diatoms and bacteria. Microb Molecular Biology Rev 76(3):667–684

Amini S, Kolle S, Petrone L, Ahanotu O, Sunny S et al (2017) Preventing mussel adhesion using lubricant-infused materials. Science 357:668–673

Anderson PSL, Friedman M, Brazeau MD, Rayfield EJ (2011) Initial radiation of jaws demonstrated stability despite faunal and environmental change. Nature 476:206–209

Andreev PD, Sansom IJ, Li Q, Zhao W, Wang J et al (2022) The oldest gnathostome teeth. Nature 609:964–968

Annenkova NV, Lavrov DV, Belikov SI (2011) Dinoflagellates associated with freshwater sponges from the ancient Lake Baikal. Protist 162:222–236

Antell GS, Kiessling W, Aberhan M, Saupe EE (2020) Marine biodiversity and geographic distributions are independent on large scales. Curr Biol 30:115–121

Armbrust EV (2009) The life of diatoms in the world's oceans. Nature 459:185–192

Aronson HS, Zellmer AJ, Goffredi SA (2017) The specific and the exclusive microbiome of the deep-sea bone-eating snail, Rubyspira osteovora. FEMS Microb Ecol 93:fiw250

Arrigo KR, van Dijken GL (2003) Phytoplankton dynamics within 37 Antarctic polynya systems. J Geophys Res 108(C8):3271

Ascani F, Richards KJ, Firing E, Grant S, Johnson KS et al (2013) Physical and biological controls of nitrate concentrations in the upper subtropical North Pacific Ocean. Deep-Sea Res 93:119–134

Ausich WL, Bottjer DJ (2003) Sessile invertebrates. In: Briggs DEG, Crowther PR (eds) Palaeobiology. Blackwell, Malden, pp 384–386

Avila SP, Azevedo JMN, Madeira P, Cordeiro R, Melo CS et al (2020) Pliocene and late Pleistocene actinopterygian fishes from Santa Maria Island, Azores (NE Atlantic Ocean): palaeoecological and palaeobiogeographical implications. Geol Mag. https://doi.org/10.1017/S0016756820000035

Avrani S, Wurtzel O, Sharon I, Sorek R, Lindell D (2011) Genomic Island variability facilitates Prochlorococcus-virus coexistence. Nature 474:604–608

Babcock LE, Peng S-C, Brett CE, Zhu M-Y, Ahlberg P et al (2015) Global climate, sea level cycles, and biotic events in the Cambrian period. Palaeoworld 24:5–15

Bailey H, Fossette S, Bogard SJ, Shillinger GL, Swithenbank AM et al (2012) Movement patterns for a critically endangered species, the leatherback turtle (Dermochelys coriacea), linked to foraging success and population status. PLoS One 7(5):e36401

Bak RPM, Joenje M, de Jong I, Lambrechts DYM, Nieuwland G (1998) Bacterial suspension feeding by coral reef benthic organisms. Mar Ecol Prog Ser 175:285–288

Baker AC (2003) Flexibility and specificity in coral-algal symbiosis: diversity, ecology, and biogeography of Symbiodinium. Annu Rev Ecol Evol Syst 34:661–689

Baker AC, Starger CJ, McClanahan TR, Glynn PW (2004) Corals' adaptive response to climate change. Nature 430:741

Baliński A, Sun Y, Dzik J (2013) Traces of marine nematodes from 470 million years old early Ordovician rocks in China. Nematology 15:567–574

Ball EE, Hayward DC, Saint R, Miller DJ (2004) A simple plan—cnidarians and the origins of developmental mechanisms. Nat Rev Genet 5:567–577

Barbara GM, Mitchell JG (2003) Bacterial tracking of motile algae. FEMS Microb 44:79–87

Barbeitos MS, Romano SL, Lasker HR (2010) Repeated loss of coloniality and symbiosis in scleractinian corals. PNAS 107(26):11877–11882

Bar-On YM, Philips R, Milo R (2018) The biomass distribution on Earth. PNAS 115(25):6506–6511

Basterretxea G, Font-Muñoz JS, Tuval I (2020) Phytoplankton orientation in a Turbulent Ocean: a microscale perspective. Front Mar Sci 7:185

Bault V, Balseiro D, Monnet C, Crônier C (2022) Post-Ordovician trilobite diversity and evolutionary faunas. Earth Sci Rev 230:104035

Baumiller TK (2008) Crinoid ecological morphology. Annu Rev Earth Planet Sci 36:221–249

Baumiller TK, Messing CG (2007) Stalked crinoid locomotion, and its ecological and evolutionary implications. Palaeontol Electron 10:1

Beardall J, Allen D, Bragg J, Finkel ZV, Flynn KJ et al (2009) Allometry and stoichiometry of unicellular, colonial and multicellular phytoplankton. New Phytol 181:295–309

Beaufort L, Lancelot Y, Camberlin P, Cayre O, Vincent E et al (1997) Insolation cycles as a major control of equatorial Indian Ocean primary production. Science 278:1451–1454

Bec B, Collos Y, Souchu P, Vaquer A, Lautier J et al (2011) Distribution of picophytoplankton and nanoplankton along an anthropogenic eutrophication gradient in French Mediterranean coastal lagoon. Aquat Microb Ecol 63:29–45

Behrenfeld M, Boss ES (2014) Resurrecting the ecological underpinnings of ocean plankton blooms. Annu Rev Mar Sci 6:167–194

Béjà O, Aravind L, Koonin EV, Suzuki MT, Hadd A et al (2000) Bacterial rhodopsin: evidence for a new type of phototrophy in the sea. Science 289:1902–1906

Bellwood DR, Goatley CHR, Bellwood O (2017) The evolution of fishes and corals on reefs: form, function and interdependence. Biol Rev 92(2):878–901

Belmonte G, Miglietta A, Rubino F, Boero F (1997) Morphological convergence of resting stages of planktonic organisms: a review. Hydrobiologia 355:159–165

Benincà E, Huisman J, Heerkloss R, Jöhnk KD, Branco P et al (2008) Chaos in a long-term experiment with a plankton community. Nature 451:822–824

Benson RBJ, Evans M, Druckenmiller PS (2012) High diversity, low disparity and small body size in plesiosaurs (Reptilia, Sauropterygia) from the Triassic-Jurassic boundary. PLoS One 7(3):e31838

Benton MJ (2018) Hyperthermal-driven mass extinctions: killing models during the Permian-Triassic mass extinction. Phil Trans R Soc A 376:20170076

Benton MJ, Wu F (2022) Triassic Revolution. Front Earh Sci 10:899541

Berbee ML, James TY, Strullu-Derrien C (2017) Early diverging fungi: diversity and impact at the dawn of terrestrial life. Ann Rev Microbiol 71:41–60

Bernal D, Dickson KA, Shadwick RE, Graham JB (2001) Analysis of the evolutionary convergence for high performance swimming in lamnid sharks and tunas. Comparative Biochem and Physiol Part A 120:695–726

Bernardino AF, Levin LA, Thurber AR, Smith CR (2012) Comparative composition, diversity and trophic ecology of sediment macrofauna at vents, seeps and organic falls. PLoS One 7(4):e33515

Bertrand EM, Allen AE (2012) Influence of vitamin B auxotrophy on nitrogen metabolism in eukaryotic phytoplankton. Front Microbiol 3:375

Bethke CM, Ding D, Jin Q, Sanford RA (2008) Origin of microbiological zoning in groundwater flows. Geology 36(9):739–742

Bever GS, Lyson TR, Field DJ, Bhullar AS (2015) Evolutionary origin of the turtle skull. Nature 525:239–242

Bianucci G, Marx FG, Collareta A, Di Stefano A, Landini W et al (2019) Rise of the titans: baleen whales became giants earlier than thought. Biol Lett 15:20190175

Biddanda B, Ogdahl M, Cotner J (2001) Dominance of bacterial metabolism in oligotrophic relative to eutrophic waters. Limnol Oceanogr 46(3):730–739

Biller SJ, Berube PM, Lindell D, Chisholm SW (2014) Prochlorococcus: the structure and function of collective diversity. Nat Rev Microbiol 13(1):13–27

Blackburn N, Fenchel T, Mitchell J (1998) Microscale nutrient patches in planktonic habitats shown by chemotactic bacteria. Science 282:2254–2256

Blieck A (1985) Paléoenvironments de Hétérostracés, Vertébrés agnathes ordoviciens à Dévoniens. Bull Mous Nat Hist Nat 4(7):143–155

Boenigk J, Arndt H (2000) Comparative studies on the feeding behaviour of two heterotrophic nano-flagellates: the filter-feeding choanoflagellate Monosiga ovata and the raptorial-feeding kineto-plastid Rhynchomonas nasuta. Aquat Microb Ecol 22:243–249

Boero F, Bouillon J, Gravili C, Miglietta MP, Parsons T, Piraino S (2008) Gelatinous plankton: irregularities rule the world (sometimes). Mar Ecol Prog Ser 356:299–310

Boessenecker RW, Ehret DJ, Long DJ, Churchill M, Martin E, Boessenecker SJ (2019) The early Pliocene extinction of the mega-toothed shark Otodus megalodon: a view from the eastern North Pacific. Peer J 7:e6088

Bond DPG, Grasby (2020) Late Ordovician mass extinction caused by volcanism, warming, and anoxia, not cooling and glaciation. Geology 48(8):777–781

Bone Q, Carré C, Chang P (2003) Tunicate feeding filters. J Mar Biol Assoc UK 83:907–919

Bonnet S, Webb EA, Panzeca C, Karl DM, Capone DG et al (2010) Vitamin B12 excretion by culture of the marine cyanobacteria Crocosphaera and Synechococcus. Limnol Oceanogr 55(5):1959–1964

Born EW, Böcher J (2001) The ecology of Greenland. Atuakkiorfik Education, Nuuk

Bowman JP (2007) Bioactive compound synthetic capacity and ecological significance of marine bacterial genus Pseudoalteromonas. Mar Drugs 5:220–241

Boyce DG, Frank KT, Leggett WC (2015) From mice to elephants: overturning the 'one size fits all' paradigm in marine plankton food chains. Ecol Lett 18:504–515. https://doi.org/10.1111/ele.12434

Boyle J, Ryan MJ (2017) New information on Titanichthys (Placodermi, Arthrodira) from the Cleveland shale (upper Devonian) of Ohia, USA. J Paleontol 91(2):318–336

Bracco A, Provenzale A, Scheuring I (2000) Mesoscale vortices and the paradox of the plankton. Proc R Soc Lond B 267:1795–1800

Bradley DC (2011) Secular trends in the geologic record and the supercontinent cycle. Earth Sci Rev 108:16–33

Brasier M, Green O, Shields G (1997) Ediacarian sponge spicule clusters from southwestern Mongolia and the origins of the Cambrian fauna. Geology 25(4):303–306

Brayard A, Krumenacker LJ, Botting JP, Jenks JF, Bylunf KG et al (2017) Unexpected early Triassic marine ecosystem and the rise of the modern evolutionary fauna. Sci Adv 3:e1602159

Brazeau MD, Friedman M (2015) The origin and early phylogenetic history of jawed vertebrates. Nature 520:490–497

Breitbart M (2012) Marine viruses: truth or dare. Annu Rev Mar Sci 4:425–448

Brey T, Clarke A (1993) Population dynamics of marine benthic invertebrates in Antarctic and subantarctic environments: are there unique adaptations? Antarct Sci 5(3):253–266

Brodeur RD, Ruzicka JJ, Steele JH (2011): Investigating alternating trophic pathways through gelatinous zooplankton and Planktivorus fishes in an upwelling ecosystem using end-to-end models. In: Omori K, Guo X, Yoshie N, Fuji N, Handoh IC, et al. (eds.): Interdisciplinary studies on environmental chemistry—marine environmental modeling and analysis. Terrapur, pp 57-63

Brown JH, Gupta VK, Li B-L, Milne BT, Restrepo C, West GB (2002) The fractal nature of nature: power laws, ecological complexity and biodiversity. Philos Trans R Soc Lond B 357:619–626

Brum JR, Hurwitz BL, Schofield O, Ducklow HW, Sullivan MB (2016) Seasonal time bombs: dominant temperate viruses affect Southern Ocean microbial dynamics. The ISME J 10:437–449

Buatois LA, Narbonne GM, Mángano MG, Carmona NB, Myrow P (2014) Ediacaran matground ecology persisted into the earliest Cambrian. Nat Commun 5:3544. https://doi.org/10.1038/ncomms4544

Buatois LA, Mángano MG, Olea RA, Wilson MA (2016) Decoupled evolution of soft and hard substrate communities during the Cambrian explosion and great Ordovician biodiversification event. PNAS 113(25):6945–6948

Buatois LA, Almond J, Mángano MG, Jensen S, Germs GJB (2018) Sediment disturbance by Ediacaran bulldozers and the roots of the Cambrian explosion. Sci Rep 8:4514

Buchan A, LeCleir GR, Gulvik CA, González JM (2014) Master recyclers: features and functions of bacteria associated with phytoplankton blooms. Nat Rev Microbiol 12:686–698

Budd GE, Jackson ISC (2016) Ecological innovations in the Cambrian and the origin of the crown group phyla. Philos Trans R Soc B 371:20150287

Buitenhuis ET, Li WKW, Vaulot D, Lomas MW, Landry MR et al (2012) Picophytoplankton biomass distribution in the global ocean. Earth Syst Sci Data 4:37–46

Bunse C, Pinhassi J (2017) Marine Bacterioplankton seasonal succession dynamics. Trends Microbiol 25(6):494–505

Burgess SD, Bowring SA (2015) High-precision geochronology confirms voluminous magmatism before, during and after Earth's most severe extinction. Sci Adv 1(7):e1500470

Butler A (1998) Acquisition and utilization of transitional metal ions by marine organisms. Science 281:207–210

Butman CA (1987) Larval settlement of soft-sediment invertebrates: the spatial scales of pattern explained by active habitat selection and the emerging role of hydrodynamical processes. Oceanogr Mar Biol Annu Rev 25:113–165

Cairns DK, Gaston AJ, Huettmann F (2008) Endothermy, ectothermy and the global structure of marine vertebrate communities. Mar Ecol Prog Ser 356:239–250

Callieri C (2008) Picophytoplankton in freshwater ecosystems: the importance of small-sized phototrophs. Fr Rev 1(1):1–28

Callieri C, Stockner JG (2002) Freshwater autotrophic picoplankton: a review. J Limnol 61(1):1–14

Canfield DE, Poulton SW, Narbonne GM (2007) Late-Neoproterozoic deep-ocean oxygenation and the rise of animal life. Science 315:92–95

Cao C, Love GD, Hays LE, Wang W, Shen S, Summons RE (2009) Biogeochemical evidence for euxinic oceans and ecological disturbance presaging the end-Permian mass extinction event. Earth Planetary Sci Let 281:188–201

Capobianco A, Friedman M (2019) Vicariance and dispersal in southern hemisphere freshwater fish clades: a palaeontological perspective. Biol Rev 94:662–699

Capriolo M, Marzoli A, Aradi LE, Callegaro S, Dal Corso J et al (2020) Deep CO2 in the end-Triassic Central Atlantic Magmatic Province. Nat Commun 11:1670. https://doi.org/10.1038/s41467-020-15325-6

Caron DA, Michaels AF, Swanberg NR, Howse FA (1995) Primary productivity by symbionts-bearing planktonic sarcodines (Acantharia, Radiolaria, foraminifera) in surface waters near Bermuda. J Plankton Res 17(1):101–129

Carpenter RC, Hackney JM, Adey WH (1991) Measurements of primary productivity and nitrogenase activity of coral reef algae in a chamber incorporating oscillatory flow. Limnol Oceanogr 36(1):40–49

Carrick HJ, Aldridge FJ, Schelske CL (1993) Wind influences phytoplankton biomass and composition in a shallow, productive lake. Limnol Oceanogr 38(6):1179–1192

Cartwright P (2003) Developmental insights into the origin of complex colonial hydrozoans. Integr Comp Biol 43:82–86

Cartwright P, Halgedahl SL, Hendricks JR, Jarrard RD, Marques AC et al (2007) Exceptionally preserved jellyfishes from the middle Cambrian. PLoS One 10:e1121

Carver CE, Mallet AL, Vercaemer B (2006) Biological synopsis of the colonial tunicates, Botryllus schlosseri and Botrylloides violaceus. Can Manuscript Rep of Fisheries and Aquatic Sci No 2747

Cascales-Miñana B, Diez JB, Gerrienne P, Cleal CJ (2015) A palaeobotanical perspective on the great end-Permian biotic crisis. Hist Biol 28:1066–1074. https://doi.org/10.1080/08912963.2015.1103237

Cathles LM, Erendi AHJ, Barrie T (1997) How Long can a hydrothermal system be sustained by a single intrusive event? Econ Geol 92:766–771

Cavin L, Guinot G (2014) Coelacanths as "almost living fossils". Front Ecol Evol 2:49

Cermeño P, Falkowski PG (2009) Controls on diatom biogeography in the ocean. Science 325:1539–1541

Chen D, Tucker ME (2003) The Frasnian-Famennian mass extinction: insights from high-resolution sequence stratigraphy and cyclostratigraphy in South China. Palaeogeogr Palaeoclimatol Palaeoecol 193:87–111

Chen X, Wei W, Wang J, Li H, Sun J et al (2019) Tide driven microbial dynamics through virus-host interactions in the estuarine ecosystem. Water Res 160:118–129

Chen H, Parry LA, Vinther J, Zhai D, Hou X, Ma X (2020) A Cambrian crown annelid reconciles phylogenomics and the fossil record. Nature 583:249–252

Cheriton OM, McManus MA, Stacey MT, Steinbuck JV (2009) Physical and biological controls on the maintenance and dissipation of a thin phytoplankton layer. Mar Ecol Prog Ser 378:55–69

Chisholm SW (1992) Phytoplankton size. In: Falkowski PG, Woodhead AD (eds) Primary productivity and biogeochemical cycles in the sea. Plenum Press, New York

Chun L, Rieppel O, Long C, Fraser NC (2016) The earliest herbivorous marine reptile and its remarkable jaw apparatus. Sci Adv 2:e1501659

Ciotti ÁM, Lewis MR, Cullen JJ (2002) Assessment of the relationships between dominant cell size in natural phytoplankton communities and the spectral shape of absorption coefficient. Limnol Oceanogr 47(2):404–417

Clack JA (2012) Gaining ground—the origin and evolution of Tetrapods. Indiana University Press, Bloomington

Clark CT, Horstmann L, de Vernal A, Jensen AM, Misarti N (2019) Pacific walrus diet across 4000 years of changing sea ice conditions. Quat Res:1–17

Clarkson E, Levi-Setti R, Horváth G (2006) The eyes of trilobites: the oldest preserved visual system. Arthropod Struct Dev 35:247–259

Clarkson MO, Wood RA, Poulton SW, Richoz S, Newton RJ et al (2016) Dynamic anoxic ferruginous conditions during the end-Permian mass extinction and recovery. Nat Commun 7. https://doi.org/10.1038/ncomms12236

Clayton S, Nagai T, Follows MJ (2014) Fine scale phytoplankton community structure across the Kuroshio front. J Plankton Res 36(4):1017–1030

Cloern JE, Dufford R (2005) Phytoplankton community ecology: principles applied in San Francisco Bay. Mar Ecol Prog Ser 285:11–28

Coatham SJ, Vinther J, Rayfield EJ, Klug C (2020) Was the Devonian placoderm Titanichthys a suspension feeder? R Soc Open Sci 7:200272

Colin SP, Costello JH, Klos E (2003) In situ swimming and feeding behaviour of eight co-occurring hydromedusae. Mar Ecol Prog Ser 253:305–309

Collard F, Gilbert B, Eppe G, Roos L, Compère P et al (2017) Morphology of the filtration apparatus of three planktivorous fishes and relation with ingested anthropogenic particles. Mar Pollut Bull 116(1–2):182–191

Collette JH, Hagadorn JW, Lacelle MA (2010) Dead in their tracks—Cambrian arthropods and their traces from intertidal sandstones of Quebec and Wisconsin. PALAIOS 25:475–486

Colley NJ, Nilsson D-E (2016) Photoreception in phytoplankton. Integr Comp Biol 56:764–775. https://doi.org/10.1093/icb/icw037

Collin R, Janis CM (1997) Morphological constraints on tetrapod feeding mechanisms: why where there no suspension-feeding marine reptiles? In: Callaway JM, Nicholls EL (eds) Ancient Reptiles. Academic, San Diego, pp 451–465

Coma R, Ribes M, Gili JM, Zabala M (1998) An energetic approach to the study of life-history traits of two modular colonial benthic invertebrates. Mar Ecol Prog Ser 162:89–103

Conley KR, Gemmell BJ, Bouquet J-M, Thompson EM, Sutherland KR (2017) A self-cleaning biological filter: how appendicularians mechanically control particle adhesion and removal. Limnol Oceanogr 63(2):927–938

Conley KR, Lombard F, Sutherland KR (2018) Mammoth grazers on the ocean's minuteness: a review of selective feeding using mucous meshes. Proc R Soc B 285:20180056

Conway Morris S, Caron J-B (2014) A primitive fish from the Cambrian of North America. Nature 512:419–422

Costello MJ, Chaudhary C (2017) Marine biodiversity, biogeography, Deep-Sea gradients, and conservation. Curr Biol 27:R511–R527

Craig J, Jamieson AJ, Bagley PM, Priede IG (2011) Naturally occurring bioluminescence on the deep-sea floor. J Mar Syst 88:563–567

Cunningham JA, Liu AG, Bengtson S, Donoghue PCJ (2016) The origin of animals: can molecular clocks and the fossil record be reconciled? BioEssays 39(1):1–12

Currie WJS, Roff JC (2006) Plankton are not passive tracers: plankton in a turbulent environment. J Geophys Res 111:C05S07

Cury P, Bakun A, Crawford RJM, Jarre A, Quiñones RA et al (2000) Small pelagics in upwelling systems: patterns of interaction and structural changes in "wasp-waist" ecosystems. ICES J of Marine Sci 57:603–618

Cusick KD, Widder EA (2014) Intensity differences in bioluminescent dinoflagellates impact foraging efficiency in a nocturnal predator. Bull Mar Sci 90(3):797–811

Czypionka T, Vargas CA, Silva N, Daneri G, González HE, Iriarte JL (2011) Importance of mixotrophic nanoplankton in Aysén Fjord (southern Chile) during austral winter. Cont Shelf Res 31:216–224

D'Hondt S, Rutherford S, Spivack AJ (2002) Metabolic activity of subsurface life in Deep-Sea sediments. Science 295:2067–2070

d'Ovidio F, De Monte S, Alvain S, Dandonneau Y, Lévy M (2010) Fluid dynamical niches of phytoplankton types. PNAS 107(43):18366–18370

da Silva RC, Sedor FA (2017) Mesosaurid swim traces. Front Ecol Evol 5:22

Dafforn KA, Lewis JA, Johnston EL (2011) Antifouling strategies: history and regulation, ecological impacts and mitigation. Mar Pollut Bull 62:453–465

Dal Corso J, Bernardi M, Sun Y, Song H, Seyfullah LJ et al (2020) Extinction and dawn of the modern world in the Carnian (late Triassic). Sci Adv 6:eaba0099

Daley AC, Paterson JR, Edgecombe GD, García-Bellido DC, Jago JB (2013) New anatomical information on Anomalocaris from the Cambrian Emu Bay shale of South Australia and a reassessment of its inferred predatory habits. Palaeontology 56:971–990. https://doi.org/10.1111/pala.12029

Danielopol DL (1976) The distribution of the Fauna in the interstitial habitats of riverine sediments of the Danube and the Piesting (Austria). Int J Speleol 8:23–51

Daniels CJ, Poulton AJ, Esposito M, Paulsen ML, Bellerby R et al (2015) Phytoplankton dynamics in contrasting early stages North Atlantic spring blooms: composition, succession, and potential drivers. Biogeosciences 12:2395–2409

Danise S, Twichett RJ, Matts K (2014) Ecological succession of a Jurassic shallow-water ichthyosaur fall. Nat Commun 5:4789. https://doi.org/10.1038/ncomms5789

Danovaro R, Dell'Anno A, Corinaldesi C, Magagnini M, Noble R et al (2008) Major viral impact on the functioning of benthic deep-sea ecosystems. Nature 454:1084–1087

Danovaro R, Dell'Anno A, Pusceddu A, Gambi C, Heiner I, Kristensen RM (2010) The first metazoa living in permanently anoxic conditions. BMC Biol 8:30

Danovaro R, Snelgrove PVR, Tyler P (2014) Challenging the paradigms of deep-sea ecology. Trends Ecol Evol 29(8):465–475

Danovaro R, Corinaklesi C, Dell'Anno A, Snelgrove PVR (2017) The deep-sea under global change. Curr Biol 27:R461–R465

Davis MP, Fielitz C (2010) Estimating divergence times of lizardfishes and their allies (Euteleostei: Aulopiformes) and the timing of deep-sea adaptations. Mol Phylogenet Evol 57:1194–1208

Davis MP, Holcroft NI, Wiley EO, Sparks JS, Smith WL (2014) Species-specific bioluminescence facilitates speciation in the deep sea. Mar Biol 161:1139–1148

Davis MP, Sparks JS, Smith WL (2016) Repeated and widespread evolution of bioluminescence in marine fishes. PLoS One 11(6):e0155154

de Busserolles F, Cortesi F, Helvik JV, Davies WIL, Templin RM et al (2017) Pushing the limits of photoreception in twilight conditions: the rod-like cone retina of the deep-sea pearlsides. Sci Adv 3:eaao4709

de Graaf IEM, Gleeson T, van Beek LPH, Sutanudjaja EH, Bierkens MFP (2019) Environmental flow limits to global groundwater pumping. Nature 574:90–94

de Lucas Pardo MA, Bakker M, van Kessel T, Cozzoli F, Winterwerp JC (2013) Erodibility of soft freshwater sediments in Markermeer: the role of bioturbation by meiobenthic fauna. Ocean Dyn 63:1137–1150

de Vargas C, Adie S, Henry N, Develle J, Mahé F et al (2015) Eukaryotic plankton diversity in the sunlit ocean. Science 348:1261605

Decelle J, Carradec Q, Pochon X, Henry N, Romac S et al (2018) Worldwide occurrence and activity of the reef-building coral. Symbiont Symbiodinium in the Open Ocean. Curr Biol 28:3625–3533

Derelle E, Ferraz C, Escande M-L, Eychenié S, Cooke R et al (2008) Life-cycle and genome of OtV5, a large DNA virus of the pelagic marine unicellular Green alga Ostreococcus tauri. PLoS One 3(5):e2250

Dickes TD (1991) The emergence of concurrent high-resolution physical and bio-optical measurements in the upper ocean and their application. Rev Geophys 29(3):383–413

Diedrich CG (2013a) Shallow marine Sauropterygian reptile biodiversity and change in the Bad Sulza formation (Illyrian, middle Triassic) of Central Germany, and a contribution to the evolution of Nothosaurus in the Germanic Basin. In: Tanner LH, Spielmann DS, Lucar SG (eds) The Triassic system, New Mexico Mus of Nat Hist and Sci bull, vol 61, pp 132–158

Diedrich CG (2013b) Evolution of white and megatooth sharks, and evidence for early predation on seals, sirenians, and whales. Nat Sci 5(11):1203–1218

Distel DL, Beaudoin DJ, Morrill W (2002) Coexistence of multiple Proteobacterial endosymbionts in the gills of the Wood-boring bivalve Lyrodus pedicellatus (Bivalvia: Teredinidae). Appl and Environ Microb 68(12):6292–6299

Dobbs FC, Guckert JB (1988) Callianassa trilobata (Crustacea: Thalassinidea) influences abundance of meiofauna and biomass, composition, and physiological state of microbial communities within its burrow. Marine Ecology—Progress Series 45:59–79

Doebeli M, Ispolatov I (2014) Chaos and unpredictability in evolution. Evolution 68(5):1365–1373

Doeller JE, Gaschen BK, Parrino V, Kraus DW (1999) Chemolithoheterotrophy in a metazoan tissue: sulphide supports cellular work in ciliated mussel gills. J Exp Biol 202:1933–1961

Donoghue PCJ, Keating JN (2014) Early vertebrate evolution. Palaeontology 57(5):879–893

dos Reis M, Thawomwattana Y, Angelis K, Telford MJ, Donoghue PCJ, Yang Z (2015) Uncertainty in the timing of origin of animals and the limits of precision in molecular timescales. Curr Biol 25:2939–2950

Doubell MJ, Prairie JC, Yamazaki H (2014) Millimeter scale profiles of chlorophyll fluorescence: deciphering the microscale spatial structure of phytoplankton. Deep-Sea Res II 101:207–215

Droser ML, Bottjer DJ, Sheehan PM, McGhee GR Jr (2000) Decoupling of taxonomic and ecologic severity of Phanerozoic marine mass extinctions. Geology 28(8):675–678

Dumont HJ (1994) The distribution and ecology of the fresh- and brackish-water medusa of the world. Hydrobiologia 272:1–12

Dunn CW (2005) Complex Colony-level organization of the deep-sea Siphonophore Bargmannia elongata (Cnidaria, hydrozoa) is directionally asymmetric and arises by the subdivision of pro-buds. Dev Dyn 234:835–845

Dunn CW, Wagner GP (2006) The evolution of colony-level development in the Siphonophora (Cnidaria: hydrozoa). Dev Genes Evol 216:743–754

Dunn CW, Puch PR, Haddock SHD (2005) Molecular Phylogenetics of the Siphonophora (Cnidaria), with implications for the evolution of functional specialization. Syst Biol 54(6):916–935

Duplisea DE, Drgas A (1999) Sensitivity of a benthic. Metazoan, biomass size spectrum to differences in sediment granulometry. Mar Ecol Prog Ser 177:73–81

Dupret V, Sanchez S, Goujet D, Tafforeau P, Ahlberg PE (2014) A primitive placoderm sheds light on the origin of the jawed vertebrate face. Nature 507:500–503

Dzik J (2007) The Verdun syndrome: simultaneous origin of protective Armour and infaunal shelters at the Precambrian-Cambrian transition. In: Vickers-Rich P, Komarower P (eds) The rise and fall of the Ediacaran biota. Geological Society, London. Special Publication 286, pp 405–414

Edwards KF, Thomas MK, Klausmeier CA, Litchman E (2012) Allometric scaling and taxonomic variation in nutrient utilization traits and maximum growth rate of phytoplankton. Limnol Oceanogr 57(2):554–566

Eiler A (2006) Evidence for the ubiquity of Mixotrophic bacteria in the Upper Ocean: implications and consequences. Appl Environ Microb 72(12):7431–7437

Eiting TP, Smith GR (2007) Miocene salmon (Oncorhynchus) from Western North America: gill Raker evolution correlated with plankton productivity in the eastern Pacific. Palaeogeogr Palaeoclimatol Palaeoecol 249:412–424

Ekdale EG, Deméré TA (2022) Neurovascular evidence for a co-occurrence of teeth and baleen in an Oligocene mysticete and the transition to filter-feeding in baleen whales. Zool J Linnean Soc 194:395–415

Ellegaard M, Moestrup Ø, Andersen TJ, Lundholm N (2016) Long-term survival of haptophyte and prasinophyte resting stages in marine sediments. Eur J Phycol 51:328–337

Erpenbeck D, Weier T, de Voogd NJ, Wörheide G, Sutcliffe P et al (2011) Insights into the evolution of freshwater sponges (Porifera: Demospongiae: Spongillina): barcoding and phylogenetic data from Lake Tanganyika endemics indicate multiple invasions and unsettle existing taxonomy. Mol Phylogenet Evol 61:231–236

Erwin DH (2015) Early metazoan life: divergence, environment and ecology. Philos Trans R Soc B 370:20150036

Estes JA, Tinker MT, Williams TM, Diak DF (1998) Killer whales predation on sea otters linking oceanic and nearshore ecosystems. Science 282:473–476

Estes JA, Danner EM, Doak DF, Konar B, Springer AM et al (2004) Complex trophic relations in kelp forest ecosystems. Bull Mar Sci 74(3):621–638

Estes JA, Heithaus M, McCauley DJ, Rasher DB, Worm B (2016) Megafaunal impacts on structure and function of ocean ecosystems. Annu Rev Environ Resour 41:83–116

Estrada M, Berdalet E (1997) Phytoplankton in a turbulent world. Sci Mar 61(1):125–140

Evans MA, MacIntyre S, Kling GW (2008) Internal wave effects on photosynthesis: experiments. Theory, and modelling. Limnol Oceanogr 53(1):339–353

Evans C, Pearce I, Brussaard CPD (2009) Viral-mediated lysis of microbes and carbon release in the sub-Antarctic and polar frontal zones of the Australian Southern Ocean. Environ Microbiol 11(11):2924–2934

Ezard THG, Aze T, Pearson PN, Purvis A (2011) Interplay between changing climate and species' ecology drives macroevolutionary dynamics. Science 332:349–351

Fahnenstiel GL, Scavia D, Lang GA, Saylor JH, Miller GS, Schwab DJ (1988) Impact of inertial period internal waves on fixed-depth primary production estimates. J Plankton Res 10(1):77–87

Falkowski PG, Katz ME, Knoll AH, Quigg A, Raven JA et al (2004) The evolution of modern eukaryotic phytoplankton. Science 305:354–360

Fang J, Zhang L, Bazylinski DA (2010) Deep-sea piezosphere and piezophiles: geomicrobiology and biogeochemistry. Trends Microbiol 18(9):413–422

Fautin DG (2009) Structural diversity, systematics, and evolution of cnidae. Toxicon 54:1054–1064

Fedonkin MA, Michelson EL (2002) Middle Proterozoic (1.5 Ga) Horodyskia moniliformis Yochelson and Fedonkin, the oldest known tissue-grade colonial Eucaryote. Smithson Contrib Paleobiol 48:1–29

Fenchel T, Finlay BJ (2006) The diversity of microbes: resurgence of the phenotype. Phil Trans R Soc A 361:1965–1973

Feng X, Tang K-H, Blankenship RE, Tang YJ (2010) Metabolic flux analysis of the mixotrophic metabolisms in the green sulfur bacterium Chlorobaculum tepidum. J Biol Chem 285:39544–39550

Fennel K (2001) The generation of phytoplankton patchiness by mesoscale current patterns. Ocean Dyn 52:58–70

Ferrón HG, Botella H (2017) Squamation and ecology of thelodonts. PLoS One 12(2):e0172781

Figueroa RI, Rengefors K (2006) Life cycle and sexuality of the freshwater Raphidophyte Gonyostomum semen (Raphidophyceae). J Phycol 42:859–871

Figueroa RI, Bravo I, Garcés E (2006) Multiple routes of sexuality in Alexandrinum taylori (Dinophyceae) in culture. J Phycol 42:1028–1039

Figueroa RI, Bravo I, Garcés E (2008) The significance of sexual versus asexual cyst formation in the life cycle of the noxious dinoflagellate Alexandrinum peruvianum. Harmful Algae 7:653–663

Finkel ZV, Beardall J, Flynn KJ, Quigg A, Rees TAV, Raven JA (2010) Phytoplankton in a changing world: cell size and elemental stoichiometry. J Plankton Res 32(1):119–137

Finn RN, Rønnestad I, van der Meeren T, Fyhn HJ (2002) Fuel and metabolic scaling during the early life stages of the Atlantic cod Gadus morhua. Mar Ecol Prog Ser 243:217–234

Fischer V, Benson RBJ, Zverkov NG, Soul LC, Arkhangelsky MS et al (2017) Plasticity and convergence in the evolution of short-necked plesiosaurs. Curr Biol 27:1667–1676

Fish FE, Hui CA (1991) Dolphin swimming—a review. Mammal Rev 21(4):181–195

Flombaum P, Gallegos JL, Gordillo RA, Rincón J, Zabala LL et al (2013) Present and future global distributions of the marine cyanobacteria Prochlorococcus and Synechococcus. PNAS 110(24):9824–9829

Flores BM, Holmgren M (2021) White-Sand savannas expand at the Core of the Amazon after Forest wildfires. Ecosystems 24:1624–1637

Flynn KJ, Stoecker DK, Mitra A, Raven JA, Glibert PM et al (2013) Misuse of the phytoplankton—zooplankton dichotomy: the need to assign organisms as mixotrophs within plankton functional types. J Plankton Res 35(1):3–11

Fordyce RE, Barnes LG (1994) The evolutionary history of whales and dolphins. Annu Rev Earth Planet Sci 22:419–455

Forsström L, Roiha T, Rautio M (2013) Response of microbial food web to increased allochthonous DOM in an oligotrophic subarctic lake. Aquat Microb Ecol 68:171–184

Fortuny J, Marcé-Nogué J, de Esteban-Trivigno S, Gili L, Galobart A (2011) Temnospondyli bite club: ecomorphological patterns of the most diverse group of early tetrapods. J Evol Biol 24:2040–2054

Frank KT, Petrie B, Choi JS, Leggett WC (2005) Trophic cascades in a formerly cod-dominated ecosystem. Science 308:1621–1623

Freiwald A, Fosså JH, Grehan A, Koslow T, Roberts M (2004) Cold-water coral reefs. UNEP WMC Bioseries, 22

Frenken T, Wolinska J, Tao Y, Rohrlack T, Agha R (2020) Infection of filamentous phytoplankton by fungal parasites enhances herbivory in pelagic food webs. Limnol Oceanogr 65:2618–2626

Friedman M (2012) Parallel evolution trajectories underlie the origin of giant suspension feeding whales and bony fishes. Proc R Soc B 279:944–951

Friedman M (2015) The early evolution of ray-finned fishes. Palaeontology 58(2):213–228

Friedman M, Sallan LC (2012) Five hundred million years of extinction and recovery: a Phanerozoic survey of large-scale diversity pattern in fishes. Palaeontology 55(4):707–742

Friedman M, Shimada K, Martin LD, Everhart MJ, Liston J et al (2010) 100-million-year dynasty of Giant Planktivorous bony fishes in the Mesozoic seas. Science 327:990–993

Friedrichs MAM, Hofmann EE (2001) Physical control of biological processes in the central equatorial Pacific Ocean. Deep-Sea Res I 48:1023–1069

Frisch AJ, Ireland M, Rizzary JR, Lönnstedt OM, Magnenat KA et al (2016) Reassessing the trophic role of reef sharks as apex predators on coral reefs. Coral Reefs 35:35/459–35/472

Fuchs HL, Franks PJS (2010) Plankton community properties determined by nutrients and size-selective feeding. Mar Ecol Prog Ser 413:1–15

Fuhrman JA, Cram JA, Needham DM (2015) Marine microbial community dynamics and their ecological interpretation. Nat Rev Microbiol 13:133–146

Fuji T, Jamieson AJ, Solan M, Bagley PM, Priede IG (2010) A large aggregation of Liparids at 7703 meters and a reappraisal of the abundance and diversity of Hadal fishes. BioSci 60(7):506–515

Gabriele M, Bellot A, Gallotti D, Brunetti R (1999) Sublittoral hard substrate communities of the northern Adriatic Sea. Cah Biol Mar 40:65–76

Gage JD, Tyler PA (1999) Deep sea biology. Cambridge University Press, Cambridge

Gai Z, Yu X, Zhu M (2017) The evolution of the zygomatic Bone from Agnatha to Tetrapoda. Anat Rec 300:16–29

Gambi C, Vanreusel A, Danovaro R (2003) Biodiversity of nematode assemblages from deep-sea sediments of the Atacama slope and trench (South Pacific Ocean). Deep-Sea Res I 50:103–117

Garcia-Pichel F, Belnap J, Neuer S, Schanz F (2003) Estimates of global cyanobacterial biomass and its distribution. Algol Stud 109:213–227

Garrison T, Ellis R (2016) Oceanography. Cengage Learning, Boston

Gavelis GS, Wakeman KC, Tillmann U, Ripken C, Mitaral S et al (2017) Microbial arms race: ballistic "nematocysts" in dinoflagellates represent a new extreme in organelle complexity. Sci Adv 3:e1602552

Geider RJ, Delucia EH, Falkowski PG, Finzi AC, Grime JP et al (2001) Primary productivity of planet earth: biological determinants and physical constraints in terrestrial and aquatic habitats. Glob Chang Biol 7:849–882

Geisler JH, McGowen MR, Yang G, Gatesy J (2011) A supermatrix analysis of genomic, morphological, and palaeontological data from crown Cetacea. BMC Evol Biol 11:112

Gemmell BJ, Costello JH, Colin SP, Stewart CJ, Dabiri JO et al (2013) Passive energy recapture in jellyfish contributes to propulsive advantage over other metazoans. PNAS 110(44):17904–17909

Ghiglione J-F, Galand PE, Pommier T, Pedrós-Alió C, Maas EE et al (2012) Pole-to-pole biogeography of surface and deep marine bacterial communities. PNAS 103(43):17633–17638

Gilbert F, Aller RC, Hulth S (2003) The influence of macrofaunal burrow spacing and diffuse scaling on sedimentary nitrification and denitrification: an experimental simulation and model approach. J Mar Res 61(1):101–125

Gilbert JA, Steele JA, Caporaso JG, Steinbrück L, Reeder J et al (2012) Defining seasonal marine microbial community dynamics. ISME J 6:298–308

Giles S, Xu GH, Near TJ, Friedman M (2017) Early members of 'living fossil' lineage imply later origin of modern ray-finned fishes. Nature 549:265–268

Ginger ML, Portman N, McKean PG (2008) Swimming with protists: perception, motility and flagellum assembly. Nature Microb 6:838–850

Gleiss AC, Jorgensen SJ, Liebsch N, Sala JE, Norman B et al (2011) Convergent evolution in locomotory patterns of flying and swimming animals. Nat Commun 2:352. https://doi.org/10.1038/ncomms1350

Gobler CJ, Deonarine S, Leigh-Bell J, Gastrich MD, Anderson OR, Wilhelm SW (2004) Ecology of phytoplankton communities dominated by Aureococcus anophagefferens: the role of viruses, nutrients, and microzooplankton grazing. Harmful Algae 3:471–483

Goffredi SK, Johnson SB, Vrijenhoek RC (2007) Genetic diversity and potential function of microbial symbionts associated with newly discovered species of Osedax Polychaete Worms. Appl and Environ Microbiol 73(7):2314–2323

Goldberg WM (2013) The biology of reefs and reef organisms. The University of Chicago Press, Chicago

Goldbogen JA, Pyenson ND, Shadwick RE (2007) Big gulps require high drag for fin whale lunge feeding. Mar Ecol Prog Ser 349:289–301

Goldbogen JA, Calambokidis J, Croll DA, Harvey JT, Newton KM et al (2008) Foraging behaviour of humpback whales: kinematic and respiratory patterns suggest a high cost for a lunge. J Exp Biol 211:3712–3719

Goldbogen JA, Calambokidis J, Piferrer E, Potvin J, Pyenson ND et al (2011) Mechanics, hydrodynamics and energetics of blue whale lunge feeding: efficiency dependence on krill density. J Exp Biol 214:131–140

Gómez-Consarnau L, Akram N, Lindell K, Pedersen A, Neutze R et al (2010) Proteorhodopsin phototrophy promotes survival of marine bacteria during starvation. PLoS Biol 8(4):e1000358

Gorelick SM, Zheng C (2015) Global change and the groundwater management challenge. Water Ressour Res 51:3031–3051

Gorsky G, Chrétiennot-Dinet MJ, Blanchot J, Palazzoli I (1999) Picoplankton and nanoplankton aggregation by appendicularians: Fecal pellet contents of Megalocerus huxleyi in the equatorial Pacific. J Geophys Res 104(C2):3381–3390

Goudemand N, Orchard MJ, Urdy S, Bucher H, Tafforeau P (2011) Synchrotron-aided reconstruction of the conodont feeding apparatus and implications for the mouth of the first vertebrates. PNAS 108(21):8720–8724

Grady JM, Maitner BS, Winter AS, Kaschner K, Tittensor DP et al (2019) Metabolic asymmetry and the global diversity of marine predators. Science 363:eaa4220

Green RH, Lowe RJ, Buckley ML (2018) Hydrodynamics of a tidally forced coral reef atoll. J Geophys Res Oceans 123:7084–7101

Greer AT, Woodson CB, Guigand CM, Cowen RK (2016) Larval fishes utilize Batesian mimicry as a survival strategy in the plankton. Mar Ecol Prog Ser 551:1–12

Grob C, Ulloa O, Claustre H, Huot Y, Alacón G, Marie D (2007) Contribution of picoplankton to the total particulate organic carbon concentration in the eastern South Pacific. Biogeosciences 4:837–852

Groh S (2014) Patterns of diversification in osteostracan evolution. Examensarbete Univ, Uppsala

Grossman A (2016) Nutrient acquisition: the generation of bioactive vitamin B12 by microalgae. Curr Biol 26:R319–R337

Guardiola FA, Cuesta A, Meseguer J, Esteban MA (2012) Risks of using antifouling biocides in aquaculture. Int Mol Sci 13:1541–1560

Guinot G, Cavin L (2015) 'Fish' (Actinopterygii and Elasmobranchii) diversification patterns through deep time. Biol Rev 91(4):950–981

Gurung K, Field KJ, Batterman SA, Goddéris Y, Donnadieu Y et al (2022) Climate windows of opportunity for plant expansion during the Phanerozoic. Nat Commun 13:4530

Guy LS, Habecker LB, Oxwang G (2014a) Giant green anemones consume seabird nestlings on the Oregon coast. Mar Ornithol 42:1–2

Guy L, Saw JH, Ettema TJG (2014b) The archaeal legacy of eukaryotes: a Phylogenomic perspective. Cold Spring Harb Perspect Biol 4(6):a016022

Haddock SHD (2007) Comparative feeding behaviour of planktonic ctenophores. Integr Comp Biol 47(6):847–853

Haddock SHD, Moline MA, Case JF (2010) Bioluminescence in the sea. Annu Rev Mar Sci 2:443–493

Hammer AC, Pitchford JW (2005) The role of mixotrophy in plankton bloom dynamics, and the consequences for productivity. ICES J Marine Sci 62:833–840

Hannisdal B, Peters SE (2011) Phanerozoic earth system evolution and marine biodiversity. Science 334:1121–1124

Harfoot MBJ, Newbold T, Tittensor DP, Emmott S, Hutton J et al (2014) Emergent global patterns of ecosystem structure and function from a mechanistic general ecosystem model. PLoS Biol 12(4):e1001841

Härnström K, Ellegaard M, Andersen TJ, Godhe A (2011) Hundred years of genetic structure in a sediment revived diatom population. PNAS 108(10):4252–4257

Harper DAT, Lefebvre B, Percival IG, Servais T (2023) The Ordovician system: key concepts, events and its distribution across Europe. In: Harper DAT, Lefebvre B, Percival IG, Servais T (eds) A global synthesis of the Ordovician system: part I. Geol Soc, London, Spec Pub 532, pp 1–11

Hartmann M, Zubkov MV, Scanlan DJ, Lepère C (2013) In situ interactions between photosynthetic picoeukaryotes and bacterioplankton in the Atlantic Ocean: evidence for mixotrophy. Environ Microb Rep 5(6):835–840

Hasting A (2004) Transients: the key to long-term ecological understanding? Trends Ecol Evol 19(1):39–45

Hatcher BG (1990) Coral reef primary productivity: a hierarchy of pattern and process. TREE 5(5):149–155

Hays GC, Doyle TK, Houghton JDR (2018) A paradigm shift in the trophic importance of jellyfish? Trend Ecol Evol 33(11):874–884

He W-H, Shi GR, Twitchett RJ, Zhang Y, Zhang K-X et al (2015) Late Permian marine ecosystem collapse began in deeper waters: evidence from brachiopod diversity and body size changes. Geobiology 13:123–138

Heaslip SG, Iverson SJ, Bowen WD, James MC (2012) Jellyfish support high energy intake to Leatherback Sea turtles (Dermochelys coriacea): video evidence from animal-borne cameras. PLoS One 7(3):e33259

Hehenberger E, Imanian B, Burki F, Keeling PJ (2014) Evidence for the retention of two evolutionary distinct plastids in dinoflagellates with diatom endosymbionts. Genome Biol 6(9):2321–2334

Helfman GS, Collette BB, Facey DE, Bowen BW (2009) The diversity of fishes. Wiley-Blackwell, Chichester

Helliwell KE, Lawrence AD, Holzer A, Kudahl UJ, Sasso S et al (2016) Cyanobacteria and eukaryotic algae use different chemical variants of vitamin B12. Curr Biol 26:999–1008

Henschke N, Everett JD, Richardson AJ, Suthers IM (2016) Rethinking the role of Salps in the ocean. Trends Ecol Evol 31(9):720–733

Henson SA, Thomas AC (2007a) Phytoplankton scales of variability in the California current system: 1. Interannual and cross-shelf variability. J Geophys Res 112:C07017

Henson SA, Thomas AC (2007b) Phytoplankton scales of variability in the California current system: 2. Latitudinal variability. J Geophys Res 112:C07018

Herrero A, Stavans J, Flores E (2016) The multicellular nature of filamentous heterocyst-forming cyanobacteria. FEMS Microb Rev fuw029(40):831–854

Hewson I, Steele JA, Capone DG, Fuhrman JA (2006) Temporal and spatial scales of variation in bacterioplankton assemblages of oligotrophic surface waters. Mar Ecol Prog Ser 311:67–77

Heydari E, Arzani N, Hassanzadeh J (2008) Mantle plume: the invisible serial killer—application to the Permian-Triassic boundary mass extinction. Palaeogeogr Palaeoclimatol Palaeoecol 264:147–162

Hoecker-Martínez MS, Smyth WD (2012) Trapping of gyrotactic organisms in an unstable shear layer. Cont Shelf Res 36:8–18

Hoffmann LJ, Peeken I, Lochte K, Assmy P, Veldhuis M (2006) Different reactions of Southern Ocean phytoplankton size classes to iron fertilization. Limnol Oceanogr 31(3):1217–1229

Holt RD (2009) Bringing the Hutchinsonian niche into the 21st century: ecological and evolutionary perspectives. PNAS 106(2):19659–19665

Holzer M, Primeau FW (2013) Global teleconnection in the oceanic phosphorus cycle: patterns, paths, and timescale. J Geophys Res 118:1775–1796

Hopkins MJ (2013) Decoupling of taxonomic diversity and morphological disparity during decline of the Cambrian trilobite family Pterocephaliidae. J Evol Biol 26:1665–1676

Horne EPW, Platt T (1984) The dominant space and time scales of variability in the physical and biological fields on continental shelves. Rapp P-v Réun Cons Int Explor Mer 183:8–19

Houde ED (1994) Differences between marine and freshwater fish larvae: implications for recruitment. ICES J Mar Sci 51:91–97

Houssaye A (2009) "Pachyostosis" in aquatic amniotes: a review. Integr Zool 4:325–340

Hu M, Zhang X, Fan C-M, Zheng Y (2020) Lineage dynamics of the endosymbiotic cell type in the soft coral Xenia. Nature 582:534–538

Huff WA, Dronov AV, Sell B, Kanygin AV, Gonta TV (2014) Traces of explosive volcanic eruptions in the upper Ordovician of the Siberian platform. Estonian J Earth Sci 63(4):244–250

Hughes RN (2005) Lessons in modularity: the evolutionary ecology of colonial invertebrates. Sci Mar 69(1):169–179

Hughes NC (2007) The evolution of trilobite body patterning. Annu Rev Earth Planet 35:402–434

Huisman J, Weissing FJ (2001) Fundamental unpredictability in multispecies competition. Am Nat 157(5):488–494

Huisman J, Thi NNP, Karl DM, Sommeijer B (2006) Reduced mixing generates oscillations and chaos in the oceanic deep chlorophyll maximum. Nature 439:322–325

Humphreys WF (2006) Aquifers: the ultimate groundwater-dependent ecosystems. Aust J Botany 54:115–132

Humphries S, Ruxton GD (2002) Why did some ichthyosaurs have such large eyes? J Exp Biol 205:439–441

Hutchings JA (2000) Collapse and recovery of marine fishes. Nature 406:882–885

Huusgaard RS, Vismann B, Kühl M, Macnaugton M, Colmander V et al (2012) The potent respiratory system of Osedax mucofloris (Siboglinidae, Annelida)—a prerequisite for the origin of Bone-eating Osedax? PLoS One 7(4):e35975

Iannicelli M (2017) Solving the mystery of endless life between conodonts and lampreys, plus a reason for final extinction of the conodonts. J Oceanogr Marine Res. https://doi.org/10.4172/2572-3103

Iglesias-Rodríguez MD, Brown CW, Doney SC, Kleypas J, Kolber D et al (2002) Representing key phytoplankton functional groups in ocean carbon cycle models: Coccolithophorids. Glob Biogeochem Cycles 16(4):1100

Inagaki F, Hinrichs K-U, Kubo Y, Bowles MW, Heuer VB et al (2015) Exploring deep microbial life in coal-bearing sediment down to ≈2.5 km below the ocean floor. Science 349:420–424

Irigolen X, Köevjer TA, Røstad A, Martinez U, Boyra G et al (2014) Large mesopelagic fishes biomass and trophic efficiency in the open ocean. Nat Commun 5:3271. https://doi.org/10.1038/ncomms4271

Irwin AJ, Finkel ZV, Schofield OME, Falkowski PG (2006) Scaling-up from nutrient physiology to the size-structure of phytoplankton communities. J Plankton Res 28(5):459–471

Iversen KB, Primigerio R, Larsen A, Egge JK, Peters F et al (2010) Effects of small-scale turbulence on lower trophic levels under different nutrient conditions. J Plankton Res 32(2):197–208

Jackson JA, Baker CS, Vant M, Steel DJ, Medrano-González L, Palumbi SR (2009) Big and slow: phylogenetic estimates of molecular evolution of baleen whales (suborder Mysticeti). Mar Biol Evol 26(11):2427–2440

Jacobson DM (1999) A brief history of dinoflagellate feeding research. J Eukaryot Microbiol 46(4):376–381

Jacobson DM, Anderson DM (1996) Widespread phagocytosis of ciliates and other protists by marine mixotrophic and heterotrophic thecate dinoflagellates. J Phycol 32:279–285

Jamieson AJ, Gebruk A, Fujii T, Solan M (2011) Functional effects of the hadal sea cucumber Elpidia atakama (Echinodermata: Holothuroidea, Elasipodida) reflect small-scale patterns of resource availability. Mar Biol 158:2695–2703

Jansson M, Blomqvist P, Jonsson A, Bergström A-K (1996) Nutrient limitation of bacterioplankton, autotrophic and mixotrophic phytoplankton, and heterotrophic nanoflagellates in Lake Örträsket. Limnol Oceanogr 41(7):1552–1559

Janvier P (2007) Living primitive fishes and fishes from deep time. In: McKenzie DJ, Farrell AP, Brauner CJ (eds) Primitive Fishes. Elsevier, Amsterdam, pp 1–54

Jarnagin ST, Kerfoot WC, Swan BK (2004) Zooplankton life cycles: direct documentation of pelagic births and deaths relative to diapausing egg production. Limnol Oceanogr 49(4):1317–1332

Jékely G (2009) Evolution of phototaxis. Philos Trans R Soc B 364:2795–2808

Jenkyns HC (2010) Geochemistry of oceanic events. Geochem Geophys Geosyst 11:Q03004

Jeong HJ, Yoo YD, Kim JS, Seong KA, Kang NS, Kim TH (2010) Growth, feeding and ecological roles of the Mixotrophic and heterotrophic dinoflagellates in marine planktonic food webs. Ocean Sci 45(2):65–91

Jessup CM, Kassen R, Forde SE, Kerr B, Buckling A et al (2004) Big questions, small worlds: microbial model systems in ecology. Trends Ecol Evol 19(4):189–197

Jiang D-Y, Motani R, Tintori A, Rieppel O, Ji C et al (2020) Evidence supporting predation of 4-m marine reptile by Triassic Megapredator. iScience 23:101347

Jiménez J (1997) Oceanic turbulence at millimetre scales. Sci Mar 61:47–56

Johnson KB, Shanks AL (2003) Low rates of predation on planktonic marine invertebrate larvae. Mar Ecol Prog Ser 248:125–139

Johnson ZI, Zinser ER, Coe A, McNully NP, Woodward EMS, Chisholm SW (2006) Niche partitioning among Prochlorococcus ecotypes along ocean-scale environmental gradients. Science 311:1737–1740

Johnson GD, Paxton JR, Sutton TT, Satoh TP, Sado T et al (2009) Deep-sea mystery solved: astonishing larval transformations and extreme sexual dimorphism unite three fish families. Biol Lett 5:235–239

Jue N, Batta-Lona PG, Trusiak S, Obergfell C, Bucklin A et al (2016) Rapid evolutionary rates and unique genomic signatures discovered in the first reference genome for the Southern Ocean Salp, Salpa thomsoni (Urochordata, Thaliacea). Genome Biol Evol 8(10):3171–3186

Juva K, Flögel S, Karstensen J, Linke P, Dullo W-A (2020) Tidal dynamics control on cold-water coral growth: a high-resolution multivariable study on eastern Atlantic cold-water coral sites. Front Mar Sci 7:132

Kaiser MJ, Attrill MJ, Jennings S, Thomas DN, Barnes DKA et al (2011) Marine ecology: processes, systems, and impacts. Oxford University Press, Oxford

Kaiser SI, Aretz M, Becker RT (2015) The global Hangenberg crisis (Devonian-carboniferous transition): review of a first-order mass extinction. Geol Soc Lond Spec Publ 423:387–437

Kamenskaya OE, Melnik VF, Gooday AJ (2013) Giant Protists (xenophyophores and komokiaceans) from the clarion-Clipperton ferromanganese nodule Field (eastern Pacific). Biol Bull Rev 3(5):388–398

Kaminski MA, Setoyama E, Cetean CG (2010) The Phanerozoic diversity of agglutinated foraminifera: origination and extinction rates. Acta Palaeontol Pol 55(3):529–539

Kammerlander B, Breiner H-W, Filker S, Sommaruga R, Sonntag B, Stoeck T (2015) High diversity of protistan plankton communities in remote high mountain lakes in the European Alps and the Himalayan mountains. FEMS Microb Ecol 91:fiv010

Kang NS, Jeong HJ, Yoo YD, Yoon EY, Lee KH et al (2011) Mixotrophy in the newly described phototrophic dinoflagellate Woloszynskia cincte from Western Korean waters: feeding mechanism, prey species and effect of prey concentration. J Eukaryot Microbiol 58(2):152–170

Kapsetaki SE, Fisher RM, West SA (2016) Predation and the formation of multicellular group in algae. Evol Ecol Res 17:651–669

Katechakis A, Stibor H, Sommer U, Hansen T (2004) Feeding selectivities and food niche separation of Acartia clausi, Penilia avirostris (Crustacea) and Doliolum denticulatum (Thaliacea) in Blanes Bay (Catalan Sea, NW Mediterranean). J Plankton Res 26(6):589–603

Katija K, Troni G, Daniels J, Lance K, Sherlock RE et al (2020) Revealing enigmatic mucus structures in the deep sea using deep PIV. Nature 583:78–82

Kato S, Kobayashi C, Kakegawa T, Yamagishi A (2009) Microbial communities in iron-silica-rich microbial mats at deep-sea hydrothermal fields of the Southern Mariana Trough. Environ Biol. https://doi.org/10.1111/j.1462-2920

Katz ME, Finkel ZV, Grzebyk D, Knoll AH, Falkowski PG (2004) Evolutionary trajectories and biogeochemical impacts of marine eukaryotic phytoplankton. Annu Rev Ecol Evol Syst 35:523–556

Katz S, Klepal W, Bright M (2011) The Osedax Trophosome: organization and ultrastructure. Biol Bull 220:128–139

Kayal E, Roure B, Philippe H, Collins AG, Lavrov DV (2013) Cnidarian phylogenetic relationships as revealed by mitogenomics. BMC Evol Biol 13:5

Keane M, Semeiks J, Webb AE, Li YI, Quesada V et al (2015) Insights into the evolution of longevity from the bowhead whale genome. Cell Rep 10:112–122

Keatley BE, Bennett EM, MacDonald GK, Taranu ZE, Gregory-Eaves I (2011) Land-use legacies are important determinants of Lake eutrophication in the Anthropocene. PLoS One 6(1):e15913

Kelley PH, Hansen TA (2003) Mesozoic Marine Revolution. In: Briggs DEG, Crowther PR (eds) Palaeobiology. Blackwell, Malden, pp 94–97

Kelley NP, Motani R (2015) Trophic convergence drives morphological convergence in marine tetrapods. Biol Lett 11:20140709

Kelley NP, Pyenson ND (2015) Evolutionary innovation and ecology in marine tetrapods from the Triassic to the Anthropocene. Science 348:aaa3716

Kenrick P, Wellman CH, Schneider H, Edgecombe GD (2012) A timeline for terrestrialization: consequences for the carbon cycle in the Palaeozoic. Philos Trans R Soc B 367:519–536

Kerner M, Hohenberg H, Ertl S, Reckermann M, Spitzy A (2003) Self-organization of dissolved organic matter to micelle-like microparticles in river-water. Nature 422:150–154

Keshtacher-Liebson E, Hadar Y, Chen Y (1995) Oligotrophic bacteria enhance algal growth under iron-deficient conditions. Appl Environ Microb 61(6):2439–2441

Kestener P, Arneodo A (2004) A three-dimensional wavelet based multifractal method: about the need of revisiting the multifractal description of turbulence dissipation data. arXiv: cond-mat/0302602v2

Keupp H, Hoffmann R, Stevens K, Albersdörfer R (2016) Key innovations in Mesozoic ammonoids: the multicuspidate radula and the calcified aptychus. Palaeontology 59(6):775–791

Key T, McCarthy A, Campbell DA, Six C, Roy S, Finkel ZV (2010) Cell size trade-off govern light exploitation strategies in marine phytoplankton. Environ Microb 12(1):95–104

Kidder DL, Erwin DH (2001) Secular distribution of biogenic silica through the Phanerozoic : comparison of silica-replaced fossils and bedded Cherts at the series level. J Geol 109:509–522

Kiessling W (2005) Long-term relationships between ecological stability and biodiversity in Phanerozoic reefs. Nature 433:410–413

Kiessling W (2009) Geologic and biologic controls on the evolution of reefs. Annu Rev Ecol Evol Syst 40:173–192

Kils U (1983) Swimming and feeding of Antarctic krill, Euphausia superba—some outstanding energetics and dynamics—some unique morphological details. In: Schnack SB (edt) Of the biology of krill Euphausia superba. Berichte zur Polarforschung, Sonderheft 4, 130–155

Kim H-M, Weber JA, Lee N, Park SG, Cho YS et al (2019) The genome of the giant Nomura's jellyfish sheds light on the early evolution of active predation. BMC Biol 17:28

Kiørboe T (2011) How zooplankton feed: mechanisms, traits and trade-offs. Biol Rev 86:311–339

Kirby MX, Jackson JBC (2004) Extinction of a first-growing oyster and changing ocean circulation in Pliocene tropical America. Geology 32(12):1025–1028

Klausen TG, Nyberg B, Heiland-Hansen W (2019) The largest delta plain in Earth's history. Geology 47(5):470–474

Klawonn I, Van den Wyngaert S, Iversen MH, Walles TJW, Flintrop CM et al (2021a) Fungal parasitism on diatoms alters formation and bio-physical properties of sinking aggregates. Comm Biol 6:206

Klawonn I, Van den Wyngaert S, Parada AE, Arandia-Gorostidi N, Whitehouse MJ et al (2021b) Characterizing the "fungal shunt": parasitic fungi on diatoms affect carbon flow and bacterial communities in aquatic microbial food webs. PNAS 118(23):e210222518

Klug C, Kröger B, Kiessling W, Mullins GL, Servais T et al (2010) The Devonian nekton revolution. Lethaia 43(4):465–477

Klug C, De Baets K, Kröger B, Bell MA, Korn D, Payne JL (2014) Normal giants? Temporal and latitudinal shifts of Palaeozoic marine invertebrate gigantism and global change. Lethaia 48:267–288

Klug C, Frey L, Pohle A, De Baets K, Korn D (2017) Palaeozoic evolution of animal mouthparts. Bull GeoSci 92(4):511–524

Knauth LP, Kennedy MJ (2009) The late Precambrian greening of the earth. Nature 460:728–732

Knoflacher, M, Gebetsroither E, Köstl, M (2003) Einträge von Stickstoff aus diffusen Quellen im Innbacheinzugsgebiet. Gewässerschutz Bericht 27/2002. Land Oberösterreich, Linz

Knoll AH, Follows MJ (2016) A bottom-up perspective on ecosystem change in Mesozoic oceans. Proc R Soc B 283:20161/ETLS20170153

Knoll AH, Bambach RK, Payne JL, Pruss S, Fischer WW (2007) Paleophysiology and end-Permian mass extinction. Earth Planet Sci Lett 256:295–313

Kogure K, Wada M (2005) Impacts of macrobenthic bioturbation in marine sediment on bacterial metabolic activity. Microbes Environ 4:191–199

Koletić N, Novosel M, Rajević N, Franjević D (2014) Bryozoans are returning home: recolonization of freshwater ecosystems inferred from phylogenetic relationships. Ecol Evol 5(2):255–264

Kooyman GL, Ponganis PJ (1998) The physiological basis of diving to depth: birds and mammals. Annu Rev Physiol 60:19–32

Kopáček J, Procházková L, Stuchlík E, Blažka P (1995) The nitrogen-phosphorus relationship in mountain lakes: influence of atmospheric input, watershed, and pH. Limnol Oceanogr 40(5):930–937

Kostylev VE, Erlandsson J, Ming MY, Williams GA (2005) The relative importance of habitat complexity and surface area in assessing biodiversity: fractal application on rocky shores. Ecol Complex 2:272–286

Kremp A (2013) Diversity of dinoflagellate life cycles: facets and implications of complex strategies. In: Lewis JM, Maret F, Bradley L (eds) Biological and geological perspectives of dinoflagellates. The Micropalaeontological Society, London

Kremp A, Parrow MW (2006) Evidence for asexual resting cysts in the life cycle of the marine Peridinoid dinoflagellate, Scrippsiella hangoei. J Phycol 42:400–409

Kröger B, Vinther J, Fuchs D (2011) Cephalopod origin and evolution: a congruent picture emerging from fossils, development and molecules. BioEssays 33(8):602–613

Ksepka DT, Bertelli S, Giannini NP (2006) The phylogeny of the living and fossil Sphenisciformes (penguins). Cladistics 22:412–441

Kuklinski P (2009) Ecology of stone-encrusting organisms in the Greenland Sea—a review. Polar Res 28(2):222–237

Kvitek RG (1986) Side-scan estimates of the utilization of gray whale feeding grounds along Vancouver Island, Canada. Master Thesis, San Francisco State University

Labandeira CC (2005) Invasion of the continents: cyanobacterial crusts to tree-inhabiting arthropods. Trends Ecol Evol 20(5):253–262

Lai X, Wignall P, Zhang K (2001) Palaeoecology of the conodonts Hindeodus and Clarkina during the Permian-Triassic transitional period. Palaeogeogr Palaeoclimatol Palaeoecol 171:63–72

LaJeunesse TC, Parkinson JE, Gabrielson PW, Jeong HJ, Reimer JD et al (2018) Systematic revision of Symbiodiniaceae highlights the antiquity and diversity of coral endosymbionts. Curr Biol 28:2570–2580

Landing E, MacGabhann BA (2010) First evidence for Cambrian glaciation provided by sections in Avalonian New Brunswick and Ireland: additional data for Avalon-Gondwana separation by the earliest Palaeozoic. Palaeogeogr Palaeoclimatol Palaeoecol 285:174–185

Landry MR, Calbet A (2004) Microzooplankton production in the oceans. ICES J of Marine Sci 61:501–507

Latysheva N, Junker VL, Palmer WJ, Codd GA, Barker D (2012) The evolution of nitrogen fixation in cyanobacteria. Bioinformatics 28(5):603–606

Laurin M, Piñeiro GH (2017) A reassessment of the taxonomic position of Mesosaurs, and a surprising phylogeny of early amniotes. Front. Earth Sci 5:88

Laurin M, Soler-Gijón R (2010) Osmotic tolerance and habitat of early stegocephalians: indirect evidence from parsimony, taphonomy, palaeobiogeography, physiology and morphology. Geol Soc Lond Spec Publ 339:151–179

Laverock B, Smith CJ, Tait K, Osborn AM, Widdicombe S, Gilbert JA (2010) Bioturbating shrimp alter the structure and diversity of bacterial communities in coastal marine sediments. ISME J 4:1531–1544

Lawrence J, Töpper J, Petelenz-Kurzdiel E, Bratbak G, Larsen A et al (2018) Viruses on the menu: the appendicularian Oikopleura dioica efficiently removes viruses from seawater. Limnol Oceanogr 63(51):5244–5253

Lazzaro X (1987) A review of planktivorous fishes: their evolution, feeding behaviours, selectivities, and impacts. Hydrobiologia 146:97–167

Leander BS, Keeling PJ (2003) Morphostasis in alveolate evolution. Trend Ecol Evol 18(8):395–402

Leander BS, Esson HJ, Breglia SA (2007) Macroevolution of complex cytoskeletal systems in euglenids. BioEssays 29:987–1000

Lee S, Ford AK, Mangubjai S, Wild C, Ferse SCA (2018) Effects of sandfish (Holothuria scabra) removal on shallow-water sediments in Fiji. PeerJ 6:e4773. https://doi.org/10.7717/peerj.4773

Lehmann B, Nägler TF, Holland HD, Wille M, Mao J et al (2007) Highly metalliferous carbonaceous shale and early Cambrian seawater. Geology 35(5):403–406

Lenton TM, Daines SJ (2017) Biogeochemical transformations in the history of the ocean. Annu Rev Mar Sci 9:31–58

Lenton TM, Crouch M, Johnson M, Pires N, Dolan L (2012) First plants cooled the Ordovician. Nat Geosci 5:86–89

Lenton TM, Dahl TW, Daines SJ, Mills BJW, Ozaki K et al (2016) Earliest land plants created modern levels of atmospheric oxygen. PNAS 113(35):9704–9709

Leprieur F, Descombes P, Gaboriau T, Cowman PF, Parravicini V et al (2016) Plate tectonics drive tropical reef biodiversity dynamics. Nat Commun 7. https://doi.org/10.1038/ncomms11461

Lerosey-Aubril R, Pates S (2018) New suspension-feeding radiodont suggests evolution of microplanktivory in Cambrian macronection. Nat Commun 9:3774

Li C-W, Chen J-Y, Hua T-E (1998) Precambrian sponges with cellular structures. Science 279:879–882

Li F, Yan J, Algeo T, Wu X (2013) Paleoceanographic conditions following the end-Permian mass extinction recorded by giant ooids (Moyang, South China). Glob Planet Chang 105:102–120

Li Z, Liu L, Chen J, Teng HH (2016a) Cellular dissolution at hypha- and spore-mineral interfaces revealing unrecognized mechanisms and scales of fungal weathering. Geology 44(4):319–322

Li C, Zhu M, Chu X (2016b) Atmospheric and oceanic oxygenation and evolution of early life on earth: new contributions from China. J Earth Sci 27(2):167–169

Liegertová M, Pergner J, Kozmiková I, Fabian P, Pombinho AR et al (2015) Cubozoan genome illustrates functional diversification of opsins and photoreceptor evolution. Sci Rep 5. https://doi.org/10.1038/srep11885

Likens GE, Bormann EH (1995) Biogeochemistry of forested ecosystems. Springer, New York

Lillywhite HB, Sheehy CM, Sandfoss MR, Crow-Riddell J, Grech A (2019) Drinking by sea snakes from oceanic freshwater lenses at first rainfall ending seasonal drought. PLoS One 14(2):e0212099

Lima-Mendez G, Faust K, Henry N, Decelle J, Colin S et al (2015) Determinants of community structure in the global plankton interactome. Science 6237:1262073

Lindberg DR, Pyenson ND (2007) Things that go bump in the night: evolutionary interactions between cephalopods and cetaceans in the tertiary. Lethaia 40:335–343

Lindgren J, Caldwell MW, Konishi T, Chiappe LM (2010) Convergent evolution in aquatic Tetrapods: insights from an exceptional fossil mosasaur. PLoS One 5(8):e11998

Linley TD, Gerringer ME, Yancey PH, Drazen JC, Weinstock CL, Jamieson AJ (2016) Fishes of the hadal zone including new species, in situ observations and depth records of Liparidae. Deep-Sea Res I 114:99–110

Liston J (2013) The plasticity of gill raker characteristics in suspension feeders: implications for Pachycormiformes. In: Arratia G, Schultze H-P, Wilson MVH (eds) Mesozoic fishes 5—global diversity and evolution. Pfeil, München

Little CTS, Vrijenhoek RC (2003) Are hydrothermal vent animals living fossils? Trends Ecol Evol 18(11):582–588

Liu K, Feng Q, Shen J, Khan M, Planavsky NJ (2018) Increased productivity as a primary driver of marine anoxia in the lower Cambrian. Palaeogeogr Palaeoclimatol Palaeoecol 491:1–9

Locsei JT, Pedley TJ (2008) Bacterial tracking of motile algae assisted by algal cell's vorticity field. arXiv:0806.0744v1

Lomstein BA, Langerhuus AT, D'Hondt S, Jørgensen BB, Spivack AJ (2012) Endospore abundance, microbial growth and necromass turnover in deep sub-seafloor sediment. Nature 484:101–104

Long JA (2016) The first jaws. Science 354:280–281

Long JA, Trinajstic K, Johanson Z (2009) Devonian arthrodire embryos and the origin of internal fertilization in vertebrates. Nature 457:1124–1127

Long JA, Choo B, Clement A (2019) The evolution of fishes through geological time. In: Johanson Z, Underwood C, Richter M (eds) Evolution and development of fishes. Cambridge University Press, Cambridge, pp 3–29

Lopez GR (1988) Comparative ecology of the macrofauna of freshwater and marine muds. Limnol Oceanogr 33/4(Part 2):946–962

López-Sandoval DC, Rodriguez-Ramos T, Cermeño P, Sobrino C, Marañon E (2014) Photosynthesis and respiration in marine phytoplankton: relationship with cell size, taxonomic affiliation, and growth phase. J Marine Biol Ecol 457:151–159

Lovecchio S, Climent E, Stocker R, Durham WM (2019) Chain formation can enhance the vertical migration of phytoplankton through turbulence. Sci Adv 5:eaaw7879

Lovejoy S, Currie WJS, Tessier Y, Claereboudt MR, Bourget E et al (2001) Universal multifractals and ocean patchiness: phytoplankton, physical fields and coastal heterogeneity. J Plankton Res 23(2):117–141

Lukeš J, Leander BS, Keeling PJ (2009) Cascades of convergent evolution: the corresponding evolutionary histories of euglenozoans and dinoflagellates. PNAS 106(1):9963–9970

Lundsten L, Schlining KL, Frasier K, Johnson SB, Kuhnz LA et al (2010) Time-series of six whale-fall communities in Monterey canyon, California USA. Deep-Sea Res I 57:1573–1584

Luppold FW, Clausen C-D, Stoppel D (1994) Davon/Karbon-Grenzprofile im Bereich von Remscheid-Altenaer Sattel, Warnsteiner Sattel, Briloner Sattel und Attendorn-Elsper Doppelmulde (Rheinisches Schiefergebirge). Geol Paläont Westf 29:7–69

Lv J, Li N, Niu D-K (2008) Association between the availability of environmental resources and the atomic composition of organismal proteomes: evidence from Prochlorococcus strains living at different depths. Biochem Biophys Res Commun 375:241–246

Macdonald FA (2020) Deep-time paleoclimate proxies. AGU Adv 1:e2020AV000244

MacDougall MJ, Modesto SP, Brocklehurst N, Verrieére A, Reisz RR, Fröbisch J (2018) Commentary: a reassessment of the taxonomic position of Mesosaurs, and a surprising phylogeny of early amniotes. Front Earth Sci 6:99

MacIntyre HL, Kans TM, Gelder RJ (2000) The effect of water motion on short-term rates of photosynthesis by marine phytoplankton. Trends Plant Sci 5(1):12–17

Malfatti F, Azam F (2009) Atomic force microscopy reveals microscale networks and possible symbioses among pelagic marine bacteria. Aquat Microb Ecol 58:1–14

Malits A, Peters F, Bayer-Giraldi M, Marrasé C, Zoppini A et al (2004) Effects of small-scale turbulence on bacteria: a matter of size. Microb Ecol 48:287–299

Manconi R, Pronzato R (2008) Global diversity of sponges (Porifera: Spongillina) in freshwater. Hydrobiologia 595:27–33

Mander Ü, Kull A, Tamm V, Kuusemets V, Karjus R (1998) Impacts of climatic fluctuations and land use change on runoff and nutrient losses in rural landscapes. Landsc Urban Plan 41:229–238

Mángano MG, Buatois LA (2017) The Cambrian revolutions: trace-fossil record, timing, links and geobiological impact. Earth-Sci Rev 173:96–108

Mann KH, Lazier JRN (2006) Dynamics of marine ecosystems. Blackwell, Malden

Mann J, Sargeant BL, Watson-Capps JJ, Gibson QA, Heithaus MR et al (2008) Why do dolphins carry sponges? PLoS One 3(12):e3868

Mapstone GM (2014) Global diversity and review of Siphonophora (Cnidaria: hydrozoa). PLoS One 9(2):e87737

Marañon E, Holligan PM, Barciela R, González N, Mouriño B et al (2001) Patterns of phytoplankton size structure and productivity in contrasting open-ocean environments. Mar Ecol Prog Ser 216:43–56

Marcus NH, Boero F (1998) The importance of benthic-pelagic coupling and the forgotten role of life cycles in coastal aquatic systems. Limnol Oceanogr 43(5):763–768

Margalef R (1978) Life-forms of phytoplankton as survival alternatives in an unstable environment. Oceanol Acta 1(4):493–509

Marhaeni B, Radjasa OK, Khoeri MM, Sabdono A, Bengen DG, Sudoyo H (2011) Antifouling activity of bacterial symbionts of seagrasses against marine biofilm-forming bacteria. J Environ Prot 2:1245–1249

Marjanović D, Laurin M (2019) Phylogeny of Paleozoic limbed vertebrates reassessed through revision and expansion of the largest published relevant data matrix. PeerJ 6:e5565. https://doi.org/10.7717/peerj.5565

Markle DF, Frost L-A (1985) Comparative morphology, seasonality, and a key to planktonic fish eggs from the Nova Scotian shelf. Can J Zool 63:246–257

Marranzino AN, Webb JF (2018) Flow sensing in the deep sea: the lateral line system of stomiiform fishes. Zool J Linnean Soc 183:945–965

Marshall JEA, Lakin J, Troth I, Wallace-Johnson SM (2020) UV-B radiation was the Devonian-carboniferous boundary terrestrial extinction kill mechanism. Sci Adv 6:eaba0768

Martens J-H, Barg H, Warren MJ, Jahn D (2002) Microbial production of vitamin B12. Appl Microbiol Biotechnol 58:275–285

Martill DM, Frey E, Tischlinger H, Mäuser M, Rivera-Sylva HE, Vidovic SU (2023) A new pterodactyloid pterosaur with a unique flter-feeding apparatus from the late Jurassic of Germany. PalZ 97:383–424

Martin KL, Carter AL (2013) Brave new propagules: terrestrial embryos in Anamniotic eggs. Integr Comp Biol 53(2):233–247

Martin R, Quigg A (2012) Evolving phytoplankton stoichiometry fueled diversification of the marine biosphere. Geoscience 2:130–146

Martinez-Pérez C, Rayfield EJ, Botella H, Donoghue PCJ (2016) Translating taxonomy into the evolution of conodont feeding ecology. Geology 44(4):247–250

Martiny AC, Huang Y, Li W (2009) Occurrence of phosphate acquisition genes in Prochlorococcus cells from different ocean regions. Environ Microb 11(6):1340–1347

Marx FG, Hocking DP, Park T, Ziegler T, Evans AR, Fitzgerals EMG (2016) Suction feeding preceded filtering in baleen whale evolution. Mem Museum Victoria 75:71–82

Masuda F, Ezaki Y (2009) A great evolution of the earth-surface environment: linking the bioinvasion onto the land and the Ordovician radiation of marine organisms. Paleontological Res 13(1):3–8

Matz MV, Frank TM, Marshall NJ, Widder EA, Johnsen S (2008) Giant Deep-Sea Protist produces Bilaterian-like traces. Curr Biol 18(23):1849–1854

McCave IN, Manighetti B, Robinson SG (1995) Sortable silt and fine sediment size/composition slicing: parameters for palaeocurrent speed and palaeoceanography. Paleoceanography 10(3):593–610

McClain CR, Allen AP, Tittensor DP, Rex MA (2012) Energetics of life on the deep seafloor. PNAS 109(38):15366–15371

McGaugh SE, Gross JB, Aken B, Blin M, Borowsky R et al (2014) The cavefish genome reveals candidate genes for eye loss. Nat Commun 5:5307. https://doi.org/10.1038/ncomms6307

McGhee GR Jr, Clapham ME, Sheehan PM, Bottjer DJ, Doser ML (2013) A new ecological-severity ranking of major Phanerozoic biodiversity crises. Palaeogeogr Palaeoclimatol Palaeoecol 370:260–270

McGowan AJ, Smith AB (2007) Ammonoids across the Permian/Triassic boundary: a cladistic perspective. Palaeontology 50(3):573–590

McGowan AJ, Smith AB (2008) Are global Phanerozoic marine diversity curves truly global? A study of the relationship between regional rock records and global Phanerozoic marine diversity. Paleobiology 34(1):80–103

McMahon WJ, Davies NS (2018) Evolution of alluvial mudrock forced by early land plants. Science 359:1022–1024

McQuoid MR, Godhe A (2004) Recruitment of coastal planktonic diatoms from benthic versus pelagic cells: variations in bloom development and species composition. Limnol Oceanogr 49(4):1123–1133

Medina-Sánchez JM, Villar-Argaiz M, Carrillo P (2004) Neither with or without you: a complex algal control on bacterioplankton in a high mountain lake. Limnol Oceanogr 49(5):1722–1733

Meier KJS, Young JR, Kirsch M, Feist-Burkhardt S (2007) Evolution of different life-cycle strategies in oceanic calcareous dinoflagellates. Eur J Phycol 42(1):81–89

Mendoza-Becerril MA, Maronna MM, Pacheco MLAF, Simões MG, Leme JM et al (2016) An evolutionary comparative analysis of the medusozoan (Cnidaria) exoskeleton. Zool J Linnean Soc 178:206–225

Mentel M, Martin W (2010) Anaerobic animals from an ancient, anoxic ecological niche. BMC Biol 8:32

Mermillod-Blondin F (2011) The functional significance of bioturbation and biodeposition on biogeochemical processes at the water-sediment interface in freshwater and marine ecosystems. J N Am Benthol Soc 30(3):770–778

Mestre M, Ferrera I, Borrull E, Ortega-Retuerta E, Mbedi S et al (2017) Spatial variability of marine bacterial and archaeal communities along the particulate matter continuum. Mol Ecol 26:6827–6840

Middelburg JJ, Mueller CE, Veuger B, Larsson AI, Form A, van Oevelen D (2015) Discovery of symbiotic nitrogen fixation and chemoautotrophy in cold-water corals. Sci Rep 5. https://doi.org/10.1038/srep17962

Miller EC, Román-Palacios (2021) Evolutionary time explains the global distribution of freshwater fish diversity. Glob Ecol Biogeogr 30(3):749–763

Miller PJO, Johnson MP, Tyack PL (2004) Sperm whale behaviour indicates the use of echolocation click buzzes 'creaks' in prey capture. Proc R Soc Lond B 271:2239–2247

Miller KG, Kominz MA, Browning JV, Wright JD, Mountain GS et al (2005) The phanerozoic record of global sea-level change. Science 310:1293–1298

Mills B, Lenton TM, Watson AJ (2014) Proterozoic oxygen rise linked to shifting balance between seafloor and terrestrial weathering. PNAS 111(25):9073–9078

Mills BJW, Batterman SA, Field KJ (2017) Nutrient acquisition by symbiotic fungi governs Palaeozoic climate transition. Philos Trans R Soc B 373:20160503

Mills DB, Francis WR, Vargas S, Larsen M, Elemans CPH et al (2018) The last common ancestor of animals lacked the HIF pathway and respired in low-oxygen environments. elife 7:e31176

Mironenko AA (2020) Endoceris: suspension feeding nautiloids? Hist Biol 32(2):281–289

Mitra A, Flynn KJ, Burkholder JM, Berge T, Calbet A et al (2014) The role of mixotrophic protists in the biological carbon pump. Biogeosciences 11:995–1005

Mitra A, Flynn KJ, Tillmann U, Raven JA, Caron D et al (2016) Defining planktonic Protist functional groups on mechanisms for energy and nutrient acquisition: incorporation of diverse Myxotrophic strategies. Protist 167:106–120

Miya M, Friedman M, Satoh TP, Takashima H, Sado T et al (2013) Evolutionary origin of the Scombridae (tunas and mackerels): members of a Paleogene adaptive radiation with 14 other pelagic Fish families. PLoS One 8(9):e73535

Mladenov N, Sommaruga R, Morales-Baquero R, Laurion I, Camarero I et al (2011) Dust inputs and bacteria influence dissolved organic matter in clear alpine lakes. Nat Commun 2:405. https://doi.org/10.1038/ncomms1411

Modesto SP (2010) The postcranial skeleton of the aquatic Parareptile Mesosaurus tenuidens from the Gondwanan Permian. J Vertebr Paleontol 30(5):1378–1395

Modesto SP, Scott DM, MacDougall MJ, Sues H-D, Evans DC, Reisz RR (2015) The oldest parareptile and the early diversification of reptiles. Proc R Soc B 282:20141912

Moniruzzaman M, Wurch LL, Alexander H, Dyhrman ST, Gobler CJ, Wilhelm SW (2017) Virus-host relationships of marine single-celled eukaryotes resolved from metatranscriptomics. Nat Commun 8. https://doi.org/10.1038/ncomms16054

Moore LM (2013) More mixotrophy in the marine microbial mix. PNAS 110(21):8323–8324

Moore JK, Villareal TA (1996) Buoyancy and growth characteristics of three positively buoyant marine diatoms. Mar Ecol Prog Ser 132:203–213

Mora C, Tittensor DP, Adl S, Simpson AGB, Worm B (2011) How many species are there on earth and in the ocean? PLoS Biol 9(8):e1001127

Moreira D, López-Garcia P (2014) The rise and fall of Picobiliphytes: how assumed autotrophs turned out to be heterotrophs. BioEssays 36:468–474

Morris JJ, Johnson ZI, Szul MJ, Keller M, Zinser ER (2011) Dependence of the cyanobacterium Prochlorococcus on hydrogen peroxide scavenging microbes for growth at the Oceans's surface. PLoS One 6(2):e16805

Morris JL, Puttick MN, Clark JW, Edwards D, Kenrick P et al (2018) The timescale of early land plant evolution. PNAS 115(10):E2274–E2283

Motani R (2009) The evolution of marine reptiles. Evo Edo Outreach 2:224–235

Motani R (2010) Warm-blooded "sea dragons"? Science 328:1361–1362

Motani R, Vermeij GJ (2021) Ecophysiological steps of marine adaptation in extant and extinct non-avian tetrapods. Biol Rev 96(5):1769–1798

Motani R, Chen X-H, Jiang D-Y, Cheng L, Tintori A, Rieppel O (2015) Lunge feeding in early marine reptiles and fast evolution of marine tetrapod feeding guilds. Sci Rep 5. https://doi.org/10.1038/srep08900

Motani R, Jiang D-Y, Tintori A, Ji C, Huang J-D (2017) Pre- versus post-mass extinction divergence of Mesozoic marine reptiles dictated by time-scale dependence of evolutionary rates. Proc R Soc B 284:20170241

Mouriam MJ, Orliac MJ (2017) Infrasonic and ultrasonic hearing evolved after the emergence of modern whales. Curr Biol 27:1776–1781

Moysiuk J, Caron J-B (2019) A new hurdiid radiodont from the Burgess shale evidences the exploitation of Cambrian infaunal food sources. Proc R Soc B 286:20191079

Muehe EM, Gerhardt S, Schink B, Kappler A (2009) Ecophysiology and the energetic benefit of mixotrophic FE(II) oxidation by various strains of nitrate-reducing bacteria. FEMS Microbiol Ecol 70:335–343

Müller A, Faubert P, Hagen M, Zu Castell W, Polle A et al (2013) Volatile profiles of fungi—Chemotyping of species and ecological functions. Fungal Genet Biology 54:25–33

Murdock DJE, Dong X-P, Repetski JE, Marone F, Stampanoni M et al (2013) The origin of conodonts and of vertebrate minaralized skeletons. Nature 501:546–549

Murdock DJE, Rayfield EJ, Donoghue PCJ (2014) Functional adaptation underpinned the evolutionary assembly of the earliest vertebrate skeleton. Evol Dev 16(6):354–351

Murray AM, Cumbaa SL, Harington R, Smith GR, Rybczynski N (2009) Early Pliocene fish remains from Arctic Canada support a pre-Pleistocene dispersal of percids (Teleostei: Perciformes). Can J Earth Sci 46:557–570

Naafs BDA, Stein R, Hefter J, Khélifi N, De Schepper S, Haug GH (2010) Late Pliocene changes in the North Atlantic current. Earth Planet Sci Lett 298:434–442

Nakamura Y, Suzuki K, Suzuki S, Hiromi J (1997) Production of Oikopleura dioica (Appendicularia) following a picoplankton 'bloom' in a eutrophic coastal area. J Plankton Res 19(1):113–124

Nakamura I, Goto Y, Sato K (2015) Ocean sunfish rewarm at the surface after deep excursions to forage for siphonophores. J Anim Ecol 84:590–603

Naselli-Flores L, Barone R (2011) Fight on plankton! Or, phytoplankton shape and size as adaptive tools to get ahead in the struggle for life. Cryptogam Algol 32(2):157–204

Near TJ, Eytan RI, Dornburg A, Kuhn KL, Moore JA et al (2012) Resolution of ray-finned fish phylogeny and timing of diversification. PNAS 109(34):13698–13702

Nehring S (1993) Mechanisms for recurrent nuisance algal blooms in coastal zones: resting cyst formation as life-strategy of dinoflagellates. In: Sterr H, Hofstede J, Plag H-P (eds) Proceedings of the international coastal congress, ICC-Kiel '92. Lang, Frankfurt, pp 454–467

Nehring S (1995) Dinoflagellates resting cysts as factors in phytoplankton ecology of the North Sea. Helgoländer Meeresun 49:375–392

Nettersheim BJ, Brocks JJ, Schwelm A, Hoper JM, Not F et al (2019) Putative sponge biomarkers in unicellular Rhizaria question an early rise of animals. Nat Ecol Evol 3:577–581. https://doi.org/10.1038/s41559-019-0806-5

Nielsen JK, Shen Y, Piasecki S, Stemmerik L (2010) No abrupt change in redox condition caused the end-Permian marine ecosystem collapse in the East Greenland Basin. Earth Planet Sci Let 291:32–38

Nielsen J, Hedeholm RB, Heinemeier J, Bushnell PG, Christiansen JS et al (2016) Eye lens radiocarbon reveals centuries of longevity in the Greenland shark (Somniosus microcephalus). Science 353:702–704

Nilsson D-E, Warrant EJ, Johnsen S, Hanlon R, Shaahar N (2012) A unique advantage for giant eyes in giant squid. Curr Biol 22:683–688

Nilsson D-E, Warrant E, Johnsen S (2014) Computational visual ecology in the pelagic realm. Philos Trans R Soc B 369:20130038

Nogaro G, Mermillod-Blondin F, Valett MH, François-Carcaillet F, Gaudet J-P et al (2009) Ecosystem engineering at the sediment-water interface: bioturbation and consumer-substrate interaction. Geology 161:125–138

Norris RD (1996) Symbiosis as an evolutionary innovation in the radiation of Paleocene foraminifera. Paleobiology 22(4):461–480

Nuñez Demarco P, Meneghel M, Laurin M, Piñeiro G (2018) Was Mesosaurus a fully aquatic reptile? Front Evol 6:109

Nützel A, Lehnert O, Frýda J (2006) Origin of planktotrophy—evidence from early molluscs. Evol Dev 8(4):325–330

O'Dea A, Lessios HA, Coates AG, Eytan RI, Restrepo-Moreno SA et al (2016) Formation of the isthmus of Panama. Sci Adv 2:e1600883

Orcutt BN, Sylvan JB, Knab NJ, Edwards KJ (2011) Microbial ecology of the Dark Ocean above, at, and below the seafloor. Microb Molec Biol Rev 75(2):361–422

Østergaard F, Kristensen A, Kristensen E (1991) Effects of burrowing macrofauna on organic matter decomposition in coastal marine sediments. Symp Zool Soc Lond 63:69–88

Otoo BKA, Clack JA, Smithson TR, Bennett CE, Kearsey TI, Coates MI (2019) A fish and tetrapod fauna from Romer's gap preserved in Scottish Tournaisian floodplain deposits. Palaeontology 62(2):225–253

Ottesen EA, Young CR, Gifford SM, Eppley JM, Marin R et al (2014) Multispecies diel transcriptional oscillations in open ocean heterotrophic bacterial assemblages. Science 345:207–212

Paerl HW, Hall NS, Calandrino ES (2011) Controlling harmful cyanobacterial blooms in a world experiencing anthropogenic and climatic-induced change. Sci Total Environ 409:1739–1745

Palkovacs EP, Dion KB, Post DM, Caccone A (2008) Independent evolutionary origins of landlocked alewife populations and rapid parallel evolution of phenotypic traits. Mol Ecol 17:582–597

Park E, Hwang D-S, Lee J-S, Song J-I, Seo T-K, Won Y-J (2012) Estimation of divergence times in cnidarian evolution based on mitochondrial protein-coding genes and the fossil record. Mol Phylogenet Evol 62:329–343

Parkes RJ, Webster G, Cragg BA, Weightman AJ, Newberry CJ et al (2005) Deep sub-seafloor pro-karyotes stimulated at interfaces over geological time. Nature 436:390–394

Parnell J, Foster S (2012) Ordovician ash geochemistry and the establishment of land plants. Geochem Trans 13:7

Parry L, Tanner A, Vinther J (2014) The origin of annelids. Palaeontology 57(6):1091–1103

Paterson JR, Garcia-Bellido DC, Lee MSY, Brock GA, Jago JB, Edgecombe GD (2011) Acute vision in the giant Cambrian predator Anomalocaris and the origin of compound eyes. Nature 480:237–240

Pawlik JR, Butman CA (1993) Settlement of a marine tube worm as a function of current velocity: interacting effects of hydrodynamics and behaviour. Limnol Oceanogr 38(8):1730–1740

Pawlowsky J, Holzmann M, Berney C, Fahrni J, Gooday AJ et al (2003) The evolution of early foraminifera. PNAS 100(20):11494–11498

Peperzak L, Gäbler Schwarz S (2012) Current knowledge of the life cycles of Phaeocystis globosa and Phaeocystis Antarctica (Prymnesiophyceae). J Phycol 48:514–517

Perissinotto R, Mayzaud P, Nichols PD, Labat JP (2007) Grazing by Pyrosoma atlanticum (Tunicata, Thaliacea) in the south Indian-Ocean. Mar Ecol Prog Ser 330:1–11

Peters F, Marrasé C (2000) Effects of turbulence on plankton: an overview of experimental evidence and some theoretical considerations. Mar Ecol Prog Ser 205:291–306

Piazza V, Dragić I, Sepčić K, Faimali M, Garaventa F et al (2014) Antifouling activity of synthetic Alkylpyridinium polymers using the barnacle model. Mar Drugs 12:1959–1976

Pietsch TW, Orr JW (2007) Phylogenetic relationships of deep-sea anglerfishes of the suborder Ceratoidei (Teleostei, Lophiiformes) based on morphology. Copeia 1:1–34

Pimiento C, Balk MA (2015) Body-size trends of the extinct giant shark Carcharocles megalodon: a deep time perspective on marine apex predators. Paleobiology 41(3):479–490

Pimiento C, Griffin JN, Clements CF, Silvestro D, Varela S et al (2017) The Pliocene marine mega-fauna extinction and its impact on functional diversity. Nat Ecol Evol 1:1100–1106. https://doi.org/10.1038/s41559-017-0223-6

Pimiento C, Cantalapiedra JL, Shimada K, Field DJ, Smaers JB (2019) Evolutionary pathways toward gigantism in sharks and rays. Evolution 73(3):588–599

Piñeiro G, Ramos A, Goso C, Scarabino F, Laurin M (2012) Unusual environmental conditions preserve a Permian mesosaur-bearing Konservat-Lagerstätte from Uruguay. Acta Palaeontol Pol 57(2):299–318

Pischedda L, Poggiale JC, Cuny P, Gilbert F (2008) Imaging oxygen distribution in marine sediments. The importance of bioturbation and sediment heterogeneity. Acta Biotheor 56(1–2):123–135

Pittera J, Humily F, Thoral M, Grulois D, Garczarek L, Six C (2014) Connecting thermal physiology and latitudinal niche partitioning in marine Synechococcus. ISME J 8:1221–1236. https://doi.org/10.1038/ismej.2013.228

Platt JF, Methven DA (1992) Threshold foraging behaviour of baleen whales. Mar Ecol Prog Ser 84:205–210

Polcyn MJ, Jacobs LL, Araújo R, Schulp AS, Mateus O (2014) Physical drivers of mosasaur evolu-tion. Palaeogeogr Palaeoclimatol Palaeoecol 400:17–27

Poore AGB, Campbell AH, Coleman RA, Edgar GJ, Jormalainen V et al (2012) Global patterns in the impact of marine herbivores on benthic primary producers. Ecol Lett 15:912–922

Porada P, Lenton TM, Pohl A, Weber B, Mander L et al (2016) High potential for weathering and climate effects of non-vascular vegetation in the Late Ordovician. Nat Commun 7. https://doi.org/10.1038/ncomms12113

Pough FH, Janis CM (2019) Vertebrate Life. Sinauer, New York

Powell MG, Moore BR, Smith TJ (2015) Origination, extinction, invasion, and extirpation components of brachiopod latitudinal biodiversity gradient through the Phanerozoic eon. Palaeobiology 41(2):330–341

Priede IG, Froese R (2013) Colonization of the deep sea by fishes. J Fish Biol 83:1528–1550

Pu JP, Bowring SA, Ramezani J, Myrow P, Raub TD et al (2016) Dodging snowballs: geochronology of the Gaskiers glaciation and the first appearance of the Ediacaran biota. Geolog Soc Am 44:895–898. https://doi.org/10.1130/G38248.1

Purcell JE (2012) Jellyfish and ctenophore blooms coincide with human proliferations and environmental perturbations. Annu Rev Mar Sci 4:209–235

Purnell MA, Jones D (2012) Quantitative analysis of conodont tooth wear and damage as a test of ecological functional hypotheses. Paleobiology 38(4):605–626

Purschke G, Bleidorn C, Struck T (2014) Systematics, evolution and phylogeny of Annelida—a morphological perspective. Mem Museum Victoria 71:247–269

Puttick MN, Kriwet J, Wen W, Hu S, Thomas GH, Benton MJ (2017) Body length of bony fishes was not a selective factor during the biggest mass extinction of all time. Palaeontology 60(5):727–741

Pyenson ND, Kelley NP, Parham JF (2014) Marine tetrapod macroevolution: physical and biological drivers on 250 Ma of invasions and evolution in ocean ecosystems. Palaeogeogr Palaeoclimatol Palaeoecol 400:1–8

Qiao Y, Liu J, Wolniewicz AS, Iijima M, Shen Y et al (2022) A globally distributed durophagous marine reptile clade supports the rapid recovery of pelagic ecosystems after the Permo-Triassic mass extinction. Comm Biol 5:1242

Quigg A, Irwin AJ, Finkel ZV (2011) Evolutionary inheritance of elemental stoichiometry in phytoplankton. Proc R Soc B 278:526–534

Quirk J, Leake JR, Johnson DA, Taylor LL, Saccone L, Beerling DJ (2015) Constraining the role of early land plants in Palaeozoic weathering and global cooling. Proc R Soc B 282:20151115

Raina J-B, Lambert BS, Parks DH, Rinke C, Siboni N et al (2022) Chemotaxis shapes the microscale organization of the ocean's microbiome. Nature 605:132–138

Randle E, Sansom RS (2019) Bite marks and predation of fossil jawless fish during the rise of jawed vertebrates. Proc R Soc B 286:20191596

Raven JA, Waite AM (2004) The evolution of silification in diatoms: inescapable sinking and sinking as escape? New Phytol 162:45–61

Read BA, Kegel J, Klute MJ, Kuo A, Lefebvre SC et al (2013) Pan genome of phytoplankton Emiliania underpins its global distribution. Nature 499:209–213

Rengefors K, Gustafsson S, Stähl-Delbanco A (2004) Factors regulating the recruitment of cyanobacterial and eukaryotic phytoplankton from littoral and profundal sediments. Aquat Microb Ecol 36:213–226

Reynolds C (2006) Ecology of phytoplankton. Cambridge University Press, Cambridge

Ribalet F, Swalwell J, Clayton S, Jiménez V, Sudek S et al (2015) Light-driven synchrony of Prochlorococcus growth and mortality in the subtropical Pacific gyre. PNAS 112(26):8008–8012

Ribeiro S, Berge T, Lundholm N, Andersen TJ, Abrantes F, Ellegaard M (2011) Phytoplankton growth after a century of dormancy illuminates past resilience to catastrophic darkness. Nat Commun 2:311. https://doi.org/10.1038/ncomms1314

Ridgway SH (1972) Mammals of the sea. Thomas Publisher, Springfield

Ridgwell A, Zeebe RE (2005) The role of the global carbonate cycle in the regulation and evolution of the earth system. Mar Geol 217(2–3):339–357

Riedl R (1966) Biologie der Meereshöhlen. Paul Parey, Hamburg

Riegl B, Bruckner A, Coles SL, Renaud P, Dodge RE (2009) Coral reefs threats and conservation in the era of global change. Ann N Y Acad Sci 1162:136–186

Rier ST, Stevenson RJ (2002) Effects of light, dissolved organic carbon, and inorganic nutrients on the relationship between algae and heterotrophic bacteria in stream periphyton. Hydrobiologia 489:179–184

Rigo M, Joachimski MM (2010) Palaeoecology of late Triassic conodonts: constraints from oxygen isotopes in biogenic apatite. Acta Palaeontol Pol 55(3):471–478

Riisgård HU, Larsen PS (2010) Particle capture mechanisms in suspension-feeding invertebrates. Mar Ecol Prog Ser 428:255–293

Roberts D, Gebruk A, Levin V, Manship BAD (2000) Feeding and digestive strategies in deposit-feeding holothurians. Oceanogr Mar Biol Annu Rev 38:257–310

Robison BH (2004) Deep pelagic biology. J Exp Mar Biol Ecol 300:253–272

Robison BH, Reisenbichler KR, Hunt JH, Haddock SHD (2003) Light production by the arm tips of the deep-sea cephalopod Vampyroteuthis inferenalis. Biol Bull 205:102–109

Robison BH, Reisenbichler KR, Sherlock RE (2017) The coevolution of midwater research and ROV technology at MBARI. Oceanography 30(4):26–37

Rocha LA, Robertson DR, Roman J, Bowen BW (2005) Ecological speciation in tropical reef fishes. Proc R Soc B 272:573–579

Rodriguez J, Tintoré J, Allen JT, Blanco JM, Gomis D et al (2001) Mesoscale vertical motion and the size structure of phytoplankton in the ocean. Nature 410:360–363

Rodriguez-Valera F, Martin-Cuadrado A-B, Rodriguez-Brito B, Pašić L, Thingstad TF et al (2009) Explaining microbial population genomics through phage predation. Nat Rev Microbiol 7:828–836

Rohwer F, Thurber RV (2009) Viruses manipulate the marine environment. Nature 459:207–212

Roman J, Palumbi SR (2003) Whales before whaling in the North Atlantic. Science 301:508–510

Romano C (2021) A hiatus obscures the early evolution of modern lineages of bony fishes. Front Earth Sci 8:618853

Romano C, Koot MB, Kogan I, Brayard A, Minikh AV et al (2016) Permian-Triassic Osteichthyes (bony fishes): diversity dynamics and body size evolution. Biol Rev 91(1):106–147

Romanuk TN, Vogt RJ, Kolasa J (2006) Nutrient enrichment weakens the stabilizing effect of species richness. Oikos 114:291–302

Rosenberg R (2001) Marine benthic faunal successional stages and related sedimentary activity. Sci Mar 65:107–119

Rota-Stabelli O, Daley AC, Pisani D (2013) Molecular Timetrees reveal a Cambrian colonization of land and a new scenario for Ecdysozoan evolution. Curr Biol 23:392–398

Roux S, Brum JR, Dutilh BE, Sunagawa S, Duhaime MB et al (2016) Ecogenomics and potential biogeochemical impacts of globally abundant ocean viruses. Nature 537:689–693

Royer M, Meyer C, Royer J, Maloney K, Cardona E et al (2023) "Breath holding" as athermoregulation strategy in the deep-diving scalloped hammerhead shark. Science 380:651–655

Rubinstein CV, Garrienne P, de la Puente GS, Astini RA, Steemans P (2010) Early middle Ordovician evidence for land plants in Argentina (eastern Gondwana). New Phytol 188:365–369

Rundell RJ, Leander BS (2010) Masters of miniaturization: convergent evolution among interstitial eukaryotes. BioEssays 32:430–437

Rusconi R, Stocker R (2016) Microbes in flow. Curr Opin Microbiol 25:1–8

Ruta M, Benton MJ (2008) Calibrated diversity, tree topology and the mother of mass extinctions: the lesson of temnospondylus. Palaeontology 51(6):1261–1288

Ruxton GD, Houston DC (2003) Could tyrannosaurus rex have been a scavenger rather than a predator? Proc R Soc Lond B 270:731–733

Sahoo SK, Planavsky NJ, Jiang G, Kendal IB, Owens JD et al (2016) Oceanic oxygenation events in the anoxic Ediacaran Ocean. Geobiology 14:457–468. https://doi.org/10.1111/gbi.12182

Salinas-de-León P, Phillips B, Ebert D, Shivji M, Cerutti-Pereyra F et al (2018) Deep-sea hydrothermal vents as natural egg-case incubators at the Galapagos rift. Sci Rep 8:1788. https://doi.org/10.1038/s41598-018-20046-4

Sallan L, Friedman M, Sansom RS, Bird CM, Sansom IJ (2018) The nearshore cradle of early vertebrate diversification. Science 362:460–464

Salzburger W (2018) Understanding explosive diversification through cichlid fish genomics. Nat Rev Genet 19:705–717

Sánchez-Baracaldo P (2015) Origin of marine planktonic cyanobacteria. Sci Rep 5:17418. https://doi.org/10.1039/srep17418

Sander PM, Christian A, Clauss M, Fechner R, Gee CT et al (2011) Biology of the sauropod dinosaurs: the evolution of gigantism. Biol Rev 86:117–155

Sansom RS, Randle E, Donoghue PCJ (2015) Discriminating signal from noise in the fossil record of early vertebrates reveals cryptic evolutionary history. Proc R Soc B 282:20142245

Santini S, Jeudy S, Bartoli J, Poirot O, Lescot M et al (2013) Genome of Phaeocystis globosa virus PgV-16T highlights the common ancestry of the largest known DNA virus infecting eukaryotes. PNAS 110(26):10800–10805

Sarmiento JL, Gruber H, Brzezinski MA, Dunne JP (2003) High-latitude controls of thermocline nutrients and low latitude biological productivity. Nature 427:56–60

Sato R, Tanaka Y, Ishimaru T (2001) House production by Oikopleura dioica (Tunicata, Appendicularia) under laboratory conditions. J Plankton Res 23(4):415–423

Saunders A, Reichow M (2009) The Siberian traps and the end-Permian mass extinction: a critical review. Chin Sci Bull 54(1):20–37

Scheffer M, Rinaldi S, Huisman J, Weissing FJ (2003) Why plankton communities have no equilibrium: solutions to the paradox. Hydrobiologia 491:9–18

Scheyer TM, Romano C, Jenks J, Bucher H (2014) Early Triassic marine biotic recovery: the predators' perspective. PLoS One 9(3):e88987

Schmidt DN, Lazarus D, Young JR, Kucera M (2006) Biogeography and evolution of body size in marine plankton. Earth Sci Rev 78:239–266

Schneebeli-Hermann E (2012) Extinguishing a Permian world. Geology 40(3):287–288

Schnetz L, Butler RJ, Coates MI, Sansom IJ (2022) Skeletal and soft tissue completeness of the acanthodian fossil record. Palaeonteol 65:e12616

Schoch RR (2009) Evolution of life cycles in early amphibians. Annu Rev Earth Planet Sci 17:135–162

Schoch RR (2014) Amphibian evolution. Wiley Blackwell, Oxford

Schoemann V, Becquevort S, Stefels J, Rousseau V, Lancelot C (2005) Phaeocystis blooms in the global ocean and their controlling mechanisms: a review. J Sea Res 53:43–66

Schorr GS, Falcone EA, Moretti DJ, Andrews RD (2014) First Long-term behavioral records from Cuvier's beaked whales (Ziphius cavirostris) reveal record-breaking dives. PLoS One 9(3):e92633

Schwalbach MS, Fuhrman JA (2005) Wide-ranging abundances of aerobic anoxygenic phototrophic bacteria in the world ocean revealed by epifluorescence microscopy and quantitative PCR. Limnol Oceanogr 50(2):620–628

Scotese CR, Song H, Mills BJW, van der Meer DG (2021) Phanerozoic paleotemperatures: the earth's changing climate during the last 540 million years. Erth Sci Rev 215:103503

Scott R, Marsh R, Hays GC (2014) Ontogeny of long distance migration. Ecology 95(10):2840–2850

Sebens KP (1986) Spatial relationships among encrusting marine organisms in the New England subtidal zone. Ecol Monogr 56(1):73–96

Selosse M-A, Charpin M, Not F (2017) Mixotrophy everywhere on land and in water: the grand écart hypothesis. Ecol Lett 20:246–263

Sengupta A, Carrara F, Stocker R (2017) Phytoplankton can actively diversify their migration strategy in response to turbulent cues. Nature 543:555–558

Sepkoski JJ (1981) A factor analytic description of the Phanerozoic marine fossil record. Paleobiology 7(1):36–53

Sepkoski JJ (1998) Rates of speciation in the fossil record. Philos Trans R Soc Lond 353:315–326

Servais T, Perrier V, Danelian T, Klug C, Martin R et al (2016) The onset of the 'Ordovician plankton revolution' in the late Cambrian. Palaeogeogr Palaeoclimatol Palaeoecol 458:12–28

Seuront L (2008) Microscale complexity in the ocean: turbulence, intermittency and plankton life. Math Model Nat Phenom 3(5):1–41

Seuront L, Schmitt F, Lagadeuc Y, Schertzer D, Lovejoy S, Frontier S (1996) Multifractal analysis of phytoplankton biomass and temperature in the ocean. Geophys Res Lett 23(24):3591–3594

Seuront L, Schmitt F, Lagadeuc Y (2001) Turbulence intermittency, small-scale phytoplankton patchiness and encounter rates in plankton patchiness and encounter rates in plankton: where do we go from here? Deep-Sea Res I 48:1199–1215

Shaked Y, de Vargas C (2006) Pelagic photosymbiosis: rDNA assessment of diversity and evolution of dinoflagellate symbionts and planktonic foraminiferal hosts. Mar Ecol Prog Ser 325:59–71

Shanks AL (2009) Pelagic larval duration and dispersal distance revisited. Biol Bull 216:373–385

Shapiro OH, Fernandez VI, Garren M, Guasto JS, Debaillon-Vesque FP et al (2014) Vortical ciliary flows actively enhance mass transport in reef corals. PNAS 111(37):13391–13396

Shen K-N, Jamandre BW, Hsu C-C, Tzeng W-N, Durand J-D (2011) Plio-Pleistocene Sea level and temperature fluctuations in the northwestern Pacific promoted speciation in the globally-distributed flathead mullet Mugil cephalus. BMC Evol Biol 11:83

Shen J, Algeo TJ, Zhou L, Feng Q, Yu J, Ellwood B (2012) Volcanic perturbations of the marine environment in South China preceding the latest Permian mass extinction and their biotic effects. Geobiology 10:82–103

Sherr EB, Sherr BF (2002) Significance of predation by protists in aquatic microbial food webs. Antonie Van Leeuwenhoek 81:293–308

Shields-Zhou G, Zhu M (2013) Biogeochemical changes across the Ediacaran-Cambrian transition in South China. Precambrian Res 225:1–6

Shimada K (2019) The size of the megatooth shark, Otodus megalodon (Lamniformes: Otodontidae), revisited. Hist Biol 33:904–911. https://doi.org/10.1080/08912963.2019.1666840

Shore AJ, Wood RA, Butler IB, Zhuravlev AY, McMahon S et al (2021) Ediacaran metazoan reveals lophotrochozoan affinity and deepens root of Cambrian explosion. Sci Adv 7:eabf2933

Shu D-G, Luo H-L, Conway Morris S, Zhang X-L, Hu S-X et al (1999) Lower Cambrian vertebrates from South China. Nature 402:42–46

Sibbald MJ, Albright LJ (1988) Aggregated and free bacteria as food sources for heterotrophic microflagellates. Appl Environ Microb 54(2):613–616

Silva RR, Ferigolo J, Bajdek P, Piñeiro G (2017) The feeding habits of Mesosauridae. Front Earth Sci 5:23

Simon N, Cras A-L, Foulon E, Lemée R (2009) Diversity and evolution of marine phytoplankton. C R Biologies 332:159–170

Simon-Lledó E, Bett BJ, Huvenne VAI, Schoening T, Benoist NMA et al (2019) Megafaunal variation in the abyssal landscape of the clarion Clipperton zone. Prog Oceanogr 170:119–133

Sims DW (1999) Threshold foraging behaviour of basking sharks on zooplankton: life on an energetic knife-edge? Proc R Soc Lond B 266:1437–1443

Sims DW, Wearmouth VJ, Southall EJ, Hill JM, Moore P et al (2006) Hunt warm, rest cool: bioenergetics strategy underlying diel vertical migration of a benthic shark. J Anim Ecol 75:176–190

Slater GJ, Goldbogen JA, Pyenson ND (2017) Independent evolution of baleen whale gigantism linked to Plio-Pleistocene Ocean dynamics. Proc R Soc B 284:20170546

Smetacek V (2012) Making sense of ocean biota: how evolution and biodiversity of land organisms differ from that of the plankton. J Biosci 37(4):589–607

Smil V (2017) Energy and civilization. MIT Press, Cambridge

Smith CR, Rabouille C (2002) What controls the mixed-layer depth in deep-sea sediments? The importance of POC flux. Limnol Oceanogr 47(2):418–426

Smith JD, Roy K (2006) Selectivity during background extinction: Plio-Pleistocene scallops in California. Paleobiology 32(3):408–416

Smith AB, Stockley B (2005) The geological history of deep-sea colonization by echinoids: roles of surface productivity and deep-water ventilation. Proc R Soc B 272:865–869

Smith CR, Bernardino AF, Baco A, Hannides A, Altamira I (2014) Seven-year enrichment: macrofaunal succession in deep-sea sediments around a 30 tonne whale fall in the Northeats Pacific. Mar Ecol Prog Ser 515:135–149

Smith CR, Glover AG, Treude T, Higgs ND, Amon DJ (2015) Whale-fall ecosystems: recent insights into ecology, paleoecology, and evolution. Annu Rev Mar Sci 7:571–596

Smithwick FM, Stubbs TL (2018) Phanerozoic survivors: Actinopterygian evolution through the Permo-Triassic and Triassic-Jurassic mass extinction events. Evolution 72(2):348–362

Sobstel M, Makowska-Haftka M, Racki G (2006) Conodont ecology in the early-middle Frasnian transition on the south polish carbonate shelf. Acta Palaeontol Pol 51(4):719–746

Sommer U, Sommer F (2006) Cladocerans versus copepods: the cause of contrasting top-down controls on freshwater and marine phytoplankton. Oecologia 147:183–194

Song H, Wignall PB, Chu D, Tong J, Sun Y et al (2014) Anoxia/high temperature double whammy during the Permian-Triassic marine crisis and its aftermath. Sci Rep 4. https://doi.org/10.1038/srep04132

Song X, Ruthensteiner B, Lyu M, Liu X, Wang J, Han J (2021) Advanced Cambrian hydroid fossils (Cnidaria: hydrozoa) extend the medusozoan evolutionary history. Proc R Soc B 288:20202939

Spiekman SNF, Fraser NC, Scheyer TM (2021) A new phylogenetic hypothesis of Tanystropheidae (Diapsida, Archosauromorpha) and other "protosaurs", and its implications for the early evolution of stem archosaurs. PeerJ 9:e11143

Stanley SM (2001) Historische Geologie. Spektrum Akademischer Verlag, Heidelberg

Stanley GD (2006) Photosymbiosis and the evolution of modern coral reefs. Science 312:857–858

Stanley SM (2009) Evidence from ammonoids and condonts for multiple early Triassic mass extinctions. PNAS 106(26):15264–15267

Stark NM, Jordan CF (1978) Nutrient retention by root mat of an Amazonian rain forest. Ecology 59(3):434–437

Steeman ME, Hebsgaard MB, Fordyce RE, Ho SYW, Rabosky DL et al (2009) Radiation of extant cetaceans driven by restructuring of the oceans. Syst Biol 58(6):573–585

Stefano GB, Mantione KJ, Casares FM, Kream RM (2015) Anaerobically functioning mitochondria: evolutionary perspective on modulation of energy metabolism in Mytilus edulis. ISJ 12:22–28

Stella JS, Pratchett MS, Hutchings PA, Jones GP (2011) Coral-associated invertebrates: diversity, ecological importance and vulnerability to disturbance. Oceanogr Mar Biol Annu Rev 49:43–104

Stigall AL (2012) Speciation collapse and invasive species dynamics during the late Devonian "mass extinction". GSA Today 22(1):4–9

Stocker R, Seymour JR (2012) Ecology and physics of bacterial chemotaxis in the ocean. Microb Molecular Biol Rev 76(4):792–812

Stocking JB, Rippe JP, Reidenbach MA (2016) Structure and dynamics of turbulent boundary layer flow over healthy and algae-covered corals. Coral Reefs 35:1047–1059

Stoecker DK, Johnson MD, de Vargas C, Not F (2009) Acquired phototrophy in aquatic protists. Aquat Microb Ecol 57:279–310

Stoecker DK, Hansen PJ, Caron DA, Mitra A (2017) Mixotrophy in the marine plankton. Annu Rev Mar Sci 9:311–335

Stolarski J, Kitahara MV, Miller DJ, Cairns SD, Mazur M, Meibom A (2011) The ancient evolutionary origins of Scleractinia revealed by azooxanthellate corals. Evol Biol 11:316

Stomp M, Huisman J, Stal LJ, Matthijs HCP (2007) Colorful niches of phototrophic microorganisms shaped by vibrations of the water molecule. ISME J 1:271–282

Stowe SD, Richards JJ, Tucker AT, Thompson R, Melander C, Cavanagh J (2011) Anti-biofilm compounds derived from marine sponges. Mar Drugs 9:2010–2035

Straube N, Li C, Claes JM, Corrigan S, Naylor GJP (2015) Molecular phylogeny of Squaliformes and first occurrence of bioluminescence in sharks. BMC Evol Biol 15:162

Suchéras-Marx B, Mattioli E, Allemand P, Giraud F, Pittet B et al (2019) The colonization of the oceans by calcifying pelagic algae. Biogeosciences 16:2501–2510

Sunda WG, Hardison DR (2010) Evolutionary tradeoffs among nutrient acquisition, cell size, and grazing defense in marine phytoplankton promote ecosystem stability. Mar Ecol Prog Ser 401:63–76

Sutherland KR, Madin LP, Stocker R (2010) Filtration of submicrometer particles by pelagic tunicates. PNAS 107:15129–15134

Suttle CA (2005) Viruses in the sea. Nature 437:356–361

Suttle CA (2007) Marine viruses—major players in the global ecosystem. Nat Rev Microbiol 5:401–812

Sutton TT, Porteiro FM, Heino M, Byrkjedal I, Langhelle G et al (2008) Vertical structure, biomass and topographic associations of deep-pelagic fishes in relation to a mid-ocean ridge system. Deep-Sea Res II 55:161–184

Svendsen H (1997) Physical oceanography and marine ecosystems: some illustrative examples. Sci Mar 61:93–108

Syverson VJP, Brett CE, Gahn FJ, Baumiller TK (2018) Spinosity, regeneration, and targeting among Paleozoic crinoids and their predators. Paleobiology 44(2):290–305

Takagi H, Kimoto K, Fujiki T, Saito H, Schmidt C et al (2019) Characterizing photosymbiosis in modern planktonic foraminifera. Biogeosciences 16:3377–3396

Takahashi K, Ichikawa T, Fukugama C, Yamane M, Kakehi S et al (2015) In situ observations of doliolid blooms in a warm water filament using a video plankton recorder: bloom development, fate, and effects on biogeochemical cycles and planktonic food webs. Limnol Oceanogr 60:1763–1780

Taniguchi M, Ishitobi T, Shimada J, Takamoto N (2006) Evaluations of spatial distribution of submarine groundwater discharge. Geophys Res Lett 33:L06605

Tarasov VG, Gebruk AV, Mironov AN, Moskalev LI (2005) Deep-sea and shallow-water hydrothermal vent communities: two different phenomena? Chem Geol 224:5–39

Taylor JR, Stocker R (2012) Trade-offs of chemotactic foraging in turbulent water. Science 338:675–670

Taylor PD, Waeschenbach A (2015) Phylogeny and diversification of bryozoans. Palaeontology 58(4):585–599

Taylor JW, Hann-Soden C, Branco S, Sylvain I, Ellison CE (2015) Clonal reproduction in fungi. PNAS 112(29):8901–8908

Taylor AR, Brownlee C, Wheeler G (2017) Coccolithophore cell biology: chalking up Progress. Annu Rec Mar Sci 9:283–310

Teal LR, Bulling MT, Parker ER, Solan M (2008) Global patterns of bioturbation intensity and mixed depth of marine soft sediments. Aquat Biol 2:207–218

Teske A (2013) Marine deep sediment microbial communities. In: DeLong EF, Lory S, Stackebrandt E, Thompson F (eds) The prokaryotes—prokaryotic communities and ecophysiology. Springer, Berlin

Teske PR, Beheregaray LB (2009) Evolution of seahorses' upright posture was linked to Oligocene expansion of seagrass habitats. Biol Lett 5:521–523

Tetu SG, Breakwell K, Elbourne LDH, Holmes AJ, Gillings MR, Paulsen IT (2013) Life in the dark: metagenomic evidence that a microbial slime community is driven by inorganic nitrogen metabolism. ISME J 7:1227–1236

Thistle D, Ertman SC, Fauchald K (1991) The fauna of the HEBBLE site: patterns in standing stock and sediment-dynamic effects. Mar Geol 99:413–432

Thuesen EV, Childress JJ (1994) Oxygen consumption rates and metabolic enzyme activities of oceanic California Medusae in relation to body size and habitat depth. Biol Bull 187:84–98

Thuy B, Gale AS, Kroh KM, Numberger-Thuy LD et al (2012) Ancient origin of the modern Deep-Sea Fauna. PLoS One 7(10):e46913

Ting CS, Rocap G, King J, Chisholm SW (2002) Cyanobacterial photosynthesis in the oceans: the origins and significance of divergent light-harvesting strategies. Trends Microbiol 10(3):134–142

Treude T, Kiel S, Linke P, Peckmann J, Goedert JL (2011) Elasmobranch egg capsules associated with modern and ancient cold seeps: a nursery for marine deep-water predators. Mar Ecol Prog Ser 437:175–181

Tsuji LA, Müller J (2009) Assembling the history of the Parareptilia: phylogeny, diversification, and a new definition of the clade. Fossil Record 12(1):71–81

Tsurpalo AP, Kostina EE (2003) Feeding characteristics of three species of Intertidal Sea anemones of the South Kuril Islands. Russ J Mar Biol 29(1):31–40

Turner JT (2002) Zooplankton fecal pellets, marine scow and sinking phytoplankton blooms. Marine Microbial Ecology 27:57–102

Vacquié-Garcia J, Royer F, Dragon A-C, Viviant M, Bailleul F, Guinet C (2012) Foraging in the darkness of the Southern Ocean: influence of bioluminescence on a deep diving predator. PLoS One 7(8):e43565

Våge S, Thingstad TF (2015) Fractal hypothesis of the pelagic microbial ecosystem—can simple ecological principles Lead to self-similar complexity in the pelagic microbial food web? Front Microbiol 6:1357

Van Iten H, Marques AC, De Moraes LJ, Forancelli Pacheco MLA, Simões MG (2014) Origin and early diversification of the phylum Cnidaria Verrill: major developments in the analysis of the taxon's Proterozoic-Cambrian history. Paléo 57(4):677–690

Van Mooy BAS, Fredricks HF, Pedler BE, Dyhrman ST, Karl DM et al (2009) Phytoplankton in the ocean use non-phosphorus lipids in response to phosphorus scarcity. Nature 458:69–72

Van Mooy BAS, Krupke A, Dyhrman ST, Fredricks HF, Frischkorn KR et al (2015) Major role of planktonic phosphate reduction in the marine phosphorus redox cycle. Science 6236:783–785

van Straalen NM (2021) Evolutionary terrestrialization scenarios for soil invertebrates. Pedobio J of soil Ecol 87–88:150753

Vanderklift MA, Lavery PS, Waddington KI (2009) Intensity of herbivory on kelp by fish and sea urchins differs between inshore and offshore reefs. Mar Ecol Prog Ser 376:203–211

Vannier J, Calandra I, Gailland C, Żylińska A (2010) Priapulid worms: Pioneer horizontal burrowers at the Precambrian-Cambrian boundary. Geology 38(8):711–714

Vaškaninova V, Chen D, Tafforeau P, Johanson Z, Ekrt B et al (2020) Marginal dentition and multiple dermal jawbones as the ancestral condition of jawed vertebrates. Science 369:211–216

Veizer J (1973) Sedimentation in geological history: recycling vs. evolution or recycling with evolution. Contrib Mineral Petrol 38:261–278

Velasco-Hogan A, Deheyn DD, Koch M, Nothdurft B, Arzt E, Meyers MA (2019) On the nature of the transparent teeth of Deep-Sea dragonfish, Aristostomias scintillans. Matter 1:1–15

Velez-Juarbe J, Domning DP, Pyenson ND (2012) Iterative evolution of sympatric Seacow (Dugongidae, Sirenia) assemblages during the past ≈ 26 million years. PLoS One 7(2):e31294

Vérard C, Hochard C, Baumgartner PO, Stampfi GM (2015) Geodynamik evolution of the earth over the Phanerozoic: plate tectonic activity and plalaeoclimatic indicators. J Palaeogeogr 4(2):167–188

Verberk WCEP, Atkinson D (2013) Why polar gigantism and Palaeozoic gigantism are not equivalent: effects of oxygen and temperature on the body size of ectotherms. Funct Ecol 27:1275–1285

Verheye HM (2000) Decadal-scale trends across several marine trophic levels in the southern Benguela upwelling system off South Africa. Ambio 29(1):30–34

Vermeij GJ (1977) The Mesozoic marine revolution: evidence from snails, predators and grazers. Paleobiology 3:245–258

Vermeij GJ, Motani R (2018) Land to sea transitions in vertebrates: the dynamics of colonization. Paleobiology 44(2):237–250

Villamil J, Nuñez Demarco P, Meneghel M, Blanco RE, Jones W et al (2016) Optimal swimming speed estimates in the early Permian meosaurid Mesosaurus tenuidens (Gervais 1865) from Uruguay. Hist Biol 28(7):963–971

Villanueva R, Norman MD (2008) Biology of the planktonic stages of benthic octopuses. In: Gibson RN, Atkinson RJA, Gordon JDM (eds) Oceanography and marine biology: an annual review, vol 46. Taylor and Francis, pp 105–202

Villareal-Chiu JF, Quinn JP, McGrath JW (2012) The genes and enzymes of phosphate metabolism by bacteria, and their distribution in the marine environment. Front Microbiol 3:19

Vinther J, Stein M, Longrich NR, Harper DAT (2014) A suspension-feeding anomalocarid from the early Cambrian. Nature 307:496–499

Vogel S (1994) Life in Moving Fluids. Princeton University Press, Princeton

Völkel S, Grieshaber MK (1997) Sulphide oxidation and oxidative phosphorylation in the mitochondria of the lugworm Arenicola marina. J Exp Biol 200:83–92

von Reumont BM, Jenner RA, Wills MA, Dell'Ampio E, Pass G et al (2012) Pancrustacean phylogeny in the light of new Phylogenomic data: support for Remipedia as the possible sister group of Hexapoda. Mol Biol Evol 29(3):1031–1045

Vrijenhoek RC (2010) Genetics and evolution of deep-sea chemosynthetic bacteria and their invertebrate hosts. In: Kiel S (ed) The vent and seep biota, Topics in geobiology, vol 33. Springer Science+Business Media, pp 15–49

Vrijenhoek RC, Johnson SB, Rouse GW (2009) A remarkable diversity of bone-eating worms (Osedax; Siboglinidae; Annelida). BMC Biol 7:74

Wagner D, Kelley CD (2017) The largest sponge in the world? Mar Biodivers 47:367–368

Wagner H-J, Douglas RH, Frank TM, Roberts NW, Partridge JC (2009) A novel vertebrate eye using both refractive and reflective optics. Curr Biol 19:108–114

Waldbauer JR, Rodriguez S, Coleman ML, Chisholm SW (2012) Transcriptome and proteome dynamics of a light dark synchronized bacterial cell cycle. PLoS One 7(8):e43432

Walker RA, Hallock P, Torres JJ, Vargo GA (2011) Photosynthesis and respiration in five species of benthic foraminifera that host algal endosymbionts. J Foraminiferal Res 41(4):314–325

Walsh EA, Kirkpatrick JB, Rutherford SD, Smith DC, Sogin M, D'Hondt S (2015) Bacterial diversity and community composition from seasurface to subseafloor. ISME J 10:979–989

Wang DY-C, Kumar S, Hedges SB (1999) Divergence time estimates for the early history of animal phyla and the origin of plants, animals and fungi. Proc R Soc Lond B 266:163–171

Wang X, Gan L, Jochum KP, Schröder HC, Müller WEG (2011) The largest bio-silica structure on earth: the Giant basal spicule from the Deep-Sea glass sponge Monorhaphis chuni. Evid Based Complement Alternat Med:540987

Ward BA, Follows MJ (2016) Marine mixotrophy increases trophic transfer efficiency, mean organism size, and vertical carbon flux. PNAS 113(11):2958–2963

Warren JK (2010) Evaporites through time: tectonic, climatic and eustatic controls in marine and nonmarine deposits. Earth Sci Rev 98:217–268

Watwood SL, Miller PJO, Johnson M, Madsen PT, Tyack PL (2006) Deep-diving foraging behaviour of sperm whales (Physeter macrocephalus). J Anim Ecol 75:814–825

Wegner NC, Snodgrass OE, Dewar H, Hyde JR (2015) Whole-body endothermy in a mesopelagic fish, the opah, Lampris guttatus. Science 348:786–789

Wellman CH, Steemans P, Vecoli M (2013) Palaeogeography of Ordovician-Silurian land plants. Geol Soc Lond Mem 38:461–476

Whalen CD, Briggs DEG (2018) The Palaeozoic colonization of the water column and the rise of global nekton. Proc R Soc B 285:20180881

Whiteside JH, Olsen PE, Kent DV, Fowell SJ, Et-Touhami M (2007) Synchrony between the Central Atlantic magmatic province and the Triassic-Jurassic mass-extinction event? Palaeogeogr Palaeoclimatol Palaeoecol 24(4):345–367

Widder EA, Latz MI, Case JF (1983) Marine bioluminescence spectra measured with an optical multichannel detection system. Biol Bull 165:791–810

Widdicombe CE, Eloire D, Barbour D, Harris RE, Somerfield PJ (2019) Long-term phytoplankton community dynamics in the Western English Channel. J Plankton Res 32(5):643–655

Wienberg C, Titschak J, Freiwald A, Frank N, Lundalv T et al (2018) The giant Mauritanian cold-water coral mound province: oxygen control on coral mound formation. Quat Sci Rev 185:135–152

Williams TM (1999) The evolution of cost efficient swimming in marine mammals: limits to energetic optimization. Philos Trans R Soc Lond B 354:193–201

Willis KJ, McElwain JC (2002) The evolution of plants. Oxford University Press, Oxford

Wilson RP, Puetz K, Bost CA, Culik BM, Bannasch R et al (1993) Diel dive depth in penguins in relation to diel vertical migration of prey: whose dinner by candlelight? Mar Ecol Prog Ser 94:101–104

Wilson ST, Hawen NJ, Armbrust EV, Barone B, Björkman KM et al (2019) Kilauea lava fuels phytoplankton bloom in the North Pacific Ocean. Science 365:1040–1044

Winston JE (2010) Life in the colonies: learning the alien ways of colonial organisms. Integr Comp Biol 50(6):919–933

Witte U, Aberle N, Sand M, Wenzhöfer F (2003a) Rapid response of a deep-sea benthic community of POM enrichment: an in situ experimental study. Mar Ecol Prog Ser 251:27–36

Witte U, Wenzhöfer F, Sommer S, Boetius A, Heinz P et al (2003b) In situ experimental evidence of the fate of a phytodetritus pulse at the abyssal sea floor. Nature 424:763–766

Wolff T (1979) Macrofaunal utilization of plant remains in the deep sea. Sarsia 64:117–136

Wood R (1998) The ecological evolution of reefs. Annu Rev Evol Syst 29:179–206

Woods JD, Onken R (1982) Diurnal variation and primary production in the ocean—preliminary results of a Lagrangian ensemble model. J Plankton Res 4(3):735–756

Woodward BL, Winn JR, Fish FE (2006) Morphological specializations of baleen whales associated with hydrodynamic performance and ecological niche. J Morphol 267:1284–1294

Worden AZ, Follows MJ, Giovannoni SJ, Wilken S, Zimmerman AE, Keeling PJ (2015) Rethinking the marine carbon cycle: factoring in the multifarious lifestyles of microbes. Science 347:1257594

Worm B, Lotze HK (2006) Effects of eutrophication, grazing, and algal blooms on rocky shores. Limnol Oceanogr 51/1(Part 2):569–579

Wu B, Feng C, Zhu C, Xu W, Yuan Y et al (2021) The genomes of two Billifisches provide insights into the evolution of Endothermy in Teleosts. Mol Biol Evol 38(6):2413–2427

Xiao S, Tang Q (2018) After the boring billion and before the freezing millions: evolutionary patterns and innovations in the Toneian period. Emerg Top in Life Sci 2:161–171

Xiong C, Wang D, Wang Q, Benton MJ, Xue J et al (2013) Diversity dynamics of Silurian-early carboniferous land plants in South China. PLoS One 8(9):e75706

Xue C-X, Liu J, Lea-Smith DJ, Rowley G, Lin H et al (2020) Insights into the vertical stratification of microbial ecological roles across the deepest seawater column on earth. Microorganisms 8:1309

Yang R-D, Mao JR, Zhang W-H, Jiang L-J, Gau H (2004) Bryophyte-like fossil (Parafunaria sinensis) from early-middle Cambrian Kaili formation in Guizhou Province, China. Acta Bot Sin 46(2):180–185

Yang J, Lan T, Zhang X-G, Smith MR (2023) Protomelission is an early dasyclad alga not a Cambrian bryozoan. Nature 615:468–471

Yao L, Aretz M, Chen J, Webb GE, Wang X (2016) Global microbial carbonate proliferation after the end-Devonian mass extinction: mainly controlled by demise of skeletal bioconstructors. Sci Rep 6:39694

Yau S, Lauro FM, DeMaere MZ, Brown MV, Thomas T et al (2011) Virophage control of antarctic algal host-virus dynamics. PNAS 108(15):6163–6168

Yin Z, Zhu M, Davidson EH, Bottjer DJ, Zhao F, Tafforeau P (2015) Sponge grade body fossil with cellular resolution dating 60 Myr before the Cambrian. PNAS 112:E1453–E1460

Yin Z, Vargas K, Cunningham J, Bengtson S, Zhu M et al (2019) The early Ediacaran Caveasphaera foreshadows the evolutionary origin of animal-like embryology. Curr Biol 29:4307–4314

Yonge CM (1955) Adaptation to rock boring in Botula and Lithophaga (Lamellibranchia, Mytilidae) with a discussion on the evolution of this habit. Q J Microscop Sci 96(3):383–410

Yoshida-Takashima Y, Nunoura T, Kazama H, Noguchi T, Inoue K et al (2011) Spatial distribution of viruses associated with planktonic and attached microbial communities in hydrothermal environments. Appl Environ Microbiol 78(5):1311–1320

Yu X, Doroghazi JR, Janga SC, Zhang JK, Circello B et al (2013) Diversity and abundance of phosphonate biosynthetic genes in nature. PNAS 110(51):20759–20764

Zachos J, Pagani M, Sloan L, Thomas E, Billups K (2001) Trends, rhythms, and aberrations in global climate 65 Ma to present. Science 292:686–693

Zaffos A, Finnegan S, Peters SE (2017) Plate tectonic regulation of global marine animal diversity. PNAS 114(22):5653–5658

Zapata F, Goetz FE, Smith SA, Howison M, Siebert S et al (2015) Phylogenomic analyses support traditional relationships within Cnidaria. PLoS One 10(10):e0139068

Zehr JP, Capone DG (2020) Changing perspectives in marine nitrogen fixation. Science 368:eaay9514

Zhang Y, Liu Z, Zhao Y, Wang W, Li J, Xu J (2014) Mesoscale eddies transport deep-sea sediments. Sci Rep 4. https://doi.org/10.1038/srep05937

Zhang X-G, Smith MR, Yang J, Hou J-B (2016) Onychophoran-like musculature in a phosphatized Cambrian lobopodian. Biol Lett 12:20160492

Zhang Z, Zhang Z, Ma J, Taylor PD, Strotz LY et al (2021) Fossil evidence unveils an early Cambrian origin for Bryozoa. Nature 599:251–255

Zhao F, Hu S, Caron J-B, Zhu M, Yin Z, Lu M (2012) Spatial variation in the diversity and composition of the lower Cambrian (series 2, stage 3) Chengjiang biota, Southwest China. Palaeogeogr Palaeoclimatol Palaeoecol 346–347:54–65

Zhaxybayeva O, Doolittle WF, Papke RT, Gogarten JP (2009) Intertwined evolutionary histories of marine Synechococcus and Prochlorococcus marinus. Genome Biol Evol 1:325–339

Zhu M, Yu X, Ahlberg PE, Choo B, Lu J et al (2013) A Siluran placoderm with osteichthyan-like marginal jaw bones. Nature 502:188–193

Zhu M, Ahlberg PE, Pan Z, Zhu Y, Qiao T et al (2016) A Siluran macillate placoderm illuminates jaw evolution. Science 354:334–336

Zhu Y-A, Lu J, Zhu M (2019a) Reappraisal of the Silurian placoderm Silurolepis and insights into the dermal neck joint evolution. R Soc Open Sci 6:191181

Zhu Y-A, Ahlberg PA, Zhu M (2019b) The evolution of vertebrate dermal jaw Bone in the light of Maxillate placoderms. In: Johanson Z, Underwood C, Richter M (eds) Evolution and development of fishes. Cambridge University Press, Cambridge, pp 71–86

Ziegler B (1998) Spezielle Paläontologie Teil 3. Schweizerbart'sche Verlagsbuchhandlung, Stuttgart

Zinser ER, Lindell D, Johnson ZI, Futschik ME, Steglich C et al (2009) Choreography of the transcriptome, photophysiology, and cell cycle of a minimal photoautotroph, Prochloro-coccus. PLoS One 4(4):e5135

Zubkov M, Tarran GA (2008) High bacterivory by the smallest phytoplankton in the North Atlantic Ocean. Nature 455:224–226

Zubkov MV, Fuchs BM, Tarran GA, Burkill PH, Amann R (2003) High rate of uptake of organic nitrogen compounds by Prochlorococcus cyanobacteria as a Key to their dominance in oligotrophic oceanic waters. Appl Environ Microb 69(2):1299–1304

Zwirglmaier K, Jardiller L, Ostrowski M, Mazard S, Garczarek L et al (2008) Global phylogeography of marine Synechococcus reveals a distinct partitioning of lineages among oceanic biomes. Experimental Microb 10(1):147–161

Biological Evolutionary Lineages in Terrestrial Habitats in the Phanerozoic

10

10.1 Land Plants as Shapers of New Habitats

The familiar sight of a landscape makes it easy to forget that it is *a priori* a two-dimensional habitat. Thinking of hills and mountains, we would immediately object to this statement, but a land plant would agree with it at once. Plants can only photosynthesize the fraction of sunlight corresponding to the horizontal proportion of the total irradiated area that they cover. The effect of that limitation is intensified by spatially limited access to mineral nutrients and water in the soil. These relationships become most clearly evident when viewing a single plant in a dense cornfield stand. Structures of natural vegetation usually deviate from these limitations because of a variety of interactions between organisms, which are ultimately variations on the basic principle explained above. Unlike algae in plankton (see Sect. 9.1), land plants are spatially "fixed". Over the long term, their development at any particular site is governed by the release of mineral nutrients through geological and geochemical processes. Over the short term, however, land plants are subject to solar radiation dynamics similar to those observed with planktonic organisms (Fig. 10.1). Unlike the latter, however, they are exposed to greater temperature fluctuations and random processes of atmospheric precipitation (Fig. 10.2 cf. Fig. 9.1).

Under these framework conditions in terrestrial vegetation, associations of different species develop that—under the classical assumption of constant climatic conditions—are expected to tend towards an optimal species composition (climax community). In reality, climatic conditions do not remain stable over the long term, and sequences of changes in species composition (successions) are not predetermined. The composition of associations is hence subject to ongoing changes due to influencing factors with differing dynamic characteristics. Aside from anthropogenic influences, this involves not only climatic but also geological factors, as well as the developmental history of the various plant species in

© The Author(s), under exclusive license to Springer-Verlag GmbH, DE, part of
Springer Nature 2024
M. Knoflacher, *Relativity of Evolution*,
https://doi.org/10.1007/978-3-662-69423-7_10

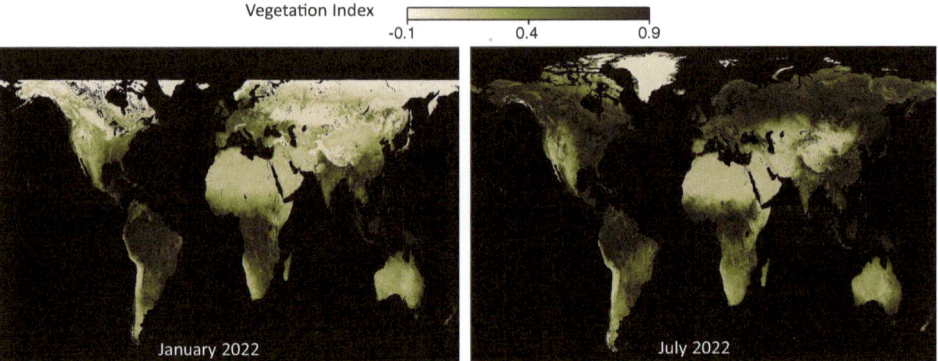

Fig. 10.1 Comparative representation of global land cover by active terrestrial vegetation (vegetation index) in January and July 2022. Credit: NASA Earth Observations (NEO)

Fig. 10.2 Energetic and
abiotic conditions for phototrophic organisms in terrestrial
systems. Explanations in
the text

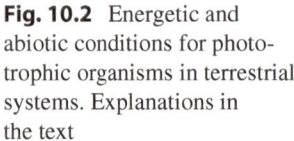

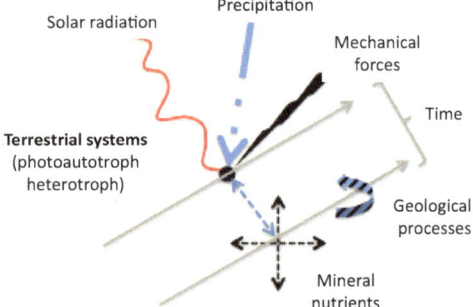

the zone of influence at any given location. Continuous interactions with microorganisms and fungi (Fig. 6.17) for utilization of mineral nutrients are easily overlooked because they are far more difficult to identify in fossil finds (Taylor and Osborn 1996; Read et al. 2000; Konhauser 2007; Bonfante and Anca 2009; Rousk et al. 2009; Dupont et al. 2012; Ogura-Tsujita et al. 2012; Feijen et al. 2018; Delaux and Schornack 2021). Nevertheless, at large spatial dimensions, recurring patterns of similar species compositions can be identified and summarized into abstract units (Walter and Breckle 1983).

Functionally, land plants have to cope with the framework conditions of dynamically and structurally extremely different atmospheric and geological systems (Fig. 8.12) (Burrough 1981; Melillo et al. 1993; Zheru et al. 2001; Ersahin et al. 2006; Senesi and Wilkinson 2008; White 2013). Morphologically, this is achieved by fractal structures of the plant body and functionally by its modular structure. This results in more potential dimensions than in oceans for differentiation of various life forms on a macroscale (Fig. 10.3). Accordingly, it is not surprising that evolution of plants on land is associated with far greater morphological and functional differentiation than in oceans. Here, it is noteworthy that, because of massive expansion of anthropogenic land use, functional differentiation is being increasingly lost (O'Loughlin et al. 2016).

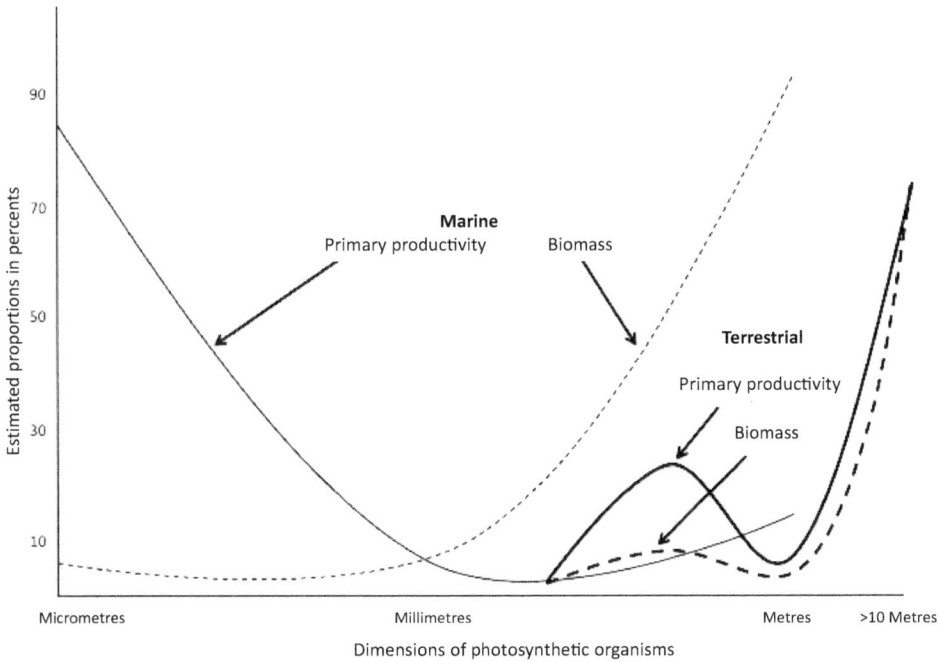

Fig. 10.3 Simplified distributions of relative proportions of current global primary production and biomass across different sizes of photosynthetic organisms in marine and terrestrial habitats. Data source: Lieth and Whittaker (1975)

As already explained (see Chap. 9), photosynthesis in marine habitats—at least since the Ediacaran—has been dominated by microscopically small microorganisms with short biomass lifetimes. Large organisms with longer life cycles—such as seagrass meadows or seaweed stands—contribute relatively little to primary production, but make up the major part of biomass in oceans (Fig. 10.3).

Contrasting with the state of affairs in in oceans, in present-day terrestrial ecosystems distributions of primary production and biomass in various habitats are almost equivalent across different plant sizes (Fig. 10.3). Functionally, therefore, they change relationships between carbon and oxygen cycles to a greater extent than marine life forms. Carbon bound in biomass can be broken down only slowly by bacteria or fungi. In the absence of fires or plant consumption by animals, biomass remains—in conjunction with geological processes—isolated from the atmosphere over millions of years (Kenrick et al. 2012). Under such conditions, oxygen accumulates in the atmosphere and also in oceans. As already mentioned, this promoted the evolution of morphologically differentiated life forms in oceans (Dahl et al. 2010; Lenton et al. 2016). Links between atmospheric oxygen concentration and fires in terrestrial ecosystems have already been discussed in Sect. 8.4.2.

Land plants establish new habitat conditions through both their structures and their physiological activities—in macroscopic dimensions for animals and other plants, and in microscopic dimensions for fungi and microorganisms. In simplified terms, spatial dimensions of these functional units are related to the plant sizes concerned and can range from fractions of a millimetre to tens of metres. On a large scale, plant associations can extend over hundreds of square kilometres (Delcourt and Delcourt 1992).

Here, the main differences between terrestrial and marine evolutionary conditions can be recognized. In terrestrial habitats, plants not only produce the food base for animal organisms, but also modify abiotic habitat conditions through their morphological structures and physiological functions (see Sect. 6.4 and Chap. 7). On land, for example, the largest biogenic structures are formed by plants alone, while in the oceans they are formed mainly by animals—such as corals—and only in shallow water zones by macroalgae and stands of seagrass.

10.2   Early Phase of Terrestrial Colonization: Many Assumptions but Few Clues

In stark contrast to marine habitats, there is no clear fossil evidence of terrestrial plants and animals until the Ordovician (Fig. 10.6). Molecular analyses suggest that colonization of land areas by plants and fungi began 600 to 700 million years ago and by animals around 510 million years ago (Heckman et al. 2001; Clarke et al. 2011). According to currently prevailing scientific opinion, colonization of land areas by multicellular plants began around 480 to 450 million years ago (Gensel 2008; Langdale 2008; Smith et al. 2010; Edwards and Kenrick 2015), accompanied by fungi and microorganisms (Heckman et al. 2001; Rota-Stabelli et al. 2013; Wellman et al. 2013; Edwards et al. 2014; Edwards and Kenrick 2015; Field et al. 2015; Knack et al. 2015; Berbee et al. 2017). This also created suitable conditions for the origin of the terrestrial evolutionary lineage of animals, which were exploited by a first wave of invertebrates to colonize land. Currently known traces of life and fossil finds suggest that the initial colonizers were millipedes (Myriapoda), which utilized plant debris, between 500 and 450 million years ago (Ward et al. 2006; Dunn 2013; Lozano-Fernandez et al. 2016; Suarez et al. 2017). Traces of utilization of plant parts by arthropods can be found as from the Devonian around 400 million years ago, followed by an increase in diversity and frequency beginning at the end of the Carboniferous around 300 million years ago (Labandeira 2013b).

Interestingly, land-dwelling isopods evolved much later from the crustacean group, with the first traces of life thought to be around 150 million years old and the first actual fossils around 110 million years old. It is possible that their late appearance on land is linked to the almost contemporaneous development of flowering plants (Broly et al. 2013; Magallón et al. 2015).

Both fungi and photosynthetic organisms such as cyanobacteria and eukaryotic algae occur in near-surface zones of oceans (Reynolds 2006; Gao et al. 2010). For unicellular life forms, there is a high probability of wind transportation to land areas (Henderson-Begg et al. 2009). Ecophysiologically, organisms in the new habitat are primarily exposed to increased stress from short- and long-wave ranges of solar radiation, as well as frequently occurring water shortages. In order to maintain their vital functions, phototrophic organisms have to cope with a variety of requirements. On the one hand, protection against the aforementioned influences requires modified cell walls and physiological processes, while on the other hand in order to maintain their metabolism cells need to interact as intensively as possible with their environment. Symbioses of cyanobacteria, other bacteria and fungi in the form of lichens are far better equipped to cope with these conflicting requirements than individual organisms (Grube and Berg 2009; Pastore et al. 2014; Insarova and Blagoveshchenskaya 2016). In view of the widespread distribution of lichens in extreme habitats even in present times, it can be assumed that they also colonized land areas after the Cryogenian (Belnap and Lange 2003; Jahren et al. 2003; Yang et al. 2004; Lalley and Viles 2008; Nash III 2008; Lakatos 2011; Henskens et al. 2012; Honegger et al. 2013; Retallack 2019). Through both their structural adaptations and exploitation of minerals from hard rock, they opened up new options for the colonization of land surfaces by eukaryotic algae and the evolution of multicellular plants. All necessary functions are performed by the groups of organisms mentioned above, which is why animals, in principle, are not required for colonization of land areas by plants (Fig. 10.4a). However, animals accelerate transformation processes by breaking down living and dead organic material. They also contribute to soil formation by mechanically mixing and transporting substrate components (Fig. 10.4b). In this scenario, it can be assumed that under suitable conditions processes repeated billions of times led to the evolution of multicellular plants. Although details of those processes are as yet unknown, fossils of the oldest forms of vegetation already show that they were not isolated functions of individual organisms (Dotzler et al. 2006; Krings and Taylor 2014; Strullu-Derrien et al. 2014; Wellman 2017). Success was achieved and secured through symbioses between various unicellular and multicellular life forms (Honegger et al. 2013; Field et al. 2015; Porada et al. 2016; Berbee et al. 2017; Mills et al. 2017).

Studies conducted in North American boreal lichen forests reveal the complexity of interactions between lichens and vascular plants. Such forests are characterized by sparse conifer stands interspersed with extensive lichen mats. In dynamic terms, they are exposed to recurrent forest fires. Because of the vegetation structure, however, trees have a low fire risk. Nonetheless, growth of trees can proliferate only to a limited extent because seed development is possible exclusively within occasional gaps in the lichen mats. Due to the specific microclimatic conditions, it is only in those gaps that roots of germinating seeds can reach mineral-containing soil so that shoots can develop (Walter and Breckle 1991).

Inhibiting effects of lichen mats on the propagation of tree vegetation are seemingly independent of climatic conditions. In 1975, during a research visit to the central Amazon region, I was able to accompany my American colleague Anthony Anderson while he studied a largely open sandy area in the heart of the rainforest.[1] Such ecosystems—known as White Sands—differ markedly from directly neighbouring rainforest in terms of severe

---

[1] The field research was enabled by a scholarship of the Austrian Science Fund.

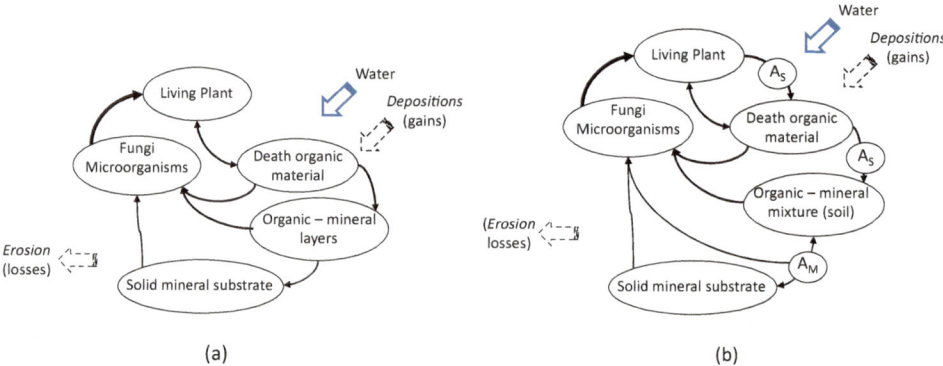

Fig. 10.4 (**a**) Schematic representation of key terrestrial processes for extraction and recovery of mineral nutrients by plants; (**b**) Extension of those processes through comminution of organic material (AS), as well as mixing and transportation of substrate (AM) by animals

soil acidification and nutrient depletion, as well as substantial daily temperature differences. The sharp vegetation boundaries between rainforest and white sand areas are striking. As with lichen forests, this is probably due to lichen mats that cover the entire area shaded by rainforest trees. This inhibits the spread of seeds from rainforest vegetation in microclimatically favourable zones. In adjoining—largely open—areas of the White Sands, only sparse, diminutive and highly specialized vegetation is able to develop (Anderson et al. 1975). Charcoal layers at a soil depth of 20–30 cm, as well as the sharp demarcation to the rainforest, suggest anthropogenic overuse in historical times—rather than natural factors—as the cause for formation of the White Sands (Adeney et al. 2016). This assessment is confirmed both by increasing evidence of pre-Columbian land use and by the observation of soil changes under the current expansion of slash-and-burn agriculture in the Amazonian region (Neves 2007; Bush et al. 2015; Flores and Holmgren 2021). The erosion-inhibiting influence of the natural vegetation could also be observed during the course of flooding, reaching a maximum level of around 25 metres above the baseline level. Although the flow of floodwaters was noticeable over several kilometres in the tributaries, no signs of erosion were observed in the floodplain forest. However, high erosion rates were observed along banks where deforested areas had been converted to pastureland.

This clearly reveals that the evolution of multicellular organisms also takes place by means of complex interactions between a wide variety of factors under randomly varying spatial and temporal conditions. However, the examples also illustrate the long-term—and unpredictable—consequences of putatively minimal changes for existing processes. Those examples provide indications of the limited potential of animals for modifying the abiotic-biotic conditions of terrestrial vegetation. As already mentioned, animals can influence the dynamics of interactions between plants, fungi and microorganisms through mechanical activities. However, they are dependent on a suitable supply of microorganisms in their digestive system and on a sufficient availability of utilizable organic material. Consequently, proposed explanations based solely on intraspecific parameters—for example genomes or morphological factors—inevitably lead to ambiguous results. The reasons for this reside primarily in the multidimensionality of interactions and not so much in inadequate knowledge of the respective effective factors.

10.3   Different Pathways for Colonization of Land Surfaces by Animals: Optional Pathways through the Subsurface

Interest in the evolution of terrestrial tetrapods from marine vertebrates overshadows the diversity of colonization processes exhibited by different marine organisms. In all cases, the basic prerequisite for a successful exodus of animals from the sea was availability of sufficient food resources from phototrophic organisms on land or in terrestrial-aquatic habitats. In the early phases of land colonization, relatively small amounts of existing photosynthetic primary production provided at the most food for small invertebrates (Labandeira 2013b; Gerrienne et al. 2016). It will never be possible to demonstrate with sufficient certainty when and which animals colonized land for the first time. Although individual fossil finds—for example of millipedes, scorpions or freshwater crustaceans (Branchiopoda) from the Silurian (Edgecombe and Giribet 2007; Sun et al. 2016; Brookfield et al. 2020; Wendruff et al. 2020)—provide evidence of the existence of such life forms, they do not represent fixed points in the evolutionary history of terrestrial life.

Faunal composition among the approximately 407-million-year-old Devonian fossils from the Rhynie and Windyfield sites in Scotland generally confirms this statement. However, that fauna is composed of an astonishingly large variety of functional life forms. In addition to detritivores—organisms living on dead plant material—species living on plant spores and fungi as well as a variety of carnivores are represented. Apparent dominance of arthropods among the fossils is presumably due to the greater durability of their exoskeletons. One indication of this is the only evidence to date of a nematode species, whose remains have been preserved solely by the integument of the plant tissue it had parasitized (Dunlop and Garwood 2017).

For early terrestrial colonization, the question of processes and pathways of emigration from the oceans remains largely unanswered. Even for aquatic life forms, the transition from salt water to fresh water may have occurred passively through large-scale separation of residual water bodies with falling sea levels, or actively by means of immigration along flowing watercourses (Fig. 10.5, a I_f). Owing to the vegetation, composed of small tracheophytes, conditions in watercourses were determined primarily by climatic and hydrogeological factors (Strother et al. 2011). As with high montane or steppe areas (Knighton 1998), floods are likely to have changed the structure and course of rivers unpredictably. Moreover, organisms had to adapt to irregularly occurring dry periods of varying duration. Regardless of this, all organisms had to cope with the physiological shift to osmotic conditions in freshwater—regarding which it should be noted that the relevant physiological processes are known only incompletely (Lee et al. 2022). Early colonization of inland

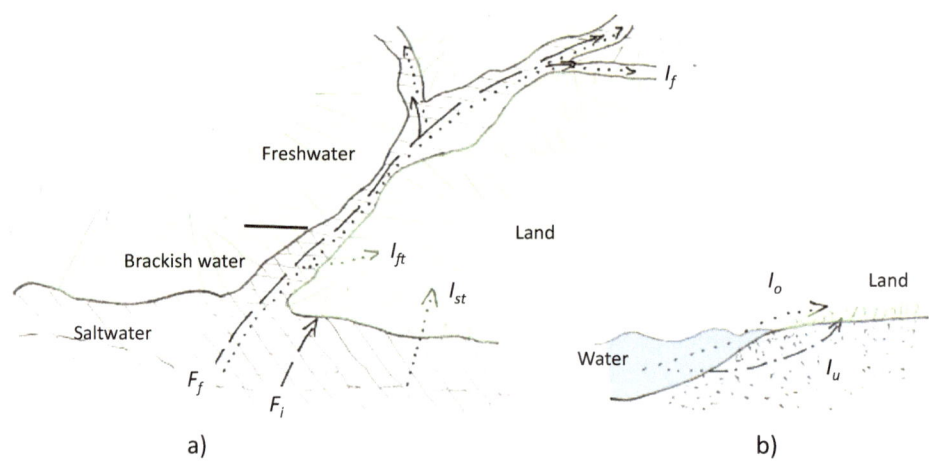

Fig. 10.5 Potential pathways for animals migrating into a terrestrial habitat: (**a**) Immigration of fish into freshwater (F_f) or intertidal zones (F_i), and of invertebrates directly onto land (I_{st}), into freshwater (I_f) or from there onto land (I_{ft}); (**b**) alternative pathways for small invertebrates above ground (I_o) or below ground, for example via the interstitial zone (Iu). Explanations in the text

waters by invertebrates in the Silurian also provided a food base for the first emigration of vertebrates—armoured, jaw-bearing fish (Arthrodira) (Friedman and Sallan 2012; Young and Lu 2020). This group of fish was completely replaced by bony fish from the Carboniferous onwards. Although various bony fish species breathe air temporarily or live in intertidal zones (Fig. 10.5a, F_i), only one group (Tetrapodomorpha) evolved into permanent land-dwelling vertebrates (Helfman et al. 2009; Friedman and Sallan 2012; Friedman 2015).

Animals, like plants, had to undergo adaptation of their physiology and morphology to the specific terrestrial environmental conditions. According to classical ideas, they simply followed the migration routes of plants along above-ground dispersal paths and successively reduced the proportion of aquatic phases in their life cycles. Small-bodied invertebrates may have followed different routes to colonize terrestrial areas (Fig. 10.5). They may have migrated through freshwater (I_{ft})—as is presumably the case for annelids, with earthworms as the best-known representatives—or directly from salt water—as is possibly the case for certain species of land snails (Erséus et al. 2020; Martinez-Redondo et al. 2022). In addition to above-ground colonization routes, animals inhabiting the marine interstitial zone (Sect. 9.2.2) can also migrate through the subsurface (Fig. 10.5b). On the one hand, interstitial zones offer protection from direct sunlight; on the other hand chemical and thermal conditions near the coast are subject to constant dynamic modification due to changes in sea level and variable currents in fresh and salt water (Taniguchi et al. 2002; Silva et al. 2006; Robinson et al. 2007). In sediments close to groundwater, both plant remains and living plants are at least temporarily accessible to small invertebrates through these pathways and offer potential for adaptation to terrestrial living conditions. As many interstitial animal groups feed on bacteria or other animals, colonization of such habitats did not directly depend on the evolution of terrestrial vegetation (Giere 2009; Rundell and Leander 2010). It is therefore possible that various groups of organisms migrated into coastal groundwater zones though the interstitial zone much earlier than previous fossil finds indicate (Fig. 10.5b).

Strong evidence for colonization of terrestrial habitats via the interstitial zone has so far been found in horned mites (Oribatida), which are members of the arthropod phylum (Schaefer et al. 2010). This colonization route can also be inferred for threadworms, flatworms and snails, given their ubiquitous distribution in a wide variety of marine and terrestrial habitats (De Ley 2006; Neusser et al. 2011). The evolution of hexapods (insects and entognaths) may also have begun via this route. Recent results of phylogenetic analyses provide evidence of a close relationship between hexapods and Remipedia—small crustaceans inhabiting the interstitial spaces of marine sediments (Regier et al. 2010; Sasaki et al. 2013; Edgecombe and Legg 2014; Schwentner et al. 2017; Giribet and Edgecombe 2019). Adaptation to terrestrial conditions is likely to have already taken place during the first phase of colonization, as there are no aquatic phases in the life cycles of the most ancient groups (class Entognatha): springtails (Collembola), "coneheads" (Protura) or two-pronged bristletails (Diplura). Insect groups with aquatic phases in their life cycles—for example dragonflies (Odonata), mayflies (Ephemeroptera) or stoneflies (Plecoptera)—first appear in the Carboniferous—about a hundred million years after the first springtails (Lancaster and Downes 2013; Misof et al. 2014).

The evolution of vertebrates from fish to land-dwelling tetrapods is often illustrated in a highly simplified manner using the examples of coelacanths (*Latimeria*) or lungfishes. Linear interpretations of evolutionary processes, in conjunction with geologically ancient fossils, easily convey the impression of an early colonization of land by vertebrates. However, various examples show that supposedly stable morphological features can change relatively quickly in response to environmental influences. In a blind, carp-like cave fish (*Cryptotora*), for example, a fixed pelvic girdle has been identified that enabled the fish to walk quadrupedally and climb waterfalls (Flammang et al. 2016). By contrast, marked reductions of the pelvic girdle are found in cetaceans due to regression of the hind limbs over the course of their evolution into fully aquatic life forms (Bejder and Hall 2002; Gatesy et al. 2013). Among fish, bichirs (*Polypterus*) show different locomotion patterns as adults, depending on whether they were either free-swimming or gravel-living with minimal water coverage as juveniles (Standen et al. 2014).

Walking on four fins has also been observed in skates, epaulette sharks (*Hemiscyllium*) and the African lungfish (*Protopterus*) (Lucifora and Vassallo 2002; Dudgeon et al. 2020; Porter et al. 2022; King et al. 2011). The—previously established—early appearance of coelacanths over 400 million years ago should accordingly be viewed as an adaptation to environmental changes (Lu et al. 2012).

Environmental changes—presumably accompanied by an expanding supply of invertebrate food and increasing structuring of inland waters by plant populations (Gibling et al. 2014; McMahon and Davies 2018)—ultimately stimulated the transition from fish-like to

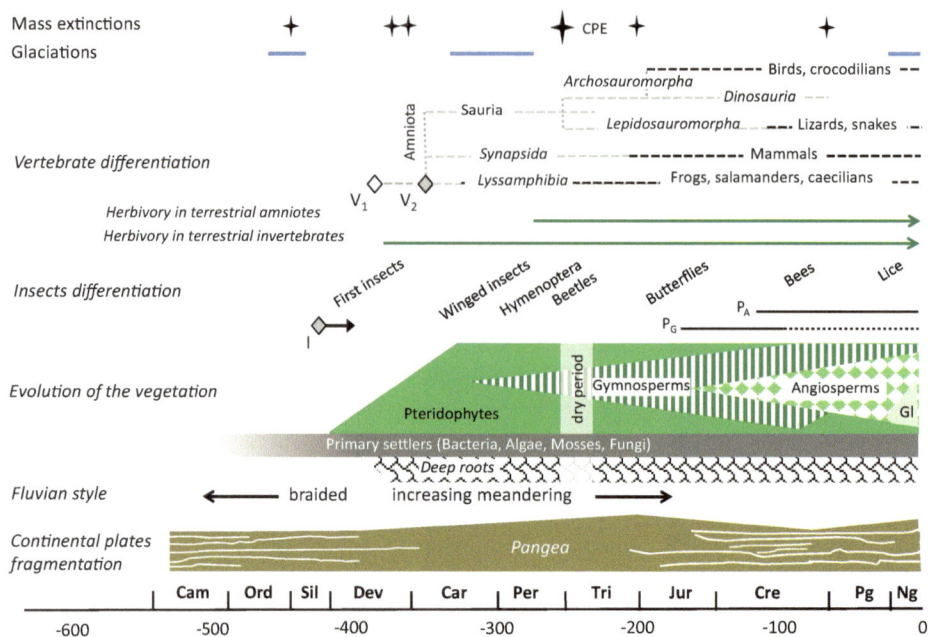

Fig. 10.6 Simplified overview of the terrestrial differentiation of plants, vertebrates and insects during the Phanerozoic. Abbreviations: *CPE* Carnian Pluvial Event, *Gl* Spread of grasslands, *I* Colonization of land areas by invertebrates, *PA* Pollination of angiosperms, *PG* Pollination of gymnosperms, *V1, V2* Phases of immigration of vertebrates into terrestrial habitats. Ages of Earth: *Dev* Devonian, Jurassic, Cambrian, Cretaceous, *Ng* Neogene, Ordovician, *Per* Permian, *Pg* Paleogene; Triassic. Literature and data sources: Anderson et al. (1999); Lenton (2001); Willis and McElwain (2002); Labandeira (2007); Ren et al. (2009); Algeo and Scheckler (1998); Fortuny et al. (2011); Kenrick et al. (2012); Retallack (2013); Misof et al. (2014); Reisz and Fröbisch (2014); Schoch (2014); Nicholson et al. (2015); Porada et al. (2016); Ross (2017); Müller et al. (2018); Close et al. (2020); Dal Corso et al. (2020); Scotese (2021); Scotese et al. (2021). Further explanations in the text

amphibian life forms. Aquatic plant structures reduced the range of movement of fish and thus opened up evolutionary opportunities for new life forms. Comparisons of eye sizes of tetrapod fish—such as Tiktaalik—indicate a primarily aquatic way of life. With the transition to amphibious life forms in the late Devonian around 380 million years ago, there is evidence not only of the evolution of feet but also of adaptation of eye dimensions to optical conditions in air (Daeschler et al. 2006; MacIver et al. 2017). Multiple environmental changes with influencing factors that have not yet been clearly identified—for example the Hangenberg Crisis (see Sect. 9.3.4)—reduced the diversity of the first land-dwelling tetrapods at the end of the Devonian (V1 in Fig. 10.6) (Kaiser et al. 2008; McGhee Jr. et al. 2013). Environmental changes during this period—formerly known as Romer's Gap[2]—stimulated differentiation into life forms with amphibian or terrestrial (amniote)

---

[2] Named after palaeontologist A. Romer.

embryonic development (Ward et al. 2006; Coates et al. 2008; Clack 2012; Schoch 2014; Pardo et al. 2017). Structure and physiology of the eggs require aquatic conditions in the former group and terrestrial conditions in the latter group (Skulan 2000). Amniotes subsequently differentiated into the group of Sauria—including dinosaurs, lizards, crocodiles and birds—and the lineage leading to mammals (Werneburg and Sánchez-Villagra 2009; Brusatte et al. 2010b; Chiari et al. 2012). Within amniotes, towards the end of the Carboniferous various species developed biomechanically stronger jaw structures and— around 315 million years ago—the capacity to eat firm plant tissue (herbivory) (Anderson et al. 2013; Reisz and Fröbisch 2014).

10.4 Terrestrial Vegetation: Differentiation Between Abiotic and Biotic Factors

The evolutionary lineages of terrestrial plants impressively demonstrate the challenges involved in conversion of organisms from aquatic habitats to life under terrestrial environmental conditions. Apart from certain evolutionary lineages, all life forms—usually presented in chronological order —are still represented in the extant flora. Contrasting with the symbiotic systems of lichens, multicellular plants have to adapt at various internal functional levels to the prevailing framework conditions when spreading into areas far removed from water. While mosses, for example, still require water for reproduction, seed plants are independent of this need and can utilize transportation of pollen by air currents or on animals to fertilize their egg cells (Culley et al. 2002; Ollerton and Coulthard 2009; Rech et al. 2016). Seeds of many plants can be spread by carriage on the fur of mammals (zoochory) or—following feeding on fruits—in the digestive systems of animals (endozoochory) (Bossuyt et al. 1999; Baltzinger et al. 2019). Dispersal of spores of mosses and ferns by snails (Boch et al. 2013) is inevitably restricted to sufficiently moist habitats. Moss plants survive temporary desiccation, while seed plants die when their water content falls below the specific minimal level (Oliver et al. 2000; Harrison and Morris 2017). On the other hand, the largely waterproof coatings of their seeds preserve their functionality over lengthy periods of drought and sometimes even in the event of fires (Holsinger 2000; Barrett 2002; Wallace et al. 2011).

The central attractor for expansion of land plants in terrestrial habitats is the—for them—unlimited availability of photosynthetically utilizable solar radiation. One indication of this is the augmentation of plant surface area by means of branching, beginning around 435 million years ago approximately 15 million years before the evolution of vascularization. Rises in photosynthetic performance are associated with increases in water and mineral metabolism, which in land plants were facilitated by the initial evolution of "true" roots around 380 million years ago. "True" roots differ from unicellular or multicellular root hairs in that they combine transportation of water and nutrients with mechanical fixation of the plant. Small—in comparison to the diameter of shoots—lowly life forms can maintain themselves solely by hydrostatic pressure and need their hair-like roots solely for exchange of substances with the substrate (Willis and McElwain 2002; Langdale 2008; Datta et al. 2011; Kenrick and Strullu-Derrien 2014). Evolution of leaves—which began around 370 million years ago—permitted the enlargement of photosynthetically active tissues and their regeneration without modification in the shoot tissues. Together with the functional development of roots, this created a multidimensional potential space for the morphological and ecophysiological differentiation of land plants.

The evolution of large stem diameters and deeply penetrating roots during the late Devonian changed the ecological conditions in and around terrestrial water bodies (Algeo and Scheckler 1998; Meyer-Berthaud et al. 2010). Physical influences of the atmosphere and solar radiation on plant populations were accordingly modified. In moving water bodies, flow conditions and sediment dynamics were modified by root structures in bank zones. Due to the stabilizing effects of roots, chaotic (braided) watercourses changed into meandering systems with wide ranges of flow velocities. In conjunction with exposed root structures, new evolutionary options for aquatic animals also emerged in inland waters (Fig. 10.6) (Abernethy and Rutherfurd 2000; Gyssels et al. 2005; De Baets et al. 2008; Algeo and Scheckler 1998; Davies and Gibling 2010; Gibling and Davies 2012; Gibling et al. 2014; MacIver et al. 2017; Hetherington and Dolan 2018; Wang et al. 2019; Hirasawa et al. 2022). It should be noted here that these conditions could only develop within the bandwidths of mechanical strength of subsoil root systems. Preconditions for their evolution were primarily present in watercourses in lowland areas with sufficiently low current velocities. Because of collapses in vegetation, flooding caused by rising sea levels, or changes in drainage conditions, such systems temporarily disappeared several times during the Phanerozoic.

Effects of the sharp increase in precipitation during the Carnian Pluvial Event (see Sect. 9.3.4) around 232 million years ago were particularly severe after the hot, dry period that lasted over 20 million years following the end of the Permian (CPE in Fig. 10.6) (Retallack et al. 1996; Retallack 2013). Among tetrapods, climate changes led to the extinction of many saurian groups. Increasing dominance of gymnosperms was spurred in tree vegetation, while among vertebrates differentiation of the Lepidosauromorpha and Archosauromorpha, with dinosaurs as the best known group (Fig. 10.6), along with flying life forms such as Pterosauria predominated (Brusatte et al. 2010a; Prentice et al. 2011; Andres 2012; Krause et al. 2019; Dal Corso et al. 2020; Benton and Wu 2022; Foffa et al. 2022). Accompanying increasing interactions with flying insects, further evolution and differentiation of flying vertebrates leading to birds and bats occurred. While the sizes of flying insects in the Carboniferous and Permian were primarily influenced by food supply and oxygen content of the atmosphere, life forms differentiated in the Mesozoic within the context of biotic conditions connected with vegetation and vertebrate fauna (Clapham and Karr 2012; Montagna et al. 2019; Nyman et al. 2019).

Extremely simplified representation of terrestrial vegetation (Fig. 10.6) serves only to provide information guidelines regarding the chronological sequences of dominant life forms. Behind collective names for various life form groups—ferns (pteridophytes), gymnosperms and angiosperms—there is a large number of different species whose systematic grouping is still under discussion. The collective term "primary settlers" serves only as a reminder of the diverse organism communities at extreme sites, which are also summarized under the collective term biological soil crusts (Belnap and Lange 2003). The representation should by no means be understood quantitatively, as many aspects—for example spatial distribution of different plant communities—are neither sufficiently known nor suitable for representation in this simple manner. Examples of interactions between plants and fungi provide the easiest means of illustrating the diversity of interactions. Mycorrhizae—interconnected networks of plant roots and fungal hyphae in a solid substrate—are among the best-known forms of co-operation between plants and fungi. They have been found in plant fossils for over 400 million years (Martin et al. 2017).

Among currently occurring plant groups, mycorrhizae are completely absent from mosses and from around 40 per cent of liverworts. In all other plant groups, over 80 per cent of species are endowed with mycorrhizae (Brundrett 2009). Endophytic fungi, which inhibit attacks by pathogenic fungi or animals by releasing toxic substances, are less well known but widespread among grasses, for example (Ownley et al. 2010; Saikkonen et al. 2013; Estrada et al. 2015). Fungi can also trigger the formation of pseudo-flowers in plants, while plants (Cano et al. 2013)—such as various orchids—can feed on fungi (Selosse and Roy 2008).

In the context of animal evolution, structural effects of vegetation—symbolized by root formation (Fig. 10.6)—are important. Indirectly, increasing height of the vegetation—and hence the generation of animal habitats—can also be inferred from this (Labandeira 2005). In contrast to oceans (see Sect. 9.3), prior to the Devonian low-slung terrestrial vegetation provided adequate habitat and feeding areas exclusively for small invertebrates. Only the above-mentioned modification by sufficiently tall plants permitted the immigration of vertebrates into terrestrial habitats (Fig. 10.6 V1 and V2). The differences give an indication of how closely animal evolution on land is linked to the evolution of vegetation.

However, a closer look at Fig. 10.6 prompts the question as to why there such long periods of time elapse between immigrations of invertebrates (I) and vertebrates and the resulting onset of feeding on living plant tissue (herbivory) (Labandeira 2007; Pearson et al. 2013; Brocklehurst et al. 2020). A major reason for this resides in the evolution of mouthparts in the context of microbiomes occurring in digestive systems for breaking down food, especially cell walls, which is not shown in that figure (Popper et al. 2011). Following the evolution of synthesis of lignins and their integration into cell walls, land plants—with the exception of mosses—increasingly developed firm tissues that were protected from decomposition by bacteria and fungi (Weng and Chapple 2010; Renault et al. 2019). More than 100 million years after the appearance of lignin synthesis in plants, the capacity for enzymatic decomposition of lignin evolved in individual groups of fungi (Floudas et al. 2012). Early animal colonizers of the mainland were able to feed far more easily on fungi, dead plant material or other animals than on living plant tissue. This gave plant evolution a relatively wide latitude up to the Permian period, although it also led to a significant increase in atmospheric oxygen content (Lenton 2001; Wallace et al. 2011). The spread of herbivory generally increased the rate of change within plant species (Niklas et al. 1983). The issue of how many plant species perished as a result of increasing herbivory remains open.

The first indications of exploitation of animals for plant reproduction have been observed in fossil finds of gymnosperms (Fig. 10.6 PC) (Ren et al. 2009; Peñalver et al. 2012; Peñalver et al. 2015). Observations on currently existing conifers reveal that, when attacked by plant parasites (*Diprion*), they can in turn attract their parasites (*Chrysonotomyia*) by releasing chemical signalling substances (Hilker et al. 2002). Gymnosperms have more complex control systems than ferns for regulating water balance. While gymnosperms employ both carbon dioxide concentration and water availability signals to regulate their stomata, ferns use only the latter possibility (Brodribb et al. 2009; Brodribb and McAdam 2011; Brodribb and McAdam 2011; McAdam and Brodribb 2013; Augusto et al. 2015). Like ferns, gymnosperms can colonize nutrient-poor sites and—by chemically inhibiting competing plants—form closed stands (Coomes et al. 2005; Fernandez et al. 2006). Conifers are also able to compensate for differences in local nutrient availability by means of direct root connections with neighbouring trees (Fraser et al. 2005).

In conjunction with climate changes in the Middle Triassic, changes in dominant vegetation during the Carnian Pluvial Event already mentioned also left their mark on the global distribution of coal deposits. Up to the end of the Palaeozoic, coal deposits were mainly found in a narrow zone around the equator. From the Middle Triassic onwards, the formation of new coal deposits shifted abruptly to northern regions between 36° and 58°. At higher latitudes of the southern hemisphere, fewer coal deposits were formed, partly due to the increasing displacement of land masses into the northern hemisphere. It was only through the generation of rainforests by angiosperm plants that coal formation also increased in the equatorial zone from around 60 million years ago (Boyce et al. 2010; Bao et al. 2023).

Increasing differentiation among angiosperms, which exploit animals most extensively for reproduction (Fig. 10.6 PA) and also for protection against herbivores, is clearly recognizable (Wing and Tiffney 1987; Olesen and Valido 2003; Tiffney 2004; Armbruster et al. 2009; Fleming et al. 2009; Thien et al. 2009; de Boer et al. 2012; van der Niet and Johnson 2012; Magallón et al. 2015; Sauquet et al. 2017). At the same time, the dependence of many angiosperm species on animal services has increased significantly. Today, around 78 per cent of angiosperm reproduction in mid-latitudes and around 94 per cent in the tropics is dependent on animal pollination of their flowers (Ollerton et al. 2011; Ratto et al. 2018). Around 90.6 per cent of all currently known plant species are angiosperms, only 0.3 per cent gymnosperms, 4.2 per cent mosses, 3 per cent ferns and 1.5 per cent liverworts (Hernández-Hernández and Wiens 2020).

Great physiological flexibility enables angiosperms to adapt quickly to changes in carbon dioxide concentrations and water availability and hence also to optimize their photosynthetic activities over the course of the day (Boyce and Zwieniecki 2012; Chen et al. 2017). Thanks to their regulatory capacities, in co-operation with bacteria, fungi and animals angiosperms can recover mineral nutrients more rapidly and comprehensively than other plant groups (Wang et al. 2018).

In regions with sufficiently high temperatures and precipitation—such as tropical rain-forests —only small fractions are lost from nutrient cycles. As a consequence, many tropical rainforests lack the thicker humus layers found in forest areas at higher latitudes. Seen from a scientific background of experience with soils in temperate zones, this fact is wrongly interpreted as an indication of inadequate functional proficiency in those ecosystems. Due to the cellular and physiological differentiation of their photosynthetic systems—for example to C4 and CAM systems (see Sect. 7.4)—in recent geological history angiosperms were able to migrate into extremely arid regions as well as to form large-scale grassland ecosystems (Gl in Fig. 10.6) (Edwards et al. 2010; Sage et al. 2012; Bouchenak-Khelladi et al. 2014). In a general way, in its turn the morphological and physiological flexibility of angiosperms promoted an increase in diversity—especially during the Cenozoic—of lichens, mosses and ferns (Page 2002; Moning et al. 2009; Laenen et al. 2014; Testo and Sundue 2016).

10.5 Terrestrial Vegetation: Heterarchical Dynamics and Structural Differentiation

Anyone who repeatedly observes a mineral soil—originally completely free of vegetation —over a lengthy period of time will form a first impression of the progress of recolonization by plants. Only some of the processes behind this phenomenon can be observed with the naked eye. Regardless of the size of the area that has been freed up—for instance due to glacial retreat or volcanic eruptions (Wood and del Moral 1987; Ohtonen et al. 1999; Kandeler et al. 2006)—microorganisms shape the initial colonization phase. Colonization by multicellular plants occurs thereafter. Depending on local climatic conditions, colonization can end with mosses—for example in the vicinity of strong water currents—or with a fully grown population of trees.

Prerequisites for the processes involved are spores and seeds that were already present on site or have been introduced. For those processes, selective factors include availability of water and nutrients, climatic conditions and influences exerted by animals (Fagan et al. 2004). Conversely, this means that multicellular plants can withstand physical and chemical influences only to a limited extent and are dependent on formation of permanent spores or seeds for their long-term development. Stored in soils, spores and seeds can retain their germination capacity for several decades to serve as "seed banks" in critical periods and will begin to grow under favourable conditions (Tsuyuzaki and Goto 2001; Holmes and Newton 2004; Brock 2011). However, specific conditions for growth after disturbance events depend on the degree of randomness in the surviving organismic communities (see Fig. 5.18). Over the medium term, this can result in counter-intuitive constellations. For example, comparative studies of beech stands and artificially afforested spruce stands in Solling (Germany) revealed similar biodiversity patterns for forest floor vegetation and invertebrates. The decisive factor here was the persistence of many species in the original beech stands in combination with new species added as a result of spruce afforestation (Ellenberg et al. 1986).

The same processes also shape global evolution of terrestrial vegetation over hundreds of millions of years. Renewed differentiation of lichens, mosses and ferns due to conditions created by flowering plants over the last hundred million years has already been mentioned. Current species diversity of palm ferns (Cycadales) can in fact be traced back to a diversification that occurred 12 million years ago and subsequent species losses over the last two million years. Details of the triggering factors are currently unknown (Nagalingum et al. 2011).

Representations of interaction processes—referred to here as sub-macroscopic processes (SM)—between vegetation and the abiotic environment are usually limited to processes taking place in soils (Paul 2007). This framework—which originated in agricultural and forestry practice in temperate zones—is by no means sufficient to describe all interactions between plants and the abiotic environment. Research, which is progressing only slowly, into functions of biocoenoses on plant surfaces has yielded the first indications of functional interactions with plants. In addition to provision of support for plant immune defences, for example, direct provision of resources to plant leaves by cyanobacteria in rainforests and support of metabolic processes on seagrasses have been demonstrated (Fürnkranz et al. 2008; Bogino et al. 2013; Costa et al. 2015). Many lichens, mosses, ferns and vascular plants never come into contact with soils and live epiphytically, mostly on vascular plants, where support from microorganisms and micrometazoans renders them self-sufficient (Zotz and Hietz 2001; Pypker et al. 2006; Hennequin et al. 2008; Calvente et al. 2011; Silvestro et al. 2013; Givnish et al. 2014). The spatial diversity and temporal dynamics of such organismic communities in their interactions with the developmental processes of macroscopic systems are extremely difficult to record scientifically. However, their ubiquitous occurrence leads to the conclusion that—in terms of functional effects— they are the most ancient ecosystems in terrestrial habitats.

Interactions between macroscopic organisms can be recognized more clearly. Prior to the Permian, plants were confronted only with relatively minor damage or loss of their tissues due to herbivorous vertebrates. The resulting room for manoeuvre—in the competition for sunlight—was exploited by some plant groups to accelerate upward growth. The oldest currently known fossils from forests of the Devonian are around 380 to 390 million years old. Trunk fragments from one site suggest tree heights of at least eight metres. Preserved root systems from a different site provide evidence of their wide range of variation. Roots of tree ferns reach maximum dimensions of only one metre in length and one centimetre in diameter, whereas branched root systems of the tree-like progymnosperms of the genus *Archaeopteris* in the late Devonian reached lengths of five to six metres, with base diameters of about 15 centimetres (Stein et al. 2007; Berry 2019; Stein et al. 2020). Rapid evolution of extremely diverse life forms of land plants in the Carboniferous led to forest stands with an average tree height of 25 metres (Niklas et al. 1983; Falcon-Lang et al. 2006; DiMichele et al. 2007; Opluštil et al. 2009; DiMichele and Falcon-Lang 2011;

Wang et al. 2012). Individual species, such as the extinct Cordaitales—classified as gymnosperms—reached tree heights of up to around 48 metres (Falcon-Lang and Bashforth 2005). This created huge three-dimensional habitats for insects, which—especially because of their distribution in swampy areas—presumably promoted the evolution of both the capacity for flight and the aquatic juvenile phases of various groups such as dragonflies (Odonata) and mayflies (Ephemeroptera) (Lancaster and Downes 2013).

Fossil evidence provides an incomplete picture of vegetation forms in the Palaeozoic, as it has yielded hardly any examples of vegetation outside of marshlands. It therefore remains unclear to what the vertebrate transitions from amphibious anamniotes to terrestrial amniotes were also stimulated by vegetation. Without reference to spatial distribution, lycopods (Lycopodiatae) achieved the greatest species diversity among terrestrial plants towards the end of the Carboniferous and into the Permian. Cordaitales disappeared during the large-scale disruptions at the end of the Permian (Willis and McElwain 2002). Feeding losses were caused solely by invertebrates, as all early amniotes were carnivorous. According to fossil evidence obtained so far, herbivorous life forms probably evolved among vertebrates only towards the end of the Carboniferous around 300 million years ago (Reisz and Fröbisch 2014). From the Devonian to the Permian, dimensions of closed land areas grew continuously and extended increasingly into high northern latitudes. Climatically, temperature differences between equatorial regions and southern latitudes increased continuously until the end of the Carboniferous. During the second half of the Carboniferous, an extensive ice sheet probably extended over the southern polar region (Boucot et al. 2013).

With the expanding distribution of megaherbivores, relative proportions of previously dominant plant groups decreased from the beginning of the Triassic onwards. On the other hand, the total number of species increased, mainly due to differentiation of conifers and to a lesser extent of palm ferns and ginkgos following the large-scale disruptions that occurred towards the end of the Triassic (Willis and McElwain 2002; Vajda and Bercovici 2014). Among the gymnosperms, some palm ferns switched from wind pollination to transference of pollen by insects in the middle Mesozoic (Fig. 10.6, BG) (Cai et al. 2018). In the Cretaceous period, around 170 million years ago, palm and tree fern species began to decline. At around the same time, the number of angiosperm species increased rapidly, accounting for around 80 per cent of all land plant species by the end of the Cretaceous period. From the middle of the Cretaceous, diversity among conifers also began to decline (Smith et al. 2010; Brown et al. 2012; Doyle 2012; Jud 2015; Magallón et al. 2015; Coiro et al. 2019; Hetherington and Dolan 2018; Li et al. 2019). Geologically, the contiguous land mass of Pangea reached its greatest extent in the Middle Jurassic and subsequently increasingly began to break up into separate continental plates. In climatic terms, temperature differences between equatorial and polar regions increased from the Triassic up to the end of the Cretaceous (Boucot et al. 2013).

The volcanic eruptions of the Deccan Traps and the asteroid impact (Chixculub crater) at the end of the Cretaceous period affected terrestrial vegetation to varying degrees. Species pollinated by insects suffered the greatest losses, while species with wind pollination were barely affected. Evergreen plant species were also more severely affected than species with regularly occurring foliage changes. In the vicinity of the asteroid impact itself, all vegetation was destroyed by fires. The first phase of recolonization was dominated by fern species, which were then increasingly replaced by angiosperms. Saprophagous communities (living on dead organic material) in the soil benefited from the vast amounts of dead organic material available, as is recognizable from relatively short-term increases in fungi and earthworms (McElwain and Punyasena 2007; Chin et al. 2013; Vajda and Bercovici 2014). The disruptive events did not interrupt the general trend towards diversification of angiosperms. As a result, large-scale forest ecosystems re-established themselves relatively quickly and were not increasingly pushed back by open herb and grass ecosystems until towards the end of the Palaeocene, a change brought about by global temperature increases. C4 grasses began to spread within grassland ecosystems around 20 million years ago. In addition to their morphological and physiological photosynthetic characteristics, they also differ from C3 grasses in that the silicate content of their cell walls is around five times greater (Stanley 2001; Strömberg 2011). Because of the open marine connections (see Sect. 8.6.3), subtropical climatic conditions were maintained in mid-latitudes until the Miocene. It was not until around five million years ago that global cooling set in with relatively sizable—regionally varying—climate dynamics (Hughes et al. 2013; Scotese 2021).

10.6   Terrestrial Fauna: Evolution under the Dynamic Conditions of Vegetation

In view of the barely discernible responses or movements of plants, it is difficult to see why this group of organisms should have influenced the evolution of animals. This notion is reinforced by the long-term success of domestication and economic utilization of plants. Within the horizon of our own experience, plants have been and can be utilized and manipulated at will. But we overlook the extremely short timespans that are comprehensible for humans. If we generously assume a knowledge-guided period of experience of two thousand years, then we have consciously observed around 0.4 seconds of a 24-hour event in the timespan of a single day. It would be utterly presumptuous to draw conclusions about events over an entire day from observations covering such a brief interval. Useful clues to events over the entire day can be provided only by remaining traces of those events that can be assigned to a specific period of time. This introduction is intended to elucidate in more detail the premise embedded in the chapter title within the context of evolutionary processes during the Phanerozoic.

It is easy to recognize the generally applicable fact that all animal life is energetically dependent on primary production by plants. However, there is a fundamental difference in structural characteristics between photosynthetic organisms in the oceans and on land. In comparison to conditions in marine habitats (see Chap. 9), on land even the smallest plants develop long-term spatial structures, only a part of which is usually suitable as food for animals. The static lifestyle of land plants precludes the evolution of passively filtering life forms among animals as seen in oceans. Moreover, because of the low density of air, edible parts are easily accessible to animals only at their above-ground body height. Any change in plant morphology in conjunction with physiological properties of tissues therefore also brings about morphological changes in animals living on them. If plants move their food-relevant parts upwards during growth, animals can reach them only by means of morphological and physiological changes. Options include, for example, increase in body size or the evolution of climbing or flying capacities. Alternatively, other food sources might also be used instead—but this would merely transfer options for utilization to other species.

The first flying insects (*Delitzschala*), which lived herbivorously on tree ferns, also evolved around 325 million years ago (Ross 2017). A connection can be presumed with increasing tree growth in the Carboniferous, as the evolution of a functionally and energetically complex flying apparatus (Iwamoto 2011) could succeed only in association with exploitation of novel food sources. In contrast to wingless hexapods, flying insects require juvenile developmental phases that are sufficiently lengthy to generate body structures and sizes suitable for flight. Juvenile life forms of hemimetabolous insects are adapted to the— often different—conditions of their habitats. Hemimetabolous insects develop their adult life forms—following one or more immature stages—when they attain the capacity to fly. These changes are clearly recognizable in dragonflies (Odonata) or mayflies (Ephemeroptera), which have aquatic juvenile forms and adult forms capable of flight (Truman 2019).

In static terms, structural effects of terrestrial vegetation have functional relationships to animals similar to those of marine coral reefs. As with the latter, structural effects depend on size ratios between animals and plants. Animals with small body sizes—for example insects—can utilize a significantly broader spectrum of structural influences than large animals such as elephants. In dynamic terms, however, there are clear differences in effects, especially in the case of non-anthropogenic disturbances. In contrast to reefs (see Sect. 9.3.4), the ecophysiological performance of plants differs significantly from that of animals. Identical abiotic disturbances can hence exert different effects on the two groups of organisms. As land plants form the basis of terrestrial food webs and also, through their structural and ecophysiological functions, enable other organisms to achieve high levels of diversity, changes in vegetation trigger far greater losses of diversity among animals. At the same time, new evolutionary options open up for animal species that remain.

Every option requires extensive adaptations, for example in morphology, metabolic physiology, sensory organ apparatus and neuronal capacities for information processing, and coordinated control of motion sequences. It also alters "sections" of the environment that are relevant for any given animal species. Animals running over the substrate have no need for information regarding structures of trunks and branches above their realm of activity. For climbing animals, however, those above-ground structures are central elements of their habitat. Rapid detection of escape routes through the three-dimensional structures of load-bearing branches can be decisive for survival in response to attacks by carnivores. For flying animals, trunks and branches are important both as take-off or land-ing sites and as obstacles to escape. Additional factors for differentiation of interactions between plants and animals arise from the evolution of specific optical, chemical or ther-mal signals in plants, for example derived from flowers or fruits.

As in marine habitats, every adaptation of herbivores inevitably triggers suitable adap-tations of carnivores for energetic reasons. In terrestrial habitats, such evolutionary changes are also influenced by plants. Their structures can make the search for prey more difficult, and they may also attract carnivores through chemical signals.

From a classical evolutionary perspective focussing on the emergence of individual species, such explanations could be interpreted as simple-minded Lamarckism. However, the question then arises as to why the evolutionary lineages of arthropods and verte-brates—given the same starting conditions and equally long evolutionary periods (dos Reis et al. 2015)—developed in the oceans in different ways than occurred on land.

However, the foregoing descriptions take only part of the interactions into account. Plants must also be adaptable in order to survive as organisms over the long term. In this context, land plants are far more exposed to the tension between abiotic and biotic factors than unicellular photosynthetic organisms in oceans. The extreme diversity of interactions between plants, microorganisms and fungi has already been discussed several times. They clearly reveal the central role of cellular information processes in interactions between organisms. Due to their immobility, plants cannot evade herbivorous animals by fleeing and must protect themselves from supercritical biomass losses either alone or in co-operation with other organisms. Like all biological interactions, these are not static but evolutionarily dynamic (Tiffney 1992).

10.7   The Slow Pendulum of Reciprocal Stimulation

Aside from fungi and microorganisms, land plants and insects have the most extensive and longest-standing interactions. Starting with the first recorded fossils, the most diverse interactions—from simple herbivory to specialized symbiotic relationships—evolved between around 300,000 plant species and over 1 million insect species (Grimaldi and Engel 2005; Schoonhoven et al. 2005; Stork 2018; Hernández-Hernández and Wiens 2020).

The example of beetles (Coleoptera) gives a first approximation to the evolution of herbivorous diets. The earliest fossilized beetles from the Permian fed solely on plant waste. The first herbivorous life forms originated in the Triassic, already accounting for the largest proportion of species found in the Jurassic and still representing the largest group of beetles today (Farrell 1998). Similar evolutionary trends can be found in the general development of herbivorous insects. The first phase—during the early Devonian—has mainly yielded traces of feeding on spore capsules (sporangia) and stems of early plants. Around 75 million years later, feeding traces on all parts of plants, which are by then tree-like, point to a wide variety of herbivorous insects (Labandeira 2007; Labandeira 2013a). According to molecular analyses, most insect life forms also evolved during this period—the Carboniferous (Condamine et al. 2016). It is remarkable in this context that in the present-day flora the evolutionarily older group of ferns shows a high frequency of leaf herbivory similar to that seen with flowering plants (Watkins Jr. and Cardelús 2012). This may be due to renewed differentiation of ferns during the Cretaceous period. Insects utilize leaves of gymnosperms —whose differentiation began in the Carboniferous—as a food source to a far lesser degree. Proportions are similarly low in C4 grasses, which evolved only a few million years ago (Turcotte et al. 2014).

Although leaf feeding is relatively rare in gymnosperms, large-scale infestation by bark beetles is in fact a recognizable phenomenon today. This group of beetles—which is not systematically cohesive—is thought to have originated in the Mesozoic era, with different hypotheses pointing to either the Jurassic or the Cretaceous period. The majority of all species—around 86 per cent—live on flowering plants in the tropics and subtropics. The lethal effects of certain bark beetles (ambrosia beetles) on conifers can be traced back to the evolution of capacities that can appear in a positive light from a human perspective. Ambrosia beetles co-operate, live in social groups and make a living by cultivating their food. These skills presumably evolved over the last 50 million years. When infesting living trees, bark beetles cooperate to overcome their defences. In differentiated social systems, various species construct tunnel systems and deposit in them spores of fungi, on which they and their larvae live (Sequeira et al. 2000; Farrell et al. 2001; Raffa 2001; Sequeira and Farrell 2001; Biedermann 2007; Jordal and Cognato 2012; Kirkendall et al. 2015).

Whereas the diversification of insects was from the very beginning directly linked to the evolution of vegetation, in the early terrestrial evolution of vertebrates the link was presumably only indirect. Changes in structures of inland waters due to the spread of large root systems during the Devonian led to the evolution of new habitat conditions for aquatic organisms. The entire life cycle could take place in niches that were structurally relatively well protected, while at the same time an already diverse invertebrate fauna alongside water bodies and on land provided adequate food resources for populations of larger vertebrates. Evidence for an increasing occurrence of foraging at or above the water surface is provided by the increasing presence of eye sockets on top of the skulls of early four-legged vertebrates (Tetrapodomorpha) (Clack 2012; Schoch 2014). Reconstructions of eye optics based on geometric dimensions of the eye sockets also indicate increasing adaptations of vision to the optical conditions in air even before limbs evolved (MacIver et al. 2017). Functional changes in the jaws, on the other hand, evolved more slowly than differentiation of limbs (Anderson et al. 2013). These indicators point to a significant influence of food availability on the evolution of terrestrial life forms among vertebrates. As already mentioned (Sect. 9.3.5) and indicated in Fig. 10.6, immigration of vertebrates into terrestrial habitats was a process that was repeated several times and associated with many failed attempts.

The first colonizers of terrestrial habitats were certainly amphibians that emerged from the water only temporarily. However, it is difficult to estimate when and how the separation between amphibians and reptiles took place because in both cases information about the eggs formed is largely lacking. This also makes it impossible to determine whether reproduction of any given fossil form involved eggs with an additional membrane (amnion) or without it. The first modality applies to Amniota (reptiles, birds and mammals), while the second typifies Anamnia—amphibians (Romer 1971). Separation between the two groups (Lissamphibia *versus* Amniota) 260 to 340 million years ago has been evaluated on the basis of anatomical features alone (Schoch 2009; Pardo et al. 2017). Moreover, fossil amphibians include various species with unusual lifestyles, for example in the stereospondyl subgroup among the amphibian temnospondyls (Ruta and Benton 2008; Schoch 2009; Laurin and Soler-Gijón 2010; Marsicano et al. 2017; Fernández-Coll et al. 2019). From the Carboniferous onwards, their fossils are often found in deposits in brackish water zones of river mouths (estuaries). Preservation conditions of the fossils clearly indicate that the animals inhabited those zones and had not been washed in. However, based on the requirements of all currently extant amphibians, various publications have assigned their original habitats to freshwater environments. Yet consistent features of the fossil sites from the Carboniferous up to the end of the Triassic, combined with morphological evolution of the group to yield crocodile-like giants with a body length of about six metres, suggest independent physiological adaptations (Steyer and Damiani 2005; Schoch 2009; Laurin and Soler-Gijón 2010; Fortuny et al. 2011; Sanchez et al. 2010; Fortuny et al. 2016; Fortuny and Steyer 2019).

Immigration into terrestrial habitats was associated with adaptative changes not only in the entire locomotor system and respiratory organs, but also in sensory physiological perception and information processing (Carrier 1987; Perry and Sander 2004; Müller et al. 2018). Very informative in this context are the evolutionary trajectories to limbless morphotypes within the squamata, in particular of snakes (see also Sects. 4.1 and 7.6.2) (Garberoglio et al. 2019; Simões et al. 2020; Mann et al. 2022; Tingle et al. 2024; Title et al. 2024).To this end, identical methods of information processing for certain phenomena evolved in systematically completely separate animal groups, for example perception of movement in insects and vertebrates (Clark and Demb 2016).

The first evidence for herbivory among tetrapods can be found in various groups of amniotes at the end of the Carboniferous. From originally relatively small life forms, species with a body weight of over 100 kilograms evolved rapidly over the course of the Permian (Reisz and Fröbisch 2014). Supposedly simple evolution from original carnivores to herbivores required fundamental changes in digestive systems and the microbiomes they contained (Mackie 2002). Because of the significantly lower specific energy content of many plant tissues (Clauss et al. 2013), a tendency towards increasing body size was promoted among leaf-eating herbivorous tetrapods. In parallel, smaller, climbing herbivores evolved, presumably feeding on more energy-rich parts of plants (Rybczynski and Reisz 2001; Fröbisch and Reisz 2009). The disruptive events at the end of the Permian led to shifts in species spectra within the large systematic groups of tetrapods (Laurin 2004; Fröbisch 2008; Fröbisch 2009; Irmis and Whiteside 2012). In some groups, giant herbivores developed towards the end of the subsequent Triassic, for example *Lisowicia* among dicynodonts, with an estimated body weight of nine tonnes (Sulej and Niedźwiedzki 2019).

During the same period, the differentiation of dinosaurs began from initially very small life forms. From them, giant herbivores—such as *Argentinosaurus* or *Brachiosaurus*—with estimated body weights of up to 50 tonnes, and perhaps even more, eventually evolved among sauropods in the subsequent Jurassic period (Sander et al. 2011). For physiological reasons, we can assume that giant life forms led semi-aquatic lifestyles—just like hippopotamuses—in inland waters and intertidal zones with mangrove-like vegetation. The animals would have benefited from sufficient food availability in these zones of maximal plant productivity. In addition, the semi-aquatic lifestyle would also have facilitated mating. On one hand, this assumption is based on the fact that the largest known terrestrial life forms among mammals—paraceratherians related to rhinoceroses—reached maximum body weights of 11 tonnes (Wang et al. 2016; Deng et al. 2021). Moreover, there have been repeated references to aquatic habitats in the tracks of sauropods, and the sometimes huge neck lengths would also have made it easier to graze on otherwise inaccessible plants in areas with many obstacles (Henderson 2004; Taylor and Wedel 2013; Brusatte et al. 2016; Xing et al. 2016; Fernández-Baldor et al. 2021).

In connection with the origin of gigantic life forms, mention is often made of Cope's Rule,[3] according to which all species continuously tend to increase in size over the course of their evolution. Aside from exaggeration in common usage, the generally applicable assertion is simply that any given species that failed to adapt to changes in its environment in a timely fashion finished up at an evolutionary dead end (Zanno and Makovicky 2012). Even among dinosaurs, however, the early decrease in body size seen in the evolutionary lineage of birds contradicts the above-mentioned "Rule" (Sereno 1999; Benson et al. 2014; Brusatte et al. 2014; Brusatte et al. 2015; Benson et al. 2018).

In fact, gaps in the fossil record can also lead to misinterpretation (Brocklehurst et al. 2013; Close et al. 2020). One example is provided by the evolution of mammals, which, according to a widespread paradigm, existed during the Mesozoic Era only as small-bodied life forms in the shadow of dinosaurs. Rich fossil finds from the last few decades paint a completely different picture. As early as the Jurassic—around 100 million years earlier than previously assumed—a diversity of life forms comparable to that seen in the Palaeogene can be found among early evolutionary lineages of mammals. Because of the absence of suitable palaeontological sites, information regarding further evolutionary developments during the Cretaceous period is currently lacking (Feduccia 1995; Luo 2007; Zhou et al. 2013; Wilson et al. 2016). For the same reasons, previous notions concerning connections between vegetation and dinosaur evolution are derived exclusively from scenarios based on a variety of assumptions (Mustoe 2007; Butler et al. 2009; Gee 2011).

---

[3] Named after American paleontologist E.D. Cope.

In connection with the spread of angiosperms, however, there is accumulating evidence for fundamental changes in global systemic evolutionary trajectories in the Early and Middle Cretaceous. Over a period of more than 35 million years (Aptian-Albian gap), a fundamental change occurred in interactions between insects and plants. Originally dominant relationships between insects and gymnosperms largely disappeared and were replaced at an increasing pace by interactions between insects—especially Hymenoptera—and angiosperms. This development was delayed, but not interrupted, by climate changes towards the end of the Cretaceous and the large-scale disruptions at the close of the Cretaceous (Labandeira et al. 2002; Rehan et al. 2013; Labandeira 2014; Nyman et al. 2019). The robustness of the systems concerned is probably due to concentration of potentials for ecosystem regulation. In structural and energetic terms, angiosperms offer a far greater diversity of factor combinations (ecological niches) above and below ground than gymnosperms. This promoted diversification of holometabolous insect groups—for example beetles, hymenopterans or butterflies—with a largely unprotected pupal stage in their ontogenetic development. Due to almost complete remodelling of the internal organs of holometabolous insects during the pupal stage, adults can feed in completely different ways compared to their juvenile stages, for example aquatic larvae of mosquitoes (Culicidae) feed on detritus, adult females on vertebrate blood and adult males on flower nectar. In combination with structural and energetic factors, the diversity of optical and chemical signals in flowers and fruits of angiosperms significantly expands the spectrum of potential interactions with animals. Because of their wide spectra of perception, insects and birds in particular can make use of the latter factors (Frentiu and Briscoe 2008; Osorio and Vorobyev 2008; Macel 2011; Patiny 2012; Kelber et al. 2017; Scholtyßek and Kelber 2017; Lindstedt et al. 2019; Volf et al. 2020). For mammals—especially bats and rodents—the new system conditions provided important triggers for further differentiation during the Cenozoic (Schipper et al. 2008; Figueirido et al. 2011).

The disruptive events at the end of the Cretaceous period not only led to extinction of all dinosaur species, with the exception of birds, but brought about other species losses among reptiles and mammals. Global cooling towards the end of the Cretaceous may have played a significant role in the differing effects on various animal groups, as a significant decline in species diversity among extinct dinosaur groups was already observable during this time (Longrich et al. 2011; Longrich et al. 2016; Sakamoto et al. 2016; Field et al. 2018; Condamine et al. 2021; Han et al. 2022). Among mammals, during the first hundred thousand years after the disruption, the number of species doubled without a significant increase in body size, occurring in parallel to diversification of land plants. Only with the increasing spread of legumes did the body sizes of mammals also increase, after a delay of some 300,000 years (Lyson et al. 2019).

As a result of interactions between plants and animals, terrestrial evolutionary processes differ significantly from processes taking place under marine conditions. Moreover, spatial and temporal coincidences of abiotic factors at different spatial dimensions modify the framework conditions for biological processes. The diversity of interactions cannot be fully captured either through fossil evidence or even in the present day. At best, large-scale and long-lasting changes can be recognized. The evolution of each individual group of organisms is also embedded in, and interacts with, such processes. Isolated analyses of the temporal sequences of life forms provide insights into the changes within the group of organisms under consideration. However, they yield barely any insights into general characteristics of evolutionary processes. *Especially in the present day, it is important to realize that evolutionary processes can never be controlled by a single species.*

To spur discussion, central global phases of terrestrial evolution have been summarized—in extremely simplified form—to show their temporal sequences and overlaps (Fig. 10.7):

- The *detritivorous system* that evolved early on are significant for terrestrial evolution as a whole. In simple terms, plant nutrients are recovered through interactions between energy conversion from organic material by animals and microbial release of mineral substances. These processes are also subject to ongoing evolutionary changes, for example through the composition of the organisms involved or changes in processes. Observations of physical activities of vertebrates in the subsurface provide indications of this (Fig. 10.7).

- In macroscopic terms, the phase of vertical growth in terrestrial vegetation ("*plant dominated*") is of central importance for the general evolution of animal organisms on land. In simple terms, land plants dominate biotic interactions during this phase. Feedback from animal organisms consistently increases over time due to the spread of herbivory—feeding on living plant material—among vertebrates. Due to a lack of feedback from megaherbivores—and a resulting increase in the fixation of carbon in plant biomass—ratios for atmospheric concentrations of oxygen and carbon dioxide shifted.

- Repercussions of megaherbivores on vegetation (Fig. 10.7, α), which began in the Permian, increased massively after the extreme environmental changes that occurred during the transition from the Permian to the Triassic (Fig. 10.7, β). The second phase is characterized by complex interactions between *gymnosperms* and *dinosaurs*.

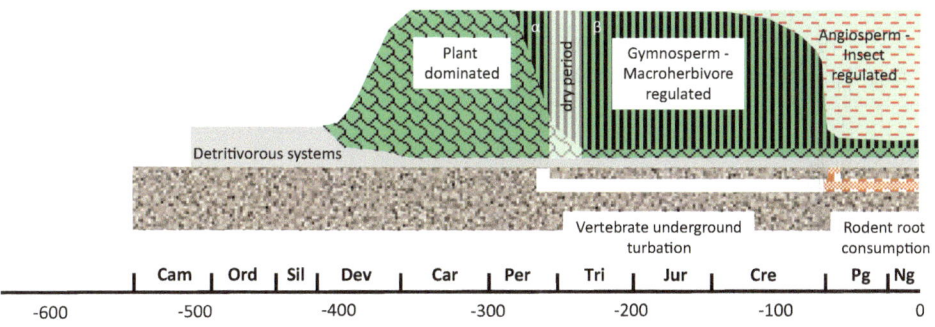

Fig. 10.7 Extremely simplified representation of terrestrial evolution through dominant phases of interactions between plant and animal organisms. Further explanations in the text. Earth ages as in Fig. 10.6

However, studies in a North American national park show that interactions between the vegetation and large herbivores are also significantly modified by carnivores—"meat-eaters"—in a non-linear fashion (Bump et al. 2009).

- As early as the Cretaceous, evolutionary processes began to be influenced by increasing interactions between angiosperms and insects, which became visible in the Cenozoic through marked diversification of various insect groups, such as hymenopterans, cole-opterans, lepidopterans and dipterans. Diversification of both insects and angiosperms subsequently stimulated evolutionary processes among birds and mammals through to primates, with repercussions for evolution of insects. The late differentiation of animal lice is illustrated as an example in Fig. 10.6.

During the millions of years that followed, a development began among shrew-sized mammals that continued relatively rapidly in two main directions: the predominantly nocturnal prosimians (Prosimii) and the predominantly diurnal monkeys and apes (Simiiformes) (Gebo 2004; Fleagle 2013; Santini et al. 2013). Until the Pliocene, differentiation and distribution of species was closely linked to climatic changes. During warm periods of the middle Eocene and early and middle Miocene, Simiiformes reached a high level of species diversity in the northern hemisphere, but disappeared again during the subsequent cooling periods. Around 7 million years ago, the family Hominidae began to differentiate among the Old World apes (members of Catarrhini), especially in Africa. Of all species identified to date, only one omnivorous life form (*Homo sapiens*) has survived to the present day (Fleagle 2013). Probably around 200,000 years ago, global dispersal of the species began (Jones et al. 1994; Hershkovitz et al. 2018).

Population growth has accelerated significantly over the past 200 years and may hit the 10 billion mark in around 40 years (Eriksson et al. 2012; Vollset et al. 2020). This is associated with the expansion of direct land utilization, which, according to current estimates, will reach the limits of availability in around 30 years (Lambin and Meyfroidt 2011). Indirect global utilization of animals and plants, together with pollution of all ecosystems through trash, waste water and exhaust gases, has significantly accelerated the loss of naturally occurring species in recent decades. Abiotic conditions on earth are also changing as a result of uncontrollable repercussions. It is accordingly possible that the last surviving hominid species will be the first animal species to trigger the next global disruption event (Barnosky et al. 2011; Barnosky et al. 2012; Rockström et al. 2023).

References

Abernethy B, Rutherfurd ID (2000) The effect of riparian tree roots on the mass-stability of river-banks. Earth Surf Process Landf 25:921–932

Adeney JM, Christensen NL, Vicentini A, Cohn-Haft M (2016) White-sand ecosystems in Amazonia. Biotropica 48(1):7–23

Algeo TJ, Scheckler SE (1998) Terrestrial-marine teleconnections in the Devonian: links between the evolution of land plants, weathering processes, and marine anoxic events. Philos Trans R Soc Lond B 353:113–130

Anderson AB, Prance GT, de Albuquerque BWP (1975) Estudos sobre a vegetação das Campinas Amazonicas – III A vegetação lenhosa da Campina da Reserva Biológica INPA-SUFRAMA (Manaus - Caracarai, km 62). Acta Amazon 5(3):225–246

Anderson JM, Anderson HM, Archangelsky S, Bamford M, Chandra S et al (1999) Patterns of Gondwana plant colonization and diversification. J Afr Earth Sci 28(1):145–167

Anderson PSL, Friedman M, Ruta M (2013) Late on the table: diversification of tetrapod mandibular biomechanics lagged behind the evolution of terrestriality. Integr Comp Biol 53:197–208. https://doi.org/10.1093/ich/ict006

Andres B (2012) The early evolutionary history and adaptive radiation of the Pterosauria. Acta Geol Sin 86(6):1356–1365

Armbruster WS, Lee J, Baldwin BG (2009) Macroevolutionary patterns of defense and pollination in Dalechampia vines: adaptation, exaptation, and evolutionary novelty. PNAS 106(43):18085–18090

Augusto L, De Schrijver A, Vesterdal L, Smolander A, Prescott C, Ranger J (2015) Influences of evergreen gymnosperm and deciduous angiosperm tree species on the functioning of temperate and boreal forests. Biol Rev 90(2):444–466

Baltzinger C, Karini S, Shukla U (2019) Plants on the move: hitch-hiking with ungulates distributes diaspores across landscapes. Front Ecol Evol 7:38

Bao X, Hu Y, Scotese CR, Li X, Guo J et al (2023) Quantifying climate conditions for the formation of coals and evaporites. Natl Sci Rev 10:nwad051

Barnosky AD, Matzke N, Tomiya S, Wogan GOU, Swartz B et al (2011) Has the Earth's sixth mass extinction already arrived? Nature 471:51–57

Barnosky AD, Hadly EA, Bascompte J, Berlow EL, Brown JH et al (2012) Approaching a state shift in Earth's biosphere. Nature 386:52–58

Barrett SCH (2002) The evolution of plant sexual diversity. Nat Rev Genet 3:274–284

Bejder L, Hall BK (2002) Limbs in whales and limblessness in other vertebrates: mechanisms of evolutionary and developmental transformation and loss. Evol Dev 4(6):445–458

Belnap J, Lange OL (eds) (2003) Biological soil crusts: structure, function, and management, Ecological studies. Springer, Berlin, p 150

Benson RBJ, Campione NE, Carrano MT, Mannion PD, Sullivan C et al (2014) Rates of dinosaur body mass evolution indicate 170 million years of sustained ecological innovation on the avian stem lineage. PLoS Biol 12(5):e1001853

Benson RBJ, Hunt G, Carrano MT, Campione N (2018) Cope's rule and the adaptive landscape of dinosaur body size evolution. Palaeontology 61(1):13–48

Benton MJ, Wu F (2022) Triassic revolution. Front Earh Sci 10:899541

Berbee ML, James TY, Strullu-Derrien C (2017) Early diverging fungi: diversity and impact at the dawn of terrestrial life. Ann Rev Microbiol 71:41–60

Berry CM (2019) Palaeobotany: the rise of the Earth's early forests. Curr Biol 29:R792–R794

Biedermann PHW (2007) Social behaviour in sib mating fungus farmers. Masterarbeit Univ, Bern

Boch S, Berlinger M, Fischer M, Knop E, Nentwig W et al (2013) Fern and bryophyte endozoochory by slugs. Oecologia 172:817–822

Bogino PC, Oliva M, Sorroche FG, Giordano W (2013) The role of bacterial biofilms and surface components in plant-bacterial associations. Int J Mol Sci 14:15838–15859

Bonfante P, Anca I-A (2009) Plants, mycorrhizal fungi, and bacteria: a network of interactions. Ann Rev Microbiol 63:363–383

Bossuyt B, Hermy M, Deckers J (1999) Migration of herbaceous plant species across ancient-recent forest ecotones in Central Belgium. J Ecol 87:628–638

Bouchenak-Khelladi Y, Muasya AM, Linder HP (2014) A revised evolutionary history of Poales: origins and diversification. Bot J of the Linnean Society 175(1):4–16

Boucot A, Xu C, Scotese CR, Morley RJ (2013) Phanerozoic paleoclimate: an atlas of lithologic indicators of climate. SEPM, Tulsa

Boyce CK, Zwieniecki MA (2012) Leaf fossil record suggests limited influence of atmospheric CO_2 on terrestrial productivity prior to angiosperm evolution. PNAS 109(26):10403–10408

Boyce CK, Lee J-E, Field TS, Brodribb TJ, Zwieniecki MA (2010) Angiosperms helped put the rain in the rainforests: the impact of plant physiological evolution of tropical biodiversity. Ann Missouri Bot Gard 97:527–540

Brock MA (2011) Persistence of seed banks in Australian temporary wetlands. Freshw Biol 56:1312–1327

Brocklehurst N, Kammerer CF, Fröbisch J (2013) The early evolution of synapsids, and the influence of sampling on their fossil record. Paleobiology 39(3):470–490

Brocklehurst N, Kammerer CF, Benson RJ (2020) The origin of tetrapod herbivory: effects on local plant diversity. Proc R Soc B287:20200124

Brodribb TJ, McAdam AM (2011) Passive origins of stomatal control in vascular plants. Science 331:582–585

Brodribb TJ, McAdam SAM, Jordan GJ, Field TS (2009) Evolution of stomatal responsiveness to CO_2 and optimization of water-use efficiency among land plants. New Phytol 183:839–847

Broly P, Deville P, Maillet S (2013) The origin of terrestrial isopods (Crustacea: isopoda: Oniscidea). Evol Ecol 27:461–476

Brookfield ME, Catlos EJ, Suarez SE (2020) Myriapod divergence time differ between molecular clock and fossil evidence: U/Pb zircon ages of the earliest fossil millipeds-bearing sediments and their significance. Hist Biol 33:1761351. https://doi.org/10.1080/10.1080/08912963.2020

Brown SAE, Scott AC, Glasspool IJ, Collison ME (2012) Cretaceous wildfires and their impact on the earth system. Cretac Res 36:162–190

Brundrett MC (2009) Mycorrhizal associations and other means of nutrition of vascular plants: understanding the global diversity of host plants by resolving conflicting information and developing reliable means of diagnosis. Plant Soil 320:37–77

Brusatte SL, Benton MJ, Desojo JB, Langer MC (2010a) The higher-level phylogeny of Archosauria (Tetrapoda: Diapsida). J Syst Palaeontol 8(1):3–47

Brusatte SL, Nesbitt SJ, Irmis RB, Butler RJ, Benson MJ, Norell MA (2010b) The origin and early radiation of dinosaurs. Earth Sci Rev 101:68–100

Brusatte SL, Lloyd GT, Wang SC, Nozell MA (2014) Gradual assembly of avian body plan culminated in rapid rates of evolution across the dinosaur-bird transition. Curr Biol 24:2386–2392

Brusatte SL, O'Connor JK, Jarvis ED (2015) The origin and diversification of birds. Curr Biol 25:R888–R898

Brusatte SL, Challands TJ, Ross DA, Wilkinson M (2016) Sauropod dinosaur trackways in a Middle Jurassic lagoon on the Isle of Skye, Scotland. Scott J Geol 52(1):1–9

Bump JK, Tischler KB, Schrank AJ, Peterson RO, Vucetich JA (2009) Large herbivores and aquatic-terrestrial links in southern boreal forests. J Anim Ecol 78:338–345

Burrough PA (1981) Fractal dimensions of landscapes and other environmental data. Nature 294:240–242

Bush MB, McMichael CH, Piperno DR, Silman MR, Barlow J et al (2015) Anthropogenic influence on Amazonian forests in pre-history: an ecological perspective. J Biogeogr 42:2277–2288

Butler RJ, Barrett PM, Kenrick P, Penn MG (2009) Diversity patterns amongst herbivorous dinosaurs and plants during the Cretaceous: implications for hypotheses of dinosaur/angiosperm co-evolution. The Natural History Museum 22:446–459

Cai C, Escalona HE, Li L, Yin Z, Huang D, Engel MS (2018) Beetle pollination of cycads in the Mesozoic. Curr Biol 28:2806–2812

Calvente A, Zappi DC, Forest F, Lohmann LG (2011) Molecular phylogeny, evolution, and biogeography of South American epiphytic cacti. Int J Plant Sci 172(7):902–914

Cano LM, Raffaele S, Haugen RH, Sainders DGO, Leonelli L et al (2013) Major transcriptome reprogramming underlies floral mimicry induced by the rust fungus *Puccinia monoica* in *Boechera stricta*. PLoS One 8(9):e75293

Carrier DR (1987) The evolution of locomotor stamina in tetrapods: circumventing a mechanical constraint. Paleobiology 13(3):326–341

Chen Z-H, Chen G, Dai F, Wang Y, Hills A et al (2017) Molecular evolution of grass stomata. Trends Plant Sci 22(2):124–169

Chiari Y, Cahais V, Galtier N, Delsuc F (2012) Phylogenomic analyses support the position of turtles as the sister group of birds and crocodiles (Archosauria). BMC Biol 10:65

Chin K, Pearson D, Ekdale AA (2013) Fossil worm Burrows reveal very early terrestrial animal activity and shed light on trophic resources after the end-cretaceous mass extinction. PLoS One 8(8):e70920

Clack JA (2012) Gaining ground—the origin and evolution of Tetrapods. Indiana University Press, Bloomington

Clapham ME, Karr JA (2012) Environmental and biotic controls in the evolutionary history of insect body size. PNAS 109(27):10927–10930

Clark DA, Demb JB (2016) Parallel computations in insect and mammalian visual motion processing. Curr Biol 26:R1062–R1072

Clarke JT, Warnock RCM, Donogue PCJ (2011) Establishing a time-scale for plant evolution. New Phytol 192:266–301

Clauss M, Steuer P, Müller DWH, Codron D, Hummel J (2013) Herbivory and body size: Allometries of diet quality and gastrointestinal physiology, and implications for herbivore ecology and dinosaur gigantism. PLoS One 8(10):e68714

Close RA, Benson RBJ, Alroy J, Carrano MT, Cleary TJ et al (2020) The apparent exponential radiation of Phanerozoic land invertebrates is an artefact of spatial sampling biases. Proc R Soc B 287:20200372

Coates MI, Ruta M, Friedman M (2008) Ever since Owen: changing perspectives on the early evolution of Tetrapods. Annu Rev Ecol Evol Sci 39:571–592

Coiro M, Doyle JA, Hilton J (2019) How deep is the conflict between molecular and fossil evidence on the age of angiosperms? New Phytol 223:83–99

Condamine FL, Clapham ME, Kergoat GJ (2016) Global patterns of insect diversification: towards a reconciliation of fossil and molecular evidence? Sci Rep 6:19208. https://doi.org/10.1038/srep19208

Condamine FL, Guinot G, Benton MJ, Currie PJ (2021) Dinosaur diversity declined well before the asteroid impact, influenced by ecological and environmental pressures. Nat Commun 12:3833

Coomes DA, Allen RB, Bentley WA, Burrows LE, Canham CD et al (2005) The hare, the tortoise and the crocodile: the ecology of angiosperm dominance, conifer persistence and fern filtering. J Ecol 93:918–935

Costa MM, Barrote I, Silva J, Olivé I, Alexandre A et al (2015) Epiphytes modulate Posidonia oceanica photosynthetic production, energetic balance, antioxidant mechanisms, and oxidative damage. Front Mar Sci 2:111

Culley TM, Weller SG, Sakai AK (2002) The evolution of wind pollination in angiosperms. Trends Ecol Evol 17(8):361–369

Daeschler EB, Shubin NH, Jenkins FA Jr (2006) A Devonian tetrapod-like fish and the evolution of the tetrapod body plan. Nature 440:757–763

Dahl TW, Hammerlund EU, Anbar AD, Bond DPG, Gili BC et al (2010) Devonian rise of atmospheric oxygen correlated to the radiations of terrestrial plants and large predatory fish. PNAS 107(42):17911–17915

Dal Corso J, Bernardi M, Sun Y, Song H, Seyfullah LJ et al (2020) Extinction and dawn of the modern world in the Carnian (Late Triassic). Sci Adv 6:eaba0099

Datta S, Kim CM, Pernas M, Pires ND, Proust H et al (2011) Root hairs: development, growth and evolution at the plant-soil interface. Plant Soil 346:1–14

Davies NS, Gibling MR (2010) Cambrian to Devonian evolution of alluvial systems: the sedimentological impact of the earliest land plants. Earth-Sci Review 98:171–200

De Baets S, Poesen J, Reubens B, Wemans K, Baerdemacker J, Muys B (2008) Root tensile strength and root distribution of typical Mediterranean plant species and their contribution to soil shear strength. Pant Soil 305:207–226

de Boer HJ, Eppinga MB, Wassen MJ, Dekker SC (2012) A critical transition in leaf evolution facilitated the Cretaceous angiosperm revolution. Nat Commun 3:1221. https://doi.org/10.1038/ncomms2217

De Ley P (2006) A quick tour of nematode diversity and the backbone of nematode phylogeny. In: Fitch DHA (ed) Wormbook, pp 1–8. https://doi.org/10.1895/wormbook.1.41.1

Delaux P-M, Schornack S (2021) Plant evolution driven by interactions with symbiotic and pathogenic microbes. Science 371:eaba6605

Delcourt PA, Delcourt HR (1992) Ecotone dynamics in space and time. In: Hansen AJ, di Castri F (eds) Landscape boundaries. Springer, Berlin, pp 19–54

Deng T, Lu X, Wang S, Flynn LJ, Sun D et al (2021) An Oligocene giant rhino provides insight into *Paraceratherium* evolution. Comm Biol 4:639

DiMichele WA, Falcon-Lang HJ (2011) Pennsylvanian 'fossil forests' in growth position (T^0 assemblahes): origin, taphonomic bias and Palaeoecological insights. J Geol Soc Lond 168:585–605

DiMichele WA, Falcon-Lang HJ, Nelson WJ, Elrick SD, Ames PR (2007) Ecological gradients within a Pennsylvanian mire forest. Geology 15(5):415–418

dos Reis M, Thawomwattana Y, Angelis K, Telford MJ, Donoghue PCJ, Yang Z (2015) Uncertainty in the timing of origin of animals and the limits of precision in molecular timescales. Curr Biol 25:2939–2950

Dotzler N, Krings M, Taylor TN, Agerer R (2006) Germination shields in *Scutellospora* (Glomeromycota: Diversisporales, Gigasporaceae) from the 400 million-year-old Rhynie chert. Mycol Prog 5:178–184. https://doi.org/10.1007/s11557-006-0511-z

Doyle JA (2012) Molecular and fossil evidence on the origin of angiosperms. Annu Rev Earth Planet Sci 40:301–326

Dudgeon CL, Corrigan S, Yang L, Allen GR, Erdmann MV et al (2020) Walking, swimming or hitching a ride? Phylogenetics and biogeography of the walking shark genus *Hemiscyllium*. Mar Freshw Res 71(9):1107–1117. https://doi.org/10.1071/MF19163

Dunlop IA, Garwood RJ (2017) Terrestrial invertebrates in the Rhynie chert ecosystem. Philos Trans R Soc B 373:20160493

Dunn CW (2013) Evolution: out of the ocean. Curr Biol 23(6):R241–R242

Dupont S, Lemetais G, Ferreira T, Cayot P, Beney L (2012) Ergosterol biosynthesis: a fungal pathway for life on land? Evolution 66(9):2961–2968

Edgecombe GD, Giribet G (2007) Evolutionary biology of centipedes (Myriapoda: Chilopoda). Annu Rev Entomol 52:151–170

Edgecombe GD, Legg DA (2014) Origins and the early evolution of arthropods. Palaeontology 57(3):457–468

Edwards D, Kenrick P (2015) The early evolution of land plants. From fossils to genomics: a commentary on Lang (1937) 'on the plant-remains from the Downtonian of England and Wales'. Philos Trans R Soc B 370:20140343

Edwards EJ, Osborne CP, Strömberg CAE, Smith SA (2010) The origins of C_4 grasslands: integrating evolutionary and ecosystem science. Science 328:587–591

Edwards D, Morris JL, Richardson JB, Kenrick P (2014) Cryptospores and cryptophytes reveal hidden diversity in early land floras. New Phytol 202:50–78

Ellenberg H, Mayer R, Schermann J (1986) Ökosystemforschung Ergebnisse des Sollingprojekts 1966–1986. Ulmer, Stuttgart

Eriksson A, Betti L, Friend AD, Lycett SJ, Singarayer JS et al (2012) Late Pleistocene climate change and the global expansion of anatomically modern humans. PNAS 109(40):16089–16094

Ersahin S, Gunal H, Kutlu T, Yetgin B, Coban S (2006) Estimating specific surface area and cation exchange capacity in soils using fractal dimension of particle-size distribution. Geoderma 136:588–597

Erséus C, Williams BW, Horn KM, Halanych KM, Santos SR et al (2020) Phylogenomic analyses reveal a palaeozoic radiation and support a freshwater origin of clitellate annelids. Zool Scr 49:614–640

Estrada C, Degner EC, Rojas EI, Wcislo WT, Van Bael SA (2015) The role of endophyte diversity in protecting plants from defoliation by leaf-cutting ants. Curr Sci 109(1):55–61

Fagan WF, Bishop JG, Schade JD (2004) Spatially structured herbivory and primary succession at Mount St. Helens: field surveys and experimental growth studies suggest a role for nutrients. Ecological Entomology 29:398–409

Falcon-Lang HJ, Bashforth AR (2005) Morphology, anatomy, and upland ecology of large cordaitalean trees from the middle Pennsylvanian of Newfoundland. Rev Palaeobot Palynol 135:223–243

Falcon-Lang HJ, Benton MJ, Braddy SJ, Davies SJ (2006) The Pennsylvanian tropical biome reconstructed from the Joggins formation of Nova Scotia, Canada. J Geol Soc Lond 163:561–576

Farrell BD (1998) "Inordinate fondness" explained: why are there so many beetles? Science 281:555–559

Farrell BD, Sequeira AS, O'Meara BC, Normark BB, Chung JH, Jordal BH (2001) The evolution of agriculture in beetles (Curculionidae: Scolytinae and Platypodinae). Evolution 55(10):2011–2027

Feduccia A (1995) Explosive evolution in tertiary birds and mammals. Science 267:637–638

Feijen FAA, Vos RA, Nuytinck J, Merckx VSFT (2018) Evolutionary dynamics of mycorrhizal symbiosis in land plant diversification. Sci Rep 8:10698. https://doi.org/10.1038/s41598-018-28920-x

Fernandez C, Lelong B, Via B, Mévy J-P, Rohles C et al (2006) Potential allelopathic effect of *Pinus halepensis* in the secondary succession: an experimental approach. Chemoecology 16:97–105

Fernández-Baldor F, Diaz-Martínez I, Huerta P, Castanera D (2021) Enigmatic tracks of solitary sauropods roaming an extensive lacustrine megatracksite in Iberia. Sci Rep 11:16939

Fernández-Coll M, Arbez T, Bernardini F, Fortuny J (2019) Cranial anatomy of the Early Triassic trematosaurine Angusaurus (Tmenospondyli: Stereospondyli): 3D endocranial insights and phylogenetic implications. J Iber Geol 45(2):269–286

Field KJ, Pressel S, Duckett JG, Riminton WR, Bidartondo MI (2015) Symbiotic options for the conquest of land. Trends Ecol Evol 30(8):477–488

Field DJ, Bercovici A, Berv JS, Dunn R, Fastovsky DE et al (2018) Early evolution of modern birds structured by global forest collapse at the end-cretaceous mass extinction. Curr Biol 28:1825–1831

Figueirido B, Janis CM, Pérez-Claros JA, De Renzi M, Palmqvist P (2011) Cenozoic climate changes influences mammalian evolutionary dynamics. PNAS 109(3):722–727

Flammang BE, Suvarnaraksha A, Markeiwicz J, Soares D (2016) Tetrapod-like pelvic girdle in a walking cavefish. Sci Rep 6:23711. https://doi.org/10.1038/srep23711

Fleagle JG (2013) Primate adaptation and evolution. Elsevier, Amsterdam

Fleming TH, Geiselman C, Kress WJ (2009) The evolution of bat pollination: a phylogenetic perspective. Ann Bot 104:1017–1043

Flores BM, Holmgren M (2021) White-sand savannas expand at the Core of the Amazon after forest wildfires. Ecosystems 24:1624–1637

Floudas D, Binder M, Riley R, Barry K, Blanchette RA et al (2012) The Paleozoic origin of enzymatic lignin decomposition reconstructed from 31 fungal genomes. Science 336:1715–1719

Foffa D, Dunne EM, Nesbitt SJ, Butler RJ, Fraser NC et al (2022) *Scleromochlus* and the early evolution of Pterosauromorpha. Nature 610:313–318

Fortuny J, Steyer JS (2019) New insights into the evolution of temnospondyls. J Iber Geol 45:247–250

Fortuny J, Marcé-Nogué J, de Esteban-Trivigno S, Gili L, Galobart A (2011) Temnospondyli bite club: ecomorphological patterns of the most diverse group of early tetrapods. J Evol Biol 24:2040–2054

Fortuny J, Marcé-Nogué J, Steyer JS, de Esteban-Trivigno S, Mujal E, Gil L (2016) Comparative 3D analyses and palaeoecology of giant early amphibians (Temnospondyli: Sterospondyli). Sci Rep 6:30387. https://doi.org/10.1038/srep30387

Fraser EC, Lieffers VJ, Landhäusser SM (2005) Age, stand density, and tree size as factors in root and basal grafting of lodgepole pine. Can J Bot 83:983–988

Frentiu FD, Briscoe AD (2008) A butterfly eye's view of birds. BioEss 30:1151–1162

Friedman M (2015) The early evolution of ray-finned fishes. Palaeontology 58(2):213–228

Friedman M, Sallan LC (2012) Five hundred million years of extinction and recovery: a Phanerozoic survey of large-scale diversity pattern in fishes. Palaeontology 55(4):707–742

Fröbisch J (2008) Global taxonomic diversity of Anomodonts (Tetrapoda, Therapsida) and the terrestrial rock record across the Permian-Triassic boundary. PLoS One 3(11):e3733

Fröbisch J (2009) Composition and similarity of global anomodont-bearing tetrapod faunas. Earth-Sci Rev 95:119–157

Fröbisch J, Reisz RR (2009) The Late Permian herbivores *Suminia* and the early evolution of arboreality in terrestrial vertebrate ecosystems. Proc R Soc B376:3611–3618

Fürnkranz M, Wanke W, Richter A, Abell G, Rasche F, Sessitsch A (2008) Nitrogen fixation by phyllosphere bacteria associated with higher plants and their colonizing epiphytes of a tropical lowland rainforest of Costa Rica. ISME J 2:561–570

Gao Z, Johnson ZI, Wang G (2010) Molecular characterization of the spatial diversity and novel lineages of mycoplankton in Hawaiian coastal waters. ISME J 4:111–120

Garberoglio FF, Apesteguía S, Simões TR, Palci A, Gómez RA et al (2019) New skulls and skeletons of the Cretaceous legged snake Najash, and the evolution of the modern snake body plan. Sci Adv 5:eaax5833

Gatesy J, Geisler JH, Chang J, Buell C, Berta A et al (2013) A phylogenetic blueprint for a modern whale. Mol Phylogenet Evol 66(2):479–506

Gebo DL (2004) A shrew-sized origin for primates. Yearb Phys Anthropol 47:40–62

Gee CT (2011) Dietary options for the sauropod dinosaurs from an integrated botanical and Paleobotanical perspective. In: Klein N, Remes K, Gee CT, Sander PM (eds) Biology of the sauropod dinosaur. Indiana Univ. Press, Bloomington, pp 34–56

Gensel PG (2008) The earliest land plants. Annu Rev Ecol Evol Syst 39:459–477

Gerrienne P, Servais T, Vecoli M (2016) Plant evolution and terrstrialization during Palaeozoic times-the phylogenetic context. Rev Palaeobot Palynol 227:4–18

Gibling MR, Davies NS (2012) Palaeozoic landscapes shaped by plant evolution. Nat Geosci 5:99–105

Gibling MR, Davies NS, Falcon-Lang HJ, Bashforth AR, DiMichele WA et al (2014) Palaeozoic co-evolution of rivers and vegetation: a synthesis of current knowledge. Proc Geol Assoc 125:524–533

Giere O (2009) Meiobenthology. Springer, Berlin

Giribet G, Edgecombe GD (2019) The phylogeny and evolutionary history of arthropods. Curr Biol 29:R592–R602

Givnish TJ, Barfuss MHJ, Van Ee B, Riina R, Schulte K et al (2014) Adaptive radiation, correlated and contingent evolution, and net species diversification in Bromeliaceae. Mol Phylogenet Evol 71:55–78

Grimaldi D, Engel MS (2005) Evolution of the insects. Cambridge University Press, Cambridge

Grube M, Berg G (2009) Microbial consortia of bacteria and fungi with focus on the lichen symbiosis. Fungal Biol Rev 23:72–85

Gyssels G, Poesen J, Bochet E, Li Y (2005) Impact of plant roots on the resistance of soils to erosion by water: a review. Prog Phys Geogr 29(2):189–217

Han F, Wang Q, Wang H, Zhu X, Zhou X et al (2022) Low dinosaur biodiversity in central China 2 million years prior to the end-Cretaceous mass extinction. PNAS 119(39):e2211234119

Harrison CJ, Morris JL (2017) The origin and early evolution of vascular plant shoots and leaves. Philos Trans R Soc B 373:20160496

Heckman DS, Geiser DM, Eidell BR, Stauffer RL, Kardos NL, Hedges SB (2001) Molecular evidence for the early colonization of land by fungi and plants. Science 293:1129–1133

Helfman GS, Collette BB, Facey DE, Bowen BW (2009) The diversity of fishes. Wiley-Blackwell, Chichester

Henderson DM (2004) Typsi punters: sauropod dinosaur pneumaticity, buoyancy and aquatic habits. Proc R Soc Lond B 271:S180–S183

Henderson-Begg SK, Hill T, Thyrhaug R, Khan M, Moffett BF (2009) Terrestrial and airborne non-bacterial ice nuclei. Atmos Sci Lett 10:215–219

Hennequin S, Schuettpelz E, Pryer KM, Ebihara A, Dubuisson J-Y (2008) Divergence times and the evolution of epiphytism in filmy ferns (Hymenophyllaceae) revisited. Int J Plant Sci 169(9):1278–1287

Henskens FL, Green TGA, Wilkins A (2012) Cyanolichens can have both cyanobacteria and green algae in a common layer as major contributors in photosynthesis. Ann Bot 110:555–563

Hernández-Hernández T, Wiens JJ (2020) Why are there so many flowering plants? A multiscale analysis of plant diversification. The Am Naturalist 195(6):948–963

Hershkovitz I, Weber GW, Quam R, Duval M, Grün R et al (2018) The earliest modern humans outside Africa. Science 359:456–459

Hetherington AJ, Dolan L (2018) Stepwise and independent origins of roots among land plants. Nature 561:235–238

Hilker M, Kohs C, Varama M, Schrank K (2002) Insect egg deposition induces Pinus sylvestris to attract egg parasitoids. J Exp Biol 205:455–461

Hirasawa T, Hu Y, Uesugi K, Hoshino M, Manabe M, Kuratani S (2022) Morphology of Palaeospondylus shows affinity to tetrapod ancestors. Nature 606:109–112

Holmes PM, Newton RJ (2004) Patterns of seed persistence in South African fynbos. Plant Ecol 172:143–158

Holsinger KE (2000) Reproductive systems and evolution in vascular plants. PNAS 97(13):7037–7042

Honegger R, Edwards D, Axe L (2013) The earliest records of internally stratified cyanobacterial and algal lichens from the Lower Devonian of the Welsh Borderland. New Phytol 197:264–275

Hughes PD, Gibbard PL, Ehlers J (2013) Timing of glaciation during the last glacial cycle: evaluating the concept of a global 'last glacial maximum' (LGM). Earth-Sci Rev 125:171–198

Insarova JD, Blagoveshchenskaya EY (2016) Lichen Symbiosis: search and recognition of partners. Biol Bull 43(5):408–418

Irmis RB, Whiteside JH (2012) Delayed recovery of non-marine tetrapods after the end-Permian mass extinction tracks global carbon cycle. Proc R Soc B 279:1310–1318

Iwamoto H (2011) Structure, function and evolution of insect flight muscle. Biophysics 7:21–28

Jahren AH, Porter S, Kuglitsch JJ (2003) Lichen metabolism identified in Early Devonian terrestrial organisms. Geology 31(2):99–102

Jones S, Martin R, Pilbeam D (eds) (1994) The Cambridge encyclopedia of human evolution. Cambridge University Press, Cambridge

Jordal BH, Cognato AI (2012) Molecular phylogeny of bark and ambrosia beetles reveals multiple origin of fungus farming during periods of global warming. BMC Evol Biol 12:133

Jud NA (2015) Fossil evidence for a herbaceous diversification of early eudicot angiosperms during the Early Cretaceous. Proc R Soc B 282:20151045

Kaiser SI, Steuber T, Becker RT (2008) Environmental change during the Late Famennian and Early Tournaisian (Late Devonian-Early Carboniferous): implications from stable isotopes and conodont biofacies in southern Europe. Geol J 43:241–260

Kandeler E, Deiglmayr K, Tscherko D, Bru D, Philippot L (2006) Abundance of narG, nir S, nirK, and nosZ genes of denitrifying bacteria during primary successions of a glacier frontier. Appl Environ Microbiol 77(9):5957–5962

Kelber A, Yovanovich C, Olsson (2017) Thresholds and noise limitations of colour vision in dim light. Philos Trans R Soc B 372:20160065

Kenrick P, Strullu-Derrien C (2014) The origin and early evolution of roots. Plant Physiol 166:570–580

Kenrick P, Wellman CH, Schneider H, Edgecombe GD (2012) A timeline for terrestrialization: consequences for the carbon cycle in the Palaeozoic. Philos Trans R Soc B 367:519–536

King HM, Shubin NH, Coates MI, Hale ME (2011) Behavioral evidence for the evolution of walking and bounding before terrestriality in sarcopterygian fishes. PNAS 108(52):21146–21151

Kirkendall LR, Biedermann PHW, Jordal BH (2015) Evolution and diversity of bark and ambrosia beetles. In: Vega FE, Hofstetter RW (eds) Bark Beetles. Elsevier, Amsterdam, pp 85–156

Knack JJ, Wilcox LW, Delaux P-M, Ané J-M, Piotrowski MJ et al (2015) Microbiomes of streptophyte algae and bryophytes suggest that a functional suite of microbiota fostered plant colonization of land. Int J Plant Sci 176(5):405–420

Knighton D (1998) Fluvial forms and processes. Arnold, London

Konhauser KO (2007) Introduction to Geomicrobiology. Blackwell, Malden

Krause DW, Sertich JJW, O'Connor PM, Rogers KC, Rogers RR (2019) The Mesozoic biogeographic history of Gondwanan terrestrial vertebrates: insights from Madagaskar's fossil record. Annu Rev Erath Plan Sci 47:519–553

Krings M, Taylor TN (2014) A mantled fungal reproductive unit from the Lower Devonian Rhynie chert that demonstrates Carboniferous "sporocarp" morphology and development. N Jb Geol Paläont (Abh) 273(2):197–205

Labandeira CC (2005) Invasion of the continents: cyanobacterial crusts to tree-inhabiting arthropods. Trends Ecol Evol 20(5):253–262

Labandeira C (2007) The origin of herbivory on land: initial patterns of plant tissue consumption by arthropods. Insect Sci 14:239–275

Labandeira CC (2013a) A paleobiologic perspective on plant-insect interactions. Curr Opin Plant Biol 16:414–421

Labandeira CC (2013b) Deep-time patterns of tissue consumption by terrestrial arthropod herbivores. Naturwissenschaften 100:355–364

Labandeira C (2014) Why did terrestrial insect diversity not increase during the angiosperm radiation? Mid-Mesozoic, plant-associated insect Lineages Harbor cues. In: Pontarotti P (ed) Evolutionary biology, genome evolution, speciation, coevolution and origin of life. Springer, Basel, pp 261–299

Labandeira CC, Johnson KR, Wilf P (2002) Impact of the terminal Cretaceous event on plant-insect associations. PNAS 99(4):2061–2066

Laenen B, Shaw B, Schneider H, Goffinet B, Paradis E et al (2014) Extant diversity of bryophytes emerged from successive post-Mesozoic diversification bursts. Nat Commun 5:5134. https://doi.org/10.1038/ncomms6134

Lakatos M (2011) Lichens and bryophytes: habitats and species. In: Lüttge U, Beck E, Bartels D (eds) Plant desiccation tolerance. Springer, Berlin, pp 65–87

Lalley JS, Viles HA (2008) Recovery of lichen-dominated soil crusts in a hyper-arid desert. Biodivers Conserv 17:1–20

Lambin EF, Meyfroidt P (2011) Global land use change, economic globalization, and the looming land scarcity. PNAS 108(9):3465–3472

Lancaster J, Downes BJ (2013) Aquatic entomology. Oxford University Press, Oxford

Langdale JA (2008) Evolution of developmental mechanisms in plants. Curr Opin Genet Dev 18:368–373

Laurin M (2004) The evolution of body size, Cope's rule and the origin of amniotes. Syst Biol 53(4):594–622

Laurin M, Soler-Gijón R (2010) Osmotic tolerance and habitat of early stegocephalians: indirect evidence from parsimony, taphonomy, palaeobiogeography, physiology and morphology. Geol Soc Lond Spec Publ 339:151–179

Lee CE, Charmantier G, Lorin-Nebel C (2022) Mechanisms of Na^+ uptake from freshwater habitats in animals. Front Physiol 13:1006113

Lenton TM (2001) The role of land plants, phosphorus weathering and fire in the rise and regulation of atmospheric oxygen. Glob Chang Biol 7:613–629

Lenton TM, Dahl TW, Daines SJ, Mills BJW, Ozaki K et al (2016) Earliest land plants created modern levels of atmospheric oxygen. PNAS 113(35):9704–9709

Li H-T, Yi T-S, Gao L-M, Ma P-F, Zhang T et al (2019) Origin of angiosperms and the puzzle of the Jurassic gap. Nature Plants 5:461–470

Lieth H, Whittaker RH (1975) Primary productivity of the biosphere. Springer, Berlin

Lindstedt C, Murphy L, Mappes J (2019) Antipredator strategies pf pupae: how to avoid predation in an immonile life stage? Philos Trans R Soc B 374:20190069

Longrich NR, Tokaryk T, Field DJ (2011) Mass extinction of birds at the Cretaceous-Paleogene (K-PG) boundary. PNAS 108(37):15253–15257

Longrich NR, Scribers J, Wills MA (2016) Severe extinction and rapid recovery of mammals across the Cretaceous-Paleogene boundary, and the effects of rarity on patterns of extinction and recovery. J Evol Biol 29:1495–1512

Lozano-Fernandez J, Carton R, Tanner AR, Puttick MN, Blaxter M et al (2016) A molecular palaeobiological exploration of arthropod terrestrialization. Philos Trans R Soc B. 371 371:20150133

Lu J, Zhu M, Long JA, Zhao W, Senden TJ et al (2012) The earliest known stem-tetrapod from the Lower Devonian of China. Nat Commun 3:1160. https://doi.org/10.1038/ncomms2170

Lucifora LO, Vassallo AI (2002) Walking in skates (Chondrichthyes, Rajidae): anatomy, behaviour and analogies to tetrapod locomotion. Biol J Linn Soc 77:35–41

Luo Z-X (2007) Transformation and diversification in early mammal evolution. Nature 450:1011–1019

Lyson TR, Miller IM, Bercovici AD, Weissenburger K, Fuentes AJ et al (2019) Exceptional continental record of biotic recovery after the Cretaceous-Paleogene mass extinction. Science 366:977–983

Macel M (2011) Attract and deter: a dual role for pyrrolizidine alkaloids in plant-insect interactions. Phytochem Rev 10:75–82

MacIver MA, Schmitz L, Mugan U, Murphey TD, Mobley CD (2017) Massive increase in visual range preceded the origin of terrestrial vertebrates. PNAS 114:E2375–E2384

Mackie RI (2002) Mutualistic fermentative digestion in the gastrointestinal tract: diversity and evolution. Integ And Comp Biol 42:319–326

Magallón S, Gómez-Acevedo S, Sánchez-Reyes LL, Hernández-Hernández T (2015) A metacalibrated time-tree documents the early rise of flowering plant phylogenetic diversity. New Phytol 207:437–453

Mann A, Pardo JD, Maddin HC (2022) Snake-like limb loss in a Carboniferous amniote. Nat Ecol Evol 6:614–621

Marsicano CA, Latimer E, Rubidge B, Smith RMH (2017) The Rhinesuchidae and early history of the Stereospondyla (Amphibia: Temnospondyli) at the end of the Palaeozoic. Zool J Linnean Soc XX:1–28

Martin FM, Uroz S, Barker DG (2017) Ancestral alliances: plant mutualistic symbioses with fungi and bacteria. Science 356:eaad4501

Martinez-Redondo GI, Guerro CS, Aristide L, Balart-Garcia P, Tonzo V, Fernández R (2022) Parallel duplication and loss of aquaporin-coding genes during the 'out of the sea' transition paved the way for animal terrestrialization. bioRxiv. https://doi.org/10.1101/2022.07.25.601387

McAdam SAM, Brodribb TJ (2013) Ancestral stomatal control results in a canalization of fern and lycophyte adaptation to drought. New Phytol 198:429–441

McElwain JC, Punyasena SW (2007) Mass extinction events and the plant fossil record. Trends Ecol Evol 22(10):548–567

McGhee GR Jr, Clapham ME, Sheehan PM, Bottjer DJ, Doser ML (2013) A new ecological-severity ranking of major Phanerozoic biodiversity crises. Palaeogeogr Palaeoclimatol Palaeoecol 370:260–270

McMahon WJ, Davies NS (2018) Evolution of alluvial mudrock forced by early land plants. Science 359:1022–1024

Melillo JM, McGuire AD, Kicklighter DW, Moore B III, Vorosmarty CJ, Schloss AL (1993) Global climate change and terrestrial net primary production. Nature 363:234–240

Meyer-Berthaud B, Soria A, Decombeix A-L (2010) The land plant cover in the Devonian: a reassessment of evolution of the tree habit. Geol Soc Lond Spec Publ 339:59–70

Mills BJW, Batterman SA, Field KJ (2017) Nutrient acquisition by symbiotic fungi governs Palaeozoic climate transition. Philos Trans R Soc B 373:20160503

Misof B, Liu S, Meusemann K, Peters RS, Donath A et al (2014) Phylogenomics resolves the timing and pattern of insect evolution. Science 356:763–767

Moning C, Werth S, Dziock F, Bässler C, Bradtka J et al (2009) Lichen diversity in temperate montane forests is influenced by forest structure more than climate. For Ecol Manag 258:745–751

Montagna M, Tong KJ, Magoga G, Strada L, Tintori A et al (2019) Recalibration of the insect evolutionary time scale using Monta San Giorgio fossils suggest survival of key lineages through the end-Permian extinction. Proc R Soc B 286:20191854

Müller J, Bickelmann C, Sobral G (2018) The evolution and fossil history of sensory perception in Amniote vertebrates. Annu Rev Earth Planet Sci 46:495–519

Mustoe GE (2007) Coevolution of cycads and dinosaurs. The Cycad Newsletter 30(1):6–9

Nagalingum NS, Marshall CR, Quental TB, Rai HS, Little DP, Mathews S (2011) Recent synchronous radiation of a living fossil. Science 334:796–799

Nash TH III (2008) Lichen Biology. Cambridge University Press, Cambridge

Neusser TP, Jörger KM, Schrödl M (2011) Cryptic species in Tropic Sands – interactive 3D anatomy, molecular phylogeny and evolution of Meiofaunal Pseudunelidae (Gastropoda, Acochlidia). PLoS One 6(8):e23313

Neves EG (2007) El Formativo que nunca terminó: la larga historia de estabilidad en las ocupaciones humanas de la Amazonía centrl. Bol de Arqueología 11:117–142

Nicholson DB, Mayhew PJ, Ross AJ (2015) Changes to the fossil record of insects through fifteen years of discovery. PLoS One 10(7):e0128554

Niklas KJ, Tiffney BH, Knoll AH (1983) Patterns in vascular land plant diversification. Nature 303:614–616

Nyman T, Onstein RE, Silvestro D, Wutke S, Taeger A et al (2019) The early wasp plucks the flower: disparate extant diversity of sawfly superfamilies (Hymenoptera: 'Symphyta') may reflect asynchronous switching to angiosperm hosts. Biol J Linn Soc 128:1–19

O'Loughlin FE, Paiva RCD, Durand M, Alsdorf DE, Bates PD (2016) A multi-sensor approach towards vegetation corrected SRTM DEM product. Remote Sens Environ 182:49–59

Ogura-Tsujita Y, Yokoyma J, Mioshi K, Yukawa T (2012) Shifts in mycorrhizal fungi during the evolution of autotrophy to mycoheterotrophy in *Cymbidium* (Orchidadeae). Am J Bot 99(7):1158–1176

Ohtonen R, Fritze H, Pennanen T, Jumpponen A, Trappe J (1999) Ecosystem properties and microbial community changes in primary succession on a glacier forefront. Oecologia 119:239–246

Olesen JM, Valido A (2003) Lizards as pollinators and seed dispersers: an island phenomenon. Trends Ecol Evol 18(4):177–181

Oliver M, Tuba Z, Mishler BD (2000) The evolution of vegetative desiccation tolerance in land plants. Plant Ecol 151:85–100

Ollerton J, Coulthard E (2009) Evolution of animal pollination. Science 326:808–809

Ollerton J, Winfree R, Tarrant S (2011) How many flowering plants are pollinated by animals? Oikos 120:321–326

Opluštil S, Pšenička J, Libertin M, Bashforth AR, Šimůnek Z et al (2009) A middle Pennsylvanian (Bolsovian) peat-forming forest preserved *in situ* in volcanic ash of the whetstone horizon in the Radnice Basin, Czech Republic. Rev Palaeobot Palynol 155:234–274

Osorio D, Vorobyev M (2008) A review of the evolution of animal colour vision and visual communication signals. Vis Res 48:2042–2051

Ownley BH, Gwinn KD, Vega FE (2010) Endophytic fungal entomopathogens with activity against plant pathogens: ecology and evolution. BioControl 55:113–128

Page CN (2002) Ecological strategies in fern evolution: a neopteridological overview. Rev Palaeobot Palynol 119:1–11

Pardo JD, Szostakiwskyj M, Ahlberg PE, Anderson JS (2017) Hidden morphological diversity among early tetrapods. Nature 546:642–645

Pastore AI, Prather CM, Gornish ES, Ryan WH, Ellis RD, Miller TE (2014) Testing the competition-colonization trade-off with a 32-year study of a saxicolous lichen community. Ecology 95(2):306–315

Patiny S (2012) Evolution of plant-pollinator relationships. Cambridge University Press, Cambridge

Paul EA (2007) Soil microbiology, ecology, and biochemistry. Academic Press, Amsterdam

Pearson MR, Benson RBJ, Upchurch P, Fröbisch J, Kammerer CF (2013) Reconstructing the diversity of early terrestrial herbivorous tetrapods. Palaeogeogr Palaeoclimatol Palaeoecol 371:42–49

Peñalver E, Labandeira CC, Barròn E, Delciòs X, Nel P et al (2012) Thrips pollination of Mesozoic gymnosperms. PNAS 109(22):8623–8628

Peñalver E, Arillo A, Pérez-de la Fuente R, Riccio ML, Delciòs X et al (2015) Long-Proboscid flies as pollinators of cretaceous gymnosperms. Curr Biol 25:1917–1923

Perry SF, Sander M (2004) Reconstructing the evolution of the respiratory apparatus in tetrapods. Respir Physiol Neurobiol 144(2–3):125–139

Popper ZA, Michel G, Hervé C, Domozych DS, Willats WGT et al (2011) Evolution and diversity of plant cell walls: from algae to flowering plants. Annu Rev Plant Biol 62:567–590

Porada P, Lenton TM, Pohl A, Weber B, Mander L et al (2016) High potential for weathering and climate effects of non-vascular vegetation in the Late Ordovician. Nat Commun 7:12113. https://doi.org/10.1038/ncomms12113

Porter ME, Hernandez AV, Gervais CR, Rummer JL (2022) Aquatic walking and swimming kinematics of neonate and juvenile epaulette sharks. Integr Comp Biol 62(6):1710–1724

Prentice KC, Ruta M, Benton MJ (2011) Evolution of morphological disparity in pterosaurs. J Syst Palaeontol 9(3):337–353

Pypker TG, Unsworth MH, Bond BJ (2006) The role of epiphytes in rainfall interception by forests in the Pacific Northwest. II. Field measurements at the branch and canopy scale. Can J For Res 36:819–832

Raffa K (2001) Mixed messages across multiple trophic levels: the ecology of bark beetle chemical communication systems. Chemoecology 11(2):49–65

Ratto F, Simmons BI, Spake R, Zamora-Gutierrez V, MacDonald MA et al (2018) Global importance of vertebrate pollinators for plant reproductive success: a meta-analysis. Front Ecol Environ 16(2):82–90

Read DJ, Duckett JG, Francis R, Ligrone R, Russell A (2000) Symbiotic fungal associations in 'lower' land plants. Philos Trans R Soc Lond B355:815–831

Rech AR, Dalsgaard B, Sandel B, Sonne J, Svenning J-C et al (2016) The macroecology of animal versus wind pollination: ecological factors are more important than historical climate stability. Plant Ecol 9(3):253–262

Regier JC, Shultz JW, Zwick A, Hussey A, Ball B et al (2010) Arthropod relationships revealed by phylogenomic analysis of nuclear protein-coding sequences. Nature 463:1079–1083

Rehan SM, Leys R, Schwarz MP (2013) First evidence for a massive extinction event affecting bees Close to the K-T boundary. PLoS One 8(10):e766783

Reisz RR, Fröbisch J (2014) The oldest Caseid synapsid from the late Pennsylvanian of Kansas, and the evolution of herbivory in terrestrial vertebrates. PLoS One 9(4):e94518

Ren D, Labandeira CC, Santiago-Blay JA, Rasnitsyn A, Shih CS et al (2009) A probable pollination mode before angiosperms: Eurasian, Long-Poboscid Scorpionflies. Science 326:840–841

Renault H, Werck-Reichhart D, Weng J-K (2019) Harnessing lignin evolution for biotechnologicsl applications. Curr Opin Biotechnol 56:105–111

Retallack GJ (2013) Permian and Triassic greenhouse crises. Gondwana Res 24:90–103

Retallack GJ (2019) Interflag sandstone laminae, a novel sedimentary structure, with implications for Ediacaran paleoenvironments. Sediment Geol 379:60–76

Retallack GJ, Veevers JJ, Morante R (1996) Global coal gap between Permian-Triassic extinction and Middle Triassic recovery of peat-forming plants. GSA Bull 108(2):195–207

Reynolds C (2006) Ecology of phytoplankton. Cambridge University Press, Cambridge

Robinson C, Gibbes B, Carey H, Li L (2007) Salt-freshwater dynamics in a subterranean estuary over a spring-neap tidal cycle. J Geophys Res 112:C09007

Rockström J, Gupta J, Qin D, Lade SJ, Abrams JF et al (2023) Safe and just earth system boundaries. Nature 619:102–111

Romer AS (1971) Vergleichende Anatomie der Wirbeltiere. Paul Parey, Hamburg

Ross A (2017) Insect evolution: the origin of wings. Curr Biol 27:R113–R115

Rota-Stabelli O, Daley AC, Pisani D (2013) Molecular Timetrees reveal a Cambrian colonization of land and a new scenario for Ecdysozoan evolution. Curr Biol 23:392–398

Rousk J, Brookes PC, Bååth E (2009) Contrasting soil pH effects on fungal and bacterial growth suggest functional redundancy in carbon mineralization. Applied and Environmental Microb 75(6):1589–1596

Rundell RJ, Leander BS (2010) Masters of miniaturization: convergent evolution among interstitial eukaryotes. BioEssays 32:430–437

Ruta M, Benton MJ (2008) Calibrated diversity, tree topology and the mother of mass extinctions: the lesson of temnospondylus. Palaeontology 51(6):1261–1288

Rybczynski N, Reisz RR (2001) Earliest evidence for efficient oral processing in a terrestrial herbivore. Nature 411:684–687

Sage RF, Sage TL, Kocacinar F (2012) Photorespiration and the evolution of C_4 photosynthesis. Annu Rev Plant Biol 63:19–47

Saikkonen K, Gundel PE, Helander M (2013) Chemical ecology mediated by fungal endophytes in grasses. J Chem Ecol 39:962–968

Sakamoto M, Benton MJ, Venditti C (2016) Dinosaurs in decline tens of millions of years before their final extinction. PNAS 113(18):5036–5040

Sanchez S, Germains D, de Ricqlés A, Abourachid A, Goussard F, Tafforeau P (2010) Limb-bone histology of temnospondyls: implications for understanding the diversification of palaeoecologies and patterns of locomotion of Permo-Triassic tetrapods. J Evol Biol 23:2076–2090

Sander PM, Christian A, Clauss M, Fechner R, Gee CT et al (2011) Biology of the sauropod dinosaurs: the evolution of gigantism. Biol Rev 86:117–155

Santini L, Rojas D, Donati G (2013) Evolving through day and night: origin and diversification of activity pattern in modern primates. Behav Ecol 26(3):789–796

Sasaki G, Ishiwata K, Machida R, Miyata T, Su Z-H (2013) Molecular phylogeny analyses support the monophyly of Hexapoda and suggests the paraphyly of Entognatha. BMC Evol Biol 13:236

Sauquet H, von Balthazar M, Magallón S, Doyle JA, Endress PK et al (2017) The ancestral flower of angiosperms and its early diversification. Nat Commun 8:16047. https://doi.org/10.1038/ncomms16047

Schaefer I, Norton RA, Scheu S, Maraun M (2010) Arthropod colonization of land—linking molecules and fossils in oribatid mites (Acari, Oribatida). Mol Phylogenet Evol 57:113–121

Schipper J, Chanson JS, Chiozza F, Cox NA, Hoffmann M et al (2008) The status of the World's land and marine mammals: diversity, threat, and knowledge. Science 122:225–230

Schoch RR (2009) Evolution of Life Cycles in early amphibians. Annu Rev Earth Planet Sci 17:135–162

Schoch RR (2014) Amphibian evolution. Wiley Blackwell, Oxford

Scholtyßek C, Kelber A (2017) Farbsehen der Tiere Ophthalmol 114:978–985

Schoonhoven LM, van Loon JJA, Dicke M (2005) Insect-plant biology. Oxford University Press, Oxford

Schwentner M, Combosch DJ, Nelson JP, Giribet G (2017) A Phylogenomic solution to the origin of insects by resolving crustacean-Hexpaod relationships. Curr Biol 27:1818–1824

Scotese CR (2021) An atlas on Phanerozoic Paleogeographic maps: the seas come in and the seas go out. Annu Rev Earth Planet Sci 49:669–718

Scotese CR, Song H, Mills BJW, van der Meer DG (2021) Phanerozoic paleotemperatures: the earth's changing climate during the last 540 million years. Erth Sci Rev 215:103503

Selosse M-A, Roy M (2008) Green plants that feed on fungi: facts and questions about mixotrophy. Trends Plant Sci 14(2):64–70

Senesi N, Wilkinson KJ (2008) Biophysical chemistry of fractal structures and processes in environmental systems. Wiley, Hoboken

Sequeira AS, Farrell BD (2001) Evolutionary origins of Gondwana interactions: how old are *Araucaria* beetle herbivores? Biol J of the Linnean Soc 74:459–474

Sequeira AS, Normark BB, Farrell BD (2000) Evolutionary assembly of the conifer fauna: distinguishing ancient from recent associations in bark beetles. Proc R Soc Lond B 267:2359–2366

Sereno PC (1999) The evolution of dinosaurs. Science 284:2137–2147

Silva G, Costa JL, de Almeida PR, Costa MJ (2006) Structure and dynamics of a benthic invertebrate community in an intertidal area of the Tagus estuary, western Portugal: a six year data series. Hydrobiologia 555:115–128

Silvestro D, Zizka G, Schulte K (2013) Disentangling the effects of key innovations on the diversification of Bromelioideae (Bromeliaceae). Evolution 68(1):163–175

Simões TR, Vernygora O, Caldwell MW, Pierce SE (2020) Megaevolutionary dynamics and the timing of evolutionary innovation in reptiles. Nat Commun 11:3322

Skulan J (2000) Has the importance of the amniote egg been overstated? Zool J of the Linnean Society 130:235–261

Smith SA, Beaulieu JM, Donoghue MJ (2010) An uncorrelated relaxed-clock analysis suggests an earlier origin for flowering plants. PNAS 107(13):5897–5902

Standen EM, Du TY, Larsson HCE (2014) Developmental plasticity and the origin of tetrapods. Nature 513:54–58

Stanley SM (2001) Historische Geologie. Spektrum Akademischer Verlag, Heidelberg

Stein WE, Mannolini F, Hernick LVA, Landing E, Berry CM (2007) Giant cladoxylopsid trees resolve the enigma of the Earth's earliest forest stumps at Gilboa. Nature 446:904–907

Stein WE, Berry CM, Morris JL, VanAller HL, Mannolini F et al (2020) Mid-Devonian Archaeopteris roots signal revolutionary change in earliest fossil forests. Curr Biol 30:1–11

Steyer JS, Damiani R (2005) A giant brachyopoid temnospondyl from the upper Triassic or lower Jurassic of Lesotho. Bull Soc Géol Fr 176(3):243–248

Stork NE (2018) How many species of insects and other terrestrial arthropods are there on earth? Annu Rev Entomol 63:31–45

Strömberg CAE (2011) Evolution of grasses and grassland ecosystems. Annu Rev Earth Planet Sci 39:517–544

Strother PK, Battison L, Brasier MD, Wellman CH (2011) Earth's earliest non-marine eukaryotes. Nature 473:505–509

Strullu-Derrien C, Kenrick P, Pressel S, Duckett JG, Rioult J-P, Strullu D-G (2014) Fungal associations in *Horneophyton ligneri* from the Rhynie Chert (c. 407 million year old) closely resemble those in extant lower land plants: novel insights into ancestral plant-fungus symbioses. New Phytol 203:964–979

Suarez SE, Brookfield ME, Catlos EJ, Stöckli DF (2017) A U-Pb zircon age constraint on the oldest-recorded air-breathing land animal. PLoS One 12(6):e0179262

Sulej T, Niedźwiedzki G (2019) An elephant-sized Late Triassic synapsid with erect limbs. Science 363:78–80

Sun X-Y, Xia X, Yang Q (2016) Dating the origin of major lineages of Branchiopoda. Palaeoworld 25(2):303–317

Taniguchi M, Burnett WC, Cable JE, Turner JV (2002) Investigation of submarine groundwater discharge. Hydrobiological Process 16:2115–2129

Taylor TN, Osborn JM (1996) The importance of fungi in shaping the paleoecosystem. Rev in Palaeobiology and Palynology 90:249–262

Taylor MP, Wedel MH (2013) Why sauropods had long necks; and giraffes have short necks. PeerJ 1:e36. https://doi.org/10.7717/peerj.36

Testo W, Sundue M (2016) A 4000-species dataset provides new insight into the evolution of ferns. Mol Phylogenet Evol 105:200–211

Thien LB, Bernhardt P, Devall MS, Chen Z-D, Luo Y-B et al (2009) Pollination biology of basal angiosperms. Am J Bot 96(1):166–182

Tiffney BH (1992) The role of vertebrate herbivory in the evolution of land plants. Palaeobotanist 41:87–97

Tiffney BH (2004) Vertebrate dispersal of seed plants through time. Annu Rev Evol Syst 35:1–29

Tingle JL, Garner KL, Astley HV (2024) Functional diversity of snake locomotor behaviors: a review of the biological literature for bioinspiration. Ann N Y Acad Sci 1533:16–37. https://doi.org/10.1111/nyas.15109

Title PO, Singhal S, Grundler MC, Costa GC, Paron RA et al (2024) The macroevolutionary singularity of snakes. Science 383:918–923

Truman JW (2019) The evolution of insect metamorphosis. Curr Biol 29:R1252–R1268

Tsuyuzaki S, Goto M (2001) Persistence of seed bank under thick volcanic deposits twenty years after eruptions of Mount Usu, Hokkaido Island, Japan. Am J Bot 88(10):1813–1817

Turcotte MM, Davies TJ, Thomsen CJM, Johnson MTJ (2014) Macroecological and macroevolutionary patterns of lead herbivory across vascular plants. Proc R Soc B 281:20140555

Vajda V, Bercovici A (2014) The global vegetation pattern across the cretaceous-Paleogene mass extinction interval: a template for other extinction events. Glob Planet Chang 122:29–49

van der Niet T, Johnson SD (2012) Phylogenetic evidence for pollinator-driven diversification of angiosperms. Trends Ecol Evol 24(6):353–361

Volf M, Wirth C, van Dam NM (2020) Localized defense induction in trees: a mosaic of leaf traits promoting variation in plant traits, communities of canopy arthropods? Am J Bot 107(4):1–4

Vollset SE, Goren E, Yuan C-W, Cao J, Smith AE et al (2020) Fertility, mortality, migration, and population scenarios for 195 countries and territories from 2017 to 2100: a forecasting analysis for the global burden of disease study. Lancet 396:1285–1306. https://doi.org/10.1016/50140-6736(20)30677-2

Wallace S, Fleming A, Wellman CH, Beerling DJ (2011) Evolutionary development of the plant spore and pollen wall. AoB Plants 2011:plr027

Walter H, Breckle S-W (1983) Ökologie der Erde Band 1, Ökologische Grundlagen in globaler Sicht. Fischer, Stuttgart

Walter H, Breckle S-W (1991) Ökologie der Erde Band 4. Gemäßigte und Arktische Zonen außerhalb Euro-Nordasiens. Fischer, Stuttgart

Wang J, Pfefferkorn HW, Zhang Y, Feng Z (2012) Permian vegetational Pompeii from Inner Mongolia and its implications for landscape paleoecology and paleobiogeography of Cathaysia. PNAS 109(13):4927–4932

Wang H, Bai B, Meng J, Wang Y (2016) Earliest known uniequivocal rhinoceotoid sheds new light on the origin of Giant rhinos and phylogeny of early rhinocerotoids. Sci Rep 6:739607

Wang C, McCormack ML, Guo D, Li J (2018) Global meta-analysis reveals different patterns of root tip adjustments by angiosperm and gymnosperm trees in response to environmental gradients. J Biogeogr 46:123–133

Wang D, Qin M, Liu L, Liu L, Zhou Y et al (2019) The most extensive Devonian fossil Forest with small Lycopsid trees bearing the earliest Stigmarian roots. Curr Biol 29:1–12

Ward P, Labandeira C, Laurin M, Berner RA (2006) Confirmation of Romer's gap as a low oxygen interval constraining the timing of initial arthropod and vertebrate terrestrialization. PNAS 103(45):16818–16822

Watkins JE Jr, Cardelús CL (2012) Ferns in an angiosperm world: cretaceous radiation into the epiphytic niche and diversification on the forest floor. Int J Plant Sci 173(6):695–710

Wellman CH (2017) Palaeoecology and palaeophytogeography of the Rhynie chert plants: further evidence from integrated analysis of in situ and dispersed spores. Philos Trans R Soc B 373:20160491

Wellman CH, Steemans P, Vecoli M (2013) Palaeogeography of Ordovician-Silurian land plants. Geol Soc Lond Mem 38:461–476

Wendruff AJ, Babcock LE, Wirkner CS, Kluessendorf J, Mikulix DG (2020) A Silurian ancestral scorpion with fossilised internal anatomy illustrating a pathway to arachnid terrestrialisation. Sci Rep 10:14

Weng J-K, Chapple C (2010) The origin and evolution of lignin biosynthesis. New Phytol 187:273–285

Werneburg I, Sánchez-Villagra MR (2009) Timing of organogenesis support nasal position of turtles in the amniote tree of life. BMC Evol Biol 9:82

White WM (2013) Geochemistry. Wiley-Blachwell, Chichester

Willis KJ, McElwain JC (2002) The evolution of plants. Oxford University Press, Oxford

Wilson GP, Ekdale EG, Hoganson JW, Calede JJ, Linden AV (2016) A large carnivorous mammal from the Late Cretaceous and the North American origin of marsupials. Nat Commun 7:13734. https://doi.org/10.1038/ncomms13734

Wing SL, Tiffney BH (1987) The reciprocal interaction of angiosperm evolution and tetrapod herbivory. Rev Palaeobot Palynol 50:179–210

Wood DM, del Moral R (1987) Mechanisms of early primary succession in subalpine habitats on Mount St. Helens. Ecology 68(4):780–790

Xing L, Li D, Falkingham PL, Lockley MG, Benton MJ et al (2016) Digit-only sauropod pes trackways from China – evidence of swimming or a preservational phenomenon? Sci Rep 6:21138

Yang R-D, Mao JR, Zhang W-H, Jiang L-J, Gau H (2004) Bryophyte-like Fossil (*Parafunaria sinensis*) from early-middle Cambrian Kaili formation in Guizhou Province, China. Acta Bot Sin 46(2):180–185

Young GC, Lu J (2020) Asia-Gondwana connections indicated by Devonian fishes from Australia: palaeogeographic considerations. J Palaeogeogr 9:8

Zanno LE, Makovicky PJ (2012) No evidence for directional evolution of body mass in herbivorous theropod dinosaurs. Proc R Soc B 280:20122526

Zheru Z, Huahai M, Cheng Q (2001) Fractal geometry of element distribution of mineral surfaces. Math Geol 33(2):217–288

Zhou CF, Wu S, Martin T, Luo Z-X (2013) A Jurassic mammaliaform and the earliest mammalian evolutionary adaptations. Nature 500:163–167

Zotz G, Hietz P (2001) The physiological ecology of vascular epiphytes: current knowledge, open questions. J Exp Bot 52(364):2067–2078

11.1 Relativity of Evolutionary Processes

Providing an answer to the final question in the previous section requires explanation in somewhat more detail. A popular scientific interpretation of evolutionary processes as the definitive fixation of the "strongest" species would render the question superfluous. However, anyone who has acquired detailed familiarity with the information provided in the preceding ten chapters will presumably respond to the question in a more differentiated manner. *A core feature of biological evolution is the capacity of organisms to cope interactively with unexpected environmental changes.* By contrast, in terms of time and space, no generally valid definition of quantitative dimensions of evolutionary processes is possible. Possible dimensions range from the planetary level down to viruses and molecules. Quantitative dimensions become more meaningful only in relation to specific species or communities of organisms. One example of this is the temporal subdivision of geological epochs in connection with emblematic collapses of biotic communities (mass extinctions) and the disruptive events behind them, which can be defined as follows (Grimaldi and Engel 2005, p 635): "Mass extinctions are the extinction of diverse and unrelated lineages, usually over *a brief period of time in geological terms* and ultimately caused by global changes in the abiotic environment." Temporal dimensions of geological epochs are on the order of millions of years, whereas effects of disruptive events can last for hundreds of thousands of years. All temporal dimensions involved far exceed those that can be grasped—let alone controlled—by humans. Accordingly, it is utterly presumptuous to talk of the beginning of an Anthropocene (Waters and Turner 2022); at best, the onset of an *Anthropogenic Disruption* can be defined.

A focus on destructive events makes it easy to forget that prerequisites for overcoming them were established long before those events occurred, by means of functional

© The Author(s), under exclusive license to Springer-Verlag GmbH, DE, part of Springer Nature 2024
M. Knoflacher, *Relativity of Evolution*,
https://doi.org/10.1007/978-3-662-69423-7_11

differentiation of organismic communities (see Fig. 5.18: Schematic sketch comparing the effects of selective random events with either high (SRE 1) or low (SRE 2) selectivity on survival rates (succession states 1) of populations of an original community of organisms (initial state). However, longer-term development of individual populations depends on whether suitable conditions—grey markers—exist for them (subsequent state 2') or not (subsequent state 2''). For example, evolution of adaptive interactions between flowering plants and insects began around 150 million years ago (Dellinger et al. 2019) and its dominant effect on the evolutionary differentiation of plants and animals emerged only after the disruptive events that occurred at the end of the Cretaceous period. Drawing on the explanations provided in Chap. 10, terrestrial habitats are currently in the *angiosperm-insect-regulated epoch* (Fig. 10.7), which continues to provide the basis for all human food.

In terrestrial habitats, the dynamics of abiotic processes dampen terrestrial vegetation—depending on its structural and functional diversity. Large damping effects develop under optimal conditions for growth of vegetation, for example in humid tropical rainforests. There, a wide variety of organisms with indecipherable interactions can develop across the scale range from microorganisms to medium-sized vertebrates. Under such conditions, there is only a low probability of biological damage to ecosystems due to mass propagation of individual species.

As the evolutionary conditions for vegetation become increasingly constrained, their dampening effects decrease, for example geographically with increasing latitude or climatically with increasing aridity. At the same time, the probability of erratic mass propagation—with corresponding effects on the entire food chain—increases, for example with migratory locusts, lemmings or reindeer (Schwerdtfeger 1968; Uboni et al. 2016). Drought favours propagation of fires, for example following lightning strikes. Due to their recurrence, morphological and physiological adaptations have developed in various animal and plant species. Examples of this are seed cases that open only after fires or the widespread development of underground storage organs in plants in African savannahs (Whelan 1995; Maurin et al. 2014).

According to fossil finds so far available, it was in such habitats that evolution of the hominine lineage of apes gave rise to the species *Homo sapiens*, which has survived to the present day (Laden and Wrangham 2005; Jones et al. 2009; Archibald et al. 2012; Davies et al. 2020). Observations of chimpanzees in Senegal suggest that early members of the hominine lineage were already able to adapt their behaviour to bush fires and to exploit advantages for foraging (Pruetz and Herzog 2017). Microstructural examination and isotope analyses of teeth indicate that around 2.5 million years ago the genus *Homo* consumed a broader range of foods than other hominine genera (Ungar and Sponheimer 2011; Cerling et al. 2013; Sponheimer et al. 2013).

It is possible that the search for underground plant parts stimulated the special evolutionary development of the genus *Homo*, as it required not only physical capacities for exploration but also cognitive abilities for finding food sources that were not visible above ground. This would also render a special feature of hominid evolution—a persistent tendency to overcome systemic barriers—more comprehensible (Wadley 2013; Whiten and

Erdal 2012). Usually, descriptions of this tendency are limited to consideration of technological developments—as evidenced by artefacts—ranging from simple stone tools up to nuclear technology and supercomputers (Schlote 2002; Foley and Lahr 2007; Johnson 2017; Marshall 2017; Service 2018). Closely linked to this are increasingly manipulative interventions in the developmental lines of potentially useful organisms, from domestication to genetic modification of microorganisms, plants and animals. The reciprocal influences of both lines of development are recognizable both from the changes in agricultural production methods over the millennia and in the development of medical methods to directly combat microbial pathogens (Semmelweis 1861; Mazoyer and Roudart 1997; Winkle 1997; Knoflacher 2017).

11.2   Evolutionary Risks Arising from Self-referential Feedback Loops

While individual cognitive achievements have yielded astonishing insights throughout history, evident limits to society's capacity to recognize and take action exist. Although directly and personally perceptible effects of an action may trigger the same responses from almost everybody, this uniformity is lost in more complex cause-effect relationships. This is increasingly the case as triggering factors and effects become increasingly separated in terms of time and space, or if individual triggers have different effects in the context of different influences. The problem is exacerbated if long-term favourable effects can be achieved only at the price of short-term individual detrimental effects. While such individual restrictions may still be acceptable in small social groups, the willingness to comply dwindles with increasing size of societies and with increasing realization of their technological potential (Ostrom 1999; Demuth 2019).

A critical dilemma of human evolution becomes recognizable in conjunction with the core tendency to overcome systemic barriers. Large majorities accept and individually support advantageous and rapidly appreciable solutions or convincingly formulated promises concerning their future realization. By contrast, indications of uncertainties and long-term detrimental effects are perceived as unsettling. This results in the societal trajectory of compulsion towards permanent growth, combined with continued accumulation of unresolved problems (Riedl and Delpos 1996). In light of previous scientific and technological successes, any doubts about human capacities to find solutions are swept aside—even regarding the mundane fact that unlimited growth is simply not feasible in a finite system such as planet Earth.

In conjunction with increasing change in abiotic environmental factors, selective furthering of beneficial organisms also permits continued growth of the global human population. However, this simply augments the demands imposed by a single species. The paradigm of yield maximization can be used to justify and implement displacement or destruction of undesirable organisms—without taking into account the complexity of interactions involved. Because of marked spatial differences in the composition of organismic communities and their interactions, problems are exacerbated by increasing global standardization of anthropogenic management methods. Above all, the associated effects can be seen in decreases affecting the most diverse groups of organisms (Gibbons et al. 2000; Seibold et al. 2019; Irwin 2022; Panziera et al. 2022; Rigal et al. 2023).

Societal reactions to increasing evidence of global climate change provide copious examples. Methodologically, the evidence is based on inference of trends from historical measurements of various climate parameters or various indicators, such as width of tree rings. Characteristics and temporal changes of the measured parameters are founded on the influences of various abiotic, biotic and anthropogenic factors and interactions between them. Estimation of future climatic developments is hence associated with marked uncertainties. Such preconditions offer considerable leeway for interpretations by human societies, ranging from complete denial of any changes to declaration of monocausal relationships between individual influencing factors and resulting outcomes (McCright and Dunlap 2011). Discussions triggered by such uncertainty alone make it difficult to identify reasonable and sustainable countermeasures. In view of the chaotic nature of climate-related processes and increasing evidence for change, more precautionary measures are needed to deal with unforeseeable crises. Regardless of this, implementation of targeted measures to contain known critical anthropogenic influencing factors, along with increased research efforts to identify critical processes that have so far been insufficiently recognized or have remained completely unrecognized, is essential.

Instead of taking such an approach, human society is engendering a causal mindset that promises to solve the problem by changing a few "control variables". This alternative procedure ignores the fact that reducing greenhouse gas emissions alone would require a global shift away from the fiscal paradigm of continuous growth. To be precise, associated measures require, for example, changes in land use and economic processes, as well as associated transport processes, to minimize energy consumption. In addition, this would have to be closely linked to increased integration into economic processes of all those capable of earning, to ensure adequate income across the board. A central guiding principle for this would be lifelong provision of safeguards for the structural and material framework conditions for self-determined lives of all members of society, regardless of current earning capacity. Socially acceptable implementation of such measures would take decades or even centuries. Decentralized energy supply for buildings could be realized somewhat more rapidly through blanket use of irradiated outdoor surfaces for energetic conversion of solar radiation.

Examples of "solar power plants" in open countryside reveal the consequences of ignoring the growth component in the climate debate. Agricultural land in Central Europe is available for other uses only because actual agricultural production is no longer profitable. Cost-effective transport, uncontrolled expansion of plantations into previously unexploited areas such as rainforests and social exploitation in all sectors of supply chains pave the way for end prices well below the real cost of ecologically and socially acceptable production.

Associated global challenges for changing the growth paradigm are confronted with numerous—from a long-term perspective irrational—social obstacles. Ideologically based interests of a religious, nationalistic or economic nature are engaged in an ongoing competition to achieve their specific goals on a global basis (Norenzayan 2013; Oka and Kuijt 2014; Neal 2015). There are no perceivable approaches to changing the growth paradigm In these overt power games. Withdrawal of individual groups or states from this paradigm is immediately interpreted as weakness and exploited to reinforce the growth endeavours of other societies. Promising approaches for regionally differentiated solutions through integration of cultural and technical knowledge potential (Rozzi et al. 2018) are simply threatened by failure because in currently dominant societies unethical behaviour is promoted for profit.

It is hence unsurprising that political targets are becoming increasingly unrealistic. For example, the European Union has been propagating the goal of "climate neutrality" by 2050 (EU 2020). For comparison, the Organization of the Petroleum Exporting Countries (OPEC) expects energy consumption in OECD countries[1] to fall by only 4 per cent by 2045 (OPEC 2020). Critical minds will notice several points here. The goal of "climate neutrality" does not translate directly into changes in energy consumption, as various loopholes exist. In the simplest case, greenhouse gas emissions are "offset" by assumed compensation—for example reforestation. Afforestation of fallow land seemingly yields particularly favourable outcomes. But such areas are not a form of natural vegetation, but rather zones that have been anthropogenically devastated—for example through slash-and-burn cultivation. In more complicated cases, the goal of climate neutrality is purportedly achieved by substituting other forms of energy for fossil fuels, such as solar radiation, wind or hydropower, biomass or nuclear energy. The first two forms of energy mentioned have a highly irregular momentum of their own that is independent of consumption. They can therefore fulfil current supply requirements only by using sufficiently large storage capacities. Widespread and comprehensive utilization of the other two forms of energy is usually associated with massive changes. In terms of energy technology, they have the advantage of being storable prior to conversion. They can therefore be easily integrated into current supply structures. Nuclear energy is categorized by its advocates as climate-neutral because of minimal release of greenhouse gases during energy conversion. But propositions do not take account of high emissions in the upstream phases of plant installation or extraction and processing of raw materials. Potential risks of plant operation and subsequent storage of waste for thousands of years are also "down-calculated" by appropriate assumptions and simulation models (Orient 2014).

A completely different starting-point for criticism is comparison of values between the EU and all OECD countries. The EU represents only part of the OECD countries, with a current share of around 39 per cent of their economic output (OPEC 2020). The comparison employed is hence not meant to be strictly numerical; it is intended merely to encourage assessment in a global context of the relative importance of measures in the EU. Global

---

[1] Organisation for Economic Co-operation and Development.

distributions of both economic output and population size—along with political power relationships— are subject to constant dynamic change. For example, in terms of global economic productivity, the share of all OECD countries is expected to fall from 43 per cent to 31 per cent by 2045, while the share of Asian non-OECD countries is expected to rise from 37 per cent to 52 per cent. Based on these trends, the OPEC scenario estimates an increase in global energy consumption of around 25 per cent by 2045.

Fundamental weaknesses in the currently dominant social model came to light with the unexpected outbreak of the SARS-CoV-2 (Covid-19) pandemic in 2019/2020. Differing responses provide evidence for effects of social conditions that cannot be directly controlled. With the SARS-CoV-2, people are both carriers and victims of a danger that our senses cannot perceive. When a new virus variant appears for the first time, society lacks experience of the real risks associated with these infections. Due to the resulting uncertainties, social leadership levels responded extremely rapidly with a diffuse spectrum of actions, ranging from socially responsible decisions to ice-cold measures to expand power or enrich individual actors. Provision of social services was ensured primarily by underpaid professional groups. And representatives of the privileged classes praised this in Sunday sermons. However, the "normality" of the current social model was never questioned. This attitude is understandable from an individual perspective, as it would also cast horrendous overpayments in a critical light. Socially, however, this is highly problematic because drastic inequalities destabilize social systems and ultimately encourage the establishment of authoritarian structures.

In a general way, in the crisis situation weaknesses in the principles of efficiency and monofactorial maximization became clearly apparent. Both principles conflict with the evolutionarily successful principles of redundancy and multifactorial optimization. Monopolization of manufacturing processes and global supply chains continue to make it difficult to manage the pandemic through vaccination. Over the long term, such structural conditions will hinder adaptations to the challenges of climate change propagated in political statements. Instead, many decision-makers see further economic growth as the sole solution to overcoming the economic and social crisis that has arisen. This response became starkly evident—at least in the context of global tourism—both after the outbreak of SARS-CoV in 2003 and following the global economic crisis in 2009 (Gössling et al. 2020).

Systemic limitations in recording the incidence of infection during the Covid-19 pandemic were barely recognized by the general public. In view of everyday experiences with information available "at the touch of a button", this comment may seem anachronistic. However, in a systemic context, the significance of information depends exclusively on its relationship to the processes observed. Global financial transactions can be captured most completely through data, as the processes themselves take place in the virtual space of digital information processing. However, all processes outside the virtual world must be recorded, measured and converted into data prior to digital processing. The easiest way to achieve this is with activities that are constantly accompanied by electronic devices such as smartphones. In processes that are highly dynamic—such as development of infections—the timespan between recording and data input significantly influences the quality of the processed information. A similar correlation exists in the recording of spatially heterogeneous processes. Here, it is the spatial density of data acquisition and not the virtual spatial resolution of the computer models employed that determines the accuracy of the information.

In the physical world, however, a wide variety of processes with diverse properties and numerous interactions take place in parallel. It is therefore impossible to capture all processes and transfer them into electronic data. The requirements can be met only approximately. Scientifically, usually only a few parameters are measured as accurately as possible and digitally recorded. Because the vast majority of all real processes remain unobserved, only incomplete information about complex systems—such as environmental systems—can be obtained from recorded data. However, this approach is seemingly attractive because the data can be easily processed electronically and models can be used to generate a variety of visualizations of the processes assessed. However, this approach produces virtual results that can be manipulated through both the selection of variables for measurement and the algorithms applied. Such approaches are ideal for justifying the enforcement of large-scale global interests. For whatever reason, the United Nations favours these solutions for ensuring sustainable global development (Ekins et al. 2019). Compared to laboratory measurements, relatively imprecise but integrative measurements of environmental parameters are regarded as methodologically outdated. However, the scientific challenges of such methods reside in systemic derivation of observed variables. Based on analyses of content along with temporal and spatial characteristics of the processes to be observed, indicators that are as easy to record as possible must be derived and dedicated recording procedures developed. The indicators can then also be recorded by sufficiently trained laypeople, as in the English River Habitat Survey or similar undertakings (Raven et al. 2000). If such methods provide a basis for development of sustainable regional strategies. They require implementation, cooperation with the local population for their formulation and realization. However, realization of such approaches is increasingly failing due to a lack of young scientists—because the scientific disciplines required for this are barely represented at universities any more. But it is also failing due to a lack of scientific support, because this means that the limits to growth are becoming more quickly apparent.

Perhaps this makes it clear that human society is capable of continuously overcoming evolutionary barriers, but has not yet been able to generate an adequate substitute for social rules for global, sustainably effective and socially balanced self-restraint. Human society is pursuing an evolutionary risk strategy based on paradigms such as efficiency, one-dimensionality, limitless growth, diversity reduction or quantitative rationality. This is at the expense of an evolutionary risk-robust strategy, recognizable for example by characteristics such as redundancy, adaptive optimization, qualitative multidimensionality, limited growth, heterogeneous regionally differentiated systems, social balance or risk competence. The latter term encompasses both a social awareness of the limits of human knowledge and controllability, as well as an adaptive capacity to act and reach decisions under unpredictable conditions (Gigerenzer 2022).

References

Archibald S, Staver AC, Levin SA (2012) Evolution of human-driven fire regimes in Africa. PNAS 109(3):847–852

Cerling TE, Manthi FK, Mbua EN, Leakey LN, Leakey MG et al (2013) Stable isotope-based diet reconstructions of Turkana Basin hominins. PNAS 110(26):10501–10506

Davies TJ, Daru BH, Bezeng BS, Charles-Dominique T, Hempson GP et al (2020) Savanna tree evolutionary ages inform the reconstruction of the palaeoenvironment of our hominin ancestors. Sci Rep 10:12430. https://doi.org/10.1038/s41598-020-69378-0

Dellinger A, Chartier M, Fernández-Fernández D, Penneys DS, Alvear M et al (2019) Beyond buzz-pollination – departures from an adaptive plateau lead to new pollination syndromes. New Phytol 221:1136–1149

Demuth B (2019) Floating coast. Norton and Company, New York

Ekins P, Gupta J, Boileau P (eds) (2019) Global environmental outlook GEO-6 healthy planet, healthy people. UN Environment, Cambridge University Press

EU (2020) What is the European green deal? European Commission

Foley R, Lahr MM (2007) On stony ground. Evol Anthropol 12:109–122

Gibbons JW, Scott DE, Ryan TJ, Buhlmann KA, Tuberville TD et al (2000) The global decline of reptiles, Déjà Vu Amphibians. BioSci 50(8):653–666

Gigerenzer G (2022) Risiko. Pantheon, München

Gössling S, Scott D, Hall CM (2020) Pandemics, tourism and global. Change: a rapid assessment of Covid-19. J Sustain Tour 29(1):1–20

Grimaldi D, Engel MS (2005) Evolution of the insects. Cambridge University Press, Cambridge

Irwin A (2022) No tree left behind. Nature 609:24–27

Johnson NF (2017) To slow or not? Challenges in subsecond networks. Science 2017:801–802

Jones S, Martin R, Pilbeam D (eds) (2009) Human Evolution. Cambridge University Press, Cambridge

Knoflacher M (Hg) (2017) Herausforderungen der evolutionären Komplexität. LIT, Wien

Laden G, Wrangham R (2005) The rise of the hominids as an adaptive shift in fallback foods: plant underground storage organs (USOs) and australopith origins. J Hum Evol 49(4):482–498

Marshall E (2017) Tweak makes U.S. nukes more precise – and deadlier. Science 355:1252–1253

Maurin O, Davies TJ, Burrows JE, Daru BH, Yessoufou K et al (2014) Savanna fire and the origins of the 'underground forests' of Africa. New Phytol 204:201–214

Mazoyer M, Roudart L (1997) Histoire des agricultures du Monde. Seuil, Paris

McCright AM, Dunlap RE (2011) The politicization of climate change and polarization in the American public's views of global warming, 2001-2010. Sociol Q 52:155–194

Neal L (2015) A concise history of international finance. Cambridge University Press, Cambridge

Norenzayan A (2013) Big gods. Princeton university Press, Princeton

Oka R, Kuijt I (2014) Greed is bad, neutral, and good: a historical perspective on excessive accumulation and consumption. Econ Anthropol 1:30–48

OPEC (2020) World oil outlook 2045. OPEC

Orient JM (2014) Fukushima and reflections on radiation as a terror weapon. J Am Physicians Surg 19(2):48–55

Ostrom E (1999) Collective action and the evolution of social norms. Workshop in political theory and policy analysis. Indiana University

Panziera D, Requier F, Chantawannakul P, Pirk CWW, Blacquière T (2022) The diversity decline in wild and managed honey bee populations urges for an integrated conservation approach. Front Ecol Evol 10:767950

Pruetz JD, Herzog NM (2017) Savanna chimpanzees at Fongoli, Senegal, navigate a fire landscape. Curr Anthropol 38:S337–S350

Raven PJ, Holmes NTH, Naura M, Dawson FH (2000) Using river habitat survey for environmental assessment and catchment planning in the U.K. Hydrobiologia 422/423:359–367

Riedl R, Delpos M (eds) (1996) Die Ursachen des Wachstums. Kremayr and Scheriau, Wien

Rigal S, Dakos V, Alonso H, Aunins A, Benkö Z et al (2023) Farmland practices are driving bird population decline across Europe. PNAS 120(21):e2216573120

Rozzi R, May RH Jr, Chapin FS III, Massardo F, Gavin MC et al (2018) From biocultural homogenization to biocultural conservation. Springer Nature, Cham

Schlote K-H (2002) Chronologie der Naturwissenschaften. Harri Deutsch, Frankfurt am Main

Schwerdtfeger F (1968) Demökologie. Paul Parey, Hamburg

Seibold S, Gossner MM, Simons NK, Blüthgen N, Müller J et al (2019) Arthropod decline in grasslands and forests is associated with landscape-level drivers. Nature 574:671–674

Semmelweis IP (1861) Die Ätiologie, der Begriff und die Prophylaxe des Kindbettfiebers. EOD Reprint, Wien

Service RF (2018) Design for U.S. exascale computer takes shape. Science 359:617–618

Sponheimer M, Alemseged Z, Cerling TE, Grine FE, Kimbel WH et al (2013) Isotopic evidence of early hominin diets. PNAS 110(26):10513–10518

Uboni A, Horstkotte T, Kaarlejärvi E, Sévêque A, Stammler F et al (2016) Long-term trends and role of climate in the population dynamics of Eurasian reindeer. PLoS One 11(6):e0158359

Ungar PS, Sponheimer M (2011) The diets of early hominins. Science 334:190–193

Wadley L (2013) Recognizing complex cognition through innovative technology in Stone Age and palaeolithic sites. Camb Archaeol J 23(2):163–183

Waters CN, Turner SD (2022) Defining the onset of the Anthropocene. Science 378:706–708

Whelan RJ (1995) The ecology of fire. Cambridge University Press, Cambridge

Whiten A, Erdal D (2012) The human socio-cognitive niche and its evolutionary origins. Philos Trans R Soc B 367:2119–2129

Winkle S (1997) Geißeln der Menschheit; Kulturgeschichte der Seuchen. Artemis and Winkler, Düsseldorf

Index